Fundamental Equations of Statics

Cartesian Vector

$$\mathbf{A} = A_x\mathbf{i} + A_y\mathbf{j} + A_z\mathbf{k}$$

Magnitude

$$A = \sqrt{A_x^2 + A_y^2 + A_z^2}$$

Directions

$$\mathbf{u}_A = \frac{\mathbf{A}}{A} = \frac{A_x}{A}\mathbf{i} + \frac{A_y}{A}\mathbf{j} + \frac{A_z}{A}\mathbf{k}$$
$$= \cos\alpha\mathbf{i} + \cos\beta\mathbf{j} + \cos\gamma\mathbf{k}$$
$$\cos^2\alpha + \cos^2\beta + \cos^2\gamma = 1$$

Dot Product

$$\mathbf{A}\cdot\mathbf{B} = AB\cos\theta$$
$$= A_xB_x + A_yB_y + A_zB_z$$

Cross Product

$$\mathbf{C} = \mathbf{A}\times\mathbf{B} = \begin{vmatrix} \mathbf{i} & \mathbf{j} & \mathbf{k} \\ A_x & A_y & A_z \\ B_x & B_y & B_z \end{vmatrix}$$

Cartesian Position Vector

$$\mathbf{r} = (x_2 - x_1)\mathbf{i} + (y_2 - y_1)\mathbf{j} + (z_2 - z_1)\mathbf{k}$$

Cartesian Force Vector

$$\mathbf{F} = F\mathbf{u} = F\left(\frac{\mathbf{r}}{r}\right)$$

Moment of a Force

$$M_o = Fd$$
$$\mathbf{M}_o = \mathbf{r}\times\mathbf{F} = \begin{vmatrix} \mathbf{i} & \mathbf{j} & \mathbf{k} \\ r_x & r_y & r_z \\ F_x & F_y & F_z \end{vmatrix}$$

Moment of a Force About a Specified Axis

$$M_a = \mathbf{u}\cdot\mathbf{r}\times\mathbf{F} = \begin{vmatrix} u_x & u_y & u_z \\ r_x & r_y & r_z \\ F_x & F_y & F_z \end{vmatrix}$$

Simplification of a Force and Couple System

$$\mathbf{F}_R = \Sigma\mathbf{F}$$
$$(\mathbf{M}_R)_O = \Sigma\mathbf{M} + \Sigma\mathbf{M}_O$$

Equilibrium

Particle

$$\Sigma F_x = 0,\ \Sigma F_y = 0,\ \Sigma F_z = 0$$

Rigid Body-Two Dimensions

$$\Sigma F_x = 0,\ \Sigma F_y = 0,\ \Sigma M_O = 0$$

Rigid Body-Three Dimensions

$$\Sigma F_x = 0,\ \Sigma F_y = 0,\ \Sigma F_z = 0$$
$$\Sigma M_{x'} = 0,\ \Sigma M_{y'} = 0,\ \Sigma M_{z'} = 0$$

Friction

Static (maximum) $F_s = \mu_s N$

Kinetic $F_k = \mu_k N$

Center of Gravity

Particles or Discrete Parts

$$\bar{r} = \frac{\Sigma\tilde{r}W}{\Sigma W}$$

Body

$$\bar{r} = \frac{\int \tilde{r}\,dW}{\int dW}$$

Area and Mass Moments of Inertia

$$I = \int r^2\,dA \qquad I = \int r^2\,dm$$

Parallel-Axis Theorem

$$I = \bar{I} + Ad^2 \qquad I = \bar{I} + md^2$$

Radius of Gyration

$$k = \sqrt{\frac{I}{A}} \qquad k = \sqrt{\frac{I}{m}}$$

Virtual Work

$$\delta U = 0$$

SI Prefixes

Multiple	*Exponential Form*	*Prefix*	*SI Symbol*
1 000 000 000	10^9	giga	G
1 000 000	10^6	mega	M
1 000	10^3	kilo	k
Submultiple			
0.001	10^{-3}	milli	m
0.000 001	10^{-6}	micro	μ
0.000 000 001	10^{-9}	nano	n

Conversion Factors (FPS) to (SI)

Quantity	*Unit of Measurement (FPS)*	*Equals*	*Unit of Measurement (SI)*
Force	lb		4.448 N
Mass	slug		14.59 kg
Length	ft		0.3048 m

Conversion Factors (FPS)

1 ft = 12 in. (inches)
1 mi. (mile) = 5280 ft
1 kip (kilopound) = 1000 lb
1 ton = 2000 lb

FUNDAMENTALS OF ENGINEERING STATICS

CUSTOM EDITION FOR BINGHAMTON UNIVERSITY

PEARSON ALWAYS LEARNING

Fundamentals of Engineering Statics

Custom Edition for Binghamton University

Taken from:
Introduction to Solid Mechanics, Third Edition
by Irving H. Shames and James M. Pitarresi

Engineering Mechanics: Statics, Thirteenth Edition
by R. C. Hibbeler

Cover Art: Courtesy of Photodisc/Getty Images.

Taken from:

Engineering Mechanics: Statics, Thirteenth Edition
by R. C. Hibbeler

Published by Pearson
Boston, Massachusetts 02116

Introduction to Solid Mechanics, Third Edition
by Irving H. Shames

Published by Pearson
Boston, Massachusetts 02116

This special edition published in cooperation with Pearson Learning Solutions.

Pearson Learning Solutions, 501 Boylston Street, Suite 900, Boston, MA 02116
A Pearson Education Company
www.pearsoned.com

Printed in the United States of America

2 3 4 5 6 7 8 9 10 V092 17 16 15 14 13

000200010271667037

ML

PEARSON

ISBN 10: 1-256-80883-0
ISBN 13: 978-1-256-80883-1

To the Student

With the hope that this work will stimulate
an interest in Engineering Mechanics
and provide an acceptable guide to its understanding.

PREFACE

The main purpose of this book is to provide the student with a clear and thorough presentation of the theory and application of engineering mechanics. To achieve this objective, this work has been shaped by the comments and suggestions of hundreds of reviewers in the teaching profession, as well as many of the author's students.

New to this Edition

New Problems. There are approximately 35% or about 410 new problems in this edition. These new problems relate to applications in many different fields of engineering. Also, a significant increase in algebraic type problems has been added, so that a generalized solution can be obtained.

Additional Fundamental Problems. These problem sets serve as extended example problems since their solutions are given in the back of the book. Additional problems have been added, especially in the areas of frames and machines, and in friction.

Expanded Solutions. Some of the fundamental problems now have more detailed solutions, including some artwork, for better clarification. Also, some of the more difficult problems have additional hints along with its answer when given in the back of the book.

Updated Photos. The relevance of knowing the subject matter is reflected by the realistic applications depicted by the many photos placed throughout the book. In this edition 20 new or updated photos are included. These, along with all the others, are generally used to explain how the relevant principles of mechanics apply to real-world situations. In some sections they are incorporated into the example problems, or to show how to model then draw the free-body diagram of an actual object.

New & Revised Example Problems. Throughout the book examples have been altered or enhanced in an attempt to help clarify concepts for students. Where appropriate new examples have been added in order to emphasize important concepts that were needed.

New Conceptual Problems. The conceptual problems given at the end of many of the problem sets are intended to engage the students in thinking through a real-life situation as depicted in a photo. They can be assigned either as individual or team projects after the students have developed some expertise in the subject matter.

Hallmark Features

Besides the new features mentioned above, other outstanding features that define the contents of the text include the following.

Organization and Approach. Each chapter is organized into well-defined sections that contain an explanation of specific topics, illustrative example problems, and a set of homework problems. The topics within each section are placed into subgroups defined by boldface titles. The purpose of this is to present a structured method for introducing each new definition or concept and to make the book convenient for later reference and review.

Chapter Contents. Each chapter begins with an illustration demonstrating a broad-range application of the material within the chapter. A bulleted list of the chapter contents is provided to give a general overview of the material that will be covered.

Emphasis on Free-Body Diagrams. Drawing a free-body diagram is particularly important when solving problems, and for this reason this step is strongly emphasized throughout the book. In particular, special sections and examples are devoted to show how to draw free-body diagrams. Specific homework problems have also been added to develop this practice.

Procedures for Analysis. A general procedure for analyzing any mechanical problem is presented at the end of the first chapter. Then this procedure is customized to relate to specific types of problems that are covered throughout the book. This unique feature provides the student with a logical and orderly method to follow when applying the theory. The example problems are solved using this outlined method in order to clarify its numerical application. Realize, however, that once the relevant principles have been mastered and enough confidence and judgment have been obtained, the student can then develop his or her own procedures for solving problems.

Important Points. This feature provides a review or summary of the most important concepts in a section and highlights the most significant points that should be realized when applying the theory to solve problems.

Fundamental Problems. These problem sets are selectively located just after most of the example problems. They provide students with simple applications of the concepts, and therefore, the chance to develop their problem-solving skills before attempting to solve any of the standard problems that follow. In addition, they can be used for preparing for exams, and they can be used at a later time when preparing for the Fundamentals in Engineering Exam.

Conceptual Understanding. Through the use of photographs placed throughout the book, theory is applied in a simplified way in order to illustrate some of its more important conceptual features and instill the physical meaning of many of the terms used in the equations. These simplified applications increase interest in the subject matter and better prepare the student to understand the examples and solve problems.

Homework Problems. Apart from the Fundamental and Conceptual type problems mentioned previously, other types of problems contained in the book include the following:

- **Free-Body Diagram Problems.** Some sections of the book contain introductory problems that only require drawing the free-body diagram for the specific problems within a problem set. These assignments will impress upon the student the importance of mastering this skill as a requirement for a complete solution of any equilibrium problem.

- **General Analysis and Design Problems.** The majority of problems in the book depict realistic situations encountered in engineering practice. Some of these problems come from actual products used in industry. It is hoped that this realism will both stimulate the student's interest in engineering mechanics and provide a means for developing the skill to reduce any such problem from its physical description to a model or symbolic representation to which the principles of mechanics may be applied.

Throughout the book, there is an approximate balance of problems using either SI or FPS units. Furthermore, in any set, an attempt has been made to arrange the problems in order of increasing difficulty except for the end of chapter review problems, which are presented in random order.

- **Computer Problems.** An effort has been made to include some problems that may be solved using a numerical procedure executed on either a desktop computer or a programmable pocket calculator. The intent here is to broaden the student's capacity for using other forms of mathematical analysis without sacrificing the time needed to focus on the application of the principles of mechanics. Problems of this type, which either can or must be solved using numerical procedures, are identified by a "square" symbol (■) preceding the problem number.

The many homework problems in this edition, have been placed into two different categories. Problems that are simply indicated by a problem number have an answer and in some cases an additional numerical result given in the back of the book. An asterisk (*) before every fourth problem number indicates a problem without an answer.

Accuracy. As with the previous editions, apart from the author, the accuracy of the text and problem solutions has been thoroughly checked by four other parties: Scott Hendricks, Virginia Polytechnic Institute and State University; Karim Nohra, University of South Florida; Kurt Norlin, Laurel Tech Integrated Publishing Services; and finally Kai Beng, a practicing engineer, who in addition to accuracy review provided suggestions for problem development.

Contents

The book is divided into 11 chapters, in which the principles are first applied to simple, then to more complicated situations. In a general sense, each principle is applied first to a particle, then a rigid body subjected to a coplanar system of forces, and finally to three-dimensional force systems acting on a rigid body.

Chapter 1 begins with an introduction to mechanics and a discussion of units. The vector properties of a concurrent force system are introduced in Chapter 2. This theory is then applied to the equilibrium of a particle in Chapter 3. Chapter 4 contains a general discussion of both concentrated and distributed force systems and the methods used to simplify them. The principles of rigid-body equilibrium are developed in Chapter 5 and then applied to specific problems involving the equilibrium of trusses, frames, and machines in Chapter 6, and to the analysis of internal forces in beams and cables in Chapter 7. Applications to problems involving frictional forces are discussed in Chapter 8, and topics related to the center of gravity and centroid are treated in Chapter 9. If time permits, sections involving more advanced topics, indicated by stars (★), may be covered. Most of these topics are included in Chapter 10 (area and mass moments of inertia) and Chapter 11 (virtual work and potential energy). Note that this material also provides a suitable reference for basic principles when it is discussed in more advanced courses. Finally, Appendix A provides a review and list of mathematical formulas needed to solve the problems in the book.

Alternative Coverage. At the discretion of the instructor, some of the material may be presented in a different sequence with no loss of continuity. For example, it is possible to introduce the concept of a force and all the necessary methods of vector analysis by first covering Chapter 2 and Section 4.2 (the cross product). Then after covering the rest of Chapter 4 (force and moment systems), the equilibrium methods of Chapters 3 and 5 can be discussed.

Acknowledgments

The author has endeavored to write this book so that it will appeal to both the student and instructor. Through the years, many people have helped in its development, and I will always be grateful for their valued suggestions and comments. Specifically, I wish to thank all the individuals who have contributed their comments relative to preparing the thirteenth edition of this work, and in particular, O. Barton, Jr. of the U.S. Naval Academy, and K. Cook-Chennault at Rutgers, the State University of New Jersey.

There are a few other people that I also feel deserve particular recognition. These include comments sent to me by H. Kuhlman and G. Benson. A long-time friend and associate, Kai Beng Yap, was of great help to me in preparing and checking problem solutions. A special note of thanks also goes to Kurt Norlin of Laurel Tech Integrated Publishing Services in this regard. During the production process I am thankful for the assistance of Rose Kernan, my production editor for many years, and to my wife,

Conny, and daughter, Mary Ann, who have helped with the proofreading and typing needed to prepare the manuscript for publication.

Lastly, many thanks are extended to all my students and to members of the teaching profession who have freely taken the time to e-mail me their suggestions and comments. Since this list is too long to mention, it is hoped that those who have given help in this manner will accept this anonymous recognition.

I would greatly appreciate hearing from you if at any time you have any comments, suggestions, or problems related to any matters regarding this edition.

Russell Charles Hibbeler
hibbeler@bellsouth.net

your work...

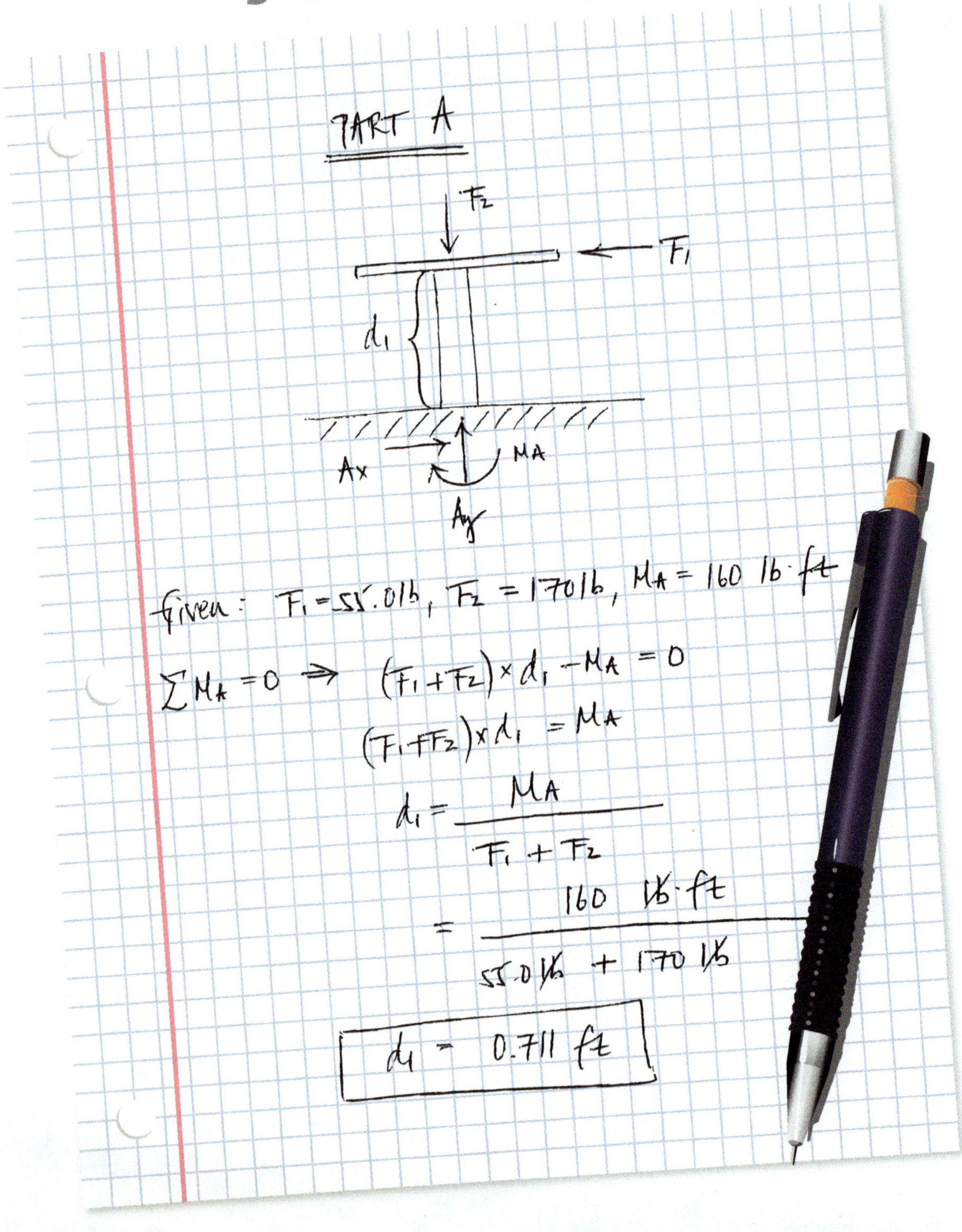
PART A
F2
F1
d1
Ax
MA
Ay
Given: F1 = 55.0 lb, F2 = 170 lb, MA = 160 lb·ft
ΣMA = 0 ⇒ (F1 + F2) × d1 − MA = 0
(F1 + F2) × d1 = MA
d1 = MA / (F1 + F2)
= 160 lb·ft / (55.0 lb + 170 lb)
d1 = 0.711 ft

Resources for Instructors

- **MasteringEngineering.** This online Tutorial Homework program allows you to integrate dynamic homework with automatic grading and adaptive tutoring. MasteringEngineering allows you to easily track the performance of your entire class on an assignment-by-assignment basis, or the detailed work of an individual student.

- **Instructor's Solutions Manual.** This supplement provides complete solutions supported by problem statements and problem figures. The thirteenth edition manual was revised to improve readability and was triple accuracy checked. The Instructor's Solutions Manual is available on Pearson Higher Education website: www.pearsonhighered.com.

- **Instructor's Resource.** Visual resources to accompany the text are located on the Pearson Higher Education website: www.pearsonhighered.com. If you are in need of a login and password for this site, please contact your local Pearson representative. Visual resources include all art from the text, available in PowerPoint slide and JPEG format.

- **Video Solutions.** Developed by Professor Edward Berger, University of Virginia, video solutions are located on the Companion Website for the text and offer step-by-step solution walkthroughs of representative homework problems from each section of the text. Make efficient use of class time and office hours by showing students the complete and concise problem-solving approaches that they can access any time and view at their own pace. The videos are designed to be a flexible resource to be used however each instructor and student prefers. A valuable tutorial resource, the videos are also helpful for student self-evaluation as students can pause the videos to check their understanding and work alongside the video. Access the videos at www.pearsonhighered.com/hibbeler/ and follow the links for the *Engineering Mechanics: Statics*, Thirteenth Edition text.

Resources for Students

- **MasteringEngineering.** Tutorial homework problems emulate the instructor's office-hour environment, guiding students through engineering concepts with self-paced individualized coaching. These in-depth tutorial homework problems are designed to coach students with feedback specific to their errors and optional hints that break problems down into simpler steps.

- **Statics Study Pack.** This supplement contains chapter-by-chapter study materials, a Free-Body Diagram Workbook and access to the Companion Website where additional tutorial resources are located.

- **Companion Website.** The Companion Website, located at www.pearsonhighered.com/hibbeler/, includes opportunities for practice and review including:

  - **Video Solutions**—Complete, step-by-step solution walkthroughs of representative homework problems from each section. Videos offer fully worked solutions that show every step of representative homework problems—this helps students make vital connections between concepts.

- **Statics Practice Problems Workbook.** This workbook contains additional worked problems. The problems are partially solved and are designed to help guide students through difficult topics.

Ordering Options

The *Statics Study Pack* and MasteringEngineering resources are available as stand-alone items for student purchase and are also available packaged with the texts. The ISBN for each valuepack is as follows:

- *Engineering Mechanics: Statics* with Study Pack: ISBN: 0133027996
- *Engineering Mechanics: Statics* with MasteringEngineering Student Access Card: ISBN: 0133009548

Custom Solutions

Please contact your local Pearson Sales Representative for more details about custom options or visit

www.pearsonlearningsolutions.com, keyword: Hibbeler.

CONTENTS

2
Force Vectors 17

3
Equilibrium of a Particle 85

4 Force System Resultants 117

5 Equilibrium of a Rigid Body 199

6
Structural Analysis 263

Chapter 7 Taken from: *Introduction to Solid Mechanics*, Third Edition by Irving H. Shames and James M. Pitarresi

7
Section Forces in Beams 331

8
Friction 379

9
Center of Gravity and Centroid 441

10 [illegible]

CREDITS

Chaper opening images are credited as follows:

Chapter 1, © Roel Meijer / Alamy
Chapter 2, Atli Mar/Getty Images
Chapter 3, John Foxx/Getty Images
Chapter 4, Huntstock/Getty Images
Chapter 5, © Bettmann/CORBIS
Chapter 6, © Martin Jenkinson / Alamy
Chapter 8, © INSADCO Photography / Alamy
Chapter 9, © Corbis Premium RF / Alamy
Chapter 10, © RubberBall / Alamy

All other photos provided by the author.

FUNDAMENTALS OF ENGINEERING STATICS

CUSTOM EDITION FOR BINGHAMTON UNIVERSITY

Chapter 2

This electric transmission tower is stabilized by cables that exert forces on the tower at their points of connection. In this chapter we will show how to express these forces as Cartesian vectors, and then determined their resultant.

Force Vectors

CHAPTER OBJECTIVES

- To show how to add forces and resolve them into components using the Parallelogram Law.
- To express force and position in Cartesian vector form and explain how to determine the vector's magnitude and direction.
- To introduce the dot product in order to determine the angle between two vectors or the projection of one vector onto another.

2.1 Scalars and Vectors

All physical quantities in engineering mechanics are measured using either scalars or vectors.

Scalar. A *scalar* is any positive or negative physical quantity that can be completely specified by its *magnitude*. Examples of scalar quantities include length, mass, and time.

Vector. A *vector* is any physical quantity that requires both a *magnitude* and a *direction* for its complete description. Examples of vectors encountered in statics are force, position, and moment. A vector is shown graphically by an arrow. The length of the arrow represents the *magnitude* of the vector, and the angle θ between the vector and a fixed axis defines the *direction of its line of action*. The head or tip of the arrow indicates the *sense of direction* of the vector, Fig. 2–1.

In print, vector quantities are represented by boldface letters such as **A**, and the magnitude of a vector is italicized, A. For handwritten work, it is often convenient to denote a vector quantity by simply drawing an arrow above it, $\vec{A}$.

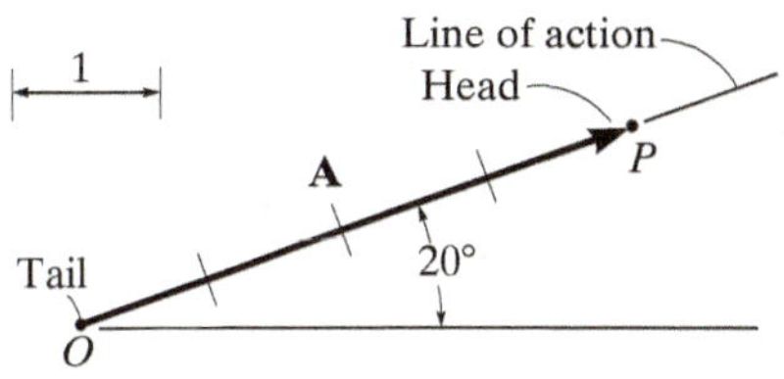

Fig. 2–1

2.2 Vector Operations

Multiplication and Division of a Vector by a Scalar. If a vector is multiplied by a positive scalar, its magnitude is increased by that amount. Multiplying by a negative scalar will also change the directional sense of the vector. Graphic examples of these operations are shown in Fig. 2–2.

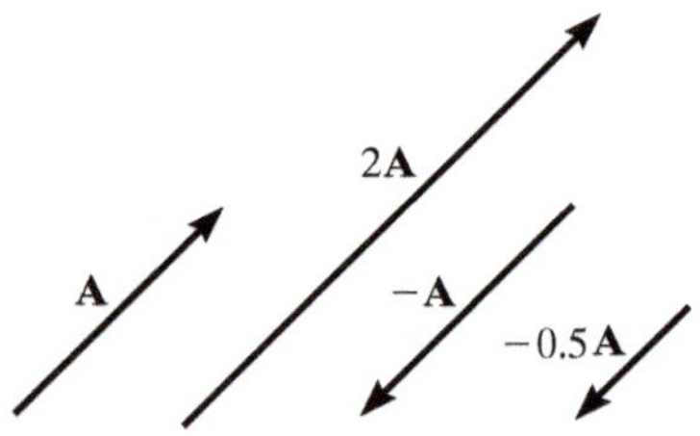

Scalar multiplication and division

Fig. 2–2

Vector Addition. All vector quantities obey the *parallelogram law of addition*. To illustrate, the two "*component*" *vectors* **A** and **B** in Fig. 2–3*a* are added to form a "*resultant*" *vector* **R** = **A** + **B** using the following procedure:

- First join the tails of the components at a point to make them concurrent, Fig. 2–3*b*.
- From the head of **B**, draw a line parallel to **A**. Draw another line from the head of **A** that is parallel to **B**. These two lines intersect at point *P* to form the adjacent sides of a parallelogram.
- The diagonal of this parallelogram that extends to *P* forms **R**, which then represents the resultant vector **R** = **A** + **B**, Fig. 2–3*c*.

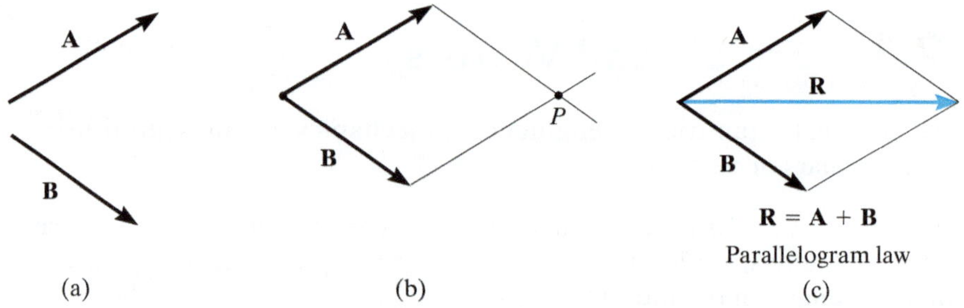

Fig. 2–3

We can also add **B** to **A**, Fig. 2–4*a*, using the *triangle rule*, which is a special case of the parallelogram law, whereby vector **B** is added to vector **A** in a "head-to-tail" fashion, i.e., by connecting the head of **A** to the tail of **B**, Fig. 2–4*b*. The resultant **R** extends from the tail of **A** to the head of **B**. In a similar manner, **R** can also be obtained by adding **A** to **B**, Fig. 2–4*c*. By comparison, it is seen that vector addition is commutative; in other words, the vectors can be added in either order, i.e., **R** = **A** + **B** = **B** + **A**.

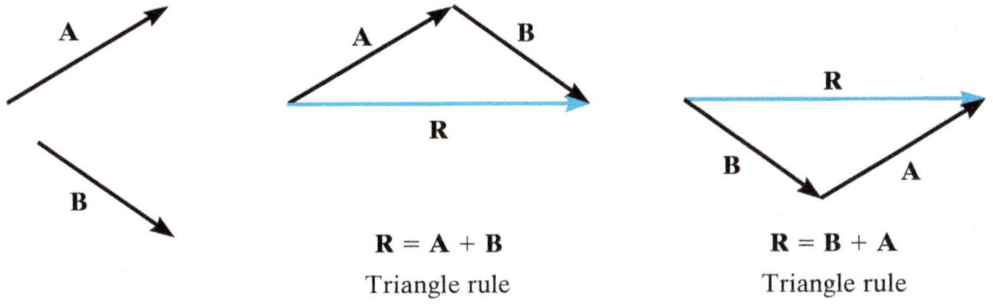

(a) (b) (c)

Fig. 2–4

As a special case, if the two vectors **A** and **B** are *collinear*, i.e., both have the same line of action, the parallelogram law reduces to an *algebraic* or *scalar addition* $R = A + B$, as shown in Fig. 2–5.

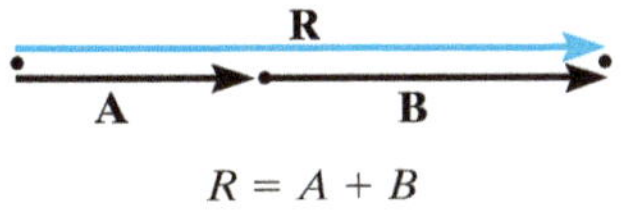

Addition of collinear vectors

Fig. 2–5

Vector Subtraction. The resultant of the *difference* between two vectors **A** and **B** of the same type may be expressed as

$$\mathbf{R}' = \mathbf{A} - \mathbf{B} = \mathbf{A} + (-\mathbf{B})$$

This vector sum is shown graphically in Fig. 2–6. Subtraction is therefore defined as a special case of addition, so the rules of vector addition also apply to vector subtraction.

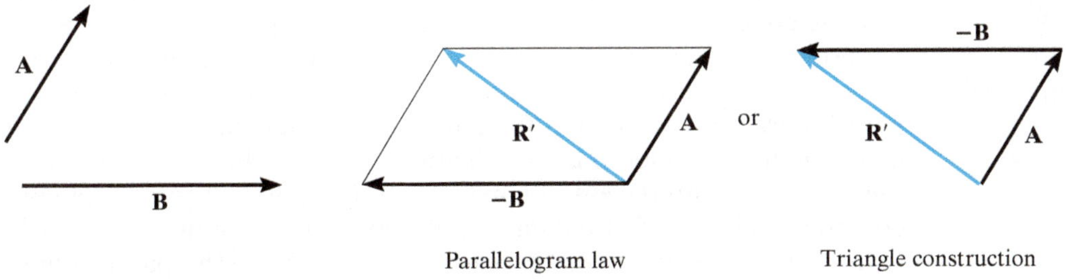

Vector subtraction

Fig. 2–6

2.3 Vector Addition of Forces

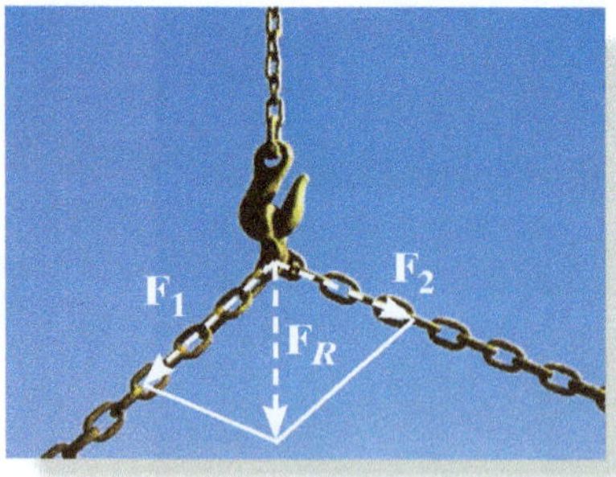

The parallelogram law must be used to determine the resultant of the two forces acting on the hook.

Experimental evidence has shown that a force is a vector quantity since it has a specified magnitude, direction, and sense and it adds according to the parallelogram law. Two common problems in statics involve either finding the resultant force, knowing its components, or resolving a known force into two components. We will now describe how each of these problems is solved using the parallelogram law.

Finding a Resultant Force. The two component forces $\mathbf{F}_1$ and $\mathbf{F}_2$ acting on the pin in Fig. 2–7*a* can be added together to form the resultant force $\mathbf{F}_R = \mathbf{F}_1 + \mathbf{F}_2$, as shown in Fig. 2–7*b*. From this construction, or using the triangle rule, Fig. 2–7*c*, we can apply the law of cosines or the law of sines to the triangle in order to obtain the magnitude of the resultant force and its direction.

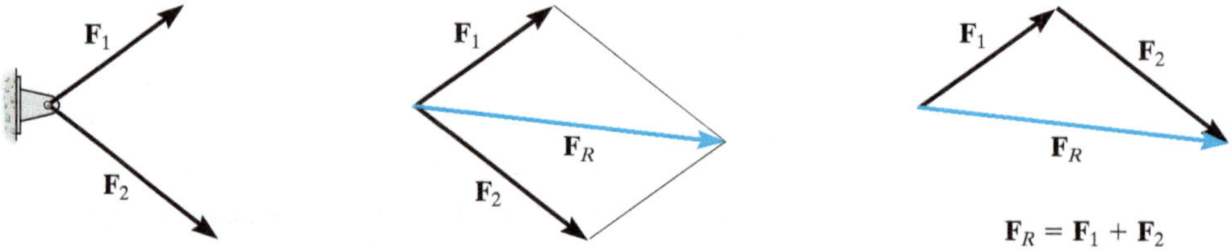

(a) (b) (c)

Fig. 2–7

Using the parallelogram law the supporting force $\mathbf{F}$ can be resolved into components acting along the u and v axes.

Finding the Components of a Force. Sometimes it is necessary to resolve a force into two *components* in order to study its pulling or pushing effect in two specific directions. For example, in Fig. 2–8*a*, $\mathbf{F}$ is to be resolved into two components along the two members, defined by the u and v axes. In order to determine the magnitude of each component, a parallelogram is constructed first, by drawing lines starting from the tip of $\mathbf{F}$, one line parallel to u, and the other line parallel to v. These lines then intersect with the v and u axes, forming a parallelogram. The force components $\mathbf{F}_u$ and $\mathbf{F}_v$ are then established by simply joining the tail of $\mathbf{F}$ to the intersection points on the u and v axes, Fig. 2–8*b*. This parallelogram can then be reduced to a triangle, which represents the triangle rule, Fig. 2–8*c*. From this, the law of sines can then be applied to determine the unknown magnitudes of the components.

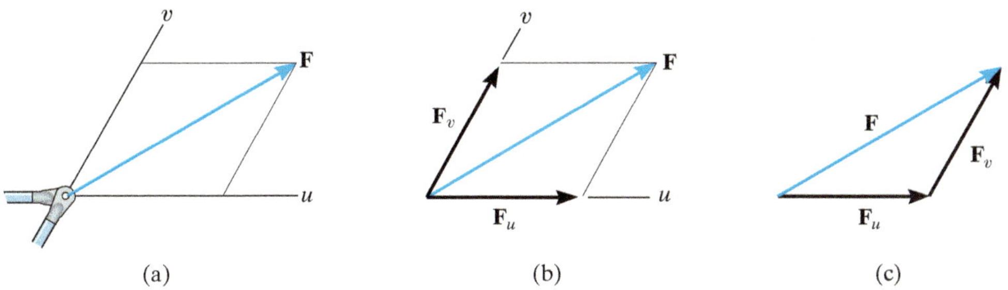

Fig. 2–8

Addition of Several Forces. If more than two forces are to be added, successive applications of the parallelogram law can be carried out in order to obtain the resultant force. For example, if three forces $\mathbf{F}_1$, $\mathbf{F}_2$, $\mathbf{F}_3$ act at a point O, Fig. 2–9, the resultant of any two of the forces is found, say, $\mathbf{F}_1 + \mathbf{F}_2$—and then this resultant is added to the third force, yielding the resultant of all three forces; i.e., $\mathbf{F}_R = (\mathbf{F}_1 + \mathbf{F}_2) + \mathbf{F}_3$. Using the parallelogram law to add more than two forces, as shown here, often requires extensive geometric and trigonometric calculation to determine the numerical values for the magnitude and direction of the resultant. Instead, problems of this type are easily solved by using the "rectangular-component method," which is explained in Sec. 2.4.

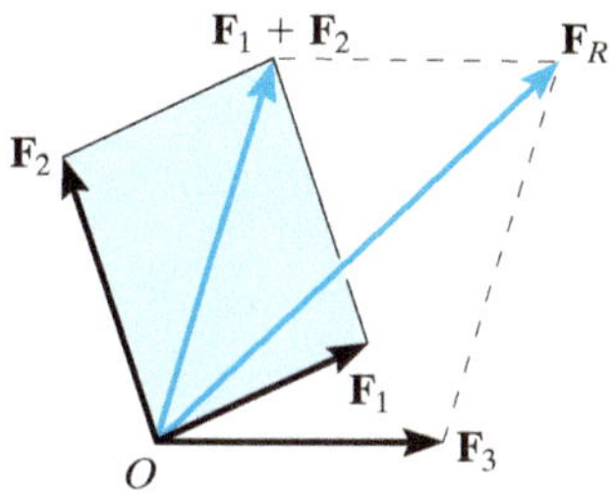

Fig. 2–9

The resultant force $\mathbf{F}_R$ on the hook requires the addition of $\mathbf{F}_1 + \mathbf{F}_2$, then this resultant is added to $\mathbf{F}_3$.

2

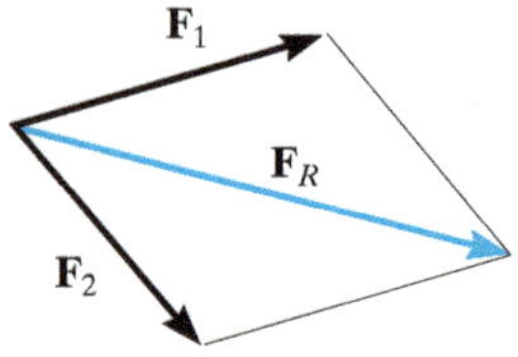

(a)

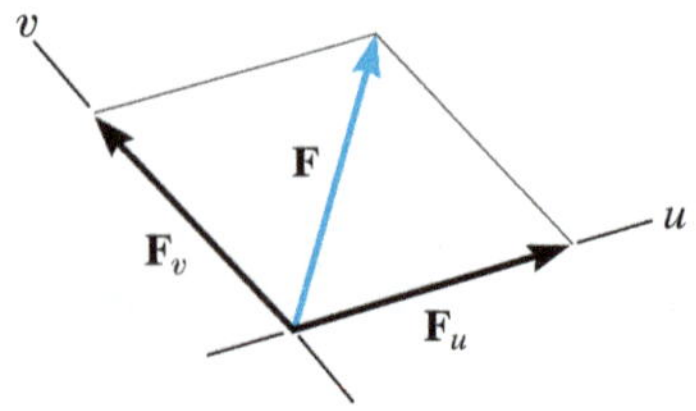

(b)

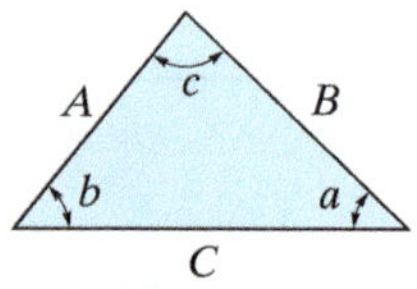

Cosine law:

$$C = \sqrt{A^2 + B^2 - 2AB \cos c}$$

Sine law:

$$\frac{A}{\sin a} = \frac{B}{\sin b} = \frac{C}{\sin c}$$

(c)

Fig. 2–10

Procedure for Analysis

Problems that involve the addition of two forces can be solved as follows:

Parallelogram Law.

- Two "component" forces $\mathbf{F}_1$ and $\mathbf{F}_2$ in Fig. 2–10*a* add according to the parallelogram law, yielding a *resultant* force $\mathbf{F}_R$ that forms the diagonal of the parallelogram.
- If a force **F** is to be resolved into *components* along two axes u and v, Fig. 2–10*b*, then start at the head of force **F** and construct lines parallel to the axes, thereby forming the parallelogram. The sides of the parallelogram represent the components, $\mathbf{F}_u$ and $\mathbf{F}_v$.
- Label all the known and unknown force magnitudes and the angles on the sketch and identify the two unknowns as the magnitude and direction of $\mathbf{F}_R$, or the magnitudes of its components.

Trigonometry.

- Redraw a half portion of the parallelogram to illustrate the triangular head-to-tail addition of the components.
- From this triangle, the magnitude of the resultant force can be determined using the law of cosines, and its direction is determined from the law of sines. The magnitudes of two force components are determined from the law of sines. The formulas are given in Fig. 2–10*c*.

Important Points

- A scalar is a positive or negative number.
- A vector is a quantity that has a magnitude, direction, and sense.
- Multiplication or division of a vector by a scalar will change the magnitude of the vector. The sense of the vector will change if the scalar is negative.
- As a special case, if the vectors are collinear, the resultant is formed by an algebraic or scalar addition.

EXAMPLE 2.1

The screw eye in Fig. 2–11*a* is subjected to two forces, $\mathbf{F}_1$ and $\mathbf{F}_2$. Determine the magnitude and direction of the resultant force.

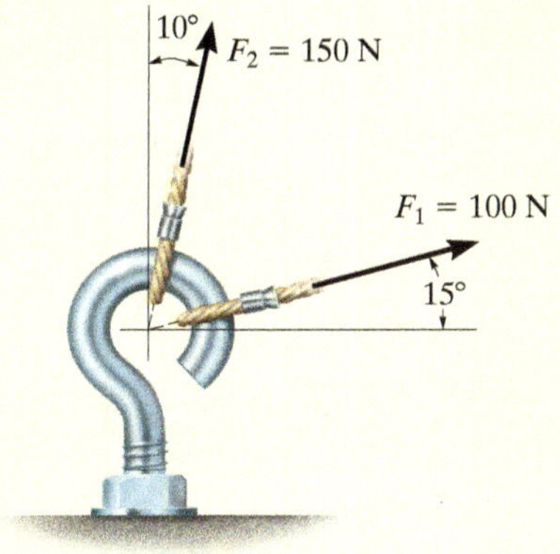

(a)

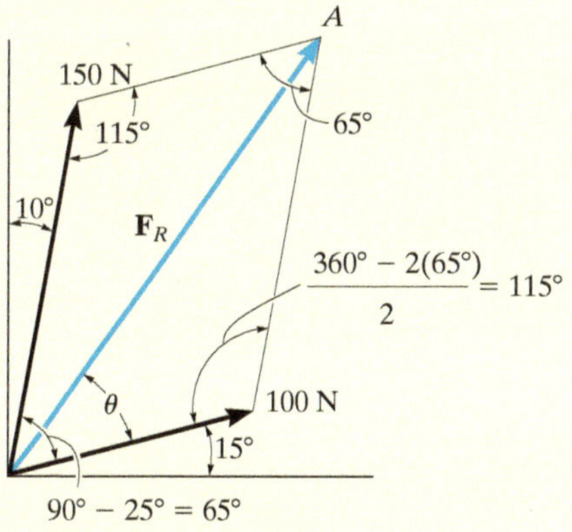

(b)

SOLUTION

Parallelogram Law. The parallelogram is formed by drawing a line from the head of $\mathbf{F}_1$ that is parallel to $\mathbf{F}_2$, and another line from the head of $\mathbf{F}_2$ that is parallel to $\mathbf{F}_1$. The resultant force $\mathbf{F}_R$ extends to where these lines intersect at point A, Fig. 2–11*b*. The two unknowns are the magnitude of $\mathbf{F}_R$ and the angle θ (theta).

Trigonometry. From the parallelogram, the vector triangle is constructed, Fig. 2–11*c*. Using the law of cosines

$$F_R = \sqrt{(100\text{ N})^2 + (150\text{ N})^2 - 2(100\text{ N})(150\text{ N})\cos 115°}$$

$$= \sqrt{10\,000 + 22\,500 - 30\,000(-0.4226)} = 212.6\text{ N}$$

$$= 213\text{ N} \qquad \textit{Ans.}$$

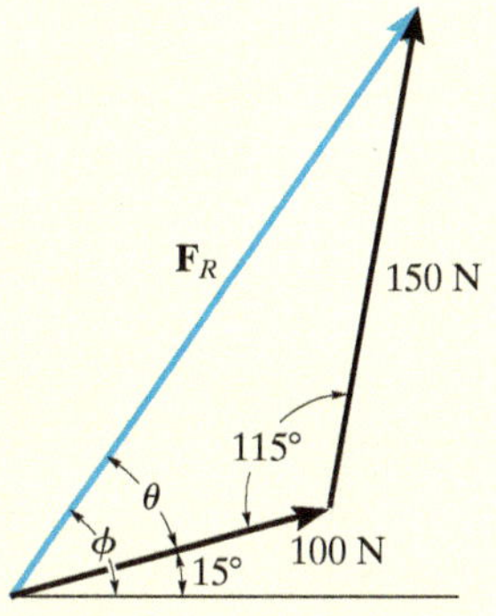

(c)

Fig. 2–11

Applying the law of sines to determine θ,

$$\frac{150\text{ N}}{\sin\theta} = \frac{212.6\text{ N}}{\sin 115°} \qquad \sin\theta = \frac{150\text{ N}}{212.6\text{ N}}(\sin 115°)$$

$$\theta = 39.8°$$

Thus, the direction ϕ (phi) of $\mathbf{F}_R$, measured from the horizontal, is

$$\phi = 39.8° + 15.0° = 54.8° \qquad \textit{Ans.}$$

NOTE: The results seem reasonable, since Fig. 2–11*b* shows $\mathbf{F}_R$ to have a magnitude larger than its components and a direction that is between them.

2

EXAMPLE 2.2

Resolve the horizontal 600-lb force in Fig. 2–12*a* into components acting along the *u* and *v* axes and determine the magnitudes of these components.

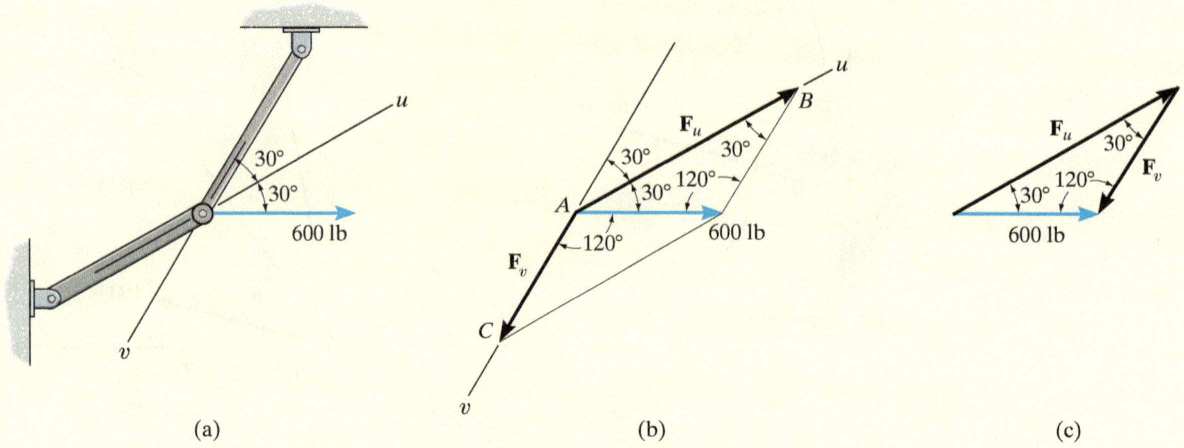

Fig. 2–12

SOLUTION

The parallelogram is constructed by extending a line from the *head* of the 600-lb force parallel to the *v* axis until it intersects the *u* axis at point *B*, Fig. 2–12*b*. The arrow from *A* to *B* represents $\mathbf{F}_u$. Similarly, the line extended from the head of the 600-lb force drawn parallel to the *u* axis intersects the *v* axis at point *C*, which gives $\mathbf{F}_v$.

The vector addition using the triangle rule is shown in Fig. 2–12*c*. The two unknowns are the magnitudes of $\mathbf{F}_u$ and $\mathbf{F}_v$. Applying the law of sines,

$$\frac{F_u}{\sin 120^\circ} = \frac{600 \text{ lb}}{\sin 30^\circ}$$

$$F_u = 1039 \text{ lb} \qquad \textit{Ans.}$$

$$\frac{F_v}{\sin 30^\circ} = \frac{600 \text{ lb}}{\sin 30^\circ}$$

$$F_v = 600 \text{ lb} \qquad \textit{Ans.}$$

NOTE: The result for F_u shows that sometimes a component can have a greater magnitude than the resultant.

EXAMPLE 2.3

Determine the magnitude of the component force **F** in Fig. 2–13*a* and the magnitude of the resultant force $\mathbf{F}_R$ if $\mathbf{F}_R$ is directed along the positive *y* axis.

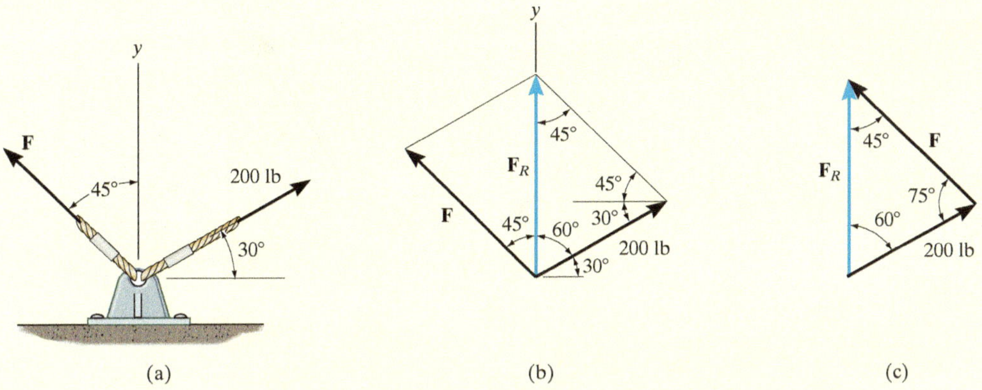

Fig. 2–13

SOLUTION

The parallelogram law of addition is shown in Fig. 2–13*b*, and the triangle rule is shown in Fig. 2–13*c*. The magnitudes of $\mathbf{F}_R$ and **F** are the two unknowns. They can be determined by applying the law of sines.

$$\frac{F}{\sin 60^\circ} = \frac{200 \text{ lb}}{\sin 45^\circ}$$

$$F = 245 \text{ lb} \qquad \textit{Ans.}$$

$$\frac{F_R}{\sin 75^\circ} = \frac{200 \text{ lb}}{\sin 45^\circ}$$

$$F_R = 273 \text{ lb} \qquad \textit{Ans.}$$

2

EXAMPLE 2.4

It is required that the resultant force acting on the eyebolt in Fig. 2–14*a* be directed along the positive *x* axis and that $\mathbf{F}_2$ have a *minimum* magnitude. Determine this magnitude, the angle θ, and the corresponding resultant force.

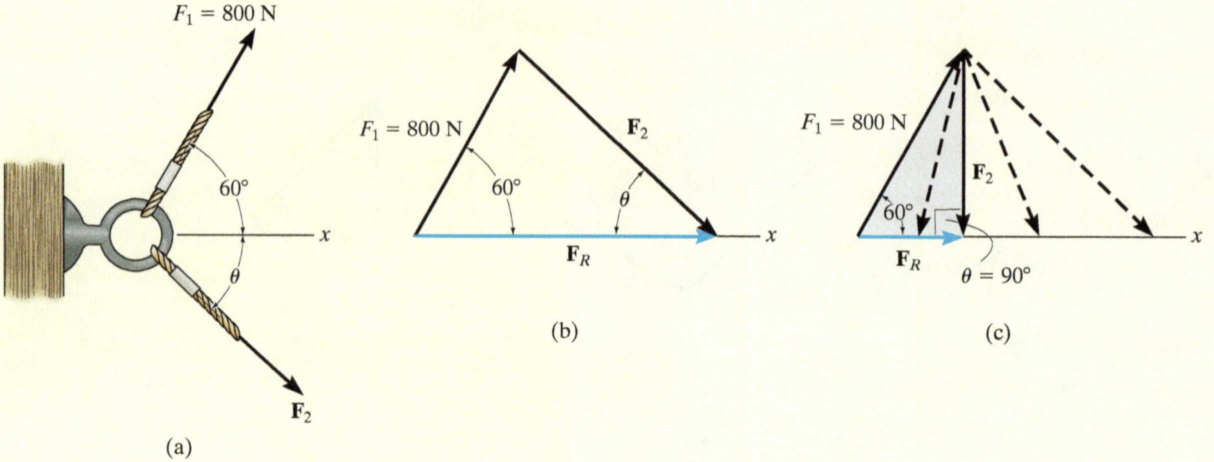

Fig. 2–14

SOLUTION

The triangle rule for $\mathbf{F}_R = \mathbf{F}_1 + \mathbf{F}_2$ is shown in Fig. 2–14*b*. Since the magnitudes (lengths) of $\mathbf{F}_R$ and $\mathbf{F}_2$ are not specified, then $\mathbf{F}_2$ can actually be any vector that has its head touching the line of action of $\mathbf{F}_R$, Fig. 2–14*c*. However, as shown, the magnitude of $\mathbf{F}_2$ is a *minimum* or the shortest length when its line of action is *perpendicular* to the line of action of $\mathbf{F}_R$, that is, when

$$\theta = 90^\circ \qquad \textit{Ans.}$$

Since the vector addition now forms the shaded right triangle, the two unknown magnitudes can be obtained by trigonometry.

$$F_R = (800\text{ N})\cos 60^\circ = 400\text{ N} \qquad \textit{Ans.}$$

$$F_2 = (800\text{ N})\sin 60^\circ = 693\text{ N} \qquad \textit{Ans.}$$

It is strongly suggested that you test yourself on the solutions to these examples, by covering them over and then trying to draw the parallelogram law, and thinking about how the sine and cosine laws are used to determine the unknowns. Then before solving any of the problems, try and solve some of the Fundamental Problems given on the next page. The solutions and answers to these are given in the back of the book. Doing this throughout the book will help immensely in developing your problem-solving skills.

FUNDAMENTAL PROBLEMS*

F2–1. Determine the magnitude of the resultant force acting on the screw eye and its direction measured clockwise from the x axis.

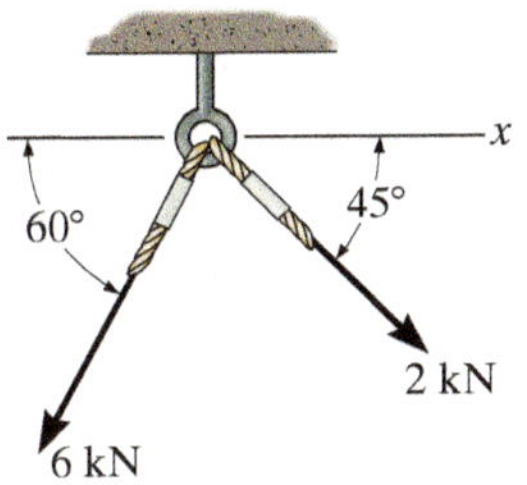

F2–1

F2–2. Two forces act on the hook. Determine the magnitude of the resultant force.

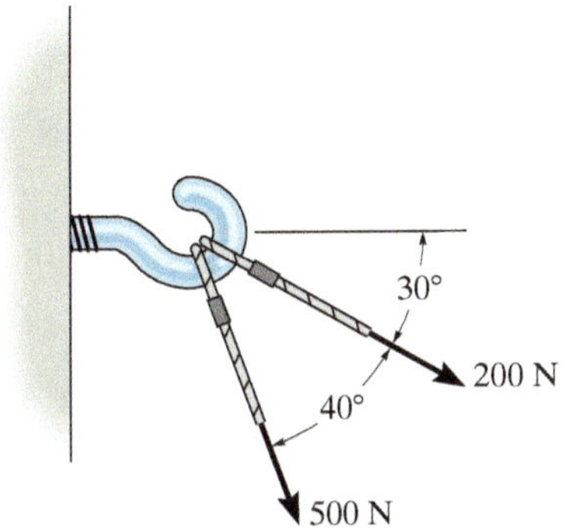

F2–2

F2–3. Determine the magnitude of the resultant force and its direction measured counterclockwise from the positive x axis.

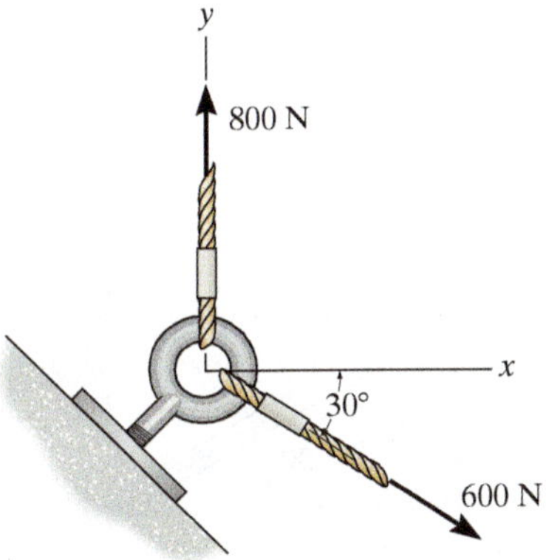

F2–3

F2–4. Resolve the 30-lb force into components along the u and v axes, and determine the magnitude of each of these components.

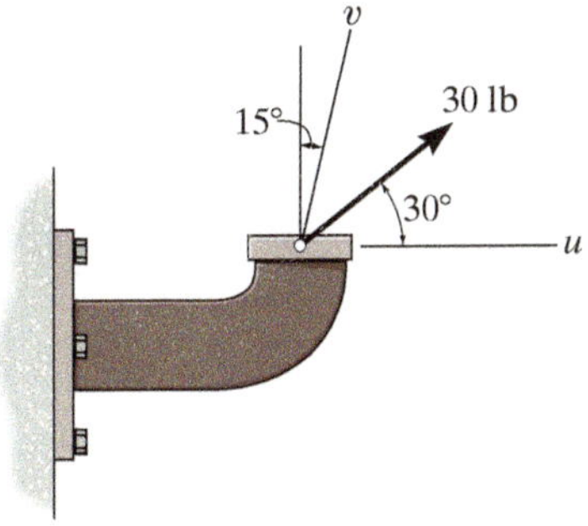

F2–4

F2–5. The force $F = 450$ lb acts on the frame. Resolve this force into components acting along members AB and AC, and determine the magnitude of each component.

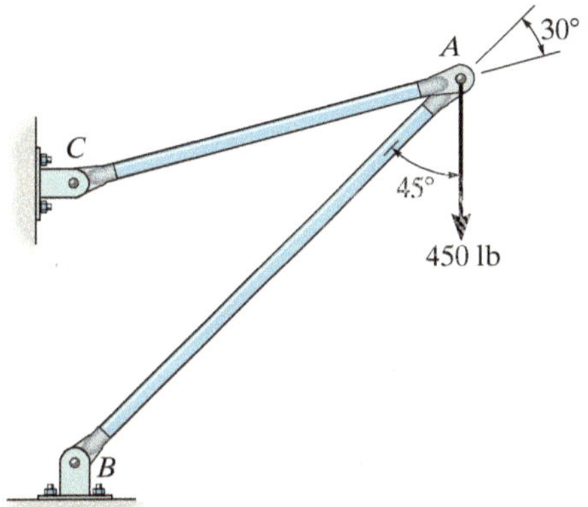

F2–5

F2–6. If force **F** is to have a component along the u axis of $F_u = 6$ kN, determine the magnitude of **F** and the magnitude of its component $\mathbf{F}_v$ along the v axis.

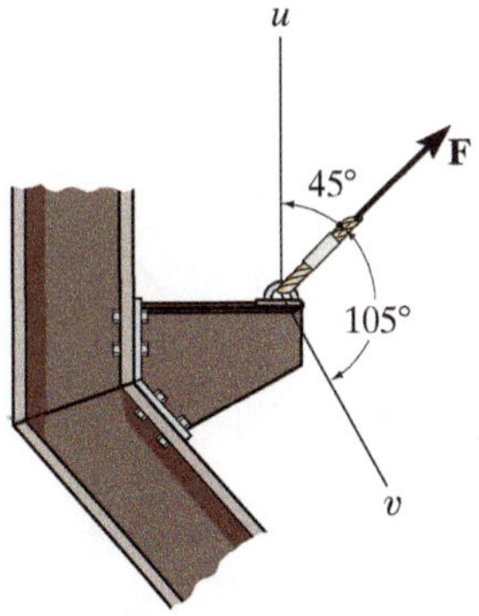

F2–6

* Partial solutions and answers to all Fundamental Problems are given in the back of the book.

PROBLEMS

2

2–1. Determine the magnitude of the resultant force $\mathbf{F}_R = \mathbf{F}_1 + \mathbf{F}_2$ and its direction, measured counterclockwise from the positive x axis.

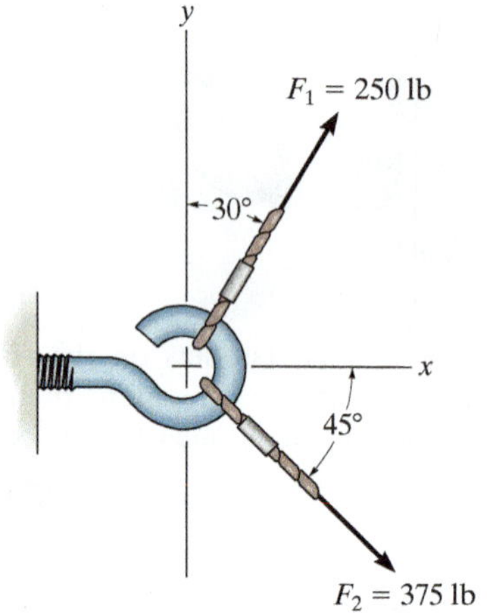

Prob. 2–1

2–2. If $\theta = 60°$ and $F = 450$ N, determine the magnitude of the resultant force and its direction, measured counterclockwise from the positive x axis.

2–3. If the magnitude of the resultant force is to be 500 N, directed along the positive y axis, determine the magnitude of force $\mathbf{F}$ and its direction θ.

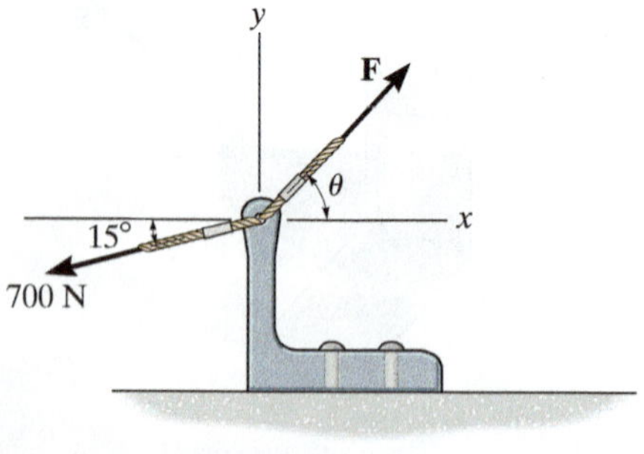

Probs. 2–2/3

***2–4.** Determine the magnitude of the resultant force $\mathbf{F}_R = \mathbf{F}_1 + \mathbf{F}_2$ and its direction, measured clockwise from the positive u axis.

2–5. Resolve the force $\mathbf{F}_1$ into components acting along the u and v axes and determine the magnitudes of the components.

2–6. Resolve the force $\mathbf{F}_2$ into components acting along the u and v axes and determine the magnitudes of the components.

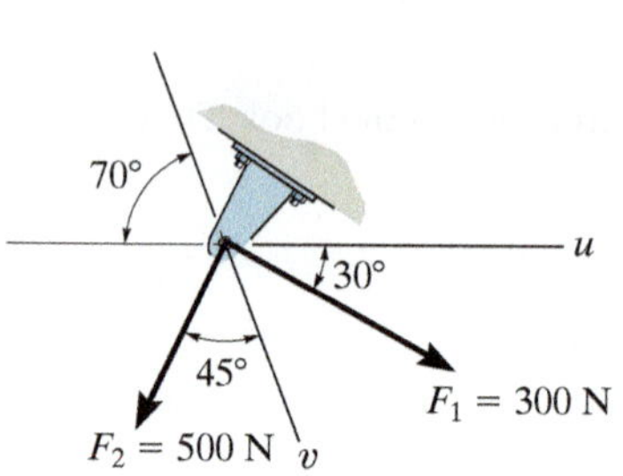

Probs. 2–4/5/6

2–7. The vertical force $\mathbf{F}$ acts downward at A on the two-membered frame. Determine the magnitudes of the two components of $\mathbf{F}$ directed along the axes of AB and AC. Set $F = 500$ N.

***2–8.** Solve Prob. 2–7 with $F = 350$ lb.

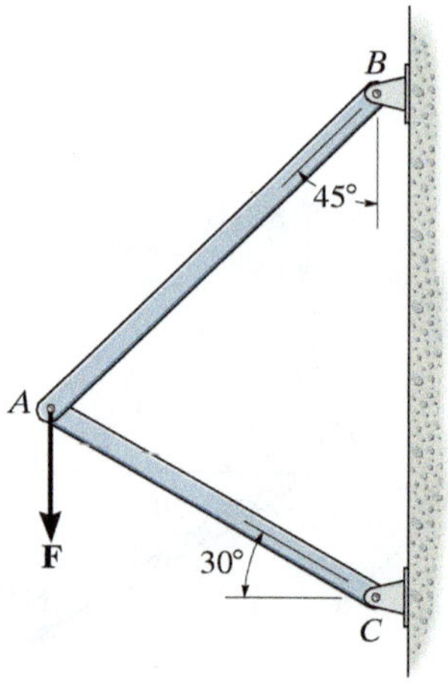

Probs. 2–7/8

2–9. Resolve $\mathbf{F}_1$ into components along the u and v axes and determine the magnitudes of these components.

2–10. Resolve $\mathbf{F}_2$ into components along the u and v axes and determine the magnitudes of these components.

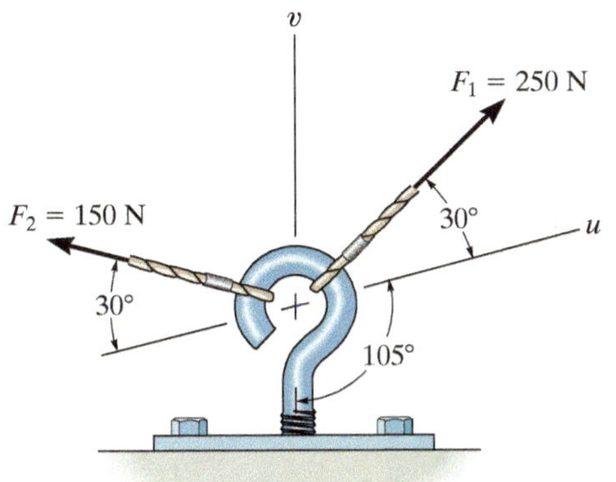

Probs. 2–9/10

2–11. The force acting on the gear tooth is $F = 20$ lb. Resolve this force into two components acting along the lines aa and bb.

***2–12.** The component of force **F** acting along line aa is required to be 30 lb. Determine the magnitude of **F** and its component along line bb.

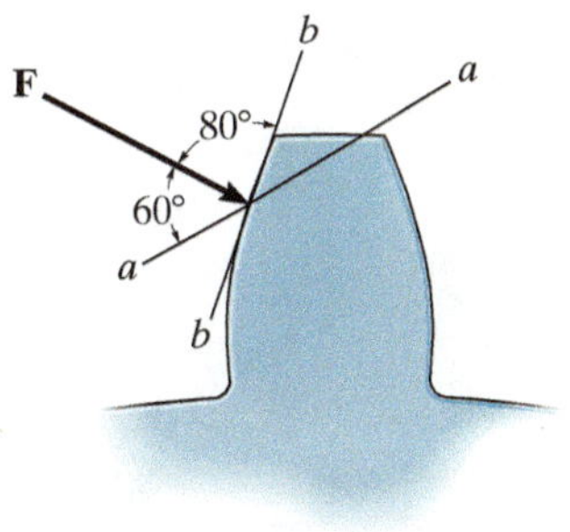

Probs. 2–11/12

2–13. Force **F** acts on the frame such that its component acting along member AB is 650 lb, directed from B towards A, and the component acting along member BC is 500 lb, directed from B towards C. Determine the magnitude of **F** and its direction θ. Set $\phi = 60°$.

2–14. Force **F** acts on the frame such that its component acting along member AB is 650 lb, directed from B towards A. Determine the required angle ϕ $(0° \leq \phi \leq 90°)$ and the component acting along member BC. Set $F = 850$ lb and $\theta = 30°$.

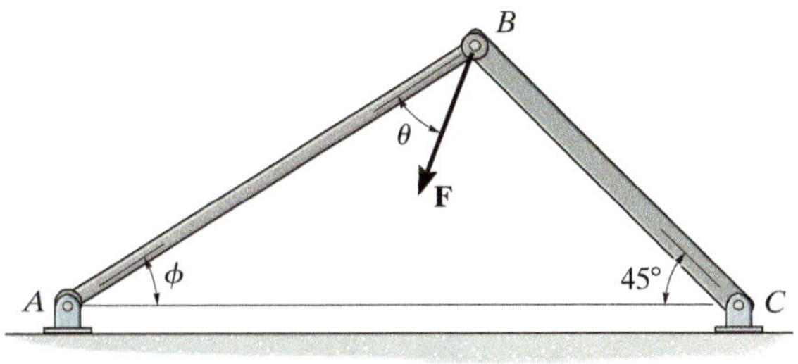

Probs. 2–13/14

2–15. The plate is subjected to the two forces at A and B as shown. If $\theta = 60°$, determine the magnitude of the resultant of these two forces and its direction measured clockwise from the horizontal.

***2–16.** Determine the angle θ for connecting member A to the plate so that the resultant force of $\mathbf{F}_A$ and $\mathbf{F}_B$ is directed horizontally to the right. Also, what is the magnitude of the resultant force?

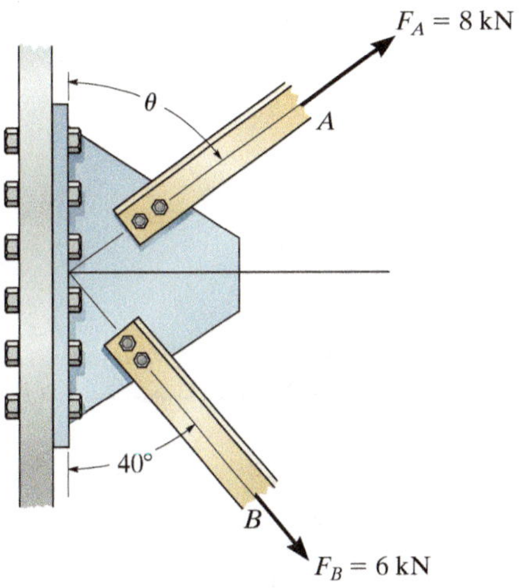

Probs. 2–15/16

2–17. Determine the design angle θ ($0° \le \theta \le 90°$) for strut AB so that the 400-lb horizontal force has a component of 500 lb directed from A towards C. What is the component of force acting along member AB? Take $\phi = 40°$.

2–18. Determine the design angle ϕ ($0° \le \phi \le 90°$) between struts AB and AC so that the 400-lb horizontal force has a component of 600 lb which acts up to the left, in the same direction as from B towards A. Take $\theta = 30°$.

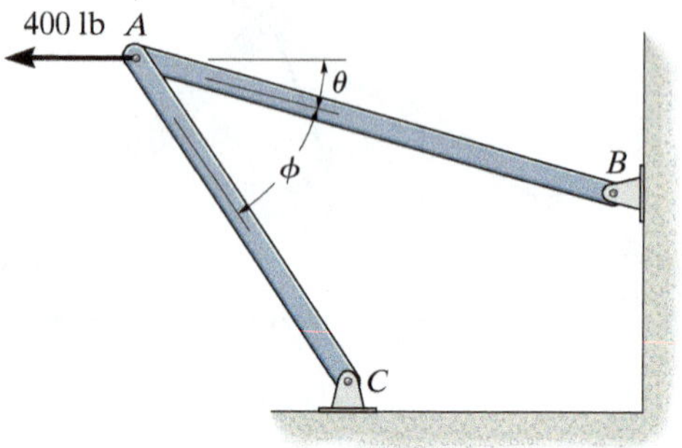

Probs. 2–17/18

2–19. Determine the magnitude and direction of the resultant $\mathbf{F}_R = \mathbf{F}_1 + \mathbf{F}_2 + \mathbf{F}_3$ of the three forces by first finding the resultant $\mathbf{F}' = \mathbf{F}_1 + \mathbf{F}_2$ and then forming $\mathbf{F}_R = \mathbf{F}' + \mathbf{F}_3$.

***2–20.** Determine the magnitude and direction of the resultant $\mathbf{F}_R = \mathbf{F}_1 + \mathbf{F}_2 + \mathbf{F}_3$ of the three forces by first finding the resultant $\mathbf{F}' = \mathbf{F}_2 + \mathbf{F}_3$ and then forming $\mathbf{F}_R = \mathbf{F}' + \mathbf{F}_1$.

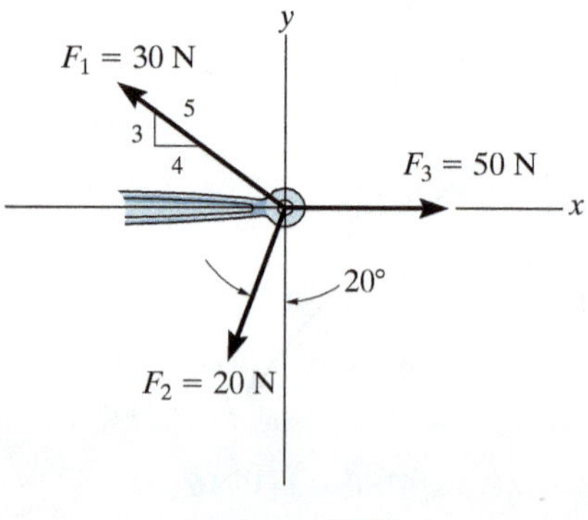

Probs. 2–19/20

2–21. Two forces act on the screw eye. If $F_1 = 400$ N and $F_2 = 600$ N, determine the angle θ ($0° \le \theta \le 180°$) between them, so that the resultant force has a magnitude of $F_R = 800$ N.

2–22. Two forces $\mathbf{F}_1$ and $\mathbf{F}_2$ act on the screw eye. If their lines of action are at an angle θ apart and the magnitude of each force is $F_1 = F_2 = F$, determine the magnitude of the resultant force $\mathbf{F}_R$ and the angle between $\mathbf{F}_R$ and $\mathbf{F}_1$.

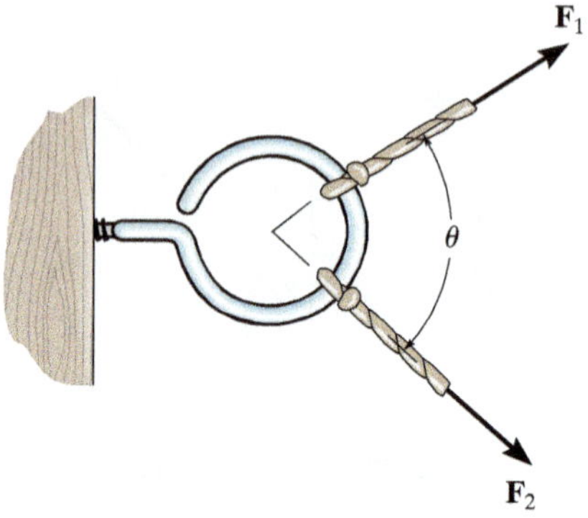

Probs. 2–21/22

2–23. Two forces act on the screw eye. If $F = 600$ N, determine the magnitude of the resultant force and the angle θ if the resultant force is directed vertically upward.

***2–24.** Two forces are applied at the end of a screw eye in order to remove the post. Determine the angle θ ($0° \le \theta \le 90°$) and the magnitude of force $\mathbf{F}$ so that the resultant force acting on the post is directed vertically upward and has a magnitude of 750 N.

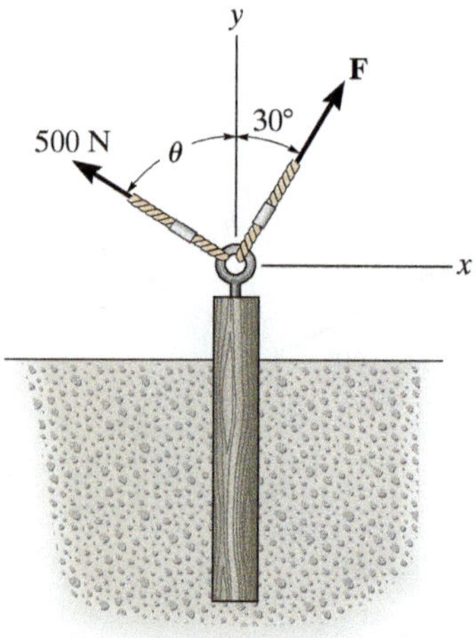

Probs. 2–23/24

2–25. The chisel exerts a force of 20 lb on the wood dowel rod which is turning in a lathe. Resolve this force into components acting (a) along the *n* and *t* axes and (b) along the *x* and *y* axes.

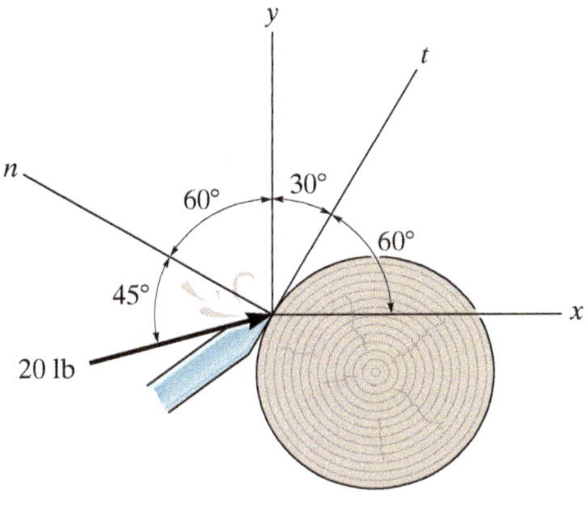

Prob. 2–25

2–26. The beam is to be hoisted using two chains. Determine the magnitudes of forces $\mathbf{F}_A$ and $\mathbf{F}_B$ acting on each chain in order to develop a resultant force of 600 N directed along the positive *y* axis. Set $\theta = 45°$.

2–27. The beam is to be hoisted using two chains. If the resultant force is to be 600 N directed along the positive *y* axis, determine the magnitudes of forces $\mathbf{F}_A$ and $\mathbf{F}_B$ acting on each chain and the angle θ of $\mathbf{F}_B$ so that the magnitude of $\mathbf{F}_B$ is a *minimum*. $\mathbf{F}_A$ acts at 30° from the *y* axis, as shown.

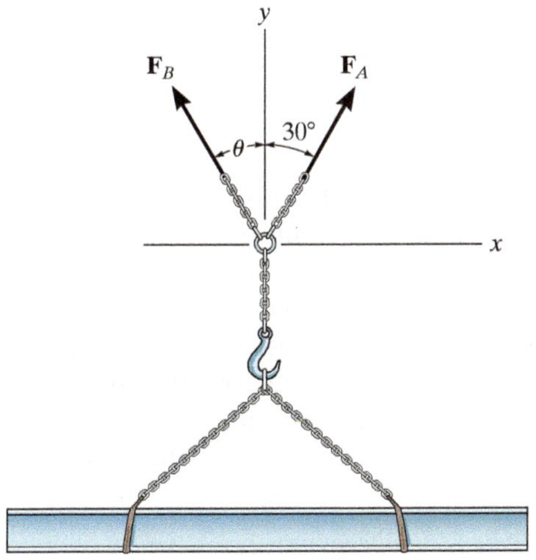

Probs. 2–26/27

***2–28.** If the resultant force of the two tugboats is 3 kN, directed along the positive *x* axis, determine the required magnitude of force $\mathbf{F}_B$ and its direction θ.

2–29. If $F_B = 3$ kN and $\theta = 45°$, determine the magnitude of the resultant force of the two tugboats and its direction measured clockwise from the positive *x* axis.

2–30. If the resultant force of the two tugboats is required to be directed towards the positive *x* axis, and $\mathbf{F}_B$ is to be a minimum, determine the magnitude of $\mathbf{F}_R$ and $\mathbf{F}_B$ and the angle θ.

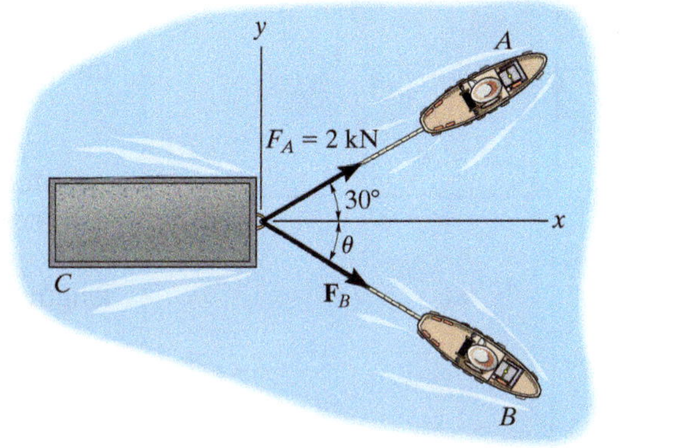

Probs. 2–28/29/30

2–31. Three chains act on the bracket such that they create a resultant force having a magnitude of 500 lb. If two of the chains are subjected to known forces, as shown, determine the angle θ of the third chain measured clockwise from the positive *x* axis, so that the magnitude of force $\mathbf{F}$ in this chain is a *minimum*. All forces lie in the *x*–*y* plane. What is the magnitude of $\mathbf{F}$? *Hint*: First find the resultant of the two known forces. Force $\mathbf{F}$ acts in this direction.

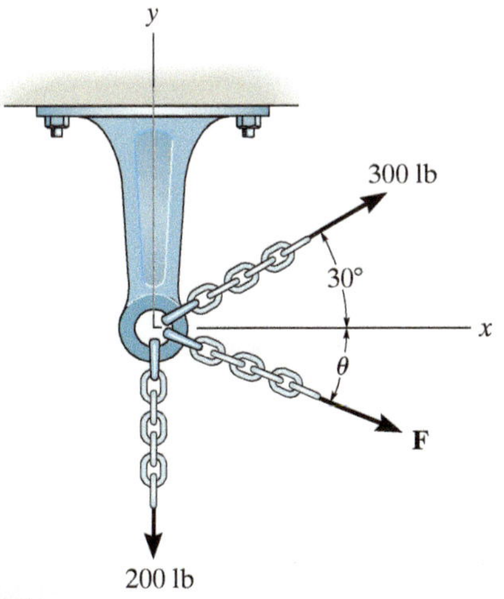

Prob. 2–31

2.4 Addition of a System of Coplanar Forces

When a force is resolved into two components along the x and y axes, the components are then called *rectangular components.* For analytical work we can represent these components in one of two ways, using either scalar notation or Cartesian vector notation.

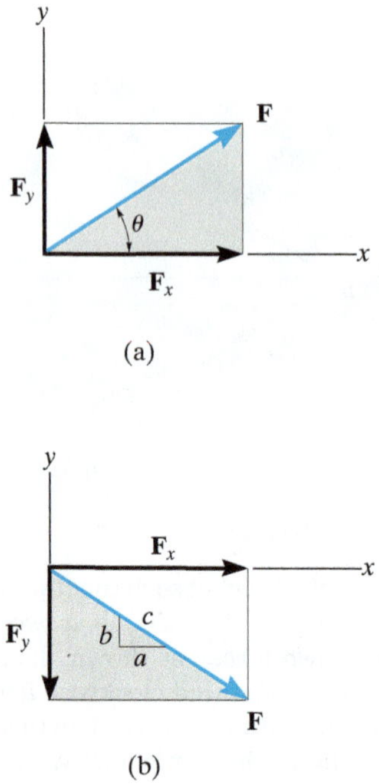

Fig. 2–15

Scalar Notation. The rectangular components of force $\mathbf{F}$ shown in Fig. 2–15*a* are found using the parallelogram law, so that $\mathbf{F} = \mathbf{F}_x + \mathbf{F}_y$. Because these components form a right triangle, they can be determined from

$$F_x = F\cos\theta \qquad \text{and} \qquad F_y = F\sin\theta$$

Instead of using the angle θ, however, the direction of $\mathbf{F}$ can also be defined using a small "slope" triangle, as in the example shown in Fig. 2–15*b*. Since this triangle and the larger shaded triangle are similar, the proportional length of the sides gives

$$\frac{F_x}{F} = \frac{a}{c}$$

or

$$F_x = F\left(\frac{a}{c}\right)$$

and

$$\frac{F_y}{F} = \frac{b}{c}$$

or

$$F_y = -F\left(\frac{b}{c}\right)$$

Here the y component is a *negative scalar* since $\mathbf{F}_y$ is directed along the negative y axis.

It is important to keep in mind that this positive and negative scalar notation is to be used only for computational purposes, not for graphical representations in figures. Throughout the book, the *head of a vector arrow* in *any figure* indicates the sense of the vector *graphically*; algebraic signs are not used for this purpose. Thus, the vectors in Figs. 2–15*a* and 2–15*b* are designated by using boldface (vector) notation.* Whenever italic symbols are written near vector arrows in figures, they indicate the *magnitude* of the vector, which is *always* a *positive* quantity.

*Negative signs are used only in figures with boldface notation when showing equal but opposite pairs of vectors, as in Fig. 2–2.

Cartesian Vector Notation. It is also possible to represent the x and y components of a force in terms of Cartesian unit vectors **i** and **j**. They are called unit vectors because they have a dimensionless magnitude of 1, and so they can be used to designate the *directions* of the x and y axes, respectively, Fig. 2–16.*

Since the *magnitude* of each component of **F** is *always a positive quantity*, which is represented by the (positive) scalars F_x and F_y, then we can express **F** as a *Cartesian vector*,

$$\mathbf{F} = F_x\mathbf{i} + F_y\mathbf{j}$$

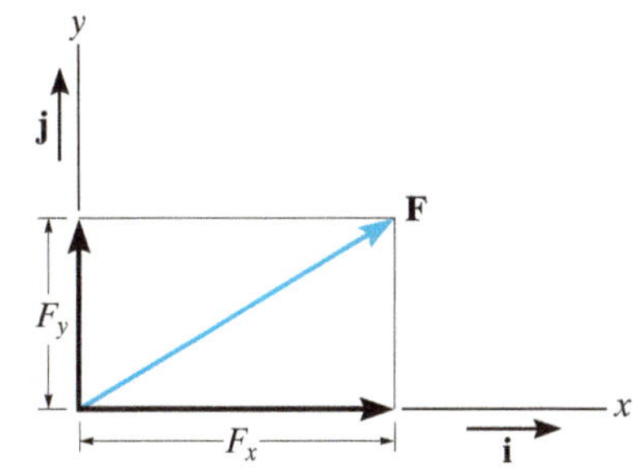

Fig. 2–16

Coplanar Force Resultants. We can use either of the two methods just described to determine the resultant of several *coplanar forces*. To do this, each force is first resolved into its x and y components, and then the respective components are added using *scalar algebra* since they are collinear. The resultant force is then formed by adding the resultant components using the parallelogram law. For example, consider the three concurrent forces in Fig. 2–17*a*, which have x and y components shown in Fig. 2–17*b*. Using *Cartesian vector notation*, each force is first represented as a Cartesian vector, i.e.,

$$\mathbf{F}_1 = F_{1x}\mathbf{i} + F_{1y}\mathbf{j}$$
$$\mathbf{F}_2 = -F_{2x}\mathbf{i} + F_{2y}\mathbf{j}$$
$$\mathbf{F}_3 = F_{3x}\mathbf{i} - F_{3y}\mathbf{j}$$

The vector resultant is therefore

$$\begin{aligned}\mathbf{F}_R &= \mathbf{F}_1 + \mathbf{F}_2 + \mathbf{F}_3\\ &= F_{1x}\mathbf{i} + F_{1y}\mathbf{j} - F_{2x}\mathbf{i} + F_{2y}\mathbf{j} + F_{3x}\mathbf{i} - F_{3y}\mathbf{j}\\ &= (F_{1x} - F_{2x} + F_{3x})\mathbf{i} + (F_{1y} + F_{2y} - F_{3y})\mathbf{j}\\ &= (F_{Rx})\mathbf{i} + (F_{Ry})\mathbf{j}\end{aligned}$$

If *scalar notation* is used, then from Fig. 2–17*b*, we have

$$(\overset{+}{\rightarrow}) \qquad (F_R)_x = F_{1x} - F_{2x} + F_{3x}$$
$$(+\uparrow) \qquad (F_R)_y = F_{1y} + F_{2y} - F_{3y}$$

These are the *same* results as the **i** and **j** components of $\mathbf{F}_R$ determined above.

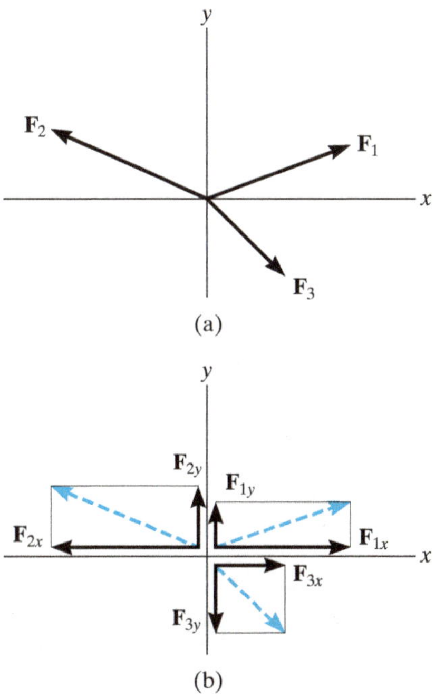

Fig. 2–17

* For handwritten work, unit vectors are usually indicated using a circumflex, e.g., $\hat{i}$ and $\hat{j}$. Also, realize that F_x and F_y in Fig. 2–16 represent the *magnitudes* of the components, which are *always positive scalars*. The directions are defined by **i** and **j**. If instead we used scalar notation, then F_x and F_y could be positive or negative scalars, since they would account for *both* the magnitude and direction of the components.

2

We can represent the components of the resultant force of any number of coplanar forces symbolically by the algebraic sum of the x and y components of all the forces, i.e.,

$$(F_R)_x = \Sigma F_x$$
$$(F_R)_y = \Sigma F_y \qquad (2\text{–}1)$$

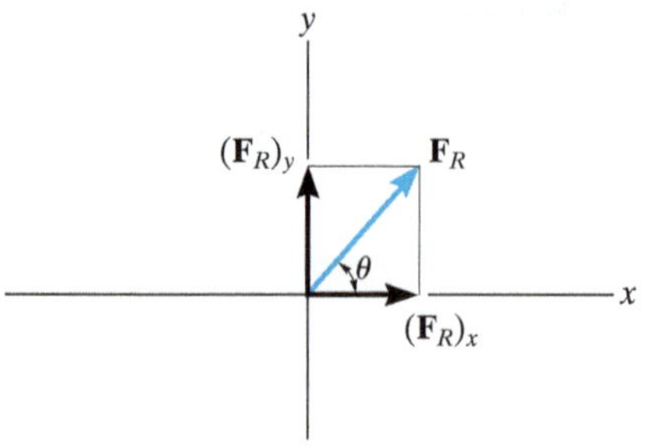

(c)

Fig. 2–17 (cont.)

Once these components are determined, they may be sketched along the x and y axes with their proper sense of direction, and the resultant force can be determined from vector addition, as shown in Fig. 2–17c. From this sketch, the magnitude of $\mathbf{F}_R$ is then found from the Pythagorean theorem; that is,

$$F_R = \sqrt{(F_R)_x^2 + (F_R)_y^2}$$

Also, the angle θ, which specifies the direction of the resultant force, is determined from trigonometry:

$$\theta = \tan^{-1}\left|\frac{(F_R)_y}{(F_R)_x}\right|$$

The above concepts are illustrated numerically in the examples which follow.

The resultant force of the three cable forces acting on the post can be determined by adding algebraically the separate x and y components of each cable force. This resultant $\mathbf{F}_R$ produces the *same pulling effect* on the post as all three cables.

Important Points

- The resultant of several coplanar forces can easily be determined if an x, y coordinate system is established and the forces are resolved along the axes.
- The direction of each force is specified by the angle its line of action makes with one of the axes, or by a slope triangle.
- The orientation of the x and y axes is arbitrary, and their positive direction can be specified by the Cartesian unit vectors **i** and **j**.
- The x and y components of the *resultant force* are simply the algebraic addition of the components of all the coplanar forces.
- The magnitude of the resultant force is determined from the Pythagorean theorem, and when the resultant components are sketched on the x and y axes, Fig. 2–17c, the direction θ can be determined from trigonometry.

EXAMPLE 2.5

Determine the x and y components of $\mathbf{F}_1$ and $\mathbf{F}_2$ acting on the boom shown in Fig. 2–18a. Express each force as a Cartesian vector.

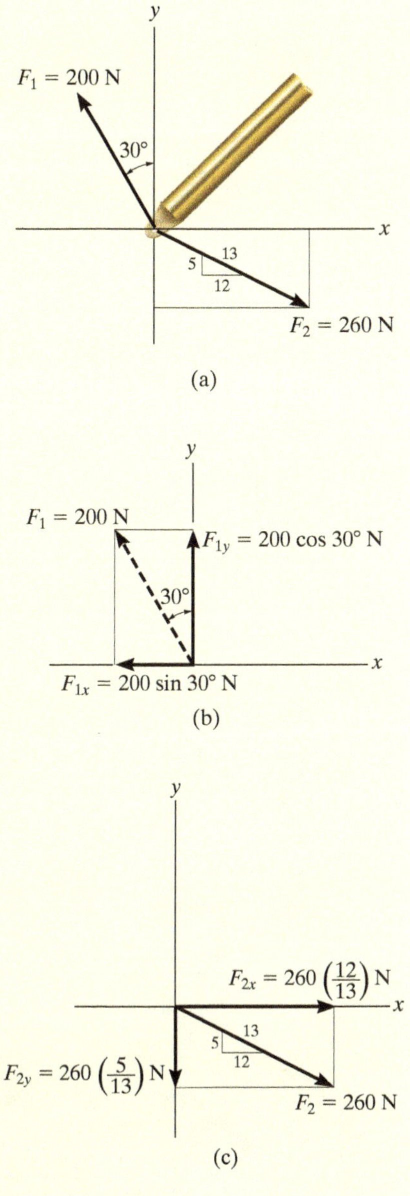

Fig. 2–18

SOLUTION

Scalar Notation. By the parallelogram law, $\mathbf{F}_1$ is resolved into x and y components, Fig. 2–18b. Since $\mathbf{F}_{1x}$ acts in the $-x$ direction, and $\mathbf{F}_{1y}$ acts in the $+y$ direction, we have

$$F_{1x} = -200 \sin 30° \text{ N} = -100 \text{ N} = 100 \text{ N} \leftarrow \qquad \textit{Ans.}$$

$$F_{1y} = 200 \cos 30° \text{ N} = 173 \text{ N} = 173 \text{ N} \uparrow \qquad \textit{Ans.}$$

The force $\mathbf{F}_2$ is resolved into its x and y components, as shown in Fig. 2–18c. Here the *slope* of the line of action for the force is indicated. From this "slope triangle" we could obtain the angle θ, e.g., $\theta = \tan^{-1}(\frac{5}{12})$, and then proceed to determine the magnitudes of the components in the same manner as for $\mathbf{F}_1$. The easier method, however, consists of using proportional parts of similar triangles, i.e.,

$$\frac{F_{2x}}{260 \text{ N}} = \frac{12}{13} \qquad F_{2x} = 260 \text{ N}\left(\frac{12}{13}\right) = 240 \text{ N}$$

Similarly,

$$F_{2y} = 260 \text{ N}\left(\frac{5}{13}\right) = 100 \text{ N}$$

Notice how the magnitude of the *horizontal component*, $\mathbf{F}_{2x}$, was obtained by multiplying the force magnitude by the ratio of the *horizontal leg* of the slope triangle divided by the hypotenuse; whereas the magnitude of the *vertical component*, F_{2y}, was obtained by multiplying the force magnitude by the ratio of the *vertical leg* divided by the hypotenuse. Hence, using scalar notation to represent these components, we have

$$F_{2x} = 240 \text{ N} = 240 \text{ N} \rightarrow \qquad \textit{Ans.}$$

$$F_{2y} = -100 \text{ N} = 100 \text{ N} \downarrow \qquad \textit{Ans.}$$

Cartesian Vector Notation. Having determined the magnitudes and directions of the components of each force, we can express each force as a Cartesian vector.

$$\mathbf{F}_1 = \{-100\mathbf{i} + 173\mathbf{j}\} \text{N} \qquad \textit{Ans.}$$

$$\mathbf{F}_2 = \{240\mathbf{i} - 100\mathbf{j}\} \text{N} \qquad \textit{Ans.}$$

EXAMPLE 2.6

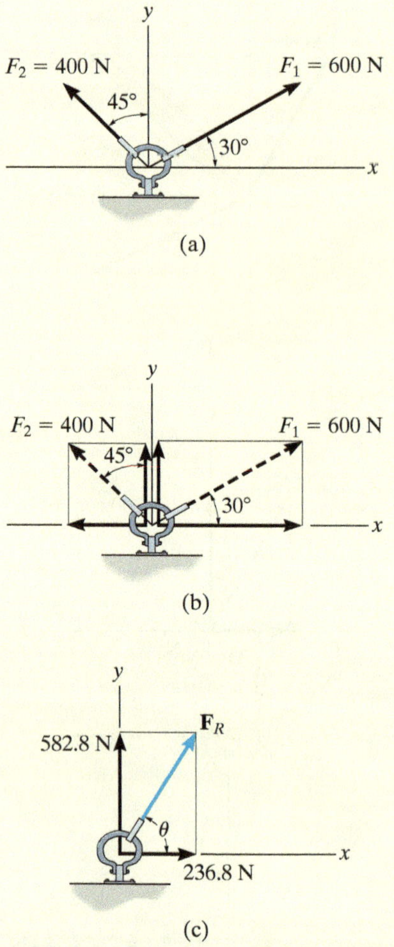

Fig. 2–19

The link in Fig. 2–19*a* is subjected to two forces $\mathbf{F}_1$ and $\mathbf{F}_2$. Determine the magnitude and direction of the resultant force.

SOLUTION I

Scalar Notation. First we resolve each force into its x and y components, Fig. 2–19*b*, then we sum these components algebraically.

$$\overset{+}{\rightarrow} (F_R)_x = \Sigma F_x; \qquad (F_R)_x = 600 \cos 30° \text{ N} - 400 \sin 45° \text{ N} = 236.8 \text{ N} \rightarrow$$

$$+\uparrow (F_R)_y = \Sigma F_y; \qquad (F_R)_y = 600 \sin 30° \text{ N} + 400 \cos 45° \text{ N} = 582.8 \text{ N} \uparrow$$

The resultant force, shown in Fig. 2–18*c*, has a *magnitude* of

$$F_R = \sqrt{(236.8 \text{ N})^2 + (582.8 \text{ N})^2} = 629 \text{ N} \qquad \textit{Ans.}$$

From the vector addition,

$$\theta = \tan^{-1}\left(\frac{582.8 \text{ N}}{236.8 \text{ N}}\right) = 67.9° \qquad \textit{Ans.}$$

SOLUTION II

Cartesian Vector Notation. From Fig. 2–19*b*, each force is first expressed as a Cartesian vector.

$$\mathbf{F}_1 = \{600 \cos 30°\mathbf{i} + 600 \sin 30°\mathbf{j}\} \text{ N}$$
$$\mathbf{F}_2 = \{-400 \sin 45°\mathbf{i} + 400 \cos 45°\mathbf{j}\} \text{ N}$$

Then,

$$\mathbf{F}_R = \mathbf{F}_1 + \mathbf{F}_2 = (600 \cos 30° \text{ N} - 400 \sin 45° \text{ N})\mathbf{i} + (600 \sin 30° \text{ N} + 400 \cos 45° \text{ N})\mathbf{j} = \{236.8\mathbf{i} + 582.8\mathbf{j}\} \text{ N}$$

The magnitude and direction of $\mathbf{F}_R$ are determined in the same manner as before.

NOTE: Comparing the two methods of solution, notice that the use of scalar notation is more efficient since the components can be found *directly*, without first having to express each force as a Cartesian vector before adding the components. Later, however, we will show that Cartesian vector analysis is very beneficial for solving three-dimensional problems.

EXAMPLE 2.7

The end of the boom O in Fig. 2–20a is subjected to three concurrent and coplanar forces. Determine the magnitude and direction of the resultant force.

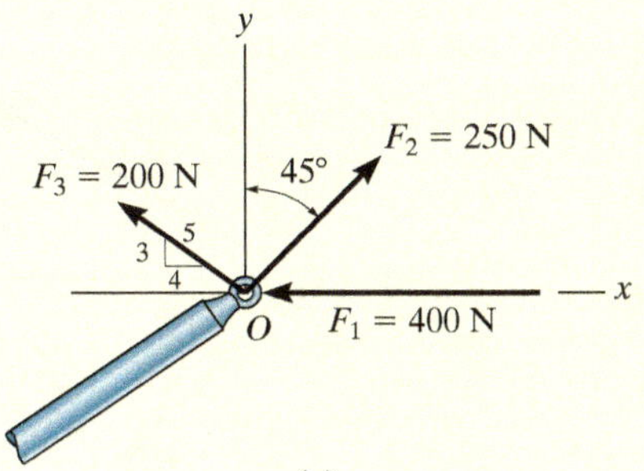

(a)

SOLUTION

Each force is resolved into its x and y components, Fig. 2–20b. Summing the x components, we have

$$\overset{+}{\rightarrow} (F_R)_x = \Sigma F_x; \qquad (F_R)_x = -400 \text{ N} + 250 \sin 45° \text{ N} - 200\left(\tfrac{4}{5}\right) \text{ N}$$

$$= -383.2 \text{ N} = 383.2 \text{ N} \leftarrow$$

The negative sign indicates that F_{Rx} acts to the left, i.e., in the negative x direction, as noted by the small arrow. Obviously, this occurs because F_1 and F_3 in Fig. 2–20b contribute a greater pull to the left than F_2 which pulls to the right. Summing the y components yields

$$+\uparrow (F_R)_y = \Sigma F_y; \qquad (F_R)_y = 250 \cos 45° \text{ N} + 200\left(\tfrac{3}{5}\right) \text{ N}$$

$$= 296.8 \text{ N} \uparrow$$

The resultant force, shown in Fig. 2–20c, has a *magnitude* of

$$F_R = \sqrt{(-383.2 \text{ N})^2 + (296.8 \text{ N})^2}$$

$$= 485 \text{ N} \qquad \textit{Ans.}$$

From the vector addition in Fig. 2–20c, the direction angle θ is

$$\theta = \tan^{-1}\left(\frac{296.8}{383.2}\right) = 37.8° \qquad \textit{Ans.}$$

NOTE: Application of this method is more convenient, compared to using two applications of the parallelogram law, first to add $\mathbf{F}_1$ and $\mathbf{F}_2$ then adding $\mathbf{F}_3$ to this resultant.

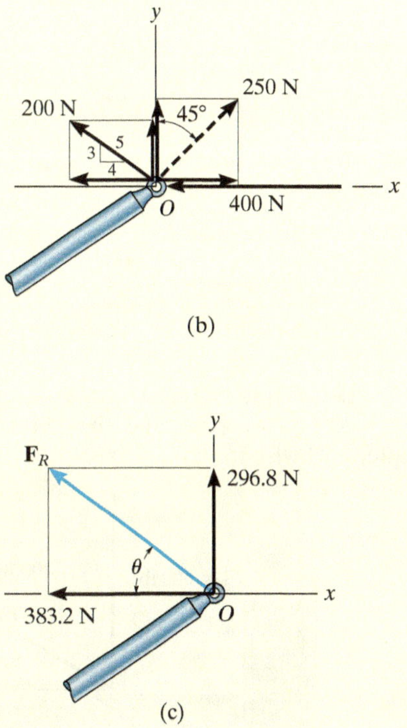

Fig. 2–20

2

FUNDAMENTAL PROBLEMS

F2–7. Resolve each force acting on the post into its x and y components.

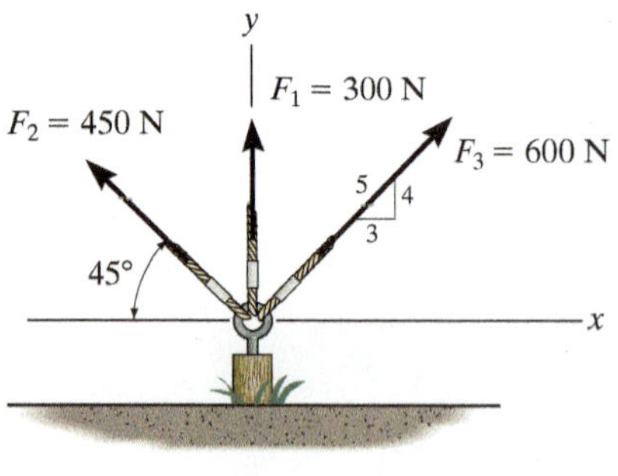

F2–7

F2–8. Determine the magnitude and direction of the resultant force.

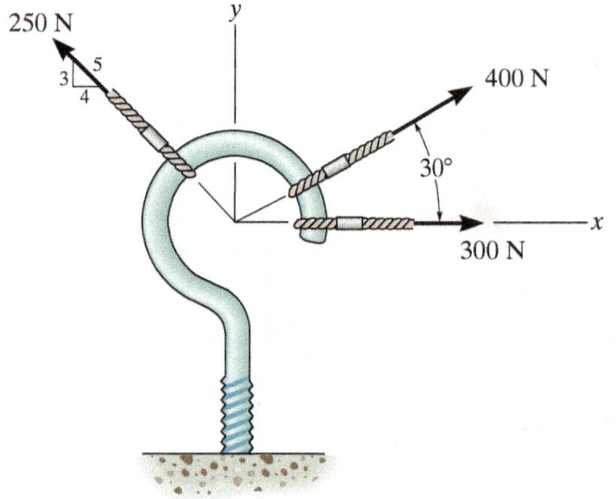

F2–8

F2–9. Determine the magnitude of the resultant force acting on the corbel and its direction θ measured counterclockwise from the x axis.

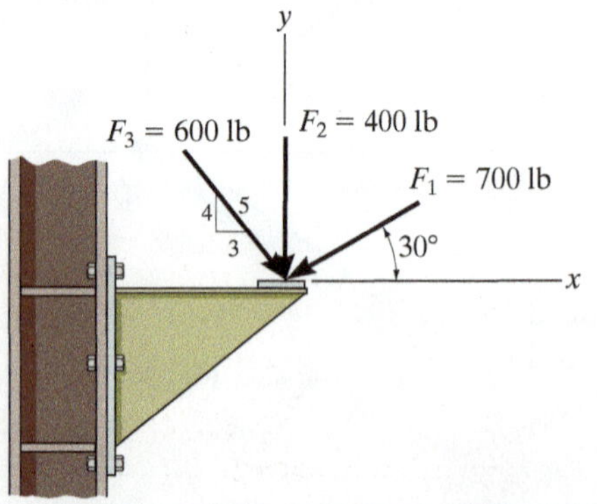

F2–9

F2–10. If the resultant force acting on the bracket is to be 750 N directed along the positive x axis, determine the magnitude of **F** and its direction θ.

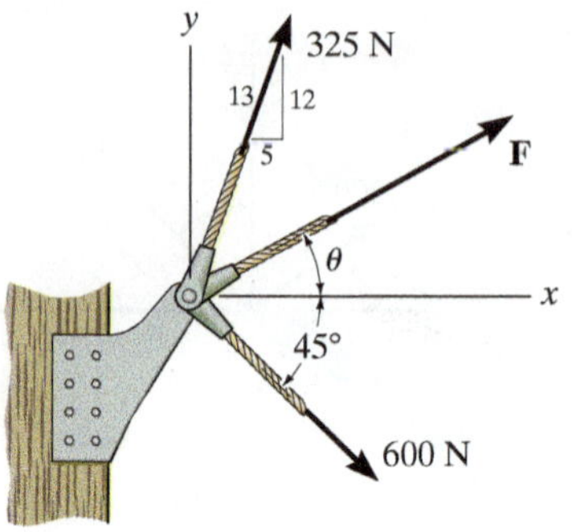

F2–10

F2–11. If the magnitude of the resultant force acting on the bracket is to be 80 lb directed along the u axis, determine the magnitude of **F** and its direction θ.

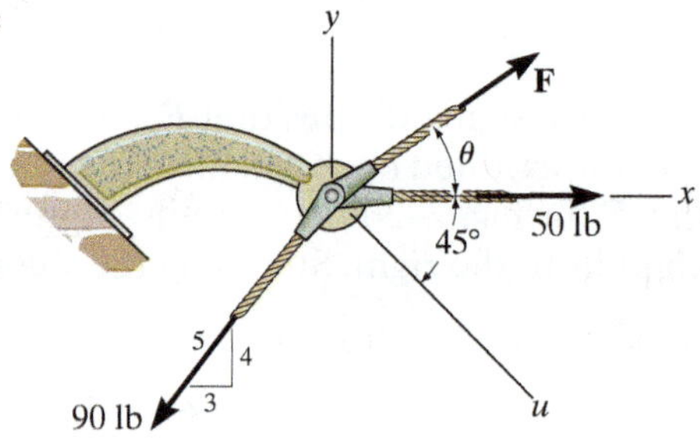

F2–11

F2–12. Determine the magnitude of the resultant force and its direction θ measured counterclockwise from the positive x axis.

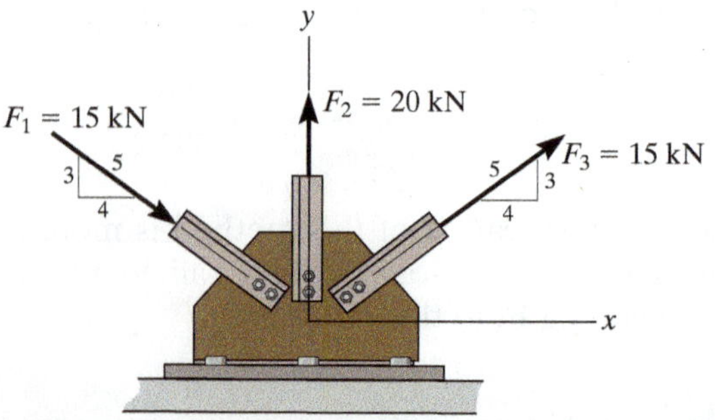

F2–12

PROBLEMS

*2–32. Determine the x and y components of the 800-lb force.

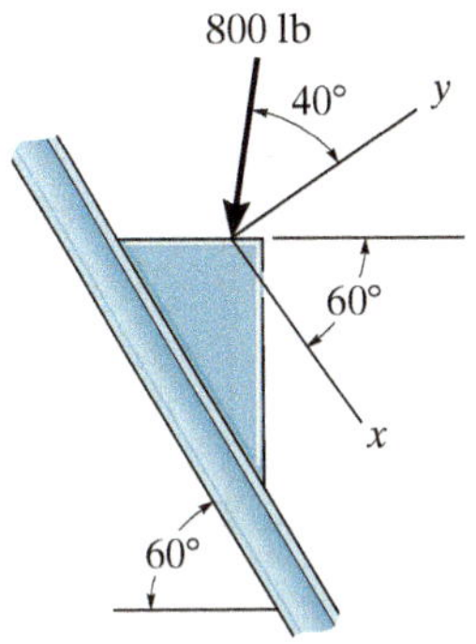

Prob. 2–32

2–33. Determine the magnitude of the resultant force and its direction, measured counterclockwise from the positive x axis.

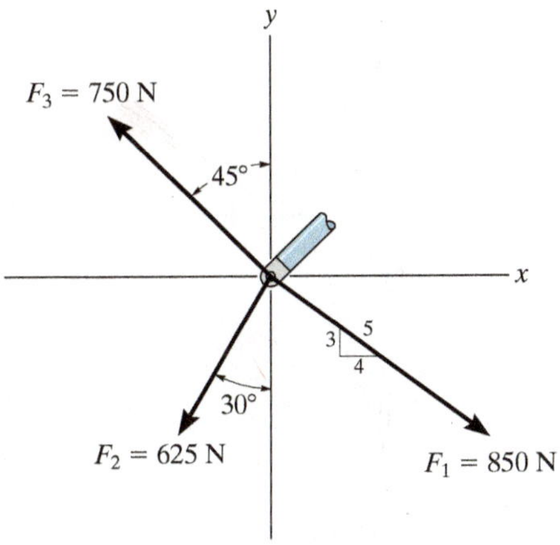

Prob. 2–33

2–34. Resolve $\mathbf{F}_1$ and $\mathbf{F}_2$ into their x and y components.

2–35. Determine the magnitude of the resultant force and its direction measured counterclockwise from the positive x axis.

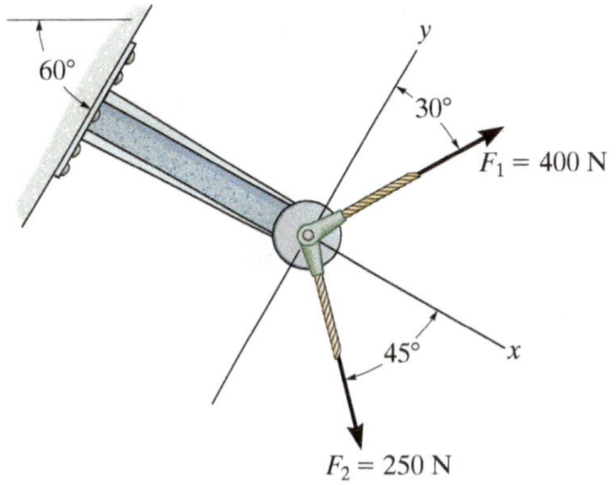

Probs. 2–34/35

*2–36. Resolve each force acting on the gusset plate into its x and y components, and express each force as a Cartesian vector.

2–37. Determine the magnitude of the resultant force acting on the plate and its direction, measured counterclockwise from the positive x axis.

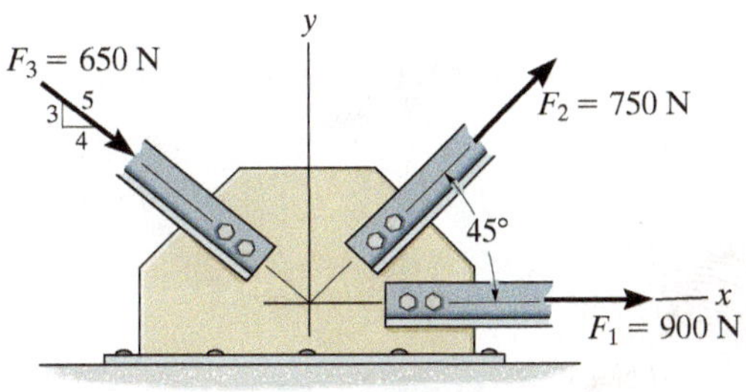

Probs. 2–36/37

2–38. Express each of the three forces acting on the column in Cartesian vector form and compute the magnitude of the resultant force.

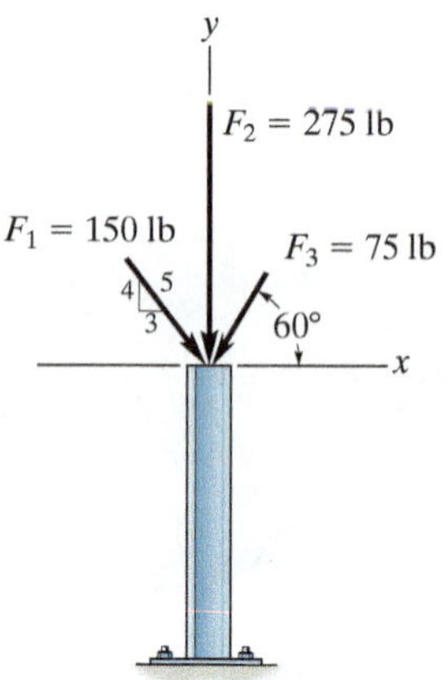

Prob. 2–38

2–41. Determine the magnitude of the resultant force and its direction measured counterclockwise from the positive x axis.

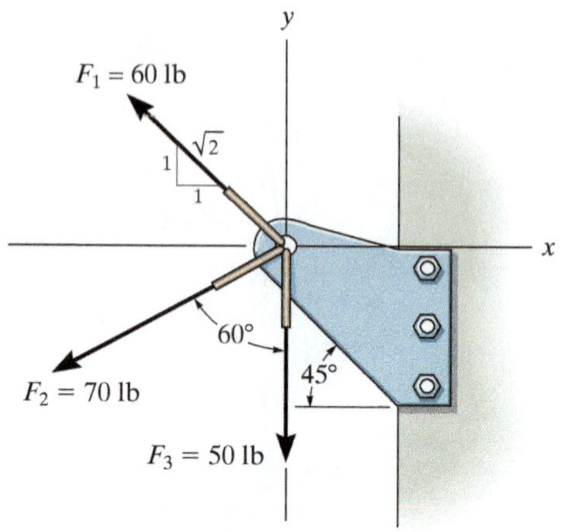

Prob. 2–41

2–39. Resolve each force acting on the support into its x and y components, and express each force as a Cartesian vector.

***2–40.** Determine the magnitude of the resultant force and its direction θ, measured counterclockwise from the positive x axis.

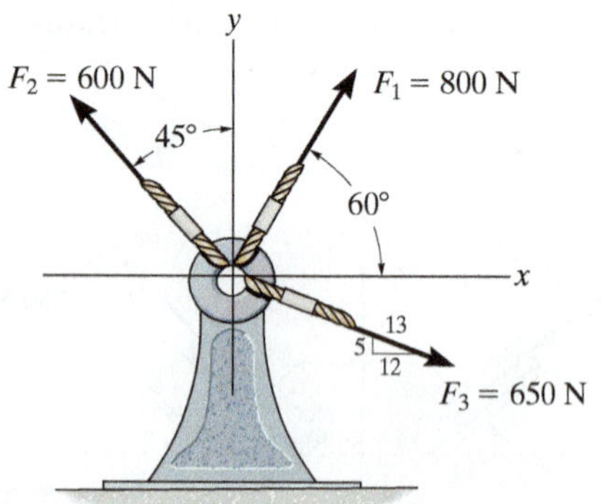

Probs. 2–39/40

2–42. Determine the magnitude and orientation θ of $\mathbf{F}_B$ so that the resultant force is directed along the positive y axis and has a magnitude of 1500 N.

2–43. Determine the magnitude and orientation, measured counterclockwise from the positive y axis, of the resultant force acting on the bracket, if $F_B = 600$ N and $\theta = 20°$.

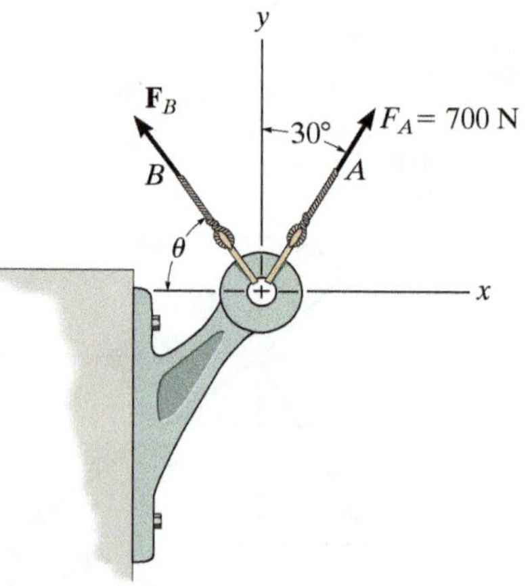

Probs. 2–42/43

*2–44. The magnitude of the resultant force acting on the bracket is to be 400 N. Determine the magnitude of $\mathbf{F}_1$ if $\phi = 30^\circ$.

2–45. If the resultant force acting on the bracket is to be directed along the positive u axis, and the magnitude of $\mathbf{F}_1$ is required to be *minimum*, determine the magnitudes of the resultant force and $\mathbf{F}_1$.

2–46. If the magnitude of the resultant force acting on the bracket is 600 N, directed along the positive u axis, determine the magnitude of $\mathbf{F}$ and its direction ϕ.

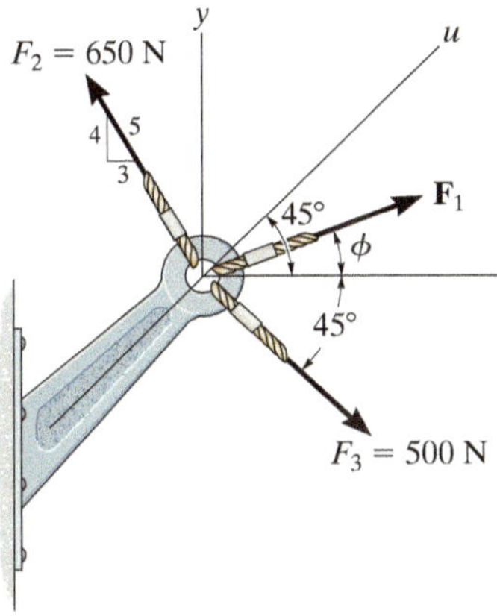

Probs. 2–44/45/46

2–47. Determine the magnitude and direction θ of the resultant force $\mathbf{F}_R$. Express the result in terms of the magnitudes of the components $\mathbf{F}_1$ and $\mathbf{F}_2$ and the angle ϕ.

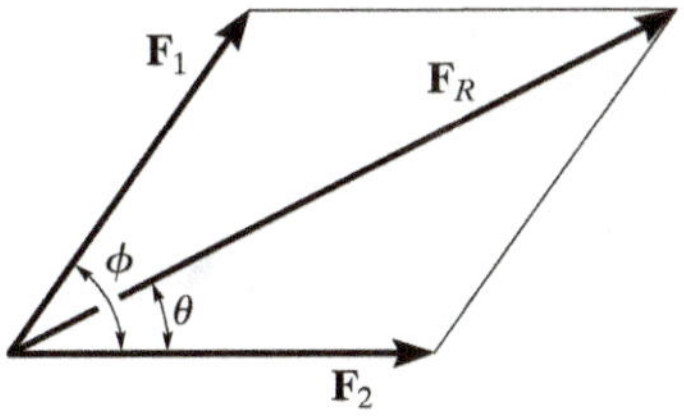

Prob. 2–47

*2–48. If $F_1 = 600$ N and $\phi = 30^\circ$, determine the magnitude of the resultant force acting on the eyebolt and its direction, measured clockwise from the positive x axis.

2–49. If the magnitude of the resultant force acting on the eyebolt is 600 N and its direction measured clockwise from the positive x axis is $\theta = 30^\circ$, determine the magnitude of $\mathbf{F}_1$ and the angle ϕ.

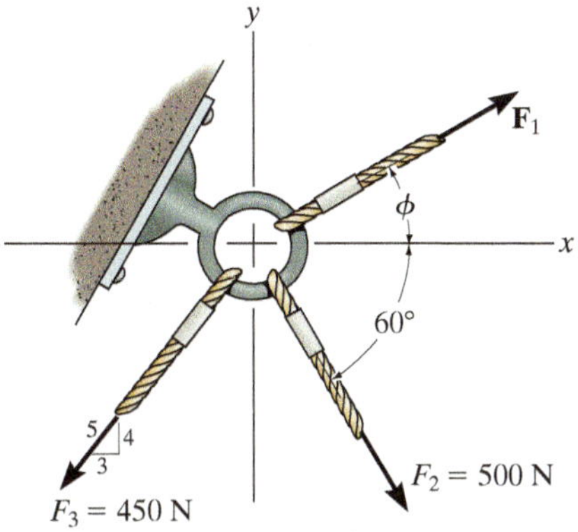

Probs. 2–48/49

2–50. Determine the magnitude of $\mathbf{F}_1$ and its direction θ so that the resultant force is directed vertically upward and has a magnitude of 800 N.

2–51. Determine the magnitude and direction measured counterclockwise from the positive x axis of the resultant force of the three forces acting on the ring A. Take $F_1 = 500$ N and $\theta = 20^\circ$.

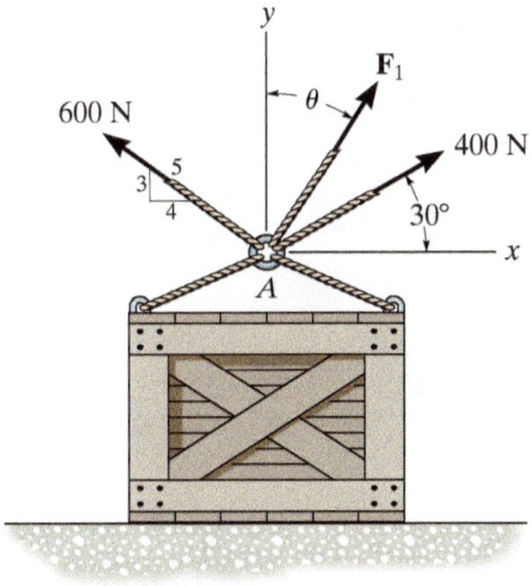

Probs. 2–50/51

*2–52. Determine the magnitude of force **F** so that the resultant $\mathbf{F}_R$ of the three forces is as small as possible. What is the minimum magnitude of $\mathbf{F}_R$?

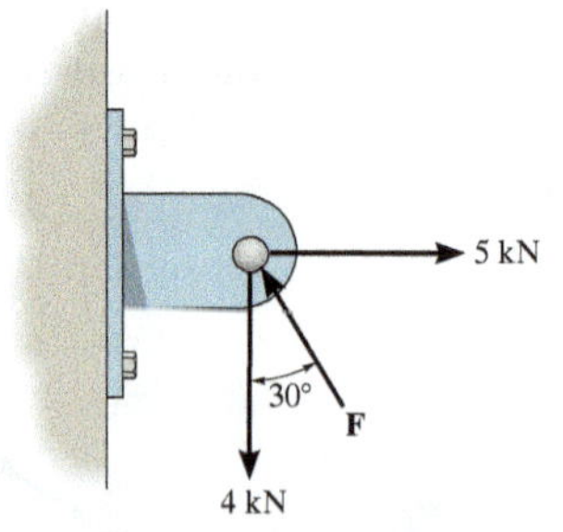

Prob. 2–52

2–53. Determine the magnitude of force **F** so that the resultant force of the three forces is as small as possible. What is the magnitude of the resultant force?

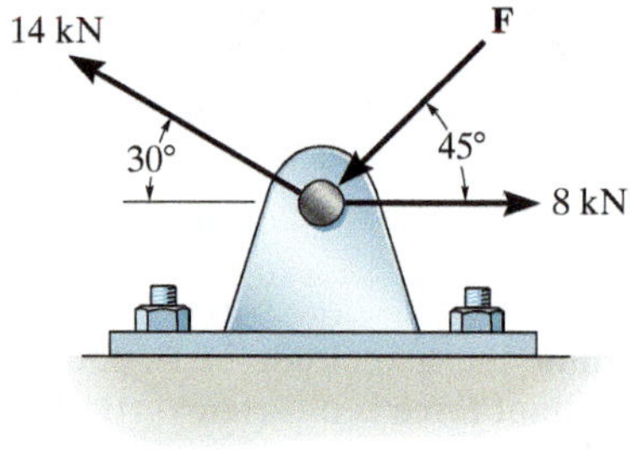

Prob. 2–53

2–54. Three forces act on the bracket. Determine the magnitude and direction θ of $\mathbf{F}_1$ so that the resultant force is directed along the positive x' axis and has a magnitude of 1 kN.

2–55. If $F_1 = 300$ N and $\theta = 20°$, determine the magnitude and direction, measured counterclockwise from the x' axis, of the resultant force of the three forces acting on the bracket.

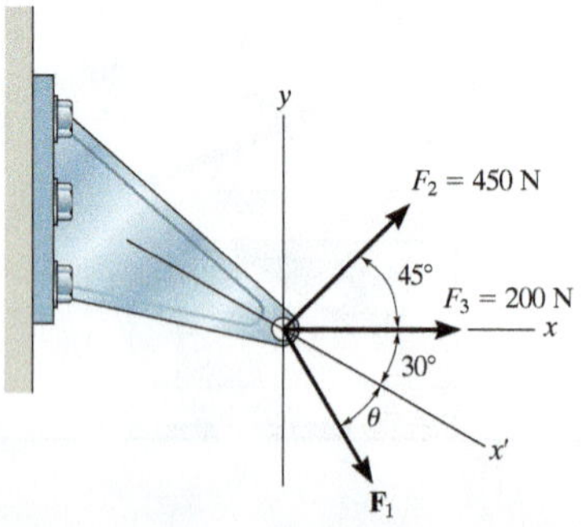

Probs. 2–54/55

*2–56. Three forces act on the bracket. Determine the magnitude and direction θ of $\mathbf{F}_2$ so that the resultant force is directed along the positive u axis and has a magnitude of 50 lb.

2–57. If $F_2 = 150$ lb and $\theta = 55°$, determine the magnitude and direction measured clockwise from the positive x axis, of the resultant force of the three forces acting on the bracket.

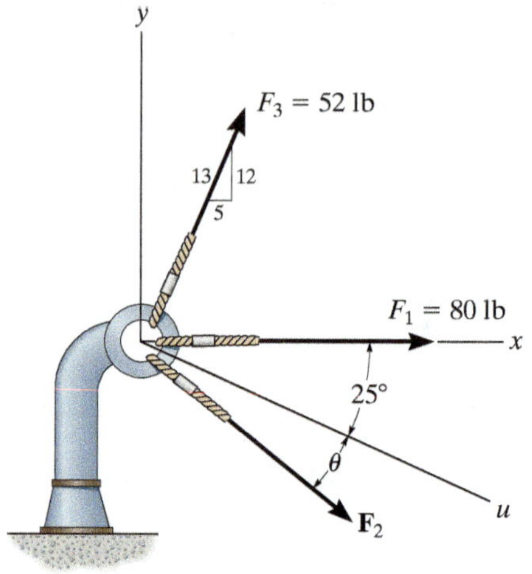

Probs. 2–56/57

2–58. If the magnitude of the resultant force acting on the bracket is to be 450 N directed along the positive u axis, determine the magnitude of $\mathbf{F}_1$ and its direction ϕ.

2–59. If the resultant force acting on the bracket is required to be a minimum, determine the magnitude of $\mathbf{F}_1$ and the resultant force. Set $\phi = 30°$.

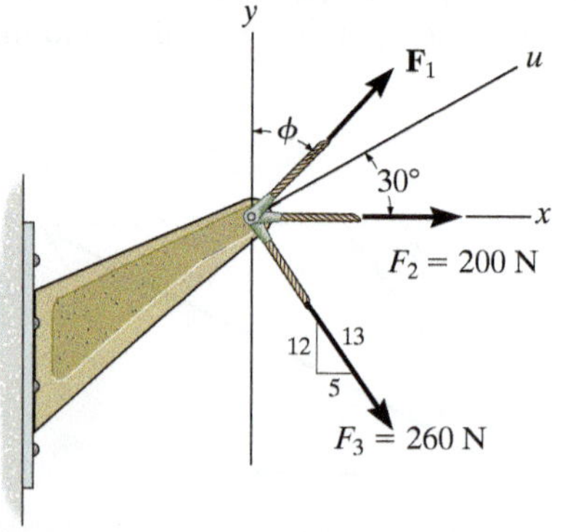

Probs. 2–58/59

2.5 Cartesian Vectors

The operations of vector algebra, when applied to solving problems in *three dimensions*, are greatly simplified if the vectors are first represented in Cartesian vector form. In this section we will present a general method for doing this; then in the next section we will use this method for finding the resultant force of a system of concurrent forces.

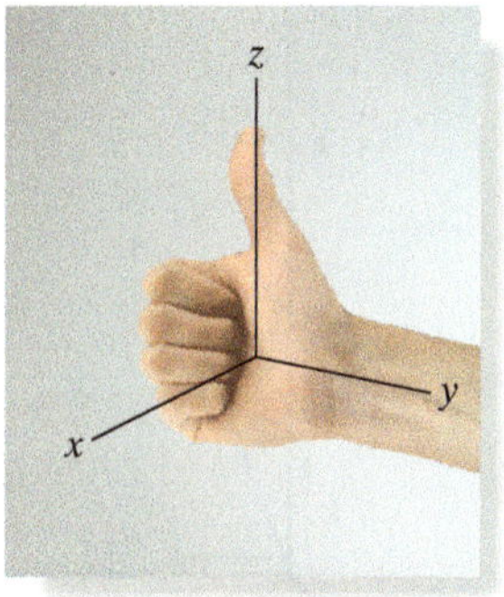

Fig. 2–21

Right-Handed Coordinate System. We will use a right-handed coordinate system to develop the theory of vector algebra that follows. A rectangular coordinate system is said to be *right-handed* if the thumb of the right hand points in the direction of the positive z axis when the right-hand fingers are curled about this axis and directed from the positive x towards the positive y axis, Fig. 2–21.

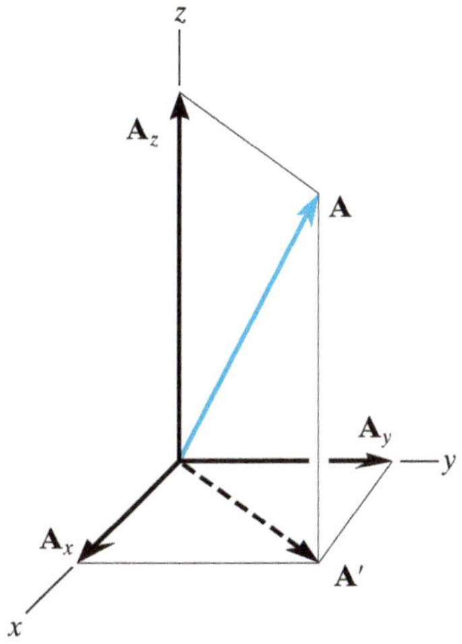

Fig. 2–22

Rectangular Components of a Vector. A vector **A** may have one, two, or three rectangular components along the x, y, z coordinate axes, depending on how the vector is oriented relative to the axes. In general, though, when **A** is directed within an octant of the x, y, z frame, Fig. 2–22, then by two successive applications of the parallelogram law, we may resolve the vector into components as $\mathbf{A} = \mathbf{A}' + \mathbf{A}_z$ and then $\mathbf{A}' = \mathbf{A}_x + \mathbf{A}_y$. Combining these equations, to eliminate $\mathbf{A}'$, **A** is represented by the vector sum of its *three* rectangular components,

$$\mathbf{A} = \mathbf{A}_x + \mathbf{A}_y + \mathbf{A}_z \tag{2–2}$$

Cartesian Unit Vectors. In three dimensions, the set of Cartesian unit vectors, **i**, **j**, **k**, is used to designate the directions of the x, y, z axes, respectively. As stated in Sec. 2.4, the *sense* (or arrowhead) of these vectors will be represented analytically by a plus or minus sign, depending on whether they are directed along the positive or negative x, y, or z axes. The positive Cartesian unit vectors are shown in Fig. 2–23.

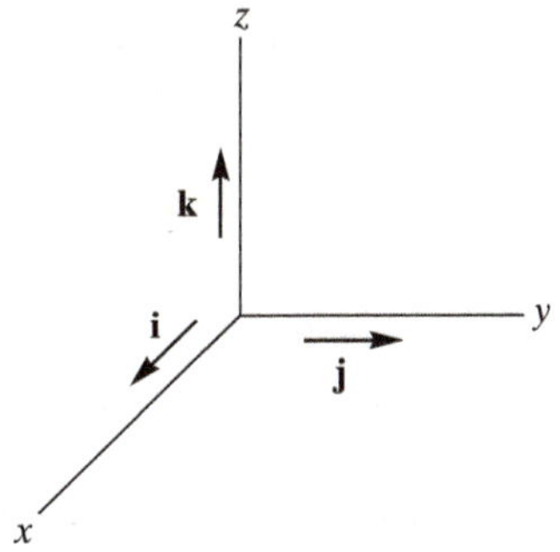

Fig. 2–23

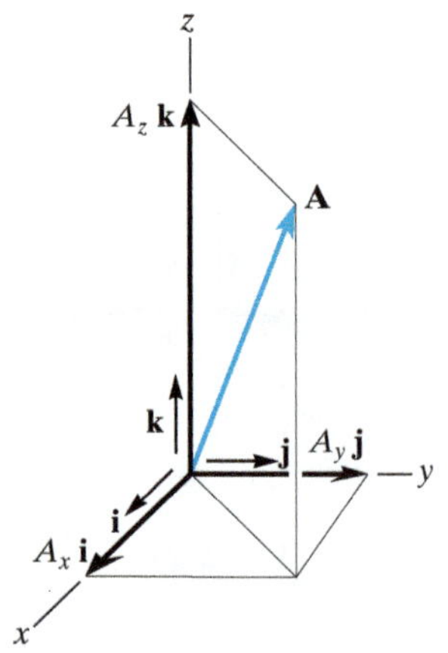

Fig. 2–24

Cartesian Vector Representation.

Since the three components of **A** in Eq. 2–2 act in the positive **i**, **j**, and **k** directions, Fig. 2–24, we can write **A** in Cartesian vector form as

$$\mathbf{A} = A_x\mathbf{i} + A_y\mathbf{j} + A_z\mathbf{k} \tag{2–3}$$

There is a distinct advantage to writing vectors in this manner. Separating the *magnitude* and *direction* of each *component vector* will simplify the operations of vector algebra, particularly in three dimensions.

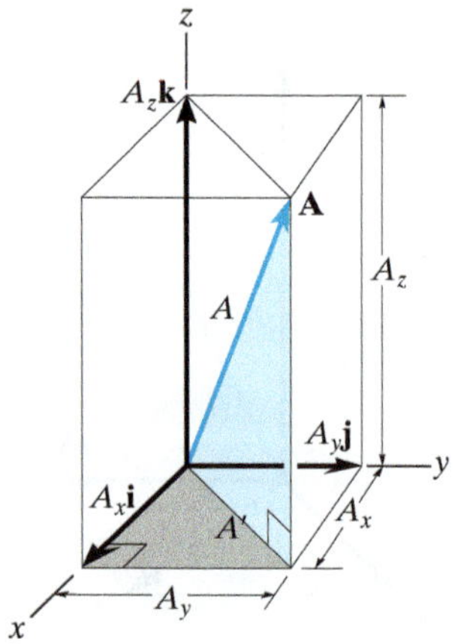

Fig. 2–25

Magnitude of a Cartesian Vector.

It is always possible to obtain the magnitude of **A** provided it is expressed in Cartesian vector form. As shown in Fig. 2–25, from the blue right triangle, $A = \sqrt{A'^2 + A_z^2}$, and from the gray right triangle, $A' = \sqrt{A_x^2 + A_y^2}$. Combining these equations to eliminate A' yields

$$A = \sqrt{A_x^2 + A_y^2 + A_z^2} \tag{2–4}$$

Hence, the magnitude of **A** *is equal to the positive square root of the sum of the squares of its components.*

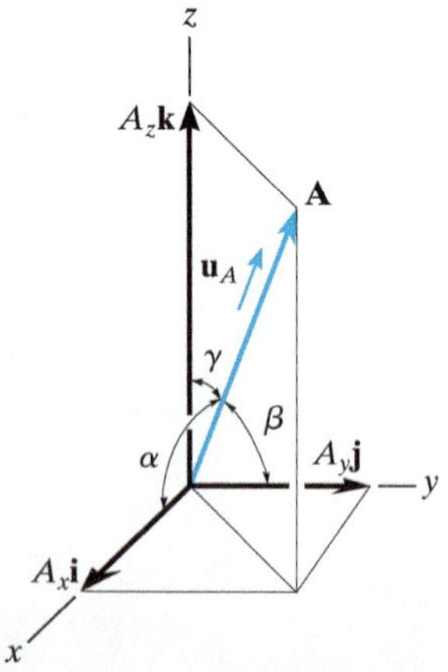

Fig. 2–26

Direction of a Cartesian Vector.

We will define the *direction* of **A** by the *coordinate direction angles* α (alpha), β (beta), and γ (gamma), measured between the *tail* of **A** and the *positive* x, y, z axes provided they are located at the tail of **A**, Fig. 2–26. Note that regardless of where **A** is directed, each of these angles will be between 0° and 180°.

To determine α, β, and γ, consider the projection of **A** onto the x, y, z axes, Fig. 2–27. Referring to the blue colored right triangles shown in each figure, we have

$$\cos\alpha = \frac{A_x}{A} \qquad \cos\beta = \frac{A_y}{A} \qquad \cos\gamma = \frac{A_z}{A} \tag{2–5}$$

These numbers are known as the *direction cosines* of **A**. Once they have been obtained, the coordinate direction angles α, β, γ can then be determined from the inverse cosines.

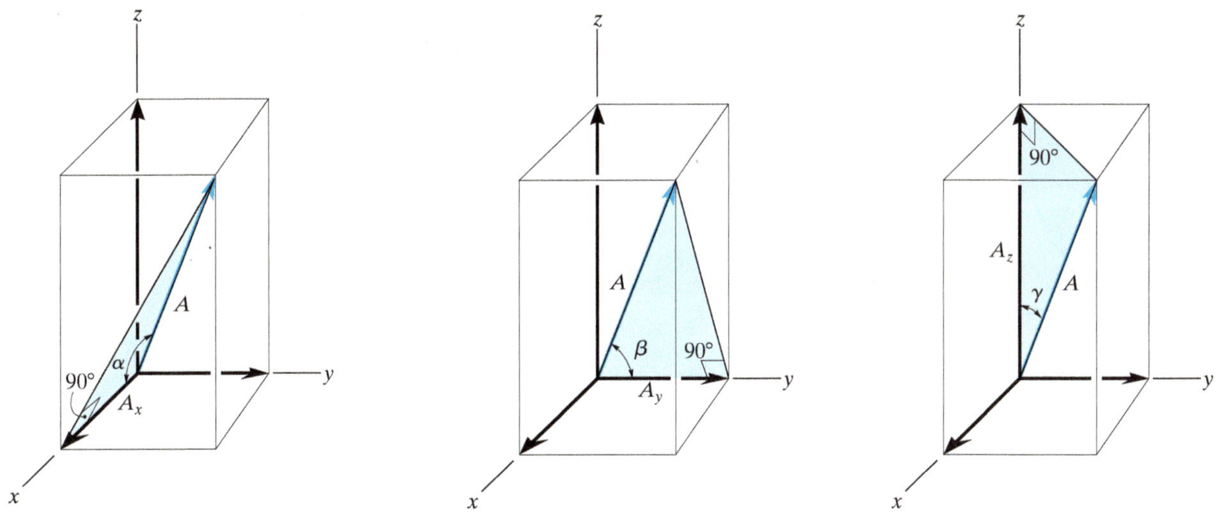

Fig. 2–27

An easy way of obtaining these direction cosines is to form a unit vector $\mathbf{u}_A$ in the direction of A, Fig. 2–26. If $\mathbf{A}$ is expressed in Cartesian vector form, $\mathbf{A} = A_x\mathbf{i} + A_y\mathbf{j} + A_z\mathbf{k}$, then $\mathbf{u}_A$ will have a magnitude of one and be dimensionless provided $\mathbf{A}$ is divided by its magnitude, i.e.,

$$\mathbf{u}_A = \frac{\mathbf{A}}{A} = \frac{A_x}{A}\mathbf{i} + \frac{A_y}{A}\mathbf{j} + \frac{A_z}{A}\mathbf{k} \qquad (2\text{–}6)$$

where $A = \sqrt{A_x^2 + A_y^2 + A_z^2}$. By comparison with Eqs. 2–5, it is seen that *the* $\mathbf{i}, \mathbf{j}, \mathbf{k}$ *components* of $\mathbf{u}_A$ *represent the direction cosines of* $\mathbf{A}$, i.e.,

$$\mathbf{u}_A = \cos\alpha\,\mathbf{i} + \cos\beta\,\mathbf{j} + \cos\gamma\,\mathbf{k} \qquad (2\text{–}7)$$

Since the magnitude of a vector is equal to the positive square root of the sum of the squares of the magnitudes of its components, and $\mathbf{u}_A$ has a magnitude of one, then from the above equation an important relation among the direction cosines can be formulated as

$$\cos^2\alpha + \cos^2\beta + \cos^2\gamma = 1 \qquad (2\text{–}8)$$

Here we can see that if only *two* of the coordinate angles are known, the third angle can be found using this equation.

Finally, if the magnitude and coordinate direction angles of $\mathbf{A}$ are known, then $\mathbf{A}$ may be expressed in Cartesian vector form as

$$\begin{aligned}\mathbf{A} &= A\mathbf{u}_A \\ &= A\cos\alpha\,\mathbf{i} + A\cos\beta\,\mathbf{j} + A\cos\gamma\,\mathbf{k} \\ &= A_x\mathbf{i} + A_y\mathbf{j} + A_z\mathbf{k}\end{aligned} \qquad (2\text{–}9)$$

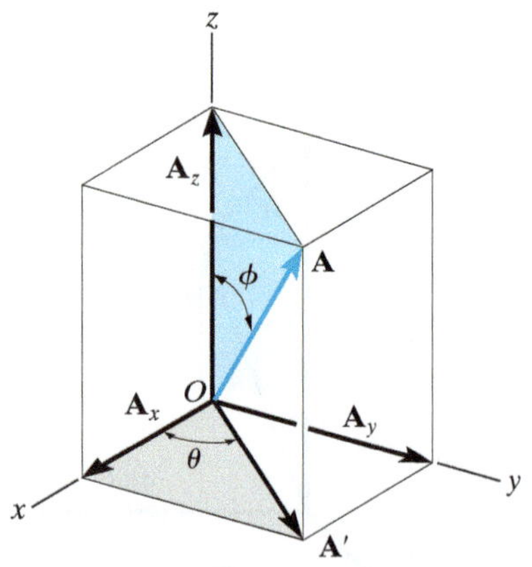

Fig. 2–28

Sometimes, the direction of **A** can be specified using two angles, θ and ϕ (phi), such as shown in Fig. 2–28. The components of **A** can then be determined by applying trigonometry first to the blue right triangle, which yields

$$A_z = A \cos \phi$$

and

$$A' = A \sin \phi$$

Now applying trigonometry to the gray shaded right triangle,

$$A_x = A' \cos \theta = A \sin \phi \cos \theta$$

$$A_y = A' \sin \theta = A \sin \phi \sin \theta$$

Therefore **A** written in Cartesian vector form becomes

$$\mathbf{A} = A \sin \phi \cos \theta \, \mathbf{i} + A \sin \phi \sin \theta \, \mathbf{j} + A \cos \phi \, \mathbf{k}$$

You should not memorize this equation, rather it is important to understand how the components were determined using trigonometry.

2.6 Addition of Cartesian Vectors

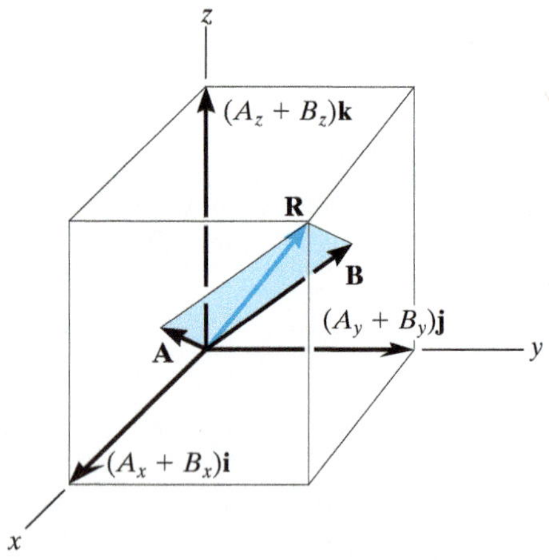

Fig. 2–29

The addition (or subtraction) of two or more vectors is greatly simplified if the vectors are expressed in terms of their Cartesian components. For example, if $\mathbf{A} = A_x\mathbf{i} + A_y\mathbf{j} + A_z\mathbf{k}$ and $\mathbf{B} = B_x\mathbf{i} + B_y\mathbf{j} + B_z\mathbf{k}$, Fig. 2–29, then the resultant vector, **R**, has components which are the scalar sums of the **i**, **j**, **k** components of **A** and **B**, i.e.,

$$\mathbf{R} = \mathbf{A} + \mathbf{B} = (A_x + B_x)\mathbf{i} + (A_y + B_y)\mathbf{j} + (A_z + B_z)\mathbf{k}$$

If this is generalized and applied to a system of several concurrent forces, then the force resultant is the vector sum of all the forces in the system and can be written as

$$\mathbf{F}_R = \Sigma \mathbf{F} = \Sigma F_x\mathbf{i} + \Sigma F_y\mathbf{j} + \Sigma F_z\mathbf{k} \qquad (2\text{–}10)$$

Here ΣF_x, ΣF_y, and ΣF_z represent the algebraic sums of the respective x, y, z or **i**, **j**, **k** components of each force in the system.

Important Points

- Cartesian vector analysis is often used to solve problems in three dimensions.
- The positive directions of the x, y, z axes are defined by the Cartesian unit vectors **i**, **j**, **k**, respectively.
- The *magnitude* of a Cartesian vector is $A = \sqrt{A_x^2 + A_y^2 + A_z^2}$.
- The *direction* of a Cartesian vector is specified using coordinate direction angles α, β, γ which the tail of the vector makes with the positive x, y, z axes, respectively. The components of the unit vector $\mathbf{u}_A = \mathbf{A}/A$ represent the direction cosines of α, β, γ. Only two of the angles α, β, γ have to be specified. The third angle is determined from the relationship $\cos^2\alpha + \cos^2\beta + \cos^2\gamma = 1$.
- Sometimes the direction of a vector is defined using the two angles θ and ϕ as in Fig. 2–28. In this case the vector components are obtained by vector resolution using trigonometry.
- To find the *resultant* of a concurrent force system, express each force as a Cartesian vector and add the **i**, **j**, **k** components of all the forces in the system.

EXAMPLE 2.8

Express the force **F** shown in Fig. 2–30 as a Cartesian vector.

SOLUTION

Since only two coordinate direction angles are specified, the third angle α must be determined from Eq. 2–8; i.e.,

$$\cos^2\alpha + \cos^2\beta + \cos^2\gamma = 1$$

$$\cos^2\alpha + \cos^2 60^\circ + \cos^2 45^\circ = 1$$

$$\cos\alpha = \sqrt{1 - (0.5)^2 - (0.707)^2} = \pm 0.5$$

Hence, two possibilities exist, namely,

$$\alpha = \cos^{-1}(0.5) = 60^\circ \quad \text{or} \quad \alpha = \cos^{-1}(-0.5) = 120^\circ$$

By inspection it is necessary that $\alpha = 60^\circ$, since $\mathbf{F}_x$ must be in the $+x$ direction.

Using Eq. 2–9, with $F = 200$ N, we have

$$\mathbf{F} = F\cos\alpha\,\mathbf{i} + F\cos\beta\,\mathbf{j} + F\cos\gamma\,\mathbf{k}$$
$$= (200\cos 60^\circ\ \text{N})\mathbf{i} + (200\cos 60^\circ\ \text{N})\mathbf{j} + (200\cos 45^\circ\ \text{N})\mathbf{k}$$
$$= \{100.0\mathbf{i} + 100.0\mathbf{j} + 141.4\mathbf{k}\}\ \text{N} \qquad \textit{Ans.}$$

Show that indeed the magnitude of $F = 200$ N.

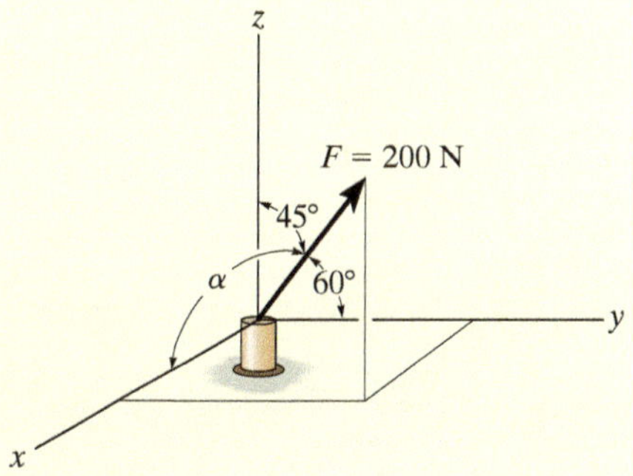

Fig. 2–30

2

EXAMPLE 2.9

Determine the magnitude and the coordinate direction angles of the resultant force acting on the ring in Fig. 2–31*a*.

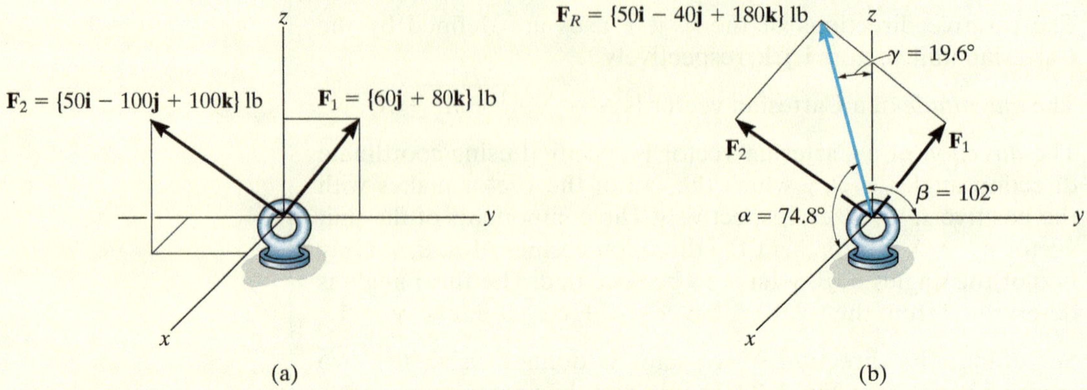

Fig. 2–31

SOLUTION

Since each force is represented in Cartesian vector form, the resultant force, shown in Fig. 2–31*b*, is

$$\mathbf{F}_R = \Sigma\mathbf{F} = \mathbf{F}_1 + \mathbf{F}_2 = \{60\mathbf{j} + 80\mathbf{k}\}\text{ lb} + \{50\mathbf{i} - 100\mathbf{j} + 100\mathbf{k}\}\text{ lb}$$
$$= \{50\mathbf{i} - 40\mathbf{j} + 180\mathbf{k}\}\text{ lb}$$

The magnitude of $\mathbf{F}_R$ is

$$F_R = \sqrt{(50\text{ lb})^2 + (-40\text{ lb})^2 + (180\text{ lb})^2} = 191.0\text{ lb}$$
$$= 191\text{ lb} \qquad \textit{Ans.}$$

The coordinate direction angles α, β, γ are determined from the components of the unit vector acting in the direction of $\mathbf{F}_R$.

$$\mathbf{u}_{F_R} = \frac{\mathbf{F}_R}{F_R} = \frac{50}{191.0}\mathbf{i} - \frac{40}{191.0}\mathbf{j} + \frac{180}{191.0}\mathbf{k}$$
$$= 0.2617\mathbf{i} - 0.2094\mathbf{j} + 0.9422\mathbf{k}$$

so that

$$\cos\alpha = 0.2617 \qquad \alpha = 74.8^\circ \qquad \textit{Ans.}$$
$$\cos\beta = -0.2094 \qquad \beta = 102^\circ \qquad \textit{Ans.}$$
$$\cos\gamma = 0.9422 \qquad \gamma = 19.6^\circ \qquad \textit{Ans.}$$

These angles are shown in Fig. 2–31*b*.

NOTE: In particular, notice that $\beta > 90^\circ$ since the **j** component of $\mathbf{u}_{F_R}$ is negative. This seems reasonable considering how $\mathbf{F}_1$ and $\mathbf{F}_2$ add according to the parallelogram law.

EXAMPLE 2.10

Express the force **F** shown in Fig. 2–32*a* as a Cartesian vector.

SOLUTION

The angles of 60° and 45° defining the direction of **F** are *not* coordinate direction angles. Two successive applications of the parallelogram law are needed to resolve **F** into its *x, y, z* components. First $\mathbf{F} = \mathbf{F}' + \mathbf{F}_z$, then $\mathbf{F}' = \mathbf{F}_x + \mathbf{F}_y$, Fig. 2–32*b*. By trigonometry, the magnitudes of the components are

$$F_z = 100 \sin 60° \text{ lb} = 86.6 \text{ lb}$$

$$F' = 100 \cos 60° \text{ lb} = 50 \text{ lb}$$

$$F_x = F' \cos 45° = 50 \cos 45° \text{ lb} = 35.4 \text{ lb}$$

$$F_y = F' \sin 45° = 50 \sin 45° \text{ lb} = 35.4 \text{ lb}$$

Realizing that $\mathbf{F}_y$ has a direction defined by $-\mathbf{j}$, we have

$$\mathbf{F} = \{35.4\mathbf{i} - 35.4\mathbf{j} + 86.6\mathbf{k}\} \text{ lb}$$ *Ans.*

To show that the magnitude of this vector is indeed 100 lb, apply Eq. 2–4,

$$F = \sqrt{F_x^2 + F_y^2 + F_z^2}$$

$$= \sqrt{(35.4)^2 + (35.4)^2 + (86.6)^2} = 100 \text{ lb}$$

If needed, the coordinate direction angles of **F** can be determined from the components of the unit vector acting in the direction of **F**. Hence,

$$\mathbf{u} = \frac{\mathbf{F}}{F} = \frac{F_x}{F}\mathbf{i} + \frac{F_y}{F}\mathbf{j} + \frac{F_z}{F}\mathbf{k}$$

$$= \frac{35.4}{100}\mathbf{i} - \frac{35.4}{100}\mathbf{j} + \frac{86.6}{100}\mathbf{k}$$

$$= 0.354\mathbf{i} - 0.354\mathbf{j} + 0.866\mathbf{k}$$

so that

$$\alpha = \cos^{-1}(0.354) = 69.3°$$

$$\beta = \cos^{-1}(-0.354) = 111°$$

$$\gamma = \cos^{-1}(0.866) = 30.0°$$

These results are shown in Fig. 2–32*c*.

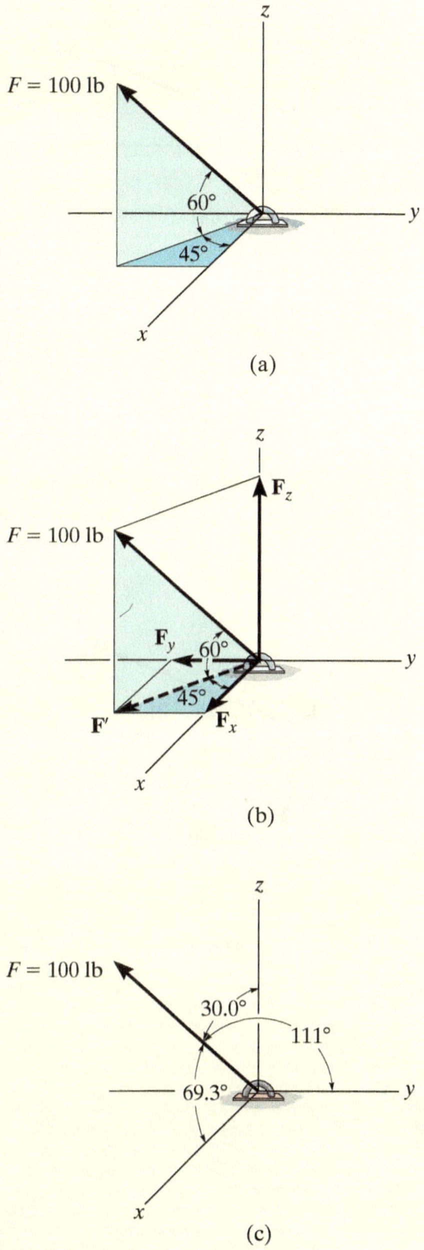

Fig. 2–32

EXAMPLE 2.11

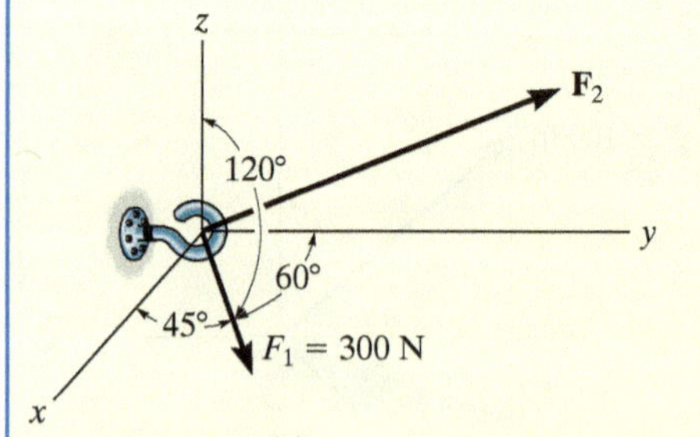

(a)

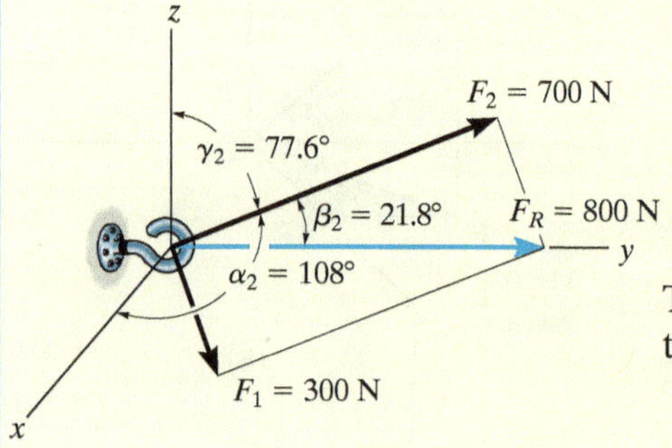

(b)

Fig. 2–33

Two forces act on the hook shown in Fig. 2–33*a*. Specify the magnitude of $\mathbf{F}_2$ and its coordinate direction angles so that the resultant force $\mathbf{F}_R$ acts along the positive *y* axis and has a magnitude of 800 N.

SOLUTION

To solve this problem, the resultant force $\mathbf{F}_R$ and its two components, $\mathbf{F}_1$ and $\mathbf{F}_2$, will each be expressed in Cartesian vector form. Then, as shown in Fig. 2–33*b*, it is necessary that $\mathbf{F}_R = \mathbf{F}_1 + \mathbf{F}_2$.

Applying Eq. 2–9,

$$\begin{aligned}\mathbf{F}_1 &= F_1 \cos\alpha_1\mathbf{i} + F_1\cos\beta_1\mathbf{j} + F_1\cos\gamma_1\mathbf{k}\\ &= 300\cos 45°\,\mathbf{i} + 300\cos 60°\,\mathbf{j} + 300\cos 120°\,\mathbf{k}\\ &= \{212.1\mathbf{i} + 150\mathbf{j} - 150\mathbf{k}\}\text{ N}\\ \mathbf{F}_2 &= F_{2x}\mathbf{i} + F_{2y}\mathbf{j} + F_{2z}\mathbf{k}\end{aligned}$$

Since $\mathbf{F}_R$ has a magnitude of 800 N and acts in the $+\mathbf{j}$ direction,

$$\mathbf{F}_R = (800\text{ N})(+\mathbf{j}) = \{800\mathbf{j}\}\text{ N}$$

We require

$$\begin{aligned}\mathbf{F}_R &= \mathbf{F}_1 + \mathbf{F}_2\\ 800\mathbf{j} &= 212.1\mathbf{i} + 150\mathbf{j} - 150\mathbf{k} + F_{2x}\mathbf{i} + F_{2y}\mathbf{j} + F_{2z}\mathbf{k}\\ 800\mathbf{j} &= (212.1 + F_{2x})\mathbf{i} + (150 + F_{2y})\mathbf{j} + (-150 + F_{2z})\mathbf{k}\end{aligned}$$

To satisfy this equation the $\mathbf{i}, \mathbf{j}, \mathbf{k}$ components of $\mathbf{F}_R$ must be equal to the corresponding $\mathbf{i}, \mathbf{j}, \mathbf{k}$ components of $(\mathbf{F}_1 + \mathbf{F}_2)$. Hence,

$$\begin{aligned}0 &= 212.1 + F_{2x} & F_{2x} &= -212.1\text{ N}\\ 800 &= 150 + F_{2y} & F_{2y} &= 650\text{ N}\\ 0 &= -150 + F_{2z} & F_{2z} &= 150\text{ N}\end{aligned}$$

The magnitude of $\mathbf{F}_2$ is thus

$$\begin{aligned}F_2 &= \sqrt{(-212.1\text{ N})^2 + (650\text{ N})^2 + (150\text{ N})^2}\\ &= 700\text{ N}\end{aligned}$$ *Ans.*

We can use Eq. 2–9 to determine $\alpha_2, \beta_2, \gamma_2$.

$$\cos\alpha_2 = \frac{-212.1}{700};\qquad \alpha_2 = 108°$$ *Ans.*

$$\cos\beta_2 = \frac{650}{700};\qquad \beta_2 = 21.8°$$ *Ans.*

$$\cos\gamma_2 = \frac{150}{700};\qquad \gamma_2 = 77.6°$$ *Ans.*

These results are shown in Fig. 2–33*b*.

FUNDAMENTAL PROBLEMS

F2–13. Determine the coordinate direction angles of the force.

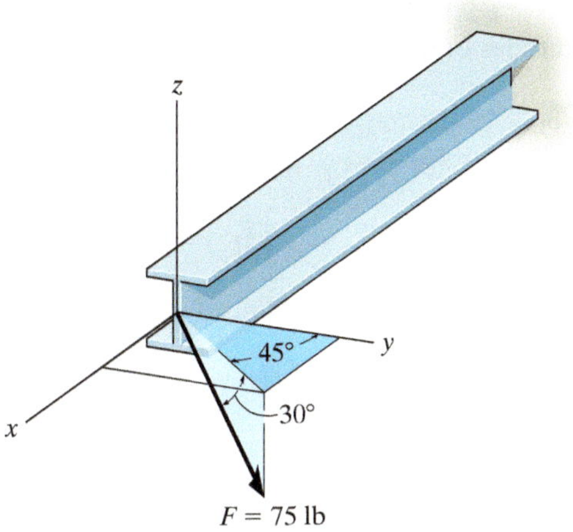

F2–13

F2–14. Express the force as a Cartesian vector.

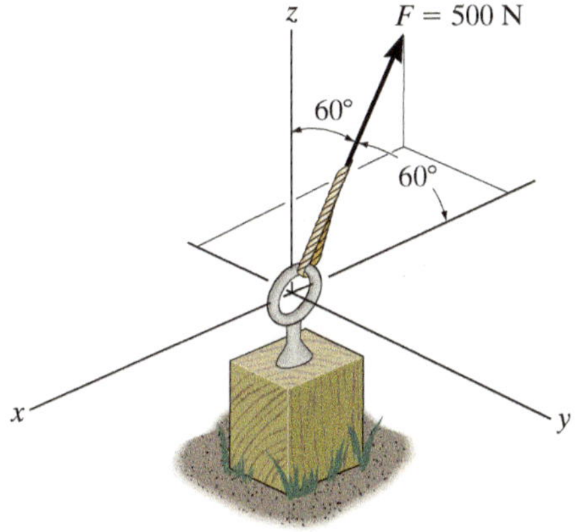

F2–14

F2–15. Express the force as a Cartesian vector.

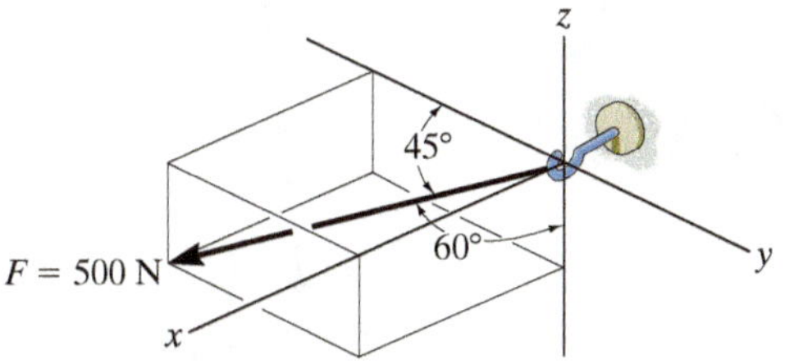

F2–15

F2–16. Express the force as a Cartesian vector.

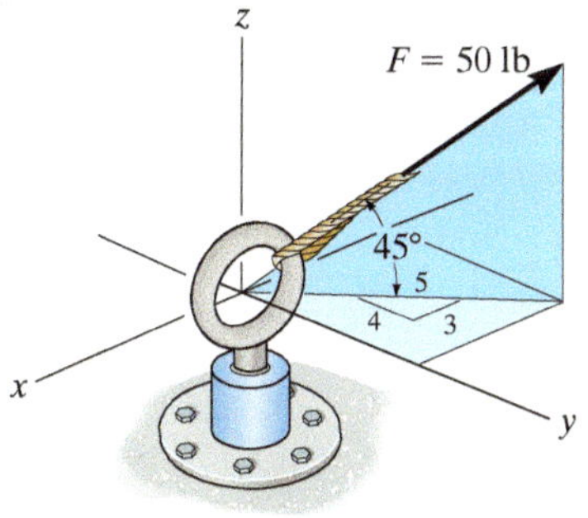

F2–16

F2–17. Express the force as a Cartesian vector.

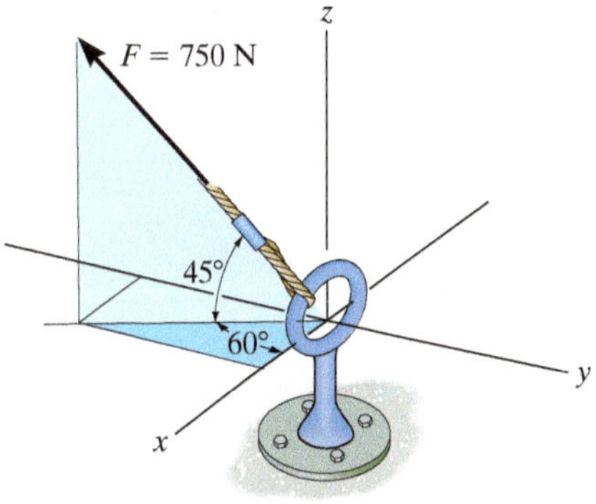

F2–17

F2–18. Determine the resultant force acting on the hook.

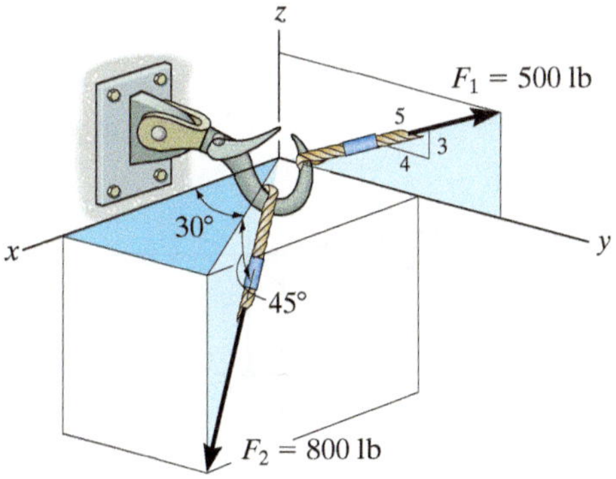

F2–18

PROBLEMS

*2–60. The stock mounted on the lathe is subjected to a force of 60 N. Determine the coordinate direction angle β and express the force as a Cartesian vector.

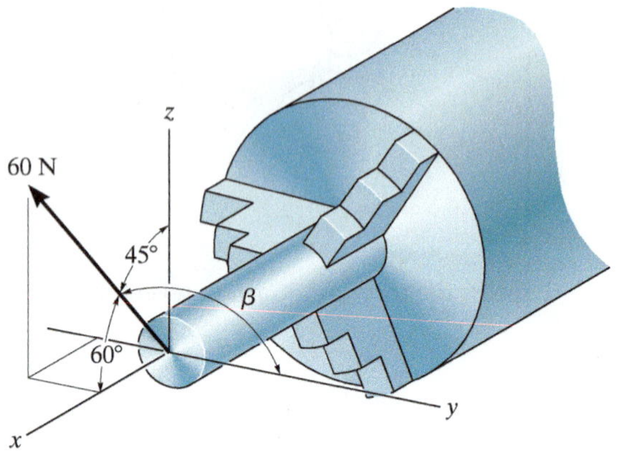

Prob. 2–60

2–61. Determine the magnitude and coordinate direction angles of the resultant force and sketch this vector on the coordinate system.

2–62. Specify the coordinate direction angles of $\mathbf{F}_1$ and $\mathbf{F}_2$ and express each force as a Cartesian vector.

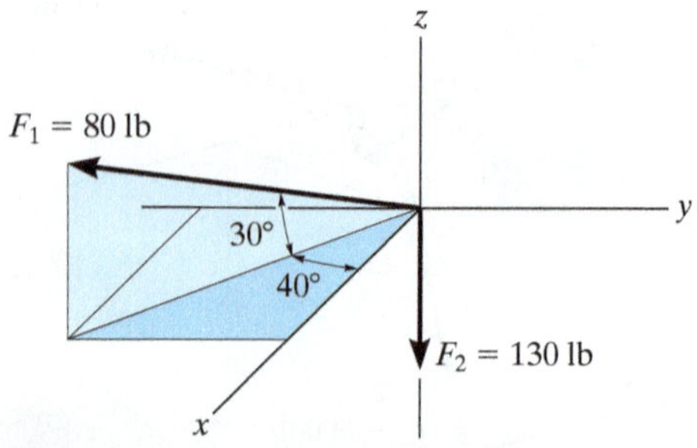

Probs. 2–61/62

2–63. The bolt is subjected to the force **F**, which has components acting along the x, y, z axes as shown. If the magnitude of **F** is 80 N, and $\alpha = 60°$ and $\gamma = 45°$, determine the magnitudes of its components.

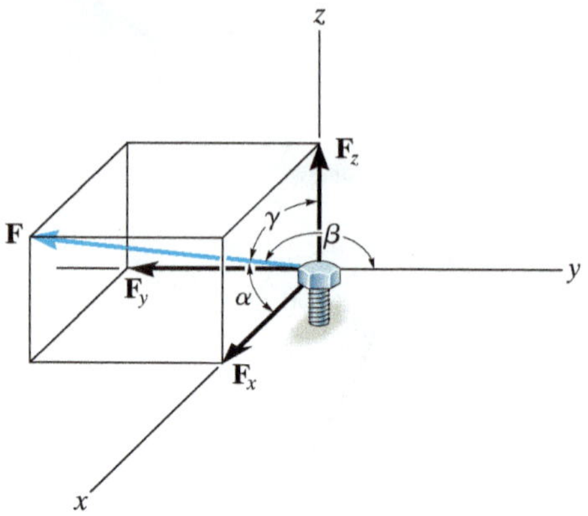

Prob. 2–63

*2–64. Determine the magnitude and coordinate direction angles of $\mathbf{F}_1 = \{60\mathbf{i} - 50\mathbf{j} + 40\mathbf{k}\}$ N and $\mathbf{F}_2 = \{-40\mathbf{i} - 85\mathbf{j} + 30\mathbf{k}\}$ N. Sketch each force on an x, y, z reference frame.

2–65. The cable at the end of the crane boom exerts a force of 250 lb on the boom as shown. Express **F** as a Cartesian vector.

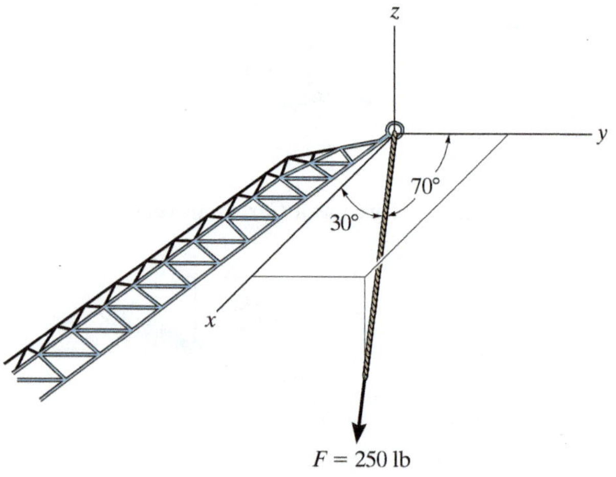

Prob. 2–65

2–66. Express each force acting on the pipe assembly in Cartesian vector form.

2–67. Determine the magnitude and the direction of the resultant force acting on the pipe assembly.

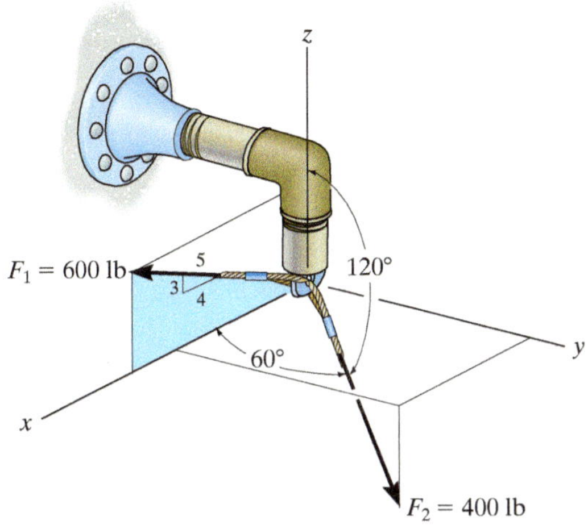

Probs. 2–66/67

***2–68.** Express each force as a Cartesian vector.

2–69. Determine the magnitude and coordinate direction angles of the resultant force acting on the hook.

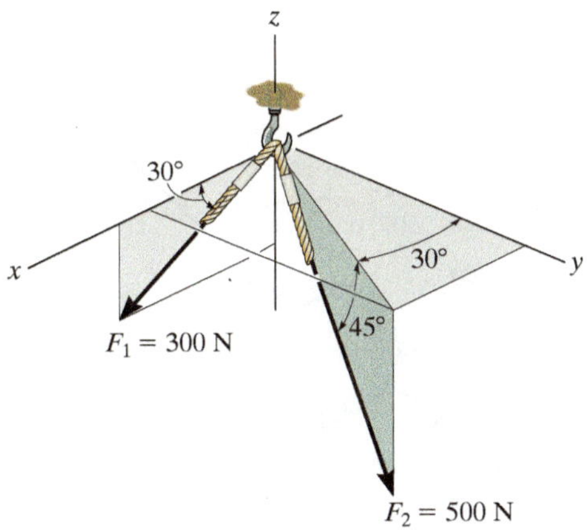

Probs. 2–68/69

2–70. The beam is subjected to the two forces shown. Express each force in Cartesian vector form and determine the magnitude and coordinate direction angles of the resultant force.

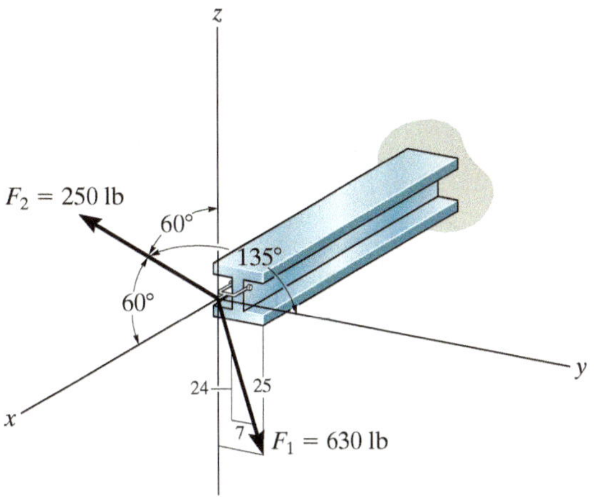

Prob. 2–70

2–71. If the resultant force acting on the bracket is directed along the positive y axis, determine the magnitude of the resultant force and the coordinate direction angles of **F** so that $\beta < 90°$.

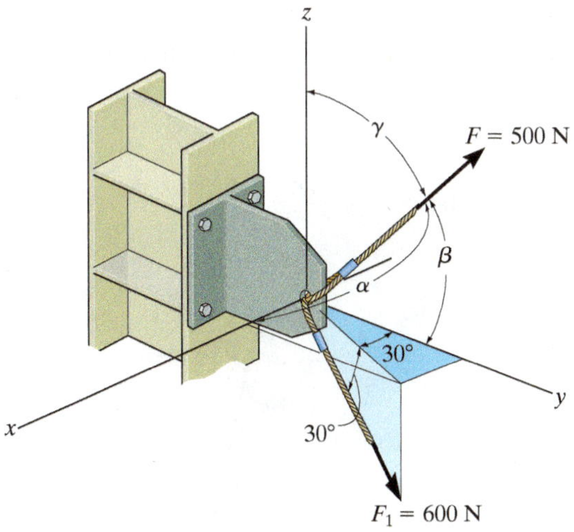

Prob. 2–71

*2–72. A force **F** is applied at the top of the tower at A. If it acts in the direction shown such that one of its components lying in the shaded y-z plane has a magnitude of 80 lb, determine its magnitude F and coordinate direction angles α, β, γ.

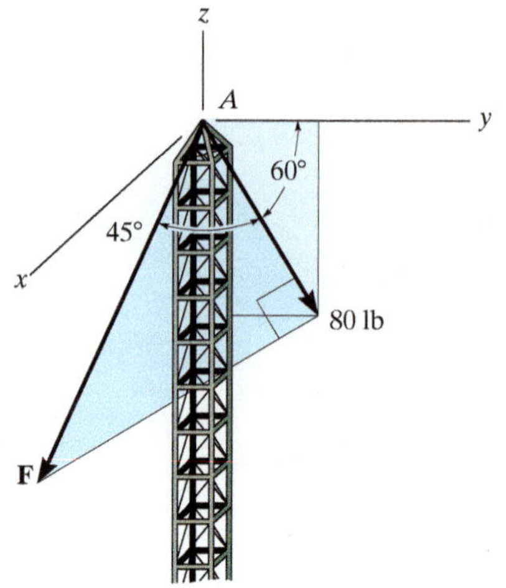

Prob. 2–72

2–73. The spur gear is subjected to the two forces caused by contact with other gears. Express each force as a Cartesian vector.

2–74. The spur gear is subjected to the two forces caused by contact with other gears. Determine the resultant of the two forces and express the result as a Cartesian vector.

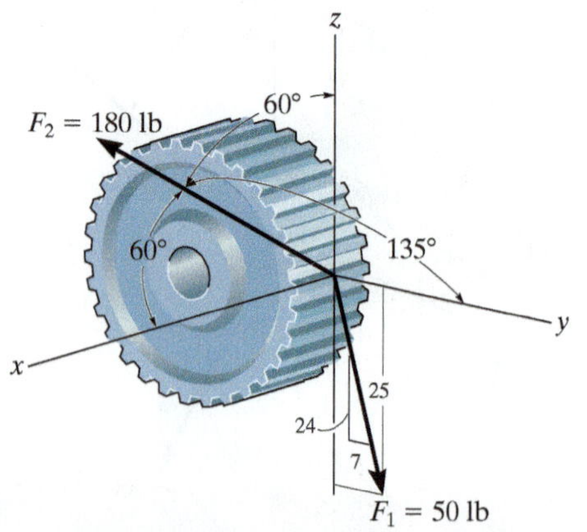

Probs. 2–73/74

2–75. Determine the coordinate direction angles of force $\mathbf{F}_1$.

*2–76. Determine the magnitude and coordinate direction angles of the resultant force acting on the eyebolt.

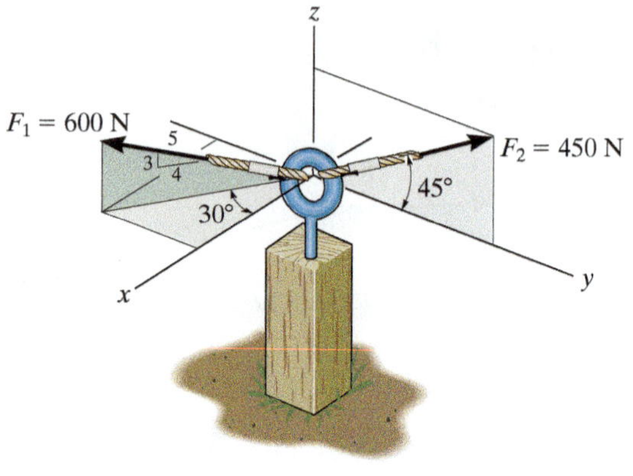

Probs. 2–75/76

2–77. The cables attached to the screw eye are subjected to the three forces shown. Express each force in Cartesian vector form and determine the magnitude and coordinate direction angles of the resultant force.

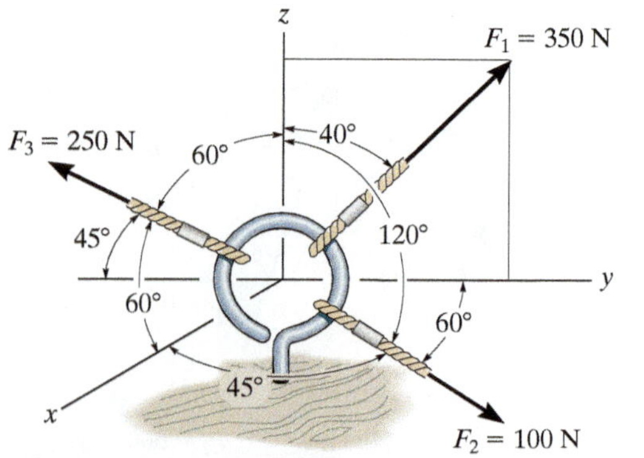

Prob. 2–77

2–78. Three forces act on the ring. If the resultant force $\mathbf{F}_R$ has a magnitude and direction as shown, determine the magnitude and the coordinate direction angles of force $\mathbf{F}_3$.

2–79. Determine the coordinate direction angles of $\mathbf{F}_1$ and $\mathbf{F}_R$.

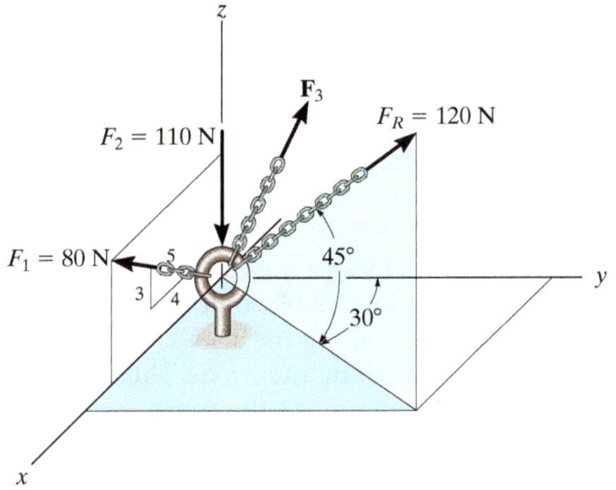

Probs. 2–78/79

*2–80. If the coordinate direction angles for $\mathbf{F}_3$ are $\alpha_3 = 120°, \beta_3 = 45°$ and $\gamma_3 = 60°$, determine the magnitude and coordinate direction angles of the resultant force acting on the eyebolt.

2–81. If the coordinate direction angles for $\mathbf{F}_3$ are $\alpha_3 = 120°$, $\beta_3 = 45°$, and $\gamma_3 = 60°$, determine the magnitude and coordinate direction angles of the resultant force acting on the eyebolt.

2–82. If the direction of the resultant force acting on the eyebolt is defined by the unit vector $\mathbf{u}_{F_R} = \cos 30°\mathbf{j} + \sin 30°\mathbf{k}$, determine the coordinate direction angles of $\mathbf{F}_3$ and the magnitude of $\mathbf{F}_R$.

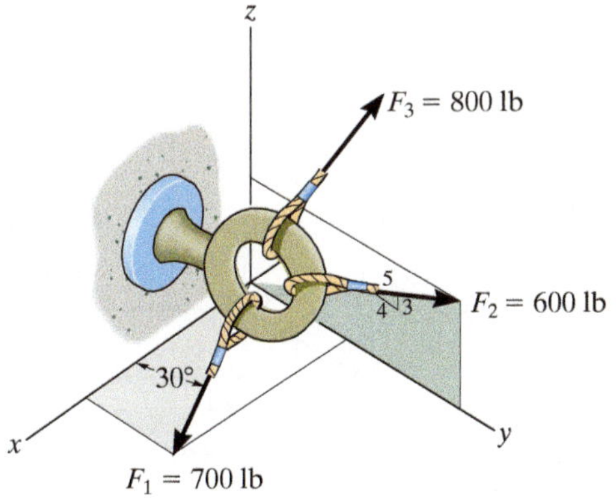

Probs. 2–80/81/82

2–83. The bracket is subjected to the two forces shown. Express each force in Cartesian vector form and then determine the resultant force $\mathbf{F}_R$. Find the magnitude and coordinate direction angles of the resultant force.

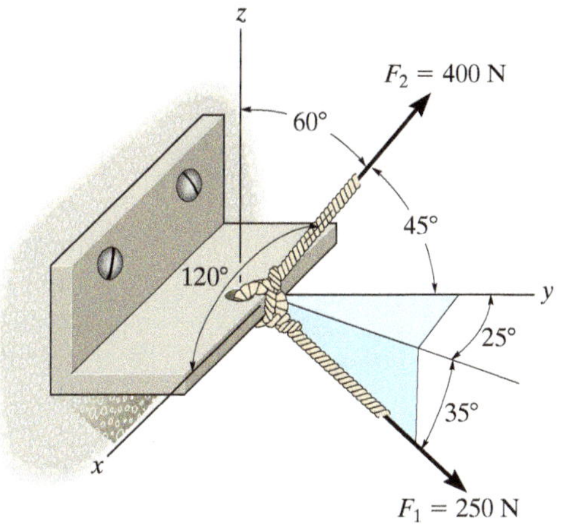

Prob. 2–83

*2–84. The pole is subjected to the force **F**, which has components acting along the x, y, z axes as shown. If the magnitude of **F** is 3 kN, $\beta = 30°$, and $\gamma = 75°$, determine the magnitudes of its three components.

2–85. The pole is subjected to the force **F** which has components $F_x = 1.5$ kN and $F_z = 1.25$ kN. If $\beta = 75°$, determine the magnitudes of **F** and $\mathbf{F}_y$.

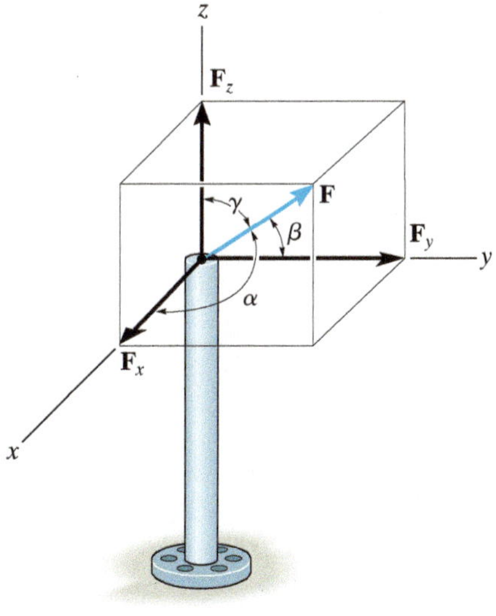

Probs. 2–84/85

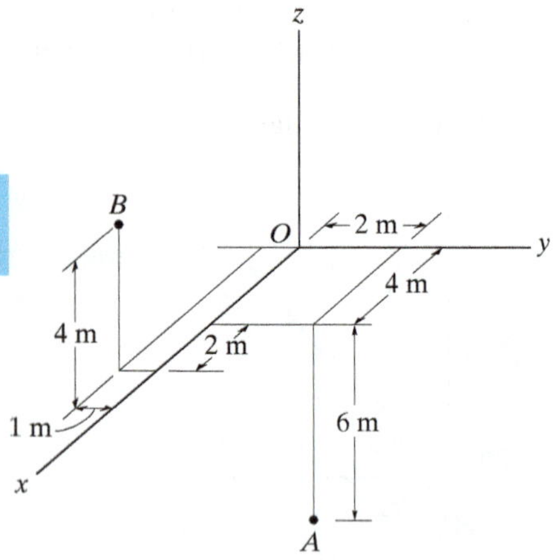

Fig. 2–34

2.7 Position Vectors

In this section we will introduce the concept of a position vector. It will be shown that this vector is of importance in formulating a Cartesian force vector directed between two points in space.

x, y, z Coordinates. Throughout the book we will use a *right-handed* coordinate system to reference the location of points in space. We will also use the convention followed in many technical books, which requires the positive z axis to be directed *upward* (the zenith direction) so that it measures the height of an object or the altitude of a point. The x, y axes then lie in the horizontal plane, Fig. 2–34. Points in space are located relative to the origin of coordinates, O, by successive measurements along the x, y, z axes. For example, the coordinates of point A are obtained by starting at O and measuring $x_A = +4$ m along the x axis, then $y_A = +2$ m along the y axis, and finally $z_A = -6$ m along the z axis. Thus, $A(4\text{ m}, 2\text{ m}, -6\text{ m})$. In a similar manner, measurements along the x, y, z axes from O to B yield the coordinates of B, i.e., $B(6\text{ m}, -1\text{ m}, 4\text{ m})$.

Position Vector. A *position vector* $\mathbf{r}$ is defined as a fixed vector which locates a point in space relative to another point. For example, if $\mathbf{r}$ extends from the origin of coordinates, O, to point $P(x, y, z)$, Fig. 2–35*a*, then $\mathbf{r}$ can be expressed in Cartesian vector form as

$$\mathbf{r} = x\mathbf{i} + y\mathbf{j} + z\mathbf{k}$$

Note how the head-to-tail vector addition of the three components yields vector $\mathbf{r}$, Fig. 2–35*b*. Starting at the origin O, one "travels" x in the $+\mathbf{i}$ direction, then y in the $+\mathbf{j}$ direction, and finally z in the $+\mathbf{k}$ direction to arrive at point $P(x, y, z)$.

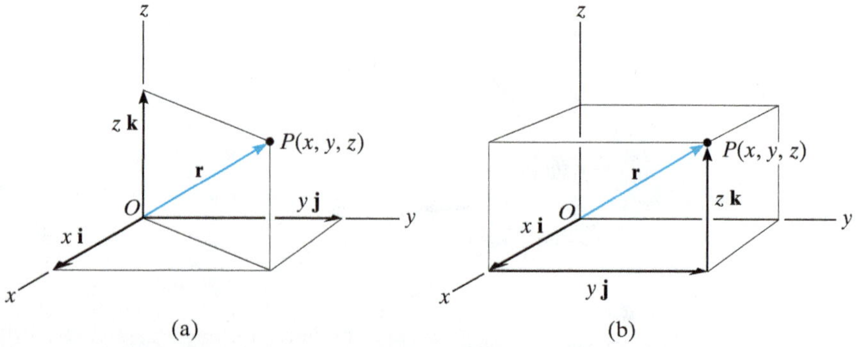

Fig. 2–35

In the more general case, the position vector may be directed from point A to point B in space, Fig. 2–36*a*. This vector is also designated by the symbol **r**. As a matter of convention, we will *sometimes* refer to this vector with *two subscripts* to indicate from and to the point where it is directed. Thus, **r** can also be designated as $\mathbf{r}_{AB}$. Also, note that $\mathbf{r}_A$ and $\mathbf{r}_B$ in Fig. 2–36*a* are referenced with only one subscript since they extend from the origin of coordinates.

From Fig. 2–36*a*, by the head-to-tail vector addition, using the triangle rule, we require

$$\mathbf{r}_A + \mathbf{r} = \mathbf{r}_B$$

Solving for **r** and expressing $\mathbf{r}_A$ and $\mathbf{r}_B$ in Cartesian vector form yields

$$\mathbf{r} = \mathbf{r}_B - \mathbf{r}_A = (x_B\mathbf{i} + y_B\mathbf{j} + z_B\mathbf{k}) - (x_A\mathbf{i} + y_A\mathbf{j} + z_A\mathbf{k})$$

or

$$\boxed{\mathbf{r} = (x_B - x_A)\mathbf{i} + (y_B - y_A)\mathbf{j} + (z_B - z_A)\mathbf{k}} \quad (2\text{–}11)$$

Thus, the **i**, **j**, **k** *components of the position vector* **r** *may be formed by taking the coordinates of the tail of the vector $A(x_A, y_A, z_A)$ and subtracting them from the corresponding coordinates of the head $B(x_B, y_B, z_B)$.* We can also form these components *directly*, Fig. 2–36*b*, by starting at A and moving through a distance of $(x_B - x_A)$ along the positive x axis (+**i**), then $(y_B - y_A)$ along the positive y axis (+**j**), and finally $(z_B - z_A)$ along the positive z axis (+**k**) to get to B.

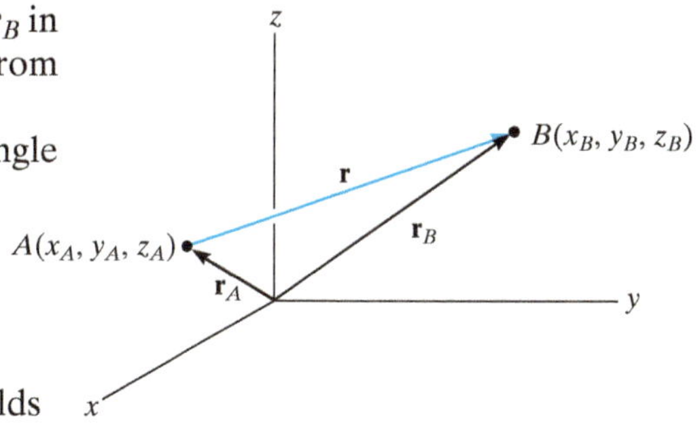

(a)

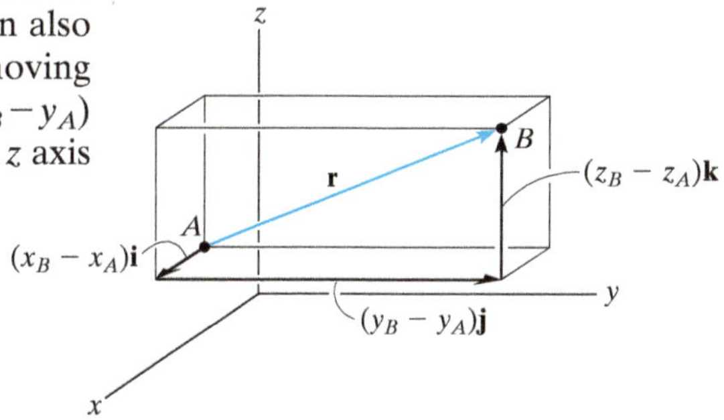

(b)

Fig. 2–36

If an x, y, z coordinate system is established, then the coordinates of two points A and B on the cable can be determined. From this the position vector **r** acting along the cable can be formulated. Its magnitude represents the distance from A to B, and its unit vector, $\mathbf{u} = \mathbf{r}/r$, gives the direction defined by α, β, γ.

EXAMPLE 2.12

An elastic rubber band is attached to points A and B as shown in Fig. 2–37*a*. Determine its length and its direction measured from A toward B.

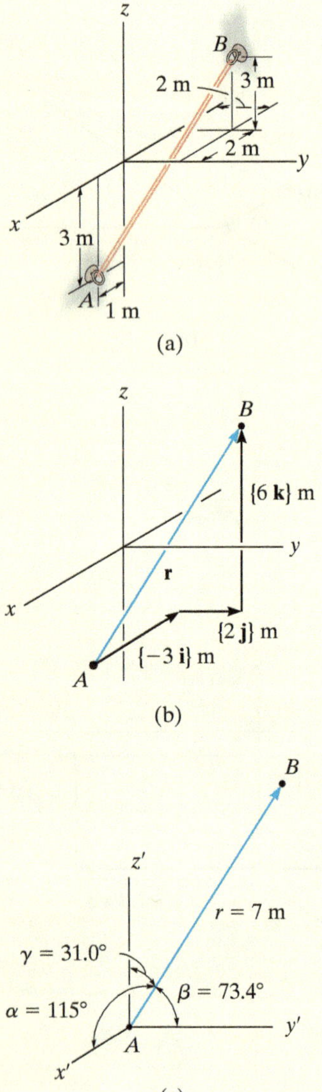

Fig. 2–37

SOLUTION

We first establish a position vector from A to B, Fig. 2–37*b*. In accordance with Eq. 2–11, the coordinates of the tail $A(1\text{ m}, 0, -3\text{ m})$ are subtracted from the coordinates of the head $B(-2\text{ m}, 2\text{ m}, 3\text{ m})$, which yields

$$\mathbf{r} = [-2\text{ m} - 1\text{ m}]\mathbf{i} + [2\text{ m} - 0]\mathbf{j} + [3\text{ m} - (-3\text{ m})]\mathbf{k}$$
$$= \{-3\mathbf{i} + 2\mathbf{j} + 6\mathbf{k}\}\text{ m}$$

These components of $\mathbf{r}$ can also be determined *directly* by realizing that they represent the direction and distance one must travel along each axis in order to move from A to B, i.e., along the x axis $\{-3\mathbf{i}\}$ m, along the y axis $\{2\mathbf{j}\}$ m, and finally along the z axis $\{6\mathbf{k}\}$ m.

The length of the rubber band is therefore

$$r = \sqrt{(-3\text{ m})^2 + (2\text{ m})^2 + (6\text{ m})^2} = 7\text{ m} \qquad \textit{Ans.}$$

Formulating a unit vector in the direction of $\mathbf{r}$, we have

$$\mathbf{u} = \frac{\mathbf{r}}{r} = -\frac{3}{7}\mathbf{i} + \frac{2}{7}\mathbf{j} + \frac{6}{7}\mathbf{k}$$

The components of this unit vector give the coordinate direction angles

$$\alpha = \cos^{-1}\left(-\frac{3}{7}\right) = 115° \qquad \textit{Ans.}$$

$$\beta = \cos^{-1}\left(\frac{2}{7}\right) = 73.4° \qquad \textit{Ans.}$$

$$\gamma = \cos^{-1}\left(\frac{6}{7}\right) = 31.0° \qquad \textit{Ans.}$$

NOTE: These angles are measured from the *positive axes* of a localized coordinate system placed at the tail of $\mathbf{r}$, as shown in Fig. 2–37*c*.

2.8 Force Vector Directed Along a Line

Quite often in three-dimensional statics problems, the direction of a force is specified by two points through which its line of action passes. Such a situation is shown in Fig. 2–38, where the force **F** is directed along the cord AB. We can formulate **F** as a Cartesian vector by realizing that it has the *same direction* and *sense* as the position vector **r** directed from point A to point B on the cord. This common direction is specified by the *unit vector* $\mathbf{u} = \mathbf{r}/r$. Hence,

$$\mathbf{F} = F\mathbf{u} = F\left(\frac{\mathbf{r}}{r}\right) = F\left(\frac{(x_B - x_A)\mathbf{i} + (y_B - y_A)\mathbf{j} + (z_B - z_A)\mathbf{k}}{\sqrt{(x_B - x_A)^2 + (y_B - y_A)^2 + (z_B - z_A)^2}}\right)$$

Although we have represented **F** symbolically in Fig. 2–38, note that it has *units of force*, unlike **r**, which has units of length.

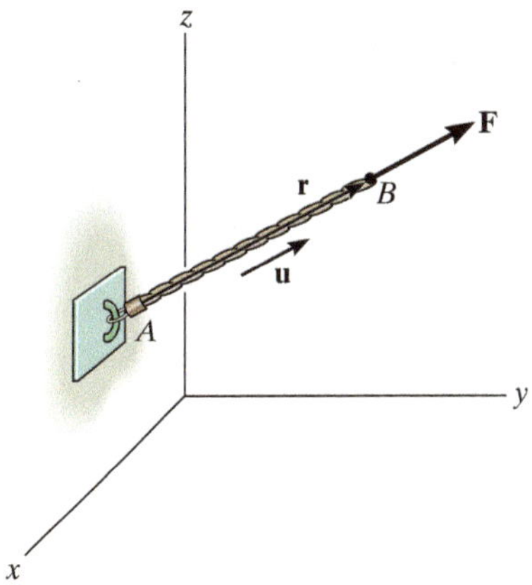

Fig. 2–38

The force **F** acting along the rope can be represented as a Cartesian vector by establishing x, y, z axes and first forming a position vector **r** along the length of the rope. Then the corresponding unit vector $\mathbf{u} = \mathbf{r}/r$ that defines the direction of both the rope and the force can be determined. Finally, the magnitude of the force is combined with its direction, $\mathbf{F} = F\mathbf{u}$.

Important Points

- A position vector locates one point in space relative to another point.
- The easiest way to formulate the components of a position vector is to determine the distance and direction that must be traveled along the x, y, z directions—going from the tail to the head of the vector.
- A force **F** acting in the direction of a position vector **r** can be represented in Cartesian form if the unit vector **u** of the position vector is determined and it is multiplied by the magnitude of the force, i.e., $\mathbf{F} = F\mathbf{u} = F(\mathbf{r}/r)$.

EXAMPLE 2.13

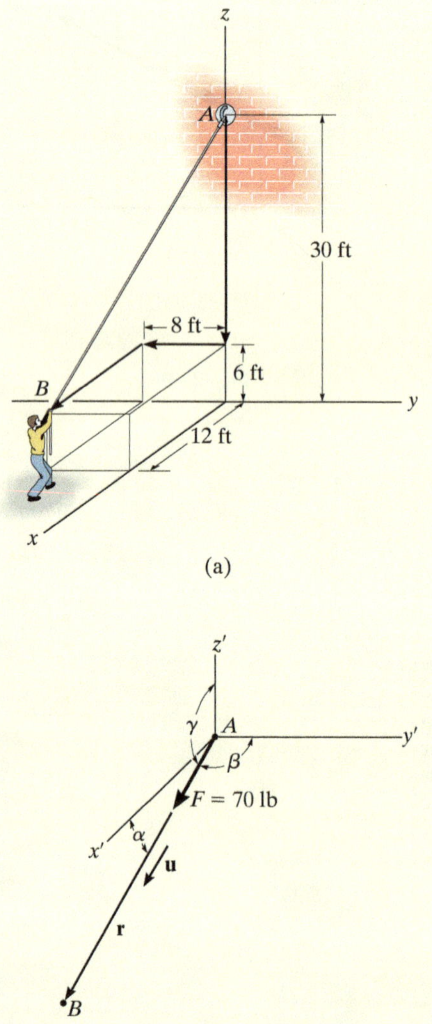

Fig. 2–39

The man shown in Fig. 2–39*a* pulls on the cord with a force of 70 lb. Represent this force acting on the support *A* as a Cartesian vector and determine its direction.

SOLUTION

Force **F** is shown in Fig. 2–39*b*. The *direction* of this vector, **u**, is determined from the position vector **r**, which extends from *A* to *B*. Rather than using the coordinates of the end points of the cord, **r** can be determined *directly* by noting in Fig. 2–39*a* that one must travel from *A* {−24**k**} ft, then {−8**j**} ft, and finally {12**i**} ft to get to *B*. Thus,

$$\mathbf{r} = \{12\mathbf{i} - 8\mathbf{j} - 24\mathbf{k}\} \text{ ft}$$

The magnitude of **r**, which represents the *length* of cord *AB*, is

$$r = \sqrt{(12 \text{ ft})^2 + (-8 \text{ ft})^2 + (-24 \text{ ft})^2} = 28 \text{ ft}$$

Forming the unit vector that defines the direction and sense of both **r** and **F**, we have

$$\mathbf{u} = \frac{\mathbf{r}}{r} = \frac{12}{28}\mathbf{i} - \frac{8}{28}\mathbf{j} - \frac{24}{28}\mathbf{k}$$

Since **F** has a *magnitude* of 70 lb and a *direction* specified by **u**, then

$$\mathbf{F} = F\mathbf{u} = 70 \text{ lb}\left(\frac{12}{28}\mathbf{i} - \frac{8}{28}\mathbf{j} - \frac{24}{28}\mathbf{k}\right)$$

$$= \{30\mathbf{i} - 20\mathbf{j} - 60\mathbf{k}\} \text{ lb} \qquad \textit{Ans.}$$

The coordinate direction angles are measured between **r** (or **F**) and the *positive axes* of a localized coordinate system with origin placed at *A*, Fig. 2–39*b*. From the components of the unit vector:

$$\alpha = \cos^{-1}\left(\frac{12}{28}\right) = 64.6° \qquad \textit{Ans.}$$

$$\beta = \cos^{-1}\left(\frac{-8}{28}\right) = 107° \qquad \textit{Ans.}$$

$$\gamma = \cos^{-1}\left(\frac{-24}{28}\right) = 149° \qquad \textit{Ans.}$$

NOTE: These results make sense when compared with the angles identified in Fig. 2–39*b*.

EXAMPLE 2.14

The force in Fig. 2–40*a* acts on the hook. Express it as a Cartesian vector.

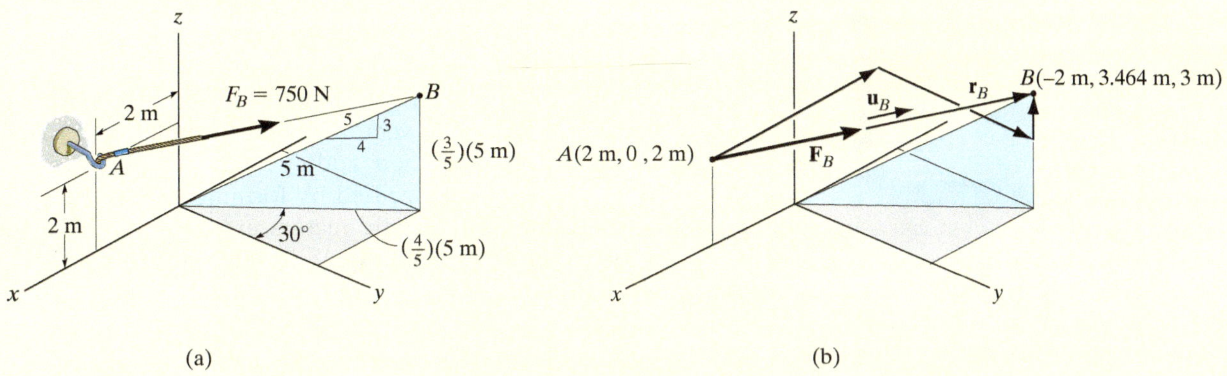

Fig. 2–40

SOLUTION

As shown in Fig. 2–40*b*, the coordinates for points *A* and *B* are

$$A(2 \text{ m}, 0, 2 \text{ m})$$

and

$$B\left[-\left(\frac{4}{5}\right)5 \sin 30° \text{ m}, \left(\frac{4}{5}\right)5 \cos 30° \text{ m}, \left(\frac{3}{5}\right) 5 \text{ m}\right]$$

or

$$B(-2 \text{ m}, 3.464 \text{ m}, 3 \text{ m})$$

Therefore, to go from *A* to *B*, one must travel $\{-4\mathbf{i}\}$ m, then $\{3.464\mathbf{j}\}$ m, and finally $\{1\mathbf{k}\}$ m. Thus,

$$\mathbf{u}_B = \left(\frac{\mathbf{r}_B}{r_B}\right) = \frac{\{-4\mathbf{i} + 3.464\mathbf{j} + 1\mathbf{k}\} \text{ m}}{\sqrt{(-4 \text{ m})^2 + (3.464 \text{ m})^2 + (1 \text{ m})^2}}$$

$$= -0.7428\mathbf{i} + 0.6433\mathbf{j} + 0.1857\mathbf{k}$$

Force $\mathbf{F}_B$ expressed as a Cartesian vector becomes

$$\mathbf{F}_B = F_B\mathbf{u}_B = (750 \text{ N})(-0.74281\mathbf{i} + 0.6433\mathbf{j} + 0.1857\mathbf{k})$$

$$= \{-557\mathbf{i} + 482\mathbf{j} + 139\mathbf{k}\} \text{ N} \qquad \textit{Ans.}$$

2

EXAMPLE 2.15

The roof is supported by cables as shown in the photo. If the cables exert forces $F_{AB} = 100$ N and $F_{AC} = 120$ N on the wall hook at A as shown in Fig. 2–41a, determine the resultant force acting at A. Express the result as a Cartesian vector.

SOLUTION

The resultant force $\mathbf{F}_R$ is shown graphically in Fig. 2–41b. We can express this force as a Cartesian vector by first formulating $\mathbf{F}_{AB}$ and $\mathbf{F}_{AC}$ as Cartesian vectors and then adding their components. The directions of $\mathbf{F}_{AB}$ and $\mathbf{F}_{AC}$ are specified by forming unit vectors $\mathbf{u}_{AB}$ and $\mathbf{u}_{AC}$ along the cables. These unit vectors are obtained from the associated position vectors $\mathbf{r}_{AB}$ and $\mathbf{r}_{AC}$. With reference to Fig. 2–41a, to go from A to B, we must travel $\{-4\mathbf{k}\}$ m, and then $\{4\mathbf{i}\}$ m. Thus,

$$\mathbf{r}_{AB} = \{4\mathbf{i} - 4\mathbf{k}\}\text{ m}$$

$$r_{AB} = \sqrt{(4\text{ m})^2 + (-4\text{ m})^2} = 5.66\text{ m}$$

$$\mathbf{F}_{AB} = F_{AB}\left(\frac{\mathbf{r}_{AB}}{r_{AB}}\right) = (100\text{ N})\left(\frac{4}{5.66}\mathbf{i} - \frac{4}{5.66}\mathbf{k}\right)$$

$$\mathbf{F}_{AB} = \{70.7\mathbf{i} - 70.7\mathbf{k}\}\text{ N}$$

To go from A to C, we must travel $\{-4\mathbf{k}\}$ m, then $\{2\mathbf{j}\}$ m, and finally $\{4\mathbf{i}\}$. Thus,

$$\mathbf{r}_{AC} = \{4\mathbf{i} + 2\mathbf{j} - 4\mathbf{k}\}\text{ m}$$

$$r_{AC} = \sqrt{(4\text{ m})^2 + (2\text{ m})^2 + (-4\text{ m})^2} = 6\text{ m}$$

$$\mathbf{F}_{AC} = F_{AC}\left(\frac{\mathbf{r}_{AC}}{r_{AC}}\right) = (120\text{ N})\left(\frac{4}{6}\mathbf{i} + \frac{2}{6}\mathbf{j} - \frac{4}{6}\mathbf{k}\right)$$

$$= \{80\mathbf{i} + 40\mathbf{j} - 80\mathbf{k}\}\text{ N}$$

The resultant force is therefore

$$\mathbf{F}_R = \mathbf{F}_{AB} + \mathbf{F}_{AC} = \{70.7\mathbf{i} - 70.7\mathbf{k}\}\text{ N} + \{80\mathbf{i} + 40\mathbf{j} - 80\mathbf{k}\}\text{ N}$$

$$= \{151\mathbf{i} + 40\mathbf{j} - 151\mathbf{k}\}\text{ N} \qquad \textit{Ans.}$$

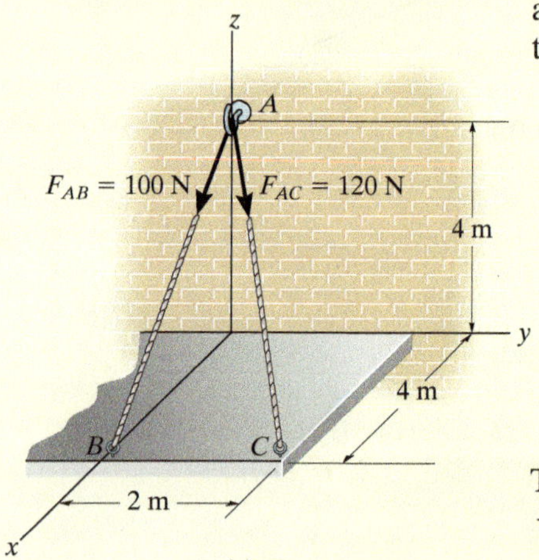

(a)

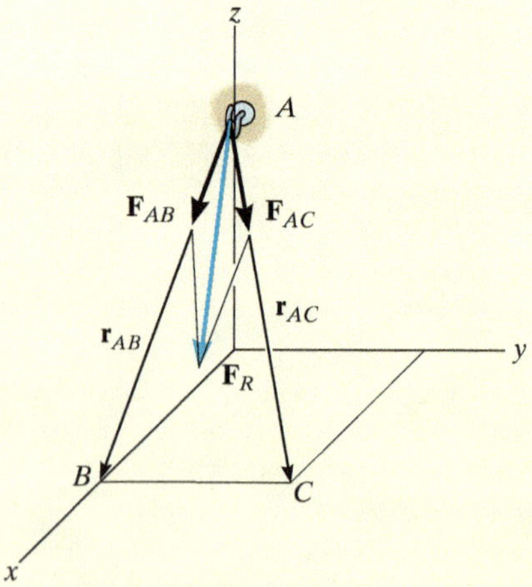

(b)

Fig. 2–41

FUNDAMENTAL PROBLEMS

F2–19. Express the position vector $\mathbf{r}_{AB}$ in Cartesian vector form, then determine its magnitude and coordinate direction angles.

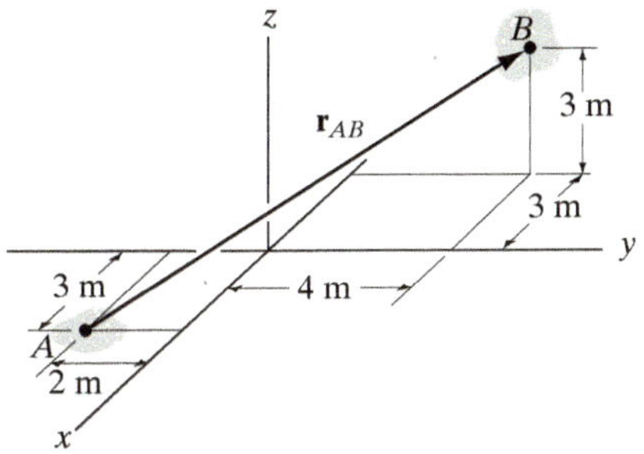

F2–19

F2–20. Determine the length of the rod and the position vector directed from A to B. What is the angle θ?

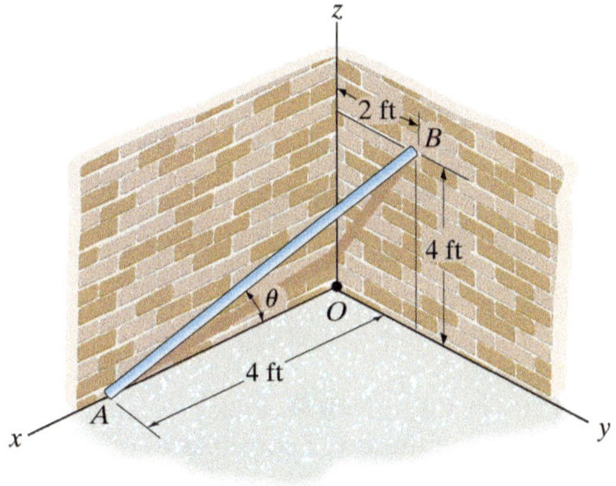

F2–20

F2–21. Express the force as a Cartesian vector.

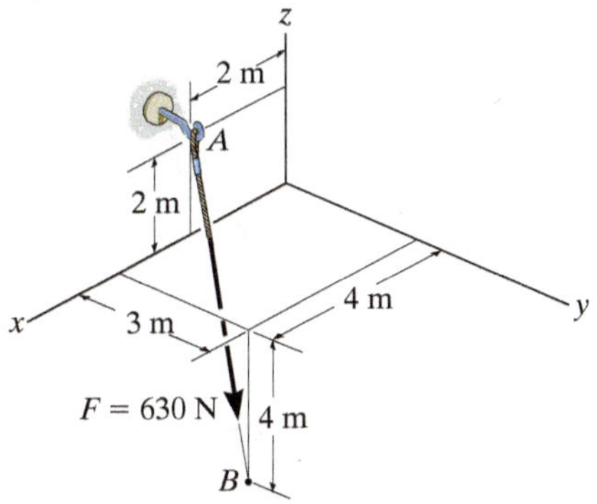

F2–21

F2–22. Express the force as a Cartesian vector.

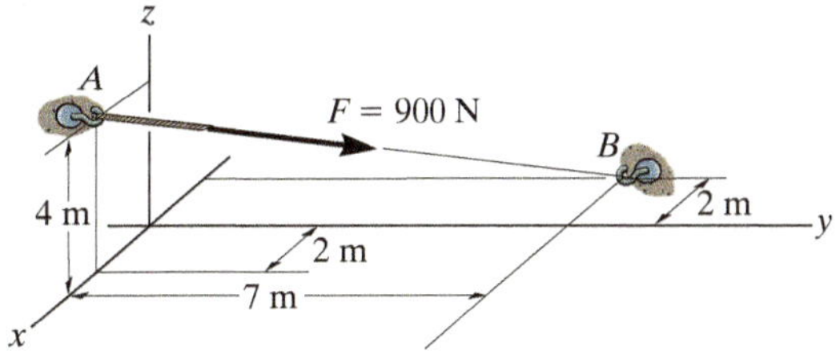

F2–22

F2–23. Determine the magnitude of the resultant force at A.

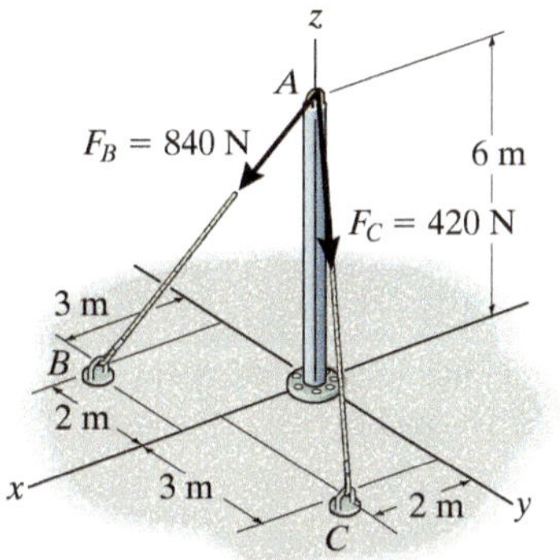

F2–23

F2–24. Determine the resultant force at A.

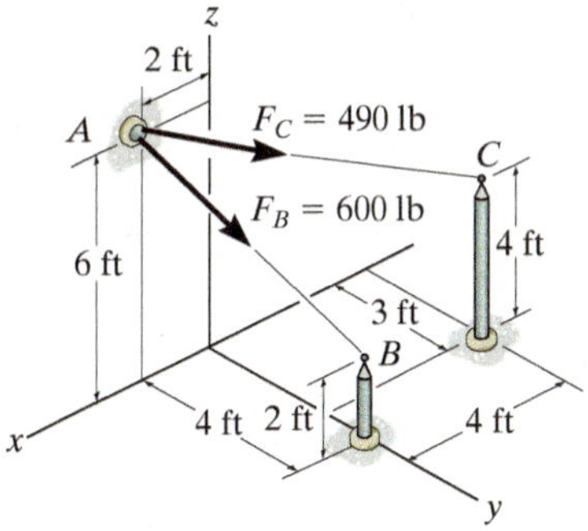

F2–24

PROBLEMS

2

2–86. Express the position vector **r** in Cartesian vector form; then determine its magnitude and coordinate direction angles.

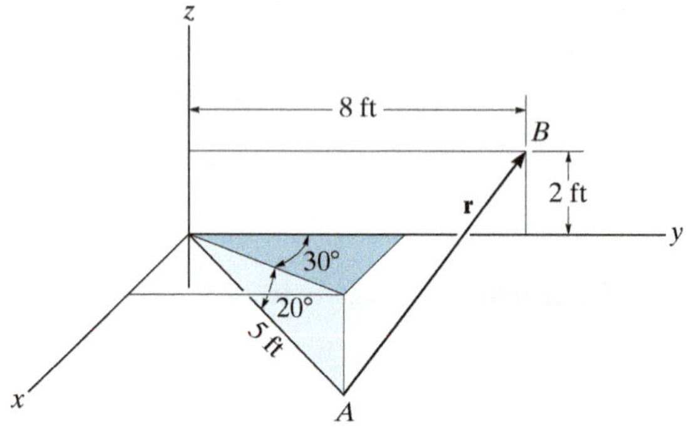

Prob. 2–86

2–87. Determine the lengths of wires *AD, BD,* and *CD*. The ring at *D* is midway between *A* and *B*.

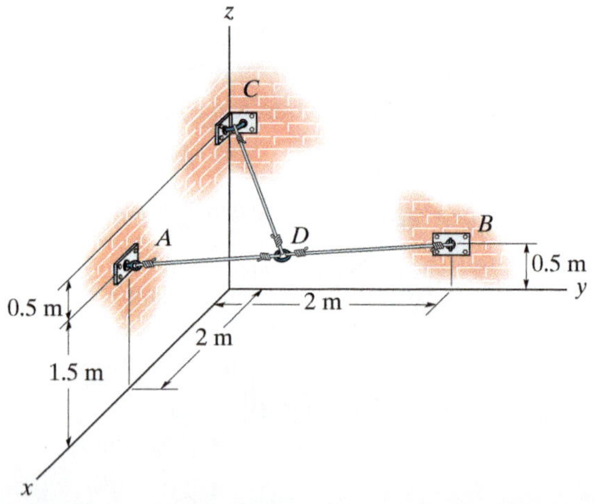

Prob. 2–87

***2–88.** Determine the length of member *AB* of the truss by first establishing a Cartesian position vector from *A* to *B* and then determining its magnitude.

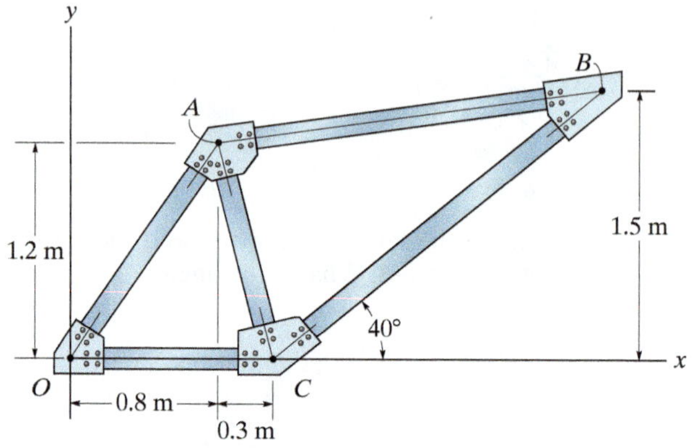

Prob. 2–88

2–89. If $\mathbf{F} = \{350\mathbf{i} - 250\mathbf{j} - 450\mathbf{k}\}$ N and cable *AB* is 9 m long, determine the *x*, *y*, *z* coordinates of point *A*.

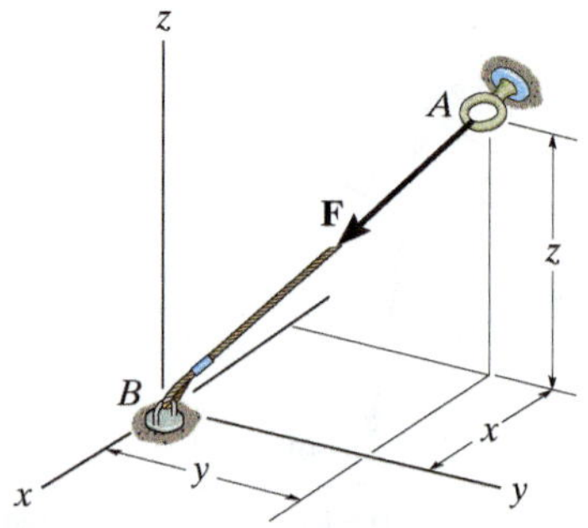

Prob. 2–89

2–90. Express $\mathbf{F}_B$ and $\mathbf{F}_C$ in Cartesian vector form.

2–91. Determine the magnitude and coordinate direction angles of the resultant force acting at A.

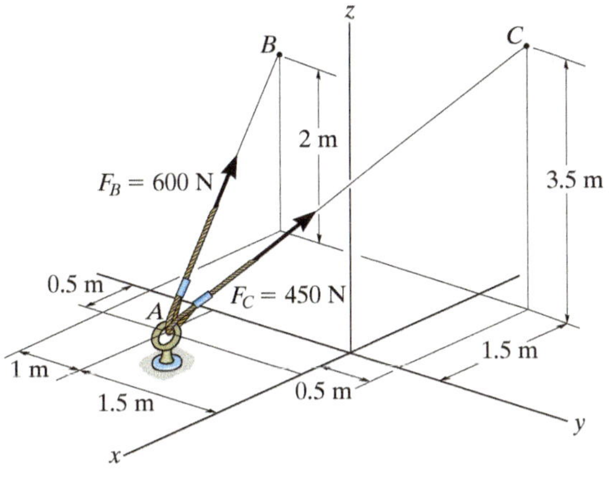

Probs. 2–90/91

2–94. The tower is held in place by three cables. If the force of each cable acting on the tower is shown, determine the magnitude and coordinate direction angles α, β, γ of the resultant force. Take $x = 20$ m, $y = 15$ m.

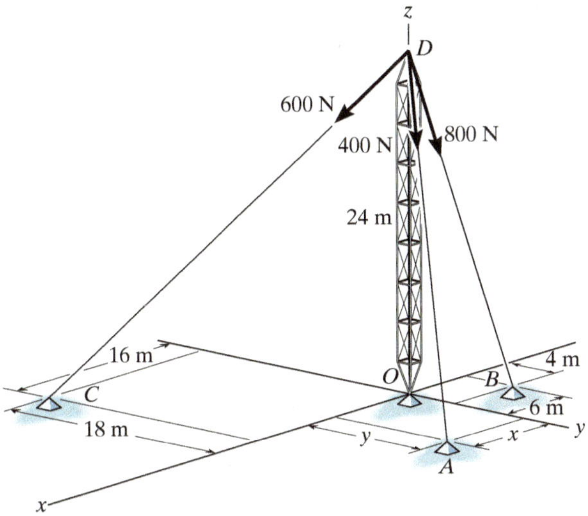

Prob. 2–94

***2–92.** If $F_B = 560$ N and $F_C = 700$ N, determine the magnitude and coordinate direction angles of the resultant force acting on the flag pole.

2–93. If $F_B = 700$ N, and $F_C = 560$ N, determine the magnitude and coordinate direction angles of the resultant force acting on the flag pole.

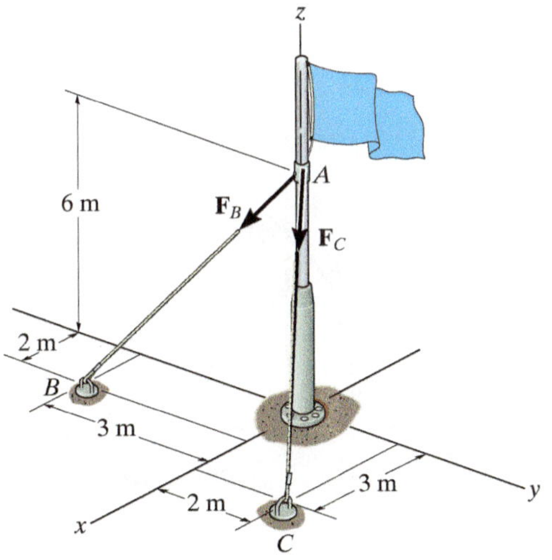

Probs. 2–92/93

2–95. At a given instant, the position of a plane at A and a train at B are measured relative to a radar antenna at O. Determine the distance d between A and B at this instant. To solve the problem, formulate a position vector, directed from A to B, and then determine its magnitude.

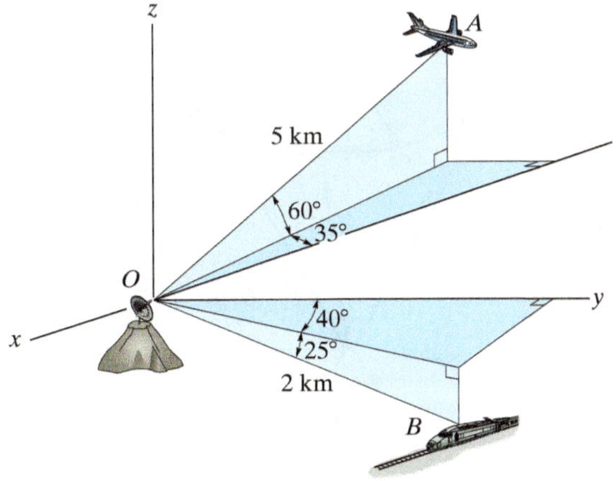

Prob. 2–95

*2–96. The man pulls on the rope at C with a force of 70 lb which causes the forces $\mathbf{F}_A$ and $\mathbf{F}_C$ at B to have this same magnitude. Express each of these two forces as Cartesian vectors.

2–97. The man pulls on the rope at C with a force of 70 lb which causes the forces $\mathbf{F}_A$ and $\mathbf{F}_C$ at B to have this same magnitude. Determine the magnitude and coordinate direction angles of the resultant force acting at B.

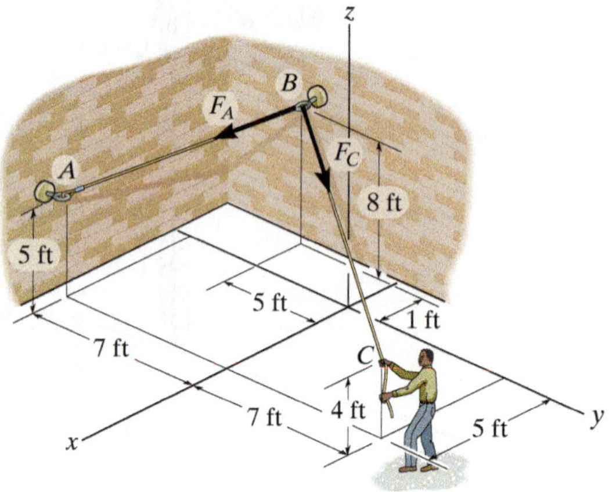

Probs. 2–96/97

2–98. The load at A creates a force of 60 lb in wire AB. Express this force as a Cartesian vector acting on A and directed toward B as shown.

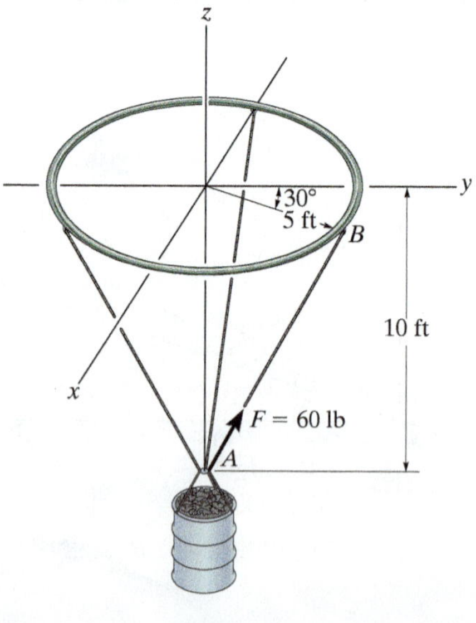

Prob. 2–98

2–99. Determine the magnitude and coordinate direction angles of the resultant force acting at point A.

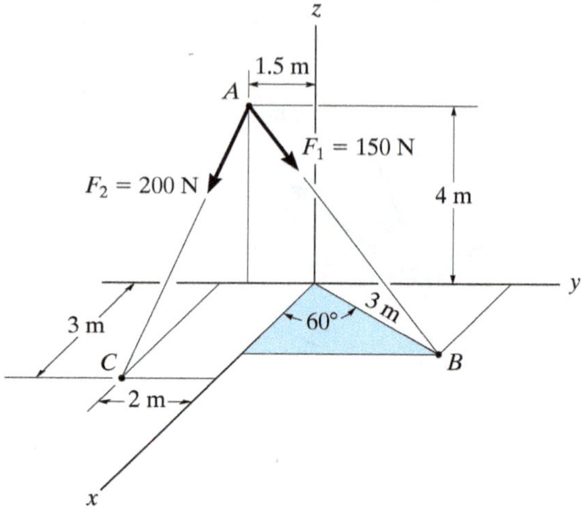

Prob. 2–99

*2–100. The guy wires are used to support the telephone pole. Represent the force in each wire in Cartesian vector form. Neglect the diameter of the pole.

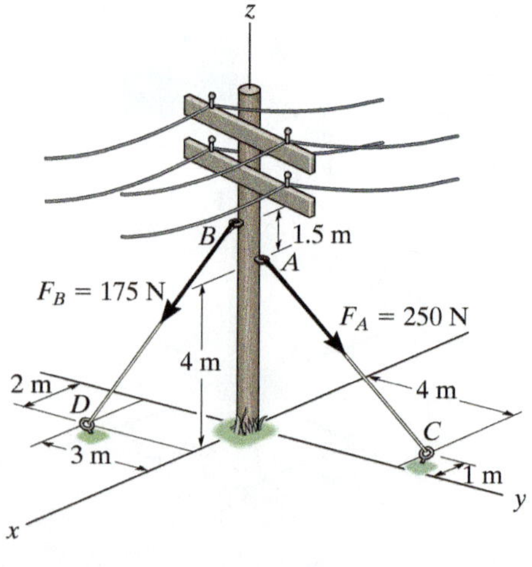

Prob. 2–100

2–101. The two mooring cables exert forces on the stern of a ship as shown. Represent each force as as Cartesian vector and determine the magnitude and coordinate direction angles of the resultant.

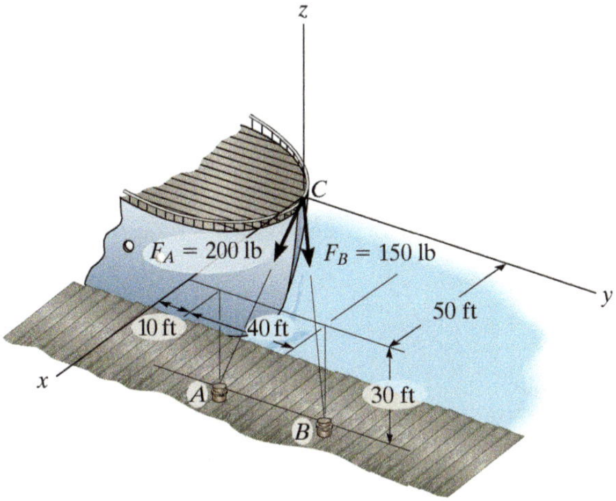

Prob. 2–101

2–102. Each of the four forces acting at E has a magnitude of 28 kN. Express each force as a Cartesian vector and determine the resultant force.

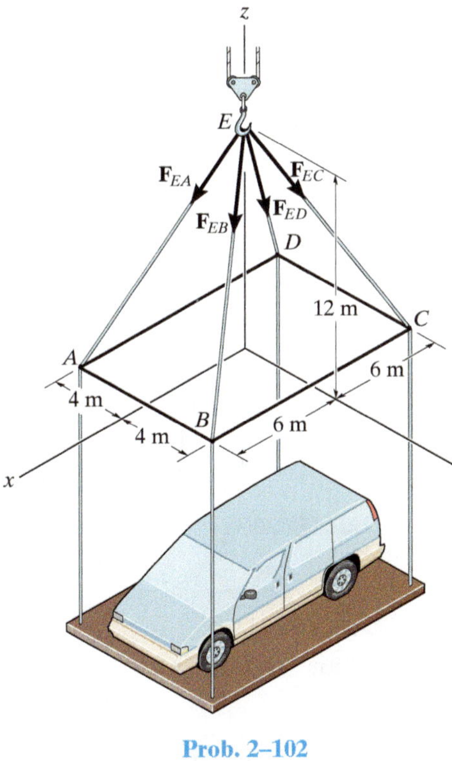

Prob. 2–102

2–103. If the force in each cable tied to the bin is 70 lb, determine the magnitude and coordinate direction angles of the resultant force.

***2–104.** If the resultant of the four forces is $\mathbf{F}_R = \{-360\mathbf{k}\}$ lb, determine the tension developed in each cable. Due to symmetry, the tension in the four cables is the same.

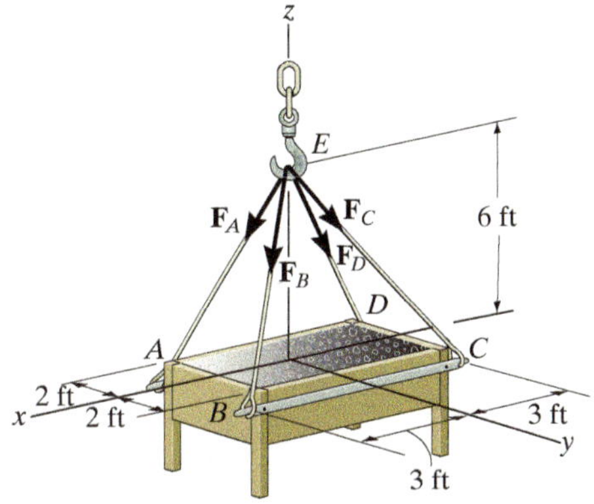

Probs. 2–103/104

2–105. The pipe is supported at its ends by a cord AB. If the cord exerts a force of $F = 12$ lb on the pipe at A, express this force as a Cartesian vector.

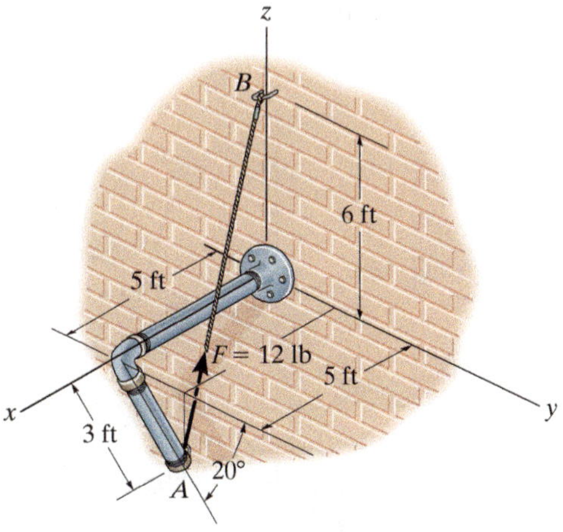

Prob. 2–105

2–106. The chandelier is supported by three chains which are concurrent at point O. If the force in each chain has a magnitude of 60 lb, express each force as a Cartesian vector and determine the magnitude and coordinate direction angles of the resultant force.

2–107. The chandelier is supported by three chains which are concurrent at point O. If the resultant force at O has a magnitude of 130 lb and is directed along the negative z axis, determine the force in each chain.

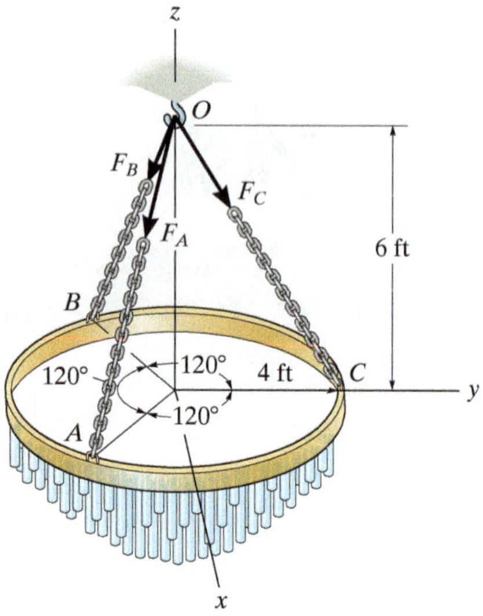

Probs. 2–106/107

***2–108.** Determine the magnitude and coordinate direction angles of the resultant force. Set $F_B = 630$ N, $F_C = 520$ N and $F_D = 750$ N, and $x = 3$ m and $z = 3.5$ m.

2–109. If the magnitude of the resultant force is 1300 N and acts along the axis of the strut, directed from point A towards O, determine the magnitudes of the three forces acting on the strut. Set $x = 0$ and $z = 5.5$ m.

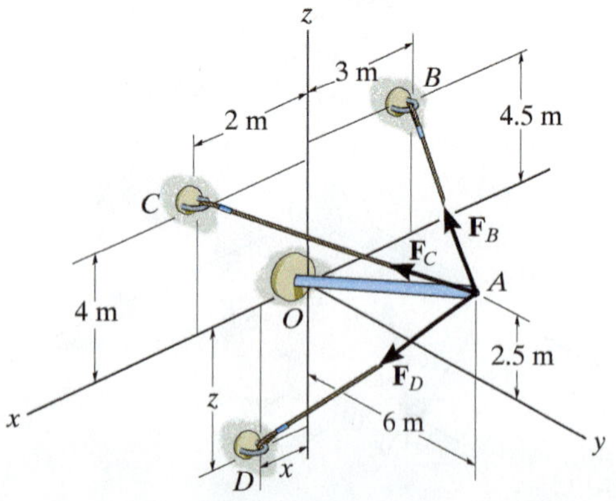

Probs. 2–108/109

2–110. The cable attached to the shear-leg derrick exerts a force on the derrick of $F = 350$ lb. Express this force as a Cartesian vector.

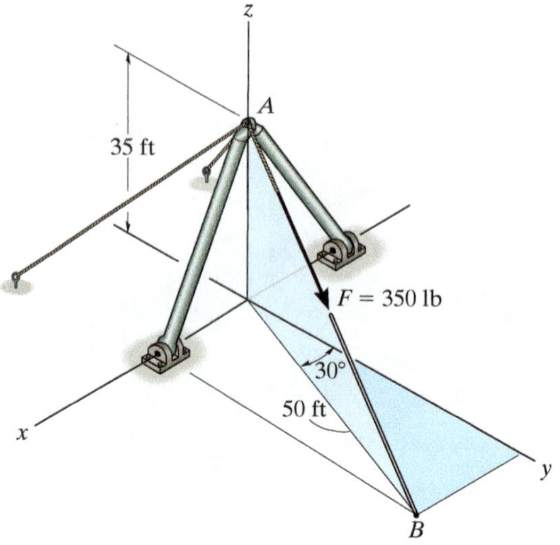

Prob. 2–110

2–111. The window is held open by chain AB. Determine the length of the chain, and express the 50-lb force acting at A along the chain as a Cartesian vector and determine its coordinate direction angles.

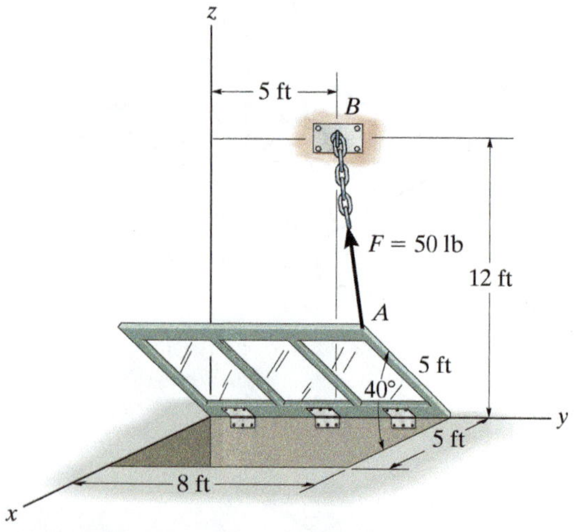

Prob. 2–111

2.9 Dot Product

Occasionally in statics one has to find the angle between two lines or the components of a force parallel and perpendicular to a line. In two dimensions, these problems can readily be solved by trigonometry since the geometry is easy to visualize. In three dimensions, however, this is often difficult, and consequently vector methods should be employed for the solution. The dot product, which defines a particular method for "multiplying" two vectors, is used to solve the above-mentioned problems.

The *dot product* of vectors **A** and **B**, written **A · B**, and read "**A** dot **B**" is defined as the product of the magnitudes of **A** and **B** and the cosine of the angle θ between their tails, Fig. 2–42. Expressed in equation form,

$$\mathbf{A} \cdot \mathbf{B} = AB \cos \theta \qquad (2\text{–}12)$$

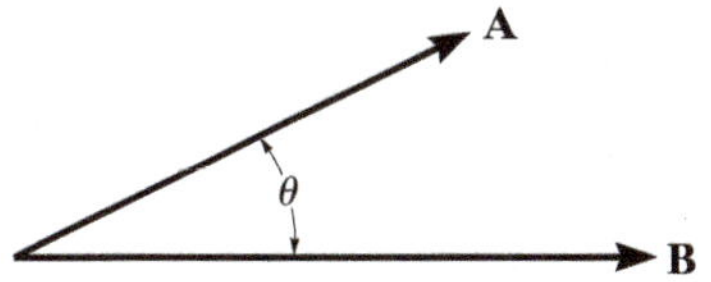

Fig. 2–42

where $0° \leq \theta \leq 180°$. The dot product is often referred to as the *scalar product* of vectors since the result is a *scalar* and not a vector.

Laws of Operation.

1. Commutative law: $\mathbf{A} \cdot \mathbf{B} = \mathbf{B} \cdot \mathbf{A}$
2. Multiplication by a scalar: $a(\mathbf{A} \cdot \mathbf{B}) = (a\mathbf{A}) \cdot \mathbf{B} = \mathbf{A} \cdot (a\mathbf{B})$
3. Distributive law: $\mathbf{A} \cdot (\mathbf{B} + \mathbf{D}) = (\mathbf{A} \cdot \mathbf{B}) + (\mathbf{A} \cdot \mathbf{D})$

It is easy to prove the first and second laws by using Eq. 2–12. The proof of the distributive law is left as an exercise (see Prob. 2–112).

Cartesian Vector Formulation. Equation 2–12 must be used to find the dot product for any two Cartesian unit vectors. For example, $\mathbf{i} \cdot \mathbf{i} = (1)(1) \cos 0° = 1$ and $\mathbf{i} \cdot \mathbf{j} = (1)(1) \cos 90° = 0$. If we want to find the dot product of two general vectors **A** and **B** that are expressed in Cartesian vector form, then we have

$$\begin{aligned}
\mathbf{A} \cdot \mathbf{B} &= (A_x\mathbf{i} + A_y\mathbf{j} + A_z\mathbf{k}) \cdot (B_x\mathbf{i} + B_y\mathbf{j} + B_z\mathbf{k}) \\
&= A_xB_x(\mathbf{i} \cdot \mathbf{i}) + A_xB_y(\mathbf{i} \cdot \mathbf{j}) + A_xB_z(\mathbf{i} \cdot \mathbf{k}) \\
&\quad + A_yB_x(\mathbf{j} \cdot \mathbf{i}) + (A_yB_y(\mathbf{j} \cdot \mathbf{j}) + A_yB_z(\mathbf{j} \cdot \mathbf{k}) \\
&\quad + A_zB_x(\mathbf{k} \cdot \mathbf{i}) + A_zB_y(\mathbf{k} \cdot \mathbf{j}) + A_zB_z(\mathbf{k} \cdot \mathbf{k})
\end{aligned}$$

Carrying out the dot-product operations, the final result becomes

$$\mathbf{A} \cdot \mathbf{B} = A_xB_x + A_yB_y + A_zB_z \qquad (2\text{–}13)$$

Thus, to determine the dot product of two Cartesian vectors, multiply their corresponding x, y, z components and sum these products algebraically. Note that the result will be either a positive or negative *scalar*.

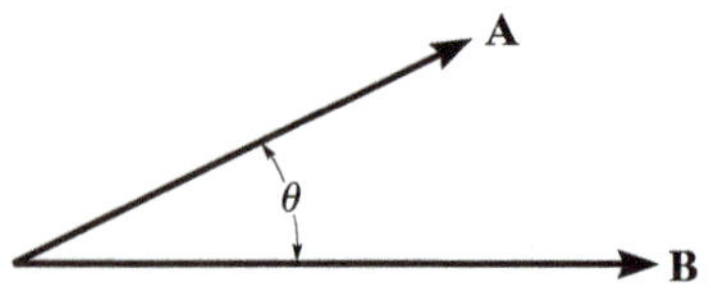

Fig. 2–42 (Repeated)

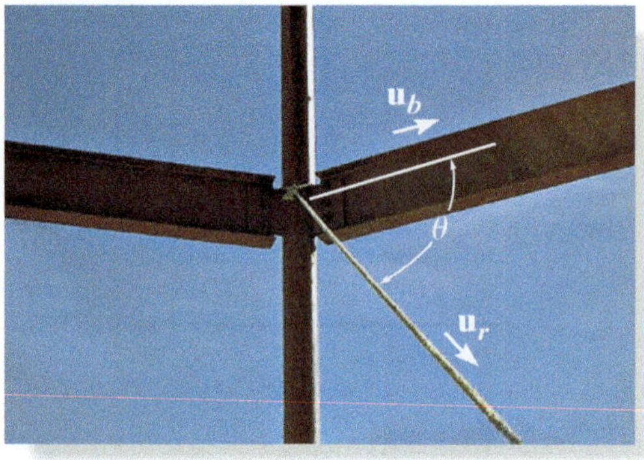

The angle θ between the rope and the beam can be determined by formulating unit vectors along the beam and rope and then using the dot product $\mathbf{u}_b \cdot \mathbf{u}_r = (1)(1)\cos\theta$.

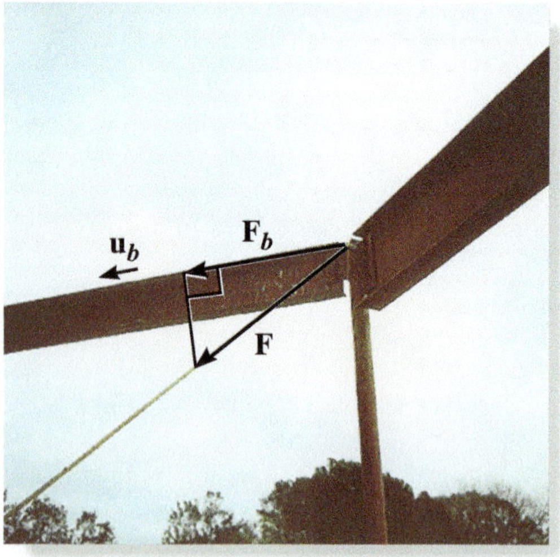

The projection of the cable force $\mathbf{F}$ along the beam can be determined by first finding the unit vector $\mathbf{u}_b$ that defines this direction. Then apply the dot product, $F_b = \mathbf{F} \cdot \mathbf{u}_b$.

Applications. The dot product has two important applications in mechanics.

- ***The angle formed between two vectors or intersecting lines.*** The angle θ between the tails of vectors $\mathbf{A}$ and $\mathbf{B}$ in Fig. 2–42 can be determined from Eq. 2–12 and written as

$$\theta = \cos^{-1}\left(\frac{\mathbf{A}\cdot\mathbf{B}}{AB}\right) \quad 0^\circ \le \theta \le 180^\circ$$

Here $\mathbf{A}\cdot\mathbf{B}$ is found from Eq. 2–13. In particular, notice that if $\mathbf{A}\cdot\mathbf{B} = 0, \theta = \cos^{-1} 0 = 90^\circ$ so that $\mathbf{A}$ will be *perpendicular* to $\mathbf{B}$.

- ***The components of a vector parallel and perpendicular to a line.*** The component of vector $\mathbf{A}$ parallel to or collinear with the line aa in Fig. 2–43 is defined by A_a where $A_a = A\cos\theta$. This component is sometimes referred to as the *projection* of $\mathbf{A}$ onto the line, since a *right angle* is formed in the construction. If the *direction* of the line is specified by the unit vector $\mathbf{u}_a$, then since $u_a = 1$, we can determine the magnitude of A_a directly from the dot product (Eq. 2–12); i.e.,

$$A_a = A\cos\theta = \mathbf{A}\cdot\mathbf{u}_a$$

Hence, the scalar projection of $\mathbf{A}$ *along a line is determined from the dot product of* $\mathbf{A}$ *and the unit vector* $\mathbf{u}_a$ *which defines the direction of the line.* Notice that if this result is positive, then $\mathbf{A}_a$ has a directional sense which is the same as $\mathbf{u}_a$, whereas if A_a is a negative scalar, then $\mathbf{A}_a$ has the opposite sense of direction to $\mathbf{u}_a$.

The component $\mathbf{A}_a$ represented as a *vector* is therefore

$$\mathbf{A}_a = A_a\,\mathbf{u}_a$$

The component of $\mathbf{A}$ that is *perpendicular* to line aa can also be obtained, Fig. 2–43. Since $\mathbf{A} = \mathbf{A}_a + \mathbf{A}_\perp$, then $\mathbf{A}_\perp = \mathbf{A} - \mathbf{A}_a$. There are two possible ways of obtaining $A_\perp$. One way would be to determine θ from the dot product, $\theta = \cos^{-1}(\mathbf{A}\cdot\mathbf{u}_A/A)$, then $A_\perp = A\sin\theta$. Alternatively, if A_a is known, then by Pythagorean's theorem we can also write $A_\perp = \sqrt{A^2 - A_a^{\,2}}$.

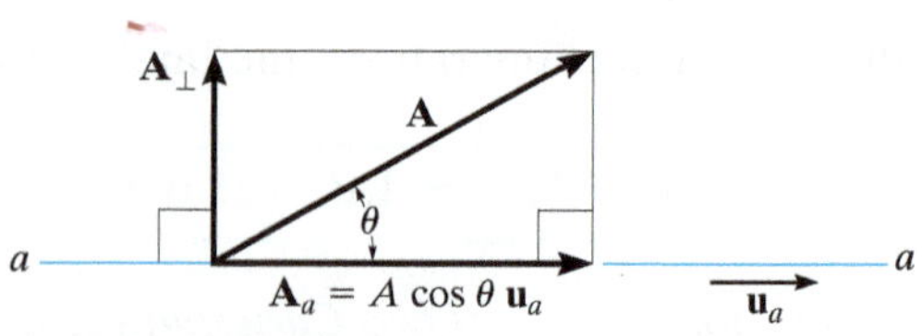

Fig. 2–43

Important Points

- The dot product is used to determine the angle between two vectors or the projection of a vector in a specified direction.
- If vectors **A** and **B** are expressed in Cartesian vector form, the dot product is determined by multiplying the respective x, y, z scalar components and algebraically adding the results, i.e., $\mathbf{A} \cdot \mathbf{B} = A_x B_x + A_y B_y + A_z B_z$.
- From the definition of the dot product, the angle formed between the tails of vectors **A** and **B** is $\theta = \cos^{-1}(\mathbf{A} \cdot \mathbf{B}/AB)$.
- The magnitude of the projection of vector **A** along a line aa whose direction is specified by $\mathbf{u}_a$ is determined from the dot product $A_a = \mathbf{A} \cdot \mathbf{u}_a$.

EXAMPLE 2.16

Determine the magnitudes of the projection of the force **F** in Fig. 2–44 onto the u and v axes.

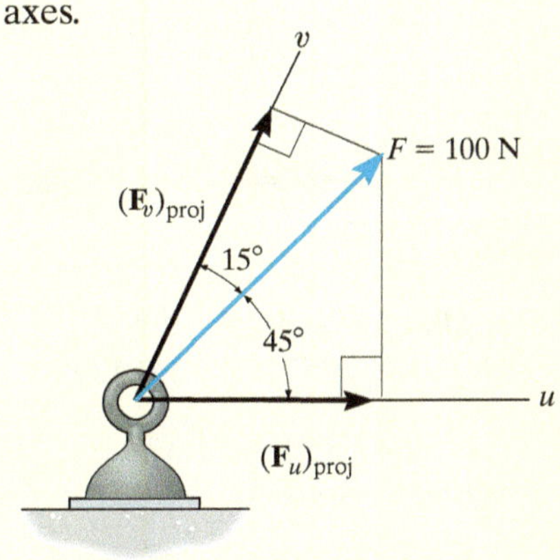

Fig. 2–44

SOLUTION

Projections of Force. The graphical representation of the *projections* is shown in Fig. 2–44. From this figure, the magnitudes of the projections of **F** onto the u and v axes can be obtained by trigonometry:

$$(F_u)_{\text{proj}} = (100\text{ N})\cos 45^\circ = 70.7\text{ N} \qquad \textit{Ans.}$$

$$(F_v)_{\text{proj}} = (100\text{ N})\cos 15^\circ = 96.6\text{ N} \qquad \textit{Ans.}$$

NOTE: These projections are not equal to the magnitudes of the components of force **F** along the u and v axes found from the parallelogram law. They will only be equal if the u and v axes are *perpendicular* to one another.

EXAMPLE 2.17

The frame shown in Fig. 2–45*a* is subjected to a horizontal force $\mathbf{F} = \{300\mathbf{j}\}$. Determine the magnitude of the components of this force parallel and perpendicular to member *AB*.

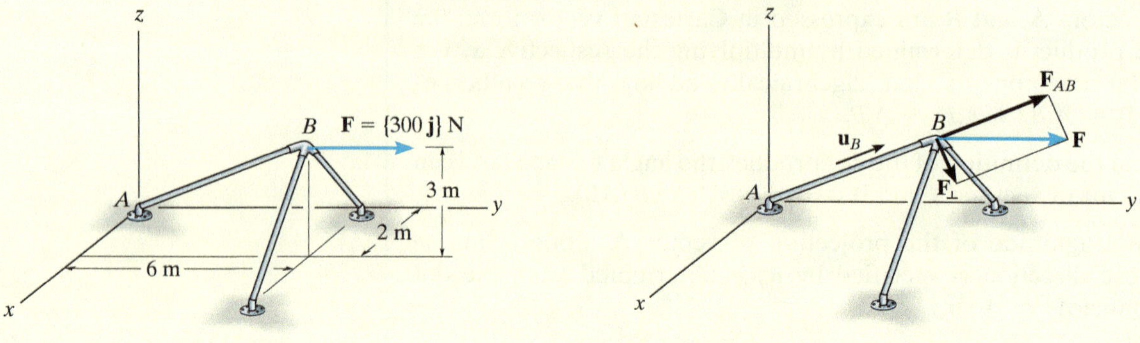

(a) (b)

Fig. 2–45

SOLUTION

The magnitude of the component of **F** along *AB* is equal to the dot product of **F** and the unit vector $\mathbf{u}_B$, which defines the direction of *AB*, Fig. 2–45*b*. Since

$$\mathbf{u}_B = \frac{\mathbf{r}_B}{r_B} = \frac{2\mathbf{i} + 6\mathbf{j} + 3\mathbf{k}}{\sqrt{(2)^2 + (6)^2 + (3)^2}} = 0.286\mathbf{i} + 0.857\mathbf{j} + 0.429\mathbf{k}$$

then

$$\begin{aligned} F_{AB} &= F\cos\theta = \mathbf{F}\cdot\mathbf{u}_B = (300\mathbf{j})\cdot(0.286\mathbf{i} + 0.857\mathbf{j} + 0.429\mathbf{k}) \\ &= (0)(0.286) + (300)(0.857) + (0)(0.429) \\ &= 257.1\text{ N} \end{aligned}$$ *Ans.*

Since the result is a positive scalar, $\mathbf{F}_{AB}$ has the same sense of direction as $\mathbf{u}_B$, Fig. 2–45*b*.

Expressing $\mathbf{F}_{AB}$ in Cartesian vector form, we have

$$\begin{aligned} \mathbf{F}_{AB} = F_{AB}\mathbf{u}_B &= (257.1\text{ N})(0.286\mathbf{i} + 0.857\mathbf{j} + 0.429\mathbf{k}) \\ &= \{73.5\mathbf{i} + 220\mathbf{j} + 110\mathbf{k}\}\text{ N} \end{aligned}$$ *Ans.*

The perpendicular component, Fig. 2–45*b*, is therefore

$$\begin{aligned} \mathbf{F}_\perp &= \mathbf{F} - \mathbf{F}_{AB} = 300\mathbf{j} - (73.5\mathbf{i} + 220\mathbf{j} + 110\mathbf{k}) \\ &= \{-73.5\mathbf{i} + 79.6\mathbf{j} - 110\mathbf{k}\}\text{ N} \end{aligned}$$

Its magnitude can be determined either from this vector or by using the Pythagorean theorem, Fig. 2–45*b*:

$$\begin{aligned} F_\perp &= \sqrt{F^2 - F_{AB}^2} = \sqrt{(300\text{ N})^2 - (257.1\text{ N})^2} \\ &= 155\text{ N} \end{aligned}$$ *Ans.*

EXAMPLE 2.18

The pipe in Fig. 2–46*a* is subjected to the force of $F = 80$ lb. Determine the angle θ between **F** and the pipe segment *BA* and the projection of **F** along this segment.

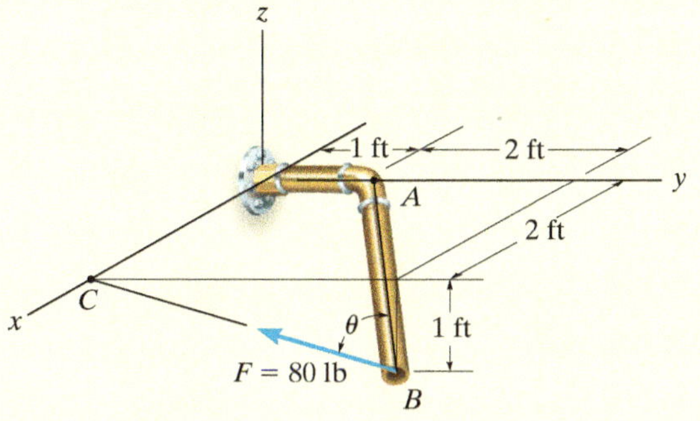

(a)

SOLUTION

Angle θ. First we will establish position vectors from *B* to *A* and *B* to *C*; Fig. 2–46*b*. Then we will determine the angle θ between the tails of these two vectors.

$$\mathbf{r}_{BA} = \{-2\mathbf{i} - 2\mathbf{j} + 1\mathbf{k}\}\text{ ft},\quad r_{BA} = 3\text{ ft}$$
$$\mathbf{r}_{BC} = \{-3\mathbf{j} + 1\mathbf{k}\}\text{ ft},\quad r_{BC} = \sqrt{10}\text{ ft}$$

Thus,

$$\cos\theta = \frac{\mathbf{r}_{BA}\cdot\mathbf{r}_{BC}}{r_{BA}r_{BC}} = \frac{(-2)(0) + (-2)(-3) + (1)(1)}{3\sqrt{10}} = 0.7379$$

$$\theta = 42.5^\circ \qquad \textit{Ans.}$$

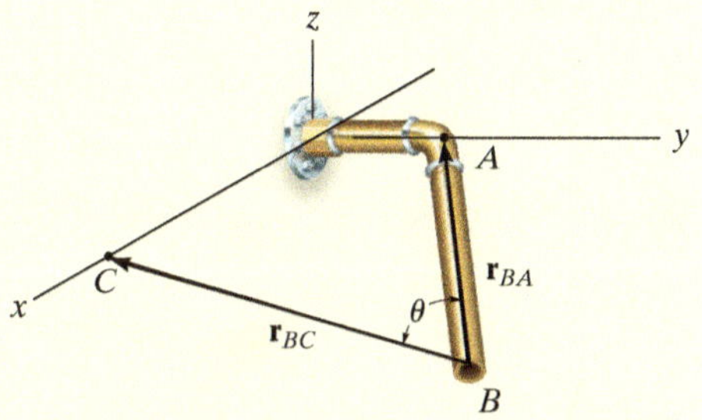

(b)

Components of F. The component of **F** along *BA* is shown in Fig. 2–46*c*. We must first formulate the unit vector along *BA* and force **F** as Cartesian vectors.

$$\mathbf{u}_{BA} = \frac{\mathbf{r}_{BA}}{r_{BA}} = \frac{(-2\mathbf{i} - 2\mathbf{j} + 1\mathbf{k})}{3} = -\frac{2}{3}\mathbf{i} - \frac{2}{3}\mathbf{j} + \frac{1}{3}\mathbf{k}$$

$$\mathbf{F} = 80\text{ lb}\left(\frac{\mathbf{r}_{BC}}{r_{BC}}\right) = 80\left(\frac{-3\mathbf{j} + 1\mathbf{k}}{\sqrt{10}}\right) = -75.89\mathbf{j} + 25.30\mathbf{k}$$

Thus,

$$F_{BA} = \mathbf{F}\cdot\mathbf{u}_{BA} = (-75.89\mathbf{j} + 25.30\mathbf{k})\cdot\left(-\frac{2}{3}\mathbf{i} - \frac{2}{3}\mathbf{j} + \frac{1}{3}\mathbf{k}\right)$$
$$= 0\left(-\frac{2}{3}\right) + (-75.89)\left(-\frac{2}{3}\right) + (25.30)\left(\frac{1}{3}\right)$$
$$= 59.0\text{ lb} \qquad \textit{Ans.}$$

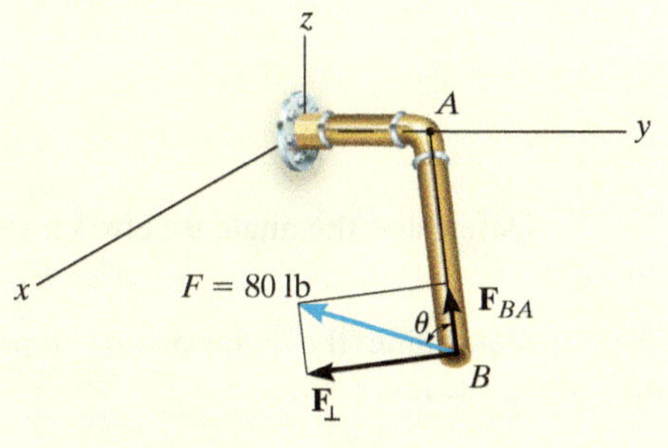

(c)

Fig. 2–46

NOTE: Since θ has been calculated, then also, $F_{BA} = F\cos\theta = 80\text{ lb}\cos 42.5^\circ = 59.0$ lb.

2

FUNDAMENTAL PROBLEMS

F2–25. Determine the angle θ between the force and the line AO.

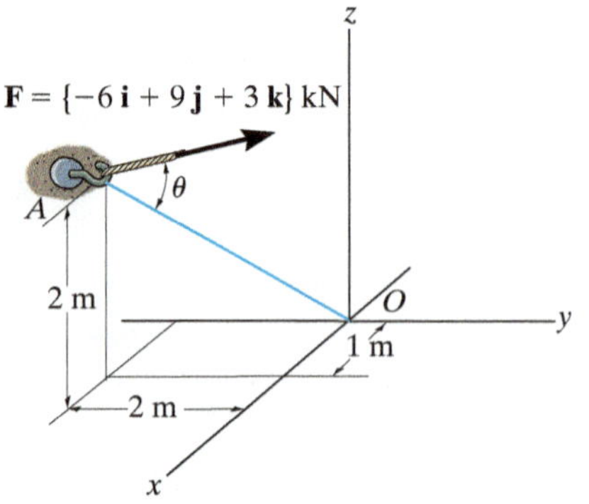

F2–25

F2–26. Determine the angle θ between the force and the line AB.

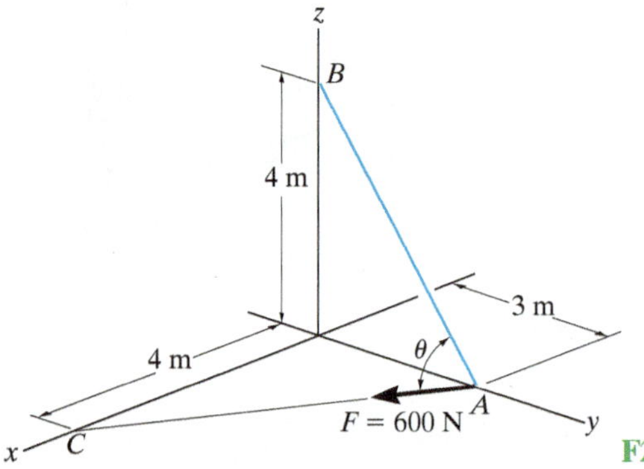

F2–26

F2–27. Determine the angle θ between the force and the line OA.

F2–28. Determine the component of projection of the force along the line OA.

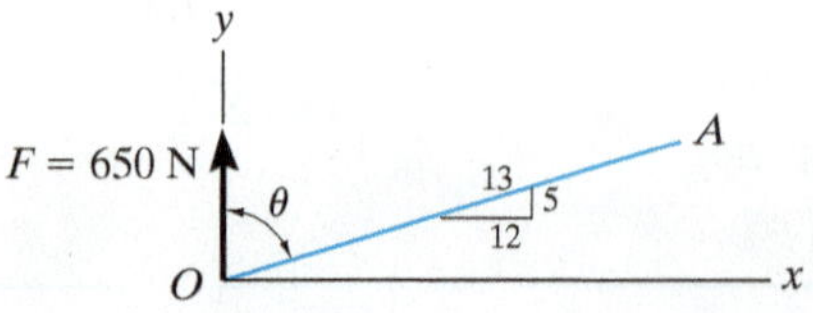

F2–27/28

F2–29. Find the magnitude of the projected component of the force along the pipe AO.

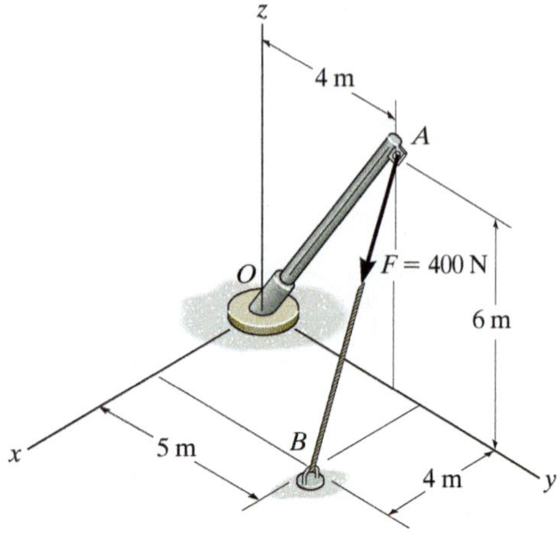

F2–29

F2–30. Determine the components of the force acting parallel and perpendicular to the axis of the pole.

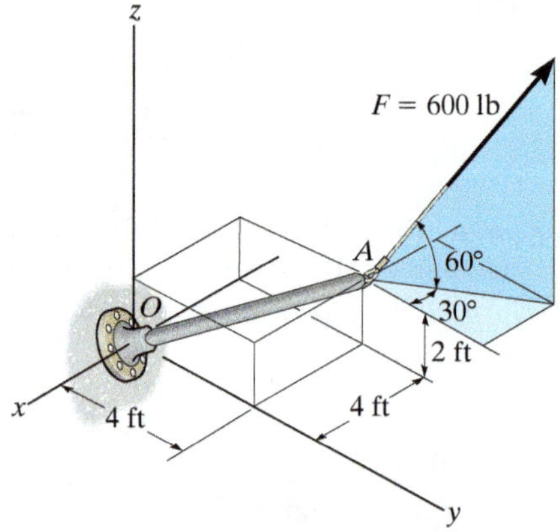

F2–30

F2–31. Determine the magnitudes of the components of force $F = 56$ N acting along and perpendicular to line AO.

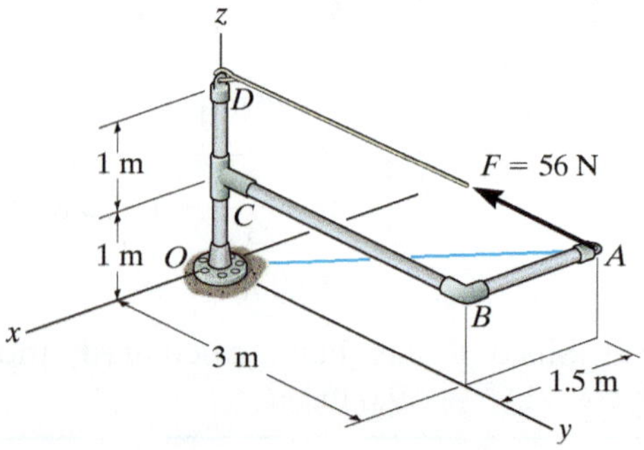

F2–31

PROBLEMS

***2–112.** Given the three vectors **A**, **B**, and **D**, show that $\mathbf{A} \cdot (\mathbf{B} + \mathbf{D}) = (\mathbf{A} \cdot \mathbf{B}) + (\mathbf{A} \cdot \mathbf{D})$.

2–113. Determine the angle θ between the edges of the sheet-metal bracket.

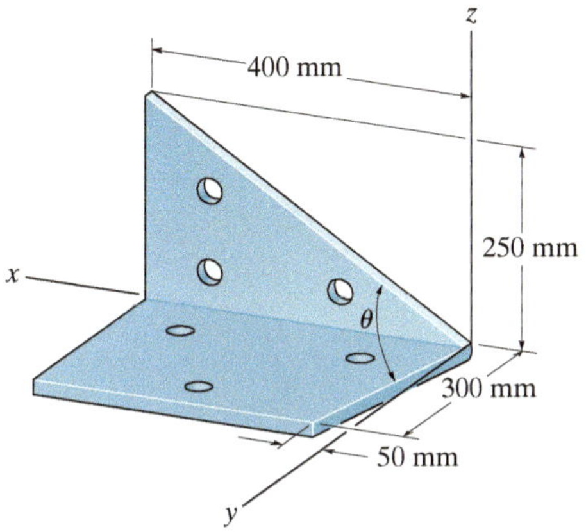

Prob. 2–113

2–114. Determine the angle θ between the sides of the triangular plate.

2–115. Determine the length of side BC of the triangular plate. Solve the problem by finding the magnitude of $\mathbf{r}_{BC}$; then check the result by first finding θ, r_{AB}, and r_{AC} and then use the cosine law.

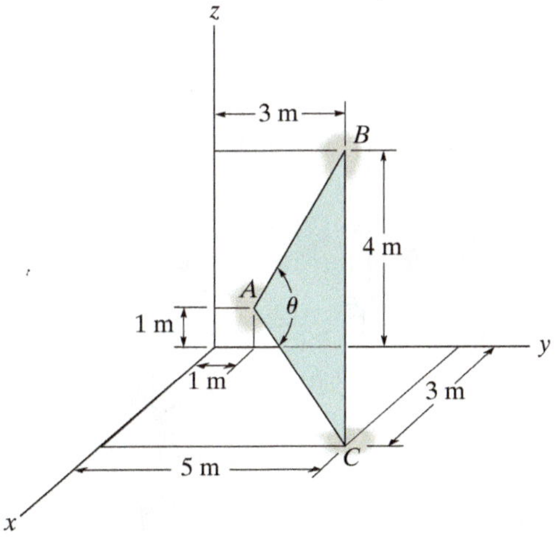

Probs. 2–114/115

***2–116.** Determine the magnitude of the projected component of force $\mathbf{F}_{AB}$ acting along the z axis.

2–117. Determine the magnitude of the projected component of force $\mathbf{F}_{AC}$ acting along the z axis.

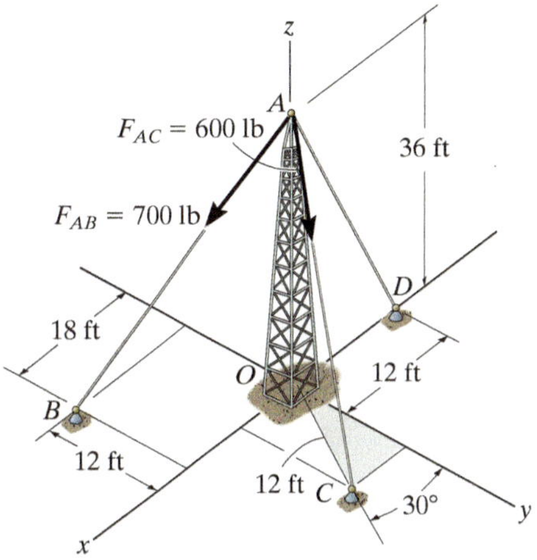

Probs. 2–116/117

2–118. Determine the projection of the force **F** along the pole.

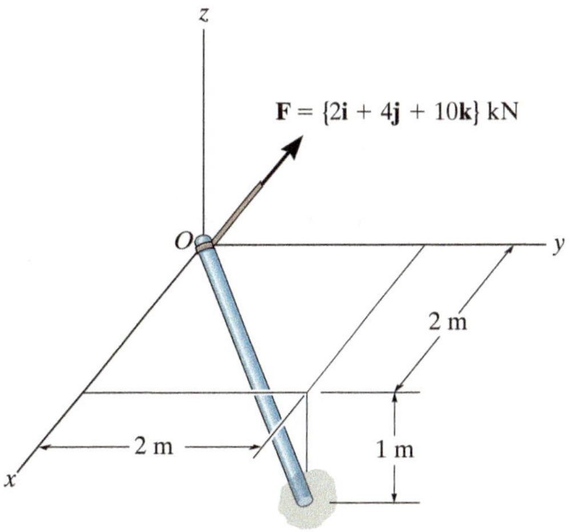

Prob. 2–118

2–119. Determine the angle θ between the y axis of the pole and the wire AB.

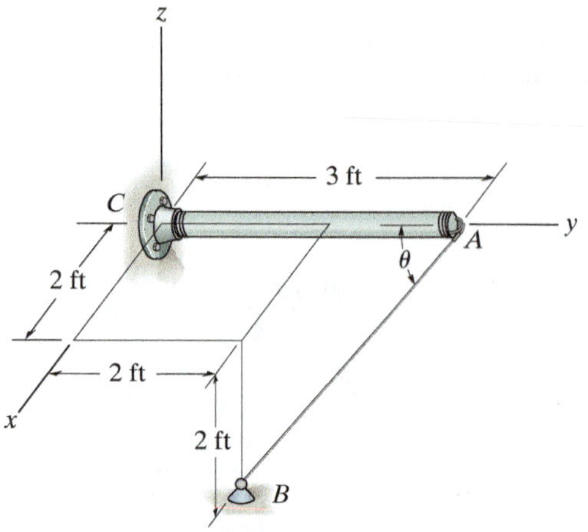

Prob. 2–119

***2–120.** Determine the magnitudes of the components of $F = 600$ N acting along and perpendicular to segment DE of the pipe assembly.

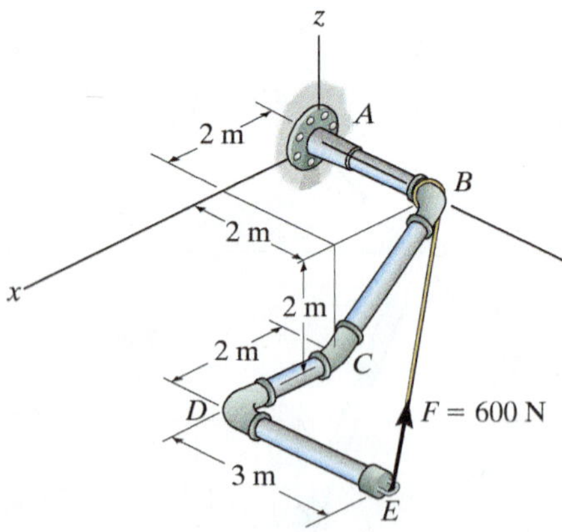

Prob. 2–120

2–121. Determine the magnitude of the projection of force $F = 600$ N along the u axis.

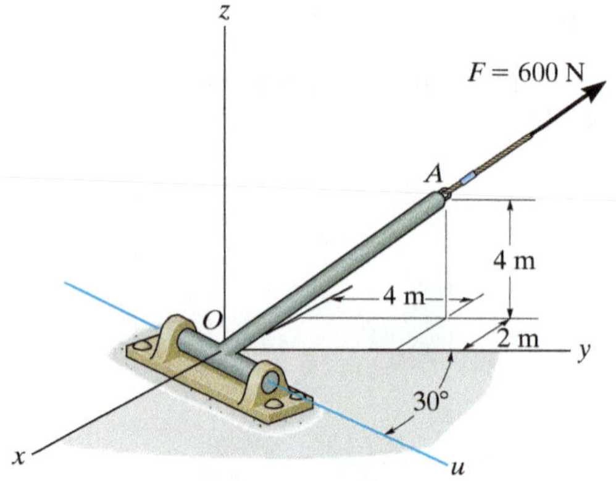

Prob. 2–121

2–122. Determine the angle θ between cables AB and AC.

2–123. Determine the angle ϕ between cable AC and strut AO.

***2–124.** Determine the projected component of force $\mathbf{F}_{AB}$ along the axis of strut AO. Express the result as a Cartesian vector.

2–125. Determine the projected component of force $\mathbf{F}_{AC}$ along the axis of strut AO. Express the result as a Cartesian vector.

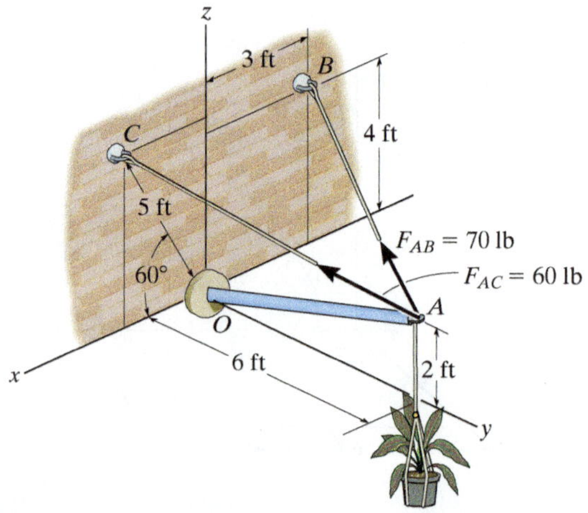

Probs. 2–122/123/124/125

2–126. Two cables exert forces on the pipe. Determine the magnitude of the projected component of $\mathbf{F}_1$ along the line of action of $\mathbf{F}_2$.

2–127. Determine the angle θ between the two cables attached to the pipe.

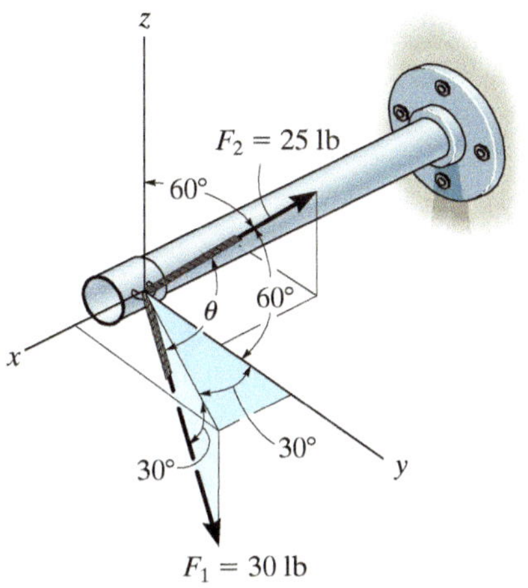

Probs. 2–126/127

***2–128.** Determine the magnitudes of the components of **F** acting along and perpendicular to segment BC of the pipe assembly.

2–129. Determine the magnitude of the projected component of **F** along AC. Express this component as a Cartesian vector.

2–130. Determine the angle θ between the pipe segments BA and BC.

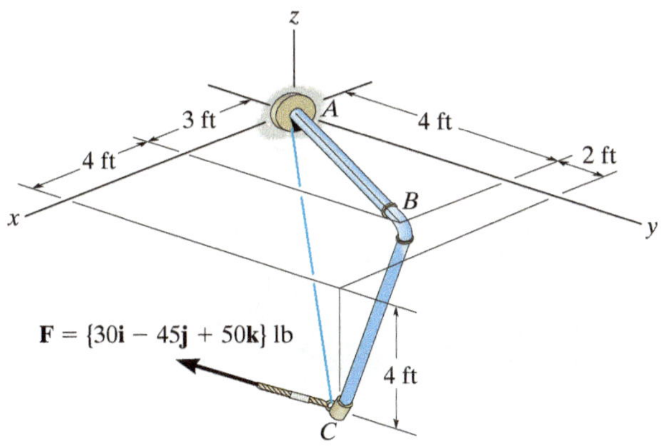

Probs. 2–128/129/130

2–131. Determine the angles θ and ϕ made between the axes OA of the flag pole and AB and AC, respectively, of each cable.

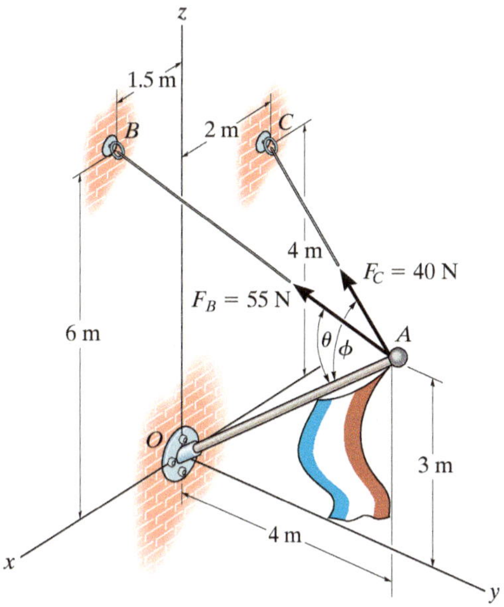

Prob. 2–131

***2–132.** The cables each exert a force of 400 N on the post. Determine the magnitude of the projected component of $\mathbf{F}_1$ along the line of action of $\mathbf{F}_2$.

2–133. Determine the angle θ between the two cables attached to the post.

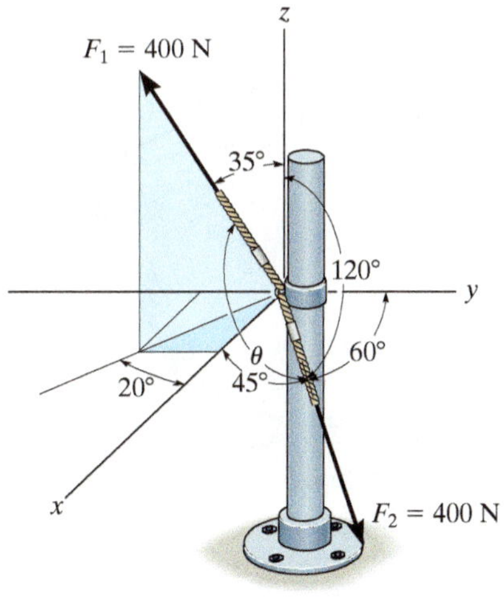

Probs. 2–132/133

2–134. Determine the magnitudes of the components of force $F = 90$ lb acting parallel and perpendicular to diagonal AB of the crate.

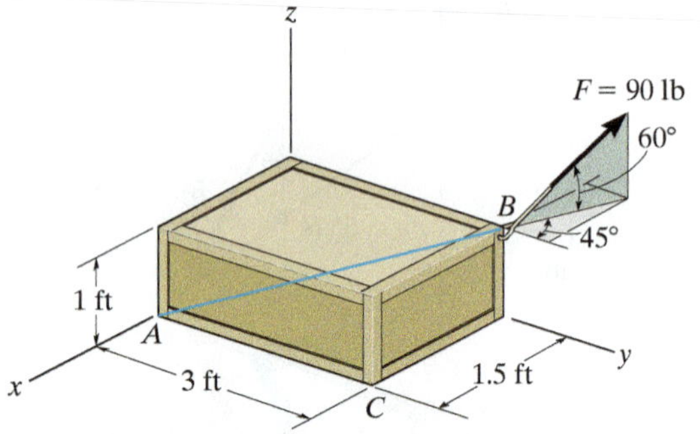

Prob. 2–134

2–135. The force $\mathbf{F} = \{25\mathbf{i} - 50\mathbf{j} + 10\mathbf{k}\}$ lb acts at the end A of the pipe assembly. Determine the magnitude of the components $\mathbf{F}_1$ and $\mathbf{F}_2$ which act along the axis of AB and perpendicular to it.

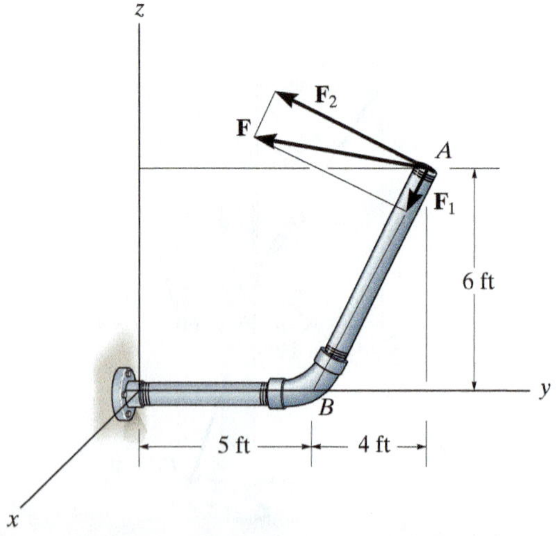

Prob. 2–135

***2–136.** Determine the components of **F** that act along rod AC and perpendicular to it. Point B is located at the midpoint of the rod.

2–137. Determine the components of **F** that act along rod AC and perpendicular to it. Point B is located 3 m along the rod from end C.

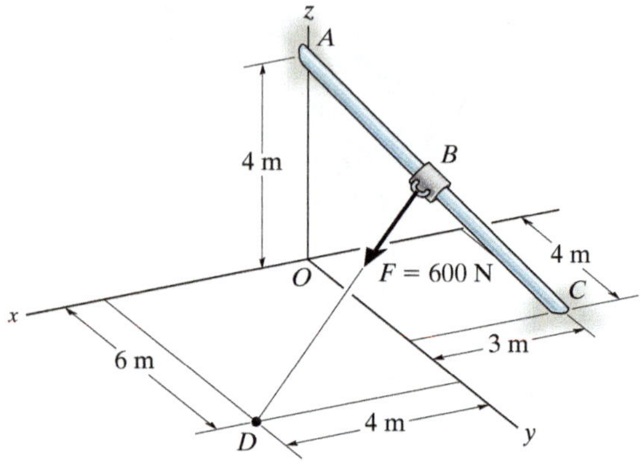

Probs. 2–136/137

2–138. Determine the magnitudes of the projected components of the force $F = 300$ N acting along the x and y axes.

2–139. Determine the magnitude of the projected component of the force $F = 300$ N acting along line OA.

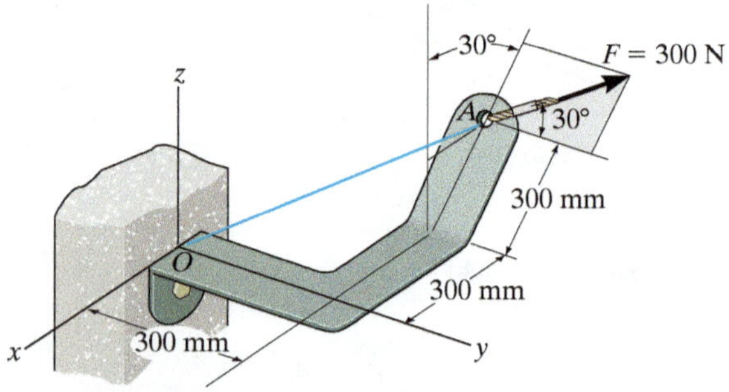

Probs. 2–138/139

CHAPTER REVIEW

A scalar is a positive or negative number; e.g., mass and temperature.

A vector has a magnitude and direction, where the arrowhead represents the sense of the vector.

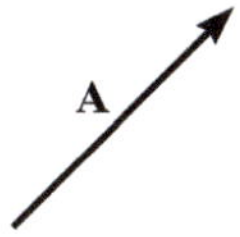

Multiplication or division of a vector by a scalar will change only the magnitude of the vector. If the scalar is negative, the sense of the vector will change so that it acts in the opposite sense.

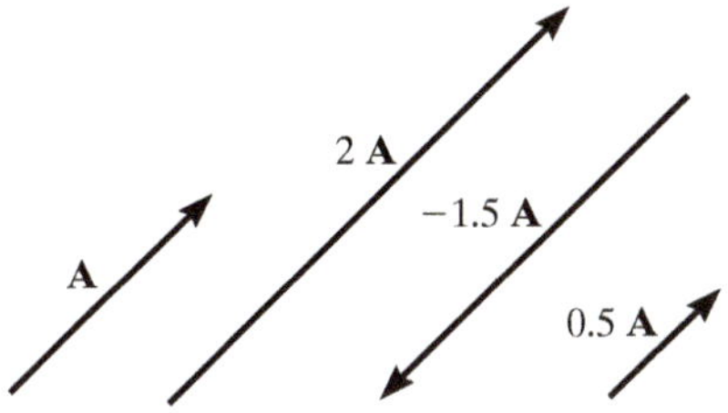

If vectors are collinear, the resultant is simply the algebraic or scalar addition.

$$R = A + B$$

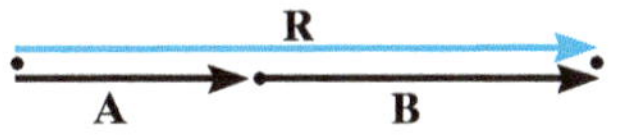

Parallelogram Law

Two forces add according to the parallelogram law. The *components* form the sides of the parallelogram and the *resultant* is the diagonal.

To find the components of a force along any two axes, extend lines from the head of the force, parallel to the axes, to form the components.

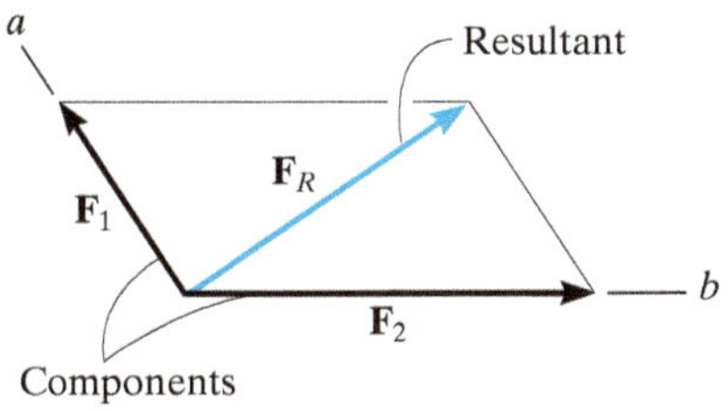

To obtain the components of the resultant, show how the forces add by tip-to-tail using the triangle rule, and then use the law of cosines and the law of sines to calculate their values.

$$F_R = \sqrt{F_1^2 + F_2^2 - 2\,F_1F_2\cos\theta_R}$$

$$\frac{F_1}{\sin\theta_1} = \frac{F_2}{\sin\theta_2} = \frac{F_R}{\sin\theta_R}$$

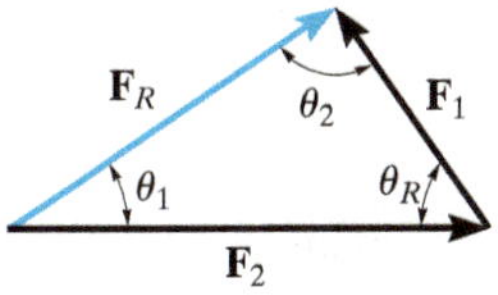

Rectangular Components: Two Dimensions

Vectors $\mathbf{F}_x$ and $\mathbf{F}_y$ are rectangular components of $\mathbf{F}$.

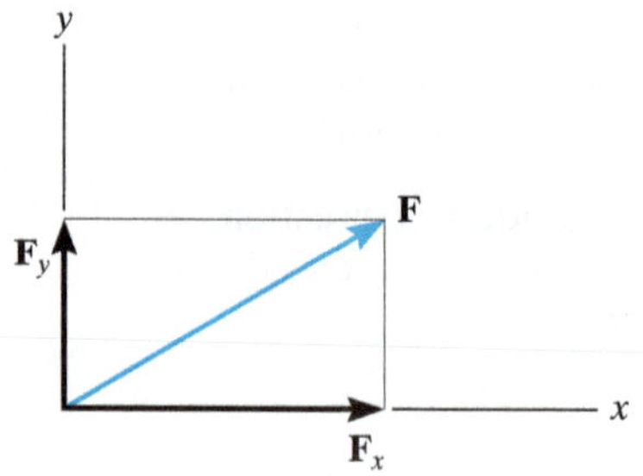

The resultant force is determined from the algebraic sum of its components.

$$(F_R)_x = \Sigma F_x$$

$$(F_R)_y = \Sigma F_y$$

$$F_R = \sqrt{(F_R)_x^2 + (F_R)_y^2}$$

$$\theta = \tan^{-1}\left|\frac{(F_R)_y}{(F_R)_x}\right|$$

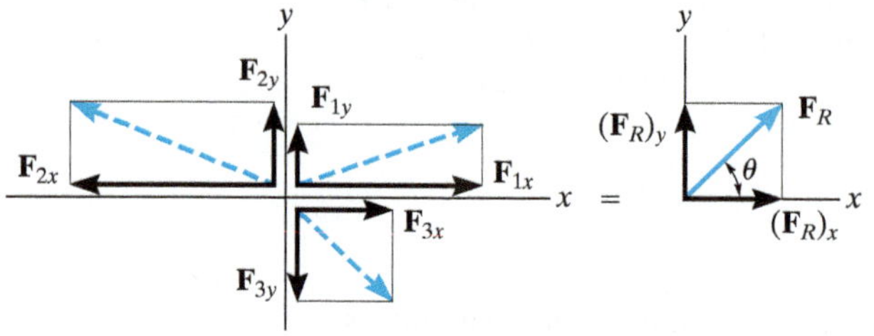

Cartesian Vectors

The unit vector $\mathbf{u}$ has a length of 1, no units, and it points in the direction of the vector $\mathbf{F}$.

$$\mathbf{u} = \frac{\mathbf{F}}{F}$$

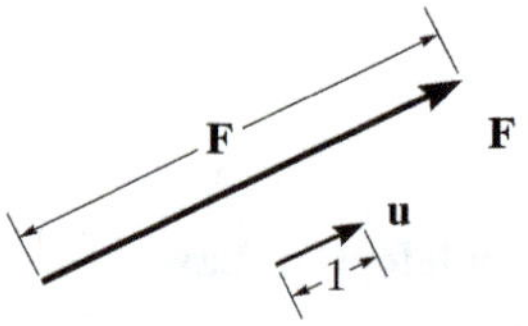

A force can be resolved into its Cartesian components along the x, y, z axes so that $\mathbf{F} = F_x\mathbf{i} + F_y\mathbf{j} + F_z\mathbf{k}$.

The magnitude of $\mathbf{F}$ is determined from the positive square root of the sum of the squares of its components.

$$F = \sqrt{F_x^2 + F_y^2 + F_z^2}$$

The coordinate direction angles α, β, γ are determined by formulating a unit vector in the direction of $\mathbf{F}$. The x, y, z components of $\mathbf{u}$ represent $\cos\alpha$, $\cos\beta$, $\cos\gamma$.

$$\mathbf{u} = \frac{\mathbf{F}}{F} = \frac{F_x}{F}\mathbf{i} + \frac{F_y}{F}\mathbf{j} + \frac{F_z}{F}\mathbf{k}$$

$$\mathbf{u} = \cos\alpha\,\mathbf{i} + \cos\beta\,\mathbf{j} + \cos\gamma\,\mathbf{k}$$

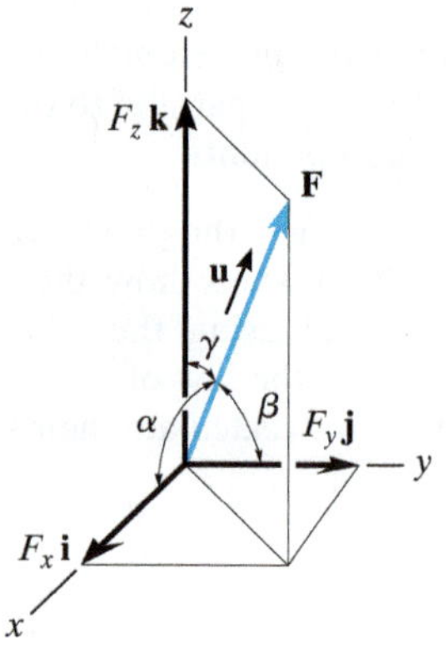

The coordinate direction angles are related so that only two of the three angles are independent of one another.

$$\cos^2 \alpha + \cos^2 \beta + \cos^2 \gamma = 1$$

To find the resultant of a concurrent force system, express each force as a Cartesian vector and add the **i**, **j**, **k** components of all the forces in the system.

$$\mathbf{F}_R = \Sigma \mathbf{F} = \Sigma F_x \mathbf{i} + \Sigma F_y \mathbf{j} + \Sigma F_z \mathbf{k}$$

Position and Force Vectors

A position vector locates one point in space relative to another. The easiest way to formulate the components of a position vector is to determine the distance and direction that one must travel along the x, y, and z directions—going from the tail to the head of the vector.

$$\mathbf{r} = (x_B - x_A)\mathbf{i} + (y_B - y_A)\mathbf{j} + (z_B - z_A)\mathbf{k}$$

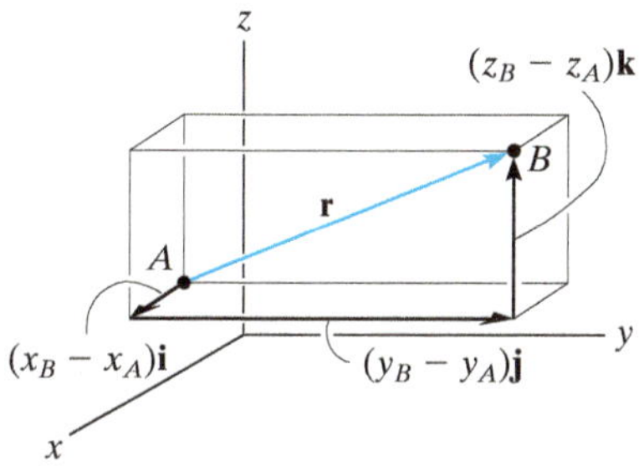

If the line of action of a force passes through points A and B, then the force acts in the same direction as the position vector **r**, which is defined by the unit vector **u**. The force can then be expressed as a Cartesian vector.

$$\mathbf{F} = F\mathbf{u} = F\left(\frac{\mathbf{r}}{r}\right)$$

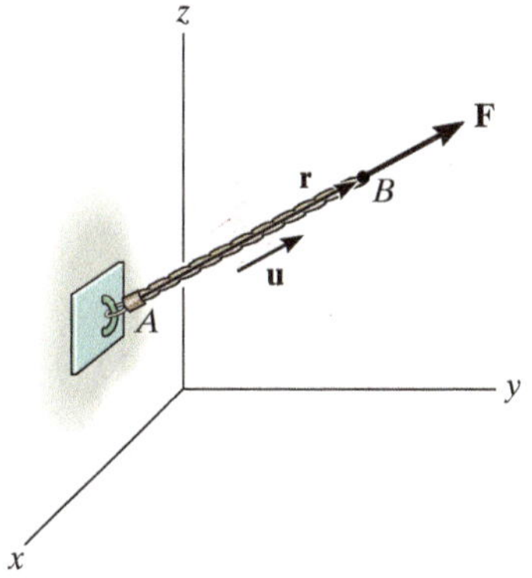

Dot Product

The dot product between two vectors **A** and **B** yields a scalar. If **A** and **B** are expressed in Cartesian vector form, then the dot product is the sum of the products of their x, y, and z components.

$$\mathbf{A} \cdot \mathbf{B} = AB \cos \theta$$
$$= A_x B_x + A_y B_y + A_z B_z$$

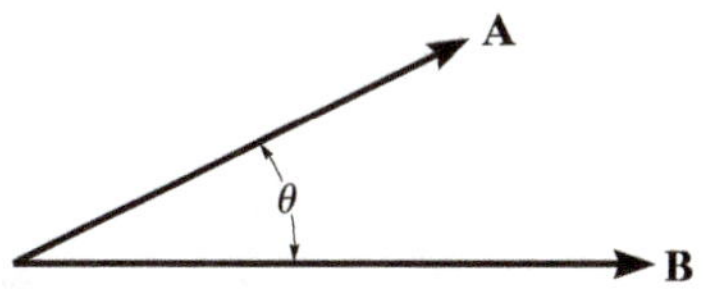

The dot product can be used to determine the angle between **A** and **B**.

$$\theta = \cos^{-1}\left(\frac{\mathbf{A} \cdot \mathbf{B}}{AB}\right)$$

The dot product is also used to determine the projected component of a vector **A** onto an axis aa defined by its unit vector $\mathbf{u}_a$.

$$\mathbf{A}_a = A \cos \theta \, \mathbf{u}_a = (\mathbf{A} \cdot \mathbf{u}_a)\mathbf{u}_a$$

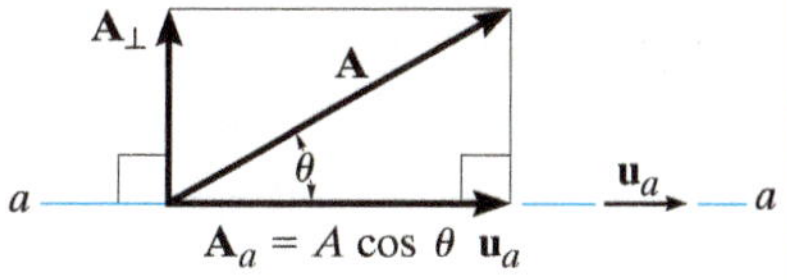

REVIEW PROBLEMS

2

*2–140. Determine the length of the conneting rod AB by first formulating a Cartesian position vector from A to B and then determining its magnitude.

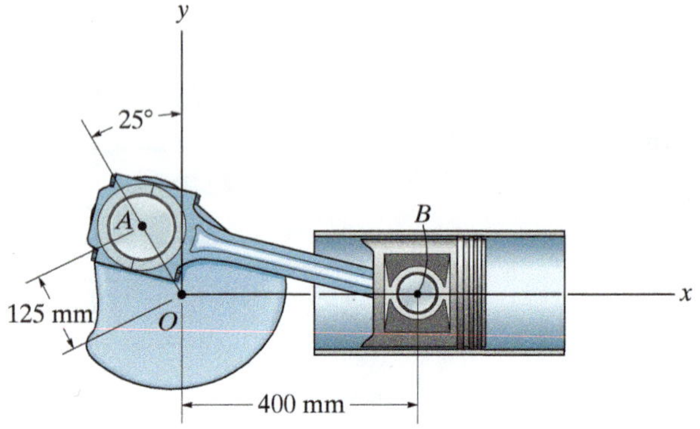

Prob. 2–140

2–141. Determine the x and y components of $\mathbf{F}_1$ and $\mathbf{F}_2$.

2–142. Determine the magnitude of the resultant force and its direction, measured counterclockwise from the positive x axis.

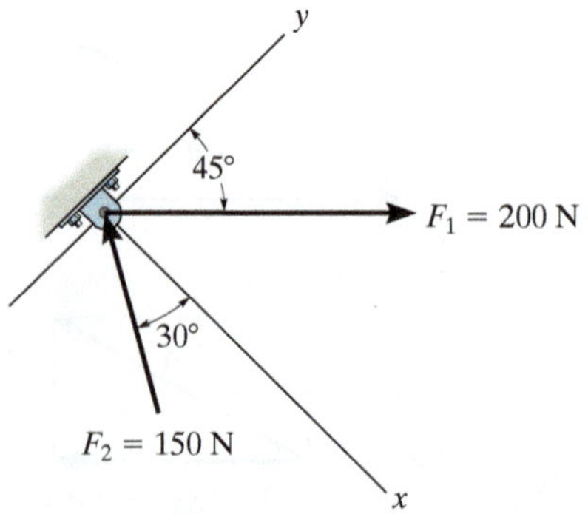

Probs. 2–141/142

2–143. Determine the x and y components of each force acting on the *gusset plate* of the bridge truss. Show that the resultant force is zero.

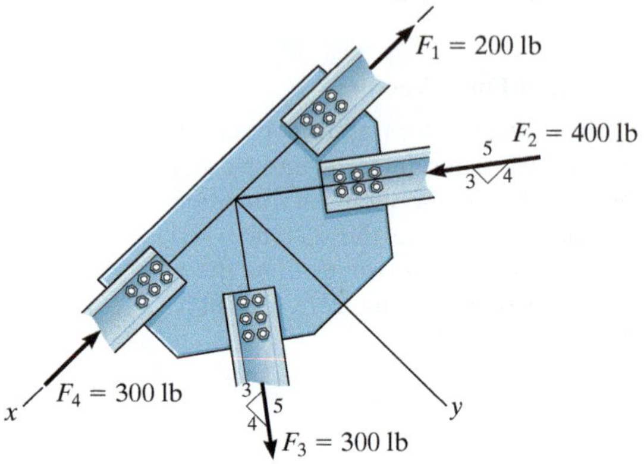

Prob. 2–143

*2–144. Express $\mathbf{F}_1$ and $\mathbf{F}_2$ as Cartesian vectors.

2–145. Determine the magnitude of the resultant force and its direction measured counterclockwise from the positive x axis.

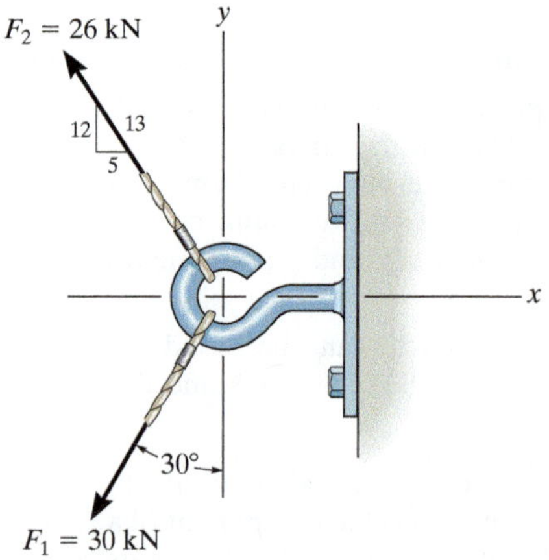

Probs. 2–144/145

2–146. The cable attached to the tractor at B exerts a force of 350 lb on the framework. Express this force as a Cartesian vector.

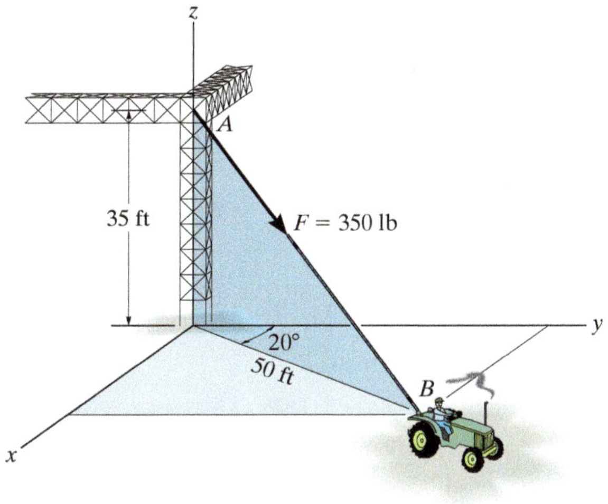

Prob. 2–146

2–147. Determine the magnitude and direction of the resultant $\mathbf{F}_R = \mathbf{F}_1 + \mathbf{F}_2 + \mathbf{F}_3$ of the three forces by first finding the resultant $\mathbf{F}' = \mathbf{F}_1 + \mathbf{F}_3$ and then forming $\mathbf{F}_R = \mathbf{F}' + \mathbf{F}_2$. Specify its direction measured counterclockwise from the positive x axis.

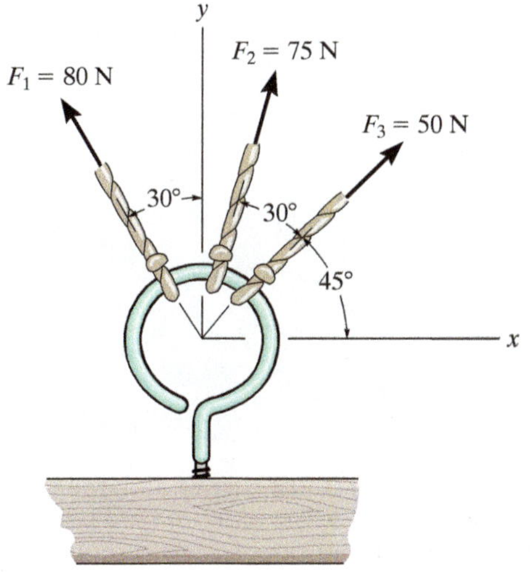

Prob. 2–147

***2–148.** If $\theta = 60°$ and $F = 20$ kN, determine the magnitude of the resultant force and its direction measured clockwise from the positive x axis.

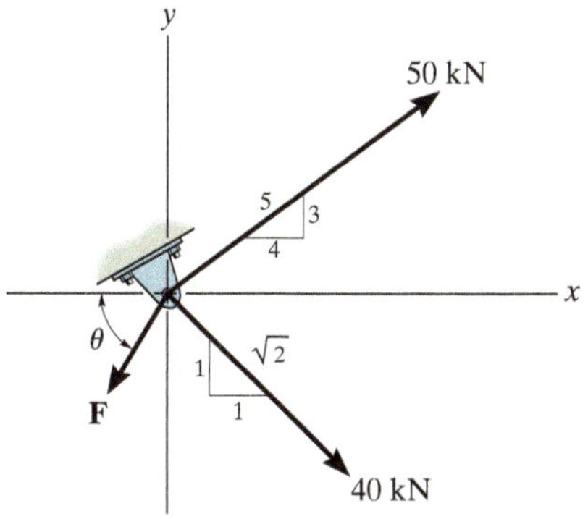

Prob. 2–148

2–149. The hinged plate is supported by the cord AB. If the force in the cord is $F = 340$ lb, express this force, directed from A toward B, as a Cartesian vector. What is the length of the cord?

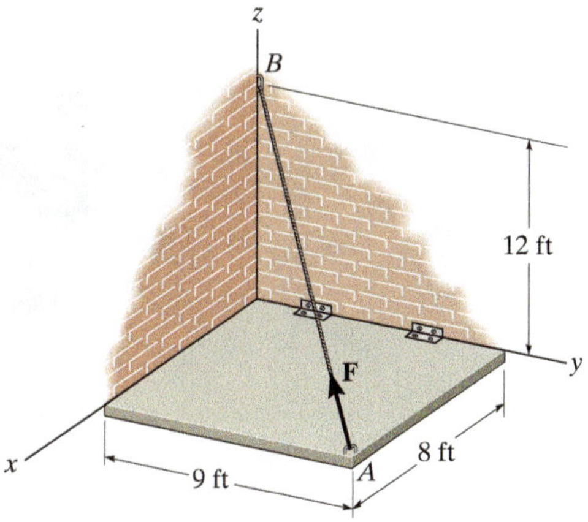

Prob. 2–149

Chapter 3

When this load is lifted at constant velocity, or is just suspended, then it is in a state of equilibrium. In this chapter we will study equilibrium for a particle and show how these ideas can be used to calculate the forces in cables used to hold suspended loads.

Equilibrium of a Particle

CHAPTER OBJECTIVES

- To introduce the concept of the free-body diagram for a particle.
- To show how to solve particle equilibrium problems using the equations of equilibrium.

3.1 Condition for the Equilibrium of a Particle

A particle is said to be in *equilibrium* if it remains at rest if originally at rest, or has a constant velocity if originally in motion. Most often, however, the term "equilibrium" or, more specifically, "static equilibrium" is used to describe an object at rest. To maintain equilibrium, it is *necessary* to satisfy Newton's first law of motion, which requires the *resultant force* acting on a particle to be equal to *zero*. This condition may be stated mathematically as

$$\Sigma \mathbf{F} = \mathbf{0} \qquad (3\text{–}1)$$

where $\Sigma \mathbf{F}$ is the vector *sum of all the forces* acting on the particle.

Not only is Eq. 3–1 a necessary condition for equilibrium, it is also a *sufficient* condition. This follows from Newton's second law of motion, which can be written as $\Sigma \mathbf{F} = m\mathbf{a}$. Since the force system satisfies Eq. 3–1, then $m\mathbf{a} = \mathbf{0}$, and therefore the particle's acceleration $\mathbf{a} = \mathbf{0}$. Consequently, the particle indeed moves with constant velocity or remains at rest.

3.2 The Free-Body Diagram

To apply the equation of equilibrium, we must account for *all* the known and unknown forces ($\Sigma\mathbf{F}$) which act *on* the particle. The best way to do this is to think of the particle as isolated and "free" from its surroundings. A drawing that shows the particle with *all* the forces that act on it is called a *free-body diagram (FBD)*.

Before presenting a formal procedure as to how to draw a free-body diagram, we will first consider two types of connections often encountered in particle equilibrium problems.

Springs. If a *linearly elastic spring* (or cord) of undeformed length l_0 is used to support a particle, the length of the spring will change in direct proportion to the force **F** acting on it, Fig. 3–1. A characteristic that defines the "elasticity" of a spring is the *spring constant* or *stiffness k*.

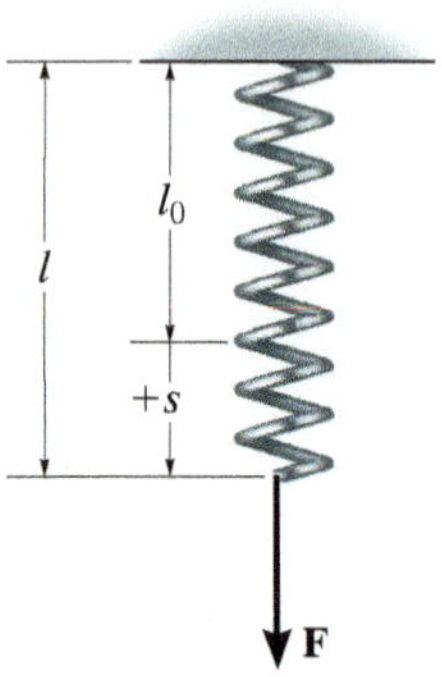

Fig. 3–1

The magnitude of force exerted on a linearly elastic spring which has a stiffness k and is deformed (elongated or compressed) a distance $s = l - l_0$, measured from its *unloaded* position, is

$$F = ks \qquad (3\text{–}2)$$

If s is positive, causing an elongation, then **F** must pull on the spring; whereas if s is negative, causing a shortening, then **F** must push on it. For example, if the spring in Fig. 3–1 has an unstretched length of 0.8 m and a stiffness $k = 500\ \text{N/m}$ and it is stretched to a length of 1 m, so that $s = l - l_0 = 1\ \text{m} - 0.8\ \text{m} = 0.2\ \text{m}$, then a force $F = ks = 500\ \text{N/m}(0.2\ \text{m}) = 100\ \text{N}$ is needed.

Cables and Pulleys. Unless otherwise stated throughout this book, except in Sec. 7.4, all cables (or cords) will be assumed to have negligible weight and they cannot stretch. Also, a cable can support *only* a tension or "pulling" force, and this force always acts in the direction of the cable. In Chapter 5, it will be shown that the tension force developed in a continuous cable which passes over a frictionless pulley must have a *constant* magnitude to keep the cable in equilibrium. Hence, for any angle θ, shown in Fig. 3–2, the cable is subjected to a constant tension T throughout its length.

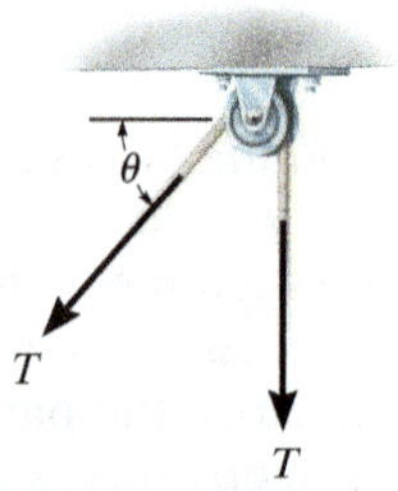

Cable is in tension

Fig. 3–2

Procedure for Drawing a Free-Body Diagram

Since we must account for *all the forces acting on the particle* when applying the equations of equilibrium, the importance of first drawing a free-body diagram cannot be overemphasized. To construct a free-body diagram, the following three steps are necessary.

Draw Outlined Shape.

Imagine the particle to be *isolated* or cut "free" from its surroundings by drawing its outlined shape.

Show All Forces.

Indicate on this sketch *all* the forces that act *on the particle*. These forces can be *active forces*, which tend to set the particle in motion, or they can be *reactive forces* which are the result of the constraints or supports that tend to prevent motion. To account for all these forces, it may be helpful to trace around the particle's boundary, carefully noting each force acting on it.

Identify Each Force.

The forces that are *known* should be labeled with their proper magnitudes and directions. Letters are used to represent the magnitudes and directions of forces that are unknown.

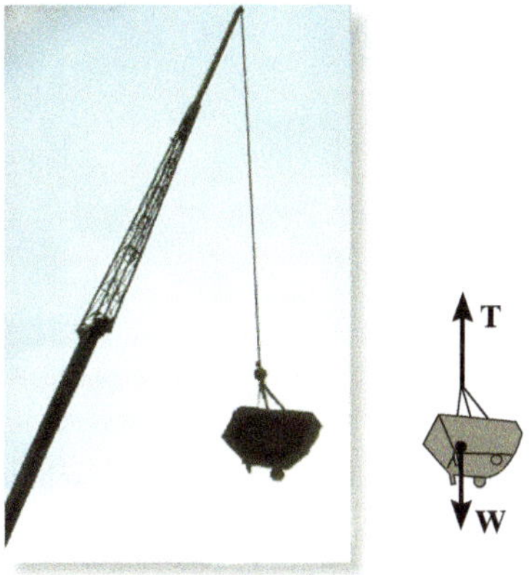

The bucket is held in equilibrium by the cable, and instinctively we know that the force in the cable must equal the weight of the bucket. By drawing a free-body diagram of the bucket we can understand why this is so. This diagram shows that there are only two forces *acting on the bucket*, namely, its weight **W** and the force **T** of the cable. For equilibrium, the resultant of these forces must be equal to zero, and so $T = W$.

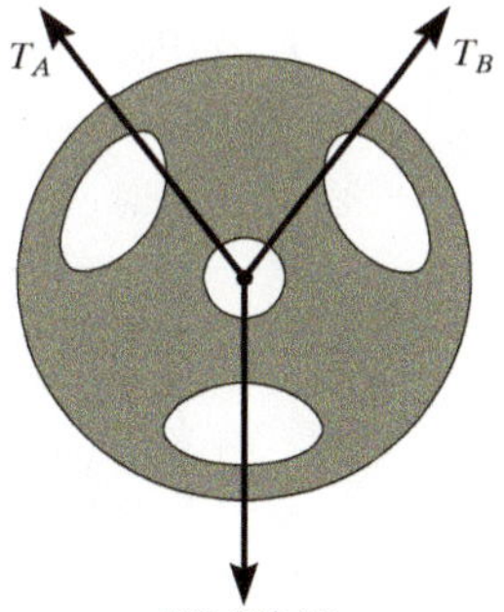

The 5-kg plate is suspended by two straps A and B. To find the force in each strap we should consider the free-body diagram of the plate. As noted, the three forces acting on it form a concurrent force system.

EXAMPLE 3.1

The sphere in Fig. 3–3*a* has a mass of 6 kg and is supported as shown. Draw a free-body diagram of the sphere, the cord *CE*, and the knot at *C*.

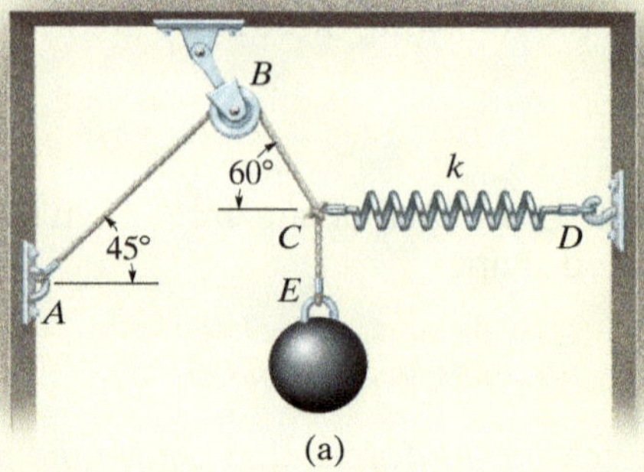

(a)

$\mathbf{F}_{CE}$ (Force of cord *CE* acting on sphere)

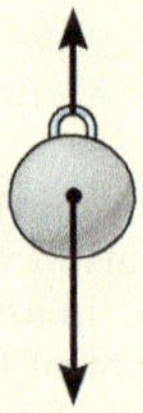

58.9 N (Weight or gravity acting on sphere)

(b)

$\mathbf{F}_{EC}$ (Force of knot acting on cord *CE*)

$\mathbf{F}_{CE}$ (Force of sphere acting on cord *CE*)

(c)

SOLUTION

Sphere. By inspection, there are only two forces acting on the sphere, namely, its weight, 6 kg (9.81 m/s^2) = 58.9 N, and the force of cord *CE*. The free-body diagram is shown in Fig. 3–3*b*.

Cord *CE*. When the cord *CE* is isolated from its surroundings, its free-body diagram shows only two forces acting on it, namely, the force of the sphere and the force of the knot, Fig. 3–3*c*. Notice that $\mathbf{F}_{CE}$ shown here is equal but opposite to that shown in Fig. 3–3*b*, a consequence of Newton's third law of action–reaction. Also, $\mathbf{F}_{CE}$ and $\mathbf{F}_{EC}$ pull on the cord and keep it in tension so that it doesn't collapse. For equilibrium, $F_{CE} = F_{EC}$.

Knot. The knot at *C* is subjected to three forces, Fig. 3–3*d*. They are caused by the cords *CBA* and *CE* and the spring *CD*. As required, the free-body diagram shows all these forces labeled with their magnitudes and directions. It is important to recognize that the weight of the sphere does not directly act on the knot. Instead, the cord *CE* subjects the knot to this force.

$\mathbf{F}_{CBA}$ (Force of cord *CBA* acting on knot)

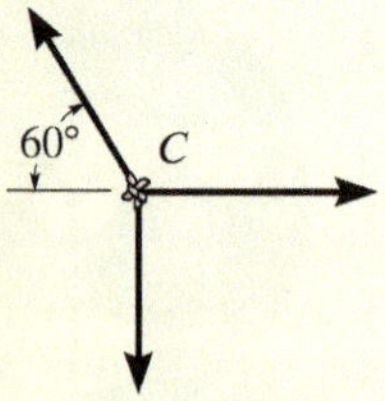

$\mathbf{F}_{CD}$ (Force of spring acting on knot)

$\mathbf{F}_{CE}$ (Force of cord *CE* acting on knot)

(d)

Fig. 3–3

3.3 Coplanar Force Systems

If a particle is subjected to a system of coplanar forces that lie in the x–y plane, as in Fig. 3–4, then each force can be resolved into its **i** and **j** components. For equilibrium, these forces must sum to produce a zero force resultant, i.e.,

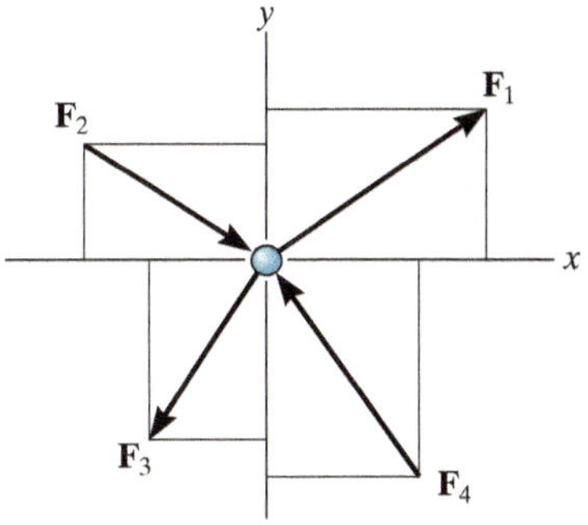

Fig. 3–4

$$\Sigma\mathbf{F} = \mathbf{0}$$
$$\Sigma F_x\mathbf{i} + \Sigma F_y\mathbf{j} = \mathbf{0}$$

For this vector equation to be satisfied, the resultant force's x and y components must both be equal to zero. Hence,

$$\Sigma F_x = 0 \qquad \Sigma F_y = 0 \tag{3–3}$$

These two equations can be solved for at most two unknowns, generally represented as angles and magnitudes of forces shown on the particle's free-body diagram.

When applying each of the two equations of equilibrium, we must account for the sense of direction of any component by using an *algebraic sign* which corresponds to the arrowhead direction of the component along the x or y axis. It is important to note that if a force has an *unknown magnitude*, then the arrowhead sense of the force on the free-body diagram can be *assumed*. Then if the *solution* yields a *negative scalar*, this indicates that the sense of the force is opposite to that which was assumed.

For example, consider the free-body diagram of the particle subjected to the two forces shown in Fig. 3–5. Here it is *assumed* that the *unknown force* **F** acts to the right to maintain equilibrium. Applying the equation of equilibrium along the x axis, we have

$$\overset{+}{\rightarrow}\Sigma F_x = 0; \qquad +F + 10\text{ N} = 0$$

Both terms are "positive" since both forces act in the positive x direction. When this equation is solved, $F = -10$ N. Here the *negative sign* indicates that **F** must act to the left to hold the particle in equilibrium, Fig. 3–5. Notice that if the $+x$ axis in Fig. 3–5 were directed to the left, both terms in the above equation would be negative, but again, after solving, $F = -10$ N, indicating that **F** would have to be directed to the left.

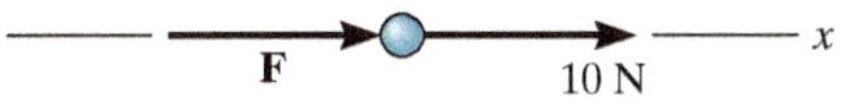

Fig. 3–5

Procedure for Analysis

Coplanar force equilibrium problems for a particle can be solved using the following procedure.

Free-Body Diagram.

- Establish the x, y axes in any suitable orientation.
- Label all the known and unknown force magnitudes and directions on the diagram.
- The sense of a force having an unknown magnitude can be assumed.

Equations of Equilibrium.

- Apply the equations of equilibrium, $\Sigma F_x = 0$ and $\Sigma F_y = 0$.
- Components are positive if they are directed along a positive axis, and negative if they are directed along a negative axis.
- If more than two unknowns exist and the problem involves a spring, apply $F = ks$ to relate the spring force to the deformation s of the spring.
- Since the magnitude of a force is always a positive quantity, then if the solution for a force yields a negative result, this indicates that its sense is the reverse of that shown on the free-body diagram.

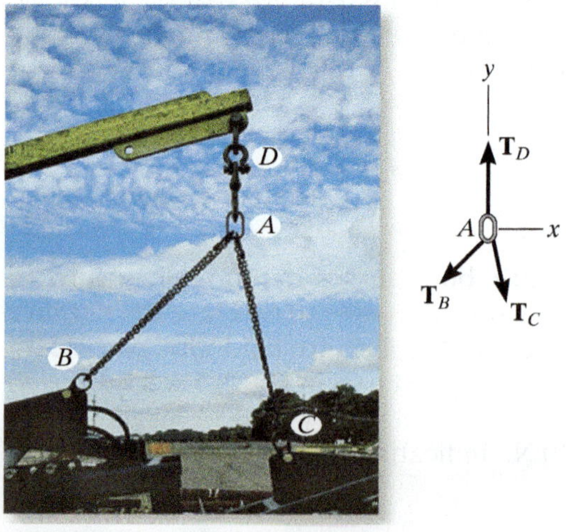

The chains exert three forces on the ring at A, as shown on its free-body diagram. The ring will not move, or will move with constant velocity, provided the summation of these forces along the x and along the y axis equals zero. If one of the three forces is known, the magnitudes of the other two forces can be obtained from the two equations of equilibrium.

EXAMPLE 3.2

Determine the tension in cables *BA* and *BC* necessary to support the 60-kg cylinder in Fig. 3–6*a*.

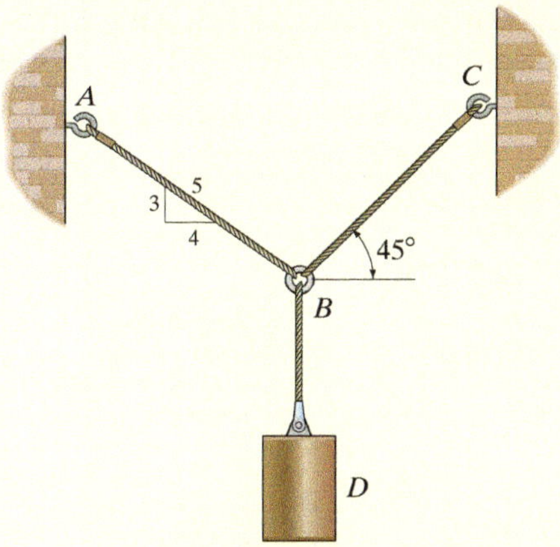

(a)

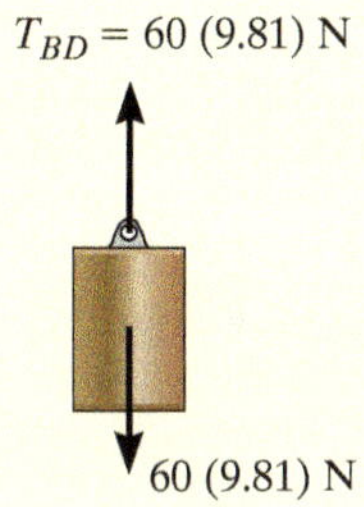

(b)

SOLUTION

Free-Body Diagram. Due to equilibrium, the weight of the cylinder causes the tension in cable *BD* to be $T_{BD} = 60(9.81)$ N, Fig. 3–6*b*. The forces in cables *BA* and *BC* can be determined by investigating the equilibrium of ring *B*. Its free-body diagram is shown in Fig. 3–6*c*. The magnitudes of $\mathbf{T}_A$ and $\mathbf{T}_C$ are unknown, but their directions are known.

Equations of Equilibrium. Applying the equations of equilibrium along the *x* and *y* axes, we have

$$\overset{+}{\rightarrow}\Sigma F_x = 0; \qquad T_C \cos 45^\circ - \left(\tfrac{4}{5}\right)T_A = 0 \qquad (1)$$

$$+\uparrow \Sigma F_y = 0; \qquad T_C \sin 45^\circ + \left(\tfrac{3}{5}\right)T_A - 60(9.81)\text{ N} = 0 \qquad (2)$$

Equation (1) can be written as $T_A = 0.8839T_C$. Substituting this into Eq. (2) yields

$$T_C \sin 45^\circ + \left(\tfrac{3}{5}\right)(0.8839T_C) - 60(9.81)\text{ N} = 0$$

so that

$$T_C = 475.66\text{ N} = 476\text{ N} \qquad \textit{Ans.}$$

Substituting this result into either Eq. (1) or Eq. (2), we get

$$T_A = 420\text{ N} \qquad \textit{Ans.}$$

NOTE: The accuracy of these results, of course, depends on the accuracy of the data, i.e., measurements of geometry and loads. For most engineering work involving a problem such as this, the data as measured to three significant figures would be sufficient.

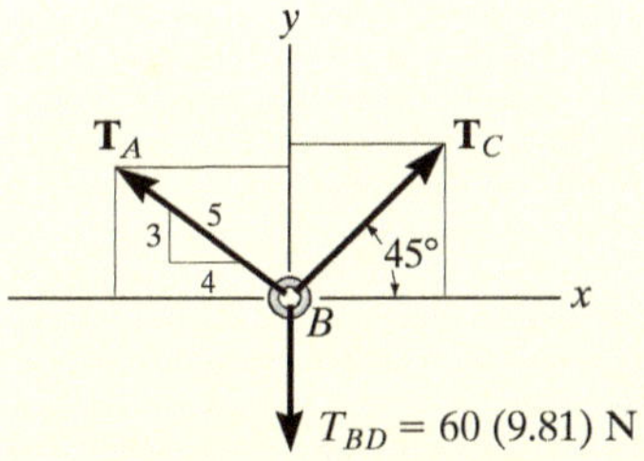

(c)

Fig. 3–6

EXAMPLE 3.3

The 200-kg crate in Fig. 3–7*a* is suspended using the ropes *AB* and *AC*. Each rope can withstand a maximum force of 10 kN before it breaks. If *AB* always remains horizontal, determine the smallest angle θ to which the crate can be suspended before one of the ropes breaks.

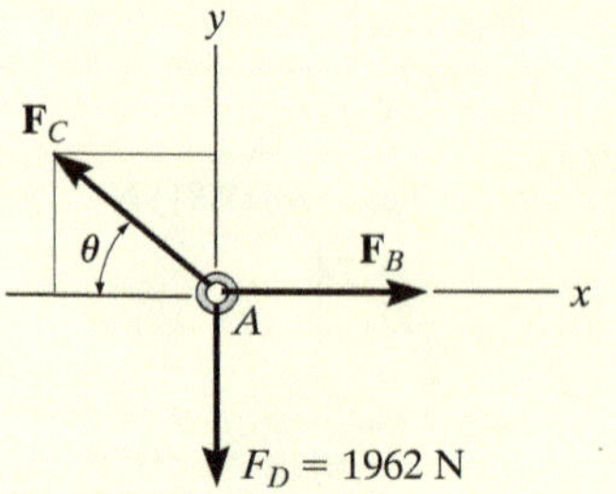

(b)

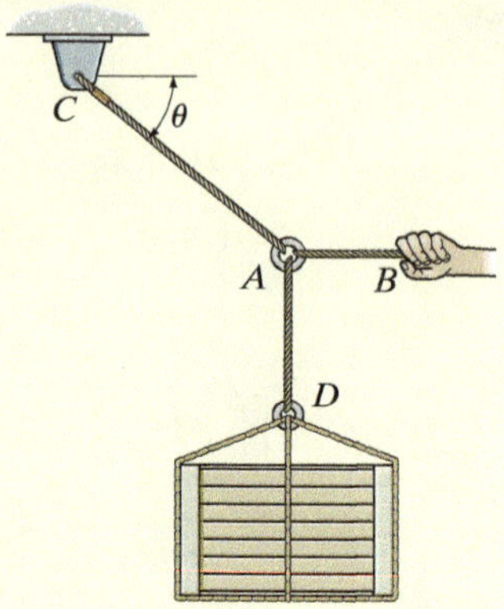

(a)

Fig. 3–7

SOLUTION

Free-Body Diagram. We will study the equilibrium of ring *A*. There are three forces acting on it, Fig. 3–7*b*. The magnitude of $\mathbf{F}_D$ is equal to the weight of the crate, i.e., $F_D = 200\,(9.81)\text{ N} = 1962\text{ N} < 10\text{ kN}$.

Equations of Equilibrium. Applying the equations of equilibrium along the *x* and *y* axes,

$$\overset{+}{\rightarrow}\Sigma F_x = 0; \qquad -F_C\cos\theta + F_B = 0; \quad F_C = \frac{F_B}{\cos\theta} \tag{1}$$

$$+\uparrow\Sigma F_y = 0; \qquad F_C\sin\theta - 1962\text{ N} = 0 \tag{2}$$

From Eq. (1), F_C is always greater than F_B since $\cos\theta \le 1$. Therefore, rope *AC* will reach the maximum tensile force of 10 kN *before* rope *AB*. Substituting $F_C = 10$ kN into Eq. (2), we get

$$[10(10^3)\text{N}]\sin\theta - 1962\text{ N} = 0$$

$$\theta = \sin^{-1}(0.1962) = 11.31^\circ = 11.3^\circ \qquad \textit{Ans.}$$

The force developed in rope *AB* can be obtained by substituting the values for θ and F_C into Eq. (1).

$$10(10^3)\text{ N} = \frac{F_B}{\cos 11.31^\circ}$$

$$F_B = 9.81\text{ kN}$$

EXAMPLE 3.4

Determine the required length of cord AC in Fig. 3–8a so that the 8-kg lamp can be suspended in the position shown. The *undeformed* length of spring AB is $l'_{AB} = 0.4$ m, and the spring has a stiffness of $k_{AB} = 300$ N/m.

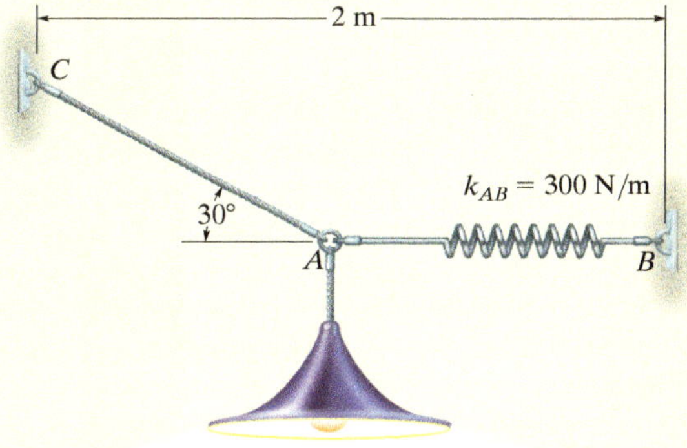

(a)

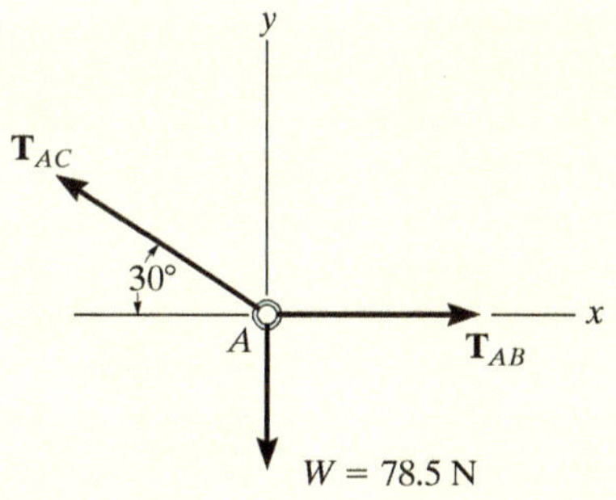

(b)

Fig. 3–8

SOLUTION

If the force in spring AB is known, the stretch of the spring can be found using $F = ks$. From the problem geometry, it is then possible to calculate the required length of AC.

Free-Body Diagram. The lamp has a weight $W = 8(9.81) = 78.5$ N and so the free-body diagram of the ring at A is shown in Fig. 3–8b.

Equations of Equilibrium. Using the x, y axes,

$$\overset{+}{\rightarrow}\Sigma F_x = 0; \qquad T_{AB} - T_{AC}\cos 30^\circ = 0$$

$$+\uparrow\Sigma F_y = 0; \qquad T_{AC}\sin 30^\circ - 78.5\text{ N} = 0$$

Solving, we obtain

$$T_{AC} = 157.0\text{ N}$$

$$T_{AB} = 135.9\text{ N}$$

The stretch of spring AB is therefore

$$T_{AB} = k_{AB}s_{AB}; \qquad 135.9\text{ N} = 300\text{ N/m}(s_{AB})$$

$$s_{AB} = 0.453\text{ m}$$

so the stretched length is

$$l_{AB} = l'_{AB} + s_{AB}$$

$$l_{AB} = 0.4\text{ m} + 0.453\text{ m} = 0.853\text{ m}$$

The horizontal distance from C to B, Fig. 3–8a, requires

$$2\text{ m} = l_{AC}\cos 30^\circ + 0.853\text{ m}$$

$$l_{AC} = 1.32\text{ m} \qquad \textit{Ans.}$$

FUNDAMENTAL PROBLEMS

All problem solutions must include an FBD.

F3–1. The crate has a weight of 550 lb. Determine the force in each supporting cable.

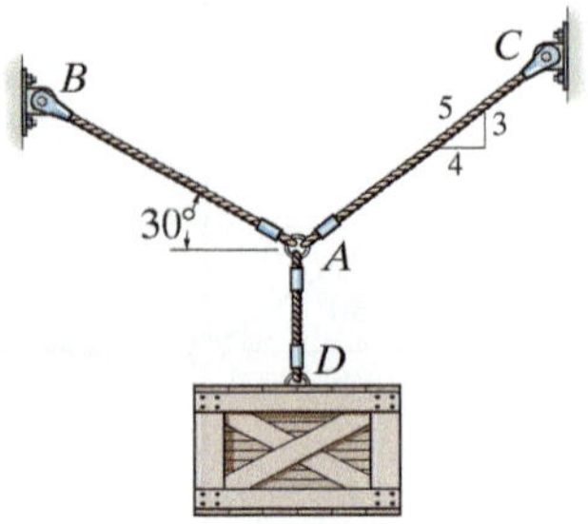

F3–1

F3–2. The beam has a weight of 700 lb. Determine the shortest cable ABC that can be used to lift it if the maximum force the cable can sustain is 1500 lb.

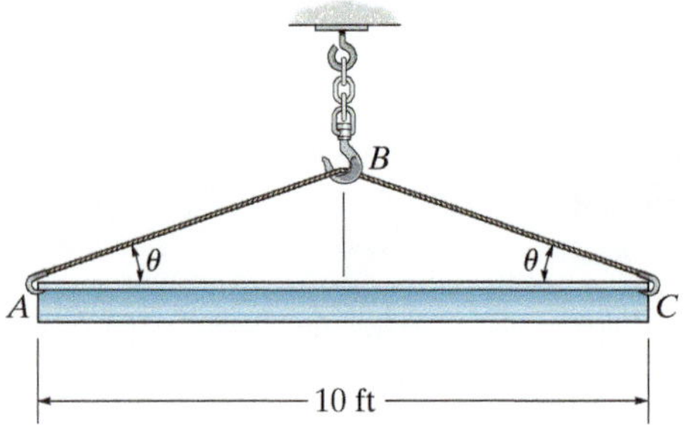

F3–2

F3–3. If the 5-kg block is suspended from the pulley B and the sag of the cord is $d = 0.15$ m, determine the force in cord ABC. Neglect the size of the pulley.

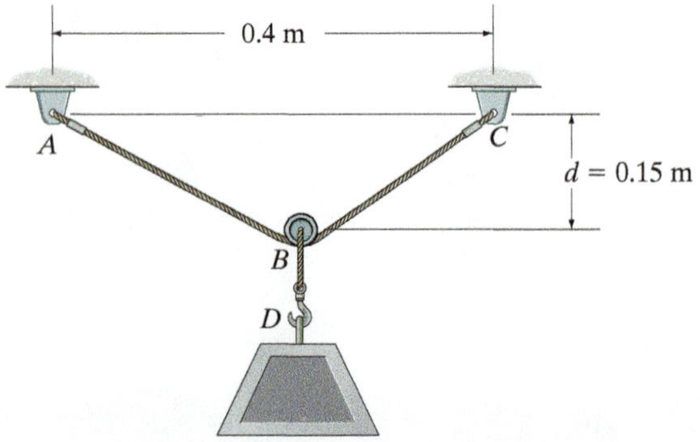

F3–3

F3–4. The block has a mass of 5 kg and rests on the smooth plane. Determine the unstretched length of the spring.

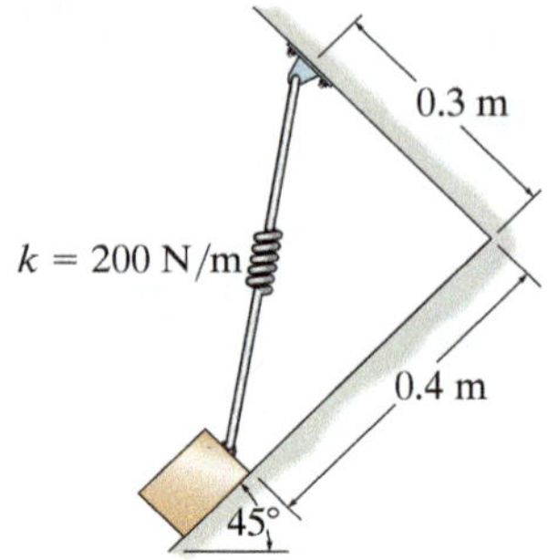

F3–4

F3–5. If the mass of cylinder C is 40 kg, determine the mass of cylinder A in order to hold the assembly in the position shown.

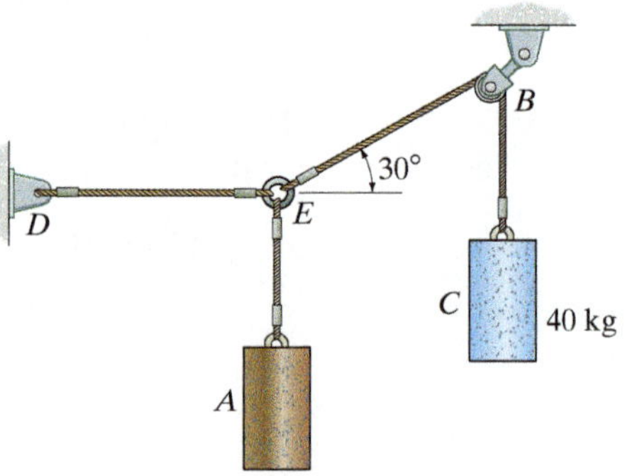

F3–5

F3–6. Determine the tension in cables AB, BC, and CD, necessary to support the 10-kg and 15-kg traffic lights at B and C, respectively. Also, find the angle θ.

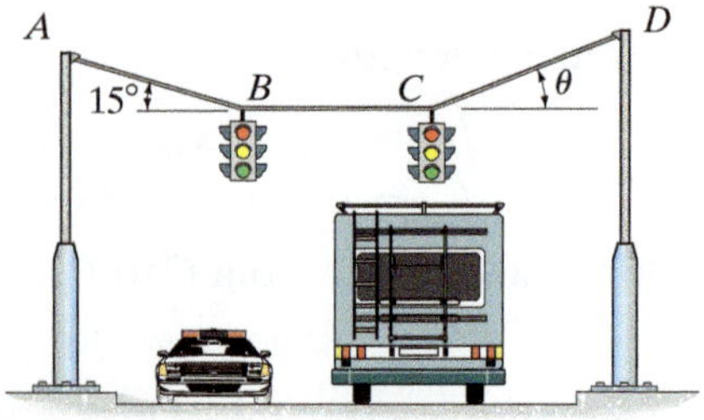

F3–6

PROBLEMS

All problem solutions must include an FBD.

3–1. The members of a truss are pin connected at joint O. Determine the magnitudes of $\mathbf{F}_1$ and $\mathbf{F}_2$ for equilibrium. Set $\theta = 60°$.

3–2. The members of a truss are pin connected at joint O. Determine the magnitude of $\mathbf{F}_1$ and its angle θ for equilibrium. Set $F_2 = 6$ kN.

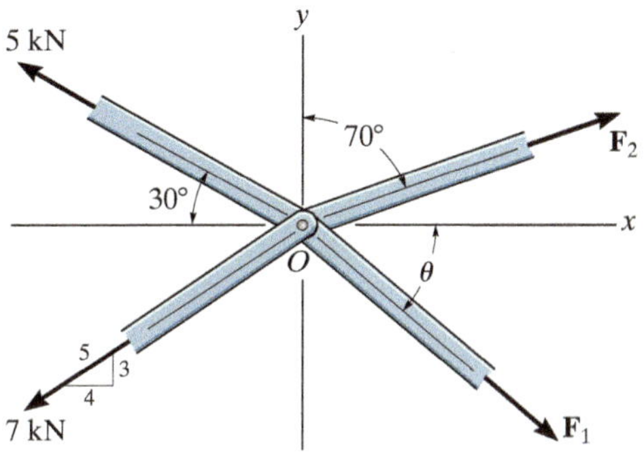

Probs. 3–1/2

3–3. The lift sling is used to hoist a container having a mass of 500 kg. Determine the force in each of the cables AB and AC as a function of θ. If the maximum tension allowed in each cable is 5 kN, determine the shortest lengths of cables AB and AC that can be used for the lift. The center of gravity of the container is located at G.

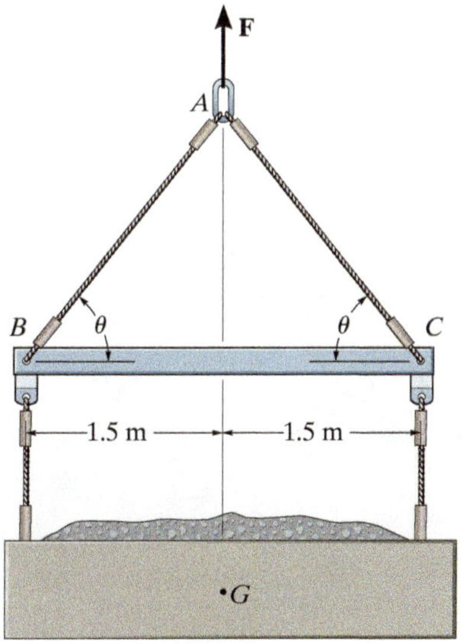

Prob. 3–3

***3–4.** Cords AB and AC can each sustain a maximum tension of 800 lb. If the drum has a weight of 900 lb, determine the smallest angle θ at which they can be attached to the drum.

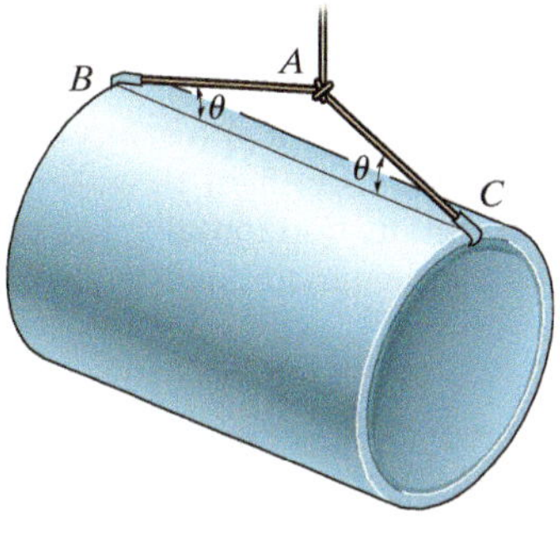

Prob. 3–4

3–5. The members of a truss are connected to the gusset plate. If the forces are concurrent at point O, determine the magnitudes of $\mathbf{F}$ and $\mathbf{T}$ for equilibrium. Take $\theta = 30°$.

3–6. The gusset plate is subjected to the forces of four members. Determine the force in member B and its proper orientation θ for equilibrium. The forces are concurrent at point O. Take $F = 12$ kN.

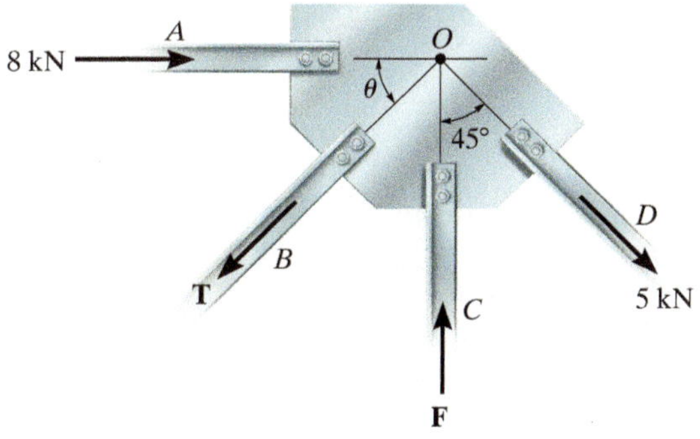

Probs. 3–5/6

3

3–7. The device shown is used to straighten the frames of wrecked autos. Determine the tension of each segment of the chain, i.e., *AB* and *BC*, if the force which the hydraulic cylinder *DB* exerts on point *B* is 3.50 kN, as shown.

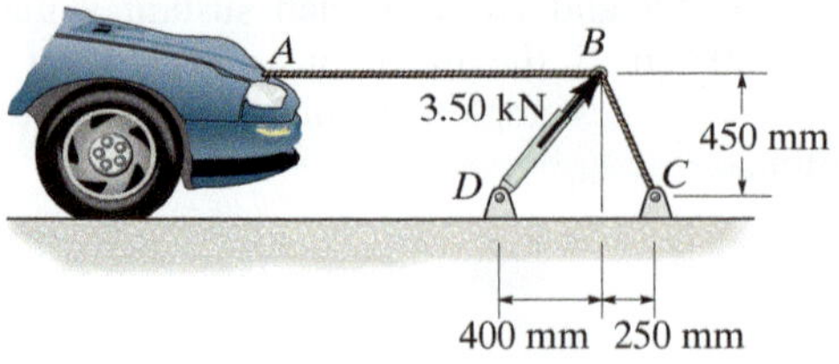

Prob. 3–7

*3–8. Two electrically charged pith balls, each having a mass of 0.2 g, are suspended from light threads of equal length. Determine the resultant horizontal force of repulsion, *F*, acting on each ball if the measured distance between them is $r = 200$ mm.

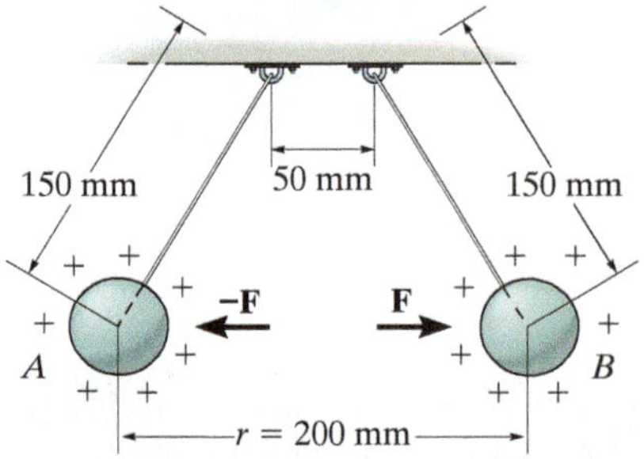

Prob. 3–8

3–9. Determine the maximum weight of the flowerpot that can be supported without exceeding a cable tension of 50 lb in either cable *AB* or *AC*.

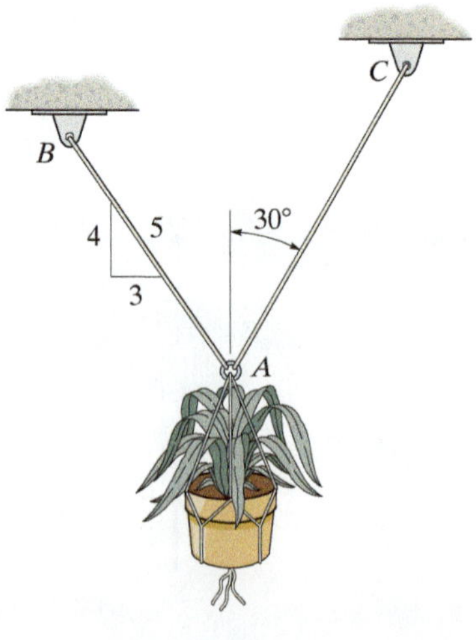

Prob. 3–9

3–10. Determine the tension developed in wires *CA* and *CB* required for equilibrium of the 10-kg cylinder. Take $\theta = 40°$.

3–11. If cable *CB* is subjected to a tension that is twice that of cable *CA*, determine the angle θ for equilibrium of the 10-kg cylinder. Also, what are the tensions in wires *CA* and *CB*?

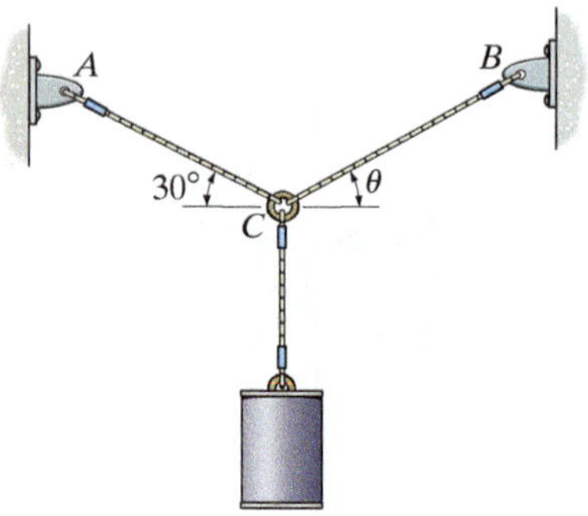

Probs. 3–10/11

*3–12. The concrete pipe elbow has a weight of 400 lb and the center of gravity is located at point *G*. Determine the force $\mathbf{F}_{AB}$ and the tension in cables *BC* and *BD* needed to support it.

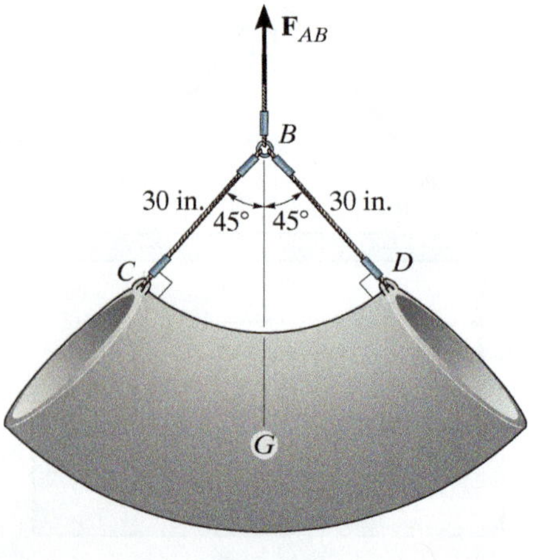

Prob. 3–12

3–13. Blocks D and F weigh 5 lb each and block E weighs 8 lb. Determine the sag s for equilibrium. Neglect the size of the pulleys.

3–14. If blocks D and F weigh 5 lb each, determine the weight of block E if the sag $s = 3$ ft. Neglect the size of the pulleys.

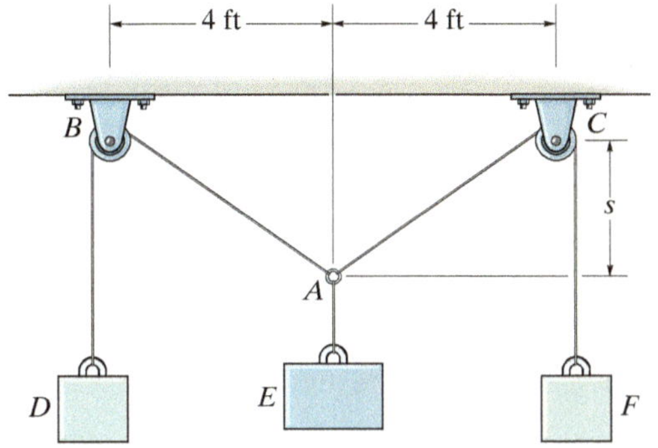

Probs. 3–13/14

*■**3–16.** The 10-lb lamp fixture is suspended from two springs, each having an unstretched length of 4 ft and stiffness of $k = 5$ lb/ft. Determine the angle θ for equilibrium.

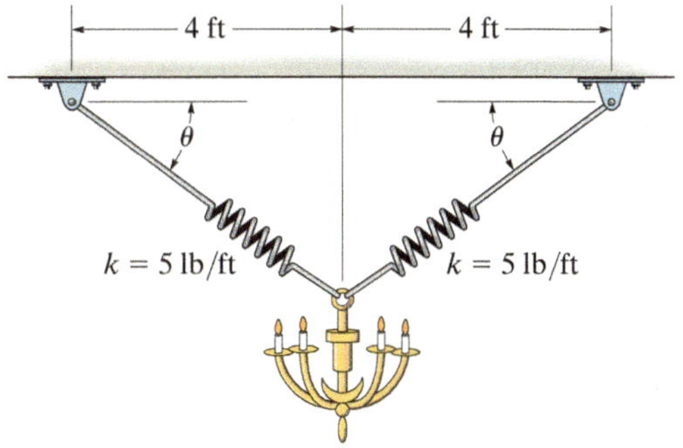

Prob. 3–16

3–15. The spring has a stiffness of $k = 800$ N/m and an unstretched length of 200 mm. Determine the force in cables BC and BD when the spring is held in the position shown.

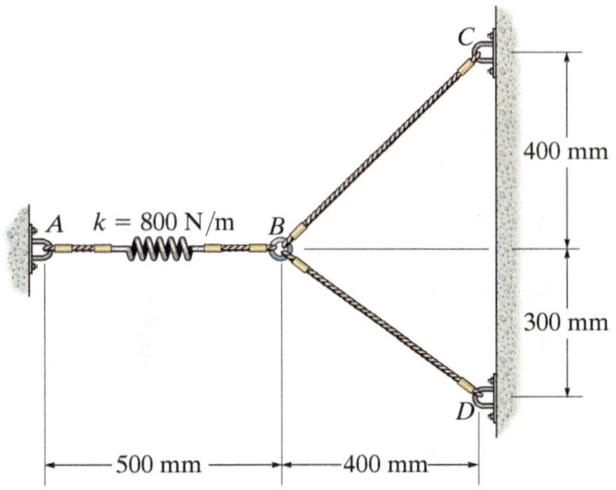

Prob. 3–15

3–17. Determine the mass of each of the two cylinders if they cause a sag of $s = 0.5$ m when suspended from the rings at A and B. Note that $s = 0$ when the cylinders are removed.

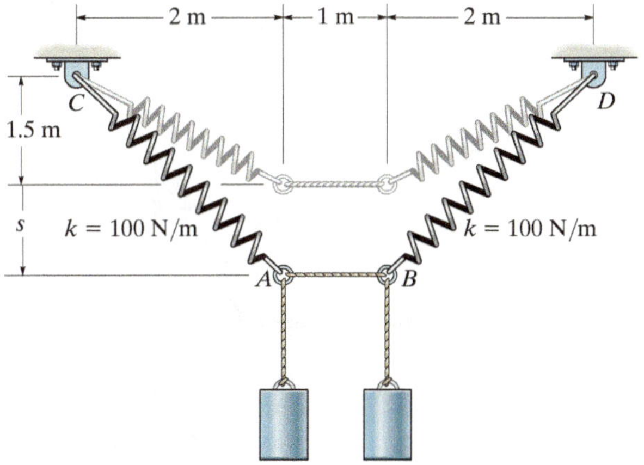

Prob. 3–17

3–18. Determine the stretch in each spring for equilibrium of the 2-kg block. The springs are shown in the equilibrium position.

3–19. The unstretched length of spring AB is 3 m. If the block is held in the equilibrium position shown, determine the mass of the block at D.

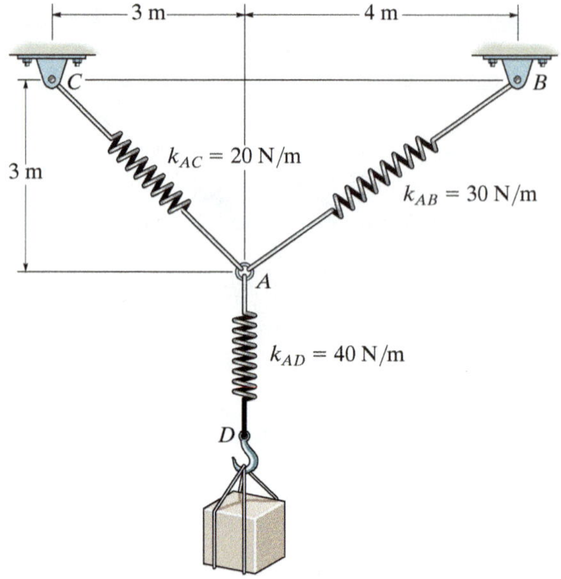

Probs. 3–18/19

***3–20.** The springs BA and BC each have a stiffness of 500 N/m and an unstretched length of 3 m. Determine the horizontal force **F** applied to the cord which is attached to the *small* ring B so that the displacement of the ring from the wall is $d = 1.5$ m.

3–21. The springs BA and BC each have a stiffness of 500 N/m and an unstretched length of 3 m. Determine the displacement d of the cord from the wall when a force $F = 175$ N is applied to the cord.

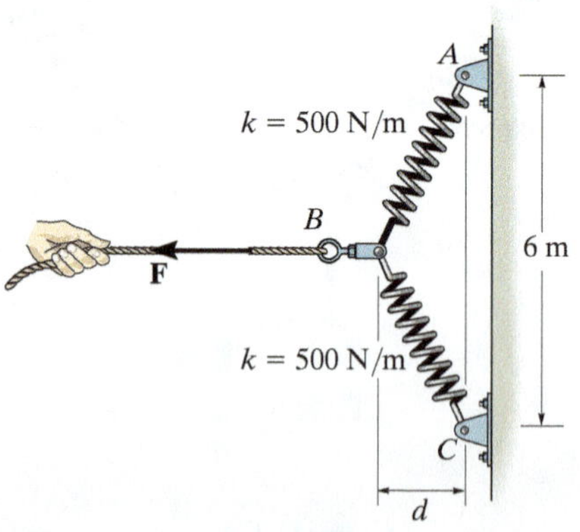

Probs. 3–20/21

■**3–22.** A vertical force $P = 10$ lb is applied to the ends of the 2-ft cord AB and spring AC. If the spring has an unstretched length of 2 ft, determine the angle θ for equilibrium. Take $k = 15$ lb/ft.

3–23. Determine the unstretched length of spring AC if a force $P = 80$ lb causes the angle $\theta = 60°$ for equilibrium. Cord AB is 2 ft long. Take $k = 50$ lb/ft.

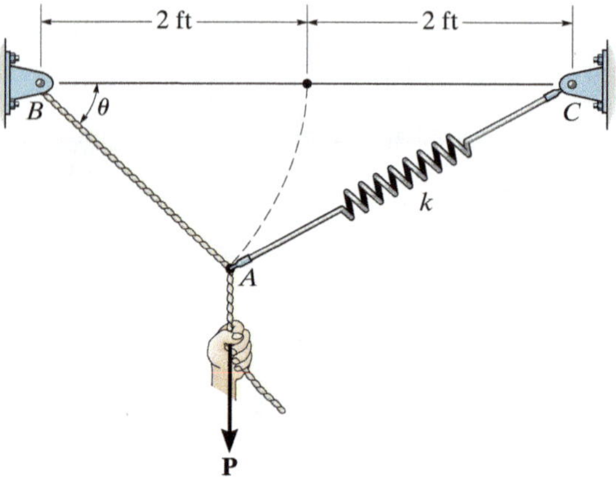

Probs. 3–22/23

***3–24.** The springs on the rope assembly are originally unstretched when $\theta = 0°$. Determine the tension in each rope when $F = 90$ lb. Neglect the size of the pulleys at B and D.

3–25. The springs on the rope assembly are originally stretched 1 ft when $\theta = 0°$. Determine the vertical force F that must be applied so that $\theta = 30°$.

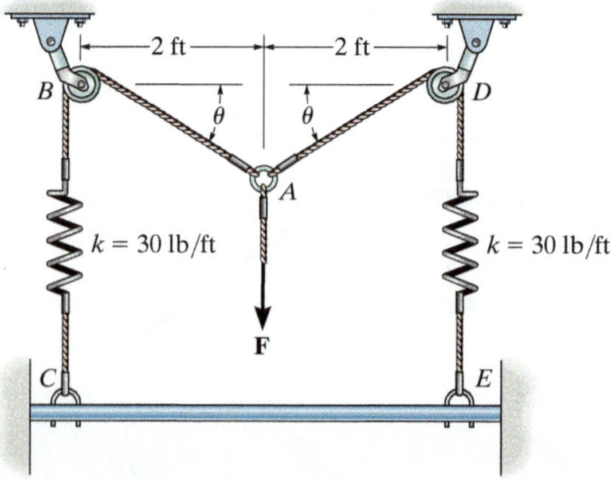

Probs. 3–24/25

3–26. The 10-lb weight A is supported by the cord AC and roller C, and by the spring that has a stiffness of $k = 10$ lb/in. If the unstretched length of the spring is 12 in. determine the distance d to where the weight is located when it is in equilibrium.

3–27. The 10-lb weight A is supported by the cord AC and roller C, and by spring AB. If the spring has an unstretched length of 8 in. and the weight is in equilibrium when $d = 4$ in., determine the stiffness k of the spring.

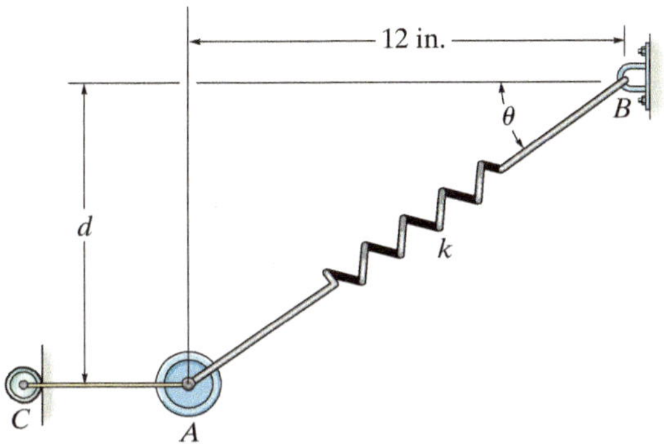

Probs. 3–26/27

3–30. A 4-kg sphere rests on the smooth parabolic surface. Determine the normal force it exerts on the surface and the mass m_B of block B needed to hold it in the equilibrium position shown.

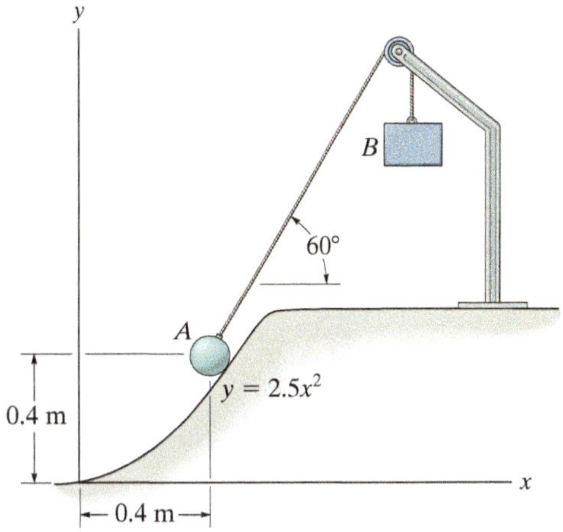

Prob. 3–30

***3–28.** Determine the tension developed in each cord required for equilibrium of the 20-kg lamp.

3–29. Determine the maximum mass of the lamp that the cord system can support so that no single cord develops a tension exceeding 400 N.

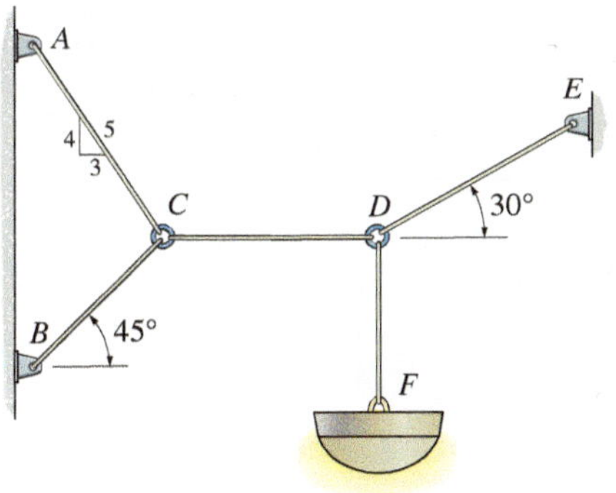

Probs. 3–28/29

3–31. If the bucket weighs 50 lb, determine the tension developed in each of the wires.

***3–32.** Determine the maximum weight of the bucket that the wire system can support so that no single wire develops a tension exceeding 100 lb.

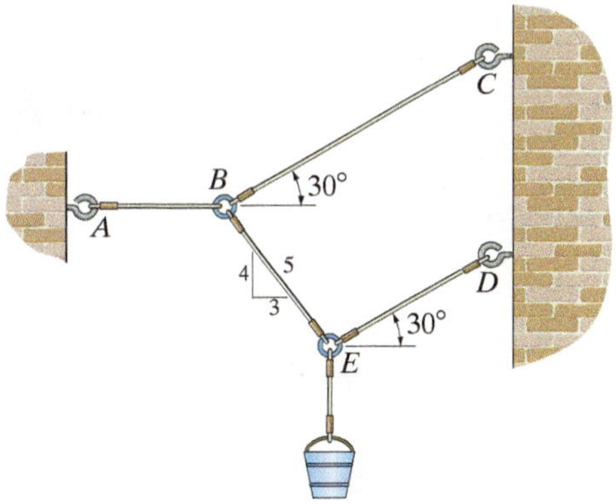

Probs. 3–31/32

3–33. Determine the tension developed in each wire which is needed to support the 50-lb flowerpot.

3–34. If the tension developed in each of the wires is not allowed to exceed 40 lb, determine the maximum weight of the flowerpot that can be safely supported.

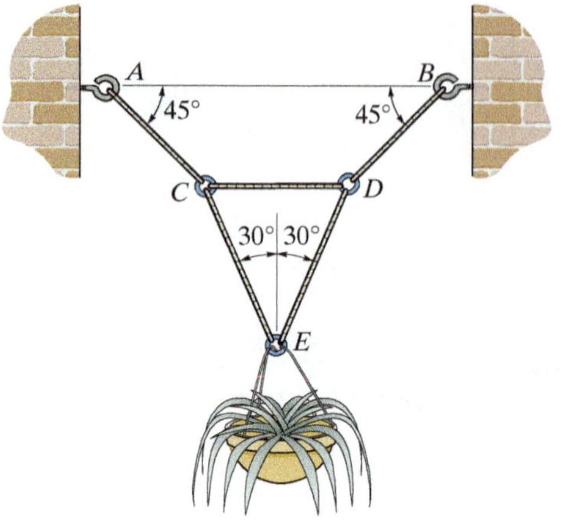

Probs. 3–33/34

3–35. Cable ABC has a length of 5 m. Determine the position x and the tension developed in ABC required for equilibrium of the 100-kg sack. Neglect the size of the pulley at B.

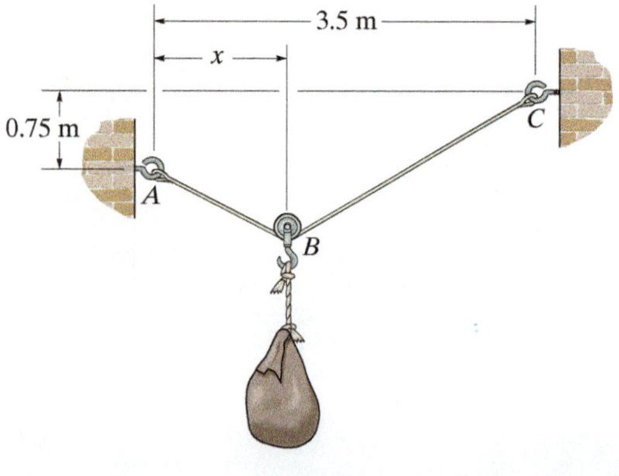

Prob. 3–35

***3–36.** The single elastic cord ABC is used to support the 40-lb load. Determine the position x and the tension in the cord that is required for equilibrium. The cord passes through the smooth ring at B and has an unstretched length of 6 ft and stiffness of $k = 50$ lb/ft.

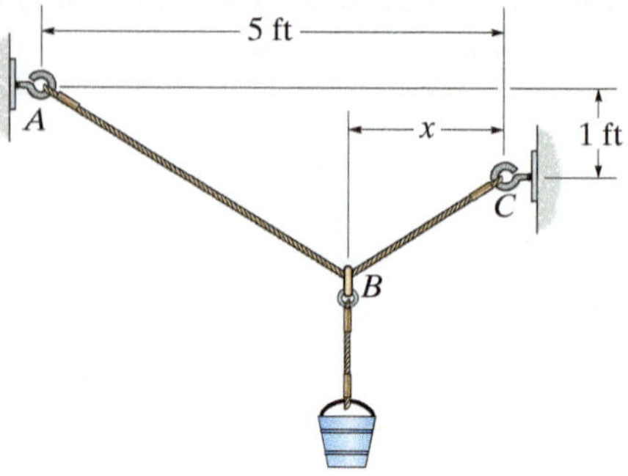

Prob. 3–36

3–37. The 200-lb uniform tank is suspended by means of a 6-ft-long cable, which is attached to the sides of the tank and passes over the small pulley located at O. If the cable can be attached at either points A and B, or C and D, determine which attachment produces the least amount of tension in the cable. What is this tension?

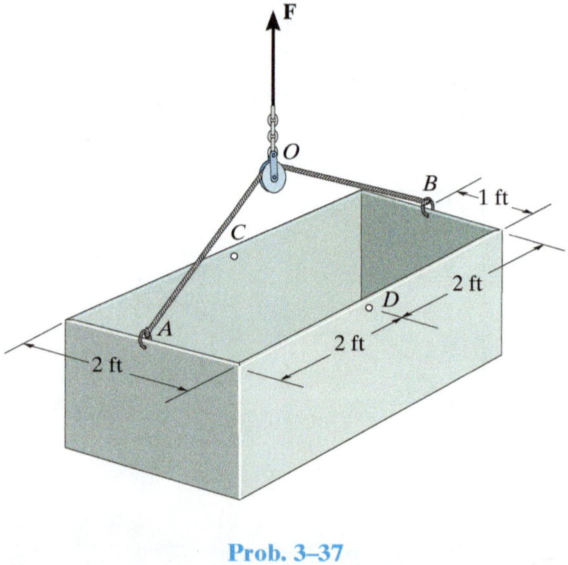

Prob. 3–37

3–38. The sling *BAC* is used to lift the 100-lb load with constant velocity. Determine the force in the sling and plot its value T (ordinate) as a function of its orientation θ where $0 \leq \theta \leq 90°$.

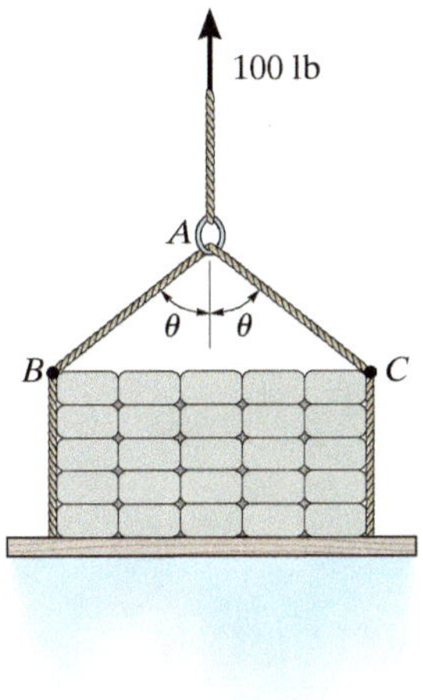

Prob. 3–38

■**3–39.** A "scale" is constructed with a 4-ft-long cord and the 10-lb block D. The cord is fixed to a pin at A and passes over two *small* pulleys. Determine the weight of the suspended block B if the system is in equilibrium when $s = 1.5$ ft.

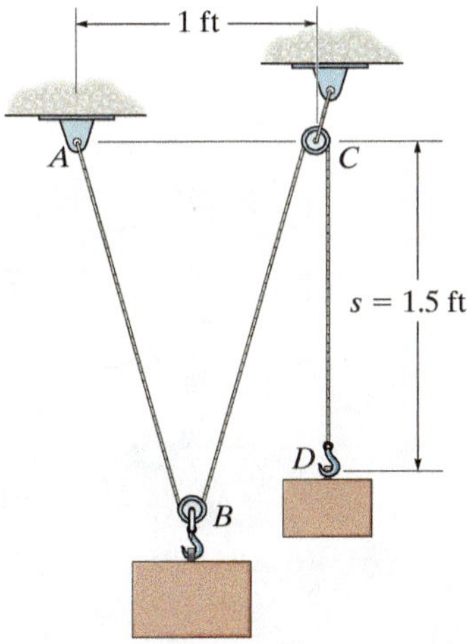

Prob. 3–39

*__3–40.__ The load has a mass of 15 kg and is lifted by the pulley system shown. Determine the force **F** in the cord as a function of the angle θ. Plot the function of force F versus the angle θ for $0 \leq \theta \leq 90°$.

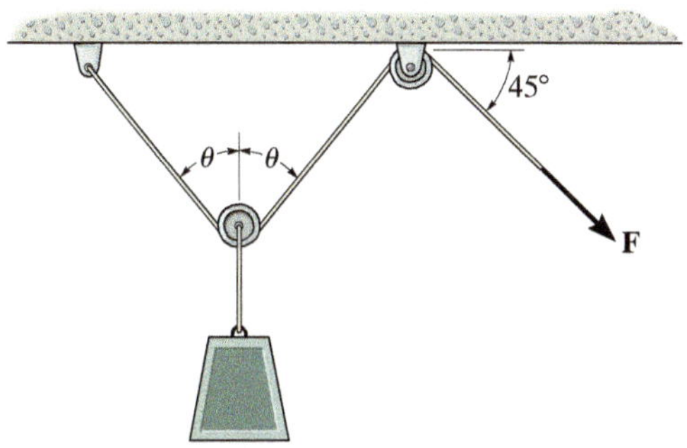

Prob. 3–40

3–41. Determine the forces in cables AC and AB needed to hold the 20-kg ball D in equlibrium. Take $F = 300$ N and $d = 1$ m.

3–42. The ball D has a mass of 20 kg. If a force of $F = 100$ N is applied horizontally to the ring at A, determine the dimension d so that the force in cable AC is zero.

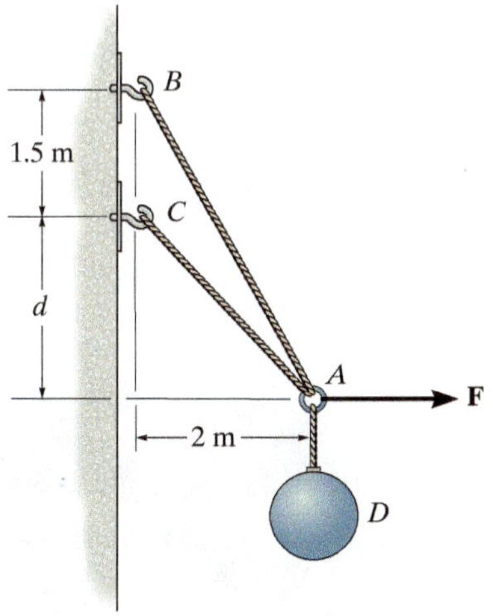

Probs. 3–41/42

CONCEPTUAL PROBLEMS

P3–1. The concrete wall panel is hoisted into position using the two cables AB and AC of equal length. Establish appropriate dimensions and use an equilibrium analysis to show that the longer the cables the less the force in each cable.

P3–1

P3–2. The hoisting cables BA and BC each have a length of 20 ft. If the maximum tension that can be supported by each cable is 900 lb, determine the maximum distance AC between them in order to lift the uniform 1200-lb truss with constant velocity.

P3–2

P3–3. The device DB is used to pull on the chain ABC to hold a door closed on the bin. If the angle between AB and BC is 30°, determine the angle between DB and BC for equilibrium.

P3–3

P3–4. Chain AB is 1-m long and chain AC is 1.2-m long. If the distance BC is 1.5 m, and AB can support a maximum force of 2 kN, whereas AC can support a maximum force of 0.8 kN, determine the largest vertical force F that can be applied to the link at A.

P3–4

3.4 Three-Dimensional Force Systems

In Section 3.1 we stated that the necessary and sufficient condition for particle equilibrium is

$$\Sigma \mathbf{F} = \mathbf{0} \qquad (3\text{–}4)$$

In the case of a three-dimensional force system, as in Fig. 3–9, we can resolve the forces into their respective **i**, **j**, **k** components, so that $\Sigma F_x\mathbf{i} + \Sigma F_y\mathbf{j} + \Sigma F_z\mathbf{k} = \mathbf{0}$. To satisfy this equation we require

$$\begin{aligned} \Sigma F_x &= 0 \\ \Sigma F_y &= 0 \\ \Sigma F_y &= 0 \end{aligned} \qquad (3\text{–}5)$$

These three equations state that the *algebraic sum* of the components of all the forces acting on the particle along each of the coordinate axes must be zero. Using them we can solve for at most three unknowns, generally represented as coordinate direction angles or magnitudes of forces shown on the particle's free-body diagram.

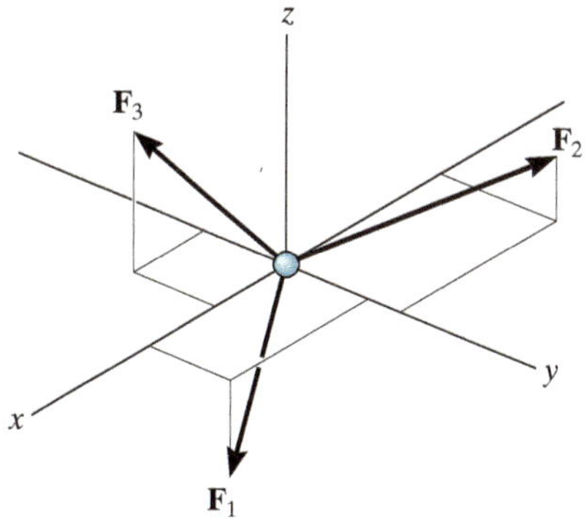

Fig. 3–9

Procedure for Analysis

Three-dimensional force equilibrium problems for a particle can be solved using the following procedure.

Free-Body Diagram.

- Establish the *x*, *y*, *z* axes in any suitable orientation.
- Label all the known and unknown force magnitudes and directions on the diagram.
- The sense of a force having an unknown magnitude can be assumed.

Equations of Equilibrium.

- Use the scalar equations of equilibrium, $\Sigma F_x = 0$, $\Sigma F_y = 0$, $\Sigma F_z = 0$, in cases where it is easy to resolve each force into its *x*, *y*, *z* components.
- If the three-dimensional geometry appears difficult, then first express each force on the free-body diagram as a Cartesian vector, substitute these vectors into $\Sigma \mathbf{F} = \mathbf{0}$, and then set the **i**, **j**, **k** components equal to zero.
- If the solution for a force yields a negative result, this indicates that its sense is the reverse of that shown on the free-body diagram.

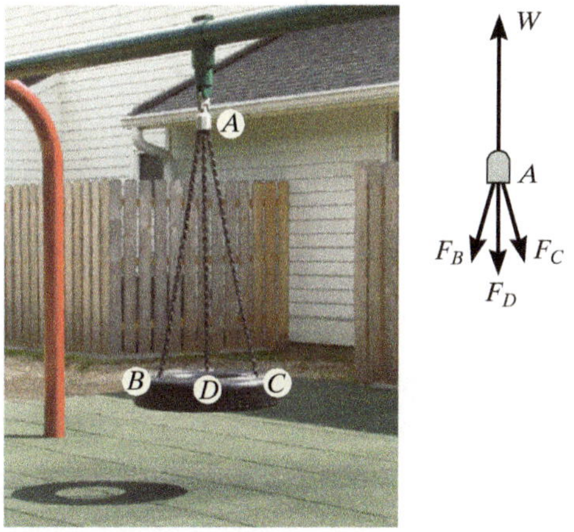

The joint at *A* is subjected to the force from the support as well as forces from each of the three chains. If the tire and any load on it have a weight *W*, then the force at the support will be **W**, and the three scalar equations of equilibrium can be applied to the free-body diagram of the joint in order to determine the chain forces, $\mathbf{F}_B$, $\mathbf{F}_C$, and $\mathbf{F}_D$.

3

EXAMPLE 3.5

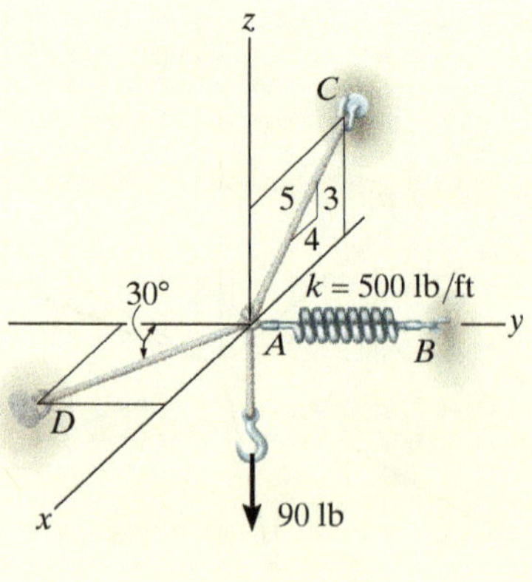

(a)

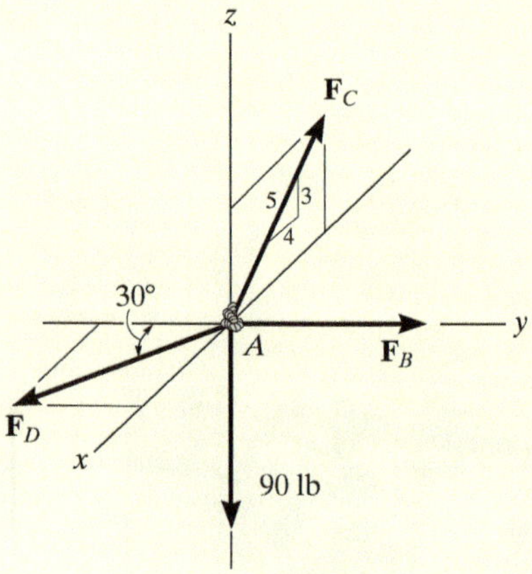

(b)

Fig. 3–10

A 90-lb load is suspended from the hook shown in Fig. 3–10*a*. If the load is supported by two cables and a spring having a stiffness $k = 500$ lb/ft, determine the force in the cables and the stretch of the spring for equilibrium. Cable *AD* lies in the *x*–*y* plane and cable *AC* lies in the *x*–*z* plane.

SOLUTION

The stretch of the spring can be determined once the force in the spring is determined.

Free-Body Diagram. The connection at *A* is chosen for the equilibrium analysis since the cable forces are concurrent at this point. The free-body diagram is shown in Fig. 3–10*b*.

Equations of Equilibrium. By inspection, each force can easily be resolved into its *x*, *y*, *z* components, and therefore the three scalar equations of equilibrium can be used. Considering components directed along each positive axis as "positive," we have

$$\Sigma F_x = 0; \qquad F_D \sin 30^\circ - \left(\tfrac{4}{5}\right)F_C = 0 \qquad (1)$$

$$\Sigma F_y = 0; \qquad -F_D \cos 30^\circ + F_B = 0 \qquad (2)$$

$$\Sigma F_z = 0; \qquad \left(\tfrac{3}{5}\right)F_C - 90 \text{ lb} = 0 \qquad (3)$$

Solving Eq. (3) for F_C, then Eq. (1) for F_D, and finally Eq. (2) for F_B, yields

$$F_C = 150 \text{ lb} \qquad \textit{Ans.}$$

$$F_D = 240 \text{ lb} \qquad \textit{Ans.}$$

$$F_B = 207.8 \text{ lb} = 208 \text{ lb} \qquad \textit{Ans.}$$

The stretch of the spring is therefore

$$F_B = ks_{AB}$$

$$207.8 \text{ lb} = (500 \text{ lb/ft})(s_{AB})$$

$$s_{AB} = 0.416 \text{ ft} \qquad \textit{Ans.}$$

NOTE: Since the results for all the cable forces are positive, each cable is in tension; that is, it pulls on point *A* as expected, Fig. 3–10*b*.

EXAMPLE 3.6

The 10-kg lamp in Fig. 3–11*a* is suspended from the three equal-length cords. Determine its smallest vertical distance *s* from the ceiling if the force developed in any cord is not allowed to exceed 50 N.

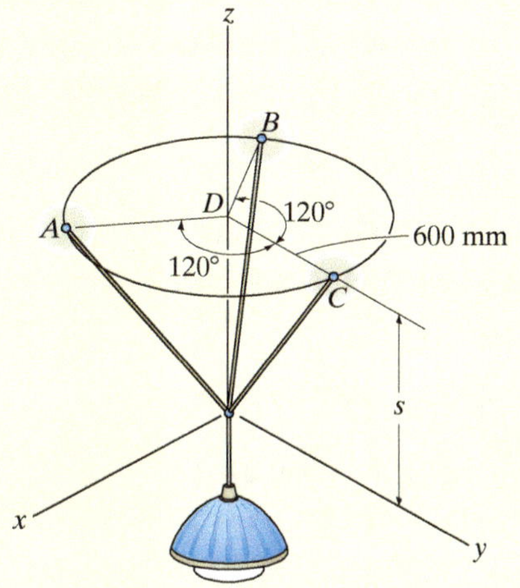

(a)

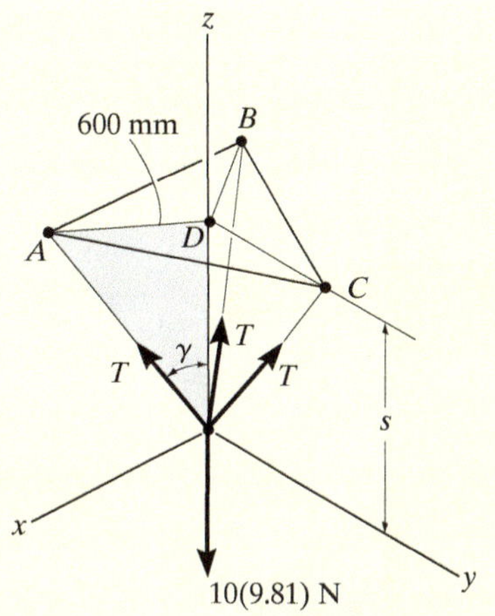

(b)

Fig. 3–11

SOLUTION

Free-Body Diagram. Due to symmetry, Fig. 3–11*b*, the distance $DA = DB = DC = 600$ mm. It follows that from $\Sigma F_x = 0$ and $\Sigma F_y = 0$, the tension T in each cord will be the same. Also, the angle between each cord and the z axis is γ.

Equation of Equilibrium. Applying the equilibrium equation along the z axis, with $T = 50$ N, we have

$$\Sigma F_z = 0; \qquad 3[(50\text{ N})\cos\gamma] - 10(9.81)\text{ N} = 0$$

$$\gamma = \cos^{-1}\frac{98.1}{150} = 49.16^\circ$$

From the shaded triangle shown in Fig. 3–11*b*,

$$\tan 49.16^\circ = \frac{600\text{ mm}}{s}$$

$$s = 519\text{ mm} \qquad \textit{Ans.}$$

EXAMPLE 3.7

Determine the force in each cable used to support the 40-lb crate shown in Fig. 3–12*a*.

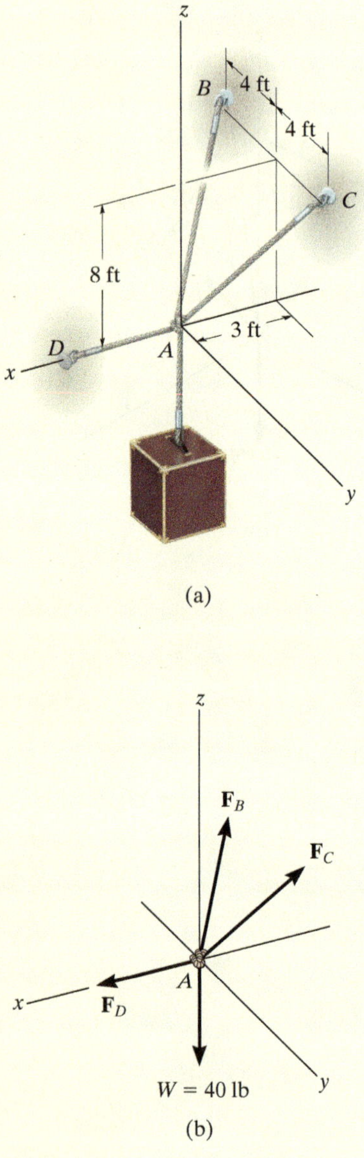

Fig. 3–12

SOLUTION

Free-Body Diagram. As shown in Fig. 3–12*b*, the free-body diagram of point *A* is considered in order to "expose" the three unknown forces in the cables.

Equations of Equilibrium. First we will express each force in Cartesian vector form. Since the coordinates of points *B* and *C* are $B(-3\text{ ft}, -4\text{ ft}, 8\text{ ft})$ and $C(-3\text{ ft}, 4\text{ ft}, 8\text{ ft})$, we have

$$\mathbf{F}_B = F_B\left[\frac{-3\mathbf{i} - 4\mathbf{j} + 8\mathbf{k}}{\sqrt{(-3)^2 + (-4)^2 + (8)^2}}\right]$$

$$= -0.318F_B\mathbf{i} - 0.424F_B\mathbf{j} + 0.848F_B\mathbf{k}$$

$$\mathbf{F}_C = F_C\left[\frac{-3\mathbf{i} + 4\mathbf{j} + 8\mathbf{k}}{\sqrt{(-3)^2 + (4)^2 + (8)^2}}\right]$$

$$= -0.318F_C\mathbf{i} + 0.424F_C\mathbf{j} + 0.848F_C\mathbf{k}$$

$$\mathbf{F}_D = F_D\mathbf{i}$$

$$\mathbf{W} = \{-40\mathbf{k}\}\text{ lb}$$

Equilibrium requires

$$\Sigma\mathbf{F} = \mathbf{0}; \qquad \mathbf{F}_B + \mathbf{F}_C + \mathbf{F}_D + \mathbf{W} = \mathbf{0}$$

$$-0.318F_B\mathbf{i} - 0.424F_B\mathbf{j} + 0.848F_B\mathbf{k}$$

$$-0.318F_C\mathbf{i} + 0.424F_C\mathbf{j} + 0.848F_C\mathbf{k} + F_D\mathbf{i} - 40\mathbf{k} = \mathbf{0}$$

Equating the respective **i**, **j**, **k** components to zero yields

$$\Sigma F_x = 0; \qquad -0.318F_B - 0.318F_C + F_D = 0 \qquad (1)$$

$$\Sigma F_y = 0; \qquad -0.424F_B + 0.424F_C = 0 \qquad (2)$$

$$\Sigma F_z = 0; \qquad 0.848F_B + 0.848F_C - 40 = 0 \qquad (3)$$

Equation (2) states that $F_B = F_C$. Thus, solving Eq. (3) for F_B and F_C and substituting the result into Eq. (1) to obtain F_D, we have

$$F_B = F_C = 23.6\text{ lb} \qquad \textit{Ans.}$$

$$F_D = 15.0\text{ lb} \qquad \textit{Ans.}$$

EXAMPLE 3.8

Determine the tension in each cord used to support the 100-kg crate shown in Fig. 3–13*a*.

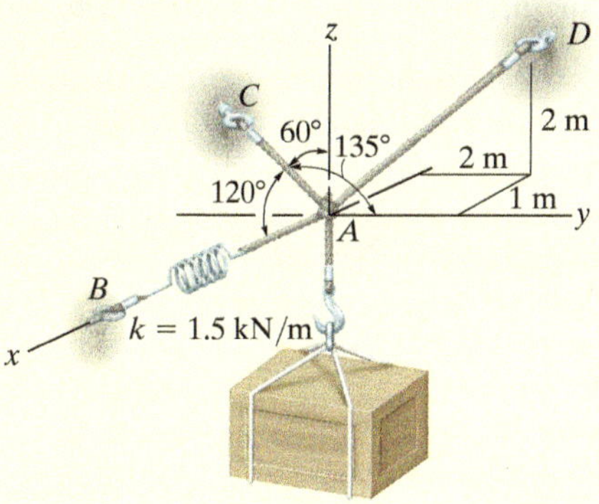

(a)

SOLUTION

Free-Body Diagram. The force in each of the cords can be determined by investigating the equilibrium of point *A*. The free-body diagram is shown in Fig. 3–13*b*. The weight of the crate is $W = 100(9.81) = 981$ N.

Equations of Equilibrium. Each force on the free-body diagram is first expressed in Cartesian vector form. Using Eq. 2–9 for $\mathbf{F}_C$ and noting point $D(-1\text{ m}, 2\text{ m}, 2\text{ m})$ for $\mathbf{F}_D$, we have

$$\mathbf{F}_B = F_B\mathbf{i}$$

$$\mathbf{F}_C = F_C\cos 120°\mathbf{i} + F_C\cos 135°\mathbf{j} + F_C\cos 60°\mathbf{k}$$

$$= -0.5F_C\mathbf{i} - 0.707F_C\mathbf{j} + 0.5F_C\mathbf{k}$$

$$\mathbf{F}_D = F_D\left[\frac{-1\mathbf{i} + 2\mathbf{j} + 2\mathbf{k}}{\sqrt{(-1)^2 + (2)^2 + (2)^2}}\right]$$

$$= -0.333F_D\mathbf{i} + 0.667F_D\mathbf{j} + 0.667F_D\mathbf{k}$$

$$\mathbf{W} = \{-981\mathbf{k}\}\text{ N}$$

Equilibrium requires

$$\Sigma\mathbf{F} = \mathbf{0}; \qquad \mathbf{F}_B + \mathbf{F}_C + \mathbf{F}_D + \mathbf{W} = \mathbf{0}$$

$$F_B\mathbf{i} - 0.5F_C\mathbf{i} - 0.707F_C\mathbf{j} + 0.5F_C\mathbf{k}$$
$$-0.333F_D\mathbf{i} + 0.667F_D\mathbf{j} + 0.667F_D\mathbf{k} - 981\mathbf{k} = \mathbf{0}$$

Equating the respective **i**, **j**, **k** components to zero,

$$\Sigma F_x = 0; \qquad F_B - 0.5F_C - 0.333F_D = 0 \qquad (1)$$

$$\Sigma F_y = 0; \qquad -0.707F_C + 0.667F_D = 0 \qquad (2)$$

$$\Sigma F_z = 0; \qquad 0.5F_C + 0.667F_D - 981 = 0 \qquad (3)$$

Solving Eq. (2) for F_D in terms of F_C and substituting this into Eq. (3) yields F_C. F_D is then determined from Eq. (2). Finally, substituting the results into Eq. (1) gives F_B. Hence,

$$F_C = 813\text{ N} \qquad \textit{Ans.}$$

$$F_D = 862\text{ N} \qquad \textit{Ans.}$$

$$F_B = 694\text{ N} \qquad \textit{Ans.}$$

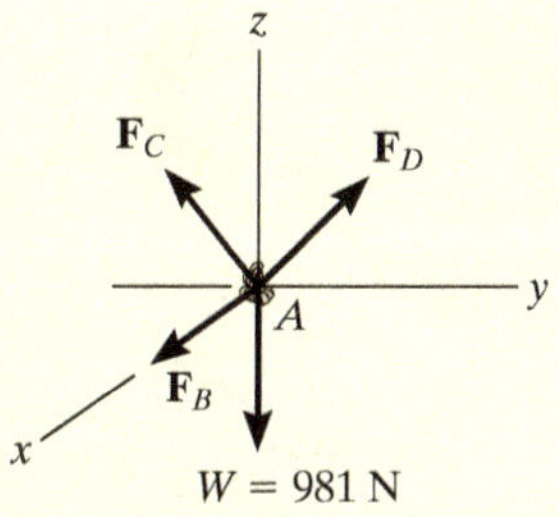

(b)

Fig. 3–13

FUNDAMENTAL PROBLEMS

All problem solutions must include an FBD.

F3–7. Determine the magnitude of forces $\mathbf{F}_1$, $\mathbf{F}_2$, $\mathbf{F}_3$, so that the particle is held in equilibrium.

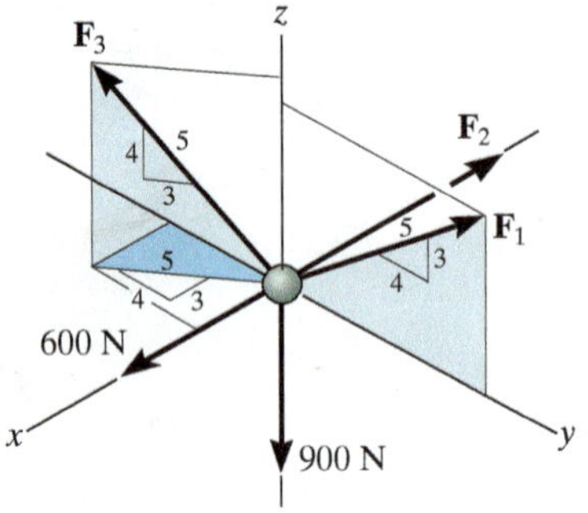

F3–7

F3–8. Determine the tension developed in cables *AB*, *AC*, and *AD*.

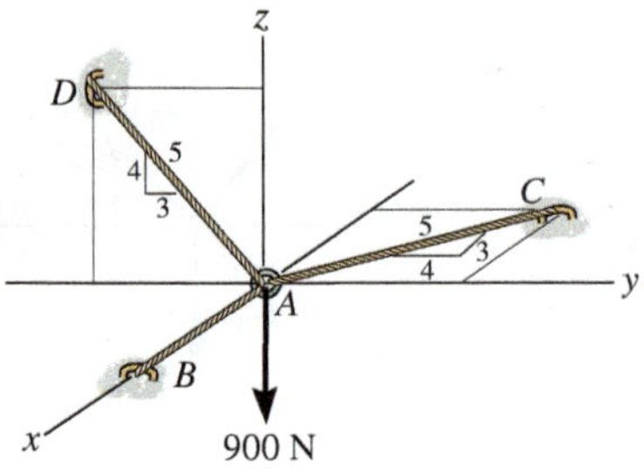

F3–8

F3–9. Determine the tension developed in cables *AB*, *AC*, and *AD*.

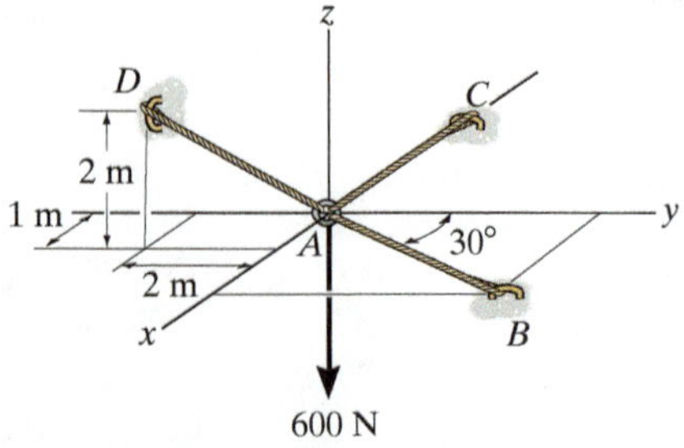

F3–9

F3–10. Determine the tension developed in cables *AB*, *AC*, and *AD*.

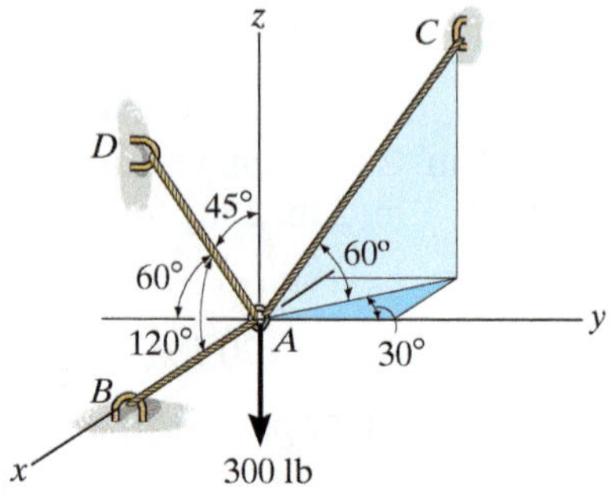

F3–10

F3–11. The 150-lb crate is supported by cables *AB*, *AC*, and *AD*. Determine the tension in these wires.

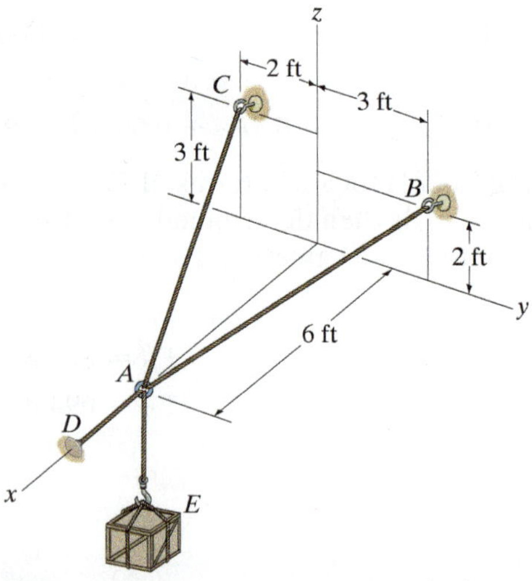

F3–11

PROBLEMS

All problem solutions must include an FBD.

3–43. Determine the magnitude and direction of the force **P** required to keep the concurrent force system in equilibrium.

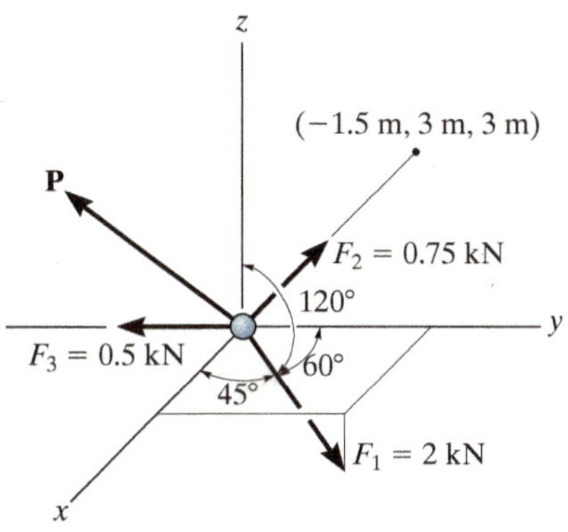

Prob. 3–43

***3–44.** If cable *AB* is subjected to a tension of 700 N, determine the tension in cables *AC* and *AD* and the magnitude of the vertical force **F**.

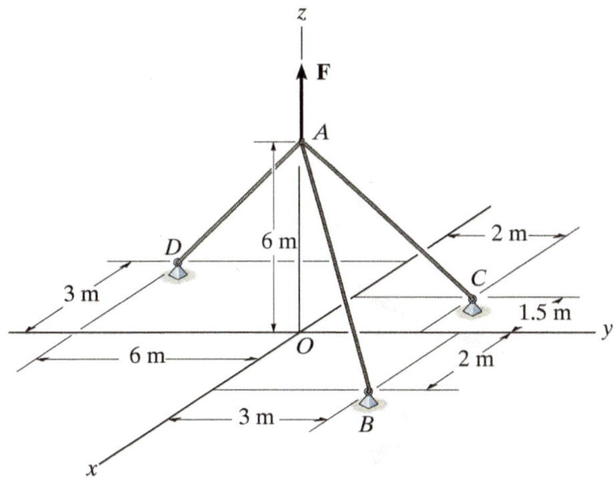

Prob. 3–44

3–45. Determine the magnitudes of $\mathbf{F}_1$, $\mathbf{F}_2$, and $\mathbf{F}_3$ for equilibrium of the particle.

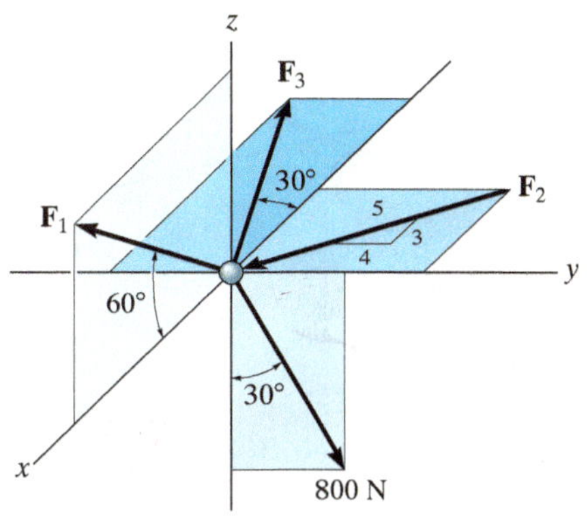

Prob. 3–45

3–46. If the bucket and its contents have a total weight of 20 lb, determine the force in the supporting cables *DA*, *DB*, and *DC*.

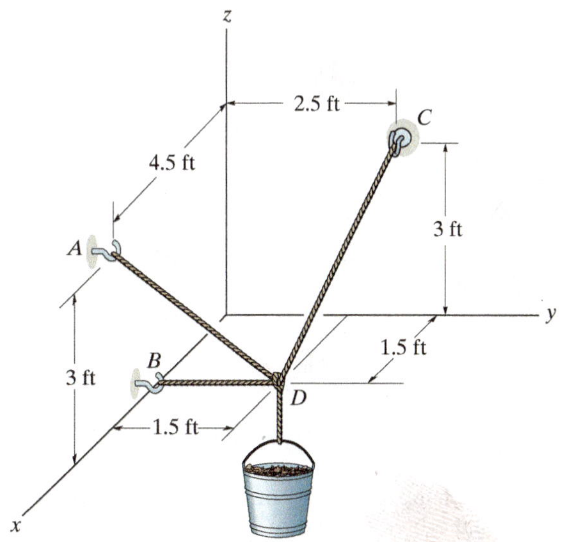

Prob. 3–46

3–47. Determine the stretch in each of the two springs required to hold the 20-kg crate in the equilibrium position shown. Each spring has an unstretched length of 2 m and a stiffness of $k = 300$ N/m.

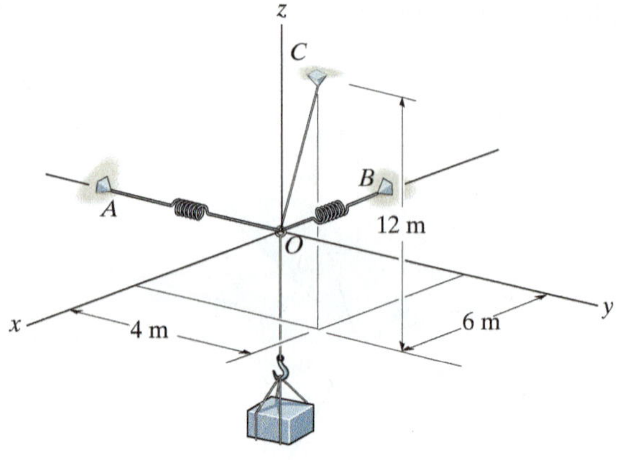

Prob. 3–47

*__3–48.__ If the balloon is subjected to a net uplift force of $F = 800$ N, determine the tension developed in ropes *AB*, *AC*, *AD*.

3–49. If each one of the ropes will break when it is subjected to a tensile force of 450 N, determine the maximum uplift force **F** the balloon can have before one of the ropes breaks.

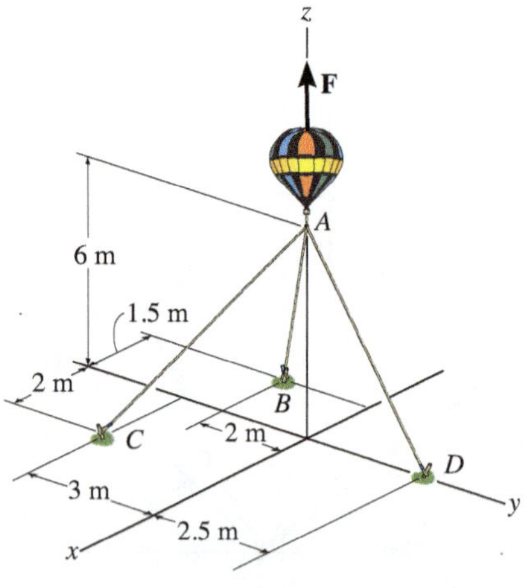

Probs. 3–48/49

■**3–50.** The lamp has a mass of 15 kg and is supported by a pole *AO* and cables *AB* and *AC*. If the force in the pole acts along its axis, determine the forces in *AO*, *AB*, and *AC* for equilibrium.

3–51. Cables *AB* and *AC* can sustain a maximum tension of 500 N, and the pole can support a maximum compression of 300 N. Determine the maximum weight of the lamp that can be supported in the position shown. The force in the pole acts along the axis of the pole.

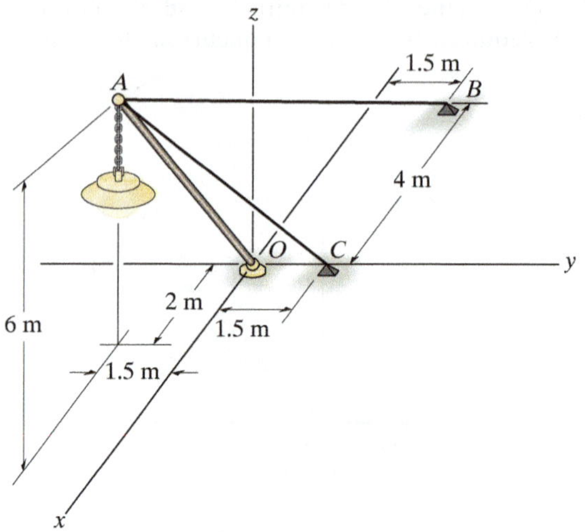

Probs. 3–50/51

*__3–52.__ The 50-kg pot is supported from *A* by the three cables. Determine the force acting in each cable for equilibrium. Take $d = 2.5$ m.

3–53. Determine the height *d* of cable *AB* so that the force in cables *AD* and *AC* is one-half as great as the force in cable *AB*. What is the force in each cable for this case? The flower pot has a mass of 50 kg.

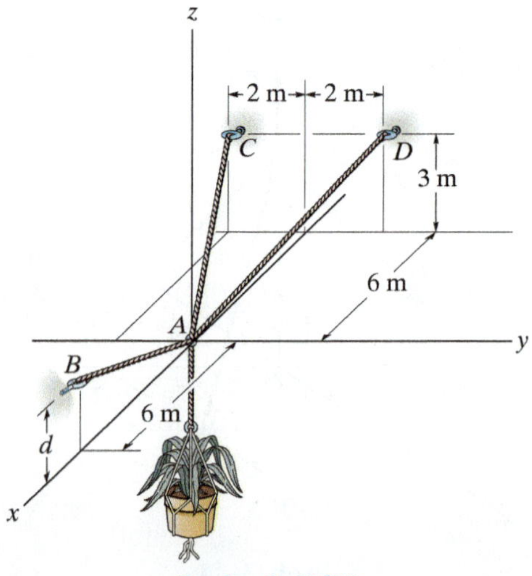

Probs. 3–52/53

3–54. Determine the tension developed in cables *AB* and *AC* and the force developed along strut *AD* for equilibrium of the 400-lb crate.

3–55. If the tension developed in each of the cables cannot exceed 300 lb, determine the largest weight of the crate that can be supported. Also, what is the force developed along strut *AD*?

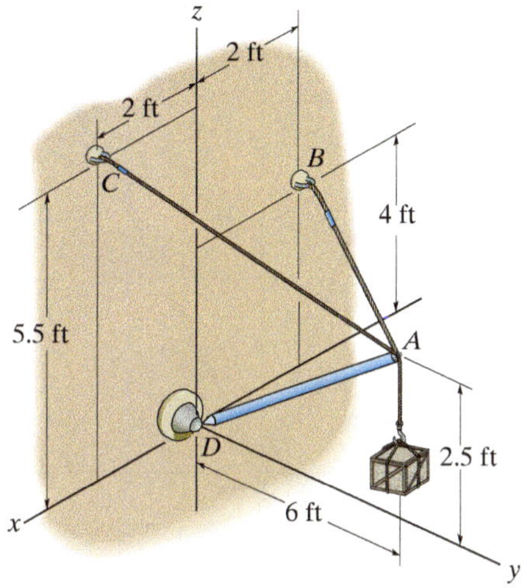

Probs. 3–54/55

*__3–56.__ Determine the force in each cable needed to support the 3500-lb platform. Set $d = 2$ ft.

3–57. Determine the force in each cable needed to support the 3500-lb platform. Set $d = 4$ ft.

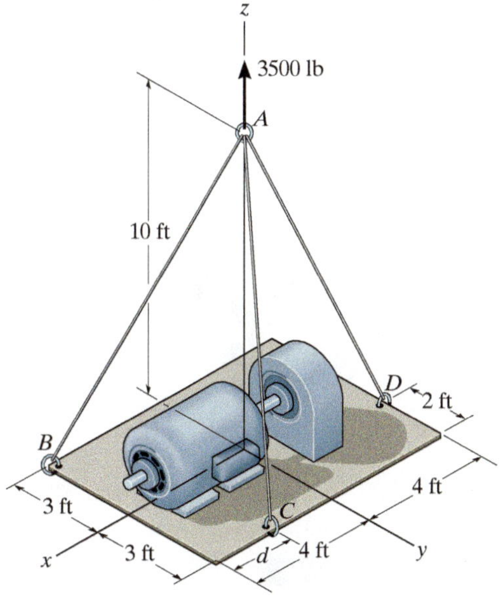

Probs. 3–56/57

3–58. Determine the tension developed in each cable for equilibrium of the 300-lb crate.

3–59. Determine the maximum weight of the crate that can be suspended from cables *AB*, *AC*, and *AD* so that the tension developed in any one of the cables does not exceed 250 lb.

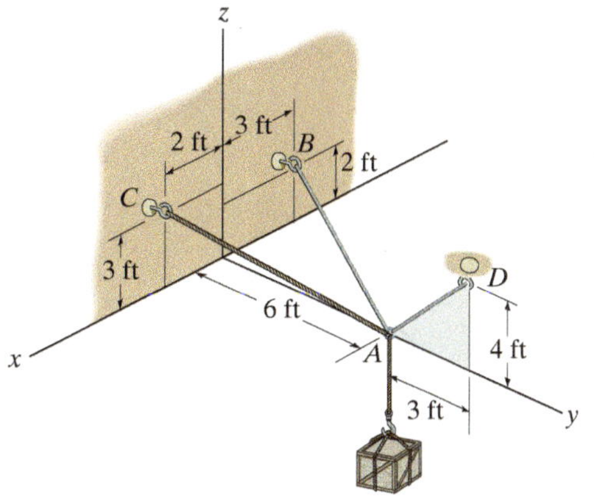

Probs. 3–58/59

*__3–60.__ The 800-lb cylinder is supported by three chains as shown. Determine the force in each chain for equilibrium. Take $d = 1$ ft.

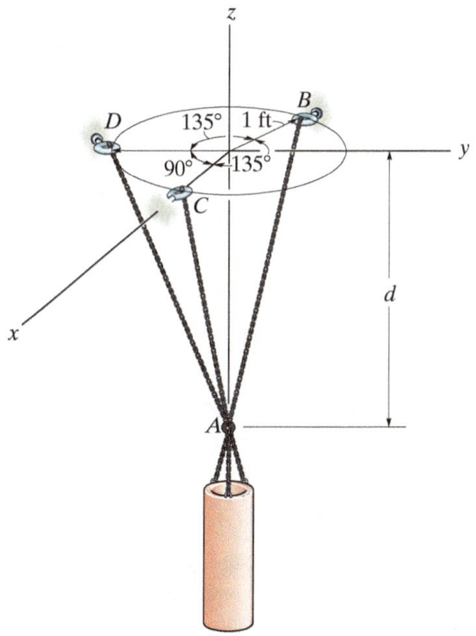

Prob. 3–60

3–61. If cable AD is tightened by a turnbuckle and develops a tension of 1300 lb, determine the tension developed in cables AB and AC and the force developed along the antenna tower AE at point A.

3–62. If the tension developed in either cable AB or AC can not exceeded 1000 lb, determine the maximum tension that can be developed in cable AD when it is tightened by the turnbuckle. Also, what is the force developed along the antenna tower at point A?

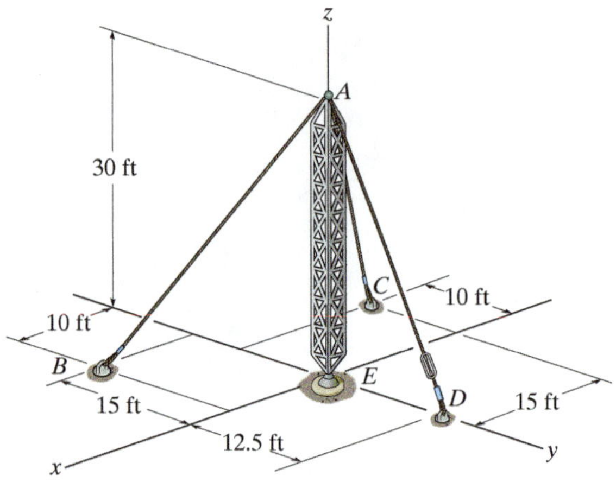

Probs. 3–61/62

3–63. The thin ring can be adjusted vertically between three equally long cables from which the 100-kg chandelier is suspended. If the ring remains in the horizontal plane and $z = 600$ mm, determine the tension in each cable.

***3–64.** The thin ring can be adjusted vertically between three equally long cables from which the 100-kg chandelier is suspended. If the ring remains in the horizontal plane and the tension in each cable is not allowed to exceed 1 kN, determine the smallest allowable distance z required for equilibrium.

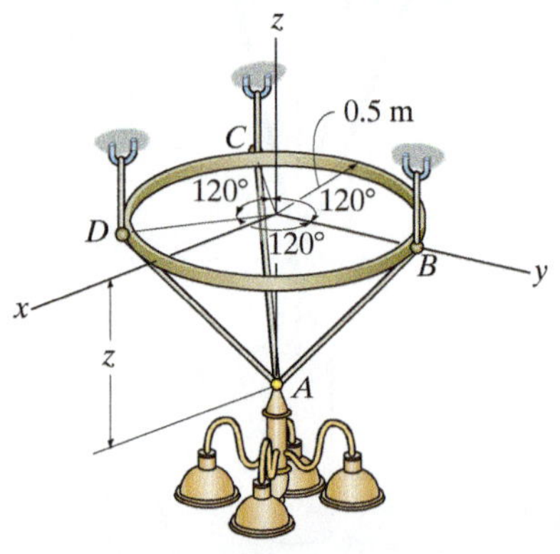

Probs. 3–63/64

3–65. The 80-lb chandelier is supported by three wires as shown. Determine the force in each wire for equilibrium.

3–66. If each wire can sustain a maximum tension of 120 lb before it fails, determine the greatest weight of the chandelier the wires will support in the position shown.

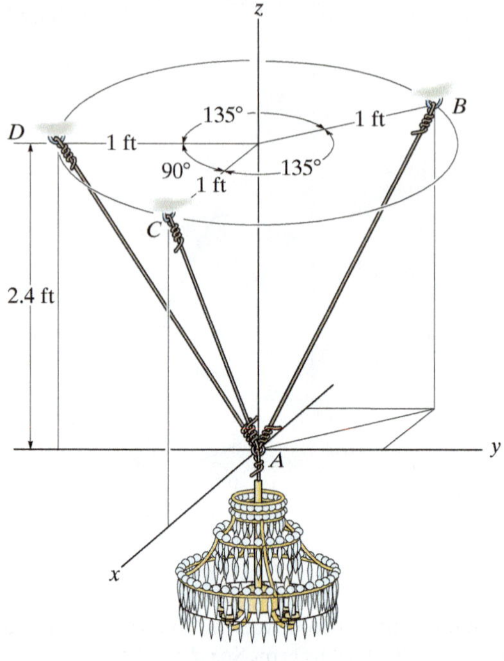

Probs. 3–65/66

■3–67. The 80-lb ball is suspended from the horizontal ring using three springs each having an unstretched length of 1.5 ft and stiffness of 50 lb/ft. Determine the vertical distance h from the ring to point A for equilibrium.

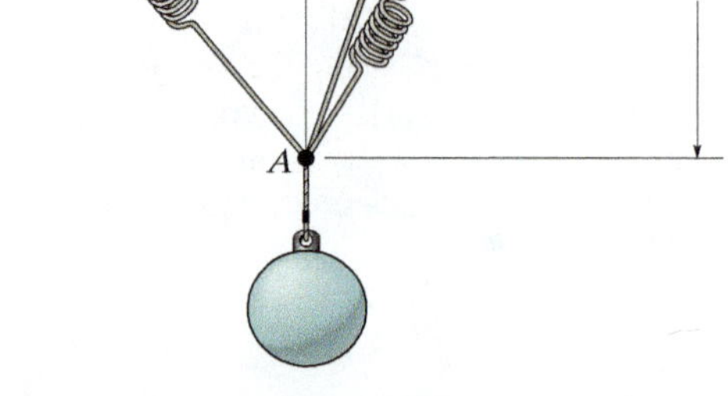

Prob. 3–67

CHAPTER REVIEW

Particle Equilibrium

When a particle is at rest or moves with constant velocity, it is in equilibrium. This requires that all the forces acting on the particle form a zero resultant force.

In order to account for all the forces that act on a particle, it is necessary to draw its free-body diagram. This diagram is an outlined shape of the particle that shows all the forces listed with their known or unknown magnitudes and directions.

$$\mathbf{F}_R = \Sigma\mathbf{F} = \mathbf{0}$$

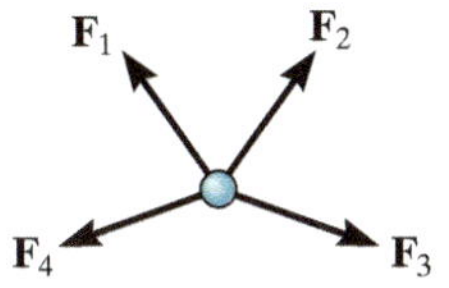

Two Dimensions

The two scalar equations of force equilibrium can be applied with reference to an established *x*, *y* coordinate system.

The tensile force developed in a *continuous cable* that passes over a frictionless pulley must have a *constant* magnitude throughout the cable to keep the cable in equilibrium.

If the problem involves a linearly elastic spring, then the stretch or compression *s* of the spring can be related to the force applied to it.

$$\Sigma F_x = 0$$
$$\Sigma F_y = 0$$

$$F = ks$$

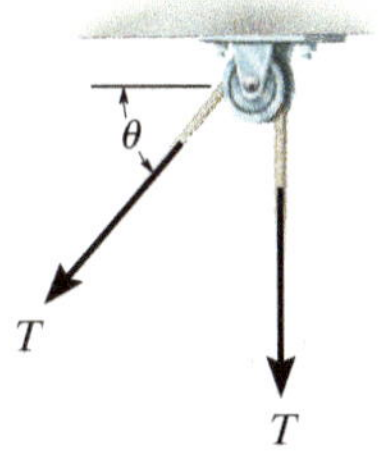

Cable is in tension

Three Dimensions

If the three-dimensional geometry is difficult to visualize, then the equilibrium equation should be applied using a Cartesian vector analysis. This requires first expressing each force on the free-body diagram as a Cartesian vector. When the forces are summed and set equal to zero, then the **i**, **j**, and **k** components are also zero.

$$\Sigma\mathbf{F} = \mathbf{0}$$

$$\Sigma F_x = 0$$
$$\Sigma F_y = 0$$
$$\Sigma F_z = 0$$

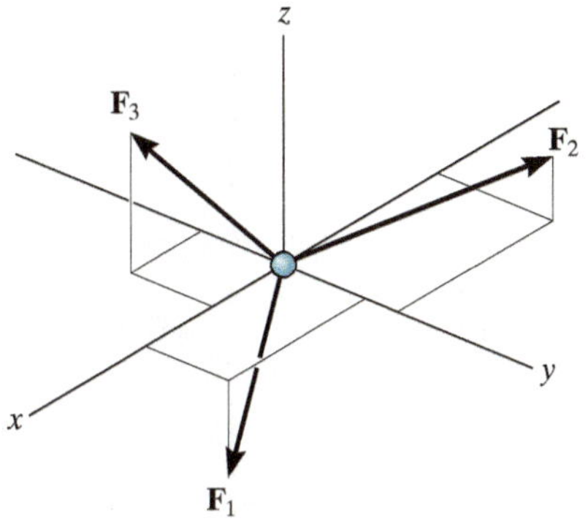

REVIEW PROBLEMS

*3–68. The pipe is held in place by the vise. If the bolt exerts a force of 50 lb on the pipe in the direction shown, determine the forces F_A and F_B that the smooth contacts at A and B exert on the pipe.

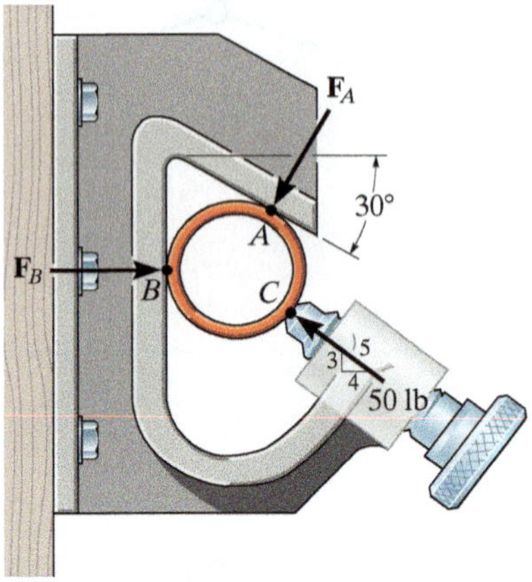

Prob. 3–68

3–69. When y is zero, the springs sustain a force of 60 lb. Determine the magnitude of the applied vertical forces **F** and −**F** required to pull point A away from point B a distance of $y = 2$ ft. The ends of cords CAD and CBD are attached to rings at C and D.

3–70. When y is zero, the springs are each stretched 1.5 ft. Determine the distance y if a force of $F = 60$ lb is applied to points A and B as shown. The ends of cords CAD and CBD are attached to rings at C and D.

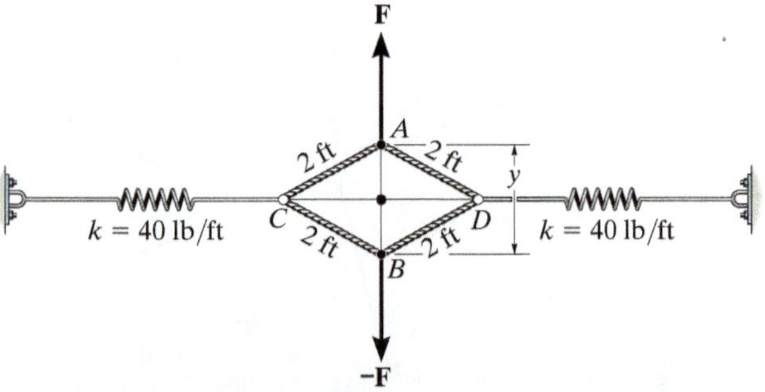

Probs. 3–69/70

3–71. Romeo tries to reach Juliet by climbing with constant velocity up a rope which is knotted at point A. Any of the three segments of the rope can sustain a maximum force of 2 kN before it breaks. Determine if Romeo, who has a mass of 65 kg, can climb the rope, and if so, can he along with Juliet, who has a mass of 60 kg, climb down with constant velocity?

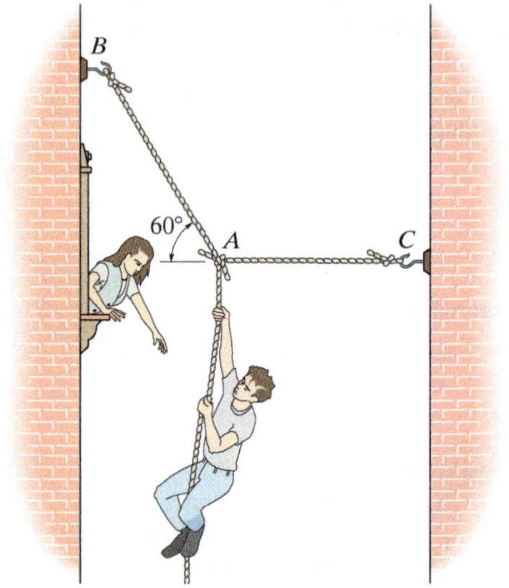

Prob. 3–71

*■3–72. Determine the magnitudes of forces $\mathbf{F}_1$, $\mathbf{F}_2$, and $\mathbf{F}_3$ necessary to hold the force $\mathbf{F} = \{-9\mathbf{i} - 8\mathbf{j} - 5\mathbf{k}\}$ kN in equilibrium.

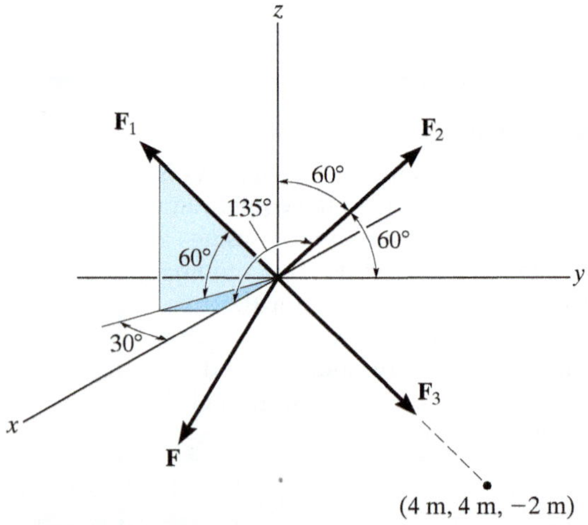

Prob. 3–72

3–73. The man attempts to pull the log at C by using the three ropes. Determine the direction θ in which he should pull on his rope with a force of 80 lb, so that he exerts a maximum force on the log. What is the force on the log for this case? Also, determine the direction in which he should pull in order to maximize the force in the rope attached to B. What is this maximum force?

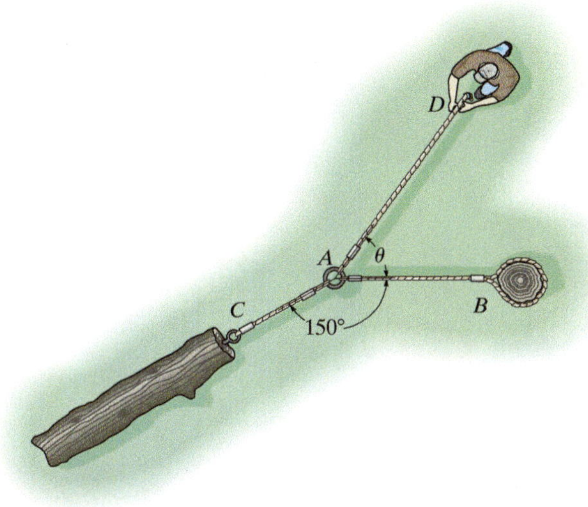

Prob. 3–73

■3–74. The ring of negligible size is subjected to a vertical force of 200 lb. Determine the required length l of cord AC such that the tension acting in AC is 160 lb. Also, what is the force acting in cord AB? *Hint:* Use the equilibrium condition to determine the required angle θ for attachment, then determine l using trigonometry applied to ΔABC.

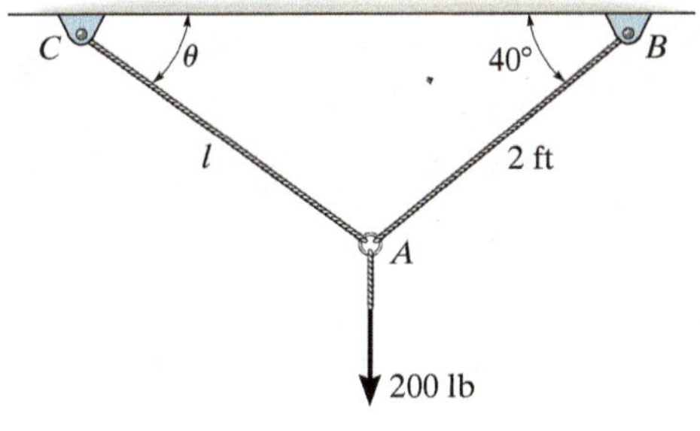

Prob. 3–74

3–75. Determine the maximum weight of the engine that can be supported without exceeding a tension of 450 lb in chain AB and 480 lb in chain AC.

Prob. 3–75

***3–76.** Determine the force in each cable needed to support the 500-lb load.

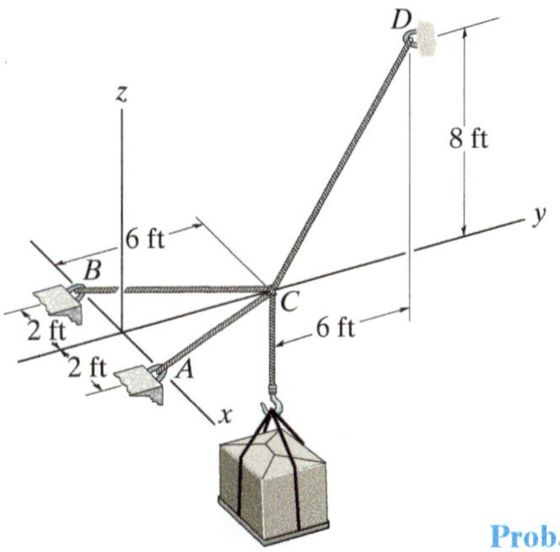

Prob. 3–76

3–77. The joint of a space frame is subjected to four member forces. Member OA lies in the $x-y$ plane and member OB lies in the $y-z$ plane. Determine the forces acting in each of the members required for equilibrium of the joint.

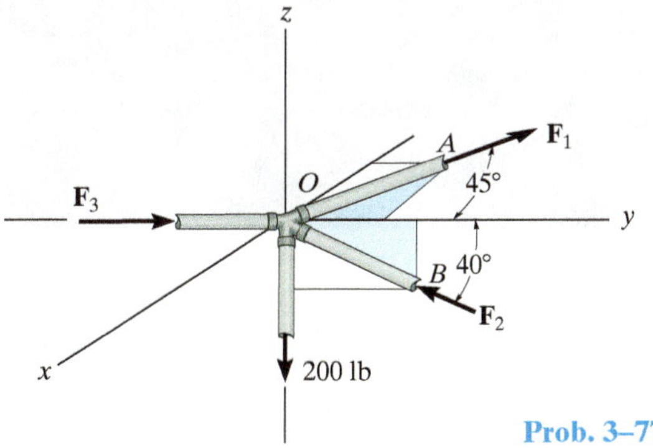

Prob. 3–77

Chapter 4

Forces applied to wrenches and wheels will produce rotation or a tendency for rotation. This effect is called a moment, and in this chapter we will study how to determine the moment of a system of forces and calculate their resultants.

Force System Resultants

CHAPTER OBJECTIVES

- To discuss the concept of the moment of a force and show how to calculate it in two and three dimensions.
- To provide a method for finding the moment of a force about a specified axis.
- To define the moment of a couple.
- To present methods for determining the resultants of nonconcurrent force systems.
- To indicate how to reduce a simple distributed loading to a resultant force having a specified location.

4.1 Moment of a Force—Scalar Formulation

When a force is applied to a body it will produce a tendency for the body to rotate about a point that is not on the line of action of the force. This tendency to rotate is sometimes called a *torque*, but most often it is called the moment of a force or simply the *moment*. For example, consider a wrench used to unscrew the bolt in Fig. 4–1*a*. If a force is applied to the handle of the wrench it will tend to turn the bolt about point O (or the z axis). The magnitude of the moment is directly proportional to the magnitude of **F** and the perpendicular distance or *moment arm d*. The larger the force or the longer the moment arm, the greater the moment or turning effect. Note that if the force **F** is applied at an angle $\theta \neq 90°$, Fig. 4–1*b*, then it will be more difficult to turn the bolt since the moment arm $d' = d \sin\theta$ will be smaller than d. If **F** is applied along the wrench, Fig. 4–1*c*, its moment arm will be zero since the line of action of **F** will intersect point O (the z axis). As a result, the moment of **F** about O is also zero and no turning can occur.

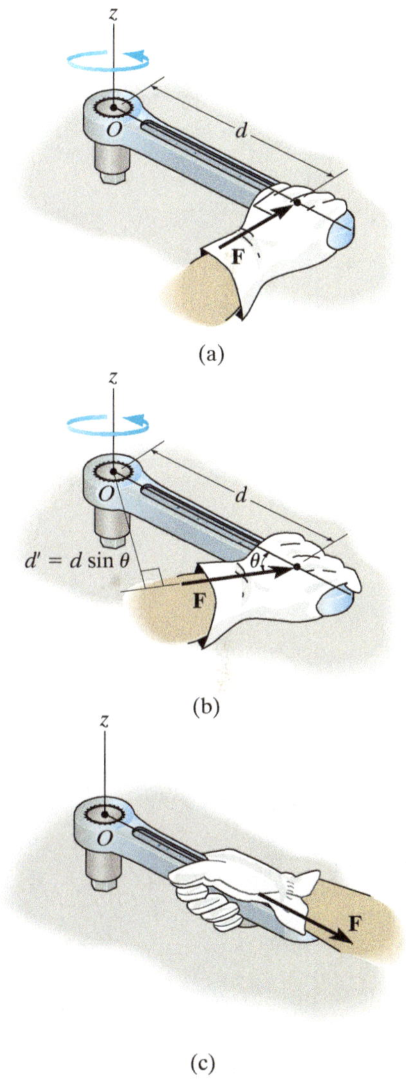

Fig. 4–1

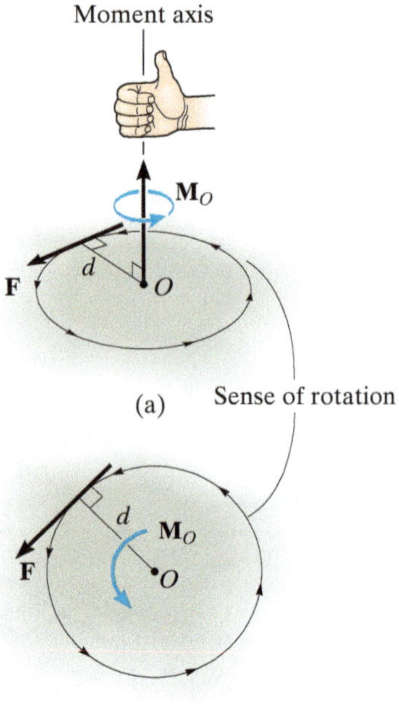

Fig. 4–2

We can generalize the above discussion and consider the force **F** and point O which lie in the shaded plane as shown in Fig. 4–2*a*. The moment $\mathbf{M}_O$ about point O, or about an axis passing through O and perpendicular to the plane, is a *vector quantity* since it has a specified magnitude and direction.

Magnitude. The magnitude of $\mathbf{M}_O$ is

$$M_O = Fd \tag{4–1}$$

where d is the *moment arm* or *perpendicular distance* from the axis at point O to the line of action of the force. Units of moment magnitude consist of force times distance, e.g., N · m or lb · ft.

Direction. The direction of $\mathbf{M}_O$ is defined by its *moment axis*, which is perpendicular to the plane that contains the force **F** and its moment arm d. The right-hand rule is used to establish the sense of direction of $\mathbf{M}_O$. According to this rule, the natural curl of the fingers of the right hand, as they are drawn towards the palm, represent the rotation, or if no movement is possible, there is a tendency for rotation caused by the moment. As this action is performed, the thumb of the right hand will give the directional sense of $\mathbf{M}_O$, Fig. 4–2*a*. Notice that the moment vector is represented three-dimensionally by a curl around an arrow. In two dimensions this vector is represented only by the curl as in Fig. 4–2*b*. Since in this case the moment will tend to cause a counterclockwise rotation, the moment vector is actually directed out of the page.

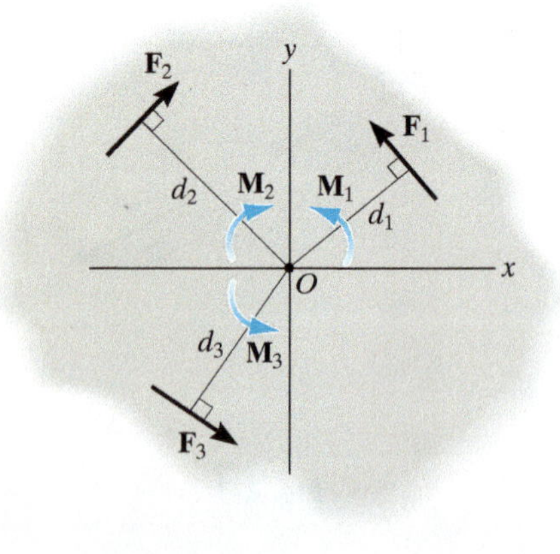

Fig. 4–3

Resultant Moment. For two-dimensional problems, where all the forces lie within the x–y plane, Fig. 4–3, the resultant moment $(\mathbf{M}_R)_o$ about point O (the z axis) can be determined by *finding the algebraic sum* of the moments caused by all the forces in the system. As a convention, we will generally consider *positive moments* as *counterclockwise* since they are directed along the positive z axis (out of the page). *Clockwise moments* will be *negative*. Doing this, the directional sense of each moment can be represented by a *plus or minus* sign. Using this sign convention, the resultant moment in Fig. 4–3 is therefore

$$\circlearrowleft +(M_R)_o = \Sigma Fd; \qquad (M_R)_o = F_1 d_1 - F_2 d_2 + F_3 d_3$$

If the numerical result of this sum is a positive scalar, $(\mathbf{M}_R)_o$ will be a counterclockwise moment (out of the page); and if the result is negative, $(\mathbf{M}_R)_o$ will be a clockwise moment (into the page).

EXAMPLE 4.1

For each case illustrated in Fig. 4–4, determine the moment of the force about point O.

SOLUTION (SCALAR ANALYSIS)

The line of action of each force is extended as a dashed line in order to establish the moment arm d. Also illustrated is the tendency of rotation of the member as caused by the force. Furthermore, the orbit of the force about O is shown as a colored curl. Thus,

Fig. 4–4*a* $\quad M_O = (100\ \text{N})(2\ \text{m}) = 200\ \text{N}\cdot\text{m} \circlearrowright$ *Ans.*

Fig. 4–4*b* $\quad M_O = (50\ \text{N})(0.75\ \text{m}) = 37.5\ \text{N}\cdot\text{m} \circlearrowright$ *Ans.*

Fig. 4–4*c* $\quad M_O = (40\ \text{lb})(4\ \text{ft} + 2\cos 30^\circ\ \text{ft}) = 229\ \text{lb}\cdot\text{ft} \circlearrowright$ *Ans.*

Fig. 4–4*d* $\quad M_O = (60\ \text{lb})(1\sin 45^\circ\ \text{ft}) = 42.4\ \text{lb}\cdot\text{ft} \circlearrowleft$ *Ans.*

Fig. 4–4*e* $\quad M_O = (7\ \text{kN})(4\ \text{m} - 1\ \text{m}) = 21.0\ \text{kN}\cdot\text{m} \circlearrowleft$ *Ans.*

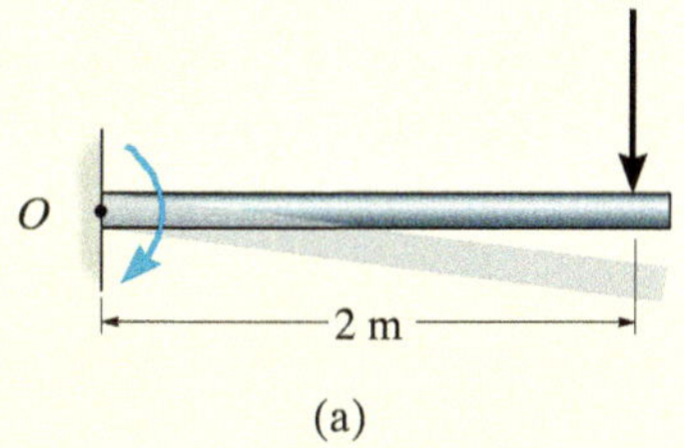

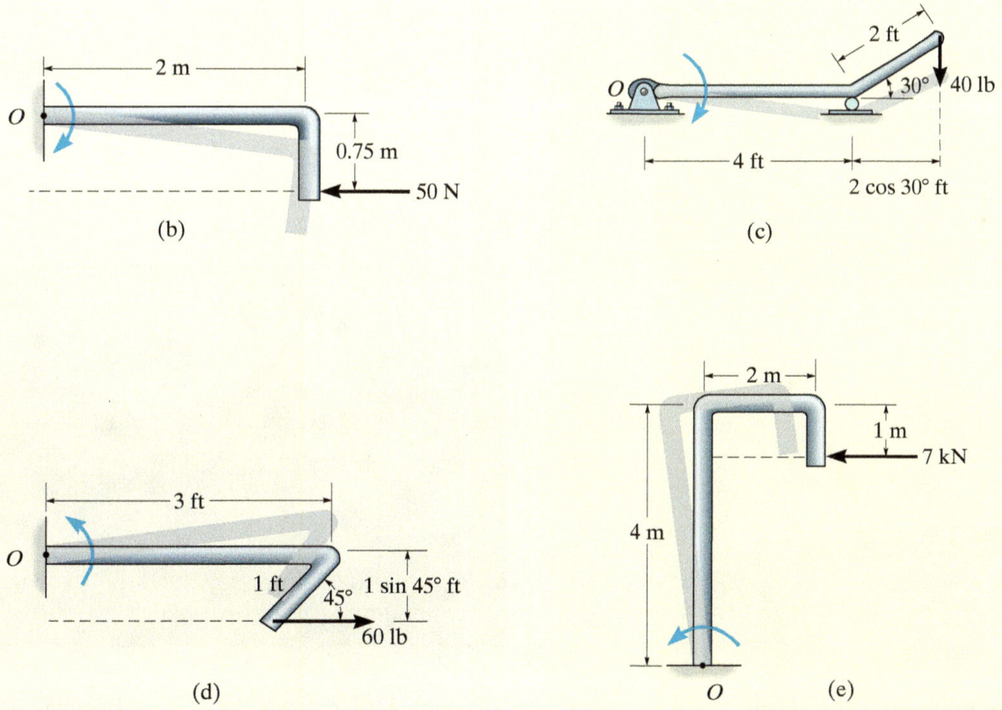

Fig. 4–4

4

EXAMPLE 4.2

Determine the resultant moment of the four forces acting on the rod shown in Fig. 4–5 about point O.

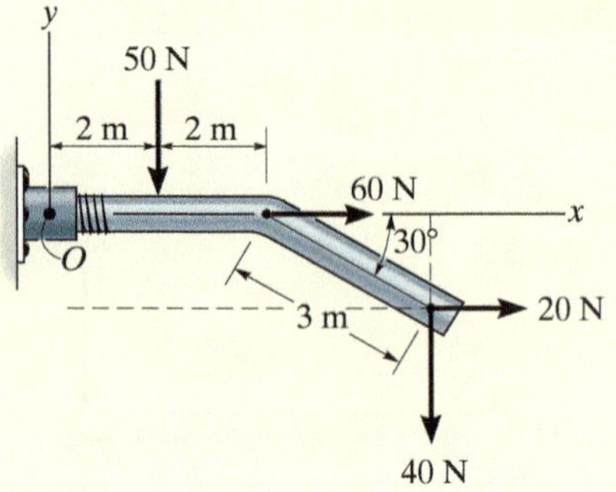

Fig. 4–5

SOLUTION

Assuming that positive moments act in the $+\mathbf{k}$ direction, i.e., counterclockwise, we have

$$\circlearrowleft + \quad (M_R)_O = \Sigma Fd;$$

$$(M_R)_O = -50\text{ N}(2\text{ m}) + 60\text{ N}(0) + 20\text{ N}(3\sin 30^\circ\text{ m})$$

$$-40\text{ N}(4\text{ m} + 3\cos 30^\circ\text{ m})$$

$$(M_R)_O = -334\text{ N}\cdot\text{m} = 334\text{ N}\cdot\text{m}\,\circlearrowright \qquad \textit{Ans.}$$

For this calculation, note how the moment-arm distances for the 20-N and 40-N forces are established from the extended (dashed) lines of action of each of these forces.

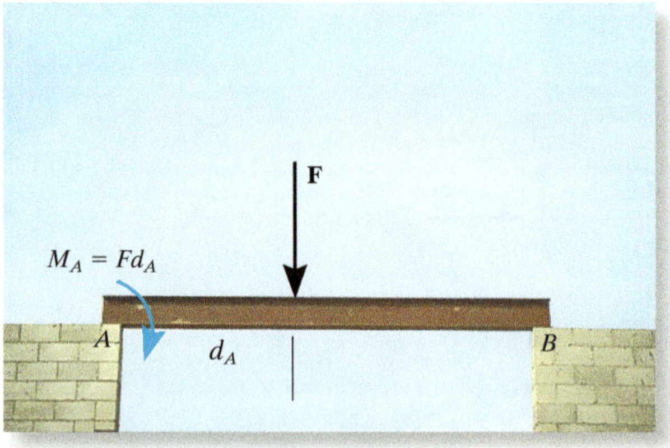

As illustrated by the example problems, the moment of a force does not always cause a rotation. For example, the force $\mathbf{F}$ tends to rotate the beam clockwise about its support at A with a moment $M_A = Fd_A$. The actual rotation would occur if the support at B were removed.

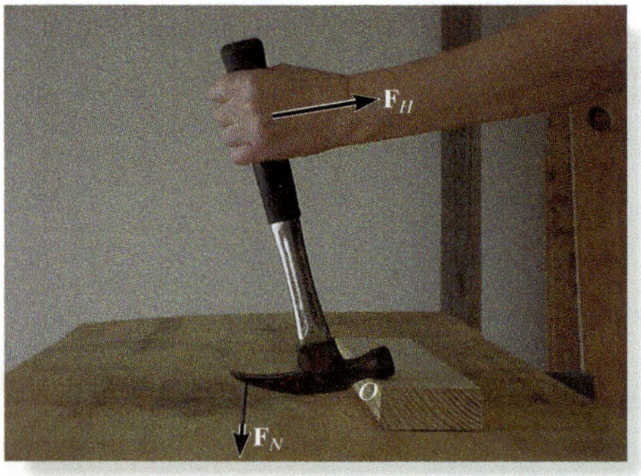

The ability to remove the nail will require the moment of $\mathbf{F}_H$ about point O to be larger than the moment of the force $\mathbf{F}_N$ about O that is needed to pull the nail out.

4.2 Cross Product

The moment of a force will be formulated using Cartesian vectors in the next section. Before doing this, however, it is first necessary to expand our knowledge of vector algebra and introduce the cross-product method of vector multiplication.

The *cross product* of two vectors **A** and **B** yields the vector **C**, which is written

$$\mathbf{C} = \mathbf{A} \times \mathbf{B} \tag{4-2}$$

and is read "**C** equals **A** cross **B**."

Magnitude. The *magnitude* of **C** is defined as the product of the magnitudes of **A** and **B** and the sine of the angle θ between their tails ($0° \leq \theta \leq 180°$). Thus, $C = AB \sin \theta$.

Direction. Vector **C** has a *direction* that is perpendicular to the plane containing **A** and **B** such that **C** is specified by the right-hand rule; i.e., curling the fingers of the right hand from vector **A** (cross) to vector **B**, the thumb points in the direction of **C**, as shown in Fig. 4–6.

Knowing both the magnitude and direction of **C**, we can write

$$\mathbf{C} = \mathbf{A} \times \mathbf{B} = (AB \sin \theta)\mathbf{u}_C \tag{4-3}$$

where the scalar $AB \sin \theta$ defines the *magnitude* of **C** and the unit vector $\mathbf{u}_C$ defines the *direction* of **C**. The terms of Eq. 4–3 are illustrated graphically in Fig. 4–6.

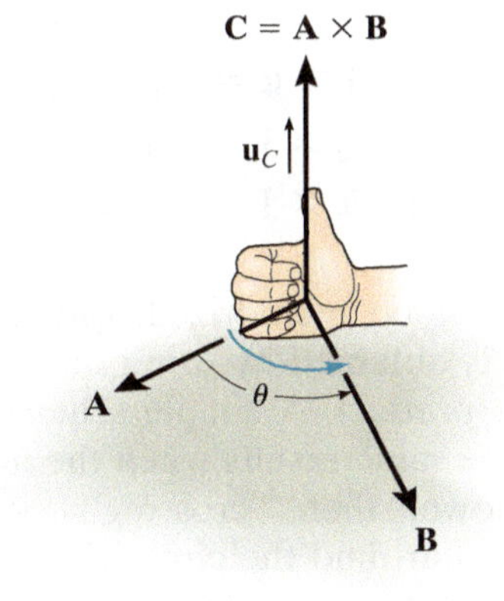

Fig. 4–6

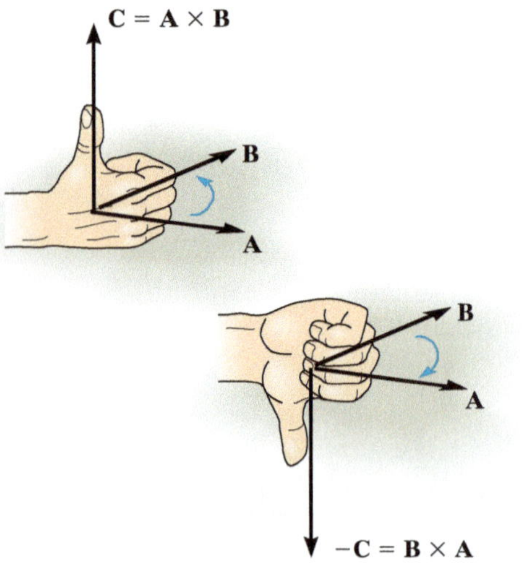

Fig. 4–7

Laws of Operation.

- The commutative law is *not* valid; i.e., $\mathbf{A} \times \mathbf{B} \neq \mathbf{B} \times \mathbf{A}$. Rather,

$$\mathbf{A} \times \mathbf{B} = -\mathbf{B} \times \mathbf{A}$$

This is shown in Fig. 4–7 by using the right-hand rule. The cross product $\mathbf{B} \times \mathbf{A}$ yields a vector that has the same magnitude but acts in the opposite direction to **C**; i.e., $\mathbf{B} \times \mathbf{A} = -\mathbf{C}$.

- If the cross product is multiplied by a scalar a, it obeys the associative law;

$$a(\mathbf{A} \times \mathbf{B}) = (a\mathbf{A}) \times \mathbf{B} = \mathbf{A} \times (a\mathbf{B}) = (\mathbf{A} \times \mathbf{B})a$$

This property is easily shown since the magnitude of the resultant vector $(|a|AB \sin \theta)$ and its direction are the same in each case.

- The vector cross product also obeys the distributive law of addition,

$$\mathbf{A} \times (\mathbf{B} + \mathbf{D}) = (\mathbf{A} \times \mathbf{B}) + (\mathbf{A} \times \mathbf{D})$$

- The proof of this identity is left as an exercise (see Prob. 4–1). It is important to note that *proper order* of the cross products must be maintained, since they are not commutative.

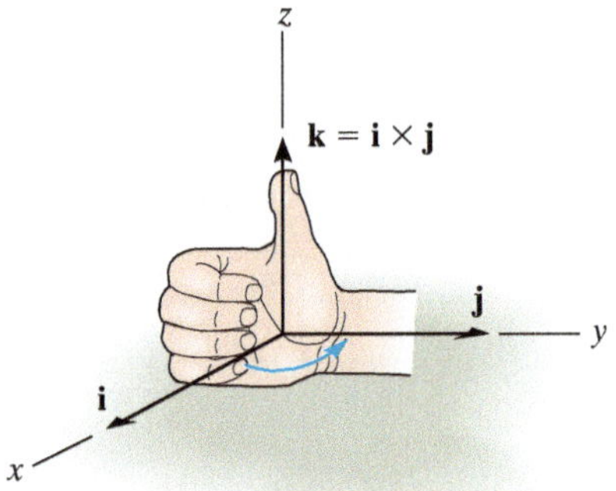

Fig. 4–8

Cartesian Vector Formulation. Equation 4–3 may be used to find the cross product of any pair of Cartesian unit vectors. For example, to find $\mathbf{i} \times \mathbf{j}$, the magnitude of the resultant vector is $(i)(j)(\sin 90^\circ) = (1)(1)(1) = 1$, and its direction is determined using the right-hand rule. As shown in Fig. 4–8, the resultant vector points in the $+\mathbf{k}$ direction. Thus, $\mathbf{i} \times \mathbf{j} = (1)\mathbf{k}$. In a similar manner,

$$\begin{array}{lll} \mathbf{i} \times \mathbf{j} = \mathbf{k} & \mathbf{i} \times \mathbf{k} = -\mathbf{j} & \mathbf{i} \times \mathbf{i} = \mathbf{0} \\ \mathbf{j} \times \mathbf{k} = \mathbf{i} & \mathbf{j} \times \mathbf{i} = -\mathbf{k} & \mathbf{j} \times \mathbf{j} = \mathbf{0} \\ \mathbf{k} \times \mathbf{i} = \mathbf{j} & \mathbf{k} \times \mathbf{j} = -\mathbf{i} & \mathbf{k} \times \mathbf{k} = \mathbf{0} \end{array}$$

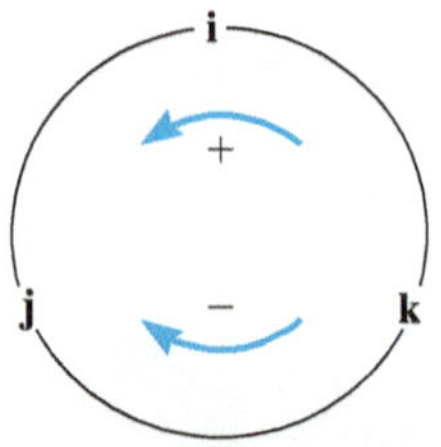

Fig. 4–9

These results should *not* be memorized; rather, it should be clearly understood how each is obtained by using the right-hand rule and the definition of the cross product. A simple scheme shown in Fig. 4–9 is helpful for obtaining the same results when the need arises. If the circle is constructed as shown, then "crossing" two unit vectors in a *counterclockwise* fashion around the circle yields the *positive* third unit vector; e.g., $\mathbf{k} \times \mathbf{i} = \mathbf{j}$. "Crossing" *clockwise*, a *negative* unit vector is obtained; e.g., $\mathbf{i} \times \mathbf{k} = -\mathbf{j}$.

Let us now consider the cross product of two general vectors **A** and **B** which are expressed in Cartesian vector form. We have

$$\begin{aligned}\mathbf{A}\times\mathbf{B} &= (A_x\mathbf{i} + A_y\mathbf{j} + A_z\mathbf{k}) \times (B_x\mathbf{i} + B_y\mathbf{j} + B_z\mathbf{k})\\ &= A_xB_x(\mathbf{i}\times\mathbf{i}) + A_xB_y(\mathbf{i}\times\mathbf{j}) + A_xB_z(\mathbf{i}\times\mathbf{k})\\ &\quad + A_yB_x(\mathbf{j}\times\mathbf{i}) + A_yB_y(\mathbf{j}\times\mathbf{j}) + A_yB_z(\mathbf{j}\times\mathbf{k})\\ &\quad + A_zB_x(\mathbf{k}\times\mathbf{i}) + A_zB_y(\mathbf{k}\times\mathbf{j}) + A_zB_z(\mathbf{k}\times\mathbf{k})\end{aligned}$$

Carrying out the cross-product operations and combining terms yields

$$\mathbf{A}\times\mathbf{B} = (A_yB_z - A_zB_y)\mathbf{i} - (A_xB_z - A_zB_x)\mathbf{j} + (A_xB_y - A_yB_x)\mathbf{k} \qquad (4\text{–}4)$$

This equation may also be written in a more compact determinant form as

$$\mathbf{A}\times\mathbf{B} = \begin{vmatrix}\mathbf{i} & \mathbf{j} & \mathbf{k}\\ A_x & A_y & A_z\\ B_x & B_y & B_z\end{vmatrix} \qquad (4\text{–}5)$$

Thus, to find the cross product of any two Cartesian vectors **A** and **B**, it is necessary to expand a determinant whose first row of elements consists of the unit vectors **i**, **j**, and **k** and whose second and third rows represent the *x, y, z* components of the two vectors **A** and **B**, respectively.*

*A determinant having three rows and three columns can be expanded using three minors, each of which is multiplied by one of the three terms in the first row. There are four elements in each minor, for example,

$$\begin{vmatrix}A_{11} & A_{12}\\ A_{21} & A_{22}\end{vmatrix}$$

By *definition*, this determinant notation represents the terms $(A_{11}A_{22} - A_{12}A_{21})$, which is simply the product of the two elements intersected by the arrow slanting downward to the right $(A_{11}A_{22})$ *minus* the product of the two elements intersected by the arrow slanting downward to the left $(A_{12}A_{21})$. For a 3 × 3 determinant, such as Eq. 4–5, the three minors can be generated in accordance with the following scheme:

For element **i**: $\begin{vmatrix}\mathbf{i} & \mathbf{j} & \mathbf{k}\\ A_x & A_y & A_z\\ B_x & B_y & B_z\end{vmatrix} = \mathbf{i}(A_yB_z - A_zB_y)$

Remember the negative sign

For element **j**: $\begin{vmatrix}\mathbf{i} & \mathbf{j} & \mathbf{k}\\ A_x & A_y & A_z\\ B_x & B_y & B_z\end{vmatrix} = -\mathbf{j}(A_xB_z - A_zB_x)$

For element **k**: $\begin{vmatrix}\mathbf{i} & \mathbf{j} & \mathbf{k}\\ A_x & A_y & A_z\\ B_x & B_y & B_z\end{vmatrix} = \mathbf{k}(A_xB_y - A_yB_x)$

Adding the results and noting that the **j** element *must include the minus sign* yields the expanded form of **A** × **B** given by Eq. 4–4.

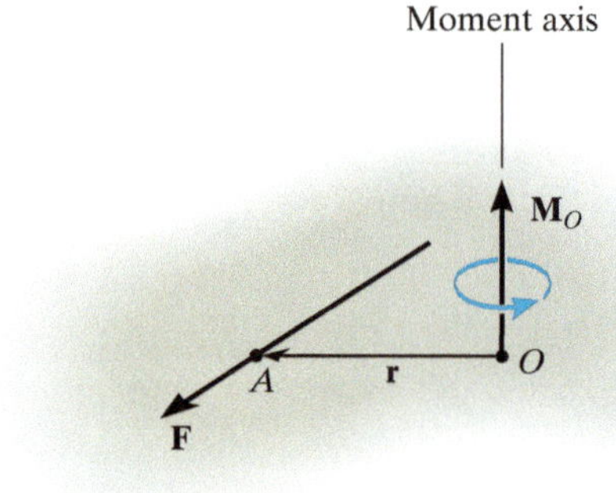

(a)

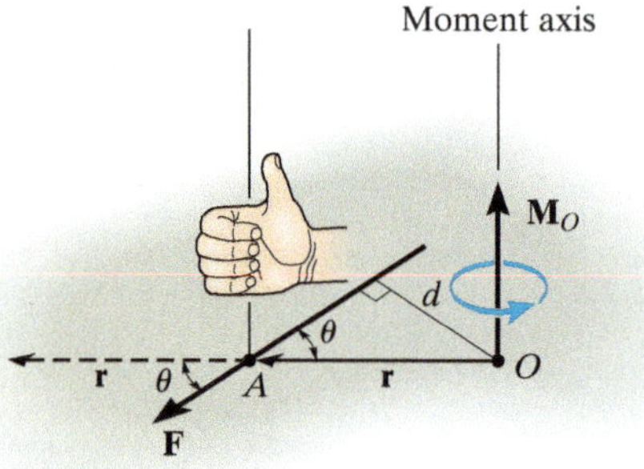

(b)

Fig. 4–10

4.3 Moment of a Force—Vector Formulation

The moment of a force **F** about point O, or actually about the moment axis passing through O and perpendicular to the plane containing O and **F**, Fig. 4–10*a*, can be expressed using the vector cross product, namely,

$$\mathbf{M}_O = \mathbf{r} \times \mathbf{F} \qquad (4\text{–}6)$$

Here **r** represents a position vector directed *from O* to *any point* on the line of action of **F**. We will now show that indeed the moment $\mathbf{M}_O$, when determined by this cross product, has the proper magnitude and direction.

Magnitude. The magnitude of the cross product is defined from Eq. 4–3 as $M_O = rF \sin \theta$, where the angle θ is measured between the *tails* of **r** and **F**. To establish this angle, **r** must be treated as a sliding vector so that θ can be constructed properly, Fig. 4–10*b*. Since the moment arm $d = r \sin \theta$, then

$$M_O = rF \sin \theta = F(r \sin \theta) = Fd$$

which agrees with Eq. 4–1.

Direction. The direction and sense of $\mathbf{M}_O$ in Eq. 4–6 are determined by the right-hand rule as it applies to the cross product. Thus, sliding **r** to the dashed position and curling the right-hand fingers from **r** toward **F**, "**r** cross **F**," the thumb is directed upward or perpendicular to the plane containing **r** and **F** and this is in the *same direction* as $\mathbf{M}_O$, the moment of the force about point O, Fig. 4–10*b*. Note that the "curl" of the fingers, like the curl around the moment vector, indicates the sense of rotation caused by the force. Since the cross product does not obey the commutative law, the order of $\mathbf{r} \times \mathbf{F}$ must be maintained to produce the correct sense of direction for $\mathbf{M}_O$.

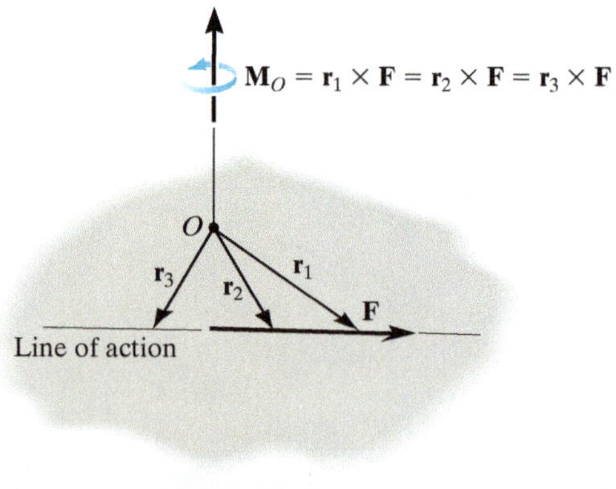

Fig. 4–11

Principle of Transmissibility. The cross product operation is often used in three dimensions since the perpendicular distance or moment arm from point O to the line of action of the force is not needed. In other words, we can use any position vector **r** measured from point O to any point on the line of action of the force **F**, Fig. 4–11. Thus,

$$\mathbf{M}_O = \mathbf{r}_1 \times \mathbf{F} = \mathbf{r}_2 \times \mathbf{F} = \mathbf{r}_3 \times \mathbf{F}$$

Since **F** can be applied at any point along its line of action and still create this *same moment* about point O, then **F** can be considered a *sliding vector*. This property is called the *principle of transmissibility* of a force.

Cartesian Vector Formulation. If we establish *x, y, z* coordinate axes, then the position vector **r** and force **F** can be expressed as Cartesian vectors, Fig. 4–12*a*. Applying Eq. 4–5 we have

$$\mathbf{M}_O = \mathbf{r} \times \mathbf{F} = \begin{vmatrix} \mathbf{i} & \mathbf{j} & \mathbf{k} \\ r_x & r_y & r_z \\ F_x & F_y & F_z \end{vmatrix} \qquad (4\text{–}7)$$

where

r_x, r_y, r_z represent the *x, y, z* components of the position vector drawn from point *O* to *any point* on the line of action of the force

F_x, F_y, F_z represent the *x, y, z* components of the force vector

If the determinant is expanded, then like Eq. 4–4 we have

$$\mathbf{M}_O = (r_yF_z - r_zF_y)\mathbf{i} - (r_xF_z - r_zF_x)\mathbf{j} + (r_xF_y - r_yF_x)\mathbf{k} \qquad (4\text{–}8)$$

The physical meaning of these three moment components becomes evident by studying Fig. 4–12*b*. For example, the **i** component of $\mathbf{M}_O$ can be determined from the moments of $\mathbf{F}_x$, $\mathbf{F}_y$, and $\mathbf{F}_z$ about the *x* axis. The component $\mathbf{F}_x$ does *not* create a moment or tendency to cause turning about the *x* axis since this force is *parallel* to the *x* axis. The line of action of $\mathbf{F}_y$ passes through point *B*, and so the magnitude of the moment of $\mathbf{F}_y$ about point *A* on the *x* axis is r_zF_y. By the right-hand rule this component acts in the *negative* **i** direction. Likewise, $\mathbf{F}_z$ passes through point *C* and so it contributes a moment component of $r_yF_z\mathbf{i}$ about the *x* axis. Thus, $(M_O)_x = (r_yF_z - r_zF_y)$ as shown in Eq. 4–8. As an exercise, establish the **j** and **k** components of $\mathbf{M}_O$ in this manner and show that indeed the expanded form of the determinant, Eq. 4–8, represents the moment of **F** about point *O*. Once $\mathbf{M}_O$ is determined, realize that it will always be *perpendicular* to the shaded plane containing vectors **r** and **F**, Fig. 4–12*a*.

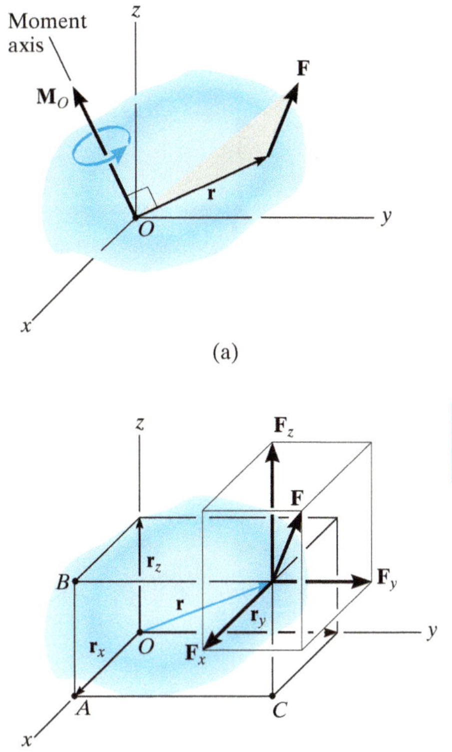

Fig. 4–12

Resultant Moment of a System of Forces. If a body is acted upon by a system of forces, Fig. 4–13, the resultant moment of the forces about point *O* can be determined by vector addition of the moment of each force. This resultant can be written symbolically as

$$(\mathbf{M}_R)_O = \Sigma(\mathbf{r} \times \mathbf{F}) \qquad (4\text{–}9)$$

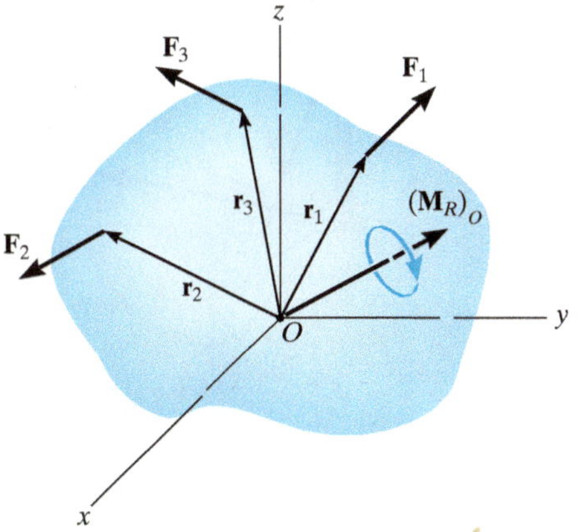

Fig. 4–13

EXAMPLE 4.3

Determine the moment produced by the force **F** in Fig. 4–14*a* about point *O*. Express the result as a Cartesian vector.

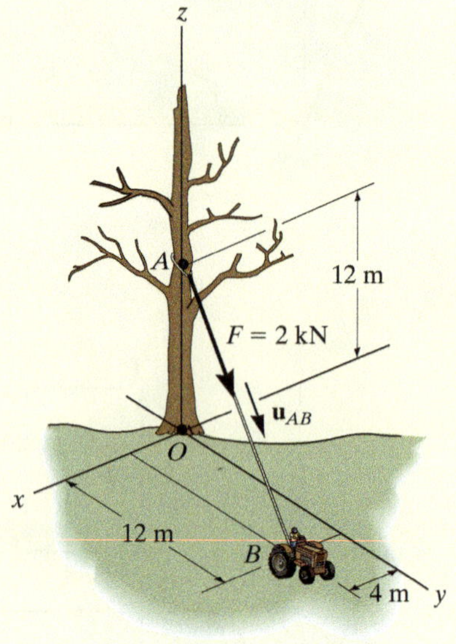

(a)

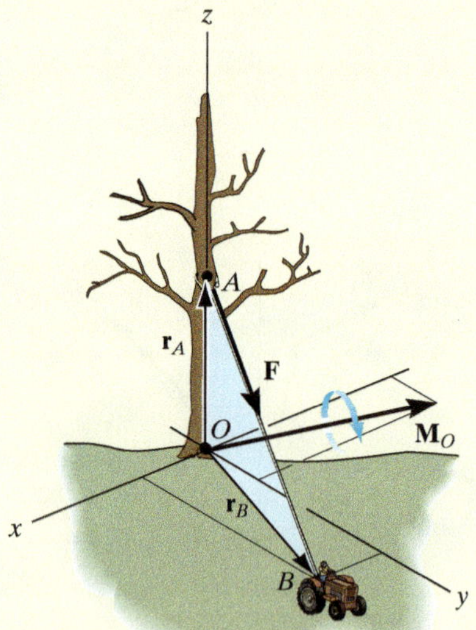

(b)

Fig. 4–14

SOLUTION

As shown in Fig. 4–14*b*, either $\mathbf{r}_A$ or $\mathbf{r}_B$ can be used to determine the moment about point *O*. These position vectors are

$$\mathbf{r}_A = \{12\mathbf{k}\}\text{ m} \quad \text{and} \quad \mathbf{r}_B = \{4\mathbf{i} + 12\mathbf{j}\}\text{ m}$$

Force **F** expressed as a Cartesian vector is

$$\mathbf{F} = F\mathbf{u}_{AB} = 2\text{ kN}\left[\frac{\{4\mathbf{i} + 12\mathbf{j} - 12\mathbf{k}\}\text{ m}}{\sqrt{(4\text{ m})^2 + (12\text{ m})^2 + (-12\text{ m})^2}}\right]$$

$$= \{0.4588\mathbf{i} + 1.376\mathbf{j} - 1.376\mathbf{k}\}\text{ kN}$$

Thus

$$\mathbf{M}_O = \mathbf{r}_A \times \mathbf{F} = \begin{vmatrix} \mathbf{i} & \mathbf{j} & \mathbf{k} \\ 0 & 0 & 12 \\ 0.4588 & 1.376 & -1.376 \end{vmatrix}$$

$$= [0(-1.376) - 12(1.376)]\mathbf{i} - [0(-1.376) - 12(0.4588)]\mathbf{j} + [0(1.376) - 0(0.4588)]\mathbf{k}$$

$$= \{-16.5\mathbf{i} + 5.51\mathbf{j}\}\text{ kN}\cdot\text{m} \qquad \textit{Ans.}$$

or

$$\mathbf{M}_O = \mathbf{r}_B \times \mathbf{F} = \begin{vmatrix} \mathbf{i} & \mathbf{j} & \mathbf{k} \\ 4 & 12 & 0 \\ 0.4588 & 1.376 & -1.376 \end{vmatrix}$$

$$= [12(-1.376) - 0(1.376)]\mathbf{i} - [4(-1.376) - 0(0.4588)]\mathbf{j} + [4(1.376) - 12(0.4588)]\mathbf{k}$$

$$= \{-16.5\mathbf{i} + 5.51\mathbf{j}\}\text{ kN}\cdot\text{m} \qquad \textit{Ans.}$$

NOTE: As shown in Fig. 4–14*b*, $\mathbf{M}_O$ acts perpendicular to the plane that contains **F**, $\mathbf{r}_A$, and $\mathbf{r}_B$. Had this problem been worked using $M_O = Fd$, notice the difficulty that would arise in obtaining the moment arm *d*.

EXAMPLE 4.4

Two forces act on the rod shown in Fig. 4–15*a*. Determine the resultant moment they create about the flange at *O*. Express the result as a Cartesian vector.

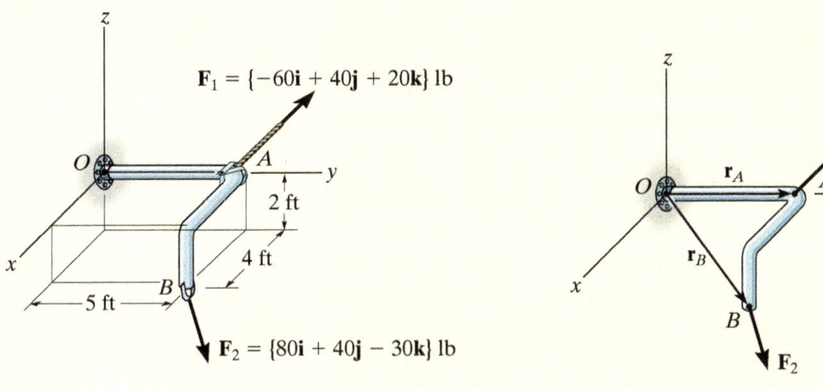

(a) (b)

SOLUTION

Position vectors are directed from point *O* to each force as shown in Fig. 4–15*b*. These vectors are

$$\mathbf{r}_A = \{5\mathbf{j}\}\ \text{ft}$$

$$\mathbf{r}_B = \{4\mathbf{i} + 5\mathbf{j} - 2\mathbf{k}\}\ \text{ft}$$

The resultant moment about *O* is therefore

$$(\mathbf{M}_R)_O = \Sigma(\mathbf{r} \times \mathbf{F})$$

$$= \mathbf{r}_A \times \mathbf{F}_1 + \mathbf{r}_B \times \mathbf{F}_2$$

$$= \begin{vmatrix} \mathbf{i} & \mathbf{j} & \mathbf{k} \\ 0 & 5 & 0 \\ -60 & 40 & 20 \end{vmatrix} + \begin{vmatrix} \mathbf{i} & \mathbf{j} & \mathbf{k} \\ 4 & 5 & -2 \\ 80 & 40 & -30 \end{vmatrix}$$

$$= [5(20) - 0(40)]\mathbf{i} - [0]\mathbf{j} + [0(40) - (5)(-60)]\mathbf{k}$$

$$+ [5(-30) - (-2)(40)]\mathbf{i} - [4(-30) - (-2)(80)]\mathbf{j} + [4(40) - 5(80)]\mathbf{k}$$

$$= \{30\mathbf{i} - 40\mathbf{j} + 60\mathbf{k}\}\ \text{lb} \cdot \text{ft}$$

Ans.

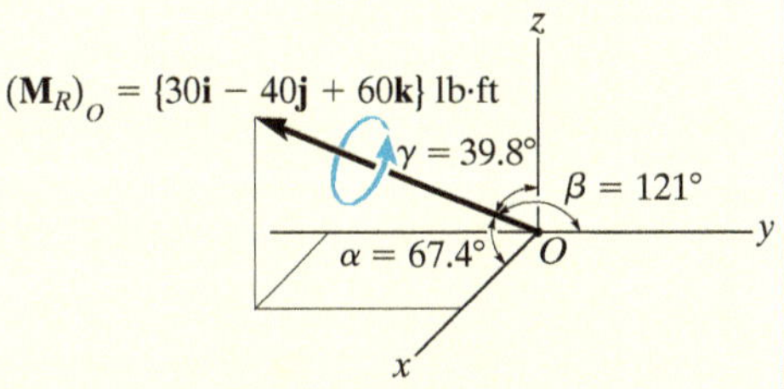

(c)

Fig. 4–15

NOTE: This result is shown in Fig. 4–15*c*. The coordinate direction angles were determined from the unit vector for $(\mathbf{M}_R)_O$. Realize that the two forces tend to cause the rod to rotate about the moment axis in the manner shown by the curl indicated on the moment vector.

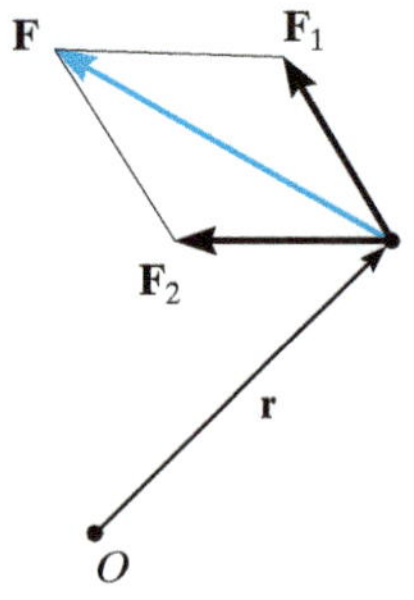

Fig. 4–16

4.4 Principle of Moments

A concept often used in mechanics is the *principle of moments*, which is sometimes referred to as *Varignon's theorem* since it was originally developed by the French mathematician Varignon (1654–1722). It states that *the moment of a force about a point is equal to the sum of the moments of the components of the force about the point.* This theorem can be proven easily using the vector cross product since the cross product obeys the *distributive law*. For example, consider the moments of the force **F** and two of its components about point O. Fig. 4–16. Since $\mathbf{F} = \mathbf{F}_1 + \mathbf{F}_2$ we have

$$\mathbf{M}_O = \mathbf{r} \times \mathbf{F} = \mathbf{r} \times (\mathbf{F}_1 + \mathbf{F}_2) = \mathbf{r} \times \mathbf{F}_1 + \mathbf{r} \times \mathbf{F}_2$$

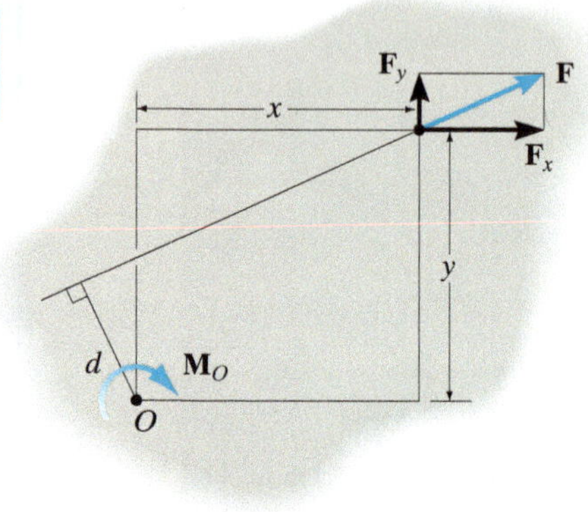

Fig. 4–17

For two-dimensional problems, Fig. 4–17, we can use the principle of moments by resolving the force into its rectangular components and then determine the moment using a scalar analysis. Thus,

$$M_O = F_x y - F_y x$$

This method is generally easier than finding the same moment using $M_O = Fd$.

The moment of the applied force **F** about point O is easy to determine if we use the principle of moments. It is simply $M_O = F_x d$.

Important Points

- The moment of a force creates the tendency of a body to turn about an axis passing through a specific point O.
- Using the right-hand rule, the sense of rotation is indicated by the curl of the fingers, and the thumb is directed along the moment axis, or line of action of the moment.
- The magnitude of the moment is determined from $M_O = Fd$, where d is called the moment arm, which represents the perpendicular or shortest distance from point O to the line of action of the force.
- In three dimensions the vector cross product is used to determine the moment, i.e., $\mathbf{M}_O = \mathbf{r} \times \mathbf{F}$. Remember that **r** is directed *from* point O *to any point* on the line of action of **F**.
- The principle of moments states that the moment of a force about a point is equal to the sum of the moments of the force's components about the point. This is a very convenient method to use in two dimensions.

EXAMPLE 4.5

Determine the moment of the force in Fig. 4–18*a* about point *O*.

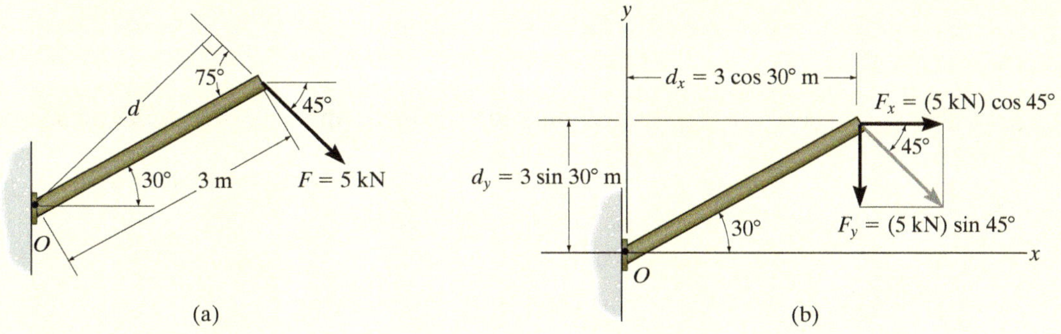

SOLUTION I

The moment arm *d* in Fig. 4–18*a* can be found from trigonometry.

$$d = (3\text{ m})\sin 75° = 2.898\text{ m}$$

Thus,

$$M_O = Fd = (5\text{ kN})(2.898\text{ m}) = 14.5\text{ kN}\cdot\text{m} \curvearrowright \qquad \textit{Ans.}$$

Since the force tends to rotate or orbit clockwise about point *O*, the moment is directed into the page.

SOLUTION II

The *x* and *y* components of the force are indicated in Fig. 4–18*b*. Considering counterclockwise moments as positive, and applying the principle of moments, we have

$$\circlearrowleft + M_O = -F_x d_y - F_y d_x$$
$$= -(5\cos 45°\text{ kN})(3\sin 30°\text{ m}) - (5\sin 45°\text{ kN})(3\cos 30°\text{ m})$$
$$= -14.5\text{ kN}\cdot\text{m} = 14.5\text{ kN}\cdot\text{m} \curvearrowright \qquad \textit{Ans.}$$

SOLUTION III

The *x* and *y* axes can be set parallel and perpendicular to the rod's axis as shown in Fig. 4–18*c*. Here $\mathbf{F}_x$ produces no moment about point *O* since its line of action passes through this point. Therefore,

$$\circlearrowleft + M_O = -F_y d_x$$
$$= -(5\sin 75°\text{ kN})(3\text{ m})$$
$$= -14.5\text{ kN}\cdot\text{m} = 14.5\text{ kN}\cdot\text{m} \curvearrowright \qquad \textit{Ans.}$$

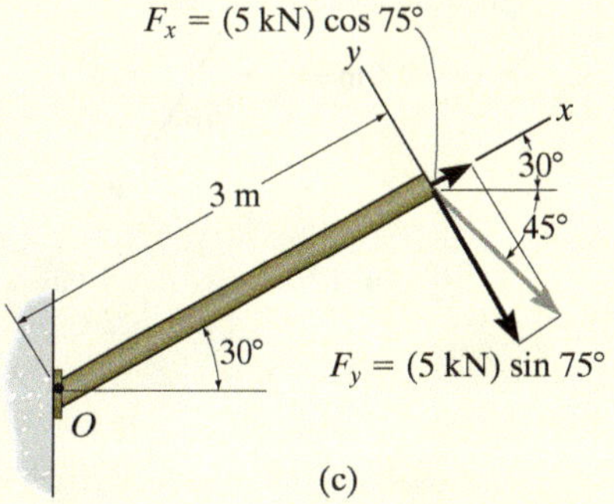

Fig. 4–18

4

EXAMPLE 4.6

Force **F** acts at the end of the angle bracket in Fig. 4–19*a*. Determine the moment of the force about point *O*.

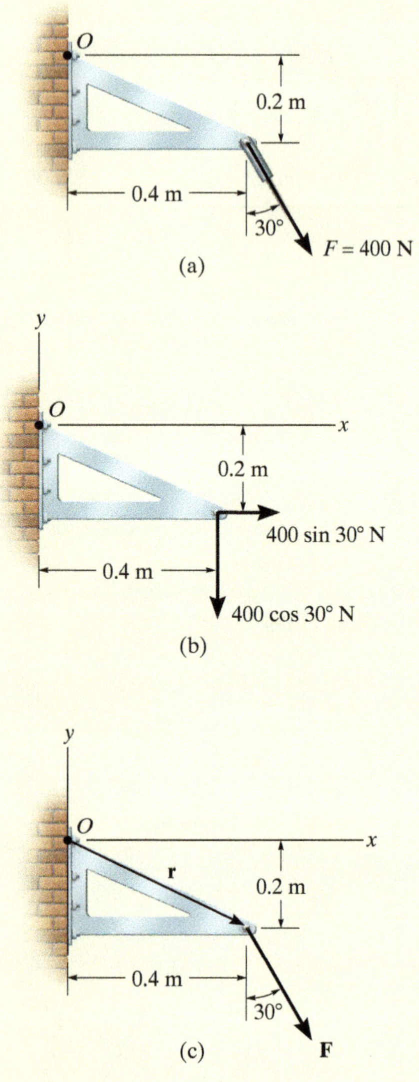

Fig. 4–19

SOLUTION I (SCALAR ANALYSIS)

The force is resolved into its *x* and *y* components, Fig. 4–19*b*, then

$$\circlearrowleft + M_O = 400 \sin 30^\circ \text{ N}(0.2 \text{ m}) - 400 \cos 30^\circ \text{ N}(0.4 \text{ m})$$
$$= -98.6 \text{ N} \cdot \text{m} = 98.6 \text{ N} \cdot \text{m} \circlearrowright$$

or

$$\mathbf{M}_O = \{-98.6\mathbf{k}\} \text{ N} \cdot \text{m}$$ *Ans.*

SOLUTION II (VECTOR ANALYSIS)

Using a Cartesian vector approach, the force and position vectors, Fig. 4–19*c*, are

$$\mathbf{r} = \{0.4\mathbf{i} - 0.2\mathbf{j}\} \text{ m}$$
$$\mathbf{F} = \{400 \sin 30^\circ \mathbf{i} - 400 \cos 30^\circ \mathbf{j}\} \text{ N}$$
$$= \{200.0\mathbf{i} - 346.4\mathbf{j}\} \text{ N}$$

The moment is therefore

$$\mathbf{M}_O = \mathbf{r} \times \mathbf{F} = \begin{vmatrix} \mathbf{i} & \mathbf{j} & \mathbf{k} \\ 0.4 & -0.2 & 0 \\ 200.0 & -346.4 & 0 \end{vmatrix}$$
$$= 0\mathbf{i} - 0\mathbf{j} + [0.4(-346.4) - (-0.2)(200.0)]\mathbf{k}$$
$$= \{-98.6\mathbf{k}\} \text{ N} \cdot \text{m}$$ *Ans.*

NOTE: It is seen that the scalar analysis (Solution I) provides a more *convenient method* for analysis than Solution II since the direction of the moment and the moment arm for each component force are easy to establish. Hence, this method is generally recommended for solving problems displayed in two dimensions, whereas a Cartesian vector analysis is generally recommended only for solving three-dimensional problems.

FUNDAMENTAL PROBLEMS

F4–1. Determine the moment of the force about point O.

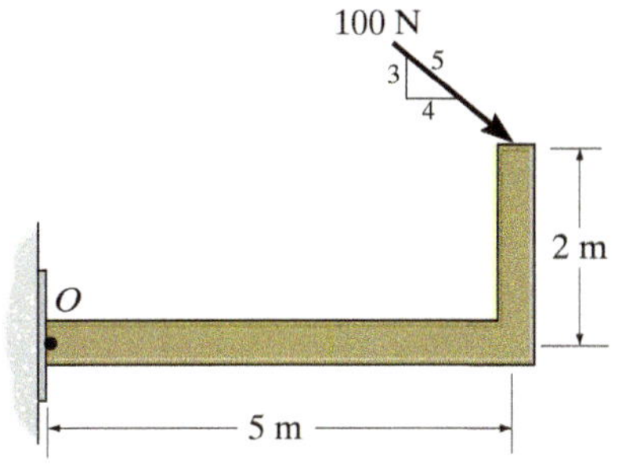

F4–1

F 4–2. Determine the moment of the force about point O.

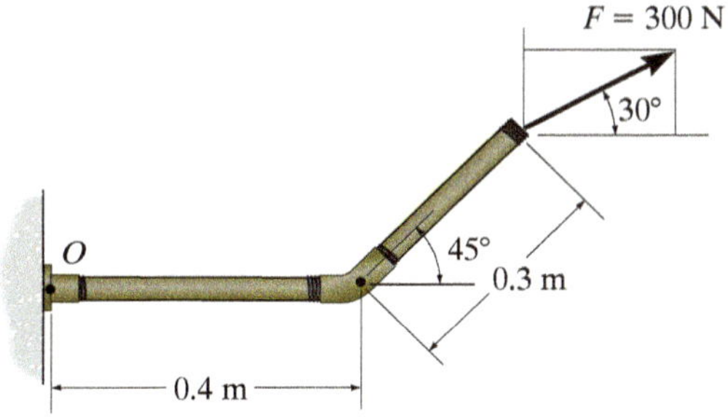

F4–2

F4–3. Determine the moment of the force about point O.

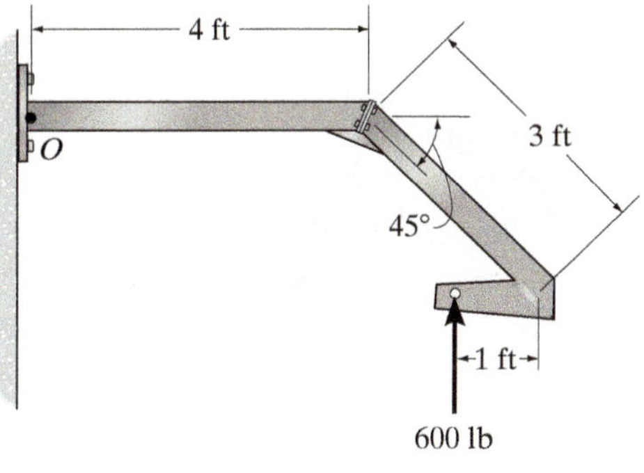

F4–3

F4–4. Determine the moment of the force about point O. Neglect the thickness of the member.

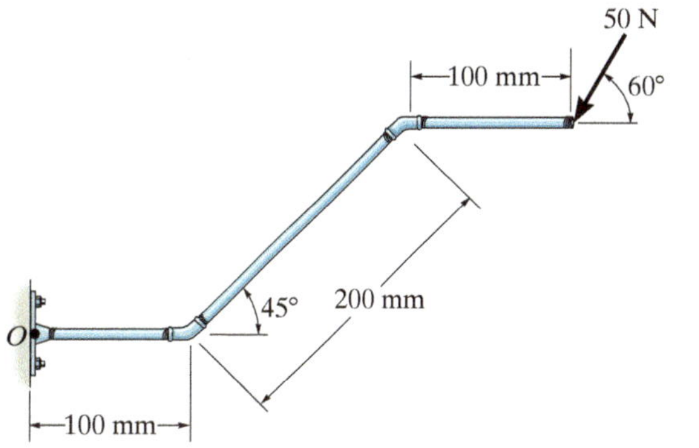

F4–4

F4–5. Determine the moment of the force about point O.

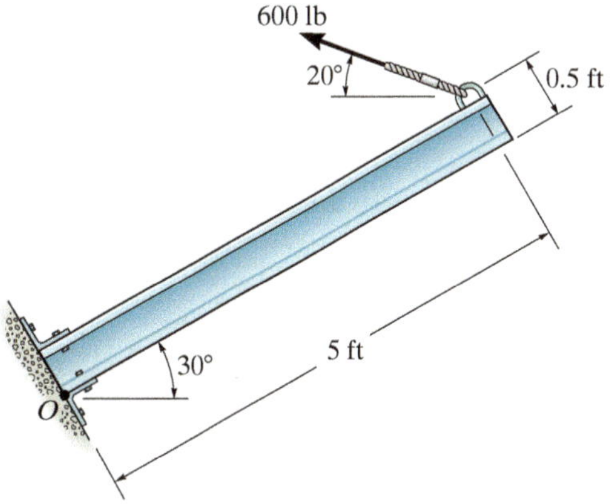

F4–5

F4–6. Determine the moment of the force about point O.

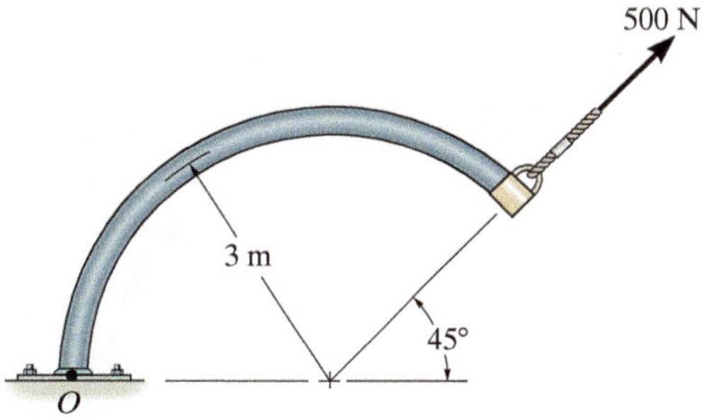

F4–6

F4–7. Determine the resultant moment produced by the forces about point O.

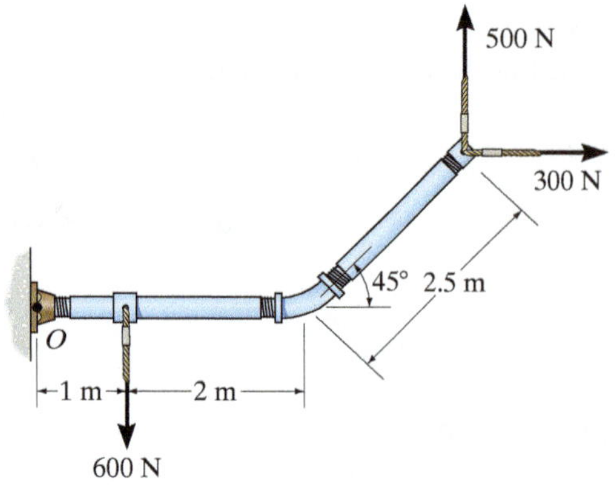

F4–7

F4–8. Determine the resultant moment produced by the forces about point O.

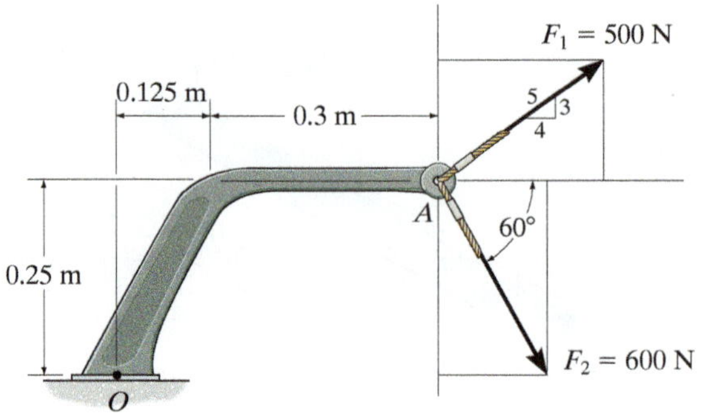

F4–8

F4–9. Determine the resultant moment produced by the forces about point O.

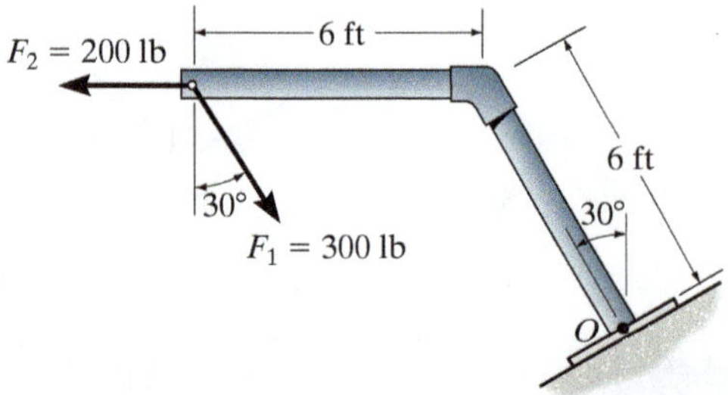

F4–9

F4–10. Determine the moment of force **F** about point O. Express the result as a Cartesian vector.

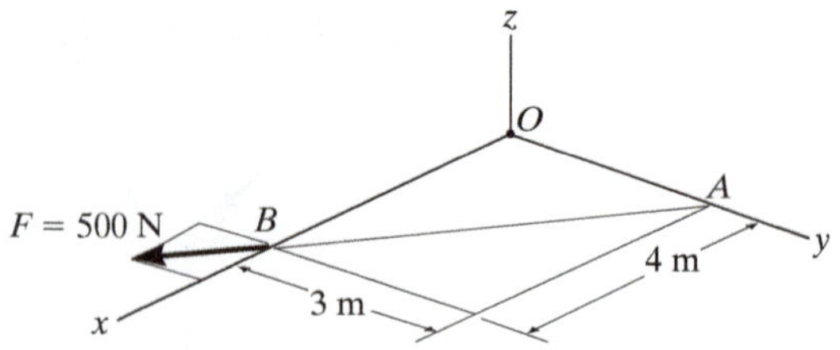

F4–10

F4–11. Determine the moment of force **F** about point O. Express the result as a Cartesian vector.

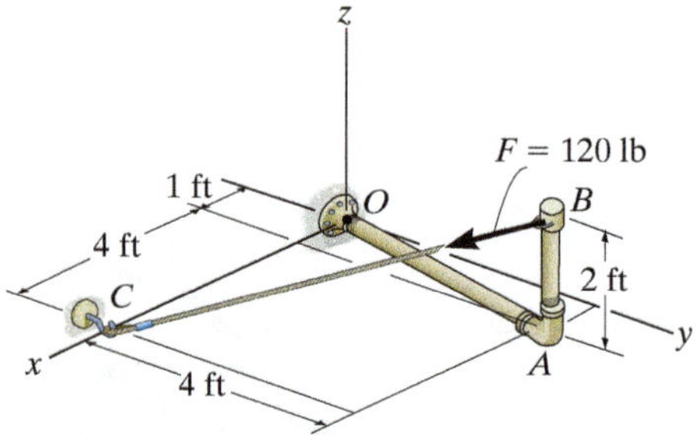

F4–11

F4–12. If $\mathbf{F}_1 = \{100\mathbf{i} - 120\mathbf{j} + 75\mathbf{k}\}$ lb and $\mathbf{F}_2 = \{-200\mathbf{i} + 250\mathbf{j} + 100\mathbf{k}\}$ lb, determine the resultant moment produced by these forces about point O. Express the result as a Cartesian vector.

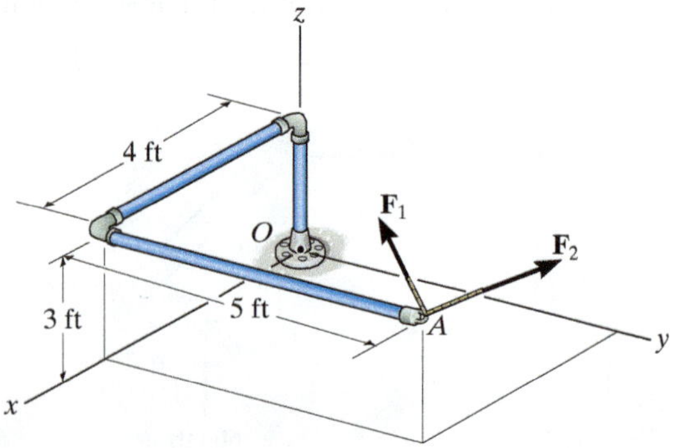

F4–12

PROBLEMS

4–1. If **A**, **B**, and **D** are given vectors, prove the distributive law for the vector cross product, i.e., $\mathbf{A} \times (\mathbf{B} + \mathbf{D}) = (\mathbf{A} \times \mathbf{B}) + (\mathbf{A} \times \mathbf{D})$.

4–2. Prove the triple scalar product identity $\mathbf{A} \cdot (\mathbf{B} \times \mathbf{C}) = (\mathbf{A} \times \mathbf{B}) \cdot \mathbf{C}$.

4–3. Given the three nonzero vectors **A**, **B**, and **C**, show that if $\mathbf{A} \cdot (\mathbf{B} \times \mathbf{C}) = 0$, the three vectors *must* lie in the same plane.

***4–4.** Determine the moment about point A of each of the three forces acting on the beam.

4–5. Determine the moment about point B of each of the three forces acting on the beam.

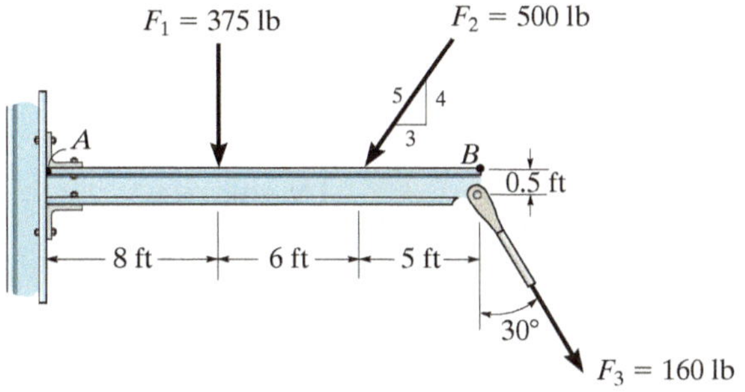

Probs. 4–4/5

4–6. The crane can be adjusted for any angle $0° \le \theta \le 90°$ and any extension $0 \le x \le 5$ m. For a suspended mass of 120 kg, determine the moment developed at A as a function of x and θ. What values of both x and θ develop the maximum possible moment at A? Compute this moment. Neglect the size of the pulley at B.

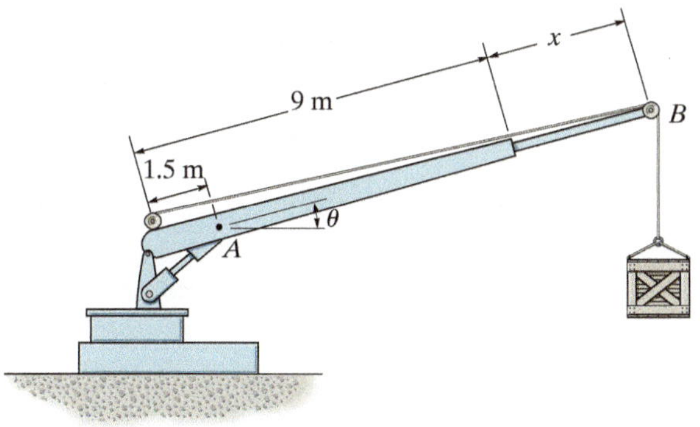

Prob. 4–6

4–7. Determine the moment of each of the three forces about point A.

***4–8.** Determine the moment of each of the three forces about point B.

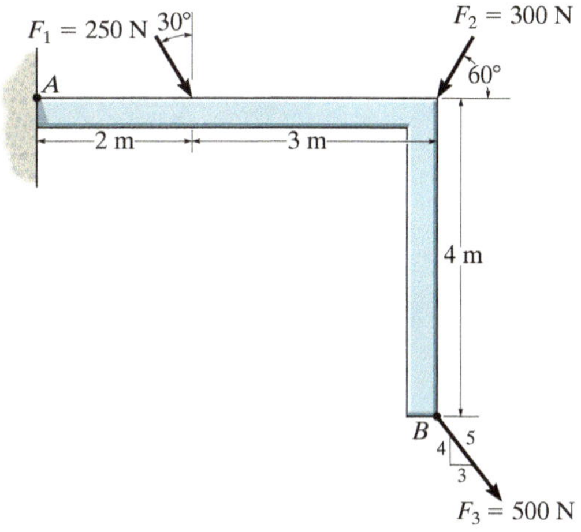

Probs. 4–7/8

4–9. Determine the moment of each force about the bolt located at A. Take $F_B = 40$ lb, $F_C = 50$ lb.

4–10. If $F_B = 30$ lb and $F_C = 45$ lb, determine the resultant moment about the bolt located at A.

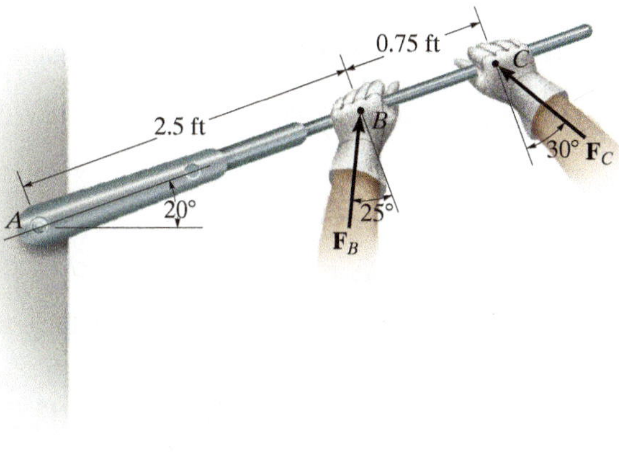

Probs. 4–9/10

4–11. The railway crossing gate consists of the 100-kg gate arm having a center of mass at G_a and the 250-kg counterweight having a center of mass at G_W. Determine the magnitude and directional sense of the resultant moment produced by the weights about point A.

***4–12.** The railway crossing gate consists of the 100-kg gate arm having a center of mass at G_a and the 250-kg counterweight having a center of mass at G_W. Determine the magnitude and directional sense of the resultant moment produced by the weights about point B.

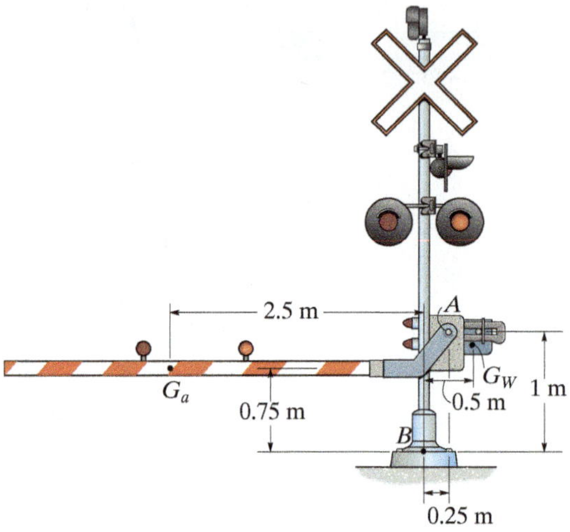

Probs. 4–11/12

***4–13.** The two boys push on the gate with forces of $F_A = 30$ lb, and $F_B = 50$ lb, as shown. Determine the moment of each force about C. Which way will the gate rotate, clockwise or counterclockwise? Neglect the thickness of the gate.

4–14. Two boys push on the gate as shown. If the boy at B exerts a force of $F_B = 30$ lb, determine the magnitude of the force F_A the boy at A must exert in order to prevent the gate from turning. Neglect the thickness of the gate.

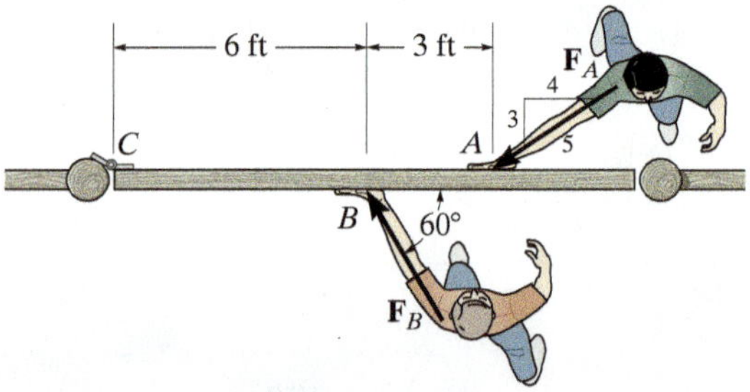

Probs. 4–13/14

4–15. The Achilles tendon force of $F_t = 650$ N is mobilized when the man tries to stand on his toes. As this is done, each of his feet is subjected to a reactive force of $N_f = 400$ N. Determine the resultant moment of $\mathbf{F}_t$ and $\mathbf{N}_f$ about the ankle joint A.

***4–16.** The Achilles tendon force $\mathbf{F}_t$ is mobilized when the man tries to stand on his toes. As this is done, each of his feet is subjected to a reactive force of $N_t = 400$ N. If the resultant moment produced by forces $\mathbf{F}_t$ and $\mathbf{N}_t$ about the ankle joint A is required to be zero, determine the magnitude of $\mathbf{F}_t$.

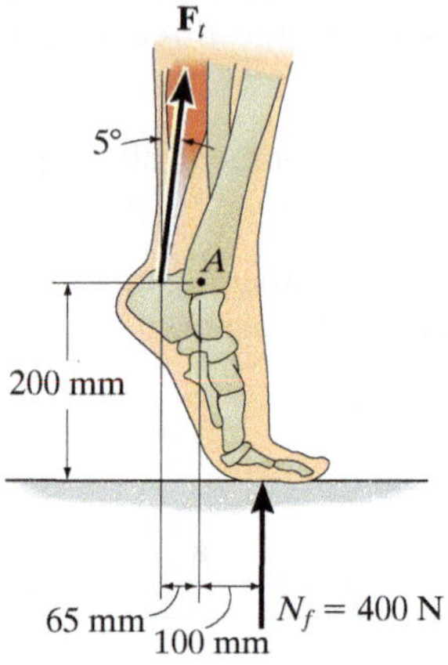

Probs. 4–15/16

4–17. The total hip replacement is subjected to a force of $F = 120$ N. Determine the moment of this force about the neck at A and the stem at B.

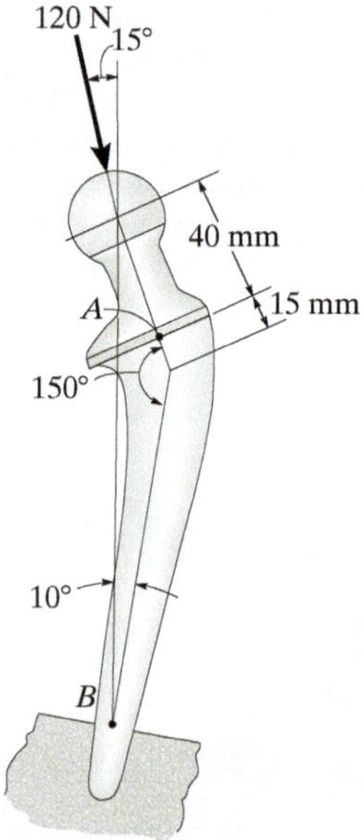

Prob. 4–17

4–18. The tower crane is used to hoist the 2-Mg load upward at constant velocity. The 1.5-Mg jib BD, 0.5-Mg jib BC, and 6-Mg counterweight C have centers of mass at G_1, G_2, and G_3, respectively. Determine the resultant moment produced by the load and the weights of the tower crane jibs about point A and about point B.

4–19. The tower crane is used to hoist a 2-Mg load upward at constant velocity. The 1.5-Mg jib BD and 0.5-Mg jib BC have centers of mass at G_1 and G_2, respectively. Determine the required mass of the counterweight C so that the resultant moment produced by the load and the weight of the tower crane jibs about point A is zero. The center of mass for the counterweight is located at G_3.

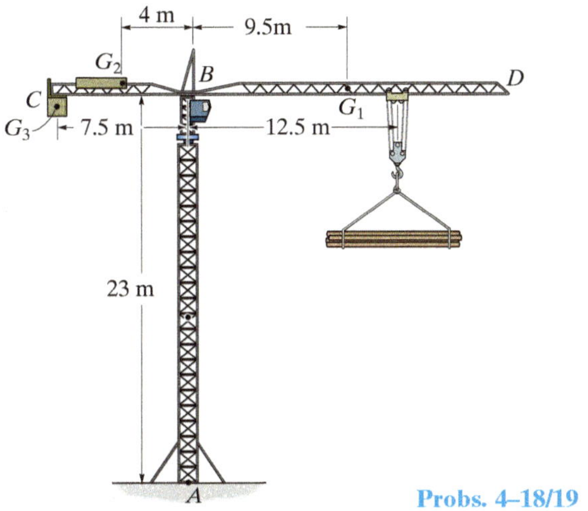

Probs. 4–18/19

***4–20.** The handle of the hammer is subjected to the force of $F = 20$ lb. Determine the moment of this force about the point A.

4–21. In order to pull out the nail at B, the force **F** exerted on the handle of the hammer must produce a clockwise moment of 500 lb · in. about point A. Determine the required magnitude of force **F**.

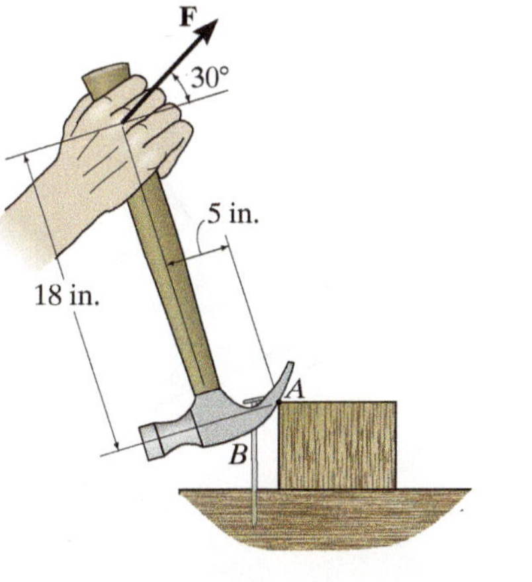

Probs. 4–20/21

4–22. The tool at A is used to hold a power lawnmower blade stationary while the nut is being loosened with the wrench. If a force of 50 N is applied to the wrench at B in the direction shown, determine the moment it creates about the nut at C. What is the magnitude of force **F** at A so that it creates the opposite moment about C?

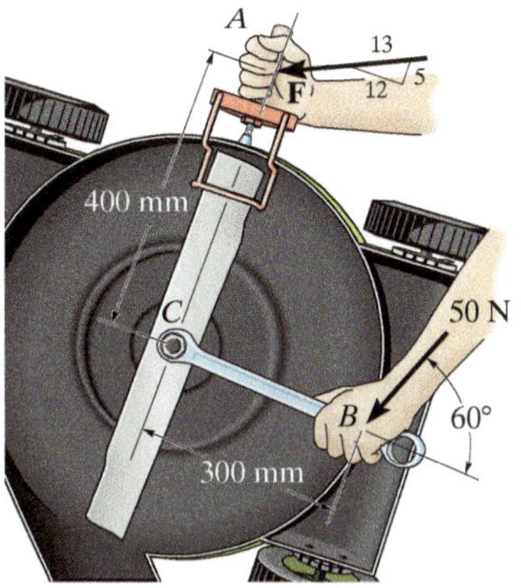

Prob. 4–22

4–23. The towline exerts a force of $P = 4$ kN at the end of the 20-m-long crane boom. If $\theta = 30°$, determine the placement x of the hook at A so that this force creates a maximum moment about point O. What is this moment?

***4–24.** The towline exerts a force of $P = 4$ kN at the end of the 20-m-long crane boom. If $x = 25$ m, determine the position θ of the boom so that this force creates a maximum moment about point O. What is this moment?

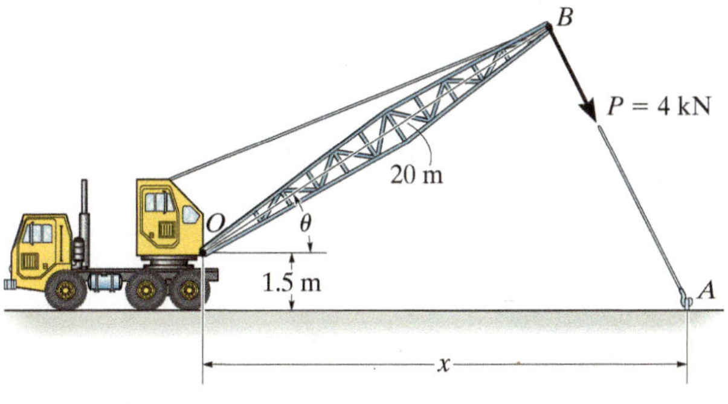

Probs. 4–23/24

4–25. If the 1500-lb boom AB, the 200-lb cage BCD, and the 175-lb man have centers of gravity located at points G_1, G_2 and G_3, respectively, determine the resultant moment produced by each weight about point A.

4–26. If the 1500-lb boom AB, the 200-lb cage BCD, and the 175-lb man have centers of gravity located at points G_1, G_2 and G_3, respectively, determine the resultant moment produced by all the weights about point A.

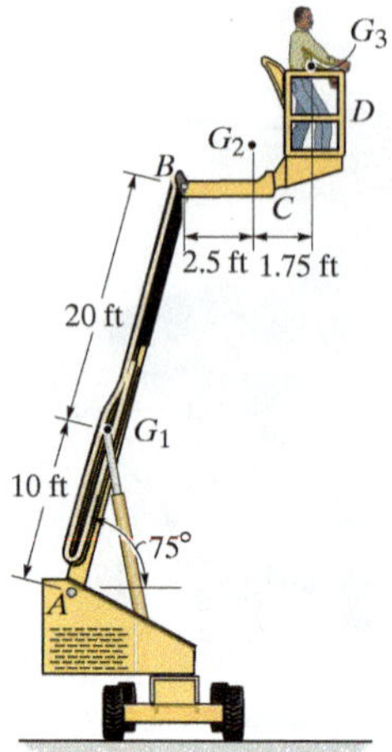

Probs. 4–25/26

4–27. The connected bar BC is used to increase the lever arm of the crescent wrench as shown. If the applied force is $F = 200$ N and $d = 300$ mm, determine the moment produced by this force about the bolt at A.

***4–28.** The connected bar BC is used to increase the lever arm of the crescent wrench as shown. If a clockwise moment of $M_A = 120$ N·m is needed to tighten the bolt at A and the force $F = 200$ N, determine the required extension d in order to develop this moment.

4–29. The connected bar BC is used to increase the lever arm of the crescent wrench as shown. If a clockwise moment of $M_A = 120$ N·m is needed to tighten the nut at A and the extension $d = 300$ mm, determine the required force **F** in order to develop this moment.

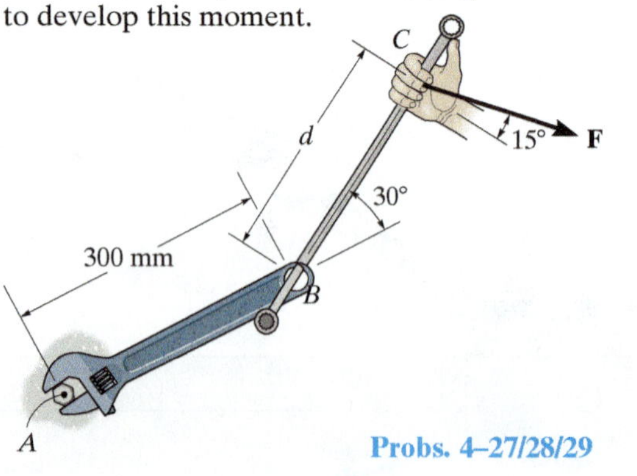

Probs. 4–27/28/29

4–30. A force **F** having a magnitude of $F = 100$ N acts along the diagonal of the parallelepiped. Determine the moment of **F** about point A, using $\mathbf{M}_A = \mathbf{r}_B \times \mathbf{F}$ and $\mathbf{M}_A = \mathbf{r}_C \times \mathbf{F}$.

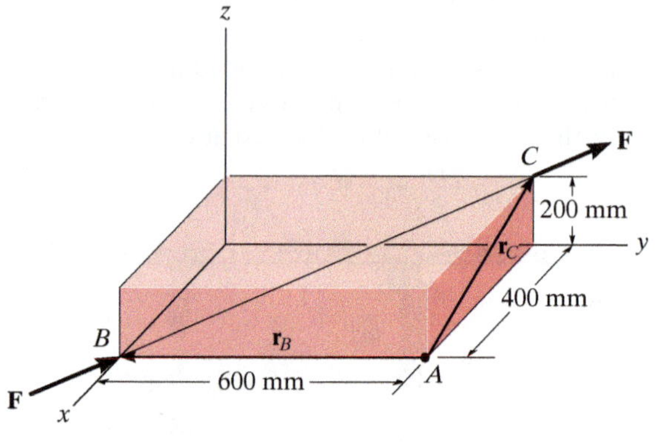

Prob. 4–30

4–31. The force $\mathbf{F} = \{600\mathbf{i} + 300\mathbf{j} - 600\mathbf{k}\}$ N acts at the end of the beam. Determine the moment of the force about point A.

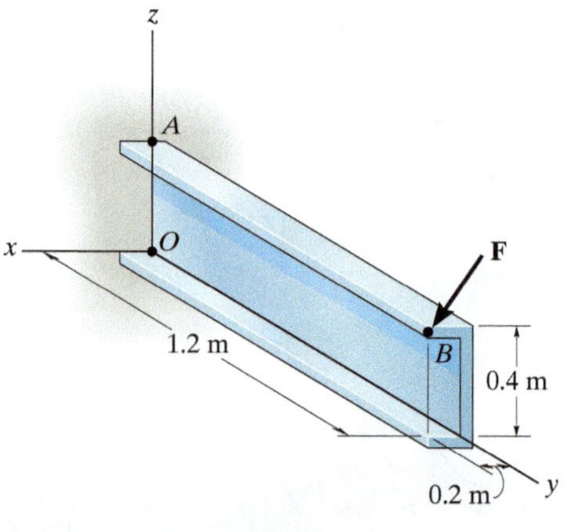

Prob. 4–31

*4–32. Determine the moment produced by force $\mathbf{F}_B$ about point O. Express the result as a Cartesian vector.

4–33. Determine the moment produced by force $\mathbf{F}_C$ about point O. Express the result as a Cartesian vector.

4–34. Determine the resultant moment produced by force $\mathbf{F}_B$ and $\mathbf{F}_C$ about point O. Express the result as a Cartesian vector.

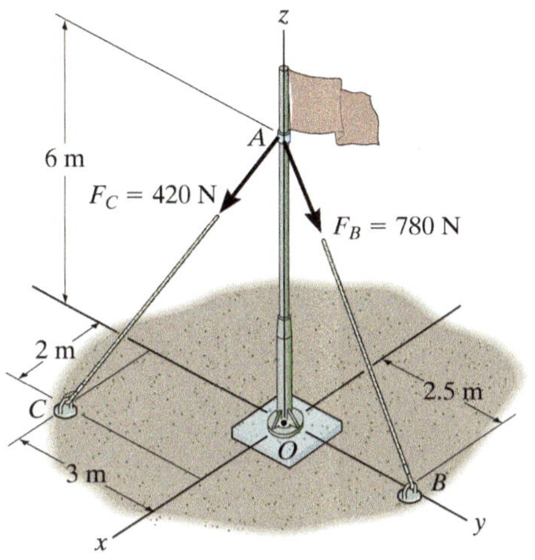

Probs. 4–32/33/34

■4–35. Using a ring collar the 75-N force can act in the vertical plane at various angles θ. Determine the magnitude of the moment it produces about point A, plot the result of M (ordinate) versus θ (abscissa) for $0° \le \theta \le 180°$, and specify the angles that give the maximum and minimum moment.

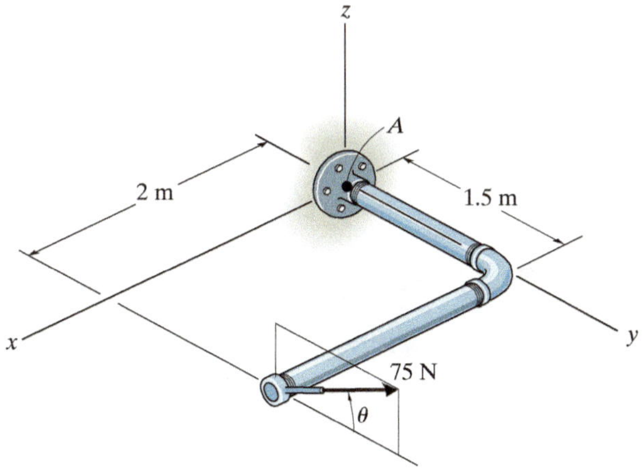

Prob. 4–35

*4–36. The curved rod lies in the x–y plane and has a radius of 3 m. If a force of $F = 80$ N acts at its end as shown, determine the moment of this force about point O.

4–37. The curved rod lies in the x–y plane and has a radius of 3 m. If a force of $F = 80$ N acts at its end as shown, determine the moment of this force about point B.

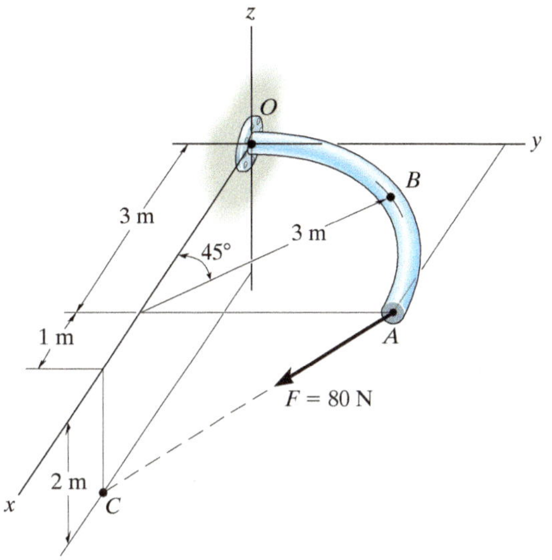

Probs. 4–36/37

4–38. Force $\mathbf{F}$ acts perpendicular to the inclined plane. Determine the moment produced by $\mathbf{F}$ about point A. Express the result as a Cartesian vector.

4–39. Force $\mathbf{F}$ acts perpendicular to the inclined plane. Determine the moment produced by $\mathbf{F}$ about point B. Express the result as a Cartesian vector.

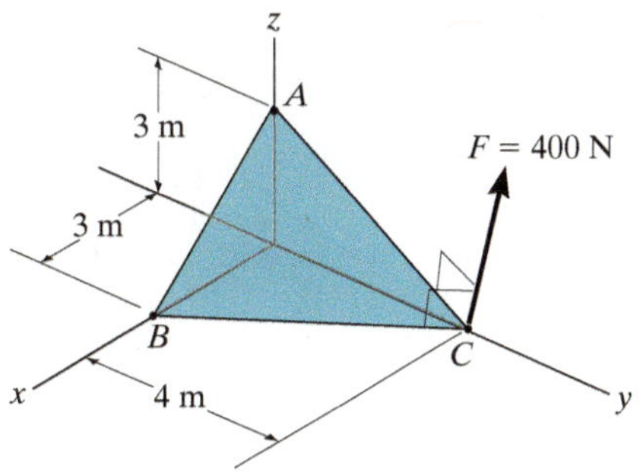

Probs. 4–38/39

***4–40.** The pipe assembly is subjected to the 80-N force. Determine the moment of this force about point A.

4–41. The pipe assembly is subjected to the 80-N force. Determine the moment of this force about point B.

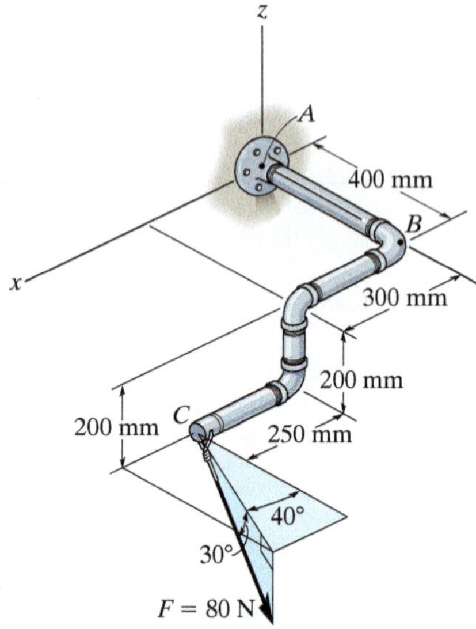

Probs. 4–40/41

4–42. Strut AB of the 1-m-diameter hatch door exerts a force of 450 N on point B. Determine the moment of this force about point O.

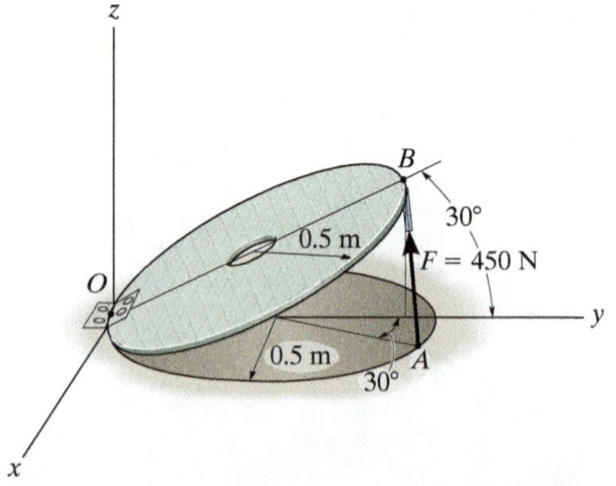

Prob. 4–42

4–43. The curved rod has a radius of 5 ft. If a force of 60 lb acts at its end as shown, determine the moment of this force about point C.

***4–44.** Determine the smallest force F that must be applied along the rope in order to cause the curved rod, which has a radius of 5 ft, to fail at the support C. This requires a moment of $M = 80\text{ lb}\cdot\text{ft}$ to be developed at C.

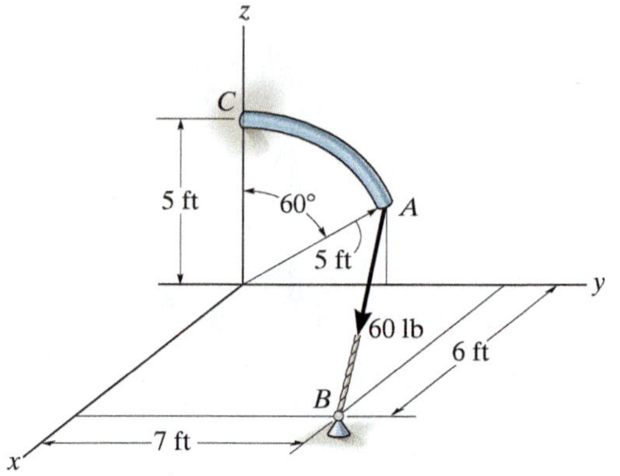

Probs. 4–43/44

4–45. A force of $\mathbf{F} = \{6\mathbf{i} - 2\mathbf{j} + 1\mathbf{k}\}$ kN produces a moment of $\mathbf{M}_O = \{4\mathbf{i} + 5\mathbf{j} - 14\mathbf{k}\}\text{ kN}\cdot\text{m}$ about the origin of coordinates, point O. If the force acts at a point having an x coordinate of $x = 1$ m, determine the y and z coordinates.

4–46. The force $\mathbf{F} = \{6\mathbf{i} + 8\mathbf{j} + 10\mathbf{k}\}$ N creates a moment about point O of $\mathbf{M}_O = \{-14\mathbf{i} + 8\mathbf{j} + 2\mathbf{k}\}\text{ N}\cdot\text{m}$. If the force passes through a point having an x coordinate of 1 m, determine the y and z coordinates of the point. Also, realizing that $M_O = Fd$, determine the perpendicular distance d from point O to the line of action of $\mathbf{F}$.

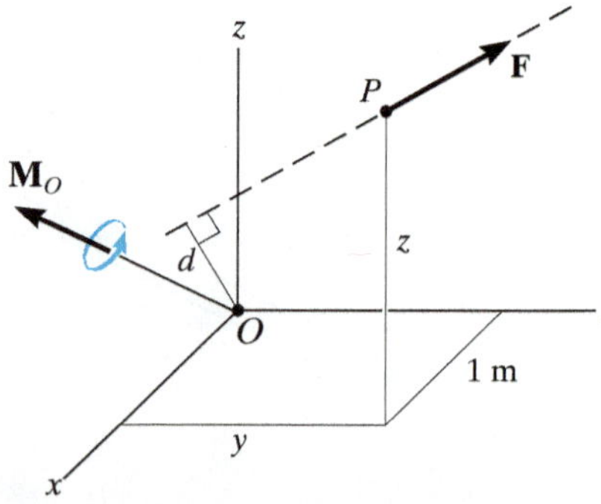

Probs. 4–45/46

4.5 Moment of a Force about a Specified Axis

Sometimes, the moment produced by a force about a *specified axis* must be determined. For example, suppose the lug nut at O on the car tire in Fig. 4–20a needs to be loosened. The force applied to the wrench will create a tendency for the wrench and the nut to rotate about the *moment axis* passing through O; however, the nut can only rotate about the y axis. Therefore, to determine the turning effect, only the y component of the moment is needed, and the total moment produced is not important. To determine this component, we can use either a scalar or vector analysis.

Scalar Analysis. To use a scalar analysis in the case of the lug nut in Fig. 4–20a, the moment arm perpendicular distance from the axis to the line of action of the force is $d_y = d\cos\theta$. Thus, the moment of $\mathbf{F}$ about the y axis is $M_y = F\,d_y = F(d\cos\theta)$. According to the right-hand rule, $\mathbf{M}_y$ is directed along the positive y axis as shown in the figure. In general, for any axis a, the moment is

$$M_a = Fd_a \tag{4–10}$$

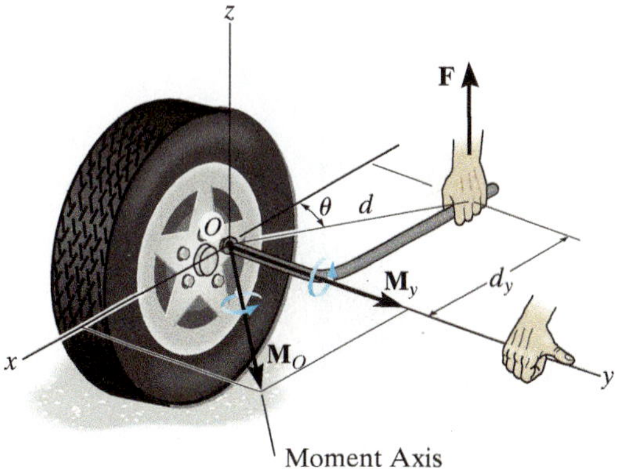

(a)

Fig. 4-20

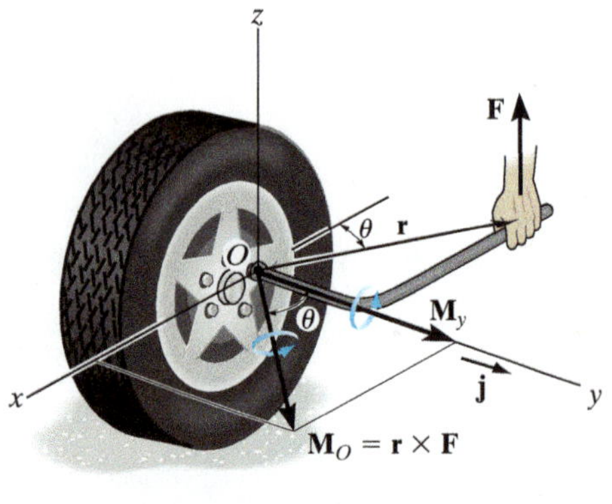

(b)

Fig. 4–20 (cont.)

Vector Analysis. To find the moment of force **F** in Fig. 4–20*b* about the *y* axis using a vector analysis, we must first determine the moment of the force about *any point O* on the *y* axis by applying Eq. 4–7, $\mathbf{M}_O = \mathbf{r} \times \mathbf{F}$. The component $\mathbf{M}_y$ along the *y* axis is the *projection* of $\mathbf{M}_O$ onto the *y* axis. It can be found using the *dot product* discussed in Chapter 2, so that $M_y = \mathbf{j} \cdot \mathbf{M}_O = \mathbf{j} \cdot (\mathbf{r} \times \mathbf{F})$, where **j** is the unit vector for the *y* axis.

We can generalize this approach by letting $\mathbf{u}_a$ be the unit vector that specifies the direction of the *a* axis shown in Fig. 4–21. Then the moment of **F** about a point *O* on the axis is $\mathbf{M}_O = \mathbf{r} \times \mathbf{F}$, and the projection of this moment onto the *a* axis is $M_a = \mathbf{u}_a \cdot (\mathbf{r} \times \mathbf{F})$. This combination is referred to as the *scalar triple product.* If the vectors are written in Cartesian form, we have

$$M_a = [u_{a_x}\mathbf{i} + u_{a_y}\mathbf{j} + u_{a_z}\mathbf{k}] \cdot \begin{vmatrix} \mathbf{i} & \mathbf{j} & \mathbf{k} \\ r_x & r_y & r_z \\ F_x & F_y & F_z \end{vmatrix}$$
$$= u_{a_x}(r_yF_z - r_zF_y) - u_{a_y}(r_xF_z - r_zF_x) + u_{a_z}(r_xF_y - r_yF_x)$$

This result can also be written in the form of a determinant, making it easier to memorize.*

$$M_a = \mathbf{u}_a \cdot (\mathbf{r} \times \mathbf{F}) = \begin{vmatrix} u_{a_x} & u_{a_y} & u_{a_z} \\ r_x & r_y & r_z \\ F_x & F_y & F_z \end{vmatrix} \qquad (4\text{–}11)$$

where

$u_{a_x}, u_{a_y}, u_{a_z}$ represent the *x, y, z* components of the unit vector defining the direction of the *a* axis

r_x, r_y, r_z represent the *x, y, z* components of the position vector extended from *any point O* on the *a* axis to *any point A* on the line of action of the force

F_x, F_y, F_z represent the *x, y, z* components of the force vector.

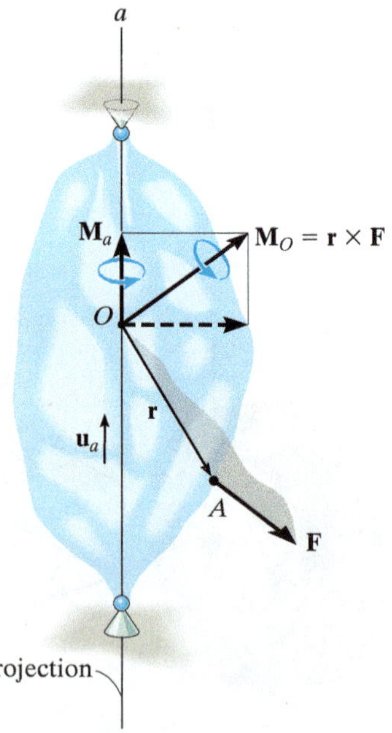

Fig. 4–21

When M_a is evaluated from Eq. 4–11, it will yield a positive or negative scalar. The sign of this scalar indicates the sense of direction of $\mathbf{M}_a$ along the *a* axis. If it is positive, then $\mathbf{M}_a$ will have the same sense as $\mathbf{u}_a$, whereas if it is negative, then $\mathbf{M}_a$ will act opposite to $\mathbf{u}_a$.

Once M_a is determined, we can then express $\mathbf{M}_a$ as a Cartesian vector, namely,

$$\mathbf{M}_a = M_a\mathbf{u}_a \qquad (4\text{–}12)$$

The examples which follow illustrate numerical applications of the above concepts.

*Take a minute to expand this determinant, to show that it will yield the above result.

Important Points

- The moment of a force about a specified axis can be determined provided the perpendicular distance d_a from the force line of action to the axis can be determined. $M_a = Fd_a$.
- If vector analysis is used, $M_a = \mathbf{u}_a \cdot (\mathbf{r} \times \mathbf{F})$, where $\mathbf{u}_a$ defines the direction of the axis and $\mathbf{r}$ is extended from *any point* on the axis to *any point* on the line of action of the force.
- If M_a is calculated as a negative scalar, then the sense of direction of $\mathbf{M}_a$ is opposite to $\mathbf{u}_a$.
- The moment $\mathbf{M}_a$ expressed as a Cartesian vector is determined from $\mathbf{M}_a = M_a\mathbf{u}_a$.

EXAMPLE 4.7

Determine the resultant moment of the three forces in Fig. 4–22 about the x axis, the y axis, and the z axis.

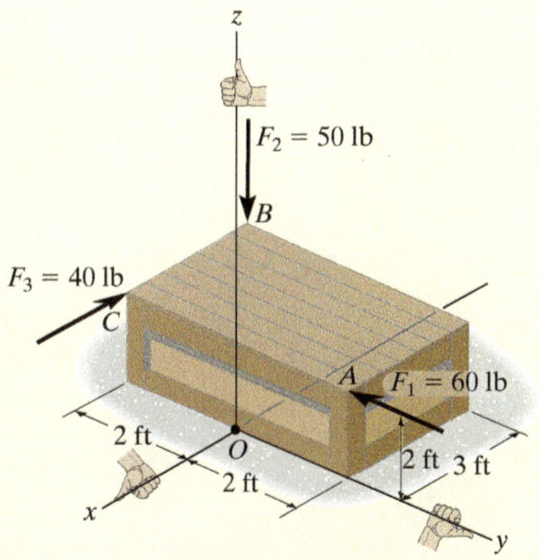

Fig. 4–22

SOLUTION

A force that is *parallel* to a coordinate axis or has a line of action that passes through the axis does *not* produce any moment or tendency for turning about that axis. Therefore, defining the positive direction of the moment of a force according to the right-hand rule, as shown in the figure, we have

$$M_x = (60\text{ lb})(2\text{ ft}) + (50\text{ lb})(2\text{ ft}) + 0 = 220\text{ lb}\cdot\text{ft} \qquad \textit{Ans.}$$

$$M_y = 0 - (50\text{ lb})(3\text{ ft}) - (40\text{ lb})(2\text{ ft}) = -230\text{ lb}\cdot\text{ft} \qquad \textit{Ans.}$$

$$M_z = 0 + 0 - (40\text{ lb})(2\text{ ft}) = -80\text{ lb}\cdot\text{ft} \qquad \textit{Ans.}$$

The negative signs indicate that $\mathbf{M}_y$ and $\mathbf{M}_z$ act in the $-y$ and $-z$ directions, respectively.

EXAMPLE 4.8

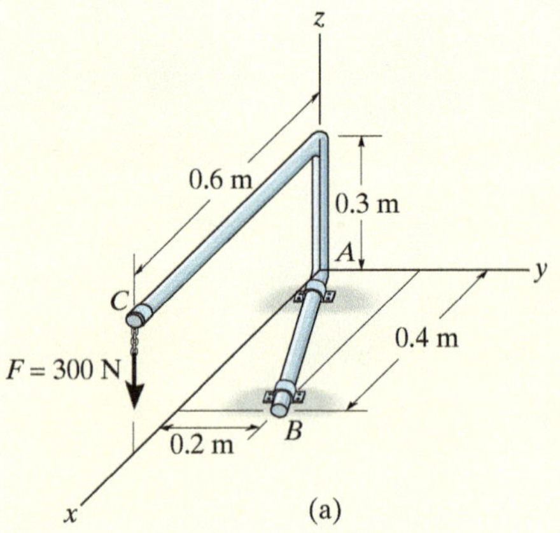

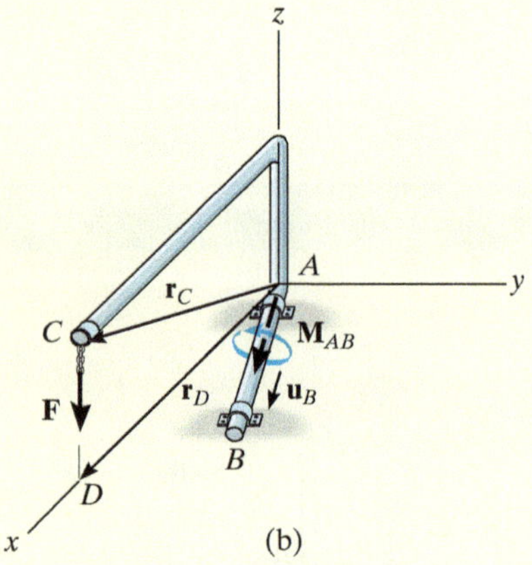

Fig. 4–23

Determine the moment $\mathbf{M}_{AB}$ produced by the force $\mathbf{F}$ in Fig. 4–23*a*, which tends to rotate the rod about the *AB* axis.

SOLUTION

A vector analysis using $M_{AB} = \mathbf{u}_B \cdot (\mathbf{r} \times \mathbf{F})$ will be considered for the solution rather than trying to find the moment arm or perpendicular distance from the line of action of $\mathbf{F}$ to the *AB* axis. Each of the terms in the equation will now be identified.

Unit vector $\mathbf{u}_B$ defines the direction of the *AB* axis of the rod, Fig. 4–23*b*, where

$$\mathbf{u}_B = \frac{\mathbf{r}_B}{r_B} = \frac{\{0.4\mathbf{i} + 0.2\mathbf{j}\}\text{ m}}{\sqrt{(0.4\text{ m})^2 + (0.2\text{ m})^2}} = 0.8944\mathbf{i} + 0.4472\mathbf{j}$$

Vector $\mathbf{r}$ is directed from *any point* on the *AB* axis to *any point* on the line of action of the force. For example, position vectors $\mathbf{r}_C$ and $\mathbf{r}_D$ are suitable, Fig. 4–23*b*. (Although not shown, $\mathbf{r}_{BC}$ or $\mathbf{r}_{BD}$ can also be used.) For simplicity, we choose $\mathbf{r}_D$, where

$$\mathbf{r}_D = \{0.6\mathbf{i}\}\text{ m}$$

The force is

$$\mathbf{F} = \{-300\mathbf{k}\}\text{ N}$$

Substituting these vectors into the determinant form and expanding, we have

$$M_{AB} = \mathbf{u}_B \cdot (\mathbf{r}_D \times \mathbf{F}) = \begin{vmatrix} 0.8944 & 0.4472 & 0 \\ 0.6 & 0 & 0 \\ 0 & 0 & -300 \end{vmatrix}$$

$$= 0.8944[0(-300) - 0(0)] - 0.4472[0.6(-300) - 0(0)] + 0[0.6(0) - 0(0)]$$

$$= 80.50\text{ N}\cdot\text{m}$$

This positive result indicates that the sense of $\mathbf{M}_{AB}$ is in the same direction as $\mathbf{u}_B$.

Expressing $\mathbf{M}_{AB}$ in Fig. 4-23*b* as a Cartesian vector yields

$$\mathbf{M}_{AB} = M_{AB}\mathbf{u}_B = (80.50\text{ N}\cdot\text{m})(0.8944\mathbf{i} + 0.4472\mathbf{j})$$
$$= \{72.0\mathbf{i} + 36.0\mathbf{j}\}\text{ N}\cdot\text{m}$$ *Ans.*

NOTE: If axis *AB* is defined using a unit vector directed from *B* toward *A*, then in the above formulation $-\mathbf{u}_B$ would have to be used. This would lead to $M_{AB} = -80.50\text{ N}\cdot\text{m}$. Consequently, $\mathbf{M}_{AB} = M_{AB}(-\mathbf{u}_B)$, and the same result would be obtained.

EXAMPLE 4.9

Determine the magnitude of the moment of force **F** about segment *OA* of the pipe assembly in Fig. 4–24*a*.

SOLUTION

The moment of **F** about the *OA* axis is determined from $M_{OA} = \mathbf{u}_{OA} \cdot (\mathbf{r} \times \mathbf{F})$, where **r** is a position vector extending from any point on the *OA* axis to any point on the line of action of **F**. As indicated in Fig. 4–24*b*, either $\mathbf{r}_{OD}$, $\mathbf{r}_{OC}$, $\mathbf{r}_{AD}$, or $\mathbf{r}_{AC}$ can be used; however, $\mathbf{r}_{OD}$ will be considered since it will simplify the calculation.

The unit vector $\mathbf{u}_{OA}$, which specifies the direction of the *OA* axis, is

$$\mathbf{u}_{OA} = \frac{\mathbf{r}_{OA}}{r_{OA}} = \frac{\{0.3\mathbf{i} + 0.4\mathbf{j}\}\text{ m}}{\sqrt{(0.3\text{ m})^2 + (0.4\text{ m})^2}} = 0.6\mathbf{i} + 0.8\mathbf{j}$$

and the position vector $\mathbf{r}_{OD}$ is

$$\mathbf{r}_{OD} = \{0.5\mathbf{i} + 0.5\mathbf{k}\}\text{ m}$$

The force **F** expressed as a Cartesian vector is

$$\mathbf{F} = F\left(\frac{\mathbf{r}_{CD}}{r_{CD}}\right)$$

$$= (300\text{ N})\left[\frac{\{0.4\mathbf{i} - 0.4\mathbf{j} + 0.2\mathbf{k}\}\text{ m}}{\sqrt{(0.4\text{ m})^2 + (-0.4\text{ m})^2 + (0.2\text{ m})^2}}\right]$$

$$= \{200\mathbf{i} - 200\mathbf{j} + 100\mathbf{k}\}\text{ N}$$

Therefore,

$$M_{OA} = \mathbf{u}_{OA} \cdot (\mathbf{r}_{OD} \times \mathbf{F})$$

$$= \begin{vmatrix} 0.6 & 0.8 & 0 \\ 0.5 & 0 & 0.5 \\ 200 & -200 & 100 \end{vmatrix}$$

$$= 0.6[0(100) - (0.5)(-200)] - 0.8[0.5(100) - (0.5)(200)] + 0$$

$$= 100\text{ N}\cdot\text{m}$$ *Ans.*

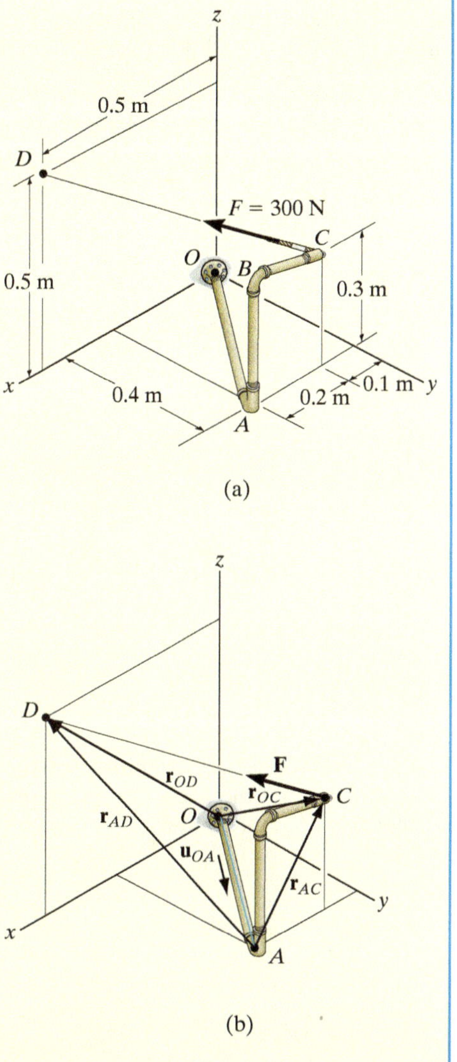

(a)

(b)

Fig. 4–24

FUNDAMENTAL PROBLEMS

F4–13. Determine the magnitude of the moment of the force $\mathbf{F} = \{300\mathbf{i} - 200\mathbf{j} + 150\mathbf{k}\}$ N about the x axis.

F4–14. Determine the magnitude of the moment of the force $\mathbf{F} = \{300\mathbf{i} - 200\mathbf{j} + 150\mathbf{k}\}$ N about the OA axis.

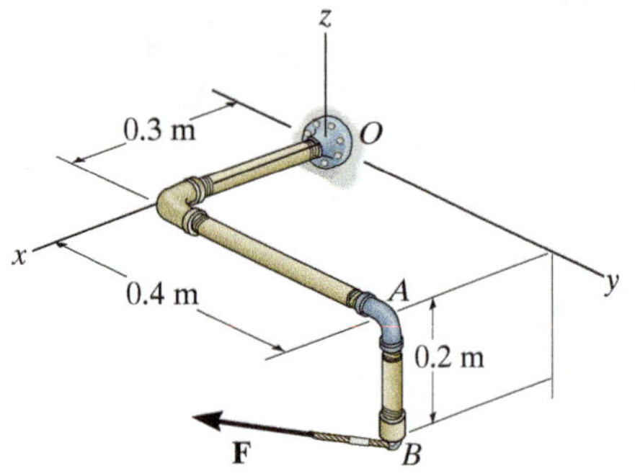

F4–13/14

F4–15. Determine the magnitude of the moment of the 200-N force about the x axis. Solve the problem using both a scalar and a vector analysis.

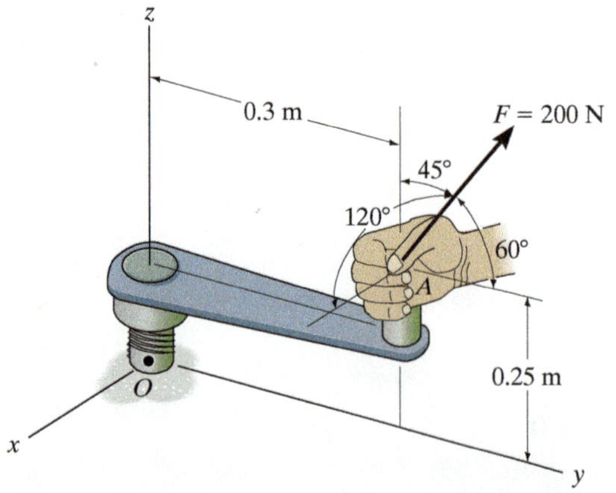

F4–15

F4–16. Determine the magnitude of the moment of the force about the y axis.

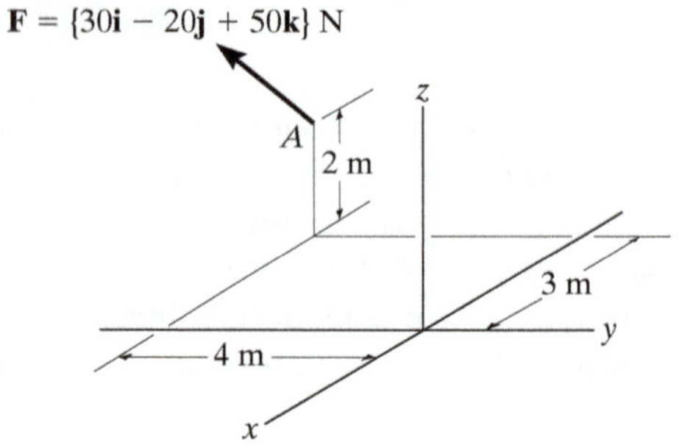

F4–16

F4–17. Determine the moment of the force $\mathbf{F} = \{50\mathbf{i} - 40\mathbf{j} + 20\mathbf{k}\}$ lb about the AB axis. Express the result as a Cartesian vector.

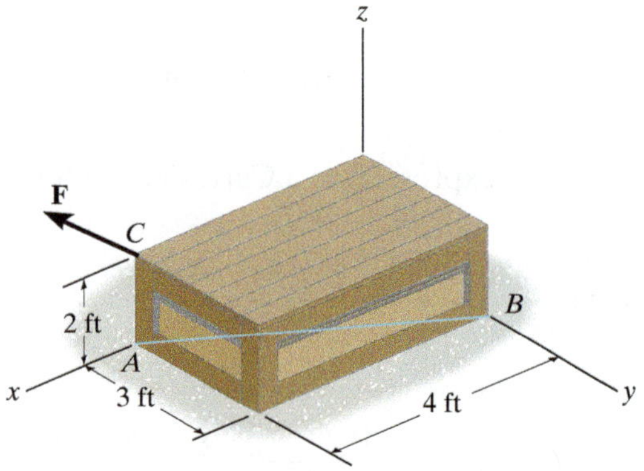

F4–17

F4–18. Determine the moment of force **F** about the x, the y, and the z axes. Solve the problem using both a scalar and a vector analysis.

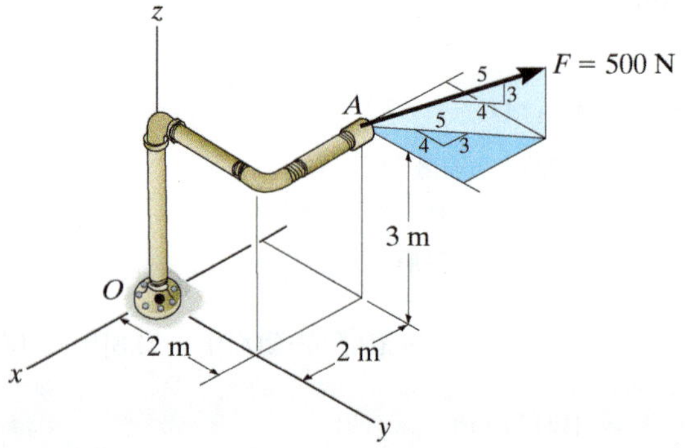

F4–18

PROBLEMS

4–47. Determine the magnitude of the moment of each of the three forces about the axis AB. Solve the problem (a) using a Cartesian vector approach and (b) using a scalar approach.

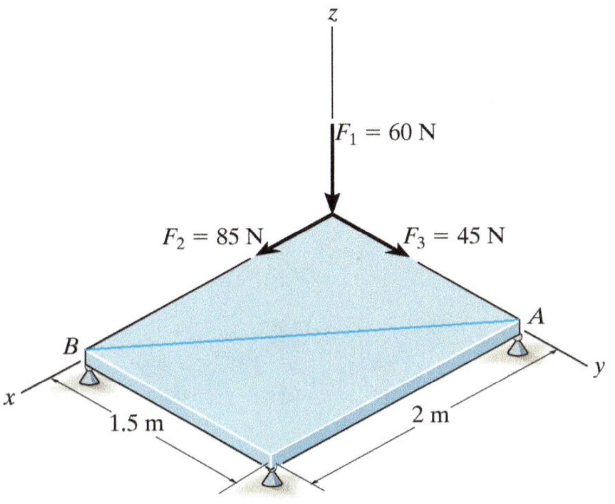

Prob. 4–47

***4–48.** The flex-headed ratchet wrench is subjected to a force of $P = 16$ lb, applied perpendicular to the handle as shown. Determine the moment or torque this imparts along the vertical axis of the bolt at A.

4–49. If a torque or moment of 80 lb · in. is required to loosen the bolt at A, determine the force P that must be applied perpendicular to the handle of the flex-headed ratchet wrench.

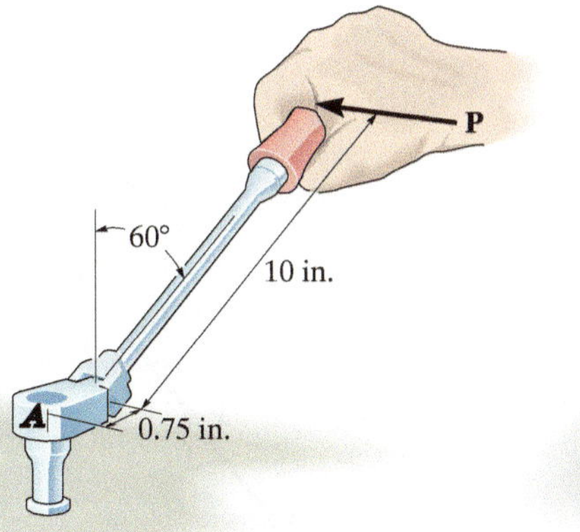

Probs. 4–48/49

4–50. The chain AB exerts a force of 20 lb on the door at B. Determine the magnitude of the moment of this force along the hinged axis x of the door.

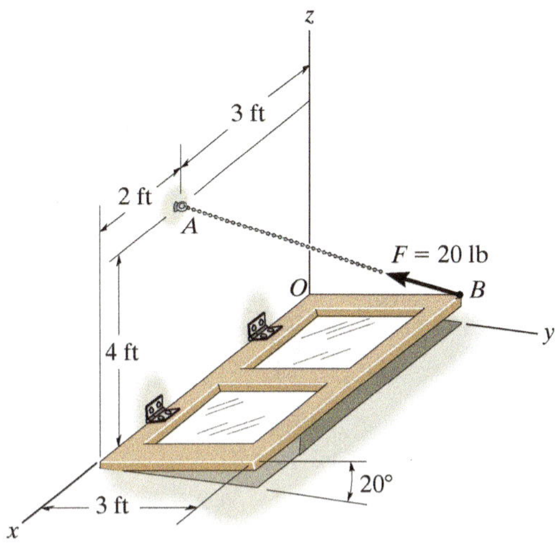

Prob. 4–50

4–51. The hood of the automobile is supported by the strut AB, which exerts a force of $F = 24$ lb on the hood. Determine the moment of this force about the hinged axis y.

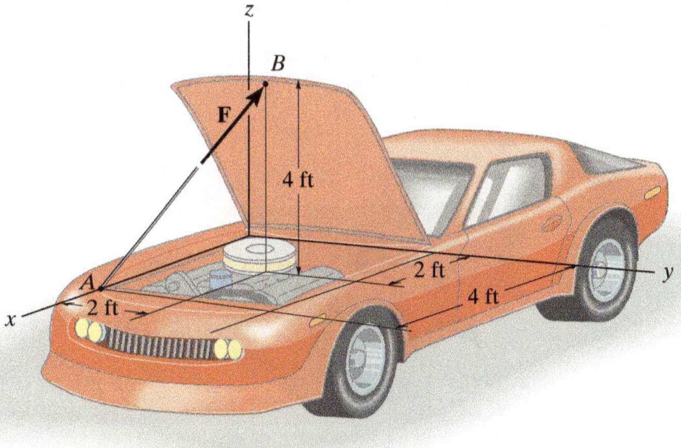

Prob. 4–51

*4–52. Determine the magnitude of the moments of the force **F** about the x, y, and z axes. Solve the problem (a) using a Cartesian vector approach and (b) using a scalar approach.

4–53. Determine the moment of the force **F** about an axis extending between A and C. Express the result as a Cartesian vector.

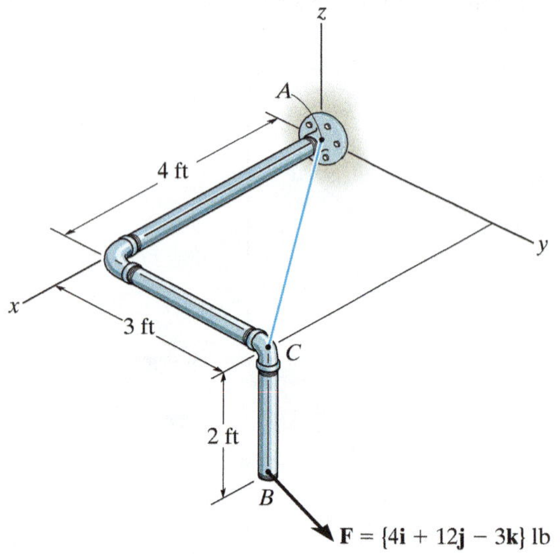

Probs. 4–52/53

4–54. The board is used to hold the end of a four-way lug wrench in the position shown when the man applies a force of $F = 100$ N. Determine the magnitude of the moment produced by this force about the x axis. Force **F** lies in a vertical plane.

4–55. The board is used to hold the end of a four-way lug wrench in position. If a torque of 30 N · m about the x axis is required to tighten the nut, determine the required magnitude of the force **F** that the man's foot must apply on the end of the wrench in order to turn it. Force **F** lies in a vertical plane.

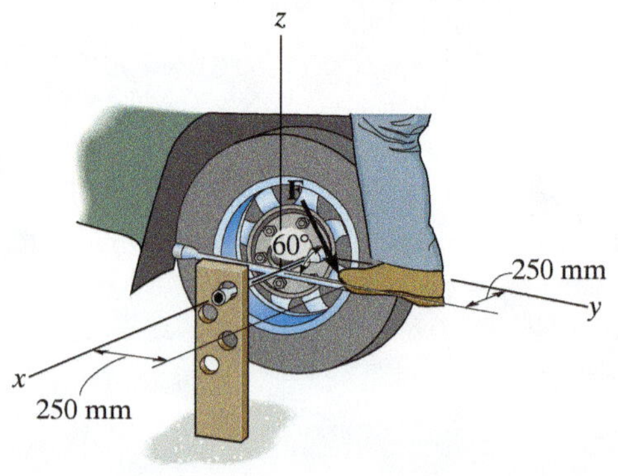

Probs. 4–54/55

*4–56. The cutting tool on the lathe exerts a force **F** on the shaft as shown. Determine the moment of this force about the y axis of the shaft.

4–57. The cutting tool on the lathe exerts a force **F** on the shaft as shown. Determine the moment of this force about the x and z axes.

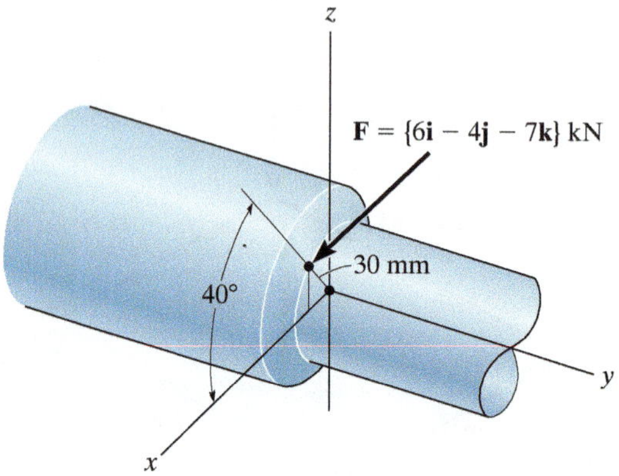

Probs. 4–56/57

4–58. If the tension in the cable is $F = 140$ lb, determine the magnitude of the moment produced by this force about the hinged axis, CD, of the panel.

4–59. Determine the magnitude of force **F** in cable AB in order to produce a moment of 500 lb · ft about the hinged axis CD, which is needed to hold the panel in the position shown.

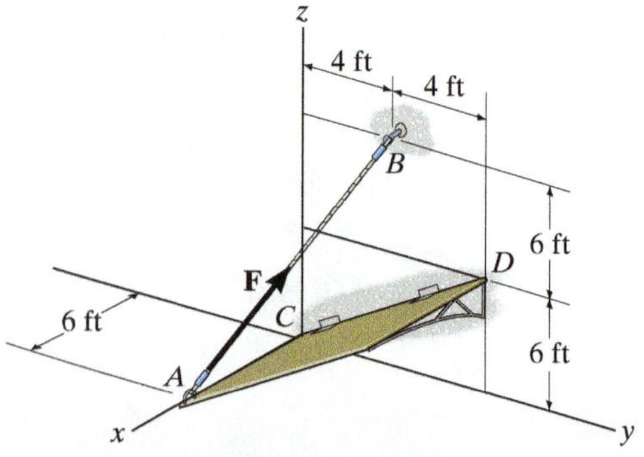

Probs. 4–58/59

***4–60.** The force of $F = 30$ N acts on the bracket as shown. Determine the moment of the force about the $a-a$ axis of the pipe. Also, determine the coordinate direction angles of F in order to produce the maximum moment about the $a-a$ axis. What is this moment?

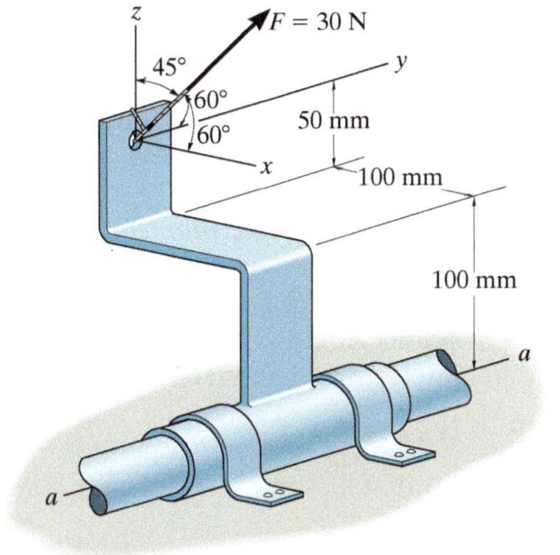

Prob. 4–60

4–61. The pipe assembly is secured on the wall by the two brackets. If the flower pot has a weight of 50 lb, determine the magnitude of the moment produced by the weight about the x, y, and z axes.

4–62. The pipe assembly is secured on the wall by the two brackets. If the flower pot has a weight of 50 lb, determine the magnitude of the moment produced by the weight about the OA axis.

4–63. The pipe assembly is secured on the wall by the two brackets. If the frictional force of both brackets can resist a maximum moment of 150 lb · ft, determine the largest weight of the flower pot that can be supported by the assembly without causing it to rotate about the OA axis.

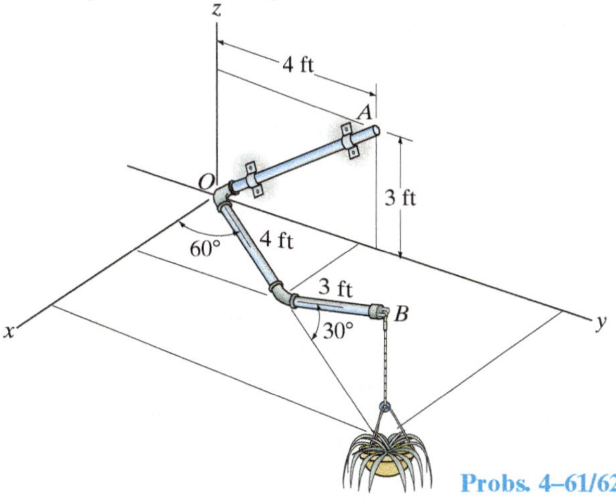

Probs. 4–61/62/63

***4–64.** The wrench A is used to hold the pipe in a stationary position while wrench B is used to tighten the elbow fitting. If $F_B = 150$ N, determine the magnitude of the moment produced by this force about the y axis. Also, what is the magnitude of force $\mathbf{F}_A$ in order to counteract this moment?

4–65. The wrench A is used to hold the pipe in a stationary position while wrench B is used to tighten the elbow fitting. Determine the magnitude of force F_B in order to develop a torque of 50 N · m about the y axis. Also, what is the required magnitude of force $\mathbf{F}_A$ in order to counteract this moment?

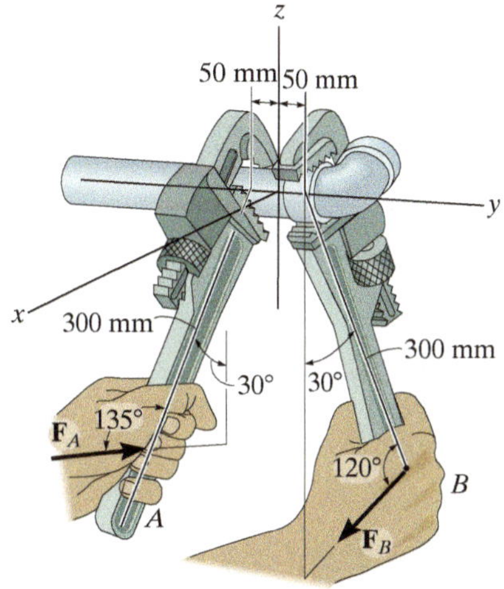

Probs. 4–64/65

4–66. The A-frame is being hoisted into an upright position by the vertical force of $F = 80$ lb. Determine the moment of this force about the y axis when the frame is in the position shown.

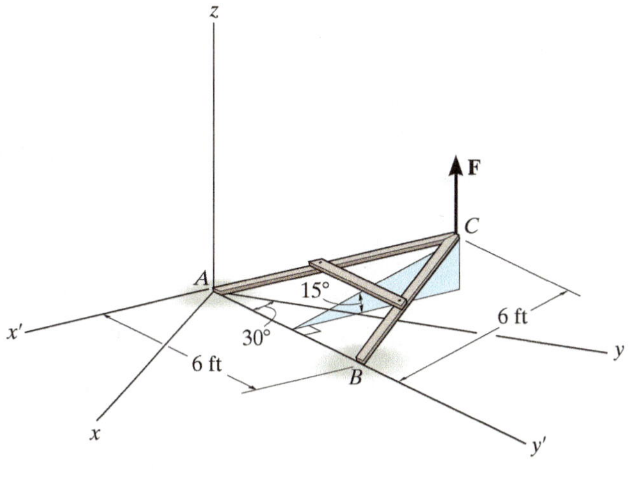

Prob. 4–66

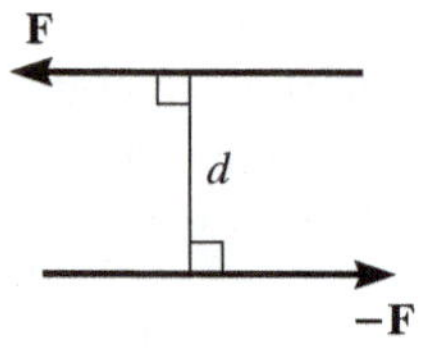

Fig. 4–25

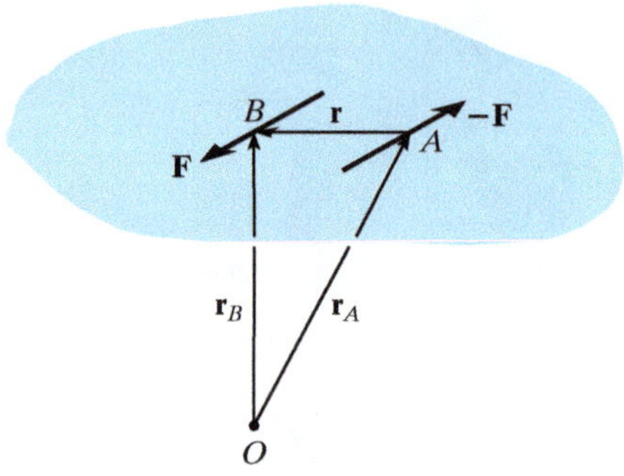

Fig. 4–26

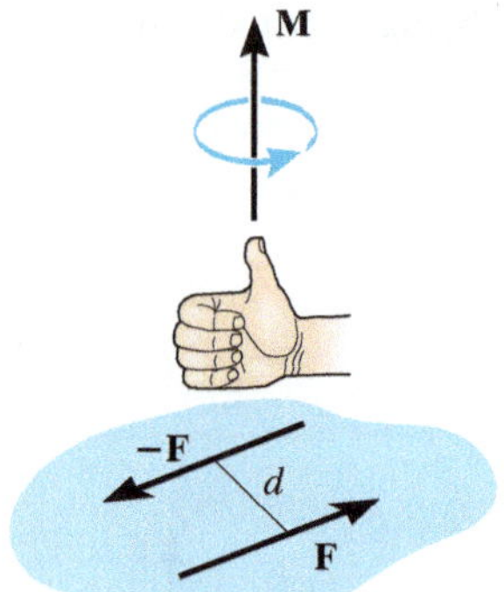

Fig. 4–27

4.6 Moment of a Couple

A *couple* is defined as two parallel forces that have the same magnitude, but opposite directions, and are separated by a perpendicular distance d, Fig. 4–25. Since the resultant force is zero, the only effect of a couple is to produce an actual rotation, or if no movement is possible, there is a tendency of rotation in a specified direction. For example, imagine that you are driving a car with both hands on the steering wheel and you are making a turn. One hand will push up on the wheel while the other hand pulls down, which causes the steering wheel to rotate.

The moment produced by a couple is called a *couple moment.* We can determine its value by finding the sum of the moments of both couple forces about *any* arbitrary point. For example, in Fig. 4–26, position vectors $\mathbf{r}_A$ and $\mathbf{r}_B$ are directed from point O to points A and B lying on the line of action of $-\mathbf{F}$ and $\mathbf{F}$. The couple moment determined about O is therefore

$$\mathbf{M} = \mathbf{r}_B \times \mathbf{F} + \mathbf{r}_A \times -\mathbf{F} = (\mathbf{r}_B - \mathbf{r}_A) \times \mathbf{F}$$

However $\mathbf{r}_B = \mathbf{r}_A + \mathbf{r}$ or $\mathbf{r} = \mathbf{r}_B - \mathbf{r}_A$, so that

$$\mathbf{M} = \mathbf{r} \times \mathbf{F} \qquad (4\text{–}13)$$

This result indicates that a couple moment is a *free vector*, i.e., it can act at *any point* since $\mathbf{M}$ depends *only* upon the position vector $\mathbf{r}$ directed *between* the forces and *not* the position vectors $\mathbf{r}_A$ and $\mathbf{r}_B$, directed from the arbitrary point O to the forces. This concept is unlike the moment of a force, which requires a definite point (or axis) about which moments are determined.

Scalar Formulation. The moment of a couple, $\mathbf{M}$, Fig. 4–27, is defined as having a *magnitude* of

$$M = Fd \qquad (4\text{–}14)$$

where F is the magnitude of one of the forces and d is the perpendicular distance or moment arm between the forces. The *direction* and sense of the couple moment are determined by the right-hand rule, where the thumb indicates this direction when the fingers are curled with the sense of rotation caused by the couple forces. In all cases, $\mathbf{M}$ will act perpendicular to the plane containing these forces.

Vector Formulation. The moment of a couple can also be expressed by the vector cross product using Eq. 4–13, i.e.,

$$\mathbf{M} = \mathbf{r} \times \mathbf{F} \qquad (4\text{–}15)$$

Application of this equation is easily remembered if one thinks of taking the moments of both forces about a point lying on the line of action of one of the forces. For example, if moments are taken about point A in Fig. 4–26, the moment of $-\mathbf{F}$ is *zero* about this point, and the moment of $\mathbf{F}$ is defined from Eq. 4–15. Therefore, in the formulation $\mathbf{r}$ is crossed with the force $\mathbf{F}$ to which it is directed.

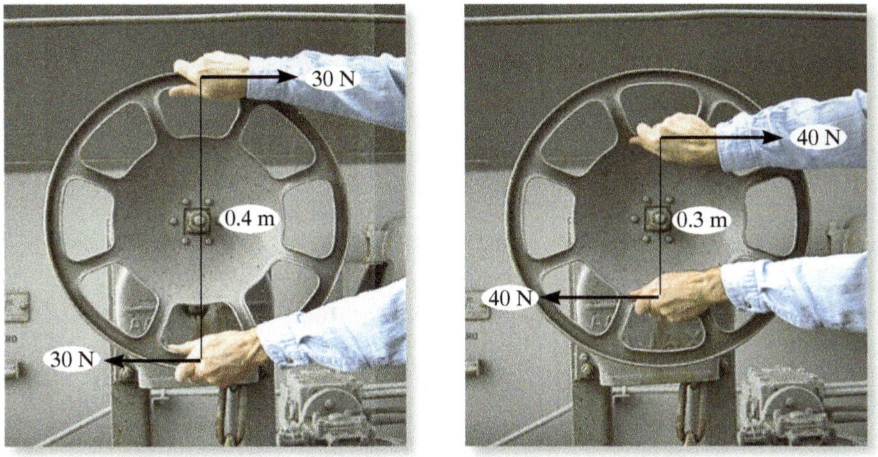

Fig. 4–28

Equivalent Couples. If two couples produce a moment with the *same magnitude and direction*, then these two couples are *equivalent*. For example, the two couples shown in Fig. 4–28 are *equivalent* because each couple moment has a magnitude of $M = 30\text{ N}(0.4\text{ m}) = 40\text{ N}(0.3\text{ m}) = 12\text{ N}\cdot\text{m}$, and each is directed into the plane of the page. Notice that larger forces are required in the second case to create the same turning effect because the hands are placed closer together. Also, if the wheel was connected to the shaft at a point other than at its center, then the wheel would still turn when each couple is applied since the $12\text{ N}\cdot\text{m}$ couple is a free vector.

Resultant Couple Moment. Since couple moments are vectors, their resultant can be determined by vector addition. For example, consider the couple moments $\mathbf{M}_1$ and $\mathbf{M}_2$ acting on the pipe in Fig. 4–29*a*. Since each couple moment is a free vector, we can join their tails at any arbitrary point and find the resultant couple moment, $\mathbf{M}_R = \mathbf{M}_1 + \mathbf{M}_2$ as shown in Fig. 4–29*b*.

If more than two couple moments act on the body, we may generalize this concept and write the vector resultant as

$$\mathbf{M}_R = \Sigma(\mathbf{r} \times \mathbf{F}) \qquad (4\text{–}16)$$

These concepts are illustrated numerically in the examples that follow. In general, problems projected in two dimensions should be solved using a scalar analysis since the moment arms and force components are easy to determine.

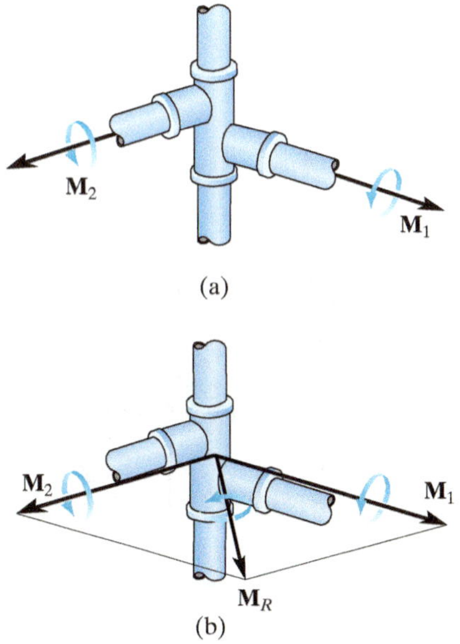

Fig. 4–29

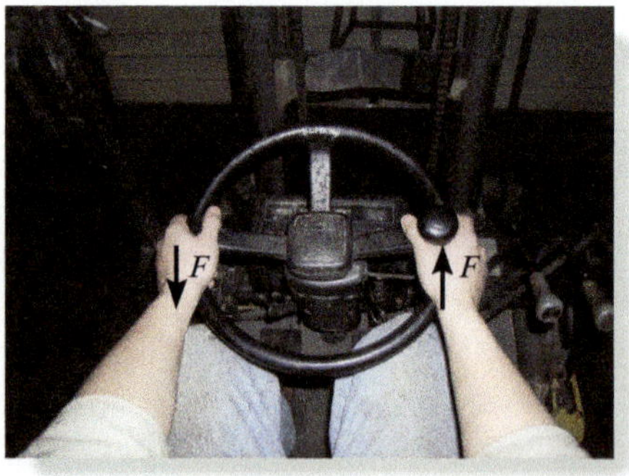

Steering wheels on vehicles have been made smaller than on older vehicles because power steering does not require the driver to apply a large couple moment to the rim of the wheel.

Important Points

- A couple moment is produced by two noncollinear forces that are equal in magnitude but opposite in direction. Its effect is to produce pure rotation, or tendency for rotation in a specified direction.
- A couple moment is a free vector, and as a result it causes the same rotational effect on a body regardless of where the couple moment is applied to the body.
- The moment of the two couple forces can be determined about *any point*. For convenience, this point is often chosen on the line of action of one of the forces in order to eliminate the moment of this force about the point.
- In three dimensions the couple moment is often determined using the vector formulation, $\mathbf{M} = \mathbf{r} \times \mathbf{F}$, where $\mathbf{r}$ is directed from *any point* on the line of action of one of the forces to *any point* on the line of action of the other force $\mathbf{F}$.
- A resultant couple moment is simply the vector sum of all the couple moments of the system.

EXAMPLE 4.10

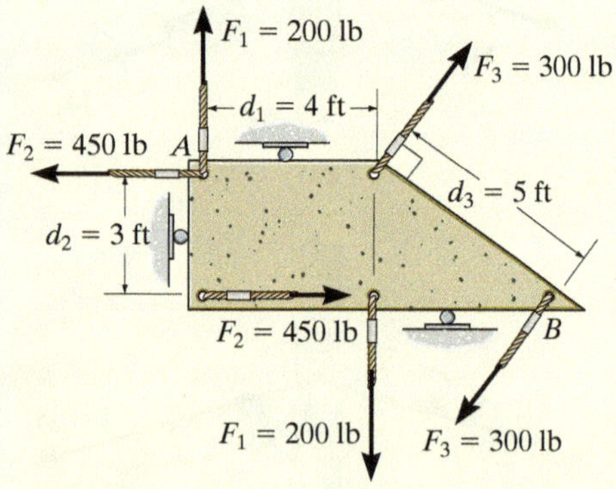

Fig. 4–30

Determine the resultant couple moment of the three couples acting on the plate in Fig. 4–30.

SOLUTION

As shown the perpendicular distances between each pair of couple forces are $d_1 = 4$ ft, $d_2 = 3$ ft, and $d_3 = 5$ ft. Considering counterclockwise couple moments as positive, we have

$$\circlearrowleft + M_R = \Sigma M; \quad M_R = -F_1 d_1 + F_2 d_2 - F_3 d_3$$

$$= -(200 \text{ lb})(4 \text{ ft}) + (450 \text{ lb})(3 \text{ ft}) - (300 \text{ lb})(5 \text{ ft})$$

$$= -950 \text{ lb}\cdot\text{ft} = 950 \text{ lb}\cdot\text{ft} \circlearrowright \qquad \textit{Ans.}$$

The negative sign indicates that $\mathbf{M}_R$ has a clockwise rotational sense.

EXAMPLE 4.11

Determine the magnitude and direction of the couple moment acting on the gear in Fig. 4–31*a*.

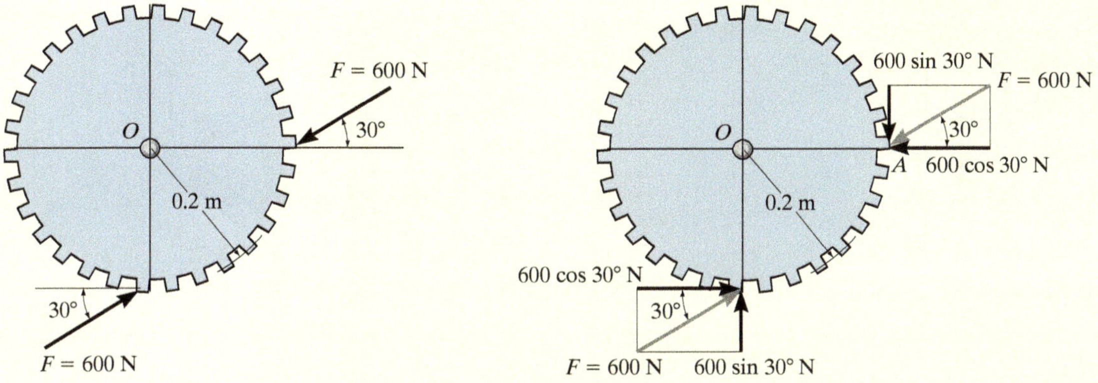

(a) (b)

SOLUTION

The easiest solution requires resolving each force into its components as shown in Fig. 4–31*b*. The couple moment can be determined by summing the moments of these force components about any point, for example, the center *O* of the gear or point *A*. If we consider counterclockwise moments as positive, we have

$$\circlearrowleft + M = \Sigma M_O;\quad M = (600 \cos 30^\circ \text{ N})(0.2 \text{ m}) - (600 \sin 30^\circ \text{ N})(0.2 \text{ m})$$

$$= 43.9 \text{ N} \cdot \text{m} \circlearrowleft \qquad \textit{Ans.}$$

or

$$\circlearrowleft + M = \Sigma M_A;\quad M = (600 \cos 30^\circ \text{ N})(0.2 \text{ m}) - (600 \sin 30^\circ \text{ N})(0.2 \text{ m})$$

$$= 43.9 \text{ N} \cdot \text{m} \circlearrowleft \qquad \textit{Ans.}$$

This positive result indicates that **M** has a counterclockwise rotational sense, so it is directed outward, perpendicular to the page.

NOTE: The same result can also be obtained using $M = Fd$, where d is the perpendicular distance between the lines of action of the couple forces, Fig. 4–31*c*. However, the computation for d is more involved. Realize that the couple moment is a free vector and can act at any point on the gear and produce the same turning effect about point *O*.

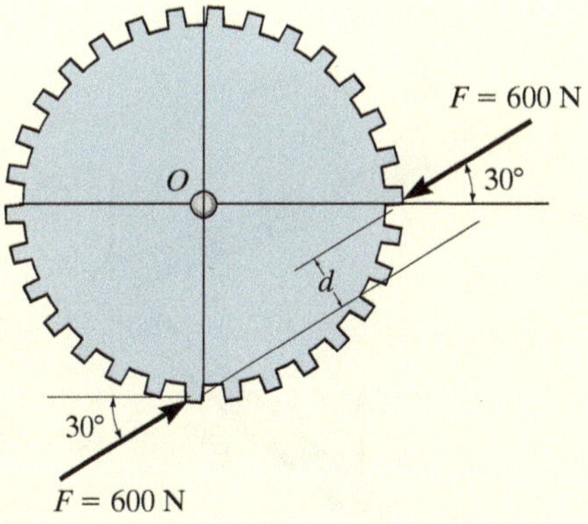

(c)

Fig. 4–31

EXAMPLE 4.12

Determine the couple moment acting on the pipe shown in Fig. 4–32*a*. Segment *AB* is directed 30° below the *x*–*y* plane.

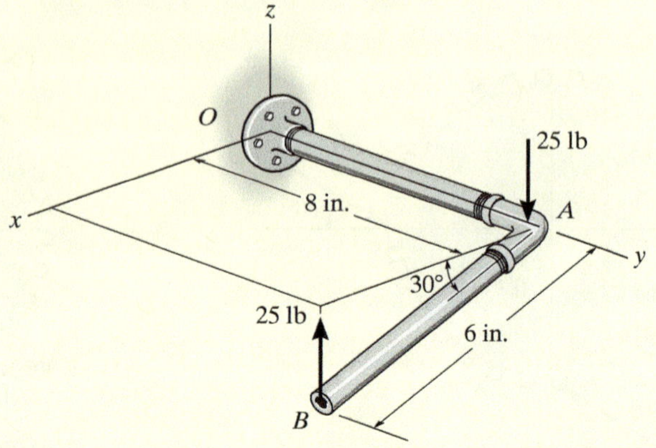

(a)

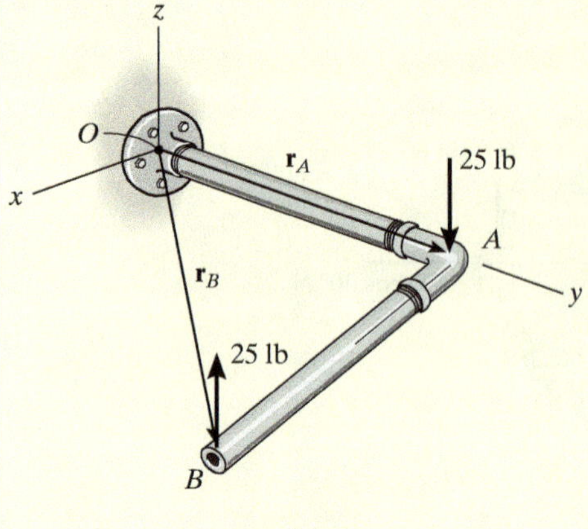

(b)

SOLUTION I *(VECTOR ANALYSIS)*

The moment of the two couple forces can be found about *any point.* If point *O* is considered, Fig. 4–32*b*, we have

$$\mathbf{M} = \mathbf{r}_A \times (-25\mathbf{k}) + \mathbf{r}_B \times (25\mathbf{k})$$
$$= (8\mathbf{j}) \times (-25\mathbf{k}) + (6 \cos 30°\mathbf{i} + 8\mathbf{j} - 6 \sin 30°\mathbf{k}) \times (25\mathbf{k})$$
$$= -200\mathbf{i} - 129.9\mathbf{j} + 200\mathbf{i}$$
$$= \{-130\mathbf{j}\} \text{ lb} \cdot \text{in.}$$

Ans.

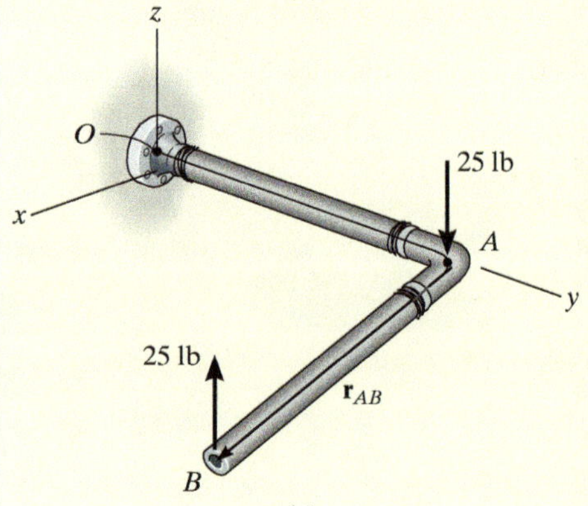

(c)

It is *easier* to take moments of the couple forces about a point lying on the line of action of one of the forces, e.g., point *A*, Fig. 4–32*c*. In this case the moment of the force at *A* is zero, so that

$$\mathbf{M} = \mathbf{r}_{AB} \times (25\mathbf{k})$$
$$= (6 \cos 30°\mathbf{i} - 6 \sin 30°\mathbf{k}) \times (25\mathbf{k})$$
$$= \{-130\mathbf{j}\} \text{ lb} \cdot \text{in.}$$

Ans.

SOLUTION II *(SCALAR ANALYSIS)*

Although this problem is shown in three dimensions, the geometry is simple enough to use the scalar equation $M = Fd$. The perpendicular distance between the lines of action of the couple forces is $d = 6 \cos 30° = 5.196$ in., Fig. 4–32*d*. Hence, taking moments of the forces about either point *A* or point *B* yields

$$M = Fd = 25 \text{ lb } (5.196 \text{ in.}) = 129.9 \text{ lb} \cdot \text{in.}$$

Applying the right-hand rule, **M** acts in the $-\mathbf{j}$ direction. Thus,

$$\mathbf{M} = \{-130\mathbf{j}\} \text{ lb} \cdot \text{in.}$$

Ans.

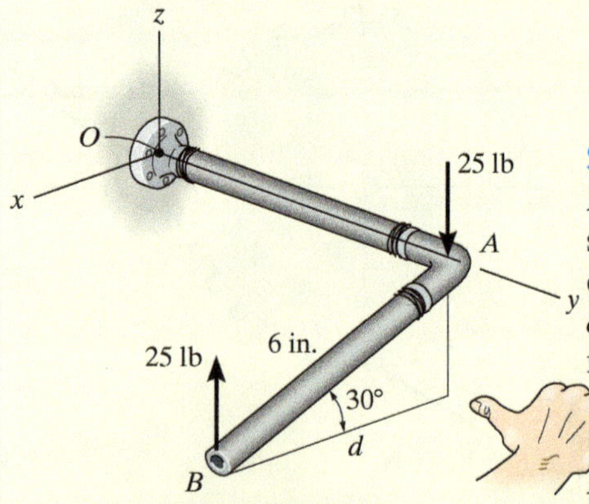

(d)

Fig. 4–32

EXAMPLE 4.13

Replace the two couples acting on the pipe column in Fig. 4–33*a* by a resultant couple moment.

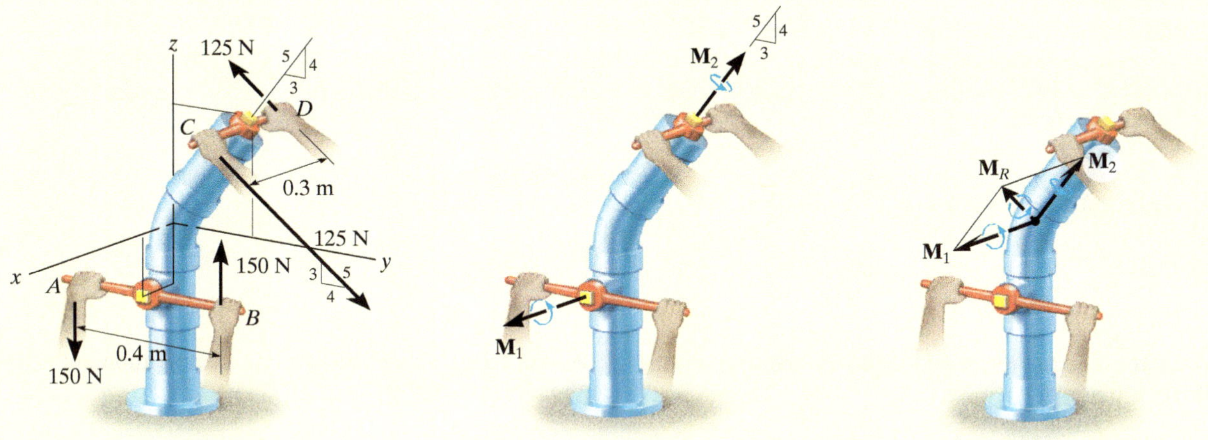

(a) (b) (c)

Fig. 4–33

SOLUTION *(VECTOR ANALYSIS)*

The couple moment $\mathbf{M}_1$, developed by the forces at A and B, can easily be determined from a scalar formulation.

$$M_1 = Fd = 150\text{ N}(0.4\text{ m}) = 60\text{ N}\cdot\text{m}$$

By the right-hand rule, $\mathbf{M}_1$ acts in the $+\mathbf{i}$ direction, Fig. 4–33*b*. Hence,

$$\mathbf{M}_1 = \{60\mathbf{i}\}\text{ N}\cdot\text{m}$$

Vector analysis will be used to determine $\mathbf{M}_2$, caused by forces at C and D. If moments are calculated about point D, Fig. 4–33*a*, $\mathbf{M}_2 = \mathbf{r}_{DC} \times \mathbf{F}_C$, then

$$\begin{aligned}\mathbf{M}_2 = \mathbf{r}_{DC} \times \mathbf{F}_C &= (0.3\mathbf{i}) \times \left[125\left(\tfrac{4}{5}\right)\mathbf{j} - 125\left(\tfrac{3}{5}\right)\mathbf{k}\right] \\ &= (0.3\mathbf{i}) \times [100\mathbf{j} - 75\mathbf{k}] = 30(\mathbf{i} \times \mathbf{j}) - 22.5(\mathbf{i} \times \mathbf{k}) \\ &= \{22.5\mathbf{j} + 30\mathbf{k}\}\text{ N}\cdot\text{m}\end{aligned}$$

Since $\mathbf{M}_1$ and $\mathbf{M}_2$ are free vectors, they may be moved to some arbitrary point and added vectorially, Fig. 4–33*c*. The resultant couple moment becomes

$$\mathbf{M}_R = \mathbf{M}_1 + \mathbf{M}_2 = \{60\mathbf{i} + 22.5\mathbf{j} + 30\mathbf{k}\}\text{ N}\cdot\text{m} \qquad \textit{Ans.}$$

4

FUNDAMENTAL PROBLEMS

F4–19. Determine the resultant couple moment acting on the beam.

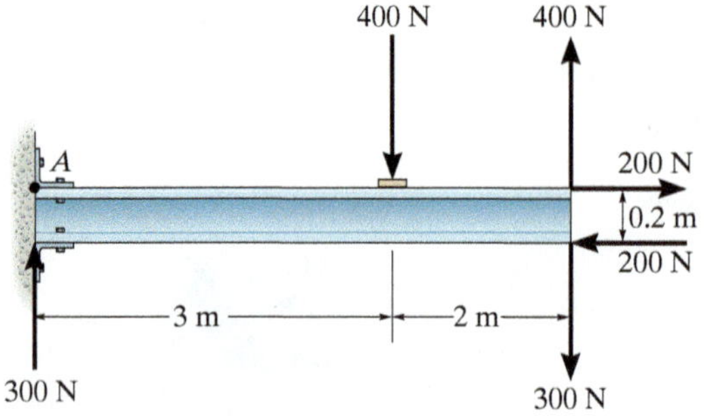

F4–19

F4–20. Determine the resultant couple moment acting on the triangular plate.

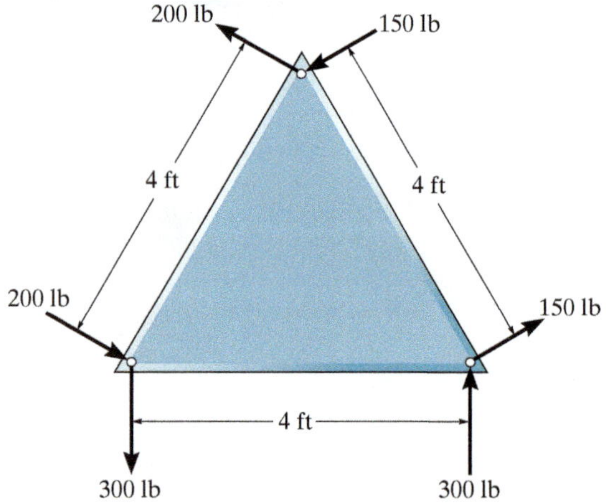

F4–20

F4–21. Determine the magnitude of **F** so that the resultant couple moment acting on the beam is 1.5 kN · m clockwise.

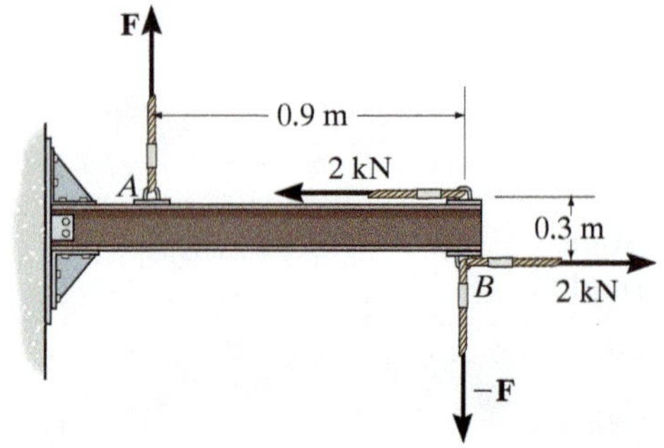

F4–21

F4–22. Determine the couple moment acting on the beam.

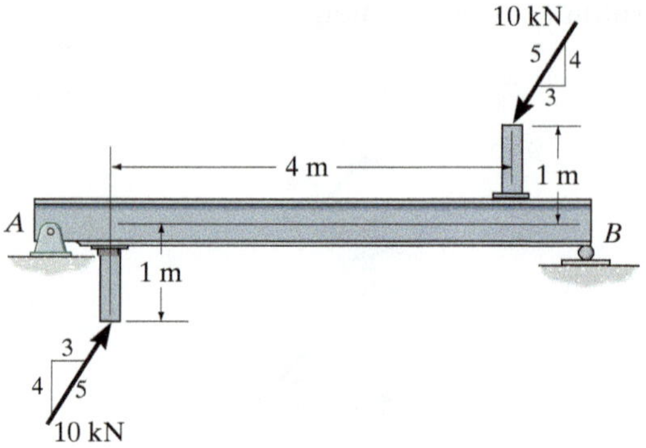

F4–22

F4–23. Determine the resultant couple moment acting on the pipe assembly.

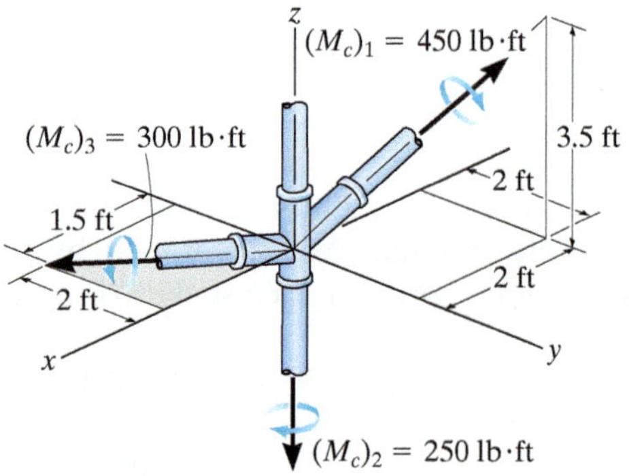

F4–23

F4–24. Determine the couple moment acting on the pipe assembly and express the result as a Cartesian vector.

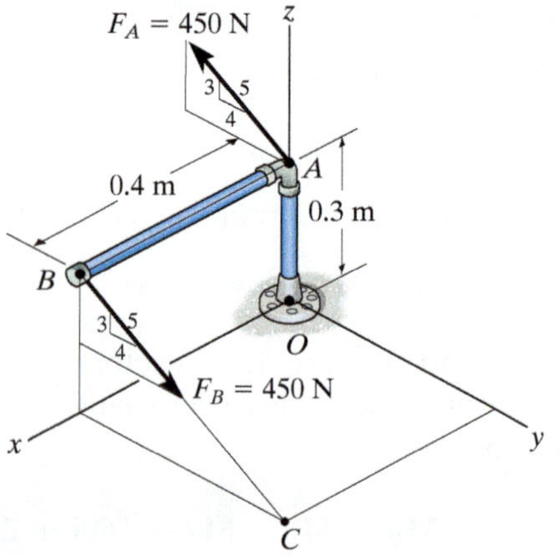

F4–24

PROBLEMS

4–67. A twist of 4 N · m is applied to the handle of the screwdriver. Resolve this couple moment into a pair of couple forces **F** exerted on the handle and **P** exerted on the blade.

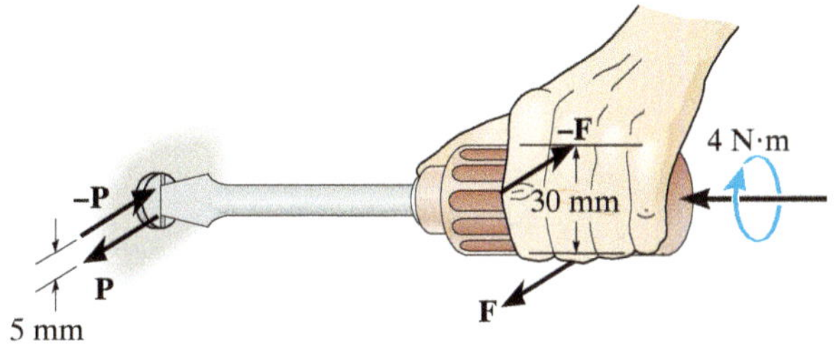

Prob. 4–67

***4–68.** The ends of the triangular plate are subjected to three couples. Determine the plate dimension d so that the resultant couple is 350 N · m clockwise.

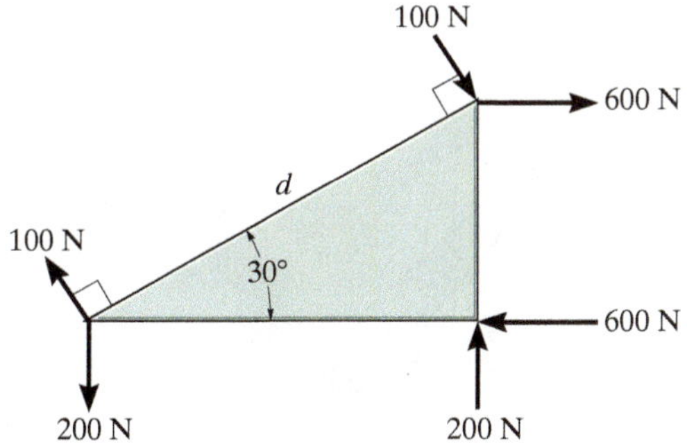

Prob. 4–68

4–69. The caster wheel is subjected to the two couples. Determine the forces **F** that the bearings create on the shaft so that the resultant couple moment on the caster is zero.

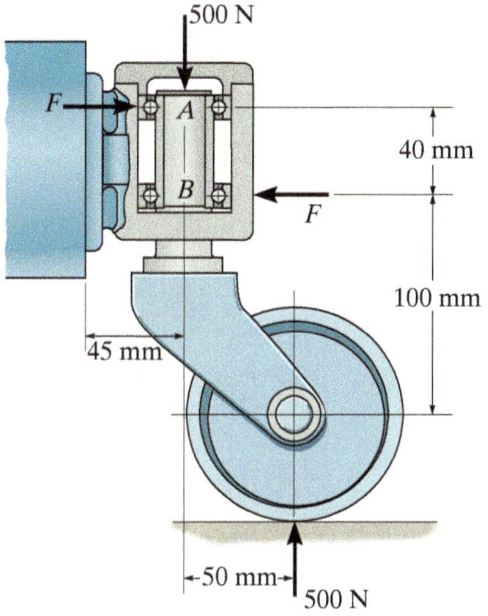

Prob. 4–69

4–70. Two couples act on the beam. If $F = 125$ lb, determine the resultant couple moment.

4–71. Two couples act on the beam. Determine the magnitude of **F** so that the resultant couple moment is 450 lb · ft, counterclockwise. Where on the beam does the resultant couple moment act?

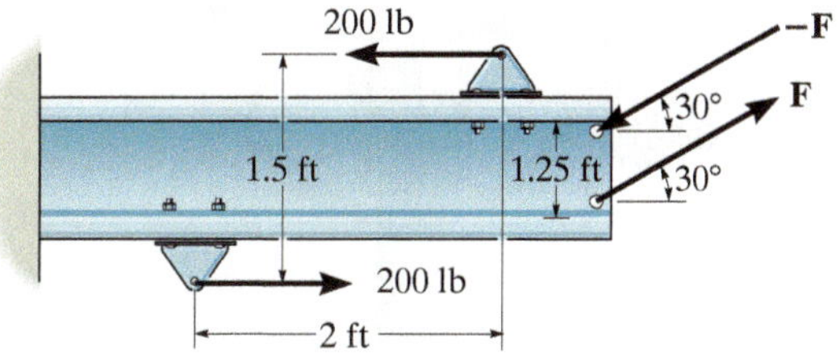

Probs. 4–70/71

*4–72. Friction on the concrete surface creates a couple moment of $M_O = 100\ \text{N}\cdot\text{m}$ on the blades of the trowel. Determine the magnitude of the couple forces so that the resultant couple moment on the trowel is zero. The forces lie in the horizontal plane and act perpendicular to the handle of the trowel.

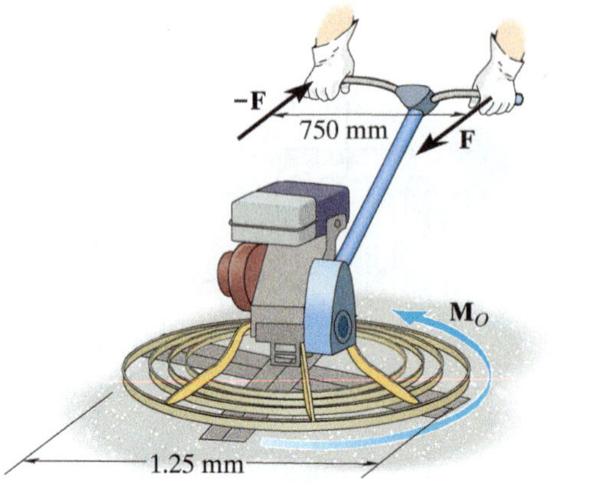

Prob. 4–72

4–73. The man tries to open the valve by applying the couple forces of $F = 75$ N to the wheel. Determine the couple moment produced.

4–74. If the valve can be opened with a couple moment of $25\ \text{N}\cdot\text{m}$, determine the required magnitude of each couple force which must be applied to the wheel.

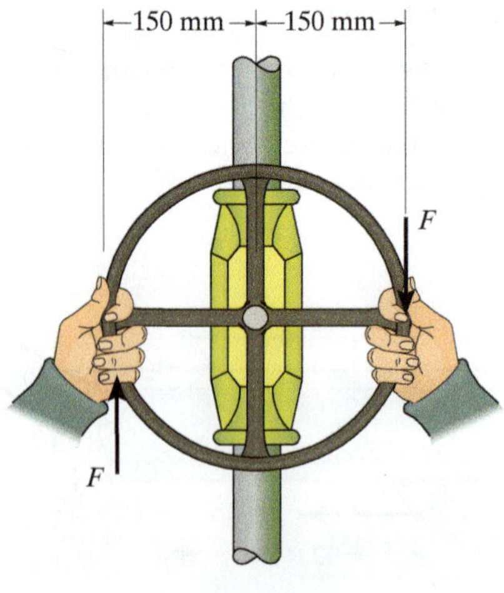

Probs. 4–73/74

4–75. When the engine of the plane is running, the vertical reaction that the ground exerts on the wheel at A is measured as 650 lb. When the engine is turned off, however, the vertical reactions at A and B are 575 lb each. The difference in readings at A is caused by a couple acting on the propeller when the engine is running. This couple tends to overturn the plane counterclockwise, which is opposite to the propeller's clockwise rotation. Determine the magnitude of this couple and the magnitude of the reaction force exerted at B when the engine is running.

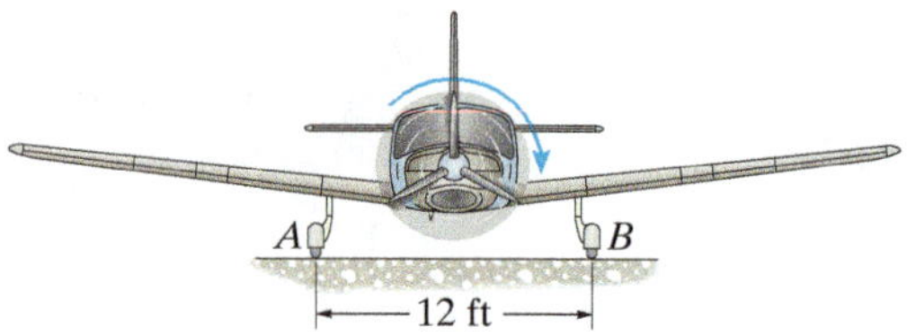

Prob. 4–75

*4–76. Determine the magnitude of the couple force **F** so that the resultant couple moment on the crank is zero.

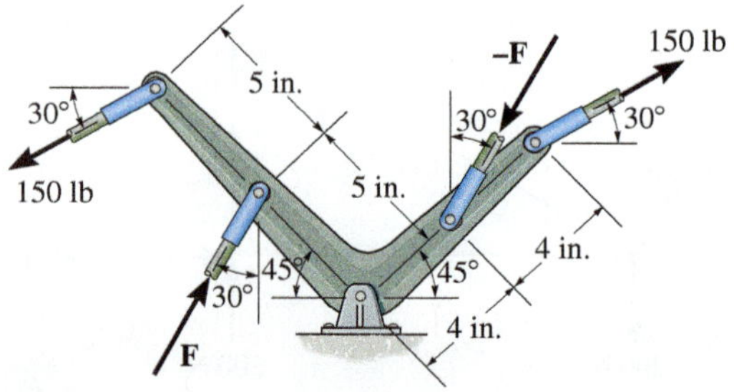

Prob. 4–76

4–77. Two couples act on the beam as shown. If $F = 150$ lb, determine the resultant couple moment.

4–78. Two couples act on the beam as shown. Determine the magnitude of **F** so that the resultant couple moment is 300 lb · ft counterclockwise. Where on the beam does the resultant couple act?

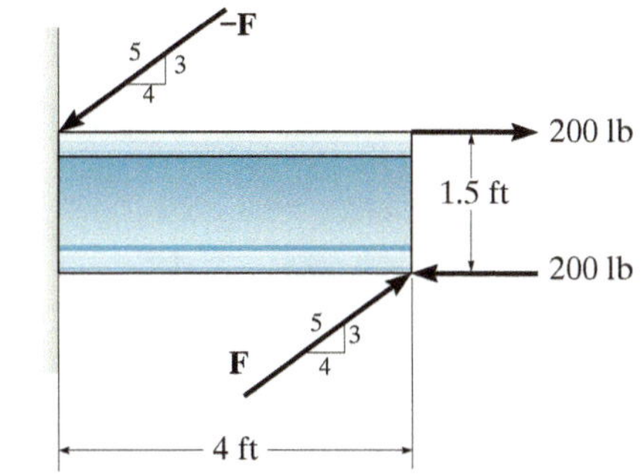

Probs. 4–77/78

4–79. If $F = 200$ lb, determine the resultant couple moment.

***4–80.** Determine the required magnitude of force **F** if the resultant couple moment on the frame is 200 lb · ft, clockwise.

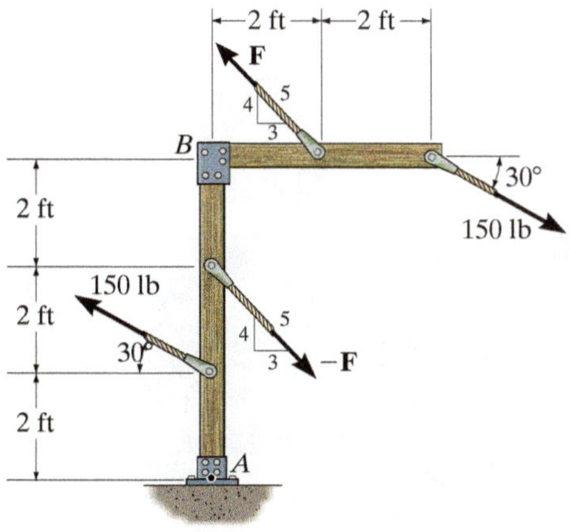

Probs. 4–79/80

4–81. Two couples act on the cantilever beam. If $F = 6$ kN, determine the resultant couple moment.

4–82. Determine the required magnitude of force **F**, if the resultant couple moment on the beam is to be zero.

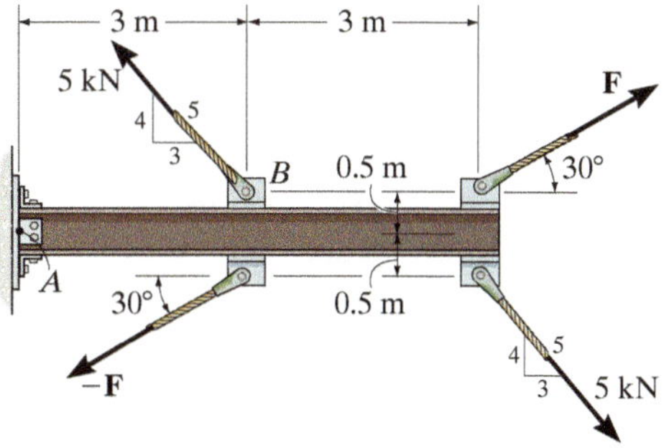

Probs. 4–81/82

4–83. Express the moment of the couple acting on the pipe assembly in Cartesian vector form. Solve the problem (a) using Eq. 4–13, and (b) summing the moment of each force about point O. Take $\mathbf{F} = \{25\mathbf{k}\}$ N.

***4–84.** If the couple moment acting on the pipe has a magnitude of 400 N · m, determine the magnitude F of the vertical force applied to each wrench.

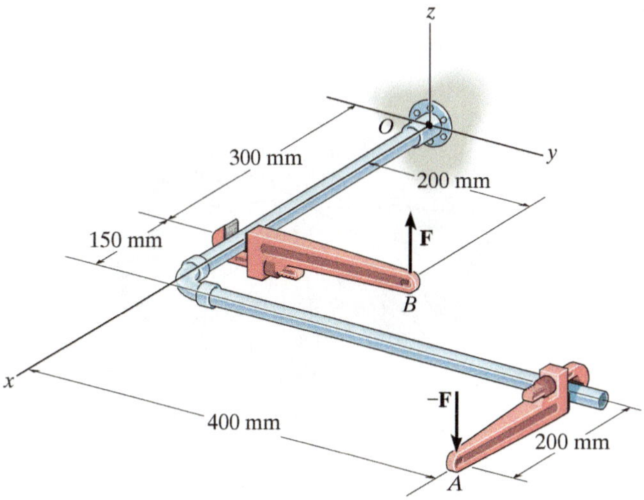

Probs. 4–83/84

4–85. The gear reducer is subjected to the couple moments shown. Determine the resultant couple moment and specify its magnitude and coordinate direction angles.

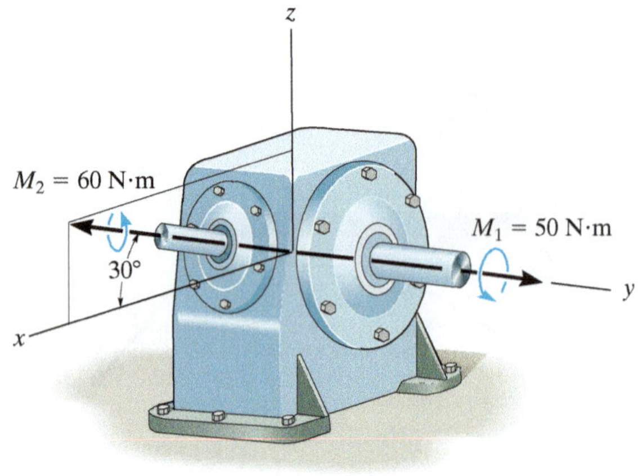

Prob. 4–85

4–86. The meshed gears are subjected to the couple moments shown. Determine the magnitude of the resultant couple moment and specify its coordinate direction angles.

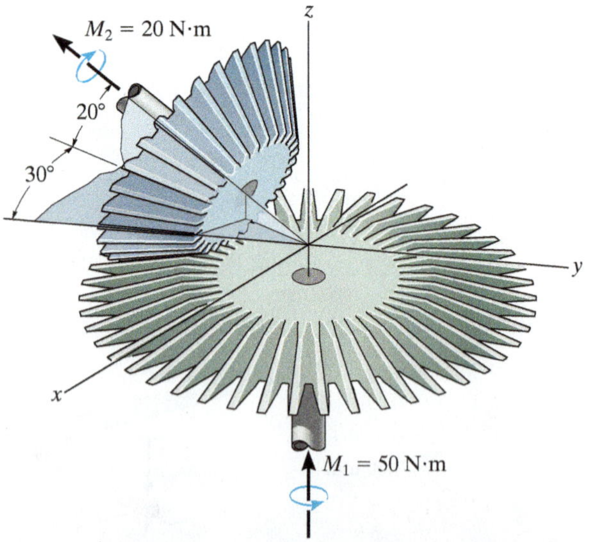

Prob. 4–86

4–87. The gear reducer is subject to the couple moments shown. Determine the resultant couple moment and specify its magnitude and coordinate direction angles.

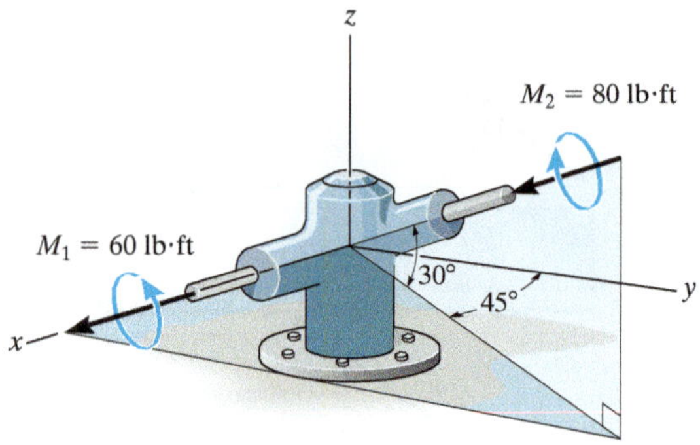

Prob. 4–87

***4–88.** A couple acts on each of the handles of the minidual valve. Determine the magnitude and coordinate direction angles of the resultant couple moment.

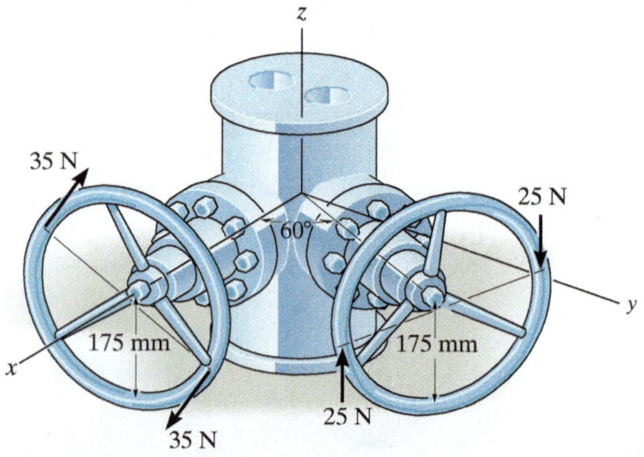

Prob. 4–88

4–89. Determine the resultant couple moment of the two couples that act on the pipe assembly. The distance from A to B is $d = 400$ mm. Express the result as a Cartesian vector.

4–90. Determine the distance d between A and B so that the resultant couple moment has a magnitude of $M_R = 20$ N·m.

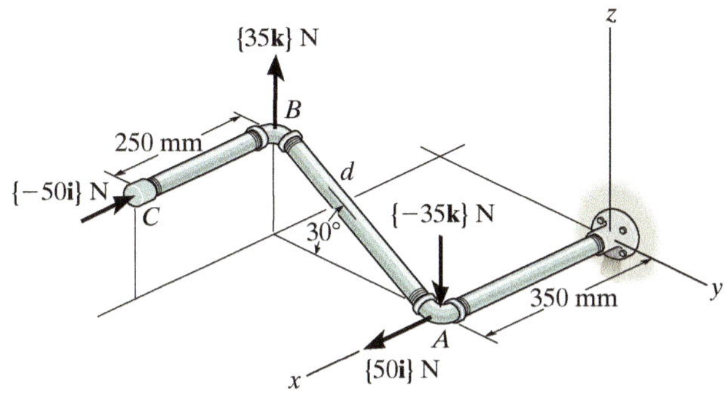

Probs. 4–89/90

4–91. If $F = 80$ N, determine the magnitude and coordinate direction angles of the couple moment. The pipe assembly lies in the x–y plane.

***4–92.** If the magnitude of the couple moment acting on the pipe assembly is 50 N·m, determine the magnitude of the couple forces applied to each wrench. The pipe assembly lies in the x–y plane.

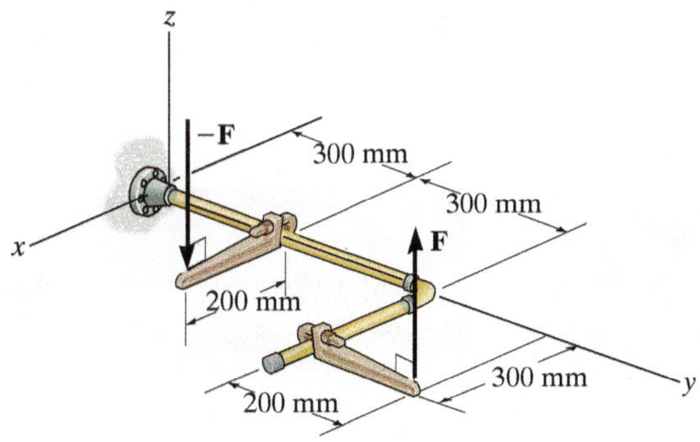

Probs. 4–91/92

4–93. If $\mathbf{F} = \{100\mathbf{k}\}$ N, determine the couple moment that acts on the assembly. Express the result as a Cartesian vector. Member BA lies in the x–y plane.

4–94. If the magnitude of the resultant couple moment is 15 N·m, determine the magnitude F of the forces applied to the wrenches.

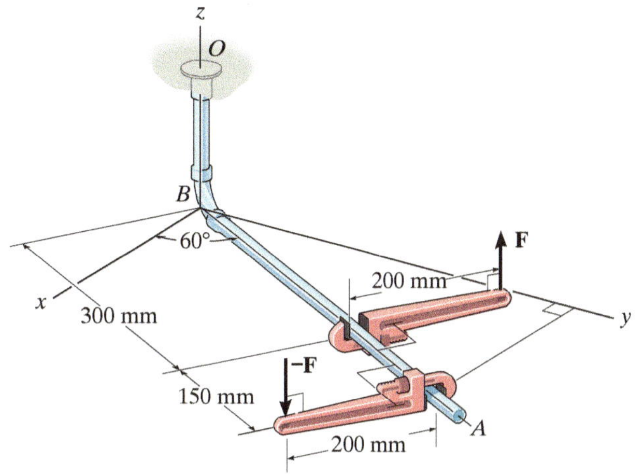

Probs. 4–93/94

4–95. If $F_1 = 100$ N, $F_2 = 120$ N and $F_3 = 80$ N, determine the magnitude and coordinate direction angles of the resultant couple moment.

***4–96.** Determine the required magnitude of $\mathbf{F}_1$, $\mathbf{F}_2$ and $\mathbf{F}_3$ so that the resultant couple moment is $(\mathbf{M}_c)_R = [50\,\mathbf{i} - 45\,\mathbf{j} - 20\,\mathbf{k}]$ N·m.

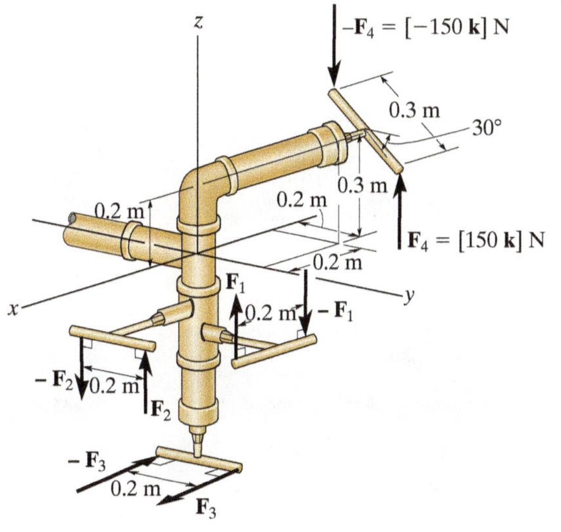

Probs. 4–95/96

4.7 Simplification of a Force and Couple System

Sometimes it is convenient to reduce a system of forces and couple moments acting on a body to a simpler form by replacing it with an *equivalent system*, consisting of a single resultant force acting at a specific point and a resultant couple moment. A system is equivalent if the *external effects* it produces on a body are the same as those caused by the original force and couple moment system. In this context, the external effects of a system refer to the *translating and rotating motion* of the body if the body is free to move, or it refers to the *reactive forces* at the supports if the body is held fixed.

For example, consider holding the stick in Fig. 4–34*a*, which is subjected to the force **F** at point A. If we attach a pair of equal but opposite forces **F** and −**F** at point B, which is *on the line of action* of **F**, Fig. 4–34*b*, we observe that −**F** at B and **F** at A will cancel each other, leaving only **F** at B, Fig. 4–34*c*. Force **F** has now been moved from A to B without modifying its *external effects* on the stick; i.e., the reaction at the grip remains the same. This demonstrates the *principle of transmissibility*, which states that a force acting on a body (stick) is a *sliding vector* since it can be applied at any point along its line of action.

We can also use the above procedure to move a force to a point that is *not* on the line of action of the force. If **F** is applied perpendicular to the stick, as in Fig. 4–35*a*, then we can attach a pair of equal but opposite forces **F** and −**F** to B, Fig. 4–35*b*. Force **F** is now applied at B, and the other two forces, **F** at A and −**F** at B, form a couple that produces the couple moment $M = Fd$, Fig. 4–35*c*. Therefore, the force **F** can be moved from A to B provided a couple moment **M** is added to maintain an equivalent system. This couple moment is determined by taking the moment of **F** about B. Since **M** is actually a *free vector*, it can act at any point on the stick. In both cases the systems are equivalent, which causes a downward force **F** and clockwise couple moment $M = Fd$ to be felt at the grip.

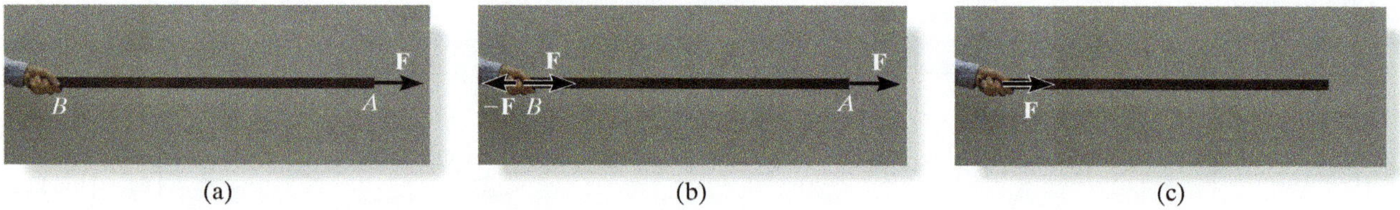

Fig. 4–34

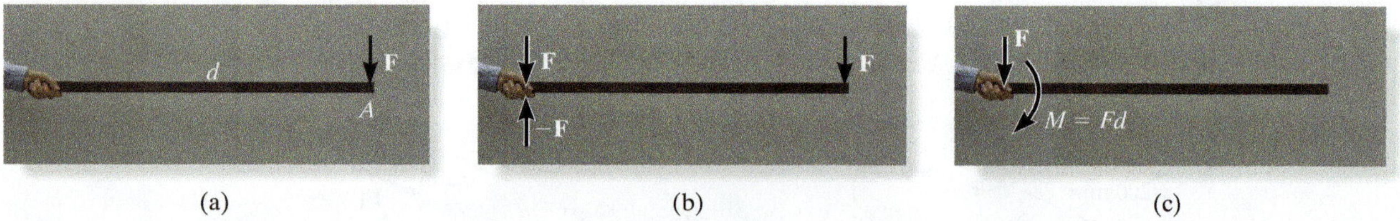

Fig. 4–35

System of Forces and Couple Moments. Using the above method, a system of several forces and couple moments acting on a body can be reduced to an equivalent single resultant force acting at a point O and a resultant couple moment. For example, in Fig. 4–36*a*, O is not on the line of action of $\mathbf{F}_1$, and so this force can be moved to point O provided a couple moment $(\mathbf{M}_O)_1 = \mathbf{r}_1 \times \mathbf{F}$ is added to the body. Similarly, the couple moment $(\mathbf{M}_O)_2 = \mathbf{r}_2 \times \mathbf{F}_2$ should be added to the body when we move $\mathbf{F}_2$ to point O. Finally, since the couple moment $\mathbf{M}$ is a free vector, it can just be moved to point O. By doing this, we obtain the equivalent system shown in Fig. 4–36*b*, which produces the same external effects (support reactions) on the body as that of the force and couple system shown in Fig. 4–36*a*. If we sum the forces and couple moments, we obtain the resultant force $\mathbf{F}_R = \mathbf{F}_1 + \mathbf{F}_2$ and the resultant couple moment $(\mathbf{M}_R)_O = \mathbf{M} + (\mathbf{M}_O)_1 + (\mathbf{M}_O)_2$, Fig. 4–36*c*.

Notice that $\mathbf{F}_R$ is independent of the location of point O since it is simply a summation of the forces. However, $(\mathbf{M}_R)_O$ depends upon this location since the moments $\mathbf{M}_1$ and $\mathbf{M}_2$ are determined using the position vectors $\mathbf{r}_1$ and $\mathbf{r}_2$, which extend from O to each force. Also note that $(\mathbf{M}_R)_O$ is a free vector and can act at *any point* on the body, although point O is generally chosen as its point of application.

We can generalize the above method of reducing a force and couple system to an equivalent resultant force $\mathbf{F}_R$ acting at point O and a resultant couple moment $(\mathbf{M}_R)_O$ by using the following two equations.

$$\begin{aligned} \mathbf{F}_R &= \Sigma \mathbf{F} \\ (\mathbf{M}_R)_O &= \Sigma \mathbf{M}_O + \Sigma \mathbf{M} \end{aligned} \tag{4–17}$$

The first equation states that the resultant force of the system is equivalent to the sum of all the forces; and the second equation states that the resultant couple moment of the system is equivalent to the sum of all the couple moments $\Sigma\mathbf{M}$ plus the moments of all the forces $\Sigma\mathbf{M}_O$ about point O. If the force system lies in the x–y plane and any couple moments are perpendicular to this plane, then the above equations reduce to the following three scalar equations.

$$\begin{aligned} (F_R)_x &= \Sigma F_x \\ (F_R)_y &= \Sigma F_y \\ (M_R)_O &= \Sigma M_O + \Sigma M \end{aligned} \tag{4–18}$$

Here the resultant force is determined from the vector sum of its two components $(F_R)_x$ and $(F_R)_y$.

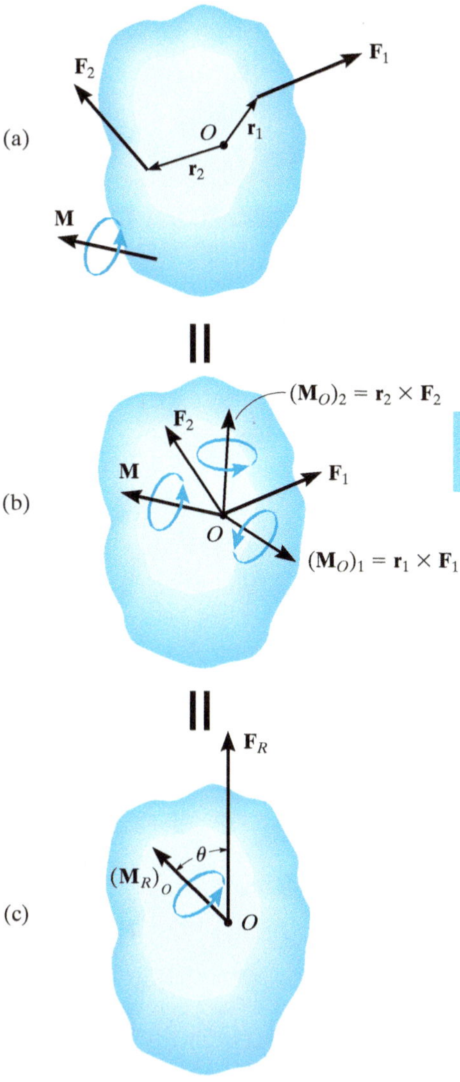

Fig. 4–36

The weights of these traffic lights can be replaced by their equivalent resultant force $W_R = W_1 + W_2$ and a couple moment $(M_R)_O = W_1 d_1 + W_2 d_2$ at the support, O. In both cases the support must provide the same resistance to translation and rotation in order to keep the member in the horizontal position.

4

Procedure for Analysis

The following points should be kept in mind when simplifying a force and couple moment system to an equivalent resultant force and couple system.

- Establish the coordinate axes with the origin located at point O and the axes having a selected orientation.

Force Summation.

- If the force system is *coplanar*, resolve each force into its x and y components. If a component is directed along the positive x or y axis, it represents a positive scalar; whereas if it is directed along the negative x or y axis, it is a negative scalar.
- In three dimensions, represent each force as a Cartesian vector before summing the forces.

Moment Summation.

- When determining the moments of a *coplanar* force system about point O, it is generally advantageous to use the principle of moments, i.e., determine the moments of the components of each force, rather than the moment of the force itself.
- In three dimensions use the vector cross product to determine the moment of each force about point O. Here the position vectors extend from O to any point on the line of action of each force.

EXAMPLE 4.14

Replace the force and couple system shown in Fig. 4–37*a* by an equivalent resultant force and couple moment acting at point *O*.

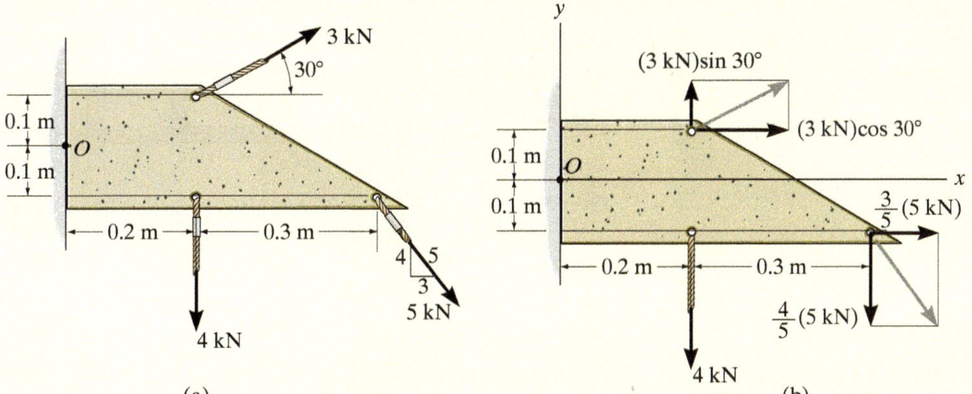

(a) (b)

SOLUTION

Force Summation. The 3 kN and 5 kN forces are resolved into their *x* and *y* components as shown in Fig. 4–37*b*. We have

$$\overset{+}{\rightarrow} (F_R)_x = \Sigma F_x; \qquad (F_R)_x = (3\text{ kN})\cos 30^\circ + \left(\tfrac{3}{5}\right)(5\text{ kN}) = 5.598\text{ kN} \rightarrow$$

$$+\uparrow (F_R)_y = \Sigma F_y; \qquad (F_R)_y = (3\text{ kN})\sin 30^\circ - \left(\tfrac{4}{5}\right)(5\text{ kN}) - 4\text{ kN} = -6.50\text{ kN} = 6.50\text{ kN}\downarrow$$

Using the Pythagorean theorem, Fig. 4–37*c*, the magnitude of $\mathbf{F}_R$ is

$$F_R = \sqrt{(F_R)_x^2 + (F_R)_y^2} = \sqrt{(5.598\text{ kN})^2 + (6.50\text{ kN})^2} = 8.58\text{ kN} \qquad \textit{Ans.}$$

Its direction θ is

$$\theta = \tan^{-1}\left(\frac{(F_R)_y}{(F_R)_x}\right) = \tan^{-1}\left(\frac{6.50\text{ kN}}{5.598\text{ kN}}\right) = 49.3^\circ \qquad \textit{Ans.}$$

Moment Summation. The moments of 3 kN and 5 kN about point *O* will be determined using their *x* and *y* components. Referring to Fig. 4–37*b*, we have

$$\circlearrowleft + (M_R)_O = \Sigma M_O;$$

$$(M_R)_O = (3\text{ kN})\sin 30^\circ(0.2\text{ m}) - (3\text{ kN})\cos 30^\circ(0.1\text{ m}) + \left(\tfrac{3}{5}\right)(5\text{ kN})\,(0.1\text{ m}) - \left(\tfrac{4}{5}\right)(5\text{ kN})\,(0.5\text{ m}) - (4\text{ kN})(0.2\text{ m})$$

$$= -2.46\text{ kN}\cdot\text{m} = 2.46\text{ kN}\cdot\text{m} \circlearrowright \qquad \textit{Ans.}$$

This clockwise moment is shown in Fig. 4–37*c*.

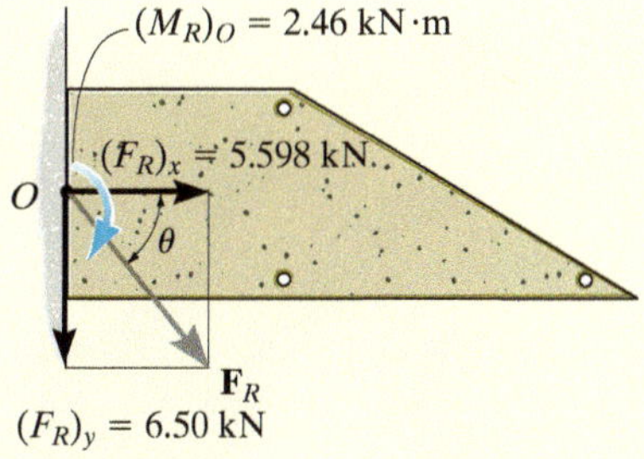

(c)

Fig. 4–37

NOTE: Realize that the resultant force and couple moment in Fig. 4–37*c* will produce the same external effects or reactions at the supports as those produced by the force system, Fig. 4–37*a*.

EXAMPLE 4.15

Replace the force and couple system acting on the member in Fig. 4–38*a* by an equivalent resultant force and couple moment acting at point *O*.

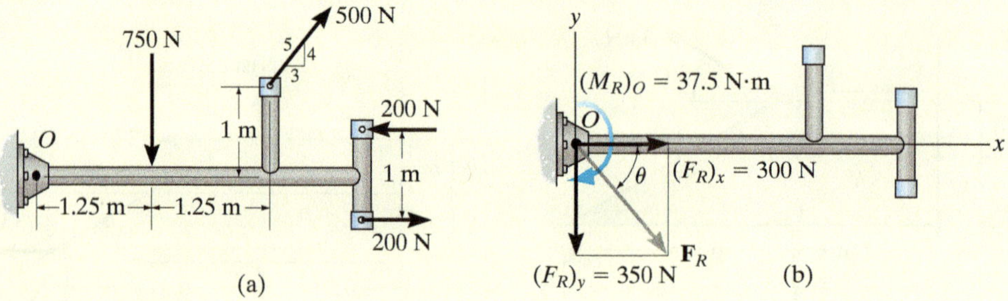

Fig. 4–38

4

SOLUTION

Force Summation. Since the couple forces of 200 N are equal but opposite, they produce a zero resultant force, and so it is not necessary to consider them in the force summation. The 500-N force is resolved into its *x* and *y* components, thus,

$$\overset{+}{\rightarrow}(F_R)_x = \Sigma F_x;\quad (F_R)_x = \left(\tfrac{3}{5}\right)(500\text{ N}) = 300\text{ N}\rightarrow$$

$$+\uparrow(F_R)_y = \Sigma F_y;\quad (F_R)_y = (500\text{ N})\left(\tfrac{4}{5}\right) - 750\text{ N} = -350\text{ N} = 350\text{ N}\downarrow$$

From Fig. 4–15*b*, the magnitude of $\mathbf{F}_R$ is

$$F_R = \sqrt{(F_R)_x^2 + (F_R)_y^2}$$

$$= \sqrt{(300\text{ N})^2 + (350\text{ N})^2} = 461\text{ N} \qquad \textit{Ans.}$$

And the angle θ is

$$\theta = \tan^{-1}\left(\frac{(F_R)_y}{(F_R)_x}\right) = \tan^{-1}\left(\frac{350\text{ N}}{300\text{ N}}\right) = 49.4^\circ \qquad \textit{Ans.}$$

Moment Summation. Since the couple moment is a free vector, it can act at any point on the member. Referring to Fig. 4–38*a*, we have

$$\circlearrowleft + (M_R)_O = \Sigma M_O + \Sigma M$$

$$(M_R)_O = (500\text{ N})\left(\tfrac{4}{5}\right)(2.5\text{ m}) - (500\text{ N})\left(\tfrac{3}{5}\right)(1\text{ m})$$

$$- (750\text{ N})(1.25\text{ m}) + 200\text{ N}\cdot\text{m}$$

$$= -37.5\text{ N}\cdot\text{m} = 37.5\text{ N}\cdot\text{m}\circlearrowright \qquad \textit{Ans.}$$

This clockwise moment is shown in Fig. 4–38*b*.

EXAMPLE 4.16

The structural member is subjected to a couple moment **M** and forces $\mathbf{F}_1$ and $\mathbf{F}_2$ in Fig. 4–39*a*. Replace this system by an equivalent resultant force and couple moment acting at its base, point *O*.

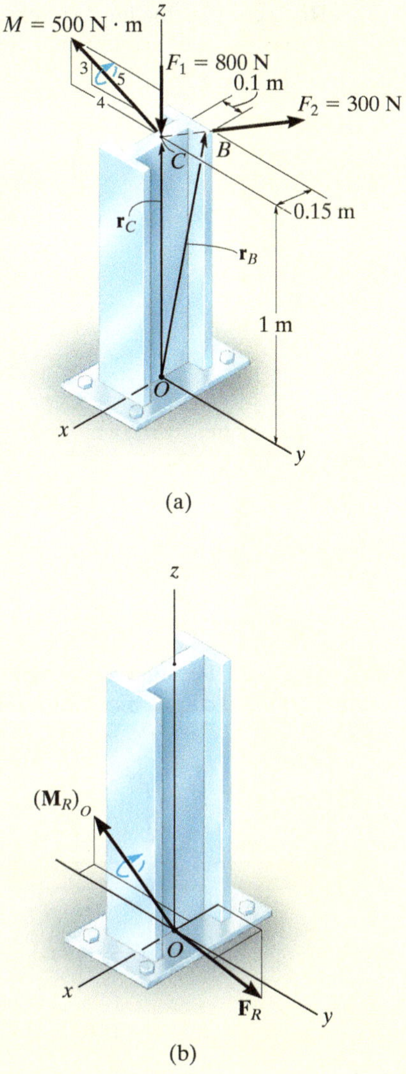

Fig. 4–39

SOLUTION (VECTOR ANALYSIS)

The three-dimensional aspects of the problem can be simplified by using a Cartesian vector analysis. Expressing the forces and couple moment as Cartesian vectors, we have

$$\mathbf{F}_1 = \{-800\mathbf{k}\}\text{ N}$$

$$\mathbf{F}_2 = (300\text{ N})\mathbf{u}_{CB}$$

$$= (300\text{ N})\left(\frac{\mathbf{r}_{CB}}{r_{CB}}\right)$$

$$= 300\text{ N}\left[\frac{\{-0.15\mathbf{i} + 0.1\mathbf{j}\}\text{ m}}{\sqrt{(-0.15\text{ m})^2 + (0.1\text{ m})^2}}\right] = \{-249.6\mathbf{i} + 166.4\mathbf{j}\}\text{ N}$$

$$\mathbf{M} = -500\left(\tfrac{4}{5}\right)\mathbf{j} + 500\left(\tfrac{3}{5}\right)\mathbf{k} = \{-400\mathbf{j} + 300\mathbf{k}\}\text{ N}\cdot\text{m}$$

Force Summation.

$$\mathbf{F}_R = \Sigma\mathbf{F}; \qquad \mathbf{F}_R = \mathbf{F}_1 + \mathbf{F}_2 = -800\mathbf{k} - 249.6\mathbf{i} + 166.4\mathbf{j}$$

$$= \{-250\mathbf{i} + 166\mathbf{j} - 800\mathbf{k}\}\text{ N} \qquad \textit{Ans.}$$

Moment Summation.

$$(\mathbf{M}_R)_o = \Sigma\mathbf{M} + \Sigma\mathbf{M}_O$$

$$(\mathbf{M}_R)_o = \mathbf{M} + \mathbf{r}_C \times \mathbf{F}_1 + \mathbf{r}_B \times \mathbf{F}_2$$

$$(\mathbf{M}_R)_o = (-400\mathbf{j} + 300\mathbf{k}) + (1\mathbf{k}) \times (-800\mathbf{k}) + \begin{vmatrix} \mathbf{i} & \mathbf{j} & \mathbf{k} \\ -0.15 & 0.1 & 1 \\ -249.6 & 166.4 & 0 \end{vmatrix}$$

$$= (-400\mathbf{j} + 300\mathbf{k}) + (\mathbf{0}) + (-166.4\mathbf{i} - 249.6\mathbf{j})$$

$$= \{-166\mathbf{i} - 650\mathbf{j} + 300\mathbf{k}\}\text{ N}\cdot\text{m} \qquad \textit{Ans.}$$

The results are shown in Fig. 4–39*b*.

4

FUNDAMENTAL PROBLEMS

F4–25. Replace the loading system by an equivalent resultant force and couple moment acting at point A.

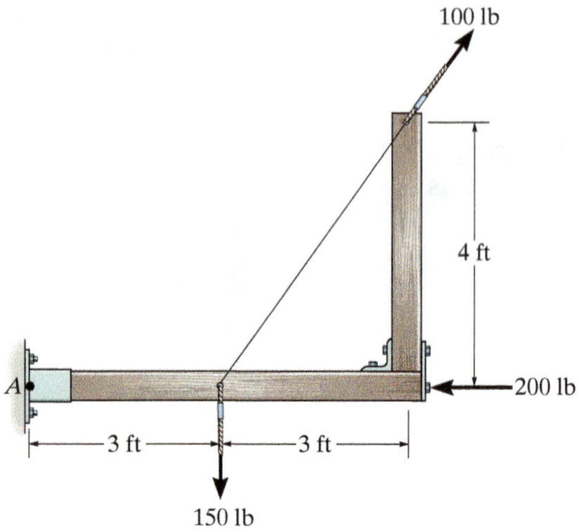

F4–25

F4–26. Replace the loading system by an equivalent resultant force and couple moment acting at point A.

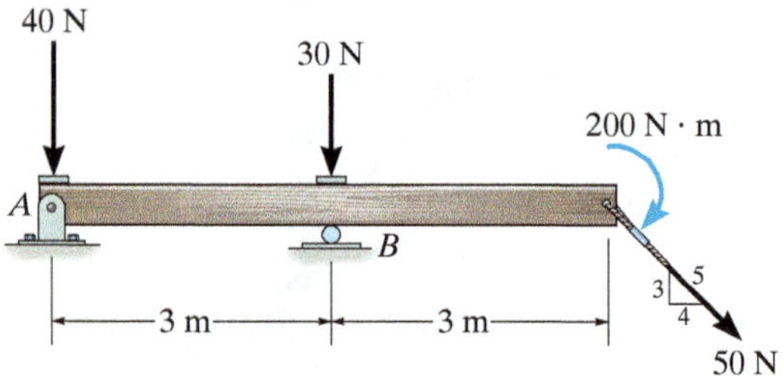

F4–26

F4–27. Replace the loading system by an equivalent resultant force and couple moment acting at point A.

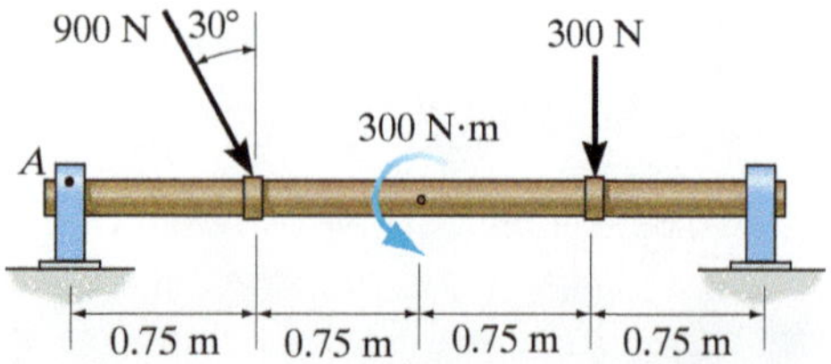

F4–27

F4–28. Replace the loading system by an equivalent resultant force and couple moment acting at point A.

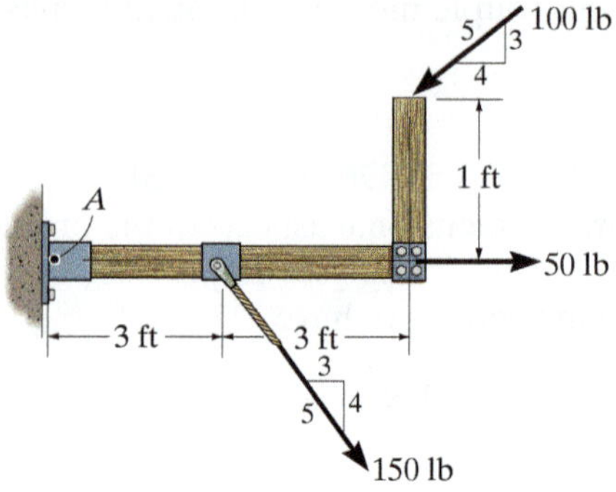

F4–28

F4–29. Replace the loading system by an equivalent resultant force and couple moment acting at point O.

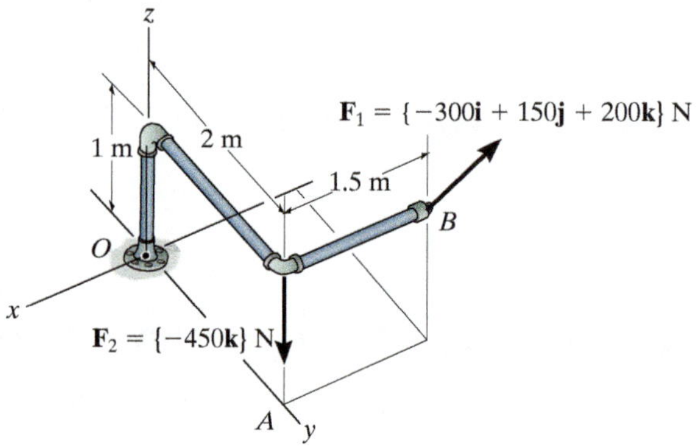

F4–29

F4–30. Replace the loading system by an equivalent resultant force and couple moment acting at point O.

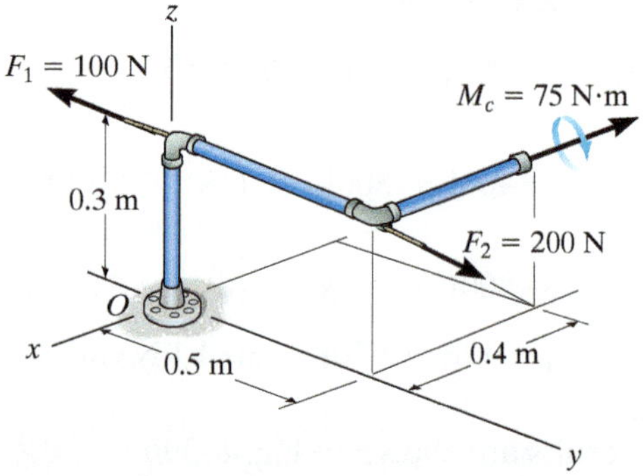

F4–30

PROBLEMS

4–97. Replace the force and couple system by an equivalent force and couple moment at point O.

4–98. Replace the force and couple system by an equivalent force and couple moment at point P.

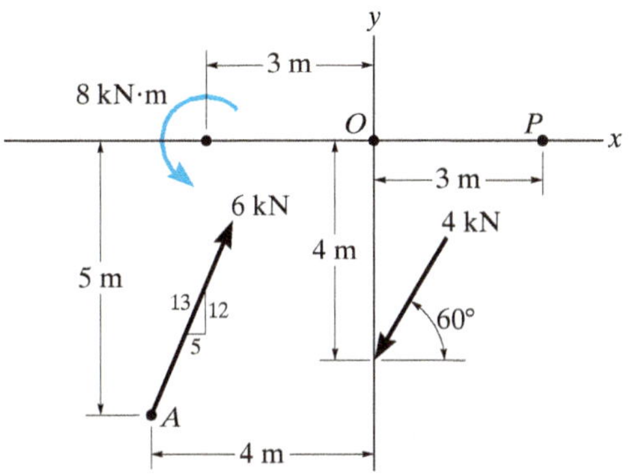

Probs. 4–97/98

4–101. Replace the force system acting on the post by a resultant force and couple moment at point O.

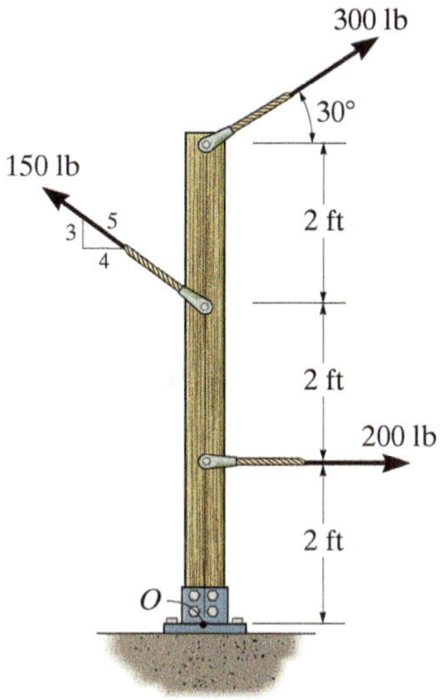

Prob. 4–101

4–99. Replace the force system acting on the beam by an equivalent force and couple moment at point A.

***4–100.** Replace the force system acting on the beam by an equivalent force and couple moment at point B.

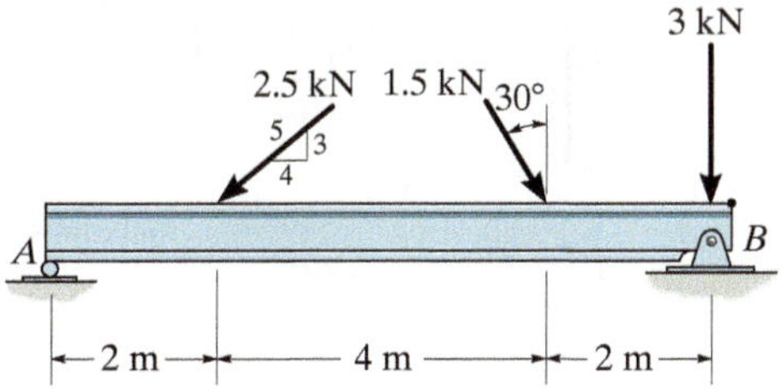

Probs. 4–99/100

4–102. Replace the two forces by an equivalent resultant force and couple moment at point O. Set $F = 20$ lb.

4–103. Replace the two forces by an equivalent resultant force and couple moment at point O. Set $F = 15$ lb.

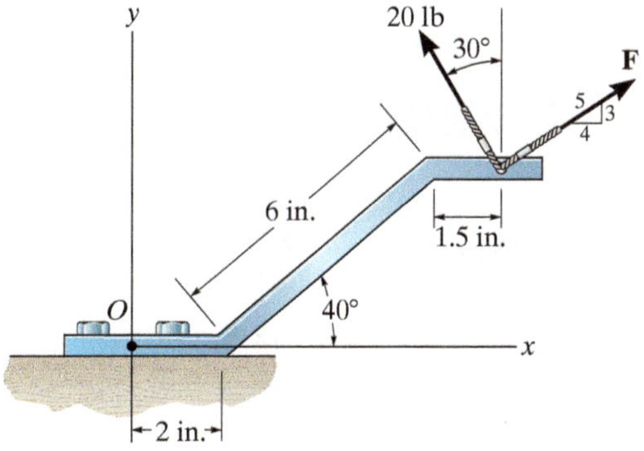

Probs. 4–102/103

***4–104.** Replace the force system acting on the crank by a resultant force, and specify where its line of action intersects BA measured from the pin at B.

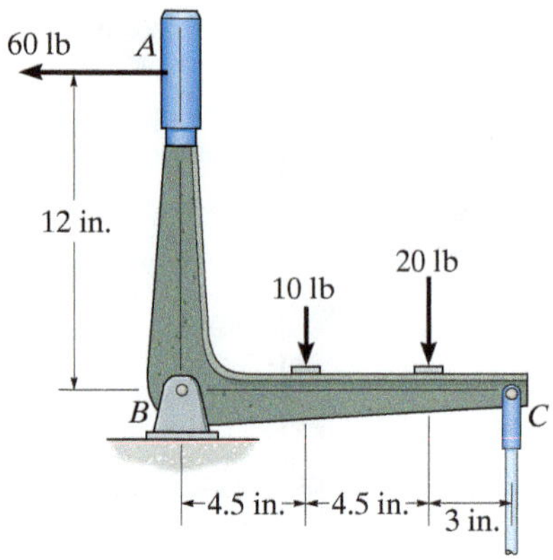

Prob. 4–104

4–105. Replace the force system acting on the frame by a resultant force and couple moment at point A.

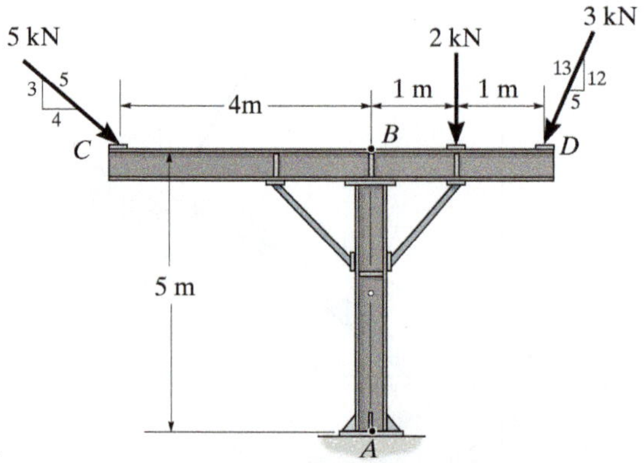

Prob. 4–105

4–106. Replace the force system acting on the bracket by a resultant force and couple moment at point A.

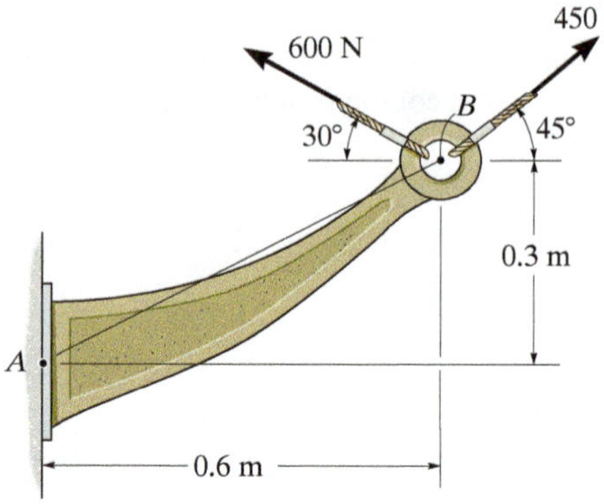

Prob. 4–106

4–107. A biomechanical model of the lumbar region of the human trunk is shown. The forces acting in the four muscle groups consist of $F_R = 35$ N for the rectus, $F_O = 45$ N for the oblique, $F_L = 23$ N for the lumbar latissimus dorsi, and $F_E = 32$ N for the erector spinae. These loadings are symmetric with respect to the y–z plane. Replace this system of parallel forces by an equivalent force and couple moment acting at the spine, point O. Express the results in Cartesian vector form.

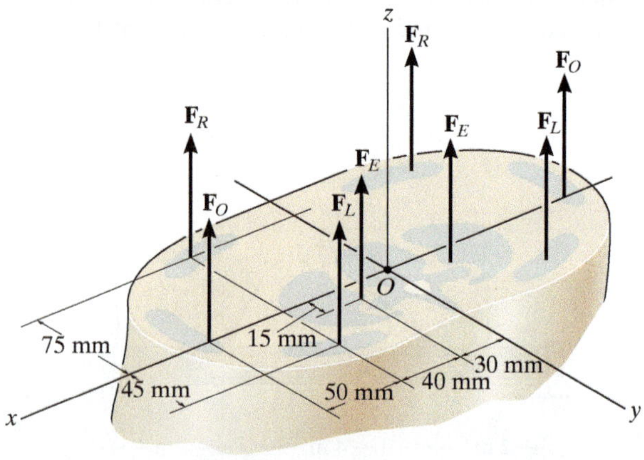

Prob. 4–107

*4–108. Replace the two forces acting on the post by a resultant force and couple moment at point O. Express the results in Cartesian vector form.

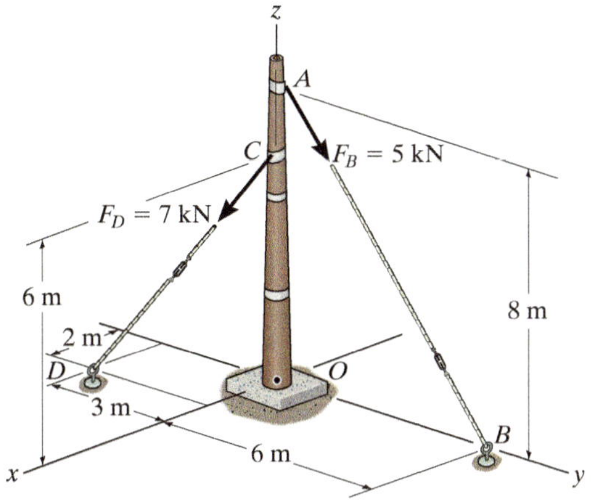

Prob. 4–108

4–109. Replace the force system by an equivalent force and couple moment at point A.

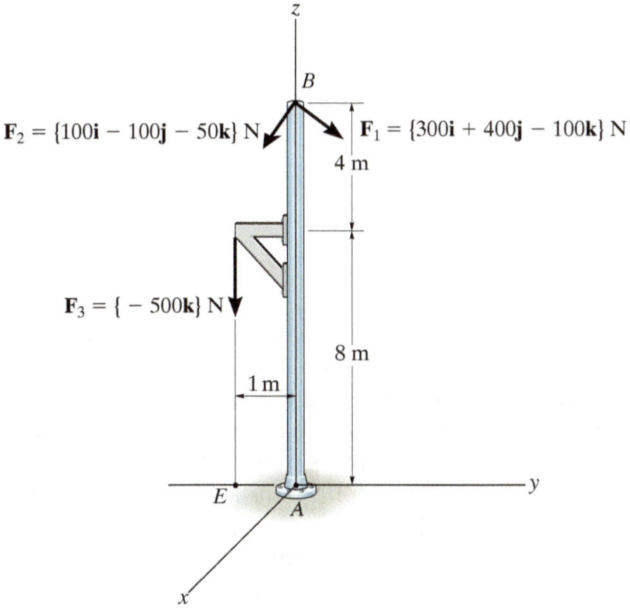

Prob. 4–109

4–110. The belt passing over the pulley is subjected to forces $\mathbf{F}_1$ and $\mathbf{F}_2$, each having a magnitude of 40 N. $\mathbf{F}_1$ acts in the $-\mathbf{k}$ direction. Replace these forces by an equivalent force and couple moment at point A. Express the result in Cartesian vector form. Set $\theta = 0°$ so that $\mathbf{F}_2$ acts in the $-\mathbf{j}$ direction.

4–111. The belt passing over the pulley is subjected to two forces $\mathbf{F}_1$ and $\mathbf{F}_2$, each having a magnitude of 40 N. $\mathbf{F}_1$ acts in the $-\mathbf{k}$ direction. Replace these forces by an equivalent force and couple moment at point A. Express the result in Cartesian vector form. Take $\theta = 45°$.

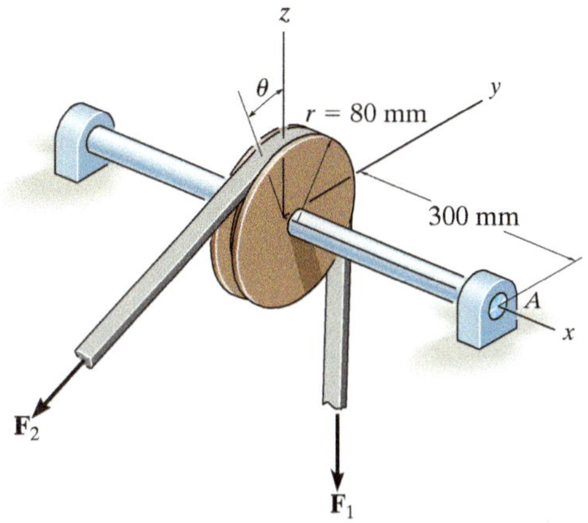

Probs. 4–110/111

*4–112. Handle forces $\mathbf{F}_1$ and $\mathbf{F}_2$ are applied to the electric drill. Replace this force system by an equivalent resultant force and couple moment acting at point O. Express the results in Cartesian vector from.

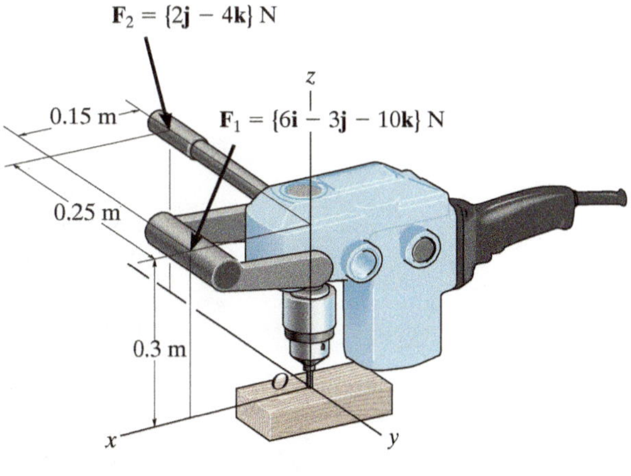

Prob. 4–112

4

4.8 Further Simplification of a Force and Couple System

In the preceding section, we developed a way to reduce a force and couple moment system acting on a rigid body into an equivalent resultant force $\mathbf{F}_R$ acting at a specific point O and a resultant couple moment $(\mathbf{M}_R)_O$. The force system can be further reduced to an equivalent single resultant force provided the lines of action of $\mathbf{F}_R$ and $(\mathbf{M}_R)_O$ are *perpendicular* to each other. Because of this condition, only concurrent, coplanar, and parallel force systems can be further simplified.

Concurrent Force System. Since a *concurrent force system* is one in which the lines of action of all the forces intersect at a common point O, Fig. 4–40*a*, then the force system produces no moment about this point. As a result, the equivalent system can be represented by a single resultant force $\mathbf{F}_R = \Sigma\mathbf{F}$ acting at O, Fig. 4–40*b*.

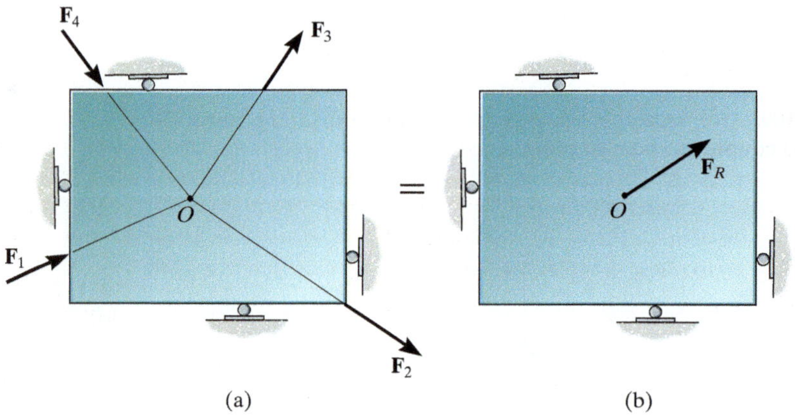

Fig. 4–40

Coplanar Force System. In the case of a *coplanar force system*, the lines of action of all the forces lie in the same plane, Fig. 4–41*a*, and so the resultant force $\mathbf{F}_R = \Sigma\mathbf{F}$ of this system also lies in this plane. Furthermore, the moment of each of the forces about any point O is directed perpendicular to this plane. Thus, the resultant moment $(\mathbf{M}_R)_O$ and resultant force $\mathbf{F}_R$ will be *mutually perpendicular*, Fig. 4–41*b*. The resultant moment can be replaced by moving the resultant force $\mathbf{F}_R$ a perpendicular or moment arm distance d away from point O such that $\mathbf{F}_R$ produces the *same moment* $(\mathbf{M}_R)_O$ about point O, Fig. 4–41*c*. This distance d can be determined from the scalar equation $(M_R)_O = F_R d = \Sigma M_O$ or $d = (M_R)_O/F_R$.

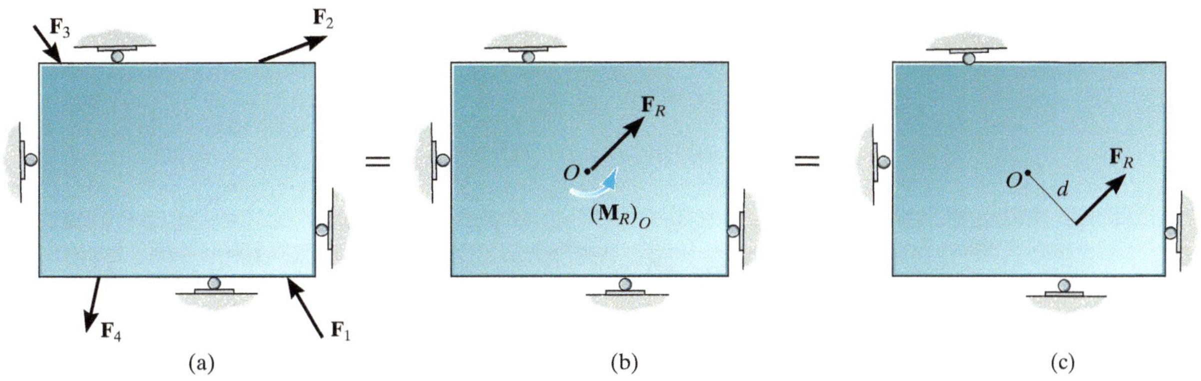

Fig. 4–41

Parallel Force System. The *parallel force system* shown in Fig. 4–42*a* consists of forces that are all parallel to the *z* axis. Thus, the resultant force $\mathbf{F}_R = \Sigma\mathbf{F}$ at point O must also be parallel to this axis, Fig. 4–42*b*. The moment produced by each force lies in the plane of the plate, and so the resultant couple moment, $(\mathbf{M}_R)_O$, will also lie in this plane, along the moment axis a since $\mathbf{F}_R$ and $(\mathbf{M}_R)_O$ are mutually perpendicular. As a result, the force system can be further reduced to an equivalent single resultant force $\mathbf{F}_R$, acting through point P located on the perpendicular b axis, Fig. 4–42*c*. The distance d along this axis from point O requires $(M_R)_O = F_R d = \Sigma M_O$ or $d = \Sigma M_O / F_R$.

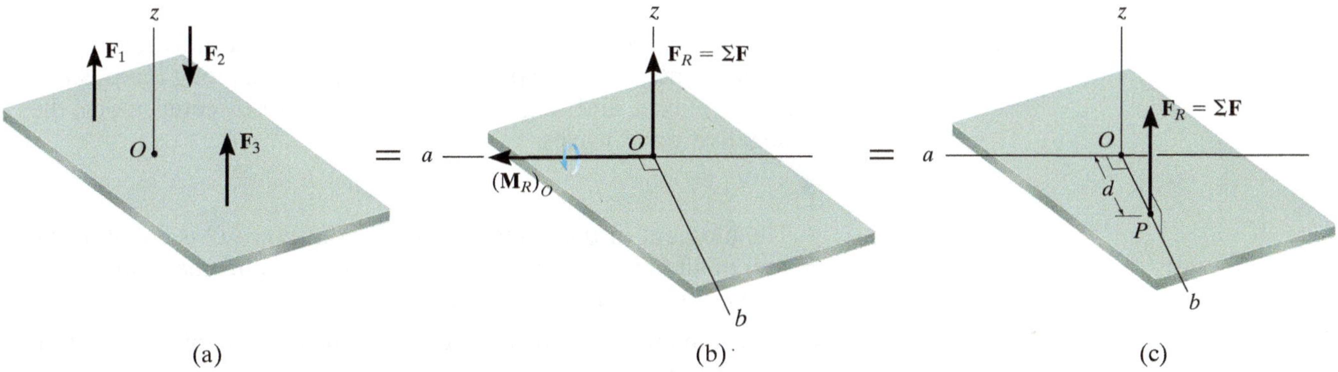

Fig. 4–42

The four cable forces are all concurrent at point O on this bridge tower. Consequently they produce no resultant moment there, only a resultant force $\mathbf{F}_R$. Note that the designers have positioned the cables so that $\mathbf{F}_R$ is directed *along* the bridge tower directly to the support, so that it does not cause any bending of the tower.

4

Procedure for Analysis

The technique used to reduce a coplanar or parallel force system to a single resultant force follows a similar procedure outlined in the previous section.

- Establish the x, y, z, axes and locate the resultant force $\mathbf{F}_R$ an arbitrary distance away from the origin of the coordinates.

Force Summation.

- The resultant force is equal to the sum of all the forces in the system.
- For a coplanar force system, resolve each force into its x and y components. Positive components are directed along the positive x and y axes, and negative components are directed along the negative x and y axes.

Moment Summation.

- The moment of the resultant force about point O is equal to the sum of all the couple moments in the system plus the moments of all the forces in the system about O.
- This moment condition is used to find the location of the resultant force from point O.

Here the weights of the traffic lights are replaced by their resultant force $W_R = W_1 + W_2$ which acts at a distance $d = (W_1d_1 + W_2d_2)/\, W_R$ from O. Both systems are equivalent.

Reduction to a Wrench. In general, a three-dimensional force and couple moment system will have an equivalent resultant force $\mathbf{F}_R$ acting at point O and a resultant couple moment $(\mathbf{M}_R)_O$ that are *not perpendicular* to one another, as shown in Fig. 4–43*a*. Although a force system such as this cannot be further reduced to an equivalent single resultant force, the resultant couple moment $(\mathbf{M}_R)_O$ can be resolved into components parallel and perpendicular to the line of action of $\mathbf{F}_R$, Fig. 4–43*a*. The perpendicular component $\mathbf{M}_\perp$ can be replaced if we move $\mathbf{F}_R$ to point P, a distance d from point O along the b axis, Fig. 4–43*b*. As seen, this axis is perpendicular to both the a axis and the line of action of $\mathbf{F}_R$. The location of P can be determined from $d = M_\perp / F_R$. Finally, because $\mathbf{M}_\parallel$ is a free vector, it can be moved to point P, Fig. 4–43*c*. This combination of a resultant force $\mathbf{F}_R$ and collinear couple moment $\mathbf{M}_\parallel$ will tend to translate and rotate the body about its axis and is referred to as a *wrench* or *screw*. A wrench is the simplest system that can represent any general force and couple moment system acting on a body.

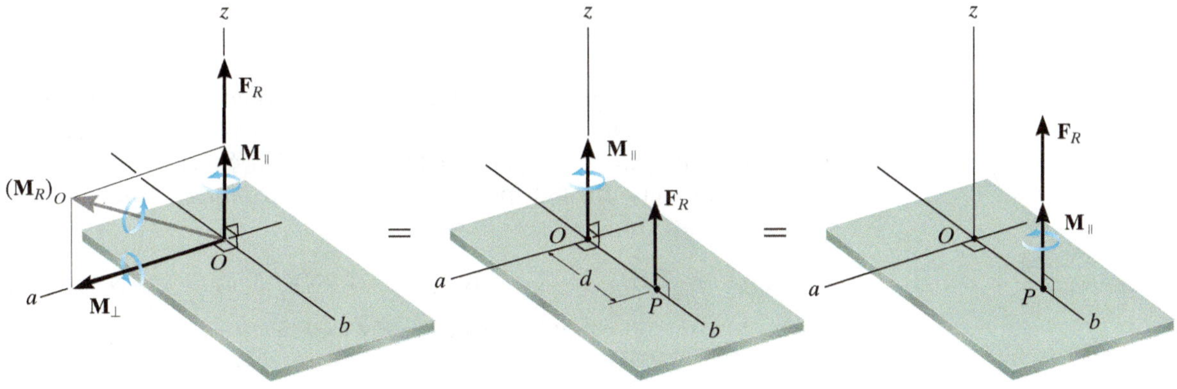

(a) (b) (c)

Fig. 4–43

EXAMPLE 4.17

Replace the force and couple moment system acting on the beam in Fig. 4–44*a* by an equivalent resultant force, and find where its line of action intersects the beam, measured from point *O*.

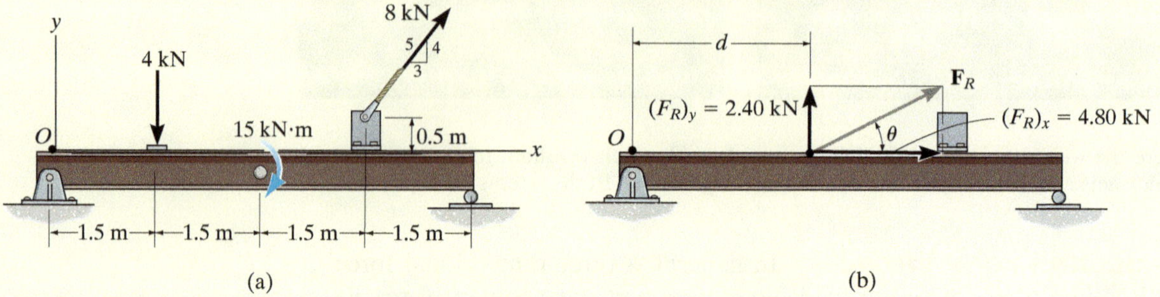

Fig. 4–44

SOLUTION

Force Summation. Summing the force components,

$$\xrightarrow{+} (F_R)_x = \Sigma F_x; \qquad (F_R)_x = 8\text{ kN}\left(\tfrac{3}{5}\right) = 4.80\text{ kN} \rightarrow$$

$$+\uparrow (F_R)_y = \Sigma F_y; \qquad (F_R)_y = -4\text{ kN} + 8\text{ kN}\left(\tfrac{4}{5}\right) = 2.40\text{ kN}\uparrow$$

From Fig. 4–44*b*, the magnitude of $\mathbf{F}_R$ is

$$F_R = \sqrt{(4.80\text{ kN})^2 + (2.40\text{ kN})^2} = 5.37\text{ kN} \qquad \textit{Ans.}$$

The angle θ is

$$\theta = \tan^{-1}\left(\frac{2.40\text{ kN}}{4.80\text{ kN}}\right) = 26.6^\circ \qquad \textit{Ans.}$$

Moment Summation. We must equate the moment of $\mathbf{F}_R$ about point *O* in Fig. 4–44*b* to the sum of the moments of the force and couple moment system about point *O* in Fig. 4–44*a*. Since the line of action of $(\mathbf{F}_R)_x$ acts through point *O*, *only* $(\mathbf{F}_R)_y$ *produces a moment* about this point. Thus,

$$\circlearrowleft + (M_R)_O = \Sigma M_O; \qquad 2.40\text{ kN}(d) = -(4\text{ kN})(1.5\text{ m}) - 15\text{ kN}\cdot\text{m}$$

$$-\left[8\text{ kN}\left(\tfrac{3}{5}\right)\right](0.5\text{ m}) + \left[8\text{ kN}\left(\tfrac{4}{5}\right)\right](4.5\text{ m})$$

$$d = 2.25\text{ m} \qquad \textit{Ans.}$$

EXAMPLE 4.18

The jib crane shown in Fig. 4–45*a* is subjected to three coplanar forces. Replace this loading by an equivalent resultant force and specify where the resultant's line of action intersects the column *AB* and boom *BC*.

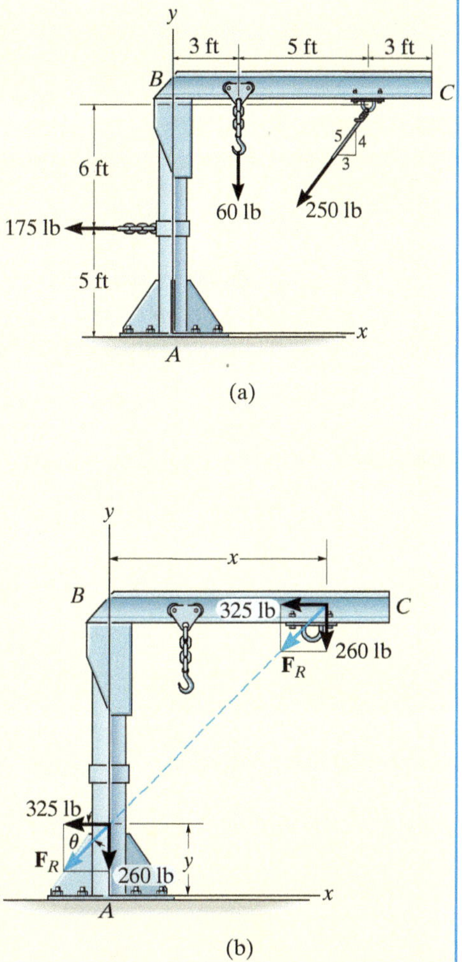

Fig. 4–45

SOLUTION

Force Summation. Resolving the 250-lb force into *x* and *y* components and summing the force components yields

$$\overset{+}{\rightarrow} (F_R)_x = \Sigma F_x; \quad (F_R)_x = -250 \text{ lb}\left(\tfrac{3}{5}\right) - 175 \text{ lb} = -325 \text{ lb} = 325 \text{ lb} \leftarrow$$

$$+\uparrow (F_R)_y = \Sigma F_y; \quad (F_R)_y = -250 \text{ lb}\left(\tfrac{4}{5}\right) - 60 \text{ lb} = -260 \text{ lb} = 260 \text{ lb}\downarrow$$

As shown by the vector addition in Fig. 4–45*b*,

$$F_R = \sqrt{(325 \text{ lb})^2 + (260 \text{ lb})^2} = 416 \text{ lb} \qquad \textit{Ans.}$$

$$\theta = \tan^{-1}\left(\frac{260 \text{ lb}}{325 \text{ lb}}\right) = 38.7° \; \theta \swarrow \qquad \textit{Ans.}$$

Moment Summation. Moments will be summed about point *A*. Assuming the line of action of $\mathbf{F}_R$ *intersects AB* at a distance *y* from *A*, Fig. 4–45*b*, we have

$$\zeta + (M_R)_A = \Sigma M_A; \qquad 325 \text{ lb } (y) + 260 \text{ lb } (0)$$

$$= 175 \text{ lb } (5 \text{ ft}) - 60 \text{ lb } (3 \text{ ft}) + 250 \text{ lb}\left(\tfrac{3}{5}\right)(11 \text{ ft}) - 250 \text{ lb}\left(\tfrac{4}{5}\right)(8 \text{ ft})$$

$$y = 2.29 \text{ ft} \qquad \textit{Ans.}$$

By the principle of transmissibility, $\mathbf{F}_R$ can be placed at a distance *x* where it intersects *BC*, Fig. 4–45*b*. In this case we have

$$\zeta + (M_R)_A = \Sigma M_A; \qquad 325 \text{ lb } (11 \text{ ft}) - 260 \text{ lb } (x)$$

$$= 175 \text{ lb } (5 \text{ ft}) - 60 \text{ lb } (3 \text{ ft}) + 250 \text{ lb}\left(\tfrac{3}{5}\right)(11 \text{ ft}) - 250 \text{ lb}\left(\tfrac{4}{5}\right)(8 \text{ ft})$$

$$x = 10.9 \text{ ft} \qquad \textit{Ans.}$$

4

EXAMPLE 4.19

The slab in Fig. 4–46*a* is subjected to four parallel forces. Determine the magnitude and direction of a resultant force equivalent to the given force system, and locate its point of application on the slab.

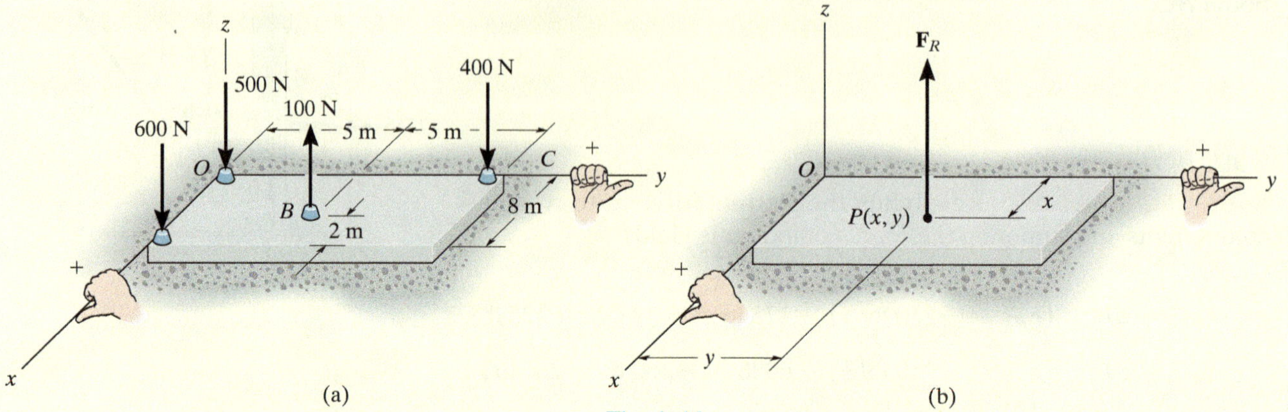

Fig. 4–46

SOLUTION *(SCALAR ANALYSIS)*

Force Summation. From Fig. 4–46*a*, the resultant force is

$+\uparrow F_R = \Sigma F;$ $\quad F_R = -600\text{ N} + 100\text{ N} - 400\text{ N} - 500\text{ N}$

$= -1400\text{ N} = 1400\text{ N}\downarrow$ *Ans.*

Moment Summation. We require the moment about the *x* axis of the resultant force, Fig. 4–46*b*, to be equal to the sum of the moments about the *x* axis of all the forces in the system, Fig. 4–46*a*. The moment arms are determined from the *y* coordinates, since these coordinates represent the *perpendicular distances* from the *x* axis to the lines of action of the forces. Using the right-hand rule, we have

$(M_R)_x = \Sigma M_x;$

$$-(1400\text{ N})y = 600\text{ N}(0) + 100\text{ N}(5\text{ m}) - 400\text{ N}(10\text{ m}) + 500\text{ N}(0)$$

$$-1400y = -3500 \qquad y = 2.50\text{ m}$$ *Ans.*

In a similar manner, a moment equation can be written about the *y* axis using moment arms defined by the *x* coordinates of each force.

$(M_R)_y = \Sigma M_y;$

$$(1400\text{ N})x = 600\text{ N}(8\text{ m}) - 100\text{ N}(6\text{ m}) + 400\text{ N}(0) + 500\text{ N}(0)$$

$$1400x = 4200$$

$$x = 3\text{ m}$$ *Ans.*

NOTE: A force of $F_R = 1400$ N placed at point $P(3.00\text{ m}, 2.50\text{ m})$ on the slab, Fig. 4–46*b*, is therefore equivalent to the parallel force system acting on the slab in Fig. 4–46*a*.

EXAMPLE 4.20

Replace the force system in Fig. 4–47*a* by an equivalent resultant force and specify its point of application on the pedestal.

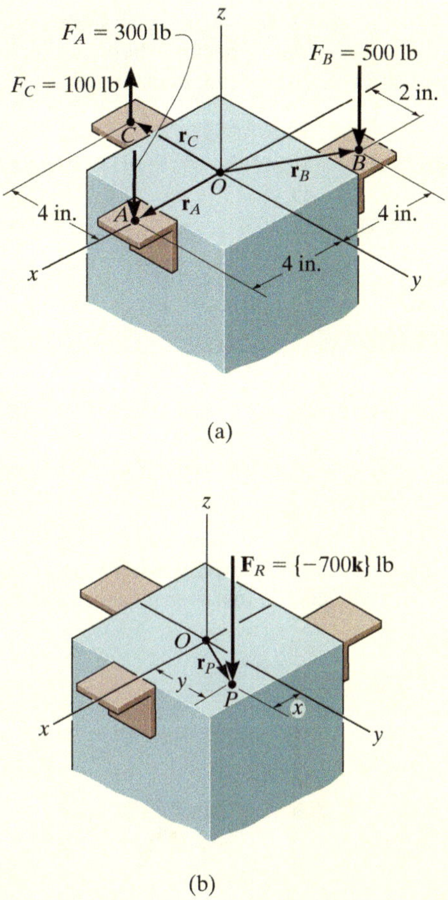

Fig. 4–47

SOLUTION

Force Summation. Here we will demonstrate a vector analysis. Summing forces,

$$\mathbf{F}_R = \Sigma\mathbf{F};\ \mathbf{F}_R = \mathbf{F}_A + \mathbf{F}_B + \mathbf{F}_C$$
$$= \{-300\mathbf{k}\}\text{ lb} + \{-500\mathbf{k}\}\text{ lb} + \{100\mathbf{k}\}\text{ lb}$$
$$= \{-700\mathbf{k}\}\text{ lb} \qquad \textit{Ans.}$$

Location. Moments will be summed about point O. The resultant force $\mathbf{F}_R$ is assumed to act through point $P\,(x, y, 0)$, Fig. 4–47*b*. Thus

$$(\mathbf{M}_R)_O = \Sigma\mathbf{M}_O;$$
$$\mathbf{r}_P \times \mathbf{F}_R = (\mathbf{r}_A \times \mathbf{F}_A) + (\mathbf{r}_B \times \mathbf{F}_B) + (\mathbf{r}_C \times \mathbf{F}_C)$$
$$(x\mathbf{i} + y\mathbf{j}) \times (-700\mathbf{k}) = [(4\mathbf{i}) \times (-300\mathbf{k})]$$
$$+ [(-4\mathbf{i} + 2\mathbf{j}) \times (-500\mathbf{k})] + [(-4\mathbf{j}) \times (100\mathbf{k})]$$
$$-700x(\mathbf{i} \times \mathbf{k}) - 700y(\mathbf{j} \times \mathbf{k}) = -1200(\mathbf{i} \times \mathbf{k}) + 2000(\mathbf{i} \times \mathbf{k})$$
$$- 1000(\mathbf{j} \times \mathbf{k}) - 400(\mathbf{j} \times \mathbf{k})$$
$$700x\mathbf{j} - 700y\mathbf{i} = 1200\mathbf{j} - 2000\mathbf{j} - 1000\mathbf{i} - 400\mathbf{i}$$

Equating the **i** and **j** components,

$$-700y = -1400 \qquad (1)$$
$$y = 2\text{ in.} \qquad \textit{Ans.}$$
$$700x = -800 \qquad (2)$$
$$x = -1.14\text{ in.} \qquad \textit{Ans.}$$

The negative sign indicates that the x coordinate of point P is negative.

NOTE: It is also possible to establish Eq. 1 and 2 directly by summing moments about the x and y axes. Using the right-hand rule, we have

$$(M_R)_x = \Sigma M_x; \qquad -700y = -100\text{ lb}(4\text{ in.}) - 500\text{ lb}(2\text{ in.})$$
$$(M_R)_y = \Sigma M_y; \qquad 700x = 300\text{ lb}(4\text{ in.}) - 500\text{ lb}(4\text{ in.})$$

4

FUNDAMENTAL PROBLEMS

F4–31. Replace the loading system by an equivalent resultant force and specify where the resultant's line of action intersects the beam measured from O.

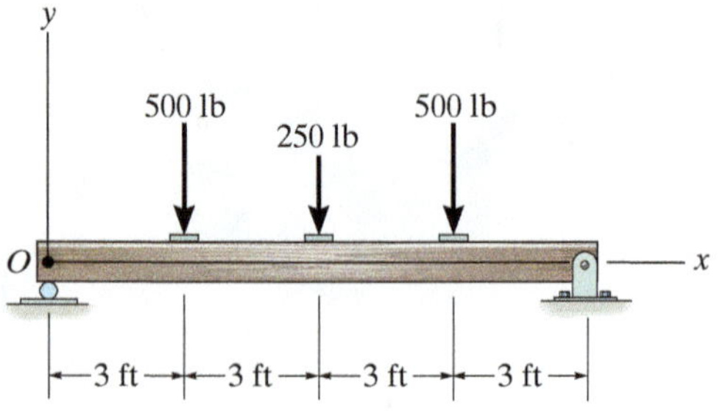

F4–31

F4–32. Replace the loading system by an equivalent resultant force and specify where the resultant's line of action intersects the member measured from A.

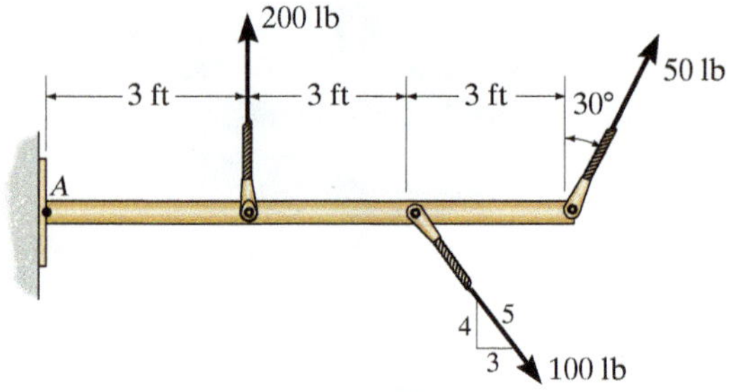

F4–32

F4–33. Replace the loading system by an equivalent resultant force and specify where the resultant's line of action intersects the horizontal segment of the member measured from A.

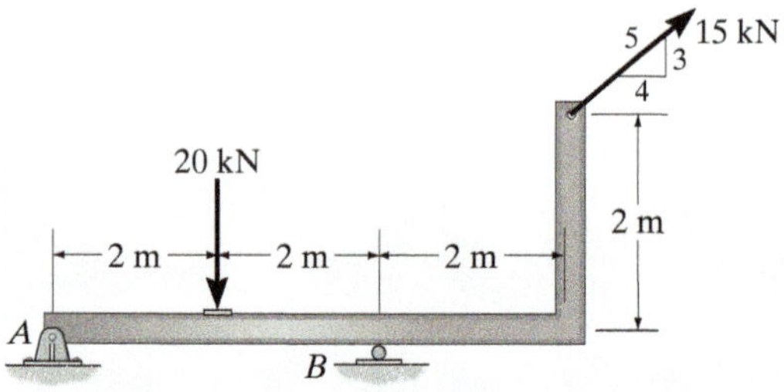

F4–33

F4–34. Replace the loading system by an equivalent resultant force and specify where the resultant's line of action intersects the member AB measured from A.

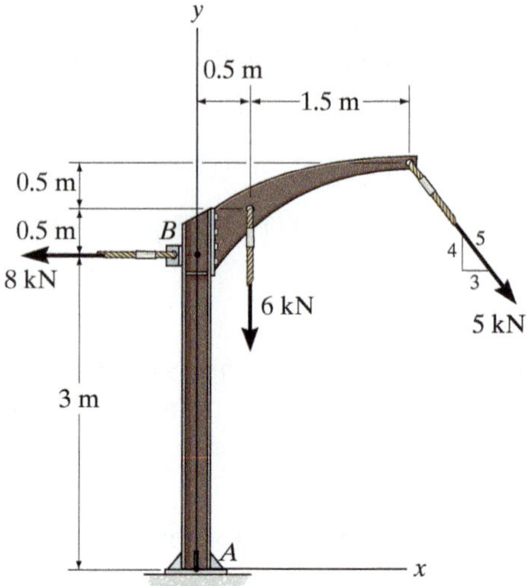

F4–34

F4–35. Replace the loading shown by an equivalent single resultant force and specify the x and y coordinates of its line of action.

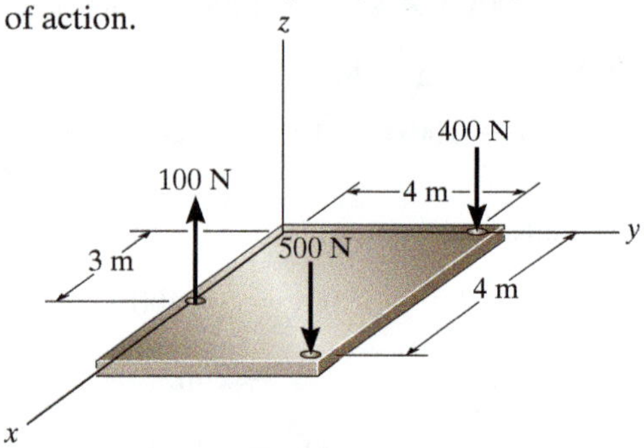

F4–35

F4–36. Replace the loading shown by an equivalent single resultant force and specify the x and y coordinates of its line of action.

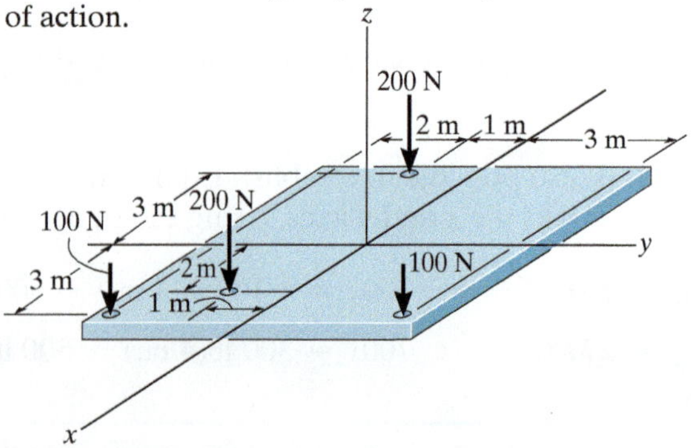

F4–36

PROBLEMS

4–113. The weights of the various components of the truck are shown. Replace this system of forces by an equivalent resultant force and specify its location measured from *B*.

4–114. The weights of the various components of the truck are shown. Replace this system of forces by an equivalent resultant force and specify its location measured from point *A*.

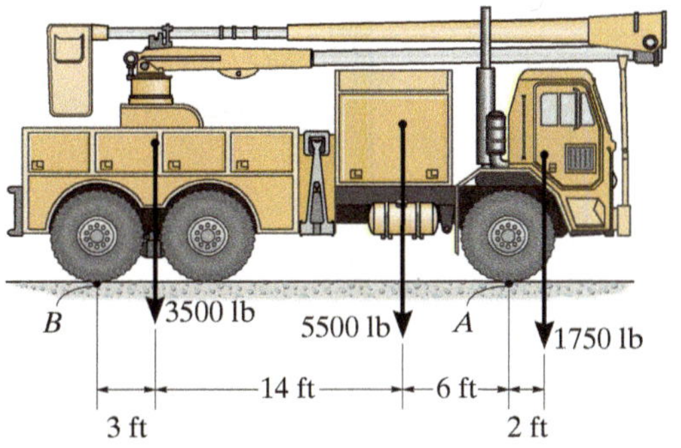

Probs. 4–113/114

4–115. Replace the three forces acting on the shaft by a single resultant force. Specify where the force acts, measured from end *A*.

***4–116.** Replace the three forces acting on the shaft by a single resultant force. Specify where the force acts, measured from end *B*.

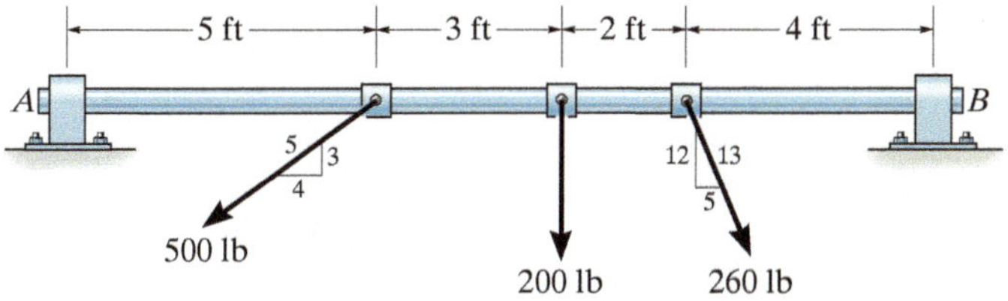

Probs. 4–115/116

4–117. Replace the loading acting on the beam by a single resultant force. Specify where the force acts, measured from end *A*.

4–118. Replace the loading acting on the beam by a single resultant force. Specify where the force acts, measured from *B*.

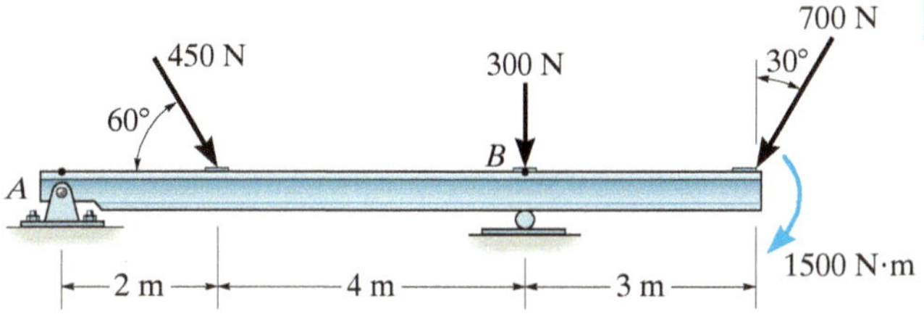

Probs. 4–117/118

4–119. Replace the force system acting on the frame by a resultant force, and specify where its line of action intersects member *AB*, measured from point *A*.

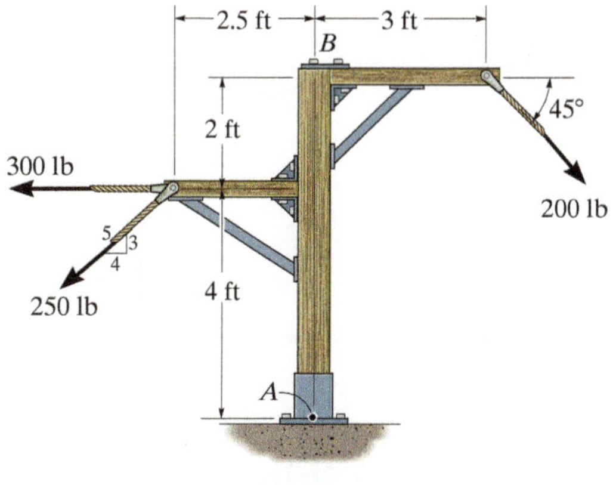

Prob. 4–119

***4–120.** Replace the loading on the frame by a single resultant force. Specify where its line of action intersects member AB, measured from A.

4–121. Replace the loading on the frame by a single resultant force. Specify where its line of action intersects member CD, measured from end C.

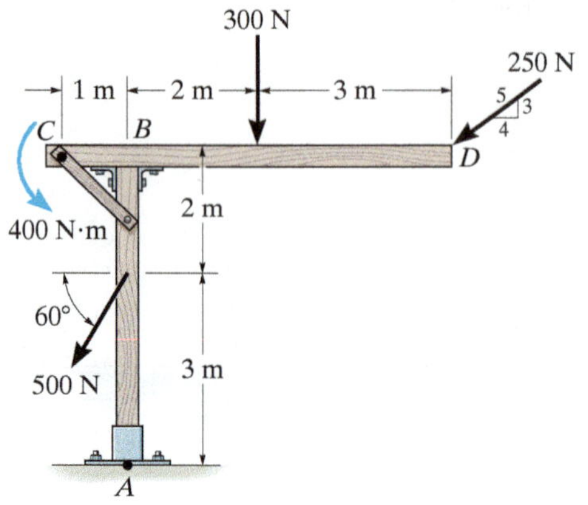

Probs. 4–120/121

4–122. Replace the force system acting on the frame by an equivalent resultant force and specify where the resultant's line of action intersects member AB, measured from point A.

4–123. Replace the force system acting on the frame by an equivalent resultant force and specify where the resultant's line of action intersects member BC, measured from point B.

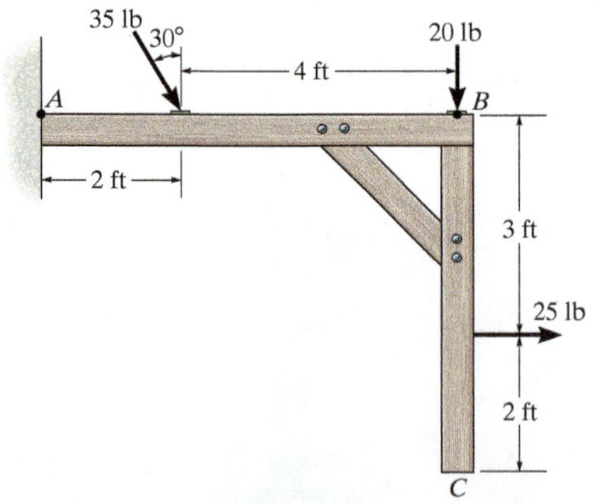

Probs. 4–122/123

***4–124.** Replace the force system acting on the post by a resultant force, and specify where its line of action intersects the post AB measured from point A.

4–125. Replace the force system acting on the post by a resultant force, and specify where its line of action intersects the post AB measured from point B.

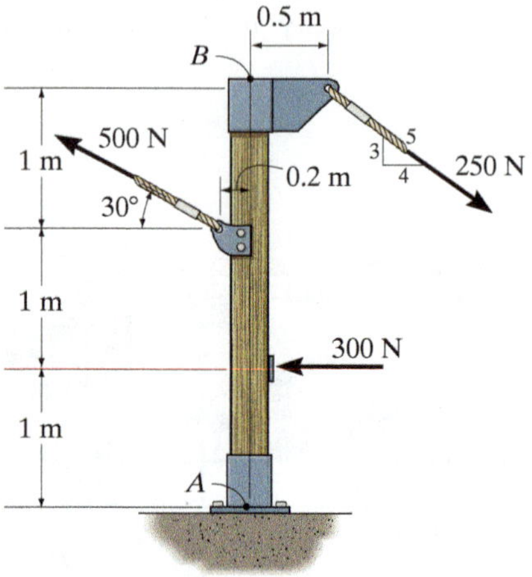

Probs. 4–124/125

4–126. Replace the loading on the frame by a single resultant force. Specify where its line of action intersects member AB, measured from A.

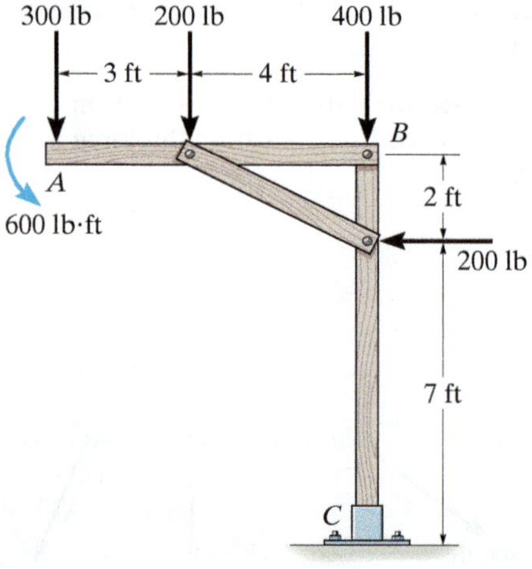

Prob. 4–126

4–127. The tube supports the four parallel forces. Determine the magnitudes of forces $\mathbf{F}_C$ and $\mathbf{F}_D$ acting at C and D so that the equivalent resultant force of the force system acts through the midpoint O of the tube.

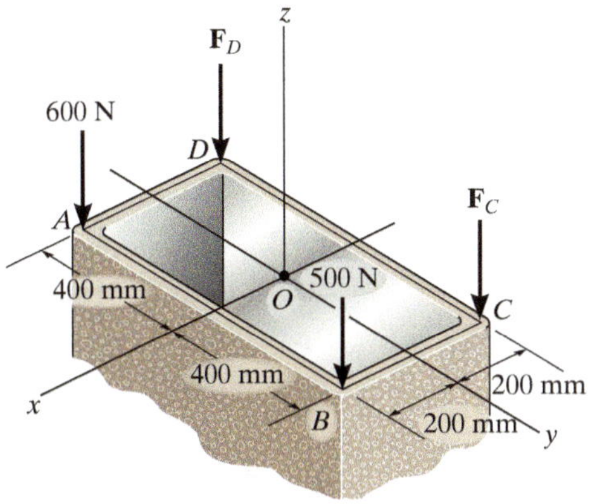

Prob. 4–127

*__4–128.__ Three parallel bolting forces act on the circular plate. Determine the resultant force, and specify its location (x, z) on the plate. $F_A = 200$ lb, $F_B = 100$ lb, and $F_C = 400$ lb.

4–129. The three parallel bolting forces act on the circular plate. If the force at A has a magnitude of $F_A = 200$ lb, determine the magnitudes of $\mathbf{F}_B$ and $\mathbf{F}_C$ so that the resultant force $\mathbf{F}_R$ of the system has a line of action that coincides with the y axis. *Hint:* This requires $\Sigma M_x = 0$ and $\Sigma M_z = 0$.

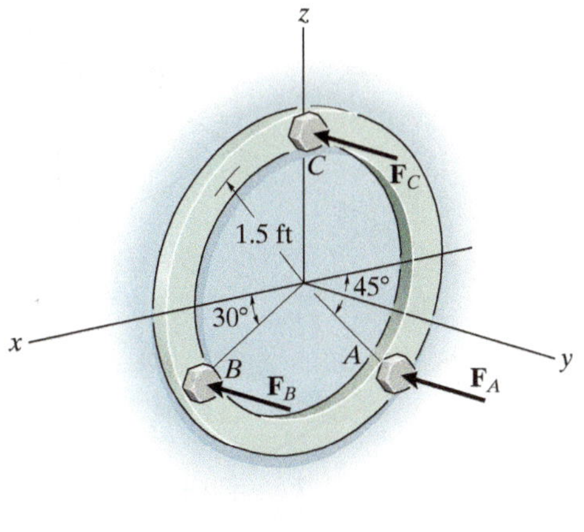

Probs. 4–128/129

4–130. The building slab is subjected to four parallel column loadings. Determine the equivalent resultant force and specify its location (x, y) on the slab. Take $F_1 = 30$ kN, $F_2 = 40$ kN.

4–131. The building slab is subjected to four parallel column loadings. Determine the equivalent resultant force and specify its location (x, y) on the slab. Take $F_1 = 20$ kN, $F_2 = 50$ kN.

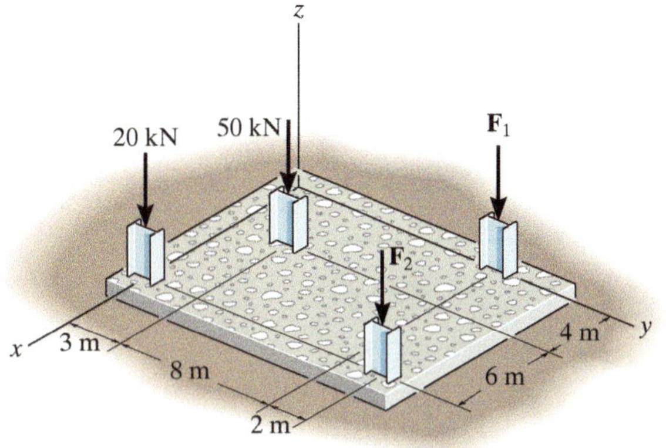

Probs. 4–130/131

*__4–132.__ If $F_A = 40$ kN and $F_B = 35$ kN, determine the magnitude of the resultant force and specify the location of its point of application (x, y) on the slab.

4–133. If the resultant force is required to act at the center of the slab, determine the magnitude of the column loadings $\mathbf{F}_A$ and $\mathbf{F}_B$ and the magnitude of the resultant force.

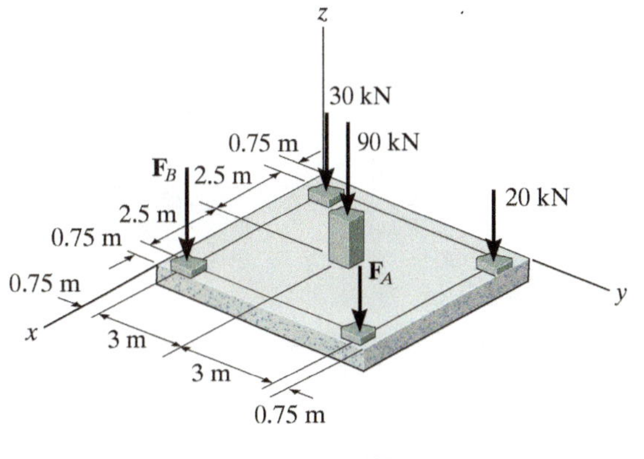

Probs. 4–132/133

4–134. Replace the two wrenches and the force, acting on the pipe assembly, by an equivalent resultant force and couple moment at point O.

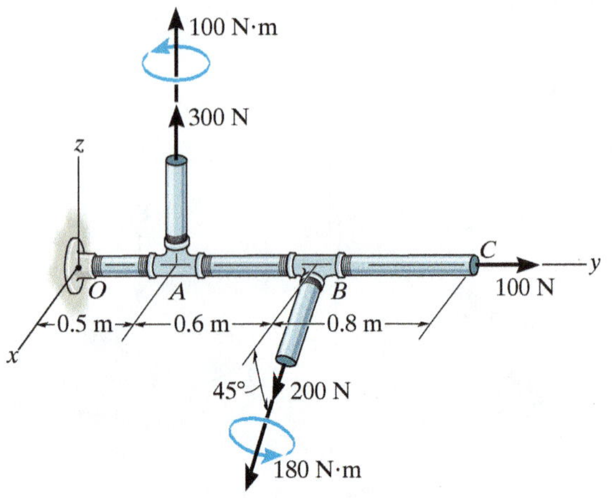

Prob. 4–134

4–135. The three forces acting on the block each have a magnitude of 10 lb. Replace this system by a wrench and specify the point where the wrench intersects the z axis, measured from point O.

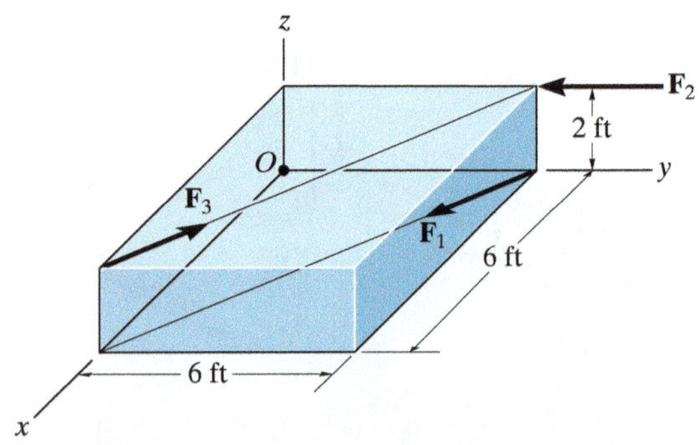

Prob. 4–135

***4–136.** Replace the force and couple moment system acting on the rectangular block by a wrench. Specify the magnitude of the force and couple moment of the wrench and where its line of action intersects the x–y plane.

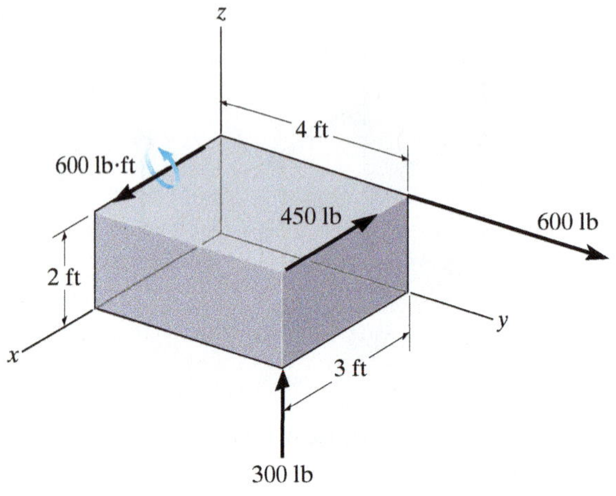

Prob. 4–136

4–137. Replace the three forces acting on the plate by a wrench. Specify the magnitude of the force and couple moment for the wrench and the point $P(x, y)$ where its line of action intersects the plate.

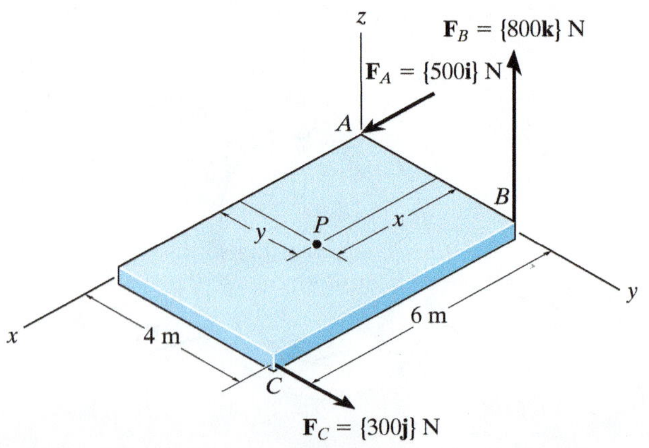

Prob. 4–137

4.9 Reduction of a Simple Distributed Loading

Sometimes, a body may be subjected to a loading that is distributed over its surface. For example, the pressure of the wind on the face of a sign, the pressure of water within a tank, or the weight of sand on the floor of a storage container, are all *distributed loadings*. The pressure exerted at each point on the surface indicates the intensity of the loading. It is measured using pascals Pa (or N/m^2) in SI units or lb/ft^2 in the U.S. Customary system.

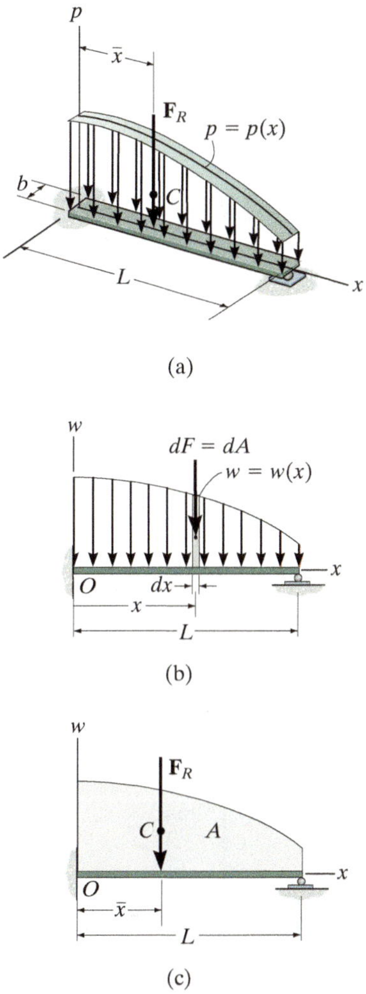

Fig. 4–48

Loading Along a Single Axis. The most common type of distributed loading encountered in engineering practice can be represented along a single axis.* For example, consider the beam (or plate) in Fig. 4–48*a* that has a constant width and is subjected to a pressure loading that varies only along the x axis. This loading can be described by the function $p = p(x)$ N/m^2. It contains only one variable x, and for this reason, we can also represent it as a *coplanar distributed load*. To do so, we multiply the loading function by the width b m of the beam, so that $w(x) = p(x)b$ N/m, Fig. 4–48*b*. Using the methods of Sec. 4.8, we can replace this coplanar parallel force system with a single equivalent resultant force $\mathbf{F}_R$ acting at a specific location on the beam, Fig. 4–48*c*.

Magnitude of Resultant Force. From Eq. 4–17 ($F_R = \Sigma F$), the magnitude of $\mathbf{F}_R$ is equivalent to the sum of all the forces in the system. In this case integration must be used since there is an infinite number of parallel forces $d\mathbf{F}$ acting on the beam, Fig. 4–48*b*. Since $d\mathbf{F}$ is acting on an element of length dx, and $w(x)$ is a force per unit length, then $dF = w(x)\,dx = dA$. In other words, the magnitude of $d\mathbf{F}$ is determined from the colored differential *area dA* under the loading curve. For the entire length L,

$$+\downarrow F_R = \Sigma F; \qquad F_R = \int_L w(x)\,dx = \int_A dA = A \qquad (4\text{–}19)$$

Therefore, the magnitude of the resultant force is equal to the area A under the loading diagram, Fig. 4–48*c*.

*The more general case of a surface loading acting on a body is considered in Sec. 9.5.

4

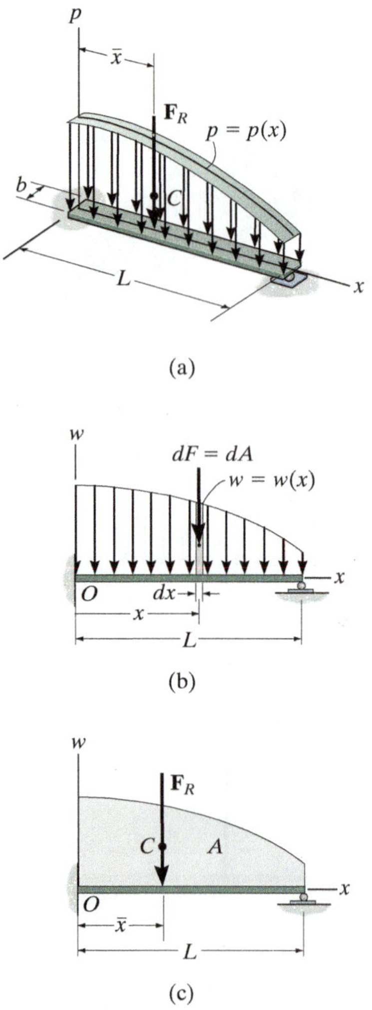

Fig. 4–48 (Repeated)

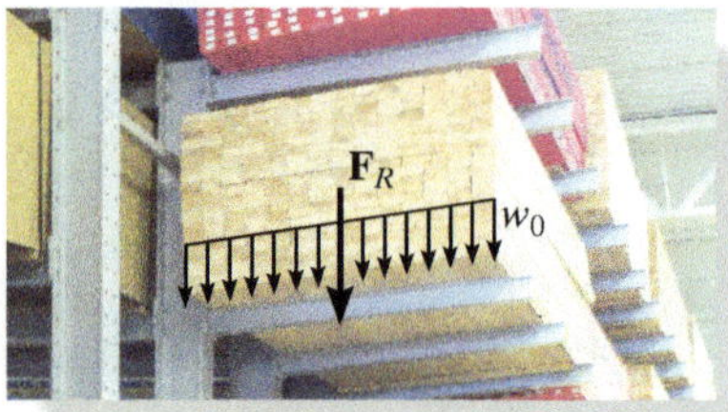

Each beam that supports this stack of lumber is subjected to a uniform loading of w_0. The resultant force is therefore equal to the area under the rectangular loading diagram. It acts through the centroid or geometric center of this area.

Location of Resultant Force. Applying Eq. 4–17 ($M_{R_O} = \Sigma M_O$), the location $\bar{x}$ of the line of action of $\mathbf{F}_R$ can be determined by equating the moments of the force resultant and the parallel force distribution about point O (the y axis). Since $d\mathbf{F}$ produces a moment of $x\,dF = xw(x)\,dx$ about O, Fig. 4–48b, then for the entire length, Fig. 4–48c,

$$\circlearrowleft + (M_R)_O = \Sigma M_O; \qquad -\bar{x}F_R = -\int_L xw(x)\,dx$$

Solving for $\bar{x}$, using Eq. 4–19, we have

$$\bar{x} = \frac{\displaystyle\int_L xw(x)\,dx}{\displaystyle\int_L w(x)\,dx} = \frac{\displaystyle\int_A x\,dA}{\displaystyle\int_A dA} \tag{4–20}$$

This coordinate $\bar{x}$, locates the geometric center or *centroid* of the *area* under the distributed loading. *In other words, the resultant force has a line of action which passes through the centroid C (geometric center) of the area under the loading diagram*, Fig. 4–48c. Detailed treatment of the integration techniques for finding the location of the centroid for areas is given in Chapter 9. In many cases, however, the distributed-loading diagram is in the shape of a rectangle, triangle, or some other simple geometric form. The centroid location for such common shapes does not have to be determined from the above equation but can be obtained directly from the tabulation given on the inside back cover.

Once $\bar{x}$ is determined, $\mathbf{F}_R$ by symmetry passes through point ($\bar{x}$, 0) on the surface of the beam, Fig. 4–48a. *Therefore, in this case the resultant force has a magnitude equal to the volume under the loading curve $p = p(x)$ and a line of action which passes through the centroid (geometric center) of this volume.*

Important Points

- Coplanar distributed loadings are defined by using a loading function $w = w(x)$ that indicates the intensity of the loading along the length of a member. This intensity is measured in N/m or lb/ft.
- The external effects caused by a coplanar distributed load acting on a body can be represented by a single resultant force.
- This resultant force is equivalent to the *area* under the loading diagram, and has a line of action that passes through the *centroid* or geometric center of this area.

EXAMPLE 4.21

Determine the magnitude and location of the equivalent resultant force acting on the shaft in Fig. 4–49*a*.

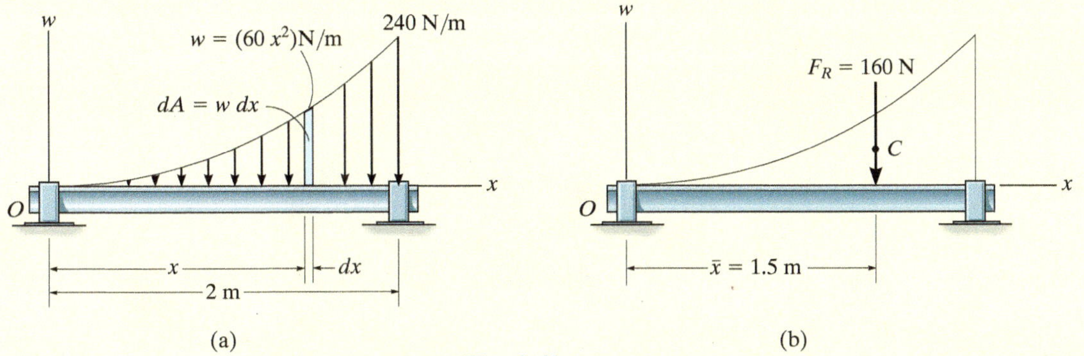

Fig. 4–49

SOLUTION

Since $w = w(x)$ is given, this problem will be solved by integration.

The differential element has an area $dA = w\,dx = 60x^2\,dx$. Applying Eq. 4–19,

$$+\downarrow F_R = \Sigma F;$$

$$F_R = \int_A dA = \int_0^{2\text{ m}} 60x^2\,dx = 60\left(\frac{x^3}{3}\right)\Bigg|_0^{2\text{ m}} = 60\left(\frac{2^3}{3} - \frac{0^3}{3}\right)$$

$$= 160\text{ N} \qquad \textit{Ans.}$$

The location $\bar{x}$ of $\mathbf{F}_R$ measured from O, Fig. 4–49*b*, is determined from Eq. 4–20.

$$\bar{x} = \frac{\int_A x\,dA}{\int_A dA} = \frac{\int_0^{2\text{ m}} x(60x^2)\,dx}{160\text{ N}} = \frac{60\left(\frac{x^4}{4}\right)\Big|_0^{2\text{ m}}}{160\text{ N}} = \frac{60\left(\frac{2^4}{4} - \frac{0^4}{4}\right)}{160\text{ N}}$$

$$= 1.5\text{ m} \qquad \textit{Ans.}$$

NOTE: These results can be checked by using the table on the inside back cover, where it is shown that formula for an exparabolic area of length a, height b, and shape shown in Fig. 4–49*a*, we have

$$A = \frac{ab}{3} = \frac{2\text{ m}(240\text{ N/m})}{3} = 160\text{ N and } \bar{x} = \frac{3}{4}a = \frac{3}{4}(2\text{ m}) = 1.5\text{ m}$$

4

EXAMPLE 4.22

A distributed loading of $p = (800x)$ Pa acts over the top surface of the beam shown in Fig. 4–50*a*. Determine the magnitude and location of the equivalent resultant force.

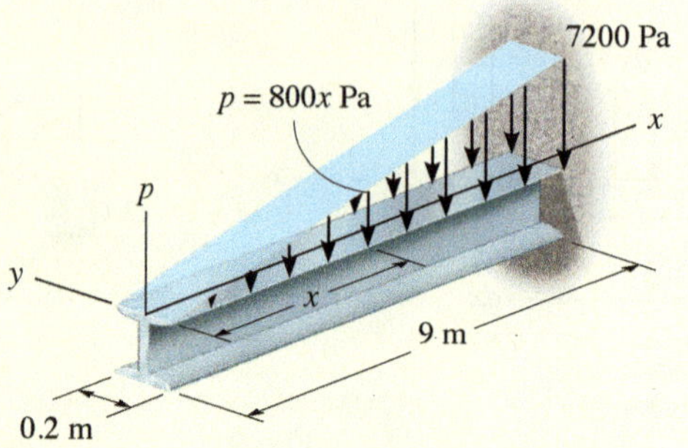

(a)

SOLUTION

Since the loading intensity is uniform along the width of the beam (the y axis), the loading can be viewed in two dimensions as shown in Fig. 4–50*b*. Here

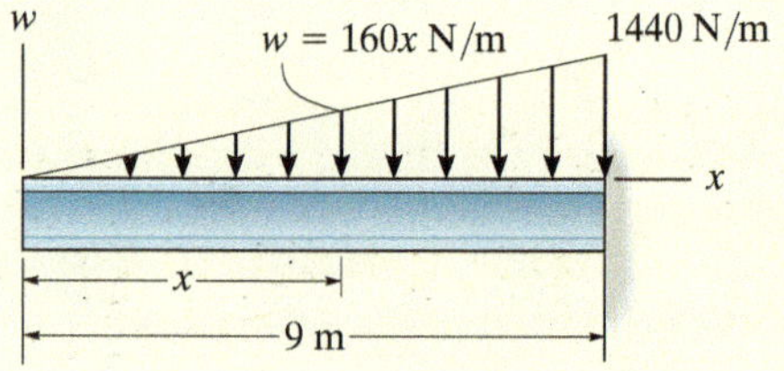

(b)

$$w = (800x \text{ N/m}^2)(0.2 \text{ m})$$
$$= (160x) \text{ N/m}$$

At $x = 9$ m, note that $w = 1440$ N/m. Although we may again apply Eqs. 4–19 and 4–20 as in the previous example, it is simpler to use the table on the inside back cover.

The magnitude of the resultant force is equivalent to the area of the triangle.

$$F_R = \tfrac{1}{2}(9 \text{ m})(1440 \text{ N/m}) = 6480 \text{ N} = 6.48 \text{ kN} \qquad \textit{Ans.}$$

The line of action of $\mathbf{F}_R$ passes through the *centroid C* of this triangle. Hence,

$$\bar{x} = 9 \text{ m} - \tfrac{1}{3}(9 \text{ m}) = 6 \text{ m} \qquad \textit{Ans.}$$

The results are shown in Fig. 4–50*c*.

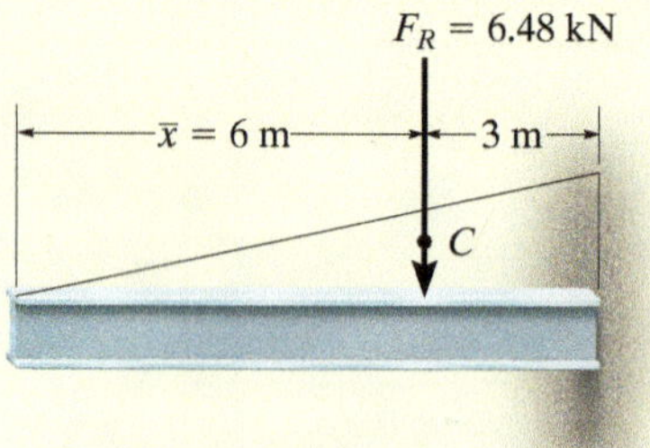

(c)

Fig. 4–50

NOTE: We may also view the resultant $\mathbf{F}_R$ as *acting* through the *centroid* of the *volume* of the loading diagram $p = p(x)$ in Fig. 4–50*a*. Hence $\mathbf{F}_R$ intersects the x–y plane at the point (6 m, 0). Furthermore, the magnitude of $\mathbf{F}_R$ is equal to the volume under the loading diagram; i.e.,

$$F_R = V = \tfrac{1}{2}(7200 \text{ N/m}^2)(9 \text{ m})(0.2 \text{ m}) = 6.48 \text{ kN} \qquad \textit{Ans.}$$

EXAMPLE 4.23

The granular material exerts the distributed loading on the beam as shown in Fig. 4–51*a*. Determine the magnitude and location of the equivalent resultant of this load.

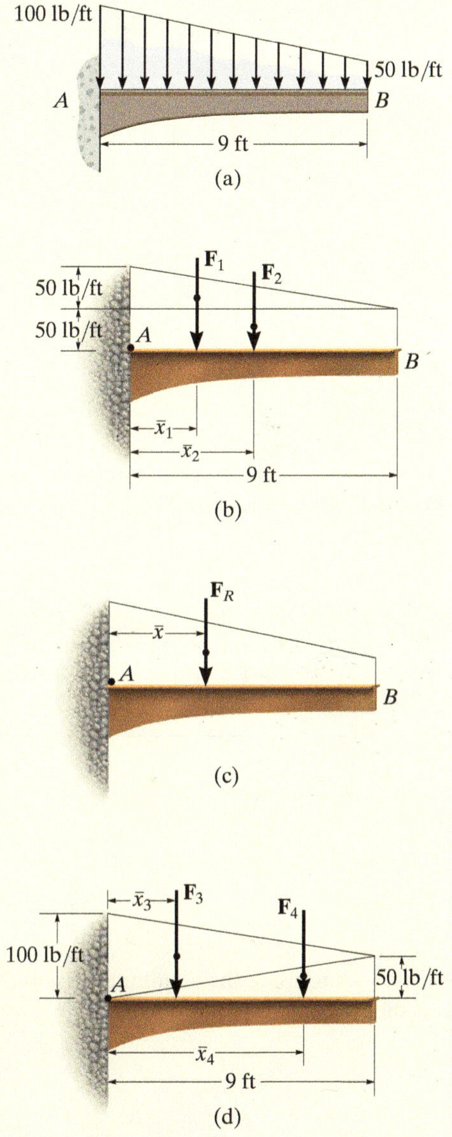

Fig. 4–51

SOLUTION

The area of the loading diagram is a *trapezoid*, and therefore the solution can be obtained directly from the area and centroid formulas for a trapezoid listed on the inside back cover. Since these formulas are not easily remembered, instead we will solve this problem by using "composite" areas. Here we will divide the trapezoidal loading into a rectangular and triangular loading as shown in Fig. 4–51*b*. The magnitude of the force represented by each of these loadings is equal to its associated *area*,

$$F_1 = \tfrac{1}{2}(9\text{ ft})(50\text{ lb/ft}) = 225\text{ lb}$$

$$F_2 = (9\text{ ft})(50\text{ lb/ft}) = 450\text{ lb}$$

The lines of action of these parallel forces act through the respective *centroids* of their associated areas and therefore intersect the beam at

$$\bar{x}_1 = \tfrac{1}{3}(9\text{ ft}) = 3\text{ ft}$$

$$\bar{x}_2 = \tfrac{1}{2}(9\text{ ft}) = 4.5\text{ ft}$$

The two parallel forces $\mathbf{F}_1$ and $\mathbf{F}_2$ can be reduced to a single resultant $\mathbf{F}_R$. The magnitude of $\mathbf{F}_R$ is

$$+\downarrow F_R = \Sigma F; \qquad F_R = 225 + 450 = 675\text{ lb} \qquad \textit{Ans.}$$

We can find the location of $\mathbf{F}_R$ with reference to point A, Fig. 4–51*b* and 4–51*c*. We require

$$\circlearrowleft + (M_R)_A = \Sigma M_A; \qquad \bar{x}(675) = 3(225) + 4.5(450)$$

$$\bar{x} = 4\text{ ft} \qquad \textit{Ans.}$$

NOTE: The trapezoidal area in Fig. 4–51*a* can also be divided into two triangular areas as shown in Fig. 4–51*d*. In this case

$$F_3 = \tfrac{1}{2}(9\text{ ft})(100\text{ lb/ft}) = 450\text{ lb}$$

$$F_4 = \tfrac{1}{2}(9\text{ ft})(50\text{ lb/ft}) = 225\text{ lb}$$

and

$$\bar{x}_3 = \tfrac{1}{3}(9\text{ ft}) = 3\text{ ft}$$

$$\bar{x}_4 = 9\text{ ft} - \tfrac{1}{3}(9\text{ ft}) = 6\text{ ft}$$

Using these results, show that again $F_R = 675$ lb and $\bar{x} = 4$ ft.

FUNDAMENTAL PROBLEMS

F4–37. Determine the resultant force and specify where it acts on the beam measured from A.

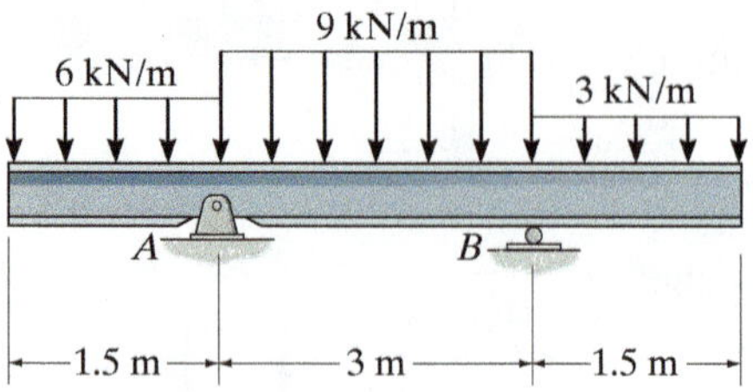

F4–37

F4–38. Determine the resultant force and specify where it acts on the beam measured from A.

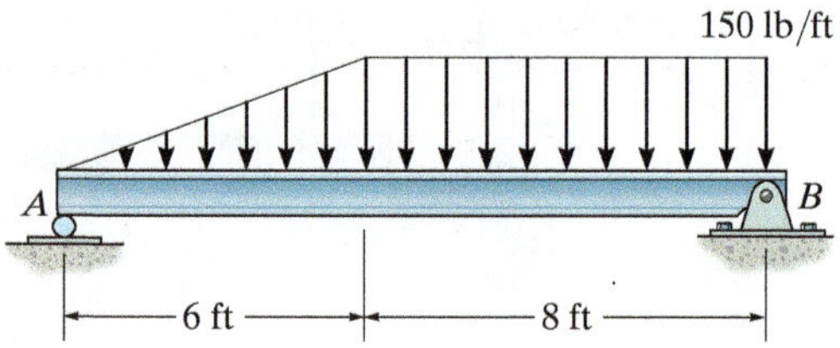

F4–38

F4–39. Determine the resultant force and specify where it acts on the beam measured from A.

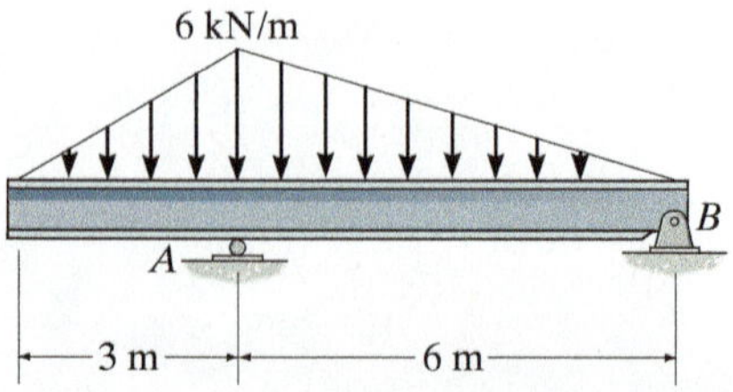

F4–39

F4–40. Determine the resultant force and specify where it acts on the beam measured from A.

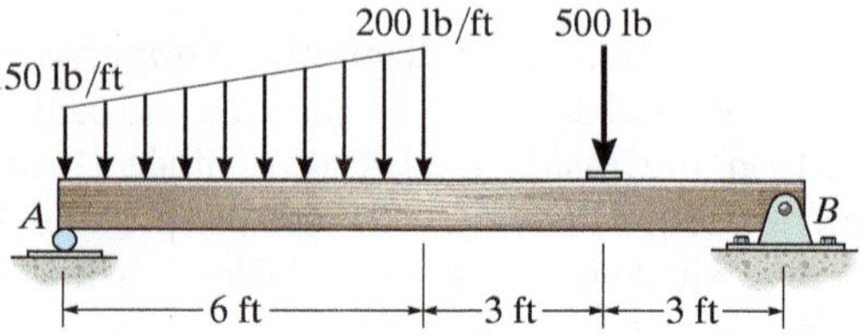

F4–40

F4–41. Determine the resultant force and specify where it acts on the beam measured from A.

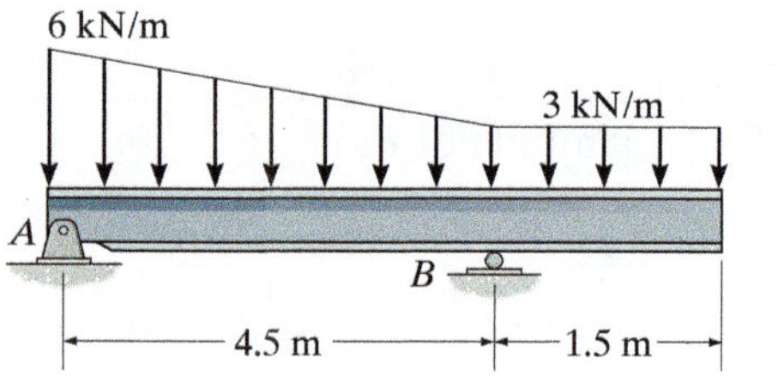

F4–41

F4–42. Determine the resultant force and specify where it acts on the beam measured from A.

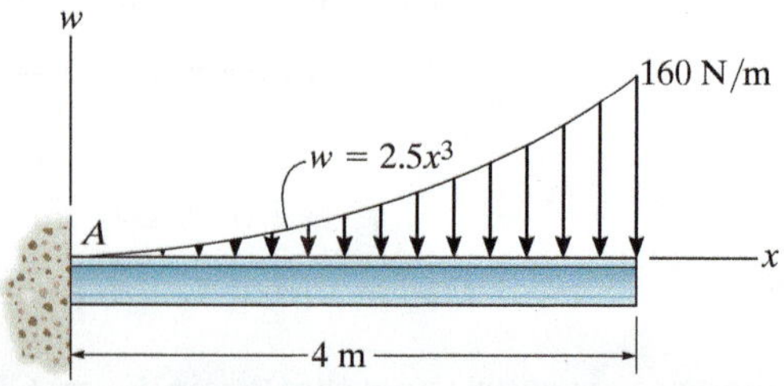

F4–42

4

PROBLEMS

4–138. The loading on the bookshelf is distributed as shown. Determine the magnitude of the equivalent resultant location, measured from point O.

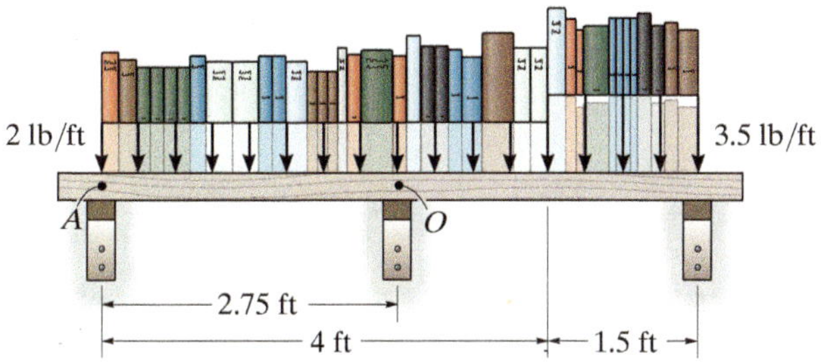

Prob. 4–138

4–139. Replace the distributed loading with an equivalent resultant force, and specify its location on the beam measured from point O.

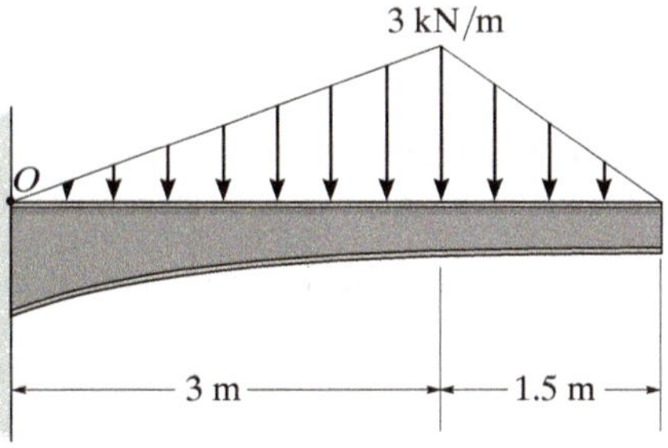

Prob. 4–139

***4–140.** Replace the loading by an equivalent force and couple moment acting at point O.

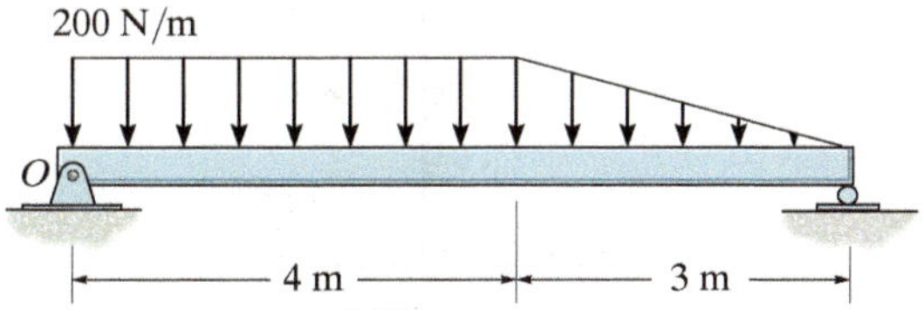

Prob. 4–140

4–141. The column is used to support the floor which exerts a force of 3000 lb on the top of the column. The effect of soil pressure along its side is distributed as shown. Replace this loading by an equivalent resultant force and specify where it acts along the column, measured from its base A.

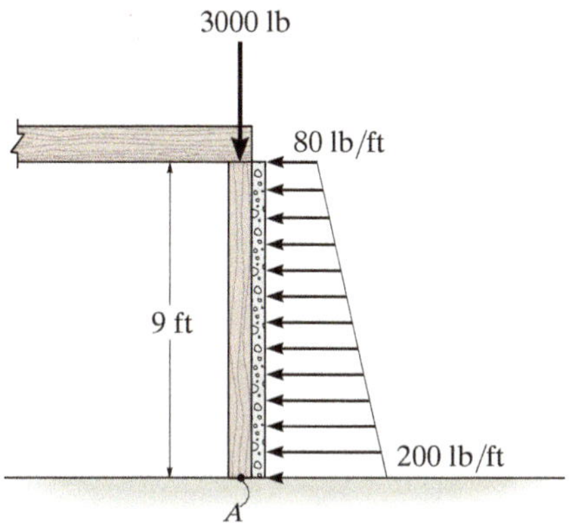

Prob. 4–141

4–142. Replace the loading by an equivalent resultant force and specify its location on the beam, measured from point B.

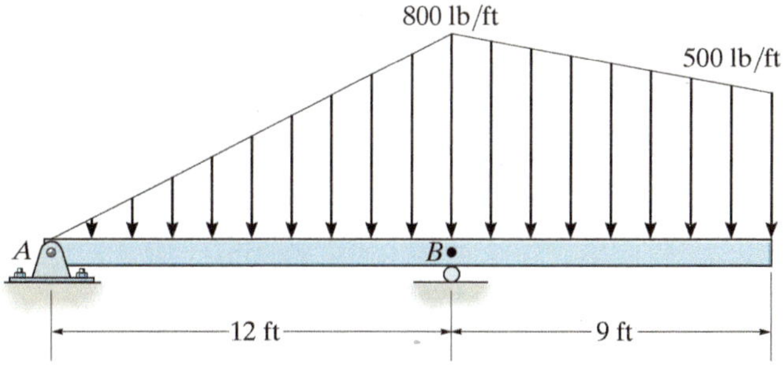

Prob. 4–142

4–143. The masonry support creates the loading distribution acting on the end of the beam. Simplify this load to a single resultant force and specify its location measured from point O.

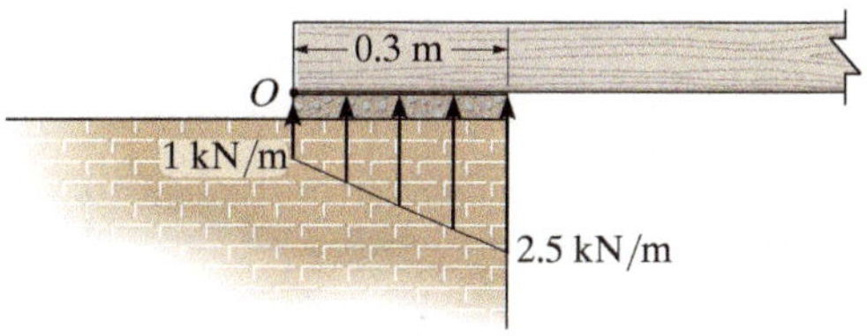

Prob. 4–143

*4–144. The distribution of soil loading on the bottom of a building slab is shown. Replace this loading by an equivalent resultant force and specify its location, measured from point O.

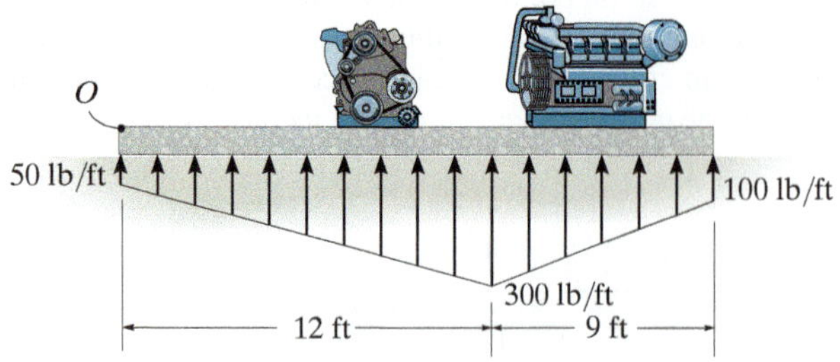

Prob. 4–144

4–145. Replace the distributed loading by an equivalent resultant force, and specify its location on the beam, measured from the pin at C.

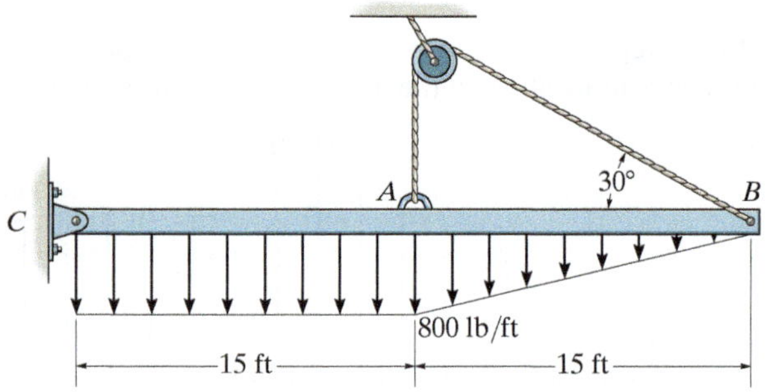

Prob. 4–145

4–146. Replace the distributed loading with an equivalent resultant force, and specify its location on the beam measured from point A.

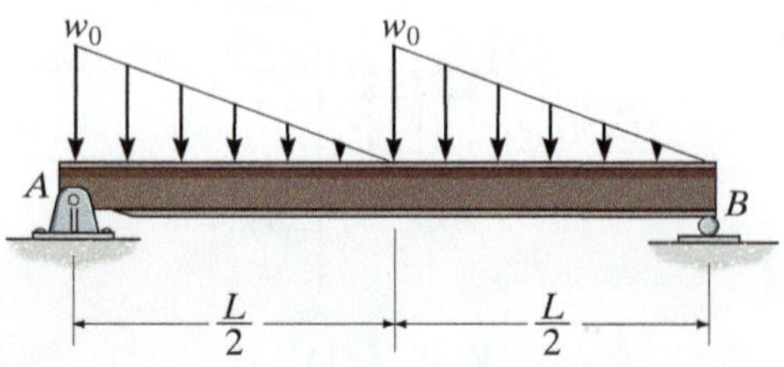

Prob. 4–146

4–147. The beam is subjected to the distributed loading. Determine the length b of the uniform load and its position a on the beam such that the resultant force and couple moment acting on the beam are zero.

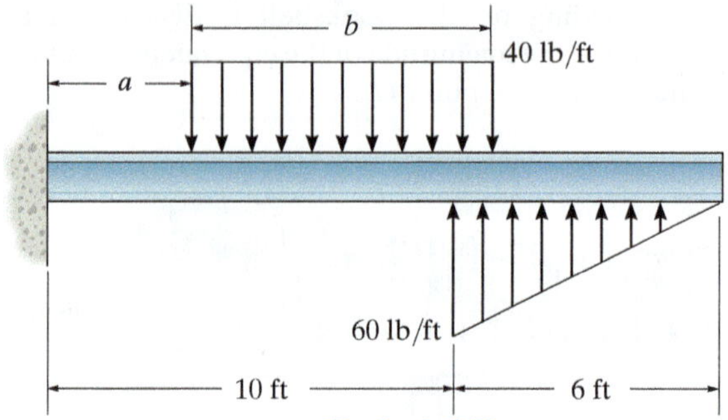

Prob. 4–147

*4–148. If the soil exerts a trapezoidal distribution of load on the bottom of the footing, determine the intensities w_1 and w_2 of this distribution needed to support the column loadings.

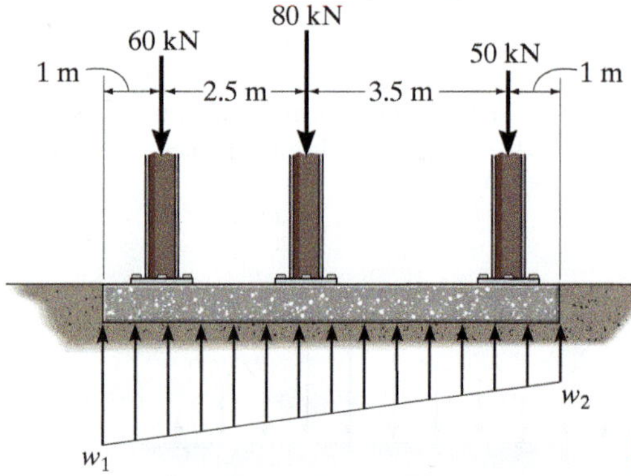

Prob. 4–148

4–149. The post is embedded into a concrete footing so that it is fixed supported. If the reaction of the concrete on the post can be approximated by the distributed loading shown, determine the intensity of w_1 and w_2 so that the resultant force and couple moment on the post due to the loadings are both zero.

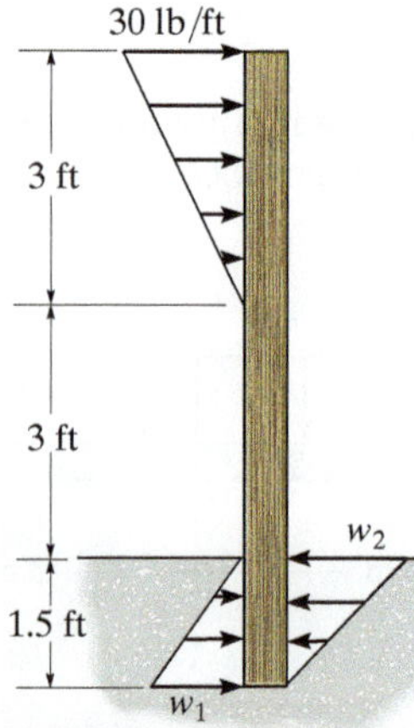

Prob. 4–149

4–150. Replace the loading by an equivalent force and couple moment acting at point O.

4–151. Replace the loading by a single resultant force, and specify the location of the force measured from point O.

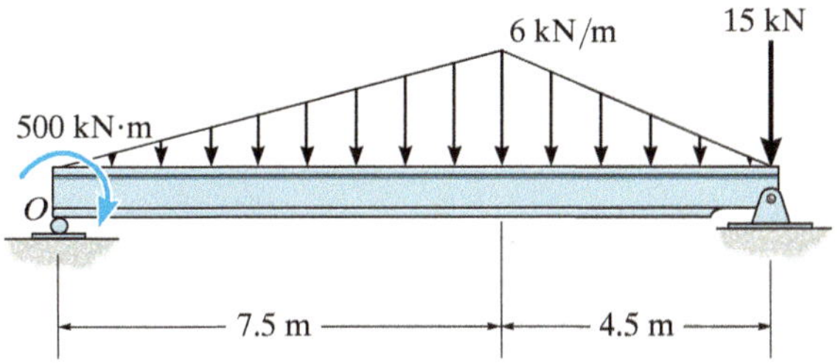

Probs. 4–150/151

4–154. Replace the distributed loading by an equivalent resultant force and specify where its line of action intersects member AB, measured from A.

4–155. Replace the distributed loading by an equivalent resultant force and specify where its line of action intersects member BC, measured from C.

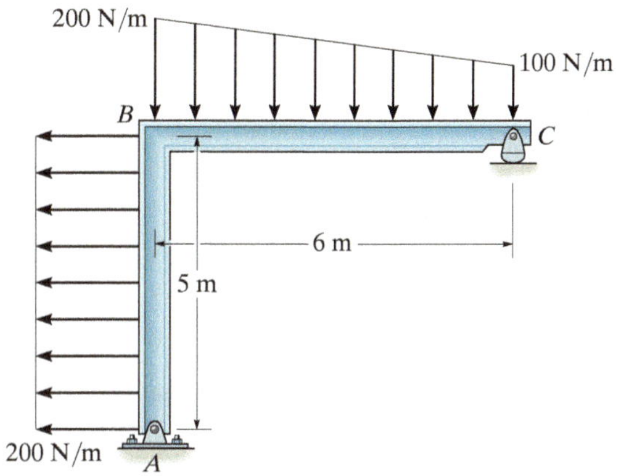

Probs. 4–154/155

***4–152.** Replace the loading by an equivalent resultant force and couple moment at point A.

4–153. Replace the loading by an equivalent resultant force and couple moment acting at point B.

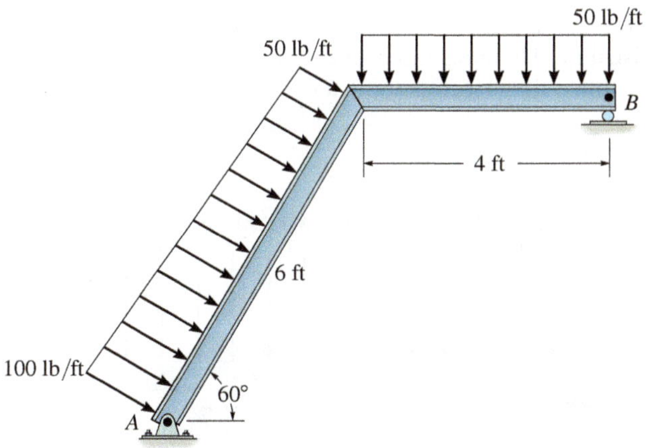

Probs. 4–152/153

***4–156.** Replace the distributed loading with an equivalent resultant force, and specify its location on the beam measured from point A.

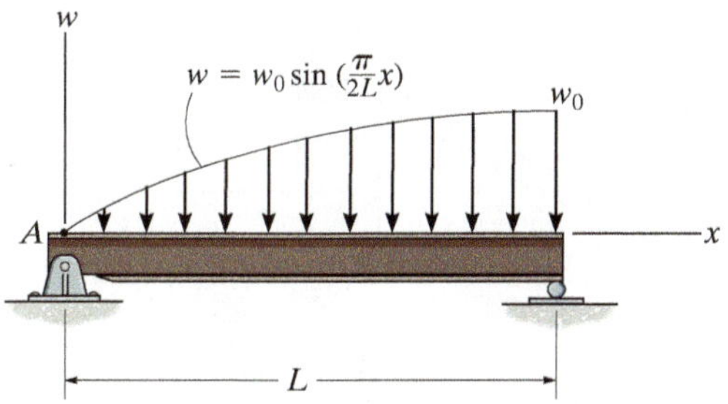

Prob. 4–156

4–157. Replace the distributed loading with an equivalent resultant force, and specify its location on the beam measured from point A.

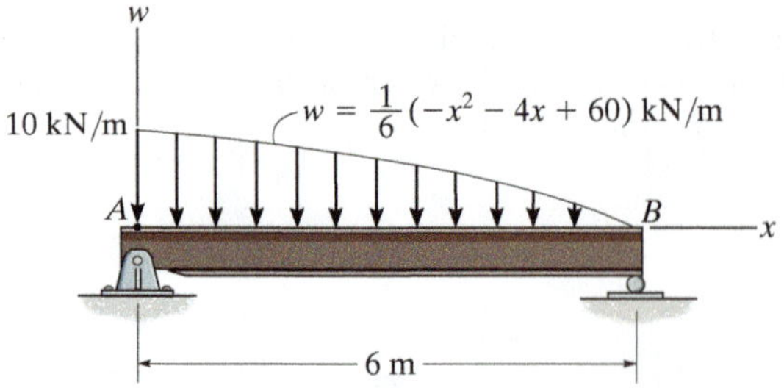

Prob. 4–157

4–158. Replace the distributed loading with an equivalent resultant force, and specify its location on the beam measured from point A.

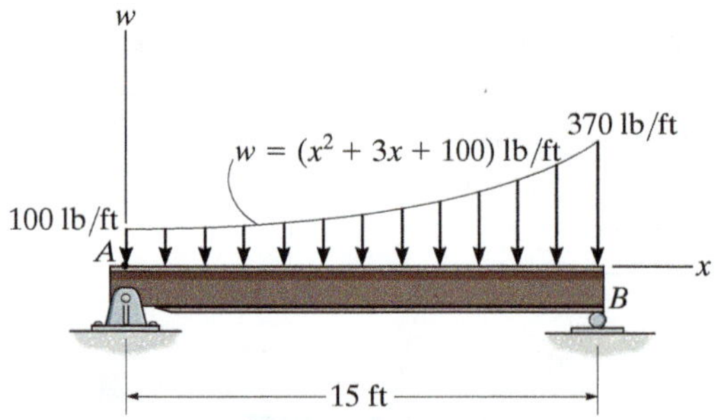

Prob. 4–158

4–159. Wet concrete exerts a pressure distribution along the wall of the form. Determine the resultant force of this distribution and specify the height h where the bracing strut should be placed so that it lies through the line of action of the resultant force. The wall has a width of 5 m.

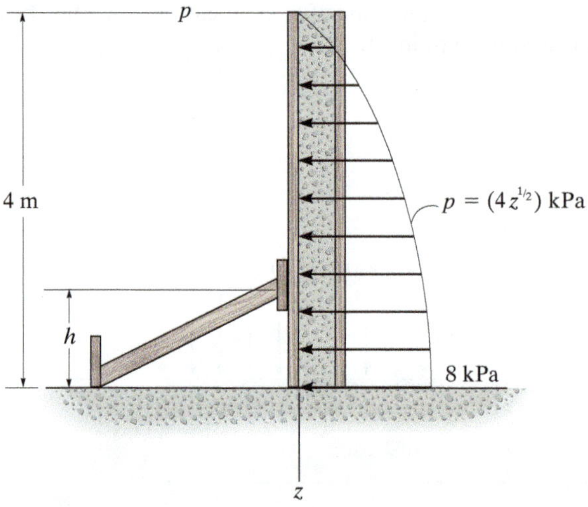

Prob. 4–159

***4–160.** Replace the loading by an equivalent force and couple moment acting at point O.

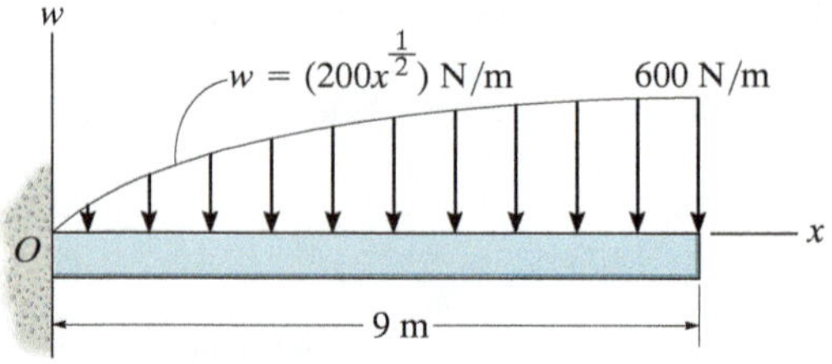

Prob. 4–160

4–161. Determine the magnitude of the equivalent resultant force of the distributed load and specify its location on the beam measured from point A.

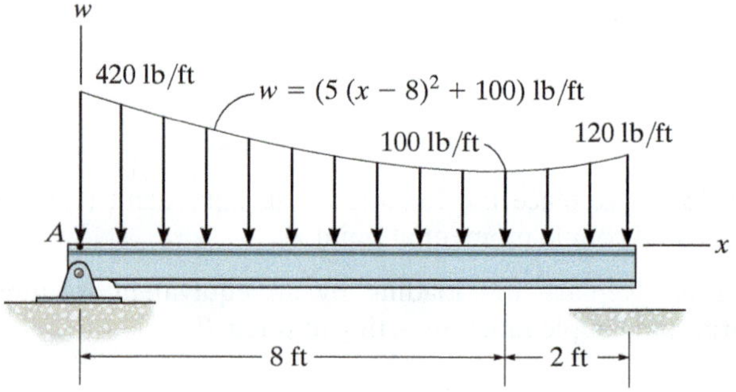

Prob. 4–161

■**4–162.** Determine the equivalent resultant force of the distributed loading and its location, measured from point A. Evaluate the integrals using a numerical method.

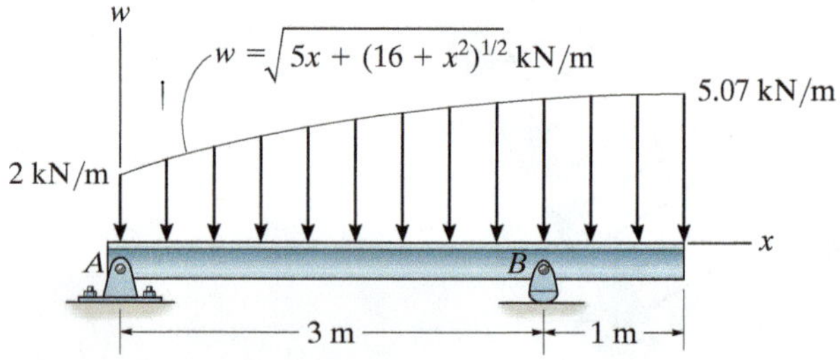

Prob. 4–162

CHAPTER REVIEW

Moment of Force—Scalar Definition

A force produces a turning effect or moment about a point O that does not lie on its line of action. In scalar form, the moment *magnitude* is the product of the force and the moment arm or perpendicular distance from point O to the line of action of the force.

The *direction* of the moment is defined using the right-hand rule. $\mathbf{M}_O$ always acts along an axis perpendicular to the plane containing $\mathbf{F}$ and d, and passes through the point O.

$$M_O = Fd$$

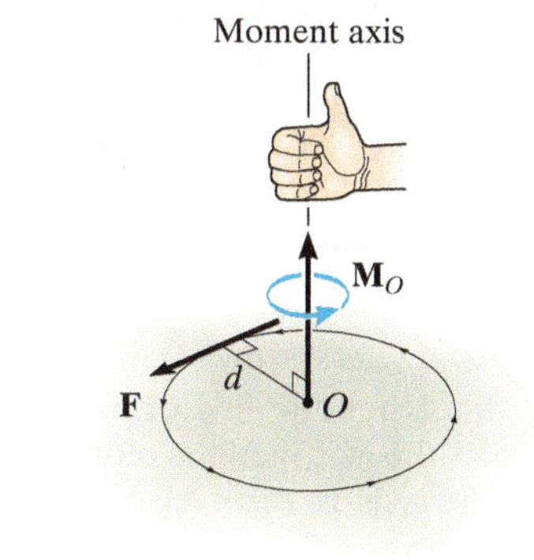

Rather than finding d, it is normally easier to resolve the force into its x and y components, determine the moment of each component about the point, and then sum the results. This is called the principle of moments.

$$M_O = Fd = F_x y - F_y x$$

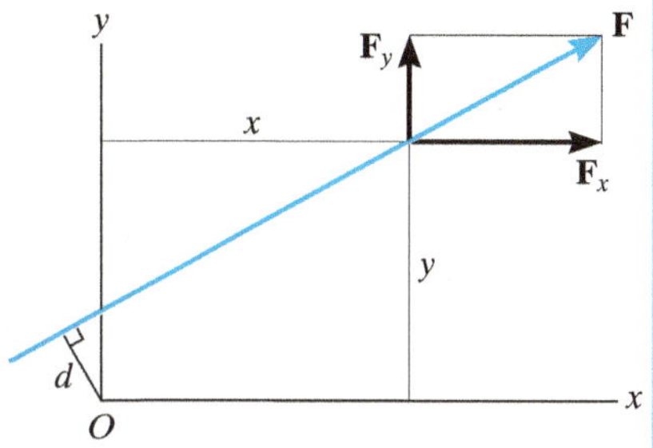

Moment of a Force—Vector Definition

Since three-dimensional geometry is generally more difficult to visualize, the vector cross product should be used to determine the moment. Here $\mathbf{M}_O = \mathbf{r} \times \mathbf{F}$, where $\mathbf{r}$ is a position vector that extends from point O to any point A, B, or C on the line of action of $\mathbf{F}$.

$$\mathbf{M}_O = \mathbf{r}_A \times \mathbf{F} = \mathbf{r}_B \times \mathbf{F} = \mathbf{r}_C \times \mathbf{F}$$

If the position vector $\mathbf{r}$ and force $\mathbf{F}$ are expressed as Cartesian vectors, then the cross product results from the expansion of a determinant.

$$\mathbf{M}_O = \mathbf{r} \times \mathbf{F} = \begin{vmatrix} \mathbf{i} & \mathbf{j} & \mathbf{k} \\ r_x & r_y & r_z \\ F_x & F_y & F_z \end{vmatrix}$$

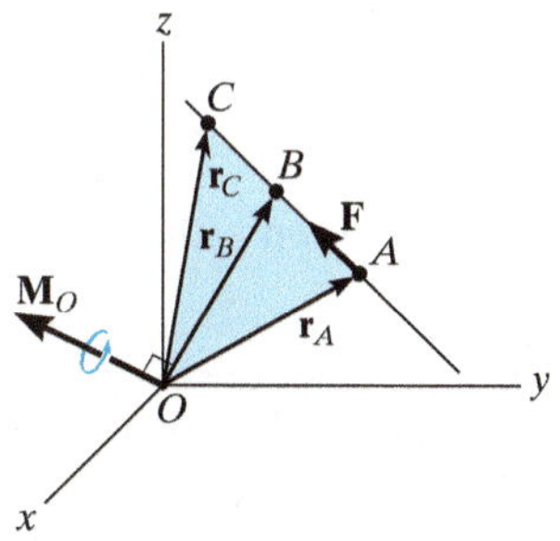

Moment about an Axis

If the moment of a force **F** is to be determined about an arbitrary axis *a*, then for a scalar solution the moment arm, or shortest distance d_a from the line of action of the force to the axis must be used. This distance is perpendicular to both the axis and the force line of action.

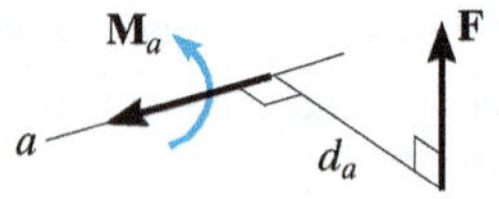

Note that when the line of action of **F** intersects the axis then the moment of **F** about the axis is zero. Also, when the line of action of **F** is parallel to the axis, the moment of **F** about the axis is zero.

In three dimensions, the scalar triple product should be used. Here $\mathbf{u}_a$ is the unit vector that specifies the direction of the axis, and **r** is a position vector that is directed from any point on the axis to any point on the line of action of the force. If M_a is calculated as a negative scalar, then the sense of direction of $\mathbf{M}_a$ is opposite to $\mathbf{u}_a$.

$$M_a = Fd_a$$

$$M_a = \mathbf{u}_a \cdot (\mathbf{r} \times \mathbf{F}) = \begin{vmatrix} u_{a_x} & u_{a_y} & u_{a_z} \\ r_x & r_y & r_z \\ F_x & F_y & F_z \end{vmatrix}$$

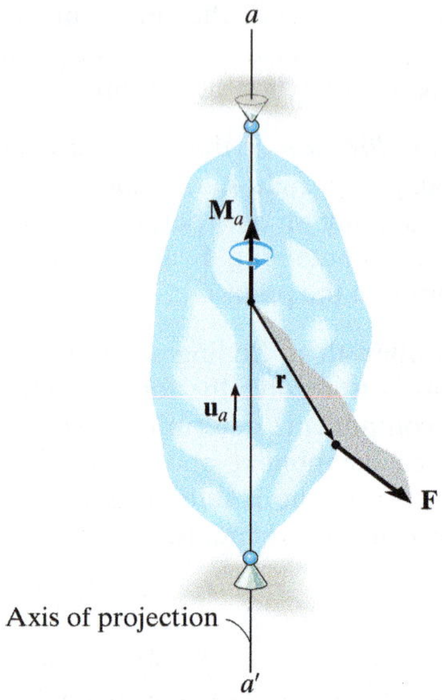

Couple Moment

A couple consists of two equal but opposite forces that act a perpendicular distance *d* apart. Couples tend to produce a rotation without translation.

$$M = Fd$$

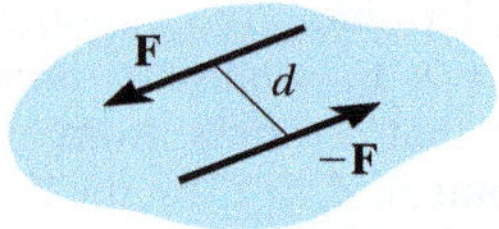

The magnitude of the couple moment is $M = Fd$, and its direction is established using the right-hand rule.

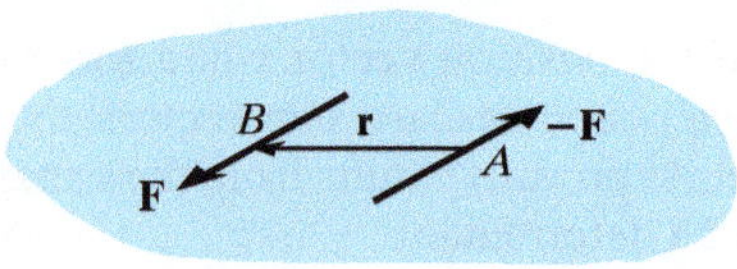

If the vector cross product is used to determine the moment of a couple, then **r** extends from any point on the line of action of one of the forces to any point on the line of action of the other force **F** that is used in the cross product.

$$\mathbf{M} = \mathbf{r} \times \mathbf{F}$$

Simplification of a Force and Couple System

Any system of forces and couples can be reduced to a single resultant force and resultant couple moment acting at a point. The resultant force is the sum of all the forces in the system, $\mathbf{F}_R = \Sigma\mathbf{F}$, and the resultant couple moment is equal to the sum of all the moments of the forces about the point and couple moments. $\mathbf{M}_{R_O} = \Sigma\mathbf{M}_O + \Sigma\mathbf{M}$.

Further simplification to a single resultant force is possible provided the force system is concurrent, coplanar, or parallel. To find the location of the resultant force from a point, it is necessary to equate the moment of the resultant force about the point to the moment of the forces and couples in the system about the same point.

If the resultant force and couple moment at a point are not perpendicular to one another, then this system can be reduced to a wrench, which consists of the resultant force and collinear couple moment.

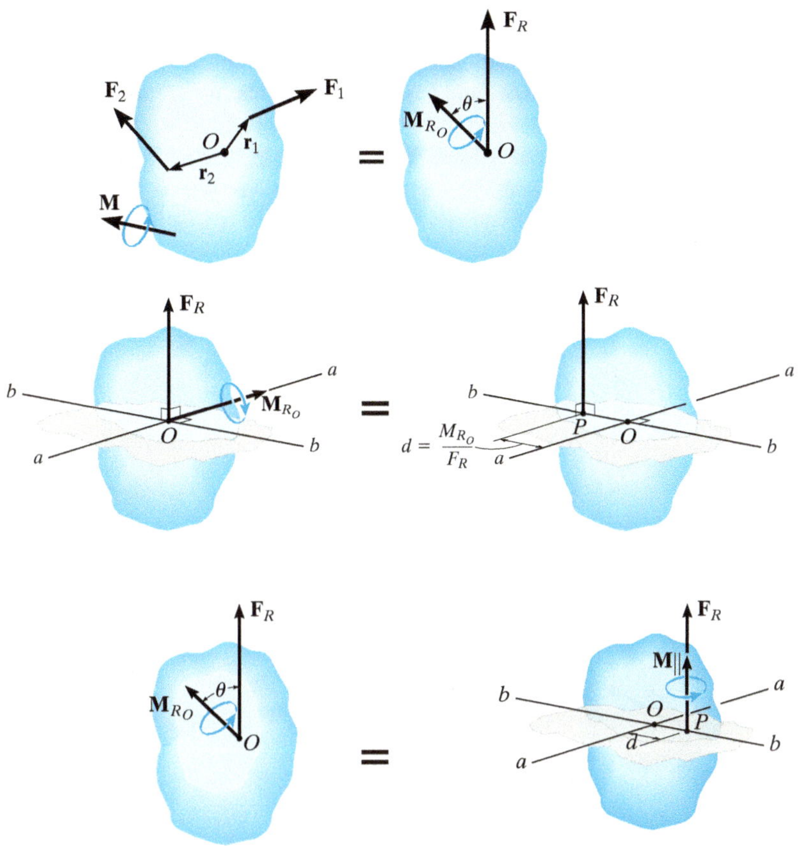

Coplanar Distributed Loading

A simple distributed loading can be represented by its resultant force, which is equivalent to the *area* under the loading curve. This resultant has a line of action that passes through the *centroid* or geometric center of the area or volume under the loading diagram.

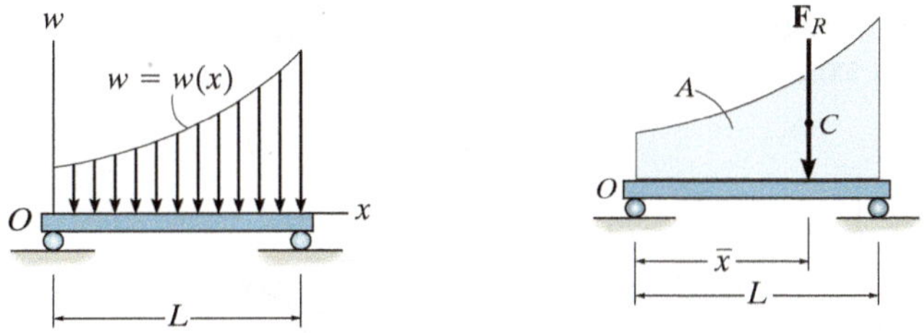

REVIEW PROBLEMS

4–163. Determine the resultant couple moment of the two couples that act on the assembly. Member OB lies in the x–z plane.

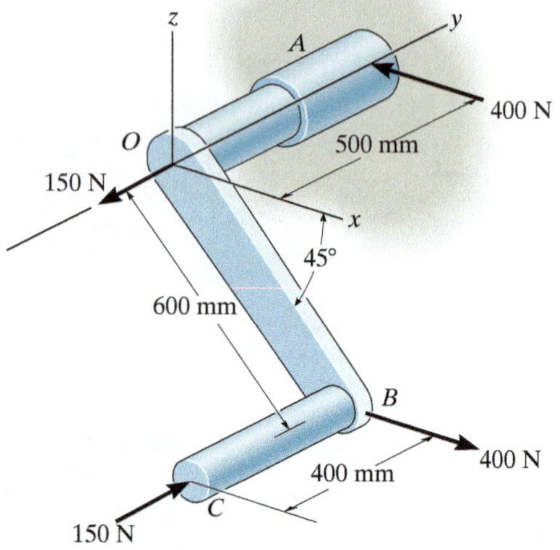

Prob. 4–163

*__4–164.__ The horizontal 30-N force acts on the handle of the wrench. What is the magnitude of the moment of this force about the z axis?

4–165. The horizontal 30-N force acts on the handle of the wrench. Determine the moment of this force about point O. Specify the coordinate direction angles α, β, γ of the moment axis.

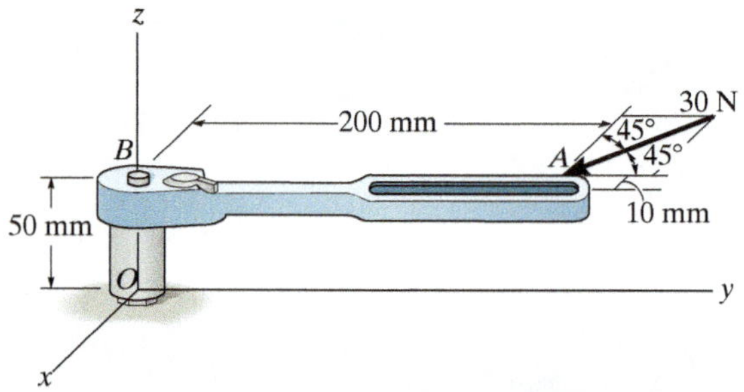

Probs. 4–164/165

4–166. The forces and couple moments that are exerted on the toe and heel plates of a snow ski are $\mathbf{F}_t = \{-50\mathbf{i} + 80\mathbf{j} - 158\mathbf{k}\}$ N, $\mathbf{M}_t = \{-6\mathbf{i} + 4\mathbf{j} + 2\mathbf{k}\}$ N·m, and $\mathbf{F}_h = \{-20\mathbf{i} + 60\mathbf{j} - 250\mathbf{k}\}$ N, $\mathbf{M}_h = \{-20\mathbf{i} + 8\mathbf{j} + 3\mathbf{k}\}$ N·m, respectively. Replace this system by an equivalent force and couple moment acting at point P. Express the results in Cartesian vector form.

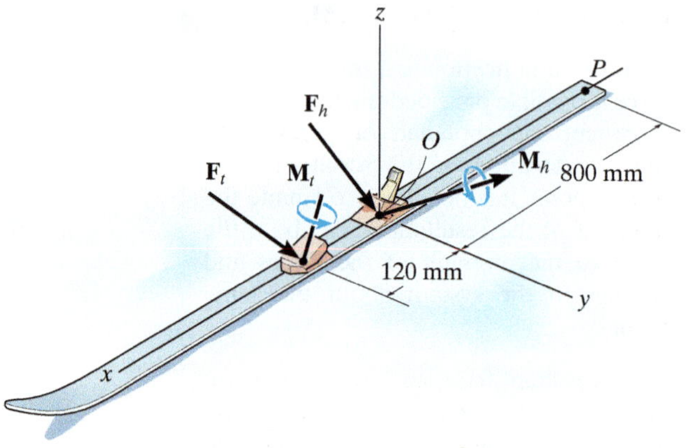

Prob. 4–166

4–167. Replace the force $\mathbf{F}$ having a magnitude of $F = 50$ lb and acting at point A by an equivalent force and couple moment at point C.

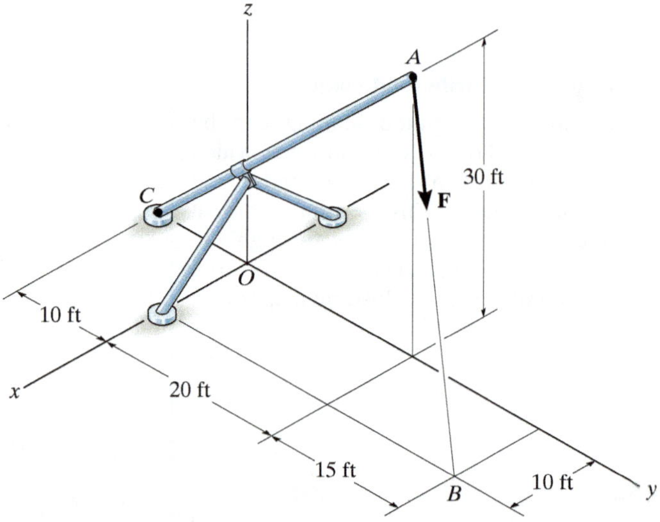

Prob. 4–167

*4–168. Determine the coordinate direction angles α, β, γ of **F**, which is applied to the end A of the pipe assembly, so that the moment of **F** about O is zero.

4–169. Determine the moment of the force **F** about point O. The force has coordinate direction angles of $\alpha = 60°$, $\beta = 120°$, $\gamma = 45°$. Express the result as a Cartesian vector.

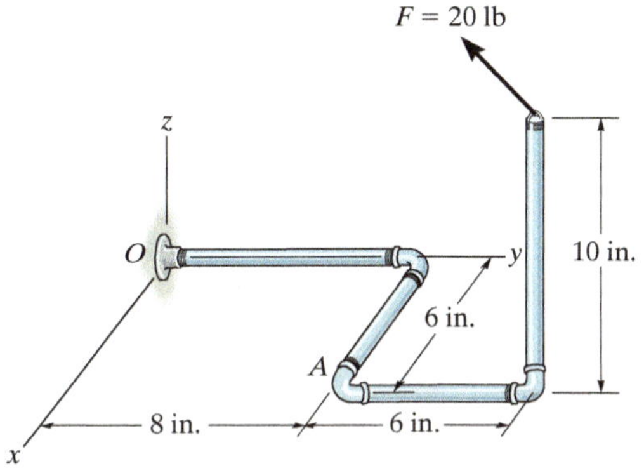

Probs. 4–168/169

4–170. Determine the moment of the force $\mathbf{F}_c$ about the door hinge at A. Express the result as a Cartesian vector.

4–171. Determine the magnitude of the moment of the force $\mathbf{F}_c$ about the hinged axis *aa* of the door.

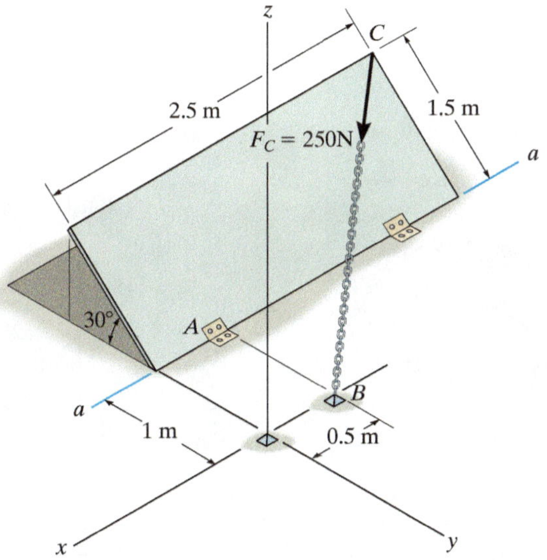

Probs. 4–170/171

*4–172. The boom has a length of 30 ft, a weight of 800 lb, and mass center at G. If the maximum moment that can be developed by the motor at A is $M = 20(10^3)$ lb · ft, determine the maximum load W, having a mass center at G', that can be lifted. Take $\theta = 30°$.

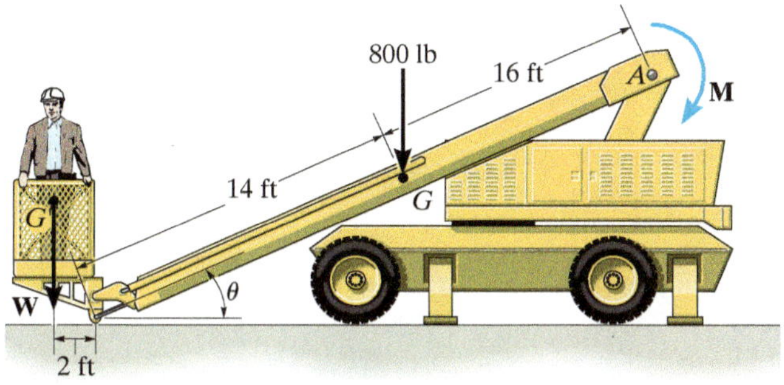

Prob. 4–172

4–173. If it takes a force of $F = 125$ lb to pull the nail out, determine the smallest vertical force **P** that must be applied to the handle of the crowbar. *Hint:* This requires the moment of **F** about point A to be equal to the moment of **P** about A. Why?

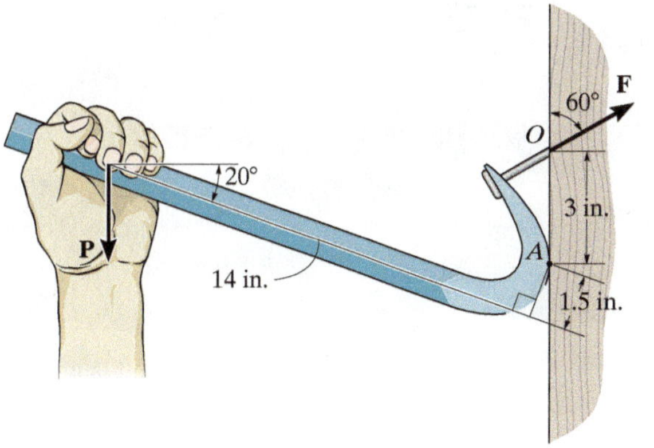

Prob. 4–173

Chapter 5

It is important to be able to determine the forces in the cables used to support this submarine to insure that they do not fail. In this chapter we will study how to apply equilibrium methods to determine the forces acting on the supports of a rigid body such as this.

Equilibrium of a Rigid Body

CHAPTER OBJECTIVES

- To develop the equations of equilibrium for a rigid body.
- To introduce the concept of the free-body diagram for a rigid body.
- To show how to solve rigid-body equilibrium problems using the equations of equilibrium.

5.1 Conditions for Rigid-Body Equilibrium

In this section, we will develop both the necessary and sufficient conditions for the equilibrium of the rigid body in Fig. 5–1*a*. As shown, this body is subjected to an external force and couple moment system that is the result of the effects of gravitational, electrical, magnetic, or contact forces caused by adjacent bodies. The internal forces caused by interactions between particles within the body are not shown in this figure because these forces occur in equal but opposite collinear pairs and hence will cancel out, a consequence of Newton's third law.

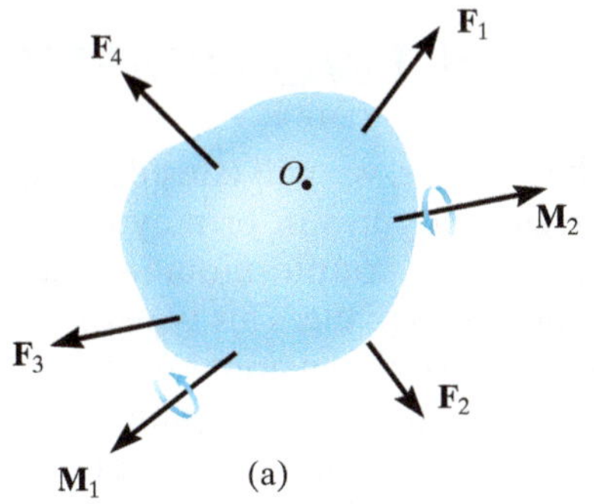

Fig. 5–1

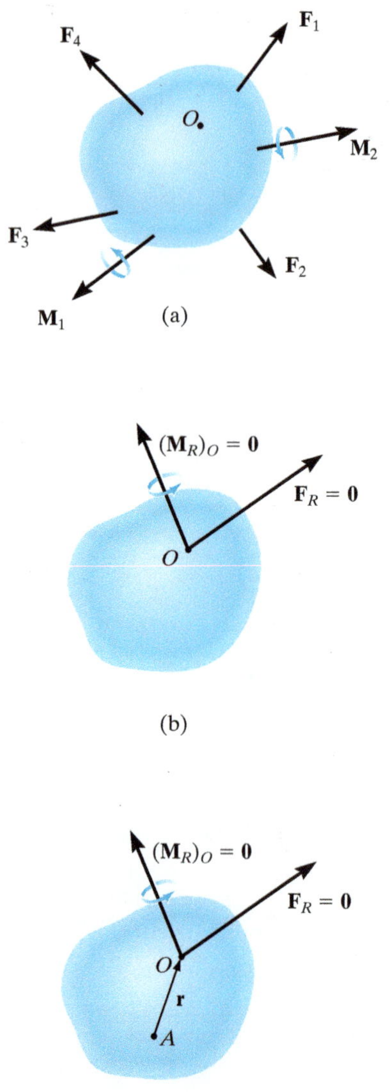

(c)

Fig. 5–1

Using the methods of the previous chapter, the force and couple moment system acting on a body can be reduced to an equivalent resultant force and resultant couple moment at any arbitrary point O on or off the body, Fig. 5–1*b*. If this resultant force and couple moment are both equal to zero, then the body is said to be in *equilibrium*. Mathematically, the equilibrium of a body is expressed as

$$\mathbf{F}_R = \Sigma \mathbf{F} = \mathbf{0}$$
$$(\mathbf{M}_R)_O = \Sigma \mathbf{M}_O = \mathbf{0} \tag{5–1}$$

The first of these equations states that the sum of the forces acting on the body is equal to *zero*. The second equation states that the sum of the moments of all the forces in the system about point O, added to all the couple moments, is equal to *zero*. These two equations are not only necessary for equilibrium, they are also sufficient. To show this, consider summing moments about some other point, such as point A in Fig. 5–1*c*. We require

$$\Sigma \mathbf{M}_A = \mathbf{r} \times \mathbf{F}_R + (\mathbf{M}_R)_O = \mathbf{0}$$

Since $\mathbf{r} \neq \mathbf{0}$, this equation is satisfied if Eqs. 5–1 are satisfied, namely $\mathbf{F}_R = \mathbf{0}$ and $(\mathbf{M}_R)_O = \mathbf{0}$.

When applying the equations of equilibrium, we will assume that the body remains rigid. In reality, however, all bodies deform when subjected to loads. Although this is the case, most engineering materials such as steel and concrete are very rigid and so their deformation is usually very small. Therefore, when applying the equations of equilibrium, we can generally assume that the body will remain *rigid* and *not deform* under the applied load without introducing any significant error. This way the direction of the applied forces and their moment arms with respect to a fixed reference remain the same both before and after the body is loaded.

EQUILIBRIUM IN TWO DIMENSIONS

In the first part of the chapter, we will consider the case where the force system acting on a rigid body lies in or may be projected onto a *single* plane and, furthermore, any couple moments acting on the body are directed perpendicular to this plane. This type of force and couple system is often referred to as a two-dimensional or *coplanar* force system. For example, the airplane in Fig. 5–2 has a plane of symmetry through its center axis, and so the loads acting on the airplane are symmetrical with respect to this plane. Thus, each of the two wing tires will support the same load $\mathbf{T}$, which is represented on the side (two-dimensional) view of the plane as $2\mathbf{T}$.

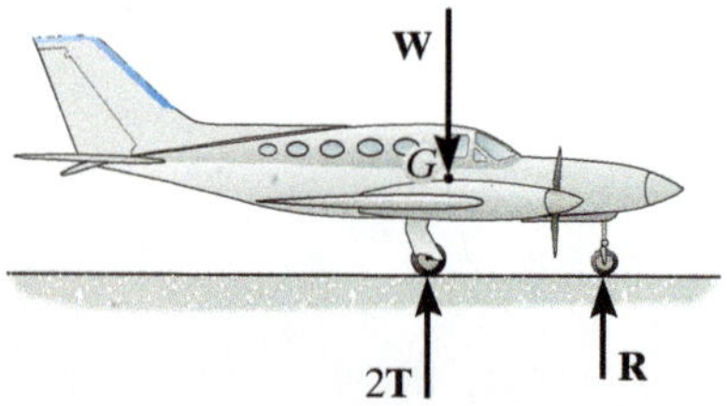

Fig. 5–2

5.2 Free-Body Diagrams

Successful application of the equations of equilibrium requires a complete specification of *all* the known and unknown external forces that act *on* the body. The best way to account for these forces is to draw a free-body diagram. This diagram is a sketch of the outlined shape of the body, which represents it as being *isolated* or "free" from its surroundings, i.e., a "free body." On this sketch it is necessary to show *all* the forces and couple moments that the surroundings exert *on the body* so that these effects can be accounted for when the equations of equilibrium are applied. *A thorough understanding of how to draw a free-body diagram is of primary importance for solving problems in mechanics.*

Support Reactions.

Before presenting a formal procedure as to how to draw a free-body diagram, we will first consider the various types of reactions that occur at supports and points of contact between bodies subjected to coplanar force systems. As a general rule,

- If a support prevents the translation of a body in a given direction, then a force is developed on the body in that direction.
- If rotation is prevented, a couple moment is exerted on the body.

For example, let us consider three ways in which a horizontal member, such as a beam, is supported at its end. One method consists of a *roller* or cylinder, Fig. 5–3*a*. Since this support only prevents the beam from *translating* in the vertical direction, the roller will only exert a *force* on the beam in this direction, Fig. 5–3*b*.

The beam can be supported in a more restrictive manner by using a *pin,* Fig. 5–3*c*. The pin passes through a hole in the beam and two leaves which are fixed to the ground. Here the pin can prevent *translation* of the beam in *any direction* ϕ, Fig. 5–3*d*, and so the pin must exert a *force* **F** on the beam in this direction. For purposes of analysis, it is generally easier to represent this resultant force **F** by its two rectangular components $\mathbf{F}_x$ and $\mathbf{F}_y$, Fig. 5–3*e*. If F_x and F_y are known, then F and ϕ can be calculated.

The most restrictive way to support the beam would be to use a *fixed support* as shown in Fig. 5–3*f*. This support will prevent both *translation and rotation* of the beam. To do this a *force and couple moment* must be developed on the beam at its point of connection, Fig. 5–3*g*. As in the case of the pin, the force is usually represented by its rectangular components $\mathbf{F}_x$ and $\mathbf{F}_y$.

Table 5–1 lists other common types of supports for bodies subjected to coplanar force systems. (In all cases the angle θ is assumed to be known.) Carefully study each of the symbols used to represent these supports and the types of reactions they exert on their contacting members.

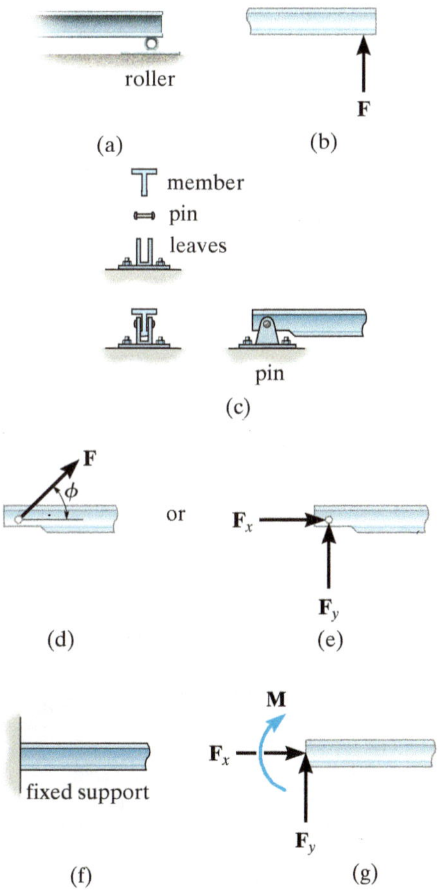

Fig. 5–3

TABLE 5–1 Supports for Rigid Bodies Subjected to Two-Dimensional Force Systems

Types of Connection	Reaction	Number of Unknowns
(1) 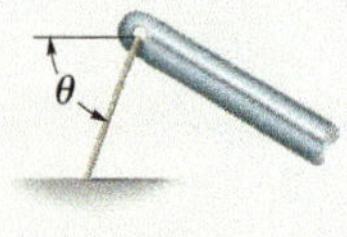 cable	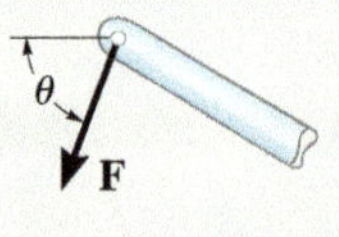	One unknown. The reaction is a tension force which acts away from the member in the direction of the cable.
(2) 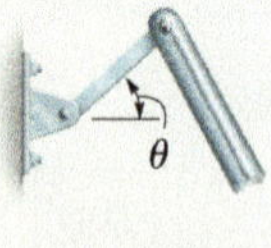 weightless link	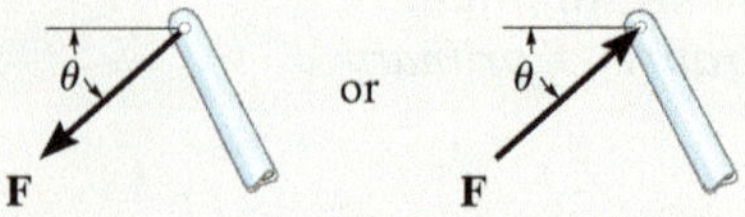	One unknown. The reaction is a force which acts along the axis of the link.
(3) 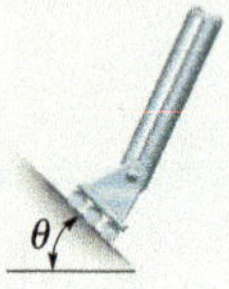 roller	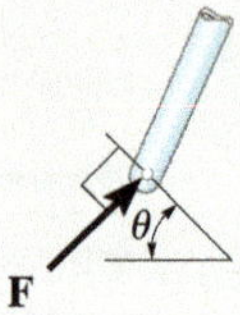	One unknown. The reaction is a force which acts perpendicular to the surface at the point of contact.
(4) 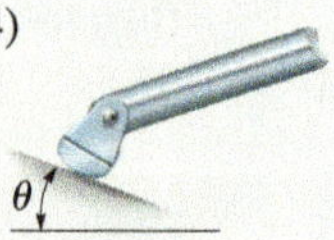 rocker	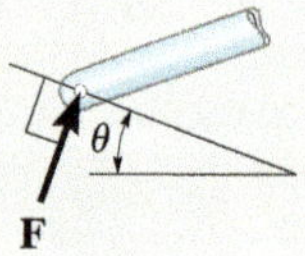	One unknown. The reaction is a force which acts perpendicular to the surface at the point of contact.
(5) 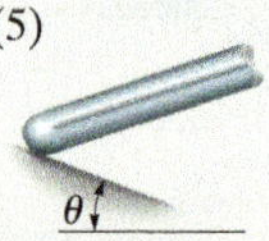 smooth contacting surface	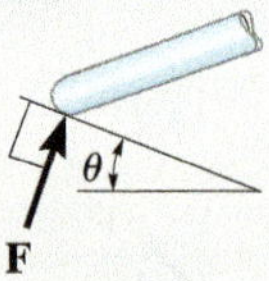	One unknown. The reaction is a force which acts perpendicular to the surface at the point of contact.
(6) 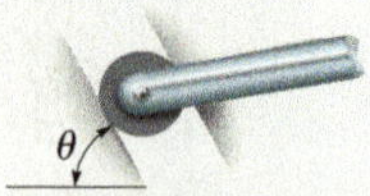 roller or pin in confined smooth slot	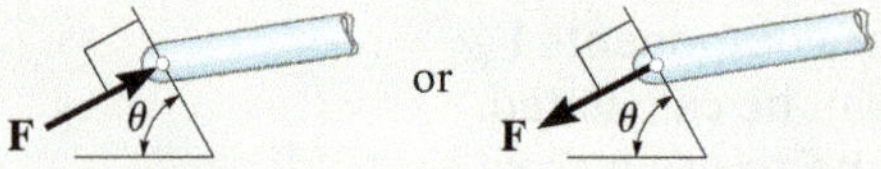	One unknown. The reaction is a force which acts perpendicular to the slot.
(7) 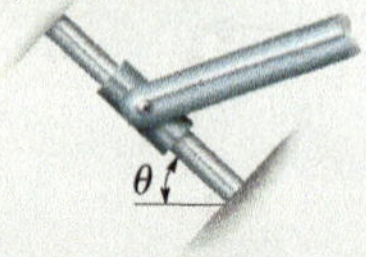 member pin connected to collar on smooth rod	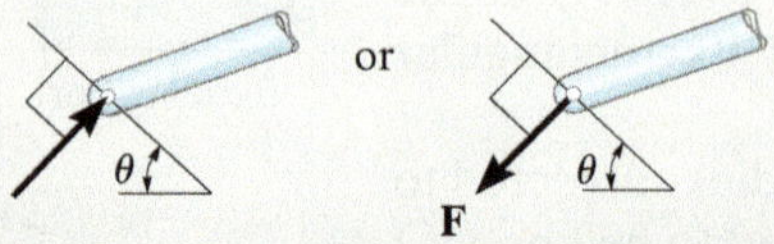	One unknown. The reaction is a force which acts perpendicular to the rod.

continued

TABLE 5–1 Continued

Types of Connection	Reaction	Number of Unknowns
(8) 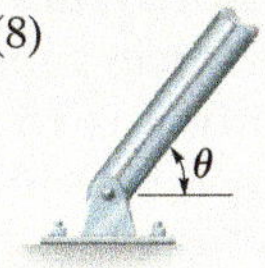 smooth pin or hinge	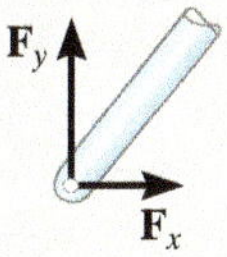 or 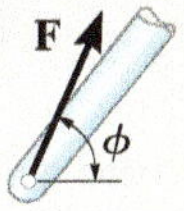	Two unknowns. The reactions are two components of force, or the magnitude and direction ϕ of the resultant force. Note that ϕ and θ are not necessarily equal [usually not, unless the rod shown is a link as in (2)].
(9) 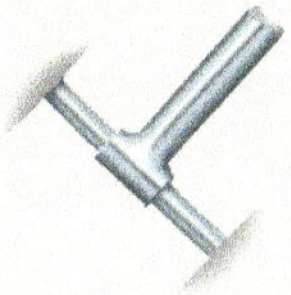 member fixed connected to collar on smooth rod	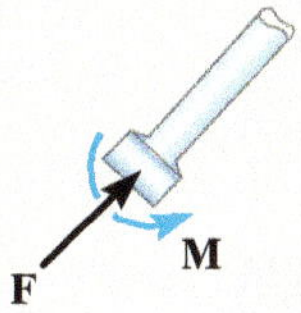	Two unknowns. The reactions are the couple moment and the force which acts perpendicular to the rod.
(10) 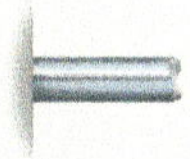 fixed support	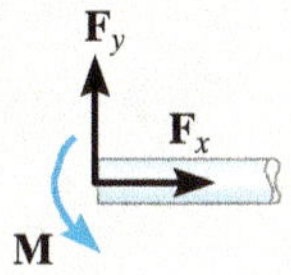 or 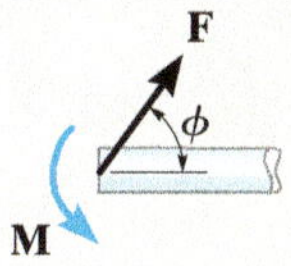	Three unknowns. The reactions are the couple moment and the two force components, or the couple moment and the magnitude and direction ϕ of the resultant force.

Typical examples of actual supports are shown in the following sequence of photos. The numbers refer to the connection types in Table 5–1.

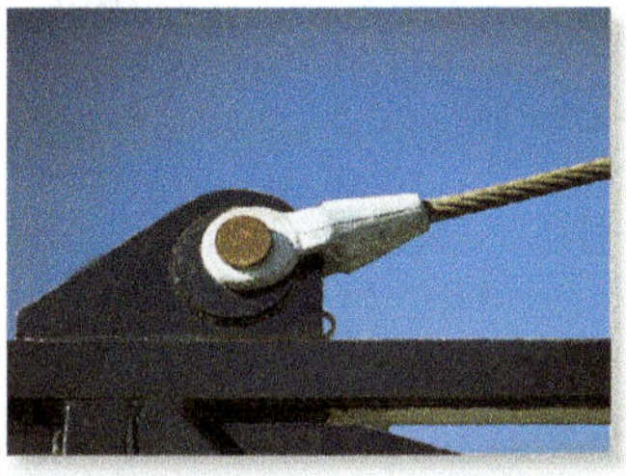

The cable exerts a force on the bracket in the direction of the cable. (1)

The rocker support for this bridge girder allows horizontal movement so the bridge is free to expand and contract due to a change in temperature. (4)

This concrete girder rests on the ledge that is assumed to act as a smooth contacting surface. (5)

This utility building is pin supported at the top of the column. (8)

The floor beams of this building are welded together and thus form fixed connections. (10)

Internal Forces. As stated in Sec. 5.1, the internal forces that act between adjacent particles in a body always occur in collinear pairs such that they have the same magnitude and act in opposite directions (Newton's third law). Since these forces cancel each other, they will not create an *external effect* on the body. It is for this reason that the internal forces should not be included on the free-body diagram if the entire body is to be considered. For example, the engine shown in Fig. 5–4*a* has a free-body diagram shown in Fig. 5–4*b*. The internal forces between all its connected parts, such as the screws and bolts, will cancel out because they form equal and opposite collinear pairs. Only the external forces $\mathbf{T}_1$ and $\mathbf{T}_2$, exerted by the chains and the engine weight **W**, are shown on the free-body diagram.

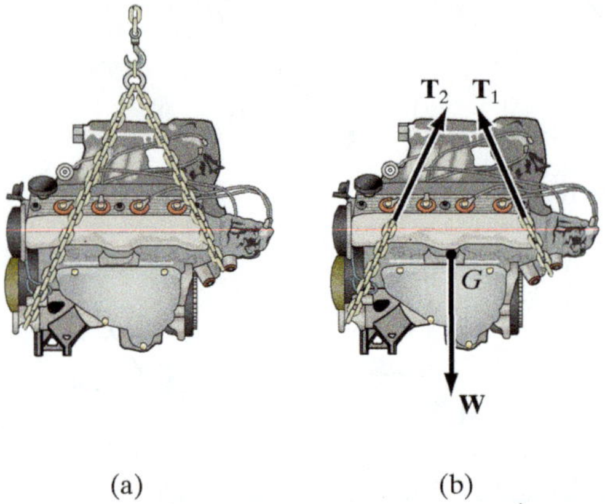

Fig. 5–4

Weight and the Center of Gravity. When a body is within a gravitational field, then each of its particles has a specified weight. It was shown in Sec. 4.8 that such a system of forces can be reduced to a single resultant force acting through a specified point. We refer to this force resultant as the *weight* **W** of the body and to the location of its point of application as the *center of gravity*. The methods used for its determination will be developed in Chapter 9.

In the examples and problems that follow, if the weight of the body is important for the analysis, this force will be reported in the problem statement. Also, when the body is *uniform* or made from the same material, the center of gravity will be located at the body's *geometric center* or *centroid*; however, if the body consists of a nonuniform distribution of material, or has an unusual shape, then the location of its center of gravity *G* will be given.

Idealized Models. When an engineer performs a force analysis of any object, he or she considers a corresponding analytical or idealized model that gives results that approximate as closely as possible the actual situation. To do this, careful choices have to be made so that selection of the type of supports, the material behavior, and the object's dimensions can be justified. This way one can feel confident that any design or analysis will yield results which can be trusted. In complex cases this process may require developing several different models of the object that must be analyzed. In any case, this selection process requires both skill and experience.

The following two cases illustrate what is required to develop a proper model. In Fig. 5–5*a*, the steel beam is to be used to support the three roof joists of a building. For a force analysis it is reasonable to assume the material (steel) is rigid since only very small deflections will occur when the beam is loaded. A bolted connection at *A* will allow for any slight rotation that occurs here when the load is applied, and so a *pin* can be considered for this support. At *B* a *roller* can be considered since this support offers no resistance to horizontal movement. Building code is used to specify the roof loading *A* so that the joist loads **F** can be calculated. These forces will be larger than any actual loading on the beam since they account for extreme loading cases and for dynamic or vibrational effects. Finally, the weight of the beam is generally neglected when it is small compared to the load the beam supports. The idealized model of the beam is therefore shown with average dimensions *a, b, c,* and *d* in Fig. 5–5*b*.

As a second case, consider the lift boom in Fig. 5–6*a*. By inspection, it is supported by a pin at *A* and by the hydraulic cylinder *BC*, which can be approximated as a weightless link. The material can be assumed rigid, and with its density known, the weight of the boom and the location of its center of gravity *G* are determined. When a design loading **P** is specified, the idealized model shown in Fig. 5–6*b* can be used for a force analysis. Average dimensions (not shown) are used to specify the location of the loads and the supports.

Idealized models of specific objects will be given in some of the examples throughout the text. It should be realized, however, that each case represents the reduction of a practical situation using simplifying assumptions like the ones illustrated here.

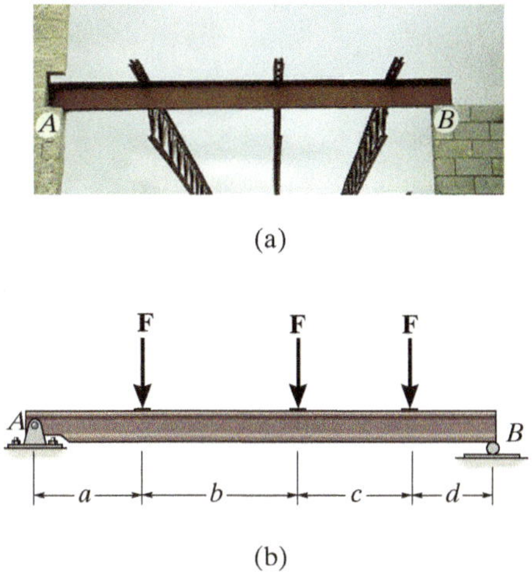

Fig. 5–5

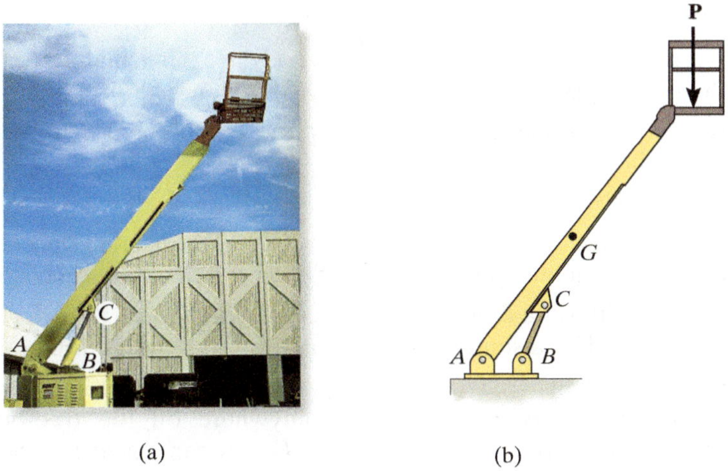

Fig. 5–6

5

Procedure for Analysis

To construct a free-body diagram for a rigid body or any group of bodies considered as a single system, the following steps should be performed:

Draw Outlined Shape.

Imagine the body to be *isolated* or cut "free" from its constraints and connections and draw (sketch) its outlined shape.

Show All Forces and Couple Moments.

Identify all the known and unknown *external forces* and couple moments that *act on the body*. Those generally encountered are due to (1) applied loadings, (2) reactions occurring at the supports or at points of contact with other bodies (see Table 5–1), and (3) the weight of the body. To account for all these effects, it may help to trace over the boundary, carefully noting each force or couple moment acting on it.

Identify Each Loading and Give Dimensions.

The forces and couple moments that are known should be labeled with their proper magnitudes and directions. Letters are used to represent the magnitudes and direction angles of forces and couple moments that are unknown. Establish an x, y coordinate system so that these unknowns, A_x, A_y, etc., can be identified. Finally, indicate the dimensions of the body necessary for calculating the moments of forces.

Important Points

- No equilibrium problem should be solved without *first drawing the free-body diagram*, so as to account for all the forces and couple moments that act on the body.
- If a support *prevents translation* of a body in a particular direction, then the support exerts a *force* on the body in that direction.
- If *rotation is prevented*, then the support exerts a *couple moment* on the body.
- Study Table 5–1.
- Internal forces are never shown on the free-body diagram since they occur in equal but opposite collinear pairs and therefore cancel out.
- The weight of a body is an external force, and its effect is represented by a single resultant force acting through the body's center of gravity G.
- *Couple moments* can be placed anywhere on the free-body diagram since they are *free vectors*. *Forces* can act at any point along their lines of action since they are *sliding vectors*.

EXAMPLE 5.1

Draw the free-body diagram of the uniform beam shown in Fig. 5–7*a*. The beam has a mass of 100 kg.

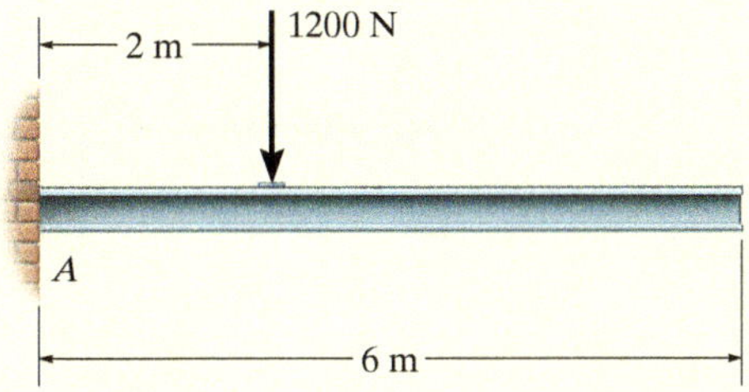

(a)

5

SOLUTION

The free-body diagram of the beam is shown in Fig. 5–7*b*. Since the support at A is fixed, the wall exerts three reactions *on the beam*, denoted as $\mathbf{A}_x$, $\mathbf{A}_y$, and $\mathbf{M}_A$. The magnitudes of these reactions are *unknown*, and their sense has been *assumed*. The weight of the beam, $W = 100(9.81)\text{ N} = 981\text{ N}$, acts through the beam's center of gravity G, which is 3 m from A since the beam is uniform.

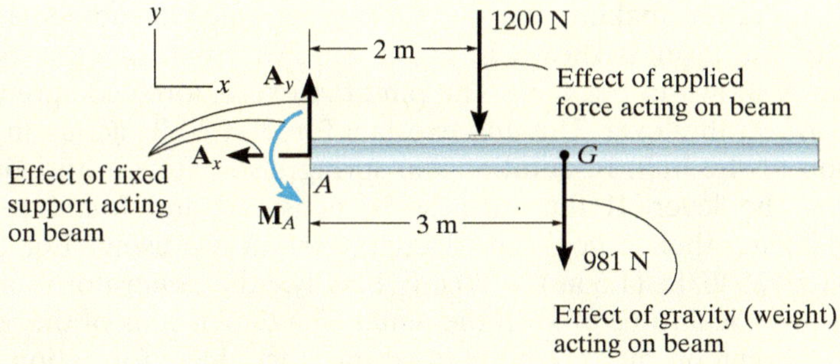

(b)

Fig. 5–7

EXAMPLE 5.2

Draw the free-body diagram of the foot lever shown in Fig. 5–8*a*. The operator applies a vertical force to the pedal so that the spring is stretched 1.5 in. and the force on the link at *B* is 20 lb.

(a)

Fig. 5–8

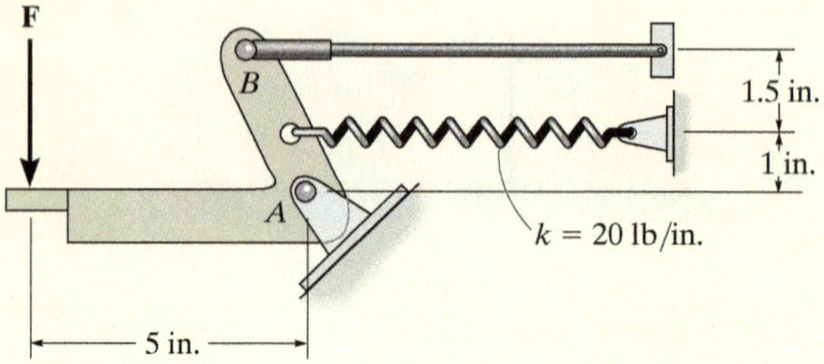

(b)

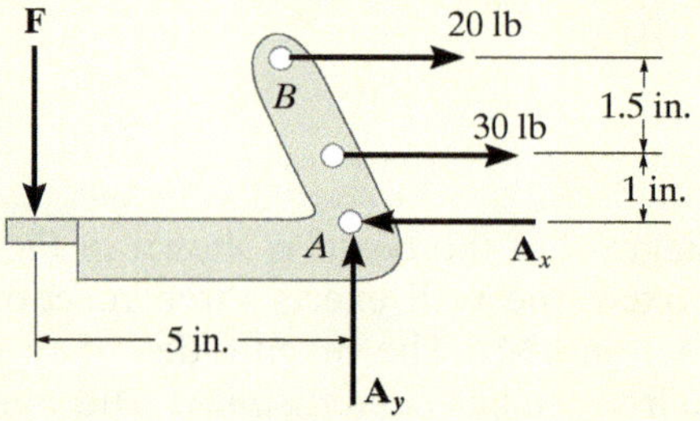

(c)

SOLUTION

By inspection of the photo the lever is loosely bolted to the frame at *A* and so this bolt acts as a pin. (See (8) in Table 5–1.) Although not shown here the link at *B* is pinned at both ends and so it is like (2) in Table 5-1. After making the proper measurements, the idealized model of the lever is shown in Fig. 5–8*b*. From this, the free-body diagram is shown in Fig. 5–8*c*. The pin at *A* exerts force components $\mathbf{A}_x$ and $\mathbf{A}_y$ on the lever. The link exerts a force of 20 lb, acting in the direction of the link. In addition the spring also exerts a horizontal force on the lever. If the stiffness is measured and found to be $k = 20\text{ lb/in.}$, then since the stretch $s = 1.5\text{ in.}$, using Eq. 3–2, $F_s = ks = 20\text{ lb/in.}\,(1.5\text{ in.}) = 30\text{ lb}$. Finally, the operator's shoe applies a vertical force of **F** on the pedal. The dimensions of the lever are also shown on the free-body diagram, since this information will be useful when calculating the moments of the forces. As usual, the senses of the unknown forces at *A* have been assumed. The correct senses will become apparent after solving the equilibrium equations.

EXAMPLE 5.3

Two smooth pipes, each having a mass of 300 kg, are supported by the forked tines of the tractor in Fig. 5–9*a*. Draw the free-body diagrams for each pipe and both pipes together.

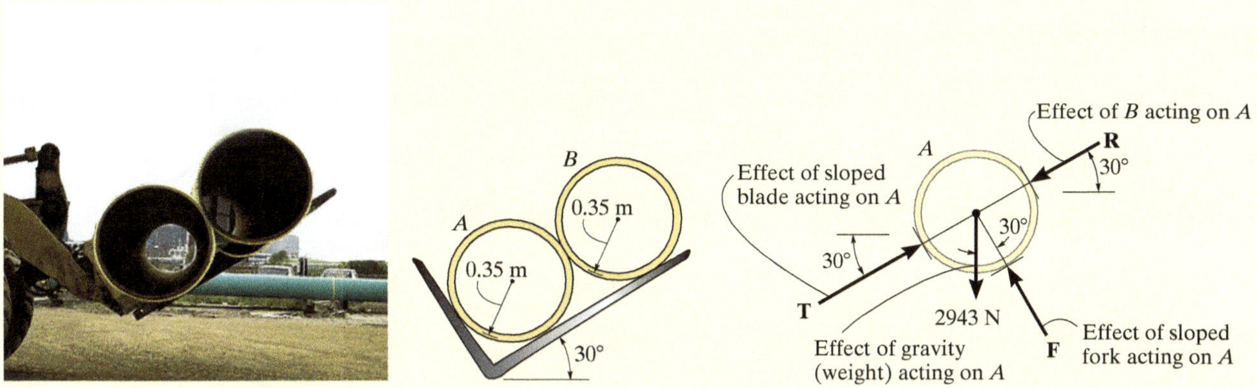

(a) (b) (c)

SOLUTION

The idealized model from which we must draw the free-body diagrams is shown in Fig. 5–9*b*. Here the pipes are identified, the dimensions have been added, and the physical situation reduced to its simplest form.

The free-body diagram for pipe *A* is shown in Fig. 5–9*c*. Its weight is $W = 300(9.81)\text{ N} = 2943\text{ N}$. Assuming all contacting surfaces are *smooth*, the reactive forces **T**, **F**, **R** act in a direction *normal* to the tangent at their surfaces of contact.

The free-body diagram of pipe *B* is shown in Fig. 5–9*d*. Can you identify each of the three forces acting *on this pipe*? In particular, note that **R**, representing the force of *A* on *B*, Fig. 5–9*d*, is equal and opposite to **R** representing the force of *B* on *A*, Fig. 5–9*c*. This is a consequence of Newton's third law of motion.

The free-body diagram of both pipes combined ("system") is shown in Fig. 5–9*e*. Here the contact force **R**, which acts between *A* and *B*, is considered as an *internal* force and hence is not shown on the free-body diagram. That is, it represents a pair of equal but opposite collinear forces which cancel each other.

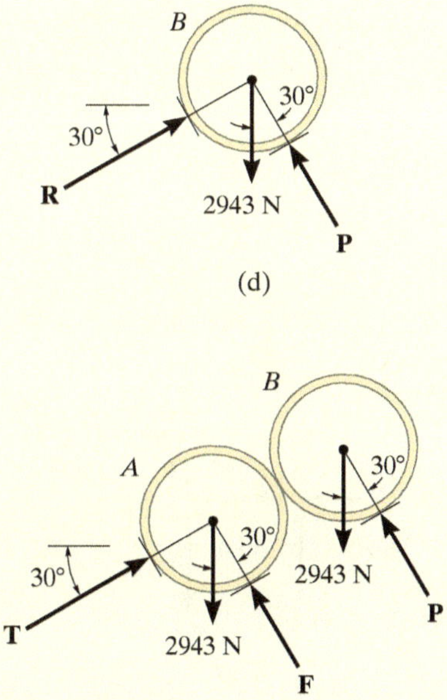

Fig. 5–9

EXAMPLE 5.4

Draw the free-body diagram of the unloaded platform that is suspended off the edge of the oil rig shown in Fig. 5–10*a*. The platform has a mass of 200 kg.

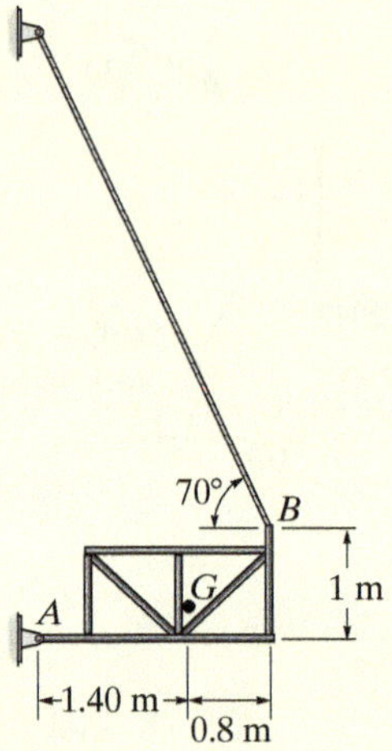

(b)

(a)

Fig. 5–10

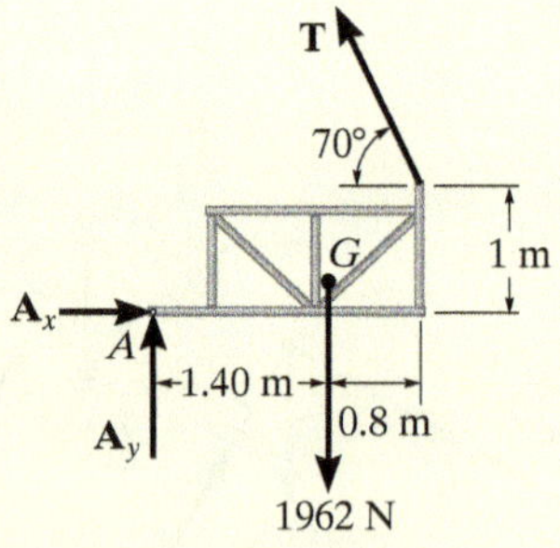

(c)

SOLUTION

The idealized model of the platform will be considered in two dimensions because by observation the loading and the dimensions are all symmetrical about a vertical plane passing through its center, Fig. 5–10*b*. The connection at *A* is considered to be a pin, and the cable supports the platform at *B*. The direction of the cable and average dimensions of the platform are listed, and the center of gravity *G* has been determined. It is from this model that we have drawn the free-body diagram shown in Fig. 5–10*c*. The platform's weight is 200(9.81) = 1962 N. The force components $\mathbf{A}_x$ and $\mathbf{A}_y$ along with the cable force **T** represent the reactions that *both* pins and *both* cables exert on the platform, Fig. 5–10*a*. As a result, half their magnitudes are developed on each side of the platform.

PROBLEMS

5–1. Draw the free-body diagram of the dumpster D of the truck, which has a mass of 2.5 Mg and a center of gravity at G. It is supported by a pin at A and a pin-connected hydraulic cylinder BC (short link). Explain the significance of each force on the diagram. (See Fig. 5–7*b*.)

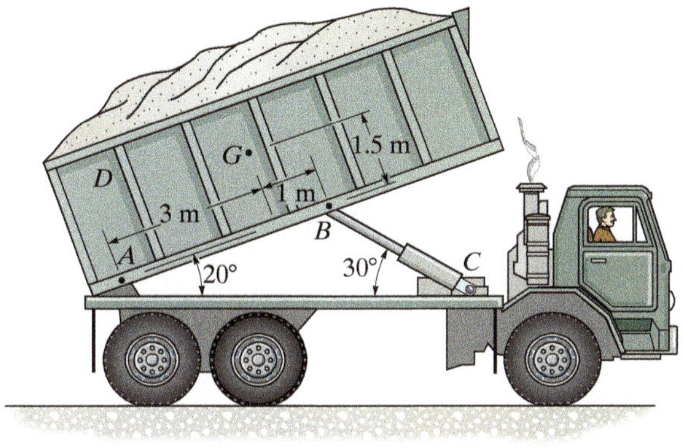

Prob. 5–1

5–2. Draw the free-body diagram of member ABC which is supported by a smooth collar at A, rocker at B, and short link CD. Explain the significance of each force acting on the diagram. (See Fig. 5–7*b*.)

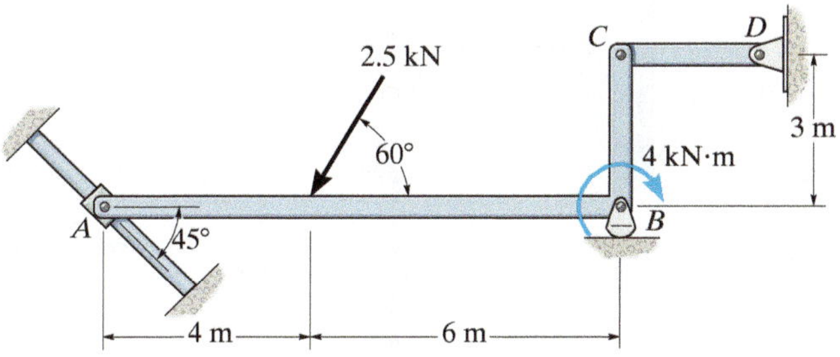

Prob. 5–2

5–3. Draw the free-body diagram of the beam which supports the 80-kg load and is supported by the pin at A and a cable which wraps around the pulley at D. Explain the significance of each force on the diagram. (See Fig. 5–7*b*.)

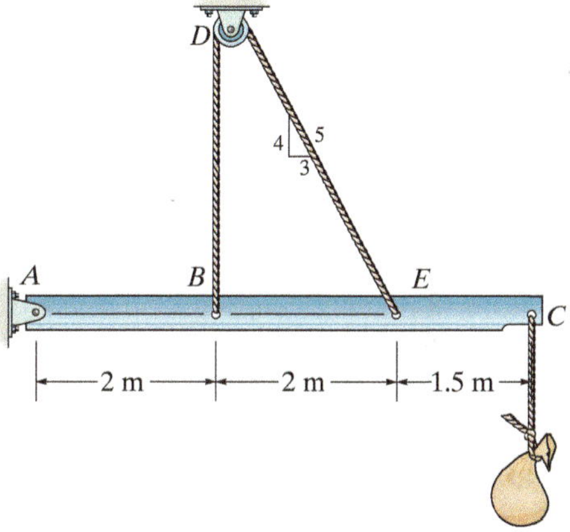

Prob. 5–3

***5–4.** Draw the free-body diagram of the hand punch, which is pinned at A and bears down on the smooth surface at B.

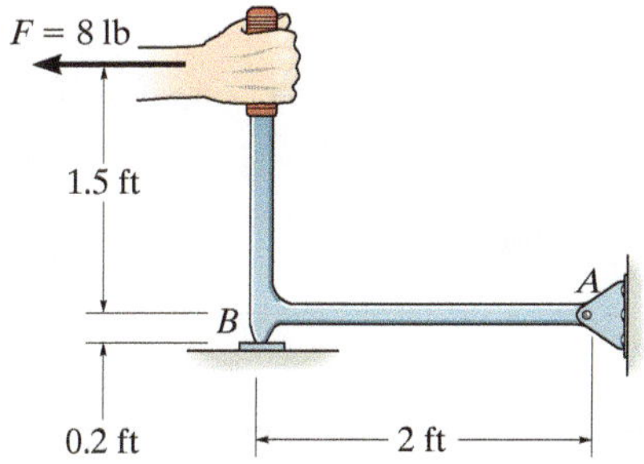

Prob. 5–4

5

5–5. Draw the free-body diagram of the uniform bar, which has a mass of 100 kg and a center of mass at G. The supports A, B, and C are smooth.

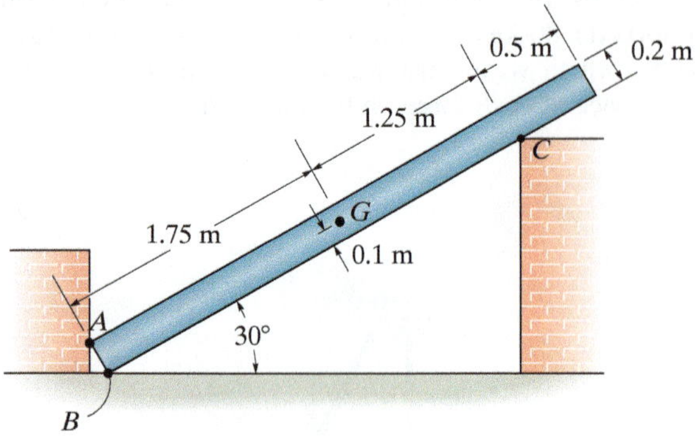

Prob. 5–5

5–6. Draw the free-body diagram of the beam, which is pin-supported at A and rests on the smooth incline at B.

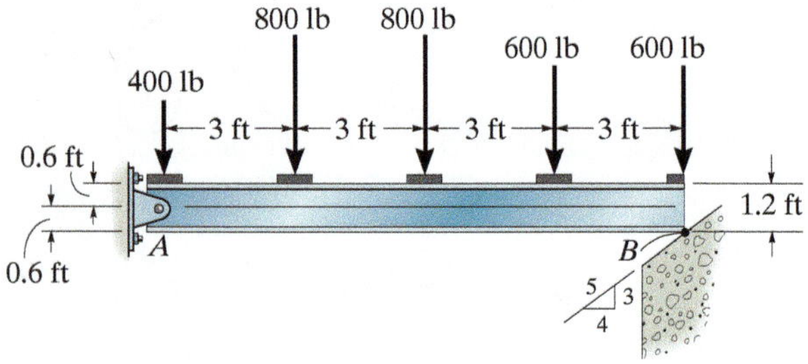

Prob. 5–6

5–7. Draw the free-body diagram of the beam, which is pin connected at A and rocker-supported at B.

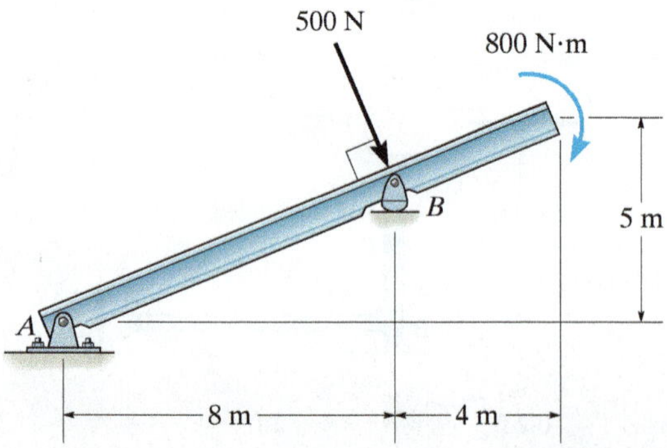

Prob. 5–7

***5–8.** Draw the free-body diagram of the bar, which has a negligible thickness and smooth points of contact at A, B, and C. Explain the significance of each force on the diagram. (See Fig. 5–7b.)

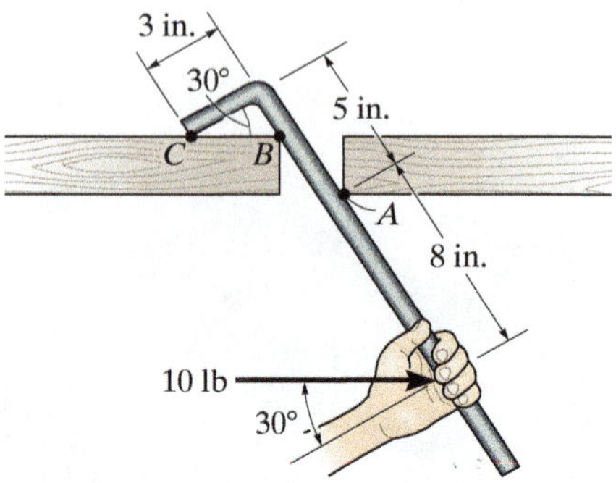

Prob. 5–8

5–9. Draw the free-body diagram of the jib crane AB, which is pin connected at A and supported by member (link) BC.

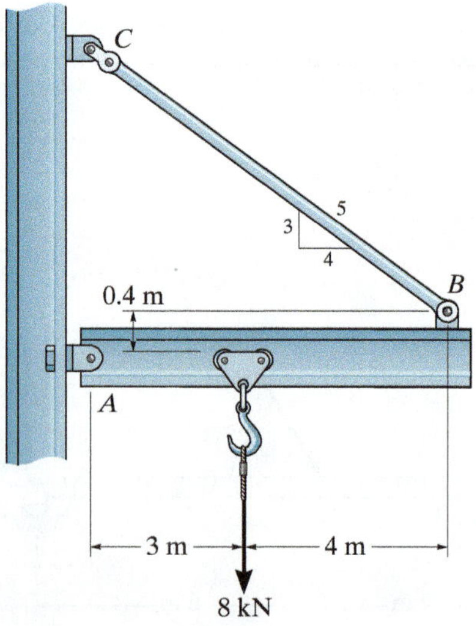

Prob. 5–9

CONCEPTUAL PROBLEMS

P5–1. Draw the free-body diagram of the uniform trash bucket which has a significant weight. It is pinned at *A* and rests against the smooth horizontal member at *B*. Show your result in side view. Label any necessary dimensions.

P5–1

P5–2. Draw the free-body diagram of the outrigger *ABC* used to support a backhoe. The pin *B* is connected to the hydraulic cylinder, which can be considered a short link (two-force member), the bearing shoe at *A* is smooth, and the outrigger is pinned to the frame at *C*.

P5–2

P5–3. Draw the free-body diagram of the wing on the passenger plane. The weights of the engine and wing are significant. The tires at *B* are smooth.

P5–3

P5–4. Draw the free-body diagrams of the wheel and member *ABC* used as part of the landing gear on a jet plane. The hydraulic cylinder *AD* acts as a two-force member, and there is a pin connection at *B*.

P5–4

5.3 Equations of Equilibrium

In Sec. 5.1 we developed the two equations which are both necessary and sufficient for the equilibrium of a rigid body, namely, $\Sigma \mathbf{F} = \mathbf{0}$ and $\Sigma \mathbf{M}_O = \mathbf{0}$. When the body is subjected to a system of forces, which all lie in the x–y plane, then the forces can be resolved into their x and y components. Consequently, the conditions for equilibrium in two dimensions are

$$\Sigma F_x = 0 \\ \Sigma F_y = 0 \\ \Sigma M_O = 0 \qquad (5\text{–}2)$$

Here ΣF_x and ΣF_y represent, respectively, the algebraic sums of the x and y components of all the forces acting on the body, and ΣM_O represents the algebraic sum of the couple moments and the moments of all the force components about the z axis, which is perpendicular to the x–y plane and passes through the arbitrary point O.

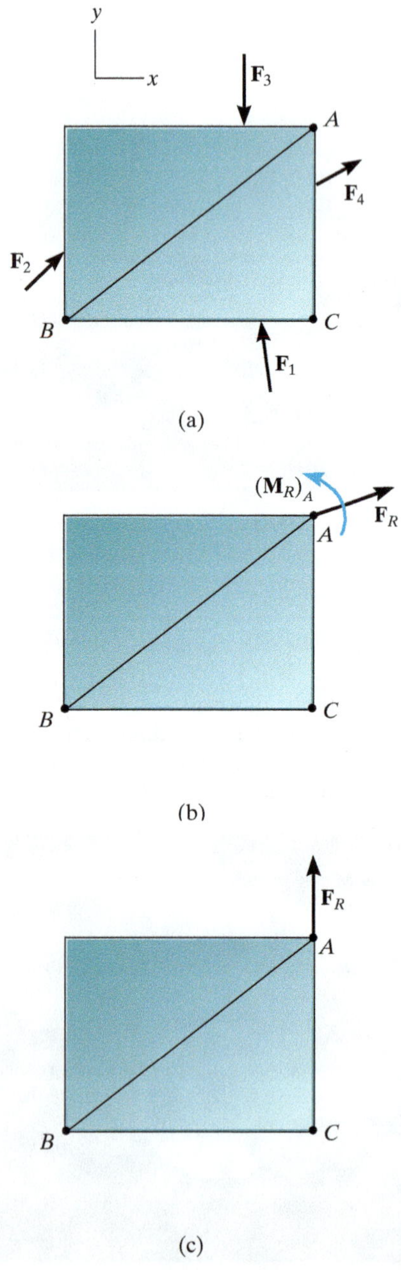

Fig. 5–11

5

Alternative Sets of Equilibrium Equations. Although Eqs. 5–2 are *most often* used for solving coplanar equilibrium problems, two *alternative* sets of three independent equilibrium equations may also be used. One such set is

$$\Sigma F_x = 0 \\ \Sigma M_A = 0 \\ \Sigma M_B = 0 \qquad (5\text{–}3)$$

When using these equations it is required that a line passing through points A and B is *not parallel* to the y axis. To prove that Eqs. 5–3 provide the *conditions* for equilibrium, consider the free-body diagram of the plate shown in Fig. 5–11*a*. Using the methods of Sec. 4.7, all the forces on the free-body diagram may be replaced by an equivalent resultant force $\mathbf{F}_R = \Sigma \mathbf{F}$, acting at point A, and a resultant couple moment $(\mathbf{M}_R)_A = \Sigma \mathbf{M}_A$, Fig. 5–11*b*. If $\Sigma M_A = 0$ is satisfied, it is necessary that $(\mathbf{M}_R)_A = \mathbf{0}$. Furthermore, in order that $\mathbf{F}_R$ satisfy $\Sigma F_x = 0$, it must have *no component* along the x axis, and therefore $\mathbf{F}_R$ must be parallel to the y axis, Fig. 5–11*c*. Finally, if it is required that $\Sigma M_B = 0$, where B does not lie on the line of action of $\mathbf{F}_R$, then $\mathbf{F}_R = \mathbf{0}$. Since Eqs. 5–3 show that both of these resultants are zero, indeed the body in Fig. 5–11*a* must be in equilibrium.

A second alternative set of equilibrium equations is

$$\begin{aligned} \Sigma M_A &= 0 \\ \Sigma M_B &= 0 \\ \Sigma M_C &= 0 \end{aligned} \tag{5–4}$$

Here it is necessary that points A, B, and C do not lie on the same line. To prove that these equations, when satisfied, ensure equilibrium, consider again the free-body diagram in Fig. 5–11*b*. If $\Sigma M_A = 0$ is to be satisfied, then $(\mathbf{M}_R)_A = \mathbf{0}$. $\Sigma M_C = 0$ is satisfied if the line of action of $\mathbf{F}_R$ passes through point C as shown in Fig. 5–11*c*. Finally, if we require $\Sigma M_B = 0$, it is necessary that $\mathbf{F}_R = \mathbf{0}$, and so the plate in Fig. 5–11*a* must then be in equilibrium.

Procedure for Analysis

Coplanar force equilibrium problems for a rigid body can be solved using the following procedure.

Free-Body Diagram.

- Establish the x, y coordinate axes in any suitable orientation.
- Draw an outlined shape of the body.
- Show all the forces and couple moments acting on the body.
- Label all the loadings and specify their directions relative to the x or y axis. The sense of a force or couple moment having an *unknown* magnitude but known line of action can be *assumed*.
- Indicate the dimensions of the body necessary for computing the moments of forces.

Equations of Equilibrium.

- Apply the moment equation of equilibrium, $\Sigma M_O = 0$, about a point (O) that lies at the intersection of the lines of action of two unknown forces. In this way, the moments of these unknowns are zero about O, and a *direct solution* for the third unknown can be determined.
- When applying the force equilibrium equations, $\Sigma F_x = 0$ and $\Sigma F_y = 0$, orient the x and y axes along lines that will provide the simplest resolution of the forces into their x and y components.
- If the solution of the equilibrium equations yields a negative scalar for a force or couple moment magnitude, this indicates that the sense is opposite to that which was assumed on the free-body diagram.

EXAMPLE 5.5

Determine the horizontal and vertical components of reaction on the beam caused by the pin at B and the rocker at A as shown in Fig. 5–12a. Neglect the weight of the beam.

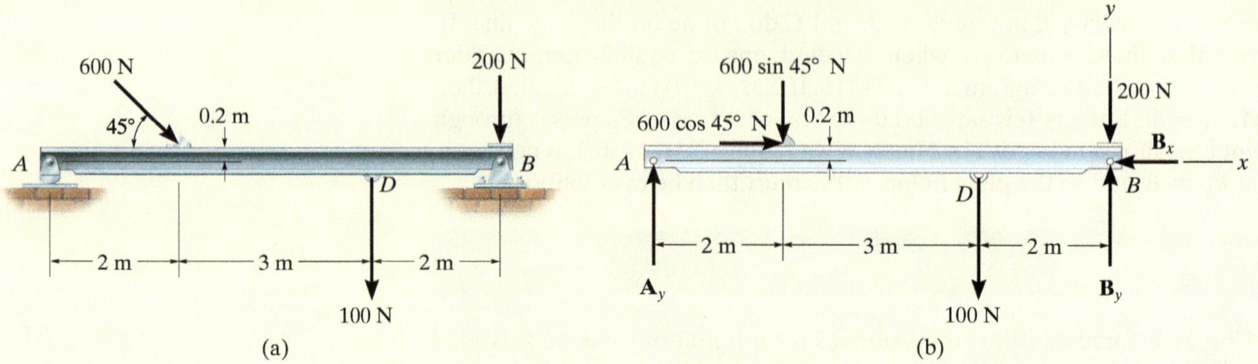

Fig. 5–12

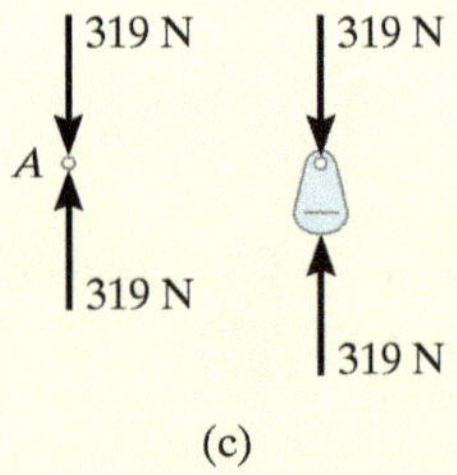

SOLUTION

Free-Body Diagram. Identify each of the forces shown on the free-body diagram of the beam, Fig. 5–12b. (See Example 5.1.) For simplicity, the 600-N force is represented by its x and y components as shown in Fig. 5–12b.

Equations of Equilibrium. Summing forces in the x direction yields

$$\xrightarrow{+} \Sigma F_x = 0; \qquad 600 \cos 45^\circ \text{ N} - B_x = 0$$

$$B_x = 424 \text{ N} \qquad \textit{Ans.}$$

A direct solution for $\mathbf{A}_y$ can be obtained by applying the moment equation $\Sigma M_B = 0$ about point B.

$$\circlearrowleft + \Sigma M_B = 0; \qquad 100 \text{ N}(2 \text{ m}) + (600 \sin 45^\circ \text{ N})(5 \text{ m})$$

$$- (600 \cos 45^\circ \text{ N})(0.2 \text{ m}) - A_y(7 \text{ m}) = 0$$

$$A_y = 319 \text{ N} \qquad \textit{Ans.}$$

Summing forces in the y direction, using this result, gives

$$+\uparrow \Sigma F_y = 0; \qquad 319 \text{ N} - 600 \sin 45^\circ \text{ N} - 100 \text{ N} - 200 \text{ N} + B_y = 0$$

$$B_y = 405 \text{ N} \qquad \textit{Ans.}$$

NOTE: Remember, the support forces in Fig. 5–12b are the result of pins that *act on the beam*. The opposite forces act on the pins. For example, Fig. 5–12c shows the equilibrium of the pin at A and the rocker.

EXAMPLE 5.6

The cord shown in Fig. 5–13*a* supports a force of 100 lb and wraps over the frictionless pulley. Determine the tension in the cord at *C* and the horizontal and vertical components of reaction at pin *A*.

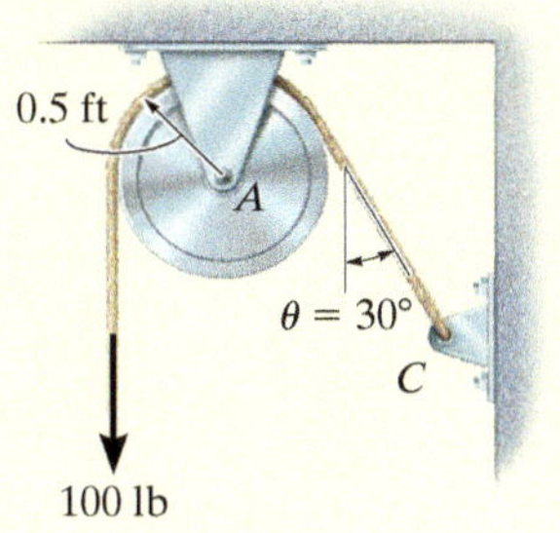

(a)

Fig. 5–13

SOLUTION

Free-Body Diagrams. The free-body diagrams of the cord and pulley are shown in Fig. 5–13*b*. Note that the principle of action, equal but opposite reaction must be carefully observed when drawing each of these diagrams: the cord exerts an unknown load distribution *p* on the pulley at the contact surface, whereas the pulley exerts an equal but opposite effect on the cord. For the solution, however, it is simpler to *combine* the free-body diagrams of the pulley and this portion of the cord, so that the distributed load becomes *internal* to this "system" and is therefore eliminated from the analysis, Fig. 5–13*c*.

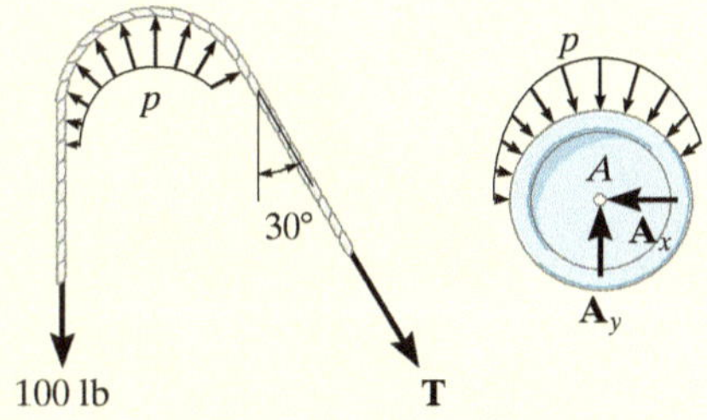

(b)

Equations of Equilibrium. Summing moments about point *A* to eliminate $\mathbf{A}_x$ and $\mathbf{A}_y$, Fig. 5–13*c*, we have

$$\circlearrowleft +\Sigma M_A = 0; \qquad 100 \text{ lb } (0.5 \text{ ft}) - T(0.5 \text{ ft}) = 0$$

$$T = 100 \text{ lb} \qquad \textit{Ans.}$$

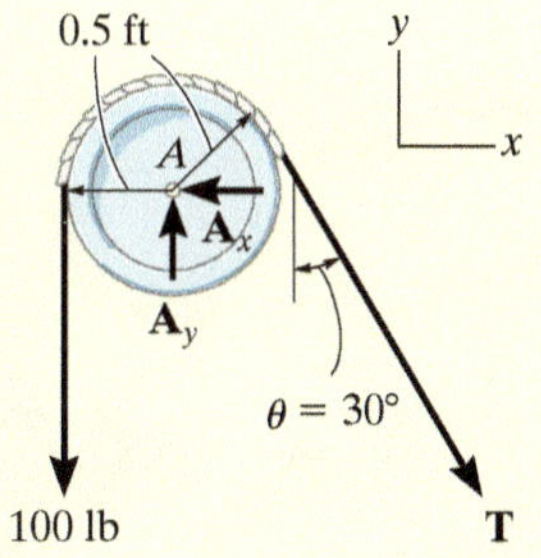

(c)

Using this result,

$$\overset{+}{\rightarrow} \Sigma F_x = 0; \qquad -A_x + 100 \sin 30° \text{ lb} = 0$$

$$A_x = 50.0 \text{ lb} \qquad \textit{Ans.}$$

$$+\uparrow \Sigma F_y = 0; \qquad A_y - 100 \text{ lb} - 100 \cos 30° \text{ lb} = 0$$

$$A_y = 187 \text{ lb} \qquad \textit{Ans.}$$

NOTE: From the moment equation, it is seen that the tension remains *constant* as the cord passes over the pulley. (This of course is true for *any angle* θ at which the cord is directed and for *any radius r* of the pulley.)

5

EXAMPLE 5.7

The member shown in Fig. 5–14*a* is pin connected at *A* and rests against a smooth support at *B*. Determine the horizontal and vertical components of reaction at the pin *A*.

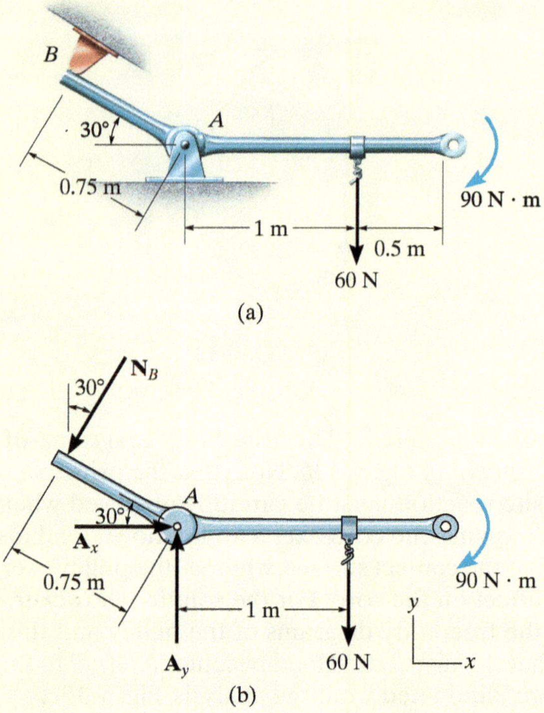

Fig. 5–14

SOLUTION

Free-Body Diagram. As shown in Fig. 5–14*b*, the reaction $\mathbf{N}_B$ is perpendicular to the member at *B*. Also, horizontal and vertical components of reaction are represented at *A*.

Equations of Equilibrium. Summing moments about *A*, we obtain a direct solution for N_B,

$$\circlearrowleft +\Sigma M_A = 0; \quad -90\ \text{N}\cdot\text{m} - 60\ \text{N}(1\ \text{m}) + N_B(0.75\ \text{m}) = 0$$

$$N_B = 200\ \text{N}$$

Using this result,

$$\overset{+}{\rightarrow} \Sigma F_x = 0; \quad A_x - 200 \sin 30^\circ\ \text{N} = 0$$

$$A_x = 100\ \text{N} \qquad \textit{Ans.}$$

$$+\uparrow \Sigma F_y = 0; \quad A_y - 200 \cos 30^\circ\ \text{N} - 60\ \text{N} = 0$$

$$A_y = 233\ \text{N} \qquad \textit{Ans.}$$

EXAMPLE 5.8

The box wrench in Fig. 5–15*a* is used to tighten the bolt at *A*. If the wrench does not turn when the load is applied to the handle, determine the torque or moment applied to the bolt and the force of the wrench on the bolt.

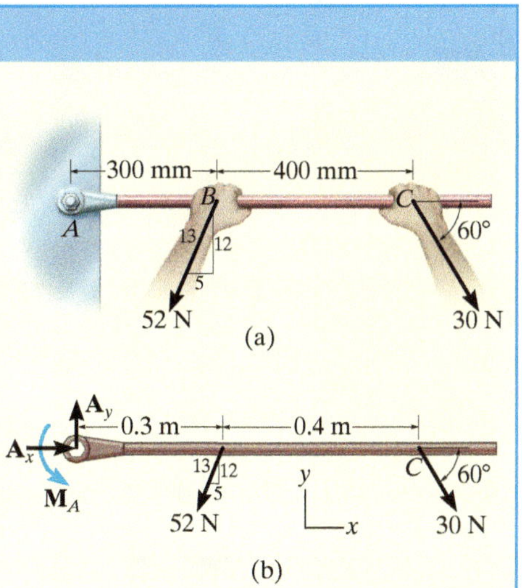

Fig. 5–15

SOLUTION

Free-Body Diagram. The free-body diagram for the wrench is shown in Fig. 5–15*b*. Since the bolt acts as a "fixed support," it exerts force components $\mathbf{A}_x$ and $\mathbf{A}_y$ and a moment $\mathbf{M}_A$ on the wrench at *A*.

Equations of Equilibrium.

$$\overset{+}{\rightarrow}\Sigma F_x = 0; \qquad A_x - 52\left(\tfrac{5}{13}\right)\text{ N} + 30\cos 60^\circ\text{ N} = 0$$

$$A_x = 5.00\text{ N} \qquad \textit{Ans.}$$

$$+\uparrow\Sigma F_y = 0; \qquad A_y - 52\left(\tfrac{12}{13}\right)\text{ N} - 30\sin 60^\circ\text{ N} = 0$$

$$A_y = 74.0\text{ N} \qquad \textit{Ans.}$$

$$\circlearrowleft+\Sigma M_A = 0; \quad M_A - \left[52\left(\tfrac{12}{13}\right)\text{N}\right](0.3\text{ m}) - (30\sin 60^\circ\text{ N})(0.7\text{ m}) = 0$$

$$M_A = 32.6\text{ N}\cdot\text{m} \qquad \textit{Ans.}$$

Note that $\mathbf{M}_A$ must be *included* in this moment summation. This couple moment is a free vector and represents the twisting resistance of the bolt on the wrench. By Newton's third law, the wrench exerts an equal but opposite moment or torque on the bolt. Furthermore, the resultant force on the wrench is

$$F_A = \sqrt{(5.00)^2 + (74.0)^2} = 74.1\text{ N} \qquad \textit{Ans.}$$

NOTE: Although only *three* independent equilibrium equations can be written for a rigid body, it is a good practice to *check* the calculations using a fourth equilibrium equation. For example, the above computations may be verified in part by summing moments about point *C*:

$$\circlearrowleft+\Sigma M_C = 0; \quad \left[52\left(\tfrac{12}{13}\right)\text{N}\right](0.4\text{ m}) + 32.6\text{ N}\cdot\text{m} - 74.0\text{ N}(0.7\text{ m}) = 0$$

$$19.2\text{ N}\cdot\text{m} + 32.6\text{ N}\cdot\text{m} - 51.8\text{ N}\cdot\text{m} = 0$$

5

EXAMPLE 5.9

Determine the horizontal and vertical components of reaction on the member at the pin A, and the normal reaction at the roller B in Fig. 5–16*a*.

SOLUTION

Free-Body Diagram. The free-body diagram is shown in Fig. 5–16*b*. The pin at A exerts two components of reaction on the member, $\mathbf{A}_x$ and $\mathbf{A}_y$.

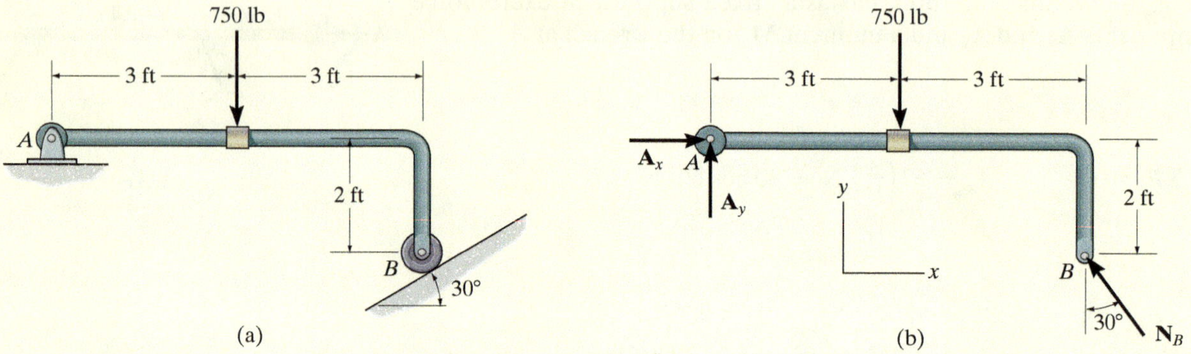

Fig. 5–16

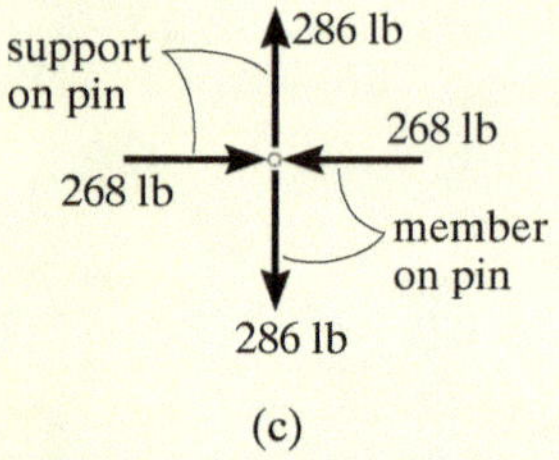

Equations of Equilibrium. The reaction N_B can be obtained *directly* by summing moments about point A, since $\mathbf{A}_x$ and $\mathbf{A}_y$ produce no moment about A.

$\circlearrowleft + \Sigma M_A = 0;$

$$[N_B \cos 30°](6 \text{ ft}) - [N_B \sin 30°](2 \text{ ft}) - 750 \text{ lb}(3 \text{ ft}) = 0$$

$$N_B = 536.2 \text{ lb} = 536 \text{ lb} \qquad \textit{Ans.}$$

Using this result,

$\overset{+}{\rightarrow} \Sigma F_x = 0; \quad A_x - (536.2 \text{ lb}) \sin 30° = 0$

$$A_x = 268 \text{ lb} \qquad \textit{Ans.}$$

$+\uparrow \Sigma F_y = 0; \quad A_y + (536.2 \text{ lb}) \cos 30° - 750 \text{ lb} = 0$

$$A_y = 286 \text{ lb} \qquad \textit{Ans.}$$

Details of the equilibrium of the pin at A are shown in Fig. 5–16*c*.

EXAMPLE 5.10

The uniform smooth rod shown in Fig. 5–17*a* is subjected to a force and couple moment. If the rod is supported at *A* by a smooth wall and at *B* and *C* either at the top or bottom by rollers, determine the reactions at these supports. Neglect the weight of the rod.

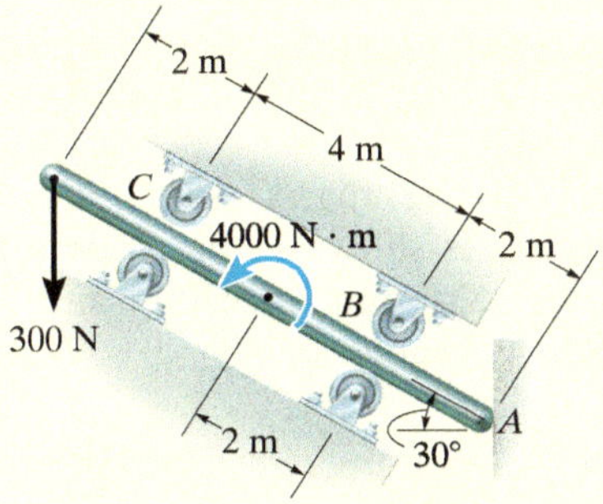

(a)

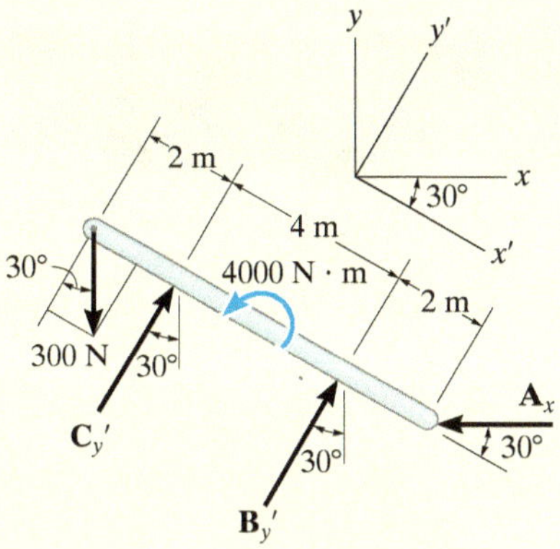

(b)

Fig. 5–17

SOLUTION

Free-Body Diagram. As shown in Fig. 5–17*b*, all the support reactions act normal to the surfaces of contact since these surfaces are smooth. The reactions at *B* and *C* are shown acting in the positive y' direction. This assumes that only the rollers located on the bottom of the rod are used for support.

Equations of Equilibrium. Using the *x*, *y* coordinate system in Fig. 5–17*b*, we have

$$\xrightarrow{+} \Sigma F_x = 0; \qquad C_{y'} \sin 30^\circ + B_{y'} \sin 30^\circ - A_x = 0 \qquad (1)$$

$$+\uparrow \Sigma F_y = 0; \qquad -300\text{ N} + C_{y'} \cos 30^\circ + B_{y'} \cos 30^\circ = 0 \qquad (2)$$

$$\circlearrowleft + \Sigma M_A = 0; \qquad -B_{y'}(2\text{ m}) + 4000\text{ N}\cdot\text{m} - C_{y'}(6\text{ m}) + (300 \cos 30^\circ\text{ N})(8\text{ m}) = 0 \qquad (3)$$

When writing the moment equation, it should be noted that the line of action of the force component 300 sin 30° N passes through point *A*, and therefore this force is not included in the moment equation.

Solving Eqs. 2 and 3 simultaneously, we obtain

$$B_{y'} = -1000.0\text{ N} = -1\text{ kN} \qquad \textit{Ans.}$$

$$C_{y'} = 1346.4\text{ N} = 1.35\text{ kN} \qquad \textit{Ans.}$$

Since $B_{y'}$ is a negative scalar, the sense of $\mathbf{B}_{y'}$ is opposite to that shown on the free-body diagram in Fig. 5–17*b*. Therefore, the top roller at *B* serves as the support rather than the bottom one. Retaining the negative sign for $B_{y'}$ (Why?) and substituting the results into Eq. 1, we obtain

$$1346.4 \sin 30^\circ\text{ N} + (-1000.0 \sin 30^\circ\text{ N}) - A_x = 0$$

$$A_x = 173\text{ N} \qquad \textit{Ans.}$$

5

EXAMPLE 5.11

(a)

The uniform truck ramp shown in Fig. 5–18*a* has a weight of 400 lb and is pinned to the body of the truck at each side and held in the position shown by the two side cables. Determine the tension in the cables.

SOLUTION

The idealized model of the ramp, which indicates all necessary dimensions and supports, is shown in Fig. 5–18*b*. Here the center of gravity is located at the midpoint since the ramp is considered to be uniform.

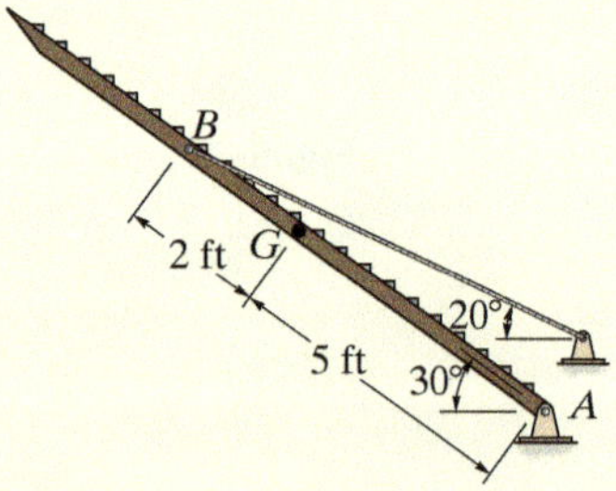

(b)

Free-Body Diagram. Working from the idealized model, the ramp's free-body diagram is shown in Fig. 5–18*c*.

Equations of Equilibrium. Summing moments about point A will yield a direct solution for the cable tension. Using the principle of moments, there are several ways of determining the moment of **T** about A. If we use x and y components, with **T** applied at B, we have

$$\zeta +\Sigma M_A = 0; \qquad -T\cos 20°(7\sin 30°\text{ ft}) + T\sin 20°(7\cos 30°\text{ ft}) + 400\text{ lb}\,(5\cos 30°\text{ ft}) = 0$$

$$T = 1425\text{ lb}$$

We can also determine the moment of **T** about A by resolving it into components along and perpendicular to the ramp at B. Then the moment of the component along the ramp will be zero about A, so that

$$\zeta +\Sigma M_A = 0; \qquad -T\sin 10°(7\text{ ft}) + 400\text{ lb}\,(5\cos 30°\text{ ft}) = 0$$

$$T = 1425\text{ lb}$$

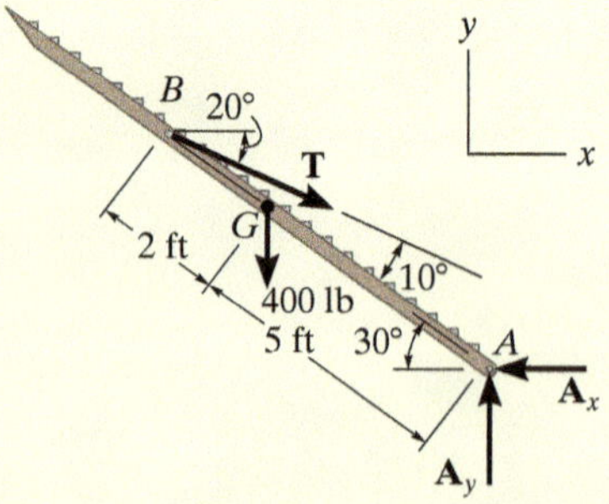

(c)

Fig. 5–18

Since there are two cables supporting the ramp,

$$T' = \frac{T}{2} = 712\text{ lb} \qquad \textit{Ans.}$$

NOTE: As an exercise, show that $A_x = 1339$ lb and $A_y = 887.4$ lb.

EXAMPLE 5.12

Determine the support reactions on the member in Fig. 5–19*a*. The collar at *A* is fixed to the member and can slide vertically along the vertical shaft.

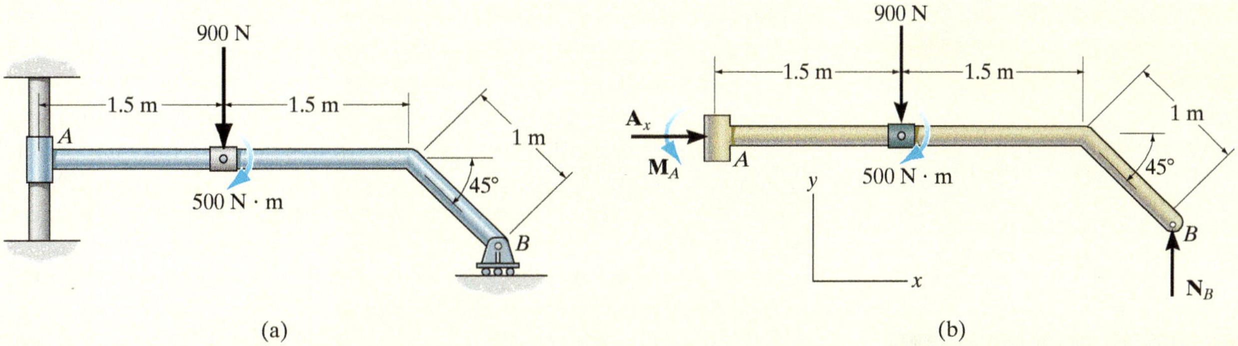

Fig. 5–19

SOLUTION

Free-Body Diagram. The free-body diagram of the member is shown in Fig. 5–19*b*. The collar exerts a horizontal force $\mathbf{A}_x$ and moment $\mathbf{M}_A$ on the member. The reaction $\mathbf{N}_B$ of the roller on the member is vertical.

Equations of Equilibrium. The forces A_x and N_B can be determined directly from the force equations of equilibrium.

$$\overset{+}{\rightarrow}\Sigma F_x = 0; \qquad A_x = 0 \qquad \textit{Ans.}$$

$$+\uparrow\Sigma F_y = 0; \qquad N_B - 900\text{ N} = 0$$

$$N_B = 900\text{ N} \qquad \textit{Ans.}$$

The moment M_A can be determined by summing moments either about point *A* or point *B*.

$$\circlearrowleft +\Sigma M_A = 0;$$

$$M_A - 900\text{ N}(1.5\text{ m}) - 500\text{ N}\cdot\text{m} + 900\text{ N}\,[3\text{ m} + (1\text{ m})\cos 45^\circ] = 0$$

$$M_A = -1486\text{ N}\cdot\text{m} = 1.49\text{ kN}\cdot\text{m}\circlearrowright \qquad \textit{Ans.}$$

or

$$\circlearrowleft +\Sigma M_B = 0; \quad M_A + 900\text{ N}\,[1.5\text{ m} + (1\text{ m})\cos 45^\circ] - 500\text{ N}\cdot\text{m} = 0$$

$$M_A = -1486\text{ N}\cdot\text{m} = 1.49\text{ kN}\cdot\text{m}\circlearrowright \qquad \textit{Ans.}$$

The negative sign indicates that $\mathbf{M}_A$ has the opposite sense of rotation to that shown on the free-body diagram.

5

5.4 Two- and Three-Force Members

The hydraulic cylinder AB is a typical example of a two-force member since it is pin connected at its ends and, provided its weight is neglected, only the pin forces act on this member.

The link used for this railroad car brake is a three-force member. Since the force $\mathbf{F}_B$ in the tie rod at B and $\mathbf{F}_C$ from the link at C are parallel, then for equilibrium the resultant force $\mathbf{F}_A$ at the pin A must also be parallel with these two forces.

The boom and bucket on this lift is a three-force member, provided its weight is neglected. Here the lines of action of the weight of the worker, $\mathbf{W}$, and the force of the two-force member (hydraulic cylinder) at B, $\mathbf{F}_B$, intersect at O. For moment equilibrium, the resultant force at the pin A, $\mathbf{F}_A$, must also be directed towards O.

The solutions to some equilibrium problems can be simplified by recognizing members that are subjected to only two or three forces.

Two-Force Members. As the name implies, a *two-force member* has forces applied at only two points on the member. An example of a two-force member is shown in Fig. 5–20*a*. To satisfy force equilibrium, $\mathbf{F}_A$ and $\mathbf{F}_B$ must be equal in magnitude, $F_A = F_B = F$, but opposite in direction ($\Sigma \mathbf{F} = \mathbf{0}$), Fig. 5–20*b*. Furthermore, moment equilibrium requires that $\mathbf{F}_A$ and $\mathbf{F}_B$ share the same line of action, which can only happen if they are directed along the line joining points A and B ($\Sigma \mathbf{M}_A = \mathbf{0}$ or $\Sigma \mathbf{M}_B = \mathbf{0}$), Fig. 5–20*c*. Therefore, for any two-force member to be in equilibrium, the two forces acting on the member *must have the same magnitude, act in opposite directions, and have the same line of action, directed along the line joining the two points where these forces act.*

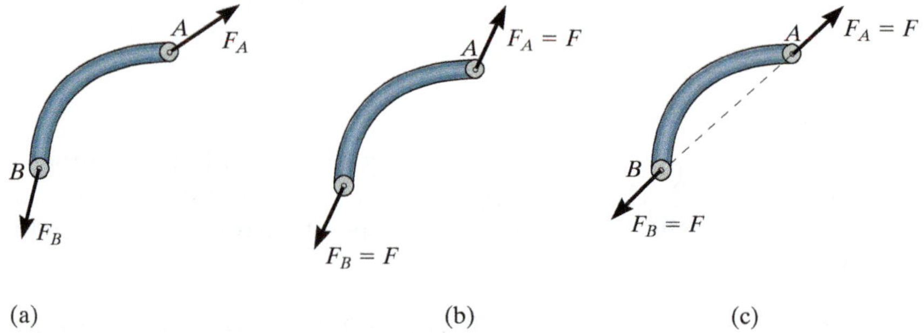

Two-force member

Fig. 5–20

Three-Force Members. If a member is subjected to only *three forces*, it is called a *three-force member*. Moment equilibrium can be satisfied only if the three forces form a *concurrent* or *parallel* force system. To illustrate, consider the member subjected to the three forces $\mathbf{F}_1$, $\mathbf{F}_2$, and $\mathbf{F}_3$, shown in Fig. 5–21*a*. If the lines of action of $\mathbf{F}_1$ and $\mathbf{F}_2$ intersect at point O, then the line of action of $\mathbf{F}_3$ must *also* pass through point O so that the forces satisfy $\Sigma \mathbf{M}_O = \mathbf{0}$. As a special case, if the three forces are all parallel, Fig. 5–21*b*, the location of the point of intersection, O, will approach infinity.

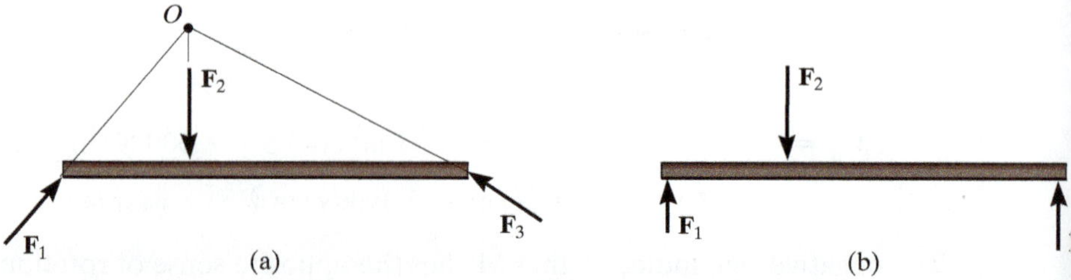

Three-force member

Fig. 5–21

EXAMPLE 5.13

The lever ABC is pin supported at A and connected to a short link BD as shown in Fig. 5–22*a*. If the weight of the members is negligible, determine the force of the pin on the lever at A.

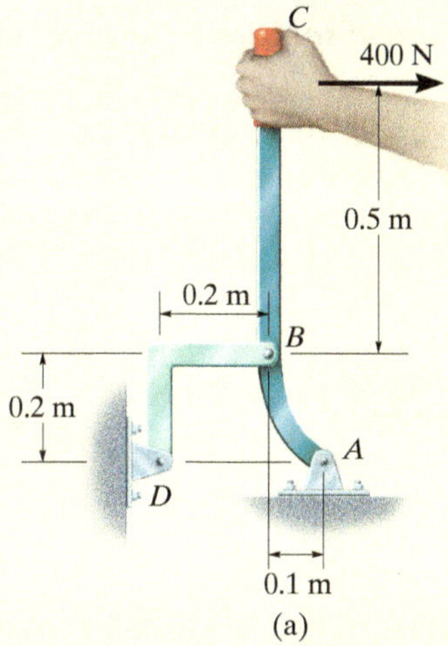

SOLUTION

Free-Body Diagrams. As shown in Fig. 5–22*b*, the short link BD is a *two-force member*, so the *resultant forces* at pins D and B must be equal, opposite, and collinear. Although the magnitude of the force is unknown, the line of action is known since it passes through B and D.

Lever ABC is a *three-force member*, and therefore, in order to satisfy moment equilibrium, the three nonparallel forces acting on it must be concurrent at O, Fig. 5–22*c*. In particular, note that the force **F** on the lever at B is equal but opposite to the force **F** acting at B on the link. Why? The distance CO must be 0.5 m since the lines of action of **F** and the 400-N force are known.

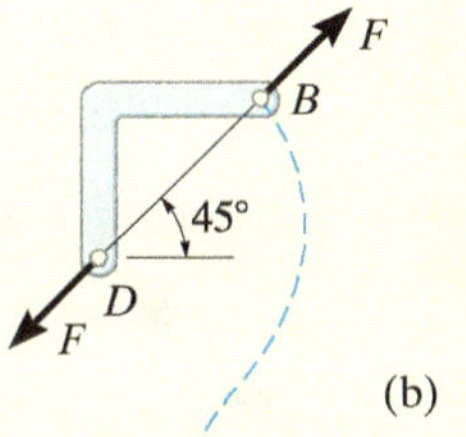

Equations of Equilibrium. By requiring the force system to be concurrent at O, since $\Sigma M_O = 0$, the angle θ which defines the line of action of $\mathbf{F}_A$ can be determined from trigonometry,

$$\theta = \tan^{-1}\left(\frac{0.7}{0.4}\right) = 60.3^\circ$$

Using the x, y axes and applying the force equilibrium equations,

$$\overset{+}{\rightarrow} \Sigma F_x = 0; \qquad F_A \cos 60.3^\circ - F \cos 45^\circ + 400 \text{ N} = 0$$

$$+\uparrow \Sigma F_y = 0; \qquad F_A \sin 60.3^\circ - F \sin 45^\circ = 0$$

Solving, we get

$$F_A = 1.07 \text{ kN} \qquad \textit{Ans.}$$

$$F = 1.32 \text{ kN}$$

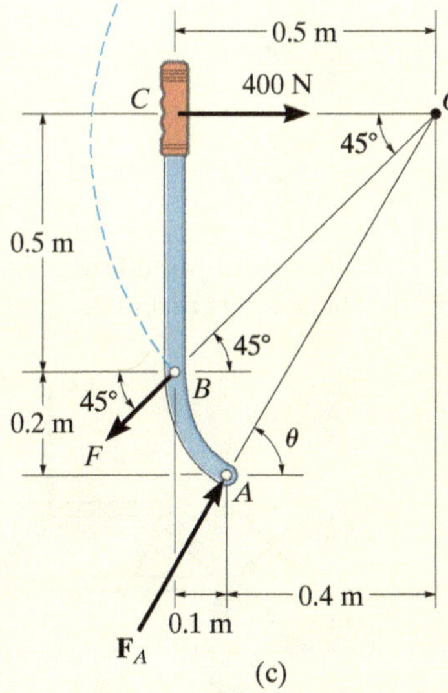

Fig. 5–22

NOTE: We can also solve this problem by representing the force at A by its two components $\mathbf{A}_x$ and $\mathbf{A}_y$ and applying $\Sigma M_A = 0$, $\Sigma F_x = 0$, $\Sigma F_y = 0$ to the lever. Once A_x and A_y are determined, we can get F_A and θ.

FUNDAMENTAL PROBLEMS

All problem solutions must include an FBD.

F5–1. Determine the horizontal and vertical components of reaction at the supports. Neglect the thickness of the beam.

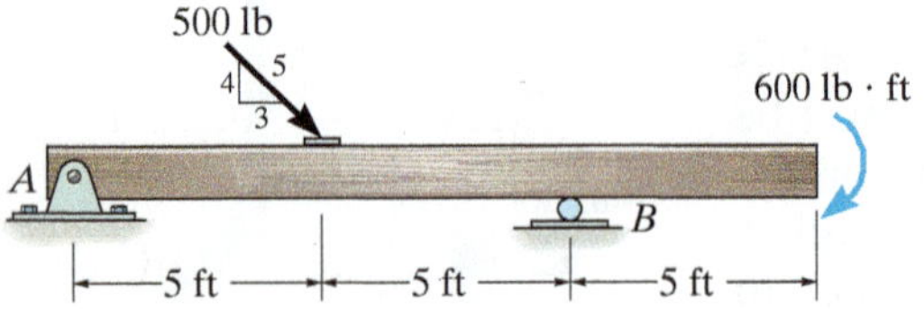

F5–1

F5–2. Determine the horizontal and vertical components of reaction at the pin A and the reaction on the beam at C.

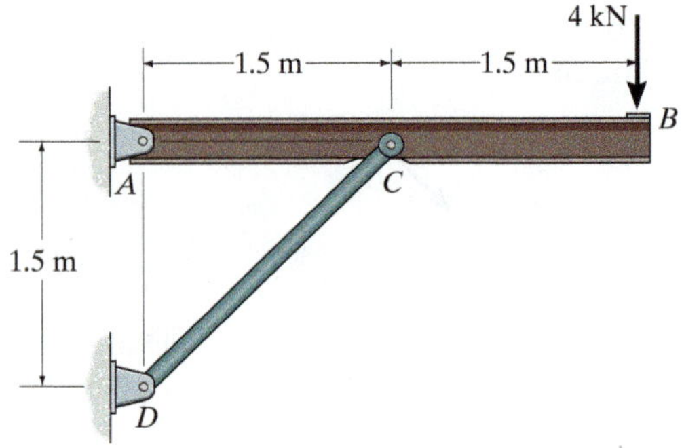

F5–2

F5–3. The truss is supported by a pin at A and a roller at B. Determine the support reactions.

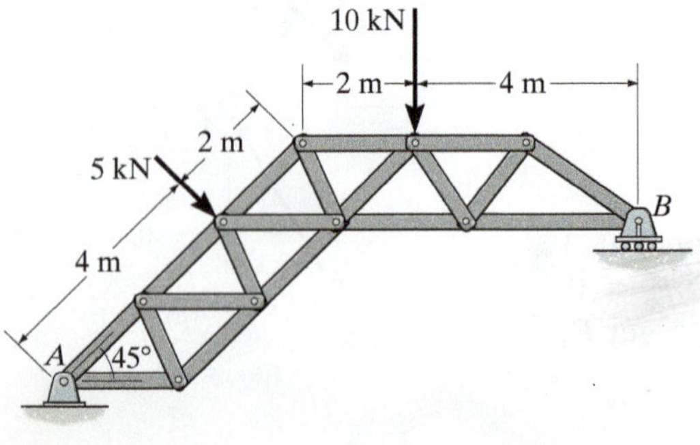

F5–3

F5–4. Determine the components of reaction at the fixed support A. Neglect the thickness of the beam.

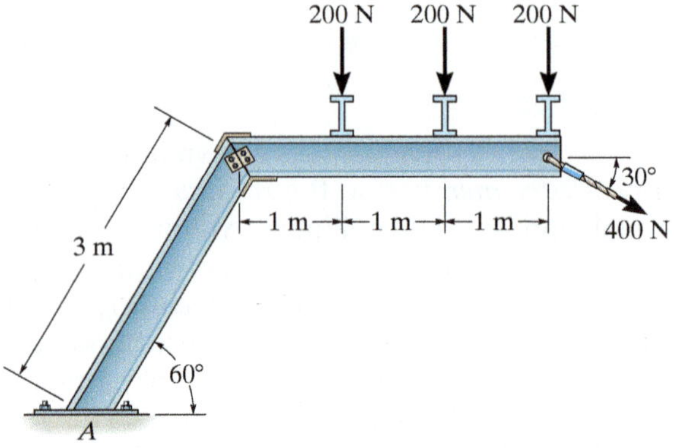

F5–4

F5–5. The 25-kg bar has a center of mass at G. If it is supported by a smooth peg at C, a roller at A, and cord AB, determine the reactions at these supports.

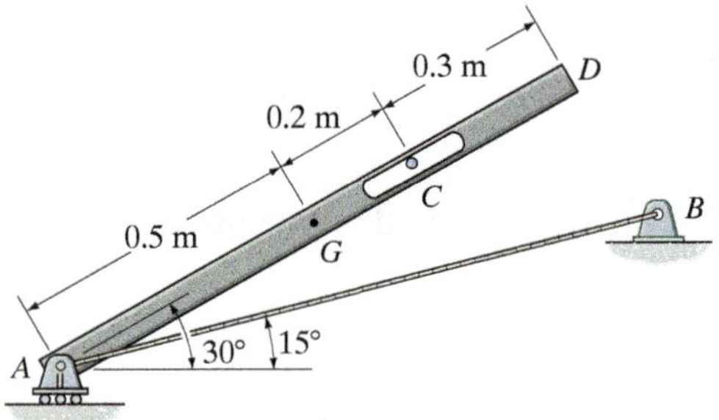

F5–5

F5–6. Determine the reactions at the smooth contact points A, B, and C on the bar.

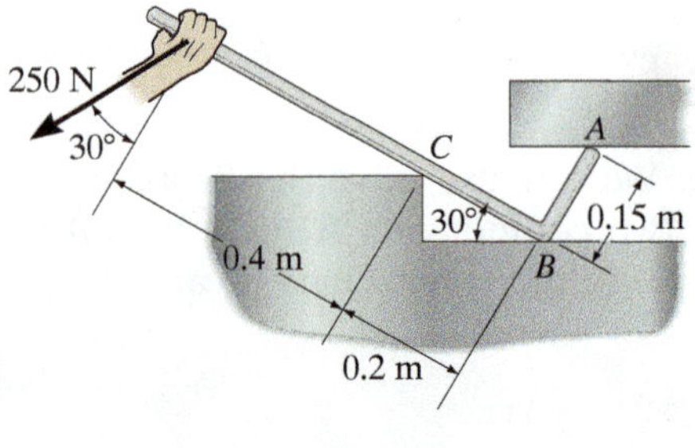

F5–6

PROBLEMS

All problem solutions must include an FBD.

5–10. Determine the horizontal and vertical components of reaction at the pin A and the reaction of the rocker B on the beam.

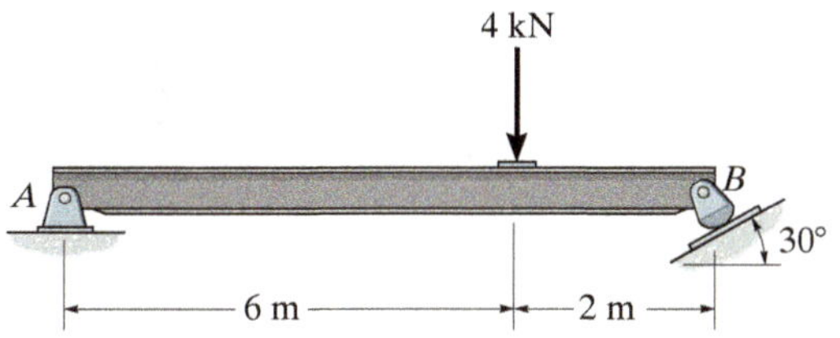

Prob. 5–10

5–11. Determine the magnitude of the reactions on the beam at A and B. Neglect the thickness of the beam.

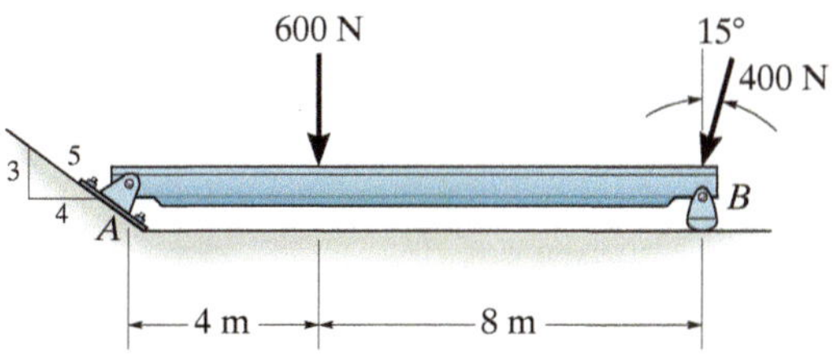

Prob. 5–11

***5–12.** Determine the components of the support reactions at the fixed support A on the cantilevered beam.

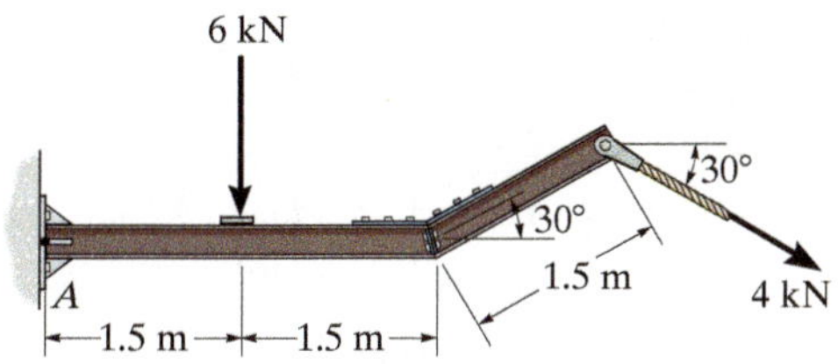

Prob. 5–12

5–13. The 75-kg gate has a center of mass located at G. If A supports only a horizontal force and B can be assumed as a pin, determine the components of reaction at these supports.

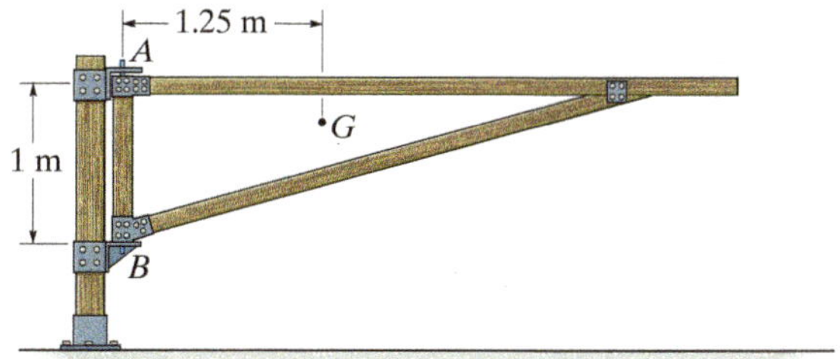

Prob. 5–13

5–14. The overhanging beam is supported by a pin at A and the two-force strut BC. Determine the horizontal and vertical components of reaction at A and the reaction at B on the beam.

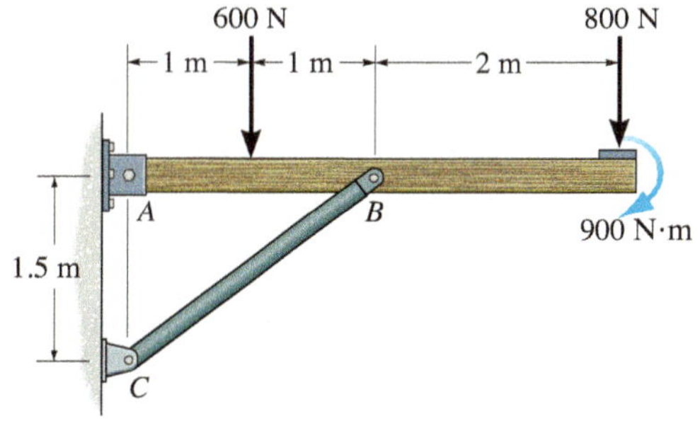

Prob. 5–14

5–15. Determine the horizontal and vertical components of reaction at the pin at A and the reaction of the roller at B on the lever.

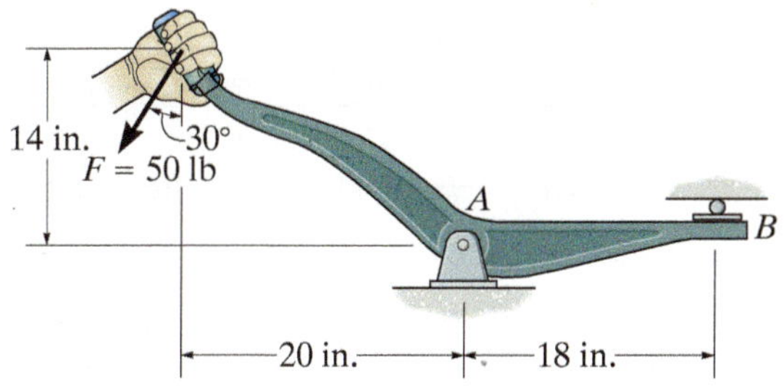

Prob. 5–15

***5–16.** Determine the components of reaction at the supports A and B on the rod.

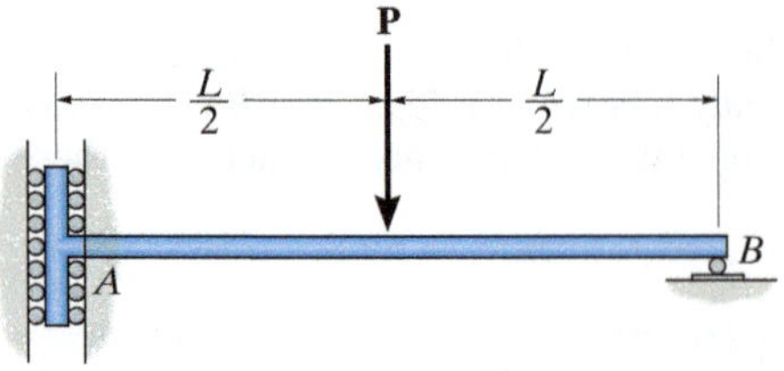

Prob. 5–16

5–17. If the wheelbarrow and its contents have a mass of 60 kg and center of mass at G, determine the magnitude of the resultant force which the man must exert on *each* of the two handles in order to hold the wheelbarrow in equilibrium.

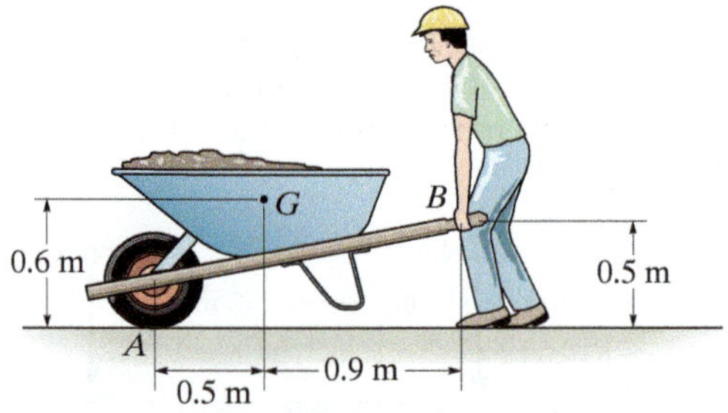

Prob. 5–17

5–18. Determine the tension in the cable and the horizontal and vertical components of reaction of the pin A. The pulley at D is frictionless and the cylinder weighs 80 lb.

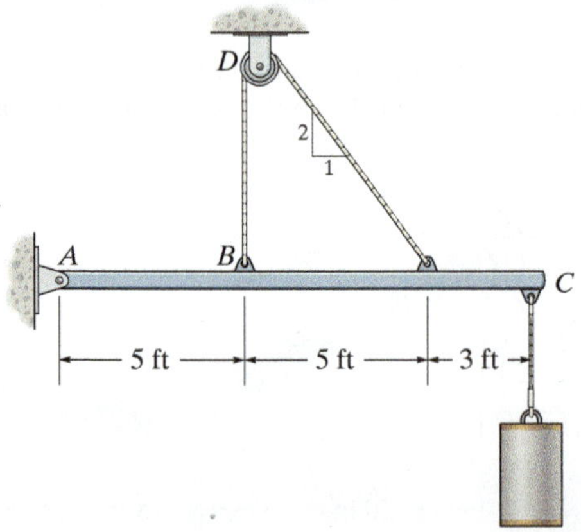

Prob. 5–18

5–19. The shelf supports the electric motor which has a mass of 15 kg and mass center at G_m. The platform upon which it rests has a mass of 4 kg and mass center at G_p. Assuming that a single bolt B holds the shelf up and the bracket bears against the smooth wall at A, determine this normal force at A and the horizontal and vertical components of reaction of the bolt on the bracket.

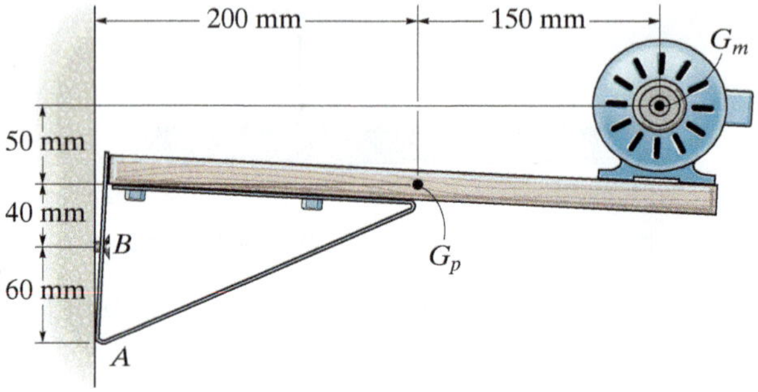

Prob. 5–19

***5–20.** The pad footing is used to support the load of 12 000 lb. Determine the intensities w_1 and w_2 of the distributed loading acting on the base of the footing for the equilibrium.

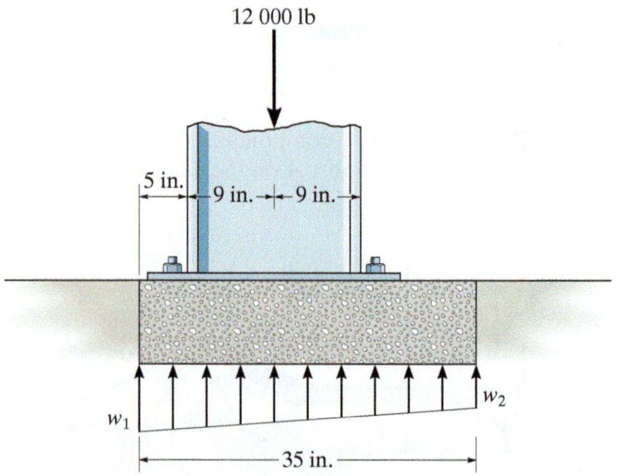

Prob. 5–20

5–21. When holding the 5-lb stone in equilibrium, the humerus H, assumed to be smooth, exerts normal forces $\mathbf{F}_C$ and $\mathbf{F}_A$ on the radius C and ulna A as shown. Determine these forces and the force $\mathbf{F}_B$ that the biceps B exerts on the radius for equilibrium. The stone has a center of mass at G. Neglect the weight of the arm.

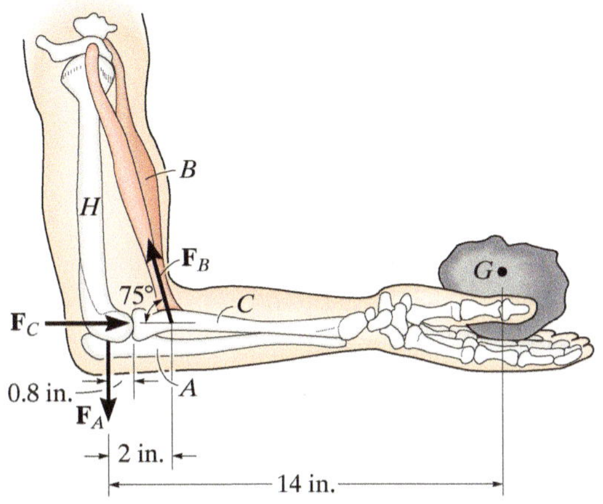

Prob. 5–21

5–22. The smooth disks D and E have a weight of 200 lb and 100 lb, respectively. If a horizontal force of $P = 200$ lb is applied to the center of disk E, determine the normal reactions at the points of contact with the ground at A, B, and C.

5–23. The smooth disks D and E have a weight of 200 lb and 100 lb, respectively. Determine the largest horizontal force P that can be applied to the center of disk E without causing the disk D to move up the incline.

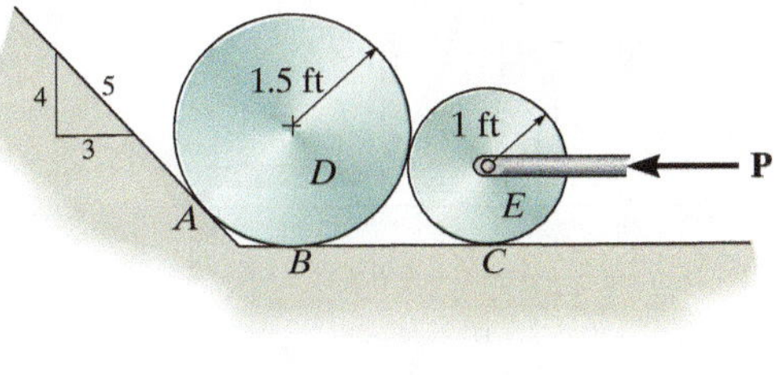

Probs. 5–22/23

***5–24.** The man is pulling a load of 8 lb with one arm held as shown. Determine the force $\mathbf{F}_H$ this exerts on the humerus bone H, and the tension developed in the biceps muscle B. Neglect the weight of the man's arm.

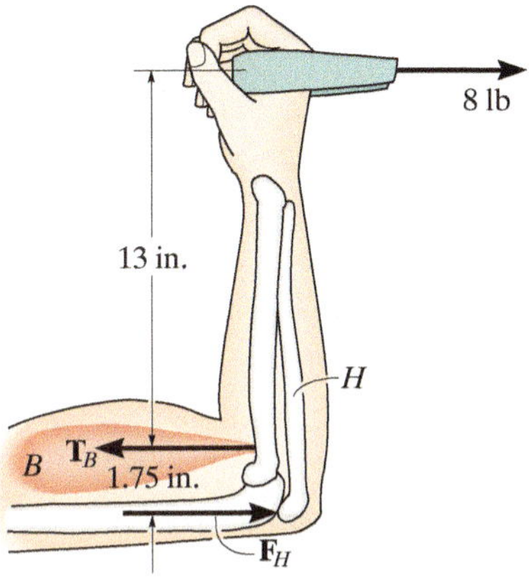

Prob. 5–24

5–25. Determine the magnitude of force at the pin A and in the cable BC needed to support the 500-lb load. Neglect the weight of the boom AB.

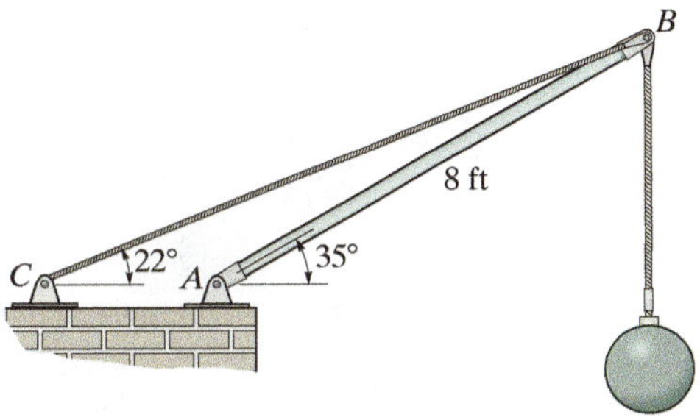

Prob. 5–25

5–26. The winch consists of a drum of radius 4 in., which is pin connected at its center C. At its outer rim is a ratchet gear having a mean radius of 6 in. The pawl AB serves as a two-force member (short link) and keeps the drum from rotating. If the suspended load is 500 lb, determine the horizontal and vertical components of reaction at the pin C.

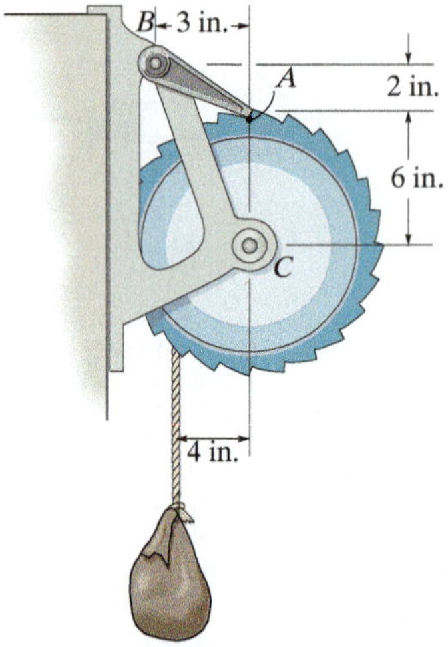

Prob. 5–26

5–27. The sports car has a mass of 1.5 Mg and mass center at G. If the front two springs each have a stiffness of $k_A = 58\ \text{kN/m}$ and the rear two springs each have a stiffness of $k_B = 65\ \text{kN/m}$, determine their compression when the car is parked on the 30° incline. Also, what friction force $\mathbf{F}_B$ must be applied to each of the rear wheels to hold the car in equilibrium? *Hint:* First determine the normal force at A and B, then determine the compression in the springs.

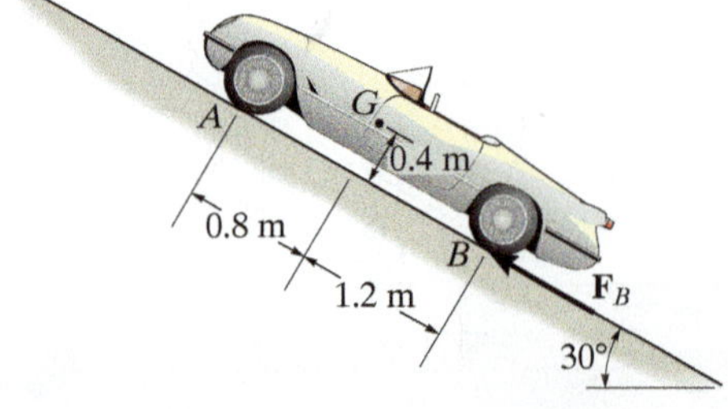

Prob. 5–27

***5–28.** The telephone pole of negligible thickness is subjected to the force of 80 lb directed as shown. It is supported by the cable BCD and can be assumed pinned at its base A. In order to provide clearance for a sidewalk right of way, where D is located, the strut CE is attached at C, as shown by the dashed lines (cable segment CD is removed). If the tension in CD' is to be twice the tension in BCD, determine the height h for placement of the strut CE.

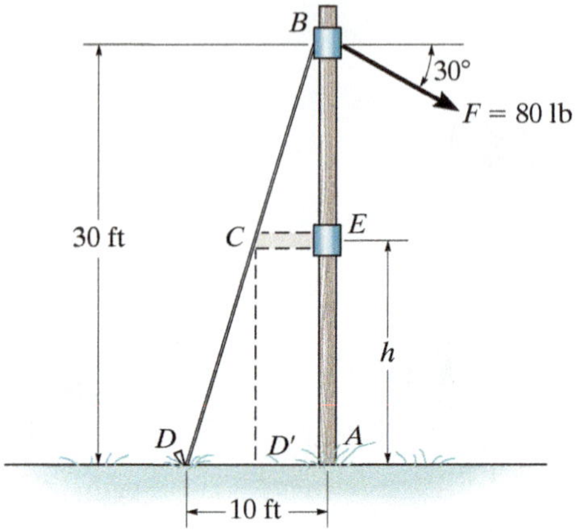

Prob. 5–28

5–29. The floor crane and the driver have a total weight of 2500 lb with a center of gravity at G. If the crane is required to lift the 500-lb drum, determine the normal reaction on *both* the wheels at A and *both* the wheels at B when the boom is in the position shown.

5–30. The floor crane and the driver have a total weight of 2500 lb with a center of gravity at G. Determine the largest weight of the drum that can be lifted without causing the crane to overturn when its boom is in the position shown.

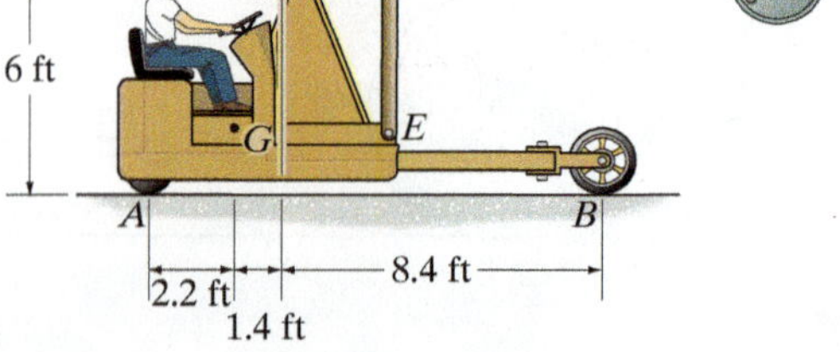

Probs. 5–29/30

5–31. The mobile crane has a weight of 120 000 lb and center of gravity at G_1; the boom has a weight of 30 000 lb and center of gravity at G_2. Determine the smallest angle of tilt θ of the boom, without causing the crane to overturn if the suspended load is $W = 40\,000$ lb. Neglect the thickness of the tracks at A and B.

***5–32.** The mobile crane has a weight of 120 000 lb and center of gravity at G_1; the boom has a weight of 30 000 lb and center of gravity at G_2. If the suspended load has a weight of $W = 16\,000$ lb, determine the normal reactions at the tracks A and B. For the calculation, neglect the thickness of the tracks and take $\theta = 30°$.

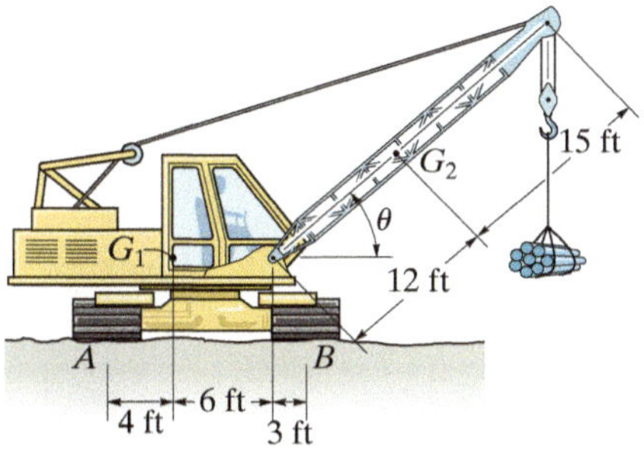

Probs. 5–31/32

5–33. The woman exercises on the rowing machine. If she exerts a holding force of $F = 200$ N on handle ABC, determine the horizontal and vertical components of reaction at pin C and the force developed along the hydraulic cylinder BD on the handle.

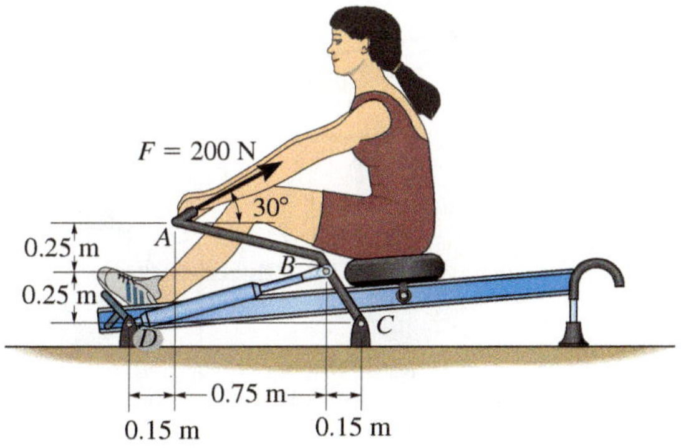

Prob. 5–33

5–34. The ramp of a ship has a weight of 200 lb and a center of gravity at G. Determine the cable force in CD needed to just start lifting the ramp, (i.e., so the reaction at B becomes zero). Also, determine the horizontal and vertical components of force at the hinge (pin) at A.

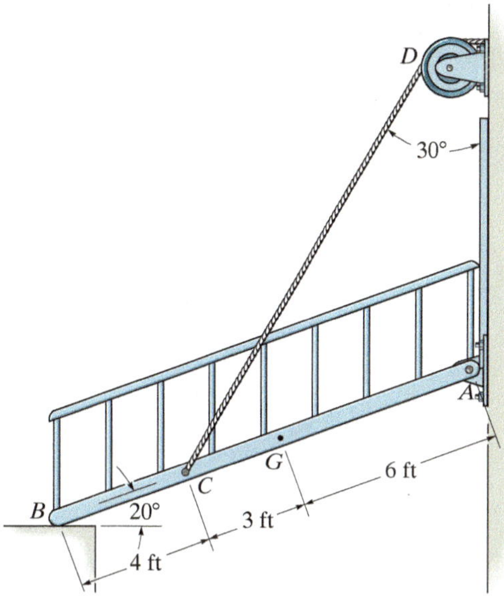

Prob. 5–34

5–35. The toggle switch consists of a cocking lever that is pinned to a fixed frame at A and held in place by the spring which has an unstretched length of 200 mm. Determine the magnitude of the resultant force at A and the normal force on the peg at B when the lever is in the position shown.

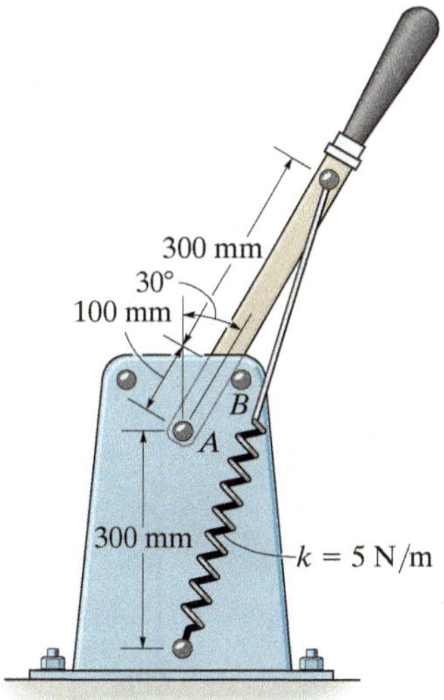

Prob. 5–35

***5–36.** The worker uses the hand truck to move material down the ramp. If the truck and its contents are held in the position shown and have a weight of 100 lb with center of gravity at G, determine the resultant normal force of both wheels on the ground A and the magnitude of the force required at the grip B.

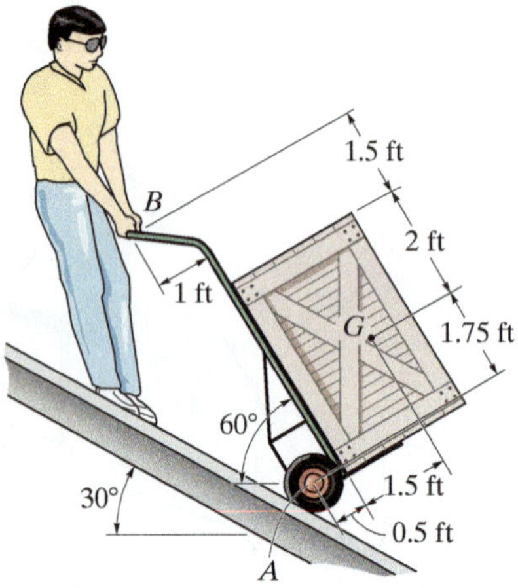

Prob. 5–36

5–37. The boom supports the two vertical loads. Neglect the size of the collars at D and B and the thickness of the boom, and compute the horizontal and vertical components of force at the pin A and the force in cable CB. Set $F_1 = 800$ N and $F_2 = 350$ N.

5–38. The boom is intended to support two vertical loads, $\mathbf{F}_1$ and $\mathbf{F}_2$. If the cable CB can sustain a maximum load of 1500 N before it fails, determine the critical loads if $F_1 = 2F_2$. Also, what is the magnitude of the maximum reaction at pin A?

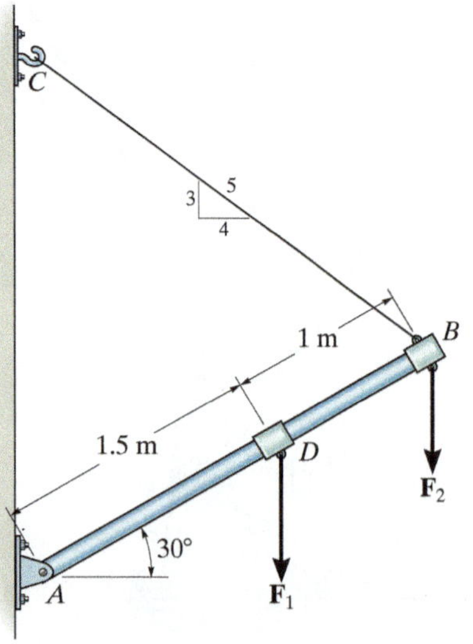

Probs. 5–37/38

5–39. The jib crane is pin connected at A and supported by a smooth collar at B. If $x = 8$ ft, determine the reactions on the jib crane at the pin A and smooth collar B. The load has a weight of 5000 lb.

***5–40.** The jib crane is pin connected at A and supported by a smooth collar at B. Determine the roller placement x of the 5000-lb load so that it gives the maximum and minimum reactions at the supports. Calculate these reactions in each case. Neglect the weight of the crane. Require 4 ft $\leq x \leq$ 10 ft.

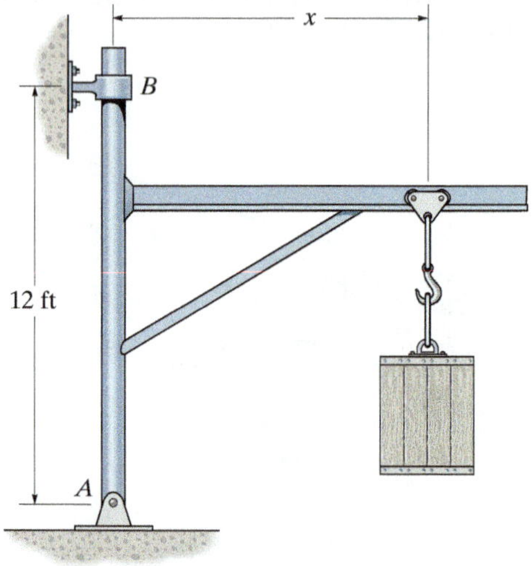

Probs. 5–39/40

5–41. The crane consists of three parts, which have weights of $W_1 = 3500$ lb, $W_2 = 900$ lb, $W_3 = 1500$ lb and centers of gravity at G_1, G_2, and G_3, respectively. Neglecting the weight of the boom, determine (a) the reactions on each of the four tires if the load is hoisted at constant velocity and has a weight of 800 lb, and (b), with the boom held in the position shown, the maximum load the crane can lift without tipping over.

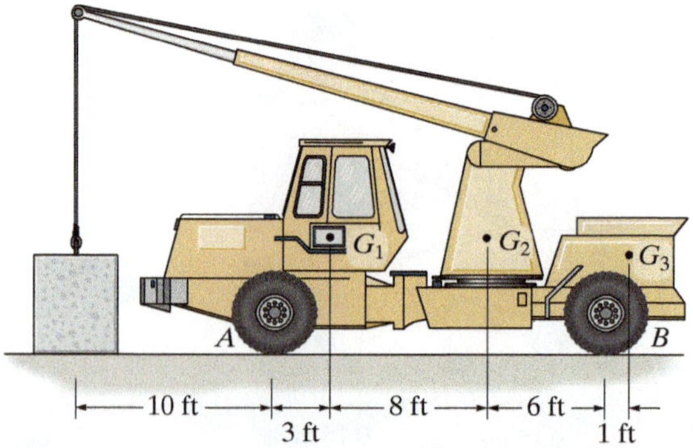

Prob. 5–41

5–42. The cantilevered jib crane is used to support the load of 780 lb. If $x = 5$ ft, determine the reactions at the supports. Note that the supports are collars that allow the crane to rotate freely about the vertical axis. The collar at B supports a force in the vertical direction, whereas the one at A does not.

5–43. The cantilevered jib crane is used to support the load of 780 lb. If the trolley T can be placed anywhere between $1.5 \text{ ft} \leq x \leq 7.5 \text{ ft}$, determine the maximum magnitude of reaction at the supports A and B. Note that the supports are collars that allow the crane to rotate freely about the vertical axis. The collar at B supports a force in the vertical direction, whereas the one at A does not.

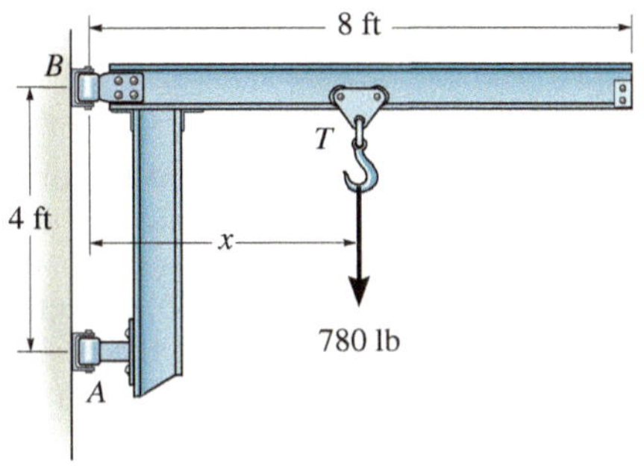

Probs. 5–42/43

***5–44.** The upper portion of the crane boom consists of the jib AB, which is supported by the pin at A, the guy line BC, and the backstay CD, each cable being separately attached to the mast at C. If the 5-kN load is supported by the hoist line, which passes over the pulley at B, determine the magnitude of the resultant force the pin exerts on the jib at A for equilibrium, the tension in the guy line BC, and the tension T in the hoist line. Neglect the weight of the jib. The pulley at B has a radius of 0.1 m.

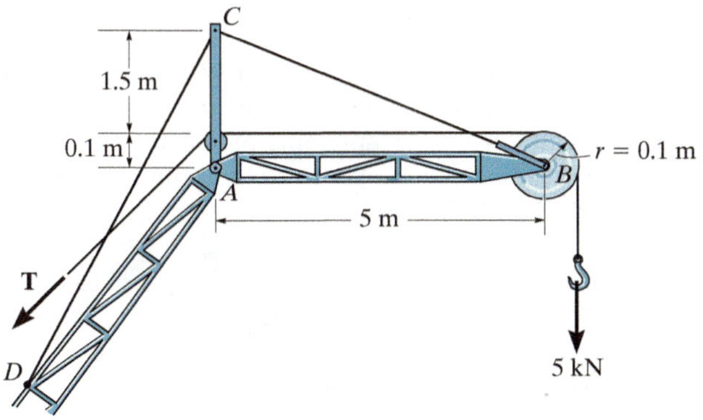

Prob. 5–44

5–45. The device is used to hold an elevator door open. If the spring has a stiffness of $k = 40$ N/m and it is compressed 0.2 m, determine the horizontal and vertical components of reaction at the pin A and the resultant force at the wheel bearing B.

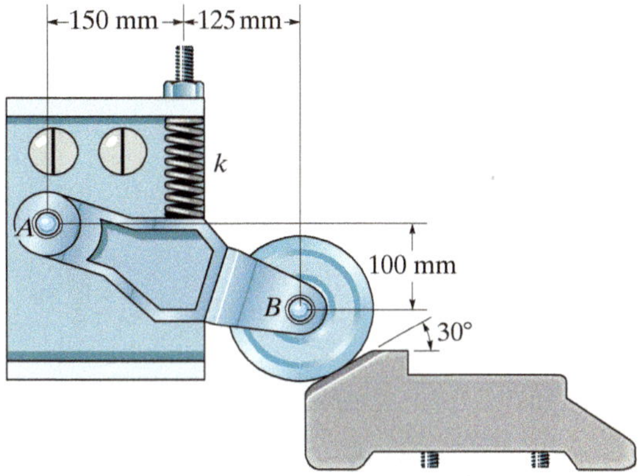

Prob. 5–45

5–46. Three uniform books, each having a weight W and length a, are stacked as shown. Determine the maximum distance d that the top book can extend out from the bottom one so the stack does not topple over.

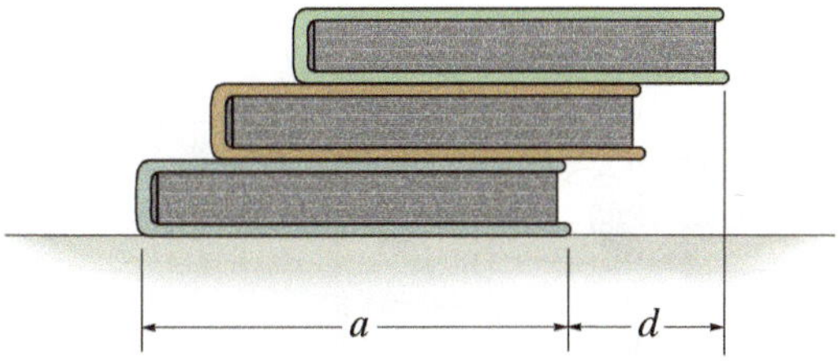

Prob. 5–46

5–47. The horizontal beam is supported by springs at its ends. Each spring has a stiffness of $k = 5\ \text{kN/m}$ and is originally unstretched when the beam is in the horizontal position. Determine the angle of tilt of the beam if a load of 800 N is applied at point C as shown.

***5–48.** The horizontal beam is supported by springs at its ends. If the stiffness of the spring at A is $k_A = 5\ \text{kN/m}$, determine the required stiffness of the spring at B so that if the beam is loaded with the 800-N force it remains in the horizontal position. The springs are originally constructed so that the beam is in the horizontal position when it is unloaded.

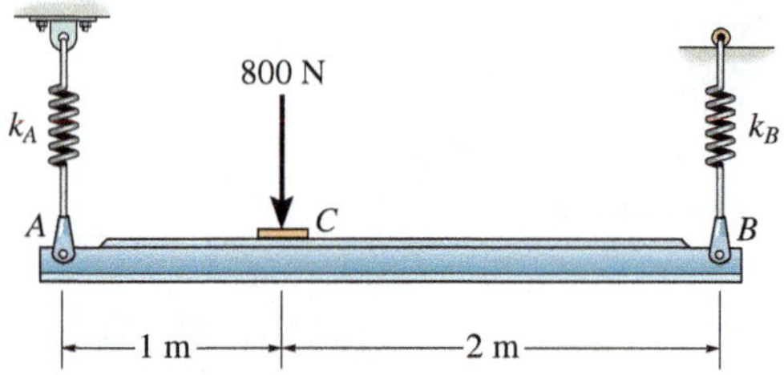

Probs. 5–47/48

5–49. The wheelbarrow and its contents have a mass of $m = 60$ kg with a center of mass at G. Determine the normal reaction on the tire and the vertical force on each hand to hold it at $\theta = 30°$. Take $a = 0.3$ m, $b = 0.45$ m, $c = 0.75$ m and $d = 0.1$ m.

5–50. The wheelbarrow and its contents have a mass m and center of mass at G. Determine the greatest angle of tilt θ without causing the wheelbarrow to tip over.

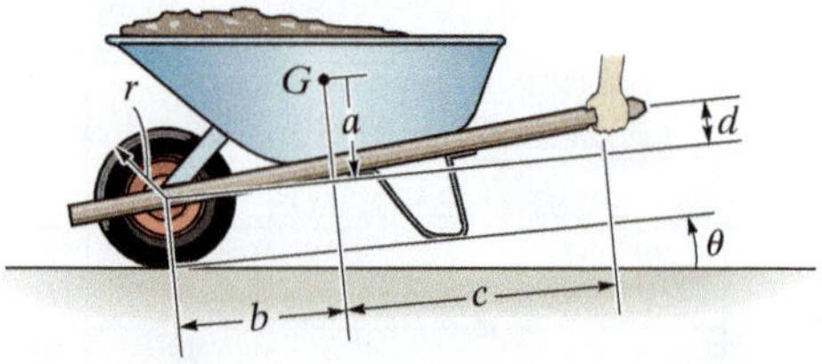

Probs. 5–49/50

5–51. The rigid beam of negligible weight is supported horizontally by two springs and a pin. If the springs are uncompressed when the load is removed, determine the force in each spring when the load **P** is applied. Also, compute the vertical deflection of end C. Assume the spring stiffness k is large enough so that only small deflections occur. *Hint:* The beam rotates about A so the deflections in the springs can be related.

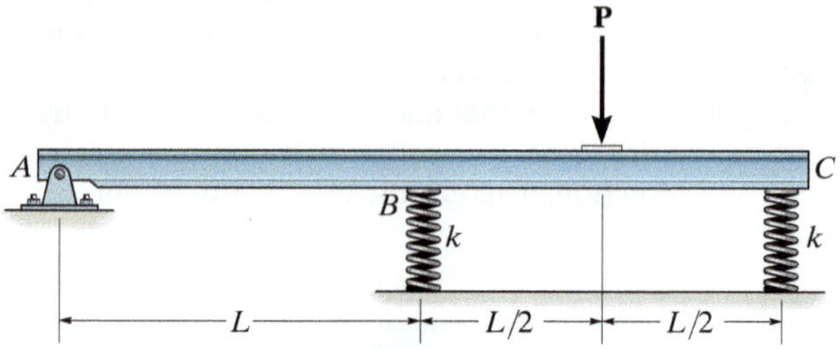

Prob. 5–51

***5–52.** A boy stands out at the end of the diving board, which is supported by two springs A and B, each having a stiffness of $k = 15\ \text{kN/m}$. In the position shown the board is horizontal. If the boy has a mass of 40 kg, determine the angle of tilt which the board makes with the horizontal after he jumps off. Neglect the weight of the board and assume it is rigid.

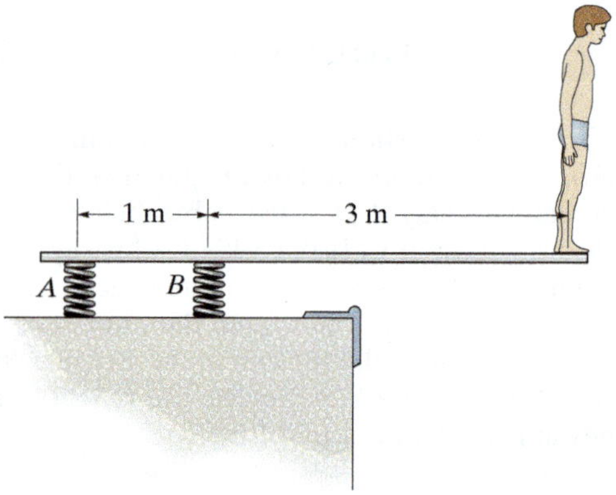

Prob. 5–52

5–53. The uniform beam has a weight W and length l and is supported by a pin at A and a cable BC. Determine the horizontal and vertical components of reaction at A and the tension in the cable necessary to hold the beam in the position shown.

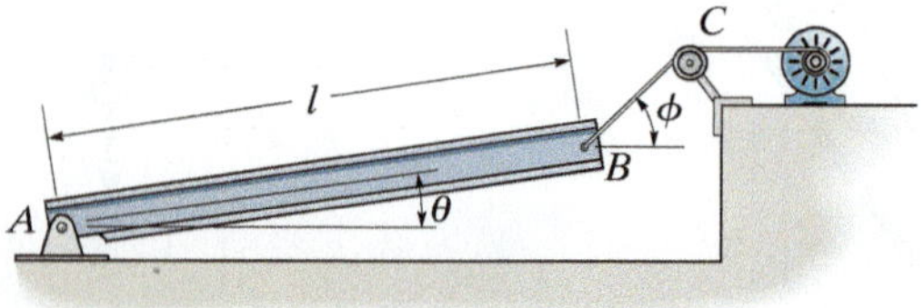

Prob. 5–53

5–54. Determine the distance d for placement of the load **P** for equilibrium of the smooth bar in the position θ as shown. Neglect the weight of the bar.

5–55. If $d = 1$ m, and $\theta = 30°$, determine the normal reaction at the smooth supports and the required distance a for the placement of the roller if $P = 600$ N. Neglect the weight of the bar.

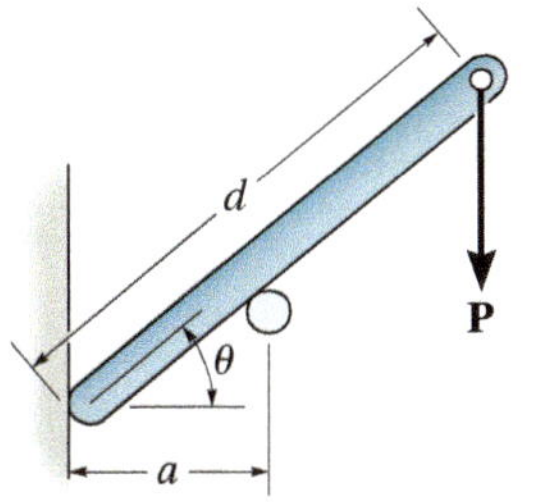

Probs. 5–54/55

***5–56.** The disk B has a mass of 20 kg and is supported on the smooth cylindrical surface by a spring having a stiffness of $k = 400$ N/m and unstretched length of $l_0 = 1$ m. The spring remains in the horizontal position since its end A is attached to the small roller guide which has negligible weight. Determine the angle θ for equilibrium of the roller.

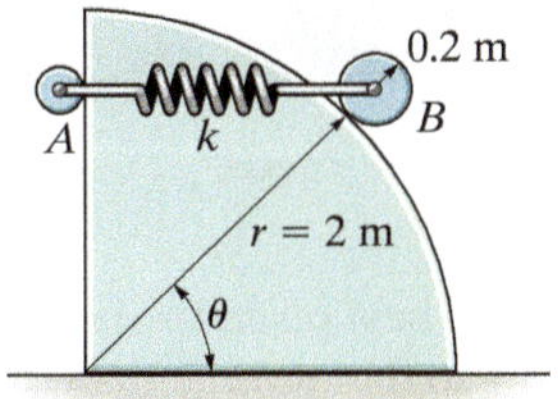

Prob. 5–56

5–57. The beam is subjected to the two concentrated loads. Assuming that the foundation exerts a linearly varying load distribution on its bottom, determine the load intensities w_1 and w_2 for equilibrium if $P = 500$ lb and $L = 12$ ft.

5–58. The beam is subjected to the two concentrated loads. Assuming that the foundation exerts a linearly varying load distribution on its bottom, determine the load intensities w_1 and w_2 for equilibrium in terms of the parameters shown.

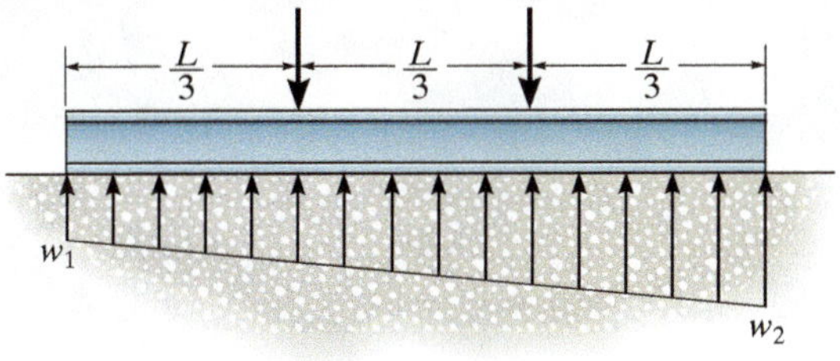

Probs. 5–57/58

5–59. The thin rod of length l is supported by the smooth tube. Determine the distance a needed for equilibrium if the applied load is **P**.

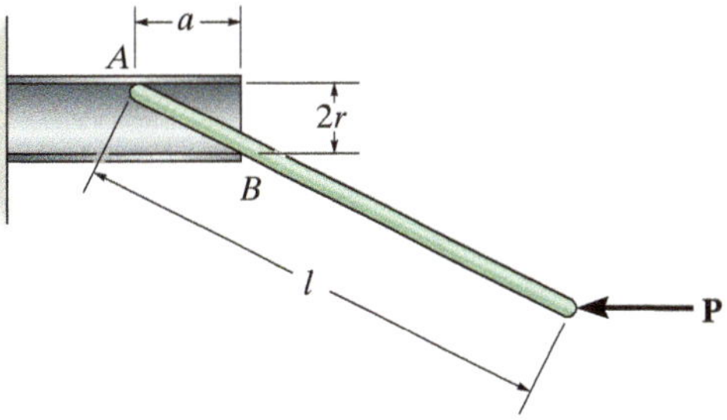

Prob. 5–59

***5–60.** The 30-N uniform rod has a length of $l = 1$ m. If $s = 1.5$ m, determine the distance h of placement at the end A along the smooth wall for equilibrium.

5–61. The uniform rod has a length l and weight W. It is supported at one end A by a smooth wall and the other end by a cord of length s which is attached to the wall as shown. Determine the placement h for equilibrium.

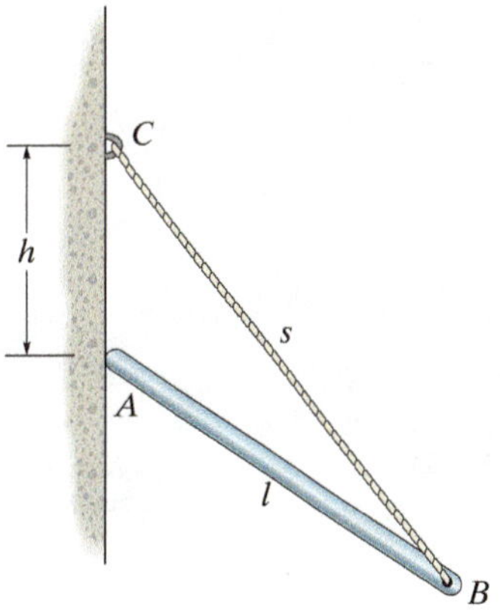

Probs. 5–60/61

CONCEPTUAL PROBLEMS

P5–5. The tie rod is used to support this overhang at the entrance of a building. If it is pin connected to the building wall at A and to the center of the overhang B, determine if the force in the rod will increase, decrease, or remain the same if (a) the support at A is moved to a lower position D, and (b) the support at B is moved to the outer position C. Explain your answer with an equilibrium analysis, using dimensions and loads. Assume the overhang is pin supported from the building wall.

P5–5

P5–6. The man attempts to pull the four wheeler up the incline and onto the trailer. From the position shown, is it more effective to pull on the rope at A, or would it be better to pull on the rope at B? Draw a free-body diagram for each case, and do an equilibrium analysis to explain your answer. Use appropriate numerical values to do your calculations.

P5–6

P5–7. Like all aircraft, this jet plane rests on three wheels. Why not use an additional wheel at the tail for better support? (Can you think of any other reason for not including this wheel?) If there was a fourth tail wheel, draw a free-body diagram of the plane from a side (2 D) view, and show why one would not be able to determine all the wheel reactions using the equations of equilibrium.

P5–7

P5–8. Where is the best place to arrange most of the logs in the wheelbarrow so that it minimizes the amount of force on the backbone of the person transporting the load? Do an equilibrium analysis to explain your answer.

P5–8

EQUILIBRIUM IN THREE DIMENSIONS

5.5 Free-Body Diagrams

The first step in solving three-dimensional equilibrium problems, as in the case of two dimensions, is to draw a free-body diagram. Before we can do this, however, it is first necessary to discuss the types of reactions that can occur at the supports.

Support Reactions. The reactive forces and couple moments acting at various types of supports and connections, when the members are viewed in three dimensions, are listed in Table 5–2. It is important to recognize the symbols used to represent each of these supports and to understand clearly how the forces and couple moments are developed. As in the two-dimensional case:

- A force is developed by a support that restricts the translation of its attached member.
- A couple moment is developed when rotation of the attached member is prevented.

For example, in Table 5–2, item (4), the ball-and-socket joint prevents any translation of the connecting member; therefore, a force must act on the member at the point of connection. This force has three components having unknown magnitudes, F_x, F_y, F_z. Provided these components are known, one can obtain the magnitude of force, $F = \sqrt{F_x^2 + F_y^2 + F_z^2}$, and the force's orientation defined by its coordinate direction angles α, β, γ, Eqs. 2–5.* Since the connecting member is allowed to rotate freely about *any* axis, no couple moment is resisted by a ball-and-socket joint.

It should be noted that the *single* bearing supports in items (5) and (7), the *single* pin (8), and the *single* hinge (9) are shown to resist both force and couple-moment components. If, however, these supports are used in conjunction with *other* bearings, pins, or hinges to hold a rigid body in equilibrium and the supports are *properly aligned* when connected to the body, then the *force reactions* at these supports *alone* are adequate for supporting the body. In other words, the couple moments become redundant and are not shown on the free-body diagram. The reason for this should become clear after studying the examples which follow.

5

* The three unknowns may also be represented as an unknown force magnitude F and two unknown coordinate direction angles. The third direction angle is obtained using the identity $\cos^2\alpha + \cos^2\beta + \cos^2\gamma = 1$, Eq. 2–8.

TABLE 5–2 Supports for Rigid Bodies Subjected to Three-Dimensional Force Systems

Types of Connection	Reaction	Number of Unknowns
(1) 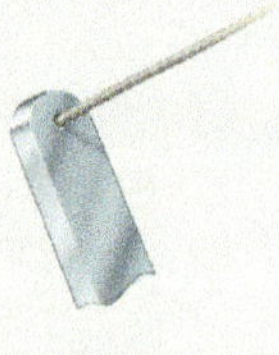 cable	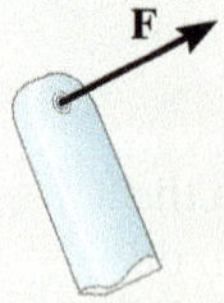	One unknown. The reaction is a force which acts away from the member in the known direction of the cable.
(2) 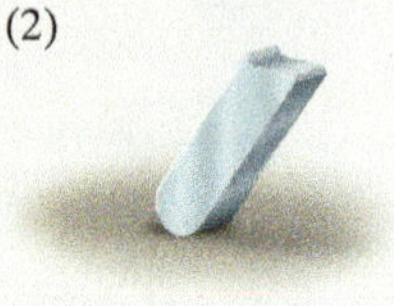 smooth surface support	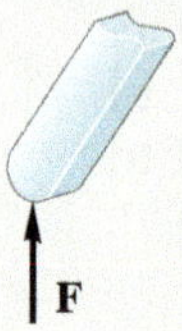	One unknown. The reaction is a force which acts perpendicular to the surface at the point of contact.
(3) 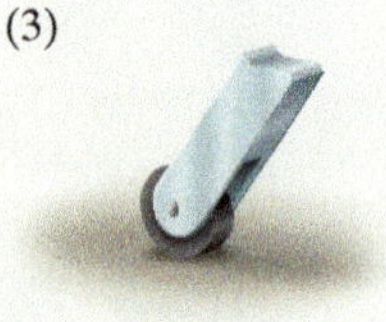 roller	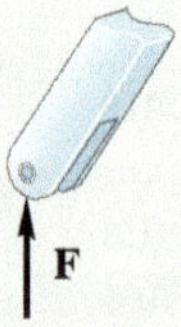	One unknown. The reaction is a force which acts perpendicular to the surface at the point of contact.
(4) 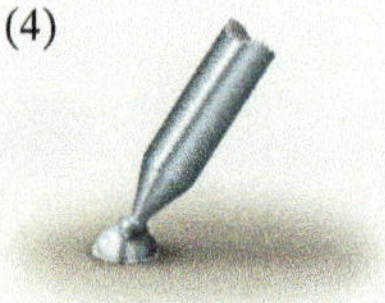 ball and socket	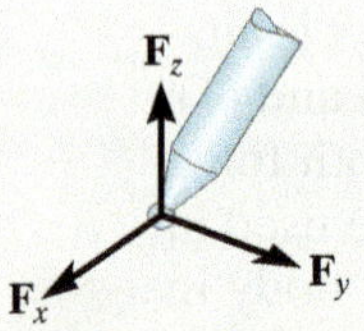	Three unknowns. The reactions are three rectangular force components.
(5)  single journal bearing	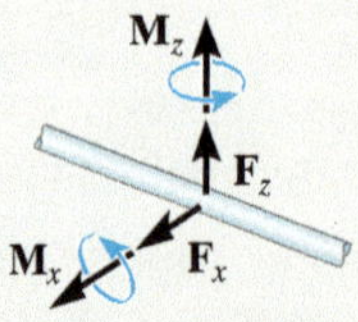	Four unknowns. The reactions are two force and two couple-moment components which act perpendicular to the shaft. Note: The couple moments are *generally not applied* if the body is supported elsewhere. See the examples.

5

continued

TABLE 5–2 Continued

<table>
<tr><th>Types of Connection</th><th>Reaction</th><th>Number of Unknowns</th></tr>
<tr><td>(6)

single journal bearing with square shaft</td><td>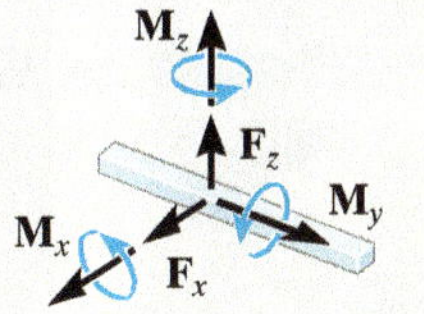
</td><td>Five unknowns. The reactions are two force and three couple-moment components. Note: The couple moments are generally not applied if the body is supported elsewhere. See the examples.</td></tr>
<tr><td>(7)
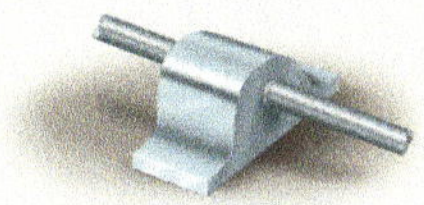
single thrust bearing</td><td>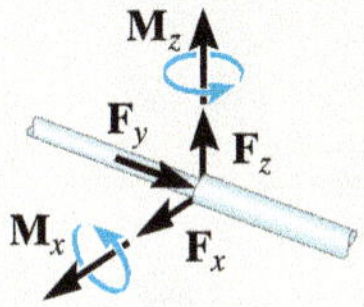
</td><td>Five unknowns. The reactions are three force and two couple-moment components. Note: The couple moments are generally not applied if the body is supported elsewhere. See the examples.</td></tr>
<tr><td>(8)
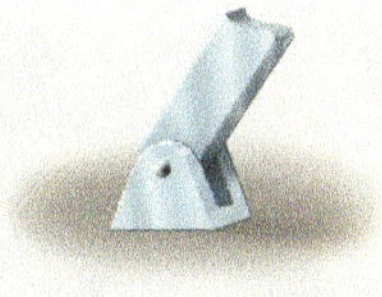
single smooth pin</td><td>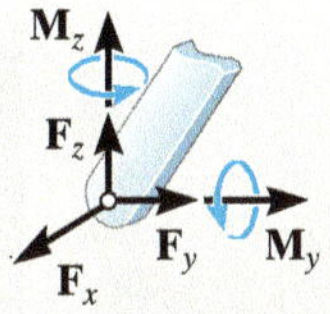
</td><td>Five unknowns. The reactions are three force and two couple-moment components. Note: The couple moments are generally not applied if the body is supported elsewhere. See the examples.</td></tr>
<tr><td>(9)

single hinge</td><td>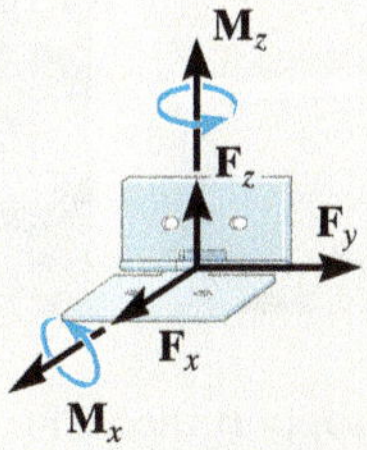
</td><td>Five unknowns. The reactions are three force and two couple-moment components. Note: The couple moments are generally not applied if the body is supported elsewhere. See the examples.</td></tr>
<tr><td>(10)

fixed support</td><td>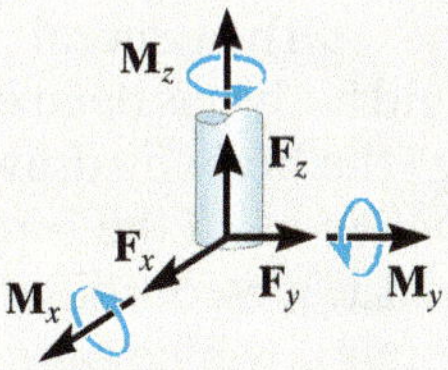
</td><td>Six unknowns. The reactions are three force and three couple-moment components.</td></tr>
</table>

Typical examples of actual supports that are referenced to Table 5–2 are shown in the following sequence of photos.

This ball-and-socket joint provides a connection for the housing of an earth grader to its frame. (4)

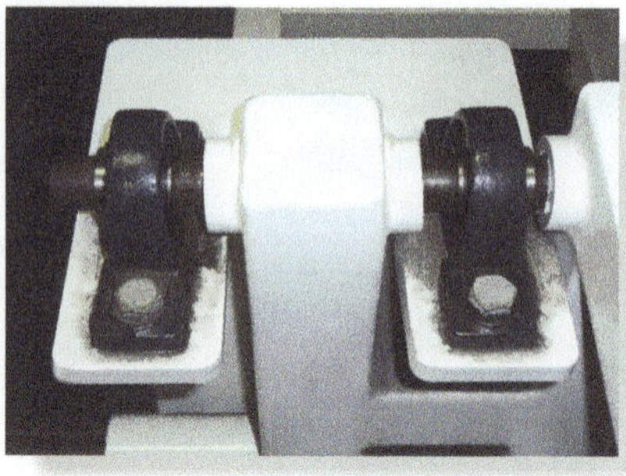

The journal bearings support the ends of the shaft. (5)

This thrust bearing is used to support the drive shaft on a machine. (7)

This pin is used to support the end of the strut used on a tractor. (8)

Free-Body Diagrams. The general procedure for establishing the free-body diagram of a rigid body has been outlined in Sec. 5.2. Essentially it requires first "isolating" the body by drawing its outlined shape. This is followed by a careful *labeling* of *all* the forces and couple moments with reference to an established x, y, z coordinate system. As a general rule, it is suggested to show the unknown components of reaction as acting on the free-body diagram in the *positive sense*. In this way, if any negative values are obtained, they will indicate that the components act in the negative coordinate directions.

EXAMPLE 5.14

Consider the two rods and plate, along with their associated free-body diagrams, shown in Fig. 5–23. The *x, y, z* axes are established on the diagram and the unknown reaction components are indicated in the *positive sense*. The weight is neglected.

SOLUTION

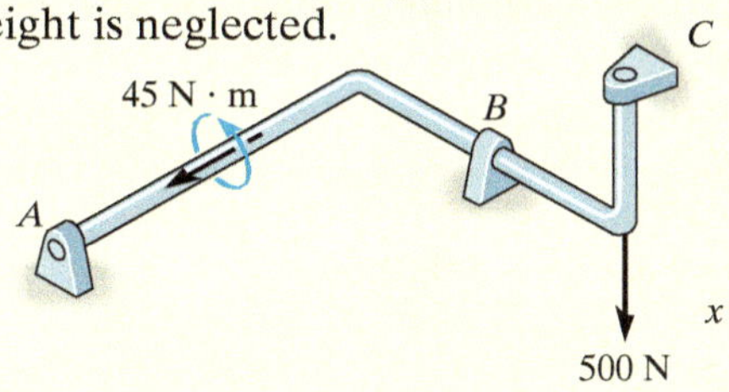

Properly aligned journal bearings at *A*, *B*, *C*.

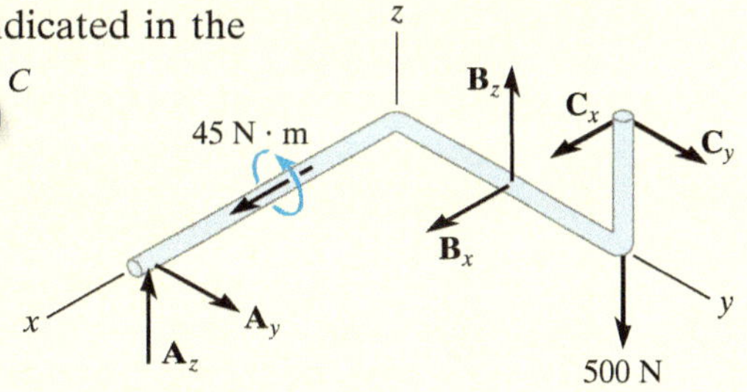

The force reactions developed by the bearings are *sufficient* for equilibrium since they prevent the shaft from rotating about each of the coordinate axes. No couple moments at each bearing are developed.

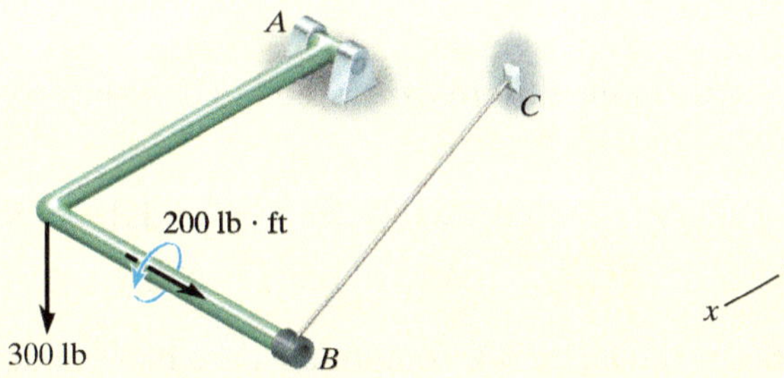

Pin at *A* and cable *BC*.

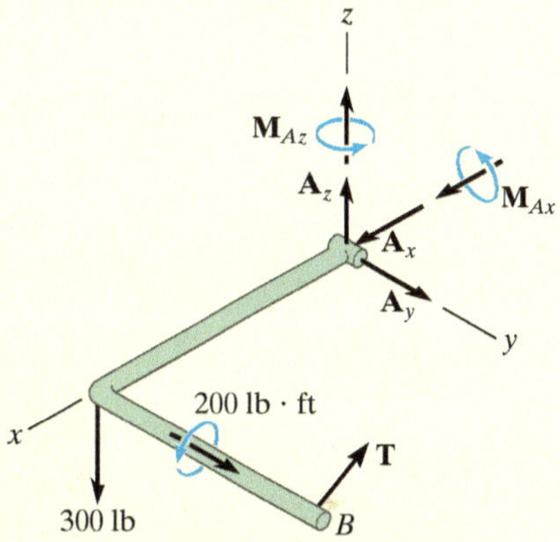

Moment components are developed by the pin on the rod to prevent rotation about the *x* and *z* axes.

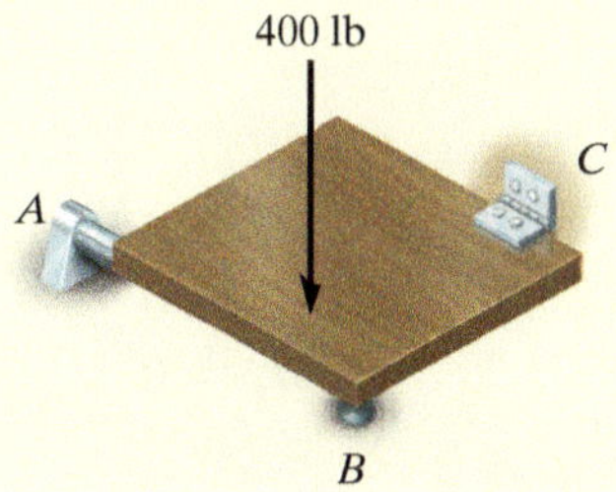

Properly aligned journal bearing at *A* and hinge at *C*. Roller at *B*.

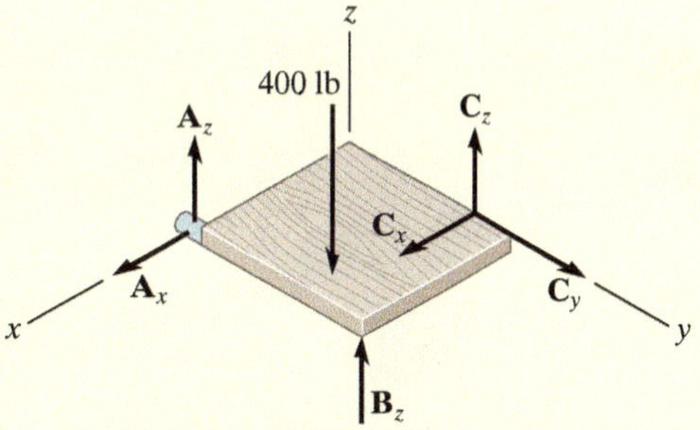

Only force reactions are developed by the bearing and hinge on the plate to prevent rotation about each coordinate axis. No moments are developed at the hinge.

Fig. 5–23

5.6 Equations of Equilibrium

As stated in Sec. 5.1, the conditions for equilibrium of a rigid body subjected to a three-dimensional force system require that both the *resultant* force and *resultant* couple moment acting on the body be equal to *zero*.

Vector Equations of Equilibrium. The two conditions for equilibrium of a rigid body may be expressed mathematically in vector form as

$$\Sigma \mathbf{F} = \mathbf{0}$$
$$\Sigma \mathbf{M}_O = \mathbf{0} \quad (5\text{–}5)$$

where $\Sigma \mathbf{F}$ is the vector sum of all the external forces acting on the body and $\Sigma \mathbf{M}_O$ is the sum of the couple moments and the moments of all the forces about any point O located either on or off the body.

Scalar Equations of Equilibrium. If all the external forces and couple moments are expressed in Cartesian vector form and substituted into Eqs. 5–5, we have

$$\Sigma \mathbf{F} = \Sigma F_x \mathbf{i} + \Sigma F_y \mathbf{j} + \Sigma F_z \mathbf{k} = \mathbf{0}$$
$$\Sigma \mathbf{M}_O = \Sigma M_x \mathbf{i} + \Sigma M_y \mathbf{j} + \Sigma M_z \mathbf{k} = \mathbf{0}$$

Since the **i**, **j**, and **k** components are independent from one another, the above equations are satisfied provided

$$\Sigma F_x = 0$$
$$\Sigma F_y = 0 \quad (5\text{–}6a)$$
$$\Sigma F_z = 0$$

and

$$\Sigma M_x = 0$$
$$\Sigma M_y = 0 \quad (5\text{–}6b)$$
$$\Sigma M_z = 0$$

These *six scalar equilibrium equations* may be used to solve for at most six unknowns shown on the free-body diagram. Equations 5–6*a* require the sum of the external force components acting in the x, y, and z directions to be zero, and Eqs. 5–6*b* require the sum of the moment components about the x, y, and z axes to be zero.

5.7 Constraints and Statical Determinacy

To ensure the equilibrium of a rigid body, it is not only necessary to satisfy the equations of equilibrium, but the body must also be properly held or constrained by its supports. Some bodies may have more supports than are necessary for equilibrium, whereas others may not have enough or the supports may be arranged in a particular manner that could cause the body to move. Each of these cases will now be discussed.

Redundant Constraints. When a body has redundant supports, that is, more supports than are necessary to hold it in equilibrium, it becomes statically indeterminate. *Statically indeterminate* means that there will be more unknown loadings on the body than equations of equilibrium available for their solution. For example, the beam in Fig. 5–24*a* and the pipe assembly in Fig. 5–24*b*, shown together with their free-body diagrams, are both statically indeterminate because of additional (or redundant) support reactions. For the beam there are five unknowns, M_A, A_x, A_y, B_y, and C_y, for which only three equilibrium equations can be written ($\Sigma F_x = 0$, $\Sigma F_y = 0$, and $\Sigma M_O = 0$, Eq. 5–2). The pipe assembly has eight unknowns, for which only six equilibrium equations can be written, Eqs. 5–6.

The additional equations needed to solve statically indeterminate problems of the type shown in Fig. 5–24 are generally obtained from the deformation conditions at the points of support. These equations involve the physical properties of the body which are studied in subjects dealing with the mechanics of deformation, such as "mechanics of materials."*

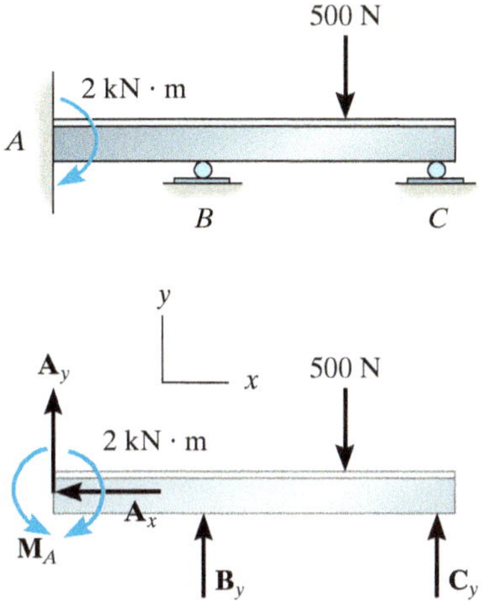

(a)

Fig. 5–24

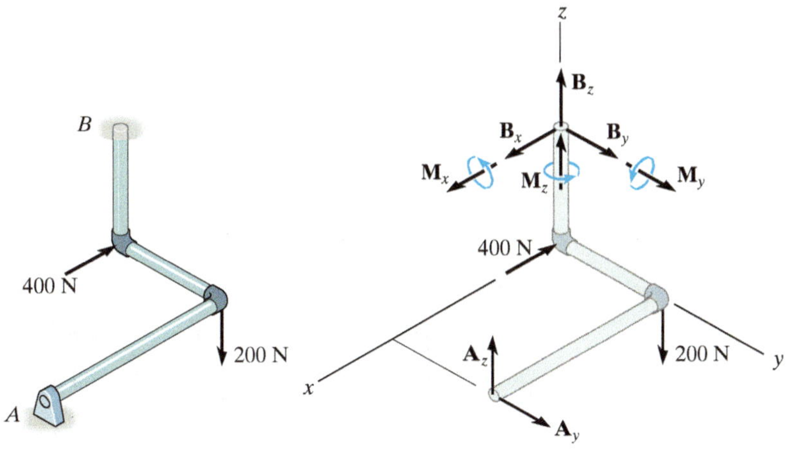

(b)

* See R. C. Hibbeler, *Mechanics of Materials*, 8th edition, Pearson Education/Prentice Hall, Inc.

5

Improper Constraints. Having the same number of unknown reactive forces as available equations of equilibrium does not always guarantee that a body will be stable when subjected to a particular loading. For example, the pin support at A and the roller support at B for the beam in Fig. 5–25*a* are placed in such a way that the lines of action of the reactive forces are *concurrent* at point A. Consequently, the applied loading **P** will cause the beam to rotate slightly about A, and so the beam is improperly constrained, $\Sigma M_A \neq 0$.

In three dimensions, a body will be improperly constrained if the lines of action of all the reactive forces intersect a common axis. For example, the reactive forces at the ball-and-socket supports at A and B in Fig. 5–25*b* all intersect the axis passing through A and B. Since the moments of these forces about A and B are all zero, then the loading **P** will rotate the member about the AB axis, $\Sigma M_{AB} \neq 0$.

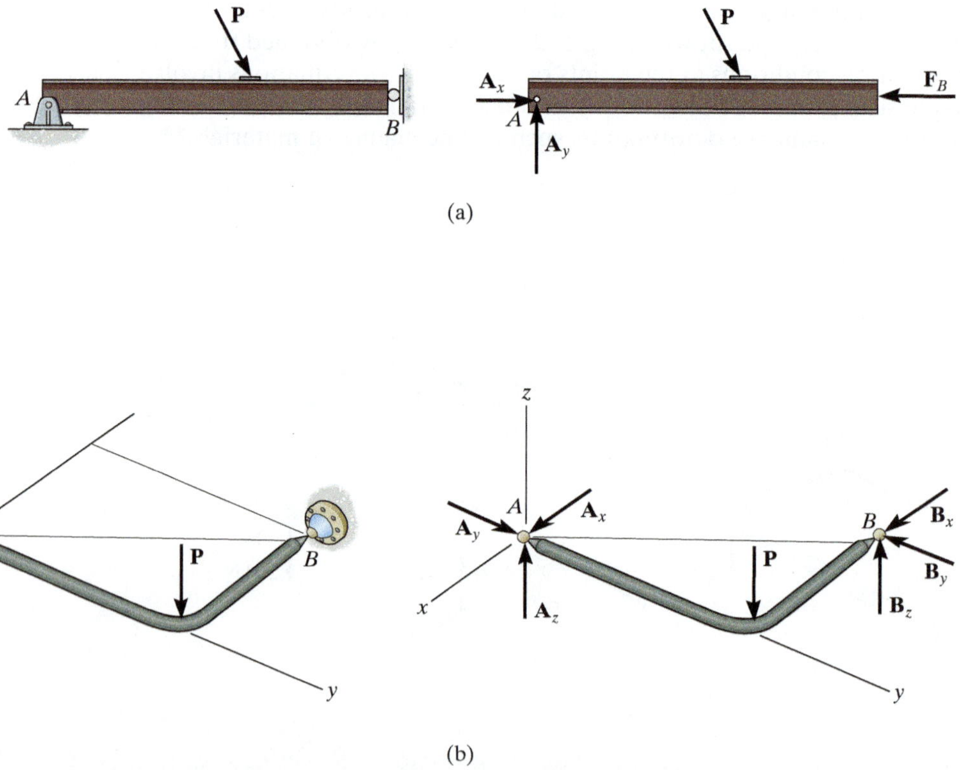

Fig. 5–25

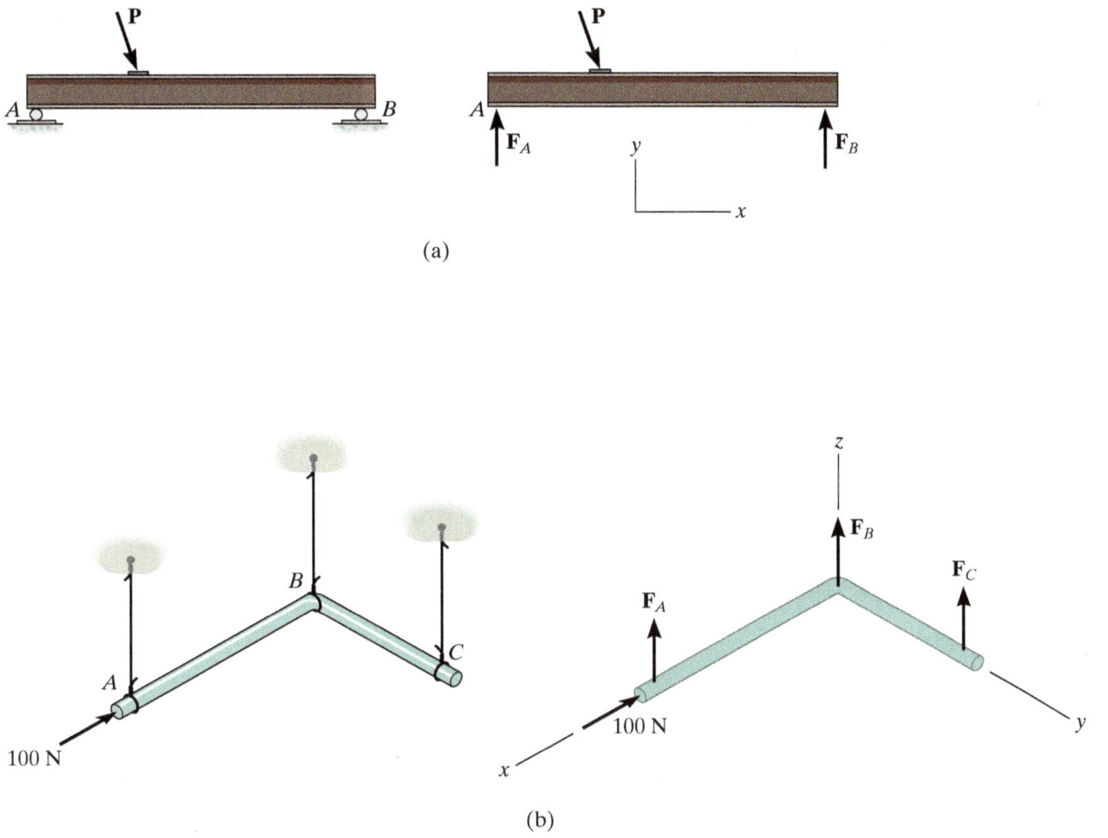

Fig. 5–26

Another way in which improper constraining leads to instability occurs when the *reactive forces* are all *parallel.* Two- and three-dimensional examples of this are shown in Fig. 5–26. In both cases, the summation of forces along the x axis will not equal zero.

In some cases, a body may have *fewer* reactive forces than equations of equilibrium that must be satisfied. The body then becomes only *partially constrained.* For example, consider member AB in Fig. 5–27a with its corresponding free-body diagram in Fig. 5–27b. Here $\Sigma F_y = 0$ will not be satisfied for the loading conditions and therefore equilibrium will not be maintained.

To summarize these points, a body is considered *improperly constrained* if all the reactive forces intersect at a common point or pass through a common axis, or if all the reactive forces are parallel. In engineering practice, these situations should be avoided at all times since they will cause an unstable condition.

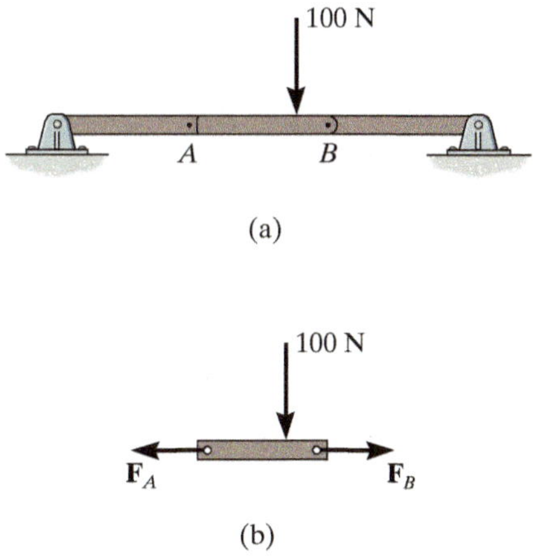

Fig. 5–27

Important Points

- Always draw the free-body diagram first when solving any equilibrium problem.
- If a support *prevents translation* of a body in a specific direction, then the support exerts a *force* on the body in that direction.
- If a support *prevents rotation about an axis*, then the support exerts a *couple moment* on the body about the axis.
- If a body is subjected to more unknown reactions than available equations of equilibrium, then the problem is *statically indeterminate*.
- A stable body requires that the lines of action of the reactive forces do not intersect a common axis and are not parallel to one another.

Procedure for Analysis

Three-dimensional equilibrium problems for a rigid body can be solved using the following procedure.

Free-Body Diagram.

- Draw an outlined shape of the body.
- Show all the forces and couple moments acting on the body.
- Establish the origin of the x, y, z axes at a convenient point and orient the axes so that they are parallel to as many of the external forces and moments as possible.
- Label all the loadings and specify their directions. In general, show all the *unknown* components having a *positive sense* along the x, y, z axes.
- Indicate the dimensions of the body necessary for computing the moments of forces.

Equations of Equilibrium.

- If the x, y, z force and moment components seem easy to determine, then apply the six scalar equations of equilibrium; otherwise use the vector equations.
- It is not necessary that the set of axes chosen for force summation coincide with the set of axes chosen for moment summation. Actually, an axis in any arbitrary direction may be chosen for summing forces and moments.
- Choose the direction of an axis for moment summation such that it intersects the lines of action of as many unknown forces as possible. Realize that the moments of forces passing through points on this axis and the moments of forces which are parallel to the axis will then be zero.
- If the solution of the equilibrium equations yields a negative scalar for a force or couple moment magnitude, it indicates that the sense is opposite to that assumed on the free-body diagram.

EXAMPLE 5.15

The homogeneous plate shown in Fig. 5–28*a* has a mass of 100 kg and is subjected to a force and couple moment along its edges. If it is supported in the horizontal plane by a roller at *A*, a ball-and-socket joint at *B*, and a cord at *C*, determine the components of reaction at these supports.

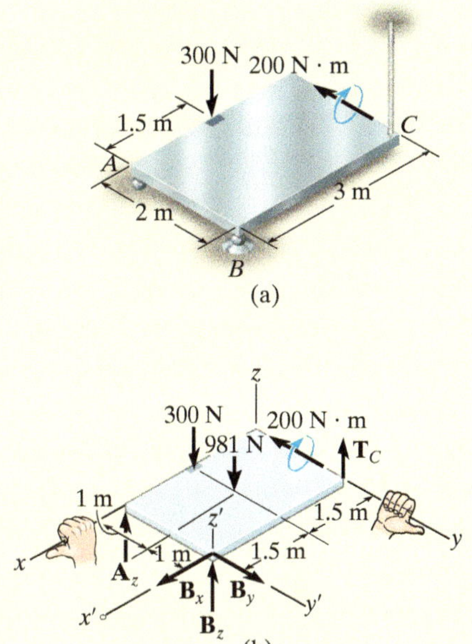

(a)

(b)

Fig. 5–28

SOLUTION (*SCALAR ANALYSIS*)

Free-Body Diagram. There are five unknown reactions acting on the plate, as shown in Fig. 5–28*b*. Each of these reactions is assumed to act in a positive coordinate direction.

Equations of Equilibrium. Since the three-dimensional geometry is rather simple, a *scalar analysis* provides a *direct solution* to this problem. A force summation along each axis yields

$\Sigma F_x = 0; \qquad B_x = 0$ *Ans.*

$\Sigma F_y = 0; \qquad B_y = 0$ *Ans.*

$\Sigma F_z = 0; \qquad A_z + B_z + T_C - 300\text{ N} - 981\text{ N} = 0 \qquad (1)$

Recall that the moment of a force about an axis is equal to the product of the force magnitude and the perpendicular distance (moment arm) from the line of action of the force to the axis. Also, forces that are parallel to an axis or pass through it create no moment about the axis. Hence, summing moments about the positive *x* and *y* axes, we have

$\Sigma M_x = 0; \qquad T_C(2\text{ m}) - 981\text{ N}(1\text{ m}) + B_z(2\text{ m}) = 0 \qquad (2)$

$\Sigma M_y = 0; \qquad 300\text{ N}(1.5\text{ m}) + 981\text{ N}(1.5\text{ m}) - B_z(3\text{ m}) - A_z(3\text{ m}) - 200\text{ N}\cdot\text{m} = 0 \qquad (3)$

The components of the force at *B* can be eliminated if moments are summed about the *x*′ and *y*′ axes. We obtain

$\Sigma M_{x'} = 0; \qquad 981\text{ N}(1\text{ m}) + 300\text{ N}(2\text{ m}) - A_z(2\text{ m}) = 0 \qquad (4)$

$\Sigma M_{y'} = 0; \qquad -300\text{ N}(1.5\text{ m}) - 981\text{ N}(1.5\text{ m}) - 200\text{ N}\cdot\text{m} + T_C(3\text{ m}) = 0 \qquad (5)$

Solving Eqs. 1 through 3 or the more convenient Eqs. 1, 4, and 5 yields

$A_z = 790\text{ N} \quad B_z = -217\text{ N} \quad T_C = 707\text{ N}$ *Ans.*

The negative sign indicates that $\mathbf{B}_z$ acts downward.

NOTE: The solution of this problem does not require a summation of moments about the *z* axis. The plate is partially constrained since the supports cannot prevent it from turning about the *z* axis if a force is applied to it in the *x*–*y* plane.

EXAMPLE 5.16

Determine the components of reaction that the ball-and-socket joint at A, the smooth journal bearing at B, and the roller support at C exert on the rod assembly in Fig. 5–29*a*.

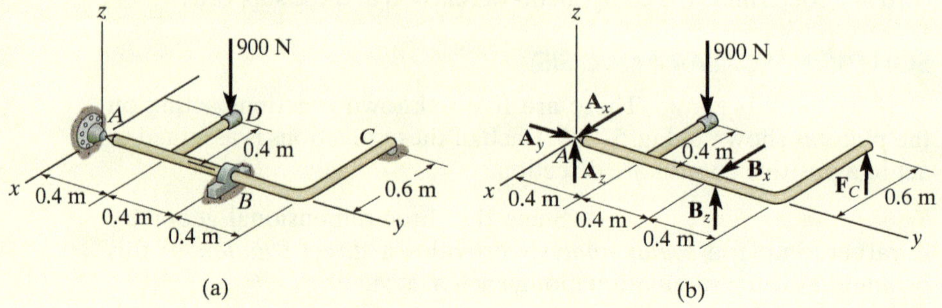

Fig. 5–29

5

SOLUTION

Free-Body Diagram. As shown on the free-body diagram, Fig. 5–29*b*, the reactive forces of the supports will prevent the assembly from rotating about each coordinate axis, and so the journal bearing at B only exerts reactive forces on the member. No couple moments are required.

Equations of Equilibrium. A direct solution for A_y can be obtained by summing forces along the y axis.

$\Sigma F_y = 0;$ $\quad A_y = 0$ *Ans.*

The force F_C can be determined directly by summing moments about the y axis.

$\Sigma M_y = 0;$ $\quad F_C(0.6\text{ m}) - 900\text{ N}(0.4\text{ m}) = 0$

$F_C = 600\text{ N}$ *Ans.*

Using this result, B_z can be determined by summing moments about the x axis.

$\Sigma M_x = 0;$ $\quad B_z(0.8\text{ m}) + 600\text{ N}(1.2\text{ m}) - 900\text{ N}(0.4\text{ m}) = 0$

$B_z = -450\text{ N}$ *Ans.*

The negative sign indicates that $\mathbf{B}_z$ acts downward. The force B_x can be found by summing moments about the z axis.

$\Sigma M_z = 0;$ $\quad -B_x(0.8\text{ m}) = 0 \quad B_x = 0$ *Ans.*

Thus,

$\Sigma F_x = 0;$ $\quad A_x + 0 = 0 \quad A_x = 0$ *Ans.*

Finally, using the results of B_z and F_C.

$\Sigma F_z = 0;$ $\quad A_z + (-450\text{ N}) + 600\text{ N} - 900\text{ N} = 0$

$A_z = 750\text{ N}$ *Ans.*

EXAMPLE 5.17

The boom is used to support the 75-lb flowerpot in Fig. 5–30*a*. Determine the tension developed in wires *AB* and *AC*.

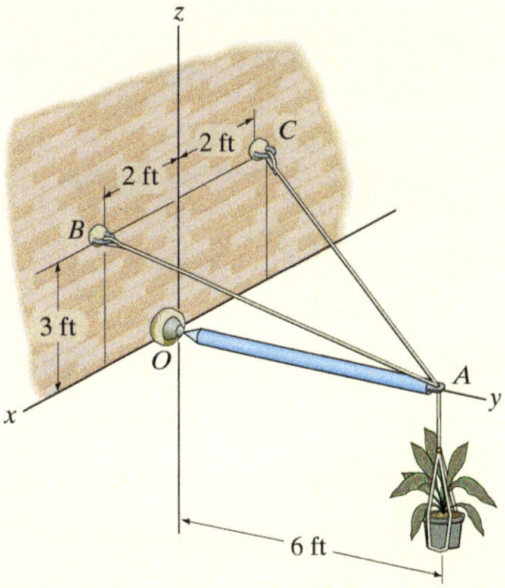

(a)

Fig. 5–30

SOLUTION

Free-Body Diagram. The free-body diagram of the boom is shown in Fig. 5–30*b*.

Equations of Equilibrium. We will use a vector analysis.

$$\mathbf{F}_{AB} = F_{AB}\left(\frac{\mathbf{r}_{AB}}{r_{AB}}\right) = F_{AB}\left(\frac{\{2\mathbf{i} - 6\mathbf{j} + 3\mathbf{k}\}\text{ ft}}{\sqrt{(2\text{ ft})^2 + (-6\text{ ft})^2 + (3\text{ ft})^2}}\right)$$

$$= \tfrac{2}{7}F_{AB}\mathbf{i} - \tfrac{6}{7}F_{AB}\mathbf{j} + \tfrac{3}{7}F_{AB}\mathbf{k}$$

$$\mathbf{F}_{AC} = F_{AC}\left(\frac{\mathbf{r}_{AC}}{r_{AC}}\right) = F_{AC}\left(\frac{\{-2\mathbf{i} - 6\mathbf{j} + 3\mathbf{k}\}\text{ ft}}{\sqrt{(-2\text{ ft})^2 + (-6\text{ ft})^2 + (3\text{ ft})^2}}\right)$$

$$= -\tfrac{2}{7}F_{AC}\mathbf{i} - \tfrac{6}{7}F_{AC}\mathbf{j} + \tfrac{3}{7}F_{AC}\mathbf{k}$$

We can eliminate the force reaction at *O* by writing the moment equation of equilibrium about point *O*.

$$\Sigma\mathbf{M}_O = \mathbf{0}; \qquad \mathbf{r}_A \times (\mathbf{F}_{AB} + \mathbf{F}_{AC} + \mathbf{W}) = \mathbf{0}$$

$$(6\mathbf{j}) \times \left[\left(\tfrac{2}{7}F_{AB}\mathbf{i} - \tfrac{6}{7}F_{AB}\mathbf{j} + \tfrac{3}{7}F_{AB}\mathbf{k}\right) + \left(-\tfrac{2}{7}F_{AC}\mathbf{i} - \tfrac{6}{7}F_{AC}\mathbf{j} + \tfrac{3}{7}F_{AC}\mathbf{k}\right) + (-75\mathbf{k})\right] = \mathbf{0}$$

$$\left(\tfrac{18}{7}F_{AB} + \tfrac{18}{7}F_{AC} - 450\right)\mathbf{i} + \left(-\tfrac{12}{7}F_{AB} + \tfrac{12}{7}F_{AC}\right)\mathbf{k} = \mathbf{0}$$

$$\Sigma M_x = 0; \qquad \tfrac{18}{7}F_{AB} + \tfrac{18}{7}F_{AC} - 450 = 0 \qquad (1)$$

$$\Sigma M_y = 0; \qquad 0 = 0$$

$$\Sigma M_z = 0; \qquad -\tfrac{12}{7}F_{AB} + \tfrac{12}{7}F_{AC} = 0 \qquad (2)$$

Solving Eqs. (1) and (2) simultaneously,

$$F_{AB} = F_{AC} = 87.5\text{ lb} \qquad \textit{Ans.}$$

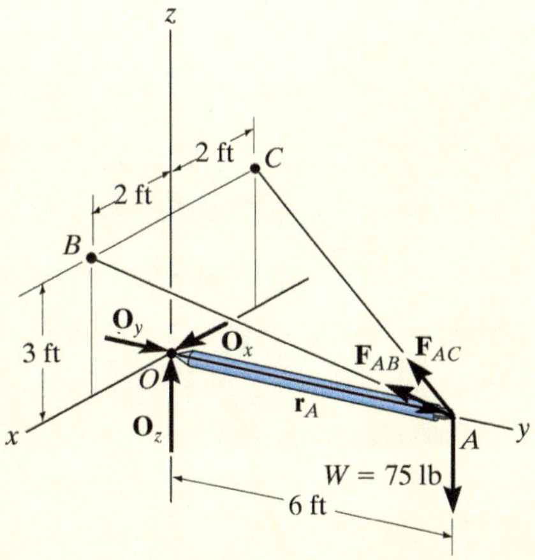

(b)

EXAMPLE 5.18

Rod AB shown in Fig. 5–31a is subjected to the 200-N force. Determine the reactions at the ball-and-socket joint A and the tension in the cables BD and BE. The collar at C is fixed to the rod.

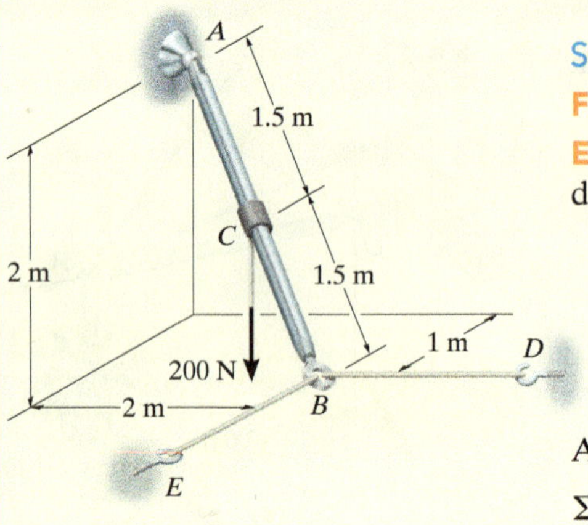

(a)

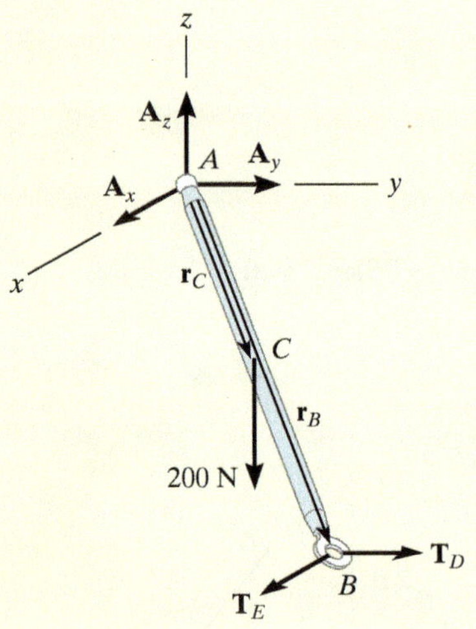

(b)

Fig. 5–31

SOLUTION (*VECTOR ANALYSIS*)

Free-Body Diagram. Fig. 5–31b.

Equations of Equilibrium. Representing each force on the free-body diagram in Cartesian vector form, we have

$$\mathbf{F}_A = A_x\mathbf{i} + A_y\mathbf{j} + A_z\mathbf{k}$$
$$\mathbf{T}_E = T_E\mathbf{i}$$
$$\mathbf{T}_D = T_D\mathbf{j}$$
$$\mathbf{F} = \{-200\mathbf{k}\}\text{ N}$$

Applying the force equation of equilibrium.

$$\Sigma\mathbf{F} = \mathbf{0};\qquad \mathbf{F}_A + \mathbf{T}_E + \mathbf{T}_D + \mathbf{F} = \mathbf{0}$$

$$(A_x + T_E)\mathbf{i} + (A_y + T_D)\mathbf{j} + (A_z - 200)\mathbf{k} = \mathbf{0}$$

$$\Sigma F_x = 0;\qquad A_x + T_E = 0 \qquad (1)$$
$$\Sigma F_y = 0;\qquad A_y + T_D = 0 \qquad (2)$$
$$\Sigma F_z = 0;\qquad A_z - 200 = 0 \qquad (3)$$

Summing moments about point A yields

$$\Sigma\mathbf{M}_A = \mathbf{0};\qquad \mathbf{r}_C \times \mathbf{F} + \mathbf{r}_B \times (\mathbf{T}_E + \mathbf{T}_D) = \mathbf{0}$$

Since $\mathbf{r}_C = \frac{1}{2}\mathbf{r}_B$, then

$$(0.5\mathbf{i} + 1\mathbf{j} - 1\mathbf{k}) \times (-200\mathbf{k}) + (1\mathbf{i} + 2\mathbf{j} - 2\mathbf{k}) \times (T_E\mathbf{i} + T_D\mathbf{j}) = \mathbf{0}$$

Expanding and rearranging terms gives

$$(2T_D - 200)\mathbf{i} + (-2T_E + 100)\mathbf{j} + (T_D - 2T_E)\mathbf{k} = \mathbf{0}$$

$$\Sigma M_x = 0;\qquad 2T_D - 200 = 0 \qquad (4)$$
$$\Sigma M_y = 0;\qquad -2T_E + 100 = 0 \qquad (5)$$
$$\Sigma M_z = 0;\qquad T_D - 2T_E = 0 \qquad (6)$$

Solving Eqs. 1 through 5, we get

$$T_D = 100\text{ N}$$ *Ans.*
$$T_E = 50\text{ N}$$ *Ans.*
$$A_x = -50\text{ N}$$ *Ans.*
$$A_y = -100\text{ N}$$ *Ans.*
$$A_z = 200\text{ N}$$ *Ans.*

NOTE: The negative sign indicates that $\mathbf{A}_x$ and $\mathbf{A}_y$ have a sense which is opposite to that shown on the free-body diagram, Fig. 5–31b.

EXAMPLE 5.19

The bent rod in Fig. 5–32*a* is supported at *A* by a journal bearing, at *D* by a ball-and-socket joint, and at *B* by means of cable *BC*. Using only *one equilibrium equation*, obtain a direct solution for the tension in cable *BC*. The bearing at *A* is capable of exerting force components only in the *z* and *y* directions since it is properly aligned on the shaft. In other words, no couple moments are required at this support.

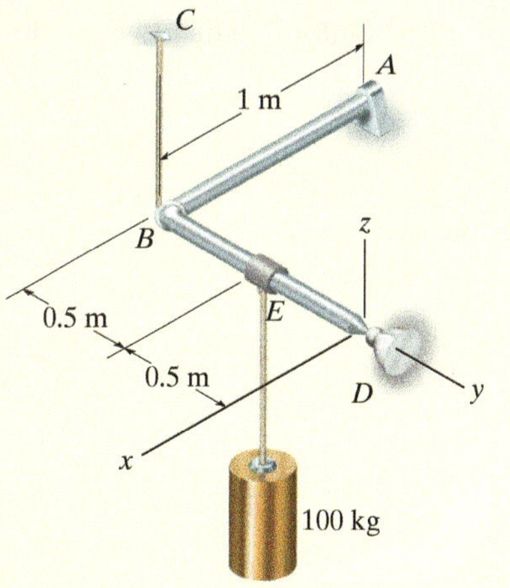

(a)

SOLUTION (*VECTOR ANALYSIS*)

Free-Body Diagram. As shown in Fig. 5–32*b*, there are six unknowns.

Equations of Equilibrium. The cable tension $\mathbf{T}_B$ may be obtained *directly* by summing moments about an axis that passes through points *D* and *A*. Why? The direction of this axis is defined by the unit vector **u**, where

$$\mathbf{u} = \frac{\mathbf{r}_{DA}}{r_{DA}} = -\frac{1}{\sqrt{2}}\mathbf{i} - \frac{1}{\sqrt{2}}\mathbf{j}$$
$$= -0.7071\mathbf{i} - 0.7071\mathbf{j}$$

Hence, the sum of the moments about this axis is zero provided

$$\Sigma M_{DA} = \mathbf{u}\cdot\Sigma(\mathbf{r}\times\mathbf{F}) = 0$$

Here **r** represents a position vector drawn from *any point* on the axis *DA* to any point on the line of action of force **F** (see Eq. 4–11). With reference to Fig. 5–32*b*, we can therefore write

$$\mathbf{u}\cdot(\mathbf{r}_B\times\mathbf{T}_B + \mathbf{r}_E\times\mathbf{W}) = \mathbf{0}$$
$$(-0.7071\mathbf{i} - 0.7071\mathbf{j})\cdot\big[(-1\mathbf{j})\times(T_B\mathbf{k}) + (-0.5\mathbf{j})\times(-981\mathbf{k})\big] = \mathbf{0}$$
$$(-0.7071\mathbf{i} - 0.7071\mathbf{j})\cdot[(-T_B + 490.5)\mathbf{i}] = \mathbf{0}$$
$$-0.7071(-T_B + 490.5) + 0 + 0 = 0$$
$$T_B = 490.5\text{ N} \qquad \textit{Ans.}$$

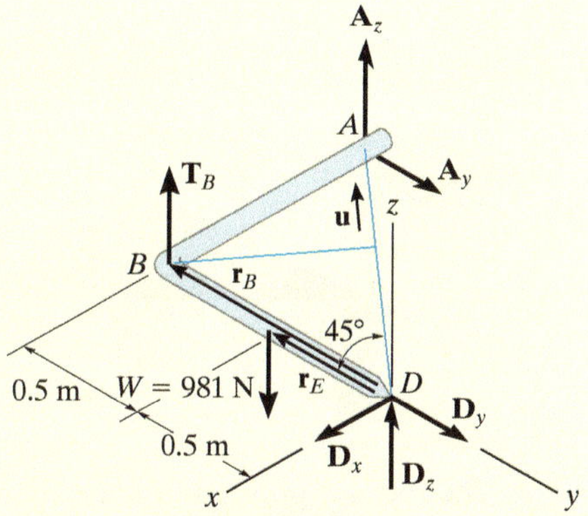

(b)

Fig. 5–32

Since the moment arms from the axis to $\mathbf{T}_B$ and **W** are easy to obtain, we can also determine this result using a scalar analysis. As shown in Fig. 5–32*b*,

$$\Sigma M_{DA} = 0;\quad T_B(1\text{ m}\sin 45^\circ) - 981\text{ N}(0.5\text{ m}\sin 45^\circ) = 0$$
$$T_B = 490.5\text{ N} \qquad \textit{Ans.}$$

5

FUNDAMENTAL PROBLEMS

All problem solutions must include an FBD.

F5–7. The uniform plate has a weight of 500 lb. Determine the tension in each of the supporting cables.

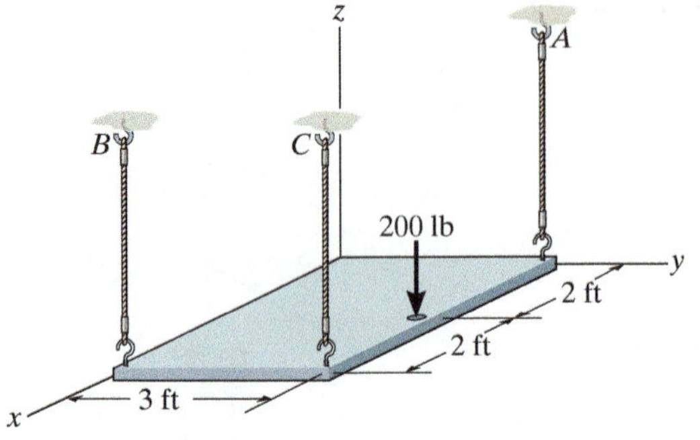

F5–7

F5–8. Determine the reactions at the roller support A, the ball-and-socket joint D, and the tension in cable BC for the plate.

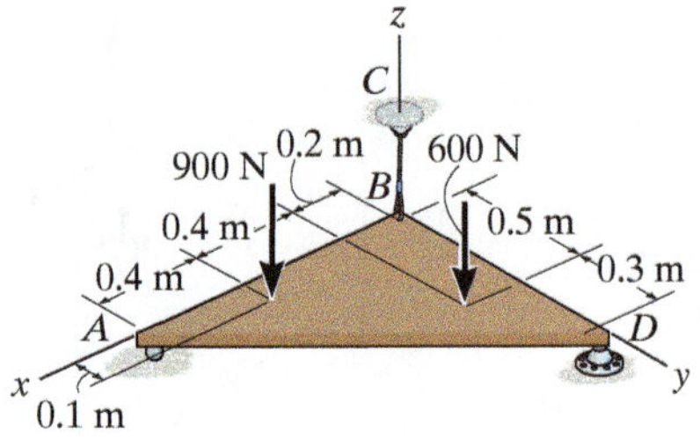

F5–8

F5–9. The rod is supported by smooth journal bearings at A, B, and C and is subjected to the two forces. Determine the reactions at these supports.

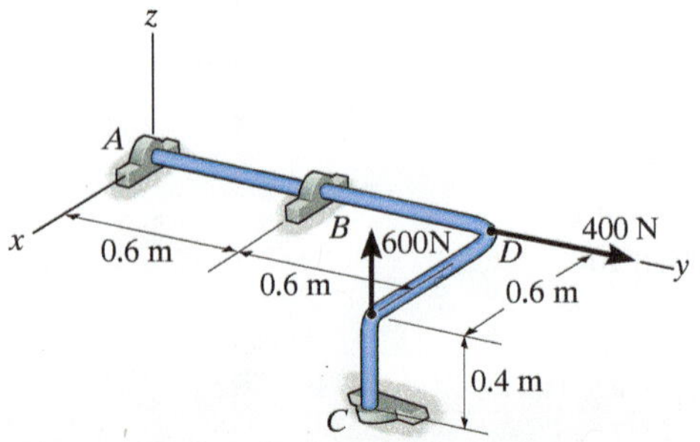

F5–9

F5–10. Determine the support reactions at the smooth journal bearings A, B, and C of the pipe assembly.

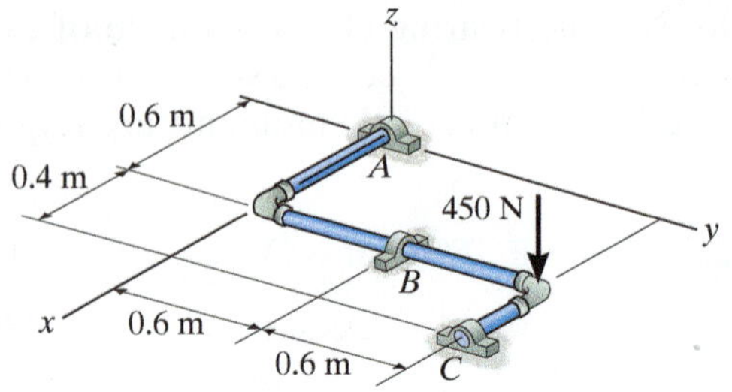

F5–10

F5–11. Determine the force developed in the short link BD, and the tension in the cords CE and CF, and the reactions of the ball-and-socket joint A on the block.

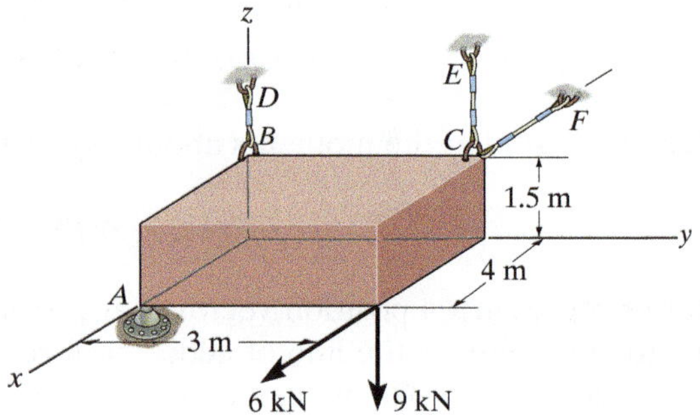

F5–11

F5–12. Determine the components of reaction that the thrust bearing A and cable BC exert on the bar.

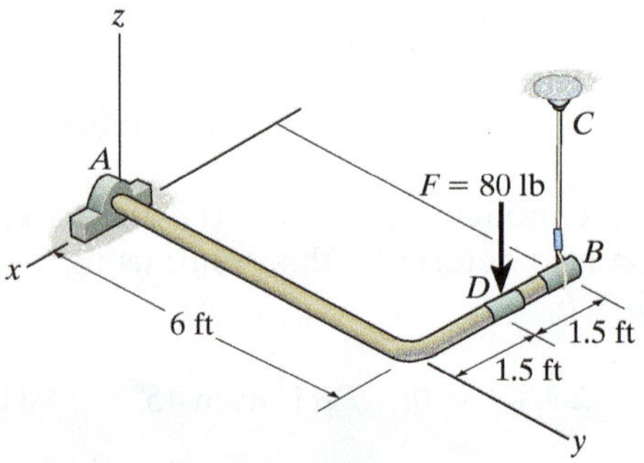

F5–12

PROBLEMS

All problem solutions must include an FBD.

5–62. The uniform load has a mass of 600 kg and is lifted using a uniform 30-kg strongback beam BAC and four ropes as shown. Determine the tension in each rope and the force that must be applied at A.

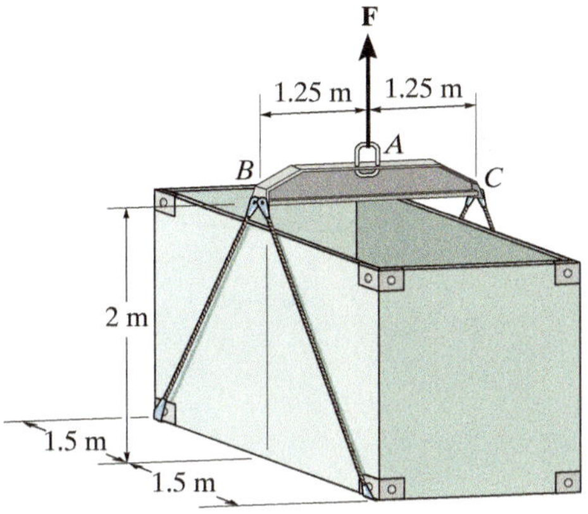

Prob. 5–62

5–63. The 50-lb mulching machine has a center of gravity at G. Determine the vertical reactions at the wheels C and B and the smooth contact point A.

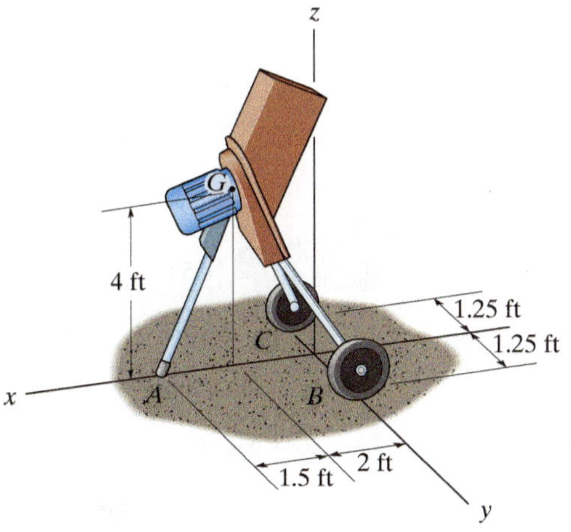

Prob. 5–63

***5–64.** The wing of the jet aircraft is subjected to a thrust of $T = 8$ kN from its engine and the resultant lift force $L = 45$ kN. If the mass of the wing is 2.1 Mg and the mass center is at G, determine the x, y, z components of reaction where the wing is fixed to the fuselage A.

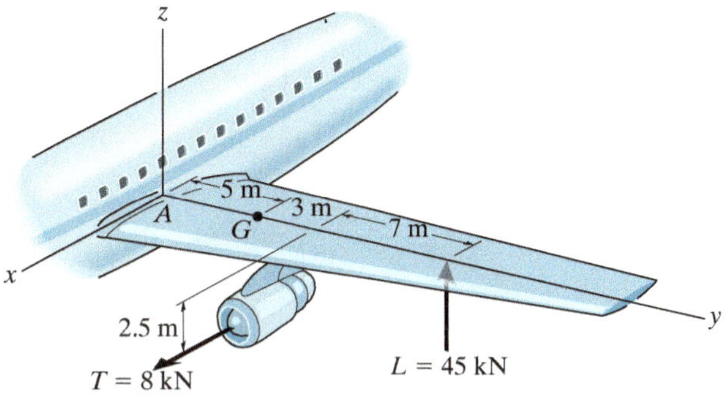

Prob. 5–64

5–65. Due to an unequal distribution of fuel in the wing tanks, the centers of gravity for the airplane fuselage A and wings B and C are located as shown. If these components have weights $W_A = 45\,000$ lb, $W_B = 8000$ lb, and $W_C = 6000$ lb, determine the normal reactions of the wheels D, E, and F on the ground.

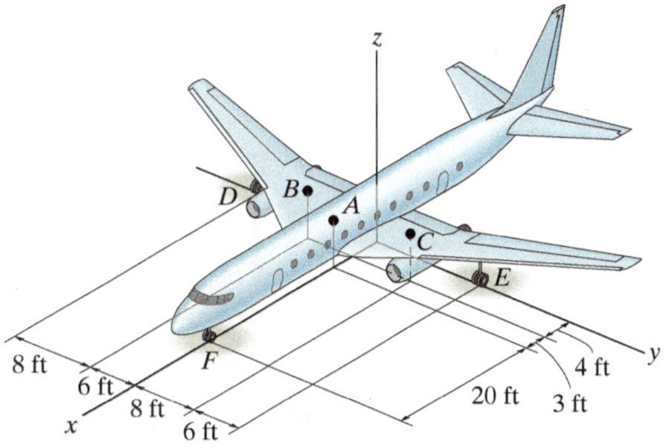

Prob. 5–65

5–66. The air-conditioning unit is hoisted to the roof of a building using the three cables. If the tensions in the cables are $T_A = 250$ lb, $T_B = 300$ lb, and $T_C = 200$ lb, determine the weight of the unit and the location (x, y) of its center of gravity G.

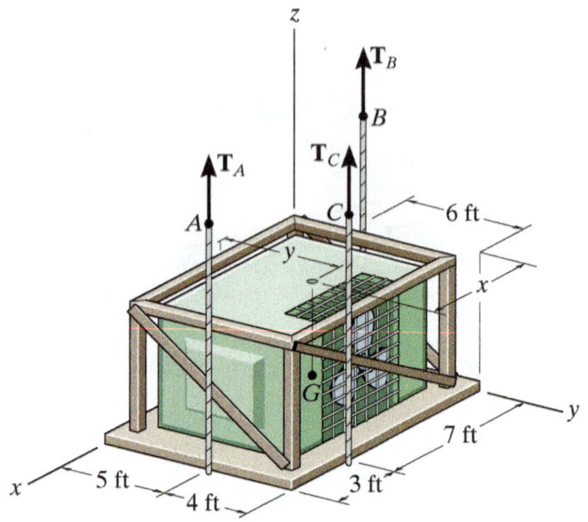

Prob. 5–66

5–67. The platform truck supports the three loadings shown. Determine the normal reactions on each of its three wheels.

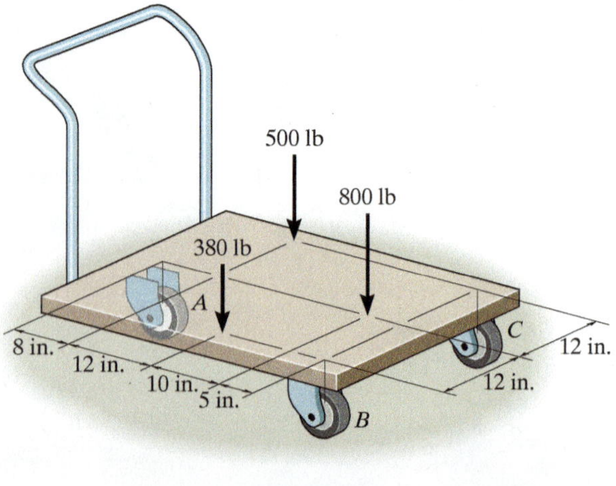

Prob. 5–67

***5–68.** Determine the force components acting on the ball-and-socket at A, the reaction at the roller B and the tension on the cord CD needed for equilibrium of the quarter circular plate.

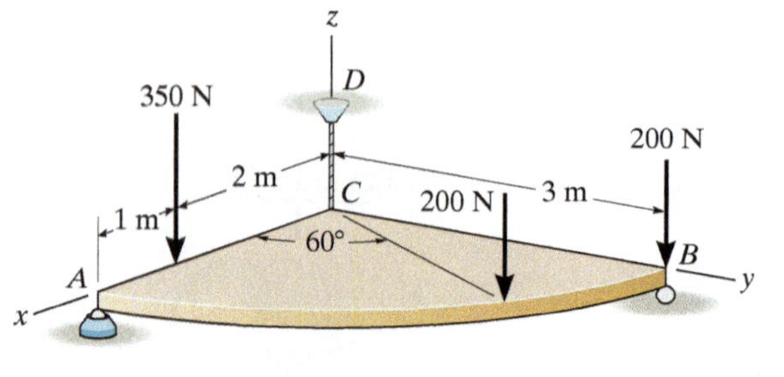

Prob. 5–68

5–69. The windlass is subjected to a load of 150 lb. Determine the horizontal force **P** needed to hold the handle in the position shown, and the components of reaction at the ball-and-socket joint A and the smooth journal bearing B. The bearing at B is in proper alignment and exerts only force reactions on the windlass.

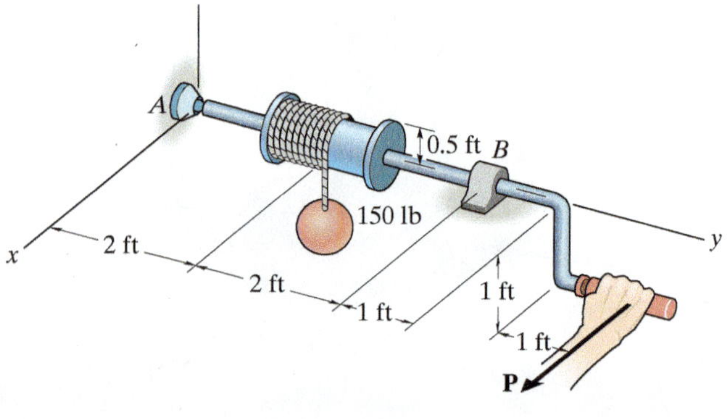

Prob. 5–69

5–70. The 100-lb door has its center of gravity at *G*. Determine the components of reaction at hinges *A* and *B* if hinge *B* resists only forces in the *x* and *y* directions and *A* resists forces in the *x*, *y*, *z* directions.

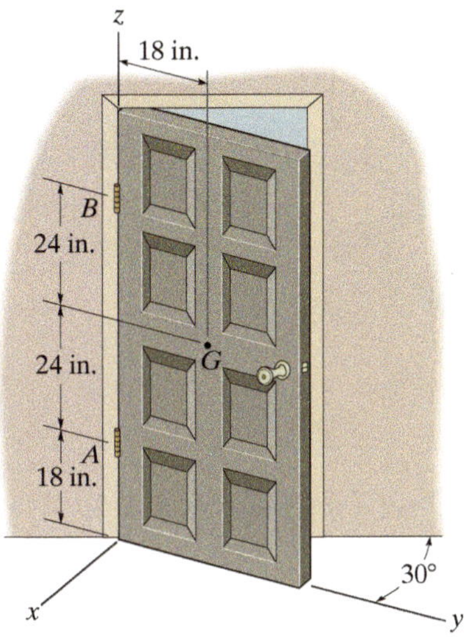

Prob. 5–70

*__5–72.__ The pole is subjected to the two forces shown. Determine the components of reaction of *A* assuming it to be a ball-and-socket joint. Also, compute the tension in each of the guy wires, *BC* and *ED*.

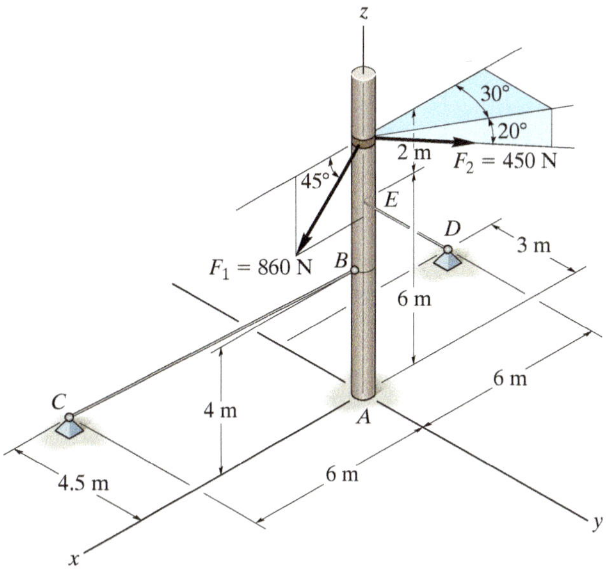

Prob. 5–72

5–73. The boom *AB* is held in equilibrium by a ball-and-socket joint *A* and a pulley and cord system as shown. Determine the *x*, *y*, *z* components of reaction at *A* and the tension in cable *DEC* if $\mathbf{F} = \{-1500\mathbf{k}\}$ lb.

5–74. The cable *CED* can sustain a maximum tension of 800 lb before it fails. Determine the greatest vertical force *F* that can be applied to the boom. Also, what are the *x*, *y*, *z* components of reaction at the ball-and-socket joint *A*?

5–71. Determine the support reactions at the smooth collar *A* and the normal reaction at the roller support *B*.

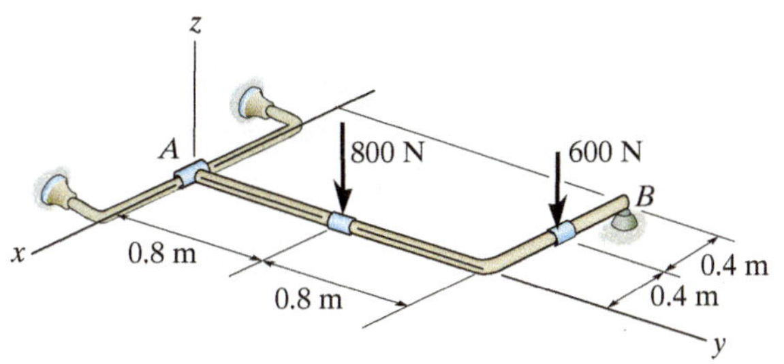

Prob. 5–71

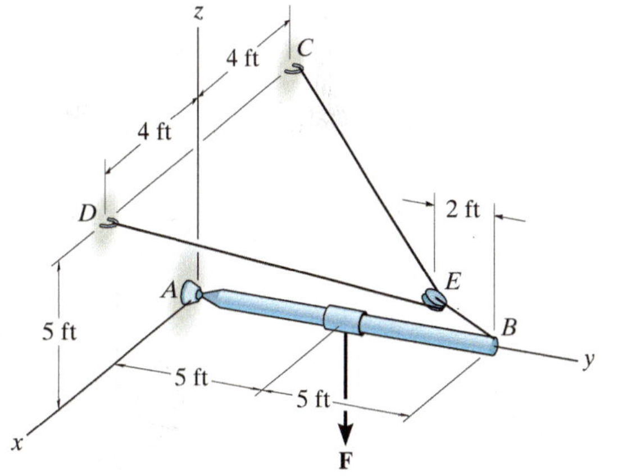

Probs. 5–73/74

5–75. If the pulleys are fixed to the shaft, determine the magnitude of tension **T** and the *x, y, z* components of reaction at the smooth thrust bearing *A* and smooth journal bearing *B*.

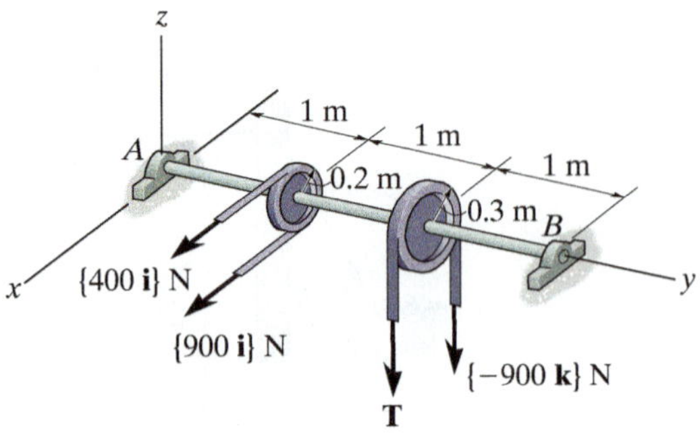

Prob. 5–75

5

***5–76.** The boom *AC* is supported at *A* by a ball-and-socket joint and by two cables *BDC* and *CE*. Cable *BDC* is continuous and passes over a pulley at *D*. Calculate the tension in the cables and the *x, y, z* components of reaction at *A* if a crate has a weight of 80 lb.

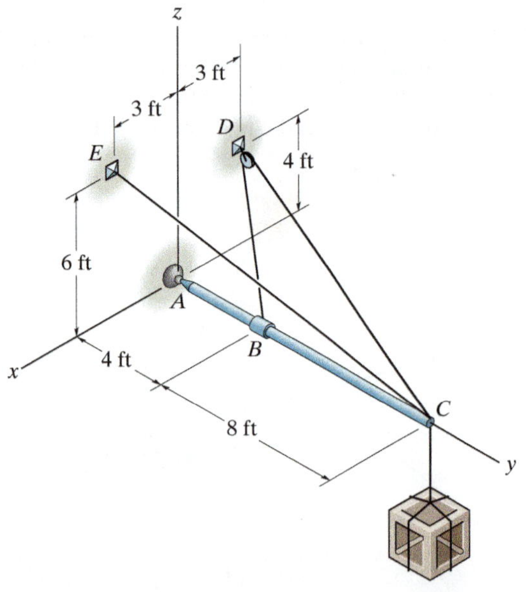

Prob. 5–76

5–77. A vertical force of 80 lb acts on the crankshaft. Determine the horizontal equilibrium force **P** that must be applied to the handle and the *x, y, z* components of reaction at the journal bearing *A* and thrust bearing *B*. The bearings are properly aligned and exert only force reactions on the shaft.

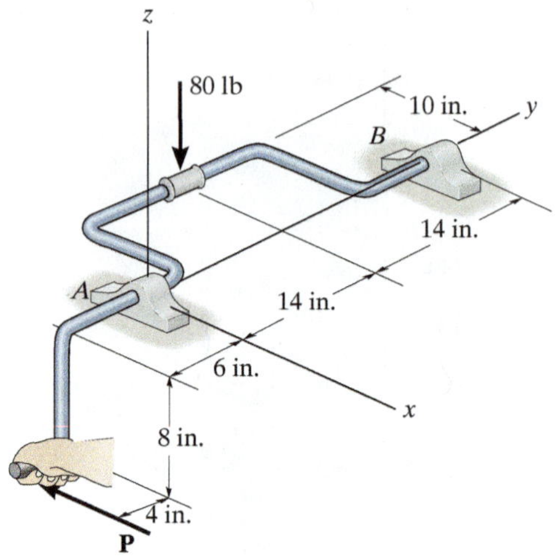

Prob. 5–77

5–78. Member *AB* is supported by a cable *BC* and at *A* by a *square* rod which fits loosely through the square hole at the end joint of the member as shown. Determine the components of reaction at *A* and the tension in the cable needed to hold the 800-lb cylinder in equilibrium.

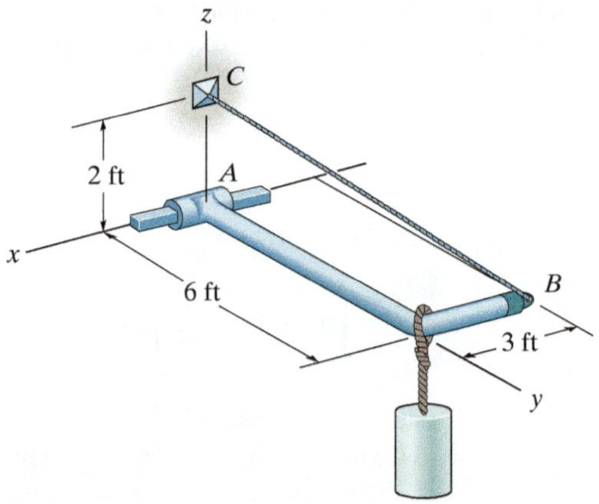

Prob. 5–78

5–79. The bent rod is supported at A, B, and C by smooth journal bearings. Compute the x, y, z components of reaction at the bearings if the rod is subjected to forces $F_1 = 300$ lb and $F_2 = 250$ lb. $\mathbf{F}_1$ lies in the y–z plane. The bearings are in proper alignment and exert only force reactions on the rod.

***5–80.** The bent rod is supported at A, B, and C by smooth journal bearings. Determine the magnitude of $\mathbf{F}_2$ which will cause the reaction $\mathbf{C}_y$ at the bearing C to be equal to zero. The bearings are in proper alignment and exert only force reactions on the rod. Set $F_1 = 300$ lb.

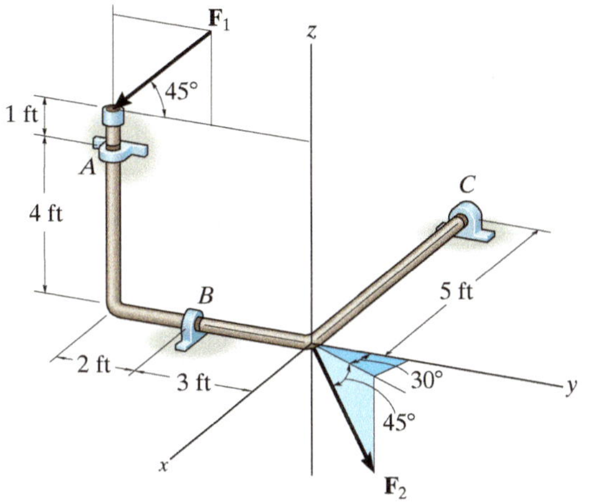

Probs. 5–79/80

5–81. The sign has a mass of 100 kg with center of mass at G. Determine the x, y, z components of reaction at the ball-and-socket joint A and the tension in wires BC and BD.

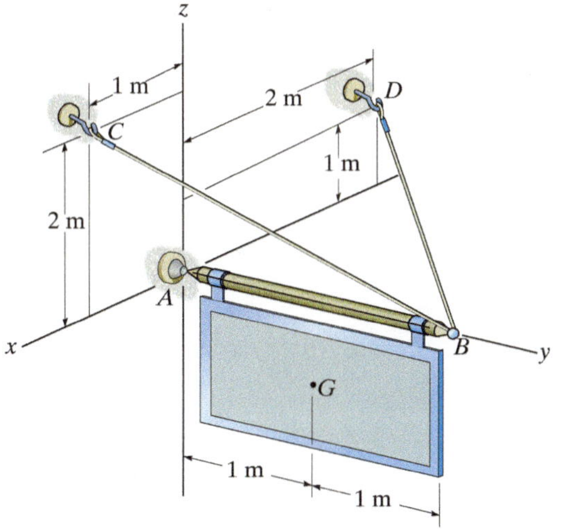

Prob. 5–81

5–82. Determine the tensions in the cables and the components of reaction acting on the smooth collar at A necessary to hold the 50-lb sign in equilibrium. The center of gravity for the sign is at G.

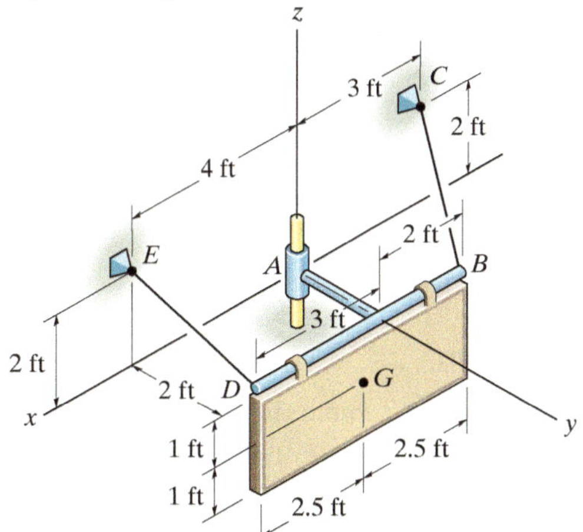

Prob. 5–82

5–83. Both pulleys are fixed to the shaft and as the shaft turns with constant angular velocity, the power of pulley A is transmitted to pulley B. Determine the horizontal tension $\mathbf{T}$ in the belt on pulley B and the x, y, z components of reaction at the journal bearing C and thrust bearing D if $\theta = 0°$. The bearings are in proper alignment and exert only force reactions on the shaft.

***5–84.** Both pulleys are fixed to the shaft and as the shaft turns with constant angular velocity, the power of pulley A is transmitted to pulley B. Determine the horizontal tension $\mathbf{T}$ in the belt on pulley B and the x, y, z components of reaction at the journal bearing C and thrust bearing D if $\theta = 45°$. The bearings are in proper alignment and exert only force reactions on the shaft.

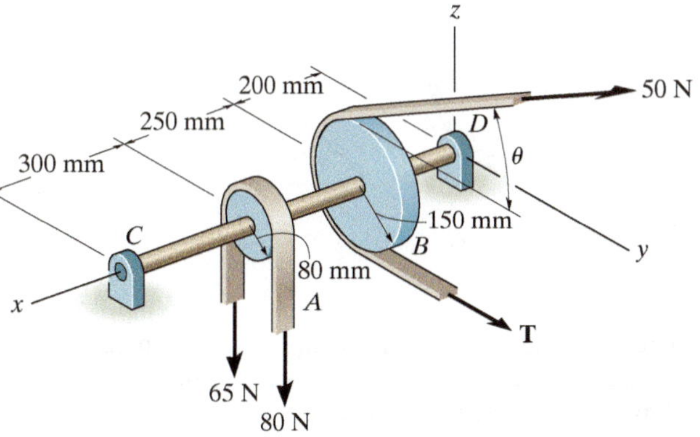

Probs. 5–83/84

5

CHAPTER REVIEW

Equilibrium

A body in equilibrium is at rest or can translate with constant velocity.

$$\Sigma \mathbf{F} = \mathbf{0}$$
$$\Sigma \mathbf{M} = \mathbf{0}$$

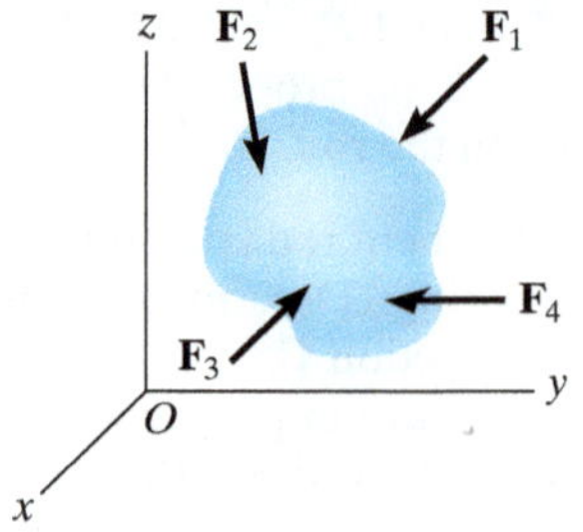

Two Dimensions

Before analyzing the equilibrium of a body, it is first necessary to draw its free-body diagram. This is an outlined shape of the body, which shows all the forces and couple moments that act on it.

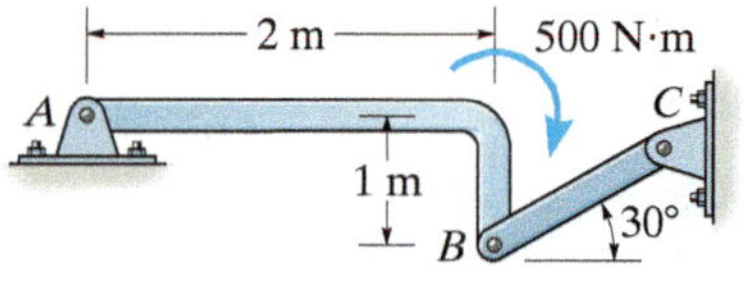

Couple moments can be placed anywhere on a free-body diagram since they are free vectors. Forces can act at any point along their line of action since they are sliding vectors.

Angles used to resolve forces, and dimensions used to take moments of the forces, should also be shown on the free-body diagram.

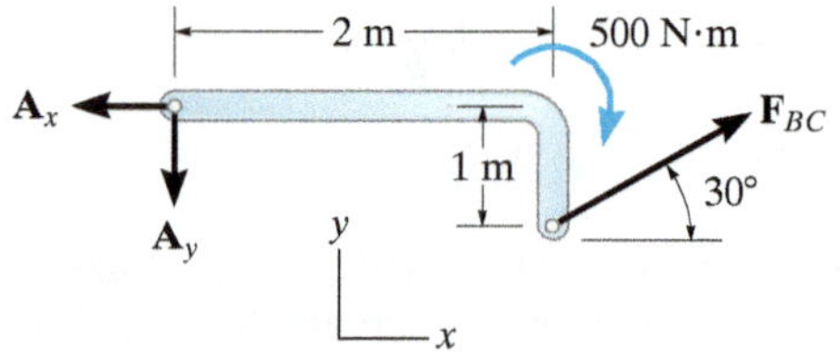

Some common types of supports and their reactions are shown below in two dimensions.

Remember that a support will exert a force on the body in a particular direction if it prevents translation of the body in that direction, and it will exert a couple moment on the body if it prevents rotation.

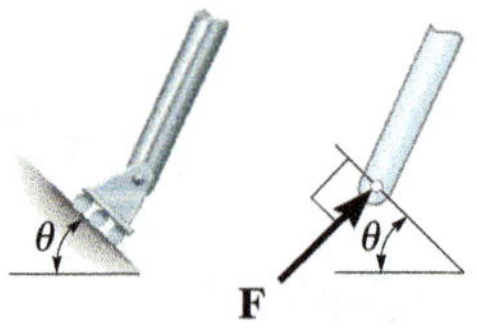

roller

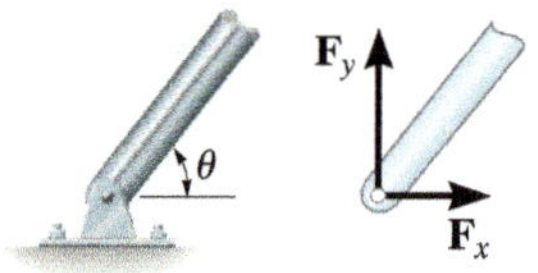

smooth pin or hinge

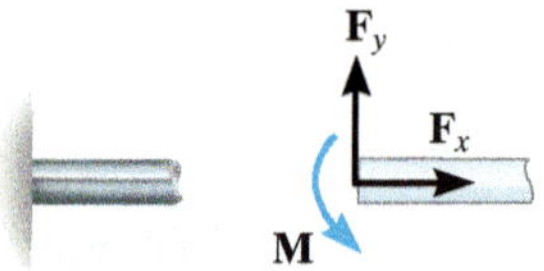

fixed support

The three scalar equations of equilibrium can be applied when solving problems in two dimensions, since the geometry is easy to visualize.

$$\Sigma F_x = 0$$
$$\Sigma F_y = 0$$
$$\Sigma M_O = 0$$

For the most direct solution, try to sum forces along an axis that will eliminate as many unknown forces as possible. Sum moments about a point A that passes through the line of action of as many unknown forces as possible.

$\Sigma F_x = 0;$

$$A_x - P_2 = 0 \quad A_x = P_2$$

$\Sigma M_A = 0;$

$$P_2 d_2 + B_y d_B - P_1 d_1 = 0$$

$$B_y = \frac{P_1 d_1 - P_2 d_2}{d_B}$$

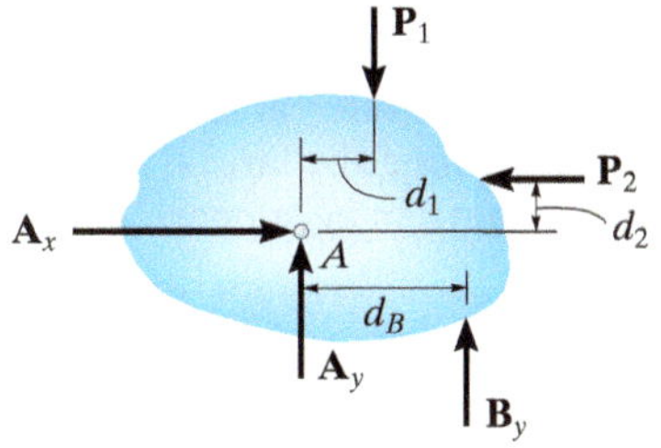

Three Dimensions

Some common types of supports and their reactions are shown here in three dimensions.

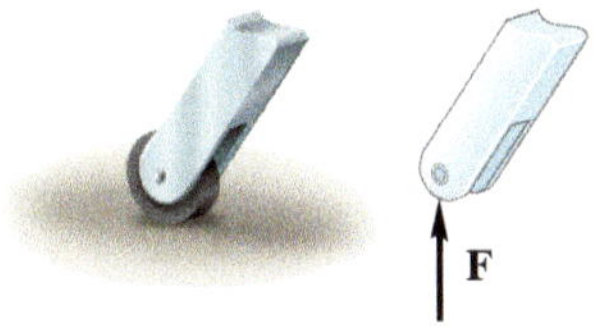

roller

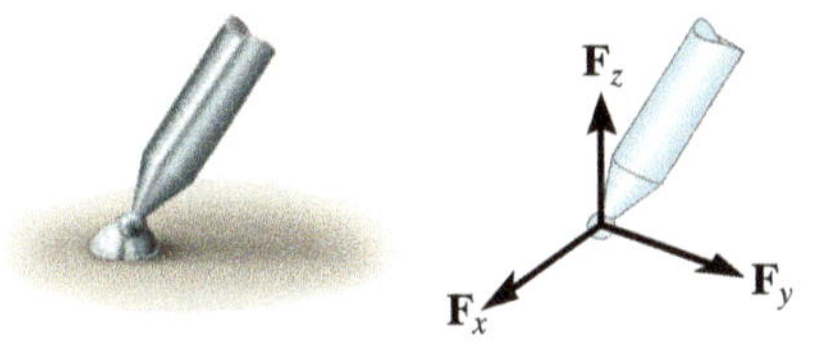

ball and socket

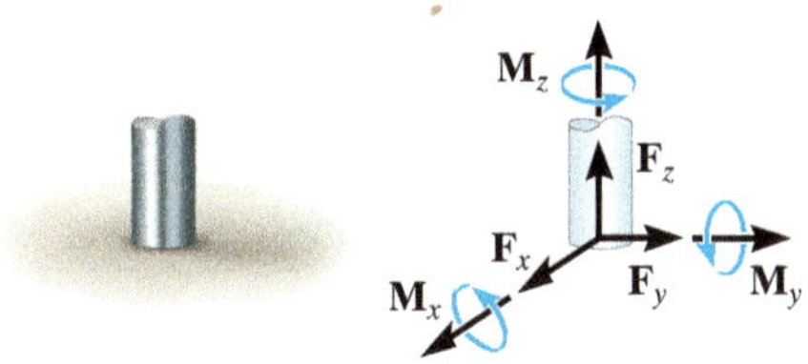

fixed support

In three dimensions, it is often advantageous to use a Cartesian vector analysis when applying the equations of equilibrium. To do this, first express each known and unknown force and couple moment shown on the free-body diagram as a Cartesian vector. Then set the force summation equal to zero. Take moments about a point O that lies on the line of action of as many unknown force components as possible. From point O direct position vectors to each force, and then use the cross product to determine the moment of each force.

$$\Sigma \mathbf{F} = \mathbf{0}$$
$$\Sigma \mathbf{M}_O = \mathbf{0}$$

The six scalar equations of equilibrium are established by setting the respective **i**, **j**, and **k** components of these force and moment summations equal to zero.

$$\Sigma F_x = 0 \qquad \Sigma M_x = 0$$
$$\Sigma F_y = 0 \qquad \Sigma M_y = 0$$
$$\Sigma F_z = 0 \qquad \Sigma M_z = 0$$

Determinacy and Stability

If a body is supported by a minimum number of constraints to ensure equilibrium, then it is statically determinate. If it has more constraints than required, then it is statically indeterminate.

To properly constrain the body, the reactions must not all be parallel to one another or concurrent.

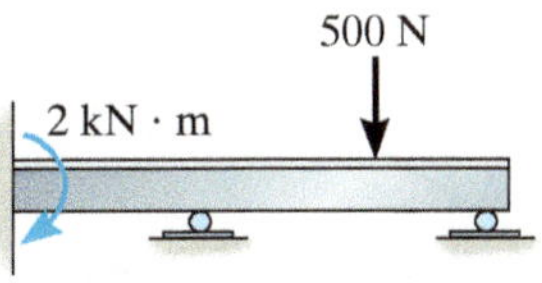

Statically indeterminate, five reactions, three equilibrium equations

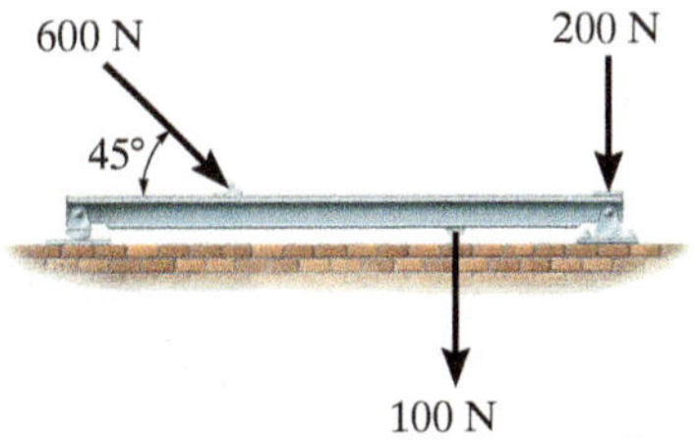

Proper constraint, statically determinate

REVIEW PROBLEMS

5–85. If the roller at B can sustain a maximum load of 3 kN, determine the largest magnitude of each of the three forces **F** that can be supported by the truss.

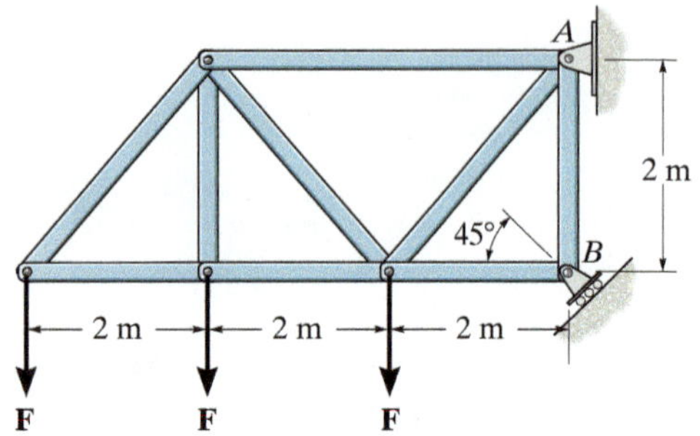

Prob. 5–85

5–86. Determine the normal reaction at the roller A and horizontal and vertical components at pin B for equilibrium of the member.

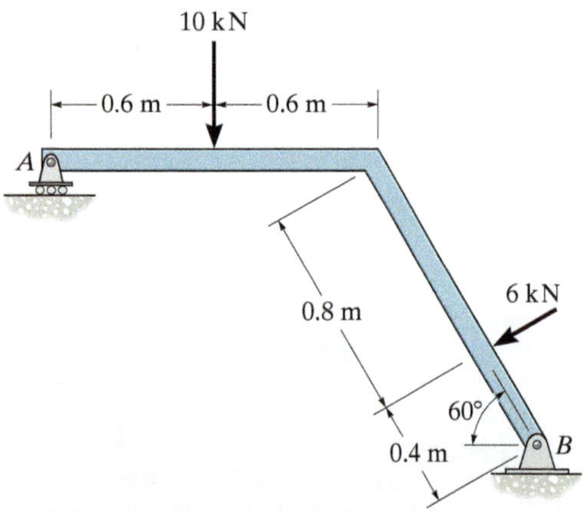

Prob. 5–86

5–87. The symmetrical shelf is subjected to a uniform load of 4 kPa. Support is provided by a bolt (or pin) located at each end A and A' and by the symmetrical brace arms, which bear against the smooth wall on both sides at B and B'. Determine the force resisted by each bolt at the wall and the normal force at B for equilibrium.

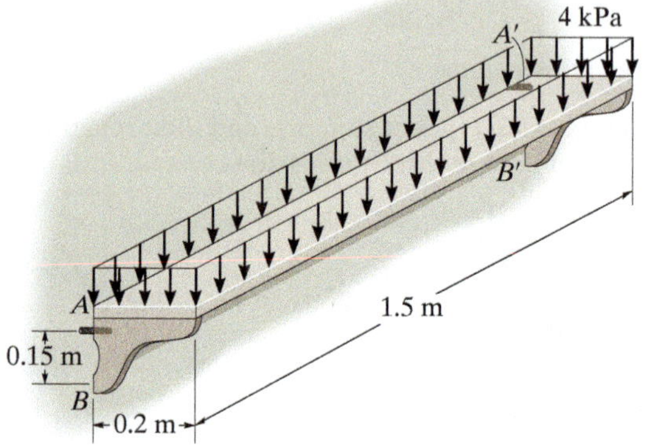

Prob. 5–87

***5–88.** Determine the x and z components of reaction at the journal bearing A and the tension in cords BC and BD necessary for equilibrium of the rod.

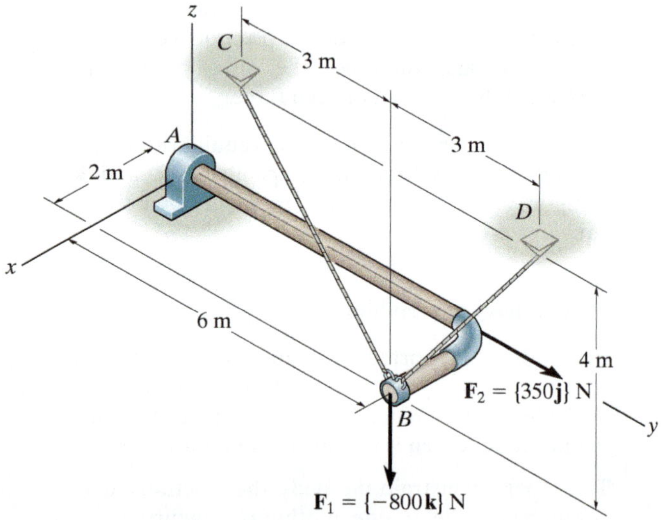

Prob. 5–88

5–89. The uniform rod of length L and weight W is supported on the smooth planes. Determine its position θ for equilibrium. Neglect the thickness of the rod.

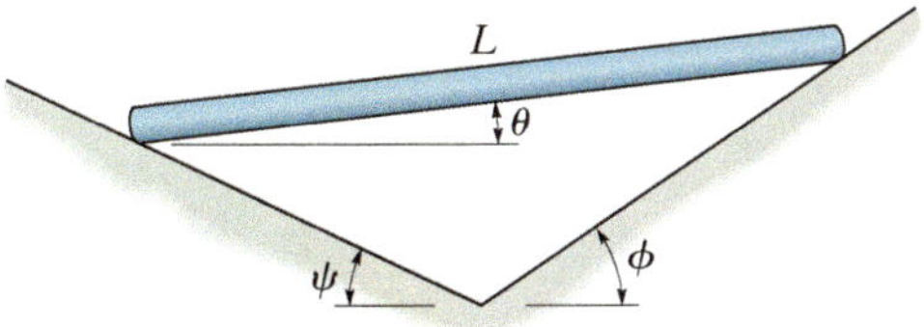

Prob. 5–89

5–90. Determine the x, y, z components of reaction at the ball supports B and C and the ball-and-socket A (not shown) for the uniformly loaded plate.

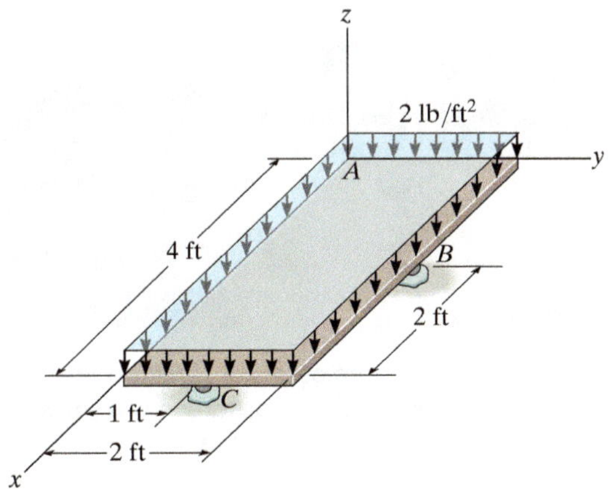

Prob. 5–90

5–91. Determine the x, y, z components of reaction at the fixed wall A. The 150-N force is parallel to the z axis and the 200-N force is parallel to the y axis.

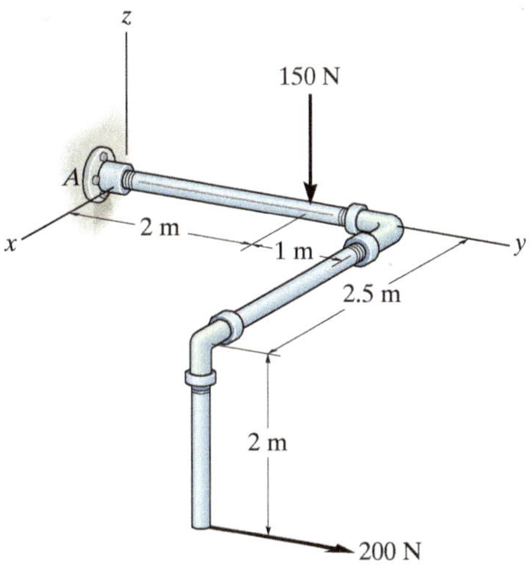

Prob. 5–91

***5–92.** Determine the reactions at the supports A and B for equilibrium of the beam.

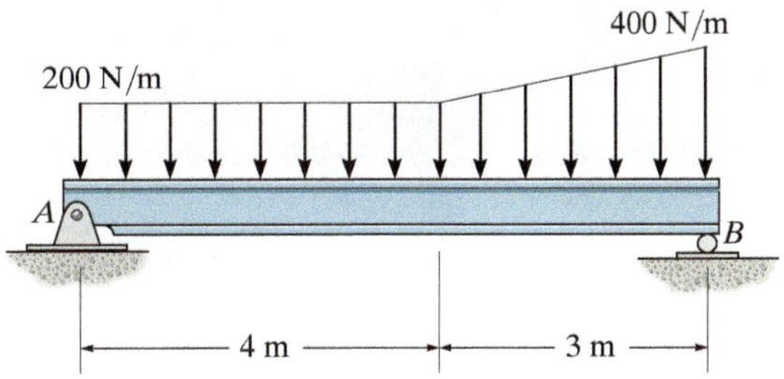

Prob. 5–92

Chapter 6

In order to design the many parts of this boom assembly it is required that we know the forces that they must support. In this chapter we will show how to analyze such structures using the equations of equilibrium.

Structural Analysis

CHAPTER OBJECTIVES

- To show how to determine the forces in the members of a truss using the method of joints and the method of sections.
- To analyze the forces acting on the members of frames and machines composed of pin-connected members.

6.1 Simple Trusses

A *truss* is a structure composed of slender members joined together at their end points. The members commonly used in construction consist of wooden struts or metal bars. In particular, *planar* trusses lie in a single plane and are often used to support roofs and bridges. The truss shown in Fig. 6–1*a* is an example of a typical roof-supporting truss. In this figure, the roof load is transmitted to the truss *at the joints* by means of a series of *purlins*. Since this loading acts in the same plane as the truss, Fig. 6–1*b*, the analysis of the forces developed in the truss members will be two-dimensional.

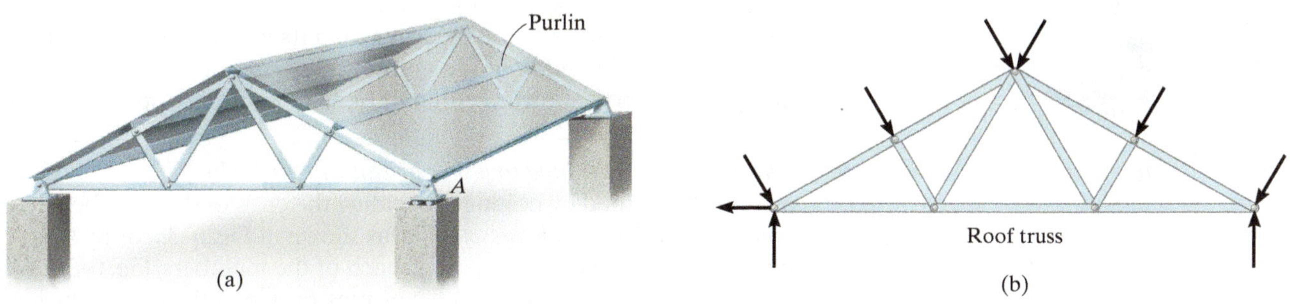

Fig. 6–1

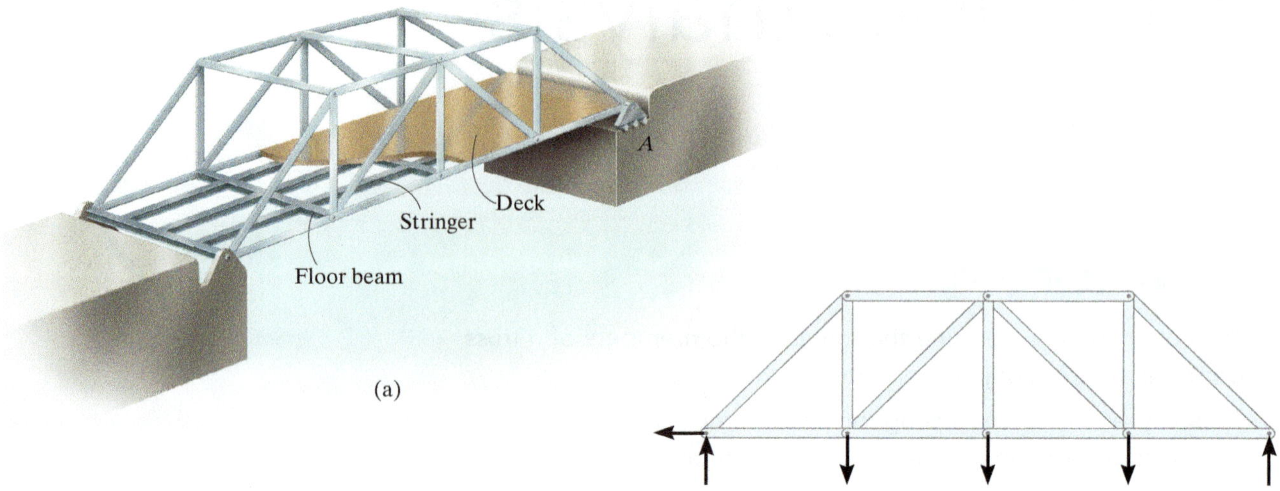

(a)

Bridge truss

(b)

Fig. 6–2

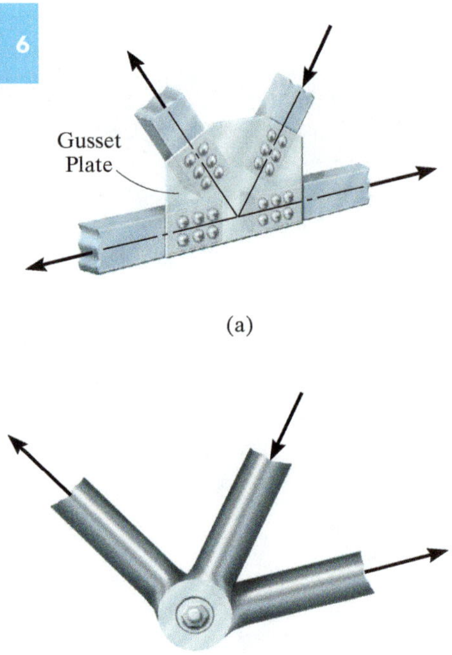

(a)

(b)

Fig. 6–3

In the case of a bridge, such as shown in Fig. 6–2*a*, the load on the deck is first transmitted to *stringers*, then to *floor beams*, and finally to the *joints* of the two supporting side trusses. Like the roof truss, the bridge truss loading is also coplanar, Fig. 6–2*b*.

When bridge or roof trusses extend over large distances, a rocker or roller is commonly used for supporting one end, for example, joint *A* in Figs. 6–1*a* and 6–2*a*. This type of support allows freedom for expansion or contraction of the members due to a change in temperature or application of loads.

Assumptions for Design. To design both the members and the connections of a truss, it is necessary first to determine the *force* developed in each member when the truss is subjected to a given loading. To do this we will make two important assumptions:

- ***All loadings are applied at the joints.*** In most situations, such as for bridge and roof trusses, this assumption is true. Frequently the weight of the members is neglected because the force supported by each member is usually much larger than its weight. However, if the weight is to be included in the analysis, it is generally satisfactory to apply it as a vertical force, with half of its magnitude applied at each end of the member.
- ***The members are joined together by smooth pins.*** The joint connections are usually formed by bolting or welding the ends of the members to a common plate, called a *gusset plate*, as shown in Fig. 6–3*a*, or by simply passing a large bolt or pin through each of the members, Fig. 6–3*b*. We can assume these connections act as pins provided the center lines of the joining members are *concurrent*, as in Fig. 6–3.

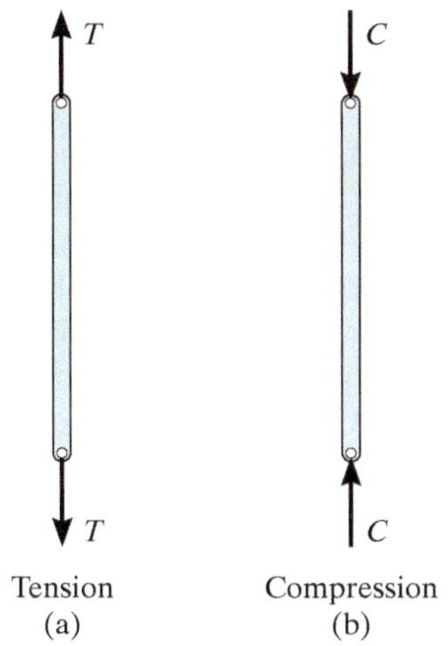

Fig. 6–4

Because of these two assumptions, *each truss member will act as a two-force member*, and therefore the force acting at each end of the member will be directed along the axis of the member. If the force tends to *elongate* the member, it is a *tensile force* (T), Fig. 6–4*a*; whereas if it tends to *shorten* the member, it is a *compressive force* (C), Fig. 6–4*b*. In the actual design of a truss it is important to state whether the nature of the force is tensile or compressive. Often, compression members must be made *thicker* than tension members because of the buckling or column effect that occurs when a member is in compression.

The use of metal gusset plates in the construction of these Warren trusses is clearly evident.

Simple Truss. If three members are pin connected at their ends, they form a *triangular truss* that will be *rigid*, Fig. 6–5. Attaching two more members and connecting these members to a new joint D forms a larger truss, Fig. 6–6. This procedure can be repeated as many times as desired to form an even larger truss. If a truss can be constructed by expanding the basic triangular truss in this way, it is called a *simple truss*.

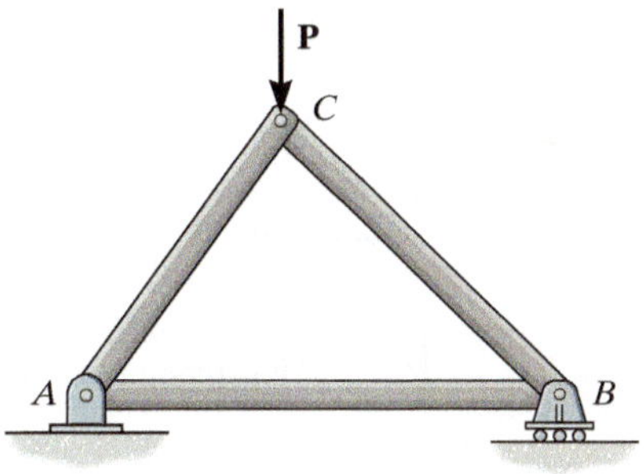

Fig. 6–5

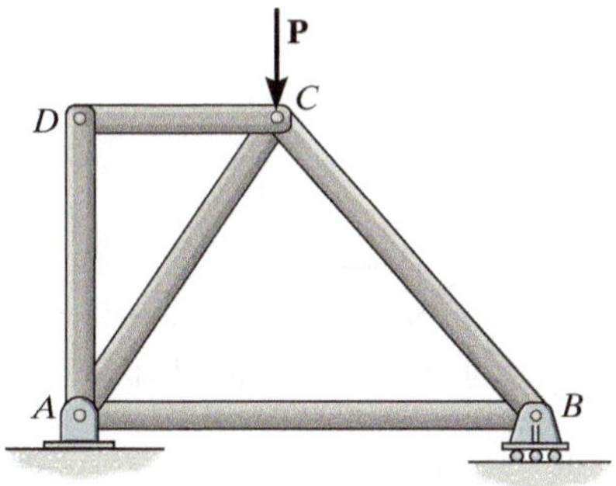

Fig. 6–6

6.2 The Method of Joints

In order to analyze or design a truss, it is necessary to determine the force in each of its members. One way to do this is to use the method of joints. This method is based on the fact that if the entire truss is in equilibrium, then each of its joints is also in equilibrium. Therefore, if the free-body diagram of each joint is drawn, the force equilibrium equations can then be used to obtain the member forces acting on each joint. Since the members of a *plane truss* are straight two-force members lying in a single plane, each joint is subjected to a force system that is *coplanar and concurrent*. As a result, only $\Sigma F_x = 0$ and $\Sigma F_y = 0$ need to be satisfied for equilibrium.

For example, consider the pin at joint B of the truss in Fig. 6–7*a*. Three forces act on the pin, namely, the 500-N force and the forces exerted by members BA and BC. The free-body diagram of the pin is shown in Fig. 6–7*b*. Here, $\mathbf{F}_{BA}$ is "pulling" on the pin, which means that member BA is in *tension;* whereas $\mathbf{F}_{BC}$ is "pushing" on the pin, and consequently member BC is in *compression*. These effects are clearly demonstrated by isolating the joint with small segments of the member connected to the pin, Fig. 6–7*c*. The pushing or pulling on these small segments indicates the effect of the member being either in compression or tension.

When using the method of joints, always start at a joint having at least one known force and at most two unknown forces, as in Fig. 6–7*b*. In this way, application of $\Sigma F_x = 0$ and $\Sigma F_y = 0$ yields two algebraic equations which can be solved for the two unknowns. When applying these equations, the correct sense of an unknown member force can be determined using one of two possible methods.

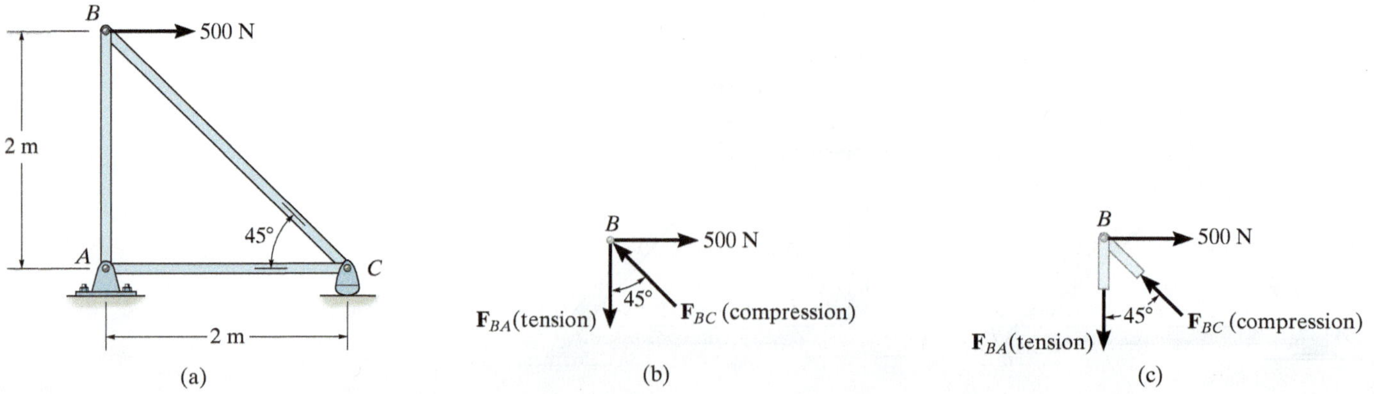

Fig. 6–7

- The *correct* sense of direction of an unknown member force can, in many cases, be determined "by inspection." For example, $\mathbf{F}_{BC}$ in Fig. 6–7*b* must push on the pin (compression) since its horizontal component, $F_{BC} \sin 45°$, must balance the 500-N force ($\Sigma F_x = 0$). Likewise, $\mathbf{F}_{BA}$ is a tensile force since it balances the vertical component, $F_{BC} \cos 45°$ ($\Sigma F_y = 0$). In more complicated cases, the sense of an unknown member force can be *assumed*; then, after applying the equilibrium equations, the assumed sense can be verified from the numerical results. A *positive* answer indicates that the sense is *correct*, whereas a *negative* answer indicates that the sense shown on the free-body diagram must be *reversed*.
- *Always assume* the *unknown member forces* acting on the joint's free-body diagram to be in *tension*; i.e., the forces "pull" on the pin. If this is done, then numerical solution of the equilibrium equations will yield *positive scalars for members in tension and negative scalars for members in compression*. Once an unknown member force is found, use its *correct* magnitude and sense (T or C) on subsequent joint free-body diagrams.

The forces in the members of this simple roof truss can be determined using the method of joints.

Procedure for Analysis

The following procedure provides a means for analyzing a truss using the method of joints.

- Draw the free-body diagram of a joint having at least one known force and at most two unknown forces. (If this joint is at one of the supports, then it may be necessary first to calculate the external reactions at the support.)
- Use one of the two methods described above for establishing the sense of an unknown force.
- Orient the x and y axes such that the forces on the free-body diagram can be easily resolved into their x and y components and then apply the two force equilibrium equations $\Sigma F_x = 0$ and $\Sigma F_y = 0$. Solve for the two unknown member forces and verify their correct sense.
- Using the calculated results, continue to analyze each of the other joints. Remember that a member in *compression* "pushes" on the joint and a member in *tension* "pulls" on the joint. Also, be sure to choose a joint having at most two unknowns and at least one known force.

6

EXAMPLE 6.1

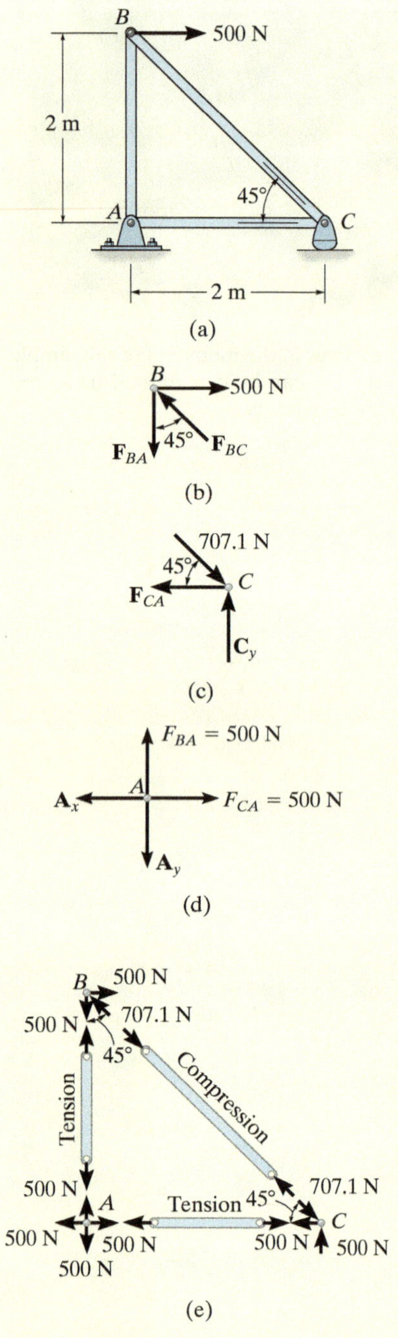

Fig. 6–8

Determine the force in each member of the truss shown in Fig. 6–8*a* and indicate whether the members are in tension or compression.

SOLUTION

Since we should have no more than two unknown forces at the joint and at least one known force acting there, we will begin our analysis at joint *B*.

Joint B. The free-body diagram of the joint at *B* is shown in Fig. 6–8*b*. Applying the equations of equilibrium, we have

$$\overset{+}{\rightarrow}\Sigma F_x = 0; \qquad 500\text{ N} - F_{BC}\sin 45^\circ = 0 \qquad F_{BC} = 707.1\text{ N (C)} \qquad \textit{Ans.}$$

$$+\uparrow\Sigma F_y = 0; \qquad F_{BC}\cos 45^\circ - F_{BA} = 0 \qquad F_{BA} = 500\text{ N (T)} \qquad \textit{Ans.}$$

Since the force in member *BC* has been calculated, we can proceed to analyze joint *C* to determine the force in member *CA* and the support reaction at the rocker.

Joint C. From the free-body diagram of joint *C*, Fig. 6–8*c*, we have

$$\overset{+}{\rightarrow}\Sigma F_x = 0; \quad -F_{CA} + 707.1\cos 45^\circ\text{ N} = 0 \qquad F_{CA} = 500\text{ N (T)} \qquad \textit{Ans.}$$

$$+\uparrow\Sigma F_y = 0; \qquad C_y - 707.1\sin 45^\circ\text{ N} = 0 \qquad C_y = 500\text{ N} \qquad \textit{Ans.}$$

Joint A. Although it is not necessary, we can determine the components of the support reactions at joint *A* using the results of F_{CA} and F_{BA}. From the free-body diagram, Fig. 6–8*d*, we have

$$\overset{+}{\rightarrow}\Sigma F_x = 0; \qquad 500\text{ N} - A_x = 0 \qquad A_x = 500\text{ N}$$

$$+\uparrow\Sigma F_y = 0; \qquad 500\text{ N} - A_y = 0 \qquad A_y = 500\text{ N}$$

NOTE: The results of the analysis are summarized in Fig. 6–8*e*. Note that the free-body diagram of each joint (or pin) shows the effects of all the connected members and external forces applied to the joint, whereas the free-body diagram of each member shows only the effects of the end joints on the member.

EXAMPLE 6.2

Determine the forces acting in all the members of the truss shown in Fig. 6–9*a*.

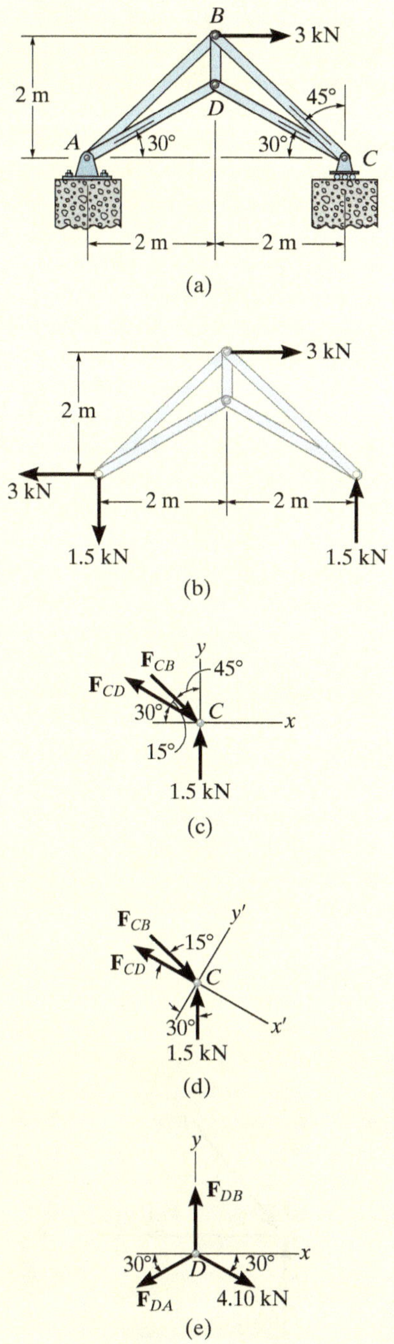

Fig. 6–9

SOLUTION

By inspection, there are more than two unknowns at each joint. Consequently, the support reactions on the truss must first be determined. Show that they have been correctly calculated on the free-body diagram in Fig. 6–9*b*. We can now begin the analysis at joint *C*. Why?

Joint *C*. From the free-body diagram, Fig. 6–9*c*,

$$\xrightarrow{+} \Sigma F_x = 0; \qquad -F_{CD} \cos 30^\circ + F_{CB} \sin 45^\circ = 0$$

$$+\uparrow \Sigma F_y = 0; \qquad 1.5 \text{ kN} + F_{CD} \sin 30^\circ - F_{CB} \cos 45^\circ = 0$$

These two equations must be solved *simultaneously* for each of the two unknowns. Note, however, that a *direct solution* for one of the unknown forces may be obtained by applying a force summation along an axis that is *perpendicular* to the direction of the other unknown force. For example, summing forces along the y' axis, which is perpendicular to the direction of $\mathbf{F}_{CD}$, Fig. 6–9*d*, yields a *direct solution* for F_{CB}.

$$+\nearrow \Sigma F_{y'} = 0; \qquad 1.5 \cos 30^\circ \text{ kN} - F_{CB} \sin 15^\circ = 0$$

$$F_{CB} = 5.019 \text{ kN} = 5.02 \text{ kN (C)} \qquad \textit{Ans.}$$

Then,

$$+\searrow \Sigma F_{x'} = 0;$$

$$-F_{CD} + 5.019 \cos 15^\circ - 1.5 \sin 30^\circ = 0; \; F_{CD} = 4.10 \text{ kN (T)} \qquad \textit{Ans.}$$

Joint *D*. We can now proceed to analyze joint *D*. The free-body diagram is shown in Fig. 6–9*e*.

$$\xrightarrow{+} \Sigma F_x = 0; \qquad -F_{DA} \cos 30^\circ + 4.10 \cos 30^\circ \text{ kN} = 0$$

$$F_{DA} = 4.10 \text{ kN} \quad \text{(T)} \qquad \textit{Ans.}$$

$$+\uparrow \Sigma F_y = 0; \qquad F_{DB} - 2(4.10 \sin 30^\circ \text{ kN}) = 0$$

$$F_{DB} = 4.10 \text{ kN} \quad \text{(T)} \qquad \textit{Ans.}$$

NOTE: The force in the last member, *BA*, can be obtained from joint *B* or joint *A*. As an exercise, draw the free-body diagram of joint *B*, sum the forces in the horizontal direction, and show that $F_{BA} = 0.776$ kN (C).

EXAMPLE 6.3

Determine the force in each member of the truss shown in Fig. 6–10*a*. Indicate whether the members are in tension or compression.

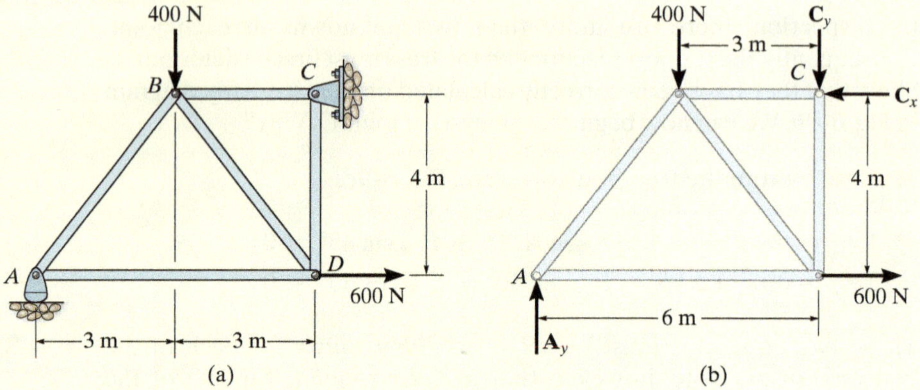

Fig. 6–10

6

SOLUTION

Support Reactions. No joint can be analyzed until the support reactions are determined, because each joint has at least three unknown forces acting on it. A free-body diagram of the entire truss is given in Fig. 6–10*b*. Applying the equations of equilibrium, we have

$$\overset{+}{\rightarrow}\Sigma F_x = 0; \qquad 600\text{ N} - C_x = 0 \qquad C_x = 600\text{ N}$$

$$\circlearrowleft + \Sigma M_C = 0; \qquad -A_y(6\text{ m}) + 400\text{ N}(3\text{ m}) + 600\text{ N}(4\text{ m}) = 0$$

$$A_y = 600\text{ N}$$

$$+\uparrow \Sigma F_y = 0; \qquad 600\text{ N} - 400\text{ N} - C_y = 0 \qquad C_y = 200\text{ N}$$

The analysis can now start at either joint *A* or *C*. The choice is arbitrary since there are one known and two unknown member forces acting on the pin at each of these joints.

Joint A. (Fig. 6–10*c*). As shown on the free-body diagram, $\mathbf{F}_{AB}$ is assumed to be compressive and $\mathbf{F}_{AD}$ is tensile. Applying the equations of equilibrium, we have

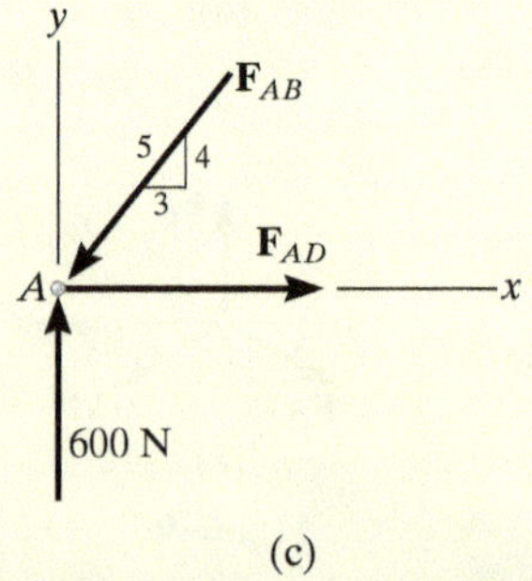

$$+\uparrow \Sigma F_y = 0; \qquad 600\text{ N} - \tfrac{4}{5}F_{AB} = 0 \qquad F_{AB} = 750\text{ N (C)} \qquad \textit{Ans.}$$

$$\overset{+}{\rightarrow}\Sigma F_x = 0; \qquad F_{AD} - \tfrac{3}{5}(750\text{ N}) = 0 \qquad F_{AD} = 450\text{ N (T)} \qquad \textit{Ans.}$$

Joint D. (Fig. 6–10*d*). Using the result for F_{AD} and summing forces in the horizontal direction, Fig. 6–10*d*, we have

$$\overset{+}{\rightarrow}\Sigma F_x = 0; \qquad -450 \text{ N} + \tfrac{3}{5}F_{DB} + 600 \text{ N} = 0 \qquad F_{DB} = -250 \text{ N}$$

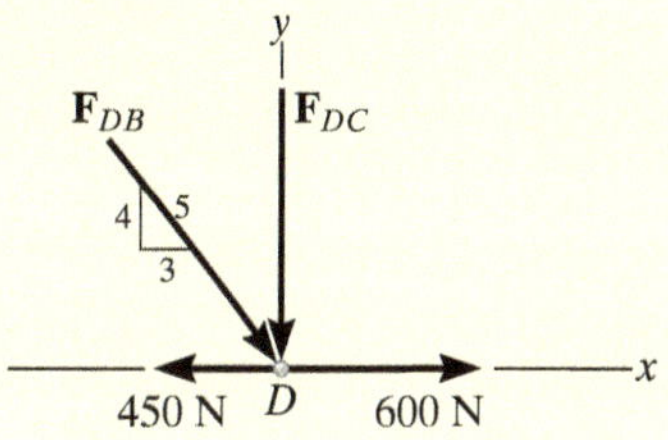

(d)

The negative sign indicates that $\mathbf{F}_{DB}$ acts in the *opposite sense* to that shown in Fig. 6–10*d*.* Hence,

$$F_{DB} = 250 \text{ N (T)} \qquad \textit{Ans.}$$

To determine $\mathbf{F}_{DC}$, we can either correct the sense of $\mathbf{F}_{DB}$ on the free-body diagram, and then apply $\Sigma F_y = 0$, or apply this equation and retain the negative sign for F_{DB}, i.e.,

$$+\uparrow\Sigma F_y = 0; \qquad -F_{DC} - \tfrac{4}{5}(-250 \text{ N}) = 0 \qquad F_{DC} = 200 \text{ N} \quad \text{(C)} \qquad \textit{Ans.}$$

Joint C. (Fig. 6–10*e*).

$$\overset{+}{\rightarrow}\Sigma F_x = 0; \qquad F_{CB} - 600 \text{ N} = 0 \qquad F_{CB} = 600 \text{ N} \quad \text{(C)} \qquad \textit{Ans.}$$

$$+\uparrow\Sigma F_y = 0; \qquad 200 \text{ N} - 200 \text{ N} \equiv 0 \quad \text{(check)}$$

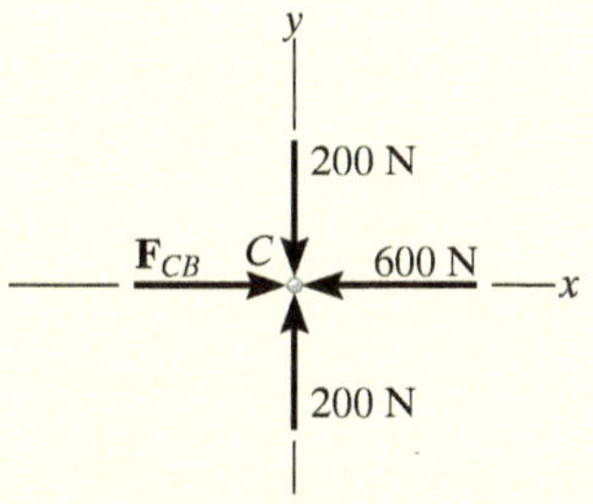

(e)

NOTE: The analysis is summarized in Fig. 6–10*f*, which shows the free-body diagram for each joint and member.

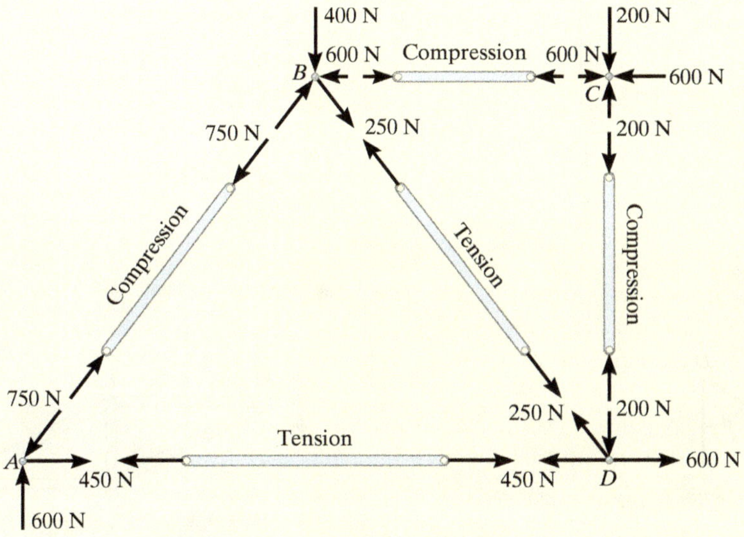

(f)

Fig. 6–10 (cont.)

*The proper sense could have been determined by inspection, prior to applying $\Sigma F_x = 0$.

6.3 Zero-Force Members

Truss analysis using the method of joints is greatly simplified if we can first identify those members which support *no loading.* These *zero-force members* are used to increase the stability of the truss during construction and to provide added support if the loading is changed.

The zero-force members of a truss can generally be found *by inspection* of each of the joints. For example, consider the truss shown in Fig. 6–11*a*. If a free-body diagram of the pin at joint A is drawn, Fig. 6–11*b*, it is seen that members AB and AF are zero-force members. (We could not have come to this conclusion if we had considered the free-body diagrams of joints F or B simply because there are five unknowns at each of these joints.) In a similar manner, consider the free-body diagram of joint D, Fig. 6–11*c*. Here again it is seen that DC and DE are zero-force members. From these observations, we can conclude that *if only two non-collinear members form a truss joint and no external load or support reaction is applied to the joint, the two members must be zero-force members.* The load on the truss in Fig. 6–11*a* is therefore supported by only five members as shown in Fig. 6–11*d*.

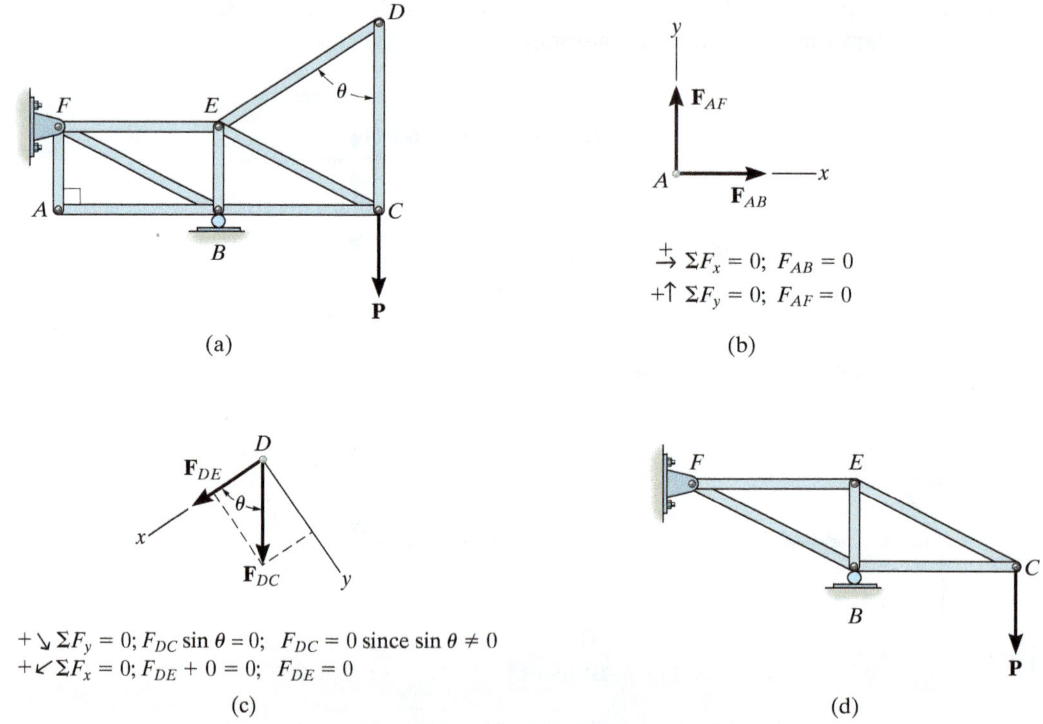

(b): $\overset{+}{\rightarrow} \Sigma F_x = 0;\; F_{AB} = 0$

$+\uparrow \Sigma F_y = 0;\; F_{AF} = 0$

(c): $+\searrow \Sigma F_y = 0;\; F_{DC}\sin\theta = 0;\; F_{DC} = 0$ since $\sin\theta \neq 0$

$+\swarrow \Sigma F_x = 0;\; F_{DE} + 0 = 0;\; F_{DE} = 0$

Fig. 6–11

Now consider the truss shown in Fig. 6–12*a*. The free-body diagram of the pin at joint D is shown in Fig. 6–12*b*. By orienting the y axis along members DC and DE and the x axis along member DA, it is seen that DA is a zero-force member. Note that this is also the case for member CA, Fig. 6–12*c*. In general then, *if three members form a truss joint for which two of the members are collinear, the third member is a zero-force member provided no external force or support reaction is applied to the joint.* The truss shown in Fig. 6–12*d* is therefore suitable for supporting the load **P**.

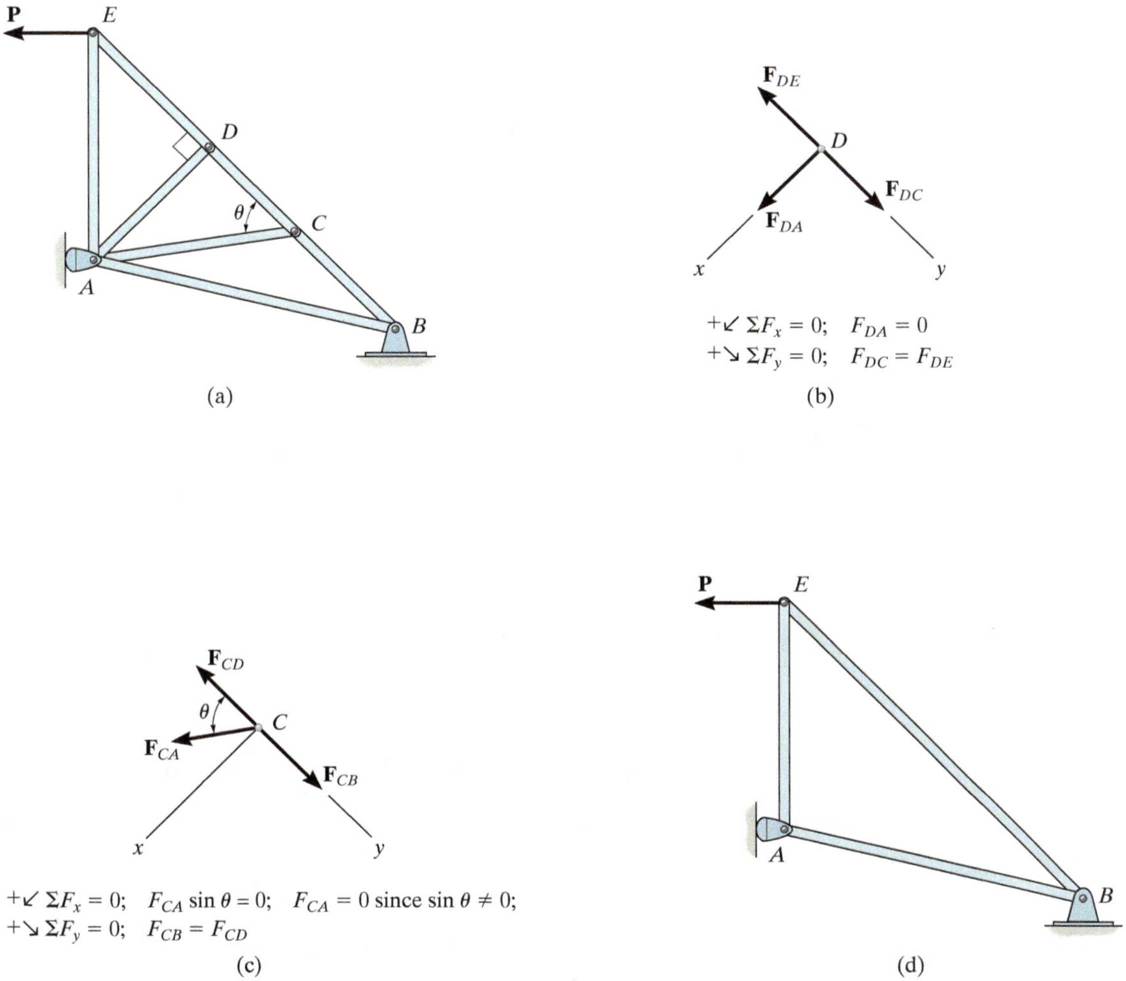

(b): $+\swarrow \Sigma F_x = 0;\quad F_{DA} = 0$

$+\searrow \Sigma F_y = 0;\quad F_{DC} = F_{DE}$

(c): $+\swarrow \Sigma F_x = 0;\quad F_{CA}\sin\theta = 0;\quad F_{CA} = 0 \text{ since } \sin\theta \neq 0;$

$+\searrow \Sigma F_y = 0;\quad F_{CB} = F_{CD}$

Fig. 6–12

EXAMPLE 6.4

Using the method of joints, determine all the zero-force members of the *Fink roof truss* shown in Fig. 6–13*a*. Assume all joints are pin connected.

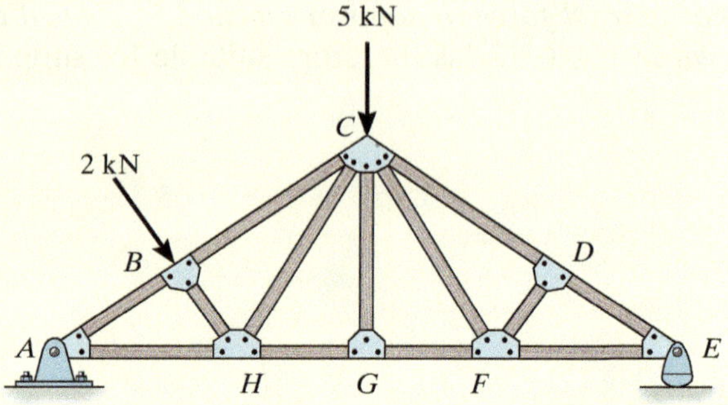

(a)

Fig. 6–13

SOLUTION

Look for joint geometries that have three members for which two are collinear. We have

Joint G. (Fig. 6–13*b*).

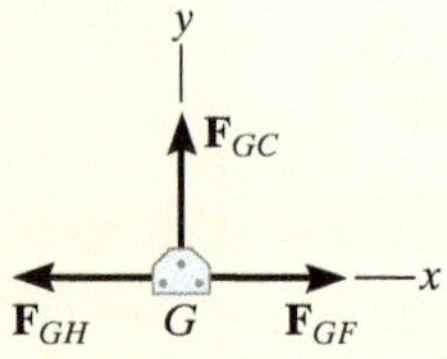

(b)

$$+\uparrow\Sigma F_y = 0; \qquad F_{GC} = 0 \qquad \textit{Ans.}$$

Realize that we could not conclude that *GC* is a zero-force member by considering joint *C*, where there are five unknowns. The fact that *GC* is a zero-force member means that the 5-kN load at *C* must be supported by members *CB*, *CH*, *CF*, and *CD*.

Joint D. (Fig. 6–13*c*).

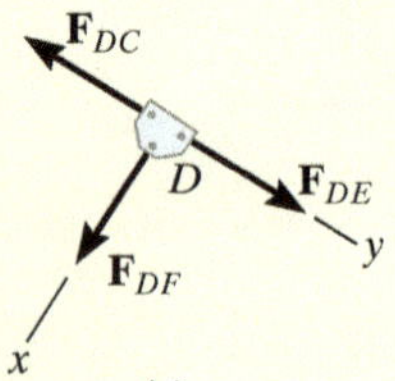

(c)

$$+\swarrow\Sigma F_x = 0; \qquad F_{DF} = 0 \qquad \textit{Ans.}$$

Joint F. (Fig. 6–13*d*).

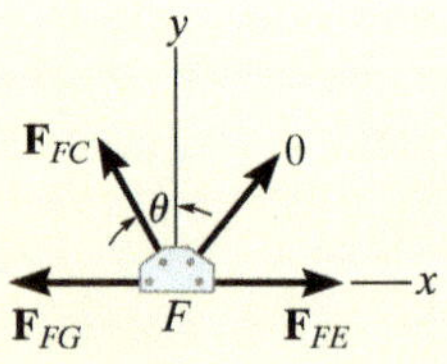

(d)

$$+\uparrow\Sigma F_y = 0; \qquad F_{FC}\cos\theta = 0 \quad \text{Since } \theta \neq 90^\circ, \quad F_{FC} = 0 \qquad \textit{Ans.}$$

NOTE: If joint *B* is analyzed, Fig. 6–13*e*,

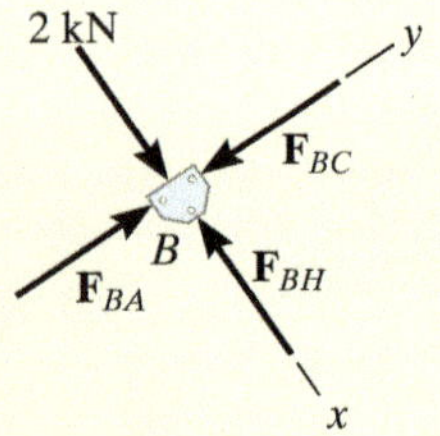

(e)

$$+\searrow\Sigma F_x = 0; \qquad 2\text{ kN} - F_{BH} = 0 \quad F_{BH} = 2\text{ kN} \quad \text{(C)}$$

Also, F_{HC} must satisfy $\Sigma F_y = 0$, Fig. 6–13*f*, and therefore *HC* is *not* a zero-force member.

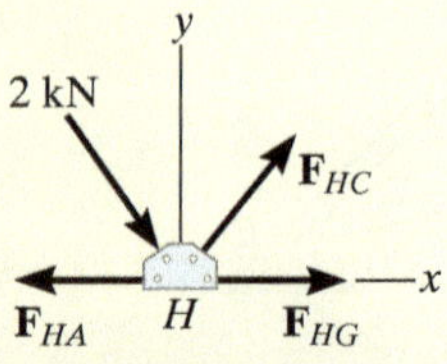

(f)

FUNDAMENTAL PROBLEMS

All problem solutions must include FBDs.

F6–1. Determine the force in each member of the truss. State if the members are in tension or compression.

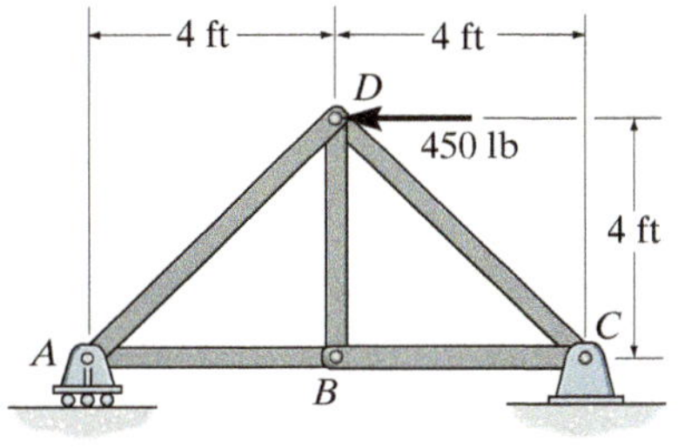

F6–1

F6–2. Determine the force in each member of the truss. State if the members are in tension or compression.

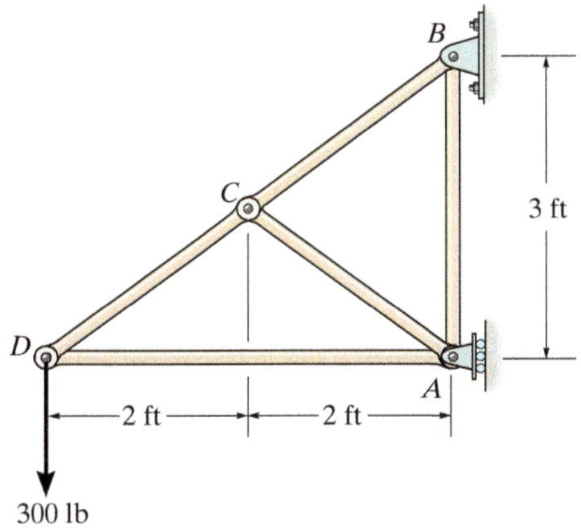

F6–2

F6–3. Determine the force in members AE and DC. State if the members are in tension or compression.

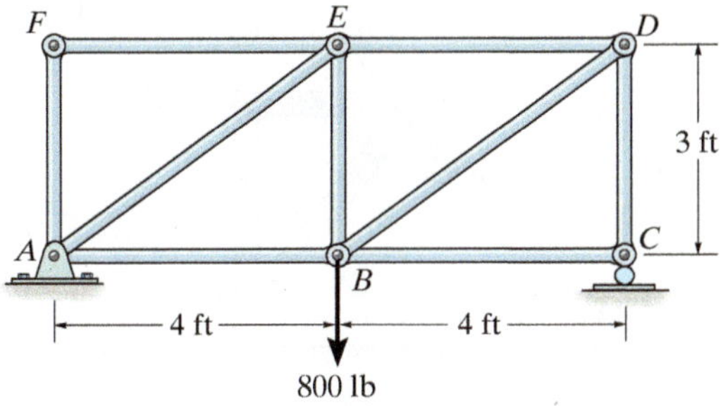

F6–3

F6–4. Determine the greatest load P that can be applied to the truss so that none of the members are subjected to a force exceeding either 2 kN in tension or 1.5 kN in compression.

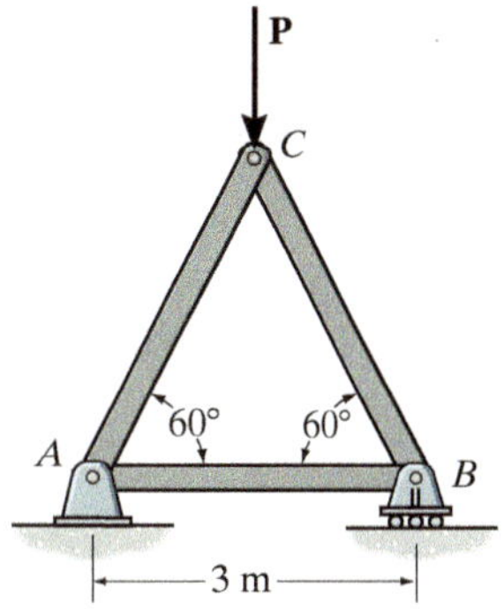

F6–4

F6–5. Identify the zero-force members in the truss.

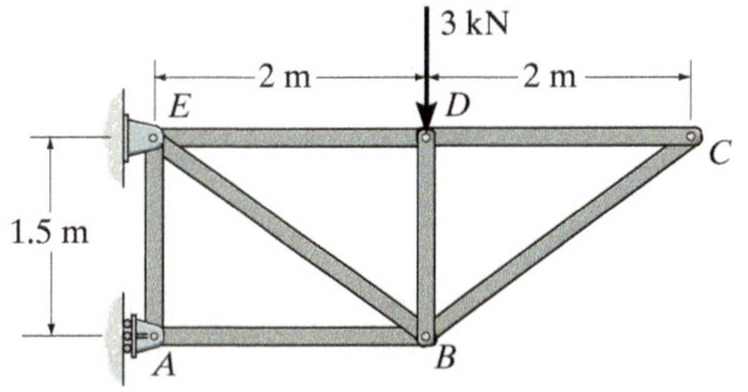

F6–5

F6–6. Determine the force in each member of the truss. State if the members are in tension or compression.

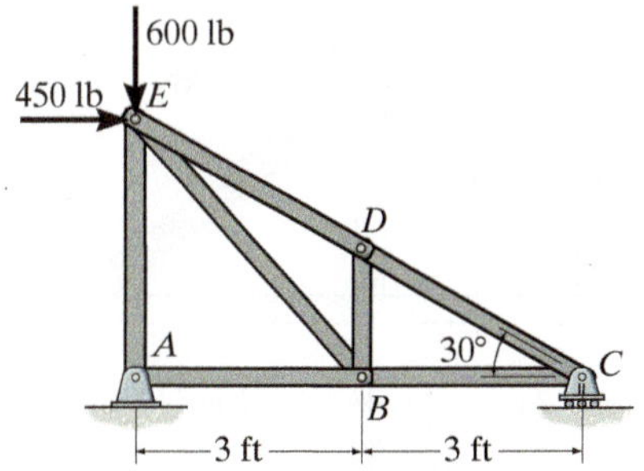

F6–6

PROBLEMS

All problem solutions must include FBDs.

6–1. Determine the force in each member of the truss and state if the members are in tension or compression. Set $P_1 = 800$ lb and $P_2 = 400$ lb.

6–2. Determine the force on each member of the truss and state if the members are in tension or compression. Set $P_1 = 500$ lb and $P_2 = 100$ lb.

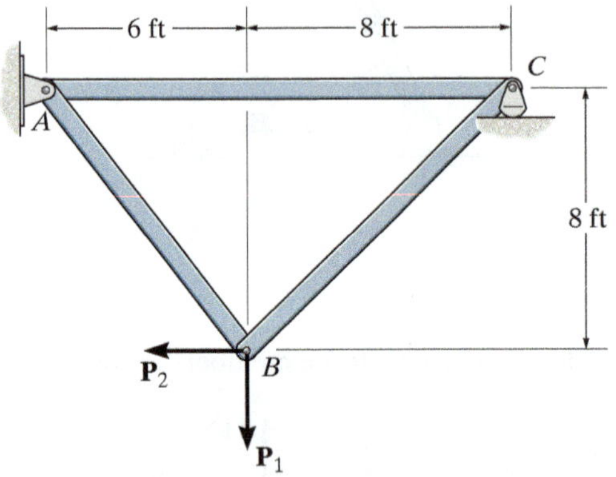

Probs. 6–1/2

6–3. Determine the force in each member of the truss, and state if the members are in tension or compression. Set $\theta = 0°$.

***6–4.** Determine the force in each member of the truss, and state if the members are in tension or compression. Set $\theta = 30°$.

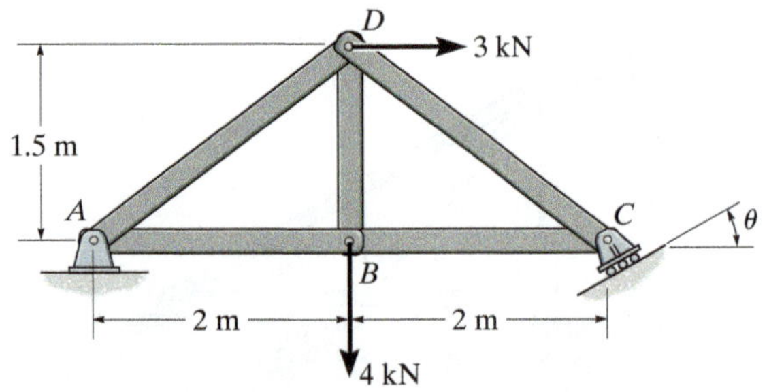

Probs. 6–3/4

6–5. Determine the force in each member of the truss, and state if the members are in tension or compression.

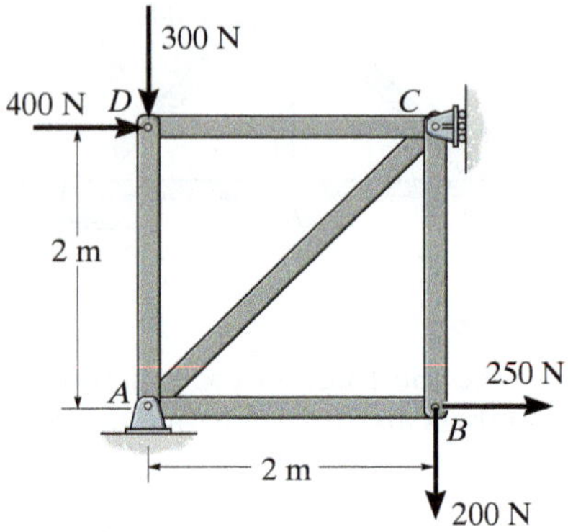

Prob. 6–5

6–6. Determine the force in each member of the truss, and state if the members are in tension or compression.

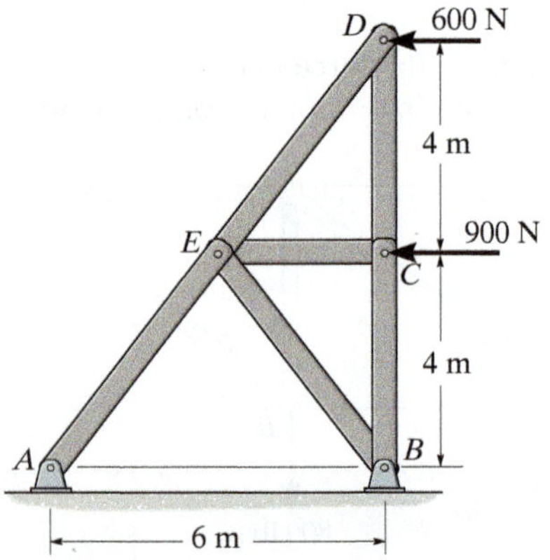

Prob. 6–6

6–7. Determine the force in each member of the *Pratt truss*, and state if the members are in tension or compression.

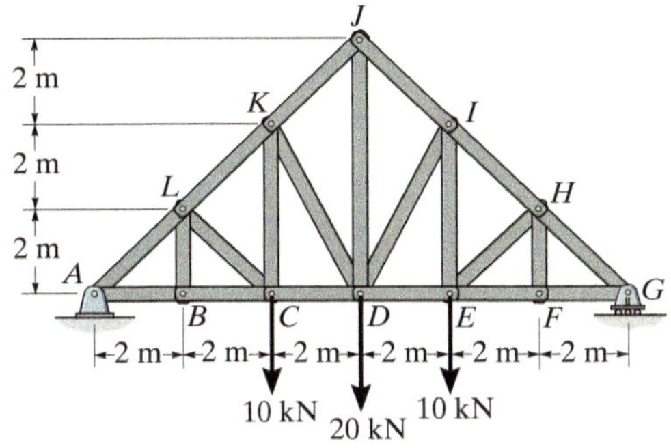

Prob. 6–7

***6–8.** Determine the force in each member of the truss, and state if the members are in tension or compression. *Hint:* The horizontal force component at *A* must be zero. Why?

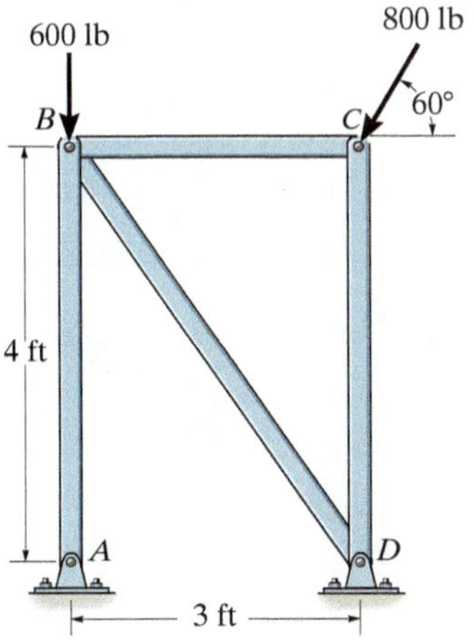

Prob. 6–8

6–9. Determine the force in each member of the truss and state if the members are in tension or compression. *Hint:* The vertical component of force at *C* must equal zero. Why?

6–10. Each member of the truss is uniform and has a mass of 8 kg/m. Remove the external loads of 6 kN and 8 kN and determine the approximate force in each member due to the weight of the truss. State if the members are in tension or compression. Solve the problem by *assuming* the weight of each member can be represented as a vertical force, half of which is applied at each end of the member.

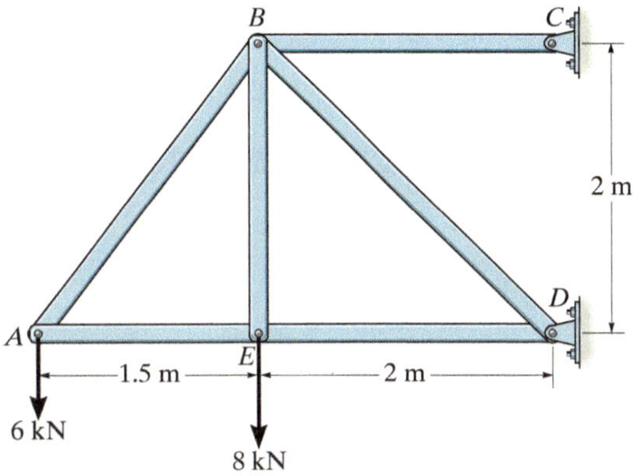

Probs. 6–9/10

6–11. Determine the force in each member of the truss and state if the members are in tension or compression.

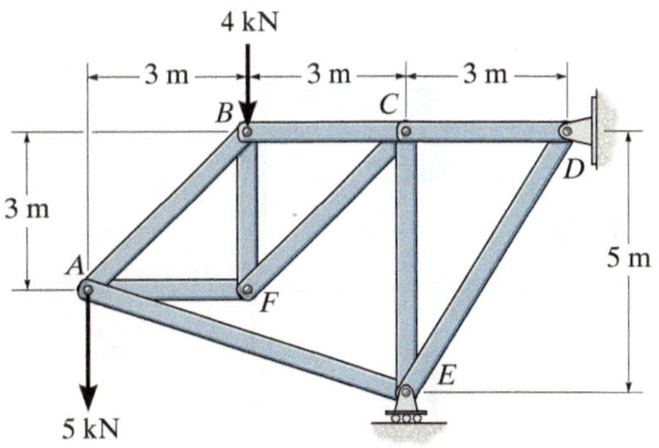

Prob. 6–11

***6–12.** Determine the force in each member of the truss and state if the members are in tension or compression. Set $P_1 = 10$ kN, $P_2 = 15$ kN.

6–13. Determine the force in each member of the truss and state if the members are in tension or compression. Set $P_1 = 0$, $P_2 = 20$ kN.

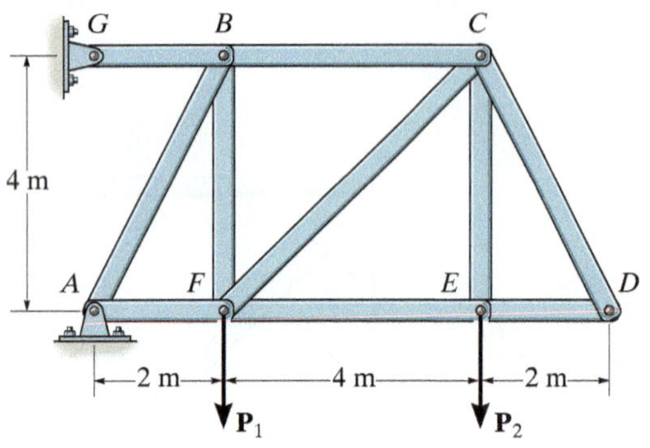

Probs. 6–12/13

***6–16.** Determine the force in each member of the truss. State whether the members are in tension or compression. Set $P = 8$ kN.

6–17. If the maximum force that any member can support is 8 kN in tension and 6 kN in compression, determine the maximum force P that can be supported at joint D.

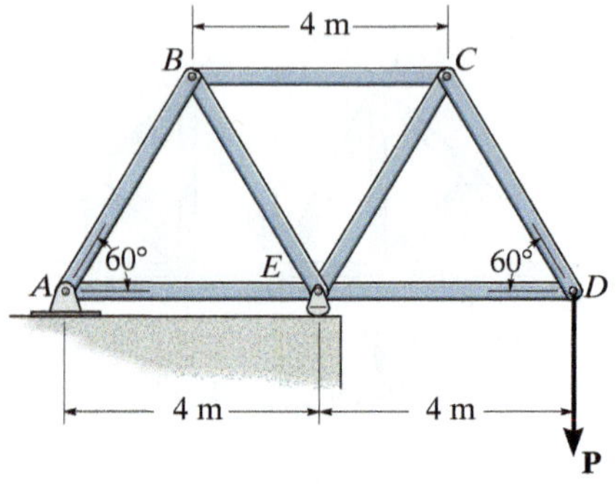

Probs. 6–16/17

6

6–14. Determine the force in each member of the truss and state if the members are in tension or compression. Set $P_1 = 100$ lb, $P_2 = 200$ lb, $P_3 = 300$ lb.

6–15. Determine the force in each member of the truss and state if the members are in tension or compression. Set $P_1 = 400$ lb, $P_2 = 400$ lb, $P_3 = 0$.

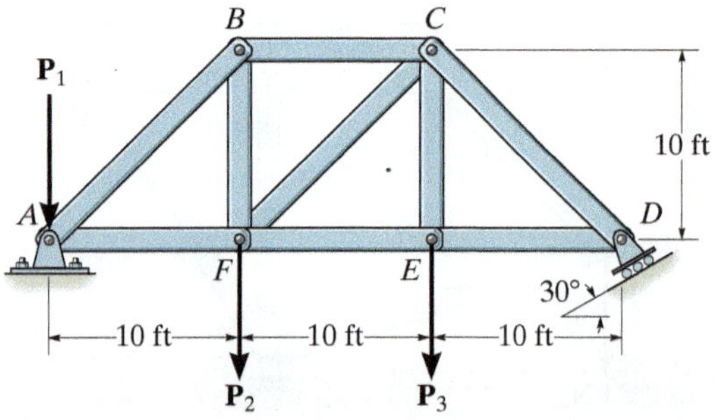

Probs. 6–14/15

6–18. Determine the force in each member of the truss and state if the members are in tension or compression. *Hint:* The resultant force at the pin E acts along member ED. Why?

6–19. Each member of the truss is uniform and has a mass of 8 kg/m. Remove the external loads of 3 kN and 2 kN and determine the approximate force in each member due to the weight of the truss. State if the members are in tension or compression. Solve the problem by *assuming* the weight of each member can be represented as a vertical force, half of which is applied at each end of the member.

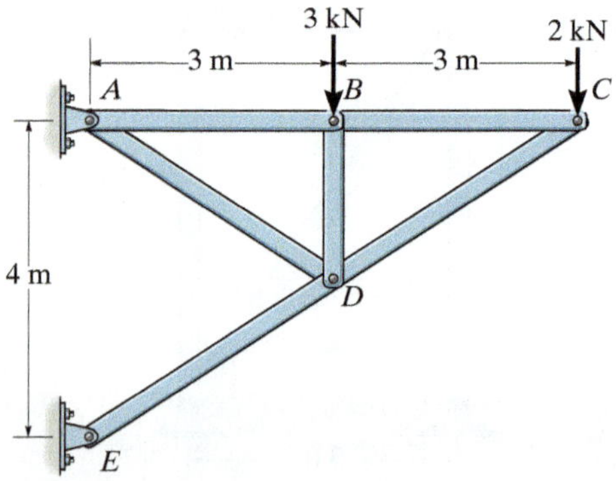

Probs. 6–18/19

*6–20. Determine the force in each member of the truss in terms of the load P, and indicate whether the members are in tension or compression.

6–21. If the maximum force that any member can support is 4 kN in tension and 3 kN in compression, determine the maximum force P that can be supported at point B. Take $d = 1$ m.

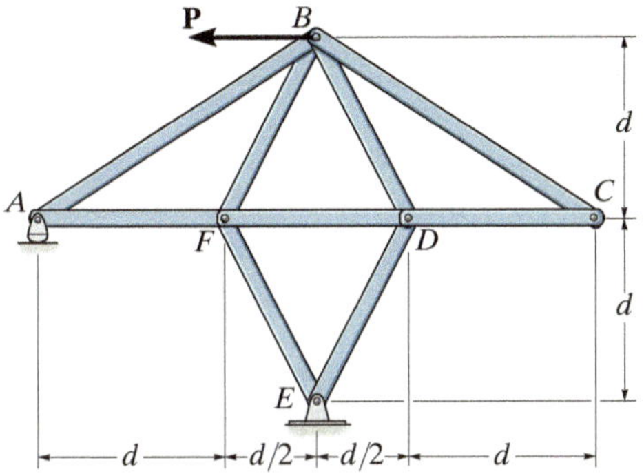

Probs. 6–20/21

6–22. Determine the force in each member of the double scissors truss in terms of the load P and state if the members are in tension or compression.

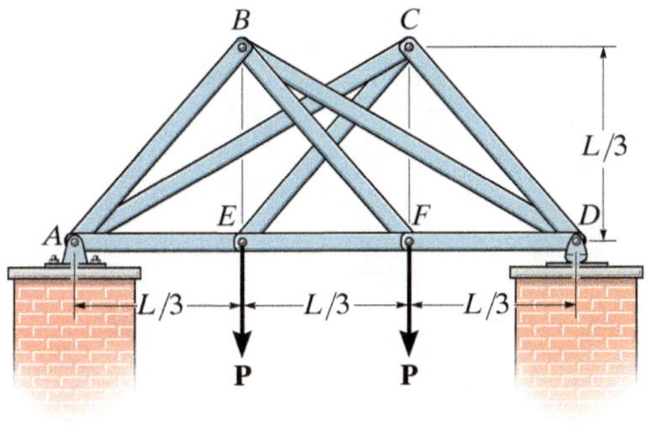

Prob. 6–22

6–23. Determine the force in each member of the truss in terms of the load P and state if the members are in tension or compression.

*6–24. Each member of the truss is uniform and has a weight W. Remove the external forces **P** and determine the approximate force in each member due to the weight of the truss. State if the members are in tension or compression. Solve the problem by *assuming* the weight of each member can be represented as a vertical force, half of which is applied at each end of the member.

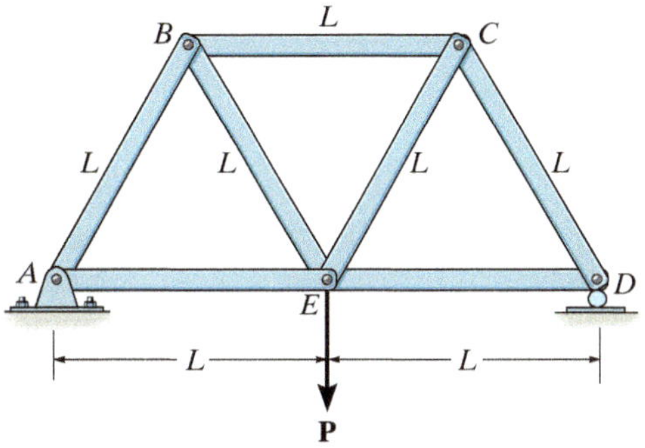

Probs. 6–23/24

6–25. Determine the force in each member of the truss in terms of the external loading and state if the members are in tension or compression.

6–26. The maximum allowable tensile force in the members of the truss is $(F_t)_{max} = 2$ kN, and the maximum allowable compressive force is $(F_c)_{max} = 1.2$ kN. Determine the maximum magnitude P of the two loads that can be applied to the truss. Take $L = 2$ m and $\theta = 30°$.

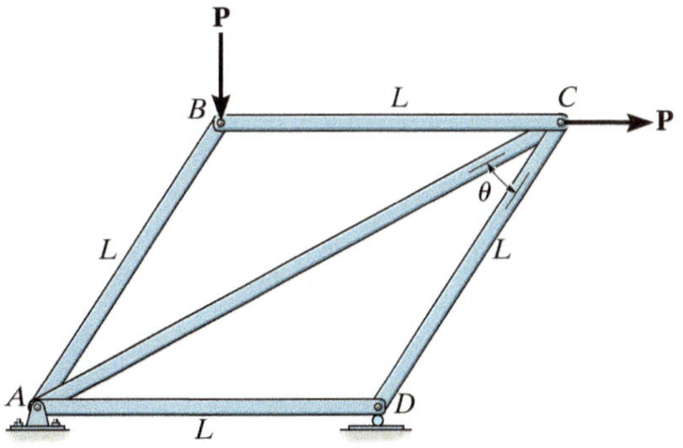

Probs. 6–25/26

6.4 The Method of Sections

When we need to find the force in only a few members of a truss, we can analyze the truss using the *method of sections*. It is based on the principle that if the truss is in equilibrium then any segment of the truss is also in equilibrium. For example, consider the two truss members shown on the left in Fig. 6–14. If the forces within the members are to be determined, then an imaginary section, indicated by the blue line, can be used to cut each member into two parts and thereby "expose" each internal force as "external" to the free-body diagrams shown on the right. Clearly, it can be seen that equilibrium requires that the member in tension (T) be subjected to a "pull," whereas the member in compression (C) is subjected to a "push."

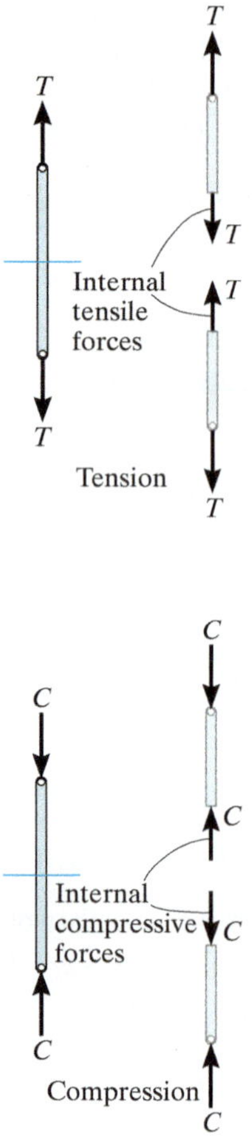

Fig. 6–14

The method of sections can also be used to "cut" or section the members of an entire truss. If the section passes through the truss and the free-body diagram of either of its two parts is drawn, we can then apply the equations of equilibrium to that part to determine the member forces at the "cut section." Since only *three* independent equilibrium equations ($\Sigma F_x = 0$, $\Sigma F_y = 0$, $\Sigma M_O = 0$) can be applied to the free-body diagram of any segment, then we should try to select a section that, in general, passes through not more than *three* members in which the forces are unknown. For example, consider the truss in Fig. 6–15*a*. If the forces in members *BC*, *GC*, and *GF* are to be determined, then section *aa* would be appropriate. The free-body diagrams of the two segments are shown in Figs. 6–15*b* and 6–15*c*. Note that the line of action of each member force is specified from the *geometry* of the truss, since the force in a member is along its axis. Also, the member forces acting on one part of the truss are equal but opposite to those acting on the other part—Newton's third law. Members *BC* and *GC* are assumed to be in *tension* since they are subjected to a "pull," whereas *GF* in *compression* since it is subjected to a "push."

The three unknown member forces $\mathbf{F}_{BC}$, $\mathbf{F}_{GC}$, and $\mathbf{F}_{GF}$ can be obtained by applying the three equilibrium equations to the free-body diagram in Fig. 6–15*b*. If, however, the free-body diagram in Fig. 6–15*c* is considered, the three support reactions $\mathbf{D}_x$, $\mathbf{D}_y$ and $\mathbf{E}_x$ will have to be known, because only three equations of equilibrium are available. (This, of course, is done in the usual manner by considering a free-body diagram of the *entire truss.*)

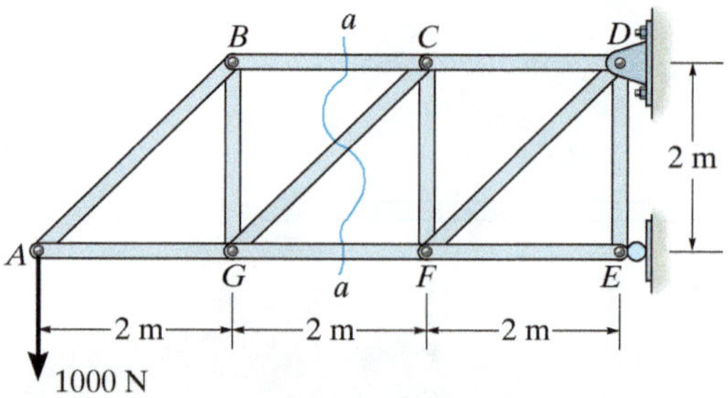

Fig. 6–15

When applying the equilibrium equations, we should carefully consider ways of writing the equations so as to yield a *direct solution* for each of the unknowns, rather than having to solve simultaneous equations. For example, using the truss segment in Fig. 6–15*b* and summing moments about C would yield a direct solution for $\mathbf{F}_{GF}$ since $\mathbf{F}_{BC}$ and $\mathbf{F}_{GC}$ create zero moment about C. Likewise, $\mathbf{F}_{BC}$ can be directly obtained by summing moments about G. Finally, $\mathbf{F}_{GC}$ can be found directly from a force summation in the vertical direction since $\mathbf{F}_{GF}$ and $\mathbf{F}_{BC}$ have no vertical components. This ability to *determine directly* the force in a particular truss member is one of the main advantages of using the method of sections.*

As in the method of joints, there are two ways in which we can determine the correct sense of an unknown member force:

- The correct sense of an unknown member force can in many cases be determined "by inspection." For example, $\mathbf{F}_{BC}$ is a tensile force as represented in Fig. 6–15*b* since moment equilibrium about G requires that $\mathbf{F}_{BC}$ create a moment opposite to that of the 1000-N force. Also, $\mathbf{F}_{GC}$ is tensile since its vertical component must balance the 1000-N force which acts downward. In more complicated cases, the sense of an unknown member force may be *assumed*. If the solution yields a *negative* scalar, it indicates that the force's sense is *opposite* to that shown on the free-body diagram.
- *Always assume* that the unknown member forces at the cut section are *tensile* forces, i.e., "pulling" on the member. By doing this, the numerical solution of the equilibrium equations will yield *positive scalars for members in tension and negative scalars for members in compression.*

The forces in selected members of this Pratt truss can readily be determined using the method of sections.

*Notice that if the method of joints were used to determine, say, the force in member GC, it would be necessary to analyze joints A, B, and G in sequence.

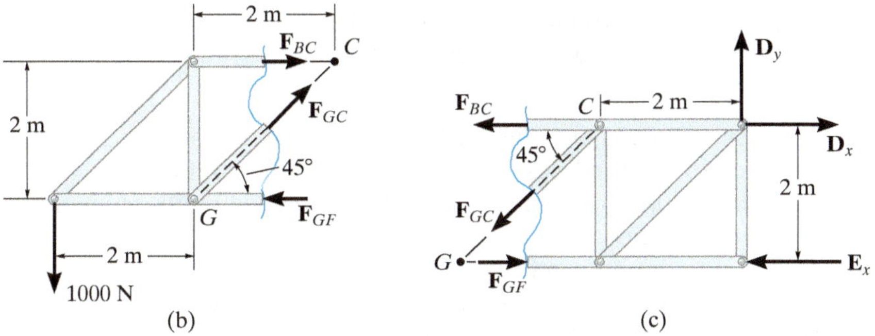

Fig. 6–15 (cont.)

Simple trusses are often used in the construction of large cranes in order to reduce the weight of the boom and tower.

Procedure for Analysis

The forces in the members of a truss may be determined by the method of sections using the following procedure.

Free-Body Diagram.

- Make a decision on how to "cut" or section the truss through the members where forces are to be determined.
- Before isolating the appropriate section, it may first be necessary to determine the truss's support reactions. If this is done then the three equilibrium equations will be available to solve for member forces at the section.
- Draw the free-body diagram of that segment of the sectioned truss which has the least number of forces acting on it.
- Use one of the two methods described above for establishing the sense of the unknown member forces.

Equations of Equilibrium.

- Moments should be summed about a point that lies at the intersection of the lines of action of two unknown forces, so that the third unknown force can be determined directly from the moment equation.
- If two of the unknown forces are *parallel*, forces may be summed *perpendicular* to the direction of these unknowns to determine *directly* the third unknown force.

EXAMPLE 6.5

Determine the force in members *GE*, *GC*, and *BC* of the truss shown in Fig. 6–16*a*. Indicate whether the members are in tension or compression.

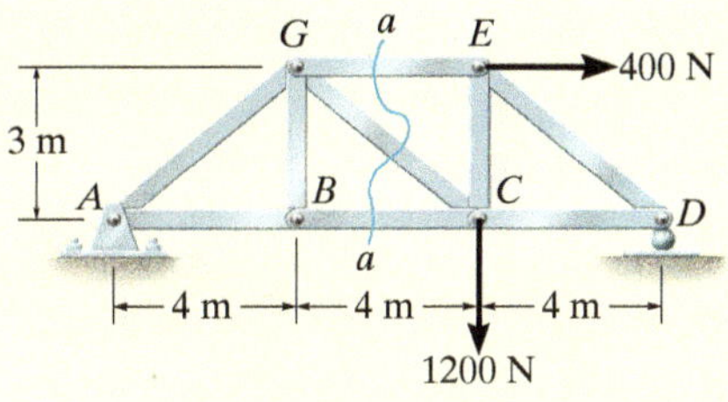

(a)

SOLUTION

Section *aa* in Fig. 6–16*a* has been chosen since it cuts through the *three* members whose forces are to be determined. In order to use the method of sections, however, it is *first* necessary to determine the external reactions at *A* or *D*. Why? A free-body diagram of the entire truss is shown in Fig. 6–16*b*. Applying the equations of equilibrium, we have

$$\xrightarrow{+} \Sigma F_x = 0; \qquad 400\text{ N} - A_x = 0 \qquad A_x = 400\text{ N}$$

$$\circlearrowleft + \Sigma M_A = 0; \qquad -1200\text{ N}(8\text{ m}) - 400\text{ N}(3\text{ m}) + D_y(12\text{ m}) = 0$$

$$D_y = 900\text{ N}$$

$$+\uparrow \Sigma F_y = 0; \qquad A_y - 1200\text{ N} + 900\text{ N} = 0 \qquad A_y = 300\text{ N}$$

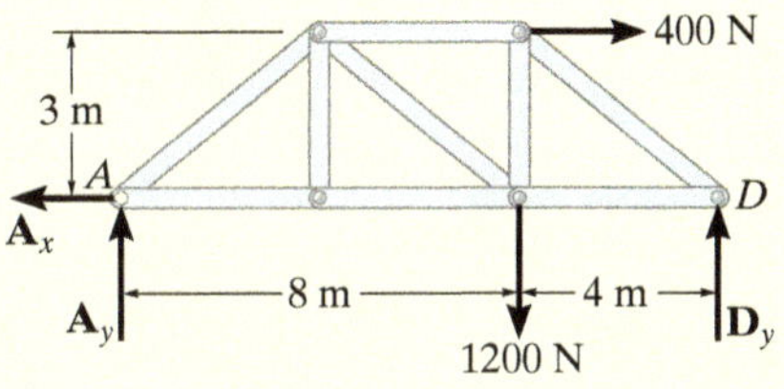

(b)

Free-Body Diagram. For the analysis the free-body diagram of the left portion of the sectioned truss will be used, since it involves the least number of forces, Fig. 6–16*c*.

Equations of Equilibrium. Summing moments about point *G* eliminates $\mathbf{F}_{GE}$ and $\mathbf{F}_{GC}$ and yields a direct solution for F_{BC}.

$$\circlearrowleft + \Sigma M_G = 0; \qquad -300\text{ N}(4\text{ m}) - 400\text{ N}(3\text{ m}) + F_{BC}(3\text{ m}) = 0$$

$$F_{BC} = 800\text{ N} \quad \text{(T)} \qquad \textit{Ans.}$$

In the same manner, by summing moments about point *C* we obtain a direct solution for F_{GE}.

$$\circlearrowleft + \Sigma M_C = 0; \qquad -300\text{ N}(8\text{ m}) + F_{GE}(3\text{ m}) = 0$$

$$F_{GE} = 800\text{ N} \quad \text{(C)} \qquad \textit{Ans.}$$

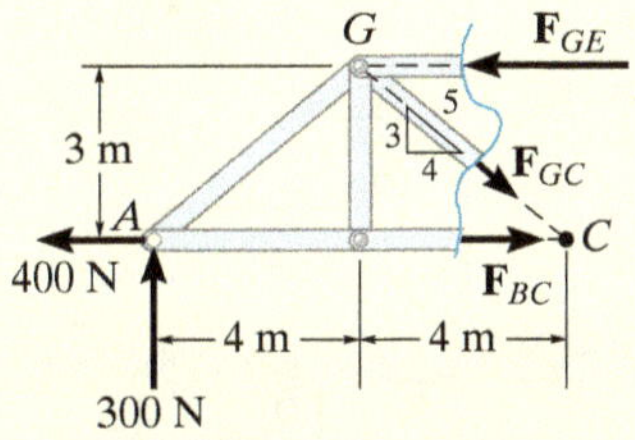

(c)

Fig. 6–16

Since $\mathbf{F}_{BC}$ and $\mathbf{F}_{GE}$ have no vertical components, summing forces in the *y* direction directly yields F_{GC}, i.e.,

$$+\uparrow \Sigma F_y = 0; \qquad 300\text{ N} - \tfrac{3}{5}F_{GC} = 0$$

$$F_{GC} = 500\text{ N} \quad \text{(T)} \qquad \textit{Ans.}$$

NOTE: Here it is possible to tell, by inspection, the proper direction for each unknown member force. For example, $\Sigma M_C = 0$ requires $\mathbf{F}_{GE}$ to be *compressive* because it must balance the moment of the 300-N force about *C*.

EXAMPLE 6.6

Determine the force in member *CF* of the truss shown in Fig. 6–17*a*. Indicate whether the member is in tension or compression. Assume each member is pin connected.

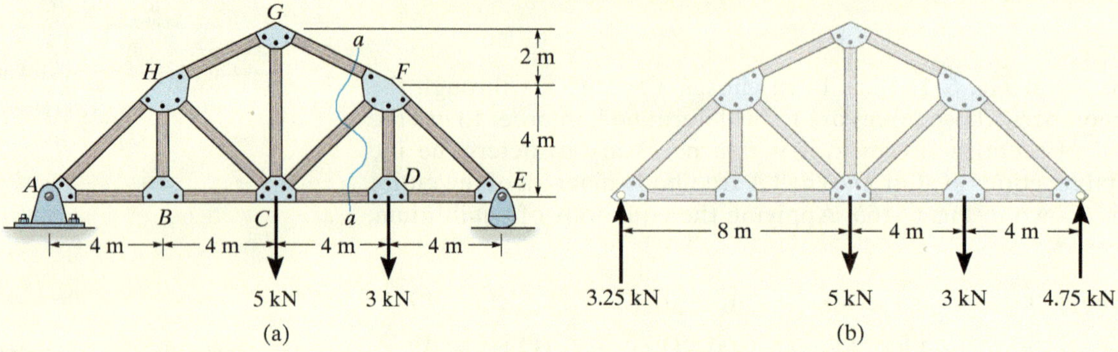

Fig. 6–17

SOLUTION

Free-Body Diagram. Section *aa* in Fig. 6–17*a* will be used since this section will "expose" the internal force in member *CF* as "external" on the free-body diagram of either the right or left portion of the truss. It is first necessary, however, to determine the support reactions on either the left or right side. Verify the results shown on the free-body diagram in Fig. 6–17*b*.

The free-body diagram of the right portion of the truss, which is the easiest to analyze, is shown in Fig. 6–17*c*. There are three unknowns, F_{FG}, F_{CF}, and F_{CD}.

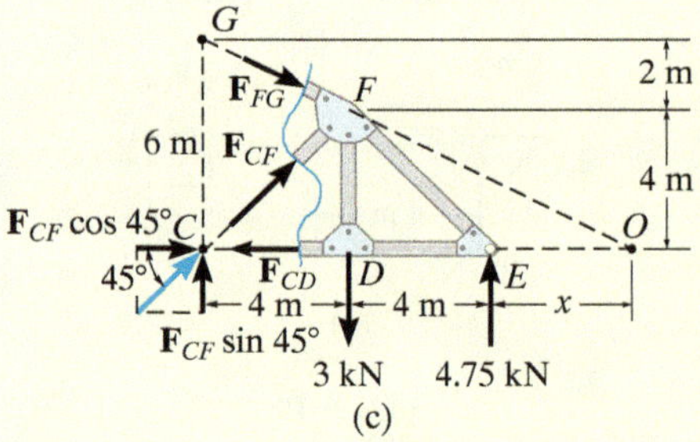

Equations of Equilibrium. We will apply the moment equation about point *O* in order to eliminate the two unknowns F_{FG} and F_{CD}. The location of point *O* measured from *E* can be determined from proportional triangles, i.e., $4/(4 + x) = 6/(8 + x)$, $x = 4$ m. Or, stated in another manner, the slope of member *GF* has a drop of 2 m to a horizontal distance of 4 m. Since *FD* is 4 m, Fig. 6–17*c*, then from *D* to *O* the distance must be 8 m.

An easy way to determine the moment of $\mathbf{F}_{CF}$ about point *O* is to use the principle of transmissibility and slide $\mathbf{F}_{CF}$ to point *C*, and then resolve $\mathbf{F}_{CF}$ into its two rectangular components. We have

$$\circlearrowleft + \Sigma M_O = 0;$$

$$-F_{CF} \sin 45°(12 \text{ m}) + (3 \text{ kN})(8 \text{ m}) - (4.75 \text{ kN})(4 \text{ m}) = 0$$

$$F_{CF} = 0.589 \text{ kN} \quad \text{(C)} \qquad \textit{Ans.}$$

EXAMPLE 6.7

Determine the force in member EB of the roof truss shown in Fig. 6–18a. Indicate whether the member is in tension or compression.

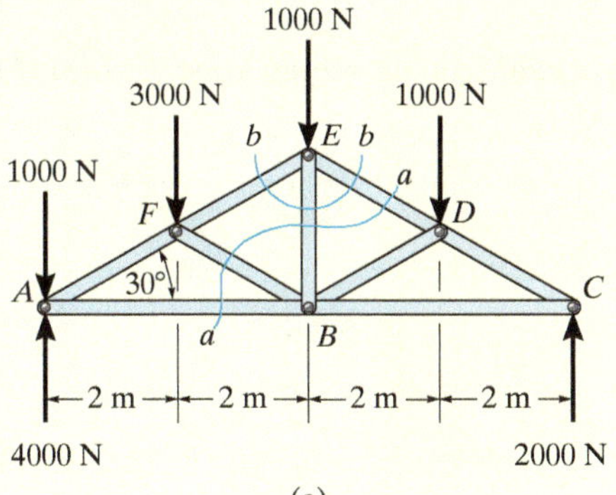

SOLUTION

Free-Body Diagrams. By the method of sections, any imaginary section that cuts through EB, Fig. 6–18a, will also have to cut through three other members for which the forces are unknown. For example, section aa cuts through ED, EB, FB, and AB. If a free-body diagram of the left side of this section is considered, Fig. 6–18b, it is possible to obtain $\mathbf{F}_{ED}$ by summing moments about B to eliminate the other three unknowns; however, $\mathbf{F}_{EB}$ cannot be determined from the remaining two equilibrium equations. One possible way of obtaining $\mathbf{F}_{EB}$ is first to determine $\mathbf{F}_{ED}$ from section aa, then use this result on section bb, Fig. 6–18a, which is shown in Fig. 6–18c. Here the force system is concurrent and our sectioned free-body diagram is the same as the free-body diagram for the joint at E.

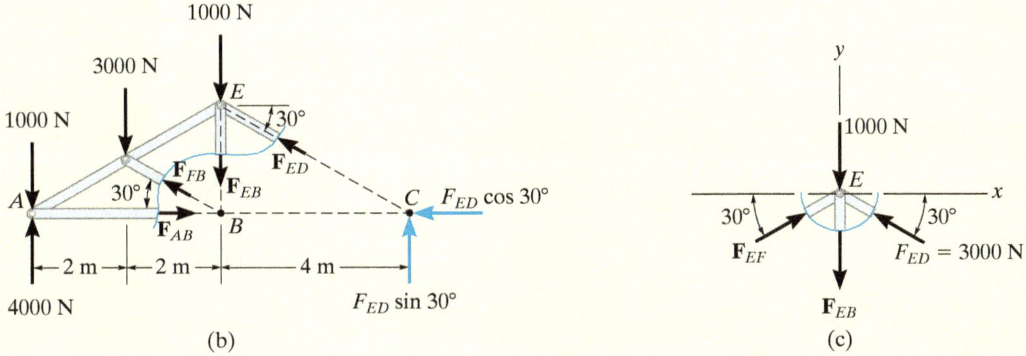

Fig. 6–18

Equations of Equilibrium. In order to determine the moment of $\mathbf{F}_{ED}$ about point B, Fig. 6–18b, we will use the principle of transmissibility and slide the force to point C and then resolve it into its rectangular components as shown. Therefore,

$$\circlearrowleft + \Sigma M_B = 0; \qquad 1000\text{ N}(4\text{ m}) + 3000\text{ N}(2\text{ m}) - 4000\text{ N}(4\text{ m}) + F_{ED}\sin 30°(4\text{ m}) = 0$$

$$F_{ED} = 3000\text{ N} \quad (\text{C})$$

Considering now the free-body diagram of section bb, Fig. 6–18c, we have

$$\overset{+}{\rightarrow} \Sigma F_x = 0; \qquad F_{EF}\cos 30° - 3000\cos 30°\text{ N} = 0$$

$$F_{EF} = 3000\text{ N} \quad (\text{C})$$

$$+\uparrow \Sigma F_y = 0; \qquad 2(3000\sin 30°\text{ N}) - 1000\text{ N} - F_{EB} = 0$$

$$F_{EB} = 2000\text{ N} \quad (\text{T})$$ *Ans.*

FUNDAMENTAL PROBLEMS

All problem solutions must include FBDs.

F6–7. Determine the force in members *BC*, *CF*, and *FE*. State if the members are in tension or compression.

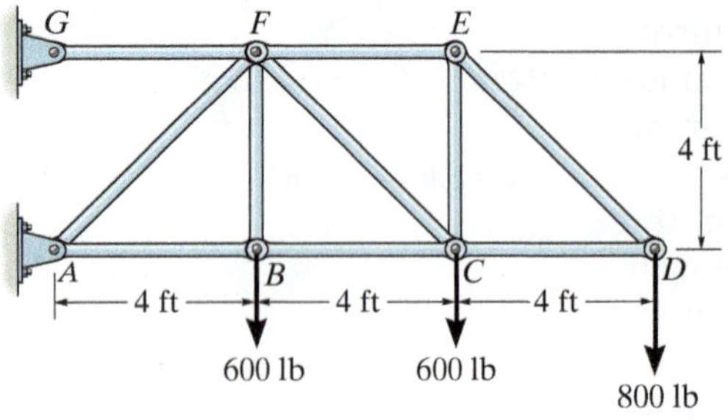

F6–7

F6–8. Determine the force in members *LK*, *KC*, and *CD* of the Pratt truss. State if the members are in tension or compression.

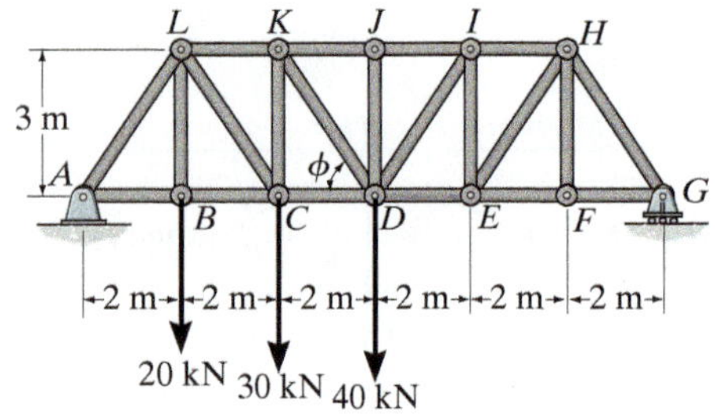

F6–8

F6–9. Determine the force in members *KJ*, *KD*, and *CD* of the Pratt truss. State if the members are in tension or compression.

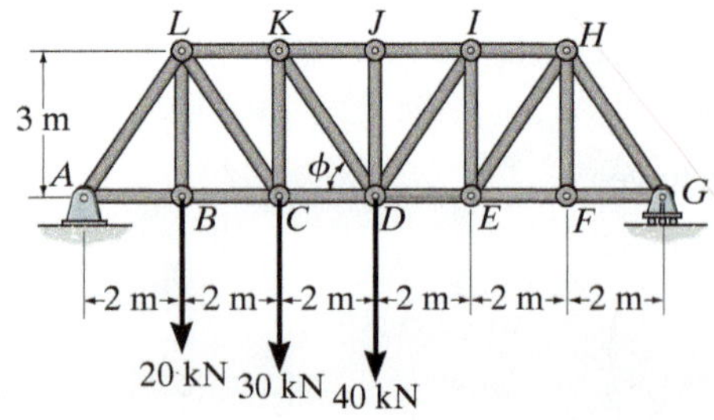

F6–9

F6–10. Determine the force in members *EF*, *CF*, and *BC* of the truss. State if the members are in tension or compression.

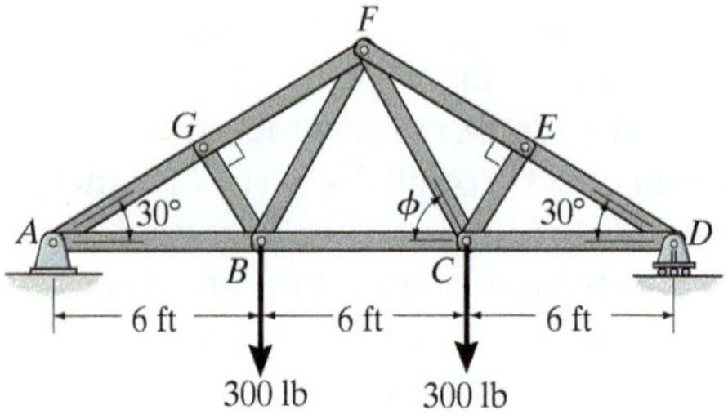

F6–10

F6–11. Determine the force in members *GF*, *GD*, and *CD* of the truss. State if the members are in tension or compression.

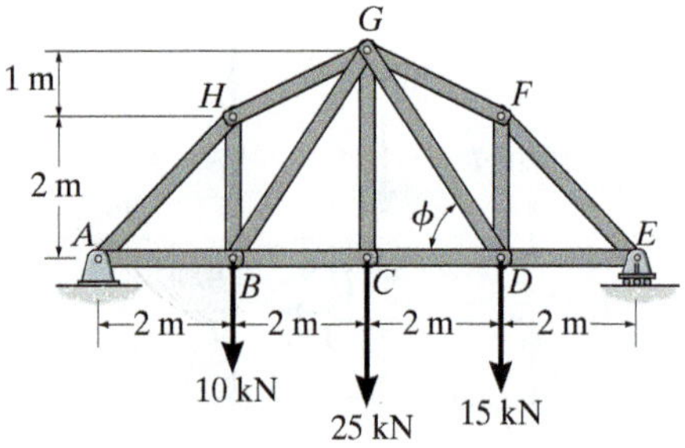

F6–11

F6–12. Determine the force in members *DC*, *HI*, and *JI* of the truss. State if the members are in tension or compression. *Suggestion:* Use the sections shown.

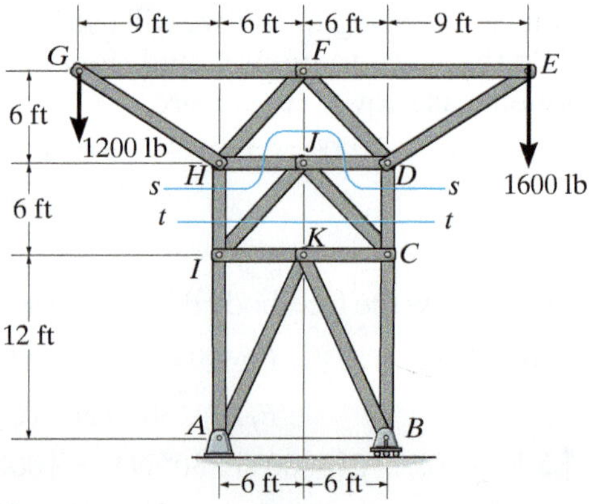

F6–12

PROBLEMS

All problem solutions must include FBDs.

6–27. Determine the force in members *HG*, *HE*, and *DE* of the truss, and state if the members are in tension or compression.

***6–28.** Determine the force in members *CD*, *HI*, and *CJ* of the truss, and state if the members are in tension or compression.

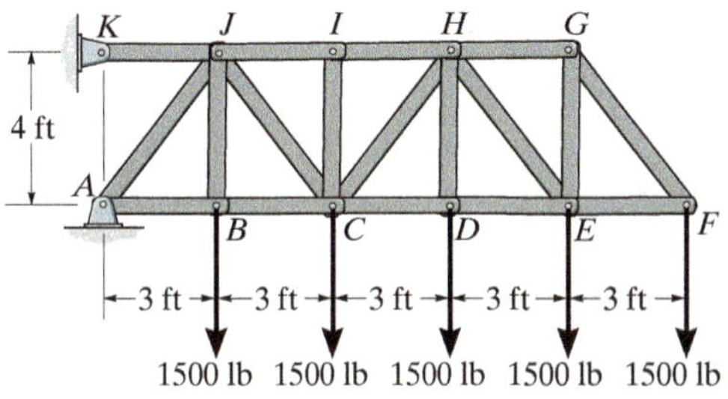

Probs. 6–27/28

6–31. Determine the force in members *CD*, *CJ*, *KJ*, and *DJ* of the truss which serves to support the deck of a bridge. State if these members are in tension or compression.

***6–32.** Determine the force in members *EI* and *JI* of the truss which serves to support the deck of a bridge. State if these members are in tension or compression.

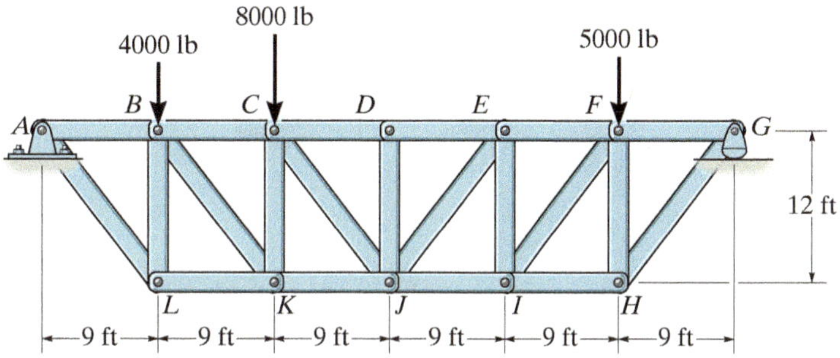

Probs. 6–31/32

6–29. Determine the force in members *GB* and *GF* of the bridge truss and state if these members are in tension or compression.

6–30. Determine the force in members *EC*, *EF*, and *FC* of the bridge truss and state if these members are in tension or compression.

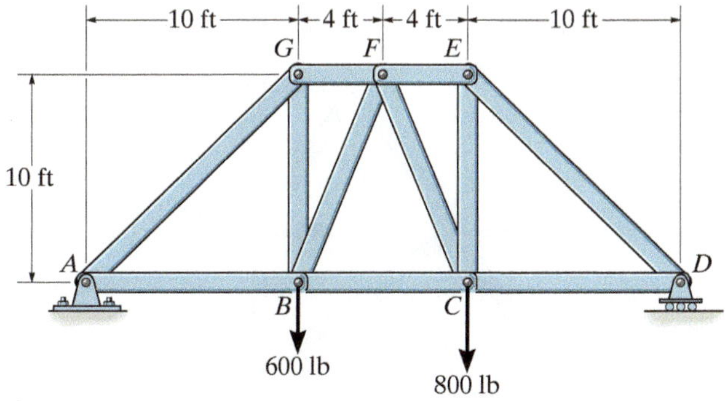

Probs. 6–29/30

6–33. Determine the force in member *GJ* of the truss and state if this member is in tension or compression.

6–34. Determine the force in member *GC* of the truss and state if this member is in tension or compression.

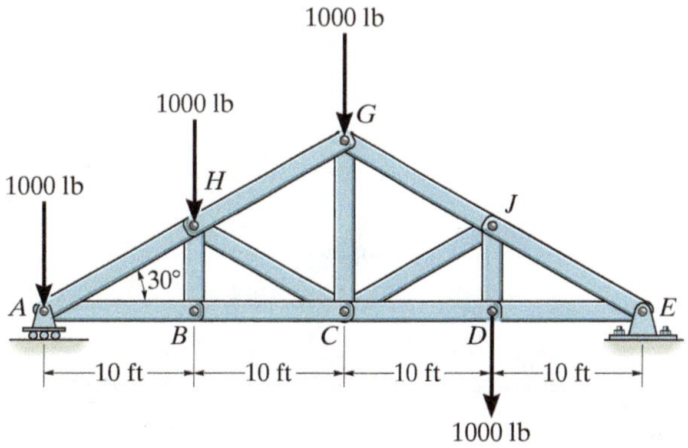

Probs. 6–33/34

6–35. Determine the force in members *BC*, *HC*, and *HG*. After the truss is sectioned use a single equation of equilibrium for the calculation of each force. State if these members are in tension or compression.

***6–36.** Determine the force in members *CD*, *CF*, and *CG* and state if these members are in tension or compression.

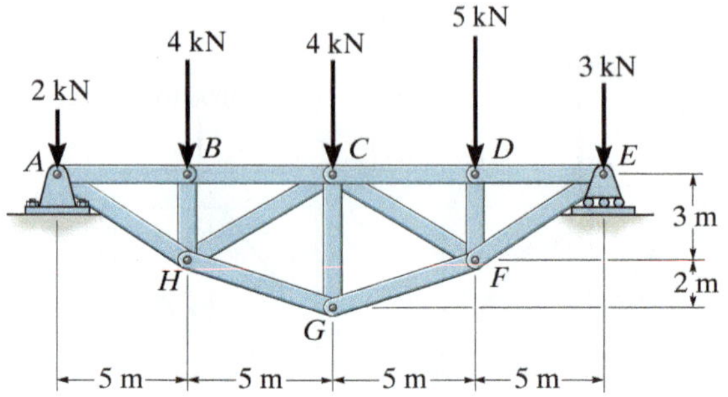

Probs. 6–35/36

6–39. Determine the force in members *IC* and *CG* of the truss and state if these members are in tension or compression. Also, indicate all zero-force members.

***6–40.** Determine the force in members *JE* and *GF* of the truss and state if these members are in tension or compression. Also, indicate all zero-force members.

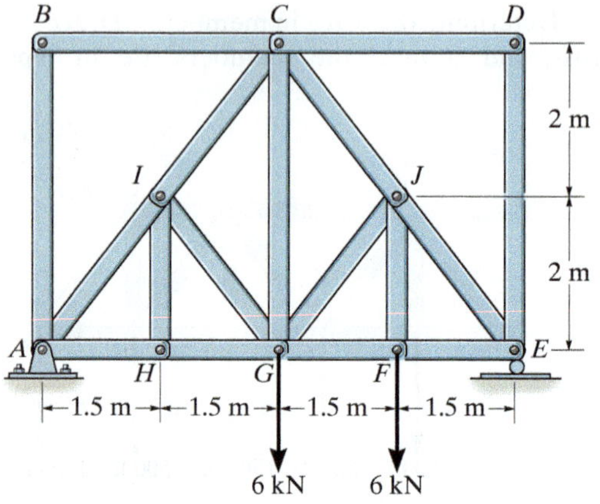

Probs. 6–39/40

6–37. Determine the force in members *GF*, *FB*, and *BC* of the *Fink truss* and state if the members are in tension or compression.

6–38. Determine the force in members *FE* and *EC* of the *Fink truss* and state if the members are in tension or compression.

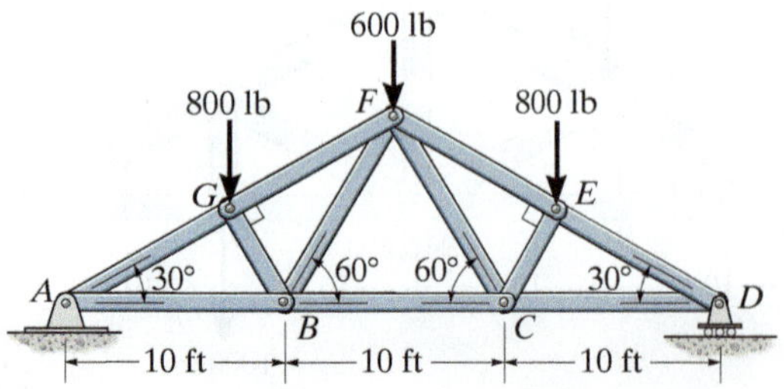

Probs. 6–37/38

6–41. Determine the force in members *FG*, *GC* and *CB* of the truss used to support the sign, and state if the members are in tension or compression.

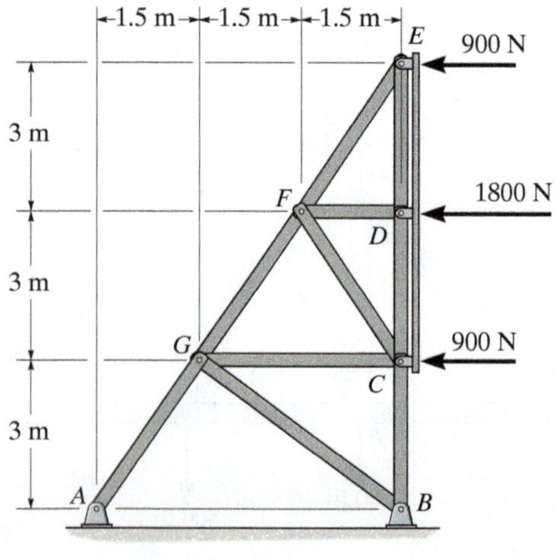

Prob. 6–41

6–42. Determine the force in members *LK*, *LC*, and *BC* of the truss, and state if the members are in tension or compression.

6–43. Determine the force in members *JI*, *JE*, and *DE* of the truss, and state if the members are in tension or compression.

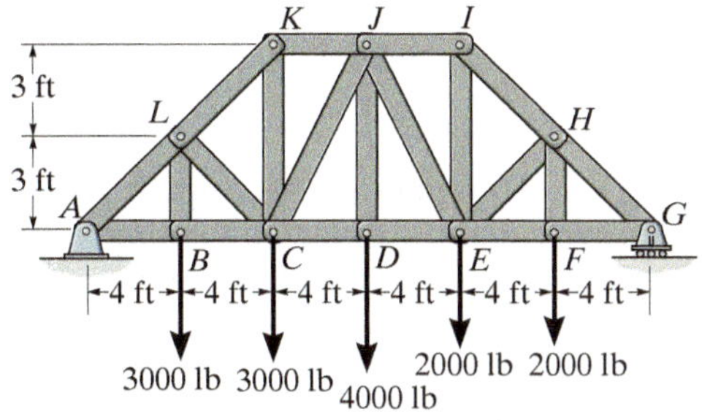

Probs. 6–42/43

***6–44.** The skewed truss carries the load shown. Determine the force in members *CB*, *BE*, and *EF* and state if these members are in tension or compression. Assume that all joints are pinned.

6–45. The skewed truss carries the load shown. Determine the force in members *AB*, *BF*, and *EF* and state if these members are in tension or compression. Assume that all joints are pinned.

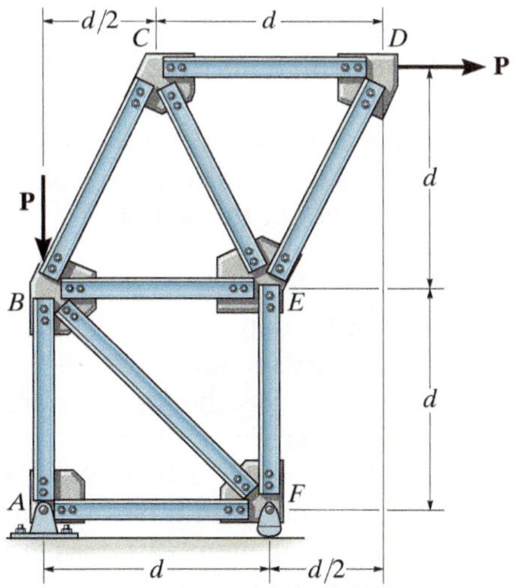

Probs. 6–44/45

6–46. Determine the force in members *CD* and *CM* of the *Baltimore bridge truss* and state if the members are in tension or compression. Also, indicate all zero-force members.

6–47. Determine the force in members *EF*, *EP*, and *LK* of the *Baltimore bridge truss* and state if the members are in tension or compression. Also, indicate all zero-force members.

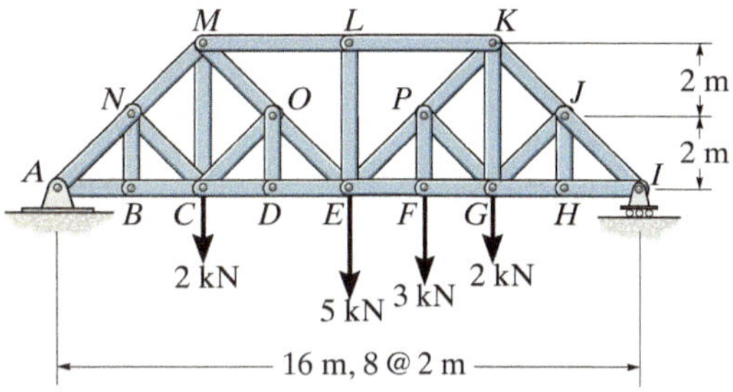

Probs. 6–46/47

***6–48.** The truss supports the vertical load of 600 N. If $L = 2$ m, determine the force on members *HG* and *HB* of the truss and state if the members are in tension or compression.

6–49. The truss supports the vertical load of 600 N. Determine the force in members *BC*, *BG*, and *HG* as the dimension *L* varies. Plot the results of *F* (ordinate with tension as positive) versus *L* (abscissa) for $0 \leq L \leq 3$ m.

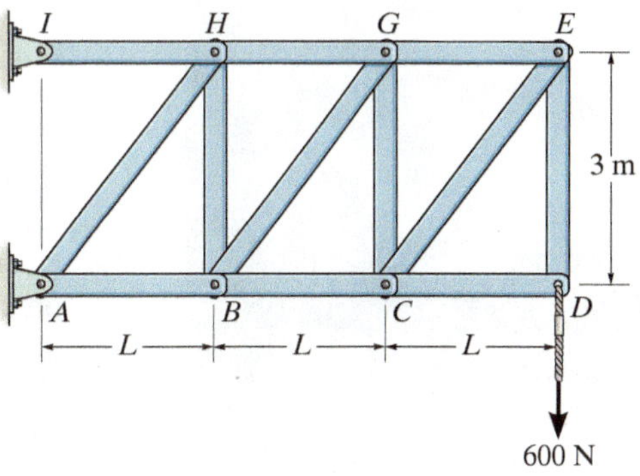

Probs. 6–48/49

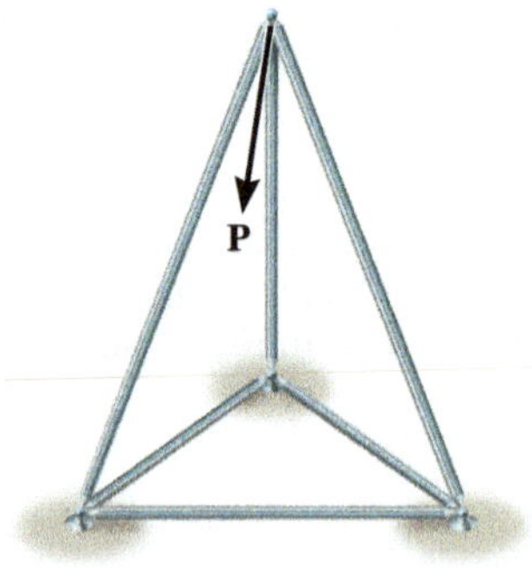

Fig. 6–19

Typical roof-supporting space truss. Notice the use of ball-and-socket joints for the connections.

For economic reasons, large electrical transmission towers are often constructed using space trusses.

*6.5 Space Trusses

A *space truss* consists of members joined together at their ends to form a stable three-dimensional structure. The simplest form of a space truss is a *tetrahedron*, formed by connecting six members together, as shown in Fig. 6–19. Any additional members added to this basic element would be redundant in supporting the force **P**. A *simple space truss* can be built from this basic tetrahedral element by adding three additional members and a joint, and continuing in this manner to form a system of multiconnected tetrahedrons.

Assumptions for Design. The members of a space truss may be treated as two-force members provided the external loading is applied at the joints and the joints consist of ball-and-socket connections. These assumptions are justified if the welded or bolted connections of the joined members intersect at a common point and the weight of the members can be neglected. In cases where the weight of a member is to be included in the analysis, it is generally satisfactory to apply it as a vertical force, half of its magnitude applied at each end of the member.

6

Procedure for Analysis

Either the method of joints or the method of sections can be used to determine the forces developed in the members of a simple space truss.

Method of Joints.

If the forces in *all* the members of the truss are to be determined, then the method of joints is most suitable for the analysis. Here it is necessary to apply the three equilibrium equations $\Sigma F_x = 0$, $\Sigma F_y = 0$, $\Sigma F_z = 0$ to the forces acting at each joint. Remember that the solution of many simultaneous equations can be avoided if the force analysis begins at a joint having at least one known force and at most three unknown forces. Also, if the three-dimensional geometry of the force system at the joint is hard to visualize, it is recommended that a Cartesian vector analysis be used for the solution.

Method of Sections.

If only a *few* member forces are to be determined, the method of sections can be used. When an imaginary section is passed through a truss and the truss is separated into two parts, the force system acting on one of the segments must satisfy the *six* equilibrium equations: $\Sigma F_x = 0$, $\Sigma F_y = 0$, $\Sigma F_z = 0$, $\Sigma M_x = 0$, $\Sigma M_y = 0$, $\Sigma M_z = 0$ (Eqs. 5–6). By proper choice of the section and axes for summing forces and moments, many of the unknown member forces in a space truss can be computed *directly*, using a single equilibrium equation.

EXAMPLE 6.8

Determine the forces acting in the members of the space truss shown in Fig. 6–20*a*. Indicate whether the members are in tension or compression.

SOLUTION

Since there are one known force and three unknown forces acting at joint A, the force analysis of the truss will begin at this joint.

Joint A. (Fig. 6–20*b*). Expressing each force acting on the free-body diagram of joint A as a Cartesian vector, we have

$$\mathbf{P} = \{-4\mathbf{j}\}\text{ kN}, \qquad \mathbf{F}_{AB} = F_{AB}\mathbf{j}, \quad \mathbf{F}_{AC} = -F_{AC}\mathbf{k},$$

$$\mathbf{F}_{AE} = F_{AE}\left(\frac{\mathbf{r}_{AE}}{r_{AE}}\right) = F_{AE}(0.577\mathbf{i} + 0.577\mathbf{j} - 0.577\mathbf{k})$$

For equilibrium,

$$\Sigma\mathbf{F} = \mathbf{0}; \qquad \mathbf{P} + \mathbf{F}_{AB} + \mathbf{F}_{AC} + \mathbf{F}_{AE} = \mathbf{0}$$

$$-4\mathbf{j} + F_{AB}\mathbf{j} - F_{AC}\mathbf{k} + 0.577F_{AE}\mathbf{i} + 0.577F_{AE}\mathbf{j} - 0.577F_{AE}\mathbf{k} = \mathbf{0}$$

$$\Sigma F_x = 0; \qquad 0.577F_{AE} = 0$$

$$\Sigma F_y = 0; \qquad -4 + F_{AB} + 0.577F_{AE} = 0$$

$$\Sigma F_z = 0; \qquad -F_{AC} - 0.577F_{AE} = 0$$

$$F_{AC} = F_{AE} = 0 \qquad \textit{Ans.}$$

$$F_{AB} = 4\text{ kN (T)} \qquad \textit{Ans.}$$

Since F_{AB} is known, joint B can be analyzed next.

Joint B. (Fig. 6–20*c*).

$$\Sigma F_x = 0; \qquad -R_B \cos 45^\circ + 0.707F_{BE} = 0$$

$$\Sigma F_y = 0; \qquad -4 + R_B \sin 45^\circ = 0$$

$$\Sigma F_z = 0; \qquad 2 + F_{BD} - 0.707F_{BE} = 0$$

$$R_B = F_{BE} = 5.66\text{ kN (T)}, \qquad F_{BD} = 2\text{ kN (C)} \qquad \textit{Ans.}$$

The *scalar* equations of equilibrium may also be applied directly to the forces acting on the free-body diagrams of joints D and C since the force components are easily determined. Show that

$$F_{DE} = F_{DC} = F_{CE} = 0 \qquad \textit{Ans.}$$

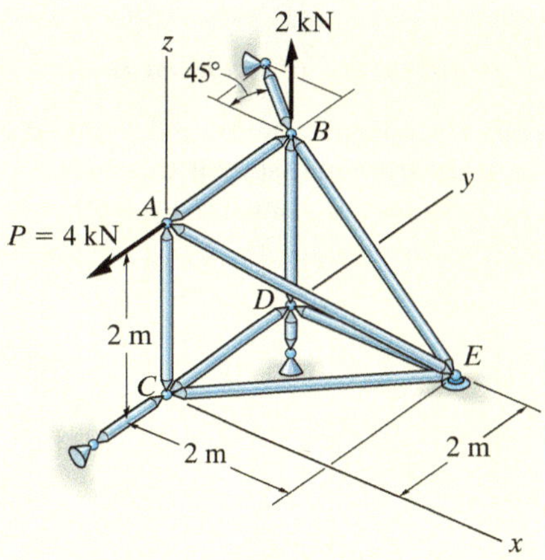

(a)

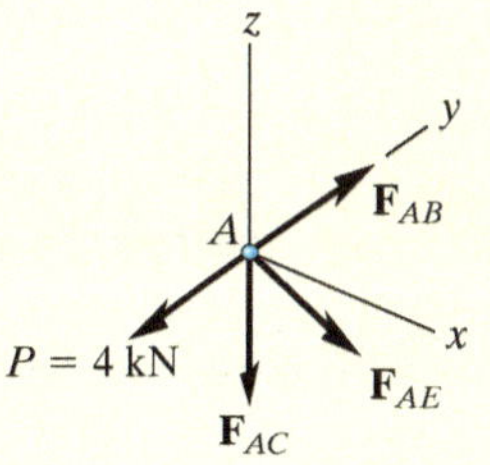

(b)

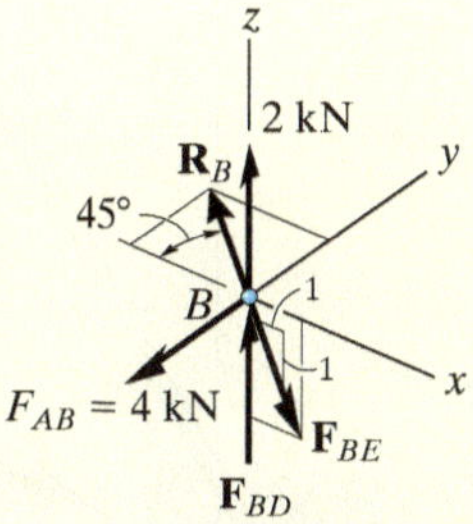

(c)

Fig. 6–20

6

PROBLEMS

All problem solutions must include FBDs.

6–50. Determine the force developed in each member of the space truss and state if the members are in tension or compression. The crate has a weight of 150 lb.

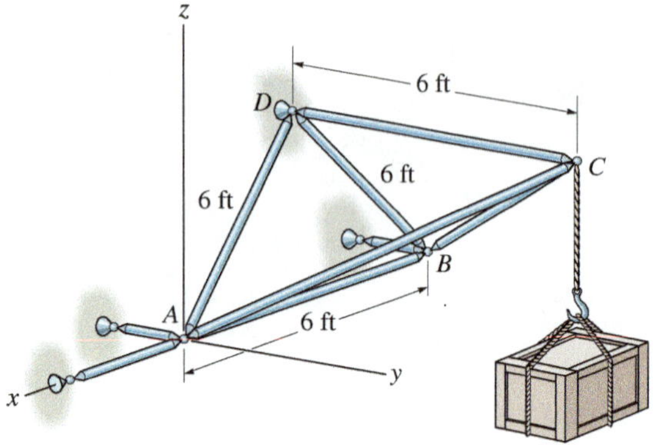

Prob. 6–50

6–51. Determine the force in each member of the space truss and state if the members are in tension or compression. *Hint:* The support reaction at E acts along member EB. Why?

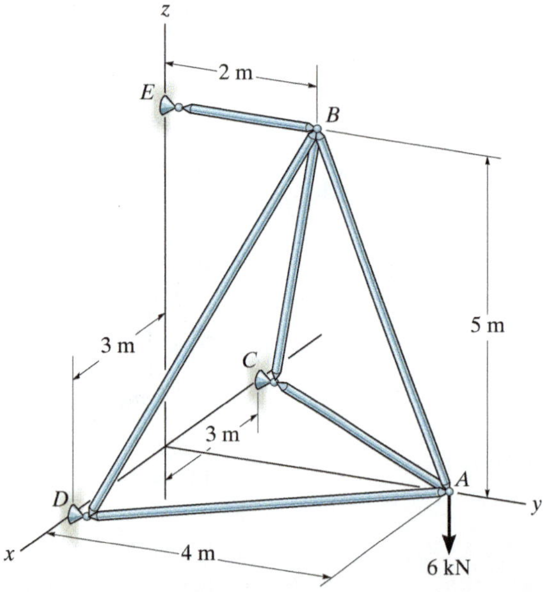

Prob. 6–51

***6–52.** Determine the force in each member of the space truss and state if the members are in tension or compression. The truss is supported by rollers at A, B, and C.

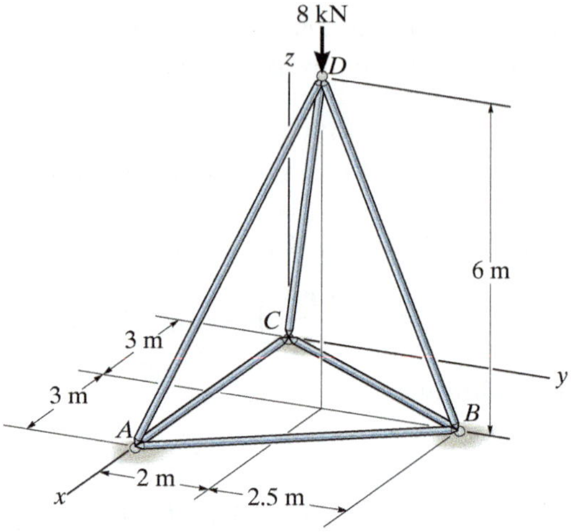

Prob. 6–52

6–53. The space truss supports a force $\mathbf{F} = [300\mathbf{i} + 400\mathbf{j} - 500\mathbf{k}]$ N. Determine the force in each member, and state if the members are in tension or compression.

6–54. The space truss supports a force $\mathbf{F} = [-400\mathbf{i} + 500\mathbf{j} + 600\mathbf{k}]$ N. Determine the force in each member, and state if the members are in tension or compression.

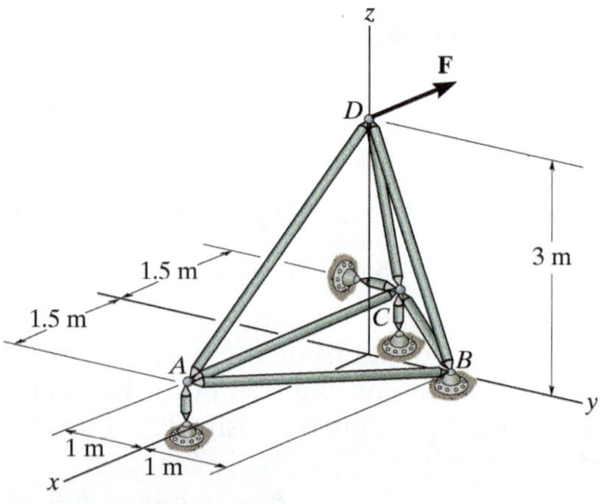

Probs. 6–53/54

6–55. Determine the force in each member of the space truss and state if the members are in tension or compression. The truss is supported by ball-and-socket joints at *C, D, E*, and *G*.

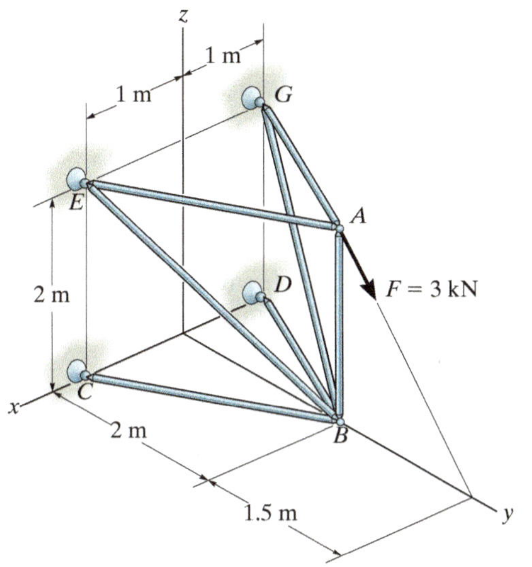

Prob. 6–55

*6–56. The space truss is used to support vertical forces at joints *B, C*, and *D*. Determine the force in each member and state if the members are in tension or compression. There is a roller at *E*, and *A* and *F* are ball-and-socket joints.

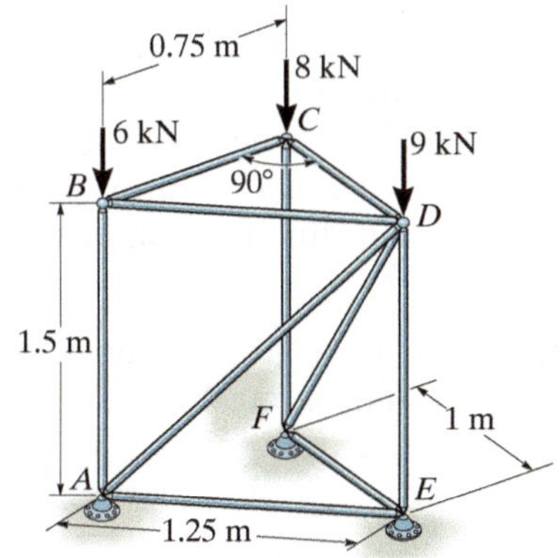

Prob. 6–56

6–57. Determine the force in members *BE, BC, BF*, and *CE* of the space truss, and state if the members are in tension or compression.

6–58. Determine the force in members *AF, AB, AD, ED, FD*, and *BD* of the space truss, and state if the members are in tension or compression.

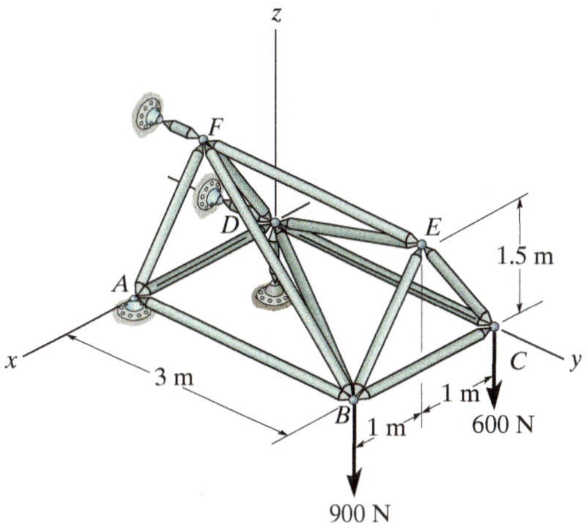

Probs. 6–57/58

6–59. The space truss is supported by a ball-and-socket joint at *D* and short links at *C* and *E*. Determine the force in each member and state if the members are in tension or compression. Take $\mathbf{F}_1 = \{-500\mathbf{k}\}$ lb and $\mathbf{F}_2 = \{400\mathbf{j}\}$ lb.

***6–60.** The space truss is supported by a ball-and-socket joint at *D* and short links at *C* and *E*. Determine the force in each member and state if the members are in tension or compression. Take $\mathbf{F}_1 = \{200\mathbf{i} + 300\mathbf{j} - 500\mathbf{k}\}$ lb and $\mathbf{F}_2 = \{400\mathbf{j}\}$ lb.

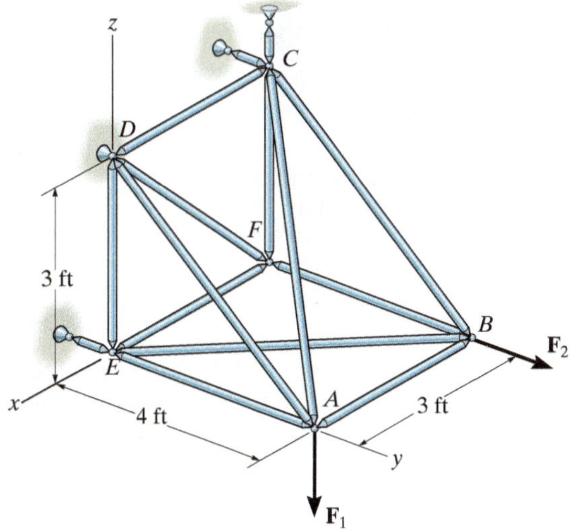

Probs. 6–59/60

6.6 Frames and Machines

This large crane is a typical example of a framework.

Common tools such as these pliers act as simple machines. Here the applied force on the handles creates a much larger force at the jaws.

Frames and machines are two types of structures which are often composed of pin-connected *multiforce members*, i.e., members that are subjected to more than two forces. *Frames* are used to support loads, whereas *machines* contain moving parts and are designed to transmit and alter the effect of forces. Provided a frame or machine contains no more supports or members than are necessary to prevent its collapse, the forces acting at the joints and supports can be determined by applying the equations of equilibrium to each of its members. Once these forces are obtained, it is then possible to *design* the size of the members, connections, and supports using the theory of mechanics of materials and an appropriate engineering design code.

Free-Body Diagrams. In order to determine the forces acting at the joints and supports of a frame or machine, the structure must be disassembled and the free-body diagrams of its parts must be drawn. The following important points *must* be observed:

- Isolate each part by drawing its *outlined shape*. Then show all the forces and/or couple moments that act on the part. Make sure to *label* or *identify* each known and unknown force and couple moment with reference to an established x, y coordinate system. Also, indicate any dimensions used for taking moments. Most often the equations of equilibrium are easier to apply if the forces are represented by their rectangular components. As usual, the sense of an unknown force or couple moment can be assumed.
- Identify all the two-force members in the structure and represent their free-body diagrams as having two equal but opposite collinear forces acting at their points of application. (See Sec. 5.4.) By recognizing the two-force members, we can avoid solving an unnecessary number of equilibrium equations.
- Forces common to *any* two *contacting* members act with equal magnitudes but opposite sense on the respective members. If the two members are treated as a *"system" of connected members*, then these forces are *"internal"* and are *not shown* on the *free-body diagram of the system*; however, if the free-body diagram of *each member* is drawn, the forces are *"external"* and *must* be shown as equal in magnitude and opposite in direction on each of the two free-body diagrams.

The following examples graphically illustrate how to draw the free-body diagrams of a dismembered frame or machine. In all cases, the weight of the members is neglected.

EXAMPLE 6.9

For the frame shown in Fig. 6–21*a*, draw the free-body diagram of (a) each member, (b) the pins at *B* and *A*, and (c) the two members connected together.

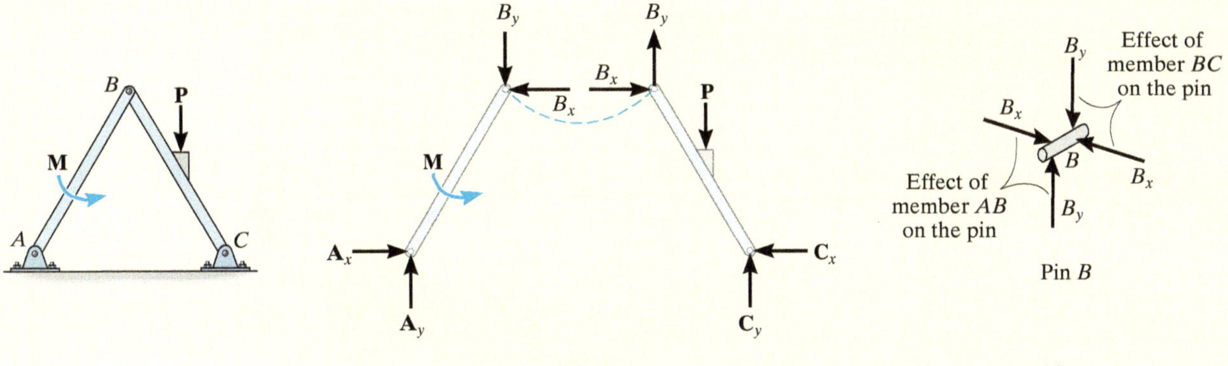

(a) (b) (c)

SOLUTION

Part (a). By inspection, members *BA* and *BC* are *not* two-force members. Instead, as shown on the free-body diagrams, Fig. 6–21*b*, *BC* is subjected to a force from each of the pins at *B* and *C* and the external force **P**. Likewise, *AB* is subjected to a force from each of the pins at *A* and *B* and the external couple moment **M**. The pin forces are represented by their *x* and *y* components.

Part (b). The pin at *B* is subjected to only *two forces*, i.e., the force of member *BC* and the force of member *AB*. For *equilibrium* these forces (or their respective components) must be equal but opposite, Fig. 6–21*c*. Realize that Newton's third law is applied between the pin and its connected members, i.e., the effect of the pin on the two members, Fig. 6–21*b*, and the equal but opposite effect of the two members on the pin, Fig. 6–21*c*. In the same manner, there are three forces on pin *A*, Fig. 6–21*d*, caused by the force components of member *AB* and each of the two pin leafs.

Part (c). The free-body diagram of both members connected together, yet removed from the supporting pins at *A* and *C*, is shown in Fig. 6–21*e*. The force components $\mathbf{B}_x$ and $\mathbf{B}_y$ are *not shown* on this diagram since they are *internal* forces (Fig. 6–21*b*) and therefore cancel out. Also, to be consistent when later applying the equilibrium equations, the unknown force components at *A* and *C* must act in the *same sense* as those shown in Fig. 6–21*b*.

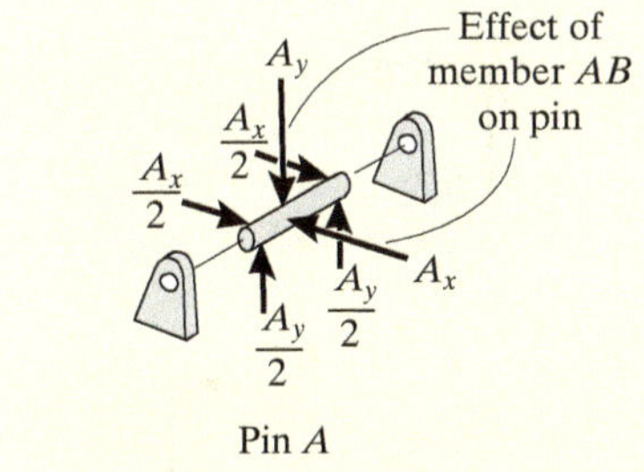

(d)

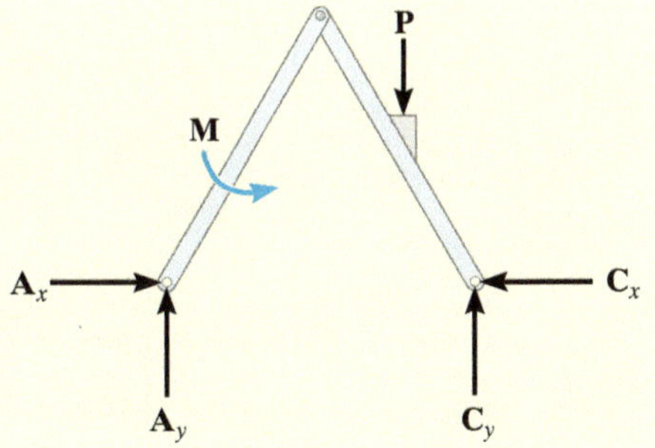

(e)

Fig. 6–21

EXAMPLE 6.10

A constant tension in the conveyor belt is maintained by using the device shown in Fig. 6–22*a*. Draw the free-body diagrams of the frame and the cylinder (or pulley) that the belt surrounds. The suspended block has a weight of *W*.

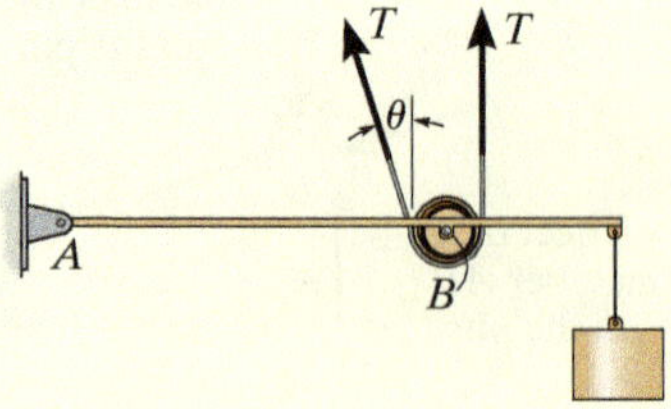

(b)

(a)

Fig. 6–22

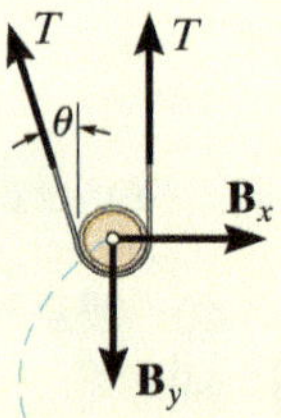

(c)

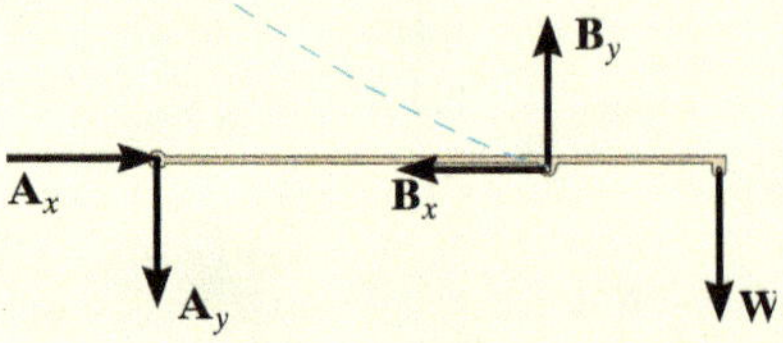

(d)

SOLUTION

The idealized model of the device is shown in Fig. 6–22*b*. Here the angle θ is assumed to be known. From this model, the free-body diagrams of the pulley and frame are shown in Figs. 6–22*c* and 6–22*d*, respectively. Note that the force components $\mathbf{B}_x$ and $\mathbf{B}_y$ that the pin at *B* exerts on the pulley must be equal but opposite to the ones acting on the frame. See Fig. 6–21*c* of Example 6.9.

EXAMPLE 6.11

For the frame shown in Fig. 6–23*a*, draw the free-body diagrams of (a) the entire frame including the pulleys and cords, (b) the frame without the pulleys and cords, and (c) each of the pulleys.

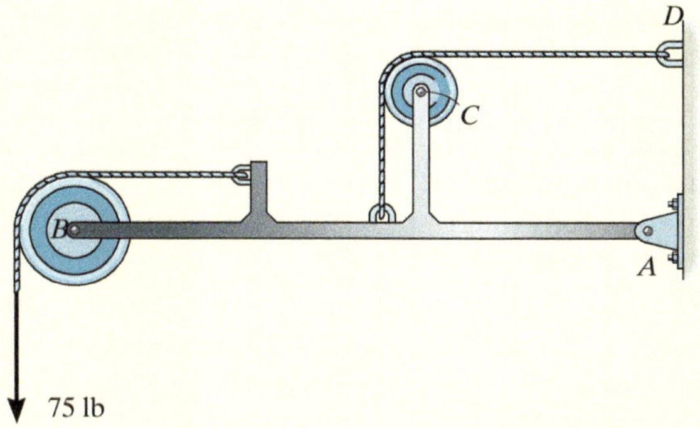

(a)

SOLUTION

Part (a). When the entire frame including the pulleys and cords is considered, the interactions at the points where the pulleys and cords are connected to the frame become pairs of *internal* forces which cancel each other and therefore are not shown on the free-body diagram, Fig. 6–23*b*.

Part (b). When the cords and pulleys are removed, their effect *on the frame* must be shown, Fig. 6–23*c*.

Part (c). The force components $\mathbf{B}_x$, $\mathbf{B}_y$, $\mathbf{C}_x$, $\mathbf{C}_y$ of the pins on the pulleys, Fig. 6–23*d*, are equal but opposite to the force components exerted by the pins on the frame, Fig. 6–23*c*. See Example 6.9.

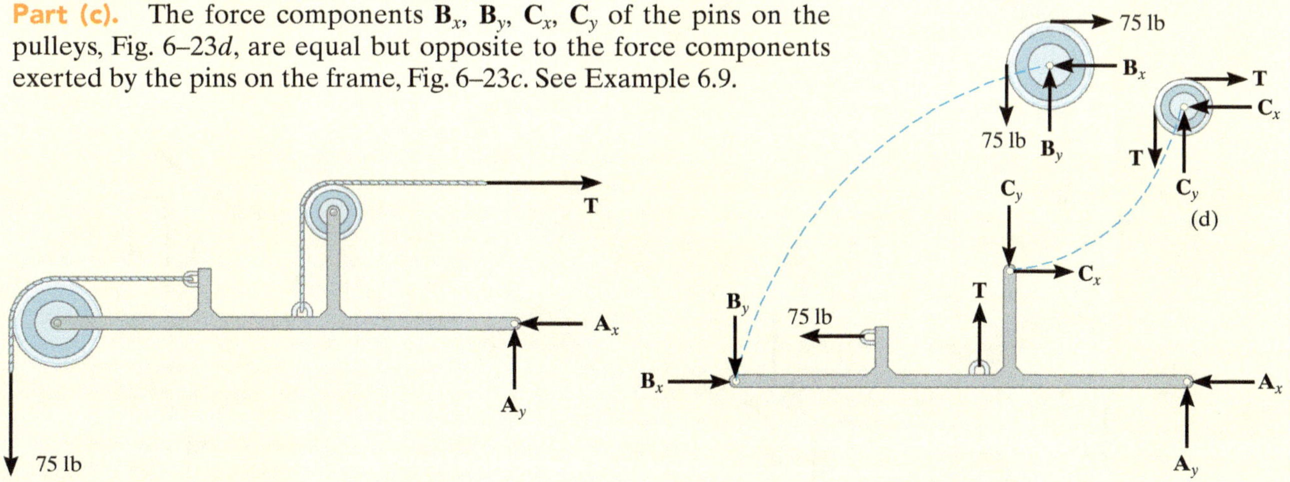

(b) (c)

Fig. 6–23

EXAMPLE 6.12

Draw the free-body diagrams of the members of the backhoe, shown in the photo, Fig. 6–24*a*. The bucket and its contents have a weight W.

(a)

Fig. 6–24

SOLUTION

The idealized model of the assembly is shown in Fig. 6–24*b*. By inspection, members AB, BC, BE, and HI are all two-force members since they are pin connected at their end points and no other forces act on them. The free-body diagrams of the bucket and the stick are shown in Fig. 6–24*c*. Note that pin C is subjected to only two forces, whereas the pin at B is subjected to three forces, Fig. 6–24*d*. The free-body diagram of the entire assembly is shown in Fig. 6–24*e*.

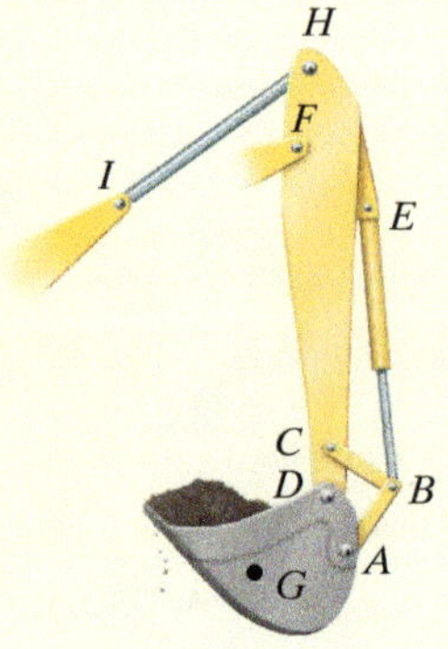

(b)

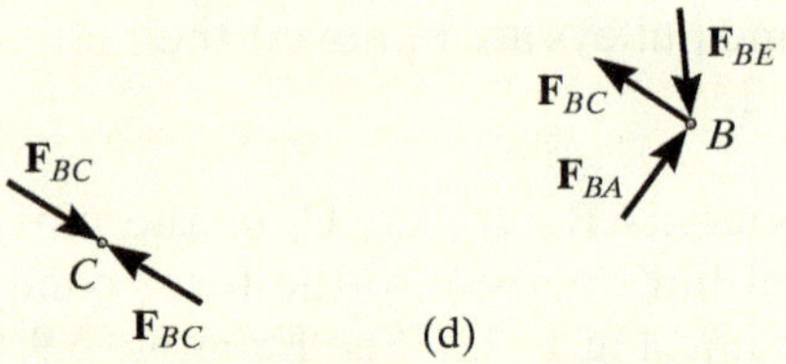

(d)

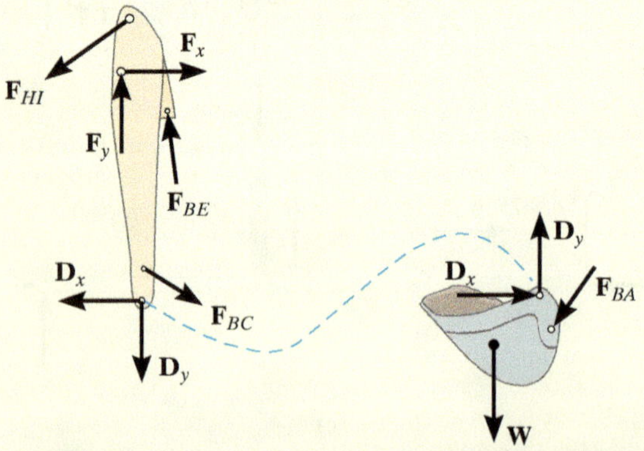

(c)

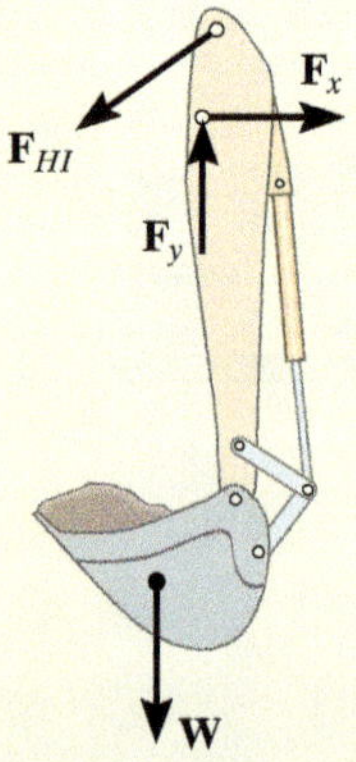

(e)

6

EXAMPLE 6.13

Draw the free-body diagram of each part of the smooth piston and link mechanism used to crush recycled cans, Fig. 6–25*a*.

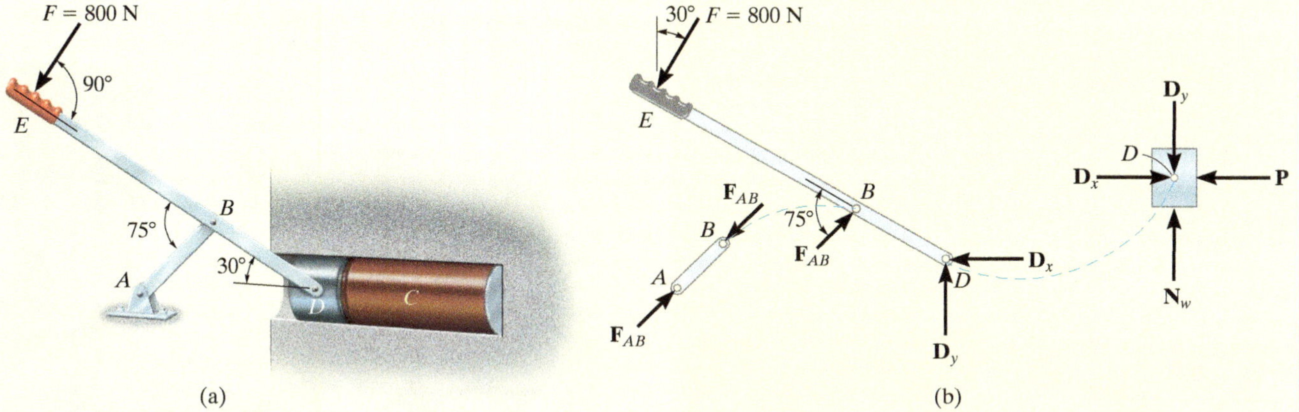

Fig. 6–25

SOLUTION

By inspection, member AB is a two-force member. The free-body diagrams of the three parts are shown in Fig. 6–25*b*. Since the pins at B and D *connect only two parts together*, the forces there are shown as equal but opposite on the separate free-body diagrams of their connected members. In particular, four components of force act on the piston: $\mathbf{D}_x$ and $\mathbf{D}_y$ represent the effect of the pin (or lever EBD), $\mathbf{N}_w$ is the *resultant force* of the wall support, and $\mathbf{P}$ is the resultant compressive force caused by the can C. The directional sense of each of the unknown forces is assumed, and the correct sense will be established after the equations of equilibrium are applied.

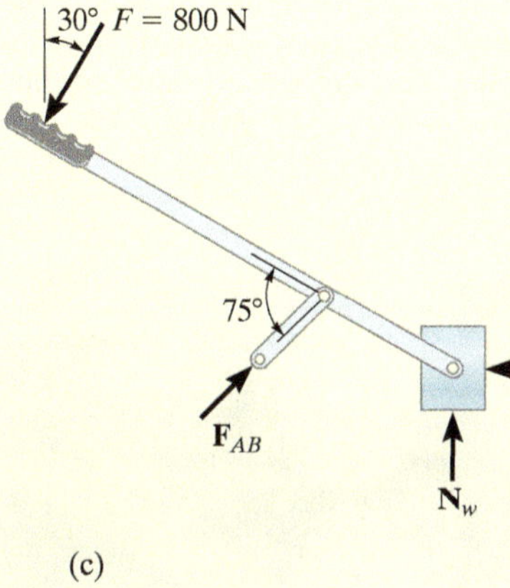

NOTE: A free-body diagram of the entire assembly is shown in Fig. 6–25*c*. Here the forces between the components are internal and are not shown on the free-body diagram.

Before proceeding, it is highly recommended that you cover the solutions to the previous examples and attempt to draw the requested free-body diagrams. When doing so, make sure the work is neat and that all the forces and couple moments are properly labeled. When finished, challenge yourself and solve the following four problems.

CONCEPTUAL PROBLEMS

P6–1. Draw the free-body diagrams of each of the crane boom segments *AB*, *BC*, and *BD*. Only the weights of *AB* and *BC* are significant. Assume *A* and *B* are pins.

P6–1

6

P6–2. Draw the free-body diagrams of the boom *ABCD* and the stick *EDFGH* of the backhoe. The weights of these two members are significant. Neglect the weights of all the other members, and assume all indicated points of connection are pins.

P6–2

P6–3. Draw the free-body diagrams of the boom *ABCDF* and the stick *FGH* of the bucket lift. Neglect the weights of the members. The bucket weighs *W*. The two–force members are *BI*, *CE*, *DE* and *GE*. Assume all indicated points of connection are pins.

P6–3

P6–4. To operate the can crusher one pushes down on the lever arm *ABC* which rotates about the fixed pin at *B*. This moves the side links *CD* downward, which causes the guide plate *E* to also move downward and thereby crush the can. Draw the free-body diagrams of the lever, side link, and guide plate. Make up some reasonable numbers and do an equilibrium analysis to show how much an applied vertical force at the handle is magnified when it is transmitted to the can. Assume all points of connection are pins and the guides for the plate are smooth.

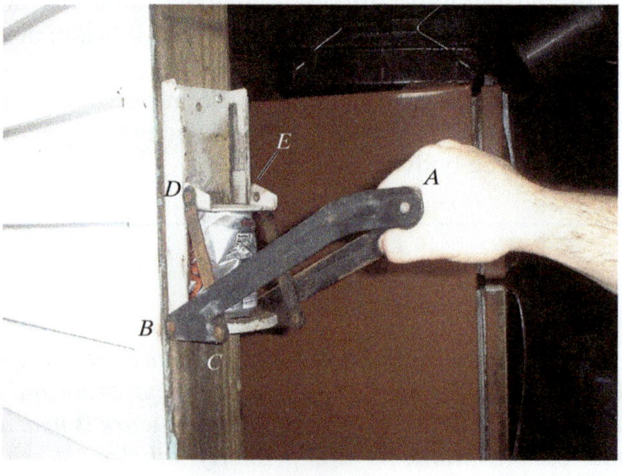

P6–4

Procedure for Analysis

The joint reactions on frames or machines (structures) composed of multiforce members can be determined using the following procedure.

Free-Body Diagram.

- Draw the free-body diagram of the entire frame or machine, a portion of it, or each of its members. The choice should be made so that it leads to the most direct solution of the problem.
- When the free-body diagram of a group of members of a frame or machine is drawn, the forces between the connected parts of this group are internal forces and are not shown on the free-body diagram of the group.
- Forces common to two members which are in contact act with equal magnitude but opposite sense on the respective free-body diagrams of the members.
- Two-force members, regardless of their shape, have equal but opposite collinear forces acting at the ends of the member.
- In many cases it is possible to tell by inspection the proper sense of the unknown forces acting on a member; however, if this seems difficult, the sense can be assumed.
- Remember that a couple moment is a free vector and can act at any point on the free-body diagram. Also, a force is a sliding vector and can act at any point along its line of action.

Equations of Equilibrium.

- Count the number of unknowns and compare it to the total number of equilibrium equations that are available. In two dimensions, there are three equilibrium equations that can be written for each member.
- Sum moments about a point that lies at the intersection of the lines of action of as many of the unknown forces as possible.
- If the solution of a force or couple moment magnitude is found to be negative, it means the sense of the force is the reverse of that shown on the free-body diagram.

EXAMPLE 6.14

Determine the tension in the cables and also the force **P** required to support the 600-N force using the frictionless pulley system shown in Fig. 6–26*a*.

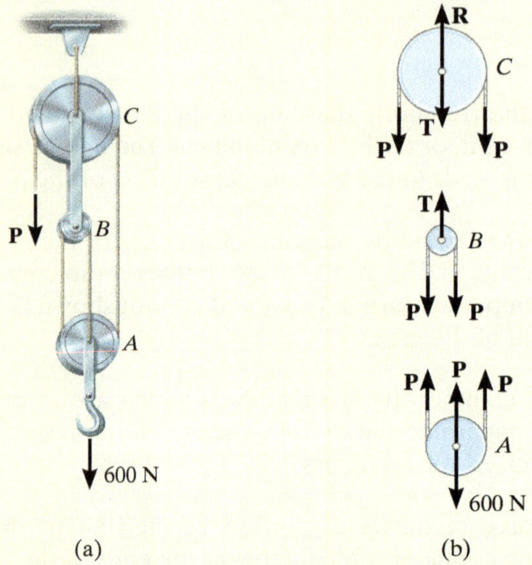

Fig. 6–26

SOLUTION

Free-Body Diagram. A free-body diagram of each pulley *including* its pin and a portion of the contacting cable is shown in Fig. 6–26*b*. Since the cable is *continuous*, it has a *constant tension* P acting throughout its length. The link connection between pulleys B and C is a two-force member, and therefore it has an unknown tension T acting on it. Notice that the *principle of action, equal but opposite reaction* must be carefully observed for forces **P** and **T** when the *separate* free-body diagrams are drawn.

Equations of Equilibrium. The three unknowns are obtained as follows:

Pulley A

$+\uparrow \Sigma F_y = 0;$ $3P - 600\text{ N} = 0$ $P = 200\text{ N}$ *Ans.*

Pulley B

$+\uparrow \Sigma F_y = 0;$ $T - 2P = 0$ $T = 400\text{ N}$ *Ans.*

Pulley C

$+\uparrow \Sigma F_y = 0;$ $R - 2P - T = 0$ $R = 800\text{ N}$ *Ans.*

EXAMPLE 6.15

A 500-kg elevator car in Fig. 6–27*a* is being hoisted by motor *A* using the pulley system shown. If the car is traveling with a constant speed, determine the force developed in the two cables. Neglect the mass of the cable and pulleys.

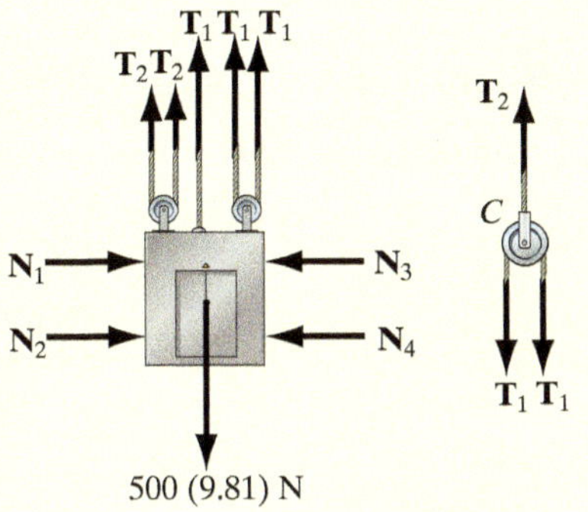

(b)

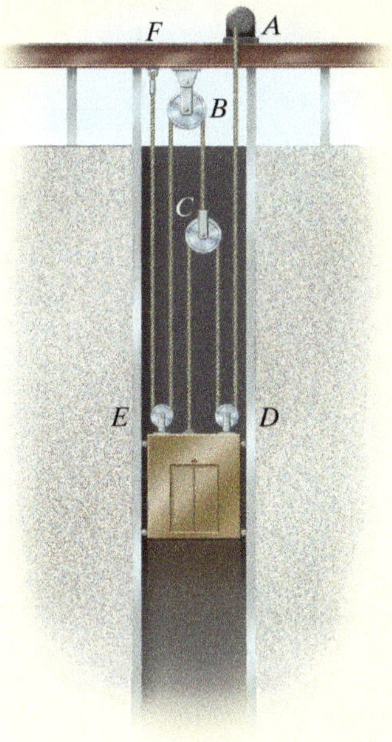

(a)

Fig. 6–27

SOLUTION

Free-Body Diagram. We can solve this problem using the free-body diagrams of the elevator car and pulley *C*, Fig. 6–27*b*. The tensile forces developed in the cables are denoted as T_1 and T_2.

Equations of Equilibrium. For pulley *C*,

$$+\uparrow \Sigma F_y = 0; \qquad T_2 - 2T_1 = 0 \quad \text{or} \quad T_2 = 2T_1 \qquad (1)$$

For the elevator car,

$$+\uparrow \Sigma F_y = 0; \qquad 3T_1 + 2T_2 - 500(9.81)\text{ N} = 0 \qquad (2)$$

Substituting Eq. (1) into Eq. (2) yields

$$3T_1 + 2(2T_1) - 500(9.81)\text{ N} = 0$$

$$T_1 = 700.71\text{ N} = 701\text{ N} \qquad \textit{Ans.}$$

Substituting this result into Eq. (1),

$$T_2 = 2(700.71)\text{ N} = 1401\text{ N} = 1.40\text{ kN} \qquad \textit{Ans.}$$

EXAMPLE 6.16

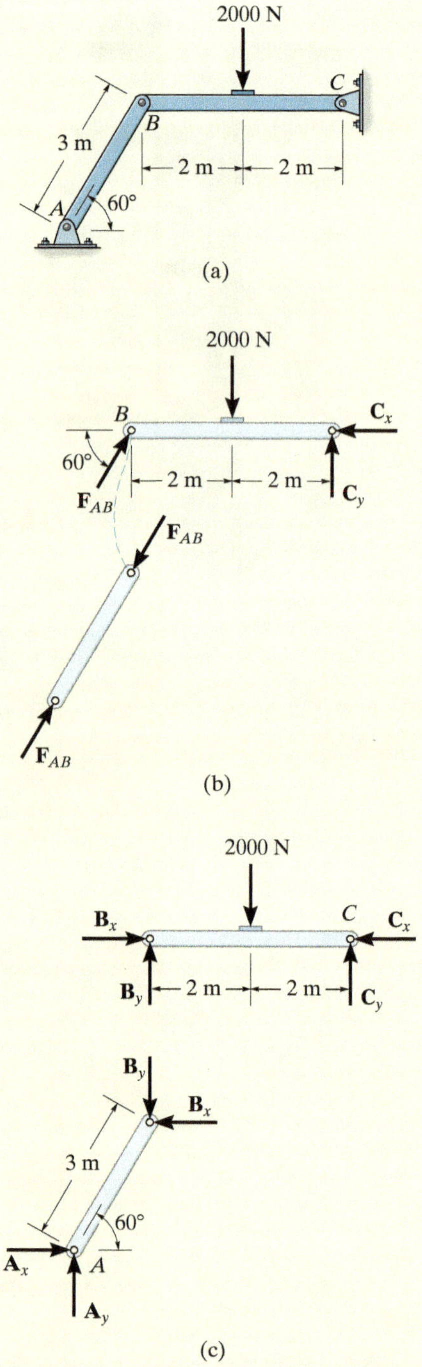

Fig. 6–28

Determine the horizontal and vertical components of force which the pin at C exerts on member BC of the frame in Fig. 6–28*a*.

SOLUTION I

Free-Body Diagrams. By inspection it can be seen that AB is a two-force member. The free-body diagrams are shown in Fig. 6–28*b*.

Equations of Equilibrium. The *three unknowns* can be determined by applying the three equations of equilibrium to member CB.

$$\circlearrowleft + \Sigma M_C = 0; \quad 2000\text{ N}(2\text{ m}) - (F_{AB}\sin 60°)(4\text{ m}) = 0 \quad F_{AB} = 1154.7\text{ N}$$

$$\xrightarrow{+} \Sigma F_x = 0; \quad 1154.7\cos 60°\text{ N} - C_x = 0 \quad C_x = 577\text{ N} \qquad \textit{Ans.}$$

$$+\uparrow \Sigma F_y = 0; \quad 1154.7\sin 60°\text{ N} - 2000\text{ N} + C_y = 0$$

$$C_y = 1000\text{ N} \qquad \textit{Ans.}$$

SOLUTION II

Free-Body Diagrams. If one does not recognize that AB is a two-force member, then more work is involved in solving this problem. The free-body diagrams are shown in Fig. 6–28*c*.

Equations of Equilibrium. The *six unknowns* are determined by applying the three equations of equilibrium to each member.

Member AB

$$\circlearrowleft + \Sigma M_A = 0; \quad B_x(3\sin 60°\text{ m}) - B_y(3\cos 60°\text{ m}) = 0 \qquad (1)$$

$$\xrightarrow{+} \Sigma F_x = 0; \quad A_x - B_x = 0 \qquad (2)$$

$$+\uparrow \Sigma F_y = 0; \quad A_y - B_y = 0 \qquad (3)$$

Member BC

$$\circlearrowleft + \Sigma M_C = 0; \quad 2000\text{ N}(2\text{ m}) - B_y(4\text{ m}) = 0 \qquad (4)$$

$$\xrightarrow{+} \Sigma F_x = 0; \quad B_x - C_x = 0 \qquad (5)$$

$$+\uparrow \Sigma F_y = 0; \quad B_y - 2000\text{ N} + C_y = 0 \qquad (6)$$

The results for C_x and C_y can be determined by solving these equations in the following sequence: 4, 1, 5, then 6. The results are

$$B_y = 1000\text{ N}$$

$$B_x = 577\text{ N}$$

$$C_x = 577\text{ N} \qquad \textit{Ans.}$$

$$C_y = 1000\text{ N} \qquad \textit{Ans.}$$

By comparison, Solution I is simpler since the requirement that F_{AB} in Fig. 6–28*b* be equal, opposite, and collinear at the ends of member AB automatically satisfies Eqs. 1, 2, and 3 above and therefore eliminates the need to write these equations. ***As a result, save yourself some time and effort by always identifying the two-force members before starting the analysis!***

6

EXAMPLE 6.17

The compound beam shown in Fig. 6–29*a* is pin connected at *B*. Determine the components of reaction at its supports. Neglect its weight and thickness.

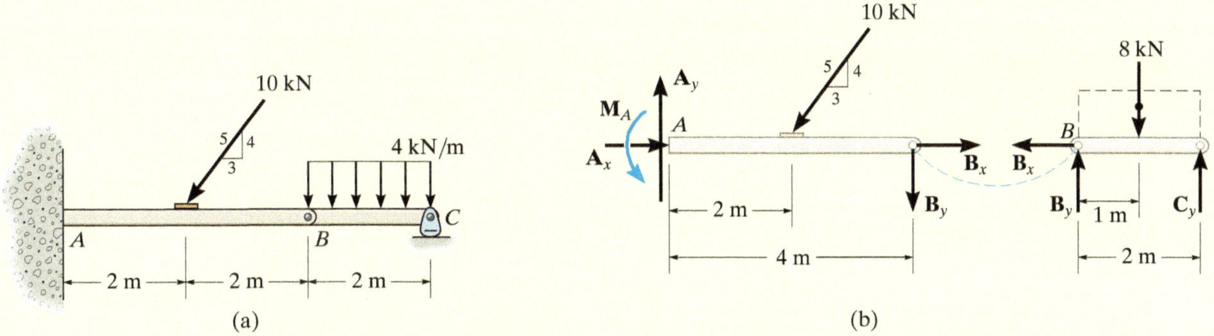

Fig. 6–29

SOLUTION

Free-Body Diagrams. By inspection, if we consider a free-body diagram of the *entire beam ABC*, there will be three unknown reactions at *A* and one at *C*. These four unknowns cannot all be obtained from the three available equations of equilibrium, and so for the solution it will become necessary to dismember the beam into its two segments, as shown in Fig. 6–29*b*.

Equations of Equilibrium. The six unknowns are determined as follows:

Segment BC

$\xleftarrow{+} \Sigma F_x = 0; \qquad B_x = 0$

$\circlearrowleft + \Sigma M_B = 0; \qquad -8\text{ kN}(1\text{ m}) + C_y(2\text{ m}) = 0$

$+\uparrow \Sigma F_y = 0; \qquad B_y - 8\text{ kN} + C_y = 0$

Segment AB

$\xrightarrow{+} \Sigma F_x = 0; \qquad A_x - (10\text{ kN})\left(\frac{3}{5}\right) + B_x = 0$

$\circlearrowleft + \Sigma M_A = 0; \qquad M_A - (10\text{ kN})\left(\frac{4}{5}\right)(2\text{ m}) - B_y(4\text{ m}) = 0$

$+\uparrow \Sigma F_y = 0; \qquad A_y - (10\text{ kN})\left(\frac{4}{5}\right) - B_y = 0$

Solving each of these equations successively, using previously calculated results, we obtain

$A_x = 6\text{ kN} \qquad A_y = 12\text{ kN} \qquad M_A = 32\text{ kN}\cdot\text{m}$ *Ans.*

$B_x = 0 \qquad B_y = 4\text{ kN}$

$C_y = 4\text{ kN}$ *Ans.*

EXAMPLE 6.18

The two planks in Fig. 6–30*a* are connected together by cable *BC* and a smooth spacer *DE*. Determine the reactions at the smooth supports *A* and *F*, and also find the force developed in the cable and spacer.

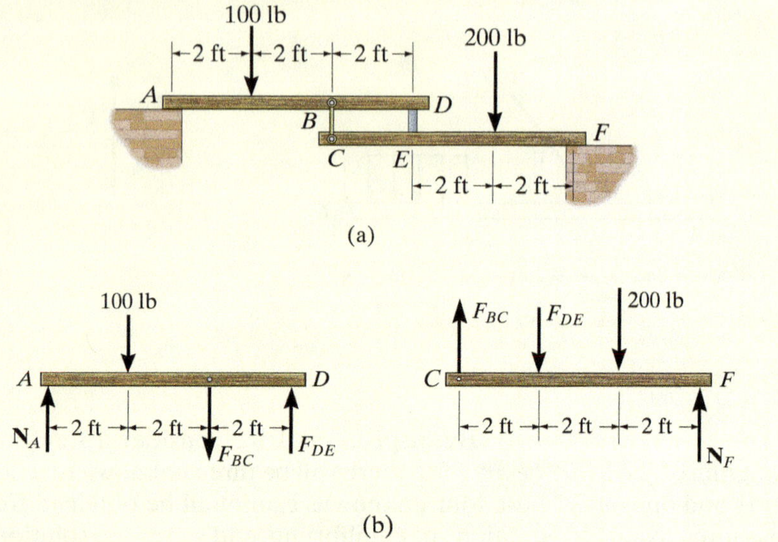

Fig. 6–30

SOLUTION

Free-Body Diagrams. The free-body diagram of each plank is shown in Fig. 6–30*b*. It is important to apply Newton's third law to the interaction forces F_{BC} and F_{DE} as shown.

Equations of Equilibrium. For plank *AD*,

$$\zeta + \Sigma M_A = 0; \qquad F_{DE}(6\text{ ft}) - F_{BC}(4\text{ ft}) - 100\text{ lb}\,(2\text{ ft}) = 0$$

For plank *CF*,

$$\zeta + \Sigma M_F = 0; \qquad F_{DE}(4\text{ ft}) - F_{BC}(6\text{ ft}) + 200\text{ lb}\,(2\text{ ft}) = 0$$

Solving simultaneously,

$$F_{DE} = 140\text{ lb} \quad F_{BC} = 160\text{ lb} \qquad \textit{Ans.}$$

Using these results, for plank *AD*,

$$+\uparrow \Sigma F_y = 0; \qquad N_A + 140\text{ lb} - 160\text{ lb} - 100\text{ lb} = 0$$

$$N_A = 120\text{ lb} \qquad \textit{Ans.}$$

And for plank *CF*,

$$+\uparrow \Sigma F_y = 0; \qquad N_F + 160\text{ lb} - 140\text{ lb} - 200\text{ lb} = 0$$

$$N_F = 180\text{ lb} \qquad \textit{Ans.}$$

NOTE: Draw the free-body diagram of the system of *both* planks and apply $\Sigma M_A = 0$ to determine N_F. Then use the free-body diagram of *CEF* to determine F_{DE} and F_{BC}.

EXAMPLE 6.19

The 75-kg man in Fig. 6–31*a* attempts to lift the 40-kg uniform beam off the roller support at B. Determine the tension developed in the cable attached to B and the normal reaction of the man on the beam when this is about to occur.

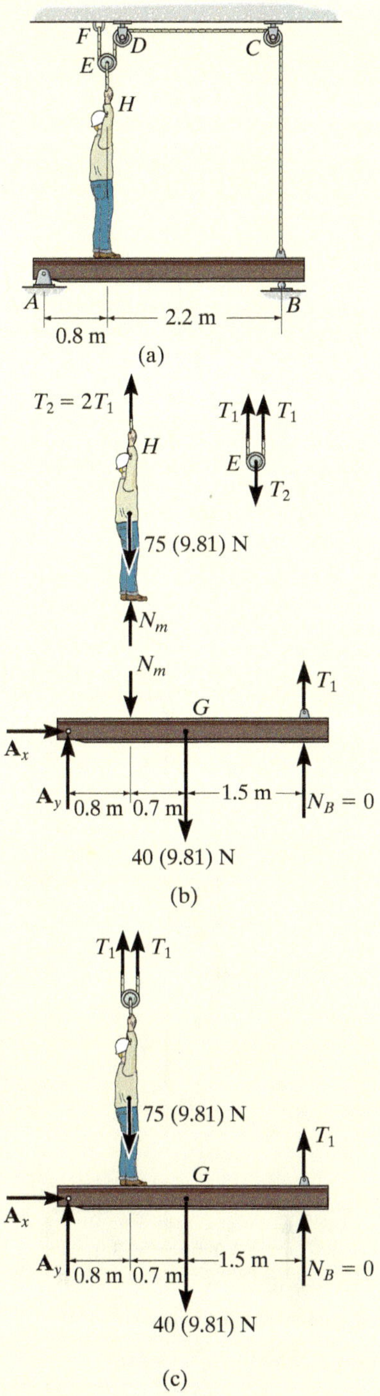

Fig. 6–31

SOLUTION

Free-Body Diagrams. The tensile force in the cable will be denoted as T_1. The free-body diagrams of the pulley E, the man, and the beam are shown in Fig. 6–31*b*. Since the man must lift the beam off the roller B then $N_B = 0$. When drawing each of these diagrams, it is very important to apply Newton's third law.

Equations of Equilibrium. Using the free-body diagram of pulley E,

$$+\uparrow \Sigma F_y = 0; \qquad 2T_1 - T_2 = 0 \quad \text{or} \quad T_2 = 2T_1 \qquad (1)$$

Referring to the free-body diagram of the man using this result,

$$+\uparrow \Sigma F_y = 0 \qquad N_m + 2T_1 - 75(9.81)\text{ N} = 0 \qquad (2)$$

Summing moments about point A on the beam,

$$\circlearrowleft + \Sigma M_A = 0; \; T_1(3\text{ m}) - N_m(0.8\text{ m}) - [40(9.81)\text{ N}]\,(1.5\text{ m}) = 0 \qquad (3)$$

Solving Eqs. 2 and 3 simultaneously for T_1 and N_m, then using Eq. (1) for T_2, we obtain

$$T_1 = 256\text{ N} \qquad N_m = 224\text{ N} \qquad T_2 = 512\text{ N}$$ *Ans.*

SOLUTION II

A direct solution for T_1 can be obtained by considering the beam, the man, and pulley E as a *single system*. The free-body diagram is shown in Fig. 6–31*c*. Thus,

$$\circlearrowleft + \Sigma M_A = 0; \qquad 2T_1(0.8\text{ m}) - [75(9.81)\text{ N}](0.8\text{ m})$$

$$- [40(9.81)\text{ N}](1.5\text{ m}) + T_1(3\text{ m}) = 0$$

$$T_1 = 256\text{ N}$$ *Ans.*

With this result Eqs. 1 and 2 can then be used to find N_m and T_2.

EXAMPLE 6.20

The smooth disk shown in Fig. 6–32*a* is pinned at *D* and has a weight of 20 lb. Neglecting the weights of the other members, determine the horizontal and vertical components of reaction at pins *B* and *D*.

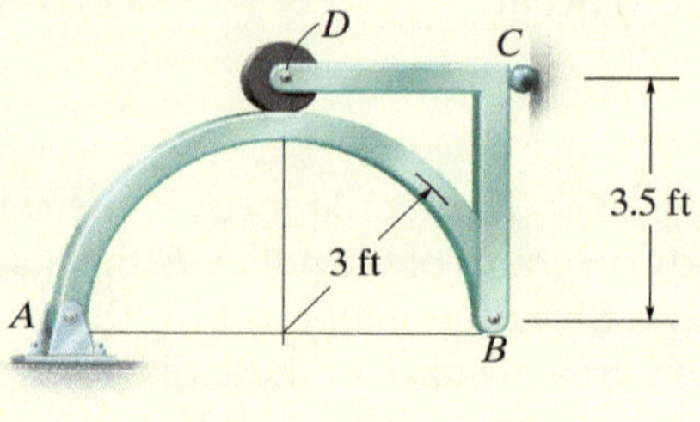

(a)

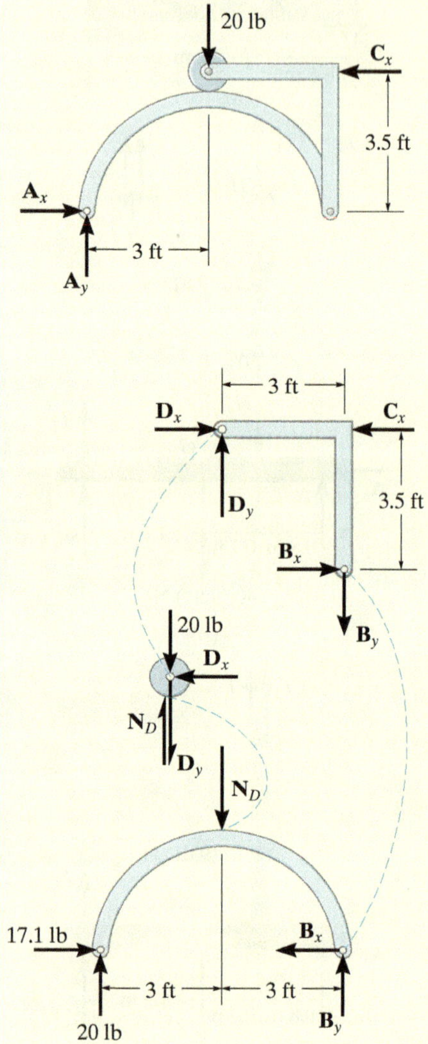

(b)

Fig. 6–32

SOLUTION

Free-Body Diagrams. The free-body diagrams of the entire frame and each of its members are shown in Fig. 6–32*b*.

Equations of Equilibrium. The eight unknowns can of course be obtained by applying the eight equilibrium equations to each member—three to member *AB*, three to member *BCD*, and two to the disk. (Moment equilibrium is automatically satisfied for the disk.) If this is done, however, all the results can be obtained only from a simultaneous solution of some of the equations. (Try it and find out.) To avoid this situation, it is best first to determine the three support reactions on the *entire* frame; then, using these results, the remaining five equilibrium equations can be applied to two other parts in order to solve successively for the other unknowns.

Entire Frame

$$\zeta+\Sigma M_A = 0; \quad -20\text{ lb}(3\text{ ft}) + C_x(3.5\text{ ft}) = 0 \qquad C_x = 17.1\text{ lb}$$

$$\overset{+}{\rightarrow}\Sigma F_x = 0; \quad A_x - 17.1\text{ lb} = 0 \qquad A_x = 17.1\text{ lb}$$

$$+\uparrow\Sigma F_y = 0; \quad A_y - 20\text{ lb} = 0 \qquad A_y = 20\text{ lb}$$

Member AB

$$\overset{+}{\rightarrow}\Sigma F_x = 0; \quad 17.1\text{ lb} - B_x = 0 \qquad B_x = 17.1\text{ lb} \qquad \textit{Ans.}$$

$$\zeta+\Sigma M_B = 0; \quad -20\text{ lb}(6\text{ ft}) + N_D(3\text{ ft}) = 0 \qquad N_D = 40\text{ lb}$$

$$+\uparrow\Sigma F_y = 0; \quad 20\text{ lb} - 40\text{ lb} + B_y = 0 \qquad B_y = 20\text{ lb} \qquad \textit{Ans.}$$

Disk

$$\overset{+}{\rightarrow}\Sigma F_x = 0; \quad D_x = 0 \qquad \textit{Ans.}$$

$$+\uparrow\Sigma F_y = 0; \quad 40\text{ lb} - 20\text{ lb} - D_y = 0 \qquad D_y = 20\text{ lb} \qquad \textit{Ans.}$$

EXAMPLE 6.21

The frame in Fig. 6–33*a* supports the 50-kg cylinder. Determine the horizontal and vertical components of reaction at *A* and the force at *C*.

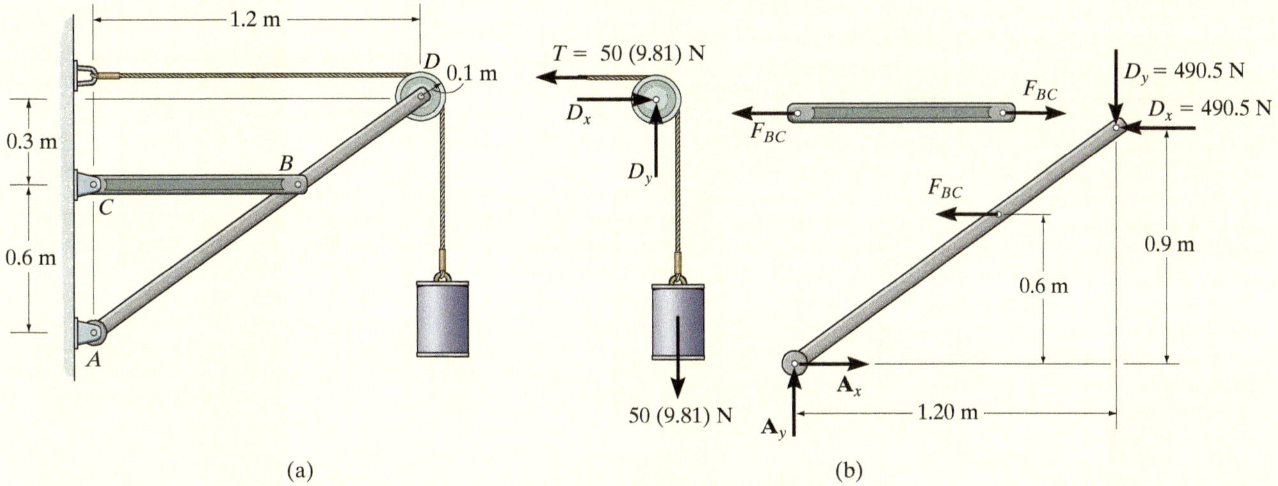

Fig. 6–33

SOLUTION

Free-Body Diagrams. The free-body diagram of pulley *D*, along with the cylinder and a portion of the cord (a system), is shown in Fig. 6–33*b*. Member *BC* is a two-force member as indicated by its free-body diagram. The free-body diagram of member *ABD* is also shown.

Equations of Equilibrium. We will begin by analyzing the equilibrium of the pulley. The moment equation of equilibrium is automatically satisfied with $T = 50(9.81)$ N, and so

$$\xrightarrow{+} \Sigma F_x = 0; \quad D_x - 50(9.81)\text{ N} = 0 \quad D_x = 490.5\text{ N}$$

$$+\uparrow \Sigma F_y = 0; \quad D_y - 50(9.81)\text{ N} = 0 \quad D_y = 490.5\text{ N} \qquad \textit{Ans.}$$

Using these results, F_{BC} can be determined by summing moments about point *A* on member *ABD*.

$$\circlearrowleft +\Sigma M_A = 0; \; F_{BC}(0.6\text{ m}) + 490.5\text{ N}(0.9\text{ m}) - 490.5\text{ N}(1.20\text{ m}) = 0$$

$$F_{BC} = 245.25\text{ N} \qquad \textit{Ans.}$$

Now A_x and A_y can be determined by summing forces.

$$\xrightarrow{+} \Sigma F_x = 0; \quad A_x - 245.25\text{ N} - 490.5\text{ N} = 0 \quad A_x = 736\text{ N} \qquad \textit{Ans.}$$

$$+\uparrow \Sigma F_y = 0; \quad A_y - 490.5\text{ N} = 0 \quad A_y = 490.5\text{ N} \qquad \textit{Ans.}$$

6

EXAMPLE 6.22

Determine the force the pins at A and B exert on the two-member frame shown in Fig. 6–34*a*.

SOLUTION I

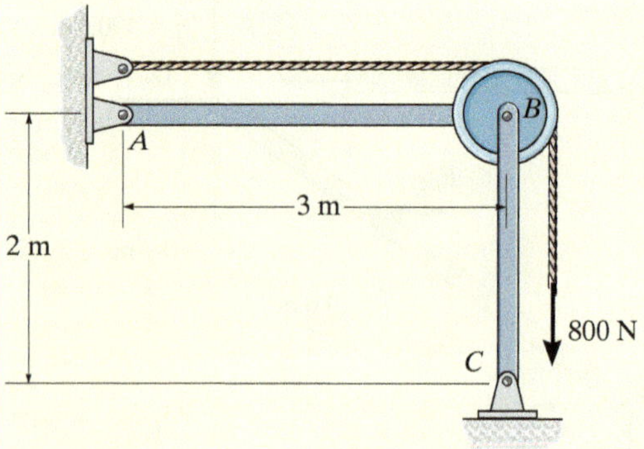

(a)

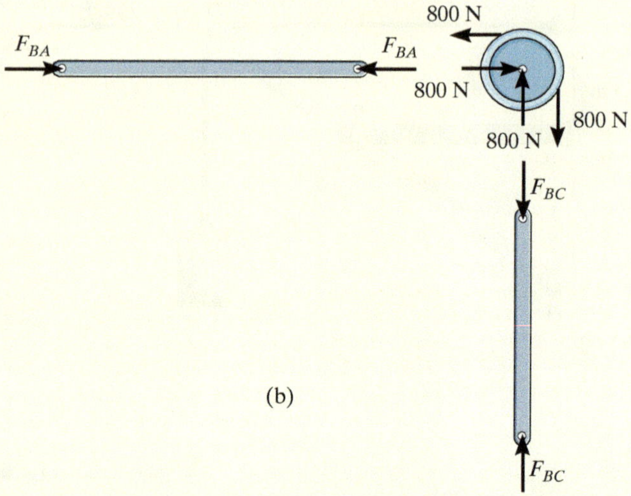

(b)

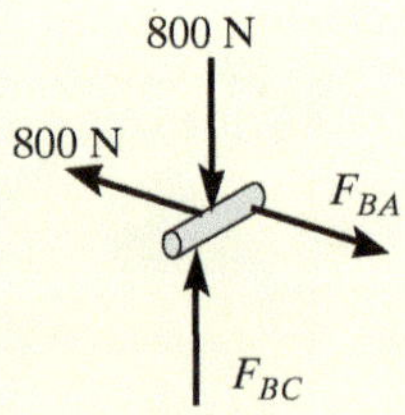

Pin B

(c)

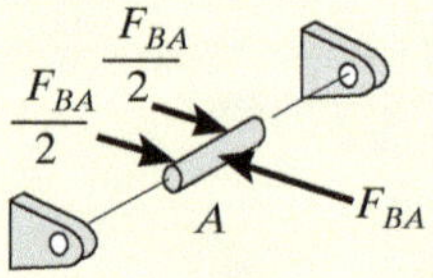

Pin A

(d)

Free-Body Diagrams. By inspection AB and BC are two-force members. Their free-body diagrams, along with that of the pulley, are shown in Fig. 6–34*b*. In order to solve this problem we must also include the free-body diagram of the pin at B because this pin connects all *three members* together, Fig. 6–34*c*.

Equations of Equilibrium: Apply the equations of force equilibrium to pin B.

$$\overset{+}{\rightarrow}\Sigma F_x = 0; \quad F_{BA} - 800\text{ N} = 0; \quad F_{BA} = 800\text{ N} \quad \textit{Ans.}$$

$$+\uparrow\Sigma F_y = 0; \quad F_{BC} - 800\text{ N} = 0; \quad F_{BC} = 800\text{ N} \quad \textit{Ans.}$$

NOTE: The free-body diagram of the pin at A, Fig. 6–34*d*, indicates how the force F_{AB} is balanced by the force $(F_{AB}/2)$ exerted on the pin by each of the two pin leaves.

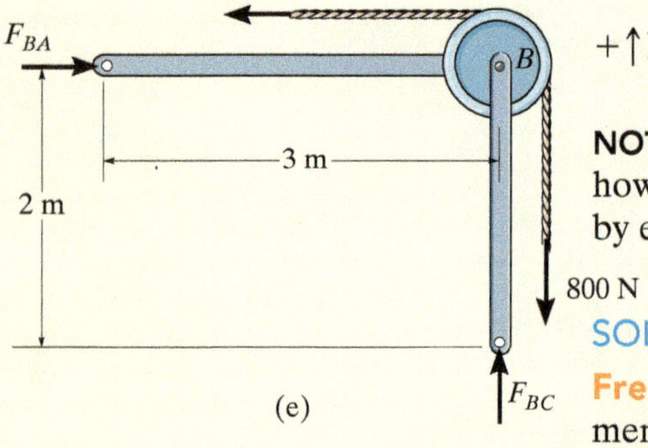

(e)

Fig. 6–34

SOLUTION II

Free-Body Diagram. If we realize that AB and BC are two-force members, then the free-body diagram of the *entire frame* produces an easier solution, Fig. 6–34*e*. The force equations of equilibrium are the same as those above. Note that moment equilibrium will be satisfied, regardless of the radius of the pulley.

FUNDAMENTAL PROBLEMS

All problem solutions must include FBDs.

F6–13. Determine the force P needed to hold the 60-lb weight in equilibrium.

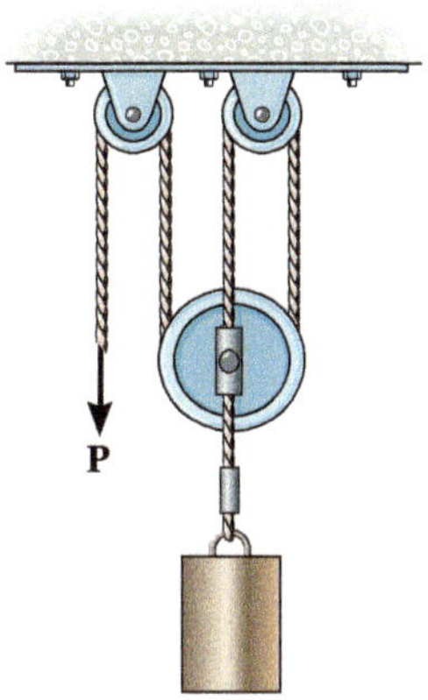

F6–13

F6–14. Determine the horizontal and vertical components of reaction at pin C.

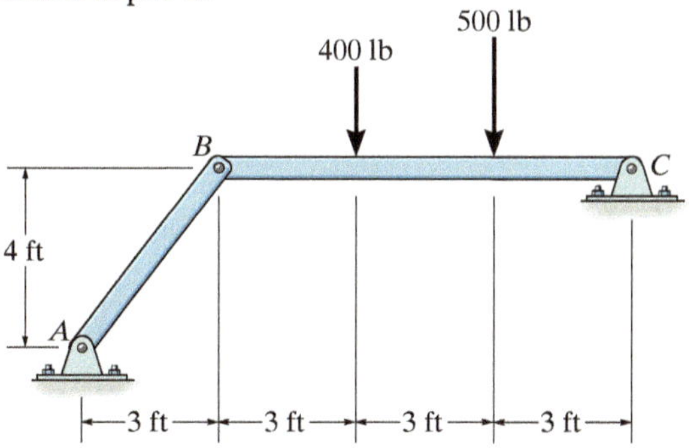

F6–14

F6–15. If a 100-N force is applied to the handles of the pliers, determine the clamping force exerted on the smooth pipe B and the magnitude of the resultant force that one of the members exerts on pin A.

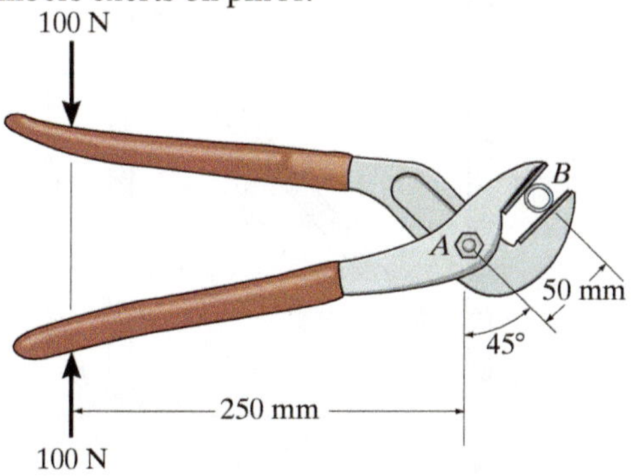

F6–15

F6–16. Determine the horizontal and vertical components of reaction at pin C.

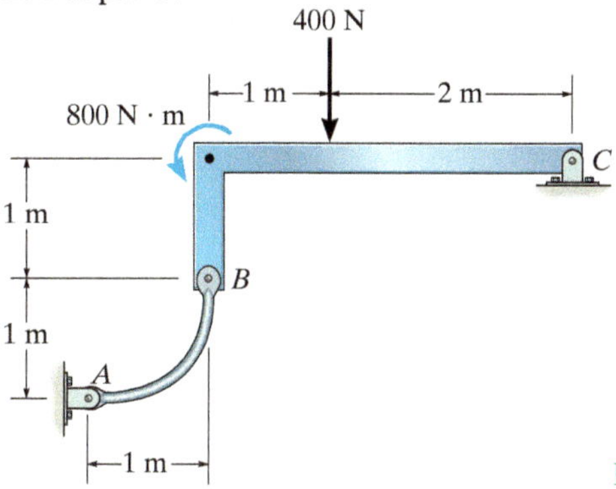

F6–16

F6–17. Determine the normal force that the 100-lb plate A exerts on the 30-lb plate B.

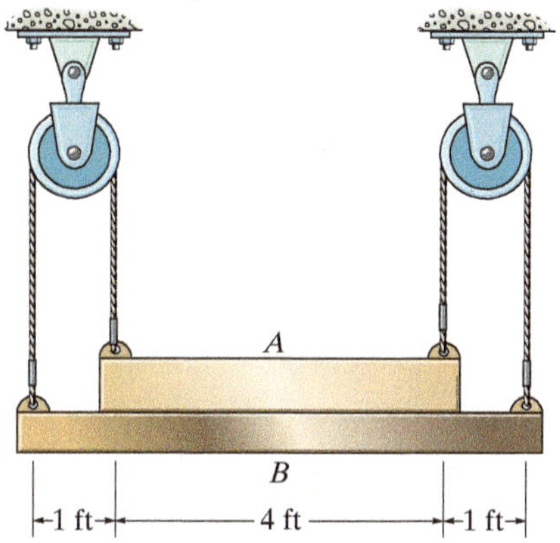

F6–17

F6–18. Determine the force P needed to lift the load. Also, determine the proper placement x of the hook for equilibrium. Neglect the weight of the beam.

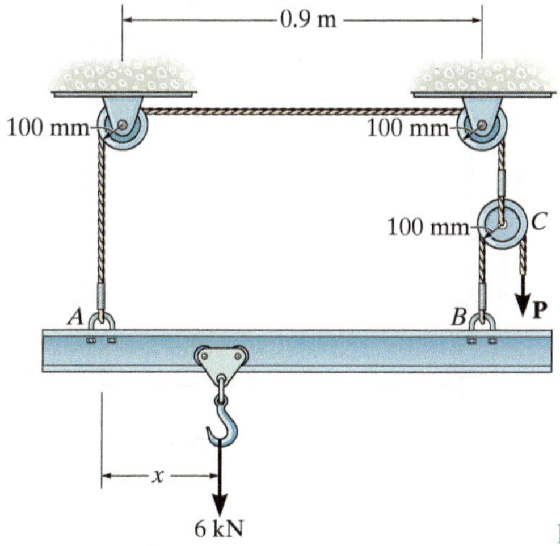

F6–18

F6–19. Determine the components of reaction at A and B.

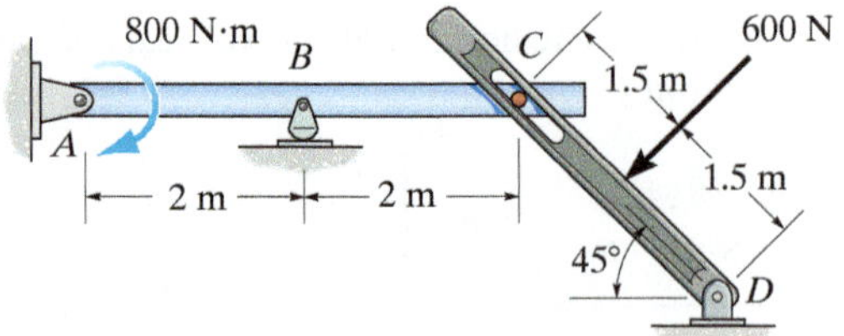

F6–19

F6–20. Determine the components of reaction at D.

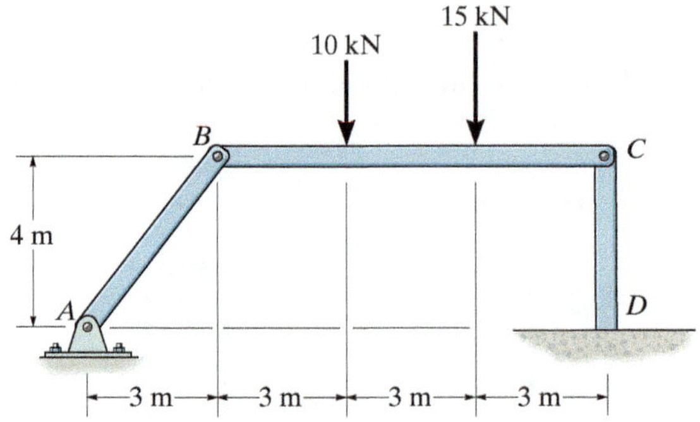

F6–20

F6–21. Determine the components of reaction at A and C.

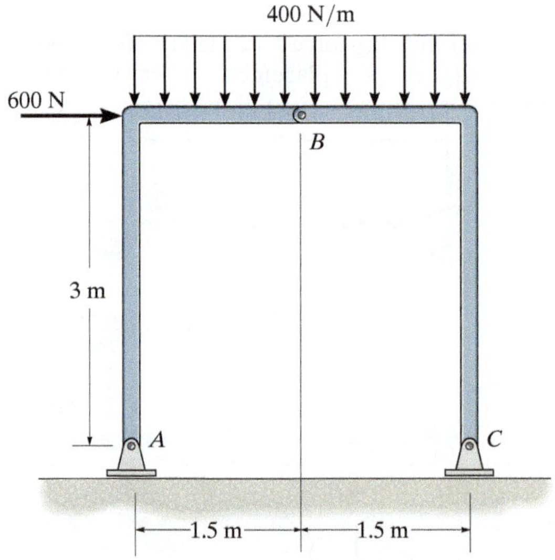

F6–21

F6–22. Determine the components of reaction at C.

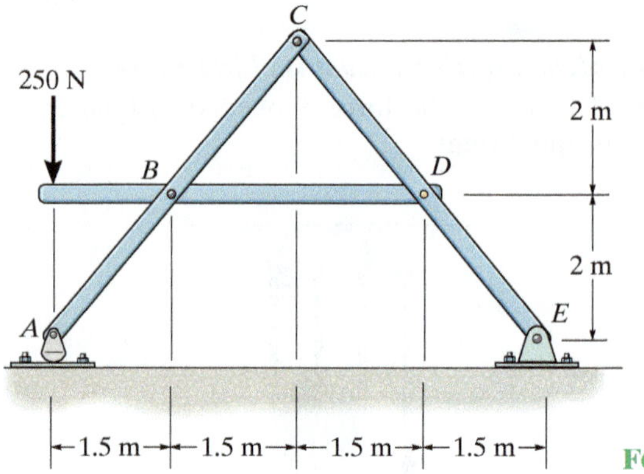

F6–22

F6–23. Determine the components of reaction at E.

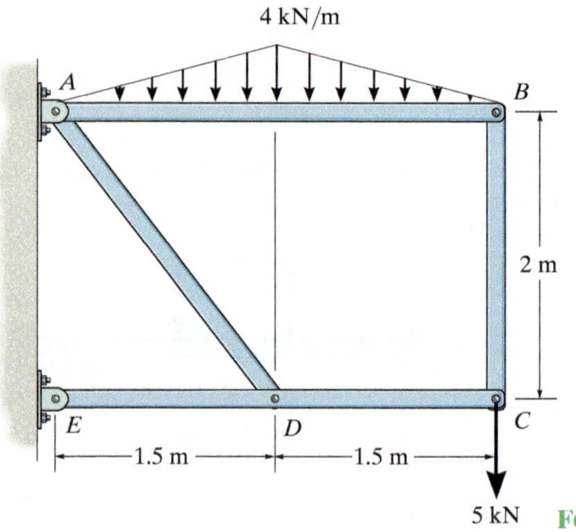

F6–23

F6–24. Determine the components of reaction at D and the components of reaction the pin at A exerts on member BA.

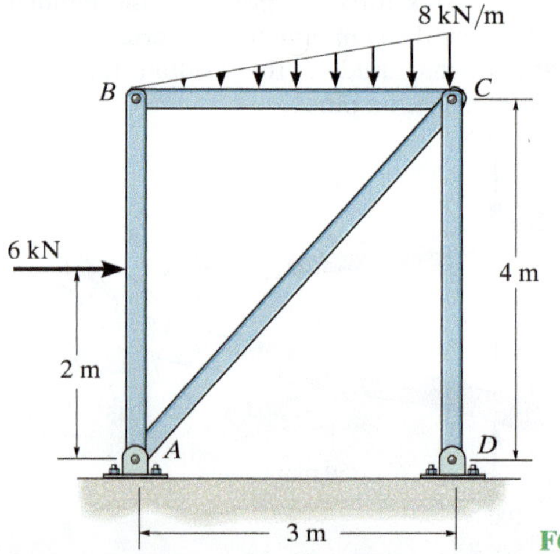

F6–24

6

PROBLEMS

All problem solutions must include FBDs.

6–61. In each case, determine the force **P** required to maintain equilibrium. The block weighs 100 lb.

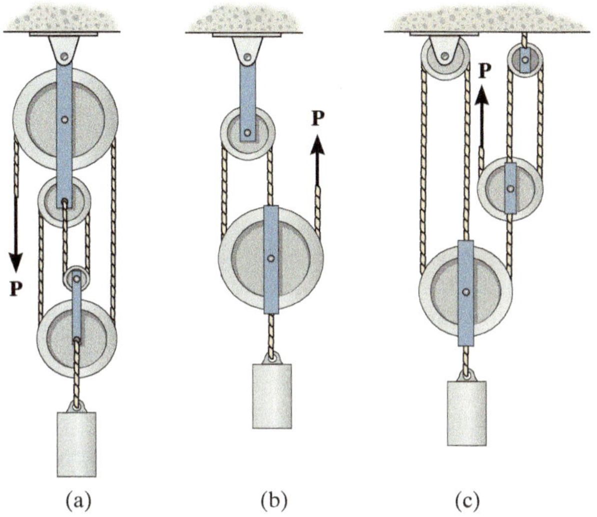

Prob. 6–61

6–62. Determine the force P on the cord, and the angle θ that the pulley-supporting link AB makes with the vertical. Neglect the mass of the pulleys and the link. The block has a weight of 200 lb and the cord is attached to the pin at B. The pulleys have radii of $r_1 = 2$ in. and $r_2 = 1$ in.

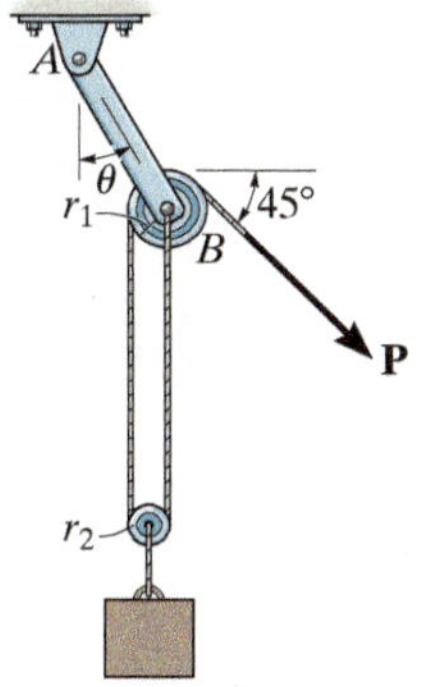

Prob. 6–62

6–63. The principles of a *differential chain block* are indicated schematically in the figure. Determine the magnitude of force **P** needed to support the 800-N force. Also, find the distance x where the cable must be attached to bar AB so the bar remains horizontal. All pulleys have a radius of 60 mm.

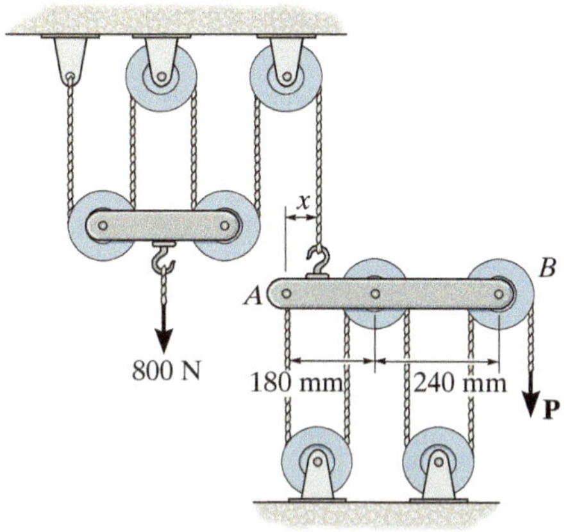

Prob. 6–63

***6–64.** Determine the force P needed to support the 20-kg mass using the *Spanish Burton rig*. Also, what are the reactions at the supporting hooks A, B, and C?

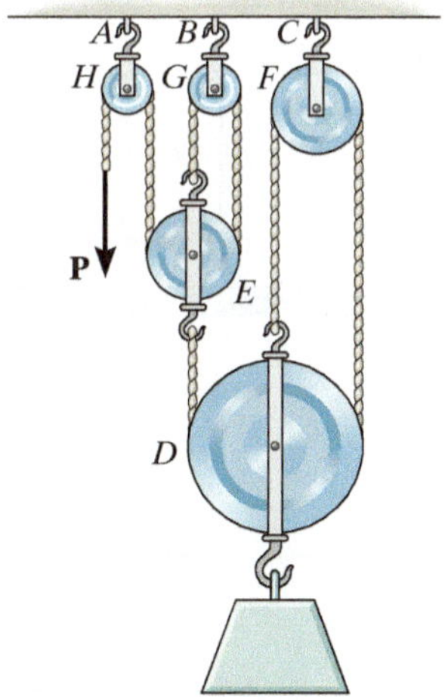

Prob. 6–64

6–65. Determine the horizontal and vertical components of force at *C* which member *ABC* exerts on member *CEF*.

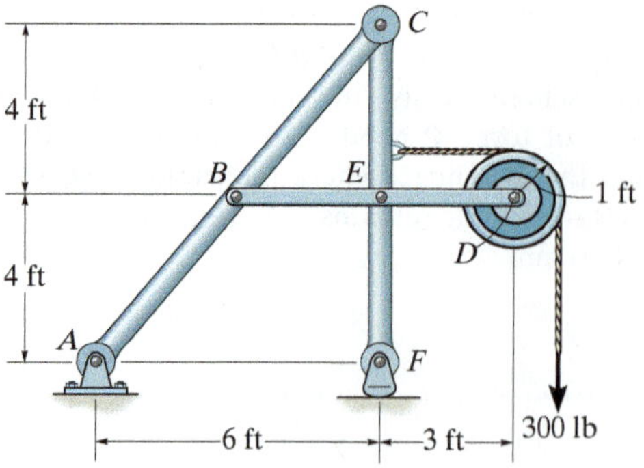

Prob. 6–65

6–66. Determine the horizontal and vertical components of force that the pins at *A*, *B*, and *C* exert on their connecting members.

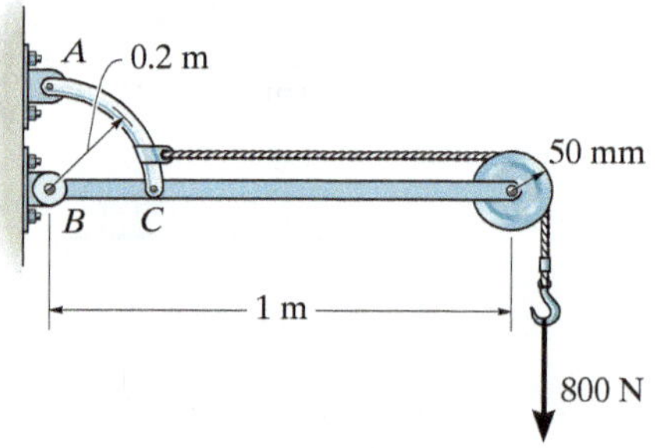

Prob. 6–66

6–67. Determine the horizontal and vertical components of force at pins *D* and *E*, and the force on the short link at *A*. The suspended cylinder has a weight of 80 lb.

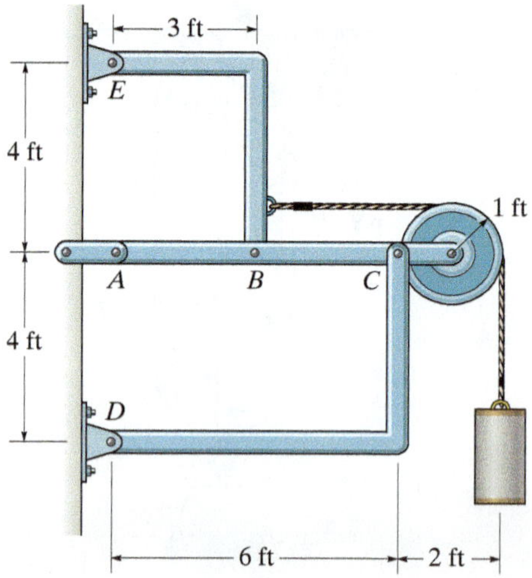

Prob. 6–67

***6–68.** Determine the greatest force *P* that can be applied to the frame if the largest force resultant acting at *A* can have a magnitude of 2 kN.

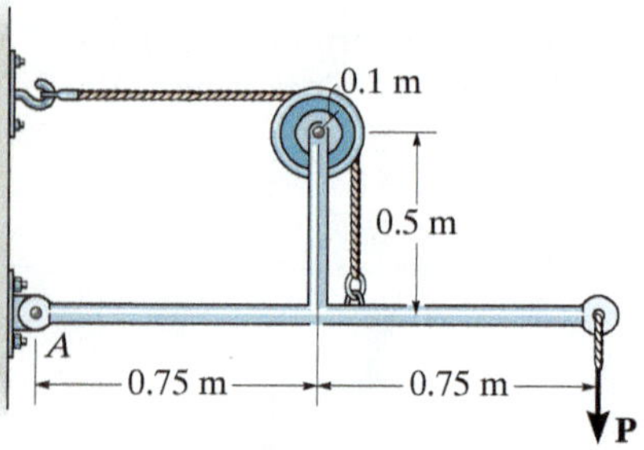

Prob. 6–68

6–69. Determine the force that the smooth roller *C* exerts on member *AB*. Also, what are the horizontal and vertical components of reaction at pin *A*? Neglect the weight of the frame and roller.

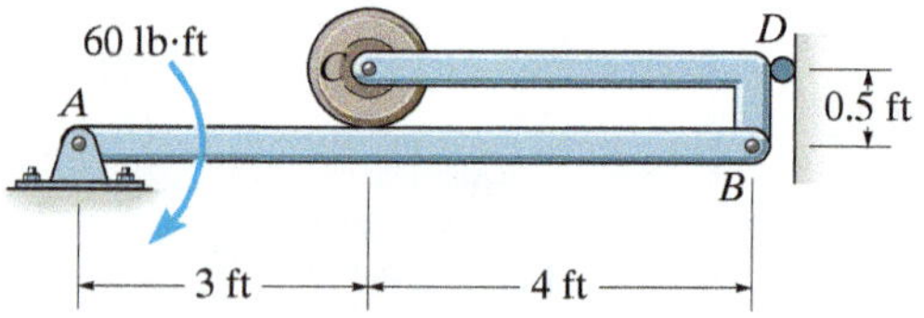

Prob. 6–69

6–70. Determine the horizontal and vertical components of force at pins *B* and *C*.

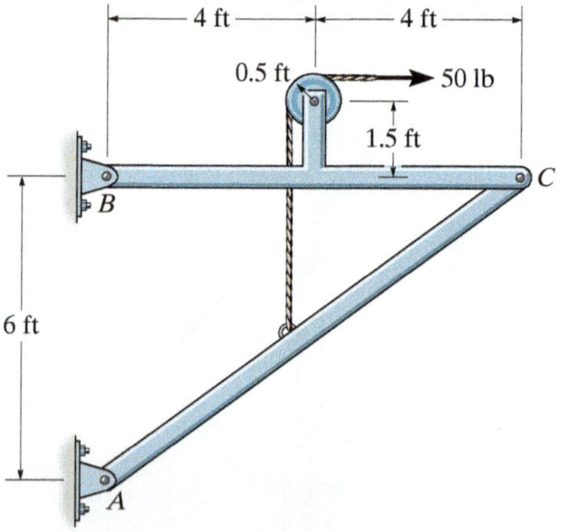

Prob. 6–70

6–71. Determine the support reactions at A, C, and E on the compound beam which is pin connected at B and D.

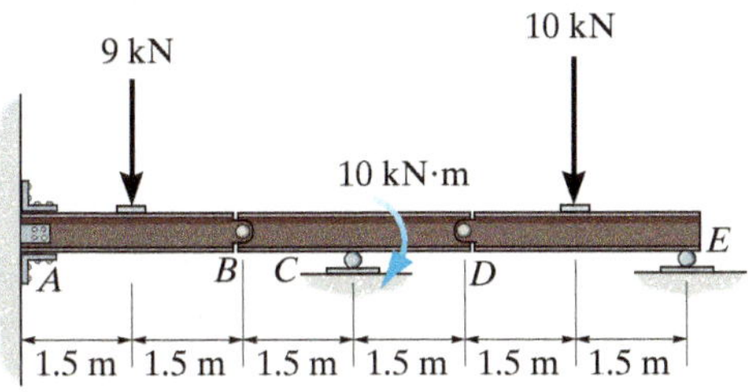

Prob. 6–71

***6–72.** Determine the horizontal and vertical components of force at pins A, B, and C, and the reactions at the fixed support D of the three-member frame.

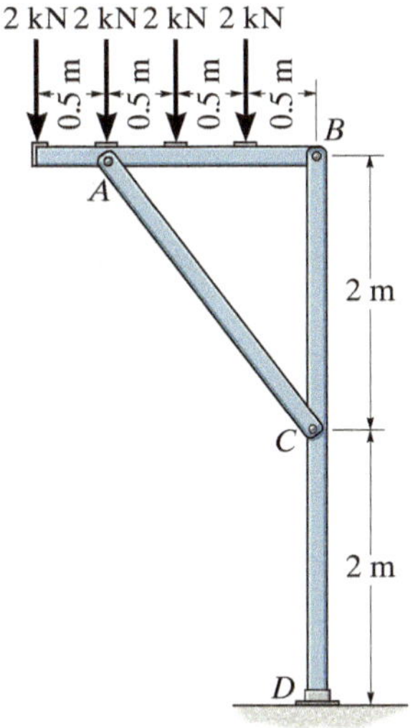

Prob. 6–72

6–73. The compound beam is fixed at A and supported by a rocker at B and C. There are hinges (pins) at D and E. Determine the reactions at the supports.

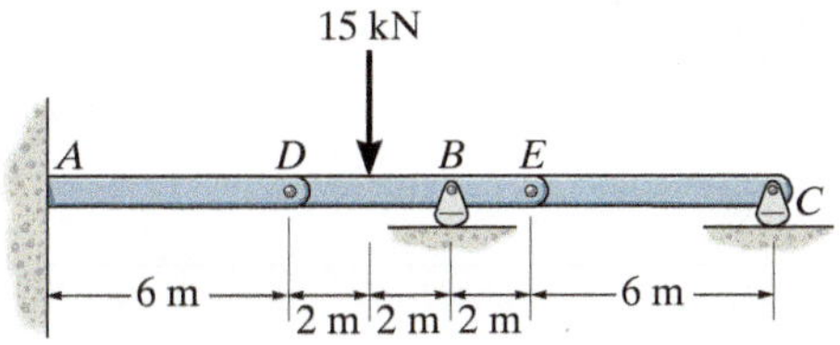

Prob. 6–73

6–74. The wall crane supports a load of 700 lb. Determine the horizontal and vertical components of reaction at the pins A and D. Also, what is the force in the cable at the winch W?

6–75. The wall crane supports a load of 700 lb. Determine the horizontal and vertical components of reaction at the pins A and D. Also, what is the force in the cable at the winch W? The jib ABC has a weight of 100 lb and member BD has a weight of 40 lb. Each member is uniform and has a center of gravity at its center.

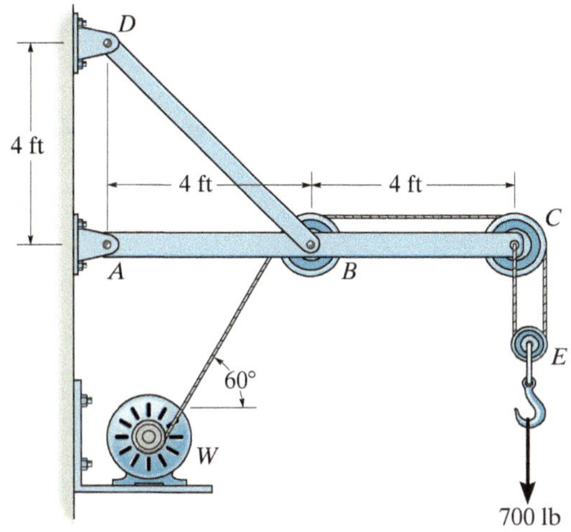

Probs. 6–74/75

***6–76.** Determine the horizontal and vertical components of force which the pins at A, B, and C exert on member ABC of the frame.

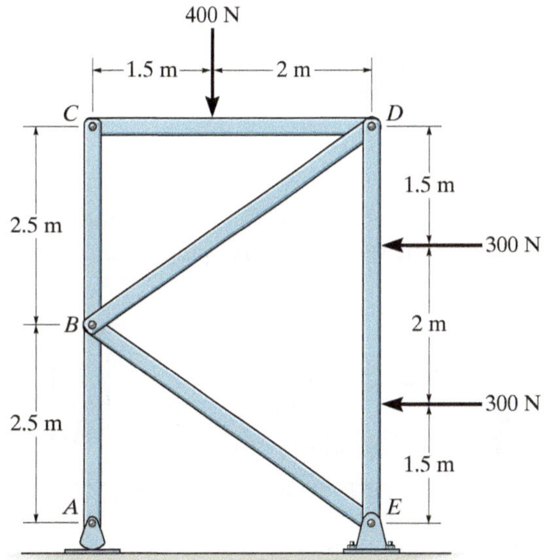

Prob. 6–76

6–77. Determine the required mass of the suspended cylinder if the tension in the chain wrapped around the freely turning gear is to be 2 kN. Also, what is the magnitude of the resultant force on pin A?

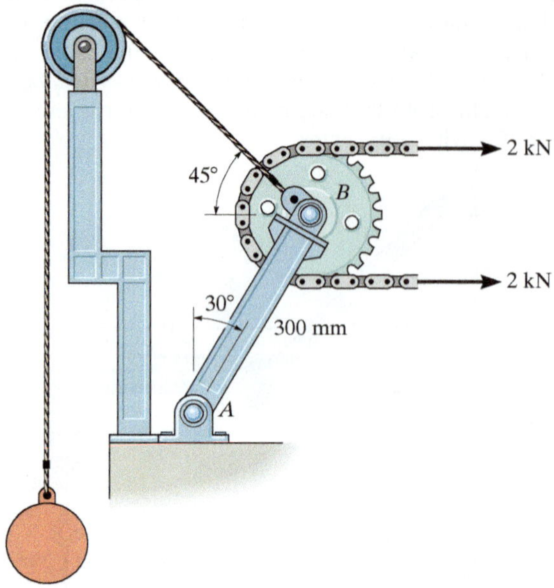

Prob. 6–77

6–78. Determine the reactions on the collar at A and the pin at C. The collar fits over a smooth rod, and rod AB is fixed connected to the collar.

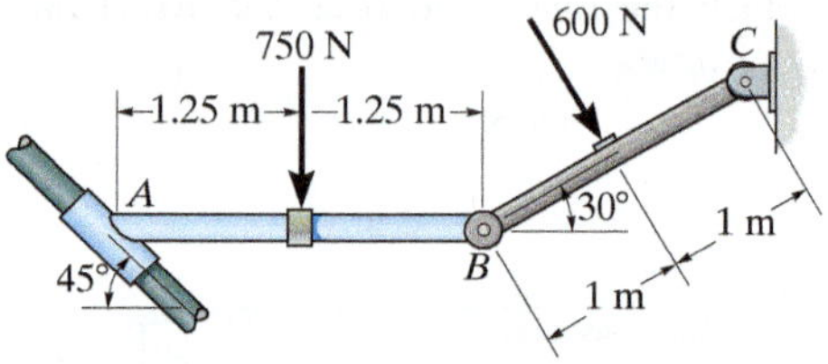

Prob. 6–78

6–79. The toggle clamp is subjected to a force **F** at the handle. Determine the vertical clamping force acting at E.

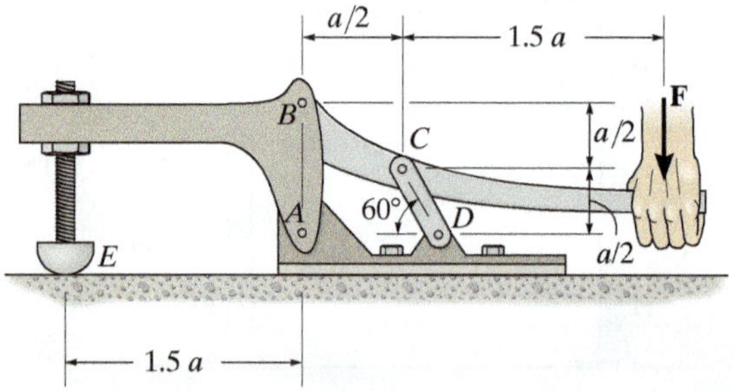

Prob. 6–79

***6–80.** When a force of 2 lb is applied to the handles of the brad squeezer, it pulls in the smooth rod AB. Determine the force **P** exerted on each of the smooth brads at C and D.

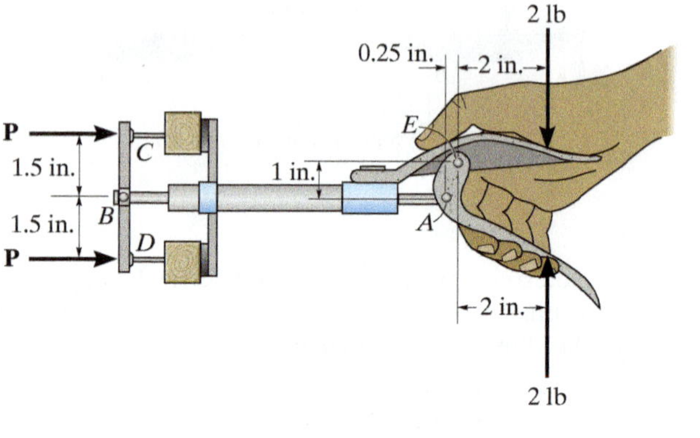

Prob. 6–80

6–81. The engine hoist is used to support the 200-kg engine. Determine the force acting in the hydraulic cylinder AB, the horizontal and vertical components of force at the pin C, and the reactions at the fixed support D.

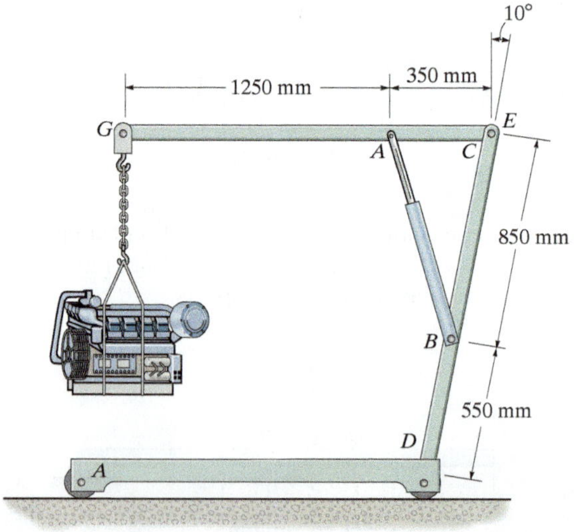

Prob. 6–81

6–82. The three power lines exert the forces shown on the pin-connected members at joints B, C, and D, which in turn are pin connected to the poles AH and EG. Determine the force in the guy cable AI and the pin reaction at the support H.

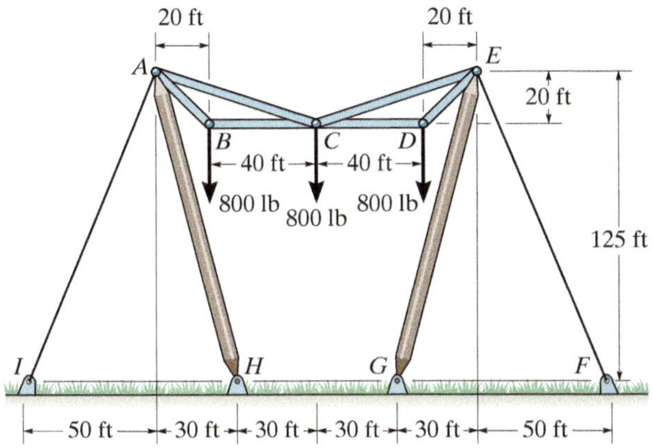

Prob. 6–82

6–83. By squeezing on the hand brake of the bicycle, the rider subjects the brake cable to a tension of 50 lb. If the caliper mechanism is pin connected to the bicycle frame at B, determine the normal force each brake pad exerts on the rim of the wheel. Is this the force that stops the wheel from turning? Explain.

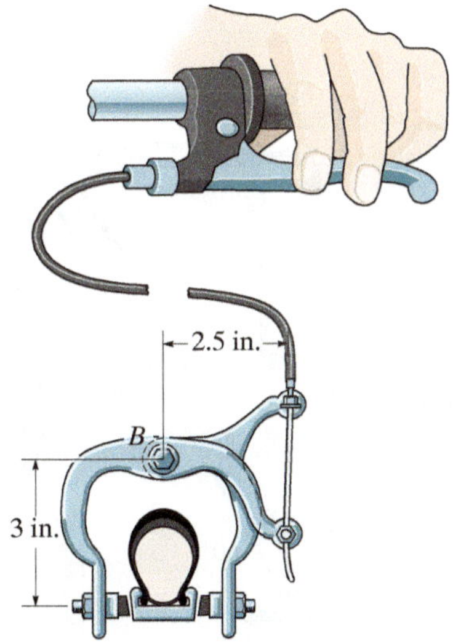

Prob. 6–83

***6–84.** Determine the required force P that must be applied at the blade of the pruning shears so that the blade exerts a normal force of 20 lb on the twig at E.

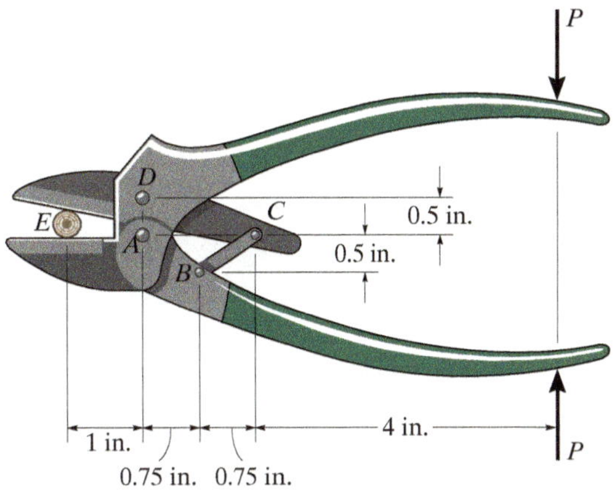

Prob. 6–84

6–85. The pruner multiplies blade-cutting power with the compound leverage mechanism. If a 20-N force is applied to the handles, determine the cutting force generated at A. Assume that the contact surface at A is smooth.

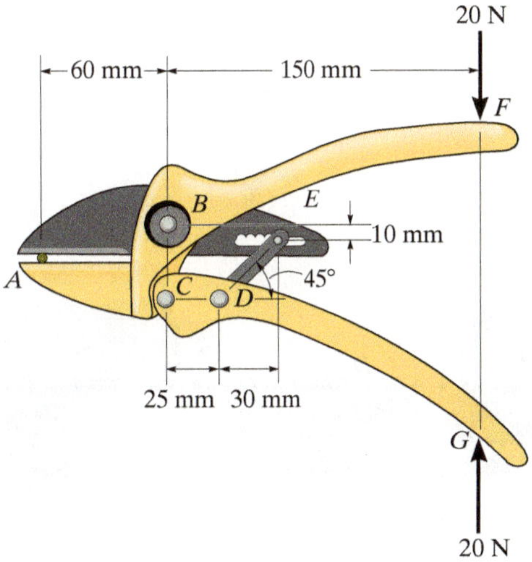

Prob. 6–85

6

6–86. The pipe cutter is clamped around the pipe P. If the wheel at A exerts a normal force of $F_A = 80$ N on the pipe, determine the normal forces of wheels B and C on the pipe. Also compute the pin reaction on the wheel at C. The three wheels each have a radius of 7 mm and the pipe has an outer radius of 10 mm.

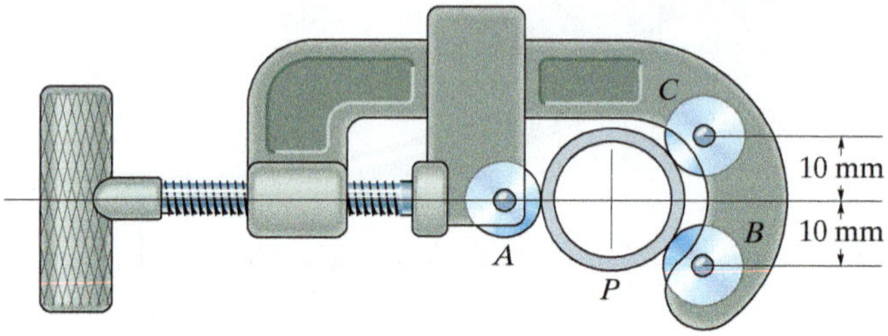

Prob. 6–86

***6–88.** Show that the weight W_1 of the counterweight at H required for equilibrium is $W_1 = (b/a)W$, and so it is independent of the placement of the load W on the platform.

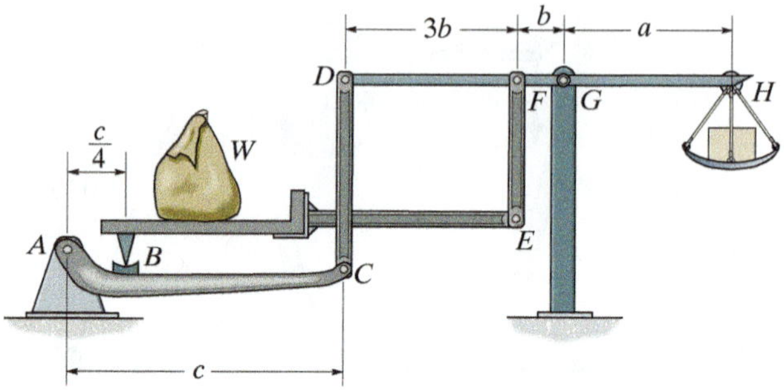

Prob. 6–88

6–87. The flat-bed trailer has a weight of 7000 lb and center of gravity at G_T. It is pin connected to the cab at D. The cab has a weight of 6000 lb and center of gravity at G_C. Determine the range of values x for the position of the 2000-lb load L so that no axle is subjected to more than 5500 lb. The load has a center of gravity at G_L.

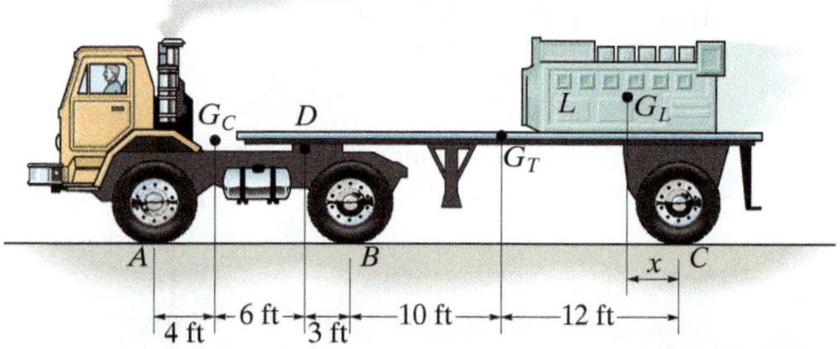

Prob. 6–87

6–89. The derrick is pin connected to the pivot at A. Determine the largest mass that can be supported by the derrick if the maximum force that can be sustained by the pin at A is 18 kN.

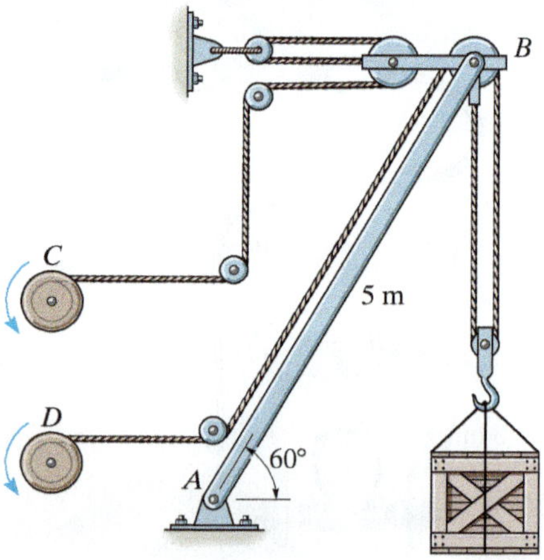

Prob. 6–89

6–90. Determine the force that the jaws J of the metal cutters exert on the smooth cable C if 100-N forces are applied to the handles. The jaws are pinned at E and A, and D and B. There is also a pin at F.

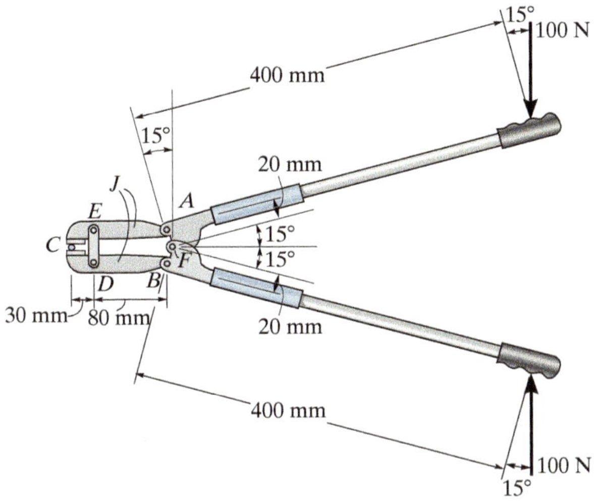

Prob. 6–90

6–91. The pumping unit is used to recover oil. When the walking beam ABC is horizontal, the force acting in the wireline at the well head is 250 lb. Determine the torque **M** which must be exerted by the motor in order to overcome this load. The horse-head C weighs 60 lb and has a center of gravity at G_C. The walking beam ABC has a weight of 130 lb and a center of gravity at G_B, and the counterweight has a weight of 200 lb and a center of gravity at G_W. The pitman, AD, is pin connected at its ends and has negligible weight.

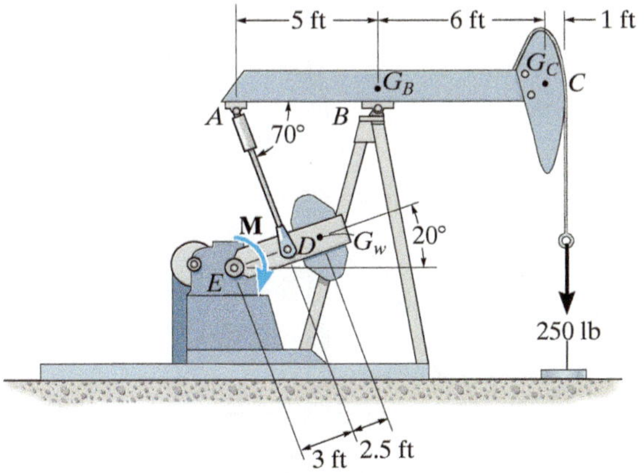

Prob. 6–91

***6–92.** The scissors lift consists of *two* sets of cross members and *two* hydraulic cylinders, DE, symmetrically located on *each side* of the platform. The platform has a uniform mass of 60 kg, with a center of gravity at G_1. The load of 85 kg, with center of gravity at G_2, is centrally located between each side of the platform. Determine the force in each of the hydraulic cylinders for equilibrium. Rollers are located at B and D.

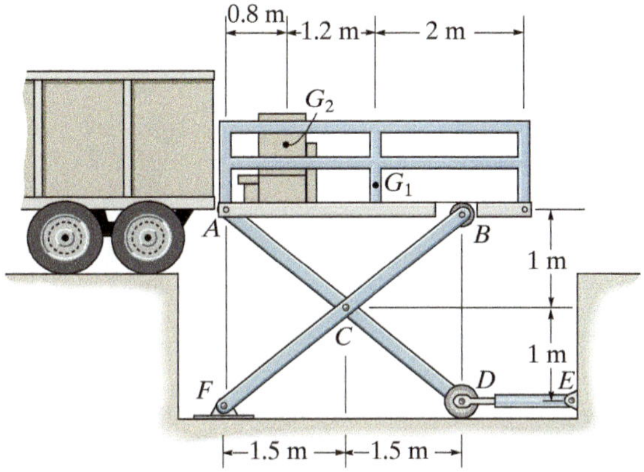

Prob. 6–92

6–93. The two disks each have a mass of 20 kg and are attached at their centers by an elastic cord that has a stiffness of $k = 2$ kN/m. Determine the stretch of the cord when the system is in equilibrium and the angle θ of the cord.

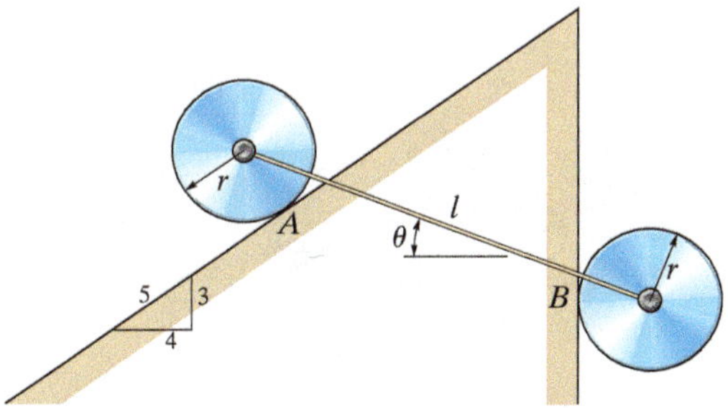

Prob. 6–93

6–94. A man having a weight of 175 lb attempts to hold himself using one of the two methods shown. Determine the total force he must exert on bar AB in each case and the normal reaction he exerts on the platform at C. Neglect the weight of the platform.

6–95. A man having a weight of 175 lb attempts to hold himself using one of the two methods shown. Determine the total force he must exert on bar AB in each case and the normal reaction he exerts on the platform at C. The platform has a weight of 30 lb.

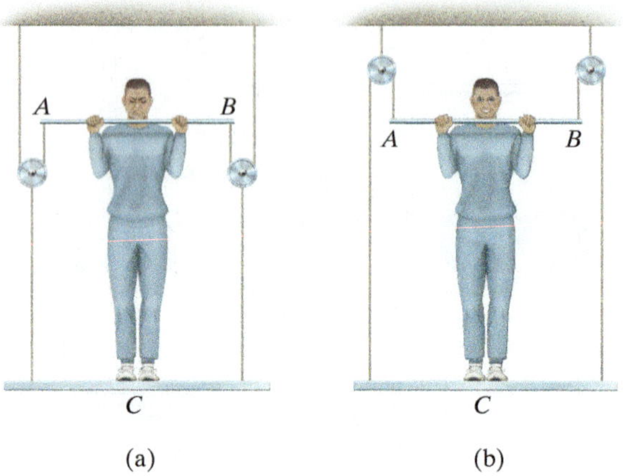

Probs. 6–94/95

*6–96. The double link grip is used to lift the beam. If the beam weighs 4 kN, determine the horizontal and vertical components of force acting on the pin at A and the horizontal and vertical components of force that the flange of the beam exerts on the jaw at B.

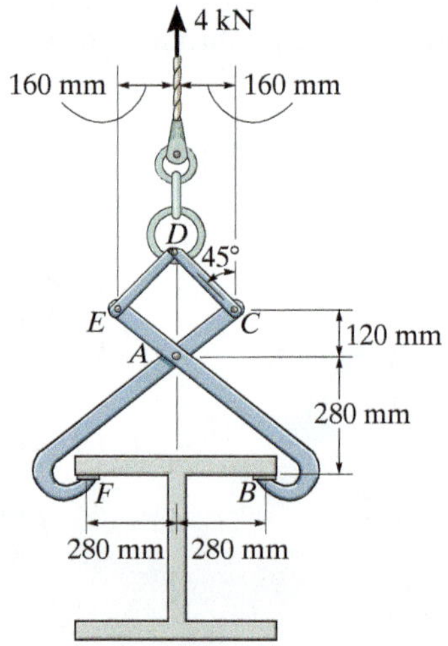

Prob. 6–96

6–97. If a force of $P = 6$ lb is applied perpendicular to the handle of the mechanism, determine the magnitude of force **F** for equilibrium. The members are pin connected at A, B, C, and D.

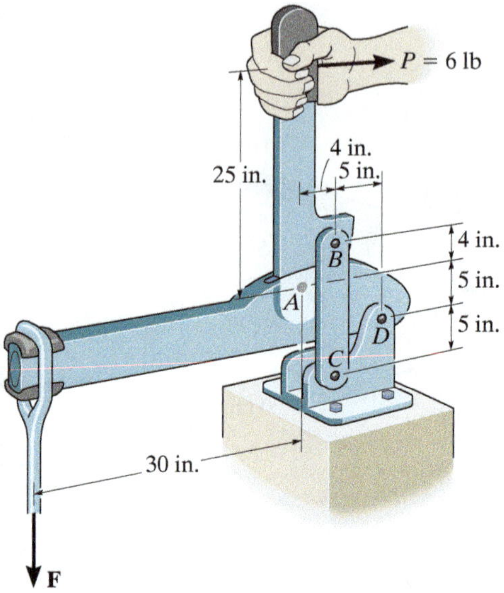

Prob. 6–97

6–98. Determine the horizontal and vertical components of force at pin B and the normal force the pin at C exerts on the smooth slot. Also, determine the moment and horizontal and vertical reactions of force at A. There is a pulley at E.

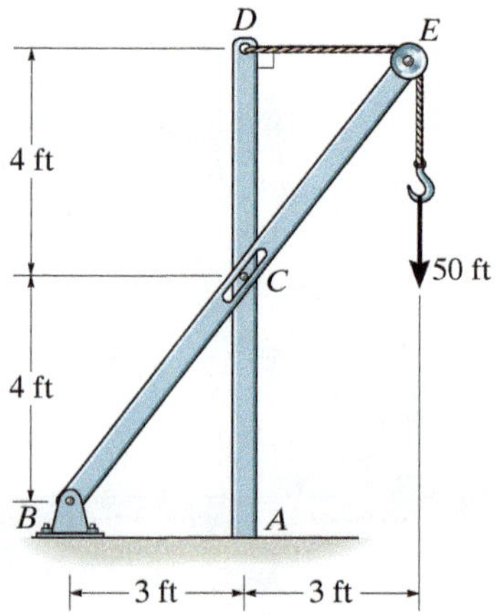

Prob. 6–98

6–99. If a clamping force of 300 N is required at A, determine the amount of force **F** that must be applied to the handle of the toggle clamp.

***6–100.** If a force of $F = 350$ N is applied to the handle of the toggle clamp, determine the resulting clamping force at A.

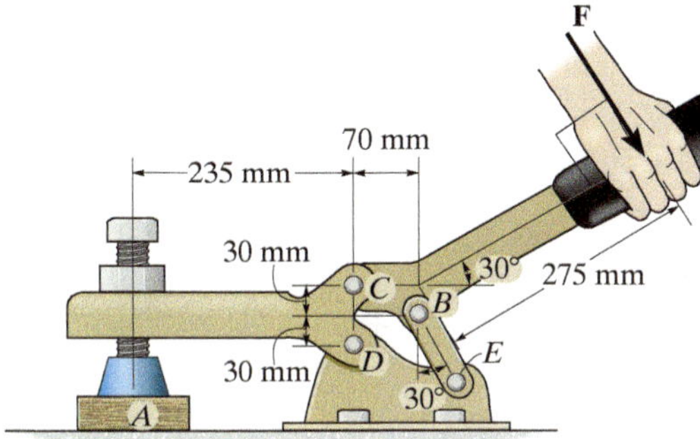

Probs. 6–99/100

6–101. If a force of 10 lb is applied to the grip of the clamp, determine the compressive force F that the wood block exerts on the clamp.

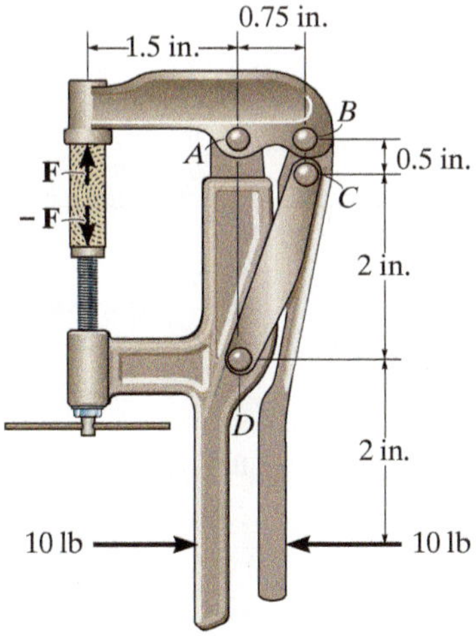

Prob. 6–101

6–102. The tractor boom supports the uniform mass of 500 kg in the bucket which has a center of mass at G. Determine the force in each hydraulic cylinder AB and CD and the resultant force at pins E and F. The load is supported equally on each side of the tractor by a similar mechanism.

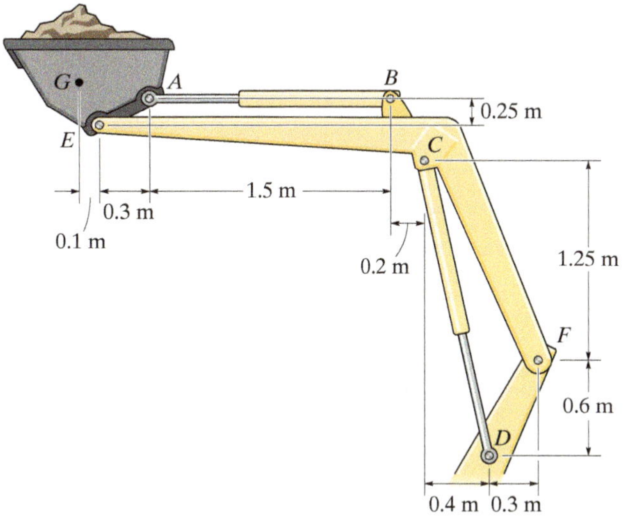

Prob. 6–102

6–103. The two-member frame supports the 200-lb cylinder and 500-lb · ft couple moment. Determine the force of the roller at B on member AC and the horizontal and vertical components of force which the pin at C exerts on member CB and the pin at A exerts on member AC. The roller C does not contact member CB.

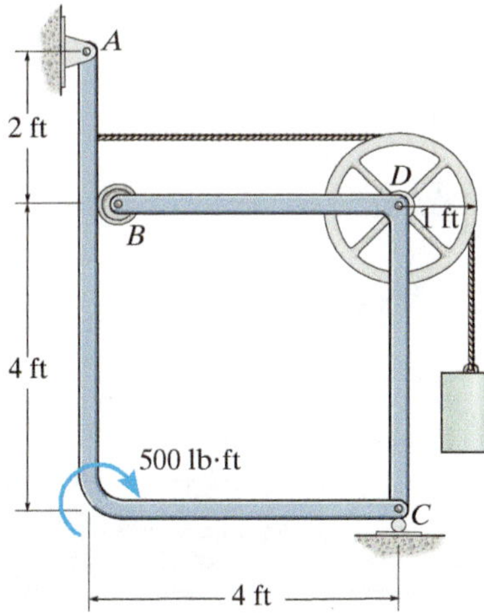

Prob. 6–103

*6–104. The mechanism is used to hide kitchen appliances under a cabinet by allowing the shelf to rotate downward. If the mixer weighs 10 lb, is centered on the shelf, and has a mass center at G, determine the stretch in the spring necessary to hold the shelf in the equilibrium position shown. There is a similar mechanism on each side of the shelf, so that each mechanism supports 5 lb of the load. The springs each have a stiffness of $k = 4$ lb/in.

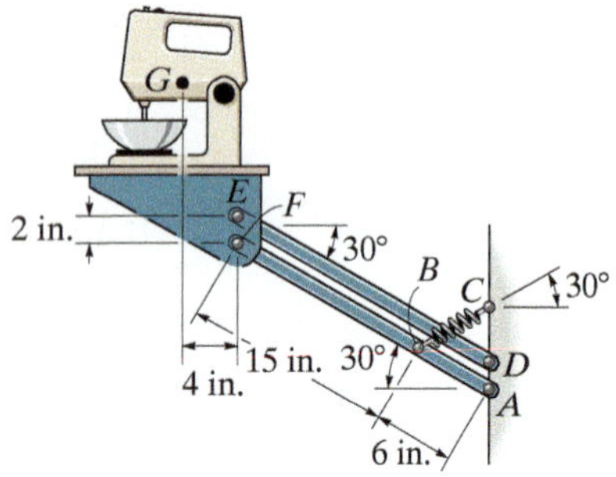

Prob. 6–104

6–106. If $d = 0.75$ ft and the spring has an unstretched length of 1 ft, determine the force F required for equilibrium.

■6–107. If a force of $F = 50$ lb is applied to the pads at A and C, determine the smallest dimension d required for equilibrium if the spring has an unstretched length of 1 ft.

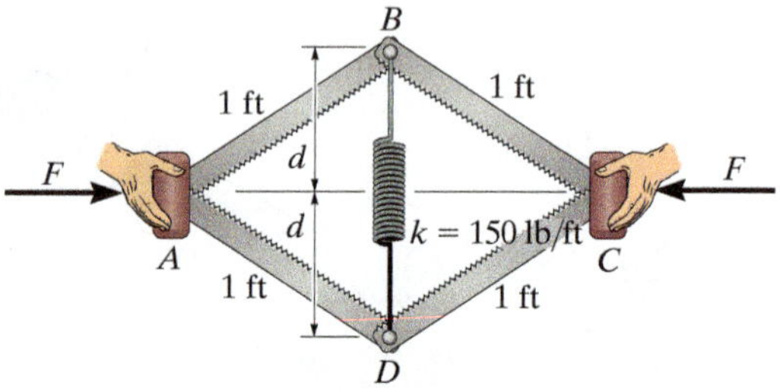

Probs. 6–106/107

6–105. The linkage for a hydraulic jack is shown. If the load on the jack is 2000 lb, determine the pressure acting on the fluid when the jack is in the position shown. All lettered points are pins. The piston at H has a cross-sectional area of $A = 2\text{ in}^2$. *Hint:* First find the force F acting along link EH. The pressure in the fluid is $p = F/A$.

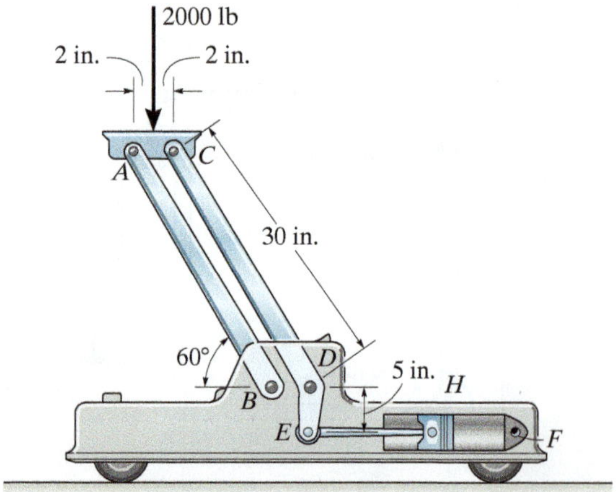

Prob. 6–105

*6–108. The hydraulic crane is used to lift the 1400-lb load. Determine the force in the hydraulic cylinder AB and the force in links AC and AD when the load is held in the position shown.

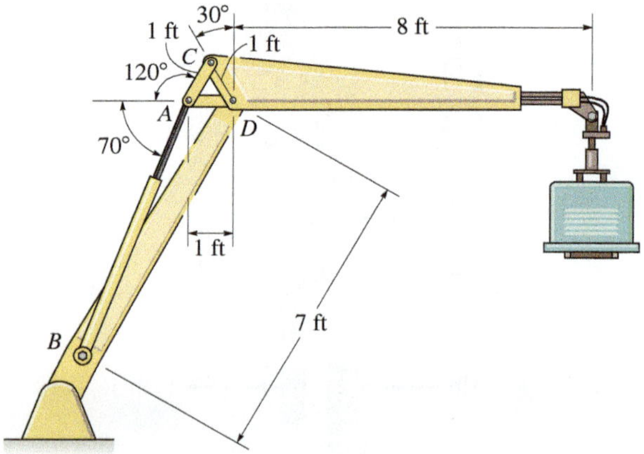

Prob. 6–108

6–109. The symmetric coil tong supports the coil which has a mass of 800 kg and center of mass at G. Determine the horizontal and vertical components of force the linkage exerts on plate $DEIJH$ at points D and E. The coil exerts only vertical reactions at K and L.

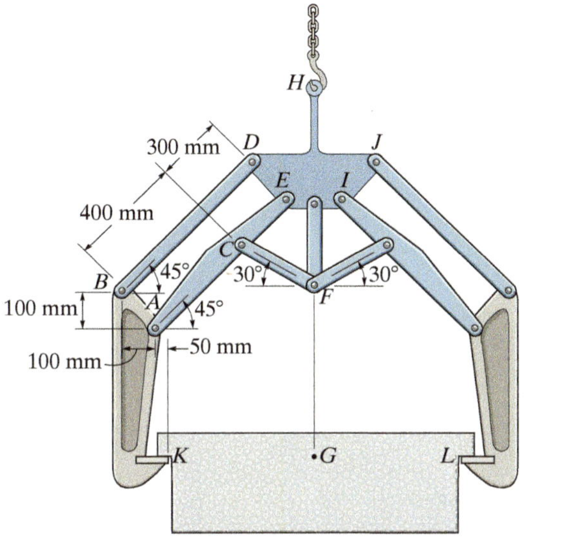

Prob. 6–109

6–110. If each of the three uniform links of the mechanism has a length $L = 3$ ft and weight of $W = 10$ lb, determine the angle θ for equilibrium. The spring has a stiffness of $k = 20$ lb/in. It always remains vertical due to the roller guide and is unstretched when $\theta = 0$.

6–111. If each of the three uniform links of the mechanism has a length L and weight W, determine the angle θ for equilibrium. The spring, which always remains vertical, is unstretched when $\theta = 0°$.

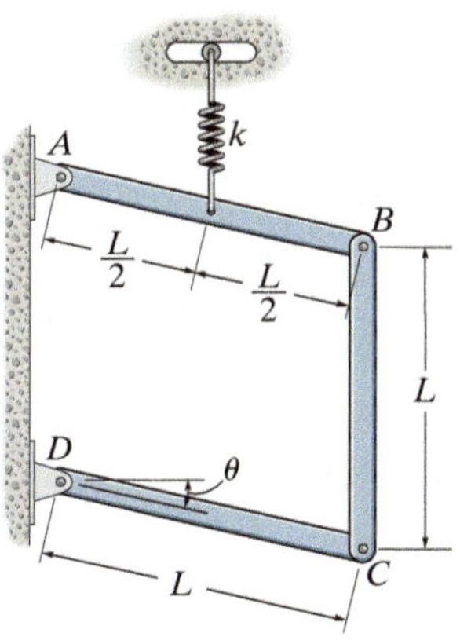

Probs. 6–110/111

*6–112. The piston C moves vertically between the two smooth walls. If the spring has a stiffness of $k = 15$ lb/in., and is unstretched when $\theta = 0°$, determine the couple **M** that must be applied to AB to hold the mechanism in equilibrium when $\theta = 30°$.

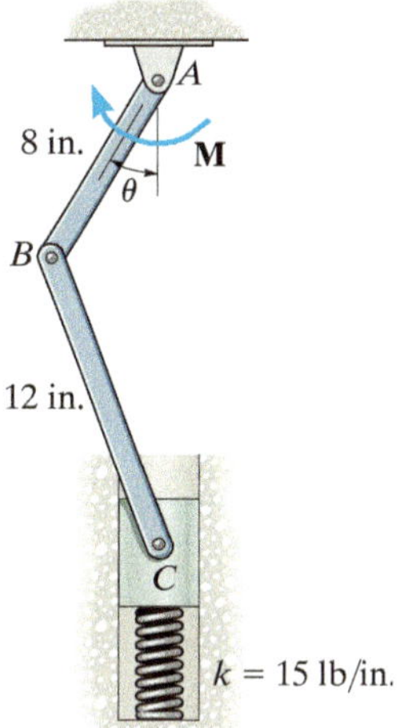

Prob. 6–112

6–113. The aircraft-hangar door opens and closes slowly by means of a motor, which draws in the cable AB. If the door is made in two sections (bifold) and each section has a uniform weight of 300 lb and height $L = 10$ ft, determine the force on the cable when $\theta = 90°$. The sections are pin connected at C and D and the bottom is attached to a roller that travels along the vertical track.

6–114. The aircraft-hangar door opens and closes slowly by means of a motor which draws in the cable AB. If the door is made in two sections (bifold) and each section has a uniform weight W and height L, determine the force in the cable as a function of the door's position θ. The sections are pin connected at C and D and the bottom is attached to a roller that travels along the vertical track.

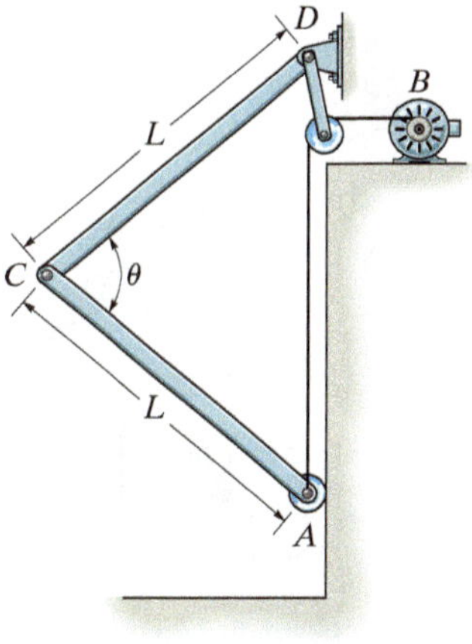

Probs. 6–113/114

6

6–115. The three pin-connected members shown in the *top view* support a downward force of 60 lb at *G*. If only vertical forces are supported at the connections *B, C, E* and pad supports *A, D, F*, determine the reactions at each pad.

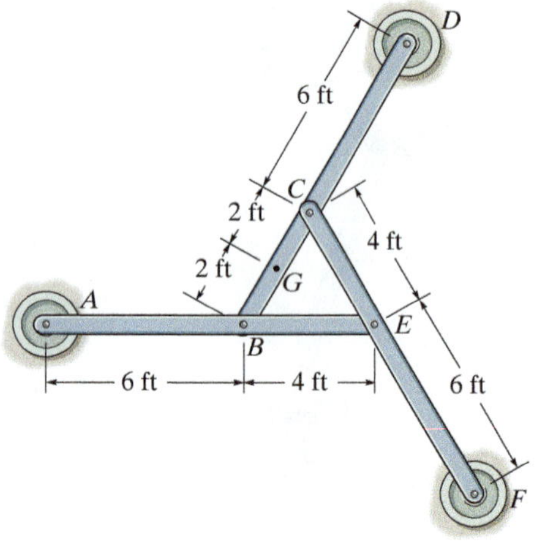

Prob. 6–115

6–117. The three-member frame is connected at its ends using ball-and-socket joints. Determine the *x, y, z* components of reaction at *B* and the tension in member *ED*. The force acting at *D* is $\mathbf{F} = \{135\mathbf{i} + 200\mathbf{j} - 180\mathbf{k}\}$ lb.

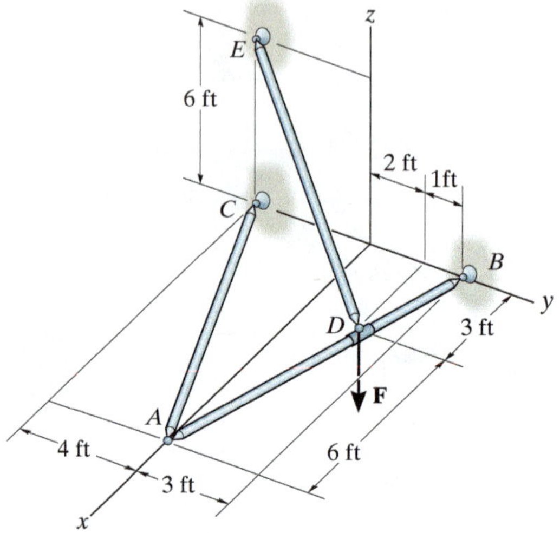

Prob. 6–117

6

***6–116.** The structure is subjected to the loading shown. Member *AD* is supported by a cable *AB* and roller at *C* and fits through a smooth circular hole at *D*. Member *ED* is supported by a roller at *D* and a pole that fits in a smooth snug circular hole at *E*. Determine the *x, y, z* components of reaction at *E* and the tension in cable *AB*.

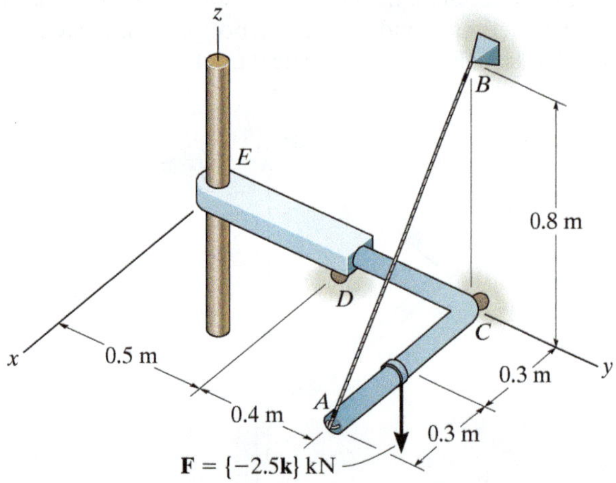

Prob. 6–116

6–118. The structure is subjected to the force of 450 lb which lies in a plane parallel to the *y*–*z* plane. Member *AB* is supported by a ball-and-socket joint at *A* and fits through a snug hole at *B*. Member *CD* is supported by a pin at *C*. Determine the *x, y, z* components of reaction at *A* and *C*.

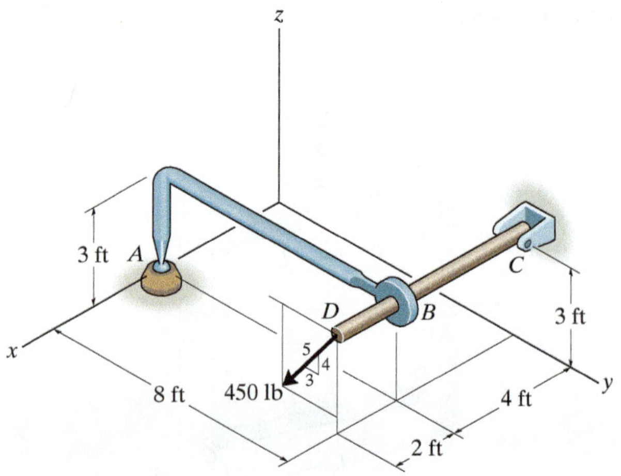

Prob. 6–118

CHAPTER REVIEW

Simple Truss

A simple truss consists of triangular elements connected together by pinned joints. The forces within its members can be determined by assuming the members are all two-force members, connected concurrently at each joint. The members are either in tension or compression, or carry no force.

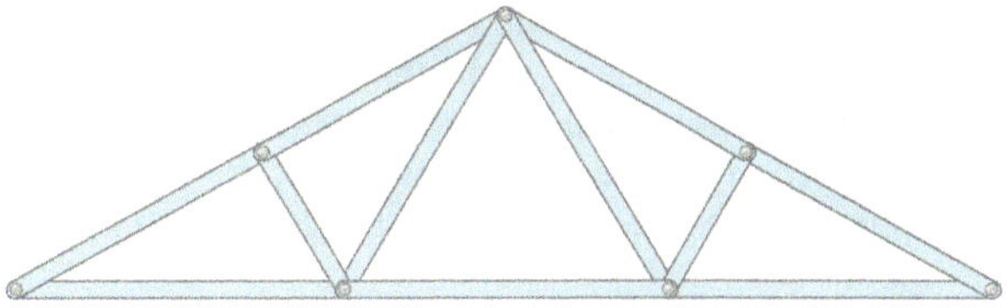

Roof truss

Method of Joints

The method of joints states that if a truss is in equilibrium, then each of its joints is also in equilibrium. For a plane truss, the concurrent force system at each joint must satisfy force equilibrium.

$$\Sigma F_x = 0$$
$$\Sigma F_y = 0$$

To obtain a numerical solution for the forces in the members, select a joint that has a free-body diagram with at most two unknown forces and one known force. (This may require first finding the reactions at the supports.)

Once a member force is determined, use its value and apply it to an adjacent joint.

Remember that forces that are found to *pull* on the joint are *tensile forces*, and those that *push* on the joint are *compressive forces*.

To avoid a simultaneous solution of two equations, set one of the coordinate axes along the line of action of one of the unknown forces and sum forces perpendicular to this axis. This will allow a direct solution for the other unknown.

The analysis can also be simplified by first identifying all the zero-force members.

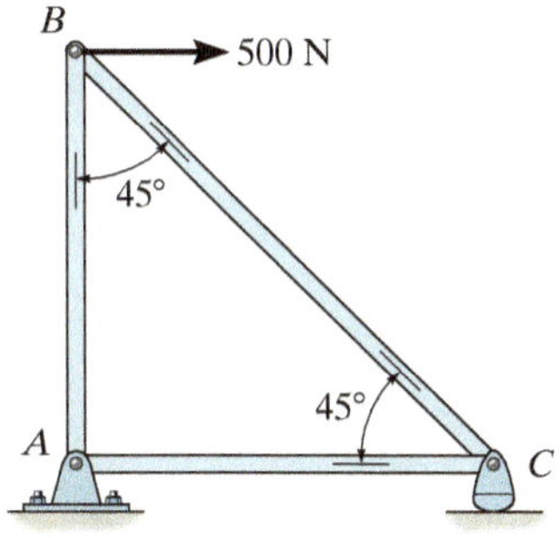

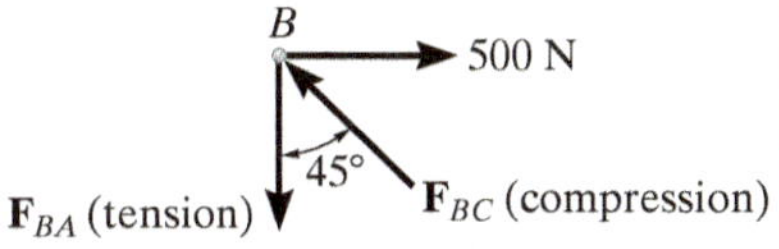

Method of Sections

The method of sections states that if a truss is in equilibrium, then each segment of the truss is also in equilibrium. Pass a section through the truss and the member whose force is to be determined. Then draw the free-body diagram of the sectioned part having the least number of forces on it.

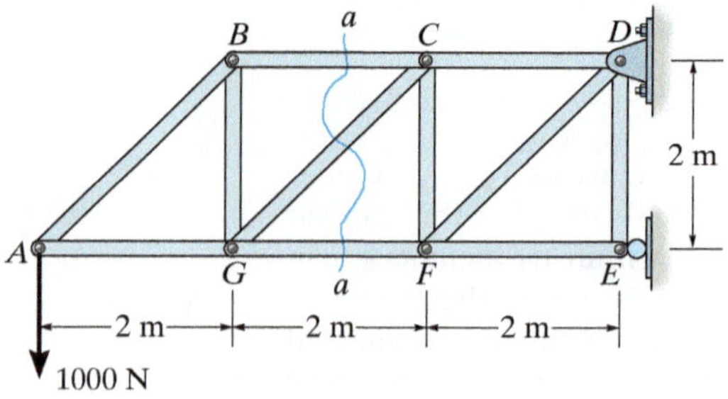

Sectioned members subjected to *pulling* are in *tension*, and those that are subjected to *pushing* are in *compression*.

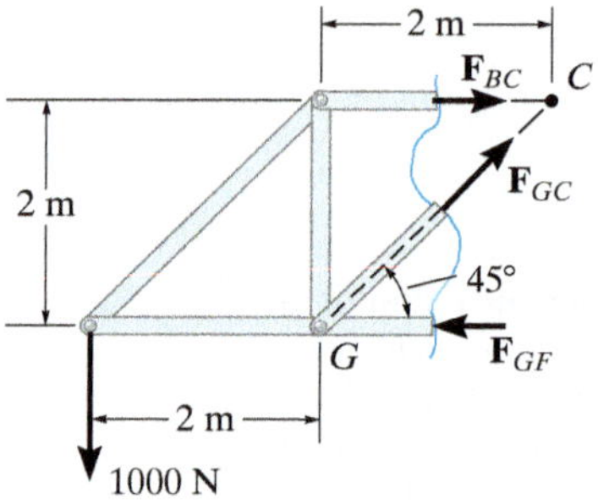

Three equations of equilibrium are available to determine the unknowns.

$$\Sigma F_x = 0$$
$$\Sigma F_y = 0$$
$$\Sigma M_O = 0$$

If possible, sum forces in a direction that is perpendicular to two of the three unknown forces. This will yield a direct solution for the third force.

$$+\uparrow \Sigma F_y = 0$$
$$-1000 \text{ N} + F_{GC} \sin 45^\circ = 0$$
$$F_{GC} = 1.41 \text{ kN (T)}$$

Sum moments about the point where the lines of action of two of the three unknown forces intersect, so that the third unknown force can be determined directly.

$$\circlearrowleft +\Sigma M_C = 0$$
$$1000 \text{ N}(4 \text{ m}) - F_{GF}(2 \text{ m}) = 0$$
$$F_{GF} = 2 \text{ kN (C)}$$

Space Truss

A space truss is a three-dimensional truss built from tetrahedral elements, and is analyzed using the same methods as for plane trusses. The joints are assumed to be ball-and-socket connections.

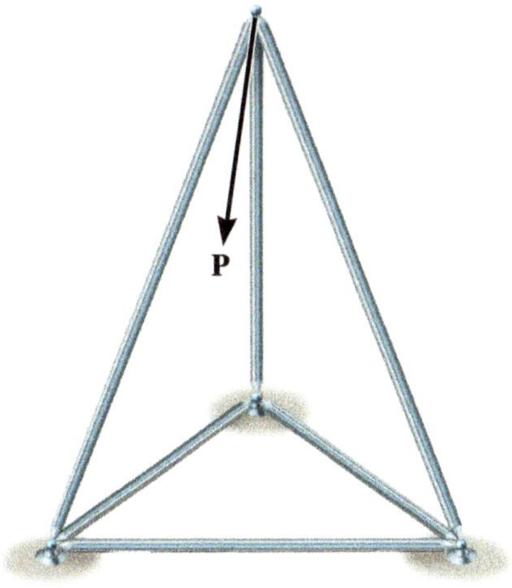

Frames and Machines

Frames and machines are structures that contain one or more multiforce members, that is, members with three or more forces or couples acting on them. Frames are designed to support loads, and machines transmit and alter the effect of forces.

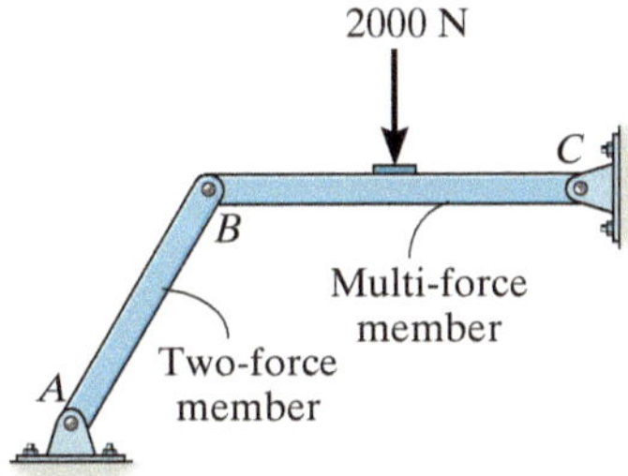

The forces acting at the joints of a frame or machine can be determined by drawing the free-body diagrams of each of its members or parts. The principle of action–reaction should be carefully observed when indicating these forces on the free-body diagram of each adjacent member or pin. For a coplanar force system, there are three equilibrium equations available for each member.

To simplify the analysis, be sure to recognize all two-force members. They have equal but opposite collinear forces at their ends.

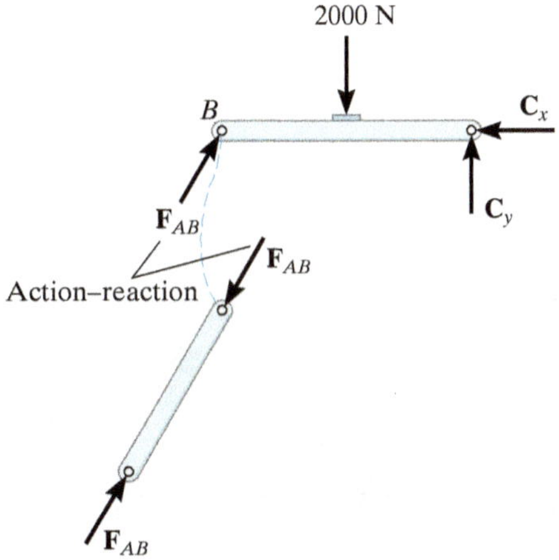

REVIEW PROBLEMS

6–119. Determine the resultant forces at pins *B* and *C* on member *ABC* of the four-member frame.

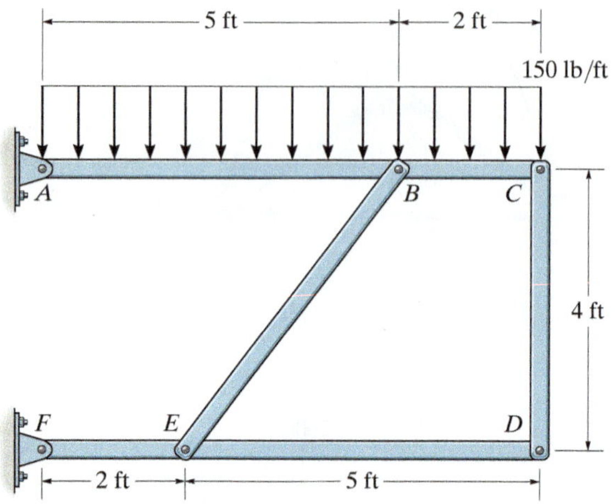

Prob. 6–119

***6–120.** Determine the force in each member of the truss and state if the members are in tension or compression.

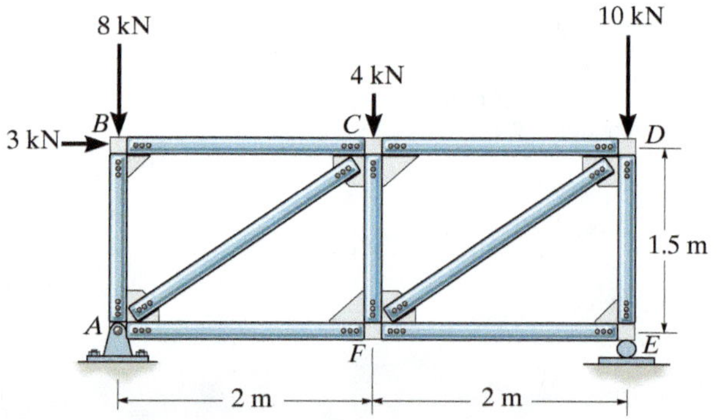

Prob. 6–120

6–121. Determine the horizontal and vertical components of force at pins *A* and *C* of the two-member frame.

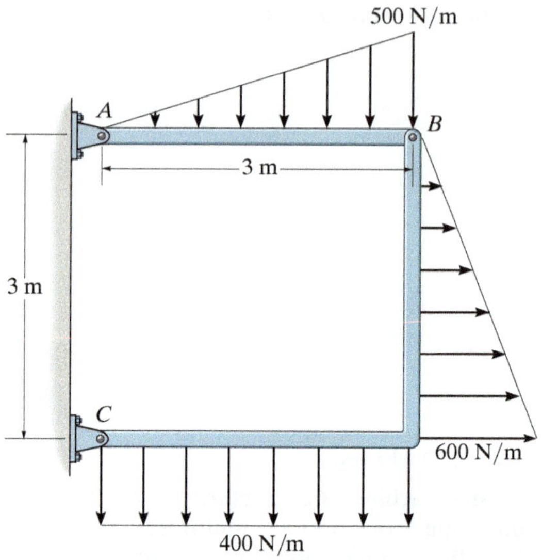

Prob. 6–121

6–122. Determine the force in members *AB*, *AD*, and *AC* of the space truss and state if the members are in tension or compression.

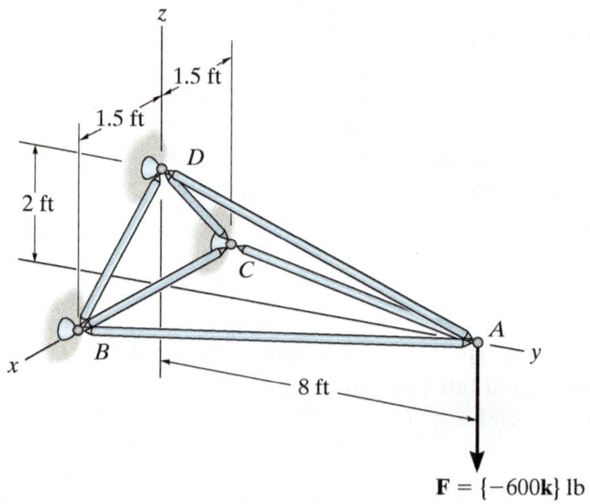

Prob. 6–122

6–123. The spring has an unstretched length of 0.3 m. Determine the mass m of each uniform link if the angle $\theta = 20^\circ$ for equilibrium.

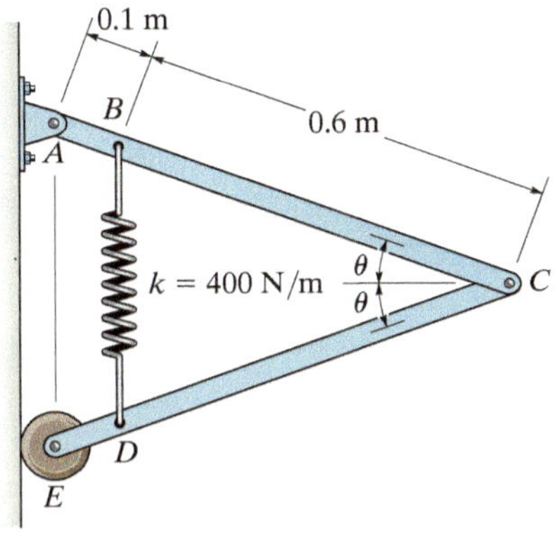

Prob. 6–123

6–125. Determine the horizontal and vertical components of force that pins A and B exert on the two-member frame. Set $F = 500$ N.

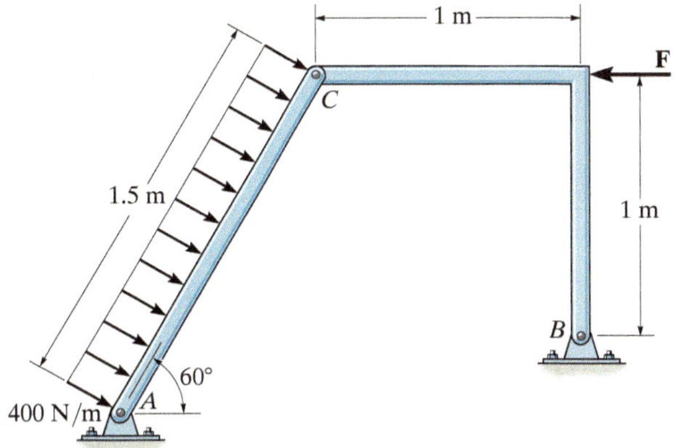

Prob. 6–125

***6–124.** Determine the horizontal and vertical components of force that the pins A and B exert on the two-member frame. Set $F = 0$.

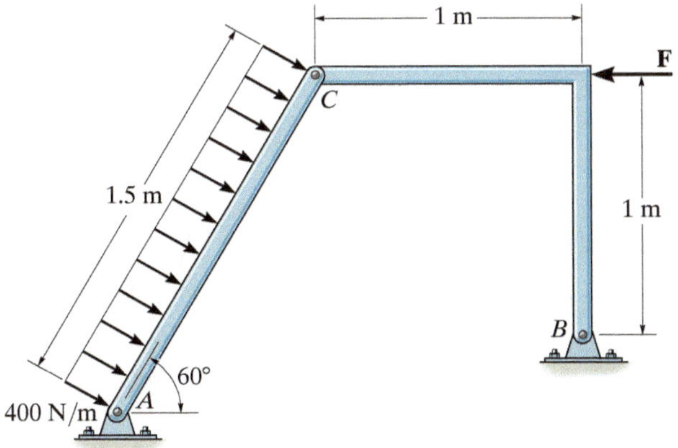

Prob. 6–124

6–126. Determine the force in each member of the truss and state if the members are in tension or compression.

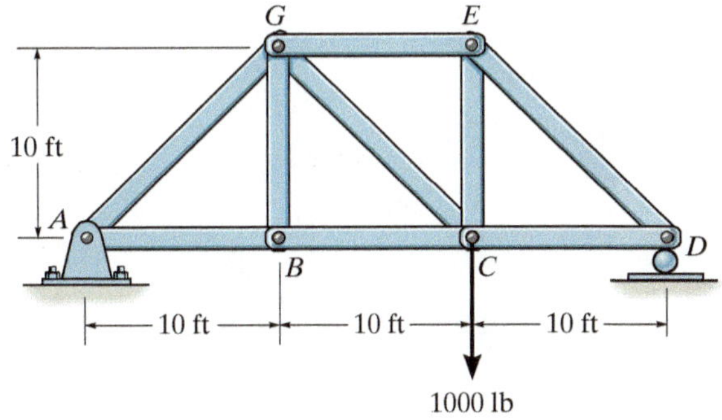

Prob. 6–126

CHAPTER 7

Section Forces in Beams

7.1 Introduction

In previous chapters, we considered a number of problems involving structural members loaded *axially* along their centroidal axes. The resultant force from the applied loads at any cross section of the member was easily established as a force normal to the cross section. Such members are referred to as rods, spars, or axial members. We now consider thin prismatic members loaded *transversely*—that is, perpendicularly to the centerline of the member. Generally, when structural members are loaded transversely we call them *beams.* Of considerable importance will be the components of the resultant force system from the applied loads acting on cross sections of the beam. Knowledge of these resultants will be used in later chapters dealing with stresses and displacements in beams. In this chapter we set our focus on developing methods for computing these force resultants.

7.2 Shear Force, Axial Force, and Bending Moment

Consider first a prismatic beam having a cross-section with a vertical axis of symmetry as shown at the upper right-hand corner of Fig. 7.1(a) (where the y axis is this axis of symmetry). A distributed loading $w(x)$ acts in the vertical plane of symmetry. In addition to this *transverse* loading, we also have an axial force P along the *longitudinal* axis of the beam [see Fig. 7.1(a)]. It will be assumed that the supporting

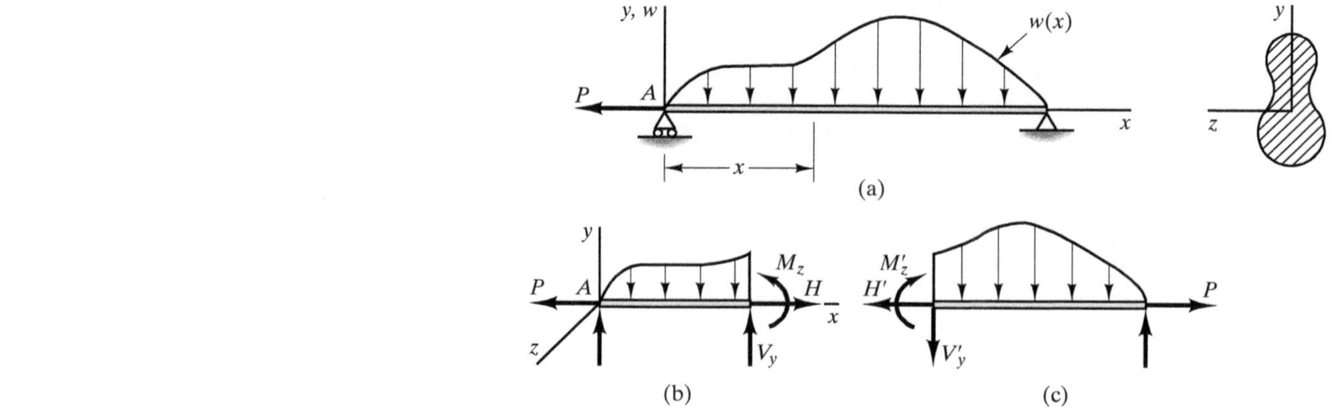

Figure 7.1. Left- and right-hand free-body diagrams exposing a section of the beam.

forces have been determined through statics. The focus of this chapter is to determine the force transmitted across any cross-sectional interface of the beam due to the applied transverse and longitudinal loads. How do we do this? We use our powerful tools of *equilibrium* and *free-body diagrams* to analyze the force resultant at any position along the beam.

We illustrate this in Fig. 7.1(b) where at a position x we make a "cut" in the beam and take a portion of the beam as a free-body so as to "expose" the cross section of the beam at x. Since the external loads acting on the beam are coplanar, it is clear that the load distribution acting on the cross section of the beam at position x must also be coplanar. We know from rigid-body mechanics that we can replace the coplanar load distribution acting on the cross-section of the beam at point x with its simplest resultant—either a single force or a single couple moment in the plane of the load distribution. If the simplest resultant is a single couple moment, it can be positioned arbitrarily. If the simplest resultant is a single force, it must have a particular line of action. However, the position of this line of action is of little interest in beam theory. *Consequently, we deliberately take the position of the resultant force to be at the centroid of the cross-section at all times.* Since we are constraining the resultant force to act through the section's centroid, we must therefore include the proper couple moment M_z to accompany the resultant force as required by static equilibrium. In addition, we decompose the resultant force into orthogonal components on the cross-section: in this case, a vertical force V_y and a horizontal force H. These quantities are shown in Fig. 7.1(b). Since these quantities are used to such a great extent in beam analysis, we have associated names with them. They are:

V_y = *shear-force* component
H = *axial-force* component
M_z = *bending-moment* component

If we had a three-dimensional load, there would have been one additional shear-force component V_z, one additional bending-moment component M_y, and a couple moment M_x along the axis of the beam. We shall call M_x the *twisting-moment* (see Fig. 7.2). We will deal with twisting of beams and shafts in preceding chapter.

Notice in Figure 7.1(c) that a second free-body diagram has been drawn that exposes the "other side" of the cross-section at position x. The shear force, axial force, and bending moment for this section have been primed in the diagram. We know from *Newton's third law* that they should be equal and opposite to the corresponding unprimed quantities in Fig. 7.1(b). We can thus choose for our computations either a left-hand or right-hand free-body diagram. But this poses somewhat of a problem for us when we come to reporting the signs of the transmitted forces and couple moment at a section. We cannot simply use the direction of a force or couple moment at the section since its sense at a section would depend on whether a left-hand or a right-hand free-body diagram was used. Therefore, to associate an unambiguous sign for shear-force, axial-force, and bending-moment at a section, a *sign convention* must be adopted. This brings us to an interesting point in our discussion. Currently, there are two different, but equally effective, sign conventions in popular use for beams. So that the student gets complete coverage of this important issue, we will describe both of them. In doing so, we will use the nomenclature of **positive** and **negative faces.** By a *positive face,* we mean a section of a body having its outward normal from the body in a positive coordinate direction. A *negative face* refers to a section of area having its outward normal from the body pointing in a negative coordinate direction.

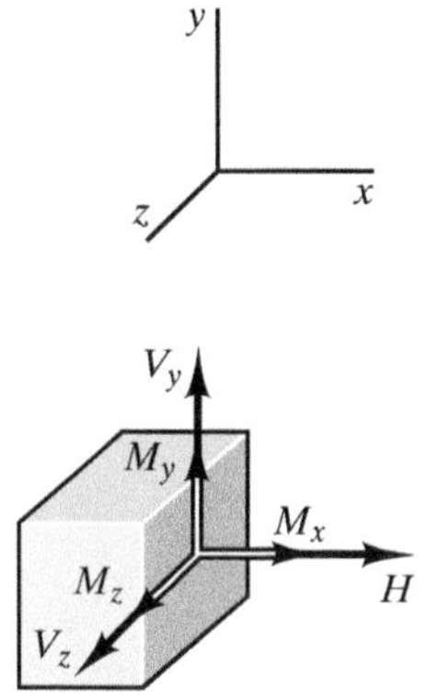

Figure 7.2. Section forces for three-dimensional loading.

For the first sign convention, called the **stress sign convention,** we adopt the following:

> The axial force, shear force, or bending moment acting on a beam cross-section is positive if it acts on a positive face and is directed in a positive coordinate direction.

We note that these components are also positive if they act on a negative face and are directed in a negative coordinate direction. Otherwise, the components are considered to be negative. The above assumes that a right-handed xyz coordinate system is used (as was shown in Fig. 7.1). Note that this is the same stress tensor sign convention that we introduced in previous chapter. It is identical to that commonly used in the theory of elasticity for the stress at a point and this is why we refer to this convention as the *stress sign convention.*

Consider Fig. 7.1. For the left-hand free-body diagram, the area vector for a beam section at a distance x from the origin points in the positive x coordinate direction. Note also that H, V_y and M_z, also point in positive directions of the xyz axes. Hence, according to the stress convention we have shown a *positive* shear force, a *positive* axial force and a *positive* bending moment at the section at x. For the right-hand free-body diagram the cross-sectional area vector points in the negative x coordinate direction. Since H', V_y' and M_z' point in negative directions of the xyz axes, these components are again positive for the section at x according to our convention. Clearly, only one sign is needed at a section for any of the section force resultants thus avoiding the aforementioned possibility of ambiguity.

For the second sign convention, called the **structural sign convention,** we adopt the following:

> An axial force or bending moment acting on a beam cross-section is positive if it acts on a positive face and is directed in a positive coordinate direction. The shear force is positive if it act in the *negative* coordinate direction on a *positive* face.

We see that the only difference between the two conventions is the sign associated with the shear force. Therefore, the axial-force and bending-moment components are also positive if they act on a negative face and are directed in a negative coordinate direction. For shear force, it is considered *negative* if it acts in the positive coordinate direction on a positive face. Again, we have assumed that a right-handed xyz coordinate system is used. Since this convention is often used in structural (civil) engineering, we refer to it as the *structural sign convention.*

Again consider Fig. 7.1 For the left-hand free-body diagram, the outward area vector for a beam section located a distance x from the origin points in the positive x direction. According to the structural sign convention we have shown a *positive* axial force, a *positive* bending moment, and a *negative* shear force at the section at x. The same is true for the right-hand free-body diagram.

As pointed out, we can solve for V_y, H, and M_z at section x using rigid-body mechanics for either a left-hand or a right-hand free-body diagram provided that we know all the external forces. The quantities V_y, H, and M_z will depend on x and so we can form equations for these quantities in terms of x. Also, it is the practice to sketch shear-force and bending-moment diagrams to convey this information for the entire beam. We shall now consider the computation of the shear force and bending moment as functions of x in the following problems. You are urged to examine them carefully. Later, we shall consider sketching these results in diagrams.

We shall use the **structural convention** in this text.

Example 7.1

We shall express the shear-force and bending-moment equations for the simply-supported beam in Fig. 7.3(a), whose weight we shall neglect. The supporting forces are seen by inspection to be 500 N each. Since there are no horizontal loads, there is no horizontal force at any cut in the beam.

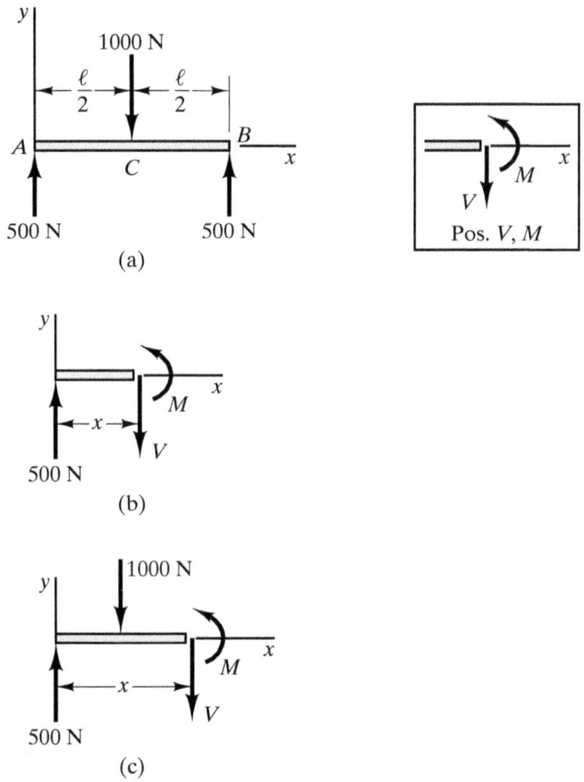

Figure 7.3. Simple beam problem requiring two domains.

To get the shear force and bending moment at a section x, we isolate the left or right side of the beam at x and employ the equations of equilibrium on the resulting free body. If x lies between A and C of the beam, then the only noninternal force present for a left-hand free body is the left supporting force [see Fig.7.3(b)]. Notice we have used directions for V and M (there is no need for subscripts in this simple problem) corresponding to the *positive* states from the point of view of the structural convention. Clearly, the *algebraic* sign we get for these quantities from equilibrium calculations where *we use the right-hand rule* will then correspond to the *convention* sign. If x is between C and B, for such a free body two external forces

Example 7.1 (Continued)

appear [see Fig. 7.3(c)]. Therefore, if the shear force and bending moment are to be expressed as a function of x, clearly *separate equations* covering the *two ranges* or *domains* are necessary.

Let us first consider the domain between A and C. It should be clear that at positions $x = 0$ and $x = l/2$ the shear force is indeterminate since at these positions there is a *point force.*[1] Therefore, in developing the shear-force equations we must *exclude* these points and thus the AC domain will be $0 < x < l/2$. Hence we can say for the AC domain on observing Fig. 7.3(b):

$\underline{0 < x < l/2}$:

$$500 - V = 0$$
$$\therefore V = 500 \text{ N}$$

We thus have positive shear force according to our structural convention for shear force. Next, consider the bending moment in the AC domain. It will be our practice in considering equilibrium to sum moments about the centroid of the cross-section at the cut section (located at a position x) in the domain of interest. In this way, the shear-force V contributes zero moment in our considerations at all times. Now examine Fig. 7.3(b). We see that the magnitude of the bending-moment M equals $(500)(x)$ N-m and that there is no problem of an indeterminacy of M at $x = 0$. That is, the value of M at $x = 0$ in this problem must be zero and nothing else. The value of M at $x = l/2$ also suffers no indeterminacy. Therefore, the domain for which M can be computed here includes $x = 0$ and $x = l/2$. Thus, from equilibrium, where we use *the right-hand rule,*we can say:

$\underline{0 \le x \le l/2}$:

$$-500x + M = 0$$
$$\therefore M = 500x \text{ N-m}$$

If we had a couple-moment C at $x = 0$, the bending moment would be zero on the left face cross section at A (i.e., at $x = 0^-$) and would be equal in magnitude to the magnitude of C on the opposite face of this cross-section (i.e., at $x = 0^+$), thereby introducing a discontinuity of M at $x = 0$, with the result that M would be indeterminate at $x = 0$.[2] Clearly, the domain for M would *exclude* $x = 0$ and we would have $0 < x \le l/2$.

[1]This will be seen more clearly when we draw shear-force diagrams.
[2]This also will be more clear when we draw the bending-moment diagrams.

Example 7.1 (Continued)

For simplicity when specifying V and M in a domain, we shall specify the limits of the domain to exclude discontinuities in *either* V and/or M. Thus, we can say for the AC domain:

$\underline{0 < x < l/2}$:

$$V = 500 \text{ N} \tag{a}$$
$$M = 500x \text{ N-m} \tag{b}$$

Now we are ready for the second domain of our problem. Thus, we can say on observing Fig. 7.3(c):

$\underline{l/2 < x < l}$:

$$500 - 1000 - V = 0$$
$$\therefore V = -500 \text{ N} \tag{c}$$
$$-500x + 1000(x - l/2) + M = 0$$
$$\therefore M = 500(l - x) \text{ N-m} \tag{d}$$

Notice from Eqs. (a) and (c) how as x passes point C, where we have a point force (1000 N), the shear force *jumps* by -1000 N, showing why we excluded the position $x = l/2$ from the domain for shear force. Notice, however, that from Eqs. (b) and (d) no such jump occurs for M at $x = l/2$. Thus we have

$\underline{0 < x < l/2}$

$$V = 500 \text{ N}$$
$$M = 500\text{x N-m}$$

$\underline{l/2 < x < l}$:

$$V = -500 \text{ N}$$
$$M = 500(l - x) \text{ N-m}$$

Example 7.2

Determine the shear-force and bending-moment equations for the simply-supported beam in Fig. 7.4. Neglect the weight of the beam.

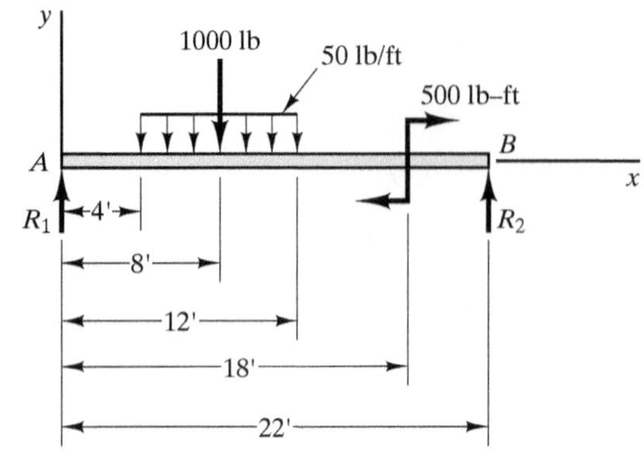

Figure 7.4. Simply supported beam.

By summing to zero moments about each end of the beam, we readily can find from statics

$$R_1 = 868 \text{ lb} \qquad R_2 = 532 \text{ lb}$$

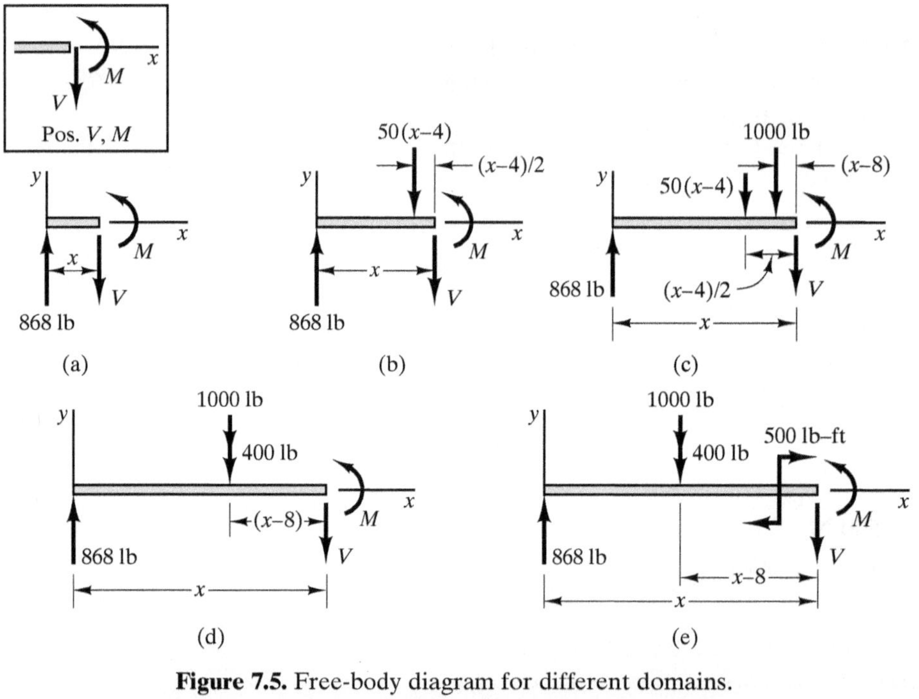

Figure 7.5. Free-body diagram for different domains.

Example 7.2 (Continued)

In Fig. 7.5(a) we have shown a free-body diagram exposing sections between the left support and the uniform load. Summing forces and taking moments about a point in the section, where we have drawn V and M as positive according to the structural convention, we get:

$\underline{0 < x \leq 4:}$

$$868 - V = 0$$
$$\therefore V = 868 \text{ lb}$$
$$-868x + M = 0$$
$$\therefore M = 868x \text{ ft-lb}$$

The next interval is between the beginning of the uniform load and the point force. Thus, observing Fig. 7.5(b), we get using the resultant of the uniform loading to the left of x:

$\underline{4 \leq x < 8:}$

$$868 - 50(x - 4) - V = 0$$
$$\therefore V = -50x + 1068 \text{ lb}$$
$$-868x + \frac{50(x - 4)^2}{2} + M = 0$$
$$\therefore M = -25x^2 + 1068x - 400 \text{ ft-lb}$$

We now consider the interval between the 1000-lb point force and the end of the uniform load. Thus, observing Fig. 7.5(c), we get:

$\underline{8 < x \leq 12:}$

$$868 - 50(x - 4) - 1000 - V = 0$$
$$\therefore V = -50x + 68 \text{ lb}$$
$$-868x + \frac{50(x - 4)^2}{2} + 1000(x - 8) + M = 0$$
$$\therefore M = -25x^2 + 68x + 7600 \text{ ft-lb}$$

The next interval is between the end of uniform loading and the point couple. We can now replace the uniform loading by its total resultant of 400 lb, as shown in Fig. 7.5(d). Thus:

$\underline{12 \leq x < 18:}$

$$868 - 400 - 1000 - V = 0$$
$$\therefore V = -532 \text{ lb}$$
$$-868x + 1400(x - 8) + M = 0$$
$$\therefore M = -532x + 11{,}200 \text{ ft-lb}$$

The last interval goes from the point couple to the right support. It is to be pointed out that the point couple does not contribute directly

Example 7.2 (Continued)

to the shear force, so we could have used the formulation for V above for the interval $18 < x < 22$. However, the couple does contribute directly to the bending moment and thus the additional interval is required. Accordingly, using Fig. 7.5(e), we get:

$18 < x < 22$:

$$V = -532 \text{ lb (as in previous interval)}$$
$$-868x + 1400(x - 8) - 500 + M = 0$$
$$\therefore M = -532x + 11{,}700 \text{ ft-lb}$$

In this example, we have carried out all the algebraic operations, as would be your natural inclination. Actually, you will be glad to know that it is more useful *not* to do this. For instance, for the range $8 < x \leq 12$, we suggest that you write the results as follows:

$8 < x \leq 12$:

$$V = -50(x - 4) - 132 \text{ lb}$$
$$M = 868x - 25(x - 4)^2 - 1000(x - 8) \text{ ft-lb}$$

We shall henceforth follow this practice.

Before we proceed further, it must be emphatically pointed out that the replacement of a distributed load, such as the uniform loading in Example 7.2, by a single resultant force is meaningful only for the *particular free* body on which the force distribution acts. For example, to compute the reactions for an *entire* beam taken as a free body, we can, if we include the weight, replace the *weight distribution* by the *total* weight at position $l/2$. For the bending moment at x, the resultant of the weight for the free body becomes wx and is *midway* at position $x/2$. In other words, in making shear-force and bending-moment equations and diagrams, we cannot replace loading distributions over the entire beam by a resultant and then proceed; there is inherent in these equations an *infinite number* of free bodies, each smaller than the beam itself, which makes the above-mentioned replacements invalid for shear-force and bending-moment considerations.

Finally, we wish to point out that for triangular loading distributions you should make good use of the well-known geometrical characteristics of the right triangle. Thus the resultant force is the area, which is 1/2 the base a times the height b. Furthermore, the resultant has a line of action which is at a distance $a/3$ from the height b, as has been indicated in Appendix IV-E

In the following example, we shall consider a cantilever beam carrying a triangular load.

Example 7.3

Determine the shear and bending-moment equations for the cantilever beam shown in Fig. 7.6 carrying a triangular load.

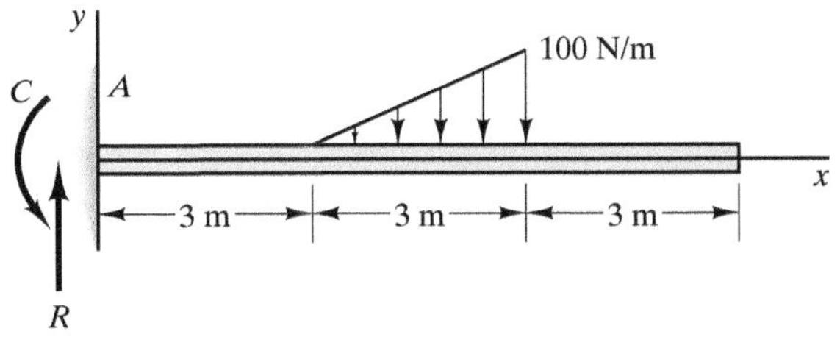

Figure 7.6. Cantilever beam with a triangular load.

The supporting force system shown by C and R is first established using as the resultant force from the triangular loading the area of the triangle with the resultant going through the centroid of the area. Thus,

$\underline{\Sigma F_y = 0}$:

$$R - \tfrac{1}{2}(3)(100) = 0$$
$$R = 150 \text{ N}$$

$\underline{\Sigma M_A = 0}$: (using the right-hand rule of statics)

$$C - [\tfrac{1}{2}(3)(100)][3 + \tfrac{2}{3}(3)] = 0$$
$$C = 750 \text{ N-m}$$

We now consider the first domain [see Fig. 7.7(a)].

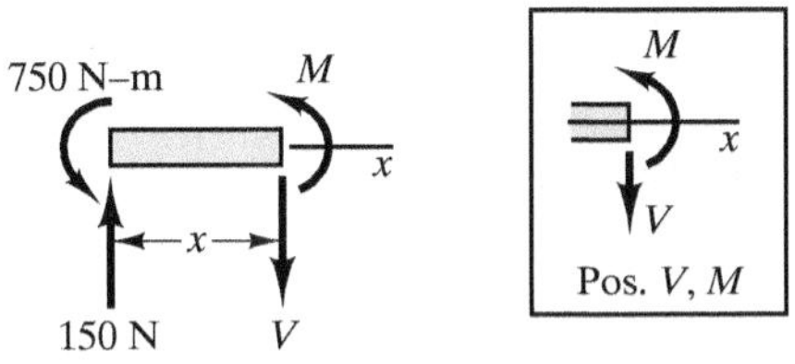

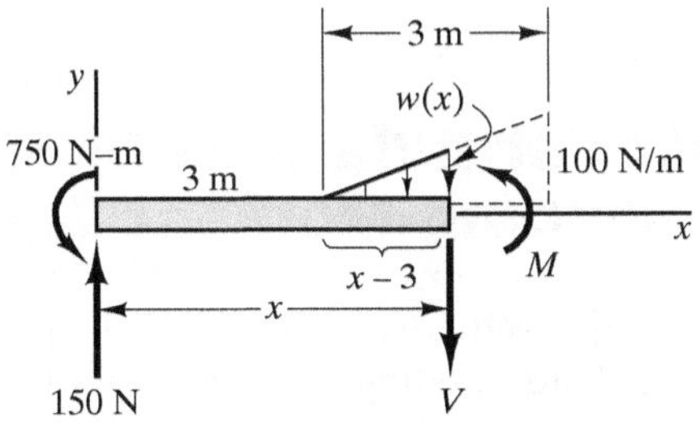

Figure 7.7. Free bodies for two domains of the beam.

Example 7.3 (Continued)

$\underline{0 < x \leq 3:}$

$$150 - V = 0$$
$$V = 150 \text{ N}$$
$$750 - (150)(x) + M = 0$$
$$M = 150x - 750 \text{ N-m}$$

For the second domain [see Fig. 7.7(b)], we need $w(x)$. From similar triangles, we can say that

$$\frac{w(x)}{100} = \frac{x - 3}{3}$$
$$\therefore\ w(x) = \frac{100}{3}(x - 3) \text{ N/m}$$

Hence:

$\underline{3 \leq x \leq 6:}$

$$150 - \frac{1}{2}(x - 3)\left[\frac{100}{3}(x - 3)\right] - V = 0$$
$$V = 150 - \frac{50}{3}(x - 3)^2 \text{ N}$$
$$750 - (150)(x) + \frac{1}{2}(x - 3)\left[\frac{100}{3}(x - 3)\right]\left[\frac{1}{3}(x - 3)\right] + M = 0$$
$$M = -750 + 150x + \frac{50}{9}(x - 3)^3 \text{ N-m}$$

Finally, it should be clear by inspection (consider right-hand free-body diagrams for $x > 6$) that the last domain is free of shear and bending moment. Thus,

$\underline{6 \leq x \leq 9:}$

$$V = M = 0$$

7.3 Direct Formulations of Shear-Force and Bending-Moment Equations

As an alternative approach to that presented in Section 7.2 one can write the shear-force and bending-moment equations in a more direct manner without drawing the simple but bothersome free-body diagrams. The idea is to take one load at a time and mentally satisfy equilibrium while adhering to the sign convention for shear and bending moment. Although this procedure takes a little time to master, it can result in significant overall savings of time. The procedure is as follows. Consider an upward force of magnitude P [see Fig. 7.8(a)]. To maintain this force in equilibrium on sections to the right of it requires a

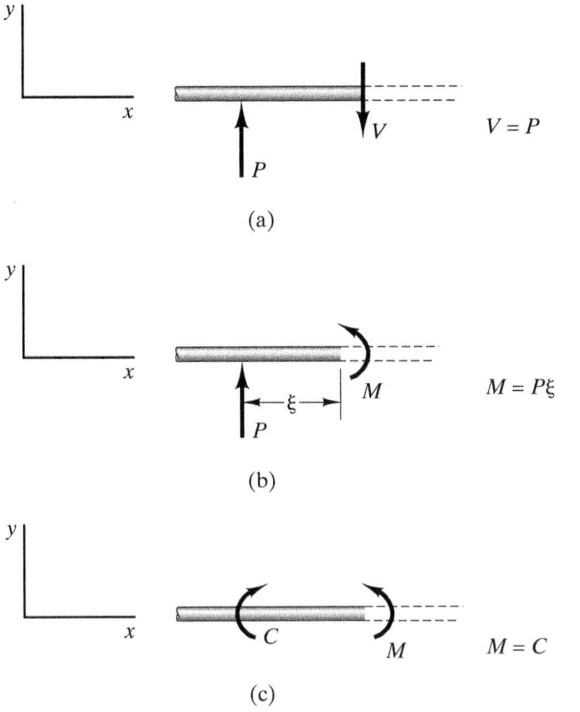

Figure 7.8. Shear forces and bending moments induced by an external force and an external couple moment.

downward shear force of the same magnitude. For our convention, this will be a positive shear force [see Fig. 7.8(a)]. Similarly, a downward force P will induce to the right of it a negative shear force according to our convention. Also, to equilibrate the upward external force at any section a distance ξ to the right of the load, a counterclockwise couple moment of magnitude $P\xi$ is needed as shown in [Fig. 7.8(b)]. The sign for this bending-moment M is clearly positive. For a downward force P, the sign would be negative. Next, a clockwise external couple moment

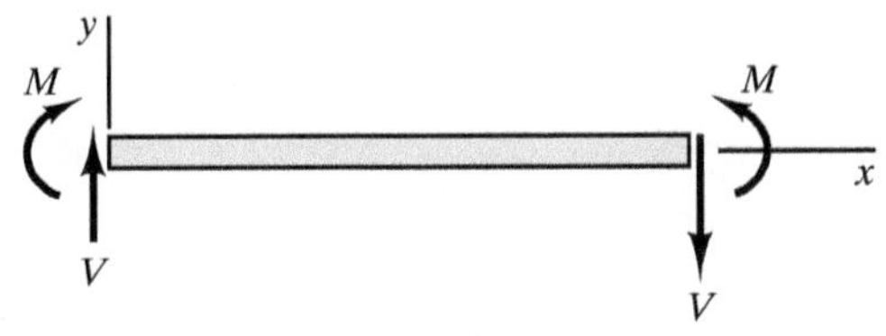

Figure 7.9. Positive shear forces and bending moments.

of magnitude C will require an opposite couple moment to the right of it [see Fig. 7.8(c)]. The induced bending moment will be positive according to our convention. A counterclockwise couple moment clearly will induce a negative bending moment.

We urge the student to use this more efficient procedure when possible for writing shear-force and bending-moment equations. Fig. 7.9 shows positive shear forces and positive bending moments at each end of a beam.

Example 7.4

Evaluate the shear-force and bending-moment equations for the beam shown in Fig. 7.10(a) *without* the aid of free-body diagrams in forming these equations.

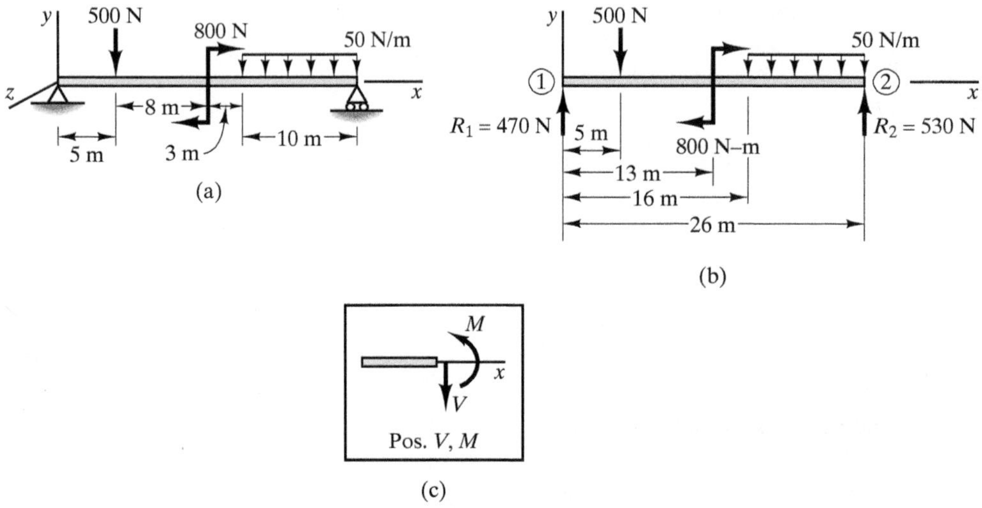

Figure 7.10. Simply supported beam problem showing positive convention for V and M.

We will however use the free-body diagram of the entire beam [Fig. 7.10(b)] to evaluate the supporting forces R_1 and R_2. Equating moments to zero at the ends of the beam, we get, using the right-hand rule of statics,

$$R_1 = 470 \text{ N} \qquad R_2 = 530 \text{ N}$$

We may now directly give the shear-force V and bending-moment M while viewing Fig. 7.10(b). Thus:

$0 < x < 5$:

$$V = R_1 = 470 \text{ N}$$
$$M = 470x \text{ N-m}$$

Example 7.4 (Continued)

$5 < x < 13$:

$$V = 470 - 500 = -30 \text{ N}$$
$$M = 470x - 500(x - 5) \text{ N-m}$$

$13 < x \leq 16$:

$$V = -30 \text{ N (same as previous interval)}$$
$$M = 470x - 500(x - 5) + 800 \text{ N-m}$$

$16 \leq x < 26$:

$$V = 470 - 500 - 50(x - 16) = -30 - 50(x - 16) \text{ N}$$
$$M = 470x - 500(x - 5) + 800 - \frac{50(x - 16)^2}{2} \text{ N-m}$$

Note we have not carried out the arithmetic for M. As indicated earlier, it is more useful to leave the bionomial expressions intact.

Example 7.5

Write the shear and bending-moment equations for the simply supported overhanging beam shown in Fig. 7.11.

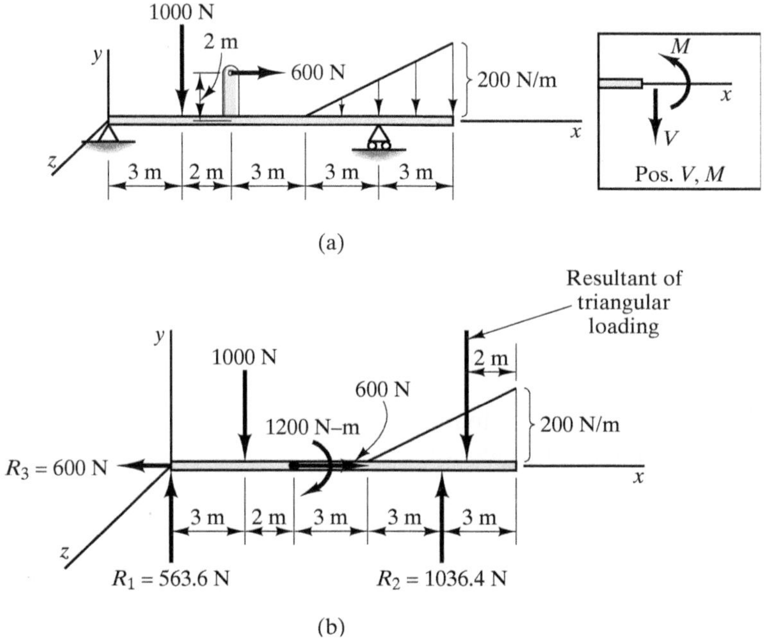

Figure 7.11. Simply supported beam with overhang.

Example 7.5 (Continued)

We shall first find the supporting force system using the free-body diagram shown in Fig. 7.11(b) and using the right-hand rule of statics.

$\Sigma M_{R_2} = 0$:

$$-R_1(11) + (1000)(8) - 1200 - \tfrac{1}{2}(6)(200)(1) = 0$$
$$R_1 = 563.6 \text{ N}$$

$\Sigma M_{R_1} = 0$:

$$R_2(11) - (1000)(3) - 1200 - \tfrac{1}{2}(6)(200)(12) = 0$$
$$R_2 = 1036.4 \text{ N}$$

$\Sigma F_x = 0$:

$$-R_3 + 600 = 0$$
$$R_3 = 600 \text{ N}$$

We now proceed with the shear-force and bending-moment equations for the various domains.

$0 < x < 3$:

$$V = 563.6 \text{ N}$$
$$M = 563.6x \text{ N-m}$$

$3 < x < 5$:

$$V = 563.6 - 1000 = -436.4 \text{ N}$$
$$M = 563.6x - 1000(x - 3) \text{ N-m}$$

$5 < x \leq 8$:

$$V = -436.4 \text{ N}$$
$$M = 563.6x - 1000(x - 3) + 1200 \text{ N-m}$$

For the next two domains we will need *w(x)* of the triangular loading, for which we refer you to Fig. 7.12. From similar triangles we can see that

$$\frac{w(x)}{200} = \frac{x - 8}{6}$$
$$w(x) = \frac{200}{6}(x - 8)$$

Example 7.5 (Continued)

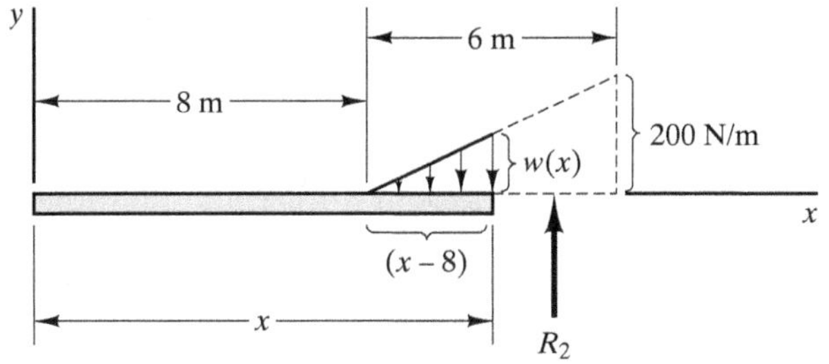

Figure 7.12. Diagram showing triangular loading contributions for domain $8 \leq x < 11$.

Now continuing:

$8 \leq x < 11$:

$$V = -436.4 - \frac{1}{2}(x-8)\left[\frac{200}{6}(x-8)\right]$$

$$= -436.4 - \frac{50}{3}(x-8)^2 \text{ N}$$

$$M = 563.6x - 1000(x-3) + 1200 - \frac{1}{2}(x-8)\left[\frac{200}{6}(x-8)\right]\left[\frac{1}{3}(x-8)\right]$$

$$= 563.6x - 1000(x-3) + 1200 - \frac{50}{9}(x-8)^3 \text{ N-m}$$

$11 < x \leq 14$:

$$V = -436.4 - \frac{50}{3}(x-8)^2 + 1036.4$$

$$= 600 - \frac{50}{3}(x-8)^2 \text{ N}$$

$$M = 563.6x - 1000\,(x-3) + 1200 - \frac{50}{9}(x-8)^3 + 1036.4\,(x-11) \text{ N-m}$$

7.4 Differential Relations for Bending Moment, Shear Force, and Load

In the design of structures, we must often be able to quickly determine the magnitude and location of peak stresses. As we will see in the next chapter, the stresses in beams are a function of (among other variables) the bending moment and shear force at a given cross section.

Consequently, the rapid determination of the maximum values of M_z and V_y in a beam is essential to the design of these structures. From the methods presented in the previous section, we have at our disposal the equations describing the shear force and bending moment at any position along the beam. These equations may be plotted or examined mathematically to determine the location and magnitude of the maximum M_z and V_y in the beam. However, this can be a time consuming exercise. In what follows, we present *differential* relationships between the applied load, shear force, and bending moment. These differential relationships and their integrals may then be used to rapidly *sketch* shear-force and bending-moment diagrams for simple but common loads, allowing for the quick assessment of critical values of M_z and V_y along the beam.

We start by examining an infinitesimal slice of the beam Δx, shown in Fig. 7.13. We adopt the standard structural convention that a *downward* loading w on the beam is *positive.* Furthermore, we shall use

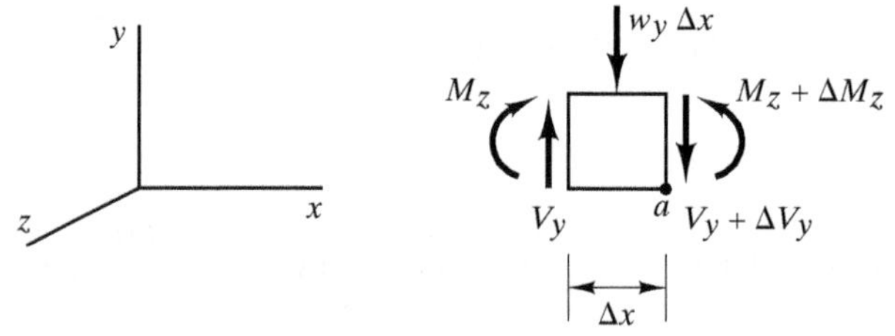

Figure 7.13. Beam element; structural convention.

a subscript for indicating the direction of the loading. Thus, we employ positive w_y for the loading intensity directed along the negative y axis. We have approximated the resultant of the loading w_y over the slice by the quantity $w_y\Delta x$ where w_y is a value of the loading intensity somewhere in the interval Δx. As Δx becomes smaller, this evaluation becomes more accurate and in the limit, as $\Delta x \to 0$, it becomes exact. We shall assume here that the weight of the beam has been included in the intensity of loading so that all force distributions acting on the element have been shown in the diagram. Again, we have employed the structural sign convention in the free-body diagram so that all forces and couple moments are shown in their positive senses in accord with the structural convention.

We now apply the equations of **equilibrium.** Thus, summing forces in the vertical direction (see Fig. 7.13), we get:

$\underline{\Sigma F_y = 0:}$

$$V_y - (V_y + \Delta V_y) - w_y\Delta x = 0$$

Taking moments about corner a of the element, we get using the right-hand rule of statics

$\underline{\Sigma M_a = 0:}$

$$\therefore -M_z - V_y\Delta x + (w_y\Delta x)(\beta\Delta x) + (M_z + \Delta M_z) = 0$$

where β is some fraction that, when multiplied by Δx, gives the proper moment arm of the force $w_y\Delta x$ about corner a. These equations can be written in the following manner after we cancel terms and divide through by Δx:

$$\frac{\Delta V_y}{\Delta x} = -w_y$$

$$\frac{\Delta M_z}{\Delta x} = V_y - w_y\beta\Delta x$$

In the limit as $\Delta x \to 0$ we get the following differential equations relating loading, shear-force, and bending moment.[3]

$$\frac{dV_y}{dx} = -w_y \tag{7.1a}$$

$$\frac{dM_z}{dx} = V_y \tag{7.1b}$$

Differentiating Eq. (7.1b) with respect to x and then replacing the right-hand side of the equation using Eq. (7.1a), we get

$$\frac{d^2M_z}{dx^2} = -w_y(x) \tag{7.2}$$

The above are differential equations that relate the distributed load w_y to the shear-force V_y and the shear force to the bending-moment M_z, respectively. We can solve these differential equations by integrating Eqs. (7.1) between two positions on the beam. Thus, we have

[3]We note that Eq. (7.1a) is not valid at the location of a point load since the rate of change of the shear with respect to x is indeterminate at such points. Furthermore, a point load induces a jump in the magnitude of the shear force. For a similar reason, Eq. (7.1b) is not valid at the location of a point couple.

$$(V_y)_2 - (V_y)_1 = -\int_1^2 w_y\,dx$$

$$\therefore\ (V_y)_2 = (V_y)_1 - \int_1^2 w_y\,dx \tag{7.3}$$

$$(M_z)_2 - (M_z)_1 = \int_1^2 V_y\,dx$$

$$\therefore\ (M_z)_2 = (M_z)_1 + \int_1^2 V_y\,dx \tag{7.4}$$

Equation (7.3) means that the change in the shear force between two points on a beam equals *minus* the area (which may be positive or negative) under the loading curve between these points, provided there is no point force present in the interval. Similarly, Eq. (7.4) indicates that the change in bending moment between two points on a beam equals the area (which may be positive or negative) of the shear-force curve between these points, provided there are no point couples applied in the interval. In sketching the diagrams for the shear-force and bending-moment diagrams, we shall make good use of Eqs. (7.1), (7.3), and (7.4).

7.5 Sketching Shear-Force and Bending-Moment Diagrams

As we pointed out earlier, we do not wish to exactly plot the shear-force and bending-moment diagrams for the beam. We are mainly interested in the locations of the peak shear force and bending moment and in the general shape of the diagrams. Therefore, we would like to simply *sketch* the diagrams showing the general shape of, and key points for the shear force and bending moment. Here is how we propose to use Eqs. (7.1) through (7.4) to sketch the diagrams. Consider a generic portion of a loading diagram that does not contain any point forces as shown in [Fig. 7.14(a)]. Directly below, we are plotting the shear-force diagram wherein at the outset only the shear at point ① (i.e., V_1) is known. The first step in sketching the shear-force curve from position ① to position ②, is to employ Eq. (7.3). The value of V_2 according to this equation equals V_1 *minus* the area under the loading curve between positions ① and ② (the area A_w). This area is **positive** so that $V_2 = V_1 - (A_w) = V_1 - A_w$. Thus there is a decrease in the value of V as we go from position ① to position ② as has been shown in the diagram. Next, we want to sketch the shape of the shear-force diagram between the points ① and ②. This curve will take on one of the following three

$$\frac{dV}{dx} = -w$$
$$V_2 = V_1 - \int_1^2 w\,dx$$

$$\frac{dM}{dx} = +V$$
$$M_2 = M_1 + \int_1^2 V\,dx$$

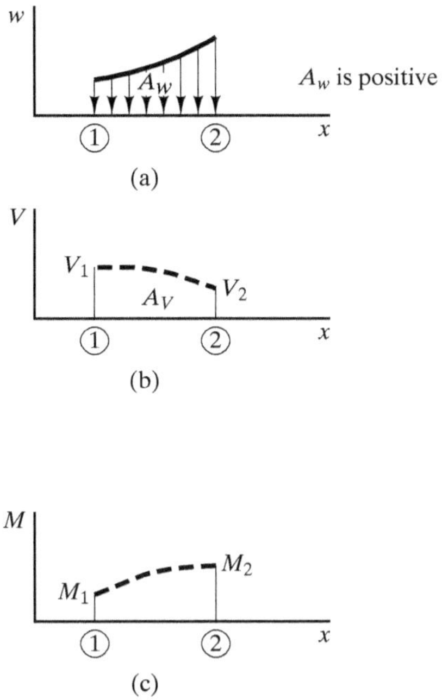

Figure 7.14. Sample construction of shear-force and bending-moment curves.

forms: a **straight line**, a curve that is **steepening**, or a curve that is **flattening**. See Fig. 7.15 for the nomenclature that we use to describe the manner in which the slope of the curve is changing as we move along the beam between the points ① and ②. The nomenclature should be clear if you imagine you are skiing or climbing along these curves. Now, we need to determine which of the three cases we have. From Eq. (7.1), we can examine the magnitude of the slope of the shear curve. That is, we examine the equation

$$\left|\frac{dV_y}{dx}\right| = |-w_y|$$

So, as we move along the beam, we can get the magnitude of the slope of the shear-force diagram from the magnitude of the loading diagram directly above. Thus, we can say that if between points ① and ②:

(1). the magnitude of loading w_y is *constant,* then the magnitude of the slope of the shear-force curve is *constant;* that is, the curve is a *straight line.*

(2). the magnitude of loading w_y is *increasing,* then the magnitude of the slope of the shear-force curve is *increasing;* that is, the curve is *steepening.*

(3). the magnitude of the loading is *decreasing,* then the magnitude of the slope of the shear-force curve is *decreasing;* that is, the curve is *flattening.*

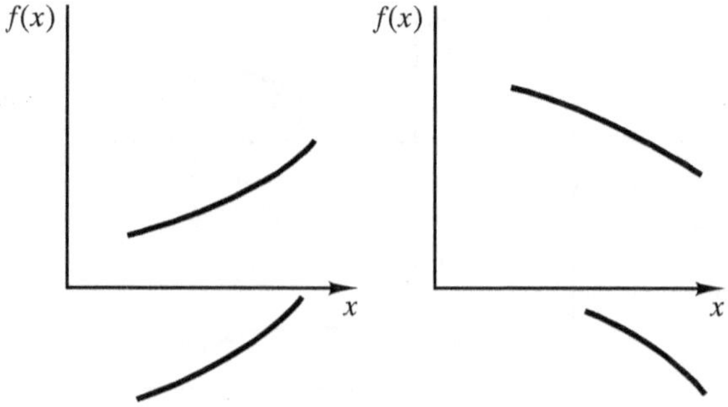

(a). $\left|\frac{df}{dx}\right|$ is *increasing*; slope is *steepening*

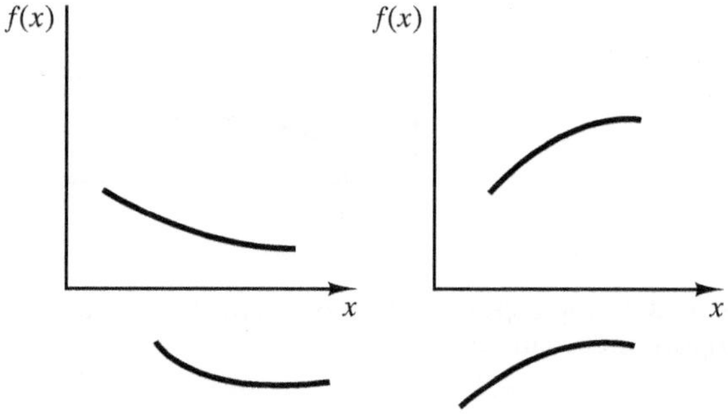

(b). $\left|\frac{df}{dx}\right|$ is *decreasing*; slope is *flattening*

Figure 7.15. Our nomenclature for the way a slope is changing as we increase x.

We see that in Fig. 7.14(a) $|w|$ is increasing. Hence, we connect point ① and ② with a *steepening* curve.

We now move to the bending-moment diagram and we will use a similar approach for sketching it as we did for the shear-force diagram. We start at point ① where M_1 is known and we assume that there are no point couples in the region from ① to ②. From the shear-force diagram in Fig. (7.14), we compute the area under the shear-force diagram between points ① and ②. We have called this area A_V. To sketch this portion of the bending-moment diagram, we use Eq. (7.4) to get the bending moment at position ②, (i.e., M_2). We have a positive area in the shear-force diagram between position ① and ② and so we must *add* this area to M_1 to get M_2. Thus, M_2 is larger than M_1 as shown in Fig. 7.14(c). Next, we must get the general shape of the bending-moment curve between points ① and ②. For this purpose, we use [Eq. 7.1(b)]. Using magnitudes, this equation can be given as follows:

$$\left|\frac{dM_z}{dx}\right| = |V_y|$$

So, as we move along the beam, we get the magnitude of the slope of the bending-moment curve from the magnitude of the shear-force curve directly above. Using the nomenclature of Fig. 7.15, we can say that if between points ① and ②:

1. the magnitude of the shear-force V_y is constant, then the magnitude of the slope of the bending-moment curve is *constant;* that is, the curve is a *straight line.*
2. the magnitude of the shear-force V_y is *increasing,* then the magnitude of the slope of bending-moment curve is *increasing;* that is, the curve is *steepening.*
3. the magnitude of the shear-force V_y is *decreasing,* then the magnitude of the slope of the bending-moment curve is *decreasing;* that is, the curve is *flattening.*

When the above procedure is used, we obtain in Fig. (7.14) a flattening curve for the bending moment between points ① and ②.

We now examine a beam problem using the approach set forth for sketching the shear-force and bending-moment diagrams for beams. In particular, note for the loading diagrams we shall insert point forces and point couples with their given or statically determined directions. It is the **loading** $\boldsymbol{w}$ that must be shown with the proper sign in accordance with the convention.

Example 7.6

Determine the shear-force and bending-moment equations for the simply supported beam shown in Fig. 7.16 and sketch the corresponding diagrams.

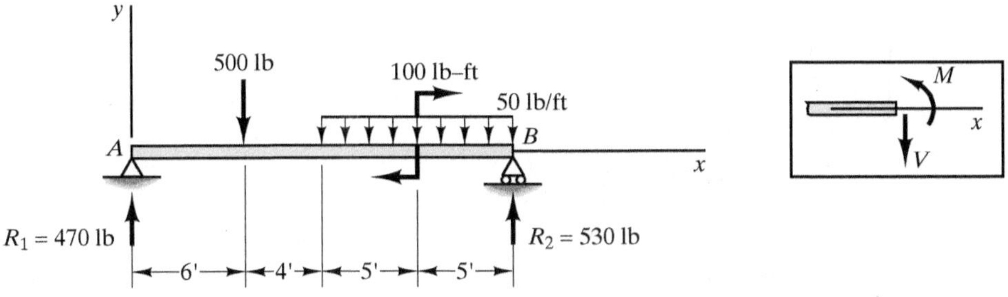

Figure 7.16. Simply supported beam.

We can directly state the shear-force and bending-moment equations as follows on observing Fig. 7.16:

Example 7.6 (Continued)

$0 < x < 6$:

$$V = R_1 = 470 \text{ lb}$$
$$M = R_1 x = 470x \text{ ft-lb}$$

$6 < x \leq 10$:

$$V = 470 - 500 = -30 \text{ lb}$$
$$M = 470x - 500(x - 6) \text{ ft-lb}$$

$10 \leq x < 15$:

$$V = 470 - 500 - 50(x - 10) \text{ lb}$$
$$M = 470x - 500(x - 6) - 50(x - 10)^2 / 2 \text{ ft-lb}$$

$15 < x < 20$:

$$V = 470 - 500 - 50(x - 10) \text{ lb}$$
$$M = 470x - 500(x - 6) - 50(x - 10)^2 / 2 + 100 \text{ ft-lb}$$

In Fig. 7.17, we show the loading diagram for the beam. Note that the *downward* 50 lb/ft loading is drawn as *positive* in the diagram in accordance with the structural convention.

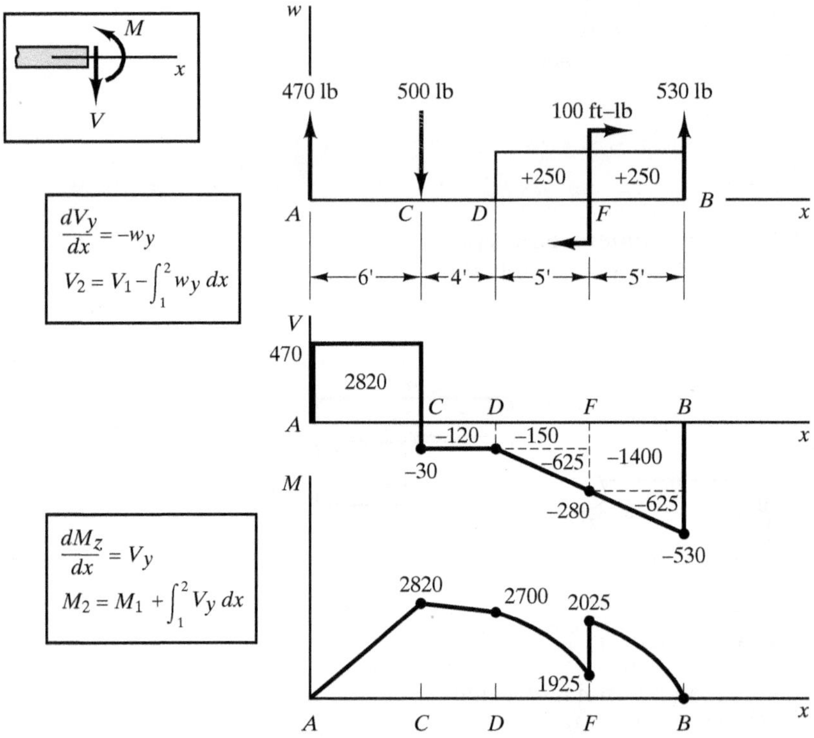

Figure 7.17. Shear and bending moment diagrams.

Example 7.6 (Continued)

Now going to the shear-force diagram, note that the upward force at the left end induces a downward shear force to the right of it. This downward shear force must be positive according to our sign convention and so the shear force jumps upward. Thus, *the jumps in shear force for point forces are in the same direction as the point forces.*[4] Since there is zero loading between A and C, the shear force is constant until we reach point C where there is a sudden downward drop equal to the 500 lb. load. From C to D, there is zero loading and thus the shear force in this region remains constant. Thus, we have horizontal lines for V from A to C and from C to D. Going to B, we have a decrease in the shear force equal to the positive area of 500 in the loading diagram between points D and B. Since w_y is positive and constant in the interval, then clearly the slope of the shear-force curve will be negative and constant in the interval (see formulas in first box). Finally, we undergo an upward jump of 530 lb. to reach a zero value at the end of the beam as must be the case if we have made no errors. Notice we have computed areas in the shear-force diagram for use later in drawing the bending-moment diagram.

Now, we proceed to the bending-moment diagram. Going from point A to point C, we increase M by 2820 ft-lb. The shear force is positive and constant in the range and hence the bending-moment curve will have a constant positive slope (see formulas in second box). From point C to point D we lose 120 ft-lb and we fill in with a straight line with a negative slope since V is negative and constant in this interval. From point D to point F, we lose some more bending moment in the amount of 775 ft-lb. Between points D and F, the shear force is negative and is *increasing in magnitude*. This means that the bending-moment curve will have a negative slope and will be *steepening*. At F, there is a clockwise couple moment and, as a result, the bending-moment curve jumps upward by 100 ft-lb. Finally, between F and B, we subtract 2025 ft-lb. We end up, as expected, with a zero bending moment at the end. Since V is negative with an increasing magnitude, the bending-moment diagram will have a negative slope and will be steepening.

7.6 Problems Requiring Equations and Diagrams

In Example 7.7 we got the equations and the diagrams of shear force and bending moment independently of each other. With simple

[4]It is important to note that while the shear force goes up or down in the direction corresponding to the direction of point loads when you are moving from left to right (i.e., in the direction of the positive x coordinate), the **opposite** occurs for the shear force when you are moving from right to left (i.e., in the **negative** x coordinate direction). Also, we suggest you use formulas in the boxes next to the curves.

loadings such as point forces, point couples, and uniform distributions, this can readily be done. Indeed, this covers many problems that occur in practice. Usually all that is needed is the labeled diagrams of the kind that we set forth in the previous example. In problems with more complex loadings we usually set forth the equations in the customary manner and then sketch the curves using the *equations* to give key values of V and M [the areas for the various curves are no longer the simple familiar ones, thus precluding advantageous use of Eqs. (7.3) and (7.4)]; the key points are then connected by curves sketched by making use of the slope relations, as we have done in Example 7.6.

Example 7.7

Shown in Fig. 7.18(a) is a cantilever beam, with a parabolic loading as well as a point force and couple moment. Shear-force and bending-moment equations and diagrams are to be formulated for the exposed portion of the beam.

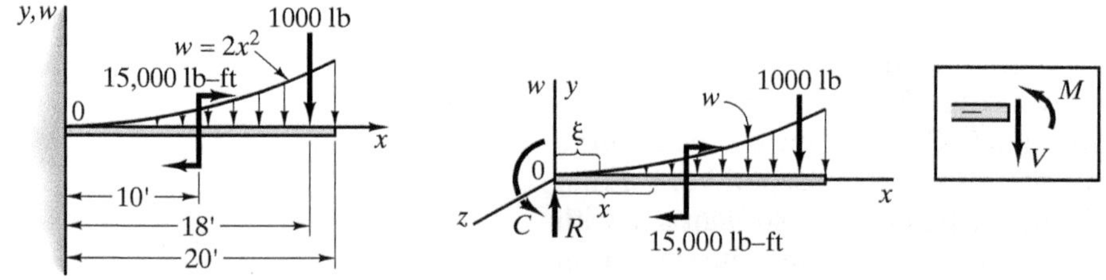

(a) (b)

Figure 7.18. Cantilever beam and its loading diagram.

We shall solve for the force and couple moment on the section of the beam adjacent to the wall using the right-hand rules of statics and we shall consider these as external supports for the beam [see Fig. 7.18(b)]. Because these quantities are to be handled as external loads, we do not apply the sign convention of shear and bending moments to them. Thus,

$\underline{\Sigma F_y = 0:}$

$$R - 1000 - \int_0^{20} 2x^2 dx = 0$$

$$\therefore R = 6333 \text{ lb}$$

$\underline{\Sigma M_0 = 0:}$

$$C - 15{,}000 - (1000)(18) - \int_0^{20} 2x^3 dx = 0$$

$$\therefore C = 113{,}000 \text{ ft-lb}$$

Example 7.7 (Continued)

We may next give the shear-force and bending-moment equations using ξ as a dummy variable [see Fig. 7.18(b)] to locate any position along the beam up to position x. We then may say:

$0 < x < 10$:

$$V = 6333 - \int_0^x 2\xi^2 \, d\xi$$

$$= -\frac{2x^3}{3} + 6333 \text{ lb} \qquad \text{(a)}$$

$$M = -113{,}000 + 6333x - \int_0^x 2\xi^2(x - \xi)\, d\xi$$

Keeping x constant during integration, we have

$$M = -\frac{x^4}{6} + 6333x - 113{,}000 \text{ ft-lb} \qquad \text{(b)}$$

$10 < x < 18$:

$$V = 6333 - \int_0^x 2\xi^2 \, d\xi$$

$$= -\frac{2x^3}{3} + 6333 \text{ lb} \qquad \text{(c)}$$

$$M = -113{,}000 + 6333x - \int_0^x 2\xi^2(x - \xi)\, d\xi + 15{,}000$$

$$= -\frac{x^4}{6} + 6333x - 98{,}000 \text{ ft-lb} \qquad \text{(d)}$$

$18 < x \leq 20$:

$$V = 6333 - \int_0^x 2\xi^2 \, d\xi + 1000$$

$$= -\frac{2x^3}{3} + 5333 \text{ lb} \qquad \text{(e)}$$

$$M = -113{,}000 + 6333x - \int_0^x 2\xi^2(x - \xi)\, d\xi + 15{,}000 - 1000(x - 18)$$

$$= -\frac{x^4}{6} + 5333x - 80{,}000 \text{ ft-lb} \qquad \text{(f)}$$

To sketch the diagrams and thus ascertain the positions of the key points, we use the above equations and the slope relations. The

Example 7.7 (Continued)

loading diagram has been redrawn (Fig. 7.19a). The shear force is 6330 lb to the right of the support at A and from Eq. (c) we see that, just to the left of where the 1000-lb load is applied at C (i.e., at $x = 18$), the shear force is 2445 lb (see Fig. 7.19b). Between A and C the loading w is positive and increasing in magnitude. Accordingly, the shear-force curve has a negative slope which is steepening. At C the 1000-lb load contributes a sudden -1000 shear-force increment. According to Eq. (e), at D (i.e., at $x = 20$), the shear force must be zero. Between C and D the loading is positive and increasing in magnitude. The slope of the shear-force diagram then is negative and is steepening. Clearly, the greatest shear force is 6330 lb at the base.

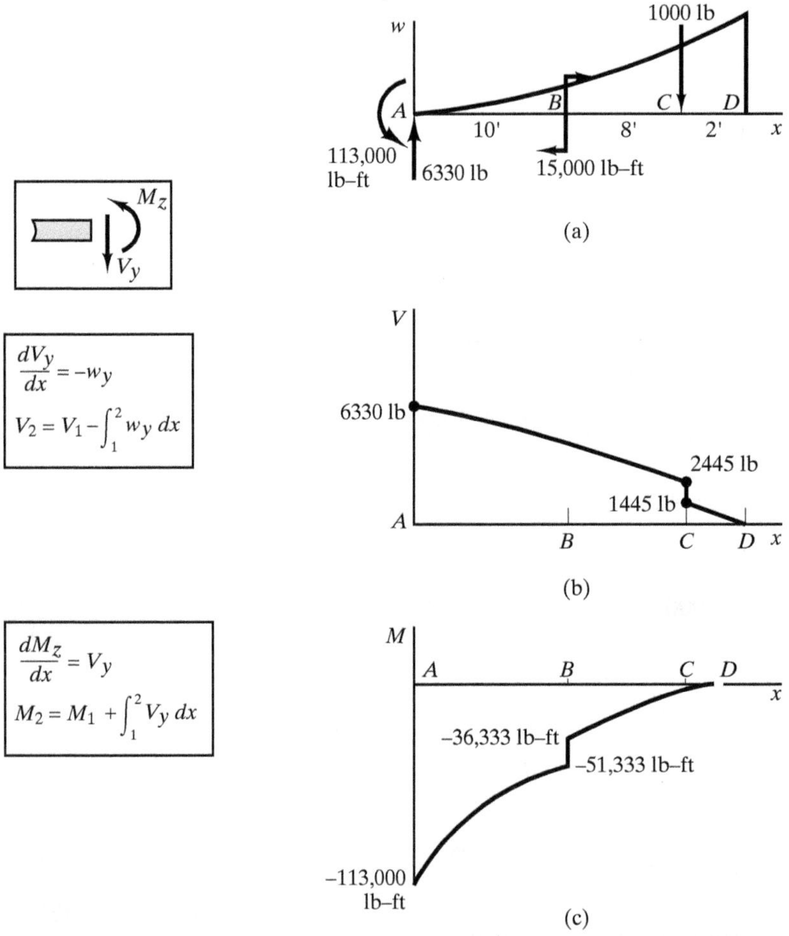

Figure 7.19. (a) Loading diagram; (b) shear-force diagram; (c) bending-moment diagram.

Example 7.7 (Continued)

As for the bending-moment diagram we have at A the instantaneous contribution of $-113{,}000$ ft-lb from the couple at the support. At B to the left of the point couple (at $x = 10$) we have from Eq. (b) the value of $-51{,}333$ ft-lb [see Fig. 7.19(c)]. Between A and B the shear force is positive and decreasing in magnitude. Accordingly, the slope of the bending moment is positive and flattening. As we cross B there is a sudden $+15{,}000$-ft-lb. increment of bending moment resulting from the applied point couple. At D (i.e., at $x = 20$) the bending moment is clearly zero. Between B and D the shear force is positive and decreasing in magnitude with a discontinuous jump at C. Accordingly, between B and D the slope of the bending-moment diagram is positive and flattening with a discontinuity in slope (a cusp) at C. The greatest bending moment, like the shear force, is at the base.

7.7 Additional Considerations

You will note that in the previous examples the key points of the shear-force and bending-moment diagrams were evaluated and labeled. From these diagrams, the maximum positive and negative shear force and bending moment in the beam were identified. We wish to point out that Eq. (7.1a) relates the *derivative* of the shear force to the load. Clearly, if we want to find the extreme values of a function (i.e., relative maximum and minimum), a first step would be to set the first derivative equal to zero. Referring back to Eq. (7.1a), this implies that locations where the load diagram passes through zero should be checked as possible locations of extreme shear force. Likewise, Eq. (7.1b) relates the derivative of the bending moment to the shear force. Again, locations where the shear force is zero (i.e., where the derivative of the bending-moment curve is zero) should be checked as possible locations of extreme bending-moment. Therefore, any position where the loading diagram or shear-force diagram passes through zero should be marked and evaluated (see Figs. 7.20 and 7.21). In the following example, we illustrate this procedure.

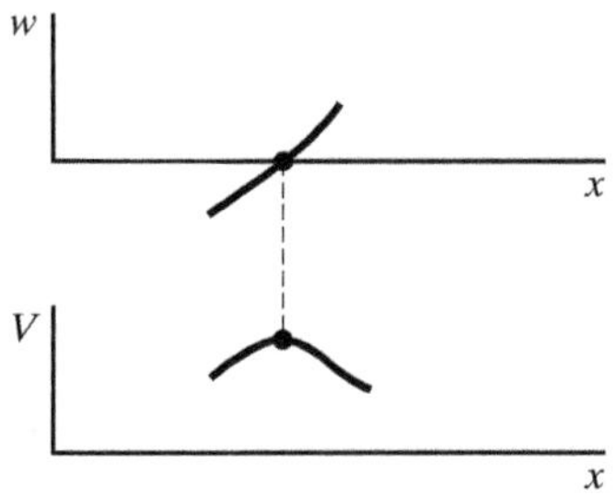

Figure 7.20. Possible maximum shear.

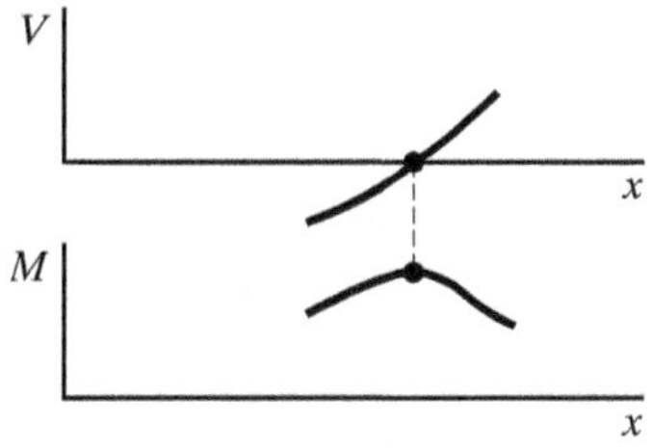

Figure 7.21. Possible maximum bending moment.

Example 7.8

Sketch the shear-force and bending-moment diagrams, being sure to locate and evaluate the bending moment having the largest value for the simply-supported beam in Fig. 7.22.

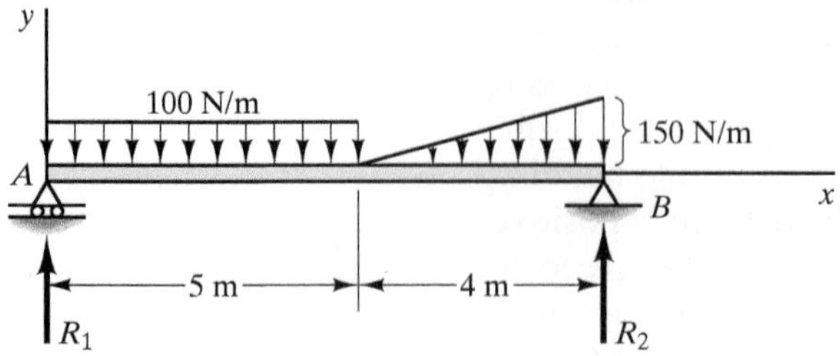

Figure 7.22. Simply-supported beam.

We shall first find the supporting forces, making use of the fact that the resultant force from the triangular loading equals the area under the loading with a line of action passing through the centroid of the triangle.

From **equilibrium** considerations we have

$\Sigma M_A = 0$:

$$-(100)(5)(2.5) - \tfrac{1}{2}(4)(150)[5 + \tfrac{2}{3}(4)] + 9(R_2) = 0$$

$$R_2 = 394.4 \text{ N}$$

$\Sigma M_B = 0$:

$$(100)(5)(6.5) + \tfrac{1}{2}(4)(150)(\tfrac{4}{3}) - 9R_1 = 0$$

$$R_1 = 405.6 \text{ N}$$

We can now sketch the loading, shear-force, and bending-moment diagrams. We do this in the usual way (see Fig. 7.23) but notice that the shear-force curve crosses the x axis at some point E at a distance d to the right of A. We must find the value of d because, as pointed out earlier, we may have a maximum bending moment at this point. Using the property of similar triangles, we can say:

$$\frac{d}{405.6} = \frac{5 - d}{94.4}$$

$$\therefore\ d = 4.056 \text{ m}$$

Example 7.8 (Continued)

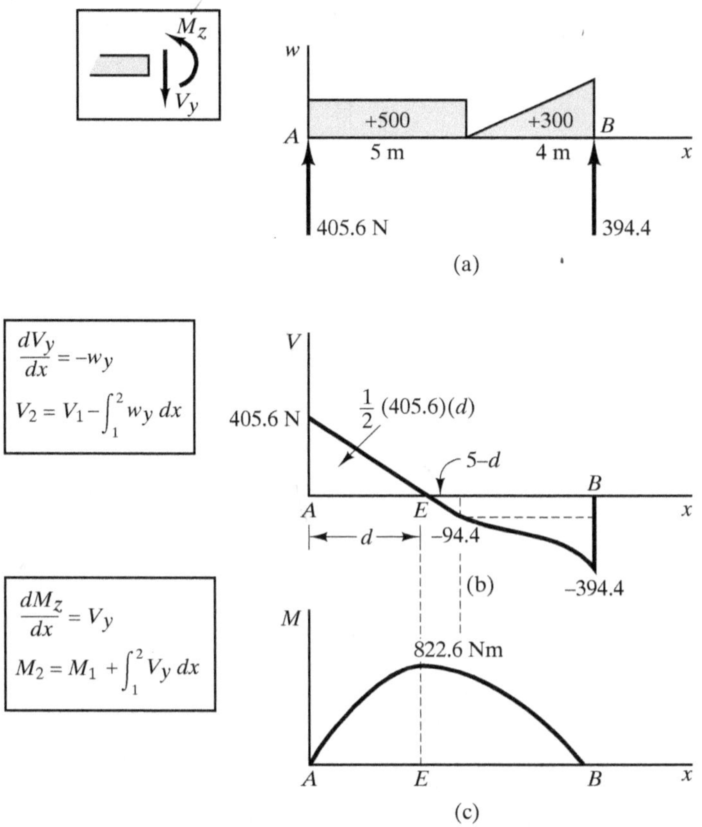

Figure 7.23. Shear-force and bending-moment diagrams showing maximum bending moment at location of zero shear force.

Now going to the bending-moment diagram it is clear that the maximum bending moment is indeed at point E and has the value

$$M_{max} = (\tfrac{1}{2})(405.6)(4.056) = 822.6 \text{ N-m}$$

Before closing this section, we would like to point out a simplification that you may wish to use in conjunction with triangular loads of the orientation shown in Fig. 7.24. If we add a triangular loading of identical shape but increasing in value with x, we can form the uniform loading of value f_o shown in Fig. 7.25. Now to maintain equivalence we insert in

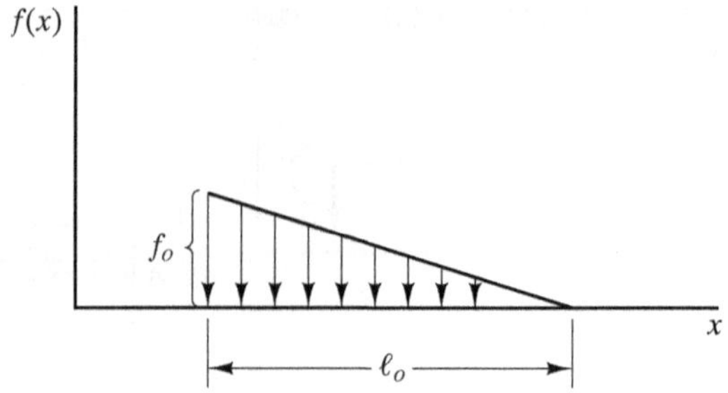

Figure 7.24. Triangular loading.

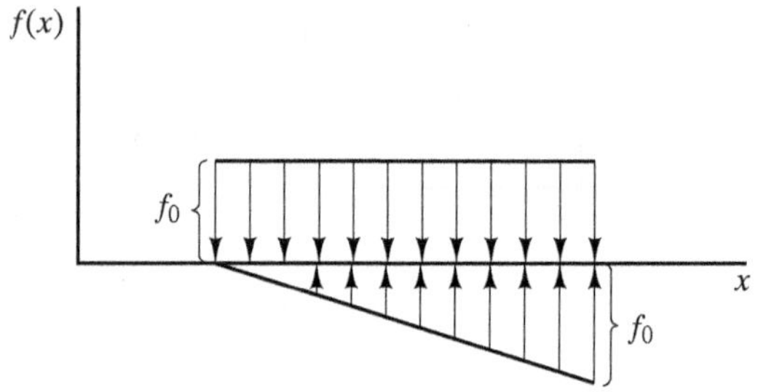

Figure 7.25. Replacement loading.

Fig. 7.25 the same triangular loading that was added in Fig. 7.25, but instead of going downward it should now be going upward, as has been shown in Fig. 7.25. We now have a uniform loading and a triangular loading, the latter in a much simpler orientation. Of course, you need not resort to this procedure; it is offered simply as a suggestion.

7.8 Closure

We have set forth in this chapter the methodology of writing equations and sketching diagrams for the shear force and bending moment at sections along a beam once the supporting forces are known. You will note that only the formulations of rigid-body statics were used in this chapter. In addition, a commonly used sign convention for these tasks was introduced. We hope that you have developed skill in writing the shear-force and bending-moment equations and also with sketching the corresponding diagrams with both speed and accuracy. We will be needing these capabilities as we move forward to the next chapters to study stresses and deformations of beams.

We finish by presenting the key results for the two sign conventions. The authors allow students to use this display on tests.

Stress Convention

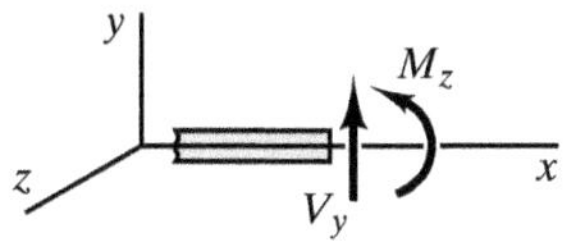

w is positive up

$$\frac{dV_y}{dx} = -w_y$$

$$V_2 = V_1 - \int_1^2 w_y dx$$

$$\frac{dM_z}{dx} = -V_y$$

$$M_2 = M_1 - \int_1^2 V_y dx$$

Structural Convention

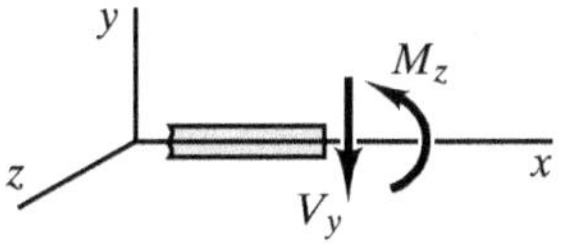

w is positive down

$$\frac{dV_y}{dx} = -w_y$$

$$V_2 = V_1 - \int_1^2 w_y dx$$

$$\frac{dM_z}{dx} = V_y$$

$$M_2 = M_1 + \int_1^2 V_y dx$$

7.9 A Look Back

In your statics course, you were introduced to a set of measures that defined the distribution of a plane area relative to a reference. These quantities were called first moments of area and second moments and products of area. In particular, the **first moment**, M_x, of an area A about the x axis is defined as

$$M_x = \iint_A y dA'$$

The centroid is the position where one could concentrate the entire area so that the first moment about an axis would be given as Ad where d is the perpendicular distance from the centroid to the axis (see Fig. 7.26). The position of the centroid is a property of the area. Finally, it follows that the first moment about a centroidal axis is always zero (since $d = 0$).

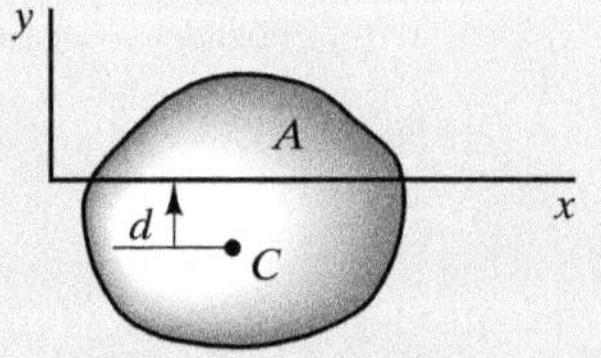

Figure 7.26. The first moment of area A about the x-axis is Ad. C is the centroid of area A.

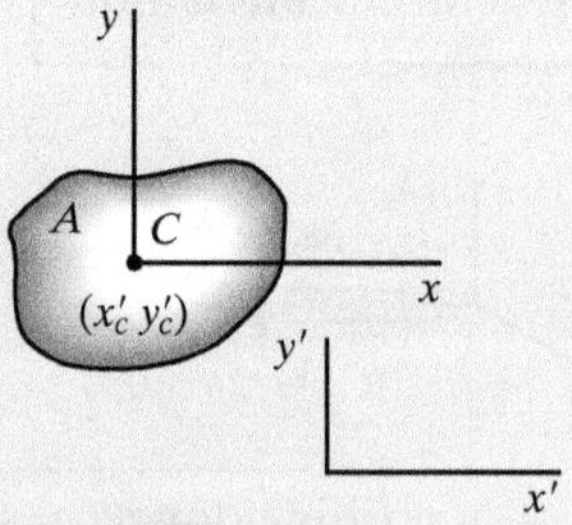

Figure 7.27.
$I_{x'x'} = I_{xx} + A(y_c')^2$
$I_{y'y'} = I_{yy} + A(x_c')^2$
$I_{x'y'} = I_{xy} + A(x_c')(y_c')$

The second moments of area are defined for axes x and y as

$$I_{xx} = \iint_A y^2 dA' \qquad I_{yy} = \iint_A x^2 dA'$$

while the product of area is defined as

$$I_{xy} = \iint_A xy dA'$$

These quantities can be obtained using the mass inertia tensor. This may be done by starting with a thin plate and then letting the thickness t go to zero while at the same time letting the density ρ go to infinity such that the product ρt becomes unity in the limit.[6]

In using the second moments and products of area, we can make good use of the so-called **parallel axis theorems** whereby the second moment of area about some axis x' equals the second moment of area about a parallel axis going through the centroid plus the product of the area times the perpendicular distance squared from the centroid to axis x'. The product of area for a pair of axes $x'y'$ (Fig. 7.27) equals the product of area about the parallel set of axes at the centroid plus the area multiplied by the coordinates of the centroid relative to the axes $x'y'$.

We shall have use of these formulations in the next chapter.

[6]See Shames, I.H., *Engineering Mechanics*, Fourth Edition, 1997, Prentice Hall Inc., Section 9.3.

Highlights (10)

We introduced conventions for the loading w_y, the shear force V_y, and the bending moment M_z at a section in order to have a representation of these quantities at the section that was independent of whether we used a right-hand or a left-hand free-body diagram to expose the section of interest. We have used in this text the structural convention. To determine the aforementioned quantities, we use the equations of statics either formally or mentally and employ **positive** section forces and moments according to our structural convention in this process. By this procedure, the algebraic sign emerging for the section forces and moments by using equations of static equilibrium would then have the proper convention sign. Remember that a positive couple moment for the right-hand rule will correspond to the case where a standard screw will advance in a positive coordinate direction when turned with the same sense as the couple moment.

We write the equations for V_y and M_z for the domains of the beam being sure to exclude positions where there is a point force or a point couple because of the indeterminacy of either V_y and/or M_z at such positions. Additionally, it is important that the student develop skill and speed in sketching shear and bending moment diagrams. The use of the *steepening* and the *flattening* terminology will be very helpful in this procedure. Remember, an increase in the magnitude of the loading function w, means a steepening slope of the corresponding shear curve and a decrease in the magnitude of the loading function w, means a flattening of the slope of the corresponding shear diagram. The very same thinking is followed when using the shear diagram to draw the bending moment diagram.

PROBLEMS

In Problems 7.1 to 7.16, use free-body diagrams as an aid.

7.1. [7.2] For the beam shown in Fig. P.7.1, what is the shear force and bending moment at the following positions?

(a). 5 ft from the left end
(b). 12 ft from the left end
(c). 5 ft from the right end

Write the shear force and bending moment as a function of x for the beam.

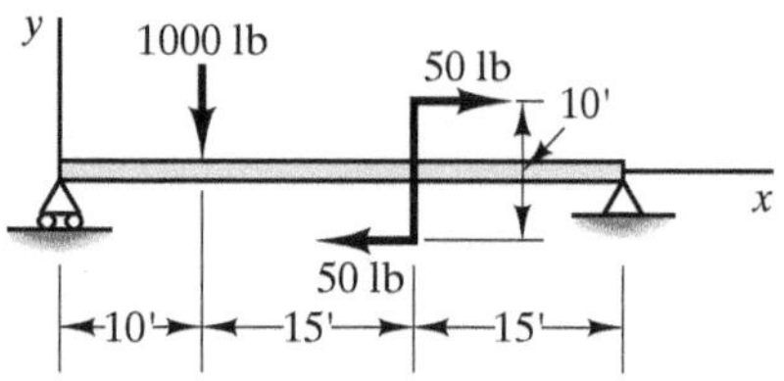

Figure P.7.1.

7.2. [7.2] A hoist can move along a beam (see Fig. P.7.2) while supporting a 10,000-N load. If the hoist starts at the left and moves from $\bar{x} = 3$ m to $\bar{x} = 12$ m, determine the shear force and bending moment at A in terms of $\bar{x}$. At what position $\bar{x}$ do we get the maximum shear force at A and the maximum bending moment at A. What are their values?

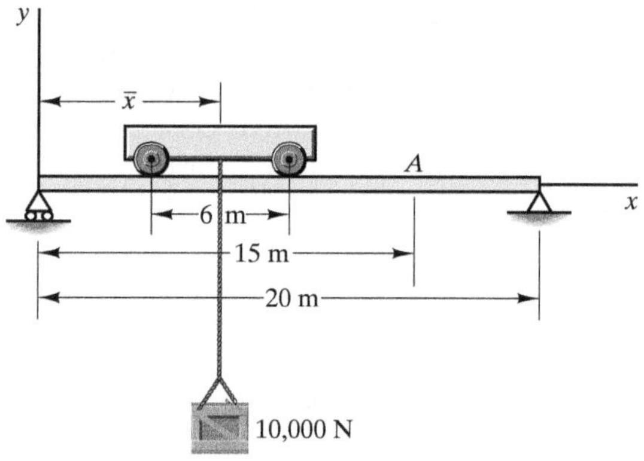

Figure P.7.2.

7.3. [7.2] Formulate the shear-force and bending-moment equations for the simply supported beam shown in Fig. P.7.3. Do not include the weight of the beam.

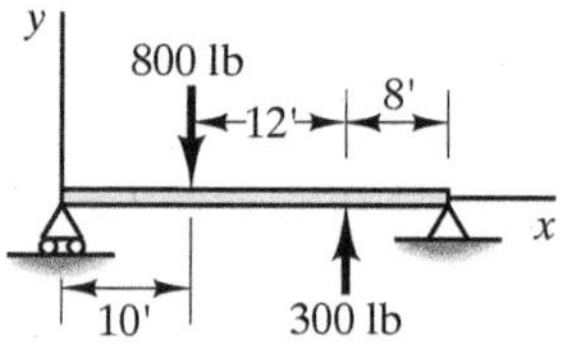

Figure P.7.3.

7.4. [7.2] Formulate the shear-force and bending-moment equations for the cantilever beam shown in Fig. P.7.4. Do not include the weight of the beam.

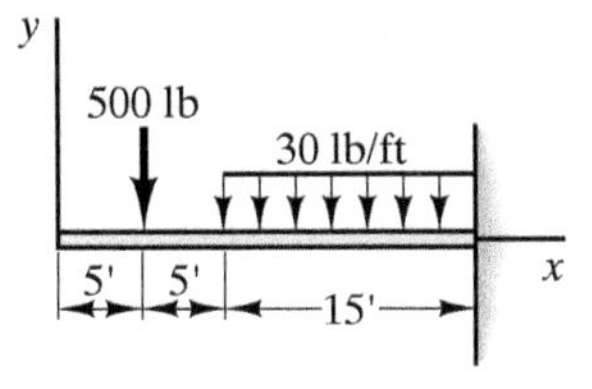

Figure P.7.4.

7.5. [7.2] Determine the shear-force and bending-moment equations for the beam in Fig. P.7.5.

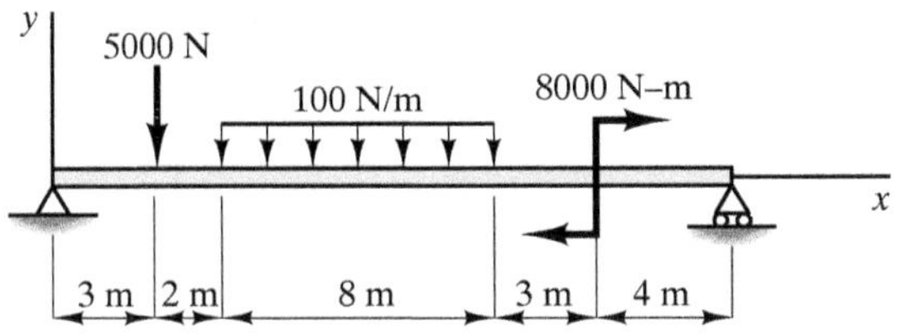

Figure P.7.5.

7.6. [7.2] *BC* Is cantilevered at *C* and pinned to *AB* at *B* (see Fig. P.7.6). Formulate the shear-force and bending-moment equations.

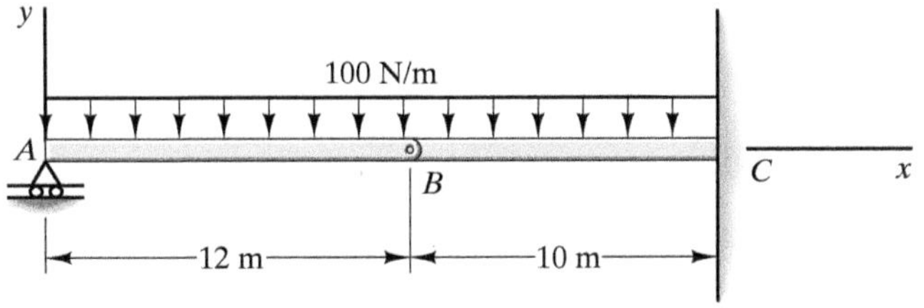

Figure P.7.6.

7.7. [7.2] Formulate shear-force and bending-moment equations for the simply supported overhanging beam (see Fig. P.7.7).

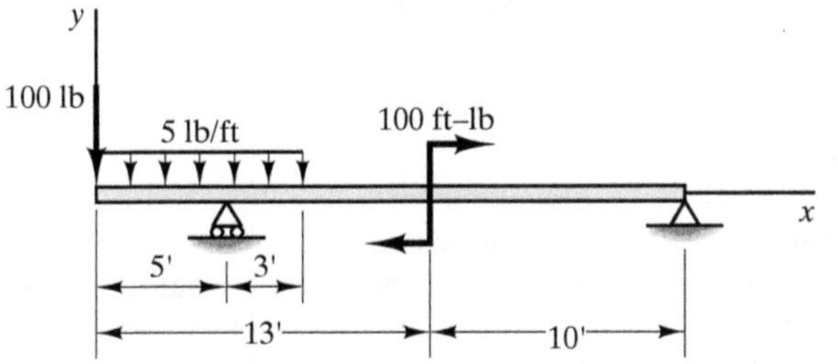

Figure P.7.7.

7.8. [7.2] A cantilever beam *AB* is pinned at point *A* to member *CA* as shown in Fig. P.7.8. Compute the shear-force and the bending-moment equations.

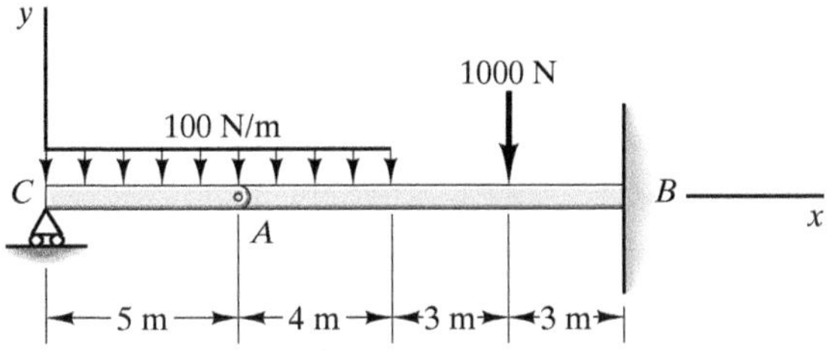

Figure P.7.8.

7.9. [7.2] Formulate the shear-force and bending-moment equations for the cantilever beam *AD* (Fig. P.7.9). *BC* is welded to beam *AD*.

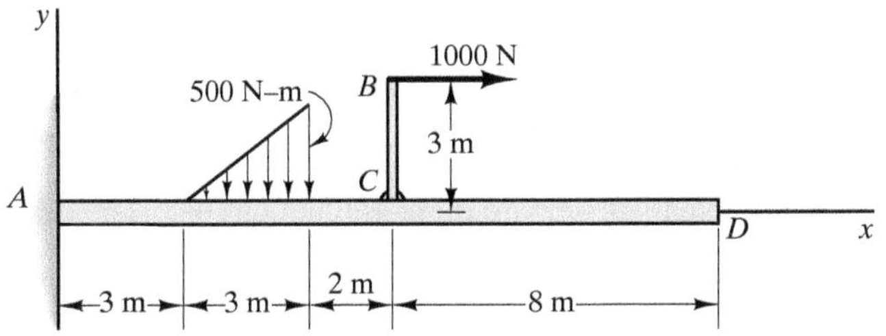

Figure P.7.9.

***7.10. [7.2]** Compute shear force, axial force, and bending moments from *A* to *C* for the bent beam shown in Fig. P.7.10 as functions of *s* along the centerline of the beam.

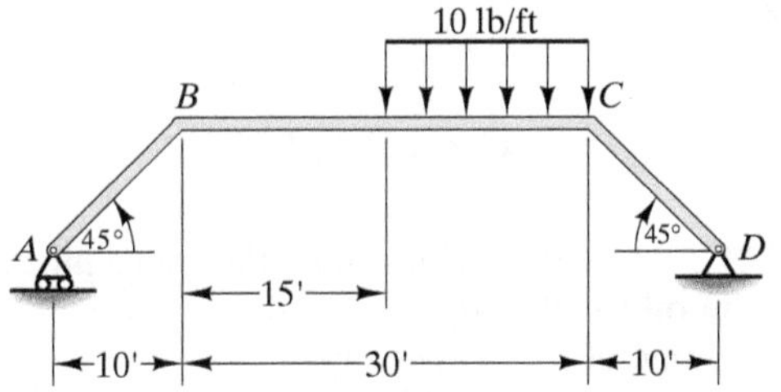

Figure P.7.10.

7.11. [7.2] Determine the shear force, bending moment, and axial force as functions of θ for the circular beam shown in Fig. P.7.11.

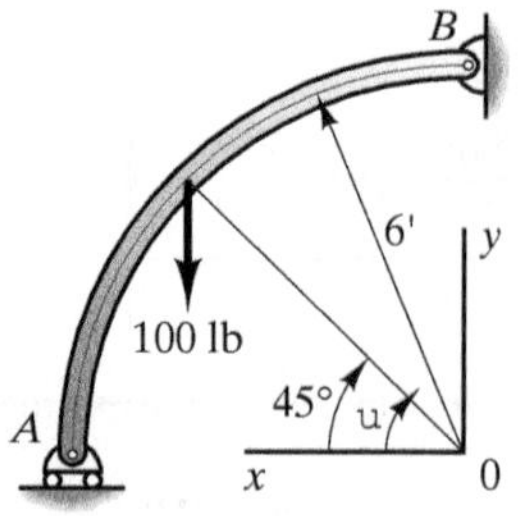

Figure P.7.11.

> Problems 7.12 to 7.16 involve three-dimensional loading for computations of shear forces, bending moments, and twisting moments over and above the V and M for the two-dimensional cases of previous problems.

***7.12. [7.2]** Shown in Fig. P.7.12 is a simply supported beam loaded in two planes. This means there will be shear-force components V_y and V_z and bending-moment components M_z and M_y. Compare these as functions of x. The beam is 40 m in length.

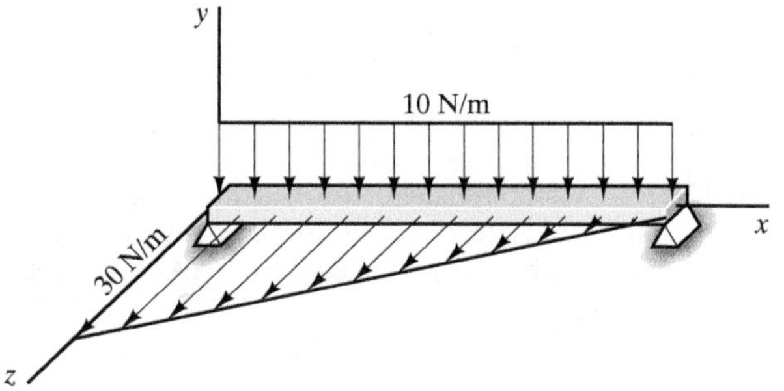

Figure P.7.12.

***7.13. [7.2]** Oil flows from a tank through a pipe AB (Fig. P.7.13). The oil weighs 40 lb/ft ^{3} and, in flowing, develops a drag on the pipe of 1 lb/ft. The pipe has an inside diameter of 3 in. and a length of 20 ft. Flow conditions are assumed the same along the entire length of the pipe. What are the shear force, bending moment, and axial force along the pipe from the oil? z is parallel to the ground and the xy plane is perpendicular to the ground.

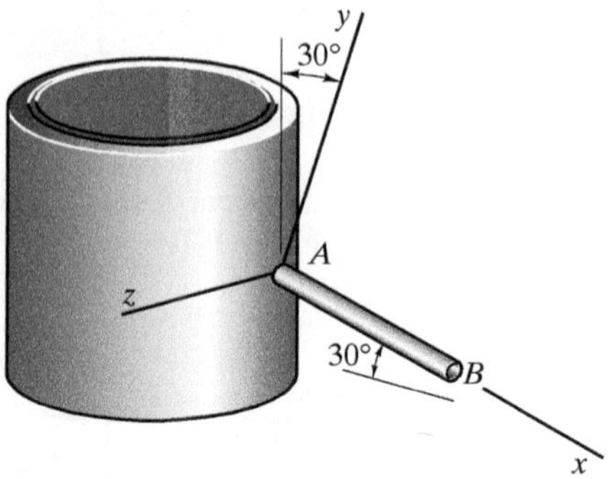

Figure P.7.13.

***7.14. [7.2]** Friction forces applied to a cantilever beam (see Fig. P.7.14) result in a uniform torque per unit length from A to B having the value 10 N-m/m while from C to D the friction forces result in an axial load distribution given by f starting from zero at C and increasing linearly until this loading intensity is 20 N/m at D. What are the various moment and force components transmitted through the beam?

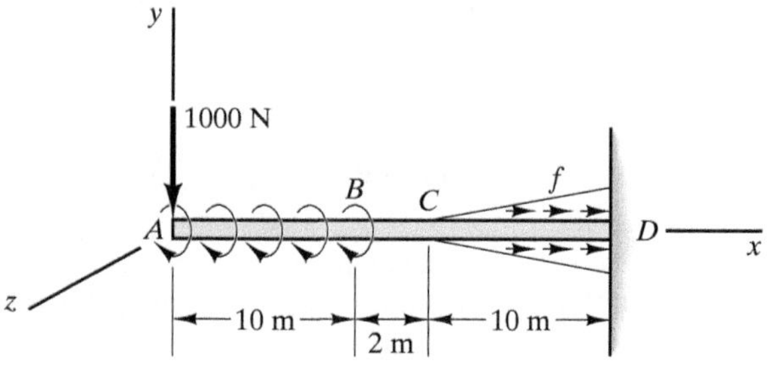

Figure P.7.14.

***7.15. [7.2]** A pipe is shown in Fig. P.7.15 closed at D. It weighs 10 lb/ft and has an inside diameter of 2 in. If it is full of water and the pressure of the water is that of the atmosphere at the entrance A, compute the shear force, axial force, and bending moment of the pipe from A to B.

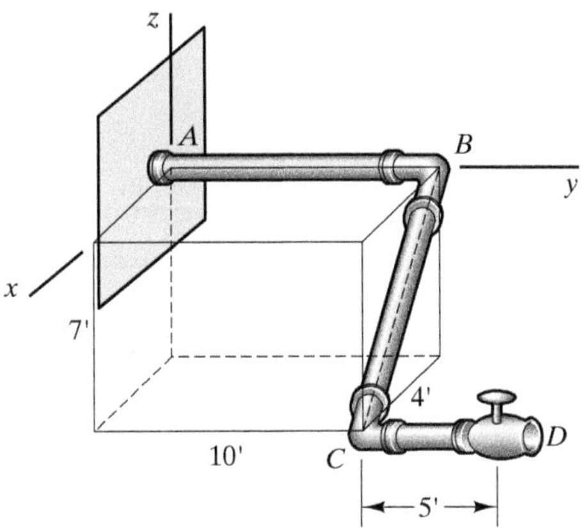

Figure P.7.15.

*****7.16. [7.2]** Consider a one-cylinder compressor as shown in Fig. P. 7.16. The pressure at the piston face is 200 psia. The bore is 3 in. and the stroke is 4 in., while the length of the connecting rod is 6 in. For the position shown, plot V_y and V_z for member AB. Also plot M_y and M_z for member AB. EB is horizontal at instant of interest.

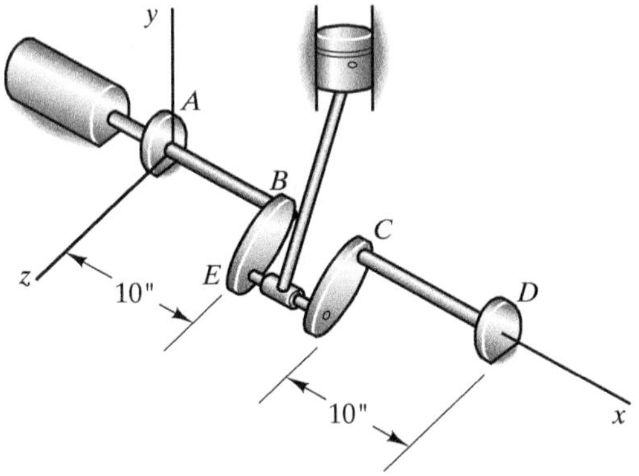

Figure P.7.16.

7.17. [7.3] Do Problem 7.3 without free-body diagrams for computing V and M.

7.18. [7.3] Do Problem 7.4 without free-body diagrams for computing V and M.

7.19. [7.3] Do Problem 7.5 without the use of free-body diagrams for computing V and M.

7.20. [7.3] Do Problem 7.6 without using free bodies for computing V and M.

7.21. [7.3] Do Problem 7.7 without using free bodies for finding V and M equations.

7.22. [7.3] Do Problem 7.8 not using free-body diagrams for getting V and M equations.

7.23. [7.3] Do Problem 7.9 without the aid of free-body diagrams for finding V and M equations.

In Problems 7.24 to 7.31 write shear-force and bending-moment equations without the aid of free-body diagrams. Then sketch shear and bending-moment diagrams labeling key points.

7.24. [7.5]

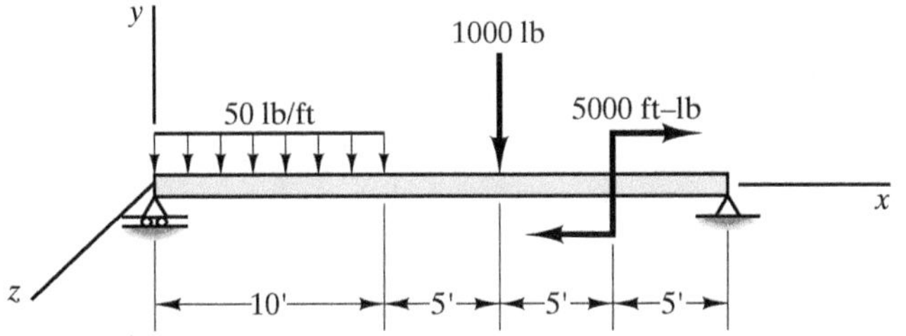

Figure P.7.24.

7.25. [7.5]

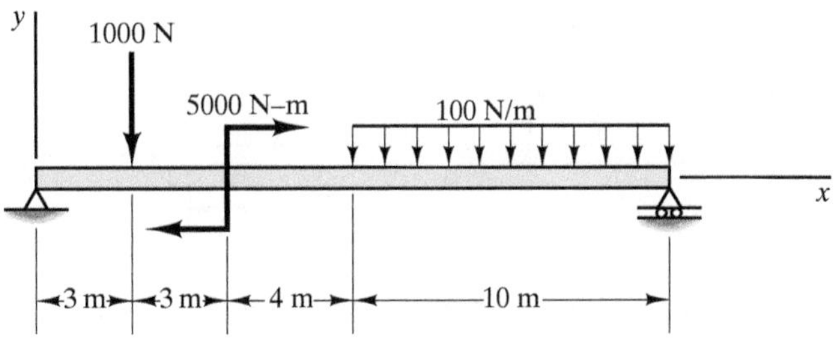

Figure P.7.25.

7.26. [7.5]

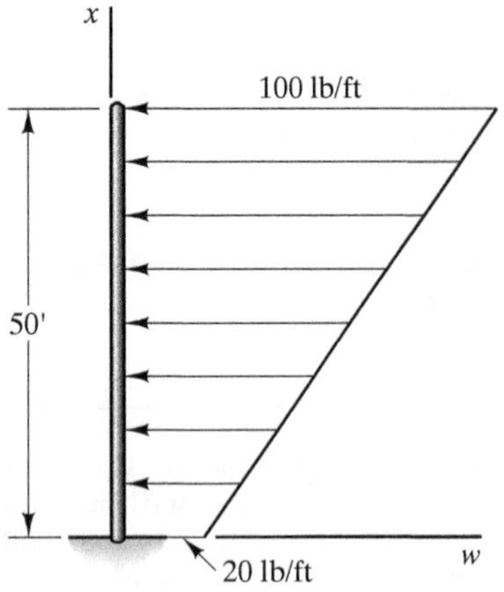

Figure P.7.26.

7.27. [7.5]

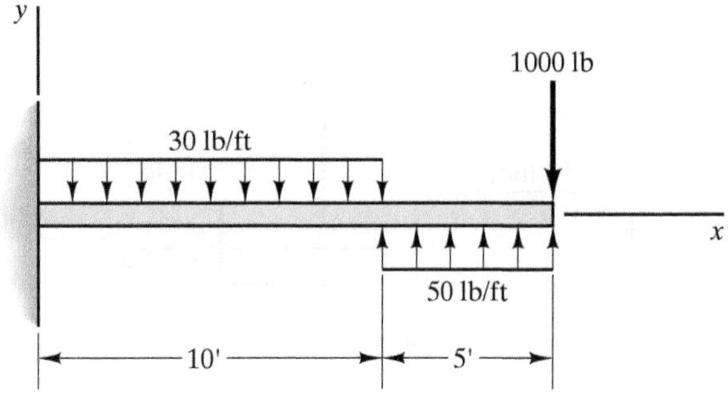

Figure P.7.27.

7.28. [7.5]

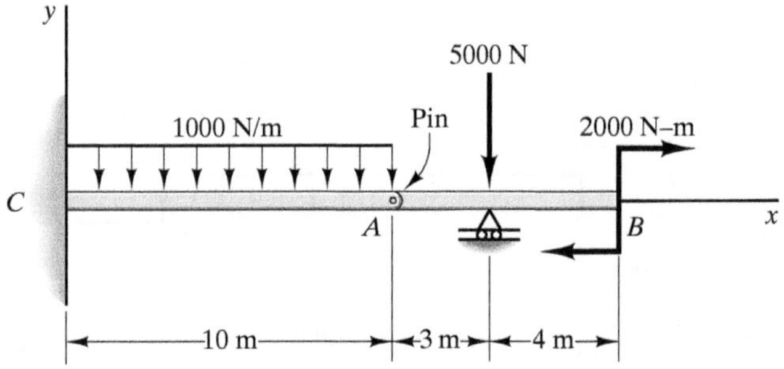

Figure P.7.28.

7.29. [7.5]

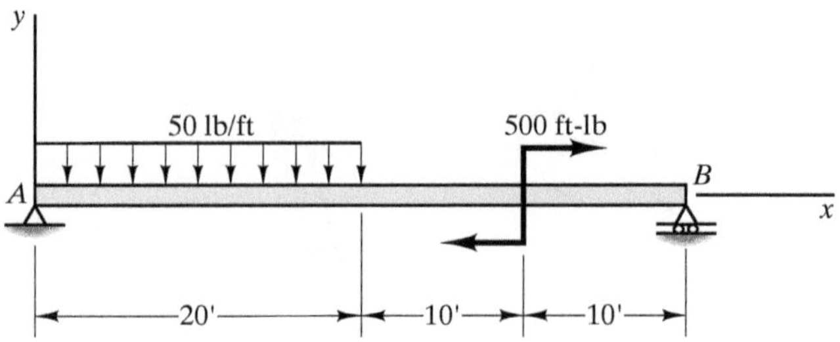

Figure P.7.29.

7.30. [7.5]

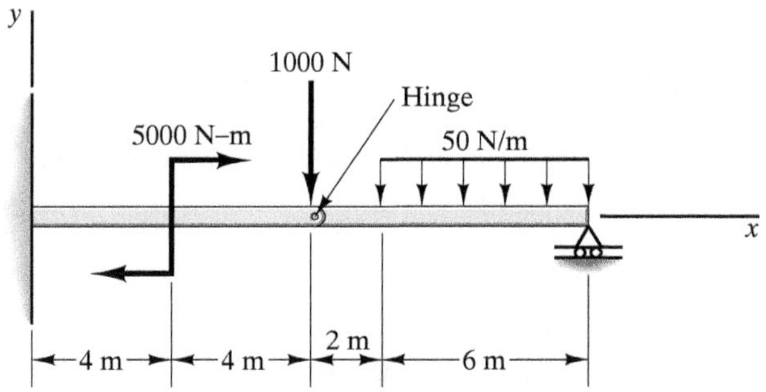

Figure P.7.30.

7.31. [7.5] Note there is a pin at B in Fig. P.7.31. The 1000-N load and 500-N-m couple-moment are applied to a rigid attachment welded to the beam $\overline{AB}$.

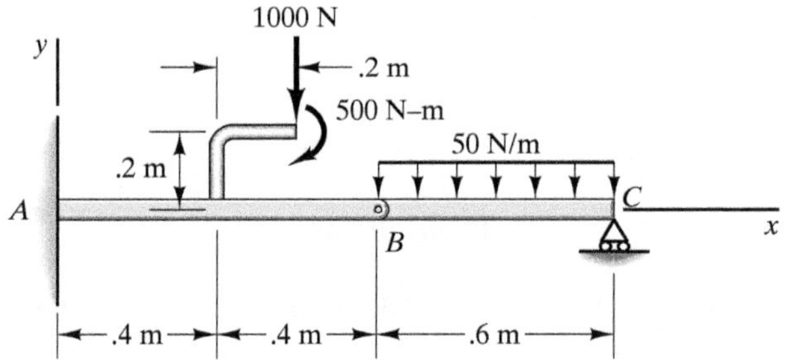

Figure P.7.31.

7.32. [5.5] For the testing of materials in bending, the "Four-Point Bend Test" is sometimes used, as shown in Fig. P.7.32. Compute the shear-force and bending-moment diagram for this configuration. List some reasons as to why this setup is of particular use. (*Hint:* What is of interest in the bending-moment distribution?)

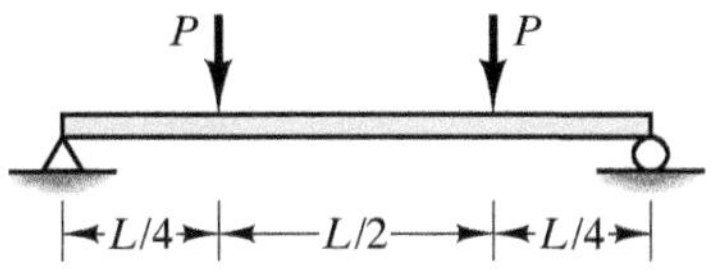

Figure P.7.32.

7.33. [7.5] A temporary elevator, shown in Fig. P.7.33, is erected at a construction site. The fully loaded elevator weighs 4500 lb and it accelerates upward uniformly from rest to a steady velocity of 2000 ft/min in 10 seconds. Draw the bending-moment diagram of the beam for the maximum load case. Take the weight of winch and motor as 500 lb and positioned as shown.

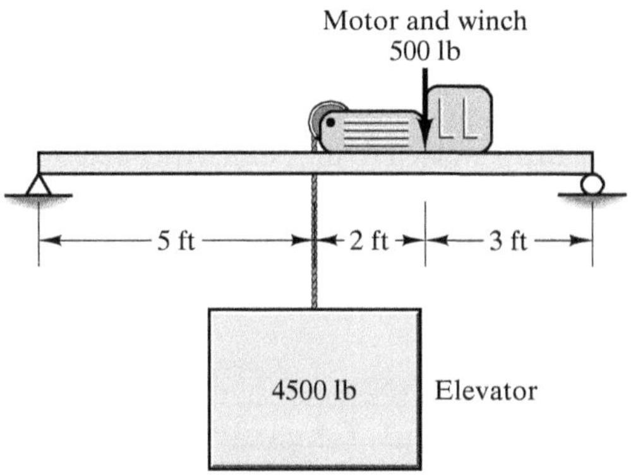

Figure P.7.33.

7.34. [7.5] Find the supporting forces for the simply supported beam in Fig. P.7.34. Then sketch the shear-force and bending-moment diagrams, labeling key points. 1K = 1000 lb.

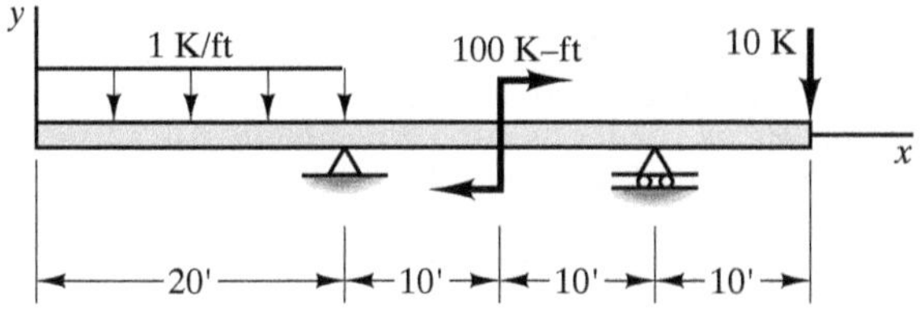

Figure P.7.34.

7.35. [7.5] A steam roller moving across the 30-ft span exerts forces on it as shown in Fig. P. 7.35. Find magnitude and location of the bending moment, and $\bar{x}$ for which this moment is an absolute maximum of all possible positions of the vehicle. (*Hint:* Sketch the bending-moment diagram for any position $\bar{x}$ and decide where M_{max} might be.)

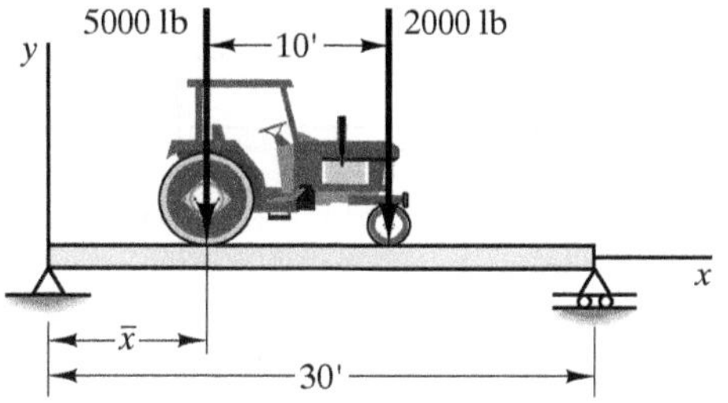

Figure P.7.35.

7.36. [7.5] A simply supported I beam is shown in Fig. P.7.36. A hole must be cut through the web to allow passage of a pipe which runs horizontally at right angles to the beam.

(a). Where, within the marked 24-ft section, would the hole least affect the moment-carrying capacity of the beam?
(b). In the same 24-ft section, where should the hole go to least affect the shear-carrying capacity of the beam?

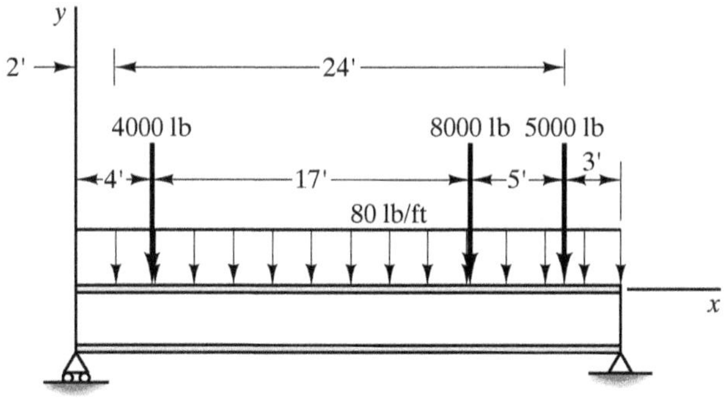

Figure P.7.36.

7.37. [7.5] In Fig. P.7.37, sketch the shear-force and bending-moment diagrams and label the key points.

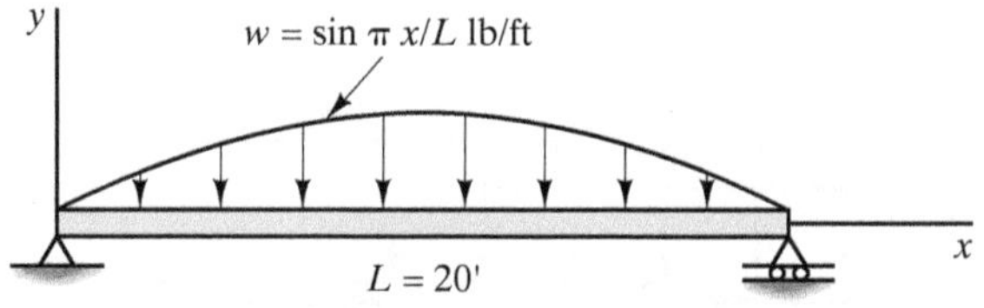

Figure P.7.37.

7.38. [7.5] After finding the supporting forces of the cantilever beam (see Fig. P.7.38) sketch the shear-force and bending-moment diagrams labeling key points.

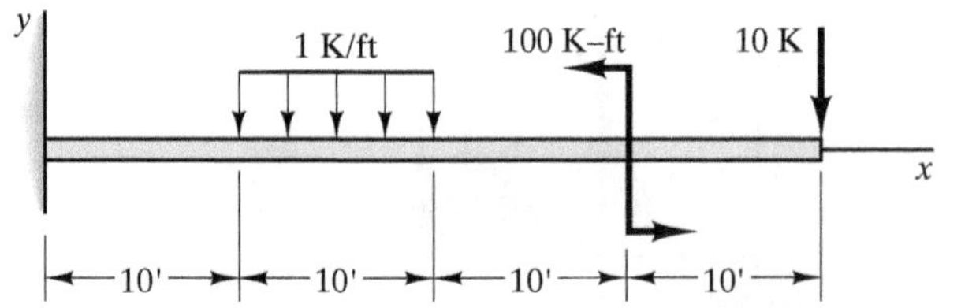

Figure P.7.38.

In Problems 7.39 to 7.42, merely sketch shear-force and bending-moment diagrams, labeling key points. Use whatever numerical calculations are necessary.

7.39. [7.8]

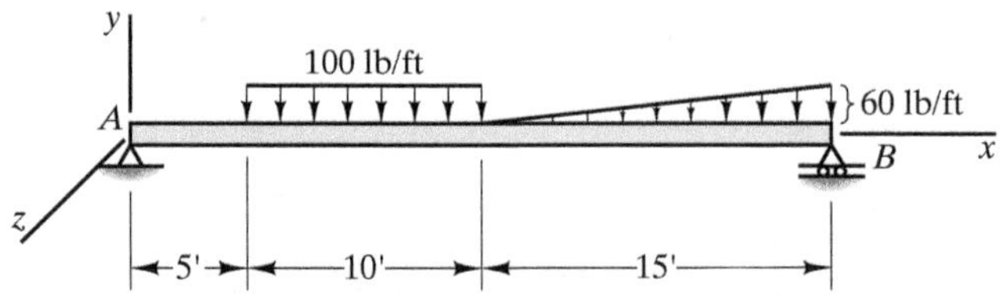

Figure P.7.39.

7.40. [7.8]

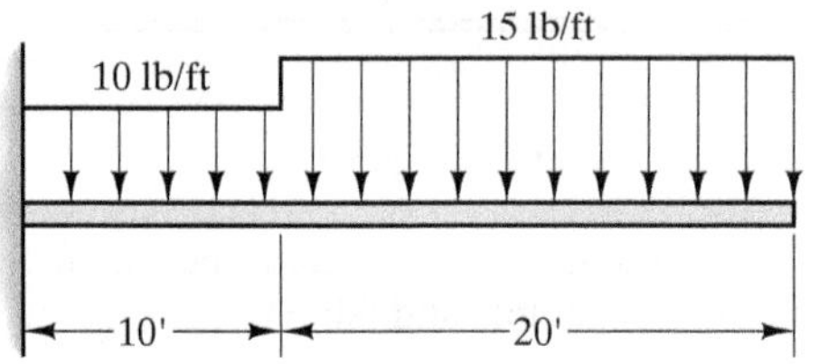

Figure P.7.40.

7.41. [7.8]

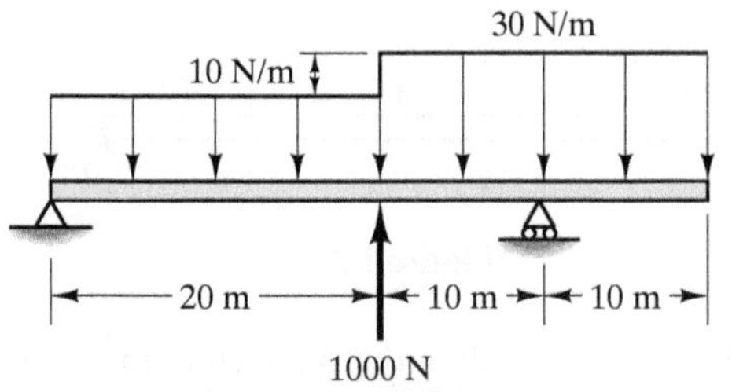

Figure P.7.41.

7.42. [7.8]

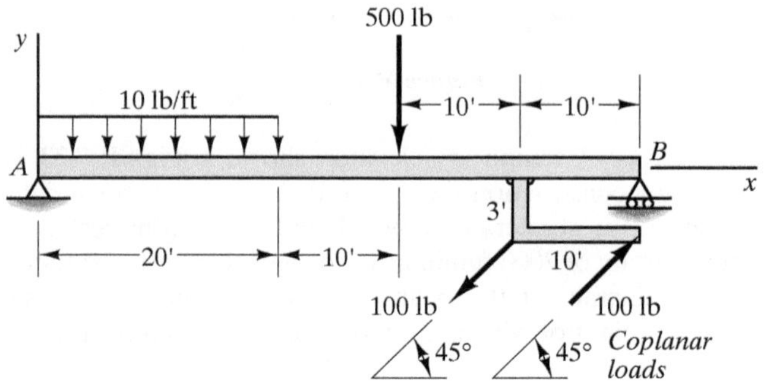

Figure P.7.42.

7.43. [7.8] A simply supported beam AB is shown in Fig. P.7.43. A bar CD is welded to the beam. After determining the supporting forces, sketch the shear-force and bending-moment diagram and determine the maximum bending moment.

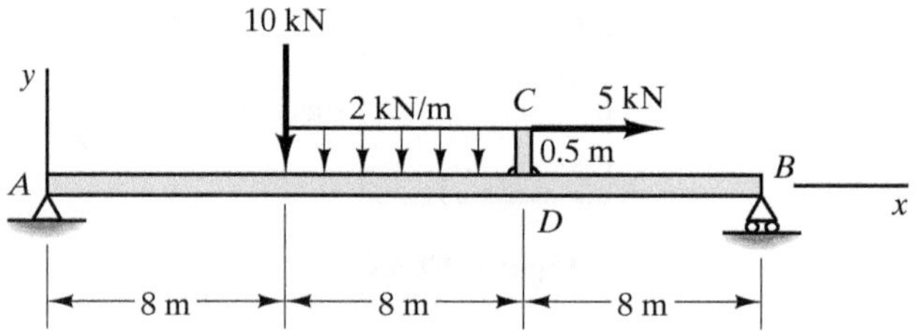

Figure P.7.43

Problems 7.44, 7.45, and 7.46 are challenging problems for the interested student looking for something to sink his or her teeth into. If available, use software for plots of variables.

∗7.44. A heavy, 50-ft long pole is to be lifted as shown in Fig. P.7.44. At what distance b should the crane attachment be located in order to impose the minimum flexural stress (caused directly by the bending moment) on its circular cross section if the weight of the pole is uniformly distributed along its length? (*Hint:* Consider the maximum bending moments in each of the two spans in terms of b. Roughly plot the magnitudes of these moments versus b. At what b will the maximum possible moment be the least value?)

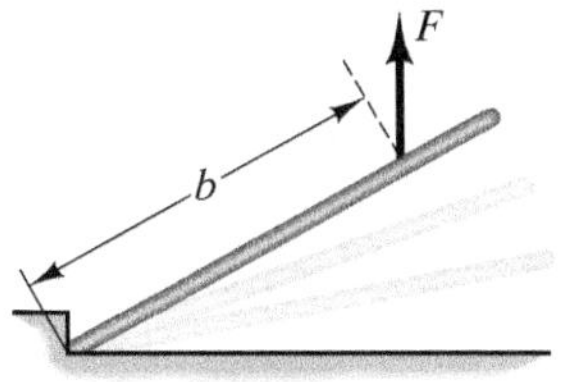

Figure P.7.44.

***7.45.** A 15-in.-o.d. (outside diameter) steel pipe is to be lifted so as to remain horizontal. For this purpose two rigidly attached yokes separated a distance a are attached to the pipe (see Fig. P.7.45). Two guy wires of length 10 ft each then connect the yokes to a hook. The pipe weighs 80 lb/ft and has a length of 20 ft. Find the proper distance a so that the maximum bending moment in the pipe is minimized. (*Hint:* Compute the maximum bending moment in each of the spans in terms of a. Roughly plot these as functions of a. At what a will the maximum possible moment be the least value?)

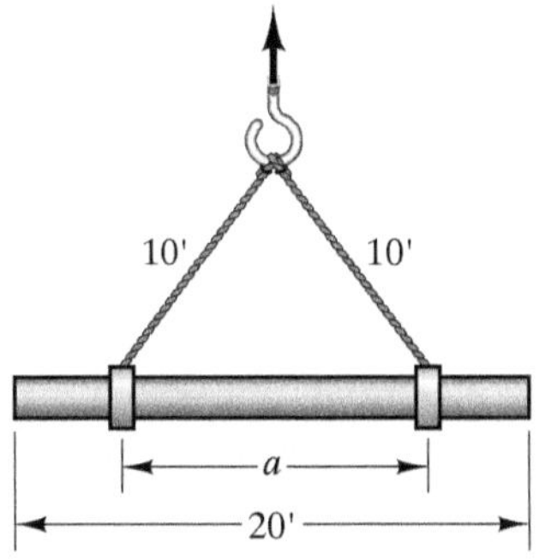

Figure P.7.45.

***7.46.** A NO PARKING sign is made of thin rigid plastic and is supported as shown in Fig. P. 7.46. If a wind blows against the sign, a uniform pressure p_0 above that of the stationary atmosphere will develop on the plate. To minimize the maximum bending moment from this pressure, at what position $\bar{x}$ should the lower support be placed? Give result as a fraction of h. (*Hint:* Compute the maximum bending moment in each of two spans in terms of $\bar{x}$ and h. Roughly plot M_{max} versus $\bar{x}$ for a given h. At what $\bar{x}$ will the maximum possible moment be the least value?)

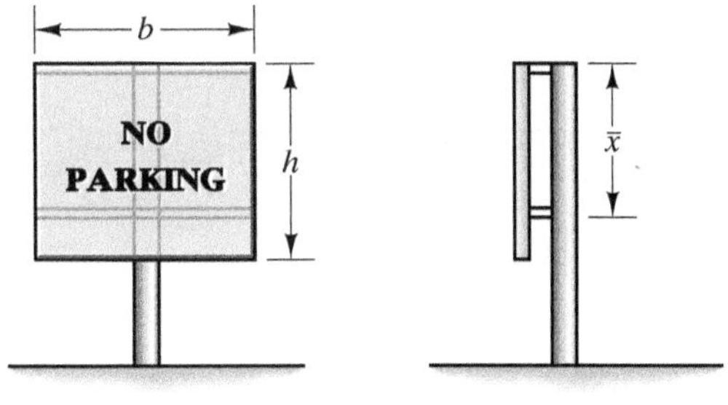

Figure P.7.46.

7.47. In Fig. P.7.47, what is the maximum negative bending moment in the region between the supports for the simply supported beam?

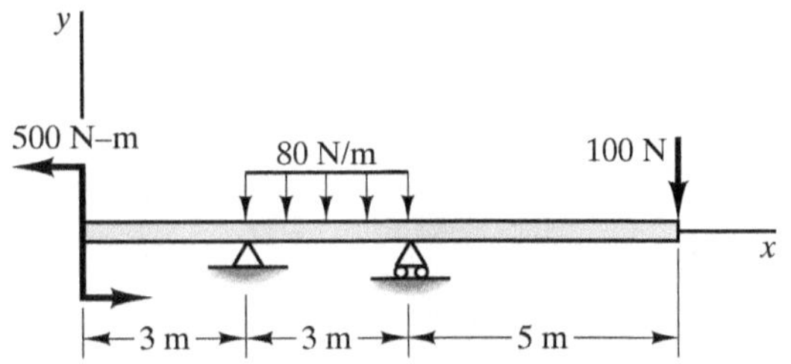

Figure P.7.47.

7.48. A 20-hp motor running at 1750 rpm drives two 1-ft diameter pulleys as shown in Fig. P.7.48. The belt tensions are given as $T_1 = 200$ lb, $T_2 = 100$ lb, and $T_3 = 3T_4$. The coupling C transmits only torque.

(a). Sketch the twisting torque diagram for M_x along the shaft AB.
(b). Sketch the shear-force diagrams for V_y and V_z.
(c). Sketch the bending-moment diagrams for M_y and M_z.

(*Hint:* As you may recall from physics the power transmitted by a shaft is torque times angular velocity in radians per unit time. Also, 1 hp = 33,000 ft-lb/min.)

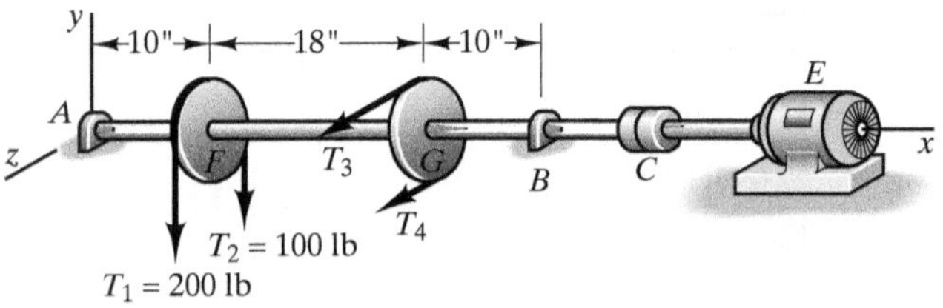

Figure P.7.48.

7.49. Shown in Fig. P.7.49 is suspension of one of the front wheels of an automobile. The pavement exerts a force of 1000 lb on the tire. The tire, brakes, and so on, weigh 100 lb; the center of gravity is taken at the center plane of the tire. Determine the force in the spring and the compression force in *CD*. Now sketch the axial force, shear force, and bending moment for member *AC*.

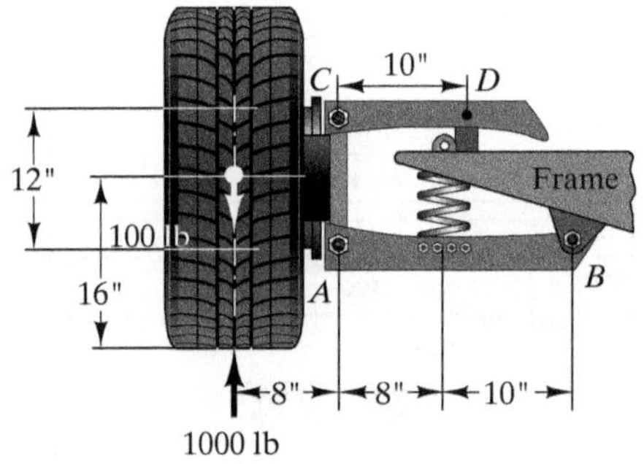

Figure P.7.49.

7.50. Sketch shear-force and bending-moment diagrams for Problem 7.5 after finding supporting forces.

7.51. Sketch shear-force and bending-moment diagrams for Problem 7.7 after finding supporting forces.

7.52. Sketch shear-force and bending-moment diagrams in Problem 7.8 after finding supporting forces and torques.

7.53. For the beam shown in Fig. P.7.53,

(a). Write the shear-force and bending-moment equations.
(b). Sketch the shear force and bending-moment diagrams labeling key points.

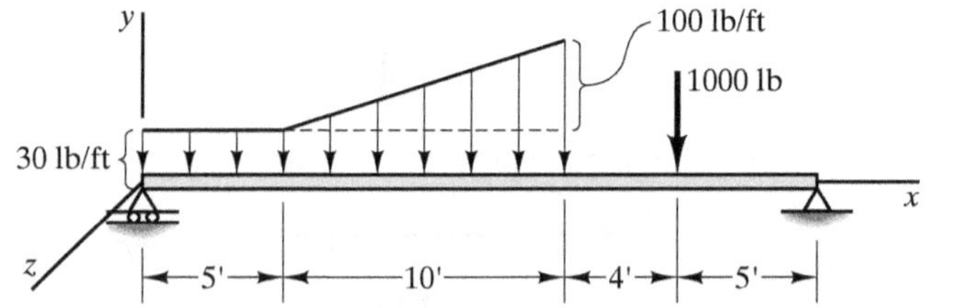

Figure P.7.53.

***7.54.** For the problem shown in Fig P.7.54:

(a). Find the supporting force system.
(b). Write the shear-force and bending-moment equations.
(c). Sketch the shear-force and bending-moment diagrams.

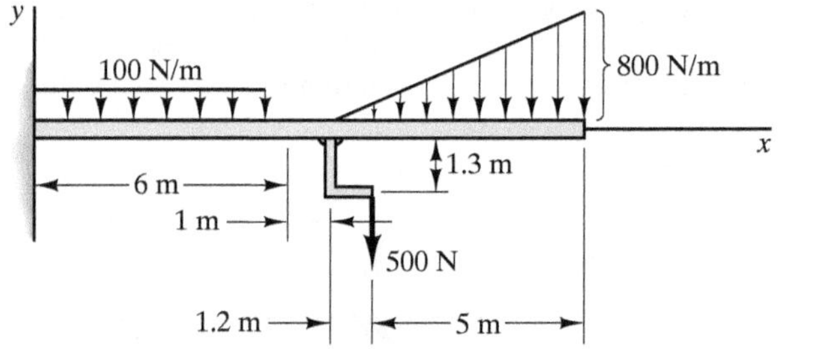

Figure P.7.54.

7.55. Write the equations and sketch the shear-force and bending-moment diagrams for the beam in Fig. P.7.55.

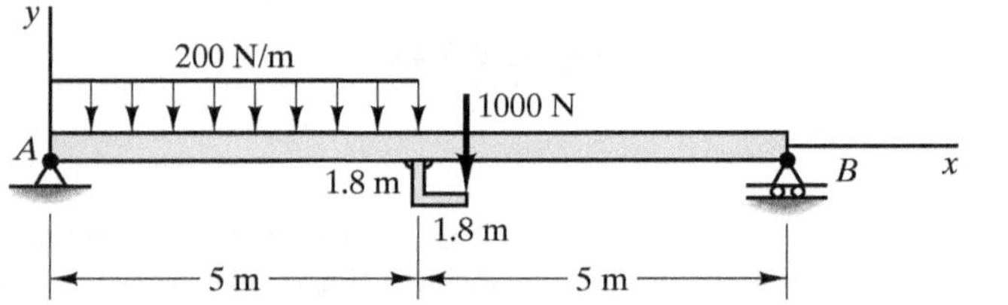

Figure P.7.55.

7.56. Given the shear-force diagram in Fig. P.7.56, draw the corresponding loading diagram. There is a clockwise couple moment of 5000 N-m at $x = 4$ m, and at $x = 0$ there is a counterclockwise couple moment at the base (it is a cantilever beam).

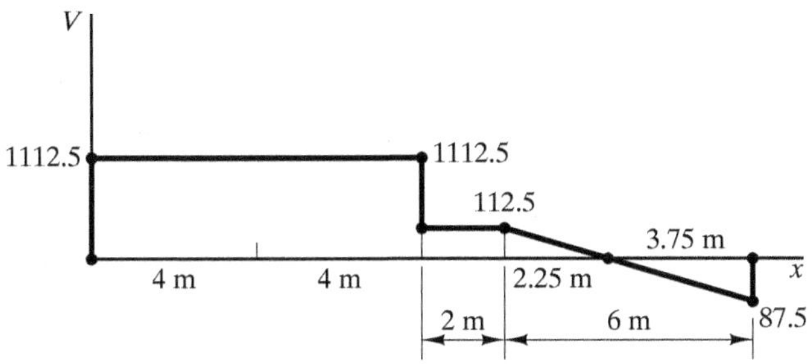

Figure P.7.56.

7.57. For the beam in Fig. P.7.57,

(a). What is the axial force on *AB*? Use three domains for this.
(b). Sketch the shear-force and bending-moment diagrams.

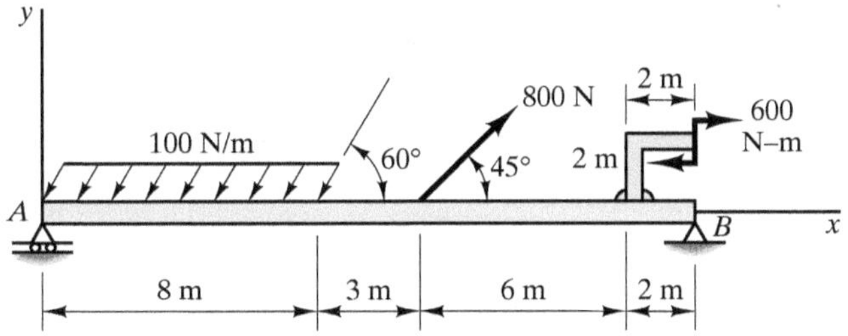

Figure P.7.57.

7.58. Determine the shear-force and bending-moment equations for the beam in Fig. P.7.58. Then sketch the diagrams using the aforementioned equations if necessary to ascertain key points in the diagrams, such as the position between the supports where $V = 0$. What is the bending moment there?

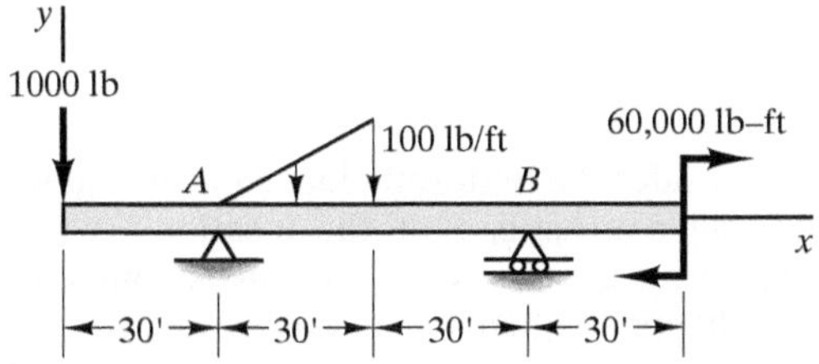

Figure P.7.58.

7.59. A cantilever beam (Fig. P.7.59) supports a parabolic and a triangular load. What are the shear-force and bending-moment equations? Sketch the shear-force and bending-moment diagrams. See the suggestion at the end of Section 7.7 regarding the triangular load.

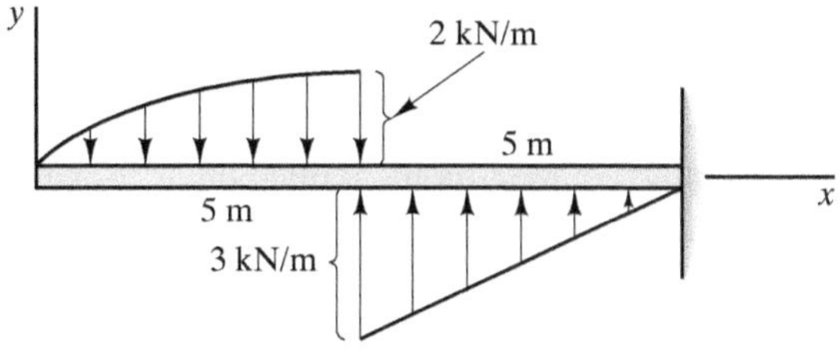

Figure P.7.59.

7.60. Formulate the shear-force and bending-moment equations and sketch the shear-force and bending-moment diagrams labeling the key points.

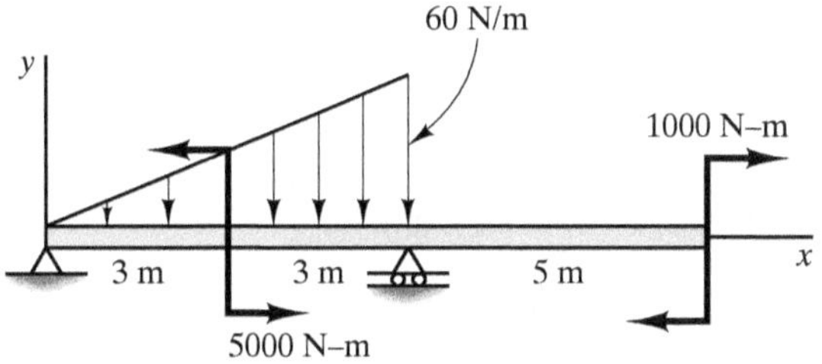

Figure P.7.60.

7.61. Formulate the shear-force and bending-moment equations for the beam in Fig. P.7.61. Sketch the shear-force and bending-moment diagrams.

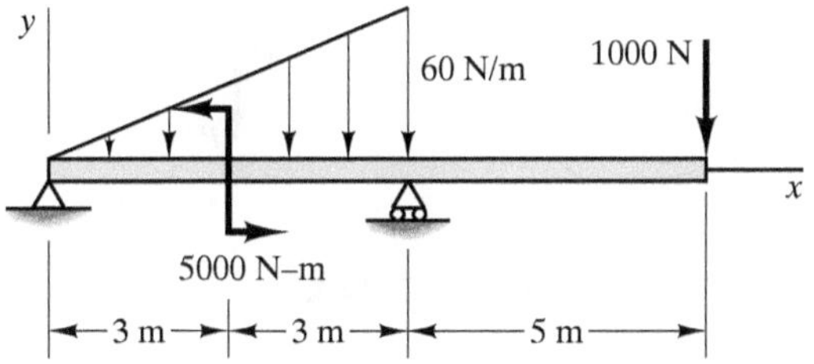

Figure P.7.61.

7.62. Sketch and label shear-force and bending-moment diagrams for the overhanging beam in Fig. P.7.62.

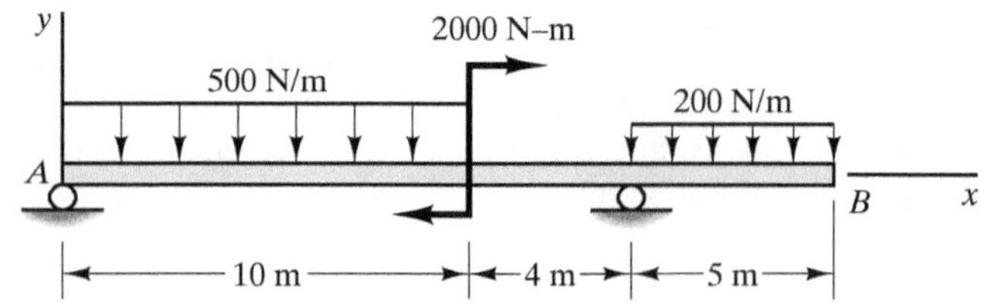

Figure P.7.62.

7.63. Given the bending-moment diagram as shown in Fig. P.7.63 sketch the corresponding shear-force diagram.

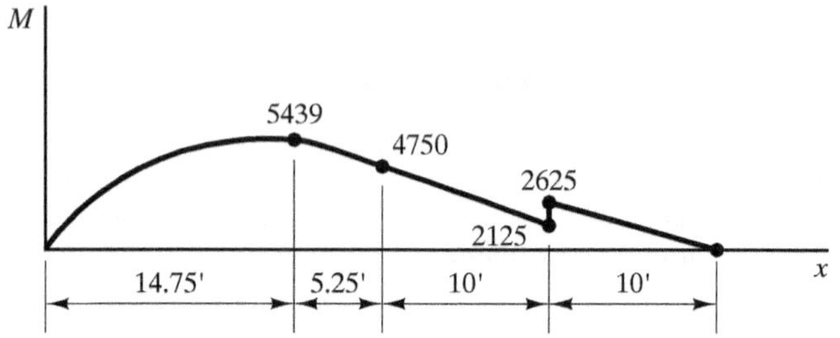

Figure P.7.63.

7.64. The structure shown in Fig. P.7.64 is proposed as a calibration fixture for an experiment. The cantilevered beam *DE* has an inverted "**T**-shaped" *rigid* bracket attached to its tip, as shown, allowing the placement of the point load *P* at any location along *ABC*. Sketch the resulting bending-moment diagrams for the beam *DE* for the load placed at *A*, *B*, and *C*.

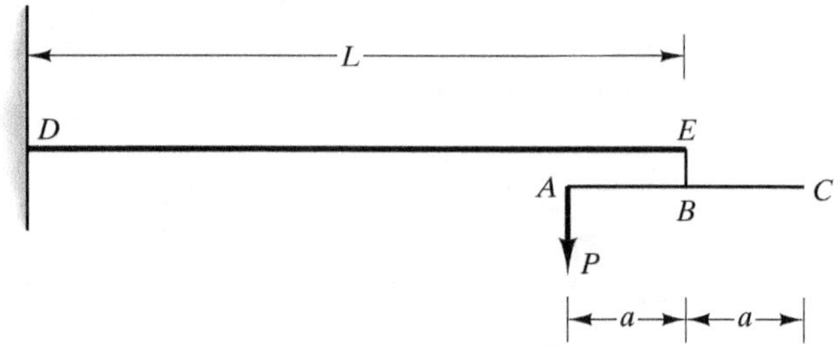

Figure P.7.64.

7.65. A design for a truss bridge is based on the top and bottom members formed into a parabolic shape as shown in Fig. P.7.65. The roadway is suspended from the truss by hangers. Why is the truss shaped this way? (*Hint:* Consider the bending moment from a uniform load.)

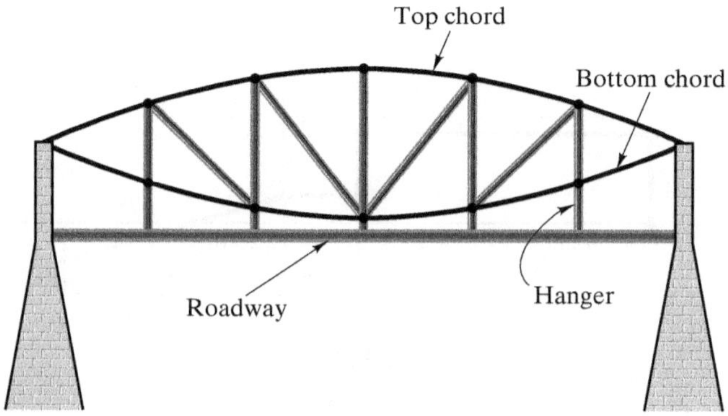

Figure P.7.65.

****7.66.** Consider the simply-supported beam shown in Fig. P.7.66. Plot the location, *(x/L)*, of the maximum bending moment in the beam versus the loading parameter α for $0 \leq \alpha \leq 10$. What should you get when $\alpha = 0$? For the range considered, what is the error associated with approximating the maximum moment location as *L/2* for $\alpha = 2$ and $\alpha = 10$?

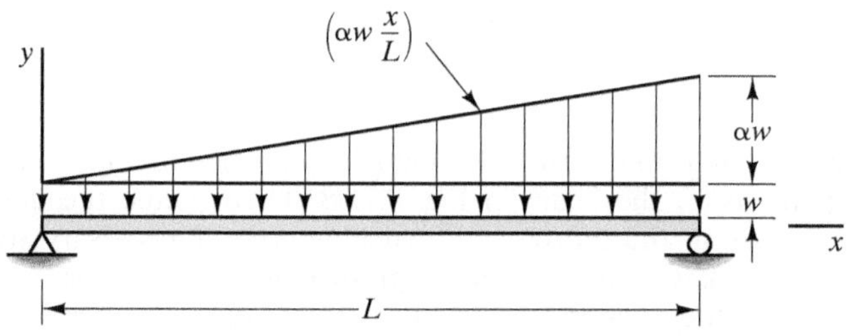

Figure P.7.66.

****7.67.** Consider the uniformly loaded beam shown in Fig. P.7.67 with an overhang. Plot the location, *(x/L)*, of the maximum positive moment versus α. Check your solution for $\alpha = 0$. What happens when $\alpha = 1$? Also, plot $M_{max}/(wL^2)$ for positive and negative bending moments versus α. At what value of α are the two magnitudes equal?

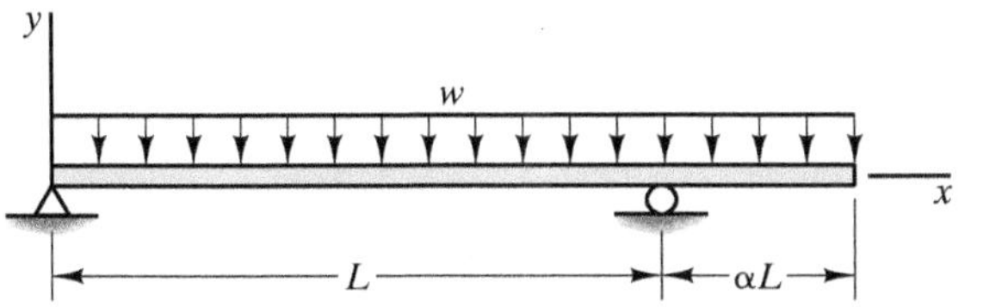

Figure P.7.67.

Project 10: Shear-Force and Bending-Moment Computation

This program is for *shear-force and bending-moment* computation for any loading, for a simply supported beam shown in Fig. P.7.68. The user defines the particular loading function for his or her program. The program should be able to handle *any loading that the user may want to insert.* Using Simpson's rule, *compute V and M at any point x.* Accept values for L, h, b, and x interactively. Also, *find the maximum shear stress at x.*

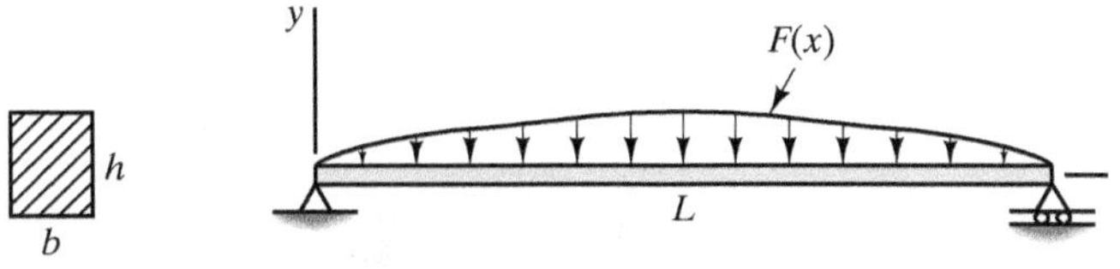

Figure P.7.68.

Work out for the following data in Fig. P.7.69:

$L = 10$ m　　$F(x) = 50$ N/m
$h = 0.1$ m　　$x = 4$ m
$b = 0.1$ m

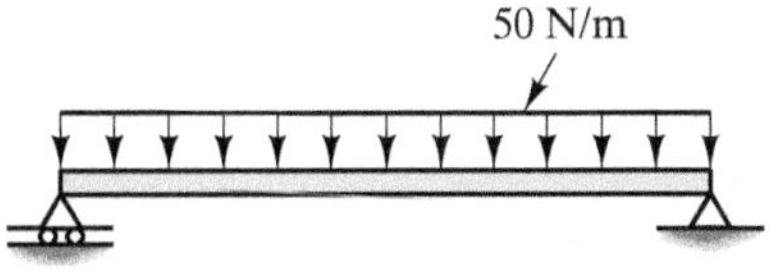

Figure P.7.69.

Chapter 8

The effective design of each brake on this railroad wheel requires that it resist the frictional forces developed between it and the wheel. In this chapter we will study dry friction, and show how to analyze friction forces for various engineering applications.

Friction

CHAPTER OBJECTIVES

- To introduce the concept of dry friction and show how to analyze the equilibrium of rigid bodies subjected to this force.
- To present specific applications of frictional force analysis on wedges, screws, belts, and bearings.
- To investigate the concept of rolling resistance.

8.1 Characteristics of Dry Friction

Friction is a force that resists the movement of two contacting surfaces that slide relative to one another. This force always acts *tangent* to the surface at the points of contact and is directed so as to oppose the possible or existing motion between the surfaces.

In this chapter, we will study the effects of *dry friction*, which is sometimes called *Coulomb friction* since its characteristics were studied extensively by C. A. Coulomb in 1781. Dry friction occurs between the contacting surfaces of bodies when there is no lubricating fluid.*

The heat generated by the abrasive action of friction can be noticed when using this grinder to sharpen a metal blade.

*Another type of friction, called fluid friction, is studied in fluid mechanics.

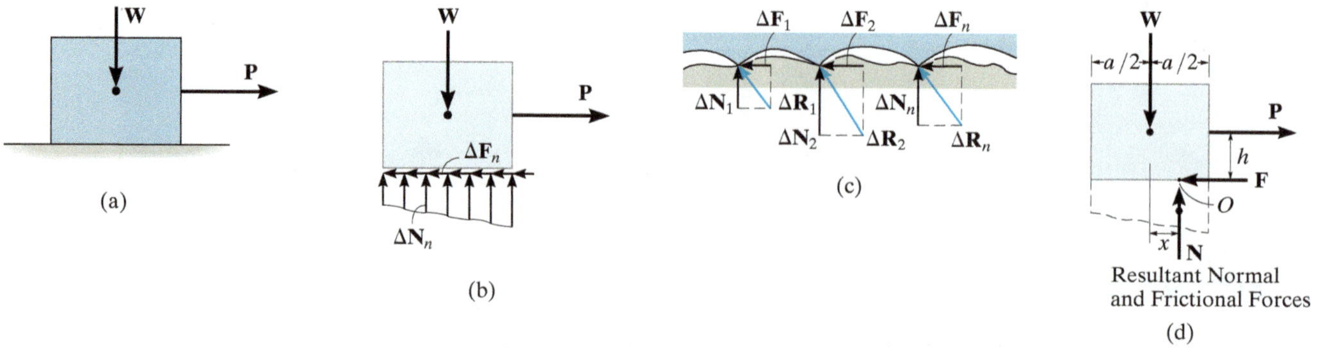

Fig. 8–1

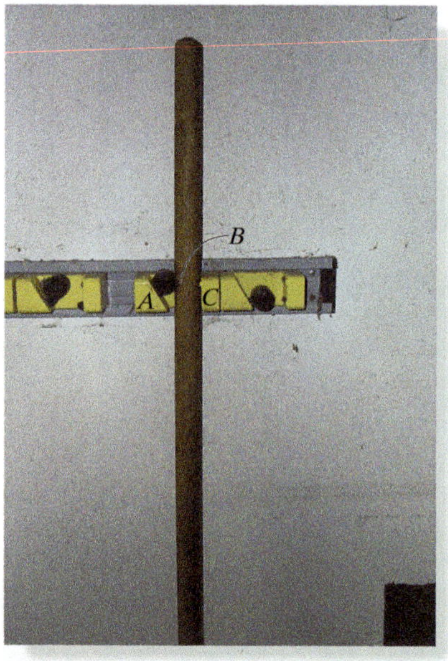

Regardless of the weight of the rake or shovel that is suspended, the device has been designed so that the small roller holds the handle in equilibrium due to frictional forces that develop at the points of contact, *A*, *B*, *C*.

Theory of Dry Friction. The theory of dry friction can be explained by considering the effects caused by pulling horizontally on a block of uniform weight **W** which is resting on a rough horizontal surface that is *nonrigid or deformable*, Fig. 8–1*a*. The upper portion of the block, however, can be considered rigid. As shown on the free-body diagram of the block, Fig. 8–1*b*, the floor exerts an uneven *distribution* of both *normal force* $\Delta \mathbf{N}_n$ and *frictional force* $\Delta \mathbf{F}_n$ along the contacting surface. For equilibrium, the normal forces must act *upward* to balance the block's weight **W**, and the frictional forces act to the left to prevent the applied force **P** from moving the block to the right. Close examination of the contacting surfaces between the floor and block reveals how these frictional and normal forces develop, Fig. 8–1*c*. It can be seen that many microscopic irregularities exist between the two surfaces and, as a result, reactive forces $\Delta \mathbf{R}_n$ are developed at each point of contact.* As shown, each reactive force contributes both a frictional component $\Delta \mathbf{F}_n$ and a normal component $\Delta \mathbf{N}_n$.

8

Equilibrium. The effect of the *distributed* normal and frictional loadings is indicated by their *resultants* **N** and **F** on the free-body diagram, Fig. 8–1*d*. Notice that **N** acts a distance x to the right of the line of action of **W**, Fig. 8–1*d*. This location, which coincides with the centroid or geometric center of the normal force distribution in Fig. 8–1*b*, is necessary in order to balance the "tipping effect" caused by **P**. For example, if **P** is applied at a height h from the surface, Fig. 8–1*d*, then moment equilibrium about point O is satisfied if $Wx = Ph$ or $x = Ph/W$.

*Besides mechanical interactions as explained here, which is referred to as a classical approach, a detailed treatment of the nature of frictional forces must also include the effects of temperature, density, cleanliness, and atomic or molecular attraction between the contacting surfaces. See J. Krim, *Scientific American*, October, 1996.

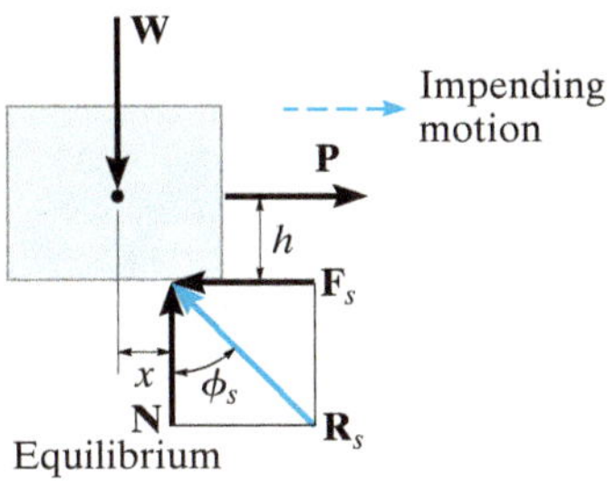

Fig. 8–1 (cont.)

Impending Motion. In cases where the surfaces of contact are rather "slippery," the frictional force **F** may *not* be great enough to balance **P**, and consequently the block will tend to slip. In other words, as P is slowly increased, F correspondingly increases until it attains a certain *maximum value* F_s, called the *limiting static frictional force*, Fig. 8–1*e*. When this value is reached, the block is in *unstable equilibrium* since any further increase in P will cause the block to move. Experimentally, it has been determined that this limiting static frictional force F_s is *directly proportional* to the resultant normal force N. Expressed mathematically,

$$F_s = \mu_s N \tag{8–1}$$

where the constant of proportionality, μ_s (mu "sub" s), is called the *coefficient of static friction.*

Thus, when the block is on the *verge of sliding*, the normal force **N** and frictional force $\mathbf{F}_s$ combine to create a resultant $\mathbf{R}_s$, Fig. 8–1*e*. The angle ϕ_s (phi "sub" s) that $\mathbf{R}_s$ makes with **N** is called the *angle of static friction.* From the figure,

$$\phi_s = \tan^{-1}\left(\frac{F_s}{N}\right) = \tan^{-1}\left(\frac{\mu_s N}{N}\right) = \tan^{-1}\mu_s$$

Typical values for μ_s are given in Table 8–1. Note that these values can vary since experimental testing was done under variable conditions of roughness and cleanliness of the contacting surfaces. For applications, therefore, it is important that both caution and judgment be exercised when selecting a coefficient of friction for a given set of conditions. When a more accurate calculation of F_s is required, the coefficient of friction should be determined directly by an experiment that involves the two materials to be used.

Table 8–1 Typical Values for μ_s

Contact Materials	Coefficient of Static Friction (μ_s)
Metal on ice	0.03–0.05
Wood on wood	0.30–0.70
Leather on wood	0.20–0.50
Leather on metal	0.30–0.60
Aluminum on aluminum	1.10–1.70

8

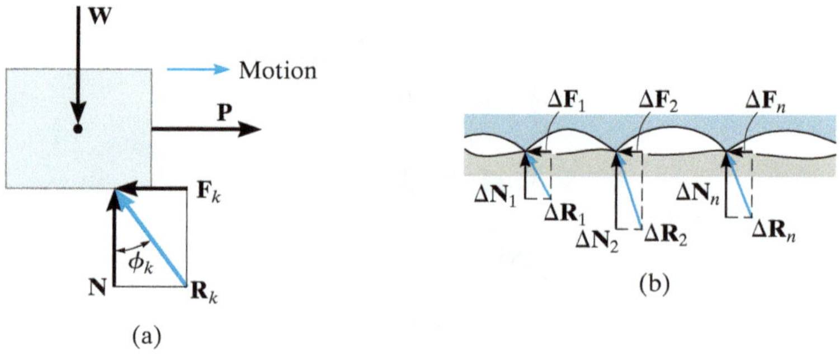

Fig. 8–2

Motion. If the magnitude of **P** acting on the block is increased so that it becomes slightly greater than F_s, the frictional force at the contacting surface will drop to a smaller value F_k, called the *kinetic frictional force*. The block will begin to slide with increasing speed, Fig. 8–2*a*. As this occurs, the block will "ride" on top of these peaks at the points of contact, as shown in Fig. 8–2*b*. The continued breakdown of the surface is the dominant mechanism creating kinetic friction.

Experiments with sliding blocks indicate that the magnitude of the kinetic friction force is directly proportional to the magnitude of the resultant normal force, expressed mathematically as

$$F_k = \mu_k N \tag{8–2}$$

Here the constant of proportionality, μ_k, is called the *coefficient of kinetic friction*. Typical values for μ_k are approximately 25 percent *smaller* than those listed in Table 8–1 for μ_s.

As shown in Fig. 8–2*a*, in this case, the resultant force at the surface of contact, $\mathbf{R}_k$, has a line of action defined by ϕ_k. This angle is referred to as the *angle of kinetic friction*, where

$$\phi_k = \tan^{-1}\left(\frac{F_k}{N}\right) = \tan^{-1}\left(\frac{\mu_k N}{N}\right) = \tan^{-1}\mu_k$$

By comparison, $\phi_s \geq \phi_k$.

8

The above effects regarding friction can be summarized by referring to the graph in Fig. 8–3, which shows the variation of the frictional force F versus the applied load P. Here the frictional force is categorized in three different ways:

- F is a *static frictional force* if equilibrium is maintained.
- F is a *limiting static frictional force* F_s when it reaches a maximum value needed to maintain equilibrium.
- F is a *kinetic frictional force* F_k when sliding occurs at the contacting surface.

Notice also from the graph that for very large values of P or for high speeds, aerodynamic effects will cause F_k and likewise μ_k to begin to decrease.

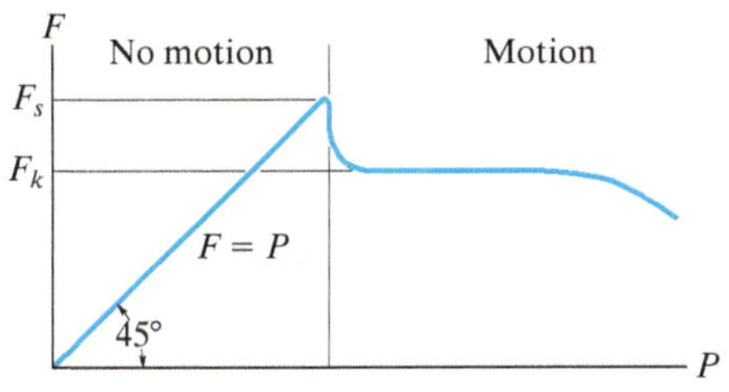

Fig. 8–3

Characteristics of Dry Friction. As a result of *experiments* that pertain to the foregoing discussion, we can state the following rules which apply to bodies subjected to dry friction.

- The frictional force acts *tangent* to the contacting surfaces in a direction *opposed* to the *motion* or tendency for motion of one surface relative to another.
- The maximum static frictional force F_s that can be developed is independent of the area of contact, provided the normal pressure is not very low nor great enough to severely deform or crush the contacting surfaces of the bodies.
- The maximum static frictional force is generally greater than the kinetic frictional force for any two surfaces of contact. However, if one of the bodies is moving with a *very low velocity* over the surface of another, F_k becomes approximately equal to F_s, i.e., $\mu_s \approx \mu_k$.
- When *slipping* at the surface of contact is *about to occur*, the maximum static frictional force is proportional to the normal force, such that $F_s = \mu_s N$.
- When *slipping* at the surface of contact is *occurring*, the kinetic frictional force is proportional to the normal force, such that $F_k = \mu_k N$.

8

8.2 Problems Involving Dry Friction

If a rigid body is in equilibrium when it is subjected to a system of forces that includes the effect of friction, the force system must satisfy not only the equations of equilibrium but *also* the laws that govern the frictional forces.

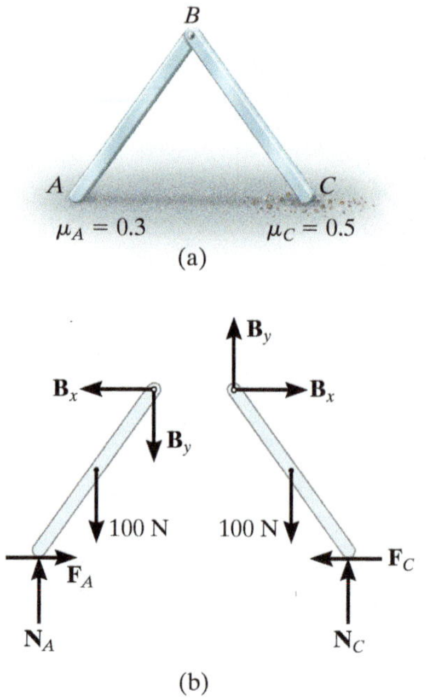

Fig. 8–4

Types of Friction Problems. In general, there are three types of mechanics problems involving dry friction. They can easily be classified once free-body diagrams are drawn and the total number of unknowns are identified and compared with the total number of available equilibrium equations.

No Apparent Impending Motion. Problems in this category are strictly equilibrium problems, which require the number of unknowns to be *equal* to the number of available equilibrium equations. Once the frictional forces are determined from the solution, however, their numerical values must be checked to be sure they satisfy the inequality $F \leq \mu_s N$; otherwise, slipping will occur and the body will not remain in equilibrium. A problem of this type is shown in Fig. 8–4*a*. Here we must determine the frictional forces at A and C to check if the equilibrium position of the two-member frame can be maintained. If the bars are uniform and have known weights of 100 N each, then the free-body diagrams are as shown in Fig. 8–4*b*. There are six unknown force components which can be determined *strictly* from the six equilibrium equations (three for each member). Once F_A, N_A, F_C, and N_C are determined, then the bars will remain in equilibrium provided $F_A \leq 0.3N_A$ and $F_C \leq 0.5N_C$ are satisfied.

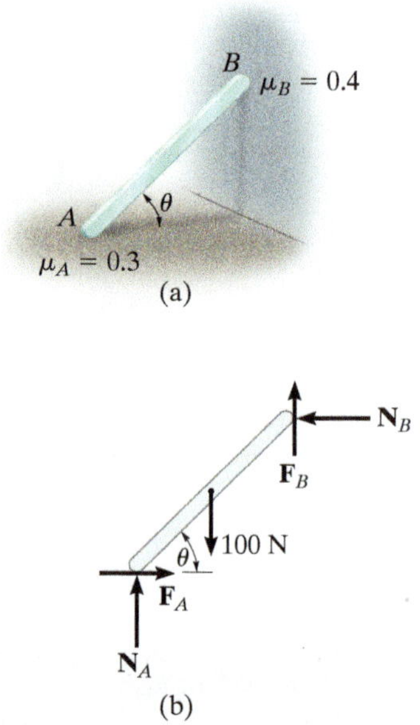

Fig. 8–5

Impending Motion at All Points of Contact. In this case the total number of unknowns will *equal* the total number of available equilibrium equations *plus* the total number of available frictional equations, $F = \mu N$. When *motion is impending* at the points of contact, then $F_s = \mu_s N$; whereas if the body is *slipping*, then $F_k = \mu_k N$. For example, consider the problem of finding the smallest angle θ at which the 100-N bar in Fig. 8–5*a* can be placed against the wall without slipping. The free-body diagram is shown in Fig. 8–5*b*. Here the *five* unknowns are determined from the *three* equilibrium equations and *two* static frictional equations which apply at *both* points of contact, so that $F_A = 0.3N_A$ and $F_B = 0.4N_B$.

Impending Motion at Some Points of Contact. Here the number of unknowns will be *less* than the number of available equilibrium equations plus the number of available frictional equations or conditional equations for tipping. As a result, several possibilities for motion or impending motion will exist and the problem will involve a determination of the kind of motion which actually occurs. For example, consider the two-member frame in Fig. 8–6*a*. In this problem we wish to determine the horizontal force P needed to cause movement. If each member has a weight of 100 N, then the free-body diagrams are as shown in Fig. 8–6*b*. There are *seven* unknowns. For a unique solution we must satisfy the *six* equilibrium equations (three for each member) and only *one* of two possible static frictional equations. This means that as P increases it will either cause slipping at A and no slipping at C, so that $F_A = 0.3N_A$ and $F_C \leq 0.5N_C$; or slipping occurs at C and no slipping at A, in which case $F_C = 0.5N_C$ and $F_A \leq 0.3N_A$. The actual situation can be determined by calculating P for each case and then choosing the case for which P is *smaller*. If in both cases the *same value* for P is calculated, which would be highly improbable, then slipping at both points occurs simultaneously; i.e., the *seven unknowns* would satisfy *eight equations*.

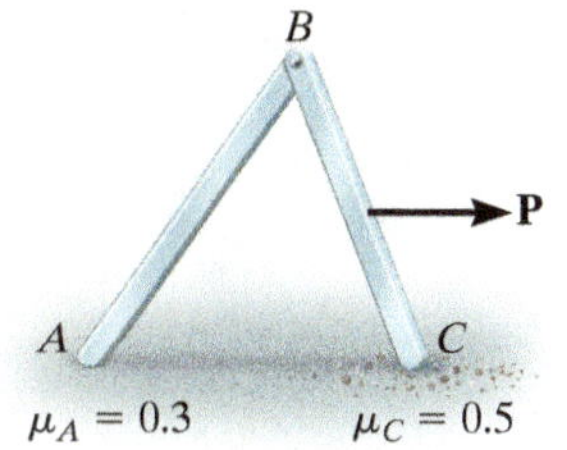

(a)

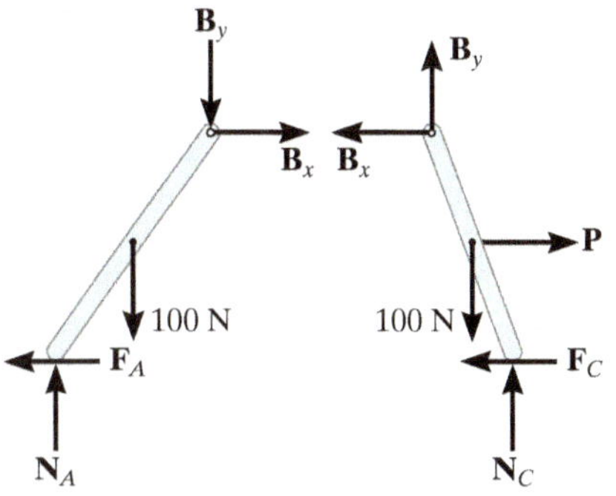

(b)

Fig. 8–6

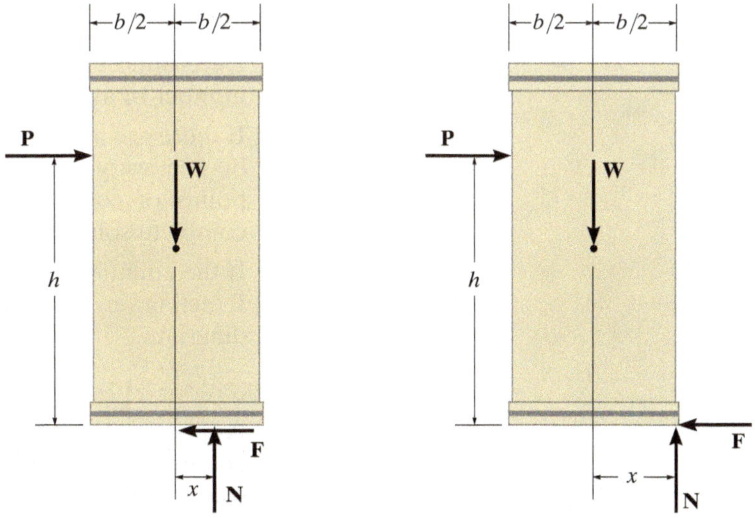

Consider pushing on the uniform crate that has a weight W and sits on the rough surface. As shown on the first free-body diagram, if the magnitude of **P** is small, the crate will remain in equilibrium. As P increases the crate will either be on the verge of slipping on the surface ($F = \mu_s N$), or if the surface is very rough (large μ_s) then the resultant normal force will shift to the corner, $x = b/2$, as shown on the second free-body diagram. At this point the crate will begin to tip over. The crate also has a greater chance of tipping if **P** is applied at a greater height h above the surface, or if its width b is smaller.

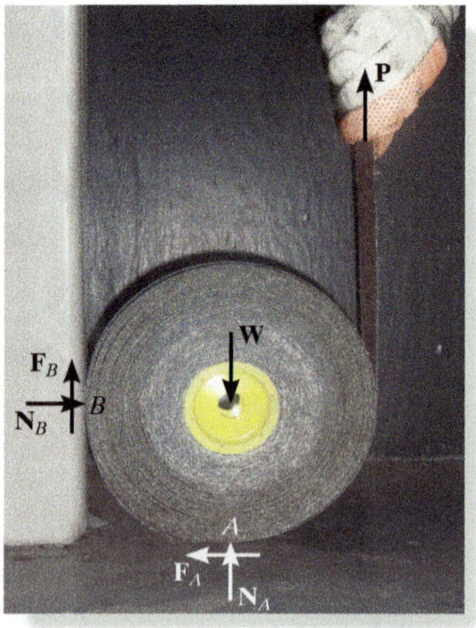

The applied vertical force **P** on this roll must be large enough to overcome the resistance of friction at the contacting surfaces A and B in order to cause rotation.

Equilibrium Versus Frictional Equations. Whenever we solve a problem such as the one in Fig. 8–4, where the friction force F is to be an "equilibrium force" and satisfies the inequality $F < \mu_s N$, then we can assume the sense of direction of F on the free-body diagram. The correct sense is made known *after* solving the equations of equilibrium for F. If F is a negative scalar the sense of **F** is the reverse of that which was assumed. This convenience of *assuming* the sense of **F** is possible because the equilibrium equations equate to zero the *components of vectors* acting in the *same direction*. However, in cases where the frictional equation $F = \mu N$ is used in the solution of a problem, as in the case shown in Fig. 8–5, then the convenience of *assuming* the sense of **F** is *lost*, since the frictional equation relates only the *magnitudes* of two *perpendicular* vectors. Consequently, **F** *must always* be shown acting with its *correct sense* on the free-body diagram, *whenever* the frictional equation is used for the solution of a problem.

Procedure for Analysis

Equilibrium problems involving dry friction can be solved using the following procedure.

Free-Body Diagrams.

- Draw the necessary free-body diagrams, and unless it is stated in the problem that impending motion or slipping occurs, *always* show the frictional forces as unknowns (i.e., *do not assume* $F = \mu N$).
- Determine the number of unknowns and compare this with the number of available equilibrium equations.
- If there are more unknowns than equations of equilibrium, it will be necessary to apply the frictional equation at some, if not all, points of contact to obtain the extra equations needed for a complete solution.
- If the equation $F = \mu N$ is to be used, it will be necessary to show **F** acting in the correct sense of direction on the free-body diagram.

Equations of Equilibrium and Friction.

- Apply the equations of equilibrium and the necessary frictional equations (or conditional equations if tipping is possible) and solve for the unknowns.
- If the problem involves a three-dimensional force system such that it becomes difficult to obtain the force components or the necessary moment arms, apply the equations of equilibrium using Cartesian vectors.

EXAMPLE 8.1

The uniform crate shown in Fig. 8–7*a* has a mass of 20 kg. If a force $P = 80$ N is applied to the crate, determine if it remains in equilibrium. The coefficient of static friction is $\mu_s = 0.3$.

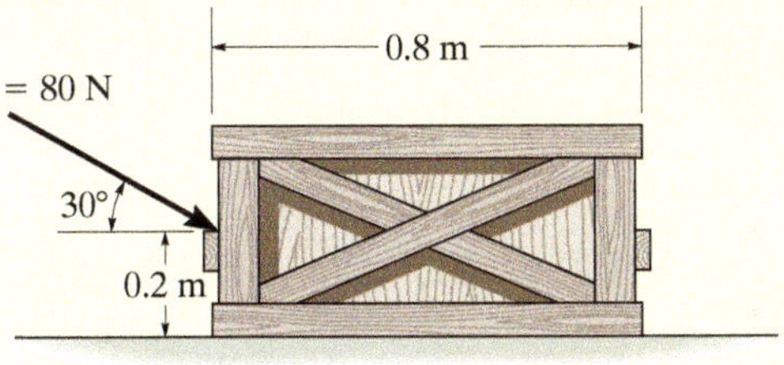

(a)

Fig. 8–7

SOLUTION

Free-Body Diagram. As shown in Fig. 8–7*b*, the *resultant* normal force $\mathbf{N}_C$ must act a distance x from the crate's center line in order to counteract the tipping effect caused by **P**. There are *three unknowns*, F, N_C, and x, which can be determined strictly from the *three* equations of equilibrium.

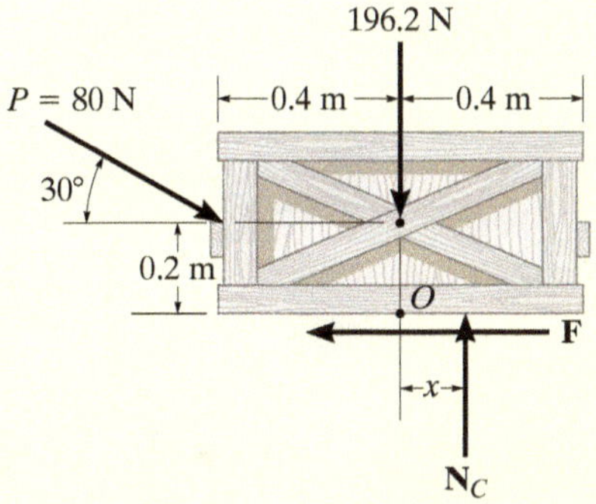

(b)

Equations of Equilibrium.

$$\overset{+}{\rightarrow}\Sigma F_x = 0; \qquad 80 \cos 30^\circ \text{ N} - F = 0$$

$$+\uparrow \Sigma F_y = 0; \qquad -80 \sin 30^\circ \text{ N} + N_C - 196.2 \text{ N} = 0$$

$$\circlearrowleft + \Sigma M_O = 0; \quad 80 \sin 30^\circ \text{ N}(0.4 \text{ m}) - 80 \cos 30^\circ \text{ N}(0.2 \text{ m}) + N_C(x) = 0$$

Solving,

$$F = 69.3 \text{ N}$$
$$N_C = 236.2 \text{ N}$$
$$x = -0.00908 \text{ m} = -9.08 \text{ mm}$$

Since x is negative it indicates the *resultant* normal force acts (slightly) to the *left* of the crate's center line. No tipping will occur since $x < 0.4$ m. Also, the *maximum* frictional force which can be developed at the surface of contact is $F_{max} = \mu_s N_C = 0.3(236.2 \text{ N}) = 70.9$ N. Since $F = 69.3 \text{ N} < 70.9$ N, the crate will *not slip*, although it is very close to doing so.

8

EXAMPLE 8.2

(a)

It is observed that when the bed of the dump truck is raised to an angle of $\theta = 25°$ the vending machines will begin to slide off the bed, Fig. 8–8*a*. Determine the static coefficient of friction between a vending machine and the surface of the truckbed.

SOLUTION

An idealized model of a vending machine resting on the truckbed is shown in Fig. 8–8*b*. The dimensions have been measured and the center of gravity has been located. We will assume that the vending machine weighs W.

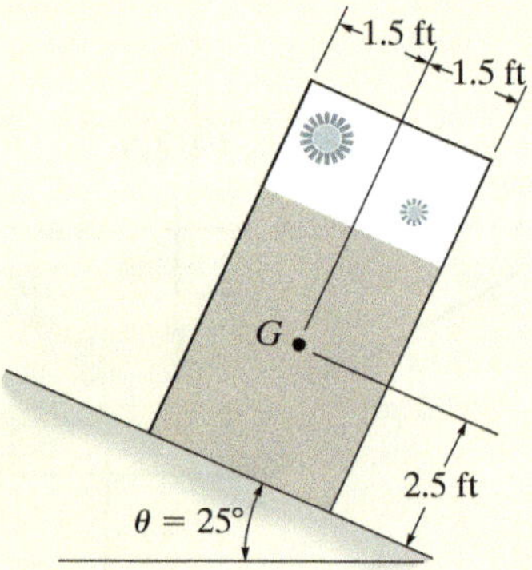

(b)

Free-Body Diagram. As shown in Fig. 8–8*c*, the dimension x is used to locate the position of the resultant normal force $\mathbf{N}$. There are four unknowns, N, F, μ_s, and x.

Equations of Equilibrium.

$$+\searrow\Sigma F_x = 0; \qquad W \sin 25° - F = 0 \qquad (1)$$

$$+\nearrow\Sigma F_y = 0; \qquad N - W \cos 25° = 0 \qquad (2)$$

$$\circlearrowleft +\Sigma M_O = 0; \quad -W \sin 25°(2.5 \text{ ft}) + W \cos 25°(x) = 0 \qquad (3)$$

Since slipping impends at $\theta = 25°$, using Eqs. 1 and 2, we have

$$F_s = \mu_s N; \qquad W \sin 25° = \mu_s(W \cos 25°)$$

$$\mu_s = \tan 25° = 0.466 \qquad \textit{Ans.}$$

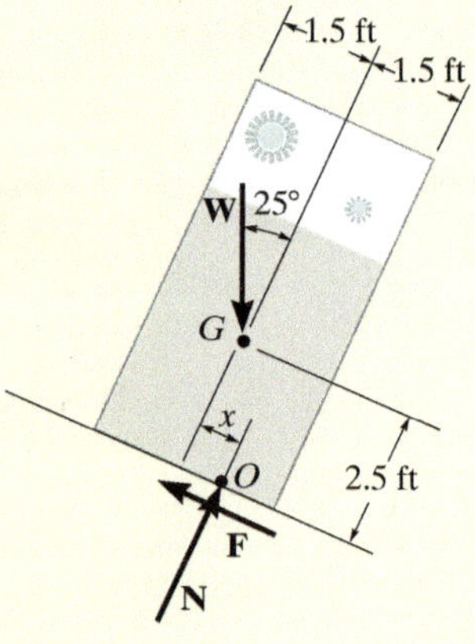

(c)

Fig. 8–8

The angle of $\theta = 25°$ is referred to as the *angle of repose*, and by comparison, it is equal to the angle of static friction, $\theta = \phi_s$. Notice from the calculation that θ is independent of the weight of the vending machine, and so knowing θ provides a convenient method for determining the coefficient of static friction.

NOTE: From Eq. 3, we find $x = 1.17$ ft. Since 1.17 ft < 1.5 ft, indeed the vending machine will slip before it can tip as observed in Fig. 8–8*a*.

EXAMPLE 8.3

The uniform 10-kg ladder in Fig. 8–9*a* rests against the smooth wall at *B*, and the end *A* rests on the rough horizontal plane for which the coefficient of static friction is $\mu_s = 0.3$. Determine the angle of inclination θ of the ladder and the normal reaction at *B* if the ladder is on the verge of slipping.

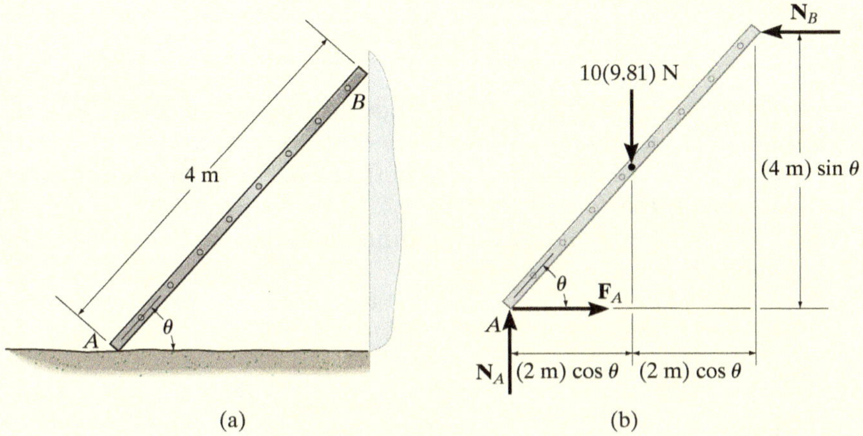

Fig. 8–9

SOLUTION

Free-Body Diagram. As shown on the free-body diagram, Fig. 8–9*b*, the frictional force $\mathbf{F}_A$ must act to the right since impending motion at *A* is to the left.

Equations of Equilibrium and Friction. Since the ladder is on the verge of slipping, then $F_A = \mu_s N_A = 0.3N_A$. By inspection, N_A can be obtained directly.

$$+\uparrow \Sigma F_y = 0; \qquad N_A - 10(9.81)\text{ N} = 0 \qquad N_A = 98.1\text{ N}$$

Using this result, $F_A = 0.3(98.1\text{ N}) = 29.43\text{ N}$. Now N_B can be found.

$$\xrightarrow{+} \Sigma F_x = 0; \qquad 29.43\text{ N} - N_B = 0$$

$$N_B = 29.43\text{ N} = 29.4\text{ N} \qquad \textit{Ans.}$$

Finally, the angle θ can be determined by summing moments about point *A*.

$$\circlearrowleft + \Sigma M_A = 0; \qquad (29.43\text{ N})(4\text{ m})\sin\theta - [10(9.81)\text{ N}](2\text{ m})\cos\theta = 0$$

$$\frac{\sin\theta}{\cos\theta} = \tan\theta = 1.6667$$

$$\theta = 59.04^\circ = 59.0^\circ \qquad \textit{Ans.}$$

8

EXAMPLE 8.4

Beam AB is subjected to a uniform load of 200 N/m and is supported at B by post BC, Fig. 8–10a. If the coefficients of static friction at B and C are $\mu_B = 0.2$ and $\mu_C = 0.5$, determine the force **P** needed to pull the post out from under the beam. Neglect the weight of the members and the thickness of the beam.

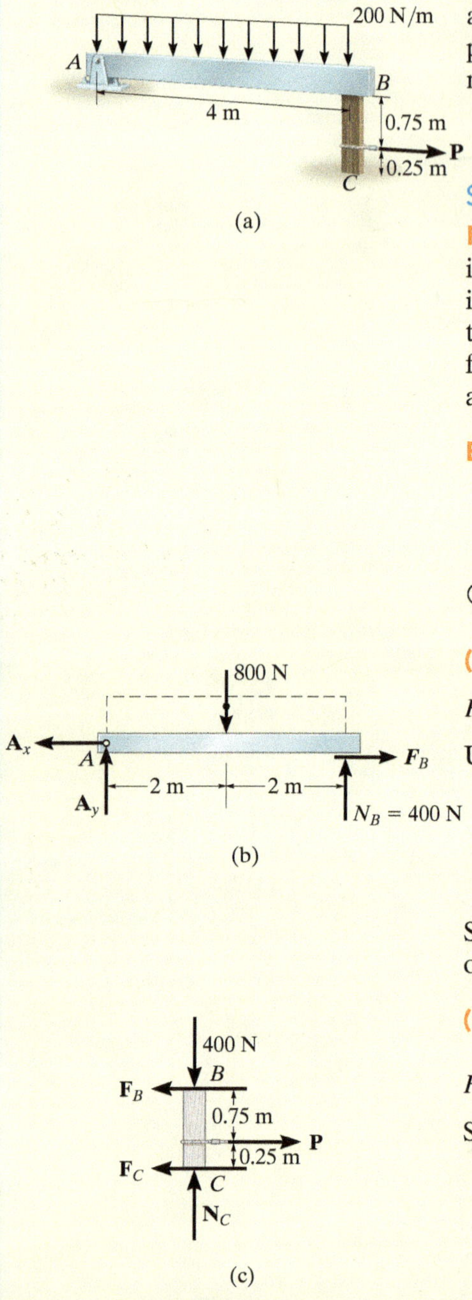

Fig. 8–10

SOLUTION

Free-Body Diagrams. The free-body diagram of the beam is shown in Fig. 8–10b. Applying $\Sigma M_A = 0$, we obtain $N_B = 400$ N. This result is shown on the free-body diagram of the post, Fig. 8–10c. Referring to this member, the *four* unknowns F_B, P, F_C, and N_C are determined from the *three* equations of equilibrium and *one* frictional equation applied either at B or C.

Equations of Equilibrium and Friction.

$$\xrightarrow{+} \Sigma F_x = 0; \qquad P - F_B - F_C = 0 \qquad (1)$$

$$+\uparrow \Sigma F_y = 0; \qquad N_C - 400\text{ N} = 0 \qquad (2)$$

$$\circlearrowleft + \Sigma M_C = 0; \qquad -P(0.25\text{ m}) + F_B(1\text{ m}) = 0 \qquad (3)$$

(Post Slips at *B* and Rotates about *C*.) This requires $F_C \le \mu_C N_C$ and

$$F_B = \mu_B N_B; \qquad F_B = 0.2(400\text{ N}) = 80\text{ N}$$

Using this result and solving Eqs. 1 through 3, we obtain

$$P = 320\text{ N}$$
$$F_C = 240\text{ N}$$
$$N_C = 400\text{ N}$$

Since $F_C = 240\text{ N} > \mu_C N_C = 0.5(400\text{ N}) = 200\text{ N}$, slipping at C occurs. Thus the other case of movement must be investigated.

(Post Slips at *C* and Rotates about *B*.) Here $F_B \le \mu_B N_B$ and

$$F_C = \mu_C N_C; \qquad F_C = 0.5 N_C \qquad (4)$$

Solving Eqs. 1 through 4 yields

$$P = 267\text{ N} \qquad \textit{Ans.}$$
$$N_C = 400\text{ N}$$
$$F_C = 200\text{ N}$$
$$F_B = 66.7\text{ N}$$

Obviously, this case occurs first since it requires a *smaller* value for P.

8

EXAMPLE 8.5

Blocks A and B have a mass of 3 kg and 9 kg, respectively, and are connected to the weightless links shown in Fig. 8–11a. Determine the largest vertical force **P** that can be applied at the pin C without causing any movement. The coefficient of static friction between the blocks and the contacting surfaces is $\mu_s = 0.3$.

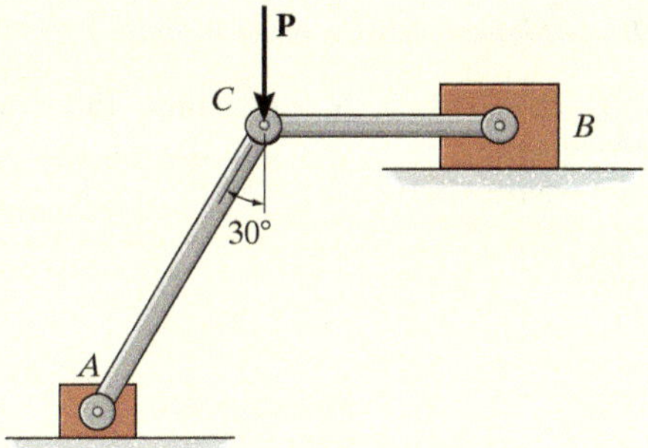

(a)

SOLUTION

Free-Body Diagram. The links are two-force members and so the free-body diagrams of pin C and blocks A and B are shown in Fig. 8–11b. Since the horizontal component of $\mathbf{F}_{AC}$ tends to move block A to the left, $\mathbf{F}_A$ must act to the right. Similarly, $\mathbf{F}_B$ must act to the left to oppose the tendency of motion of block B to the right, caused by $\mathbf{F}_{BC}$. There are seven unknowns and six available force equilibrium equations, two for the pin and two for each block, so that *only one* frictional equation is needed.

Equations of Equilibrium and Friction. The force in links AC and BC can be related to P by considering the equilibrium of pin C.

$$+\uparrow \Sigma F_y = 0; \qquad F_{AC}\cos 30^\circ - P = 0; \qquad F_{AC} = 1.155P$$

$$\overset{+}{\rightarrow} \Sigma F_x = 0; \qquad 1.155P \sin 30^\circ - F_{BC} = 0; \qquad F_{BC} = 0.5774P$$

Using the result for F_{AC}, for block A,

$$\overset{+}{\rightarrow} \Sigma F_x = 0; \qquad F_A - 1.155P \sin 30^\circ = 0; \qquad F_A = 0.5774P \qquad (1)$$

$$+\uparrow \Sigma F_y = 0; \qquad N_A - 1.155P\cos 30^\circ - 3(9.81\text{ N}) = 0;$$

$$N_A = P + 29.43\text{ N} \qquad (2)$$

Using the result for F_{BC}, for block B,

$$\overset{+}{\rightarrow} \Sigma F_x = 0; \qquad (0.5774P) - F_B = 0; \qquad F_B = 0.5774P \qquad (3)$$

$$+\uparrow \Sigma F_y = 0; \qquad N_B - 9(9.81)\text{ N} = 0; \qquad N_B = 88.29\text{ N}$$

Movement of the system may be caused by the initial slipping of *either* block A or block B. If we assume that block A slips first, then

$$F_A = \mu_s N_A = 0.3N_A \qquad (4)$$

Substituting Eqs. 1 and 2 into Eq. 4,

$$0.5774P = 0.3(P + 29.43)$$

$$P = 31.8\text{ N} \qquad \textit{Ans.}$$

Substituting this result into Eq. 3, we obtain $F_B = 18.4$ N. Since the maximum static frictional force at B is $(F_B)_{max} = \mu_s N_B = 0.3(88.29\text{ N}) = 26.5\text{ N} > F_B$, block B will not slip. Thus, the above assumption is correct. Notice that if the inequality were not satisfied, we would have to assume slipping of block B and then solve for P.

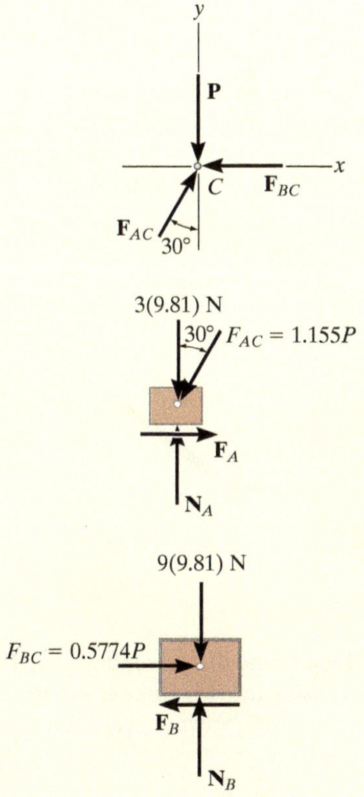

(b)

Fig. 8–11

FUNDAMENTAL PROBLEMS

All problem solutions must include FBDs.

F8–1. If $P = 200$ N, determine the friction developed between the 50-kg crate and the ground. The coefficient of static friction between the crate and the ground is $\mu_s = 0.3$.

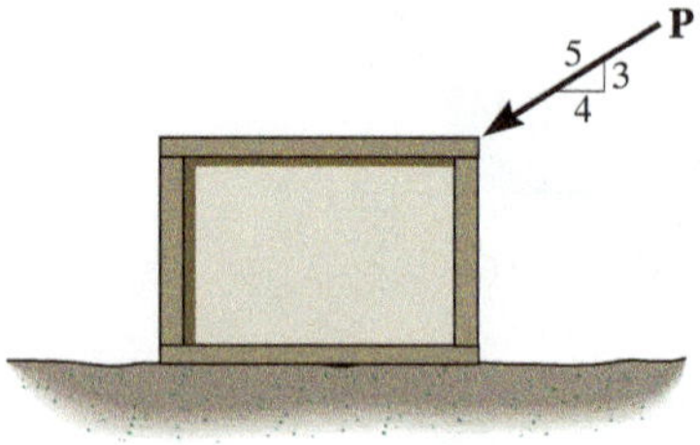

F8–1

F8–2. Determine the minimum force P to prevent the 30-kg rod AB from sliding. The contact surface at B is smooth, whereas the coefficient of static friction between the rod and the wall at A is $\mu_s = 0.2$.

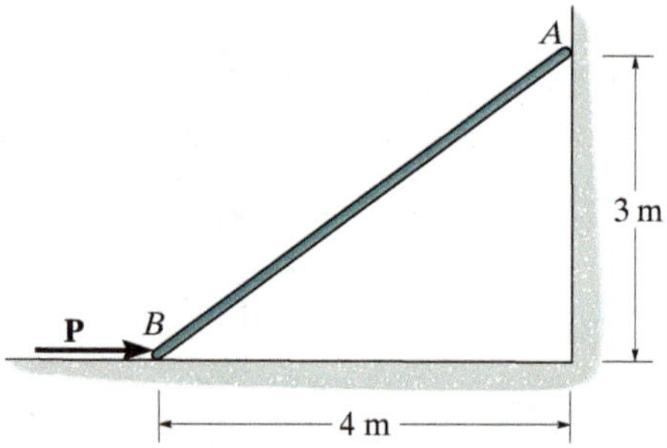

F8–2

F8–3. Determine the maximum force P that can be applied without causing the two 50-kg crates to move. The coefficient of static friction between each crate and the ground is $\mu_s = 0.25$.

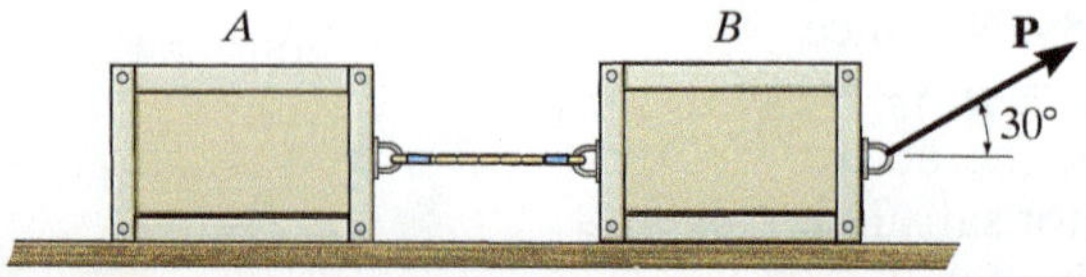

F8–3

F8–4. If the coefficient of static friction at contact points A and B is $\mu_s = 0.3$, determine the maximum force P that can be applied without causing the 100-kg spool to move.

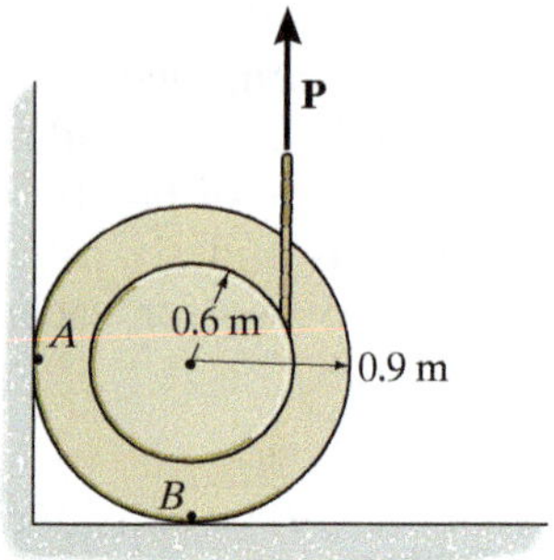

F8–4

F8–5. Determine the maximum force P that can be applied without causing movement of the 250-lb crate that has a center of gravity at G. The coefficient of static friction at the floor is $\mu_s = 0.4$.

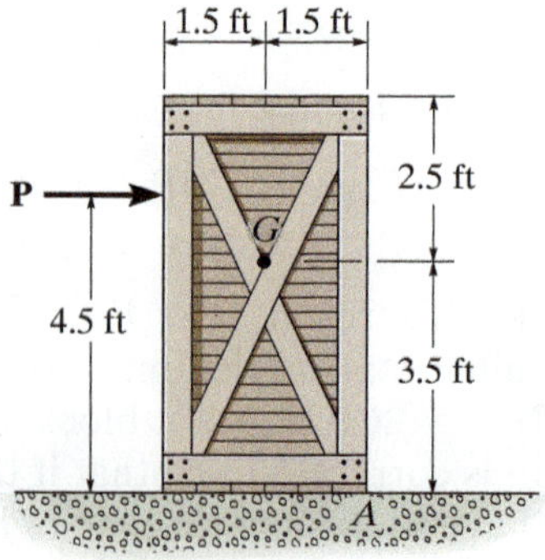

F8–5

8

F8–6. Determine the minimum coefficient of static friction between the uniform 50-kg spool and the wall so that the spool does not slip.

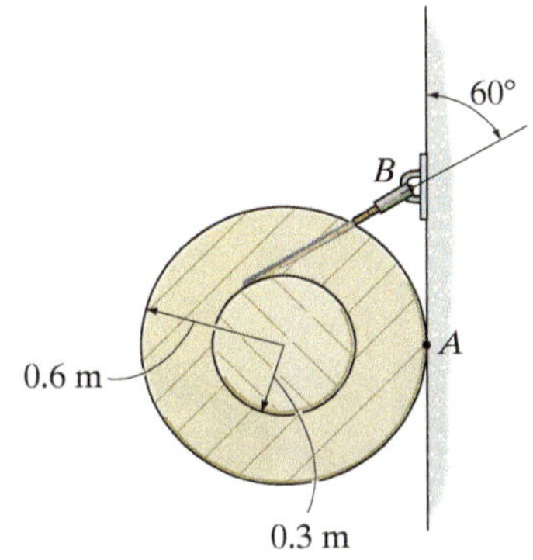

F8–6

F8–7. Blocks A, B, and C have weights of 50 N, 25 N, and 15 N, respectively. Determine the smallest horizontal force P that will cause impending motion. The coefficient of static friction between A and B is $\mu_s = 0.3$, between B and C, $\mu_s' = 0.4$, and between block C and the ground, $\mu_s'' = 0.35$.

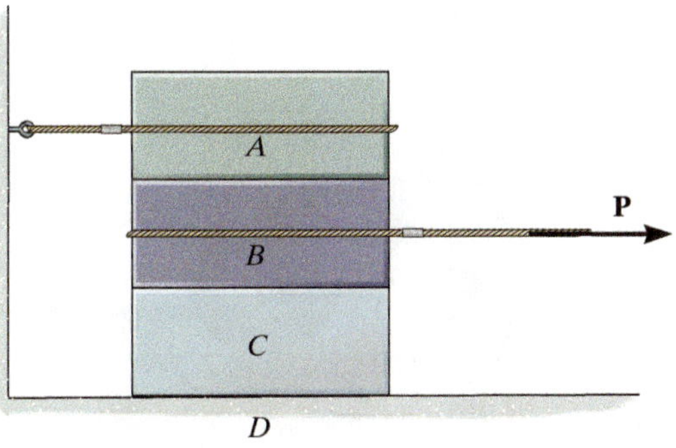

F8–7

F8–8. If the coefficient of static friction at all contacting surfaces is μ_s, determine the inclination θ at which the identical blocks, each of weight W, begin to slide.

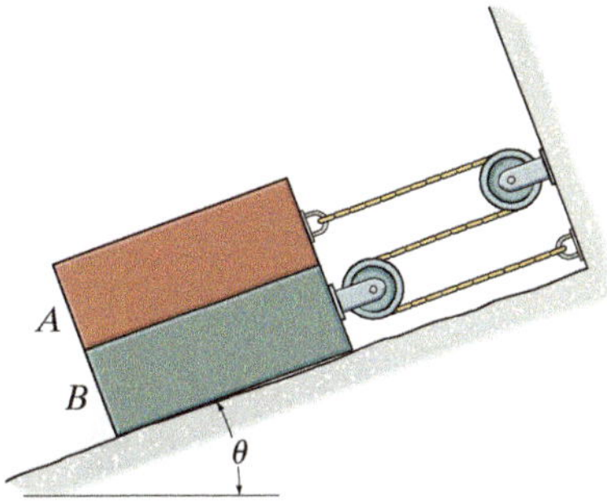

F8–8

F8–9. Blocks A and B have a mass of 7 kg and 10 kg, respectively. Using the coefficients of static friction indicated, determine the largest force P which can be applied to the cord without causing motion. There are pulleys at C and D.

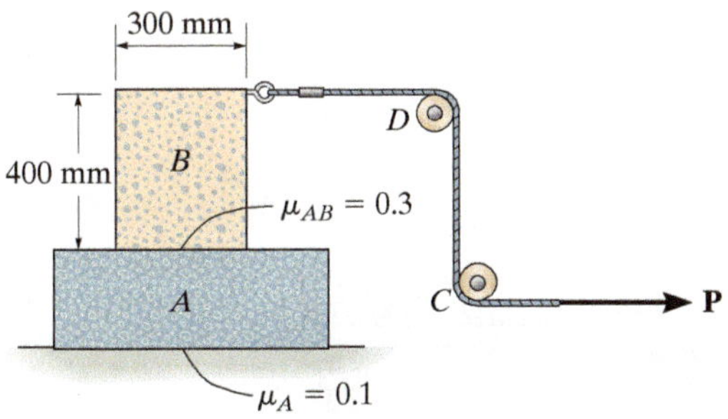

F8–9

8

PROBLEMS

All problem solutions must include FBDs.

8–1. The mine car and its contents have a total mass of 6 Mg and a center of gravity at G. If the coefficient of static friction between the wheels and the tracks is $\mu_s = 0.4$ when the wheels are locked, find the normal force acting on the front wheels at B and the rear wheels at A when the brakes at both A and B are locked. Does the car move?

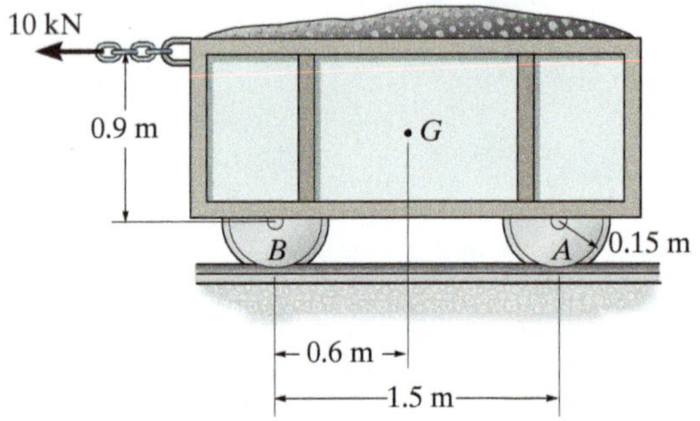

Prob. 8–1

8–2. Determine the maximum force P the connection can support so that no slipping occurs between the plates. There are four bolts used for the connection and each is tightened so that it is subjected to a tension of 4 kN. The coefficient of static friction between the plates is $\mu_s = 0.4$.

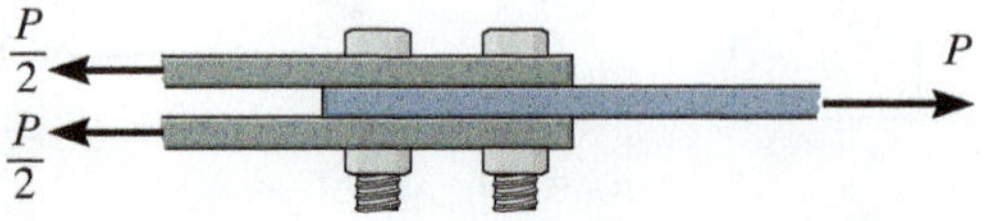

Prob. 8–2

8–3. The winch on the truck is used to hoist the garbage bin onto the bed of the truck. If the loaded bin has a weight of 8500 lb and center of gravity at G, determine the force in the cable needed to begin the lift. The coefficients of static friction at A and B are $\mu_A = 0.3$ and $\mu_B = 0.2$, respectively. Neglect the height of the support at A.

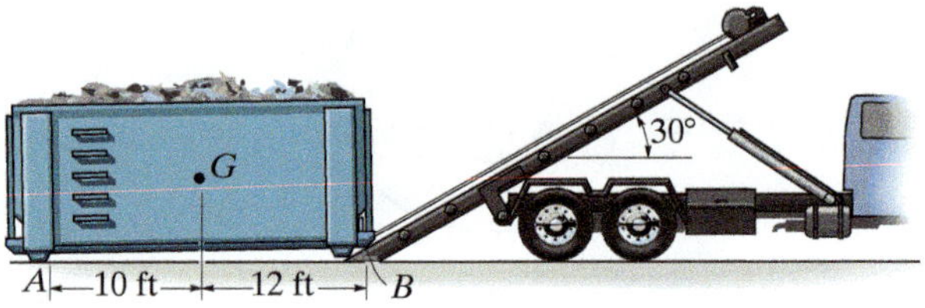

Prob. 8–3

***8–4.** The tractor has a weight of 4500 lb with center of gravity at G. The driving traction is developed at the rear wheels B, while the front wheels at A are free to roll. If the coefficient of static friction between the wheels at B and the ground is $\mu_s = 0.5$, determine if it is possible to pull at $P = 1200$ lb without causing the wheels at B to slip or the front wheels at A to lift off the ground.

Prob. 8–4

8

8–5. The ladder has a uniform weight of 80 lb and rests against the smooth wall at B. If the coefficient of static friction at A is $\mu_A = 0.4$, determine if the ladder will slip. Take $\theta = 60°$.

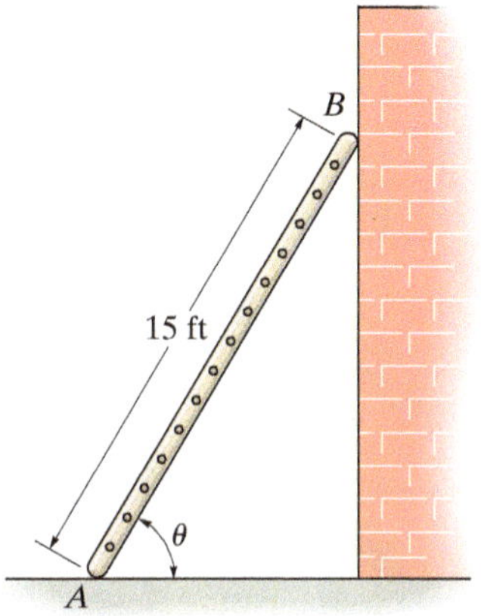

Prob. 8–5

8–6. The ladder has a uniform weight of 80 lb and rests against the wall at B. If the coefficient of static friction at A and B is $\mu = 0.4$, determine the smallest angle θ at which the ladder will not slip.

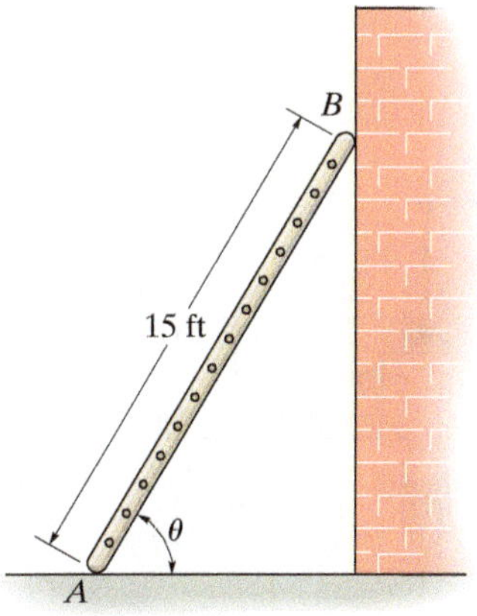

Prob. 8–6

8–7. The block brake consists of a pin-connected lever and friction block at B. The coefficient of static friction between the wheel and the lever is $\mu_s = 0.3$, and a torque of 5 N · m is applied to the wheel. Determine if the brake can hold the wheel stationary when the force applied to the lever is (a) $P = 30$ N, (b) $P = 70$ N.

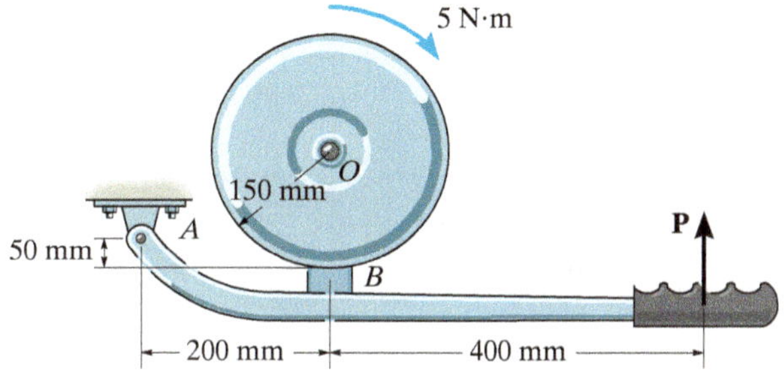

Prob. 8–7

***8–8.** The block brake consists of a pin-connected lever and friction block at B. The coefficient of static friction between the wheel and the lever is $\mu_s = 0.3$, and a torque of 5 N · m is applied to the wheel. Determine if the brake can hold the wheel stationary when the force applied to the lever is (a) $P = 30$ N, (b) $P = 70$ N.

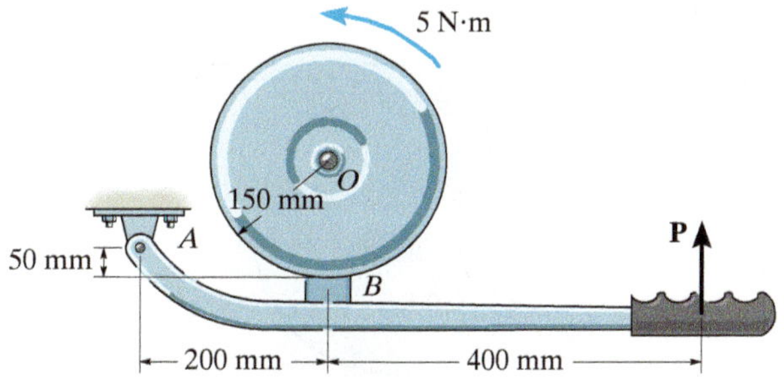

Prob. 8–8

8–9. The block brake is used to stop the wheel from rotating when the wheel is subjected to a couple moment $\mathbf{M}_0$. If the coefficient of static friction between the wheel and the block is μ_s, determine the smallest force P that should be applied.

8–10. Show that the brake in Prob. 8–9 is self-locking, i.e., $P \leq 0$, provided $b/c \leq \mu_s$.

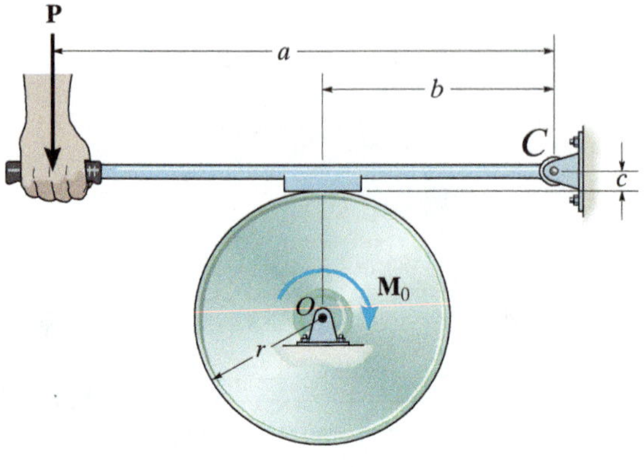

Probs. 8–9/10

***8–12.** If a torque of $M = 300\ \mathrm{N\cdot m}$ is applied to the flywheel, determine the force that must be developed in the hydraulic cylinder CD to prevent the flywheel from rotating. The coefficient of static friction between the friction pad at B and the flywheel is $\mu_s = 0.4$.

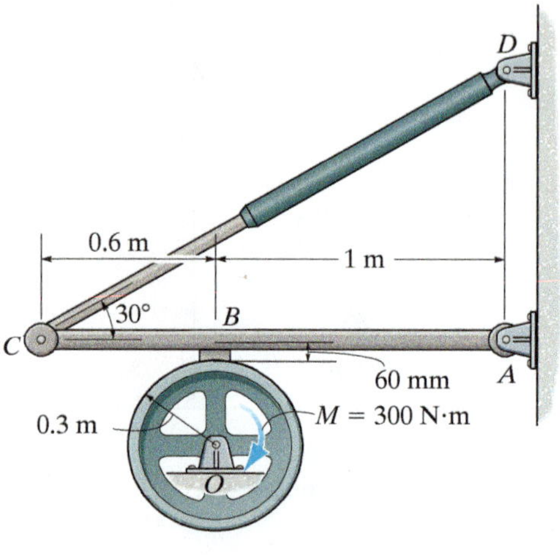

Prob. 8–12

8–11. The block brake is used to stop the wheel from rotating when the wheel is subjected to a couple moment $\mathbf{M}_0$. If the coefficient of static friction between the wheel and the block is μ_s, determine the smallest force P that should be applied.

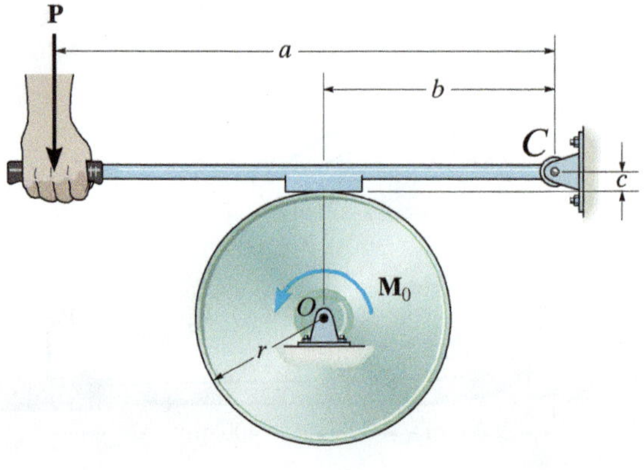

Prob. 8–11

8–13. The cam is subjected to a couple moment of $5\ \mathrm{N\cdot m}$ Determine the minimum force P that should be applied to the follower in order to hold the cam in the position shown. the coefficient of static friction between the cam and the follower is $\mu = 0.4$. The guide at A is smooth.

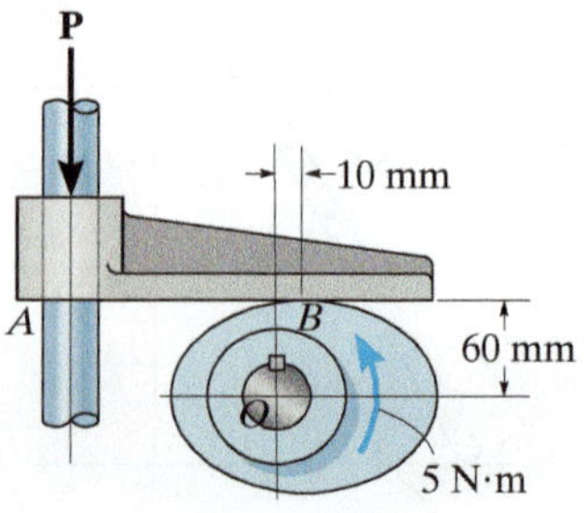

Prob. 8–13

8–14. Determine the maximum weight W the man can lift with constant velocity using the pulley system, without and then with the "leading block" or pulley at A. The man has a weight of 200 lb and the coefficient of static friction between his feet and the ground is $\mu_s = 0.6$.

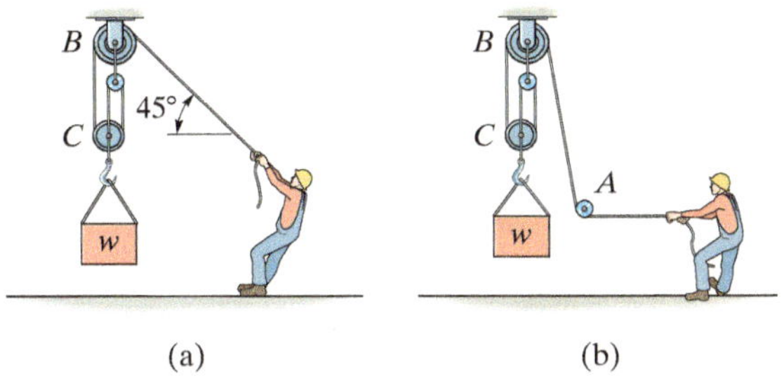

Prob. 8–14

8–15. The car has a mass of 1.6 Mg and center of mass at G. If the coefficient of static friction between the shoulder of the road and the tires is $\mu_s = 0.4$, determine the greatest slope θ the shoulder can have without causing the car to slip or tip over if the car travels along the shoulder at constant velocity.

Prob. 8–15

***8–16.** The uniform dresser has a weight of 90 lb and rests on a tile floor for which $\mu_s = 0.25$. If the man pushes on it in the horizontal direction $\theta = 0°$, determine the smallest magnitude of force $\mathbf{F}$ needed to move the dresser. Also, if the man has a weight of 150 lb, determine the smallest coefficient of static friction between his shoes and the floor so that he does not slip.

8–17. The uniform dresser has a weight of 90 lb and rests on a tile floor for which $\mu_s = 0.25$. If the man pushes on it in the direction $\theta = 30°$, determine the smallest magnitude of force $\mathbf{F}$ needed to move the dresser. Also, if the man has a weight of 150 lb, determine the smallest coefficient of static friction between his shoes and the floor so that he does not slip.

Probs. 8–16/17

8–18. The 5-kg cylinder is suspended from two equal-length cords. The end of each cord is attached to a ring of negligible mass that passes along a horizontal shaft. If the rings can be separated by the greatest distance $d = 400$ mm and still support the cylinder, determine the coefficient of static friction between each ring and the shaft.

8–19. The 5-kg cylinder is suspended from two equal-length cords. The end of each cord is attached to a ring of negligible mass that passes along a horizontal shaft. If the coefficient of static friction between each ring and the shaft is $\mu_s = 0.5$, determine the greatest distance d by which the rings can be separated and still support the cylinder.

8

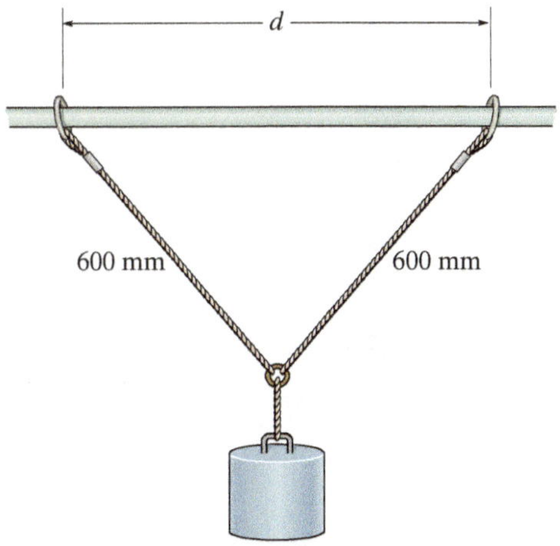

Probs. 8–18/19

*8–20. The board can be adjusted vertically by tilting it up and sliding the smooth pin A along the vertical guide G. When placed horizontally, the bottom C then bears along the edge of the guide, where $\mu_s = 0.4$. Determine the largest dimension d which will support any applied force **F** without causing the board to slip downward.

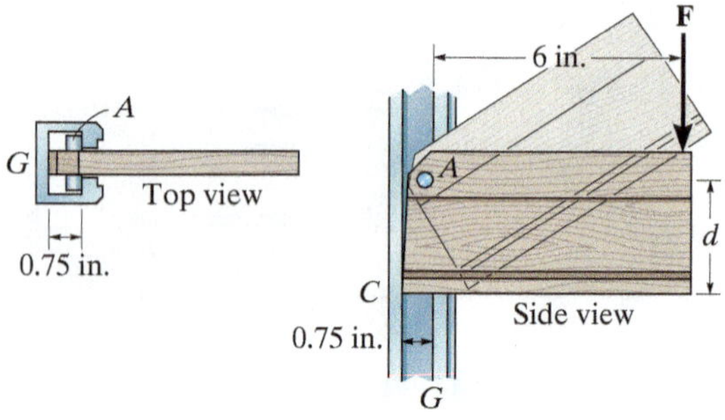

Prob. 8–20

8–21. The uniform pole has a weight W and length L. Its end B is tied to a supporting cord, and end A is placed against the wall, for which the coefficient of static friction is μ_s. Determine the largest angle θ at which the pole can be placed without slipping.

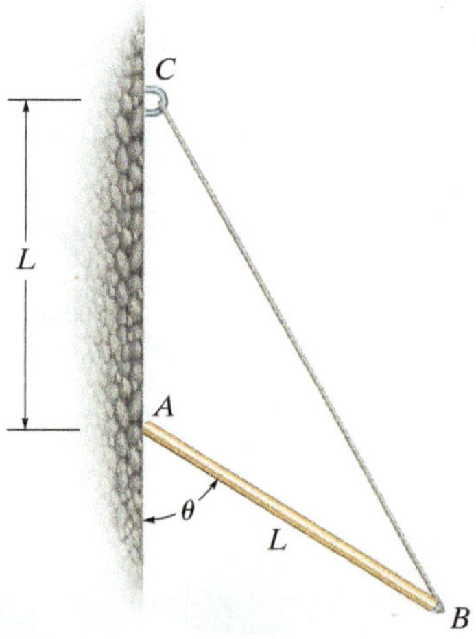

Prob. 8–21

8–22. If the clamping force is $F = 200$ N and each board has a mass of 2 kg, determine the maximum number of boards the clamp can support. The coefficient of static friction between the boards is $\mu_s = 0.3$, and the coefficient of static friction between the boards and the clamp is $\mu_s' = 0.45$.

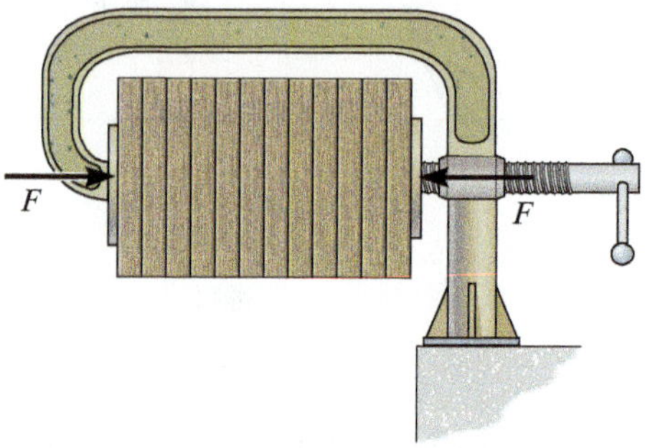

Prob. 8–22

8–23. A 35-kg disk rests on an inclined surface for which $\mu_s = 0.2$. Determine the maximum vertical force **P** that may be applied to bar AB without causing the disk to slip at C. Neglect the mass of the bar.

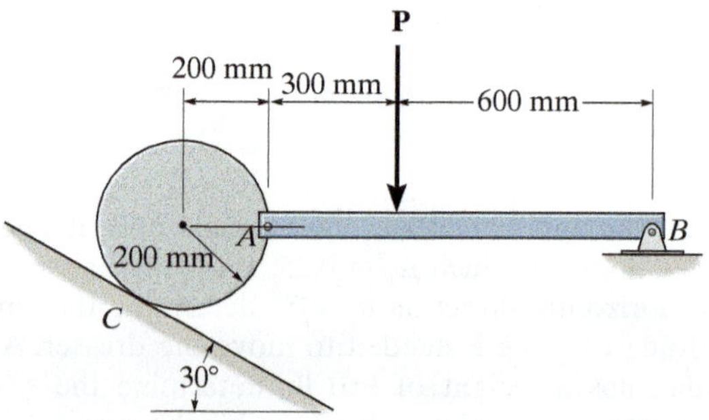

Prob. 8–23

8

***8–24.** The man has a weight of 200 lb, and the coefficient of static friction between his shoes and the floor is $\mu_s = 0.5$. Determine where he should position his center of gravity G at d in order to exert the maximum horizontal force on the door. What is this force?

Prob. 8–24

8–25. The crate has a weight of $W = 150$ lb, and the coefficients of static and kinetic friction are $\mu_s = 0.3$ and $\mu_k = 0.2$, respectively. Determine the friction force on the floor if $\theta = 30°$ and $P = 200$ lb.

8–26. The crate has a weight of $W = 350$ lb, and the coefficients of static and kinetic friction are $\mu_s = 0.3$ and $\mu_k = 0.2$, respectively. Determine the friction force on the floor if $\theta = 45°$ and $P = 100$ lb.

8–27. The crate has a weight W and the coefficient of static friction at the surface is $\mu_s = 0.3$. Determine the orientation of the cord and the smallest possible force **P** that has to be applied to the cord so that the crate is on the verge of moving.

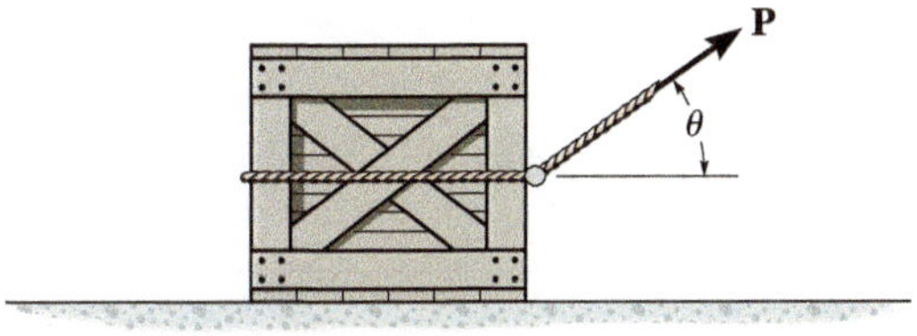

Probs. 8–25/26/27

***8–28.** If the coefficient of static friction between the man's shoes and the pole is $\mu_s = 0.6$, determine the minimum coefficient of static friction required between the belt and the pole at A in order to support the man. The man has a weight of 180 lb and a center of gravity at G.

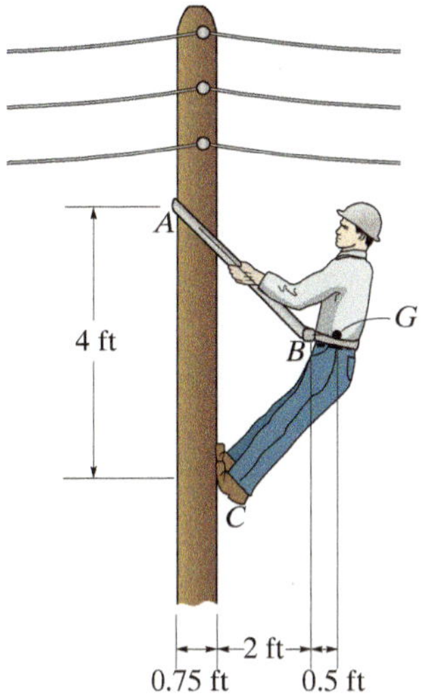

Prob. 8–28

8–29. The friction pawl is pinned at A and rests against the wheel at B. It allows freedom of movement when the wheel is rotating counterclockwise about C. Clockwise rotation is prevented due to friction of the pawl which tends to bind the wheel. If $(\mu_s)_B = 0.6$, determine the design angle θ which will prevent clockwise motion for any value of applied moment M. *Hint:* Neglect the weight of the pawl so that it becomes a two-force member.

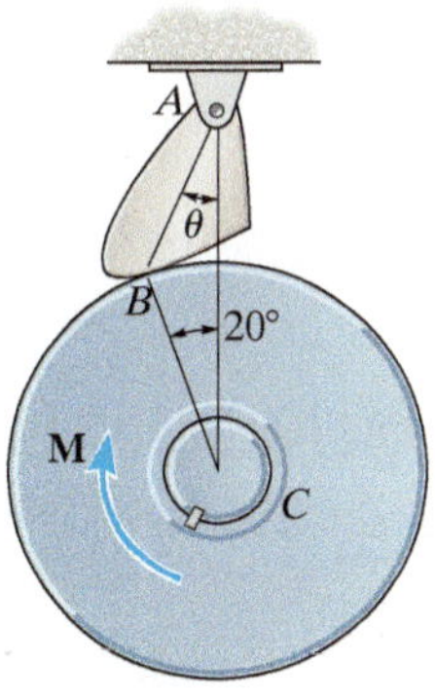

Prob. 8–29

8

8–30. If $\theta = 30°$, determine the minimum coefficient of static friction at A and B so that equilibrium of the supporting frame is maintained regardless of the mass of the cylinder C. Neglect the mass of the rods.

8–31. If the coefficient of static friction at A and B is $\mu_s = 0.6$, determine the maximum angle θ so that the frame remains in equilibrium, regardless of the mass of the cylinder. Neglect the mass of the rods.

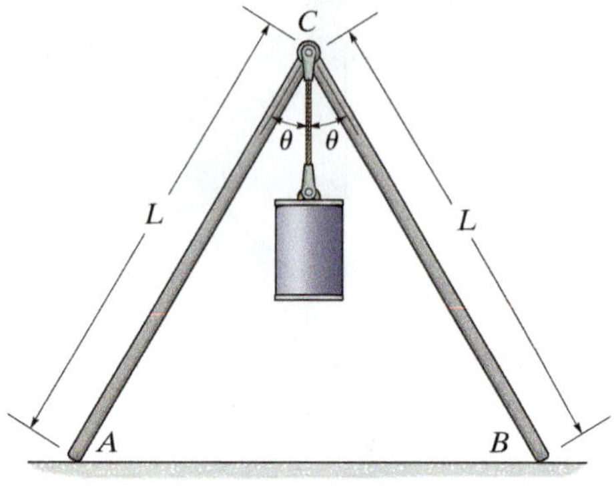

Probs. 8–30/31

*8–32. The semicylinder of mass m and radius r lies on the rough inclined plane for which $\phi = 10°$ and the coefficient of static friction is $\mu_s = 0.3$. Determine if the semicylinder slides down the plane, and if not, find the angle of tip θ of its base AB.

8–33. The semicylinder of mass m and radius r lies on the rough inclined plane. If the inclination $\phi = 15°$, determine the smallest coefficient of static friction which will prevent the semicylinder from slipping.

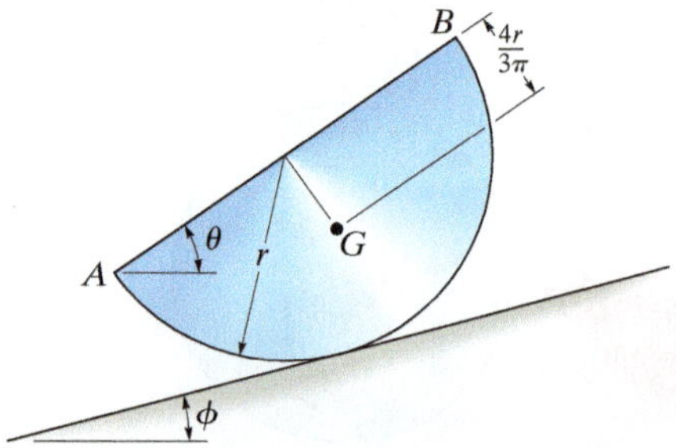

Probs. 8–32/33

8–34. The coefficient of static friction between the 150-kg crate and the ground is $\mu_s = 0.3$, while the coefficient of static friction between the 80-kg man's shoes and the ground is $\mu'_s = 0.4$. Determine if the man can move the crate.

8–35. If the coefficient of static friction between the crate and the ground in Prob. 8–34 is $\mu_s = 0.3$, determine the minimum coefficient of static friction between the man's shoes and the ground so that the man can move the crate.

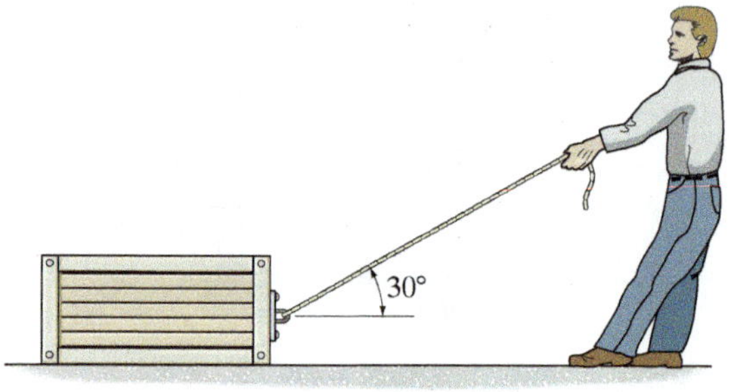

Probs. 8–34/35

*8–36. The rod has a weight W and rests against the floor and wall for which the coefficients of static friction are μ_A and μ_B, respectively. Determine the smallest value of θ for which the rod will not move.

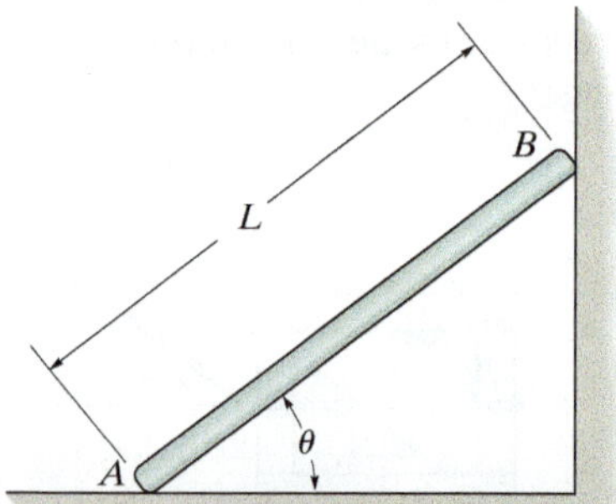

Prob. 8–36

8–37. The 80-lb boy stands on the beam and pulls on the cord with a force large enough to just cause him to slip. If $(\mu_s)_D = 0.4$ between his shoes and the beam, determine the reactions at A and B. The beam is uniform and has a weight of 100 lb. Neglect the size of the pulleys and the thickness of the beam.

8–38. The 80-lb boy stands on the beam and pulls with a force of 40 lb. If $(\mu_s)_D = 0.4$, determine the frictional force between his shoes and the beam and the reactions at A and B. The beam is uniform and has a weight of 100 lb. Neglect the size of the pulleys and the thickness of the beam.

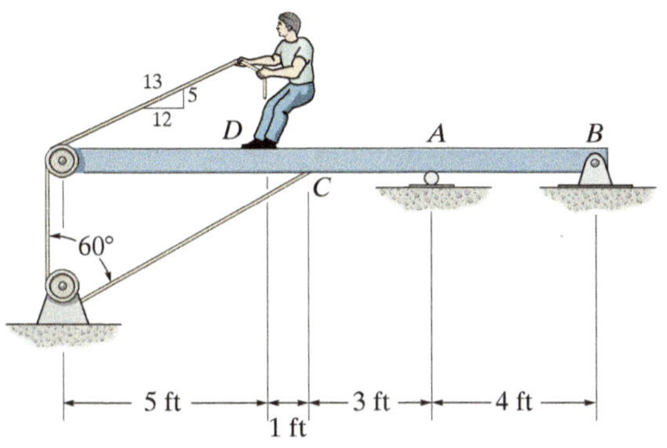

Probs. 8–37/38

8–39. Determine the smallest force the man must exert on the rope in order to move the 80-kg crate. Also, what is the angle θ at this moment? The coefficient of static friction between the crate and the floor is $\mu_s = 0.3$.

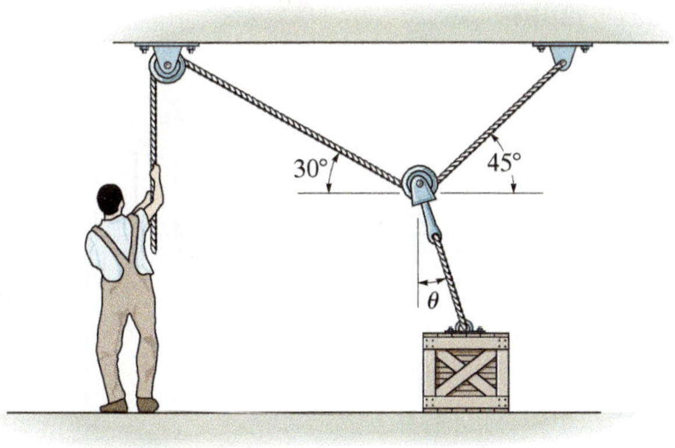

Prob. 8–39

***8–40.** Two blocks A and B have a weight of 10 lb and 6 lb, respectively. They are resting on the incline for which the coefficients of static friction are $\mu_A = 0.15$ and $\mu_B = 0.25$. Determine the incline angle θ for which both blocks begin to slide. Also find the required stretch or compression in the connecting spring for this to occur. The spring has a stiffness of $k = 2$ lb/ft.

8–41. Two blocks A and B have a weight of 10 lb and 6 lb, respectively. They are resting on the incline for which the coefficients of static friction are $\mu_A = 0.15$ and $\mu_B = 0.25$. Determine the angle θ which will cause motion of one of the blocks. What is the friction force under each of the blocks when this occurs? The spring has a stiffness of $k = 2$ lb/ft and is originally unstretched.

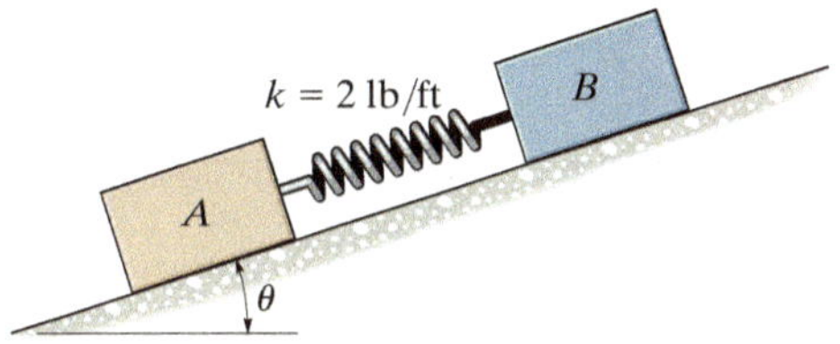

Probs. 8–40/41

8–42. The friction hook is made from a fixed frame and a cylinder of negligible weight. A piece of paper is placed between the wall and the cylinder. If $\theta = 20°$, determine the smallest coefficient of static friction μ at all points of contact so that any weight W of paper p can be held.

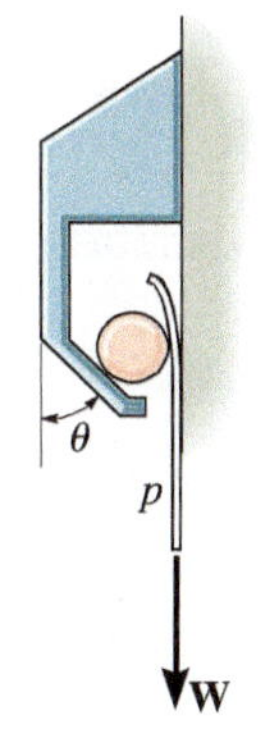

Prob. 8–42

8

8–43. The uniform rod has a mass of 10 kg and rests on the inside of the smooth ring at B and on the ground at A. If the rod is on the verge of slipping, determine the coefficient of static friction between the rod and the ground.

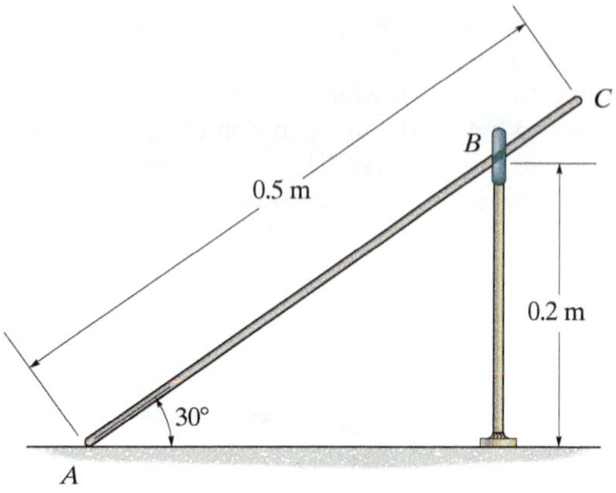

Prob. 8–43

***8–44.** The rings A and C each weigh W and rest on the rod, which has a coefficient of static friction of μ_s. If the suspended ring at B has a weight of $2W$, determine the largest distance d between A and C so that no motion occurs. Neglect the weight of the wire. The wire is smooth and has a total length of l.

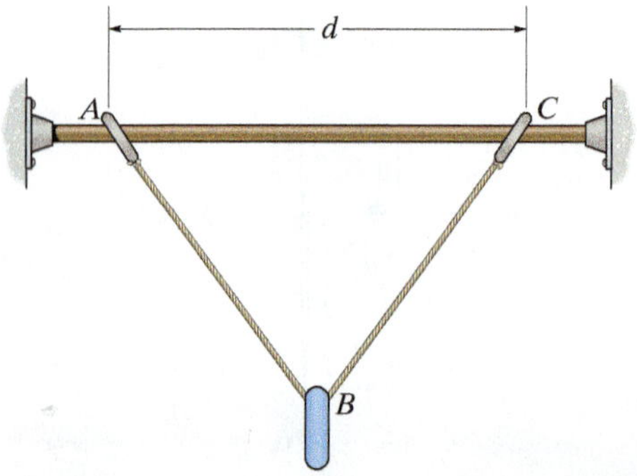

Prob. 8–44

8–45. The three bars have a weight of $W_A = 20$ lb, $W_B = 40$ lb, and $W_C = 60$ lb, respectively. If the coefficients of static friction at the surfaces of contact are as shown, determine the smallest horizontal force P needed to move block A.

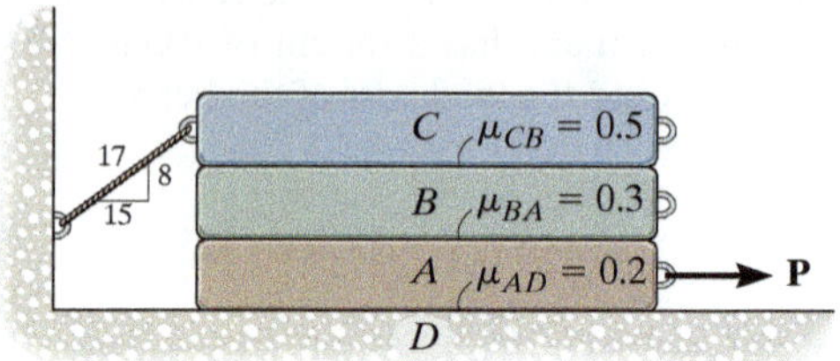

Prob. 8–45

8–46. The beam AB has a negligible mass and thickness and is subjected to a triangular distributed loading. It is supported at one end by a pin and at the other end by a post having a mass of 50 kg and negligible thickness. Determine the minimum force P needed to move the post. The coefficients of static friction at B and C are $\mu_B = 0.4$ and $\mu_C = 0.2$, respectively.

8–47. The beam AB has a negligible mass and thickness and is subjected to a triangular distributed loading. It is supported at one end by a pin and at the other end by a post having a mass of 50 kg and negligible thickness. Determine the two coefficients of static friction at B and at C so that when the magnitude of the applied force is increased to $P = 150$ N, the post slips at both B and C simultaneously.

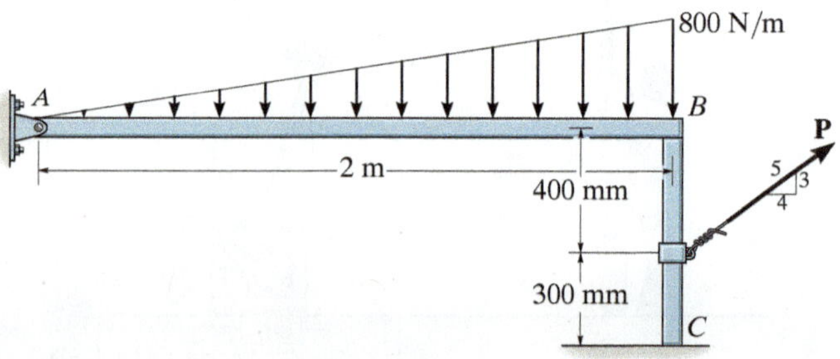

Probs. 8–46/47

***8–48.** The beam AB has a negligible mass and thickness and is subjected to a force of 200 N. It is supported at one end by a pin and at the other end by a spool having a mass of 40 kg. If a cable is wrapped around the inner core of the spool, determine the minimum cable force P needed to move the spool. The coefficients of static friction at B and D are $\mu_B = 0.4$ and $\mu_D = 0.2$, respectively.

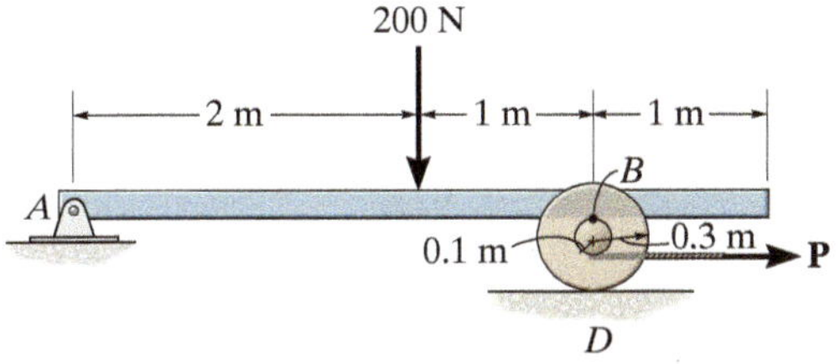

Prob. 8–48

8–49. If each box weighs 150 lb, determine the least horizontal force P that the man must exert on the top box in order to cause motion. The coefficient of static friction between the boxes is $\mu_s = 0.5$, and the coefficient of static friction between the box and the floor is $\mu_s' = 0.2$.

8–50. If each box weighs 150 lb, determine the least horizontal force P that the man must exert on the top box in order to cause motion. The coefficient of static friction between the boxes is $\mu_s = 0.65$, and the coefficient of static friction between the box and the floor is $\mu_s' = 0.35$.

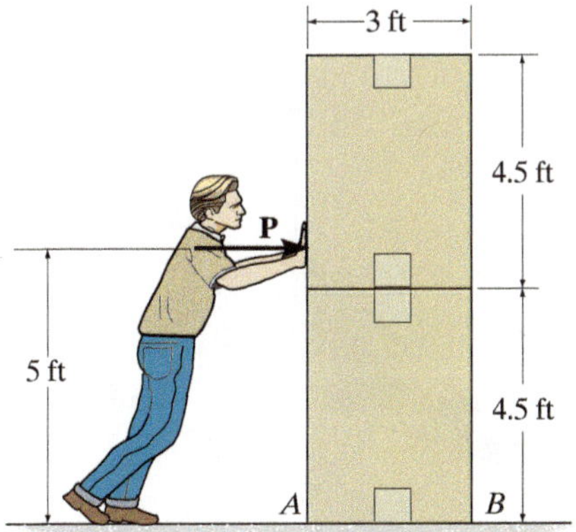

Probs. 8–49/50

8–51. The block of weight W is being pulled up the inclined plane of slope α using a force $\mathbf{P}$. If $\mathbf{P}$ acts at the angle ϕ as shown, show that for slipping to occur, $P = W\sin(\alpha + \theta)/\cos(\phi - \theta)$, where θ is the angle of friction; $\theta = \tan^{-1}\mu$.

***8–52.** Determine the angle ϕ at which $\mathbf{P}$ should act on the block so that the magnitude of $\mathbf{P}$ is as small as possible to begin pulling the block up the incline. What is the corresponding value of P? The block weighs W and the slope α is known.

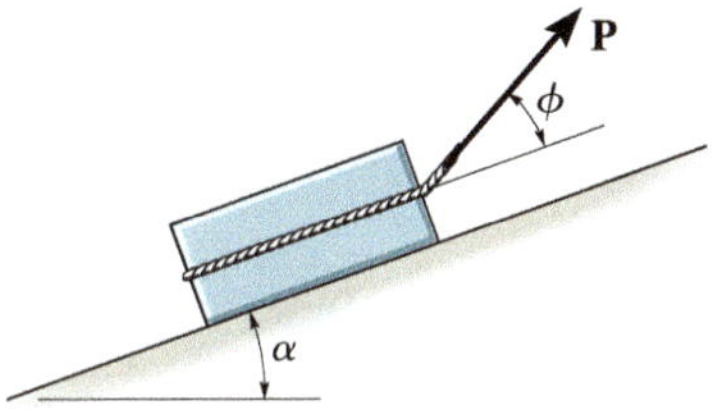

Probs. 8–51/52

8–53. The wheel weighs 20 lb and rests on a surface for which $\mu_B = 0.2$. A cord wrapped around it is attached to the top of the 30-lb homogeneous block. If the coefficient of static friction at D is $\mu_D = 0.3$, determine the smallest vertical force that can be applied tangentially to the wheel which will cause motion to impend.

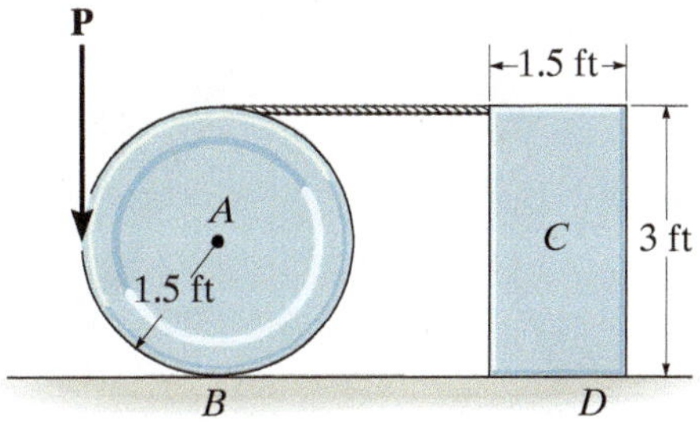

Prob. 8–53

8–54. The uniform beam has a weight W and length $4a$. It rests on the fixed rails at A and B. If the coefficient of static friction at the rails is μ_s, determine the horizontal force P, applied perpendicular to the face of the beam, which will cause the beam to move.

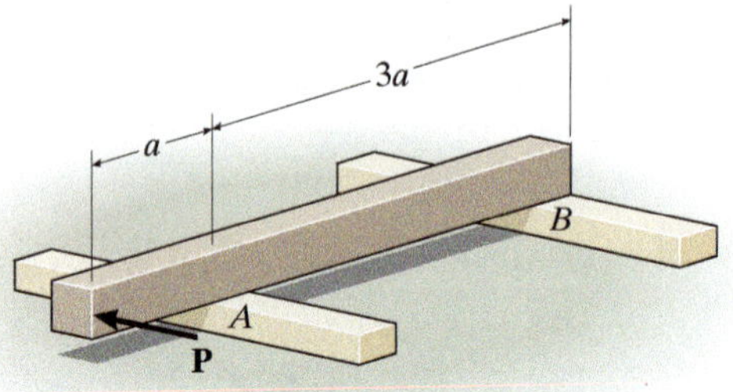

Prob. 8–54

8–55. Determine the greatest angle θ so that the ladder does not slip when it supports the 75-kg man in the position shown. The surface is rather slippery, where the coefficient of static friction at A and B is $\mu_s = 0.3$.

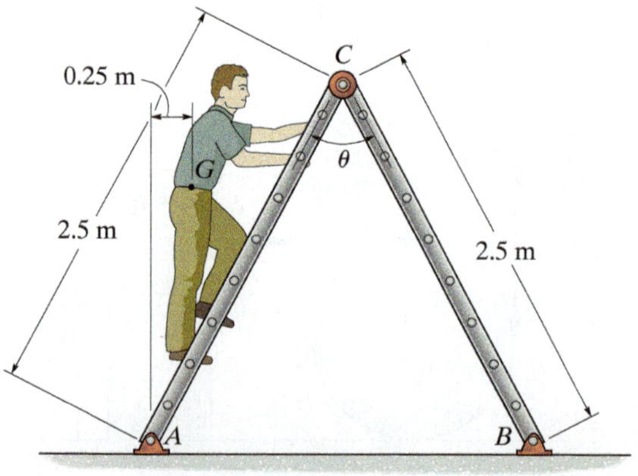

Prob. 8–55

***8–56.** The uniform 6-kg slender rod rests on the top center of the 3-kg block. If the coefficients of static friction at the points of contact are $\mu_A = 0.4$, $\mu_B = 0.6$, and $\mu_C = 0.3$, determine the largest couple moment M which can be applied to the rod without causing motion of the rod.

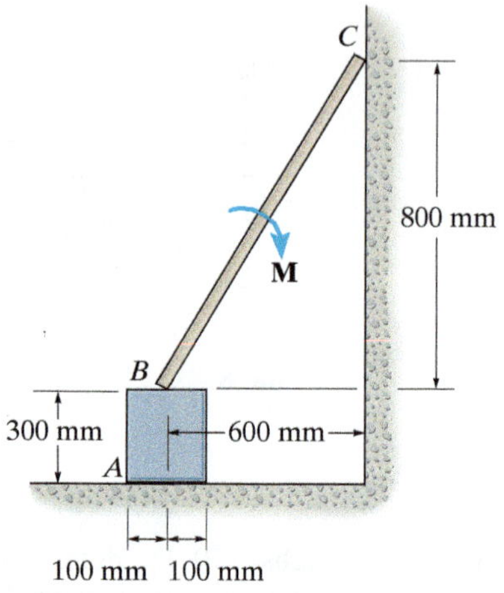

Prob. 8–56

8–57. The disk has a weight W and lies on a plane that has a coefficient of static friction μ. Determine the maximum height h to which the plane can be lifted without causing the disk to slip.

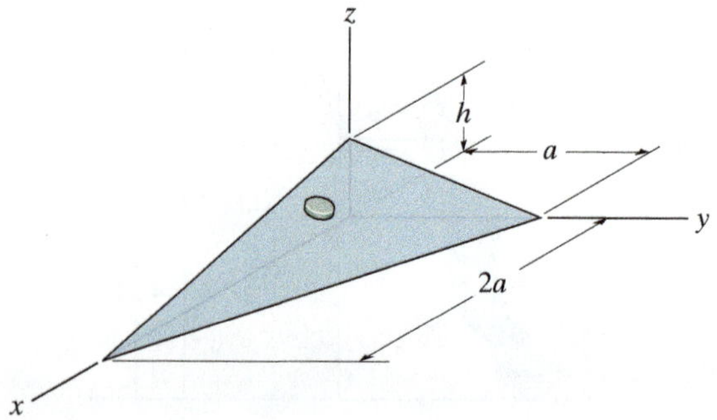

Prob. 8–57

8

CONCEPTUAL PROBLEMS

P8–1. Is it more effective to move the load forward at constant velocity with the boom fully extended as shown, or should the boom be fully retracted? Power is supplied to the rear wheels. The front wheels are free to roll. Do an equilibrium analysis to explain your answer.

P8–1

P8–2. The lug nut on the free-turning wheel is to be removed using the wrench. Which is the most effective way to apply force to the wrench? Also, why is it best to keep the car tire on the ground rather than first jacking it up? Explain your answers with an equilibrium analysis.

P8–2

P8–3. The rope is used to tow the refrigerator. Is it best to pull slightly up on the rope as shown, pull horizontally, or pull somewhat downwards? Also, is it best to attach the rope at a high position as shown, or at a lower position? Do an equilibrium analysis to explain your answer.

P8–4. The rope is used to tow the refrigerator. In order to prevent yourself from slipping while towing, is it best to pull up as shown, pull horizontally, or pull downwards on the rope? Do an equilibrium analysis to explain your answer.

P8–3/4

P8–5. Is it easier to tow the load by applying a force along the tow bar when it is in an almost horizontal position as shown, or is it better to pull on the bar when it has a steeper slope? Do an equilibrium analysis to explain your answer.

P8–5

8.3 Wedges

Wedges are often used to adjust the elevation of structural or mechanical parts. Also, they provide stability for objects such as this pipe.

A *wedge* is a simple machine that is often used to transform an applied force into much larger forces, directed at approximately right angles to the applied force. Wedges also can be used to slightly move or adjust heavy loads.

Consider, for example, the wedge shown in Fig. 8–12*a*, which is used to *lift* the block by applying a force to the wedge. Free-body diagrams of the block and wedge are shown in Fig. 8–12*b*. Here we have excluded the weight of the wedge since it is usually *small* compared to the weight **W** of the block. Also, note that the frictional forces $\mathbf{F}_1$ and $\mathbf{F}_2$ must oppose the motion of the wedge. Likewise, the frictional force $\mathbf{F}_3$ of the wall on the block must act downward so as to oppose the block's upward motion. The locations of the resultant normal forces are not important in the force analysis since neither the block nor wedge will "tip." Hence the moment equilibrium equations will not be considered. There are seven unknowns, consisting of the applied force **P**, needed to cause motion of the wedge, and six normal and frictional forces. The seven available equations consist of four force equilibrium equations, $\Sigma F_x = 0$, $\Sigma F_y = 0$ applied to the wedge and block, and three frictional equations, $F = \mu N$, applied at each surface of contact.

If the block is to be *lowered*, then the frictional forces will all act in a sense opposite to that shown in Fig. 8–12*b*. Provided the coefficient of friction is very *small* or the wedge angle θ is *large*, then the applied force **P** must act to the right to hold the block. Otherwise, **P** may have a reverse sense of direction in order to *pull* on the wedge to remove it. If **P** is *not applied* and friction forces hold the block in place, then the wedge is referred to as *self-locking*.

8

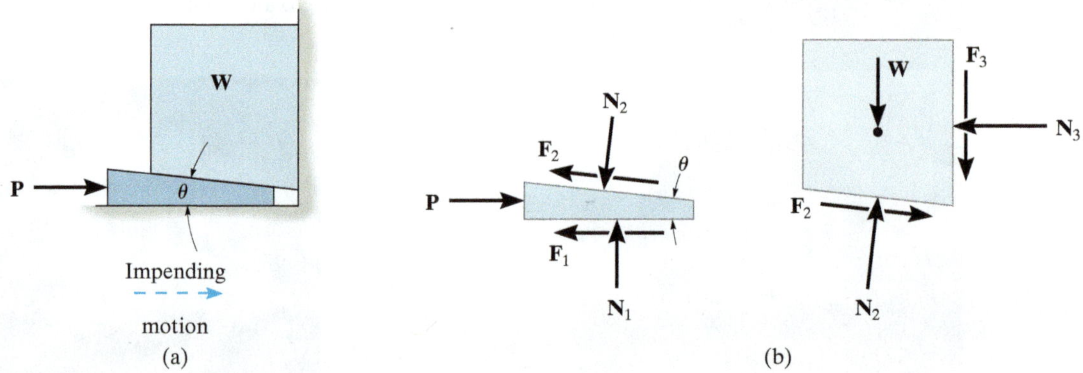

Fig. 8–12

EXAMPLE 8.6

The uniform stone in Fig. 8–13*a* has a mass of 500 kg and is held in the horizontal position using a wedge at B. If the coefficient of static friction is $\mu_s = 0.3$ at the surfaces of contact, determine the minimum force P needed to remove the wedge. Assume that the stone does not slip at A.

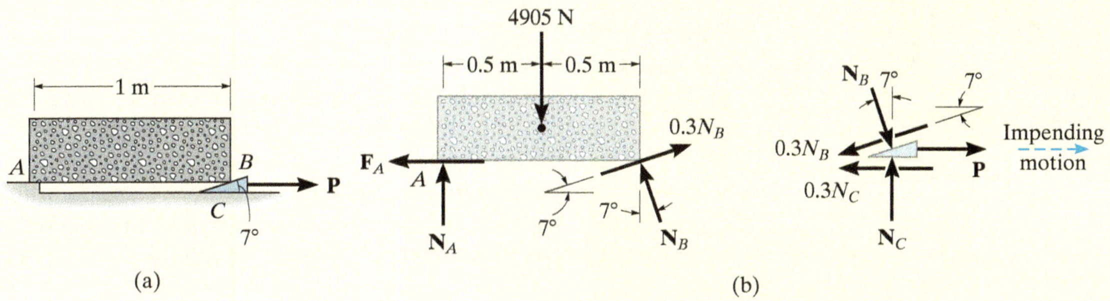

Fig. 8–13

SOLUTION

The minimum force P requires $F = \mu_s N$ at the surfaces of contact with the wedge. The free-body diagrams of the stone and wedge are shown in Fig. 8–13*b*. On the wedge the friction force opposes the impending motion, and on the stone at A, $F_A \leq \mu_s N_A$, since slipping does not occur there. There are five unknowns. Three equilibrium equations for the stone and two for the wedge are available for solution. From the free-body diagram of the stone,

$$\circlearrowleft +\Sigma M_A = 0; \quad -4905\text{ N}(0.5\text{ m}) + (N_B \cos 7°\text{ N})(1\text{ m}) + (0.3N_B \sin 7°\text{ N})(1\text{ m}) = 0$$

$$N_B = 2383.1\text{ N}$$

Using this result for the wedge, we have

$$+\uparrow \Sigma F_y = 0; \quad N_C - 2383.1 \cos 7°\text{ N} - 0.3(2383.1 \sin 7°\text{ N}) = 0$$

$$N_C = 2452.5\text{ N}$$

$$\overset{+}{\rightarrow} \Sigma F_x = 0; \quad 2383.1 \sin 7°\text{ N} - 0.3(2383.1 \cos 7°\text{ N}) + P - 0.3(2452.5\text{ N}) = 0$$

$$P = 1154.9\text{ N} = 1.15\text{ kN} \qquad \textit{Ans.}$$

NOTE: Since P is positive, indeed the wedge must be pulled out. If P were zero, the wedge would remain in place (self-locking) and the frictional forces developed at B and C would satisfy $F_B < \mu_s N_B$ and $F_C < \mu_s N_C$.

8

Square-threaded screws find applications on valves, jacks, and vises, where particularly large forces must be developed along the axis of the screw.

8.4 Frictional Forces on Screws

In most cases, screws are used as fasteners; however, in many types of machines they are incorporated to transmit power or motion from one part of the machine to another. A *square-threaded screw* is commonly used for the latter purpose, especially when large forces are applied along its axis. In this section, we will analyze the forces acting on square-threaded screws. The analysis of other types of screws, such as the V-thread, is based on these same principles.

For analysis, a square-threaded screw, as in Fig. 8–14, can be considered a cylinder having an inclined square ridge or *thread* wrapped around it. If we unwind the thread by one revolution, as shown in Fig. 8–14*b*, the slope or the *lead angle* θ is determined from $\theta = \tan^{-1}(l/2\pi r)$. Here l and $2\pi r$ are the vertical and horizontal distances between A and B, where r is the mean radius of the thread. The distance l is called the *lead* of the screw and it is equivalent to the distance the screw advances when it turns one revolution.

Upward Impending Motion. Let us now consider the case of the square-threaded screw jack in Fig. 8–15 that is subjected to upward impending motion caused by the applied torsional moment ***M**. A free-body diagram of the *entire unraveled thread h* in contact with the jack can be represented as a block as shown in Fig. 8–16*a*. The force **W** is the vertical force acting on the thread or the axial force applied to the shaft, Fig. 8–15, and M/r is the resultant horizontal force produced by the couple moment M about the axis of the shaft. The reaction **R** of the groove on the thread has both frictional and normal components, where $F = \mu_s N$. The angle of static friction is $\phi_s = \tan^{-1}(F/N) = \tan^{-1}\mu_s$. Applying the force equations of equilibrium along the horizontal and vertical axes, we have

$$\xrightarrow{+} \Sigma F_x = 0; \qquad M/r - R \sin(\theta + \phi_s) = 0$$

$$+\uparrow \Sigma F_y = 0; \qquad R \cos(\theta + \phi_s) - W = 0$$

Eliminating R from these equations, we obtain

$$M = rW \tan(\theta + \phi_s) \tag{8–3}$$

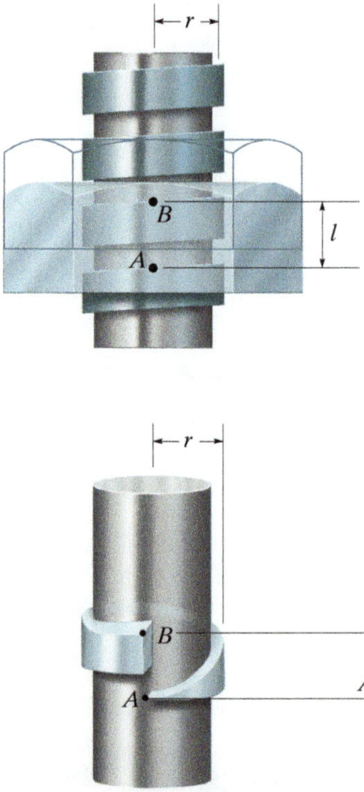

(a)

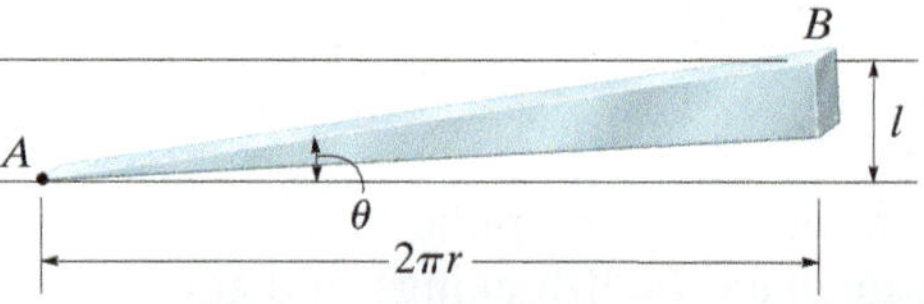

(b)

Fig. 8–14

*For applications, **M** is developed by applying a horizontal force **P** at a right angle to the end of a lever that would be fixed to the screw.

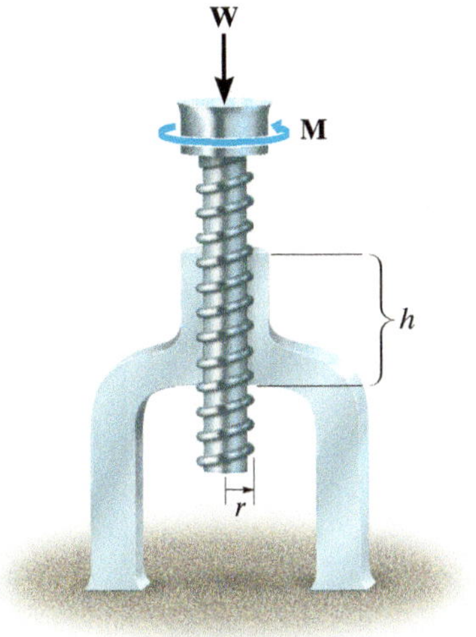

Fig. 8–15

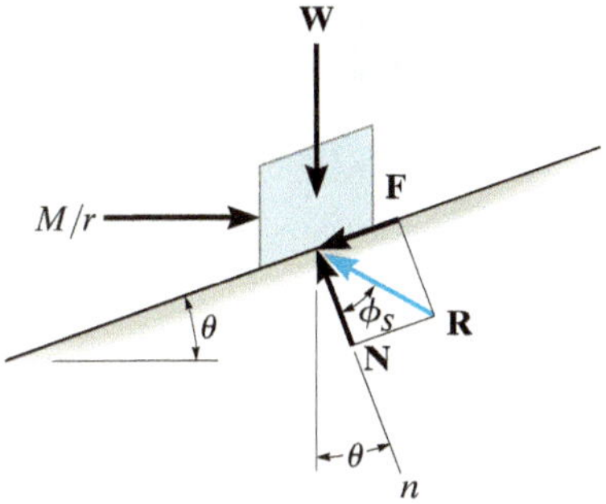

Upward screw motion

(a)

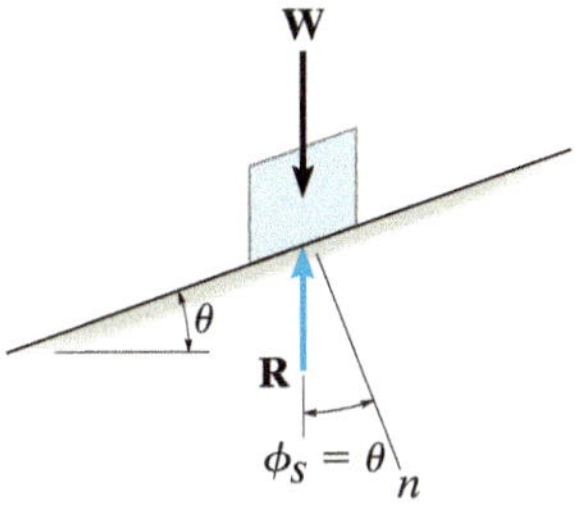

Self-locking screw ($\theta = \phi_s$)
(on the verge of rotating downward)

(b)

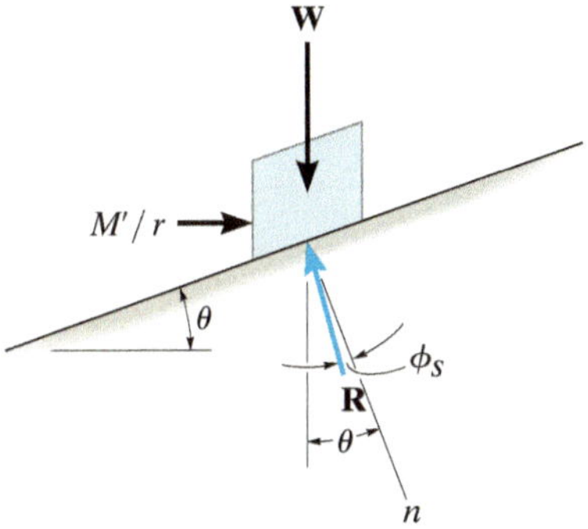

Downward screw motion ($\theta > \phi_s$)

(c)

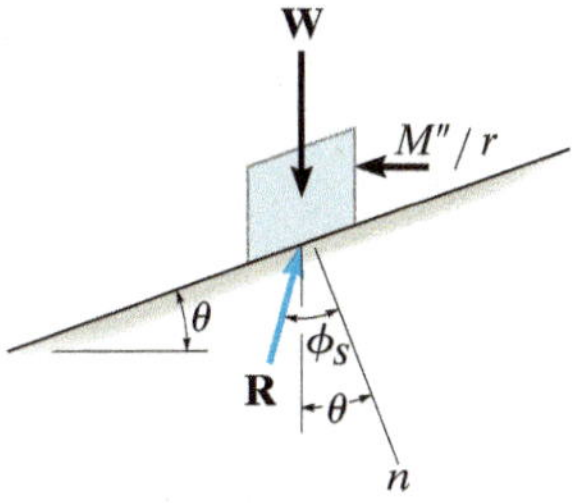

Downward screw motion ($\theta < \phi_s$)

(d)

Fig. 8–16

Self-Locking Screw. A screw is said to be *self-locking* if it remains in place under any axial load **W** when the moment **M** is removed. For this to occur, the direction of the frictional force must be reversed so that **R** acts on the other side of **N**. Here the angle of static friction ϕ_s becomes greater than or equal to θ, Fig. 8–16*d*. If $\phi_s = \theta$, Fig. 8–16*b*, then **R** will act vertically to balance **W**, and the screw will be on the verge of winding downward.

Downward Impending Motion, $(\theta > \phi_s)$. If the screw is not self-locking, it is necessary to apply a moment **M′** to prevent the screw from winding downward. Here, a horizontal force M'/r is required to push against the thread to prevent it from sliding down the plane, Fig. 8–16*c*. Using the same procedure as before, the magnitude of the moment **M′** required to prevent this unwinding is

$$M' = rW \tan(\theta - \phi_s) \quad (8\text{–}4)$$

Downward Impending Motion, $(\phi_s > \theta)$. If a screw is self-locking, a couple moment **M″** must be applied to the screw in the opposite direction to wind the screw downward ($\phi_s > \theta$). This causes a reverse horizontal force M''/r that pushes the thread down as indicated in Fig. 8–16*d*. In this case, we obtain

$$M'' = rW \tan(\phi_s - \theta) \quad (8\text{–}5)$$

If *motion of the screw* occurs, Eqs. 8–3, 8–4, and 8–5 can be applied by simply replacing ϕ_s with ϕ_k.

8

EXAMPLE 8.7

The turnbuckle shown in Fig. 8–17 has a square thread with a mean radius of 5 mm and a lead of 2 mm. If the coefficient of static friction between the screw and the turnbuckle is $\mu_s = 0.25$, determine the moment **M** that must be applied to draw the end screws closer together.

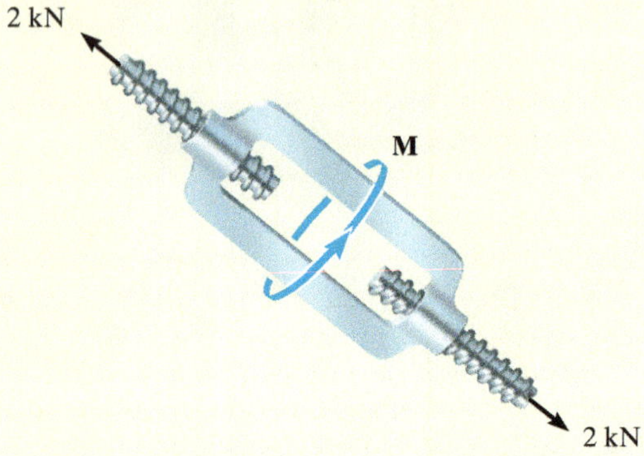

Fig. 8–17

SOLUTION

The moment can be obtained by applying Eq. 8–3. Since friction at *two screws* must be overcome, this requires

$$M = 2[rW \tan(\theta + \phi_s)] \qquad (1)$$

Here $W = 2000$ N, $\phi_s = \tan^{-1}\mu_s = \tan^{-1}(0.25) = 14.04°$, $r = 5$ mm, and $\theta = \tan^{-1}(l/2\pi r) = \tan^{-1}(2 \text{ mm}/[2\pi(5 \text{ mm})]) = 3.64°$. Substituting these values into Eq. 1 and solving gives

$$M = 2[(2000 \text{ N})(5 \text{ mm}) \tan(14.04° + 3.64°)]$$

$$= 6374.7 \text{ N}\cdot\text{mm} = 6.37 \text{ N}\cdot\text{m} \qquad \textit{Ans.}$$

NOTE: When the moment is *removed*, the turnbuckle will be self-locking; i.e., it will not unscrew since $\phi_s > \theta$.

8

PROBLEMS

8–58. Determine the largest angle θ that will cause the wedge to be self-locking regardless of the magnitude of horizontal force P applied to the blocks. The coefficient of static friction between the wedge and the blocks is $\mu_s = 0.3$. Neglect the weight of the wedge.

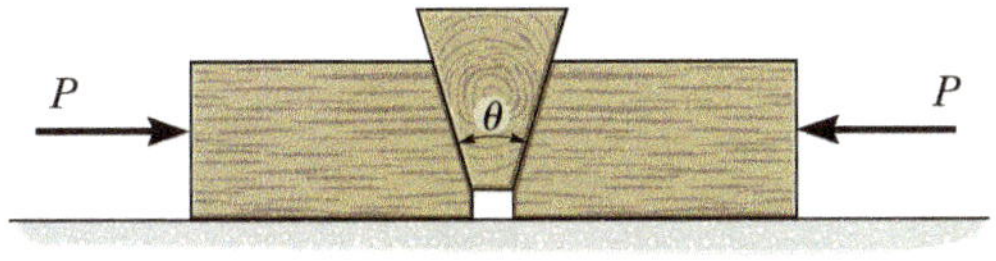

Prob. 8–58

8–59. If the beam AD is loaded as shown, determine the horizontal force P which must be applied to the wedge in order to remove it from under the beam. The coefficients of static friction at the wedge's top and bottom surfaces are $\mu_{CA} = 0.25$ and $\mu_{CB} = 0.35$, respectively. If $P = 0$, is the wedge self-locking? Neglect the weight and size of the wedge and the thickness of the beam.

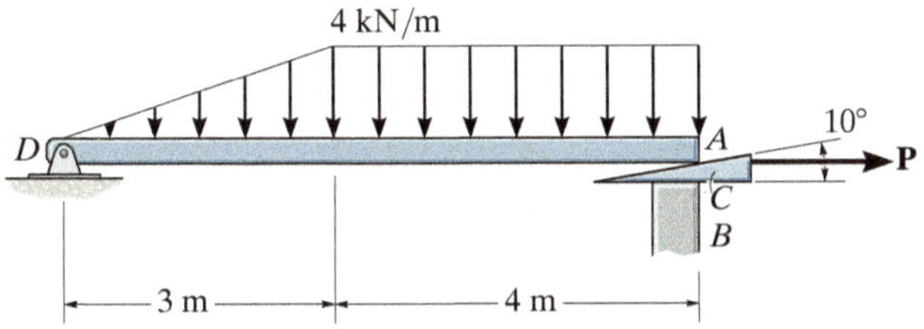

Prob. 8–59

***8–60.** The wedge has a negligible weight and a coefficient of static friction $\mu_s = 0.35$ with all contacting surfaces. Determine the largest angle θ so that it is "self-locking." This requires no slipping for any magnitude of the force **P** applied to the joint.

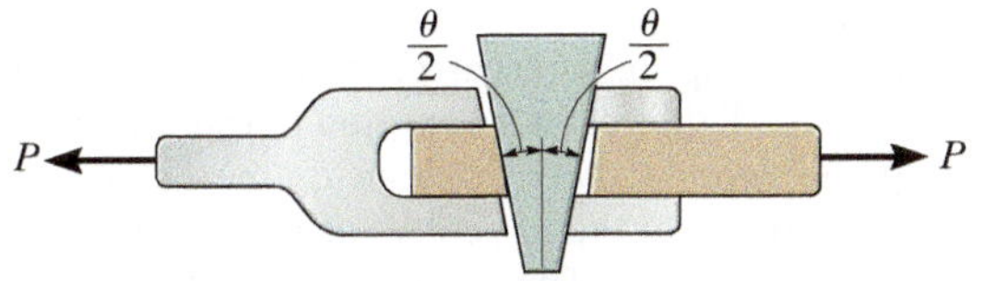

Prob. 8–60

8–61. If the spring is compressed 60 mm and the coefficient of static friction between the tapered stub S and the slider A is $\mu_{SA} = 0.5$, determine the horizontal force **P** needed to move the slider forward. The stub is free to move without friction within the fixed collar C. The coefficient of static friction between A and surface B is $\mu_{AB} = 0.4$. Neglect the weights of the slider and stub.

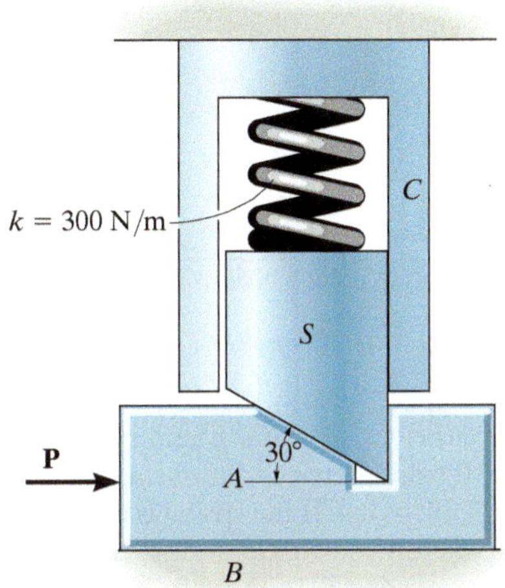

Prob. 8–61

8–62. If $P = 250$ N, determine the required minimum compression in the spring so that the wedge will not move to the right. Neglect the weight of A and B. The coefficient of static friction for all contacting surfaces is $\mu_s = 0.35$. Neglect friction at the rollers.

8–63. Determine the minimum applied force **P** required to move wedge A to the right. The spring is compressed a distance of 175 mm. Neglect the weight of A and B. The coefficient of static friction for all contacting surfaces is $\mu_s = 0.35$. Neglect friction at the rollers.

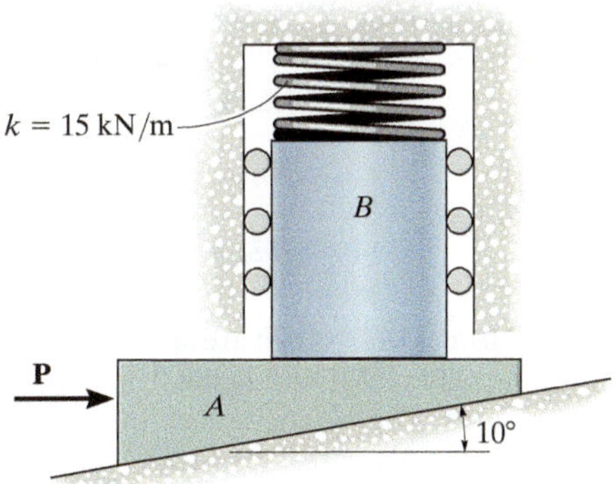

Probs. 8–62/63

8

*8–64. Determine the largest weight of the wedge that can be placed between the 8-lb cylinder and the wall without upsetting equilibrium. The coefficient of static friction at A and C is $\mu_s = 0.5$ and at B, $\mu_s' = 0.6$.

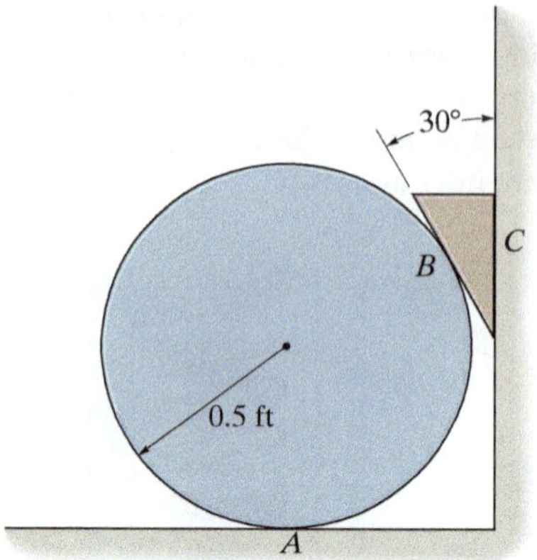

Prob. 8–64

8–65. The coefficient of static friction between wedges B and C is $\mu_s = 0.6$ and between the surfaces of contact B and A and C and D, $\mu_s' = 0.4$. If the spring is compressed 200 mm when in the position shown, determine the smallest force P needed to move wedge C to the left. Neglect the weight of the wedges.

8–66. The coefficient of static friction between the wedges B and C is $\mu_s = 0.6$ and between the surfaces of contact B and A and C and D, $\mu_s' = 0.4$. If $P = 50$ N, determine the smallest allowable compression of the spring without causing wedge C to move to the left. Neglect the weight of the wedges.

8

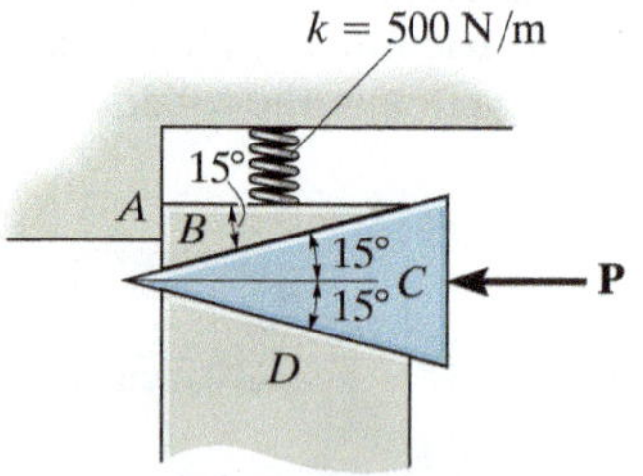

Probs. 8–65/66

8–67. If couple forces of $F = 10$ lb are applied perpendicular to the lever of the clamp at A and B, determine the clamping force on the boards. The single square-threaded screw of the clamp has a mean diameter of 1 in. and a lead of 0.25 in. The coefficient of static friction is $\mu_s = 0.3$.

*8–68. If the clamping force on the boards is 600 lb, determine the required magnitude of the couple forces that must be applied perpendicular to the lever AB of the clamp at A and B in order to loosen the screw. The single square-threaded screw has a mean diameter of 1 in. and a lead of 0.25 in. The coefficient of static friction is $\mu_s = 0.3$.

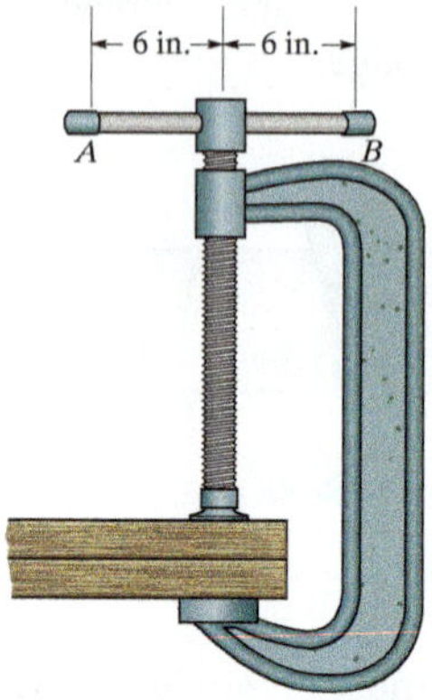

Probs. 8–67/68

8–69. The column is used to support the upper floor. If a force $F = 80$ N is applied perpendicular to the handle to tighten the screw, determine the compressive force in the column. The square-threaded screw on the jack has a coefficient of static friction of $\mu_s = 0.4$, mean diameter of 25 mm, and a lead of 3 mm.

8–70. If the force **F** is removed from the handle of the jack in Prob. 8–69, determine if the screw is self-locking.

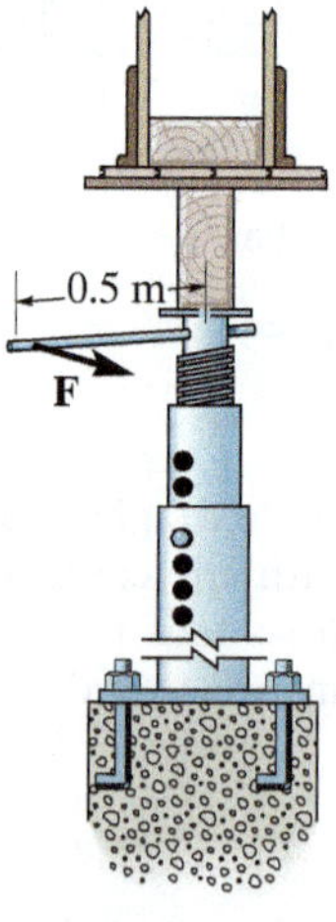

Probs. 8–69/70

8–71. If the clamping force at G is 900 N, determine the horizontal force **F** that must be applied perpendicular to the handle of the lever at E. The mean diameter and lead of both single square-threaded screws at C and D are 25 mm and 5 mm, respectively. The coefficient of static friction is $\mu_s = 0.3$.

*8–72. If a horizontal force of $F = 50$ N is applied perpendicular to the handle of the lever at E, determine the clamping force developed at G. The mean diameter and lead of the single square-threaded screw at C and D are 25 mm and 5 mm, respectively. The coefficient of static friction is $\mu_s = 0.3$.

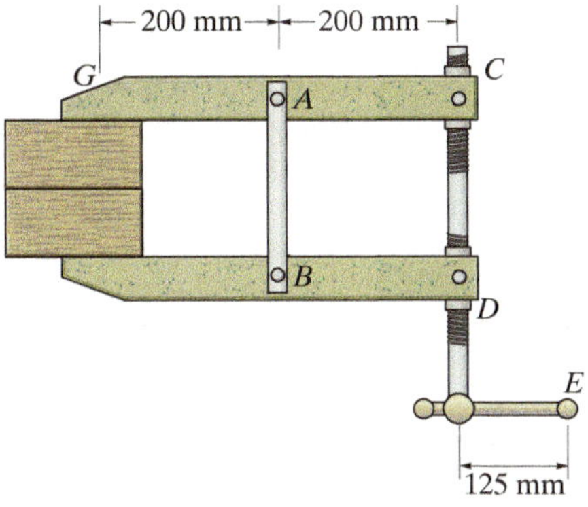

Probs. 8–71/72

8–73. A turnbuckle, similar to that shown in Fig. 8–17, is used to tension member AB of the truss. The coefficient of the static friction between the square threaded screws and the turnbuckle is $\mu_s = 0.5$. The screws have a mean radius of 6 mm and a lead of 3 mm. If a torque of $M = 10$ N · m is applied to the turnbuckle, to draw the screws closer together, determine the force in each member of the truss. No external forces act on the truss.

8–74. A turnbuckle, similar to that shown in Fig. 8–17, is used to tension member AB of the truss. The coefficient of the static friction between the square-threaded screws and the turnbuckle is $\mu_s = 0.5$. The screws have a mean radius of 6 mm and a lead of 3 mm. Determine the torque M which must be applied to the turnbuckle to draw the screws closer together, so that the compressive force of 500 N is developed in member BC.

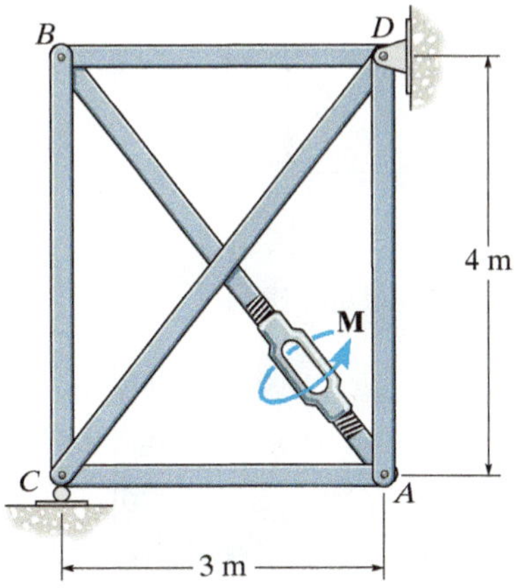

Prob. 8–74

8–75. The shaft has a square-threaded screw with a lead of 8 mm and a mean radius of 15 mm. If it is in contact with a plate gear having a mean radius of 30 mm, determine the resisting torque **M** on the plate gear which can be overcome if a torque of 7 N · m is applied to the shaft. The coefficient of static friction at the screw is $\mu_B = 0.2$. Neglect friction of the bearings located at A and B.

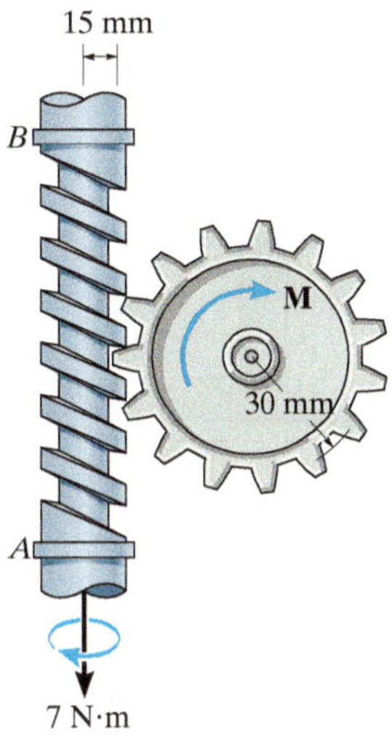

Prob. 8–75

*8–76. The square-threaded screw has a mean diameter of 20 mm and a lead of 4 mm. If the weight of the plate A is 5 lb, determine the smallest coefficient of static friction between the screw and the plate so that the plate does not travel down the screw when the plate is suspended as shown.

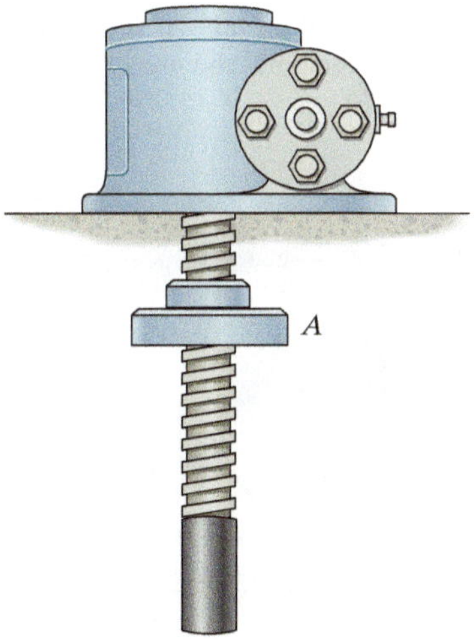

Prob. 8–76

8–77. The fixture clamp consist of a square-threaded screw having a coefficient of static friction of $\mu_s = 0.3$, mean diameter of 3 mm, and a lead of 1 mm. The five points indicated are pin connections. Determine the clamping force at the smooth blocks D and E when a torque of $M = 0.08\ \text{N}\cdot\text{m}$ is applied to the handle of the screw.

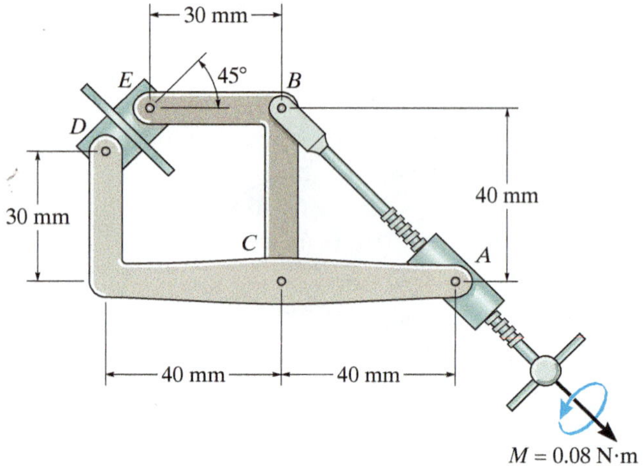

Prob. 8–77

8–78. The braking mechanism consists of two pinned arms and a square-threaded screw with left- and right-hand threads. Thus when turned, the screw draws the two arms together. If the lead of the screw is 4 mm, the mean diameter 12 mm, and the coefficient of static friction is $\mu_s = 0.35$, determine the tension in the screw when a torque of $5\ \text{N}\cdot\text{m}$ is applied to tighten the screw. If the coefficient of static friction between the brake pads A and B and the circular shaft is $\mu_s' = 0.5$, determine the maximum torque M the brake can resist.

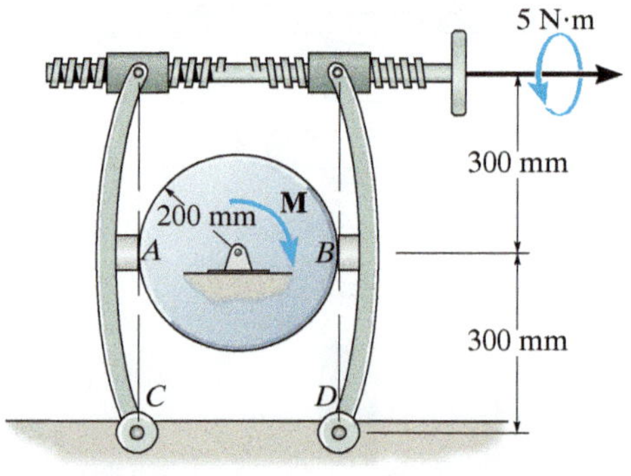

Prob. 8–78

8–79. If a horizontal force of $P = 100$ N is applied perpendicular to the handle of the lever at A, determine the compressive force **F** exerted on the material. Each single square-threaded screw has a mean diameter of 25 mm and a lead of 7.5 mm. The coefficient of static friction at all contacting surfaces of the wedges is $\mu_s = 0.2$, and the coefficient of static friction at the screw is $\mu_s' = 0.15$.

***8–80.** Determine the horizontal force **P** that must be applied perpendicular to the handle of the lever at A in order to develop a compressive force of 12 kN on the material. Each single square-threaded screw has a mean diameter of 25 mm and a lead of 7.5 mm. The coefficient of static friction at all contacting surfaces of the wedges is $\mu_s = 0.2$, and the coefficient of static friction at the screw is $\mu_s' = 0.15$.

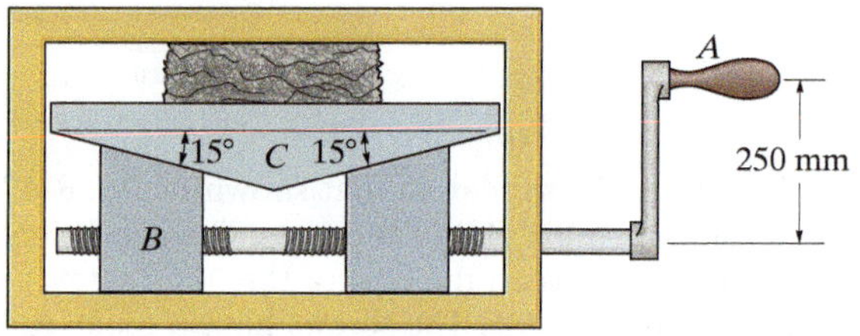

Probs. 8–79/80

8–81. Determine the clamping force on the board A if the screw of the "C" clamp is tightened with a twist of $M = 8\ \text{N}\cdot\text{m}$. The single square-threaded screw has a mean radius of 10 mm, a lead of 3 mm, and the coefficient of static friction is $\mu_s = 0.35$.

8–82. If the required clamping force at the board A is to be 50 N, determine the torque M that must be applied to the handle of the "C" clamp to tighten it down. The single square-threaded screw has a mean radius of 10 mm, a lead of 3 mm, and the coefficient of static friction is $\mu_s = 0.35$.

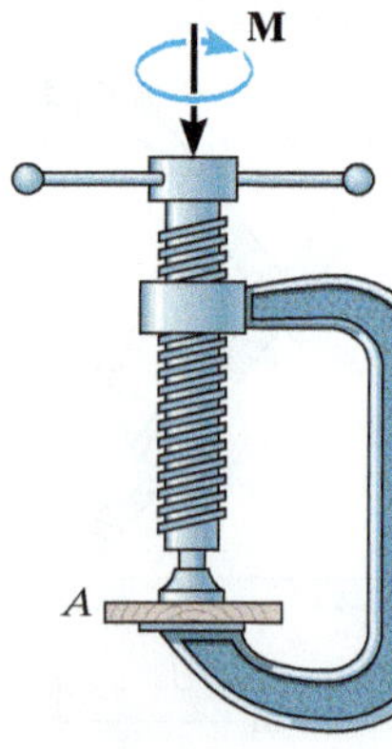

Probs. 8–81/82

8.5 Frictional Forces on Flat Belts

Whenever belt drives or band brakes are designed, it is necessary to determine the frictional forces developed between the belt and its contacting surface. In this section we will analyze the frictional forces acting on a flat belt, although the analysis of other types of belts, such as the V-belt, is based on similar principles.

Consider the flat belt shown in Fig. 8–18*a*, which passes over a fixed curved surface. The total angle of belt to surface contact in radians is β, and the coefficient of friction between the two surfaces is μ. We wish to determine the tension T_2 in the belt, which is needed to pull the belt counterclockwise over the surface, and thereby overcome both the frictional forces at the surface of contact and the tension T_1 in the other end of the belt. Obviously, $T_2 > T_1$.

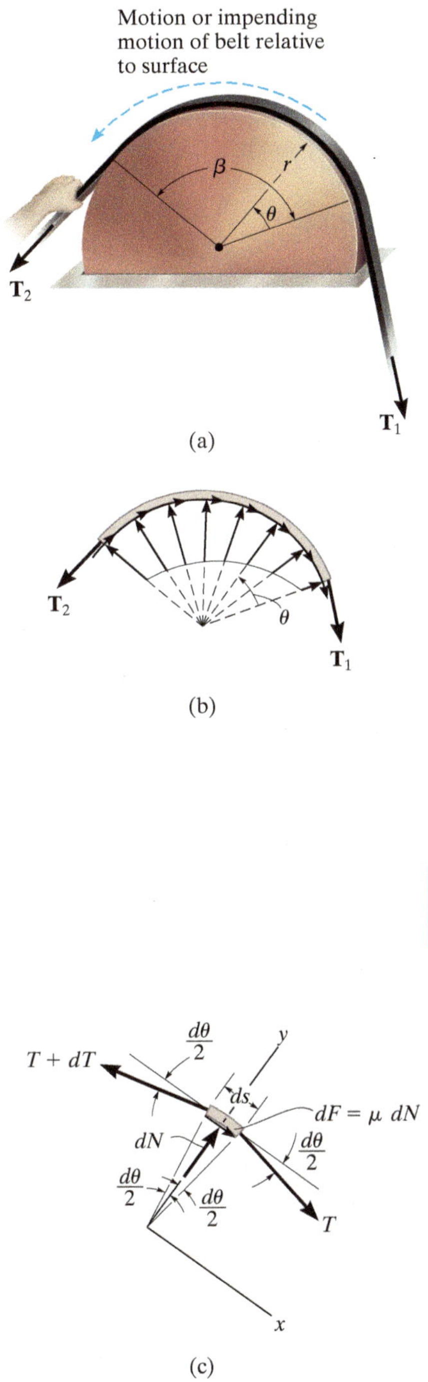

Fig. 8–18

Frictional Analysis. A free-body diagram of the belt segment in contact with the surface is shown in Fig. 8–18*b*. As shown, the normal and frictional forces, acting at different points along the belt, will vary both in magnitude and direction. Due to this *unknown* distribution, the analysis of the problem will first require a study of the forces acting on a differential element of the belt.

A free-body diagram of an element having a length ds is shown in Fig. 8–18*c*. Assuming either impending motion or motion of the belt, the magnitude of the frictional force $dF = \mu\, dN$. This force opposes the sliding motion of the belt, and so it will increase the magnitude of the tensile force acting in the belt by dT. Applying the two force equations of equilibrium, we have

$$\searrow + \Sigma F_x = 0; \qquad T\cos\left(\frac{d\theta}{2}\right) + \mu\, dN - (T + dT)\cos\left(\frac{d\theta}{2}\right) = 0$$

$$+\nearrow \Sigma F_y = 0; \qquad dN - (T + dT)\sin\left(\frac{d\theta}{2}\right) - T\sin\left(\frac{d\theta}{2}\right) = 0$$

Since $d\theta$ is of *infinitesimal size*, $\sin(d\theta/2) = d\theta/2$ and $\cos(d\theta/2) = 1$. Also, the *product* of the two infinitesimals dT and $d\theta/2$ may be neglected when compared to infinitesimals of the first order. As a result, these two equations become

$$\mu\, dN = dT$$

and

$$dN = T\, d\theta$$

Eliminating dN yields

$$\frac{dT}{T} = \mu\, d\theta$$

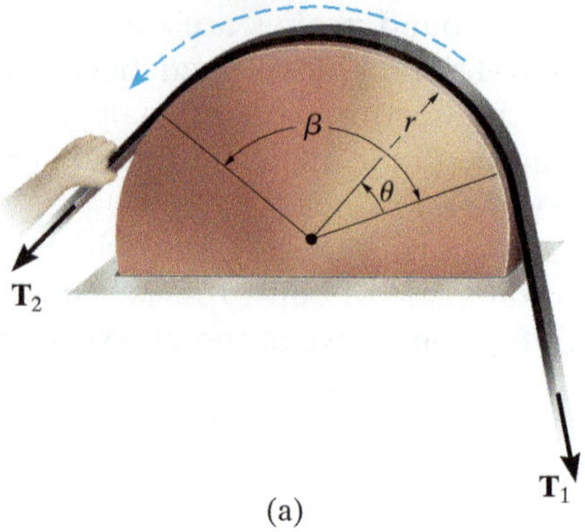

Fig. 8–18 (Repeated)

Flat or V-belts are often used to transmit the torque developed by a motor to a wheel attached to a pump, fan or blower.

8

Integrating this equation between all the points of contact that the belt makes with the drum, and noting that $T = T_1$ at $\theta = 0$ and $T = T_2$ at $\theta = \beta$, yields

$$\int_{T_1}^{T_2} \frac{dT}{T} = \mu \int_0^{\beta} d\theta$$

$$\ln \frac{T_2}{T_1} = \mu\beta$$

Solving for T_2, we obtain

$$T_2 = T_1 e^{\mu\beta} \qquad (8\text{–}6)$$

where

T_2, T_1 = belt tensions; T_1 opposes the direction of motion (or impending motion) of the belt measured relative to the surface, while T_2 acts in the direction of the relative belt motion (or impending motion); because of friction, $T_2 > T_1$

μ = coefficient of static or kinetic friction between the belt and the surface of contact

β = angle of belt to surface contact, measured in radians

e = 2.718 . . . , base of the natural logarithm

Note that T_2 is *independent* of the *radius* of the drum, and instead it is a function of the angle of belt to surface contact, β. As a result, this equation is valid for flat belts passing over any curved contacting surface.

EXAMPLE 8.8

The maximum tension that can be developed in the cord shown in Fig. 8–19*a* is 500 N. If the pulley at *A* is free to rotate and the coefficient of static friction at the fixed drums *B* and *C* is $\mu_s = 0.25$, determine the largest mass of the cylinder that can be lifted by the cord.

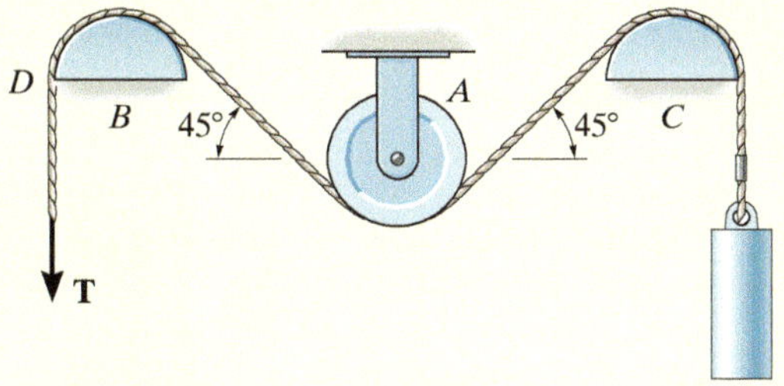

(a)

SOLUTION

Lifting the cylinder, which has a weight $W = mg$, causes the cord to move counterclockwise over the drums at *B* and *C*; hence, the maximum tension T_2 in the cord occurs at *D*. Thus, $F = T_2 = 500$ N. A section of the cord passing over the drum at *B* is shown in Fig. 8–19*b*. Since $180° = \pi$ rad the angle of contact between the drum and the cord is $\beta = (135°/180°)\pi = 3\pi/4$ rad. Using Eq. 8–6, we have

$$T_2 = T_1 e^{\mu_s \beta}; \qquad 500 \text{ N} = T_1 e^{0.25[(3/4)\pi]}$$

Hence,

$$T_1 = \frac{500 \text{ N}}{e^{0.25[(3/4)\pi]}} = \frac{500 \text{ N}}{1.80} = 277.4 \text{ N}$$

Since the pulley at *A* is free to rotate, equilibrium requires that the tension in the cord remains the *same* on both sides of the pulley.

The section of the cord passing over the drum at *C* is shown in Fig. 8–19*c*. The weight $W < 277.4$ N. Why? Applying Eq. 8–6, we obtain

$$T_2 = T_1 e^{\mu_s \beta}; \qquad 277.4 \text{ N} = W e^{0.25[(3/4)\pi]}$$

$$W = 153.9 \text{ N}$$

so that

$$m = \frac{W}{g} = \frac{153.9 \text{ N}}{9.81 \text{ m/s}^2}$$

$$= 15.7 \text{ kg} \qquad \textit{Ans.}$$

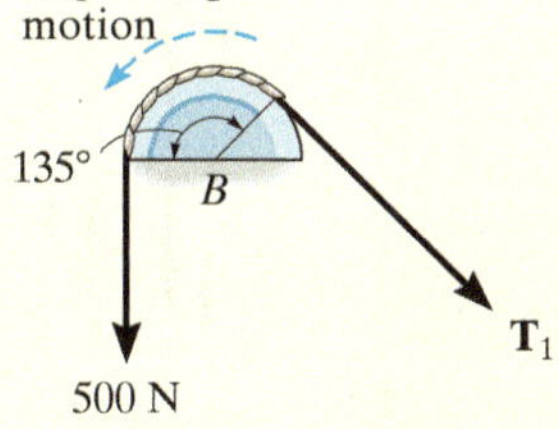

(b)

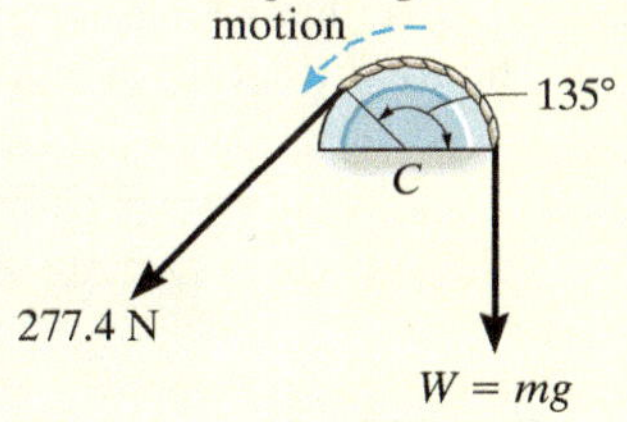

(c)

Fig. 8–19

PROBLEMS

8–83. A cylinder having a mass of 250 kg is to be supported by the cord that wraps over the pipe. Determine the smallest vertical force **F** needed to support the load if the cord passes (a) once over the pipe, $\beta = 180°$, and (b) two times over the pipe, $\beta = 540°$. Take $\mu_s = 0.2$.

***8–84.** A cylinder having a mass of 250 kg is to be supported by the cord that wraps over the pipe. Determine the largest vertical force **F** that can be applied to the cord without moving the cylinder. The cord passes (a) once over the pipe, $\beta = 180°$, and (b) two times over the pipe, $\beta = 540°$. Take $\mu_s = 0.2$.

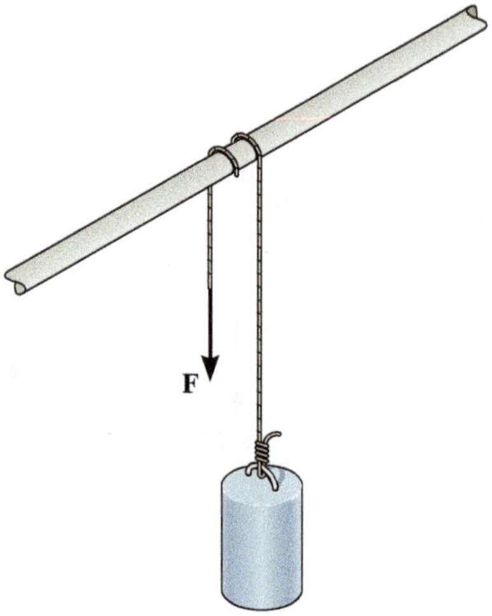

Probs. 8–83/84

8–85. A "hawser" is wrapped around a fixed "capstan" to secure a ship for docking. If the tension in the rope, caused by the ship, is 1500 lb, determine the least number of complete turns the rope must be rapped around the capstan in order to prevent slipping of the rope. The greatest horizontal force that a longshoreman can exert on the rope is 50 lb. The coefficient of static friction is $\mu_s = 0.3$.

Prob. 8–85

8–86. A force of $P = 25$ N is just sufficient to prevent the 20-kg cylinder from descending. Determine the required force **P** to begin lifting the cylinder. The rope passes over a rough peg with two and half turns.

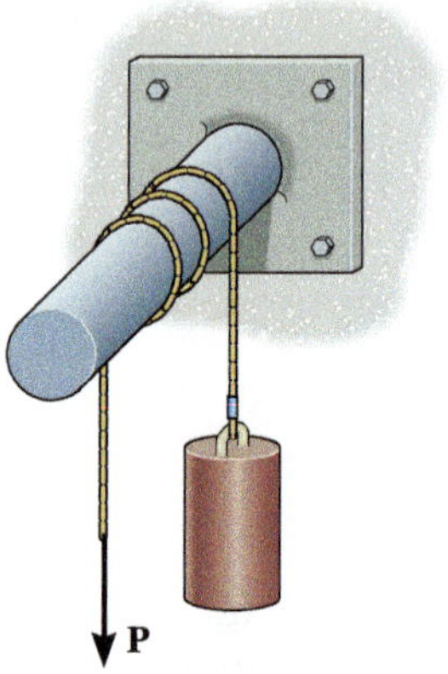

Prob. 8–86

8–87. The 20-kg cylinder A and 50-kg cylinder B are connected together using a rope that passes around a rough peg two and a half turns. If the cylinders are on the verge of moving, determine the coefficient of static friction between the rope and the peg.

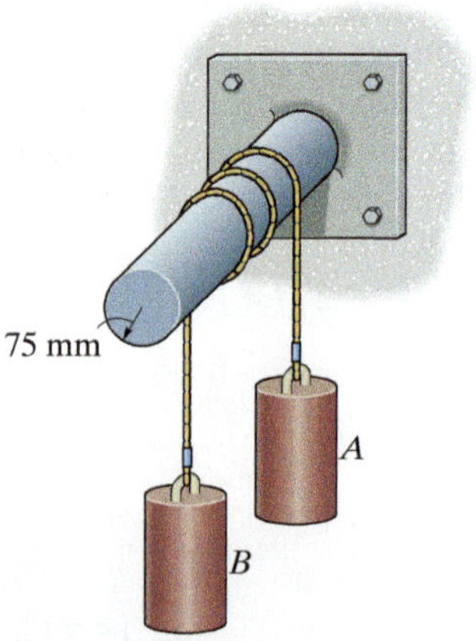

Prob. 8–87

8

***8–88.** Determine the maximum and the minimum values of weight W which may be applied without causing the 50-lb block to slip. The coefficient of static friction between the block and the plane is $\mu_s = 0.2$, and between the rope and the drum D $\mu_s' = 0.3$.

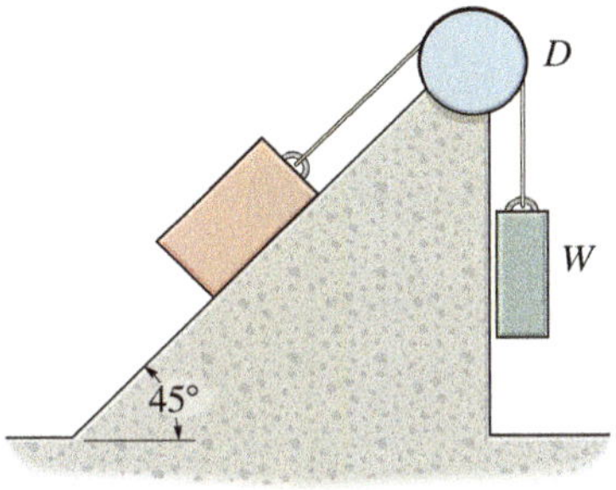

Prob. 8–88

8–89. The truck, which has a mass of 3.4 Mg, is to be lowered down the slope by a rope that is wrapped around a tree. If the wheels are free to roll and the man at A can resist a pull of 300 N, determine the minimum number of turns the rope should be wrapped around the tree to lower the truck at a constant speed. The coefficient of kinetic friction between the tree and rope is $\mu_k = 0.3$.

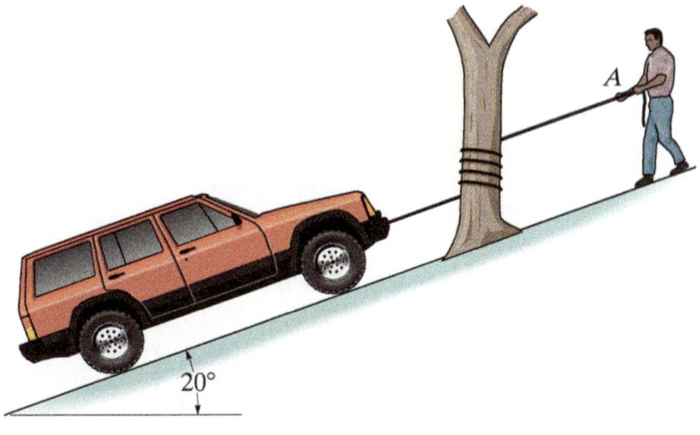

Prob. 8–89

8–90. The smooth beam is being hoisted using a rope that is wrapped around the beam and passes through a ring at A as shown. If the end of the rope is subjected to a tension $\mathbf{T}$ and the coefficient of static friction between the rope and ring is $\mu_s = 0.3$, determine the smallest angle of θ for equilibrium.

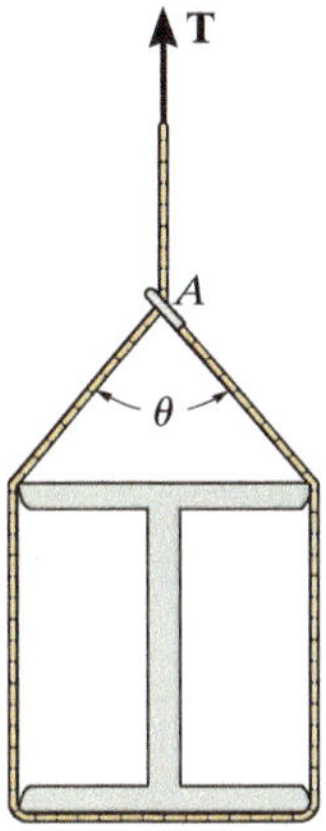

Prob. 8–90

8–91. The uniform concrete pipe has a weight of 800 lb and is unloaded slowly from the truck bed using the rope and skids shown. If the coefficient of kinetic friction between the rope and pipe is $\mu_k = 0.3$, determine the force the worker must exert on the rope to lower the pipe at constant speed. There is a pulley at B, and the pipe does not slip on the skids. The lower portion of the rope is parallel to the skids.

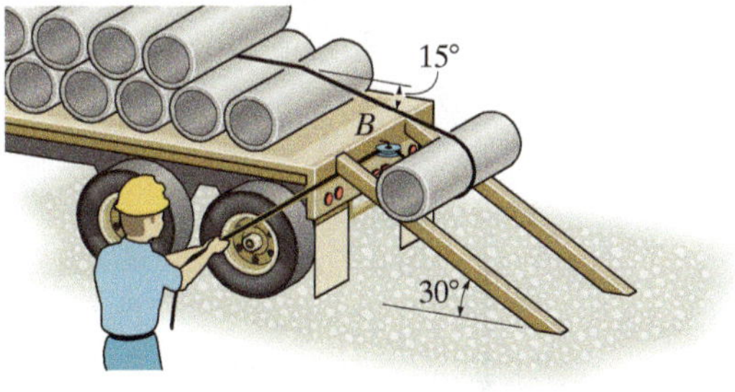

Prob. 8–91

***8–92.** The simple band brake is constructed so that the ends of the friction strap are connected to the pin at A and the lever arm at B. If the wheel is subjected to a torque of $M = 80 \text{ lb} \cdot \text{ft}$, and the minimum force $P = 20$ lb is needed to apply to the lever to hold the wheel stationary, determine the coefficient of static friction between the wheel and the band.

8–93. The simple band brake is constructed so that the ends of the friction strap are connected to the pin at A and the lever arm at B. If the wheel is subjected to a torque of $M = 80 \text{ lb} \cdot \text{ft}$, determine the smallest force P applied to the lever that is required to hold the wheel stationary. The coefficient of static friction between the strap and wheel is $\mu_s = 0.5$.

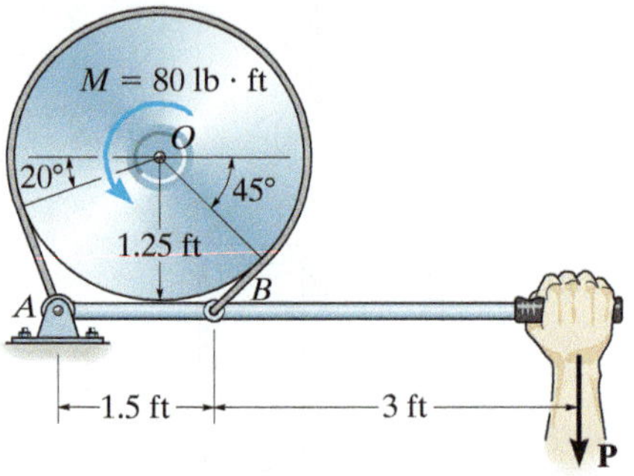

Probs. 8–92/93

8–94. A minimum force of $P = 50$ lb is required to hold the cylinder from slipping against the belt and the wall. Determine the weight of the cylinder if the coefficient of friction between the belt and cylinder is $\mu_s = 0.3$ and slipping does not occur at the wall.

8–95. The cylinder weighs 10 lb and is held in equilibrium by the belt and wall. If slipping does not occur at the wall, determine the minimum vertical force P which must be applied to the belt for equilibrium. The coefficient of static friction between the belt and the cylinder is $\mu_s = 0.25$.

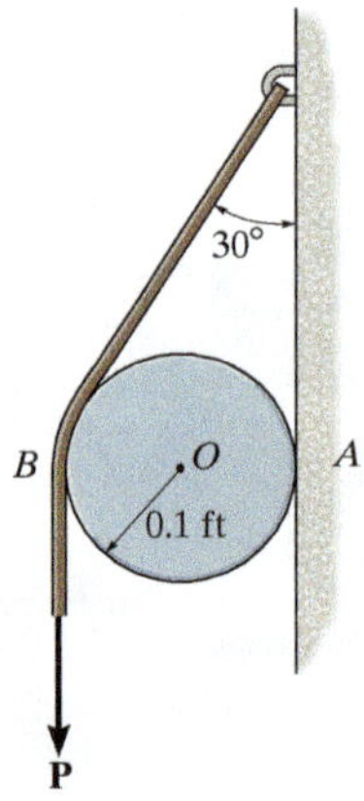

Probs. 8–94/95

***8–96.** A cord having a weight of 0.5 lb/ft and a total length of 10 ft is suspended over a peg P as shown. If the coefficient of static friction between the peg and cord is $\mu_s = 0.5$, determine the longest length h which one side of the suspended cord can have without causing motion. Neglect the size of the peg and the length of cord draped over it.

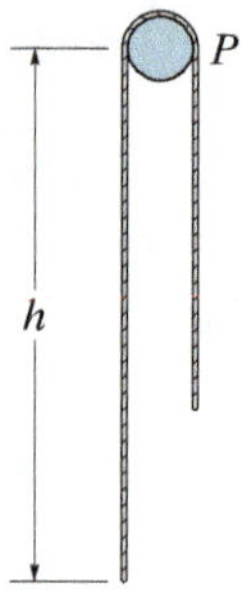

Prob. 8–96

8–97. Determine the smallest force **P** required to lift the 40-kg crate. The coefficient of static friction between the cable and each peg is $\mu_s = 0.1$.

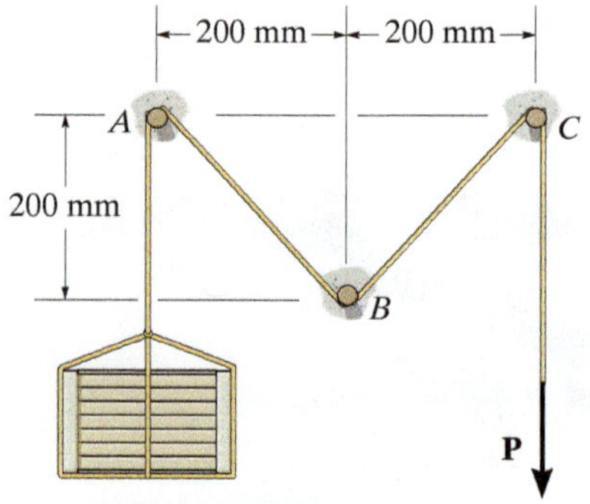

Prob. 8–97

8–98. Show that the frictional relationship between the belt tensions, the coefficient of friction μ, and the angular contacts α and β for the V-belt is $T_2 = T_1 e^{\mu\beta/\sin(\alpha/2)}$.

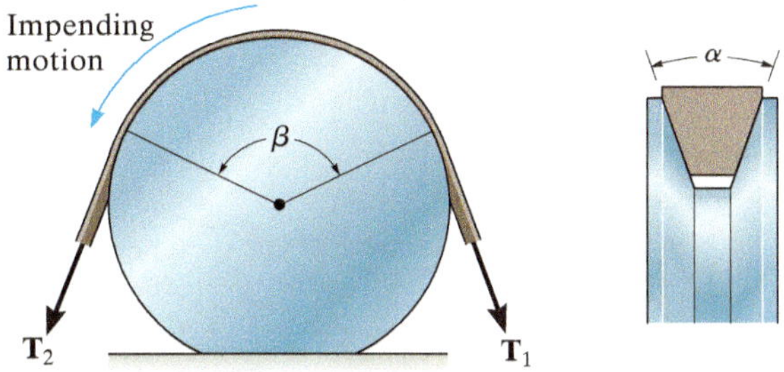

Prob. 8–98

8–99. If a force of $P = 200$ N is applied to the handle of the bell crank, determine the maximum torque **M** that can be resisted so that the flywheel does not rotate clockwise. The coefficient of static friction between the brake band and the rim of the wheel is $\mu_s = 0.3$.

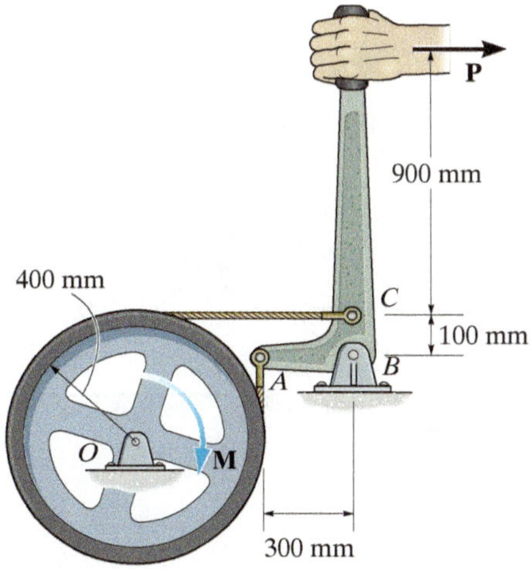

Prob. 8–99

***8–100.** A 10-kg cylinder D, which is attached to a small pulley B, is placed on the cord as shown. Determine the largest angle θ so that the cord does not slip over the peg at C. The cylinder at E also has a mass of 10 kg, and the coefficient of static friction between the cord and the peg is $\mu_s = 0.1$.

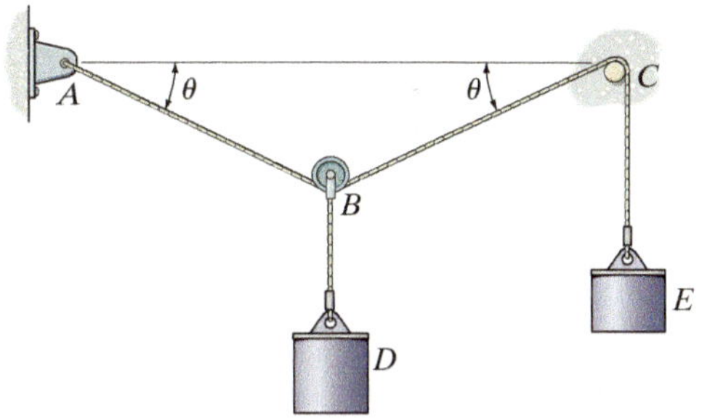

Prob. 8–100

8–101. A V-belt is used to connect the hub A of the motor to wheel B. If the belt can withstand a maximum tension of 1200 N, determine the largest mass of cylinder C that can be lifted and the corresponding torque **M** that must be supplied to A. The coefficient of static friction between the hub and the belt is $\mu_s = 0.3$, and between the wheel and the belt is $\mu_s' = 0.20$. *Hint*: See Prob. 8–98.

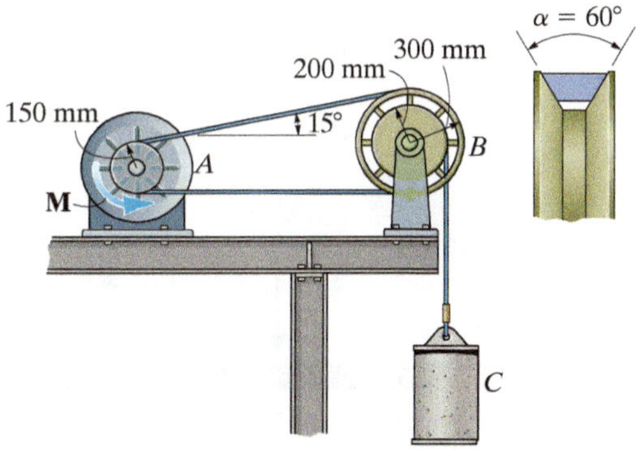

Prob. 8–101

8

8–102. The 20-kg motor has a center of gravity at G and is pin-connected at C to maintain a tension in the drive belt. Determine the smallest counterclockwise twist or torque **M** that must be supplied by the motor to turn the disk B if wheel A locks and causes the belt to slip over the disk. No slipping occurs at A. The coefficient of static friction between the belt and the disk is $\mu_s = 0.3$.

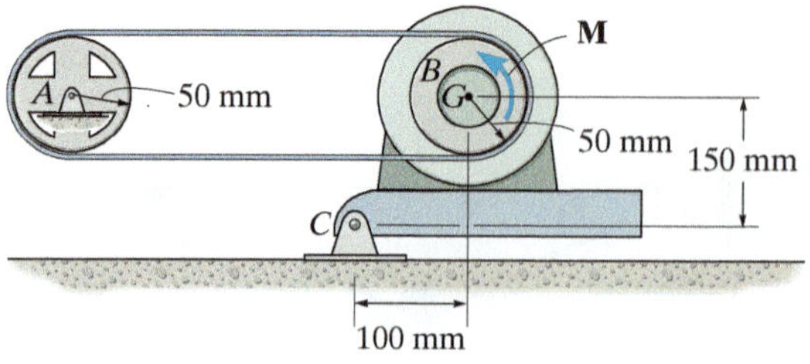

Prob. 8–102

8–103. Blocks A and B have a mass of 100 kg and 150 kg, respectively. If the coefficient of static friction between A and B and between B and C is $\mu_s = 0.25$ and between the ropes and the pegs D and E $\mu'_s = 0.5$ determine the smallest force F needed to cause motion of block B if $P = 30$ N.

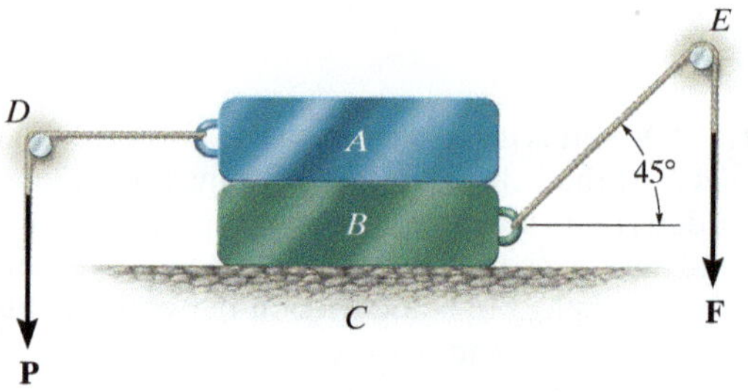

Prob. 8–103

***8–104.** Determine the minimum coefficient of static friction μ_s between the cable and the peg and the placement d of the 3-kN force for the uniform 100-kg beam to maintain equilibrium.

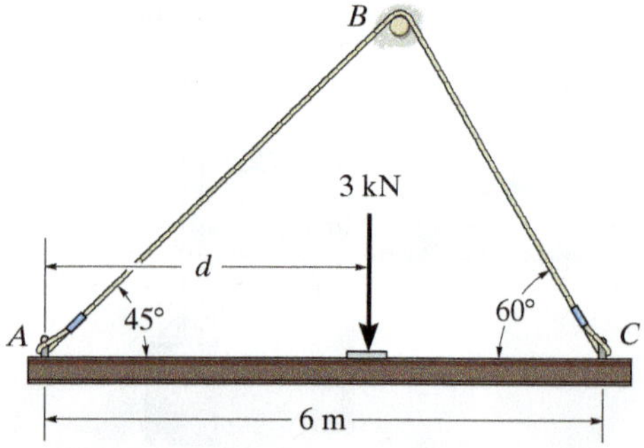

Prob. 8–104

8–105. A conveyer belt is used to transfer granular material and the frictional resistance on the top of the belt is $F = 500$ N. Determine the smallest stretch of the spring attached to the moveable axle of the idle pulley B so that the belt does not slip at the drive pulley A when the torque **M** is applied. What minimum torque **M** is required to keep the belt moving? The coefficient of static friction between the belt and the wheel at A is $\mu_s = 0.2$.

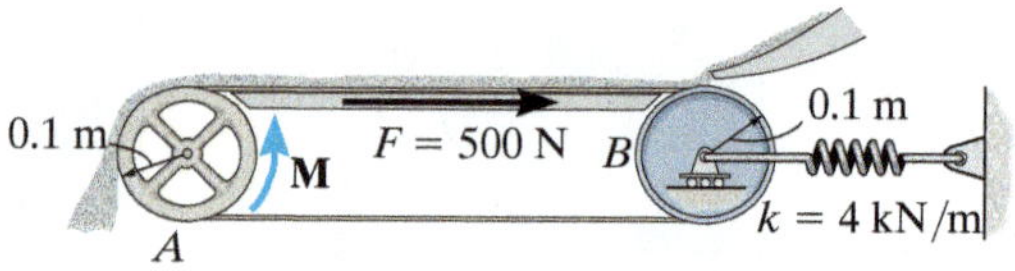

Prob. 8–105

8–106. The belt on the portable dryer wraps around the drum D, idler pulley A, and motor pulley B. If the motor can develop a maximum torque of $M = 0.80$ N·m, determine the smallest spring tension required to hold the belt from slipping. The coefficient of static friction between the belt and the drum and motor pulley is $\mu_s = 0.3$.

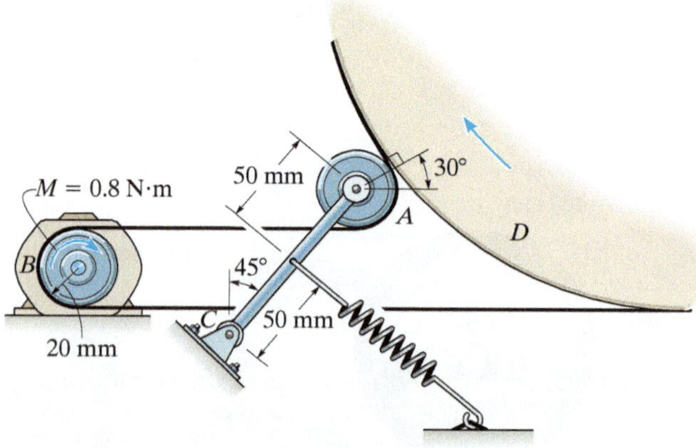

Prob. 8–106

8

*8.6 Frictional Forces on Collar Bearings, Pivot Bearings, and Disks

Pivot and *collar bearings* are commonly used in machines to support an *axial load* on a rotating shaft. Typical examples are shown in Fig. 8–20. Provided these bearings are not lubricated, or are only partially lubricated, the laws of dry friction may be applied to determine the moment needed to turn the shaft when it supports an axial force.

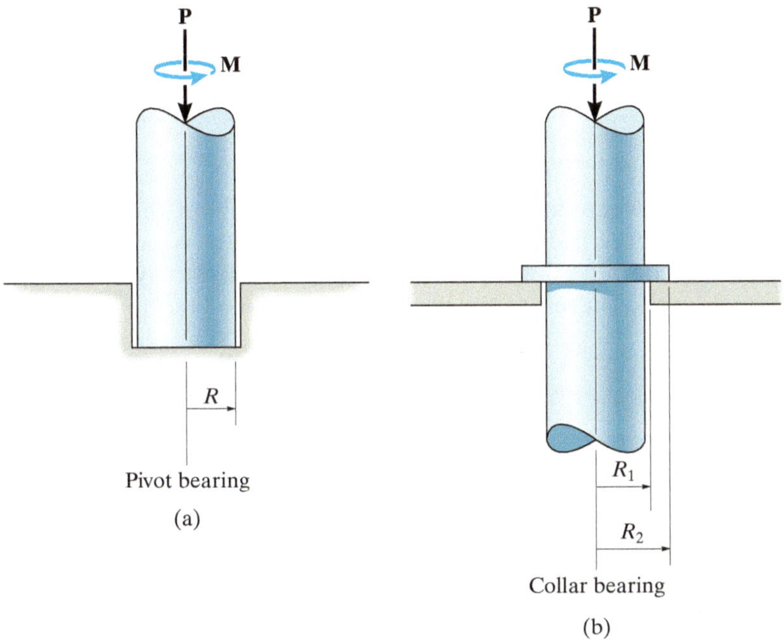

Fig. 8–20

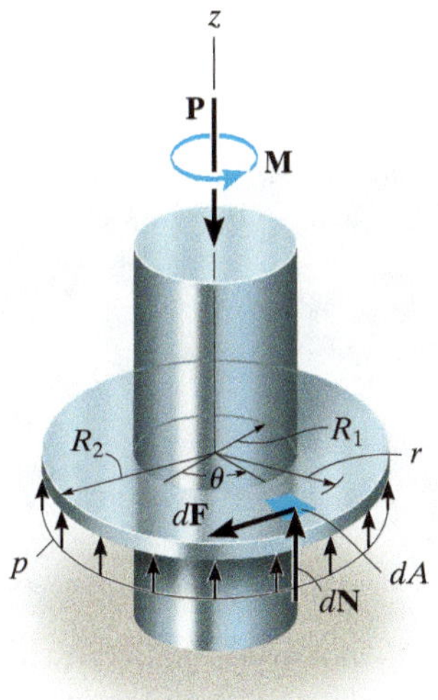

Fig. 8–21

Frictional Analysis. The collar bearing on the shaft shown in Fig. 8–21 is subjected to an axial force **P** and has a total bearing or contact area $\pi(R_2^2 - R_1^2)$. Provided the bearing is new and evenly supported, then the normal pressure p on the bearing will be *uniformly distributed* over this area. Since $\Sigma F_z = 0$, then p, measured as a force per unit area, is $p = P/\pi(R_2^2 - R_1^2)$.

The moment needed to cause impending rotation of the shaft can be determined from moment equilibrium about the z axis. A differential area element $dA = (r\,d\theta)(dr)$, shown in Fig. 8–21, is subjected to both a normal force $dN = p\,dA$ and an associated frictional force,

$$dF = \mu_s\,dN = \mu_s p\,dA = \frac{\mu_s P}{\pi(R_2^2 - R_1^2)}dA$$

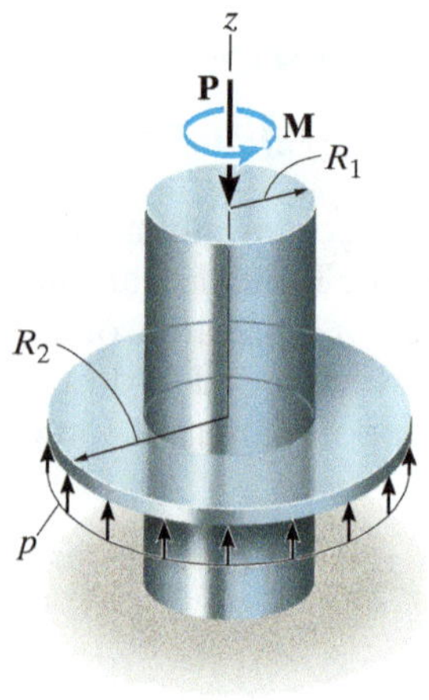

Fig. 8–21 (Repeated)

The normal force does not create a moment about the z axis of the shaft; however, the frictional force does; namely, $dM = r\,dF$. Integration is needed to compute the applied moment **M** needed to overcome all the frictional forces. Therefore, for impending rotational motion,

$$\Sigma M_z = 0; \qquad M - \int_A r\,dF = 0$$

Substituting for dF and dA and integrating over the entire bearing area yields

$$M = \int_{R_1}^{R_2}\int_0^{2\pi} r\left[\frac{\mu_s P}{\pi(R_2^2 - R_1^2)}\right](r\,d\theta\,dr) = \frac{\mu_s P}{\pi(R_2^2 - R_1^2)}\int_{R_1}^{R_2} r^2\,dr \int_0^{2\pi} d\theta$$

or

$$M = \frac{2}{3}\mu_s P\left(\frac{R_2^3 - R_1^3}{R_2^2 - R_1^2}\right) \qquad (8\text{–}7)$$

The motor that turns the disk of this sanding machine develops a torque that must overcome the frictional forces acting on the disk.

The moment developed at the end of the shaft, when it is *rotating* at constant speed, can be found by substituting μ_k for μ_s in Eq. 8–7.

In the case of a pivot bearing, Fig. 8–20*a*, then $R_2 = R$ and $R_1 = 0$, and Eq. 8–7 reduces to

$$M = \frac{2}{3}\mu_s PR \qquad (8\text{–}8)$$

Remember that Eqs. 8–7 and 8–8 apply only for bearing surfaces subjected to *constant pressure*. If the pressure is not uniform, a variation of the pressure as a function of the bearing area must be determined before integrating to obtain the moment. The following example illustrates this concept.

EXAMPLE 8.9

The uniform bar shown in Fig. 8–22*a* has a weight of 4 lb. If it is assumed that the normal pressure acting at the contacting surface varies linearly along the length of the bar as shown, determine the couple moment **M** required to rotate the bar. Assume that the bar's width is negligible in comparison to its length. The coefficient of static friction is equal to $\mu_s = 0.3$.

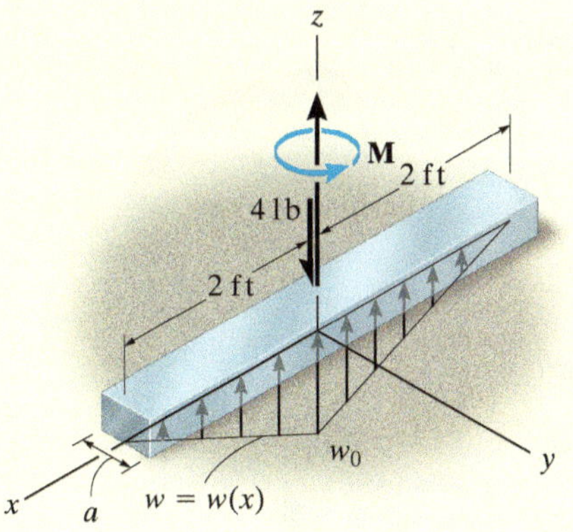

(a)

SOLUTION

A free-body diagram of the bar is shown in Fig. 8–22*b*. The intensity w_0 of the distributed load at the center ($x = 0$) is determined from vertical force equilibrium, Fig. 8–22*a*.

$$+\uparrow \Sigma F_z = 0; \qquad -4\text{ lb} + 2\left[\frac{1}{2}\left(2\text{ ft}\right)w_0\right] = 0 \qquad w_0 = 2\text{ lb/ft}$$

Since $w = 0$ at $x = 2$ ft, the distributed load expressed as a function of x is

$$w = (2\text{ lb/ft})\left(1 - \frac{x}{2\text{ ft}}\right) = 2 - x$$

The magnitude of the normal force acting on a differential segment of area having a length dx is therefore

$$dN = w\,dx = (2 - x)dx$$

The magnitude of the frictional force acting on the same element of area is

$$dF = \mu_s\,dN = 0.3(2 - x)dx$$

Hence, the moment created by this force about the z axis is

$$dM = x\,dF = 0.3(2x - x^2)dx$$

The summation of moments about the z axis of the bar is determined by integration, which yields

$$\Sigma M_z = 0; \qquad M - 2\int_0^2 (0.3)(2x - x^2)\,dx = 0$$

$$M = 0.6\left(x^2 - \frac{x^3}{3}\right)\Big|_0^2$$

$$M = 0.8\text{ lb}\cdot\text{ft} \qquad \textit{Ans.}$$

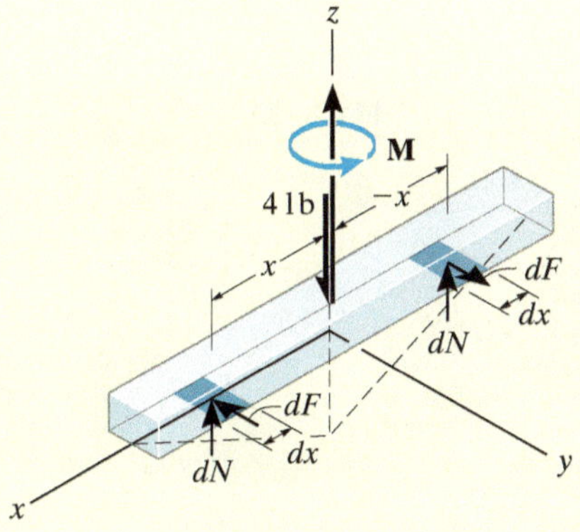

(b)

Fig. 8–22

8.7 Frictional Forces on Journal Bearings

Unwinding the cable from this spool requires overcoming friction from the supporting shaft.

When a shaft or axle is subjected to lateral loads, a *journal bearing* is commonly used for support. Provided the bearing is not lubricated, or is only partially lubricated, a reasonable analysis of the frictional resistance on the bearing can be based on the laws of dry friction.

Frictional Analysis. A typical journal-bearing support is shown in Fig. 8–23*a*. As the shaft rotates, the contact point moves up the wall of the bearing to some point A where slipping occurs. If the vertical load acting at the end of the shaft is **P**, then the bearing reactive force **R** acting at A will be equal and opposite to **P**, Fig. 8–23*b*. The moment needed to maintain constant rotation of the shaft can be found by summing moments about the z axis of the shaft; i.e.,

$$\Sigma M_z = 0; \qquad M - (R \sin \phi_k)r = 0$$

or

$$M = Rr \sin \phi_k \tag{8–9}$$

where ϕ_k is the angle of kinetic friction defined by $\tan \phi_k = F/N = \mu_k N/N = \mu_k$. In Fig. 8–23*c*, it is seen that $r \sin \phi_k = r_f$. The dashed circle with radius r_f is called the *friction circle*, and as the shaft rotates, the reaction **R** will always be tangent to it. If the bearing is partially lubricated, μ_k is small, and therefore $\sin \phi_k \approx \tan \phi_k \approx \mu_k$. Under these conditions, a reasonable *approximation* to the moment needed to overcome the frictional resistance becomes

$$M \approx Rr\mu_k \tag{8–10}$$

Notice that to minimize friction the bearing radius r should be as small as possible. In practice, however, this type of journal bearing is not suitable for long service since friction between the shaft and bearing will eventually wear down the surfaces. Instead, designers will incorporate "ball bearings" or "rollers" in journal bearings to minimize frictional losses.

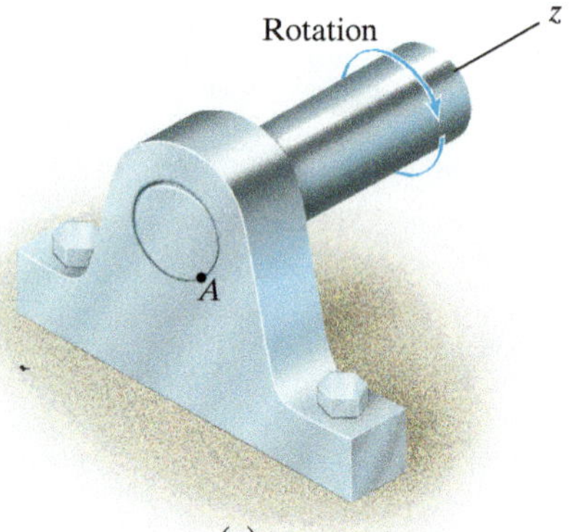

(a)

Fig. 8–23

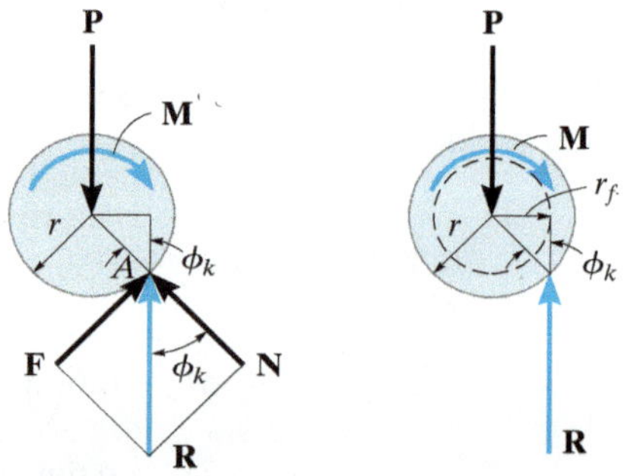

(b) (c)

8

EXAMPLE 8.10

The 100-mm-diameter pulley shown in Fig. 8–24*a* fits loosely on a 10-mm-diameter shaft for which the coefficient of static friction is $\mu_s = 0.4$. Determine the minimum tension T in the belt needed to (a) raise the 100-kg block and (b) lower the block. Assume that no slipping occurs between the belt and pulley and neglect the weight of the pulley.

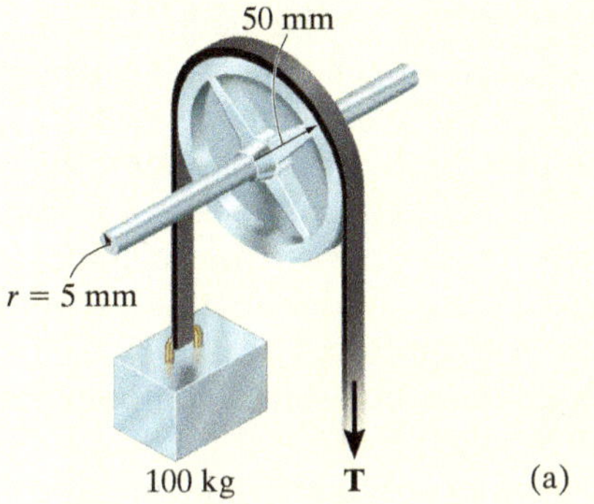

(a)

SOLUTION

Part (a). A free-body diagram of the pulley is shown in Fig. 8–24*b*. When the pulley is subjected to belt tensions of 981 N each, it makes contact with the shaft at point P_1. As the tension T is *increased*, the contact point will move around the shaft to point P_2 before motion impends. From the figure, the friction circle has a radius $r_f = r \sin \phi_s$. Using the simplification that $\sin \phi_s \approx \tan \phi_s \approx \mu_s$ then $r_f \approx r\mu_s = (5 \text{ mm})(0.4) = 2 \text{ mm}$, so that summing moments about P_2 gives

$$\circlearrowleft + \Sigma M_{P_2} = 0; \qquad 981 \text{ N}(52 \text{ mm}) - T(48 \text{ mm}) = 0$$

$$T = 1063 \text{ N} = 1.06 \text{ kN} \qquad \textit{Ans.}$$

If a more exact analysis is used, then $\phi_s = \tan^{-1} 0.4 = 21.8°$. Thus, the radius of the friction circle would be $r_f = r \sin \phi_s = 5 \sin 21.8° = 1.86 \text{ mm}$. Therefore,

$$\circlearrowleft + \Sigma M_{P_2} = 0;$$

$$981 \text{ N}(50 \text{ mm} + 1.86 \text{ mm}) - T(50 \text{ mm} - 1.86 \text{ mm}) = 0$$

$$T = 1057 \text{ N} = 1.06 \text{ kN} \qquad \textit{Ans.}$$

Part (b). When the block is lowered, the resultant force **R** acting on the shaft passes through point as shown in Fig. 8–24*c*. Summing moments about this point yields

$$\circlearrowleft + \Sigma M_{P_3} = 0; \qquad 981 \text{ N}(48 \text{ mm}) - T(52 \text{ mm}) = 0$$

$$T = 906 \text{ N} \qquad \textit{Ans.}$$

NOTE: Using the approximate analysis, the difference between raising and lowering the block is thus 157 N.

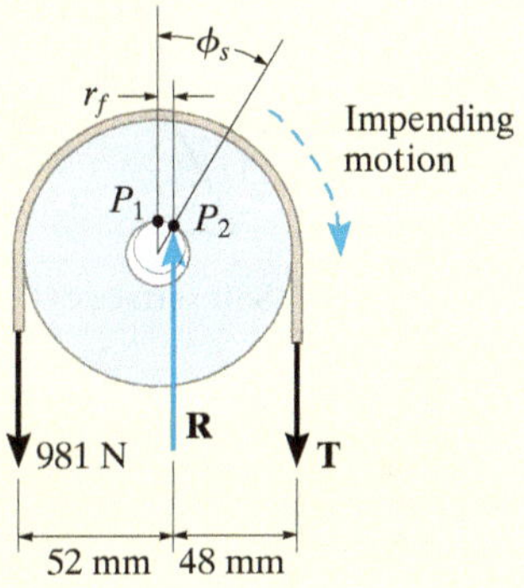

(b)

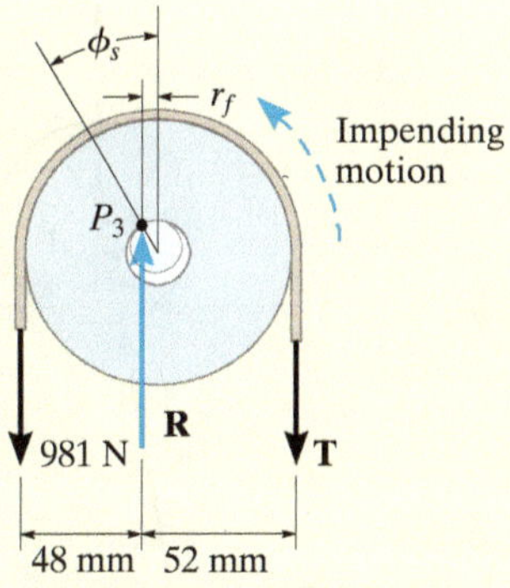

(c)

Fig. 8–24

8

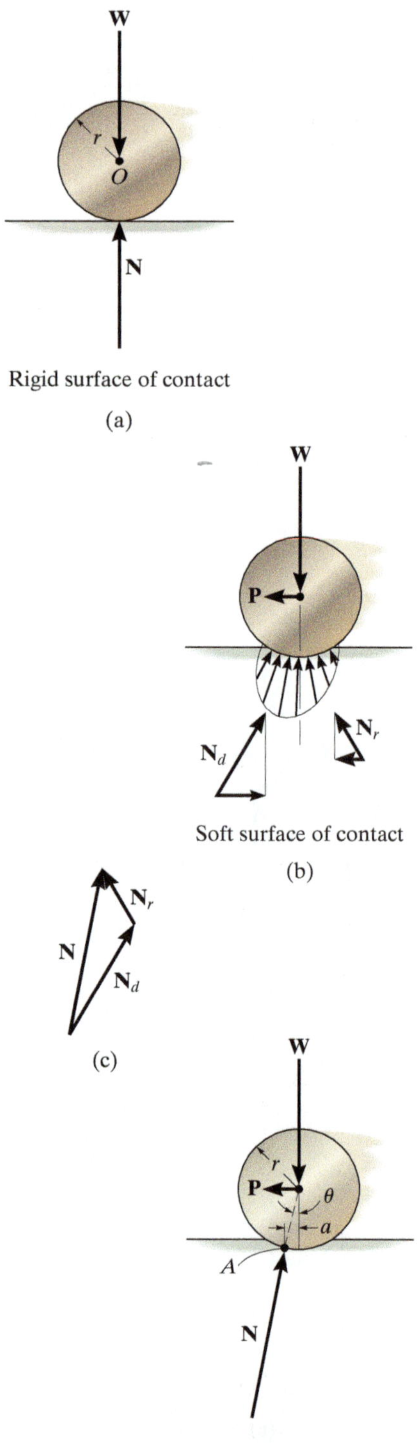

Fig. 8–25

*8.8 Rolling Resistance

When a *rigid* cylinder rolls at constant velocity along a *rigid* surface, the normal force exerted by the surface on the cylinder acts perpendicular to the tangent at the point of contact, as shown in Fig. 8–25*a*. Actually, however, no materials are perfectly rigid, and therefore the reaction of the surface on the cylinder consists of a distribution of normal pressure. For example, consider the cylinder to be made of a very hard material, and the surface on which it rolls to be relatively soft. Due to its weight, the cylinder compresses the surface underneath it, Fig. 8–25*b*. As the cylinder rolls, the surface material in front of the cylinder *retards* the motion since it is being *deformed*, whereas the material in the rear is *restored* from the deformed state and therefore tends to *push* the cylinder forward. The normal pressures acting on the cylinder in this manner are represented in Fig. 8–25*b* by their resultant forces $\mathbf{N}_d$ and $\mathbf{N}_r$. The magnitude of the force of *deformation*, $\mathbf{N}_d$, and its horizontal component is *always greater* than that of *restoration*, $\mathbf{N}_r$, and consequently a horizontal driving force **P** must be applied to the cylinder to maintain the motion. Fig. 8–25*b*.*

Rolling resistance is caused primarily by this effect, although it is also, to a lesser degree, the result of surface adhesion and relative micro-sliding between the surfaces of contact. Because the actual force **P** needed to overcome these effects is difficult to determine, a simplified method will be developed here to explain one way engineers have analyzed this phenomenon. To do this, we will consider the resultant of the *entire* normal pressure, $\mathbf{N} = \mathbf{N}_d + \mathbf{N}_r$, acting on the cylinder, Fig. 8–25*c*. As shown in Fig. 8–25*d*, this force acts at an angle θ with the vertical. To keep the cylinder in equilibrium, i.e., rolling at a constant rate, it is necessary that **N** be *concurrent* with the driving force **P** and the weight **W**. Summing moments about point A gives $Wa = P(r\cos\theta)$. Since the deformations are generally very small in relation to the cylinder's radius, $\cos\theta \approx 1$; hence,

$$Wa \approx Pr$$

or

$$P \approx \frac{Wa}{r} \qquad (8\text{–}11)$$

The distance a is termed the *coefficient of rolling resistance,* which has the dimension of length. For instance, $a \approx 0.5$ mm for a wheel rolling on a rail, both of which are made of mild steel. For hardened steel ball

*Actually, the deformation force $\mathbf{N}_d$ causes *energy* to be stored in the material as its magnitude is increased, whereas the restoration force $\mathbf{N}_r$, as its magnitude is decreased, allows some of this energy to be released. The remaining energy is *lost* since it is used to heat up the surface, and if the cylinder's weight is very large, it accounts for permanent deformation of the surface. Work must be done by the horizontal force **P** to make up for this loss.

bearings on steel, $a \approx 0.1$ mm. Experimentally, though, this factor is difficult to measure, since it depends on such parameters as the rate of rotation of the cylinder, the elastic properties of the contacting surfaces, and the surface finish. For this reason, little reliance is placed on the data for determining a. The analysis presented here does, however, indicate why a heavy load (W) offers greater resistance to motion (P) than a light load under the same conditions. Furthermore, since Wa/r is generally very small compared to $\mu_k W$, the force needed to *roll* a cylinder over the surface will be much less than that needed to *slide* it across the surface. It is for this reason that a roller or ball bearings are often used to minimize the frictional resistance between moving parts.

Rolling resistance of railroad wheels on the rails is small since steel is very stiff. By comparison, the rolling resistance of the wheels of a tractor in a wet field is very large.

EXAMPLE 8.11

A 10-kg steel wheel shown in Fig. 8–26*a* has a radius of 100 mm and rests on an inclined plane made of soft wood. If θ is increased so that the wheel begins to roll down the incline with constant velocity when $\theta = 1.2°$, determine the coefficient of rolling resistance.

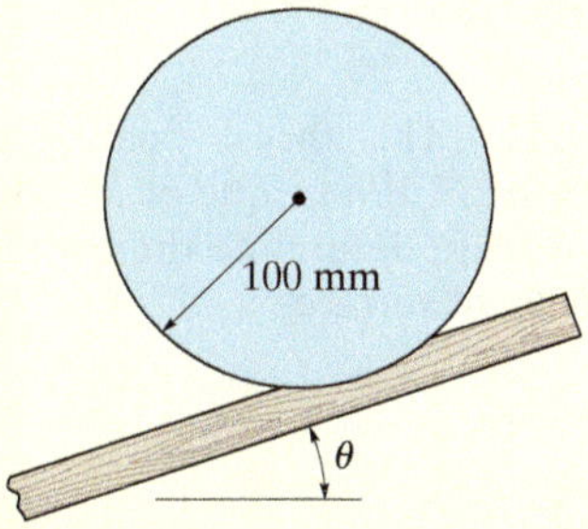

(a)

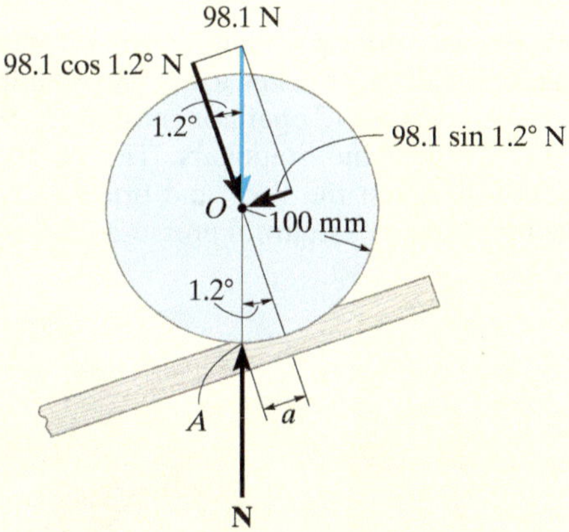

(b)

Fig. 8–26

SOLUTION

As shown on the free-body diagram, Fig. 8–26*b*, when the wheel has impending motion, the normal reaction **N** acts at point A defined by the dimension a. Resolving the weight into components parallel and perpendicular to the incline, and summing moments about point A, yields

$\circlearrowleft + \Sigma M_A = 0;$

$$-(98.1 \cos 1.2° \text{ N})(a) + (98.1 \sin 1.2° \text{ N})(100 \cos 1.2° \text{ mm}) = 0$$

Solving, we obtain

$$a = 2.09 \text{ mm} \qquad \textit{Ans.}$$

PROBLEMS

8–107. The annular ring bearing is subjected to a thrust of 800 lb. Determine the smallest required coefficient of static friction if a torque of $M = 15 \text{ lb} \cdot \text{ft}$ must be resisted to prevent the shaft from rotating.

***8–108.** The annular ring bearing is subjected to a thrust of 800 lb. If $\mu_s = 0.35$, determine the torque M that must be applied to overcome friction.

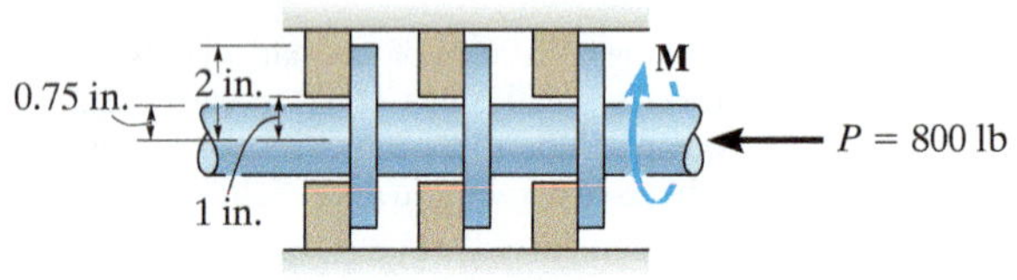

Probs. 8–107/108

8–109. The floor-polishing machine rotates at a constant angular velocity. If it has a weight of 80 lb. determine the couple forces F the operator must apply to the handles to hold the machine stationary. The coefficient of kinetic friction between the floor and brush is $\mu_k = 0.3$. Assume the brush exerts a uniform pressure on the floor.

Prob. 8–109

8–110. The shaft is supported by a thrust bearing A and a journal bearing B. Determine the torque **M** required to rotate the shaft at constant angular velocity. The coefficient of kinetic friction at the thrust bearing is $\mu_k = 0.2$. Neglect friction at B.

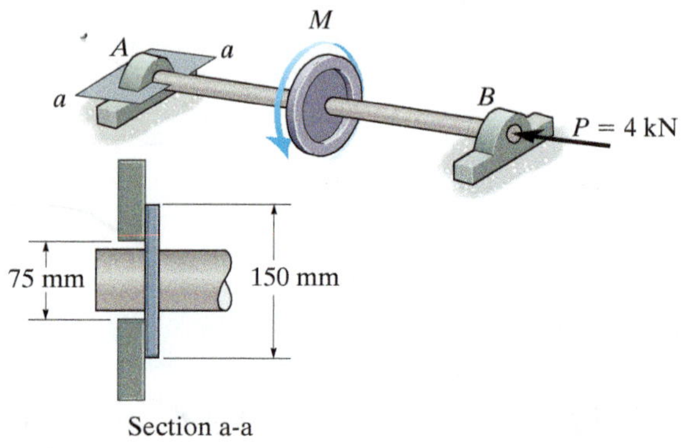

Prob. 8–110

8–111. The thrust bearing supports an axial load of $P = 6$ kN. If a torque of $M = 150 \text{ N} \cdot \text{m}$ is required to rotate the shaft, determine the coefficient of static friction at the constant surface.

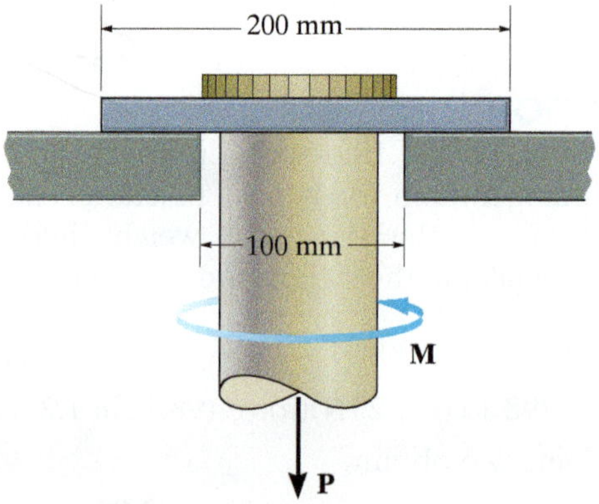

Prob. 8–111

*8–112. Assuming that the variation of pressure at the bottom of the pivot bearing is defined as $p = p_0(R_2/r)$, determine the torque M needed to overcome friction if the shaft is subjected to an axial force **P**. The coefficient of static friction is μ_s. For the solution, it is necessary to determine p_0 in terms of P and the bearing dimensions R_1 and R_2.

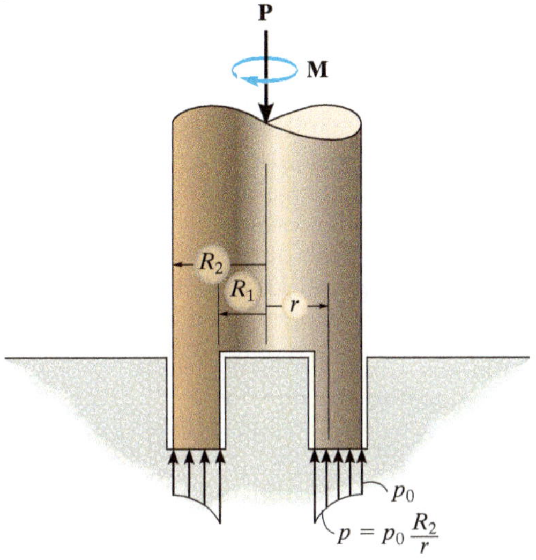

Prob. 8–112

8–113. The plate clutch consists of a flat plate A that slides over the rotating shaft S. The shaft is fixed to the driving plate gear B. If the gear C, which is in mesh with B, is subjected to a torque of $M = 0.8\ \text{N}\cdot\text{m}$, determine the smallest force P, that must be applied via the control arm, to stop the rotation. The coefficient of static friction between the plates A and D is $\mu_s = 0.4$. Assume the bearing pressure between A and D to be uniform.

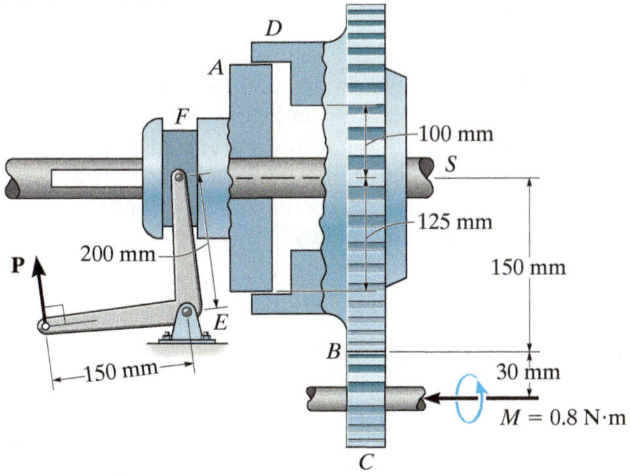

Prob. 8–113

8–114. The conical bearing is subjected to a constant pressure distribution at its surface of contact. If the coefficient of static friction is μ_s, determine the torque M required to overcome friction if the shaft supports an axial force **P**.

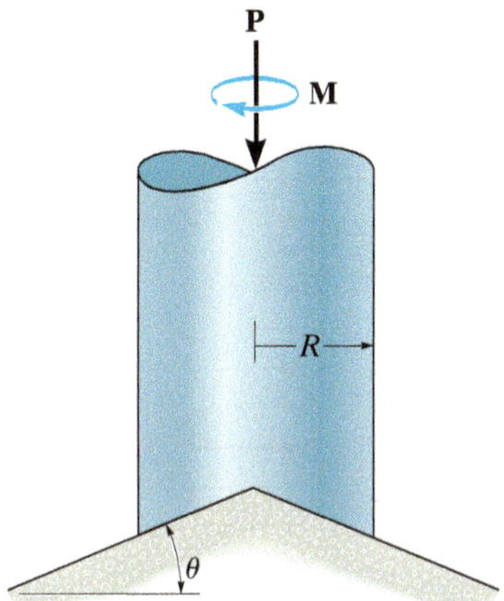

Prob. 8–114

8–115. The pivot bearing is subjected to a pressure distribution at its surface of contact which varies as shown. If the coefficient of static friction is μ, determine the torque M required to overcome friction if the shaft supports an axial force **P**.

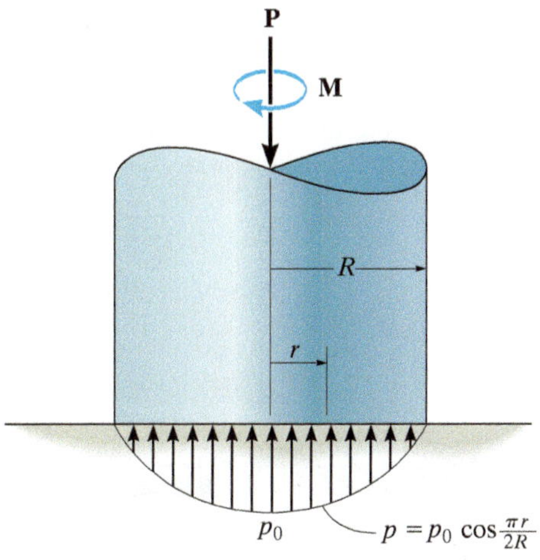

Prob. 8–115

*8–116. A 200-mm-diameter post is driven 3 m into sand for which $\mu_s = 0.3$. If the normal pressure acting *completely around the post* varies linearly with depth as shown, determine the frictional torque **M** that must be overcome to rotate the post.

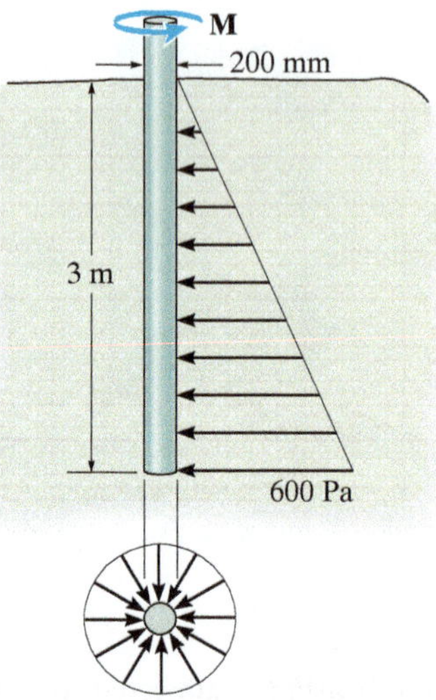

Prob. 8–116

8–117. A beam having a uniform weight W rests on the rough horizontal surface having a coefficient of static friction μ_s. If the horizontal force **P** is applied perpendicular to the beam's length, determine the location d of the point O about which the beam begins to rotate.

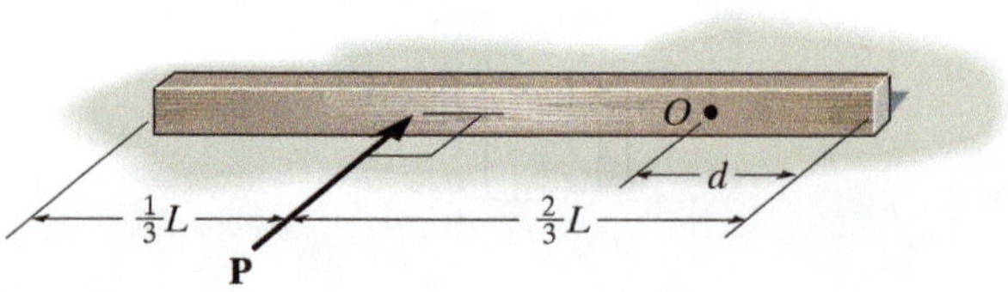

Prob. 8–117

8–118. The collar fits *loosely* around a fixed shaft that has a radius of 2 in. If the coefficient of kinetic friction between the shaft and the collar is $\mu_k = 0.3$, determine the force P on the horizontal segment of the belt so that the collar rotates *counterclockwise* with a constant angular velocity. Assume that the belt does not slip on the collar; rather, the collar slips on the shaft. Neglect the weight and thickness of the belt and collar. The radius, measured from the center of the collar to the mean thickness of the belt, is 2.25 in.

8–119. The collar fits *loosely* around a fixed shaft that has a radius of 2 in. If the coefficient of kinetic friction between the shaft and the collar is $\mu_k = 0.3$, determine the force P on the horizontal segment of the belt so that the collar rotates *clockwise* with a constant angular velocity. Assume that the belt does not slip on the collar; rather, the collar slips on the shaft. Neglect the weight and thickness of the belt and collar. The radius, measured from the center of the collar to the mean thickness of the belt, is 2.25 in.

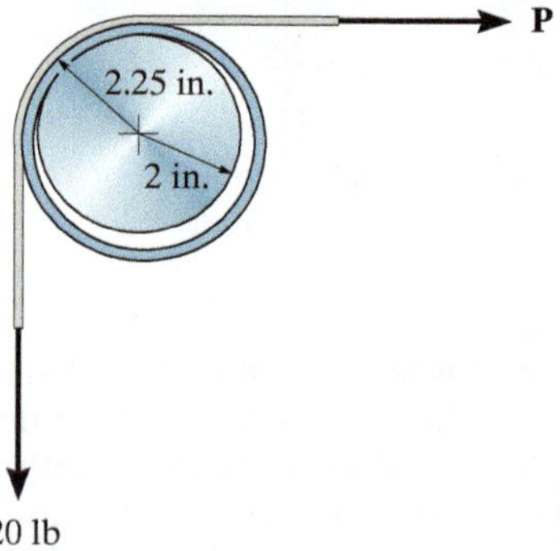

Probs. 8–118/119

*8–120. The pulley has a radius of 3 in. and fits loosely on the 0.5-in.-diameter shaft. If the loadings acting on the belt cause the pulley to rotate with constant angular velocity, determine the frictional force between the shaft and the pulley and compute the coefficient of kinetic friction. The pulley weighs 18 lb.

8–121. The pulley has a radius of 3 in. and fits loosely on the 0.5-in.-diameter shaft. If the loadings acting on the belt cause the pulley to rotate with constant angular velocity, determine the frictional force between the shaft and the pulley and compute the coefficient of kinetic friction. Neglect the weight of the pulley.

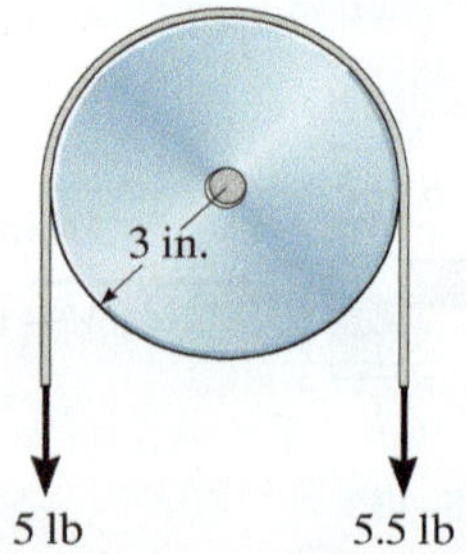

Probs. 8–120/121

8–122. Determine the tension **T** in the belt needed to overcome the tension of 200 lb created on the other side. Also, what are the normal and frictional components of force developed on the collar bushing? The coefficient of static friction is $\mu_s = 0.21$.

8–123. If a tension force $T = 215$ lb is required to pull the 200-lb force around the collar bushing, determine the coefficient of static friction at the contacting surface. The belt does not slip on the collar.

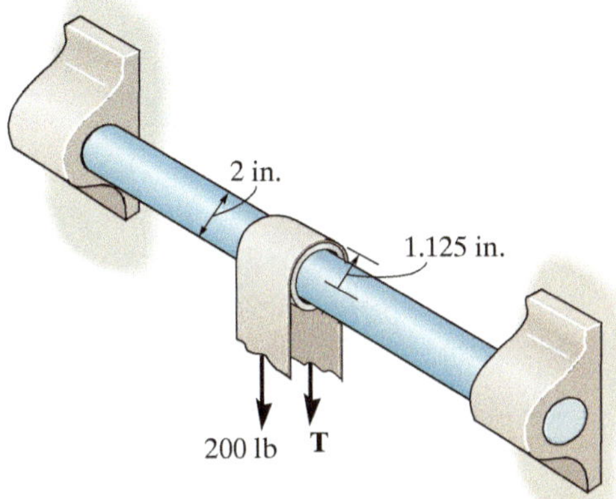

Probs. 8–122/123

*8–124. A pulley having a diameter of 80 mm and mass of 1.25 kg is supported loosely on a shaft having a diameter of 20 mm. Determine the torque M that must be applied to the pulley to cause it to rotate with constant motion. The coefficient of kinetic friction between the shaft and pulley is $\mu_k = 0.4$. Also calculate the angle θ which the normal force at the point of contact makes with the horizontal. The shaft itself cannot rotate.

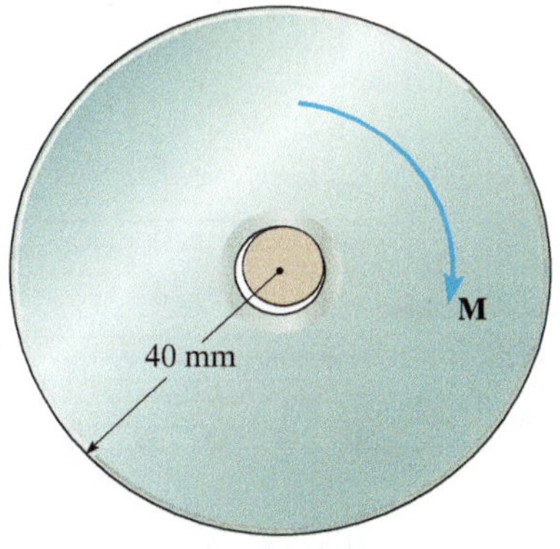

Prob. 8–124

8–125. The 5-kg skateboard rolls down the 5° slope at constant speed. If the coefficient of kinetic friction between the 12.5 mm diameter axles and the wheels is $\mu_k = 0.3$, determine the radius of the wheels. Neglect rolling resistance of the wheels on the surface. The center of mass for the skateboard is at G.

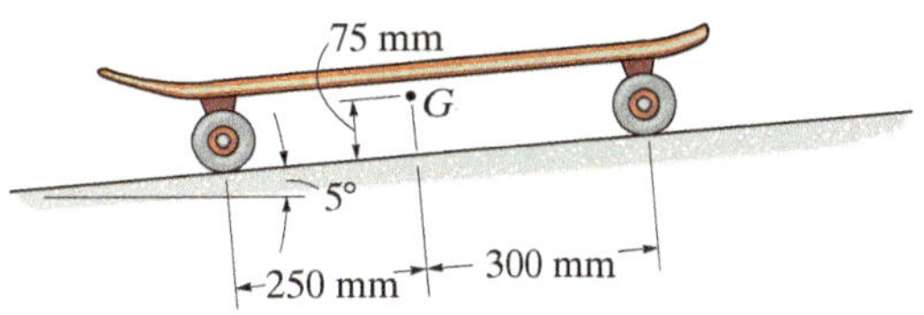

Prob. 8–125

8–126. The cart together with the load weighs 150 lb and has a center of gravity at G. If the wheels fit loosely on the 1.5-in. diameter axles, determine the horizontal force **P** required to pull the cart with constant velocity. The coefficient of kinetic friction between the axles and the wheels is $\mu_k = 0.2$. Neglect rolling resistance of the wheels on the ground.

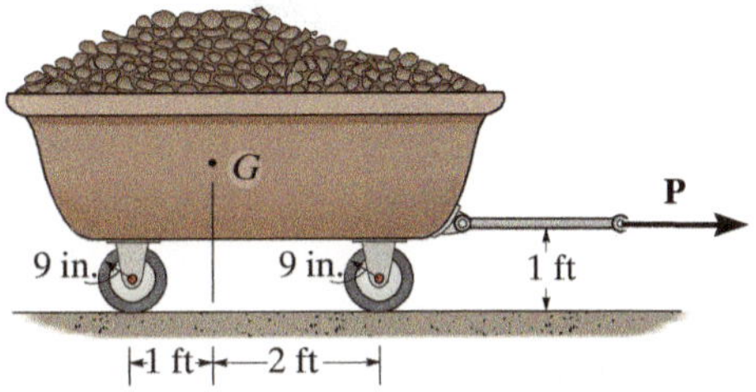

Prob. 8–126

8–127. The trailer has a total weight of 850 lb and center of gravity at G which is directly over its axle. If the axle has a diameter of 1 in., the radius of the wheel is $r = 1.5$ ft, and the coefficient of kinetic friction at the bearing is $\mu_k = 0.08$, determine the horizontal force P needed to pull the trailer.

Prob. 8–127

8

*8–128. The vehicle has a weight of 2600 lb and center of gravity at G. Determine the horizontal force $\mathbf{P}$ that must be applied to overcome the rolling resistance of the wheels. The coefficient of rolling resistance is 0.5 in. The tires have a diameter of 2.75 ft.

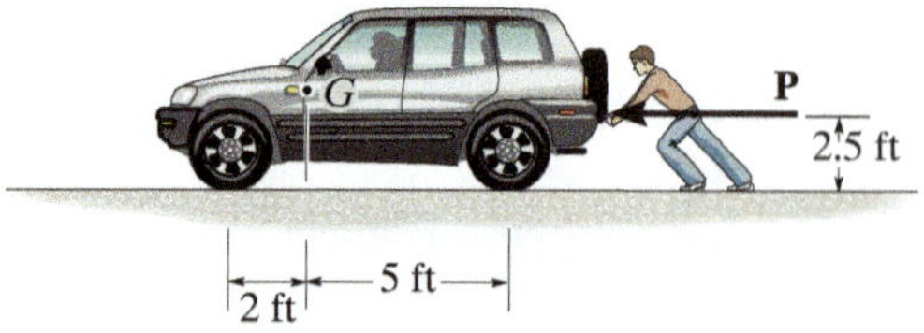

Prob. 8–128

8–129. The tractor has a weight of 16 000 lb and the coefficient of rolling resistance is $a = 2$ in. Determine the force $\mathbf{P}$ needed to overcome rolling resistance at all four wheels and push it forward.

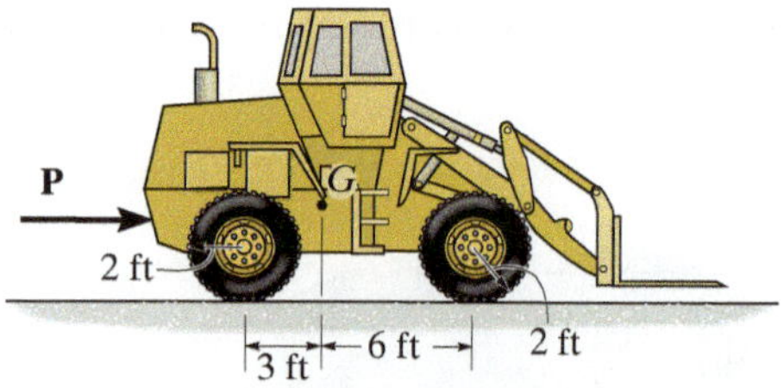

Prob. 8–129

8

8–130. The hand cart has wheels with a diameter of 80 mm. If a crate having a mass of 500 kg is placed on the cart so that each wheel carries an equal load, determine the horizontal force P that must be applied to the handle to overcome the rolling resistance. The coefficient of rolling resistance is 2 mm. Neglect the mass of the cart.

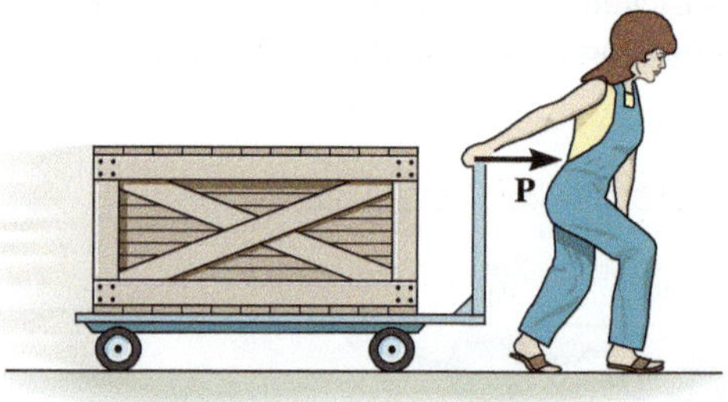

Prob. 8–130

8–131. The cylinder is subjected to a load that has a weight W. If the coefficients of rolling resistance for the cylinder's top and bottom surfaces are a_A and a_B, respectively, show that a horizontal force having a magnitude of $P = [W(a_A + a_B)]/2r$ is required to move the load and thereby roll the cylinder forward. Neglect the weight of the cylinder.

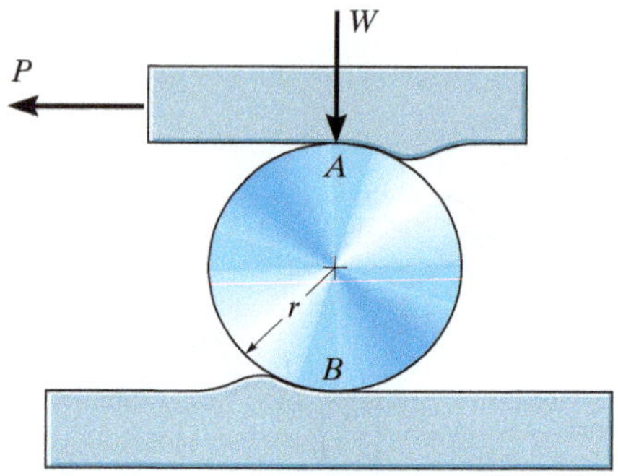

Prob. 8–131

*8–132. A large crate having a mass of 200 kg is moved along the floor using a series of 150-mm-diameter rollers for which the coefficient of rolling resistance is 3 mm at the ground and 7 mm at the bottom surface of the crate. Determine the horizontal force $\mathbf{P}$ needed to push the crate forward at a constant speed. *Hint:* Use the result of Prob. 8–131.

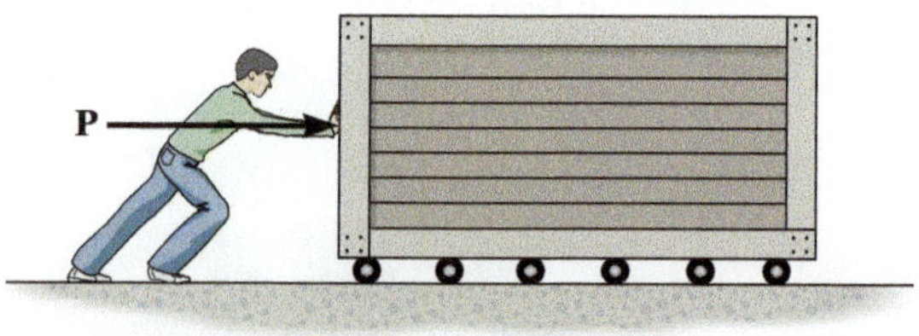

Prob. 8–132

CHAPTER REVIEW

Dry Friction

Frictional forces exist between two rough surfaces of contact. These forces act on a body so as to oppose its motion or tendency of motion.

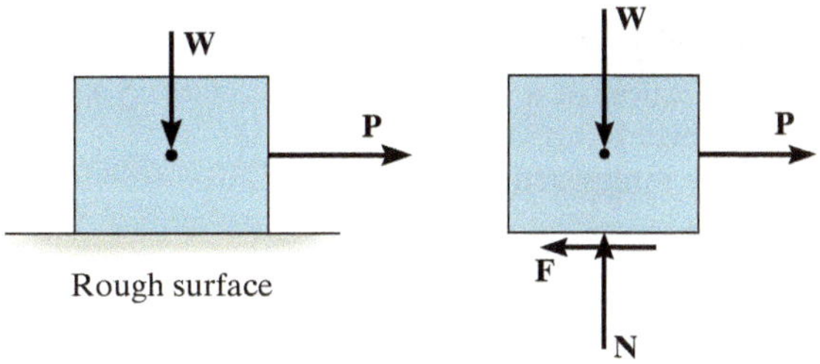

A static frictional force approaches a maximum value of $F_s = \mu_s N$, where μ_s is the *coefficient of static friction*. In this case, motion between the contacting surfaces is *impending*.

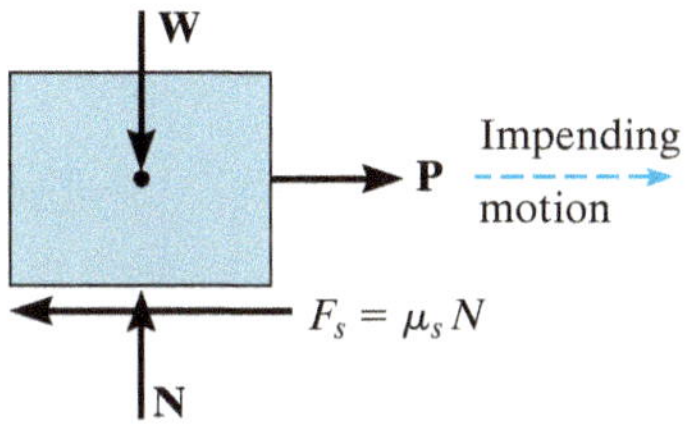

If slipping occurs, then the friction force remains essentially constant and equal to $F_k = \mu_k N$. Here μ_k is the *coefficient of kinetic friction*.

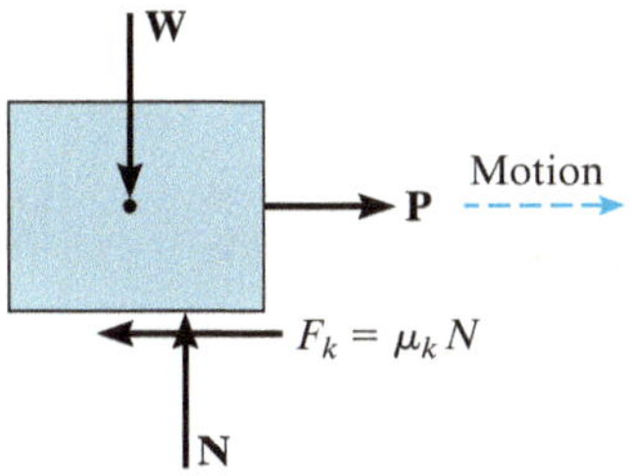

The solution of a problem involving friction requires first drawing the free-body diagram of the body. If the unknowns cannot be determined strictly from the equations of equilibrium, and the possibility of slipping occurs, then the friction equation should be applied at the appropriate points of contact in order to complete the solution.

It may also be possible for slender objects, like crates, to tip over, and this situation should also be investigated.

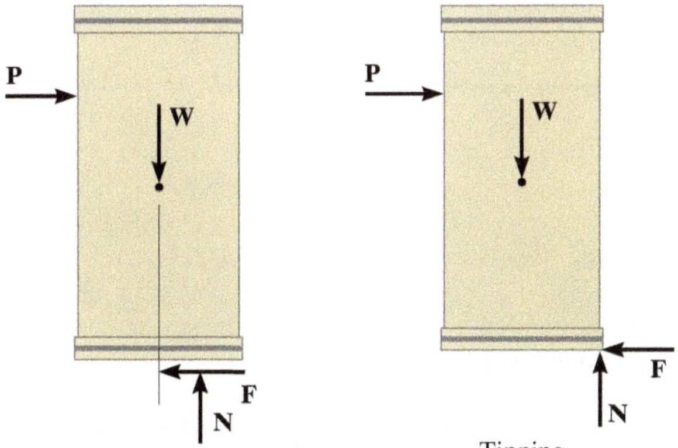

Impending slipping
$F = \mu_s N$

Tipping

Wedges

Wedges are inclined planes used to increase the application of a force. The two force equilibrium equations are used to relate the forces acting on the wedge.

An applied force **P** must push on the wedge to move it to the right.

If the coefficients of friction between the surfaces are large enough, then **P** can be removed, and the wedge will be self-locking and remain in place.

$$\Sigma F_x = 0$$
$$\Sigma F_y = 0$$

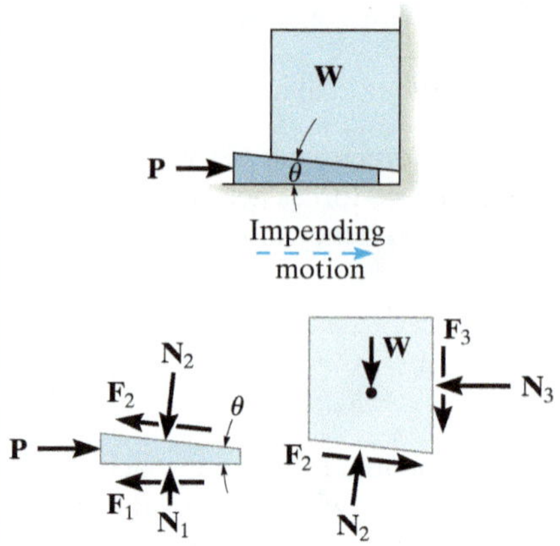

Screws

Square-threaded screws are used to move heavy loads. They represent an inclined plane, wrapped around a cylinder.

The moment needed to turn a screw depends upon the coefficient of friction and the screw's lead angle θ.

If the coefficient of friction between the surfaces is large enough, then the screw will support the load without tending to turn, i.e., it will be self-locking.

$$M = rW \tan(\theta + \phi_s)$$

Upward Impending Screw Motion

$$M' = rW \tan(\theta - \phi_s)$$

Downward Impending Screw Motion

$$\theta > \phi_s$$

$$M'' = rW \tan(\phi_s - \theta)$$

Downward Screw Motion

$$\phi_s > \theta$$

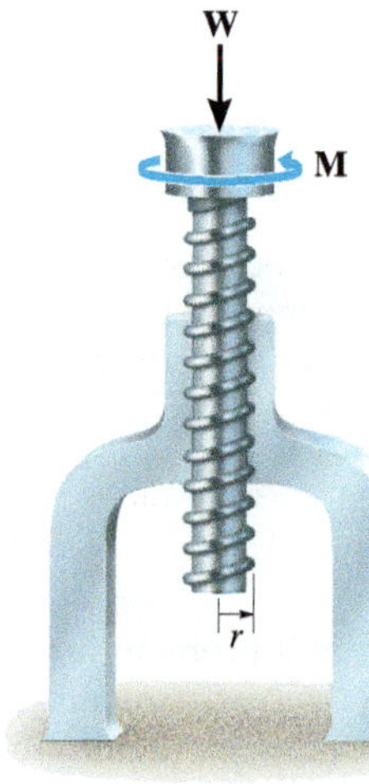

Flat Belts

The force needed to move a flat belt over a rough curved surface depends only on the angle of belt contact, β, and the coefficient of friction.

$$T_2 = T_1 e^{\mu\beta}$$

$$T_2 > T_1$$

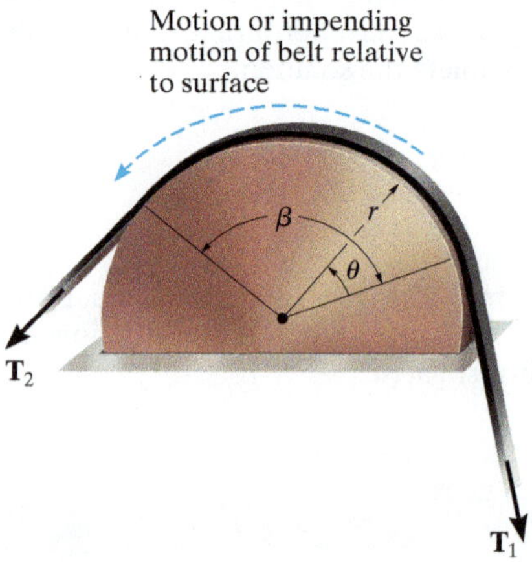

Collar Bearings and Disks

The frictional analysis of a collar bearing or disk requires looking at a differential element of the contact area. The normal force acting on this element is determined from force equilibrium along the shaft, and the moment needed to turn the shaft at a constant rate is determined from moment equilibrium about the shaft's axis.

If the pressure on the surface of a collar bearing is uniform, then integration gives the result shown.

$$M = \frac{2}{3}\mu_s P\left(\frac{R_2^3 - R_1^3}{R_2^2 - R_1^2}\right)$$

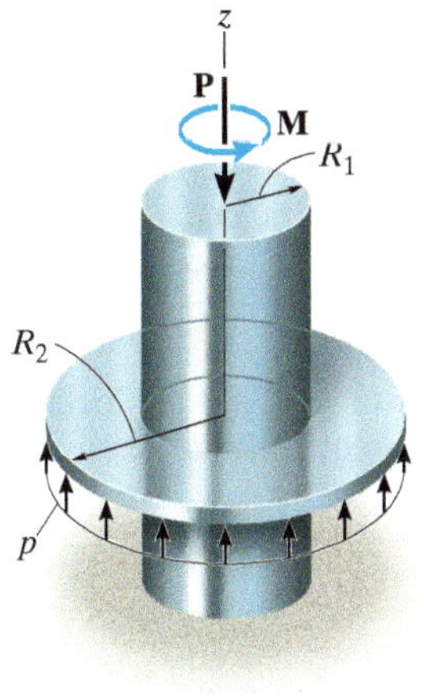

Journal Bearings

When a moment is applied to a shaft in a nonlubricated or partially lubricated journal bearing, the shaft will tend to roll up the side of the bearing until slipping occurs. This defines the radius of a friction circle, and from it the moment needed to turn the shaft can be determined.

$$M = Rr \sin \phi_k$$

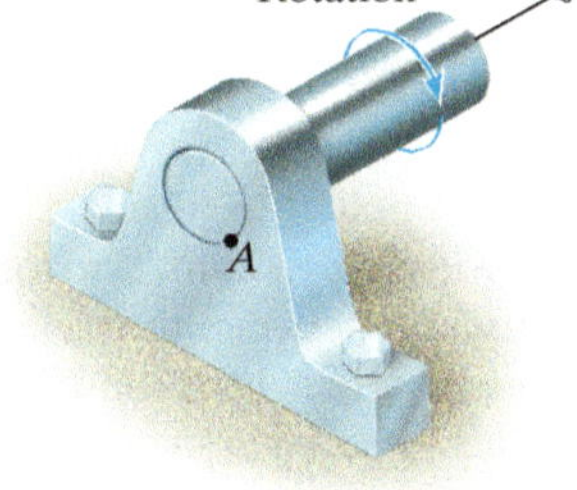

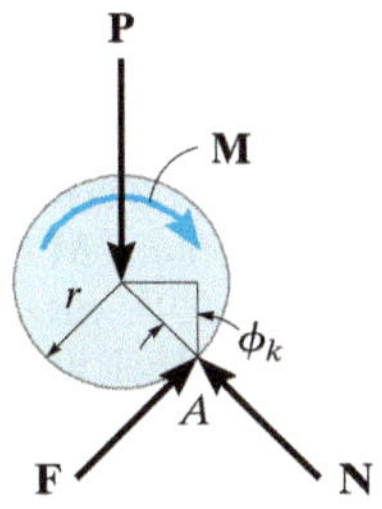

Rolling Resistance

The resistance of a wheel to rolling over a surface is caused by localized *deformation* of the two materials in contact. This causes the resultant normal force acting on the rolling body to be inclined so that it provides a component that acts in the opposite direction of the applied force **P** causing the motion. This effect is characterized using the *coefficient of rolling resistance*, *a*, which is determined from experiment.

$$P \approx \frac{Wa}{r}$$

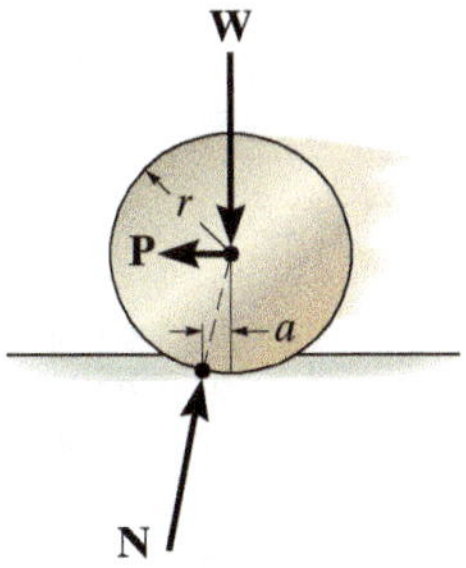

8

REVIEW PROBLEMS

8–133. The uniform 50-lb beam is supported by the rope that is attached to the end of the beam, wraps over the rough peg, and is then connected to the 100-lb block. If the coefficient of static friction between the beam and the block, and between the rope and the peg, is $\mu_s = 0.4$, determine the maximum distance that the block can be placed from A and still remain in equilibrium. Assume the block will not tip.

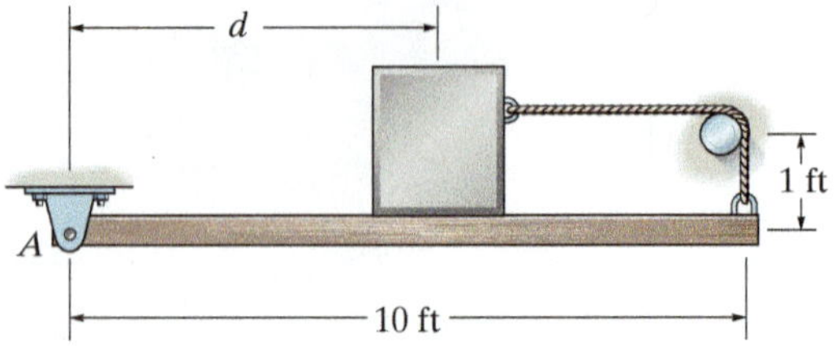

Prob. 8–133

8–134. Determine the maximum number of 50-lb packages that can be placed on the belt without causing the belt to slip at the drive wheel A, which is rotating with a constant angular velocity. Wheel B is free to rotate. Also, find the corresponding torsional moment **M** that must be supplied to wheel A. The conveyor belt is pre-tensioned with the 300-lb horizontal force. The coefficient of kinetic friction between the belt and platform P is $\mu_k = 0.2$, and the coefficient of static friction between the belt and the rim of each wheel is $\mu_s = 0.35$.

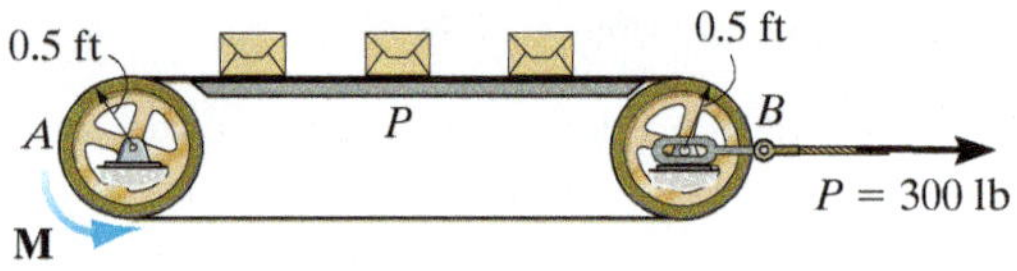

Prob. 8–134

8–135. If $P = 900$ N is applied to the handle of the bell crank, determine the maximum torque M the cone clutch can transmit. The coefficient of static friction at the contacting surface is $\mu_s = 0.3$.

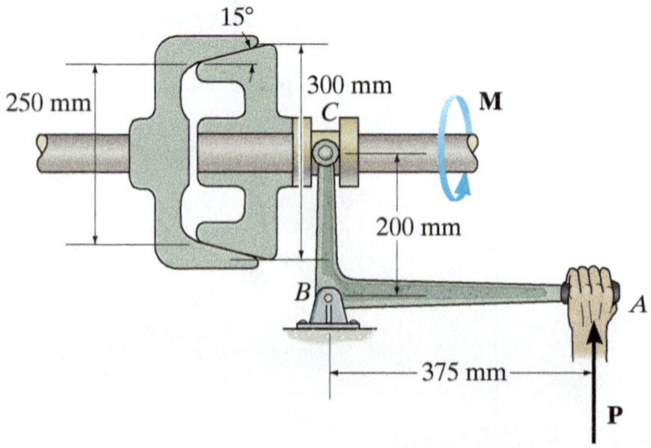

Prob. 8–135

***8–136.** The lawn roller has a mass of 80 kg. If the arm BA is held at an angle of 30° from the horizontal and the coefficient of rolling resistance for the roller is 25 mm, determine the force P needed to push the roller at constant speed. Neglect friction developed at the axle, A, and assume that the resultant force **P** acting on the handle is applied along arm BA.

Prob. 8–136

8–137. The three stone blocks have weights of $W_A = 600$ lb, $W_B = 150$ lb, and $W_C = 500$ lb. Determine the smallest horizontal force P that must be applied to block C in order to move this block. The coefficient of static friction between the blocks is $\mu_s = 0.3$, and between the floor and each block $\mu_s' = 0.5$.

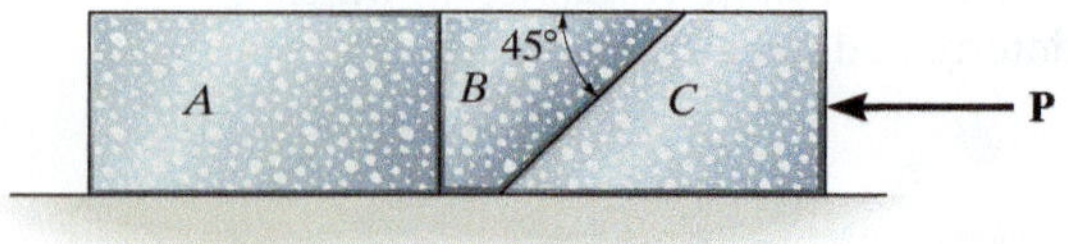

Prob. 8–137

8–138. The uniform 60-kg crate C rests uniformly on a 10-kg dolly D. If the front casters of the dolly at A are locked to prevent rolling while the casters at B are free to roll, determine the maximum force **P** that may be applied without causing motion of the crate. The coefficient of static friction between the casters and the floor is $\mu_f = 0.35$ and between the dolly and the crate, $\mu_d = 0.5$.

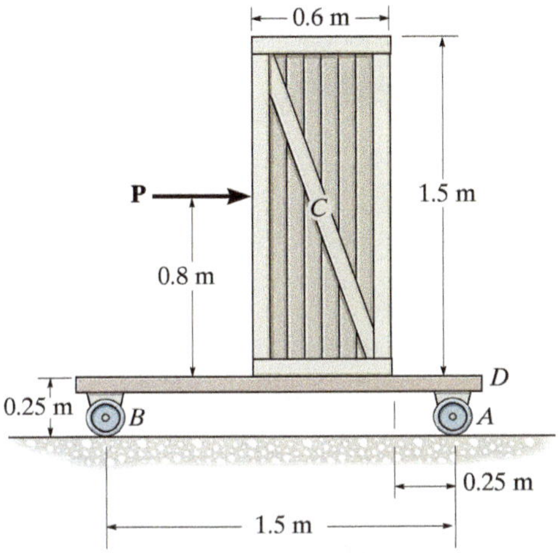

Prob. 8–138

8–139. The uniform 20-lb ladder rests on the rough floor for which the coefficient of static friction is $\mu_s = 0.8$ and against the smooth wall at B. Determine the horizontal force P the man must exert on the ladder in order to cause it to move.

***8–140.** The uniform 20-lb ladder rests on the rough floor for which the coefficient of static friction is $\mu_s = 0.4$ and against the smooth wall at B. Determine the horizontal force P the man must exert on the ladder in order to cause it to move.

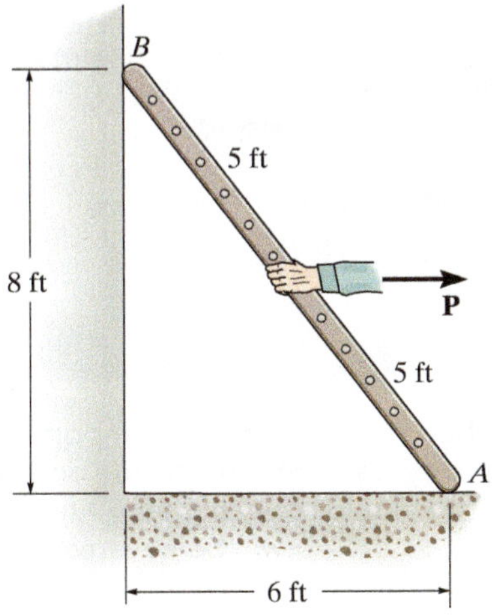

Probs. 8–139/140

8–141. The jacking mechanism consists of a link that has a square-threaded screw with a mean diameter of 0.5 in. and a lead of 0.20 in., and the coefficient of static friction is $\mu_s = 0.4$. Determine the torque M that should be applied to the screw to start lifting the 6000-lb load acting at the end of member ABC.

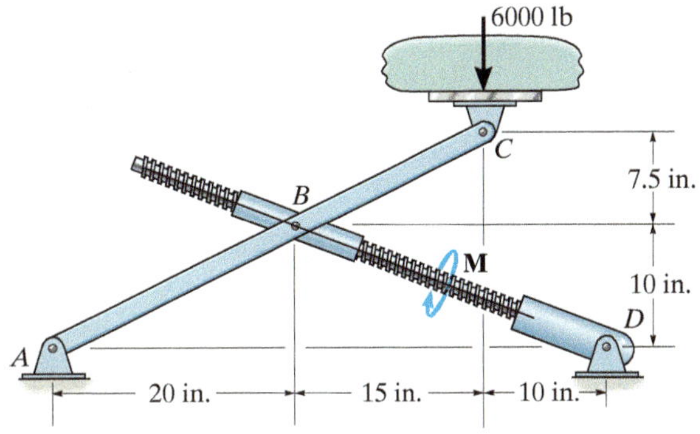

Prob. 8–141

8–142. Determine the minimum horizontal force P required to hold the crate from sliding down the plane. The crate has a mass of 50 kg and the coefficient of static friction between the crate and the plane is $\mu_s = 0.25$.

8–143. Determine the minimum force P required to push the crate up the plane. The crate has a mass of 50 kg and the coefficient of static friction between the crate and the plane is $\mu_s = 0.25$.

***8–144.** A horizontal force of $P = 100$ N is just sufficient to hold the crate from sliding down the plane, and a horizontal force of $P = 350$ N is required to just push the crate up the plane. Determine the coefficient of static friction between the plane and the crate, and find the mass of the crate.

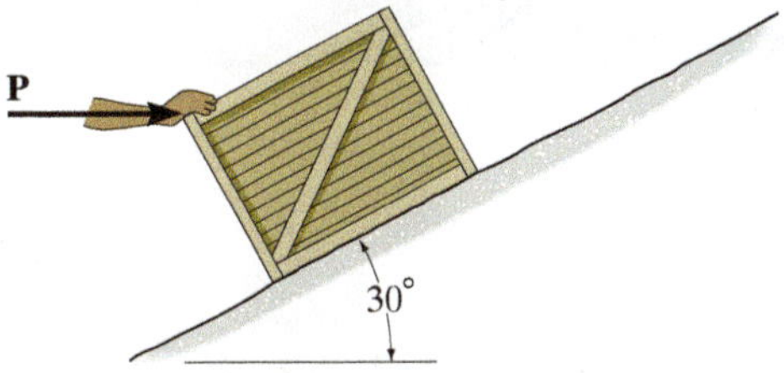

Probs. 8–142/143/144

8

Chapter 9

When a pressure tank of any shape is designed, it is important to be able to determine its center of gravity, calculate its volume and surface area, and determine the forces of the liquids they contain. All of these topics will be covered in this chapter.

Center of Gravity and Centroid

CHAPTER OBJECTIVES

- To discuss the concept of the center of gravity, center of mass, and the centroid.
- To show how to determine the location of the center of gravity and centroid for a system of discrete particles and a body of arbitrary shape.
- To use the theorems of Pappus and Guldinus for finding the surface area and volume for a body having axial symmetry.
- To present a method for finding the resultant of a general distributed loading and show how it applies to finding the resultant force of a pressure loading caused by a fluid.

9.1 Center of Gravity, Center of Mass, and the Centroid of a Body

In this section we will first show how to locate the center of gravity for a body, and then we will show that the center of mass and the centroid of a body can be developed using this same method.

Center of Gravity. A body is composed of an infinite number of particles of differential size, and so if the body is located within a gravitational field, then each of these particles will have a weight dW, Fig. 9–1*a*. These weights will form an approximately parallel force system, and the resultant of this system is the total weight of the body, which passes through a single point called the *center of gravity, G*, Fig. 9–1*b*.*

*This is true as long as the gravity field is assumed to have the same magnitude and direction everywhere. That assumption is appropriate for most engineering applications, since gravity does not vary appreciably between, for instance, the bottom and the top of a building.

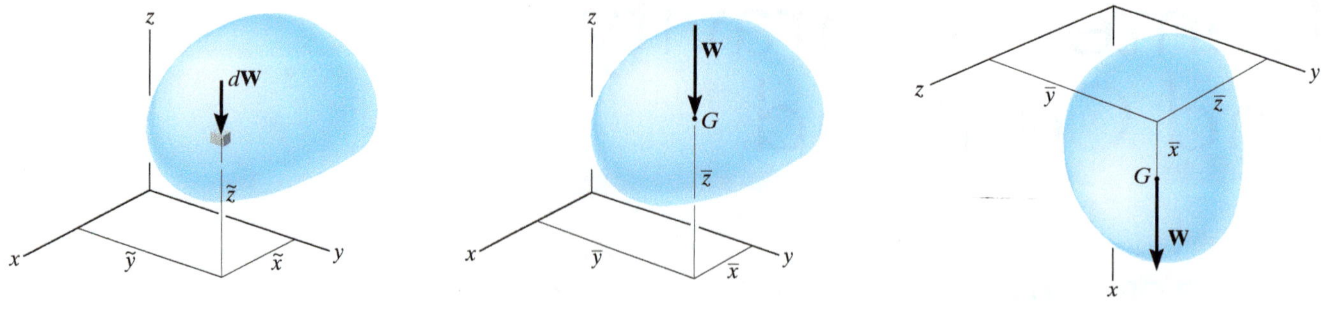

Fig. 9–1

Using the methods outlined in Sec. 4.8, the weight of the body is the sum of the weights of all of its particles, that is

$$+\downarrow F_R = \Sigma F_z; \qquad W = \int dW$$

The location of the center of gravity, measured from the y axis, is determined by equating the moment of W about the y axis, Fig. 9–1b, to the sum of the moments of the weights of the particles about this same axis. If dW is located at point $(\tilde{x}, \tilde{y}, \tilde{z})$, Fig. 9–1$a$, then

$$(M_R)_y = \Sigma M_y; \qquad \bar{x}W = \int \tilde{x}\,dW$$

Similarly, if moments are summed about the x axis,

$$(M_R)_x = \Sigma M_x; \qquad \bar{y}W = \int \tilde{y}\,dW$$

Finally, imagine that the body is fixed within the coordinate system and this system is rotated 90° about the y axis, Fig. 9–1c. Then the sum of the moments about the y axis gives

$$(M_R)_y = \Sigma M_y; \qquad \bar{z}W = \int \tilde{z}\,dW$$

Therefore, the location of the center of gravity G with respect to the x, y, z axes becomes

$$\bar{x} = \frac{\int \tilde{x}\,dW}{\int dW} \qquad \bar{y} = \frac{\int \tilde{y}\,dW}{\int dW} \qquad \bar{z} = \frac{\int \tilde{z}\,dW}{\int dW} \tag{9–1}$$

Here

$\bar{x}, \bar{y}, \bar{z}$ are the coordinates of the center of gravity G, Fig. 9–1b.
$\tilde{x}, \tilde{y}, \tilde{z}$ are the coordinates of each particle in the body, Fig. 9–1a.

9

Center of Mass of a Body. In order to study the *dynamic response* or accelerated motion of a body, it becomes important to locate the body's center of mass C_m, Fig. 9–2. This location can be determined by substituting $dW = g\,dm$ into Eqs. 9–1. Since g is constant, it cancels out, and so

$$\bar{x} = \frac{\int \tilde{x}\,dm}{\int dm} \qquad \bar{y} = \frac{\int \tilde{y}\,dm}{\int dm} \qquad \bar{z} = \frac{\int \tilde{z}\,dm}{\int dm} \tag{9–2}$$

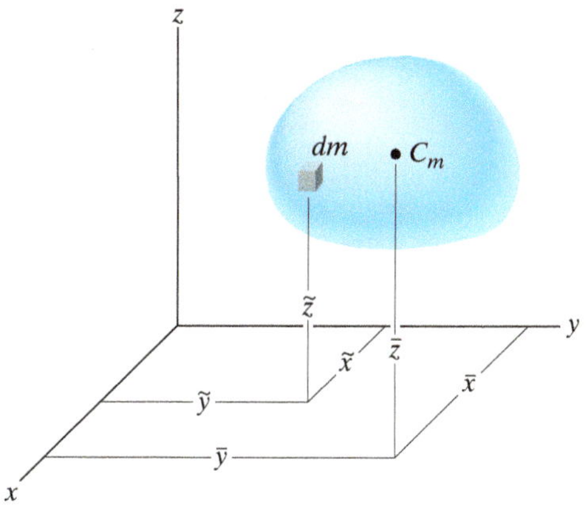

Fig. 9–2

Centroid of a Volume. If the body in Fig. 9–3 is made from a homogeneous material, then its density ρ (rho) will be constant. Therefore, a differential element of volume dV has a mass $dm = \rho\,dV$. Substituting this into Eqs. 9–2 and canceling out ρ, we obtain formulas that locate the *centroid* C or geometric center of the body; namely

$$\bar{x} = \frac{\int_V \tilde{x}\,dV}{\int_V dV} \qquad \bar{y} = \frac{\int_V \tilde{y}\,dV}{\int_V dV} \qquad \bar{z} = \frac{\int_V \tilde{z}\,dV}{\int_V dV} \tag{9–3}$$

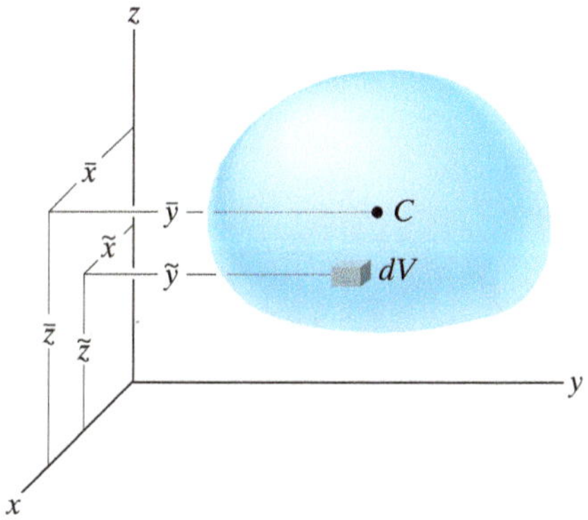

Fig. 9–3

These equations represent a balance of the moments of the volume of the body. Therefore, if the volume possesses two planes of symmetry, then its centroid must lie along the line of intersection of these two planes. For example, the cone in Fig. 9–4 has a centroid that lies on the y axis so that $\bar{x} = \bar{z} = 0$. The location $\bar{y}$ can be found using a single integration by choosing a differential element represented by a *thin disk* having a thickness dy and radius $r = z$. Its volume is $dV = \pi r^2\,dy = \pi z^2\,dy$ and its centroid is at $\tilde{x} = 0$, $\tilde{y} = y$, $\tilde{z} = 0$.

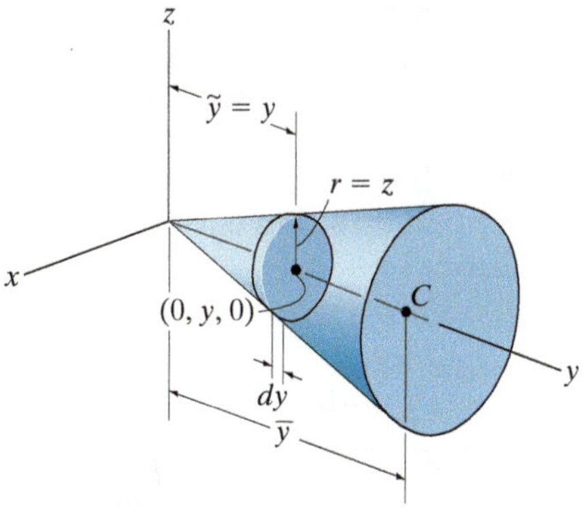

Fig. 9–4

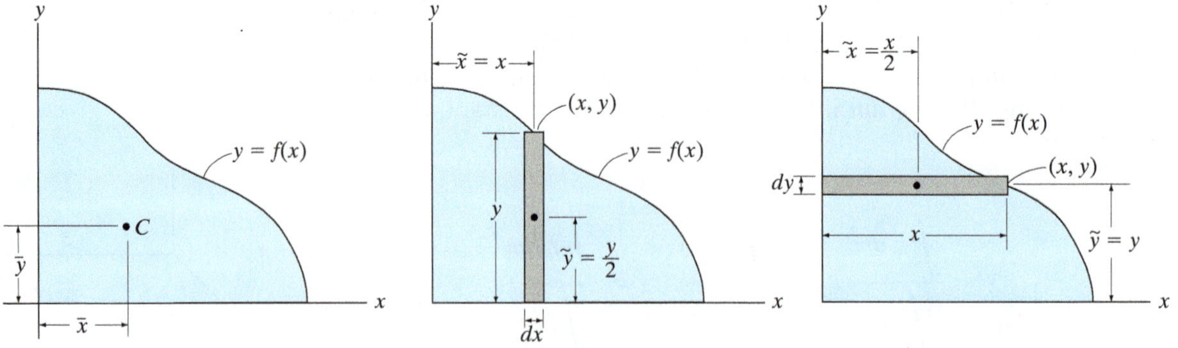

Fig. 9–5

Integration must be used to determine the location of the center of gravity of this goal post due to the curvature of the supporting member.

Centroid of an Area. If an area lies in the x–y plane and is bounded by the curve $y = f(x)$, as shown in Fig. 9–5*a*, then its centroid will be in this plane and can be determined from integrals similar to Eqs. 9–3, namely,

$$\bar{x} = \frac{\int_A \tilde{x}\, dA}{\int_A dA} \qquad \bar{y} = \frac{\int_A \tilde{y}\, dA}{\int_A dA} \tag{9–4}$$

These integrals can be evaluated by performing a *single integration* if we use a *rectangular strip* for the differential area element. For example, if a vertical strip is used, Fig. 9–5*b*, the area of the element is $dA = y\,dx$, and its centroid is located at $\tilde{x} = x$ and $\tilde{y} = y/2$. If we consider a horizontal strip, Fig. 9–5*c*, then $dA = x\,dy$, and its centroid is located at $\tilde{x} = x/2$ and $\tilde{y} = y$.

Centroid of a Line. If a line segment (or rod) lies within the x–y plane and it can be described by a thin curve $y = f(x)$, Fig. 9–6*a*, then its centroid is determined from

$$\bar{x} = \frac{\int_L \tilde{x}\, dL}{\int_L dL} \qquad \bar{y} = \frac{\int_L \tilde{y}\, dL}{\int_L dL} \tag{9–5}$$

Here, the length of the differential element is given by the Pythagorean theorem, $dL = \sqrt{(dx)^2 + (dy)^2}$, which can also be written in the form

$$dL = \sqrt{\left(\frac{dx}{dx}\right)^2 dx^2 + \left(\frac{dy}{dx}\right)^2 dx^2}$$

$$= \left(\sqrt{1 + \left(\frac{dy}{dx}\right)^2}\right) dx$$

or

$$dL = \sqrt{\left(\frac{dx}{dy}\right)^2 dy^2 + \left(\frac{dy}{dy}\right)^2 dy^2}$$

$$= \left(\sqrt{\left(\frac{dx}{dy}\right)^2 + 1}\right) dy$$

Either one of these expressions can be used; however, for application, the one that will result in a simpler integration should be selected. For example, consider the rod in Fig. 9–6*b*, defined by $y = 2x^2$. The length of the element is $dL = \sqrt{1 + (dy/dx)^2}\, dx$, and since $dy/dx = 4x$, then $dL = \sqrt{1 + (4x)^2}\, dx$. The centroid for this element is located at $\tilde{x} = x$ and $\tilde{y} = y$.

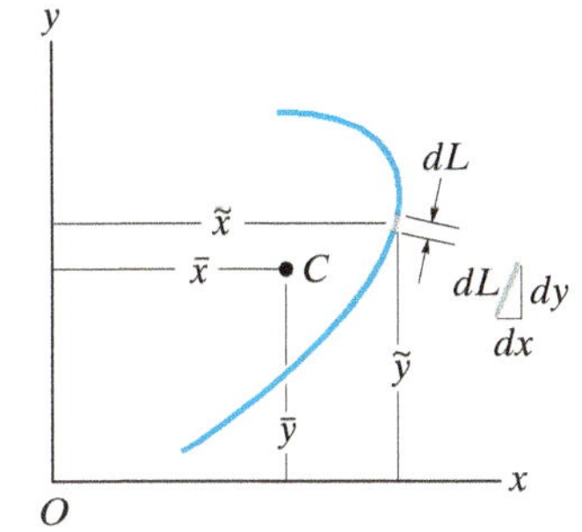

(a)

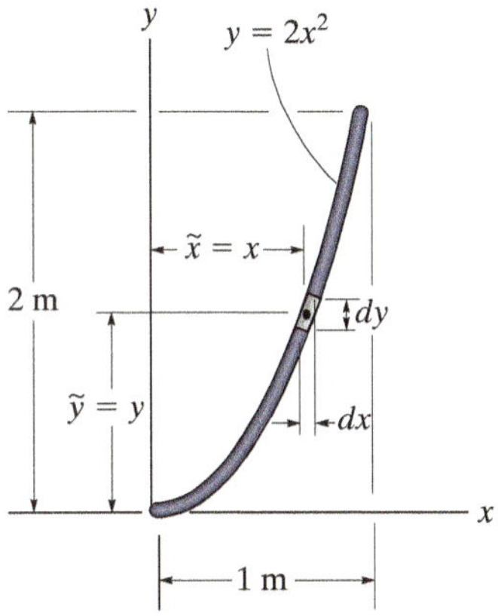

(b)

Fig. 9–6

Important Points

- The centroid represents the geometric center of a body. This point coincides with the center of mass or the center of gravity only if the material composing the body is uniform or homogeneous.
- Formulas used to locate the center of gravity or the centroid simply represent a balance between the sum of moments of all the parts of the system and the moment of the "resultant" for the system.
- In some cases the centroid is located at a point that is not on the object, as in the case of a ring, where the centroid is at its center. Also, this point will lie on any axis of symmetry for the body, Fig. 9–7.

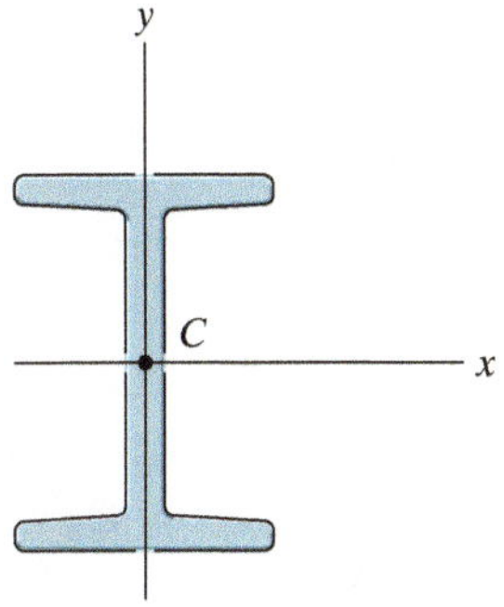

Fig. 9–7

9

Procedure for Analysis

The center of gravity or centroid of an object or shape can be determined by single integrations using the following procedure.

Differential Element.

- Select an appropriate coordinate system, specify the coordinate axes, and then choose a differential element for integration.
- For lines the element is represented by a differential line segment of length dL.
- For areas the element is generally a rectangle of area dA, having a finite length and differential width.
- For volumes the element can be a circular disk of volume dV, having a finite radius and differential thickness.
- Locate the element so that it touches the arbitrary point (x, y, z) on the curve that defines the boundary of the shape.

Size and Moment Arms.

- Express the length dL, area dA, or volume dV of the element in terms of the coordinates describing the curve.
- Express the moment arms $\tilde{x}, \tilde{y}, \tilde{z}$ for the centroid or center of gravity of the element in terms of the coordinates describing the curve.

Integrations.

- Substitute the formulations for $\tilde{x}, \tilde{y}, \tilde{z}$ and dL, dA, or dV into the appropriate equations (Eqs. 9–1 through 9–5).
- Express the function in the integrand in terms of the *same variable as the differential thickness of the element.*
- The limits of the integral are defined from the two extreme locations of the element's differential thickness, so that when the elements are "summed" or the integration performed, the entire region is covered.

EXAMPLE 9.1

Locate the centroid of the rod bent into the shape of a parabolic arc as shown in Fig. 9–8.

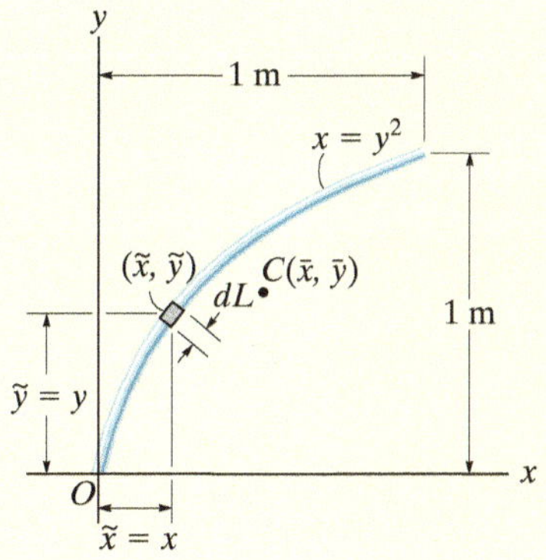

Fig. 9–8

SOLUTION

Differential Element. The differential element is shown in Fig. 9–8. It is located on the curve at the *arbitrary point* (x, y).

Area and Moment Arms. The differential element of length dL can be expressed in terms of the differentials dx and dy using the Pythagorean theorem.

$$dL = \sqrt{(dx)^2 + (dy)^2} = \sqrt{\left(\frac{dx}{dy}\right)^2 + 1}\, dy$$

Since $x = y^2$, then $dx/dy = 2y$. Therefore, expressing dL in terms of y and dy, we have

$$dL = \sqrt{(2y)^2 + 1}\, dy$$

As shown in Fig. 9–8, the centroid of the element is located at $\tilde{x} = x$, $\tilde{y} = y$.

Integrations. Applying Eq. 9–5 and using the integration formula to evaluate the integrals, we get

$$\bar{x} = \frac{\int_L \tilde{x}\, dL}{\int_L dL} = \frac{\int_0^{1\text{ m}} x\sqrt{4y^2 + 1}\, dy}{\int_0^{1\text{ m}} \sqrt{4y^2 + 1}\, dy} = \frac{\int_0^{1\text{ m}} y^2\sqrt{4y^2 + 1}\, dy}{\int_0^{1\text{ m}} \sqrt{4y^2 + 1}\, dy}$$

$$= \frac{0.6063}{1.479} = 0.410 \text{ m} \qquad \textit{Ans.}$$

$$\bar{y} = \frac{\int_L \tilde{y}\, dL}{\int_L dL} = \frac{\int_0^{1\text{ m}} y\sqrt{4y^2 + 1}\, dy}{\int_0^{1\text{ m}} \sqrt{4y^2 + 1}\, dy} = \frac{0.8484}{1.479} = 0.574 \text{ m} \qquad \textit{Ans.}$$

NOTE: These results for C seem reasonable when they are plotted on Fig. 9–8.

9

EXAMPLE 9.2

Locate the centroid of the circular wire segment shown in Fig. 9–9.

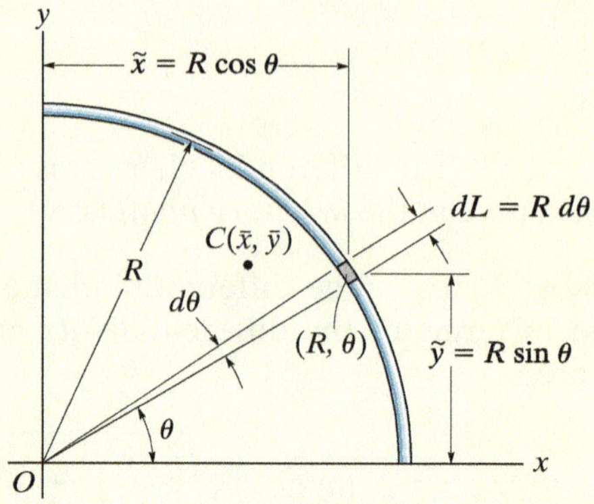

Fig. 9–9

SOLUTION

Polar coordinates will be used to solve this problem since the arc is circular.

Differential Element. A differential circular arc is selected as shown in the figure. This element lies on the curve at (R, θ).

Length and Moment Arm. The length of the differential element is $dL = R\,d\theta$, and its centroid is located at $\tilde{x} = R\cos\theta$ and $\tilde{y} = R\sin\theta$.

Integrations. Applying Eqs. 9–5 and integrating with respect to θ, we obtain

$$\bar{x} = \frac{\int_L \tilde{x}\,dL}{\int_L dL} = \frac{\int_0^{\pi/2} (R\cos\theta)R\,d\theta}{\int_0^{\pi/2} R\,d\theta} = \frac{R^2\int_0^{\pi/2} \cos\theta\,d\theta}{R\int_0^{\pi/2} d\theta} = \frac{2R}{\pi} \qquad \textit{Ans.}$$

$$\bar{y} = \frac{\int_L \tilde{y}\,dL}{\int_L dL} = \frac{\int_0^{\pi/2} (R\sin\theta)R\,d\theta}{\int_0^{\pi/2} R\,d\theta} = \frac{R^2\int_0^{\pi/2} \sin\theta\,d\theta}{R\int_0^{\pi/2} d\theta} = \frac{2R}{\pi} \qquad \textit{Ans.}$$

NOTE: As expected, the two coordinates are numerically the same due to the symmetry of the wire.

9

EXAMPLE 9.3

Determine the distance $\bar{y}$ measured from the x axis to the centroid of the area of the trianglc shown in Fig. 9–10.

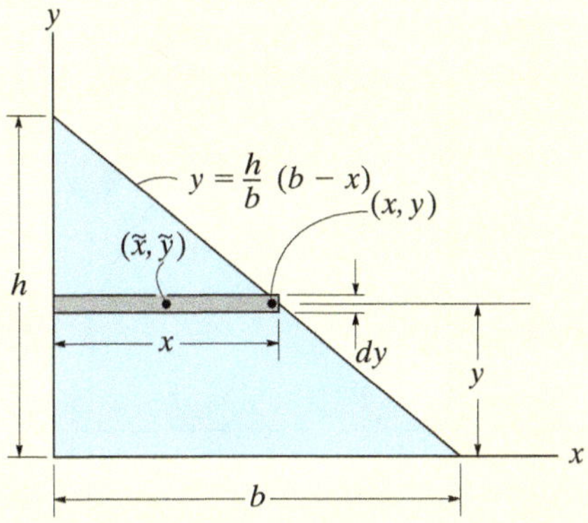

Fig. 9–10

SOLUTION

Differential Element. Consider a rectangular element having a thickness dy, and located in an arbitrary position so that it intersects the boundary at (x, y), Fig. 9–10.

Area and Moment Arms. The area of the element is $dA = x\,dy = \frac{b}{h}(h - y)\,dy$, and its centroid is located a distance $\tilde{y} = y$ from the x axis.

Integration. Applying the second of Eqs. 9–4 and integrating with respect to y yields

$$\bar{y} = \frac{\int_A \tilde{y}\,dA}{\int_A dA} = \frac{\int_0^h y\left[\frac{b}{h}(h - y)\,dy\right]}{\int_0^h \frac{b}{h}(h - y)\,dy} = \frac{\frac{1}{6}bh^2}{\frac{1}{2}bh}$$

$$= \frac{h}{3} \qquad \textit{Ans.}$$

NOTE: This result is valid for any shape of triangle. It states that the centroid is located at one-third the height, measured from the base of the triangle.

9

EXAMPLE 9.4

Locate the centroid for the area of a quarter circle shown in Fig. 9–11.

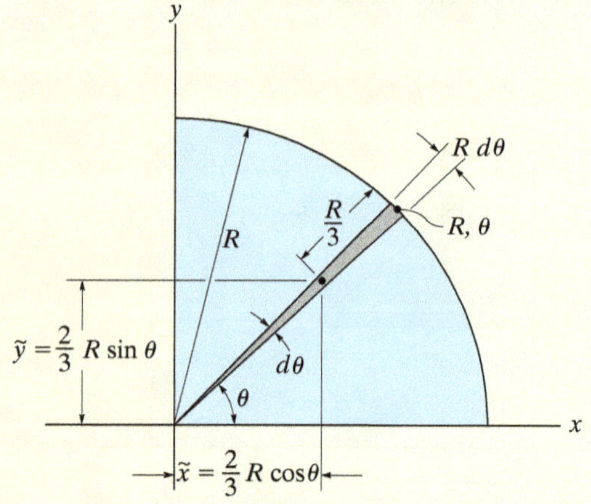

Fig. 9–11

SOLUTION

Differential Element. Polar coordinates will be used, since the boundary is circular. We choose the element in the shape of a *triangle*, Fig. 9–11. (Actually the shape is a circular sector; however, neglecting higher-order differentials, the element becomes triangular.) The element intersects the curve at point (R, θ).

Area and Moment Arms. The area of the element is

$$dA = \tfrac{1}{2}(R)(R\,d\theta) = \frac{R^2}{2}d\theta$$

and using the results of Example 9.3, the centroid of the (triangular) element is located at $\tilde{x} = \frac{2}{3}R\cos\theta$, $\tilde{y} = \frac{2}{3}R\sin\theta$.

Integrations. Applying Eqs. 9–4 and integrating with respect to θ, we obtain

$$\bar{x} = \frac{\int_A \tilde{x}\,dA}{\int_A dA} = \frac{\int_0^{\pi/2}\left(\frac{2}{3}R\cos\theta\right)\frac{R^2}{2}d\theta}{\int_0^{\pi/2}\frac{R^2}{2}d\theta} = \frac{\left(\frac{2}{3}R\right)\int_0^{\pi/2}\cos\theta\,d\theta}{\int_0^{\pi/2}d\theta} = \frac{4R}{3\pi} \qquad \textit{Ans.}$$

$$\bar{y} = \frac{\int_A \tilde{y}\,dA}{\int_A dA} = \frac{\int_0^{\pi/2}\left(\frac{2}{3}R\sin\theta\right)\frac{R^2}{2}d\theta}{\int_0^{\pi/2}\frac{R^2}{2}d\theta} = \frac{\left(\frac{2}{3}R\right)\int_0^{\pi/2}\sin\theta\,d\theta}{\int_0^{\pi/2}d\theta} = \frac{4R}{3\pi} \qquad \textit{Ans.}$$

9

EXAMPLE 9.5

Locate the centroid of the area shown in Fig. 9–12*a*.

SOLUTION I

Differential Element. A differential element of thickness dx is shown in Fig. 9–12*a*. The element intersects the curve at the *arbitrary point* (x, y), and so it has a height y.

Area and Moment Arms. The area of the element is $dA = y\,dx$, and its centroid is located at $\tilde{x} = x$, $\tilde{y} = y/2$.

Integrations. Applying Eqs. 9–4 and integrating with respect to x yields

$$\bar{x} = \frac{\int_A \tilde{x}\,dA}{\int_A dA} = \frac{\int_0^{1\text{ m}} xy\,dx}{\int_0^{1\text{ m}} y\,dx} = \frac{\int_0^{1\text{ m}} x^3\,dx}{\int_0^{1\text{ m}} x^2\,dx} = \frac{0.250}{0.333} = 0.75\text{ m} \qquad \textit{Ans.}$$

$$\bar{y} = \frac{\int_A \tilde{y}\,dA}{\int_A dA} = \frac{\int_0^{1\text{ m}} (y/2)y\,dx}{\int_0^{1\text{ m}} y\,dx} = \frac{\int_0^{1\text{ m}} (x^2/2)x^2\,dx}{\int_0^{1\text{ m}} x^2\,dx} = \frac{0.100}{0.333} = 0.3\text{ m} \qquad \textit{Ans.}$$

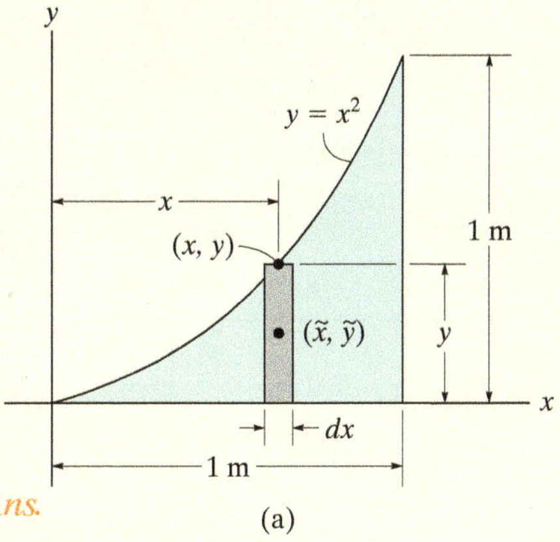

SOLUTION II

Differential Element. The differential element of thickness dy is shown in Fig. 9–12*b*. The element intersects the curve at the *arbitrary point* (x, y), and so it has a length $(1 - x)$.

Area and Moment Arms. The area of the element is $dA = (1 - x)\,dy$, and its centroid is located at

$$\tilde{x} = x + \left(\frac{1 - x}{2}\right) = \frac{1 + x}{2}, \quad \tilde{y} = y$$

Integrations. Applying Eqs. 9–4 and integrating with respect to y, we obtain

$$\bar{x} = \frac{\int_A \tilde{x}\,dA}{\int_A dA} = \frac{\int_0^{1\text{ m}} [(1 + x)/2](1 - x)\,dy}{\int_0^{1\text{ m}} (1 - x)\,dy} = \frac{\frac{1}{2}\int_0^{1\text{ m}} (1 - y)\,dy}{\int_0^{1\text{ m}} (1 - \sqrt{y})\,dy} = \frac{0.250}{0.333} = 0.75\text{ m} \qquad \textit{Ans.}$$

$$\bar{y} = \frac{\int_A \tilde{y}\,dA}{\int_A dA} = \frac{\int_0^{1\text{ m}} y(1 - x)\,dy}{\int_0^{1\text{ m}} (1 - x)\,dy} = \frac{\int_0^{1\text{ m}} (y - y^{3/2})\,dy}{\int_0^{1\text{ m}} (1 - \sqrt{y})\,dy} = \frac{0.100}{0.333} = 0.3\text{ m} \qquad \textit{Ans.}$$

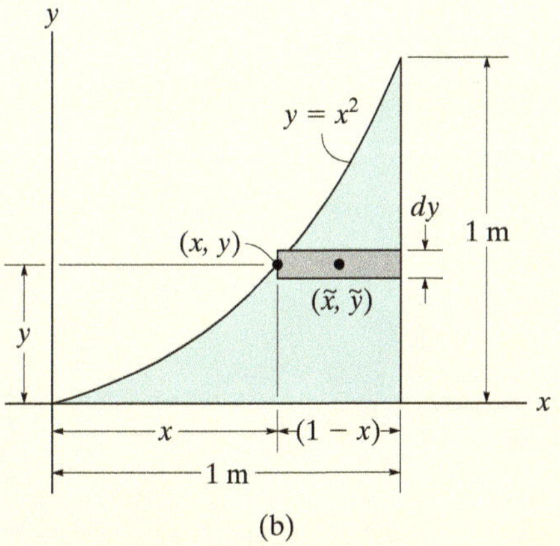

Fig. 9–12

NOTE: Plot these results and notice that they seem reasonable. Also, for this problem, elements of thickness dx offer a simpler solution.

9

EXAMPLE 9.6

Locate the centroid of the semi-elliptical area shown in Fig. 9–13*a*.

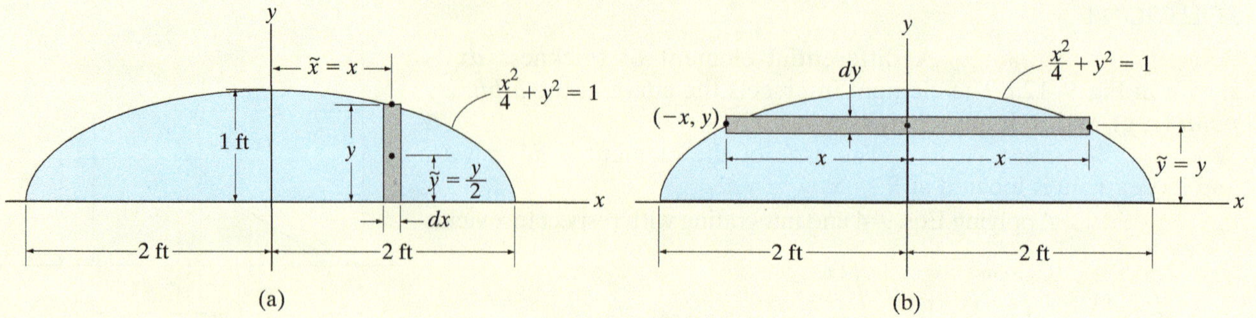

Fig. 9–13

SOLUTION I

Differential Element. The rectangular differential element parallel to the y axis shown shaded in Fig. 9–13*a* will be considered. This element has a thickness of dx and a height of y.

Area and Moment Arms. Thus, the area is $dA = y\,dx$, and its centroid is located at $\tilde{x} = x$ and $\tilde{y} = y/2$.

Integration. Since the area is symmetrical about the y axis,

$$\bar{x} = 0 \qquad \textit{Ans.}$$

Applying the second of Eqs. 9–4 with $y = \sqrt{1 - \frac{x^2}{4}}$, we have

$$\bar{y} = \frac{\int_A \tilde{y}\,dA}{\int_A dA} = \frac{\int_{-2\text{ ft}}^{2\text{ ft}} \frac{y}{2}(y\,dx)}{\int_{-2\text{ ft}}^{2\text{ ft}} y\,dx} = \frac{\frac{1}{2}\int_{-2\text{ ft}}^{2\text{ ft}} \left(1 - \frac{x^2}{4}\right)dx}{\int_{-2\text{ ft}}^{2\text{ ft}} \sqrt{1 - \frac{x^2}{4}}\,dx} = \frac{4/3}{\pi} = 0.424\text{ ft} \qquad \textit{Ans.}$$

SOLUTION II

Differential Element. The shaded rectangular differential element of thickness dy and width $2x$, parallel to the x axis, will be considered, Fig. 9–13*b*.

Area and Moment Arms. The area is $dA = 2x\,dy$, and its centroid is at $\tilde{x} = 0$ and $\tilde{y} = y$.

Integration. Applying the second of Eqs. 9–4, with $x = 2\sqrt{1 - y^2}$, we have

$$\bar{y} = \frac{\int_A \tilde{y}\,dA}{\int_A dA} = \frac{\int_0^{1\text{ ft}} y(2x\,dy)}{\int_0^{1\text{ ft}} 2x\,dy} = \frac{\int_0^{1\text{ ft}} 4y\sqrt{1 - y^2}\,dy}{\int_0^{1\text{ ft}} 4\sqrt{1 - y^2}\,dy} = \frac{4/3}{\pi}\text{ ft} = 0.424\text{ ft} \qquad \textit{Ans.}$$

9

EXAMPLE 9.7

Locate the $\bar{y}$ centroid for the paraboloid of revolution, shown in Fig. 9–14.

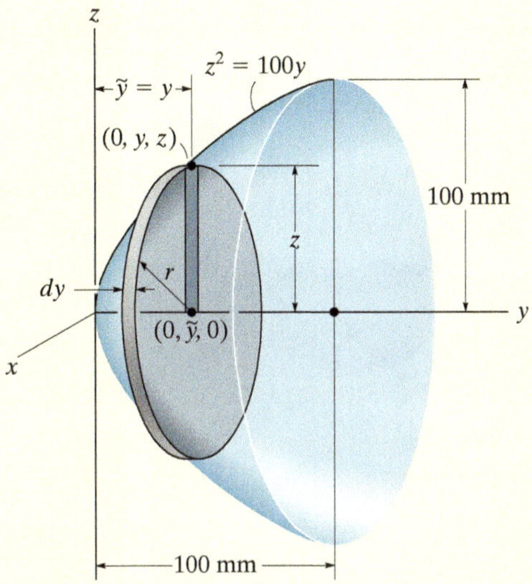

Fig. 9–14

SOLUTION

Differential Element. An element having the shape of a *thin disk* is chosen. This element has a thickness dy, it intersects the generating curve at the *arbitrary point* $(0, y, z)$, and so its radius is $r = z$.

Volume and Moment Arm. The volume of the element is $dV = (\pi z^2)\,dy$, and its centroid is located at $\tilde{y} = y$.

Integration. Applying the second of Eqs. 9–3 and integrating with respect to y yields.

$$\bar{y} = \frac{\int_V \tilde{y}\,dV}{\int_V dV} = \frac{\int_0^{100\text{ mm}} y(\pi z^2)\,dy}{\int_0^{100\text{ mm}} (\pi z^2)\,dy} = \frac{100\pi \int_0^{100\text{ mm}} y^2\,dy}{100\pi \int_0^{100\text{ mm}} y\,dy} = 66.7\text{ mm} \qquad \textit{Ans.}$$

9

EXAMPLE 9.8

Determine the location of the center of mass of the cylinder shown in Fig. 9–15 if its density varies directly with the distance from its base, i.e., $\rho = 200z\ \text{kg/m}^3$.

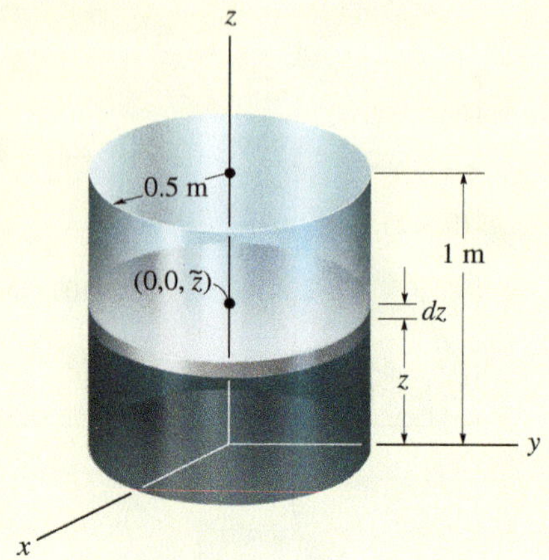

Fig. 9–15

SOLUTION

For reasons of material symmetry,

$$\bar{x} = \bar{y} = 0 \qquad \textit{Ans.}$$

Differential Element. A disk element of radius 0.5 m and thickness dz is chosen for integration, Fig. 9–15, since the *density of the entire element is constant* for a given value of z. The element is located along the z axis at the *arbitrary point* $(0, 0, z)$.

Volume and Moment Arm. The volume of the element is $dV = \pi(0.5)^2\,dz$, and its centroid is located at $\tilde{z} = z$.

Integrations. Using the third of Eqs. 9–2 with $dm = \rho\,dV$ and integrating with respect to z, noting that $\rho = 200z$, we have

$$\bar{z} = \frac{\int_V \tilde{z}\rho\,dV}{\int_V \rho\,dV} = \frac{\int_0^{1\text{ m}} z(200z)\left[\pi(0.5)^2\,dz\right]}{\int_0^{1\text{ m}} (200z)\pi(0.5)^2\,dz}$$

$$= \frac{\int_0^{1\text{ m}} z^2\,dz}{\int_0^{1\text{ m}} z\,dz} = 0.667\text{ m} \qquad \textit{Ans.}$$

9

FUNDAMENTAL PROBLEMS

F9–1. Determine the centroid $(\bar{x}, \bar{y})$ of the shaded area.

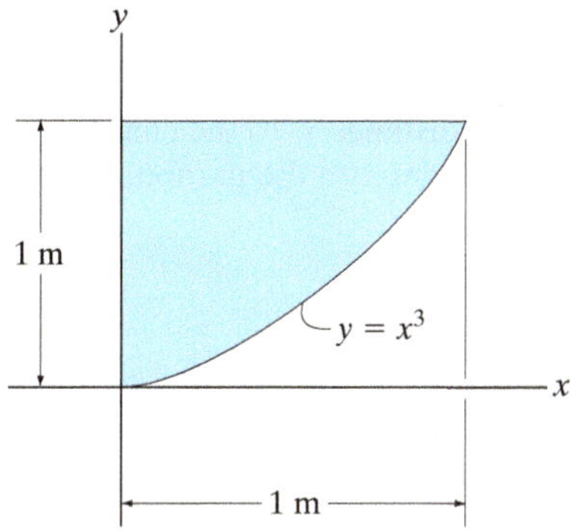

F9–1

F9–2. Determine the centroid $(\bar{x}, \bar{y})$ of the shaded area.

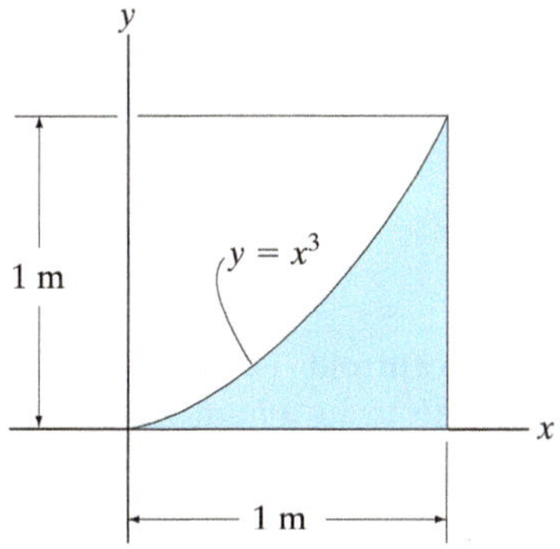

F9–2

F9–3. Determine the centroid $\bar{y}$ of the shaded area.

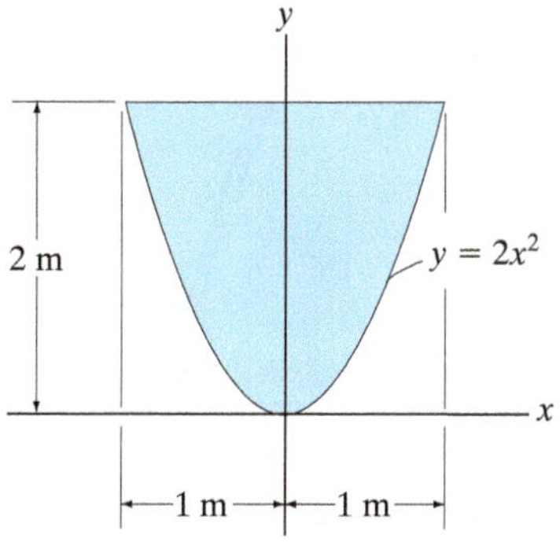

F9–3

F9–4. Locate the center mass $\bar{x}$ of the straight rod if its mass per unit length is given by $m = m_0(1 + x^2/L^2)$.

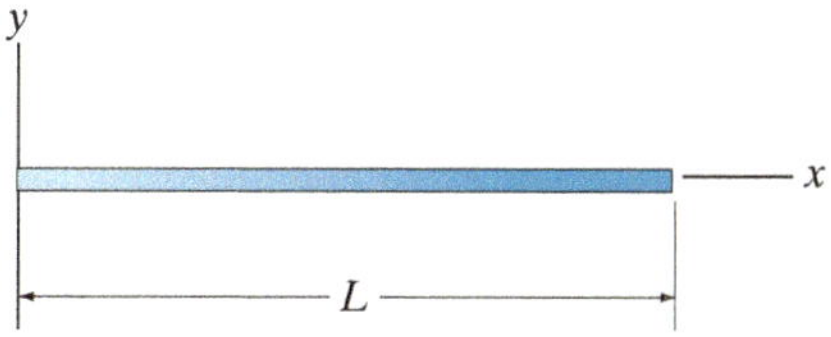

F9–4

F9–5. Locate the centroid $\bar{y}$ of the homogeneous solid formed by revolving the shaded area about the y axis.

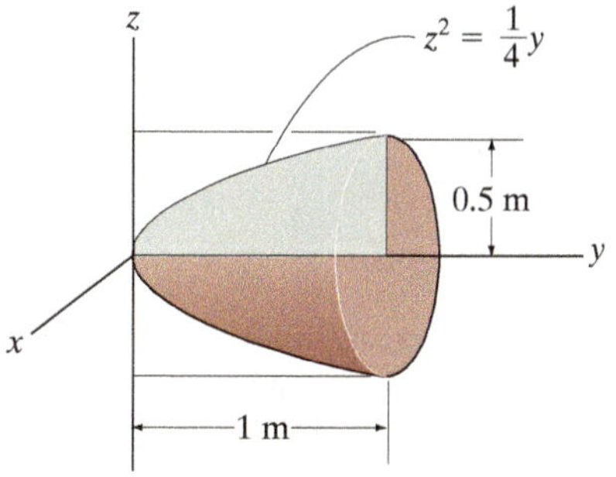

F9–5

F9–6. Locate the centroid $\bar{z}$ of the homogeneous solid formed by revolving the shaded area about the z axis.

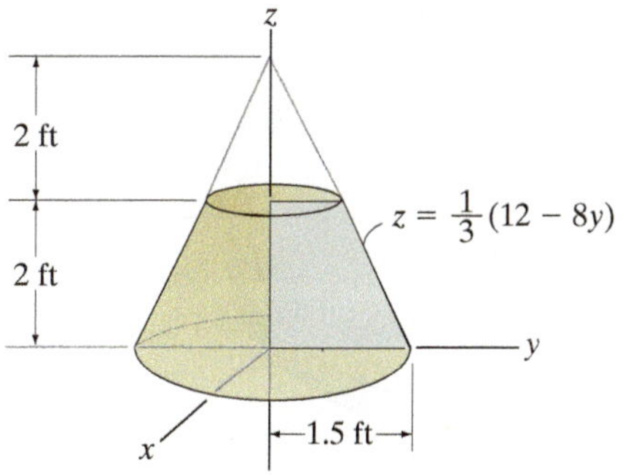

F9–6

PROBLEMS

9–1. Locate the center of mass of the homogeneous rod bent into the shape of a circular arc.

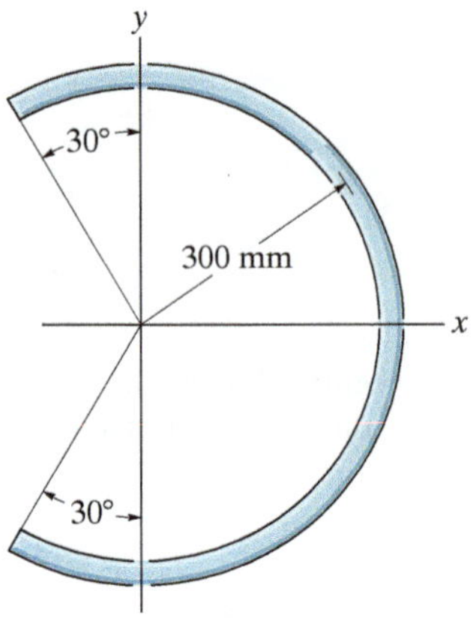

Prob. 9–1

9–2. Locate the center of gravity $\bar{x}$ of the homogeneous rod bent in the form of a semicircular arc. The rod has a weight per unit length of 0.5 lb/ft. Also, determine the horizontal reaction at the smooth support B and the x and y components of reaction at the pin A.

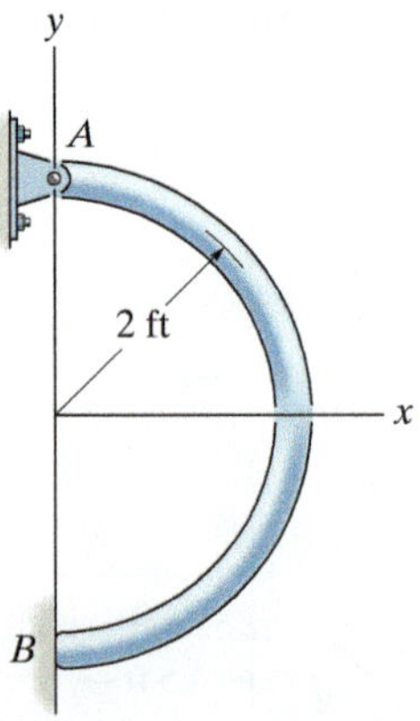

Prob. 9–2

9–3. Locate the distance $\bar{x}$ to the center of gravity of the homogeneous rod bent into the parabolic shape. If the rod has a weight per unit length of 0.5 lb/ft, determine the reactions at the fixed support O.

***9–4.** Locate the distance $\bar{y}$ to the center of gravity of the homogeneous rod bent into the parabolic shape.

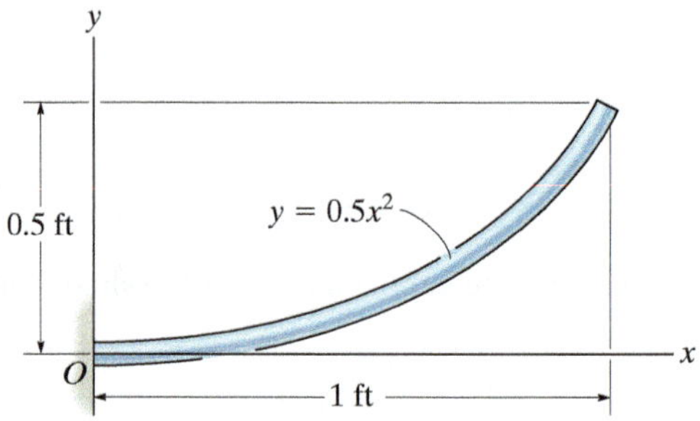

Probs. 9–3/4

9–5. Locate the centroid $(\bar{x}, \bar{y})$ of the uniform rod. Evaluate the integrals using a numerical method.

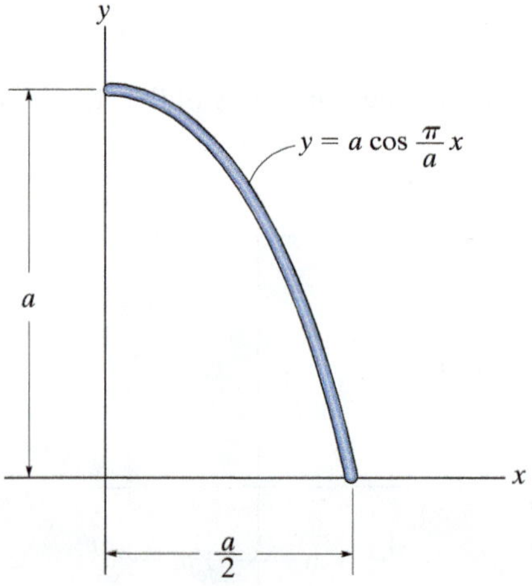

Prob. 9–5

9–6. Locate the centroid $\bar{y}$ of the area.

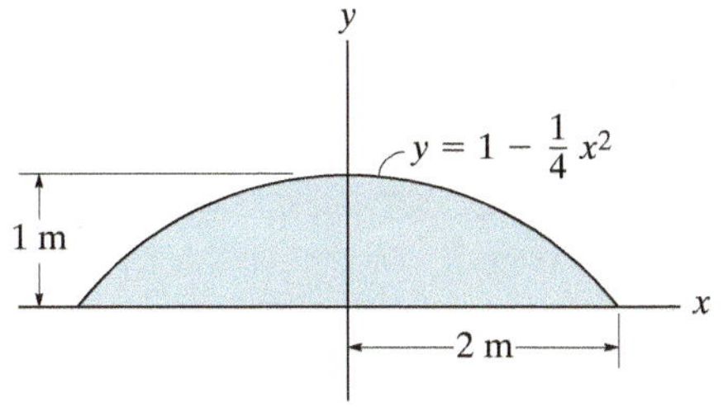

Prob. 9–6

9–7. Locate the centroid $\bar{x}$ of the parabolic area.

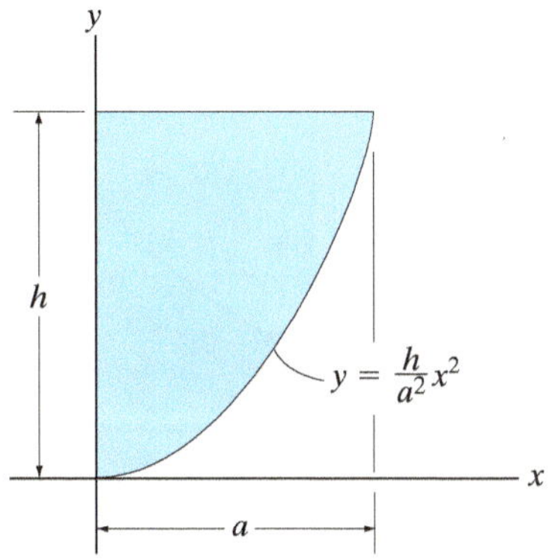

Prob. 9–7

***9–8.** Locate the centroid $\bar{y}$ of the parabolic area.

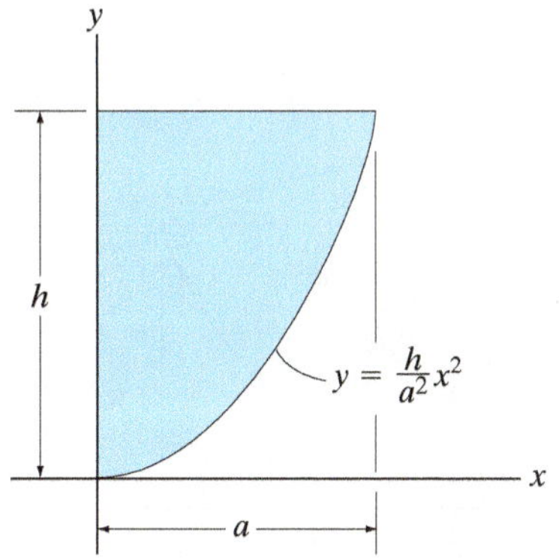

Prob. 9–8

9–9. Locate the centroid $\bar{x}$ of the area.

9–10. Locate the centroid $\bar{y}$ of the area.

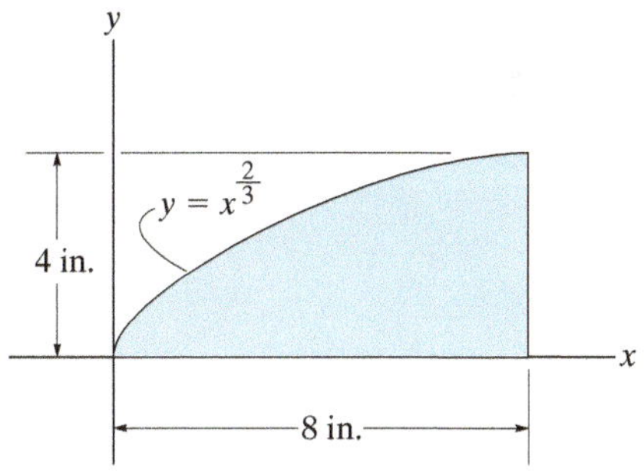

Probs. 9–9/10

9–11. Locate the centroid $\bar{x}$ of the area.

***9–12.** Locate the centroid $\bar{y}$ of the area.

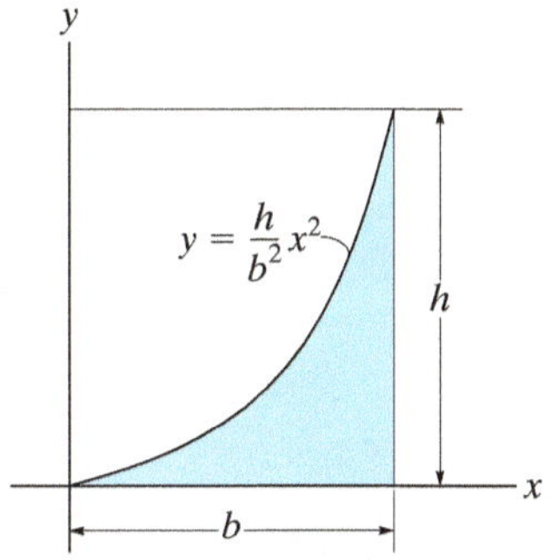

Probs. 9–11/12

9–13. Locate the centroid $\bar{x}$ of the area.

9–14. Locate the centroid $\bar{y}$ of the area.

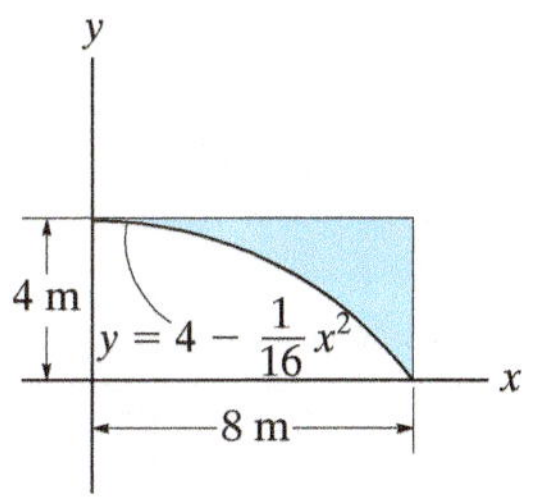

Probs. 9–13/14

9

9–15. Locate the centroid $\bar{x}$ of the area.

***9–16.** Locate the centroid $\bar{y}$ of the area.

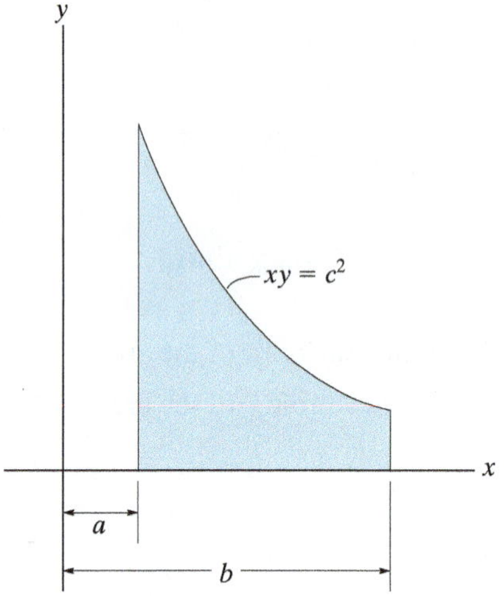

Probs. 9–15/16

9–17. Locate the centroid $\bar{x}$ of the shaded area.

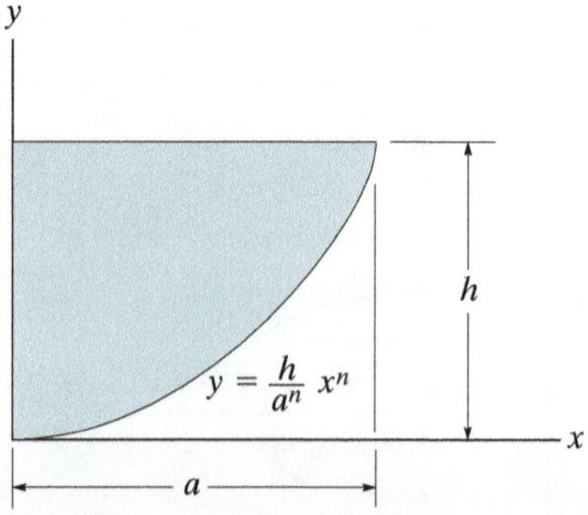

Prob. 9–17

9–18. Locate the centroid $\bar{x}$ of the area.

9–19. Locate the centroid $\bar{y}$ of the area.

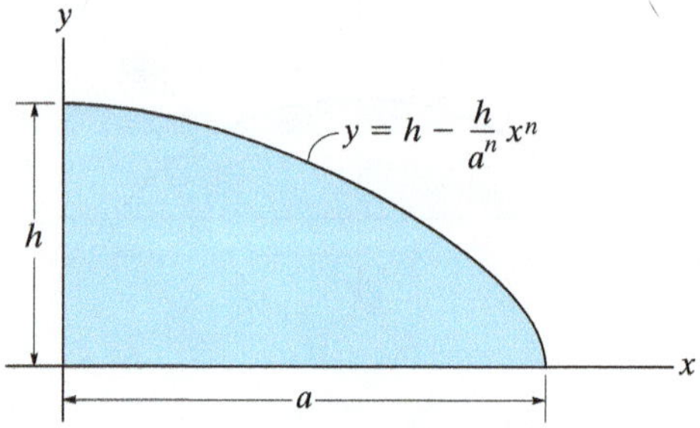

Probs. 9–18/19

***9–20.** Locate the centroid $\bar{y}$ of the shaded area.

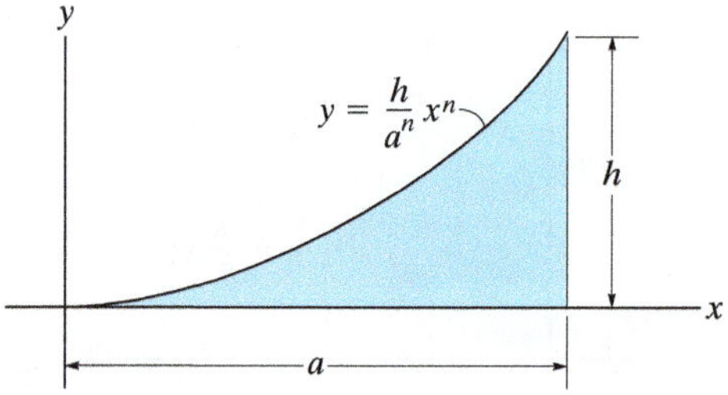

Prob. 9–20

9–21. Locate the centroid $\bar{x}$ of the area.

9–22. Locate the centroid $\bar{y}$ of the area.

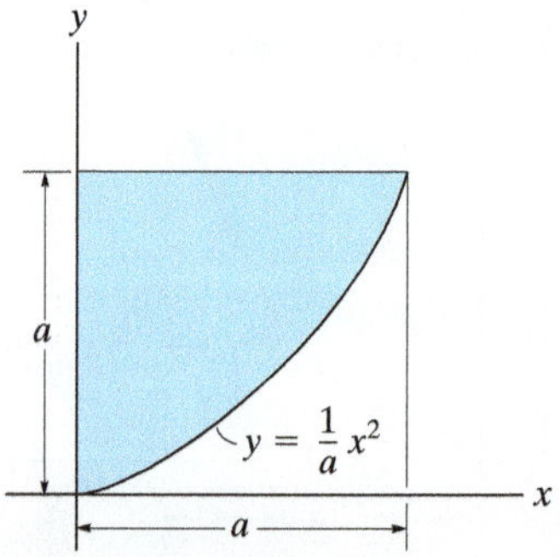

Probs. 9–21/22

9–23. Locate the centroid $\bar{x}$ of the quarter elliptical area.

***9–24.** Locate the centroid $\bar{y}$ of the quarter elliptical area.

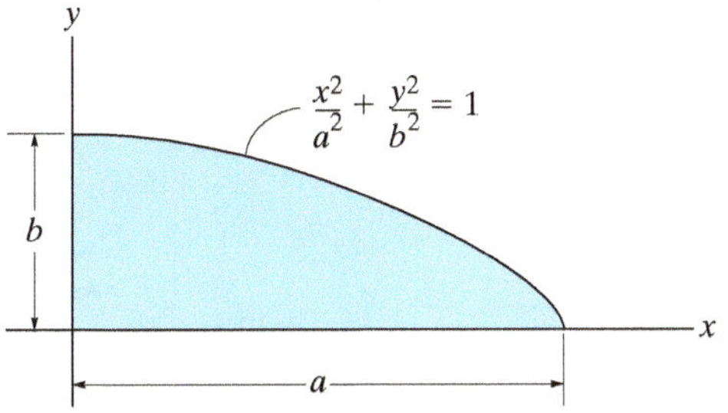

Probs. 9–23/24

9–25. The plate has a thickness of 0.25 ft and a specific weight of $\gamma = 180\ \text{lb/ft}^3$. Determine the location of its center of gravity. Also, find the tension in each of the cords used to support it.

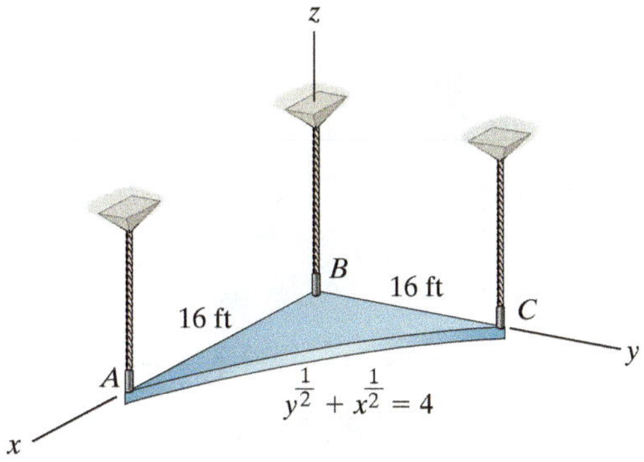

Prob. 9–25

9–26. Locate the centroid $\bar{x}$ of the area.

9–27. Locate the centroid $\bar{y}$ of the area.

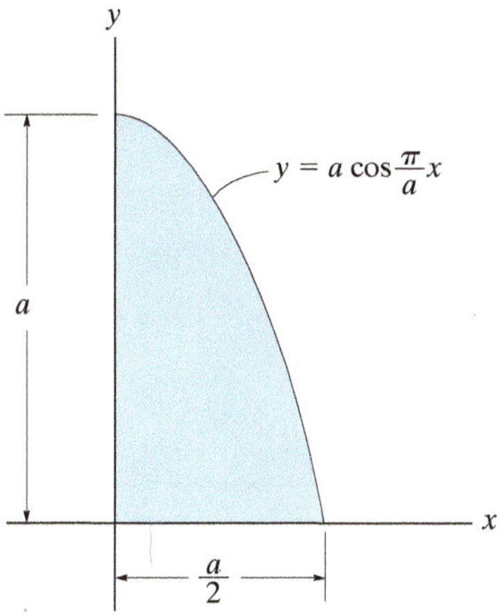

Probs. 9–26/27

***9–28.** Locate the centroid $\bar{x}$ of the area.

9–29. Locate the centroid $\bar{y}$ of the area.

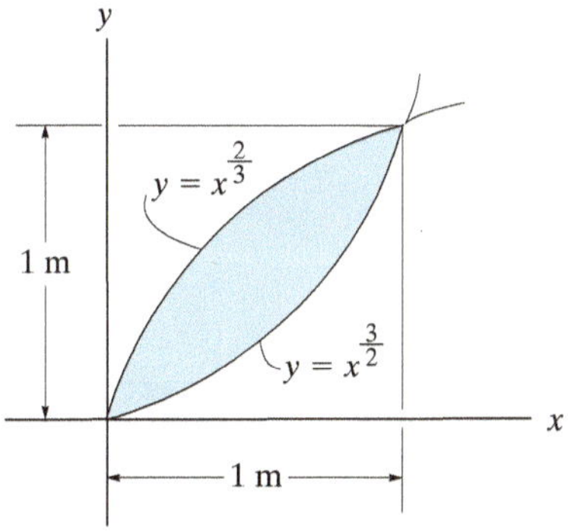

Probs. 9–28/29

9–30. Locate the centroid $\bar{x}$ of the area.

9–31. Locate the centroid $\bar{y}$ of the area.

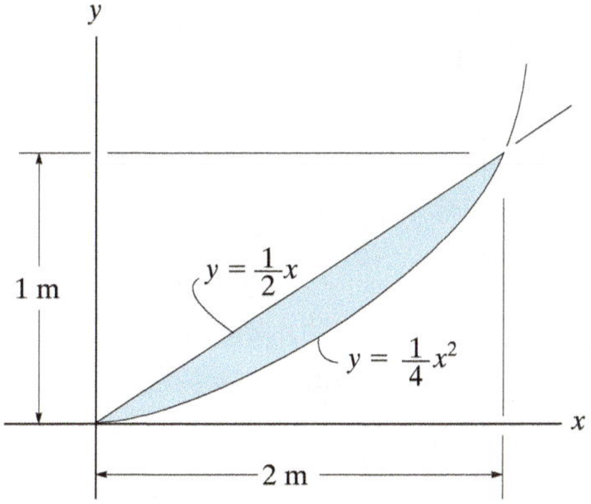

Probs. 9–30/31

9

***9–32.** Locate the centroid $\bar{x}$ of the area.

9–33. Locate the centroid $\bar{y}$ of the area.

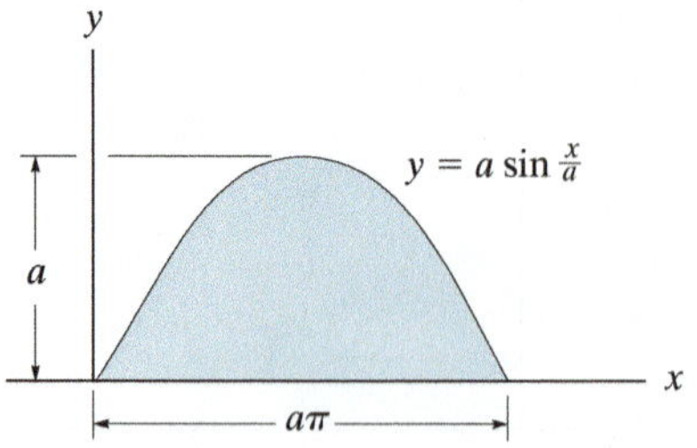

Probs. 9–32/33

9–34. The steel plate is 0.3 m thick and has a density of $7850\ \text{kg/m}^3$. Determine the location of its center of mass. Also find the reactions at the pin and roller support.

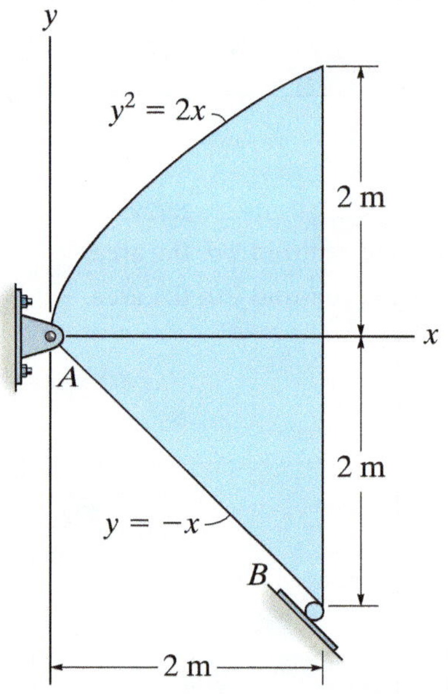

Prob. 9–34

9–35. Locate the centroid $\bar{x}$ of the shaded area.

***9–36.** Locate the centroid $\bar{y}$ of the shaded area.

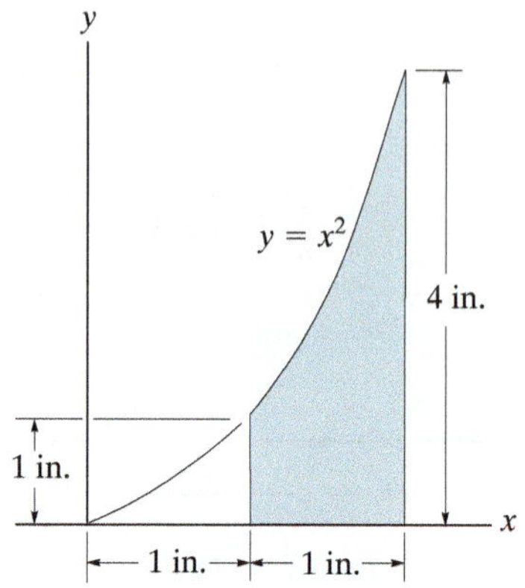

Probs. 9–35/36

9–37. If the density at any point in the quarter circular plate is defined by $\rho = \rho_0 xy$, where ρ_0 is a constant, determine the mass and locate the center of mass $(\bar{x}, \bar{y})$ of the plate. The plate has a thickness t.

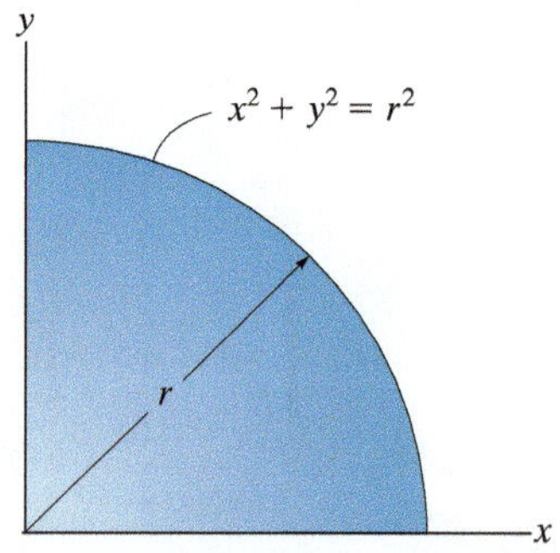

Prob. 9–37

9–38. Determine the location $\bar{r}$ of the centroid C of the cardioid, $r = a(1 - \cos\theta)$.

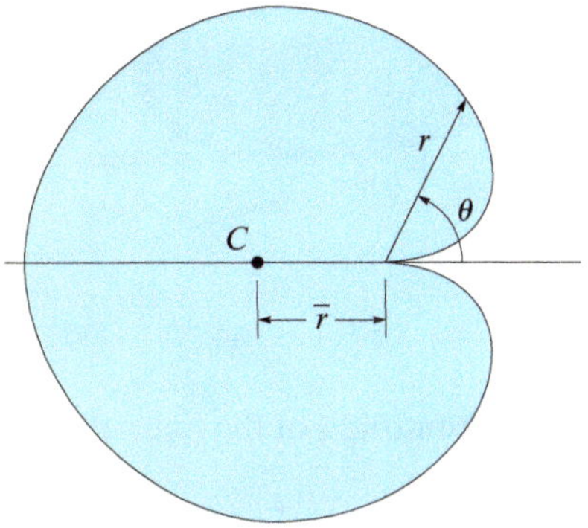

Prob. 9–38

9–39. Locate the centroid $\bar{y}$ of the paraboloid.

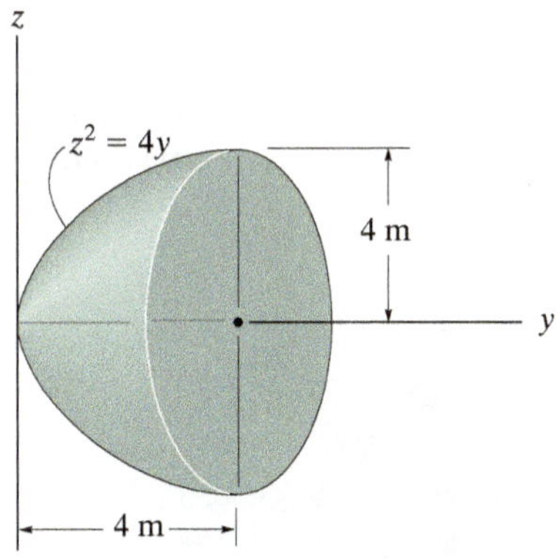

Prob. 9–39

***9–40.** Locate the center of gravity of the volume. The material is homogeneous.

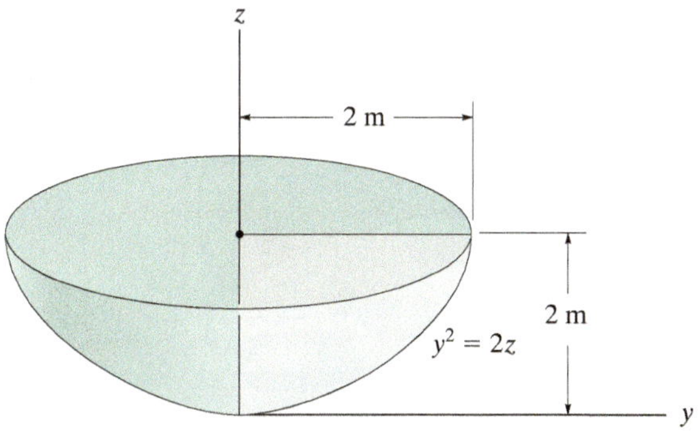

Prob. 9–40

9–41. Locate the centroid $\bar{z}$ of the hemisphere.

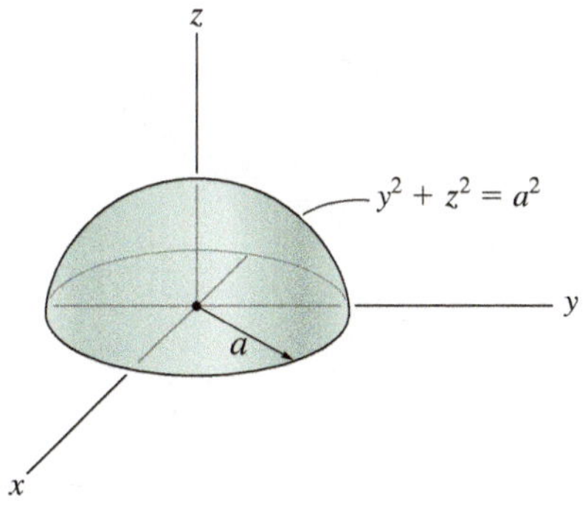

Prob. 9–41

9

9–42. Determine the centroid $\bar{y}$ of the solid.

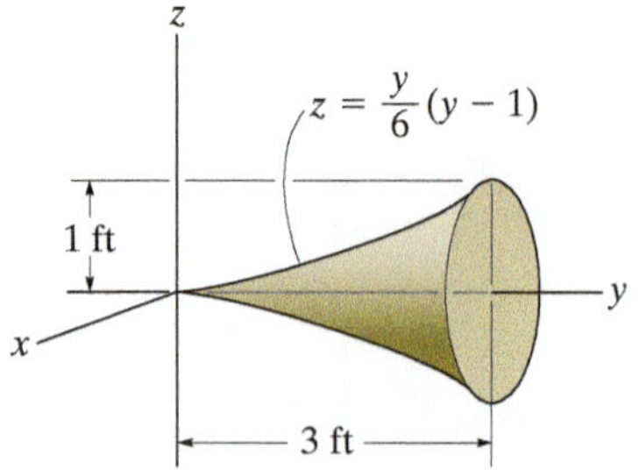

Prob. 9–42

9–43. Locate the center of gravity $\bar{z}$ of the solid.

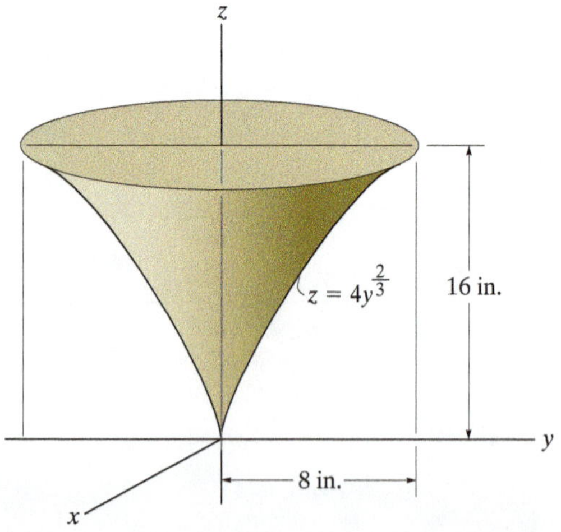

Prob. 9–43

***9–44.** Locate the centroid of the ellipsoid of revolution.

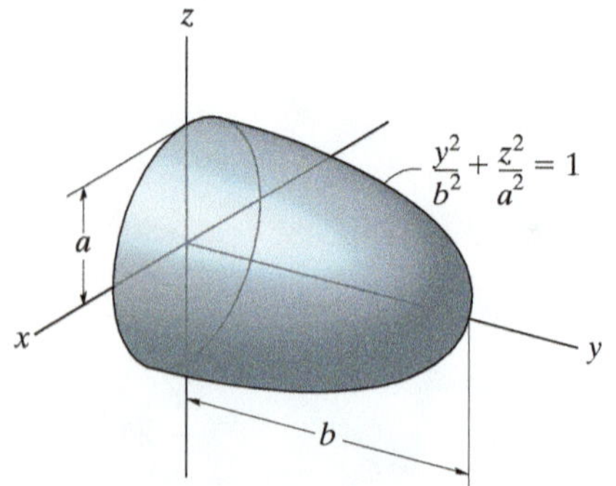

Prob. 9–44

9–45. Locate the centroid $\bar{z}$ of the right-elliptical cone.

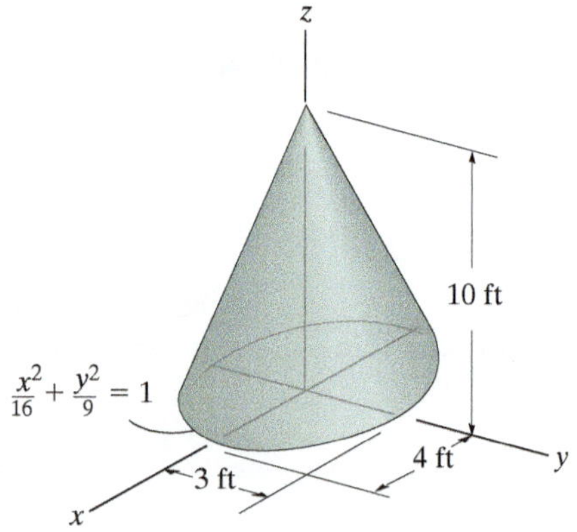

Prob. 9–45

9–46. The hemisphere of radius r is made from a stack of very thin plates such that the density varies with height, $\rho = kz$, where k is a constant. Determine its mass and the distance $\bar{z}$ to the center of mass G.

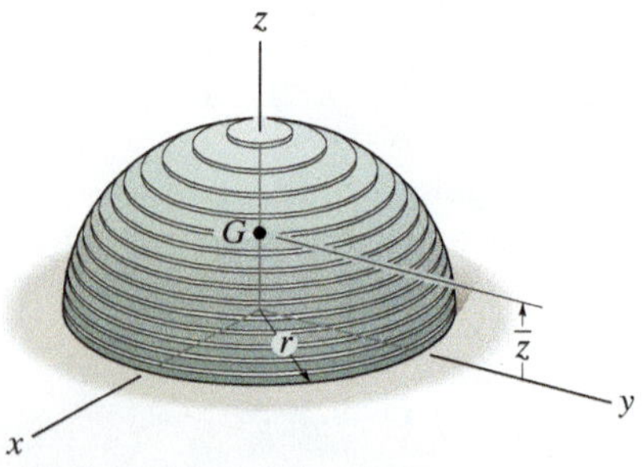

Prob. 9–46

9–47. Locate the centroid of the quarter-cone.

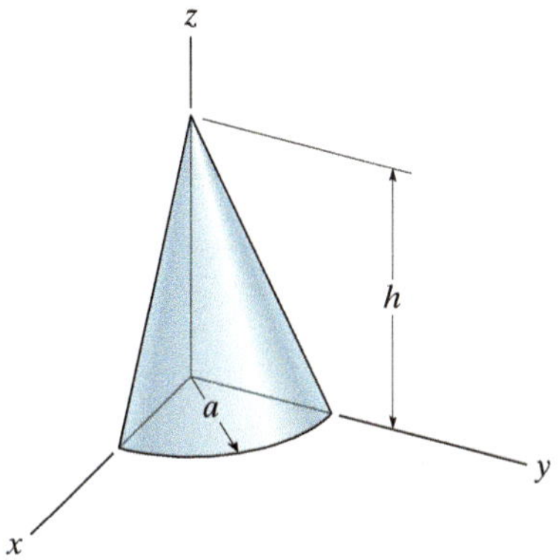

Prob. 9–47

***9–48.** Locate the centroid $\bar{z}$ of the frustum of the right-circular cone.

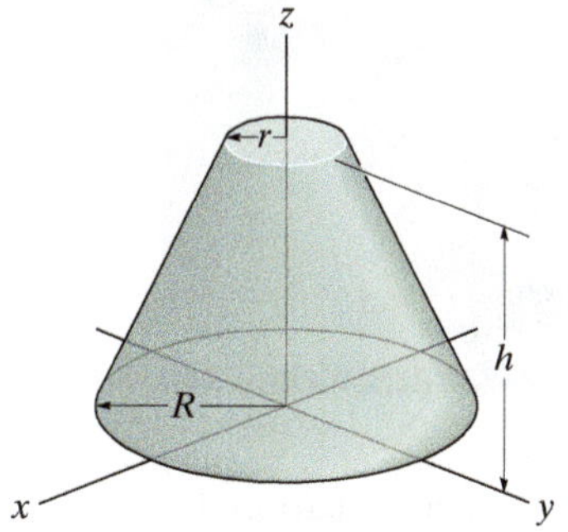

Prob. 9–48

9–49. The king's chamber of the Great Pyramid of Giza is located at its centroid. Assuming the pyramid to be a solid, prove that this point is at $\bar{z} = \frac{1}{4}h$, *Suggestion:* Use a rectangular differential plate element having a thickness dz and area $(2x)(2y)$.

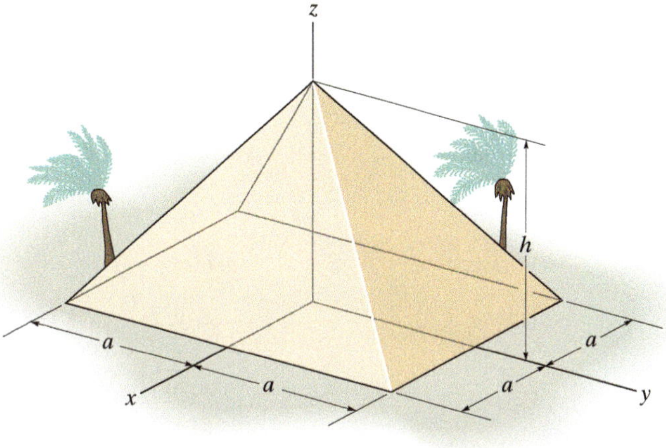

Prob. 9–49

9–50. Determine the location $\bar{z}$ of the centroid for the tetrahedron. *Suggestion:* Use a triangular "plate" element parallel to the x–y plane and of thickness dz.

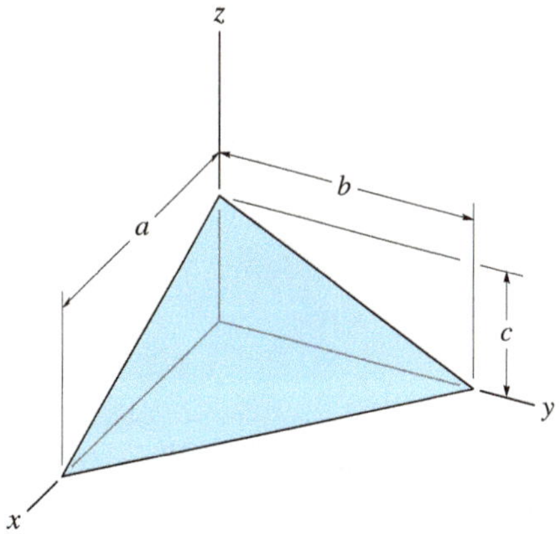

Prob. 9–50

9

9.2 Composite Bodies

A *composite body* consists of a series of connected "simpler" shaped bodies, which may be rectangular, triangular, semicircular, etc. Such a body can often be sectioned or divided into its composite parts and, provided the *weight* and location of the center of gravity of each of these parts are known, we can then eliminate the need for integration to determine the center of gravity for the entire body. The method for doing this follows the same procedure outlined in Sec. 9.1. Formulas analogous to Eqs. 9–1 result; however, rather than account for an infinite number of differential weights, we have instead a finite number of weights. Therefore,

$$\bar{x} = \frac{\Sigma \tilde{x} W}{\Sigma W} \qquad \bar{y} = \frac{\Sigma \tilde{y} W}{\Sigma W} \qquad \bar{z} = \frac{\Sigma \tilde{z} W}{\Sigma W} \tag{9–6}$$

Here

$\bar{x}, \bar{y}, \bar{z}$ represent the coordinates of the center of gravity G of the composite body.

$\tilde{x}, \tilde{y}, \tilde{z}$ represent the coordinates of the center of gravity of each composite part of the body.

ΣW is the sum of the weights of all the composite parts of the body, or simply the total weight of the body.

When the body has a *constant density or specific weight*, the center of gravity *coincides* with the centroid of the body. The centroid for composite lines, areas, and volumes can be found using relations analogous to Eqs. 9–6; however, the W's are replaced by L's, A's, and V's, respectively. Centroids for common shapes of lines, areas, shells, and volumes that often make up a composite body are given in the table on the inside back cover.

In order to determine the force required to tip over this concrete barrier it is first necessary to determine the location of its center of gravity G. Due to symmetry, G will lie on the vertical axis of symmetry.

Procedure for Analysis

The location of the center of gravity of a body or the centroid of a composite geometrical object represented by a line, area, or volume can be determined using the following procedure.

Composite Parts.

- Using a sketch, divide the body or object into a finite number of composite parts that have simpler shapes.
- If a composite body has a *hole*, or a geometric region having no material, then consider the composite body without the hole and consider the hole as an *additional* composite part having *negative* weight or size.

Moment Arms.

- Establish the coordinate axes on the sketch and determine the coordinates $\tilde{x}, \tilde{y}, \tilde{z}$ of the center of gravity or centroid of each part.

Summations.

- Determine $\bar{x}, \bar{y}, \bar{z}$ by applying the center of gravity equations, Eqs. 9–6, or the analogous centroid equations.
- If an object is *symmetrical* about an axis, the centroid of the object lies on this axis.

If desired, the calculations can be arranged in tabular form, as indicated in the following three examples.

The center of gravity of this water tank can be determined by dividing it into composite parts and applying Eqs. 9–6.

9

EXAMPLE 9.9

Locate the centroid of the wire shown in Fig. 9–16*a*.

SOLUTION

Composite Parts. The wire is divided into three segments as shown in Fig. 9–16*b*.

Moment Arms. The location of the centroid for each segment is determined and indicated in the figure. In particular, the centroid of segment ① is determined either by integration or by using the table on the inside back cover.

Summations. For convenience, the calculations can be tabulated as follows:

Segment	L (mm)	$\tilde{x}$ (mm)	$\tilde{y}$ (mm)	$\tilde{z}$ (mm)	$\tilde{x}L$ (mm^2)	$\tilde{y}L$ (mm^2)	$\tilde{z}L$ (mm^2)
1	$\pi(60) = 188.5$	60	−38.2	0	11 310	−7200	0
2	40	0	20	0	0	800	0
3	20	0	40	−10	0	800	−200
	$\Sigma L = 248.5$				$\Sigma\tilde{x}L = 11\,310$	$\Sigma\tilde{y}L = -5600$	$\Sigma\tilde{z}L = -200$

Thus,

$$\bar{x} = \frac{\Sigma\tilde{x}L}{\Sigma L} = \frac{11\,310}{248.5} = 45.5 \text{ mm} \qquad \textit{Ans.}$$

$$\bar{y} = \frac{\Sigma\tilde{y}L}{\Sigma L} = \frac{-5600}{248.5} = -22.5 \text{ mm} \qquad \textit{Ans.}$$

$$\bar{z} = \frac{\Sigma\tilde{z}L}{\Sigma L} = \frac{-200}{248.5} = -0.805 \text{ mm} \qquad \textit{Ans.}$$

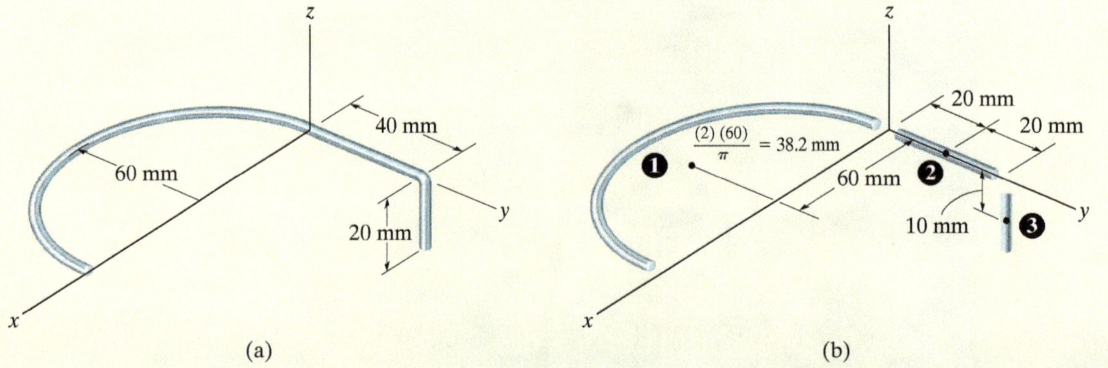

Fig. 9–16

EXAMPLE 9.10

Locate the centroid of the plate area shown in Fig. 9–17*a*.

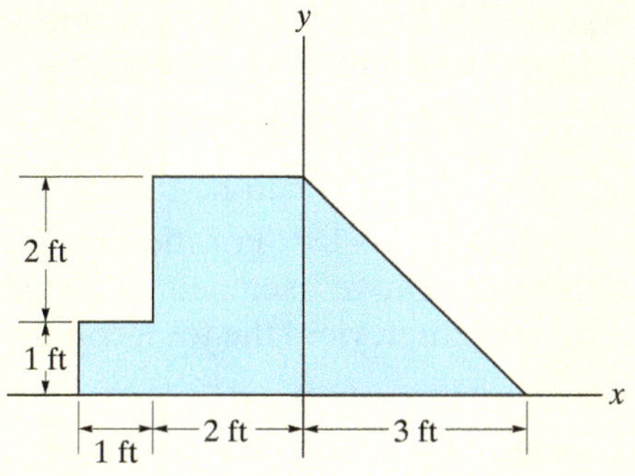

(a)

Fig. 9–17

SOLUTION

Composite Parts. The plate is divided into three segments as shown in Fig. 9–17*b*. Here the area of the small rectangle ③ is considered "negative" since it must be subtracted from the larger one ②.

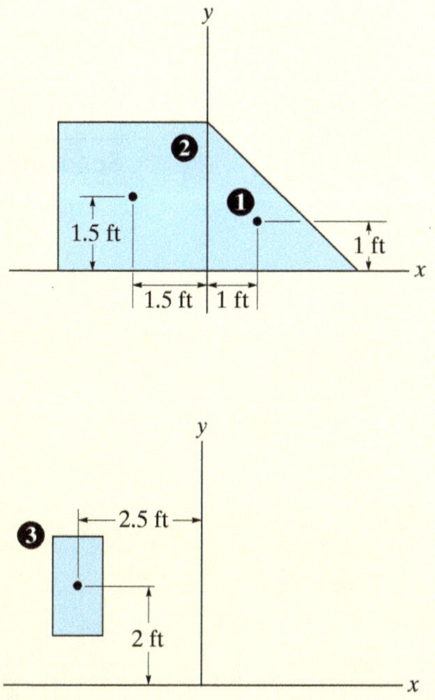

(b)

Moment Arms. The centroid of each segment is located as indicated in the figure. Note that the $\tilde{x}$ coordinates of ② and ③ are *negative.*

Summations. Taking the data from Fig. 9–17*b*, the calculations are tabulated as follows:

Segment	A (ft^2)	$\tilde{x}$ (ft)	$\tilde{y}$ (ft)	$\tilde{x}A$ (ft^3)	$\tilde{y}A$ (ft^3)
1	$\frac{1}{2}(3)(3) = 4.5$	1	1	4.5	4.5
2	$(3)(3) = 9$	-1.5	1.5	-13.5	13.5
3	$-(2)(1) = -2$	-2.5	2	5	-4
	$\Sigma A = 11.5$			$\Sigma\tilde{x}A = -4$	$\Sigma\tilde{y}A = 14$

Thus,

$$\bar{x} = \frac{\Sigma\tilde{x}A}{\Sigma A} = \frac{-4}{11.5} = -0.348 \text{ ft} \qquad \textit{Ans.}$$

$$\bar{y} = \frac{\Sigma\tilde{y}A}{\Sigma A} = \frac{14}{11.5} = 1.22 \text{ ft} \qquad \textit{Ans.}$$

NOTE: If these results are plotted in Fig. 9–17*a*, the location of point *C* seems reasonable.

EXAMPLE 9.11

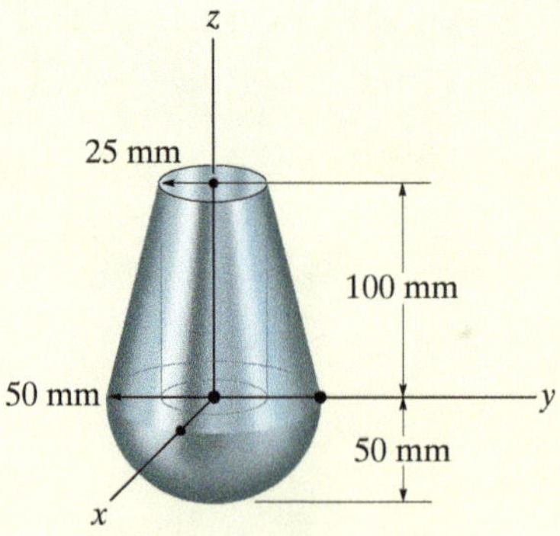

(a)

Fig. 9–18

Locate the center of mass of the assembly shown in Fig. 9–18*a*. The conical frustum has a density of $\rho_c = 8\ \text{Mg/m}^3$, and the hemisphere has a density of $\rho_h = 4\ \text{Mg/m}^3$. There is a 25-mm-radius cylindrical hole in the center of the frustum.

SOLUTION

Composite Parts. The assembly can be thought of as consisting of four segments as shown in Fig. 9–18*b*. For the calculations, ③ and ④ must be considered as "negative" segments in order that the four segments, when added together, yield the total composite shape shown in Fig. 9–18*a*.

Moment Arm. Using the table on the inside back cover, the computations for the centroid $\tilde{z}$ of each piece are shown in the figure.

Summations. Because of *symmetry*, note that

$$\bar{x} = \bar{y} = 0 \qquad \textit{Ans.}$$

Since $W = mg$, and g is constant, the third of Eqs. 9–6 becomes $\bar{z} = \Sigma\tilde{z}m/\Sigma m$. The mass of each piece can be computed from $m = \rho V$ and used for the calculations. Also, $1\ \text{Mg/m}^3 = 10^{-6}\ \text{kg/mm}^3$, so that

Segment	m (kg)	$\tilde{z}$ (mm)	$\tilde{z}m$ (kg·mm)
1	$8(10^{-6})\left(\frac{1}{3}\right)\pi(50)^2(200) = 4.189$	50	209.440
2	$4(10^{-6})\left(\frac{2}{3}\right)\pi(50)^3 = 1.047$	−18.75	−19.635
3	$-8(10^{-6})\left(\frac{1}{3}\right)\pi(25)^2(100) = -0.524$	100 + 25 = 125	−65.450
4	$-8(10^{-6})\pi(25)^2(100) = -1.571$	50	−78.540
	$\Sigma m = 3.142$		$\Sigma\tilde{z}m = 45.815$

$$\text{Thus,}\quad \tilde{z} = \frac{\Sigma\tilde{z}m}{\Sigma m} = \frac{45.815}{3.142} = 14.6\ \text{mm} \qquad \textit{Ans.}$$

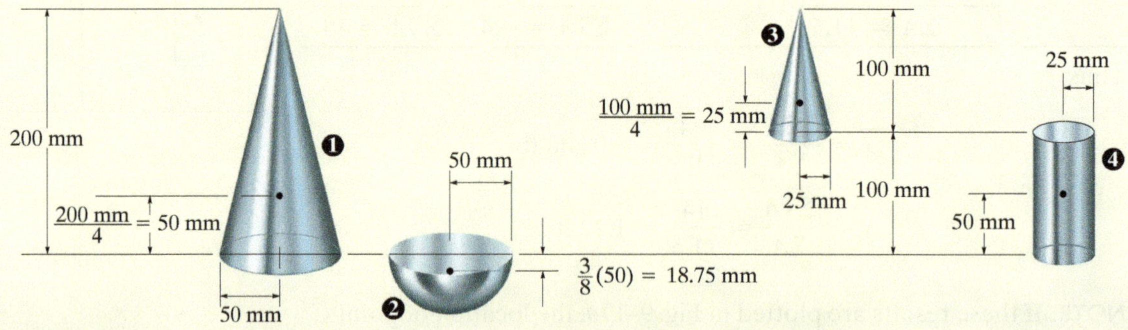

(b)

FUNDAMENTAL PROBLEMS

F9–7. Locate the centroid $(\bar{x}, \bar{y}, \bar{z})$ of the wire bent in the shape shown.

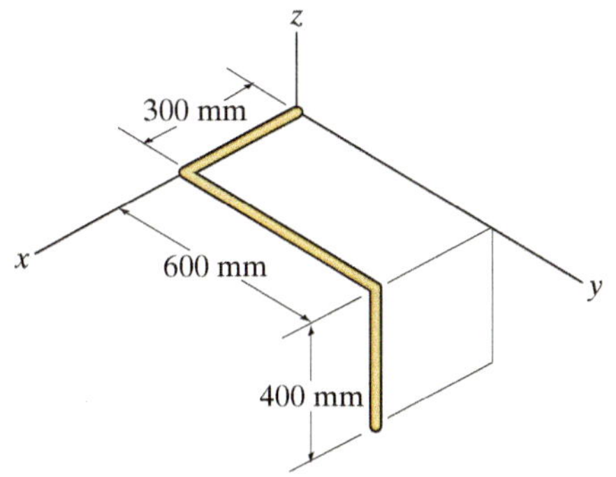

F9–7

F9–8. Locate the centroid $\bar{y}$ of the beam's cross-sectional area.

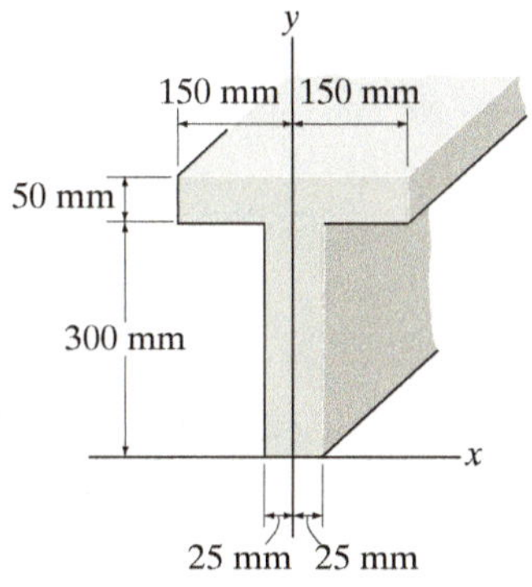

F9–8

F9–9. Locate the centroid $\bar{y}$ of the beam's cross-sectional area.

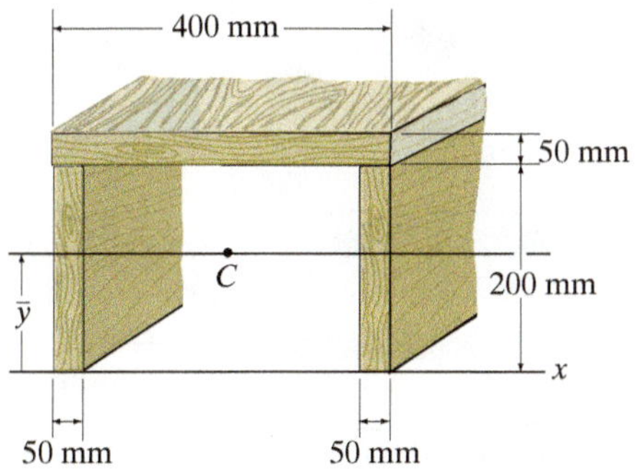

F9–9

F9–10. Locate the centroid $(\bar{x}, \bar{y})$ of the cross-sectional area.

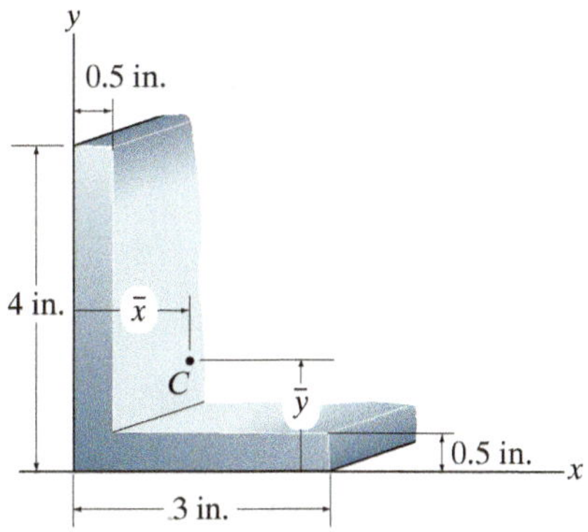

F9–10

F9–11. Locate the center of mass $(\bar{x}, \bar{y}, \bar{z})$ of the homogeneous solid block.

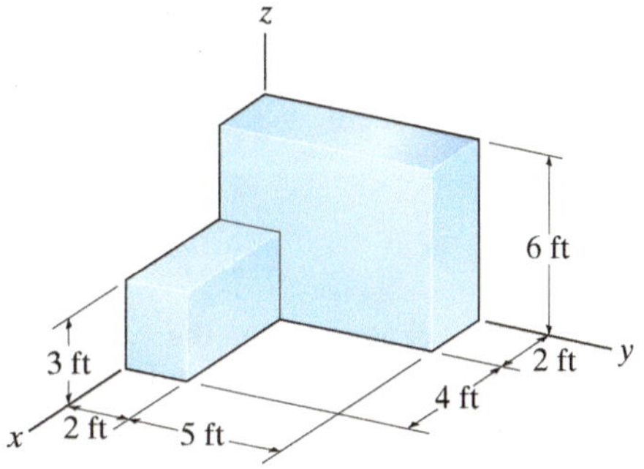

F9–11

F9–12. Determine the center of mass $(\bar{x}, \bar{y}, \bar{z})$ of the homogeneous solid block.

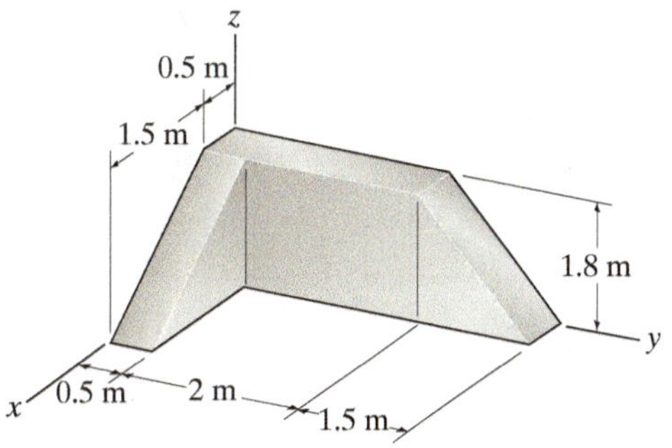

F9–12

PROBLEMS

9–51. The truss is made from five members, each having a length of 4 m and a mass of 7 kg/m. If the mass of the gusset plates at the joints and the thickness of the members can be neglected, determine the distance d to where the hoisting cable must be attached, so that the truss does not tip (rotate) when it is lifted.

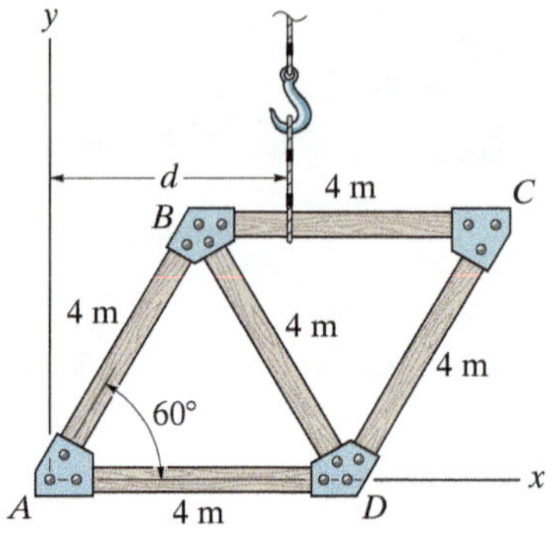

Prob. 9–51

9–53. Locate the centroid $(\bar{x}, \bar{y})$ of the cross section. All the dimensions are measured to the centerline thickness of each thin segment.

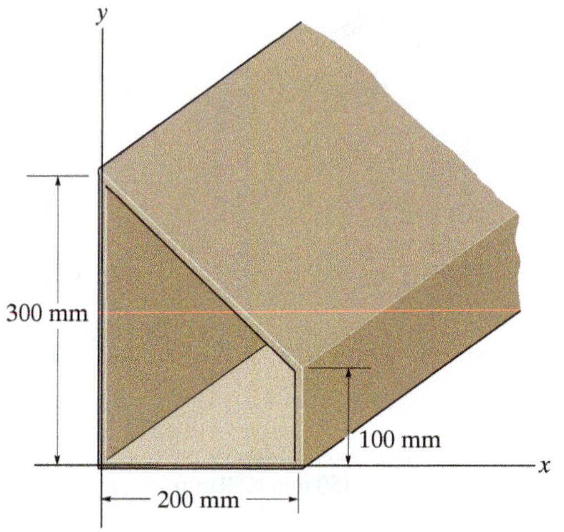

Prob. 9–53

***9–52.** Locate the centroid for the wire. Neglect the thickness of the material and slight bends at the corners.

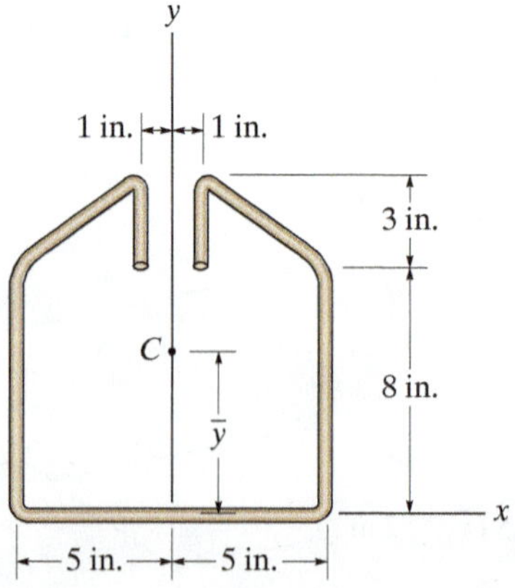

Prob. 9–52

9–54. Locate the centroid $(\bar{x}, \bar{y})$ of the metal cross section. Neglect the thickness of the material and slight bends at the corners.

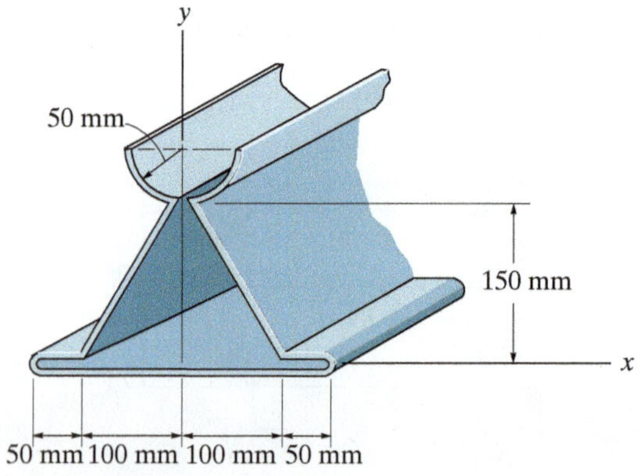

Prob. 9–54

9–55. The three members of the frame each have a weight per unit length of 4 lb/ft. Locate the position $(\bar{x}, \bar{y})$ of the center of gravity. Neglect the size of the pins at the joints and the thickness of the members. Also, calculate the reactions at the fixed support A.

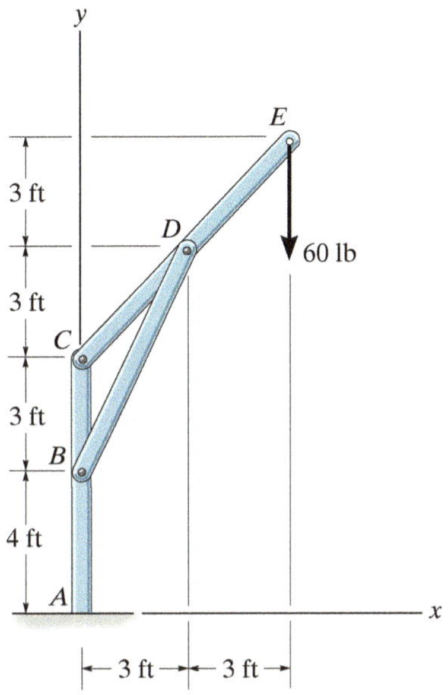

Prob. 9–55

***9–56.** The steel and aluminum plate assembly is bolted together and fastened to the wall. Each plate has a constant width in the z direction of 200 mm and thickness of 20 mm. If the density of A and B is $\rho_s = 7.85\ \text{Mg/m}^3$, and for C, $\rho_{al} = 2.71\ \text{Mg/m}^3$, determine the location $\bar{x}$ of the center of mass. Neglect the size of the bolts.

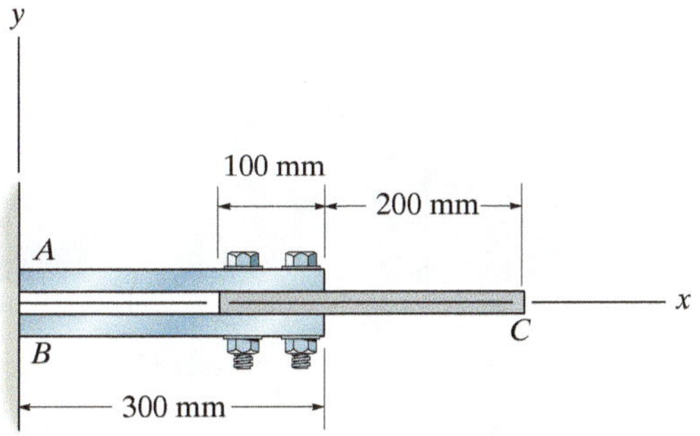

Prob. 9–56

9–57. To determine the location of the center of gravity of the automobile it is first placed in a *level position*, with the two wheels on one side resting on the scale platform P. In this position the scale records a reading of W_1. Then, one side is elevated to a convenient height c as shown. The new reading on the scale is W_2. If the automobile has a total weight of W, determine the location of its center of gravity $G(\bar{x}, \bar{y})$.

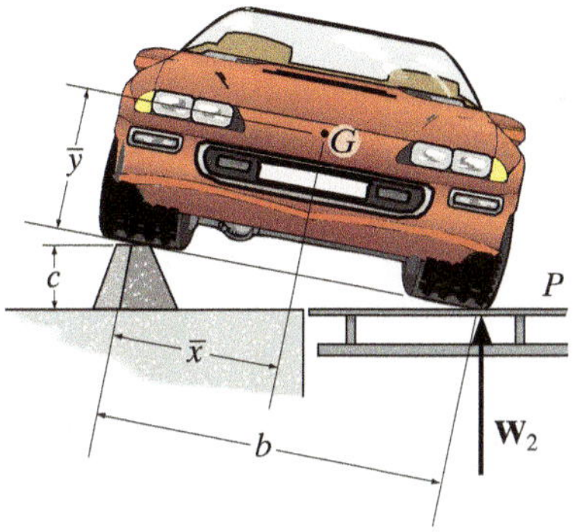

Prob. 9–57

9–58. Determine the location $\bar{y}$ of the centroidal axis $\bar{x}$–$\bar{x}$ of the beam's cross-sectional area. Neglect the size of the corner welds at A and B for the calculation.

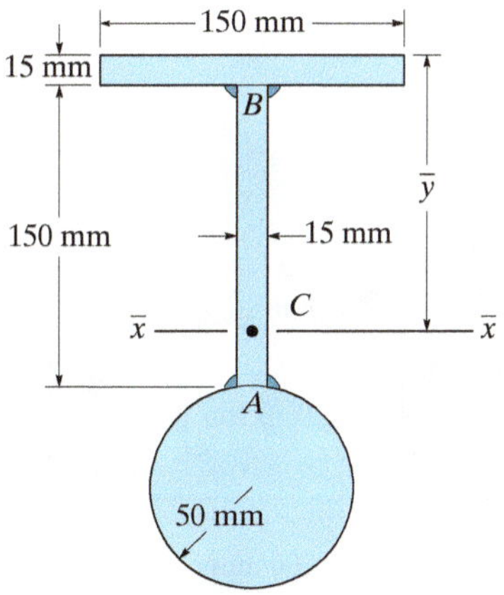

Prob. 9–58

9–59. Locate the centroid $(\bar{x}, \bar{y})$ for the angle's cross-sectional area.

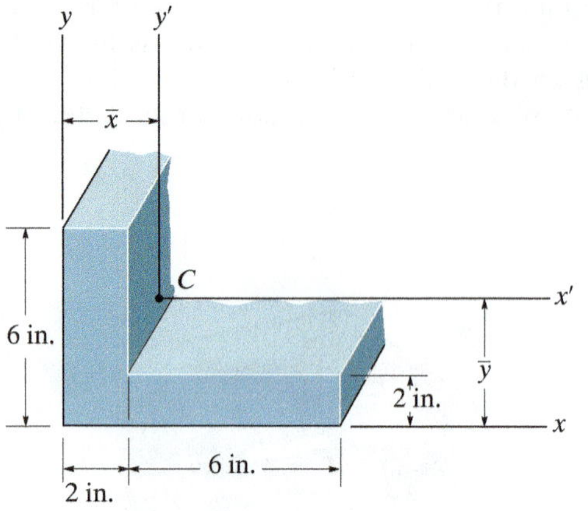

Prob. 9–59

***9–60.** Locate the centroid $\bar{y}$ of the channel's cross-sectional area.

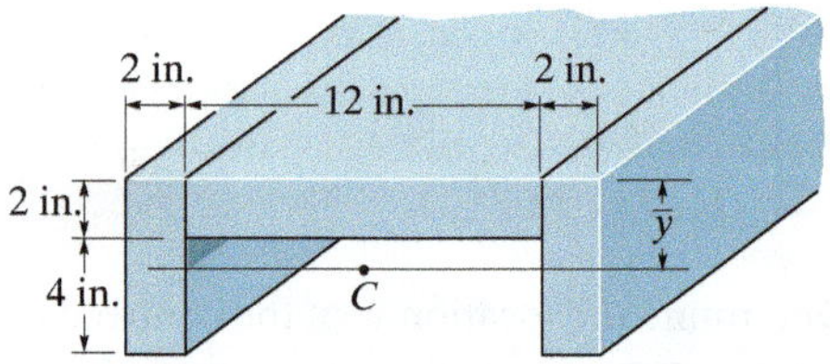

Prob. 9–60

9–61. Locate the centroid $(\bar{x}, \bar{y})$ of the member's cross-sectional area.

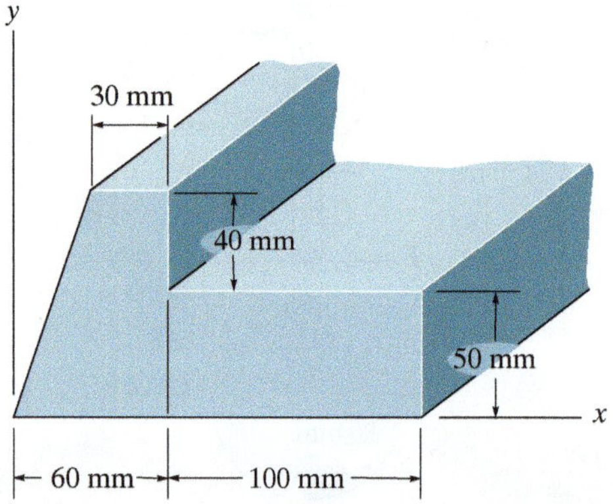

Prob. 9–61

9–62. Locate the centroid $\bar{y}$ of the bulb-tee cross section.

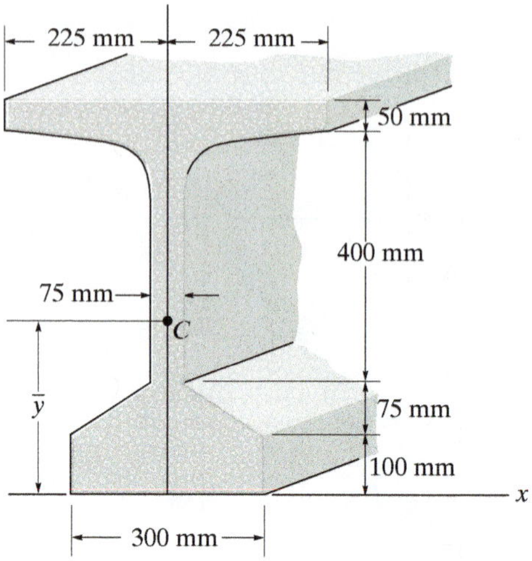

Prob. 9–62

9–63. Determine the location $(\bar{x}, \bar{y})$ of the centroid C of the cross-sectional area for the structural member constructed from two equal-sized channels welded together as shown. Assume all corners are square. Neglect the size of the welds.

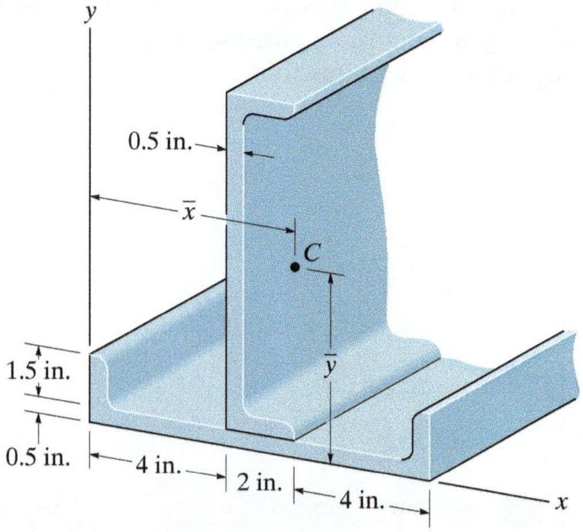

Prob. 9–63

***9–64.** Locate the centroid $\bar{y}$ of the concrete beam having the tapered cross section shown.

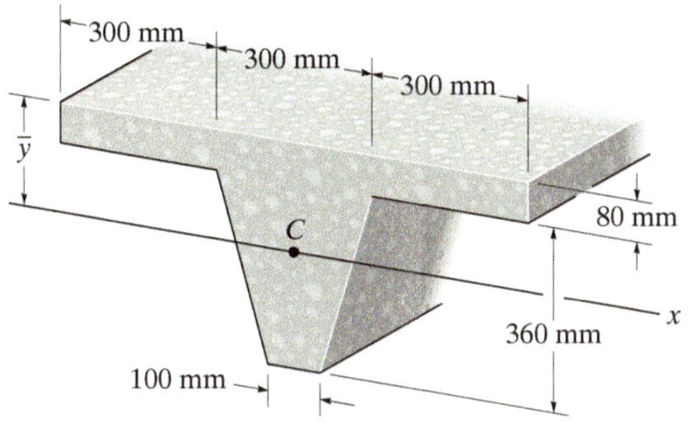

Prob. 9–64

9–65. Locate the centroid $\bar{y}$ of the beam's cross-section built up from a channel and a wide-flange beam.

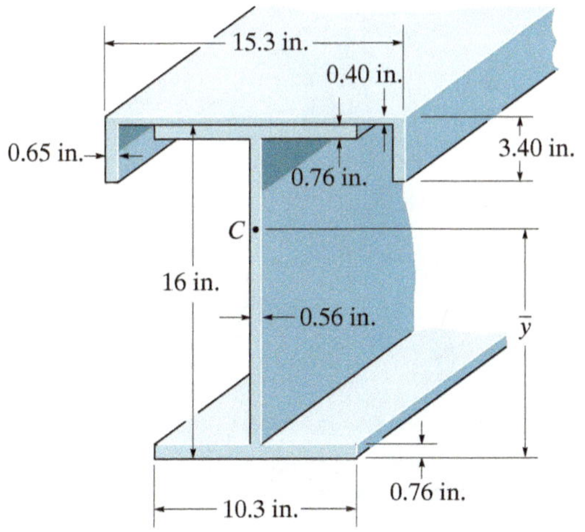

Prob. 9–65

9–66. Locate the centroid $\bar{y}$ of the cross-sectional area of the beam constructed from a channel and a plate. Assume all corners are square and neglect the size of the weld at A.

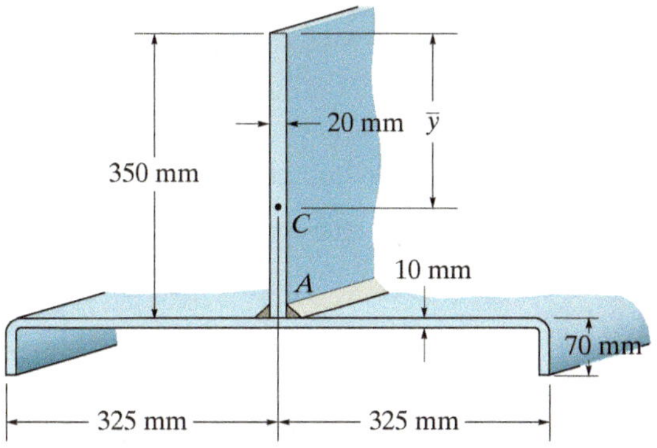

Prob. 9–66

9–67. An aluminum strut has a cross section referred to as a deep hat. Locate the centroid $\bar{y}$ of its area. Each segment has a thickness of 10 mm.

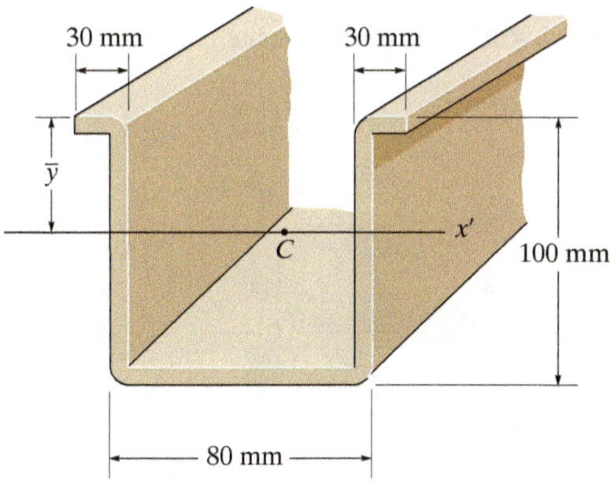

Prob. 9–67

9

***9–68.** Locate the centroid $\bar{y}$ of the beam's cross-sectional area.

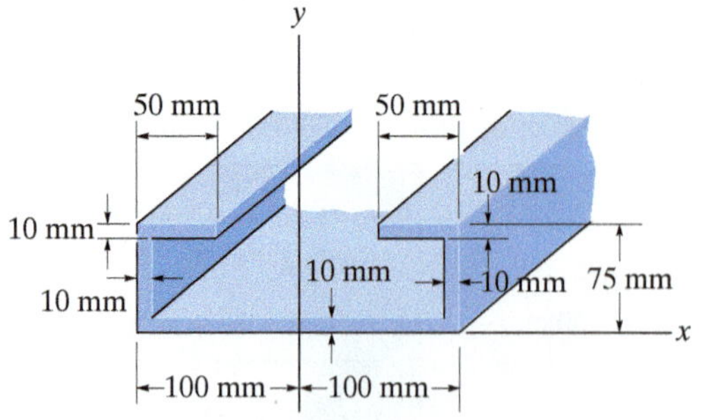

Prob. 9–68

9–69. Determine the location $\bar{x}$ of the centroid C of the shaded area that is part of a circle having a radius r.

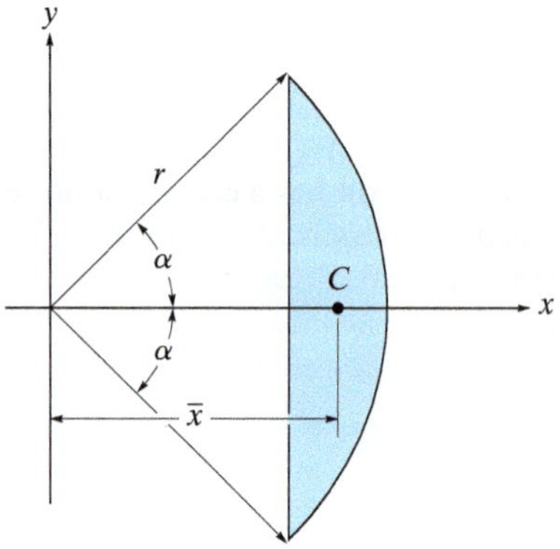

Prob. 9–69

9–70. Locate the centroid $\bar{y}$ for the cross-sectional area of the angle.

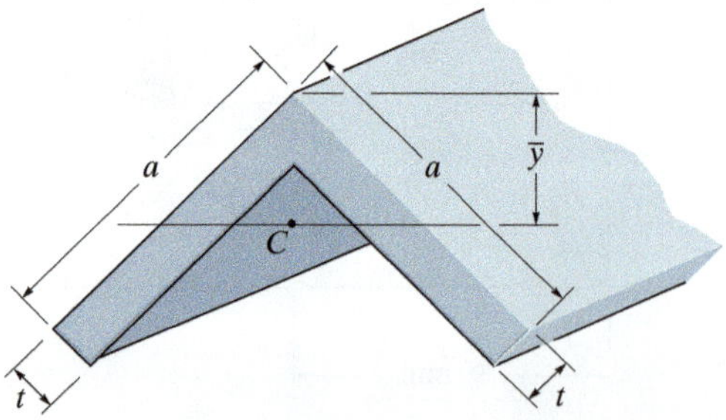

Prob. 9–70

9–71. Determine the location $\bar{y}$ of the centroid of the beam's cross-sectional area. Neglect the size of the corner welds at A and B for the calculation.

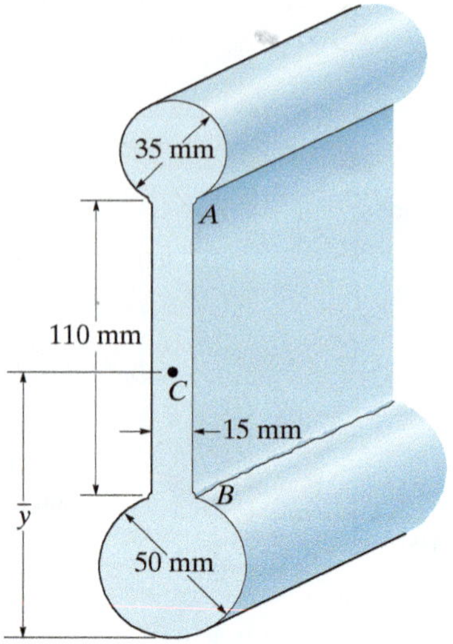

Prob. 9–71

***9–72.** A triangular plate made of homogeneous material has a constant thickness that is very small. If it is folded over as shown, determine the location $\bar{y}$ of the plate's center of gravity G.

9–73. A triangular plate made of homogeneous material has a constant thickness that is very small. If it is folded over as shown, determine the location $\bar{z}$ of the plate's center of gravity G.

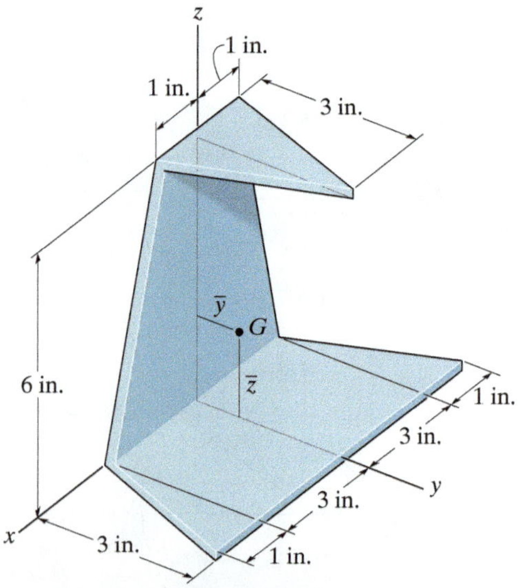

Probs. 9–72/73

9–74. The sheet metal part has the dimensions shown. Determine the location $(\bar{x}, \bar{y}, \bar{z})$ of its centroid.

9–75. The sheet metal part has a weight per unit area of $2\ \text{lb/ft}^2$ and is supported by the smooth rod and the cord at C. If the cord is cut, the part will rotate about the y axis until it reaches equilibrium. Determine the equilibrium angle of tilt, measured downward from the negative x axis, that AD makes with the $-x$ axis.

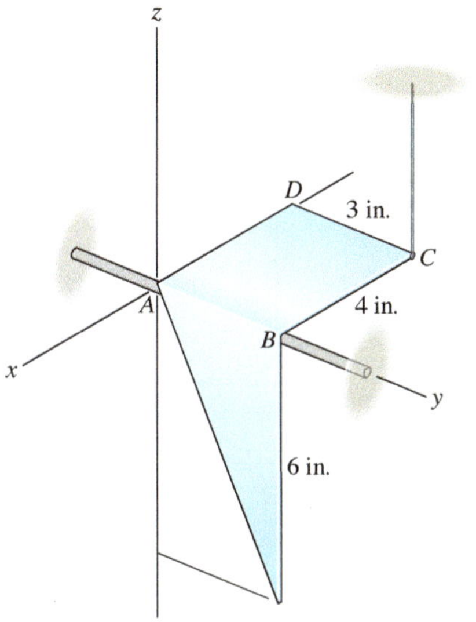

Probs. 9–74/75

***9–76.** The wooden table is made from a square board having a weight of 15 lb. Each of the legs weighs 2 lb and is 3 ft long. Determine how high its center of gravity is from the floor. Also, what is the angle, measured from the horizontal, through which its top surface can be tilted on two of its legs before it begins to overturn? Neglect the thickness of each leg.

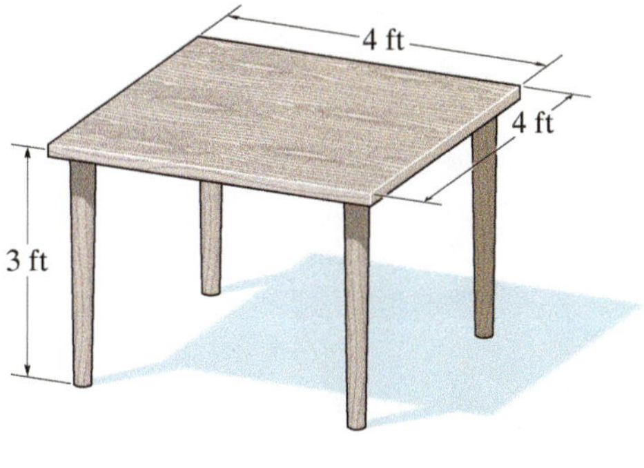

Prob. 9–76

9–77. Determine the location $(\bar{x}, \bar{y})$ of the center of gravity of the three-wheeler. The location of the center of gravity of each component and its weight are tabulated in the figure. If the three-wheeler is symmetrical with respect to the x–y plane, determine the normal reaction each of its wheels exerts on the ground.

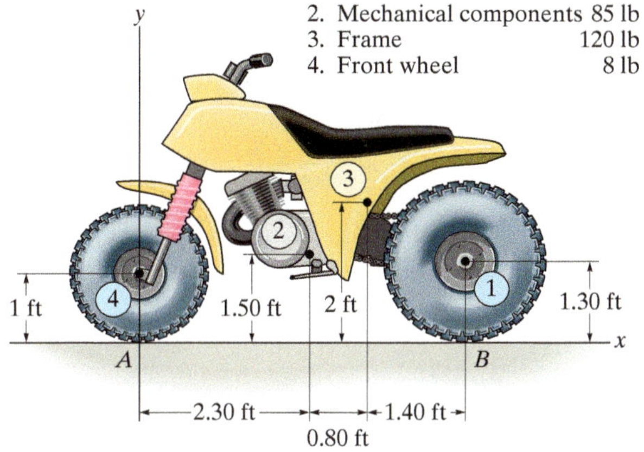

Prob. 9–77

9–78. Locate the center of gravity of the two-block assembly. The specific weights of the materials A and B are $\gamma_A = 150\ \text{lb/ft}^3$ and $\gamma_B = 400\ \text{lb/ft}^3$, respectively.

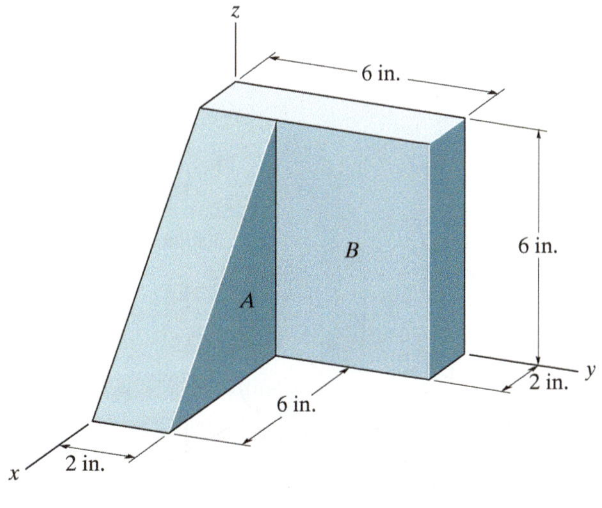

Prob. 9–78

9

9–79. Locate the center of mass of the block. Materials 1, 2, and 3 have densities of 2.70 Mg/m^3, 5.70 Mg/m^3, and 7.80 Mg/m^3, respectively.

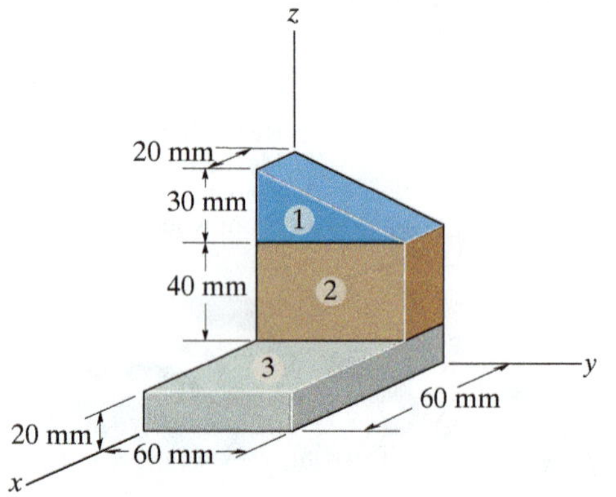

Prob. 9–79

*9–80. Locate the centroid $\bar{z}$ of the homogenous solid formed by boring a hemispherical hole into the cylinder that is capped with a cone.

9–81. Locate the center of mass $\bar{z}$ of the solid formed by boring a hemispherical hole into a cylinder that is capped with a cone. The cone and cylinder are made of materials having densities of 7.80 Mg/m^3 and 2.70 Mg/m^3, respectively.

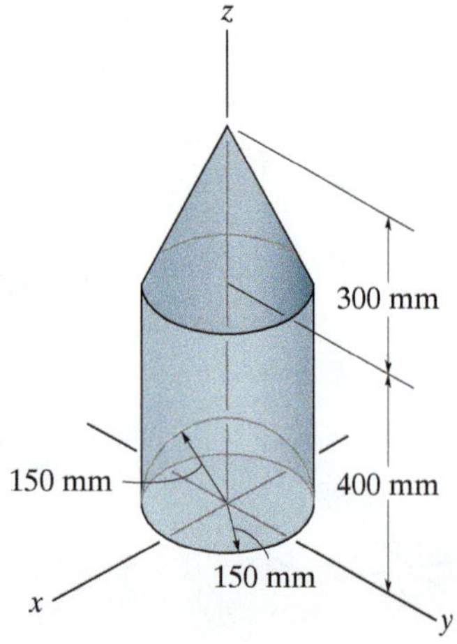

Probs. 9–80/81

9–82. Determine the distance h to which a 100-mm-diameter hole must be bored into the base of the cone so that the center of mass of the resulting shape is located at $\bar{z} = 115$ mm. The material has a density of 8 Mg/m^3.

9–83. Determine the distance $\bar{z}$ to the centroid of the shape that consists of a cone with a hole of height $h = 50$ mm bored into its base.

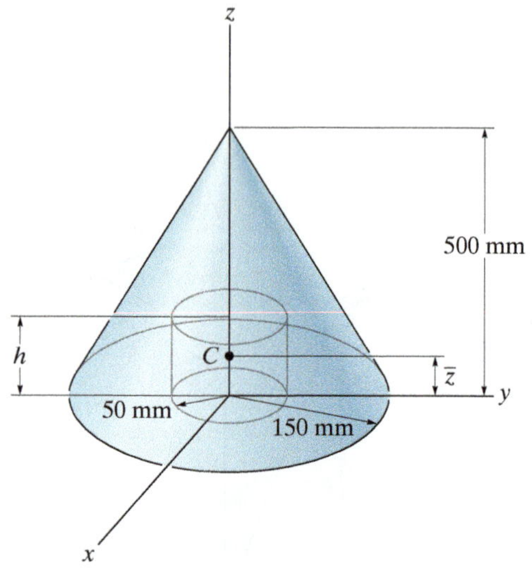

Probs. 9–82/83

*9–84. The buoy is made from two homogeneous cones each having a radius of 1.5 ft. If $h = 1.2$ ft, find the distance $\bar{z}$ to the buoy's center of gravity G.

9–85. The buoy is made from two homogeneous cones each having a radius of 1.5 ft. If it is required that the buoy's center of gravity G be located at $\bar{z} = 0.5$ ft, determine the height h of the top cone.

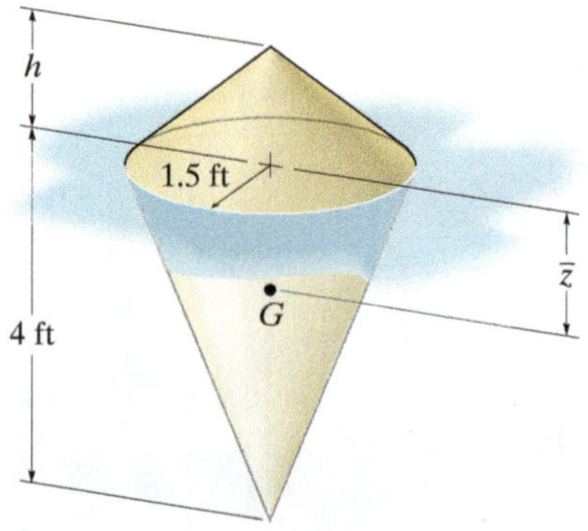

Probs. 9–84/85

9

9–86. Locate the center of mass $\bar{z}$ of the assembly. The assembly consists of a cylindrical center core, A, having a density of 7.90 Mg/m^3, and a cylindrical outer part, B, and a cone cap, C, each having a density of 2.70 Mg/m^3.

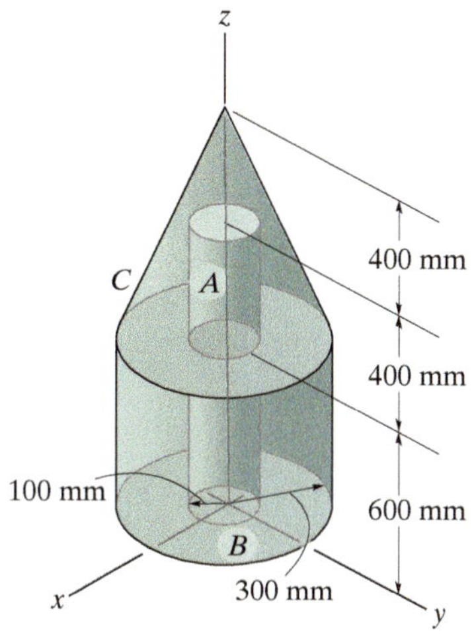

Prob. 9–86

9–87. The assembly consists of a 20-in. wooden dowel rod and a tight-fitting steel collar. Determine the distance $\bar{x}$ to its center of gravity if the specific weights of the materials are $\gamma_w = 150$ lb/ft^3 and $\gamma_{st} = 490$ lb/ft^3. The radii of the dowel and collar are shown.

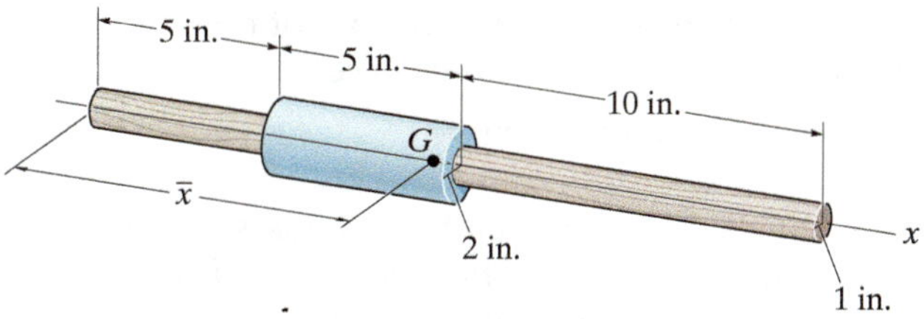

Prob. 9–87

***9–88.** A hole having a radius r is to be drilled in the center of the homogeneous block. Determine the depth h of the hole so that the center of gravity G is as low as possible.

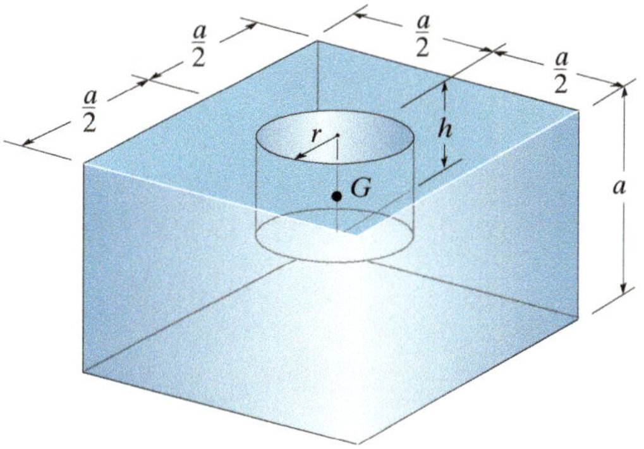

Prob. 9–88

9–89. Locate the center of mass $\bar{z}$ of the assembly. The cylinder and the cone are made from materials having densities of 5 Mg/m^3 and 9 Mg/m^3, respectively.

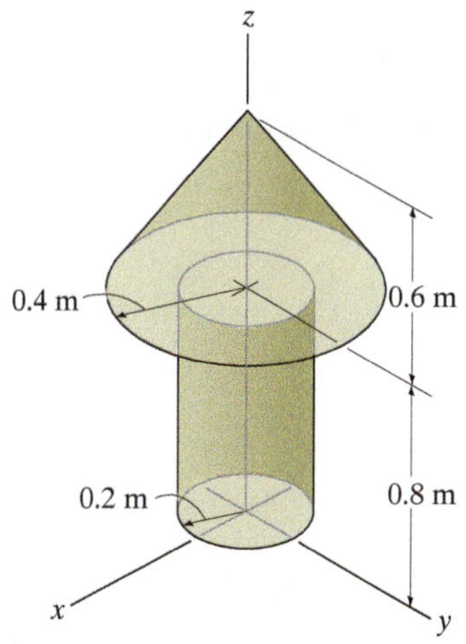

Prob. 9–89

9

*9.3 Theorems of Pappus and Guldinus

The two *theorems of Pappus and Guldinus* are used to find the surface area and volume of any body of revolution. They were first developed by Pappus of Alexandria during the fourth century A.D. and then restated at a later time by the Swiss mathematician Paul Guldin or Guldinus (1577–1643).

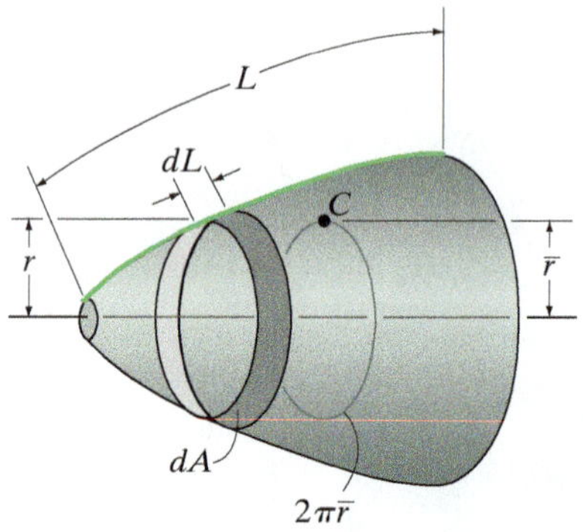

Fig. 9–19

The amount of roofing material used on this storage building can be estimated by using the first theorem of Pappus and Guldinus to determine its surface area.

Surface Area. If we revolve a *plane curve* about an axis that does not intersect the curve we will generate a *surface area of revolution*. For example, the surface area in Fig. 9–19 is formed by revolving the curve of length L about the horizontal axis. To determine this surface area, we will first consider the differential line element of length dL. If this element is revolved 2π radians about the axis, a ring having a surface area of $dA = 2\pi r\, dL$ will be generated. Thus, the surface area of the entire body is $A = 2\pi \int r\, dL$. Since $\int r\, dL = \bar{r}L$ (Eq. 9–5), then $A = 2\pi\bar{r}L$. If the curve is revolved only through an angle θ (radians), then

$$A = \theta\bar{r}L \qquad (9\text{–}7)$$

where

A = surface area of revolution

θ = angle of revolution measured in radians, $\theta \leq 2\pi$

$\bar{r}$ = perpendicular distance from the axis of revolution to the centroid of the generating curve

L = length of the generating curve

Therefore the first theorem of Pappus and Guldinus states that *the area of a surface of revolution equals the product of the length of the generating curve and the distance traveled by the centroid of the curve in generating the surface area.*

9

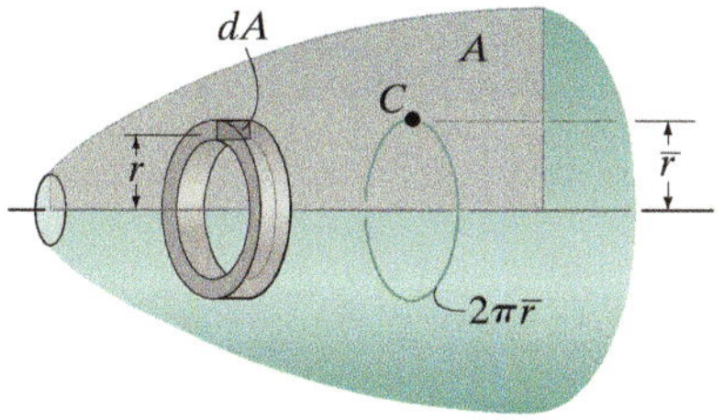

Fig. 9–20

Volume. A *volume* can be generated by revolving a *plane area* about an axis that does not intersect the area. For example, if we revolve the shaded area A in Fig. 9–20 about the horizontal axis, it generates the volume shown. This volume can be determined by first revolving the differential element of area dA 2π radians about the axis, so that a ring having the volume $dV = 2\pi r\, dA$ is generated. The entire volume is then $V = 2\pi \int r\, dA$. However, $\int r\, dA = \bar{r}A$, Eq. 9–4, so that $V = 2\pi\bar{r}A$. If the area is only revolved through an angle θ (radians), then

$$V = \theta \bar{r} A \qquad (9\text{–}8)$$

where

V = volume of revolution

θ = angle of revolution measured in radians, $\theta \leq 2\pi$

$\bar{r}$ = perpendicular distance from the axis of revolution to the centroid of the generating area

A = generating area

Therefore the second theorem of Pappus and Guldinus states that *the volume of a body of revolution equals the product of the generating area and the distance traveled by the centroid of the area in generating the volume.*

Composite Shapes. We may also apply the above two theorems to lines or areas that are composed of a series of composite parts. In this case the total surface area or volume generated is the addition of the surface areas or volumes generated by each of the composite parts. If the perpendicular distance from the axis of revolution to the centroid of each composite part is $\tilde{r}$, then

$$A = \theta \Sigma(\tilde{r}L) \qquad (9\text{–}9)$$

and

$$V = \theta \Sigma(\tilde{r}A) \qquad (9\text{–}10)$$

Application of the above theorems is illustrated numerically in the following examples.

The volume of fertilizer contained within this silo can be determined using the second theorem of Pappus and Guldinus.

9

EXAMPLE 9.12

Show that the surface area of a sphere is $A = 4\pi R^2$ and its volume is $V = \frac{4}{3}\pi R^3$.

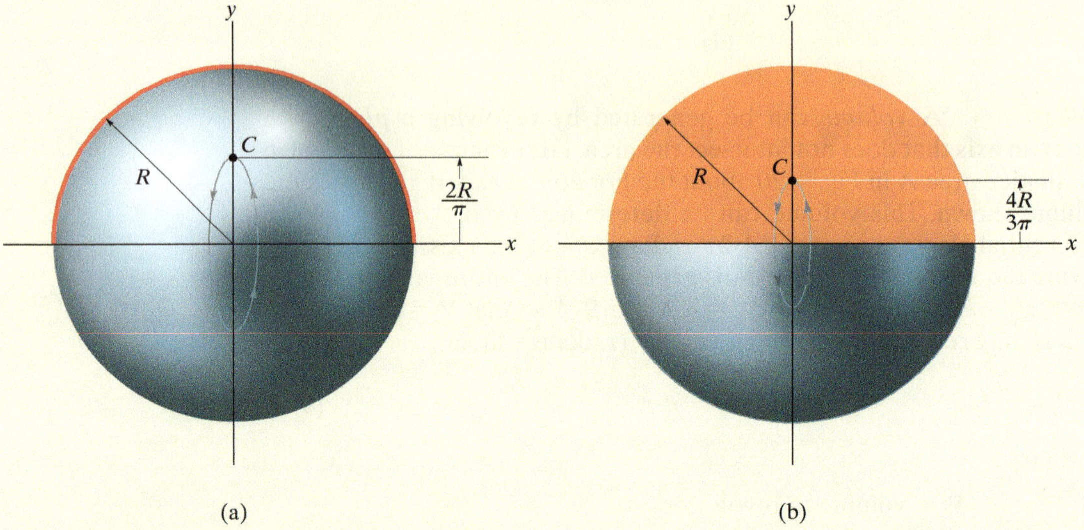

Fig. 9–21

SOLUTION

Surface Area. The surface area of the sphere in Fig. 9–21*a* is generated by revolving a semicircular *arc* about the *x* axis. Using the table on the inside back cover, it is seen that the centroid of this arc is located at a distance $\bar{r} = 2R/\pi$ from the axis of revolution (*x* axis). Since the centroid moves through an angle of $\theta = 2\pi$ rad to generate the sphere, then applying Eq. 9–7 we have

$$A = \theta \bar{r} L; \qquad A = 2\pi\left(\frac{2R}{\pi}\right)\pi R = 4\pi R^2 \qquad \textit{Ans.}$$

Volume. The volume of the sphere is generated by revolving the semicircular *area* in Fig. 9–21*b* about the *x* axis. Using the table on the inside back cover to locate the centroid of the area, i.e., $\bar{r} = 4R/3\pi$, and applying Eq. 9–8, we have

$$V = \theta \bar{r} A; \qquad V = 2\pi\left(\frac{4R}{3\pi}\right)\left(\frac{1}{2}\pi R^2\right) = \frac{4}{3}\pi R^3 \qquad \textit{Ans.}$$

9

EXAMPLE 9.13

Determine the surface area and volume of the full solid in Fig. 9–22*a*.

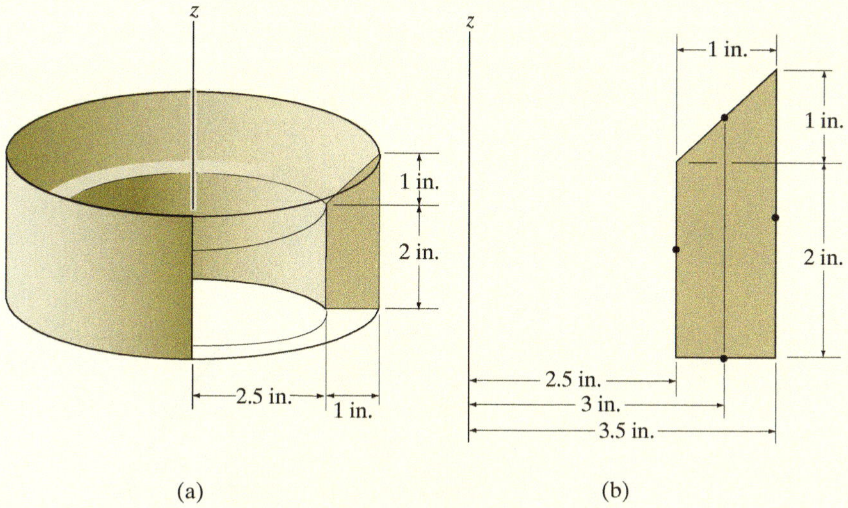

Fig. 9–22

SOLUTION

Surface Area. The surface area is generated by revolving the four line segments shown in Fig. 9–22*b* 2π radians about the *z* axis. The distances from the centroid of each segment to the *z* axis are also shown in the figure. Applying Eq. 9–7 yields

$$A = 2\pi\Sigma\tilde{r}L = 2\pi\left[(2.5\text{ in.})(2\text{ in.}) + (3\text{ in.})\left(\sqrt{(1\text{ in.})^2 + (1\text{ in.})^2}\right) + (3.5\text{ in.})(3\text{ in.}) + (3\text{ in.})(1\text{ in.})\right]$$

$$= 143\text{ in}^2 \qquad \textit{Ans.}$$

Volume. The volume of the solid is generated by revolving the two area segments shown in Fig. 9–22*c* 2π radians about the *z* axis. The distances from the centroid of each segment to the *z* axis are also shown in the figure. Applying Eq. 9–10, we have

$$V = 2\pi\Sigma\tilde{r}A = 2\pi\left\{(3.1667\text{ in.})\left[\frac{1}{2}(1\text{ in.})(1\text{ in.})\right] + (3\text{ in.})[(2\text{ in.})(1\text{ in.})\right\}$$

$$= 47.6\text{ in}^3 \qquad \textit{Ans.}$$

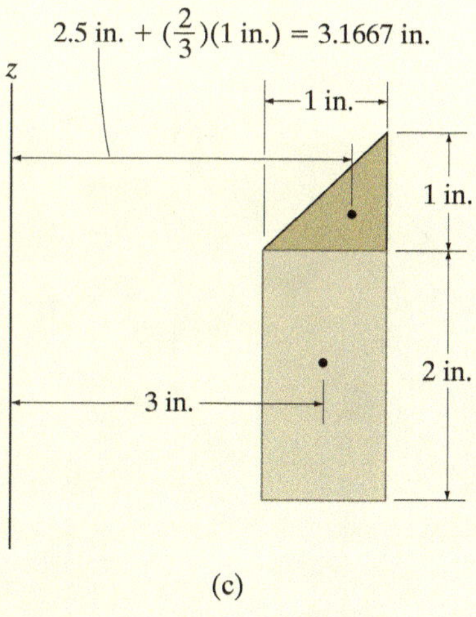

(c)

9

FUNDAMENTAL PROBLEMS

F9–13. Determine the surface area and volume of the solid formed by revolving the shaded area 360° about the z axis.

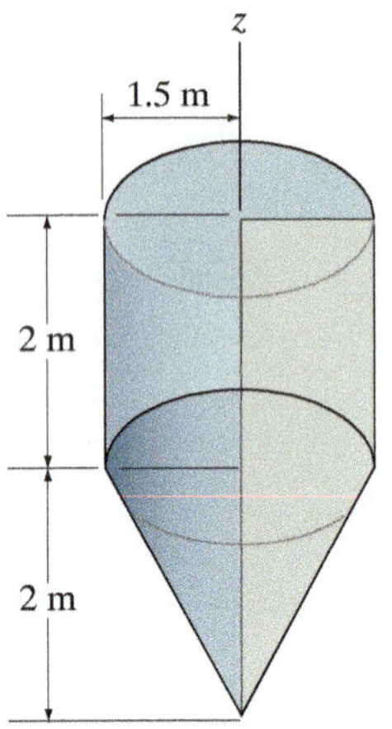

F9–13

F9–14. Determine the surface area and volume of the solid formed by revolving the shaded area 360° about the z axis.

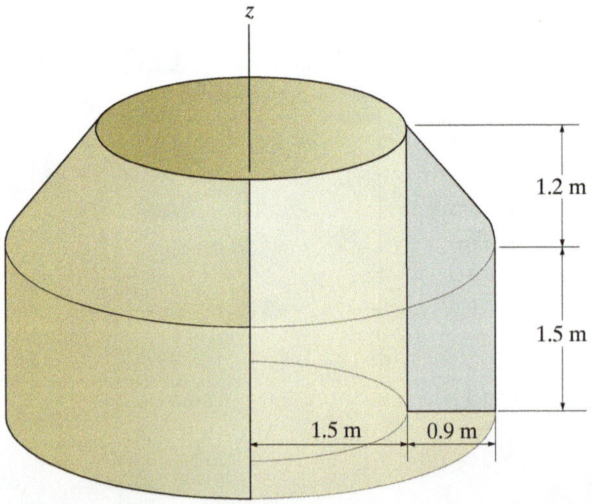

F9–14

F9–15. Determine the surface area and volume of the solid formed by revolving the shaded area 360° about the z axis.

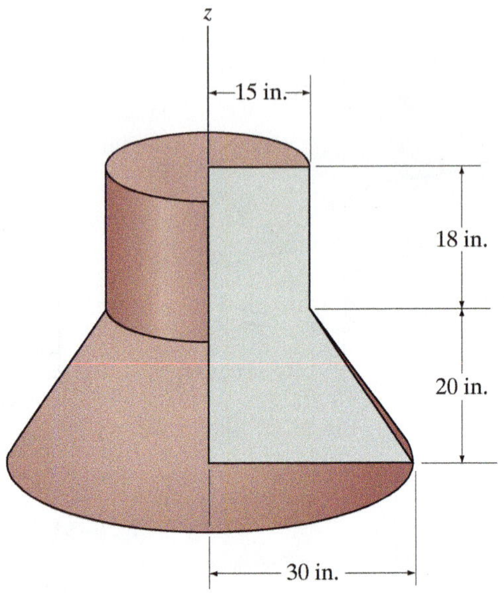

F9–15

F9–16. Determine the surface area and volume of the solid formed by revolving the shaded area 360° about the z axis.

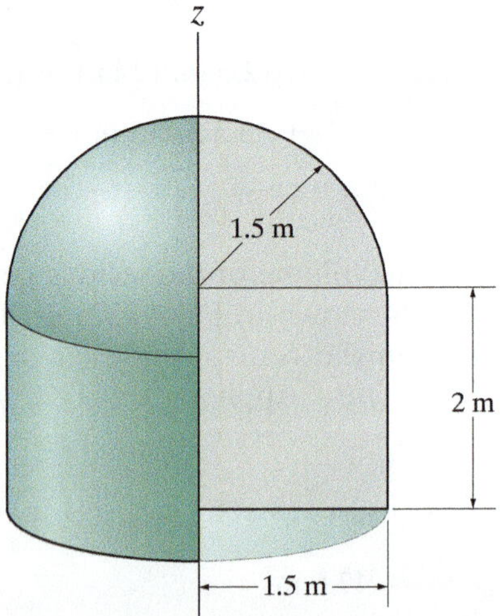

F9–16

9

PROBLEMS

9–90. Determine the outside surface area of the storage tank.

9–91. Determine the volume of the storage tank.

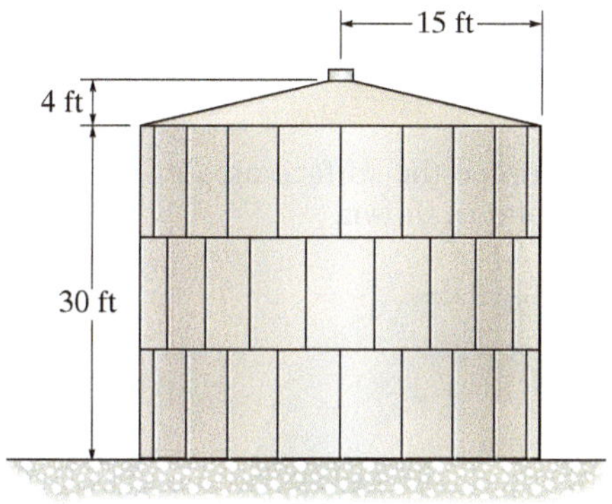

Probs. 9–90/91

9–94. The *rim* of a flywheel has the cross section *A–A* shown. Determine the volume of material needed for its construction.

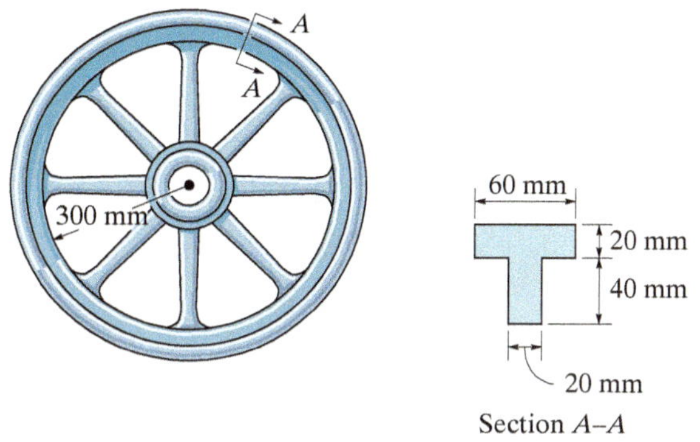

Prob. 9–94

***9–92.** Determine the outside surface area of the hopper.

9–93. The hopper is filled to its top with coal. Determine the volume of coal if the voids (air space) are 35 percent of the volume of the hopper.

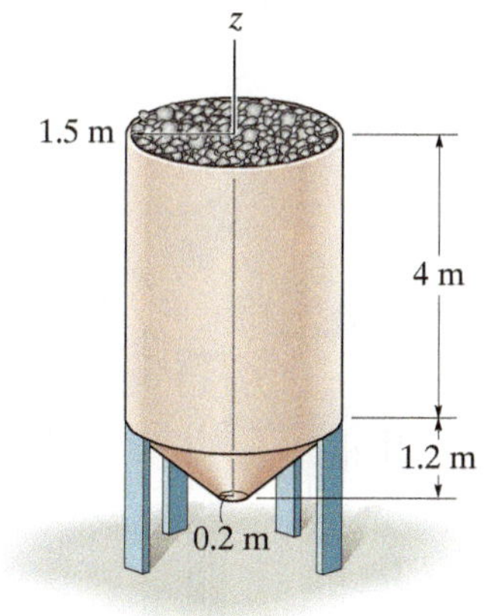

Probs. 9–92/93

9–95. Determine the surface area of the concrete sea wall, excluding its bottom.

***9–96.** A circular sea wall is made of concrete. Determine the total weight of the wall if the concrete has a specific weight of $\gamma_c = 150\ \text{lb/ft}^3$.

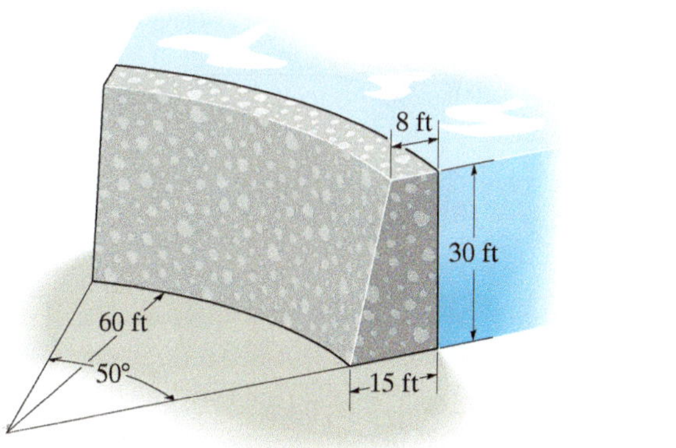

Probs. 9–95/96

9

9–97. The process tank is used to store liquids during manufacturing. Estimate the outside surface area of the tank. The tank has a flat top and the plates from which the tank is made have negligible thickness.

9–98. The process tank is used to store liquids during manufacturing. Estimate the volume of the tank. The tank has a flat top and the plates from which the tank is made have negligible thickness.

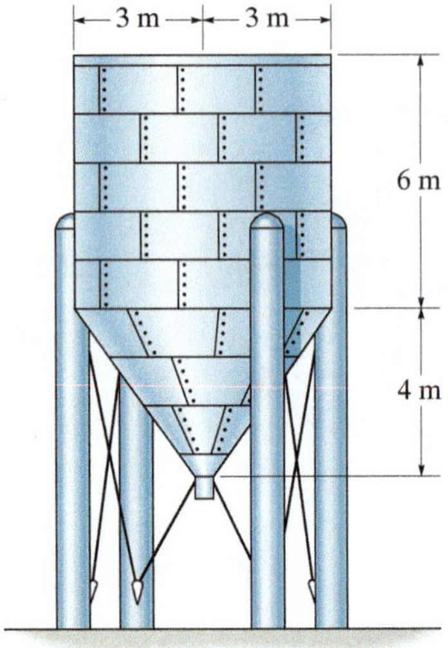

Probs. 9–97/98

9–99. The V-belt has an inner radius of 6 in. and a cross-sectional area as shown. Determine the outside surface area of the belt.

***9–100.** A V-belt has an inner radius of 6 in., and a cross-sectional area as shown. Determine the volume of material used in making the V-belt.

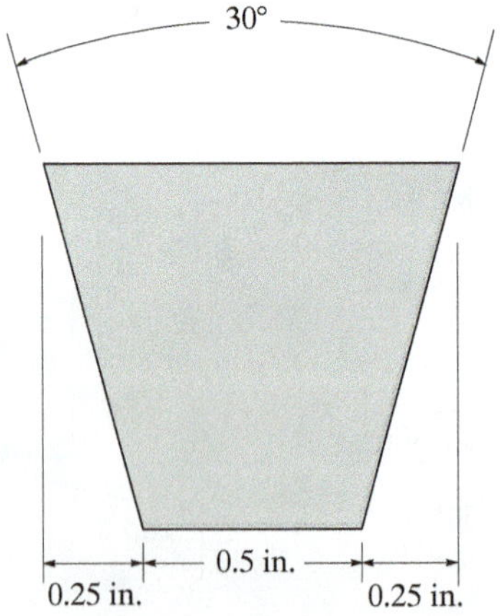

Probs. 9–99/100

9–101. Determine the surface area and the volume of the ring formed by rotating the square about the vertical axis.

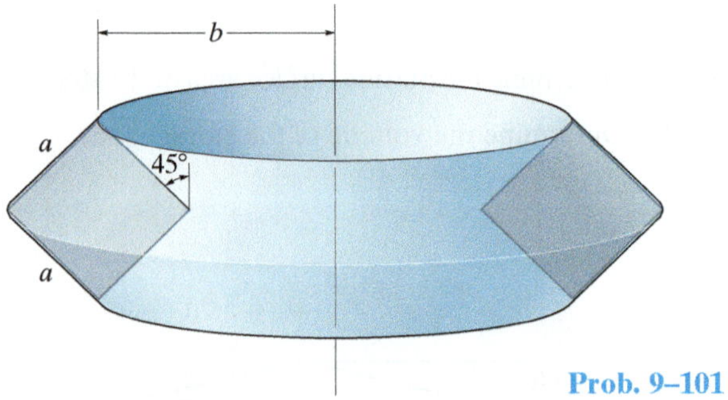

Prob. 9–101

9–102. Determine the surface area of the ring. The cross section is circular as shown.

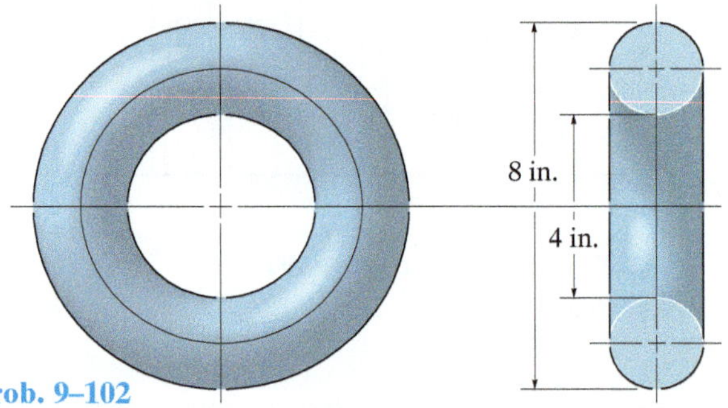

Prob. 9–102

9–103. The water-supply tank has a hemispherical bottom and cylindrical sides. Determine the weight of water in the tank when it is filled to the top at C. Take $\gamma_w = 62.4\ \text{lb/ft}^3$.

***9–104.** Determine the number of gallons of paint needed to paint the outside surface of the water-supply tank, which consists of a hemispherical bottom, cylindrical sides, and conical top. Each gallon of paint can cover $250\ \text{ft}^2$.

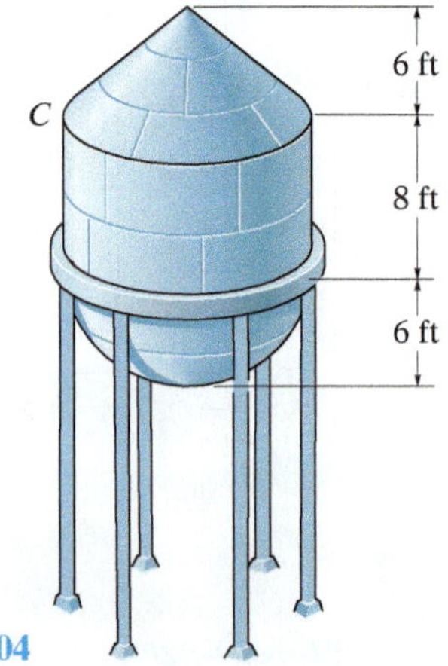

Probs. 9–103/104

9

9–105. The full circular aluminum housing is used in an automotive brake system. The cross section is shown in the figure. Determine its weight if aluminum has a specific weight of 169 lb/ft^3.

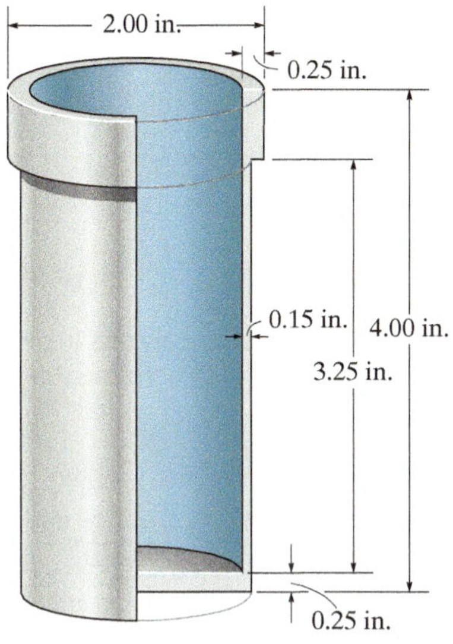

Prob. 9–105

9–106. Determine the volume of an ellipsoid formed by revolving the shaded area about the x axis using the second theorem of Pappus–Guldinus. The area and centroid y of the shaded area should first be obtained by using integration.

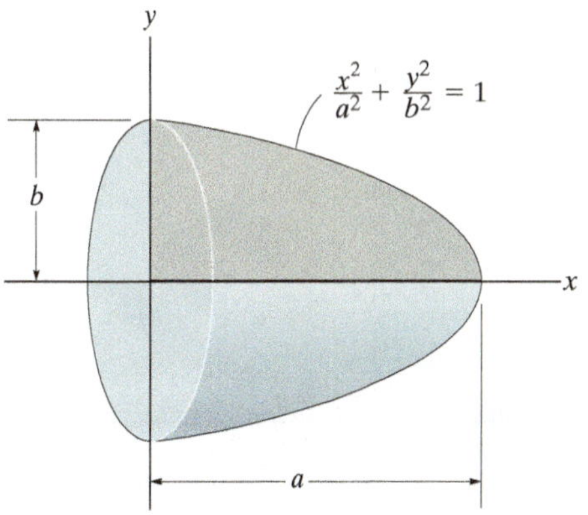

Prob. 9–106

9–107. Using integration, determine both the area and the centroidal distance $\bar{x}$ of the shaded area. Then, using the second theorem of Pappus–Guldinus, determine the volume of the solid generated by revolving the area about the y axis.

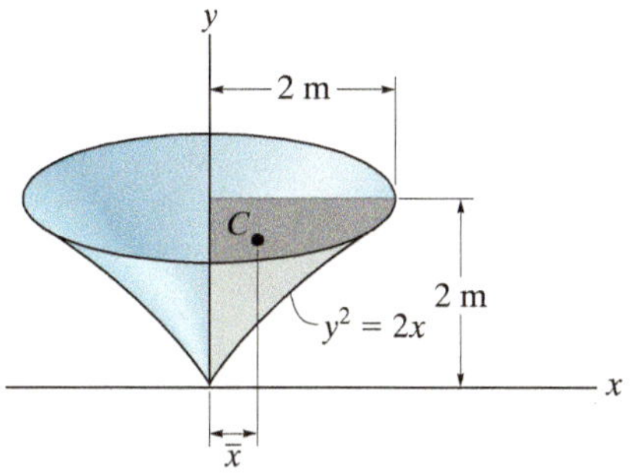

Prob. 9–107

***9–108.** Determine the height h to which liquid should be poured into the conical cup so that it contacts half the surface area on the inside of the cup.

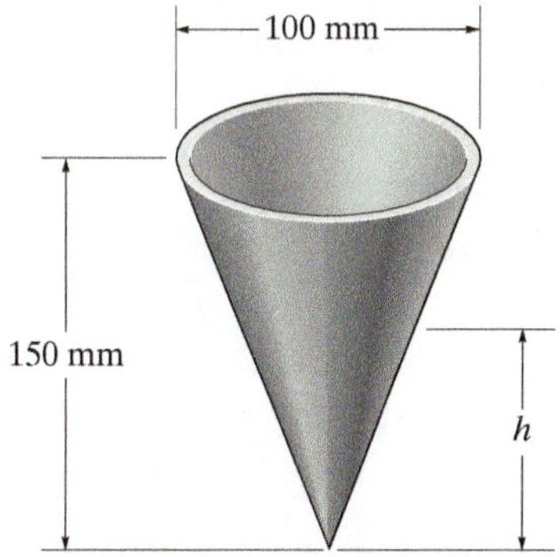

Prob. 9–108

9–109. Determine the volume of the solid formed by revolving the shaded area about the u–u axis using the second theorem of Pappus–Guldinus. The area and centroid of the area should first be obtained by using integration.

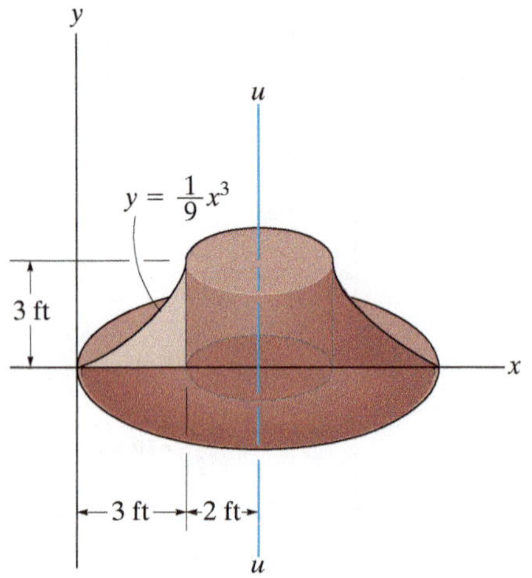

Prob. 9–109

9–110. Determine the volume of material needed to make the casting.

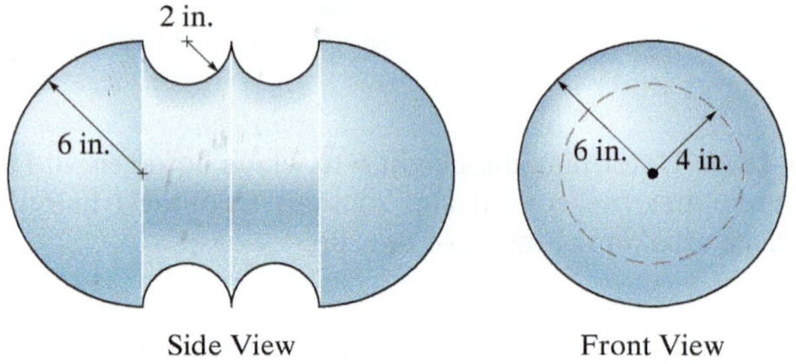

Prob. 9–110

9–111. Determine the height h to which liquid should be poured into the cup so that it contacts half the surface area on the inside of the cup. Neglect the cup's thickness for the calculation.

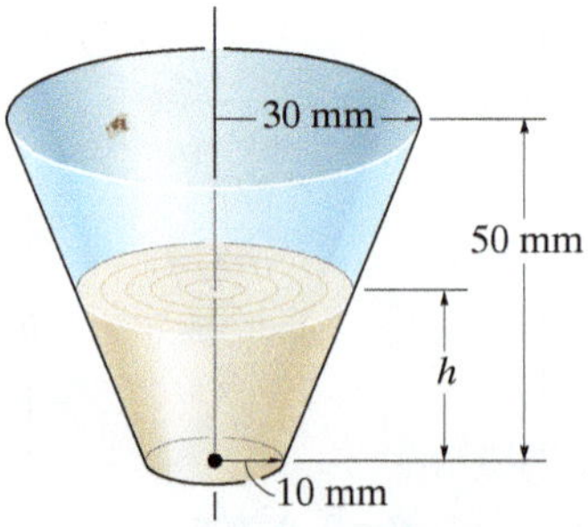

Prob. 9–111

***9–112.** The water tank has a paraboloid-shaped roof. If one liter of paint can cover 3 m^2 of the tank, determine the number of liters required to coat the roof.

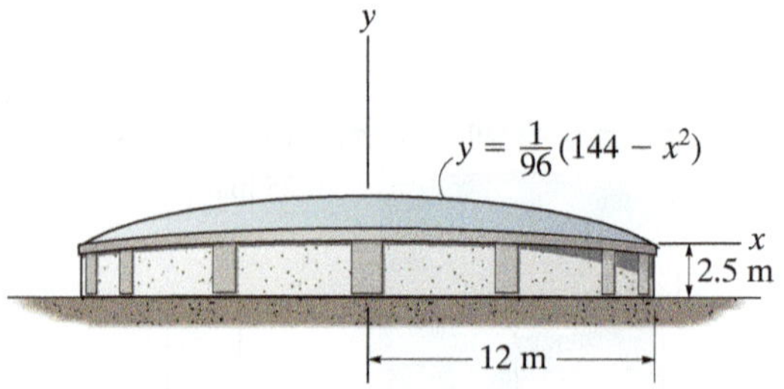

Prob. 9–112

9–113. A steel wheel has a diameter of 840 mm and a cross section as shown in the figure. Determine the total mass of the wheel if $\rho = 5\ \mathrm{Mg/m^3}$.

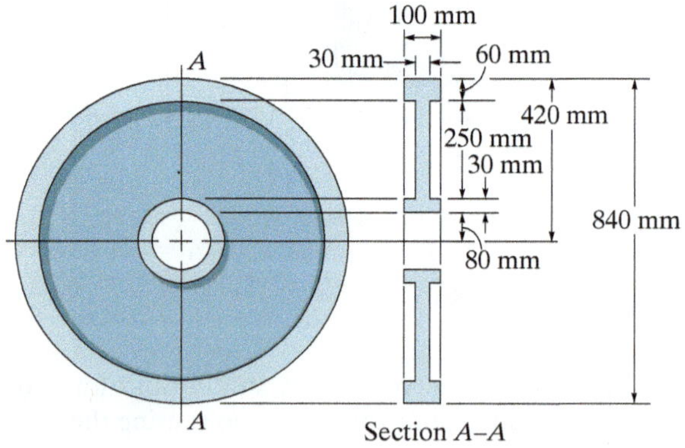

Prob. 9–113

9–114. Determine the surface area of the roof of the structure if it is formed by rotating the parabola about the y axis.

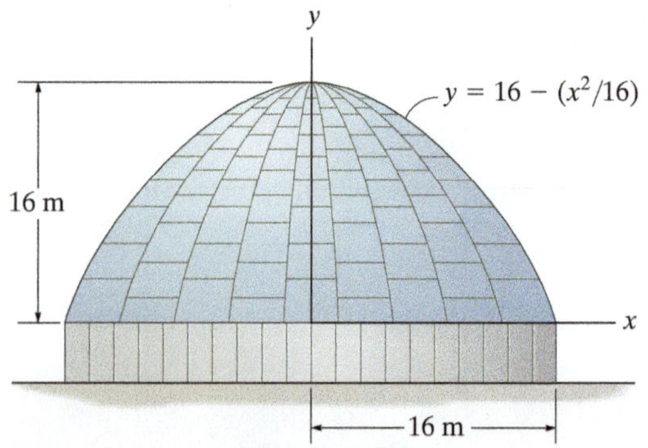

Prob. 9–114

*9.4 Resultant of a General Distributed Loading

In Sec. 4.9, we discussed the method used to simplify a two-dimensional distributed loading to a single resultant force acting at a specific point. In this section we will generalize this method to include flat surfaces that have an arbitrary shape and are subjected to a variable load distribution. Consider, for example, the flat plate shown in Fig. 9–23*a*, which is subjected to the loading defined by $p = p(x, y)$ Pa, where 1 Pa (pascal) $= 1\ \text{N/m}^2$. Knowing this function, we can determine the resultant force $\mathbf{F}_R$ acting on the plate and its location $(\bar{x}, \bar{y})$, Fig. 9–23*b*.

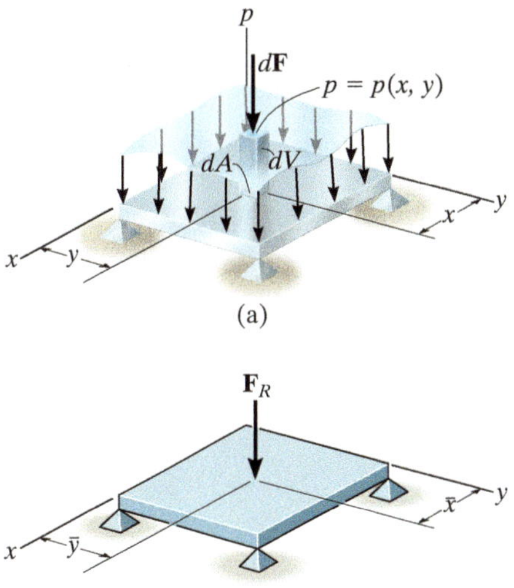

Fig. 9–23

Magnitude of Resultant Force. The force $d\mathbf{F}$ acting on the differential area dA m^2 of the plate, located at the arbitrary point (x, y), has a magnitude of $dF = [p(x, y)\ \text{N/m}^2](dA\ \text{m}^2) = [p(x, y)\ dA]$ N. Notice that $p(x, y)\ dA = dV$, the colored differential *volume element* shown in Fig. 9–23*a*. The *magnitude* of $\mathbf{F}_R$ is the sum of the differential forces acting over the plate's *entire surface area A*. Thus:

$$F_R = \Sigma F; \qquad F_R = \int_A p(x, y)\, dA = \int_V dV = V \tag{9–11}$$

This result indicates that the *magnitude of the resultant force is equal to the total volume under the distributed-loading diagram.*

The resultant of a wind loading that is distributed on the front or side walls of this building must be calculated using integration in order to design the framework that holds the building together.

Location of Resultant Force. The location $(\bar{x}, \bar{y})$ of $\mathbf{F}_R$ is determined by setting the moments of $\mathbf{F}_R$ equal to the moments of all the differential forces $d\mathbf{F}$ about the respective y and x axes: From Figs. 9–23*a* and 9–23*b*, using Eq. 9–11, this results in

$$\bar{x} = \frac{\displaystyle\int_A x p(x, y)\, dA}{\displaystyle\int_A p(x, y)\, dA} = \frac{\displaystyle\int_V x\, dV}{\displaystyle\int_V dV} \qquad \bar{y} = \frac{\displaystyle\int_A y p(x, y)\, dA}{\displaystyle\int_A p(x, y)\, dA} = \frac{\displaystyle\int_V y\, dV}{\displaystyle\int_V dV} \tag{9–12}$$

Hence, the *line of action of the resultant force passes through the geometric center or centroid of the volume under the distributed-loading diagram.*

9

*9.5 Fluid Pressure

According to Pascal's law, a fluid at rest creates a pressure p at a point that is the *same* in *all* directions. The magnitude of p, measured as a force per unit area, depends on the specific weight γ or mass density ρ of the fluid and the depth z of the point from the fluid surface.* The relationship can be expressed mathematically as

$$p = \gamma z = \rho g z \tag{9–13}$$

where g is the acceleration due to gravity. This equation is valid only for fluids that are assumed *incompressible*, as in the case of most liquids. Gases are compressible fluids, and since their density changes significantly with both pressure and temperature, Eq. 9–13 cannot be used.

To illustrate how Eq. 9–13 is applied, consider the submerged plate shown in Fig. 9–24. Three points on the plate have been specified. Since point B is at depth z_1 from the liquid surface, the *pressure* at this point has a magnitude $p_1 = \gamma z_1$. Likewise, points C and D are both at depth z_2; hence, $p_2 = \gamma z_2$. In all cases, the pressure acts *normal* to the surface area dA located at the specified point.

Using Eq. 9–13 and the results of Sec. 9.4, it is possible to determine the resultant force caused by a liquid and specify its location on the surface of a submerged plate. Three different shapes of plates will now be considered.

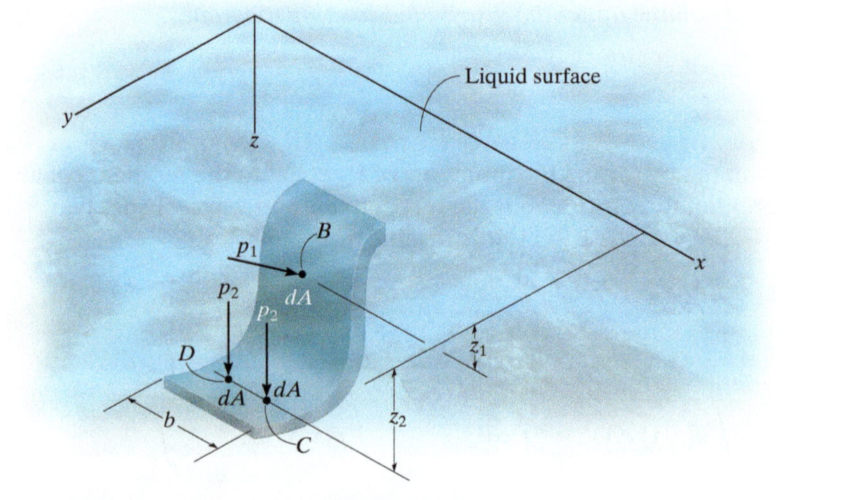

Fig. 9–24

*In particular, for water $\gamma = 62.4\ \text{lb/ft}^3$, or $\gamma = \rho g = 9810\ \text{N/m}^3$ since $\rho = 1000\ \text{kg/m}^3$ and $g = 9.81\ \text{m/s}^2$.

9

Flat Plate of Constant Width. A flat rectangular plate of constant width, which is submerged in a liquid having a specific weight γ, is shown in Fig. 9–25*a*. Since pressure varies linearly with depth, Eq. 9–13, the distribution of pressure over the plate's surface is represented by a trapezoidal volume having an intensity of $p_1 = \gamma z_1$ at depth z_1 and $p_2 = \gamma z_2$ at depth z_2. As noted in Sec. 9.4, the magnitude of the *resultant force* $\mathbf{F}_R$ is equal to the *volume* of this loading diagram and $\mathbf{F}_R$ has a *line of action* that passes through the volume's centroid C. Hence, $\mathbf{F}_R$ does *not* act at the centroid of the plate; rather, it acts at point P, called the *center of pressure*.

Since the plate has a *constant width*, the loading distribution may also be viewed in two dimensions, Fig. 9–25*b*. Here the loading intensity is measured as force/length and varies linearly from $w_1 = bp_1 = b\gamma z_1$ to $w_2 = bp_2 = b\gamma z_2$. The magnitude of $\mathbf{F}_R$ in this case equals the trapezoidal *area*, and $\mathbf{F}_R$ has a *line of action* that passes through the area's *centroid* C. For numerical applications, the area and location of the centroid for a trapezoid are tabulated on the inside back cover.

The walls of the tank must be designed to support the pressure loading of the liquid that is contained within it.

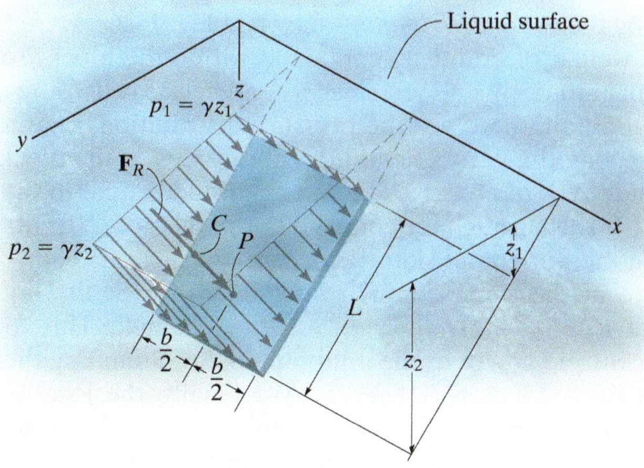

(a)

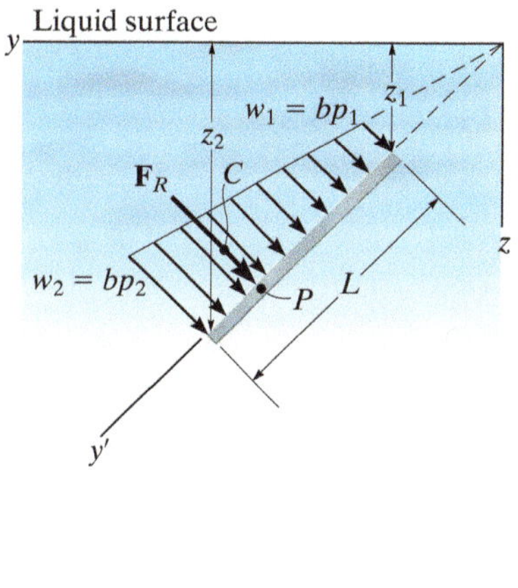

(b)

Fig. 9–25

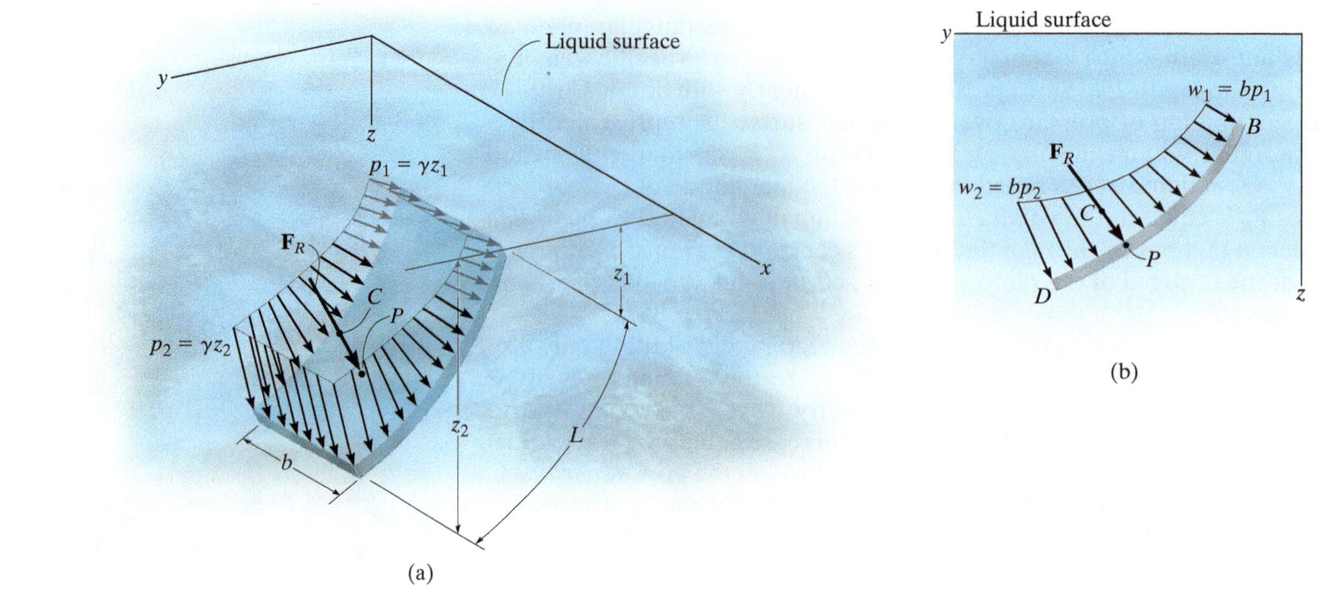

Fig. 9–26

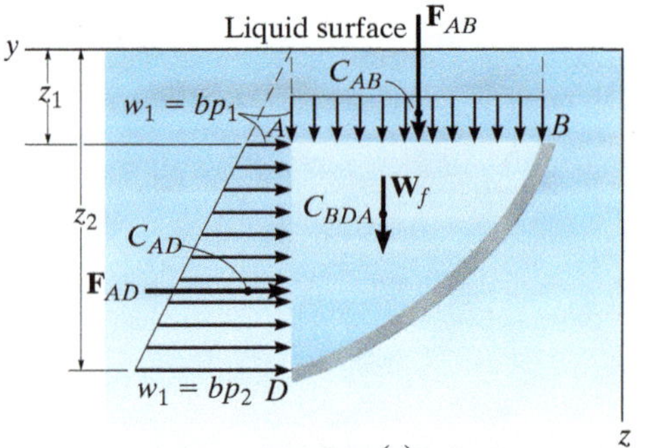

Curved Plate of Constant Width. When a submerged plate of constant width is curved, the pressure acting normal to the plate continually changes both its magnitude and direction, and therefore calculation of the magnitude of $\mathbf{F}_R$ and its location P is more difficult than for a flat plate. Three- and two-dimensional views of the loading distribution are shown in Figs. 9–26*a* and 9–26*b*, respectively. Although integration can be used to solve this problem, a simpler method exists. This method requires separate calculations for the horizontal and vertical *components* of $\mathbf{F}_R$.

For example, the distributed loading acting on the plate can be represented by the *equivalent loading* shown in Fig. 9–26*c*. Here the plate supports the weight of liquid W_f contained within the block BDA. This force has a magnitude $W_f = (\gamma b)(\text{area}_{BDA})$ and acts through the centroid of BDA. In addition, there are the pressure distributions caused by the liquid acting along the vertical and horizontal sides of the block. Along the vertical side AD, the force $\mathbf{F}_{AD}$ has a magnitude equal to the area of the trapezoid. It acts through the centroid C_{AD} of this area. The distributed loading along the horizontal side AB is *constant* since all points lying in this plane are at the same depth from the surface of the liquid. The magnitude of $\mathbf{F}_{AB}$ is simply the area of the rectangle. This force acts through the centroid C_{AB} or at the midpoint of AB. Summing these three forces yields $\mathbf{F}_R = \Sigma\mathbf{F} = \mathbf{F}_{AD} + \mathbf{F}_{AB} + \mathbf{W}_f$. Finally, the location of the center of pressure P on the plate is determined by applying $M_R = \Sigma M$, which states that the moment of the resultant force about a convenient reference point such as D or B, in Fig. 9–26*b*, is equal to the sum of the moments of the three forces in Fig. 9–26*c* about this same point.

9

Flat Plate of Variable Width. The pressure distribution acting on the surface of a submerged plate having a variable width is shown in Fig. 9–27. If we consider the force $d\mathbf{F}$ acting on the differential area strip dA, parallel to the x axis, then its magnitude is $dF = p\,dA$. Since the depth of dA is z, the pressure on the element is $p = \gamma z$. Therefore, $dF = (\gamma z)dA$ and so the resultant force becomes

$$F_R = \int dF = \gamma \int z\,dA$$

If the depth to the centroid C' of the area is $\bar{z}$, Fig. 9–27, then, $\int z\,dA = \bar{z}A$. Substituting, we have

$$F_R = \gamma \bar{z} A \tag{9–14}$$

In other words, *the magnitude of the resultant force acting on any flat plate is equal to the product of the area A of the plate and the pressure* $p = \gamma\bar{z}$ *at the depth of the area's centroid* C'. As discussed in Sec. 9.4, this force is also equivalent to the volume under the pressure distribution. Realize that its line of action passes through the centroid C of this *volume* and intersects the plate at the center of pressure P, Fig. 9–27. Notice that the location of C' does not coincide with the location of P.

The resultant force of the water pressure and its location on the elliptical back plate of this tank truck must be determined by integration.

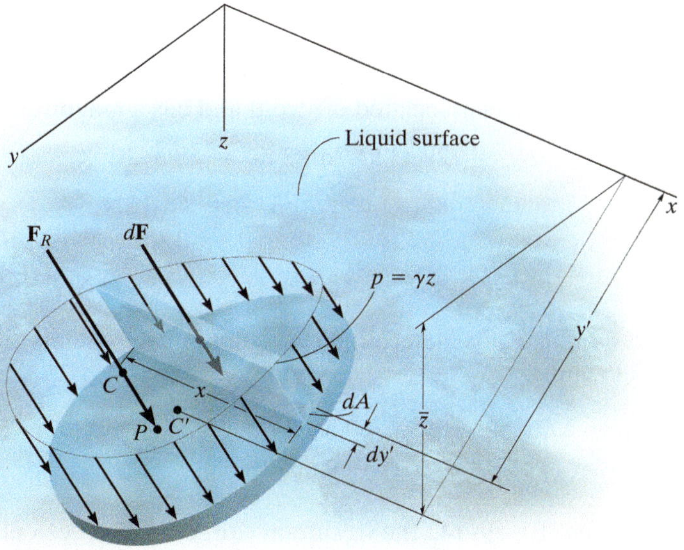

Fig. 9–27

9

EXAMPLE 9.14

Determine the magnitude and location of the resultant hydrostatic force acting on the submerged rectangular plate *AB* shown in Fig. 9–28*a*. The plate has a width of 1.5 m; $\rho_w = 1000\ \text{kg/m}^3$.

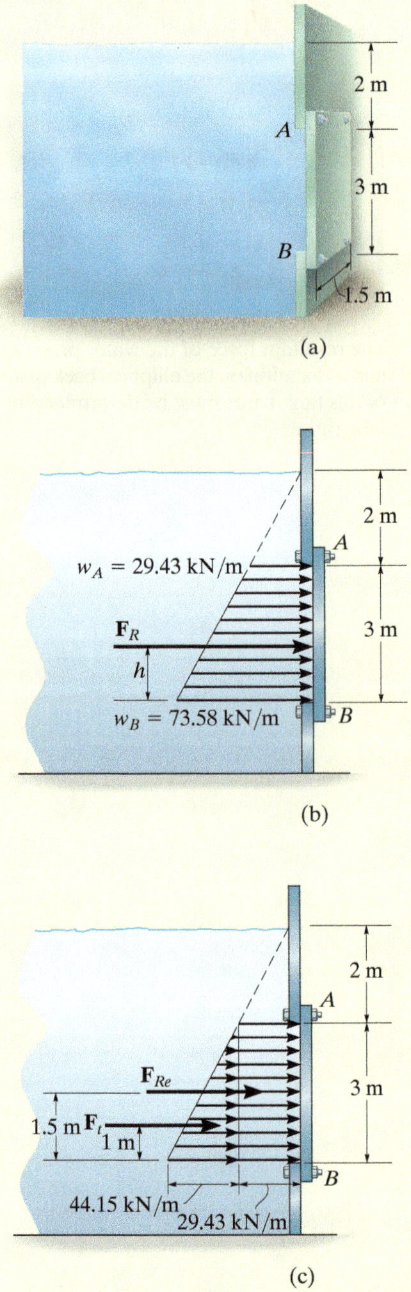

Fig. 9–28

SOLUTION I

The water pressures at depths *A* and *B* are

$$p_A = \rho_w g z_A = (1000\ \text{kg/m}^3)(9.81\ \text{m/s}^2)(2\ \text{m}) = 19.62\ \text{kPa}$$

$$p_B = \rho_w g z_B = (1000\ \text{kg/m}^3)(9.81\ \text{m/s}^2)(5\ \text{m}) = 49.05\ \text{kPa}$$

Since the plate has a constant width, the pressure loading can be viewed in two dimensions as shown in Fig. 9–28*b*. The intensities of the load at *A* and *B* are

$$w_A = bp_A = (1.5\ \text{m})(19.62\ \text{kPa}) = 29.43\ \text{kN/m}$$

$$w_B = bp_B = (1.5\ \text{m})(49.05\ \text{kPa}) = 73.58\ \text{kN/m}$$

From the table on the inside back cover, the magnitude of the resultant force $\mathbf{F}_R$ created by this distributed load is

$$F_R = \text{area of a trapezoid} = \tfrac{1}{2}(3)(29.4 + 73.6) = 154.5\ \text{kN} \qquad \textit{Ans.}$$

This force acts through the centroid of this area,

$$h = \frac{1}{3}\left(\frac{2(29.43) + 73.58}{29.43 + 73.58}\right)(3) = 1.29\ \text{m} \qquad \textit{Ans.}$$

measured upward from *B*, Fig. 9–31*b*.

SOLUTION II

The same results can be obtained by considering two components of $\mathbf{F}_R$, defined by the triangle and rectangle shown in Fig. 9–28*c*. Each force acts through its associated centroid and has a magnitude of

$$F_{Re} = (29.43\ \text{kN/m})(3\ \text{m}) = 88.3\ \text{kN}$$

$$F_t = \tfrac{1}{2}(44.15\ \text{kN/m})(3\ \text{m}) = 66.2\ \text{kN}$$

Hence,

$$F_R = F_{Re} + F_t = 88.3 + 66.2 = 154.5\ \text{kN} \qquad \textit{Ans.}$$

The location of $\mathbf{F}_R$ is determined by summing moments about *B*, Figs. 9–28*b* and *c*, i.e.,

$$\circlearrowleft +(M_R)_B = \Sigma M_B;\quad (154.5)h = 88.3(1.5) + 66.2(1)$$

$$h = 1.29\ \text{m} \qquad \textit{Ans.}$$

NOTE: Using Eq. 9–14, the resultant force can be calculated as $F_R = \gamma\bar{z}A = (9810\ \text{N/m}^3)(3.5\ \text{m})(3\ \text{m})(1.5\ \text{m}) = 154.5\ \text{kN}$.

9

EXAMPLE 9.15

Determine the magnitude of the resultant hydrostatic force acting on the surface of a seawall shaped in the form of a parabola as shown in Fig. 9–29*a*. The wall is 5 m long; $\rho_w = 1020 \text{ kg/m}^3$.

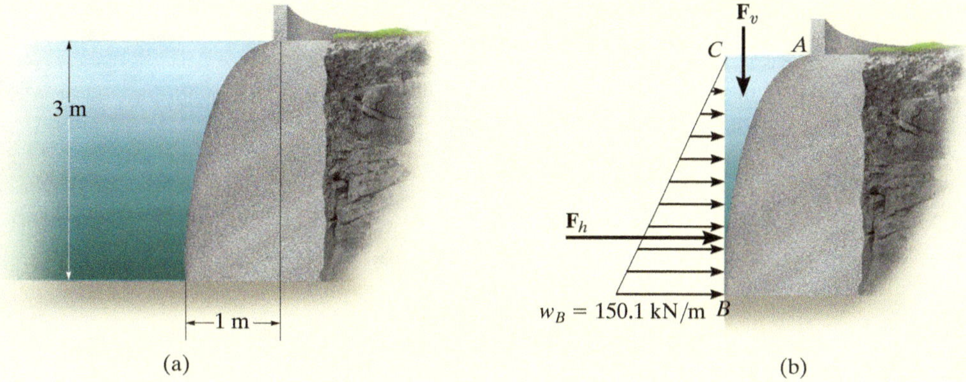

Fig. 9–29

SOLUTION

The horizontal and vertical components of the resultant force will be calculated, Fig. 9–29*b*. Since

$$p_B = \rho_w g z_B = (1020 \text{ kg/m}^3)(9.81 \text{ m/s}^2)(3 \text{ m}) = 30.02 \text{ kPa}$$

then

$$w_B = bp_B = 5 \text{ m}(30.02 \text{ kPa}) = 150.1 \text{ kN/m}$$

Thus,

$$F_h = \tfrac{1}{2}(3 \text{ m})(150.1 \text{ kN/m}) = 225.1 \text{ kN}$$

The area of the parabolic section *ABC* can be determined using the formula for a parabolic area $A = \frac{1}{3}ab$. Hence, the weight of water within this 5-m-long region is

$$F_v = (\rho_w g b)(\text{area}_{ABC})$$
$$= (1020 \text{ kg/m}^3)(9.81 \text{ m/s}^2)(5 \text{ m})\left[\tfrac{1}{3}(1 \text{ m})(3 \text{ m})\right] = 50.0 \text{ kN}$$

The resultant force is therefore

$$F_R = \sqrt{F_h^2 + F_v^2} = \sqrt{(225.1 \text{ kN})^2 + (50.0 \text{ kN})^2}$$
$$= 231 \text{ kN} \qquad \textit{Ans.}$$

9

EXAMPLE 9.16

Determine the magnitude and location of the resultant force acting on the triangular end plates of the water trough shown in Fig. 9–30*a*; $\rho_w = 1000\ \text{kg/m}^3$.

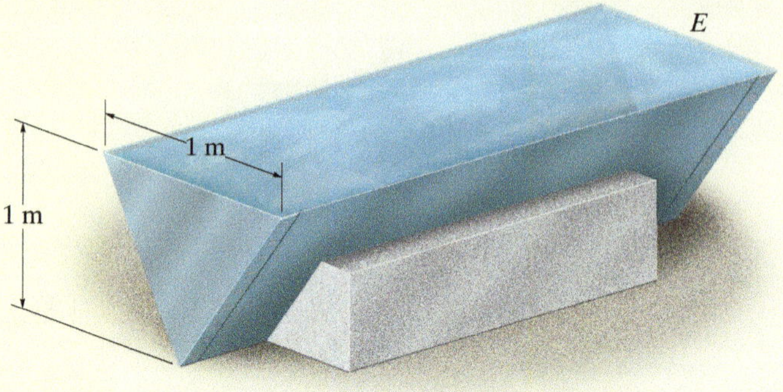

(a)

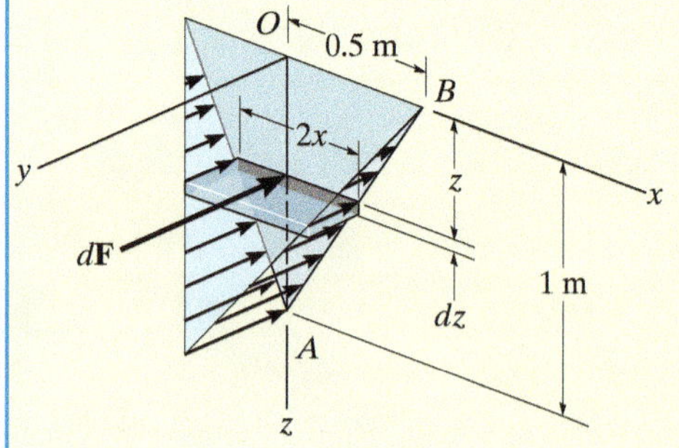

(b)

Fig. 9–30

SOLUTION

The pressure distribution acting on the end plate *E* is shown in Fig. 9–30*b*. The magnitude of the resultant force is equal to the volume of this loading distribution. We will solve the problem by integration. Choosing the differential volume element shown in the figure, we have

$$dF = dV = p\,dA = \rho_w g z(2x\,dz) = 19\,620zx\,dz$$

The equation of line *AB* is

$$x = 0.5(1 - z)$$

Hence, substituting and integrating with respect to z from $z = 0$ to $z = 1$ m yields

$$F = V = \int_V dV = \int_0^{1\text{ m}} (19\,620)z[0.5(1 - z)]\,dz$$

$$= 9810\int_0^{1\text{ m}} (z - z^2)\,dz = 1635\text{ N} = 1.64\text{ kN} \qquad \textit{Ans.}$$

This resultant passes through the *centroid of the volume.* Because of symmetry,

$$\bar{x} = 0 \qquad \textit{Ans.}$$

Since $\tilde{z} = z$ for the volume element, then

$$\bar{z} = \frac{\int_V \tilde{z}\,dV}{\int_V dV} = \frac{\int_0^{1\text{ m}} z(19\,620)z[0.5(1 - z)]\,dz}{1635} = \frac{9810\int_0^{1\text{ m}} (z^2 - z^3)\,dz}{1635}$$

$$= 0.5\text{ m} \qquad \textit{Ans.}$$

NOTE: We can also determine the resultant force by applying Eq. 9–14, $F_R = \gamma\bar{z}A = (9810\text{ N/m}^3)(\tfrac{1}{3})(1\text{ m})[\tfrac{1}{2}(1\text{ m})(1\text{ m})] = 1.64\text{ kN}$.

FUNDAMENTAL PROBLEMS

F9–17. Determine the magnitude of the hydrostatic force acting per meter length of the wall. Water has a density of $\rho = 1\ \text{Mg/m}^3$.

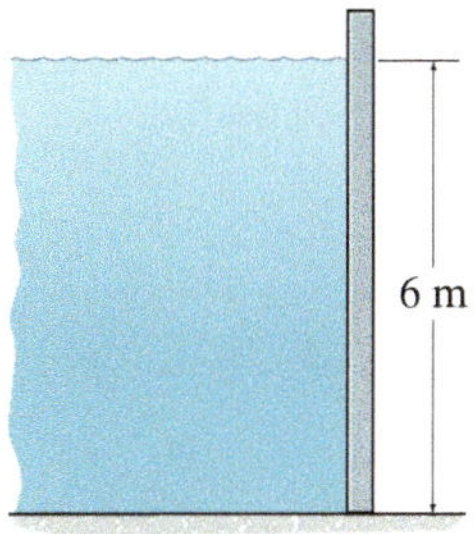

F9–17

F9–18. Determine the magnitude of the hydrostatic force acting on gate AB, which has a width of 4 ft. The specific weight of water is $\gamma = 62.4\ \text{lb/ft}^3$.

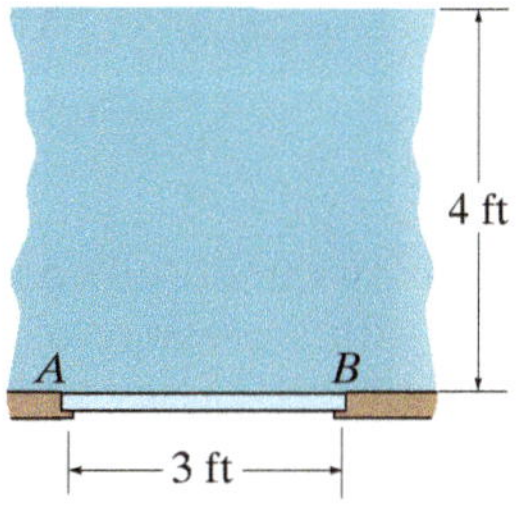

F9–18

F9–19. Determine the magnitude of the hydrostatic force acting on gate AB, which has a width of 1.5 m. Water has a density of $\rho = 1\ \text{Mg/m}^3$.

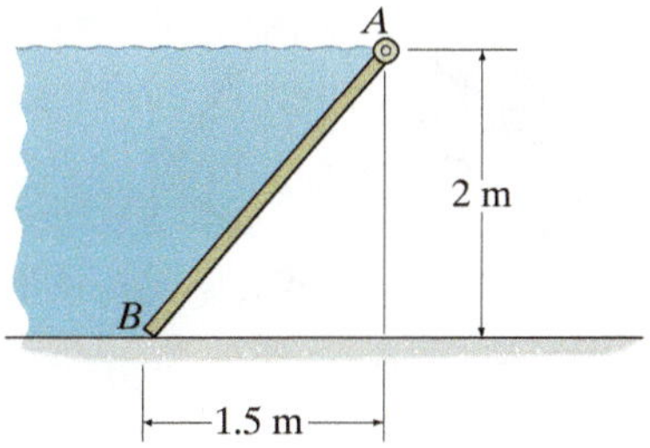

F9–19

F9–20. Determine the magnitude of the hydrostatic force acting on gate AB, which has a width of 2 m. Water has a density of $\rho = 1\ \text{Mg/m}^3$.

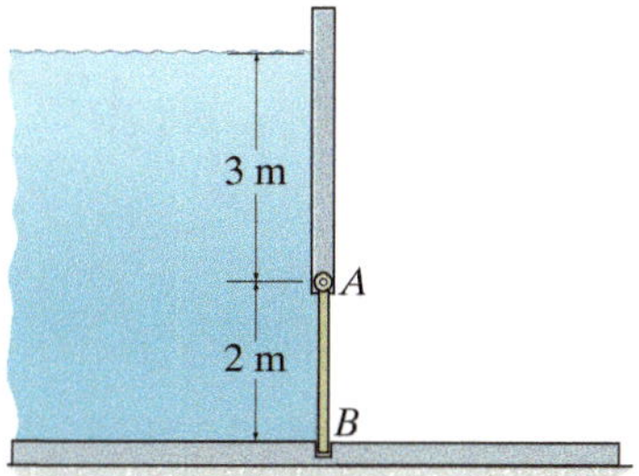

F9–20

F9–21. Determine the magnitude of the hydrostatic force acting on gate AB, which has a width of 2 ft. The specific weight of water is $\gamma = 62.4\ \text{lb/ft}^3$.

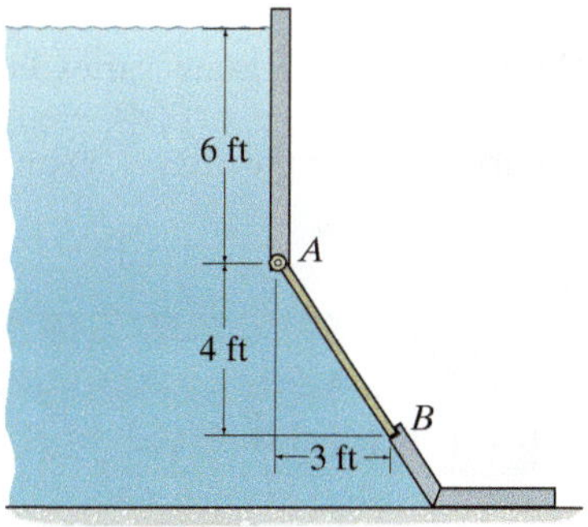

F9–21

PROBLEMS

9–115. The pressure loading on the plate is described by the function $p = 10\,[6/(x + 1) + 8]\ \text{lb/ft}^2$. Determine the magnitude of the resultant force and the coordinates $(\bar{x}, \bar{y})$ of the point where the line of action of the force intersects the plate.

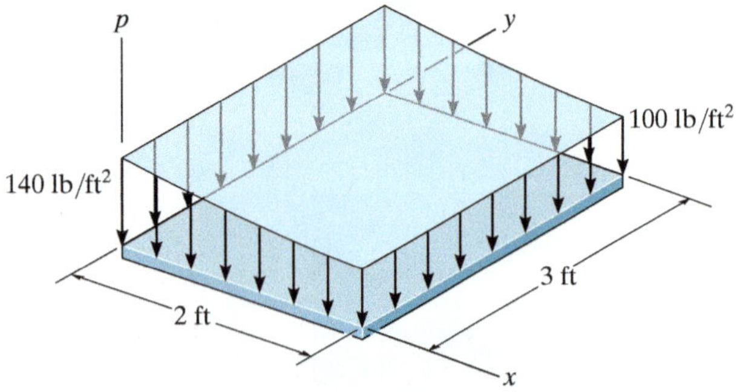

Prob. 9–115

***9–116.** The loading acting on a square plate is represented by a parabolic pressure distribution. Determine the magnitude of the resultant force and the coordinates $(\bar{x}, \bar{y})$ of the point where the line of action of the force intersects the plate. Also, what are the reactions at the rollers B and C and the ball-and-socket joint A? Neglect the weight of the plate.

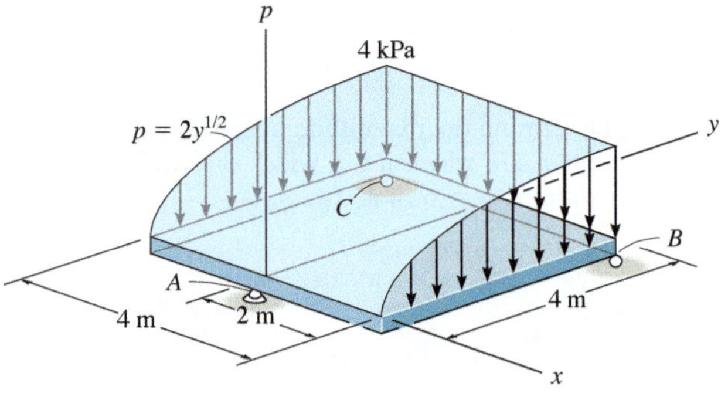

Prob. 9–116

9–117. The load over the plate varies linearly along the sides of the plate such that $p = \frac{2}{3}[x(4 - y)]$ kPa. Determine the resultant force and its position $(\bar{x}, \bar{y})$ on the plate.

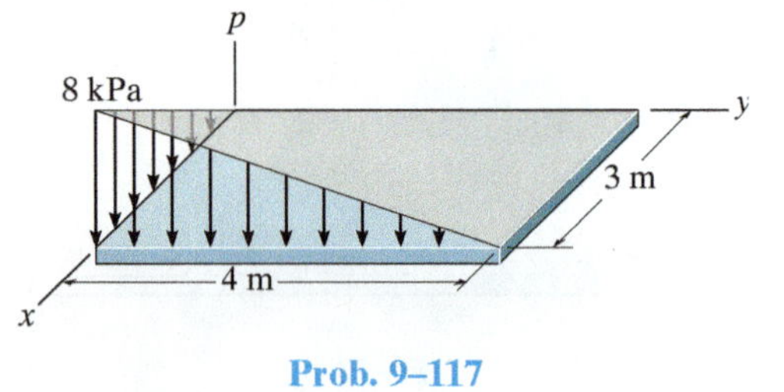

Prob. 9–117

9–118. The rectangular plate is subjected to a distributed load over its *entire surface*. The load is defined by the expression $p = p_0 \sin(\pi x/a) \sin(\pi y/b)$, where p_0 represents the pressure acting at the center of the plate. Determine the magnitude and location of the resultant force acting on the plate.

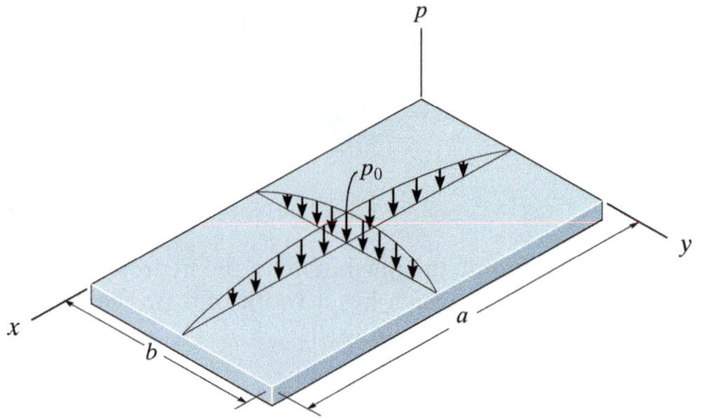

Prob. 9–118

9–119. The wind blows uniformly on the front surface of the metal building with a pressure of $30\ \text{lb/ft}^2$. Determine the resultant force it exerts on the surface and the position of this resultant.

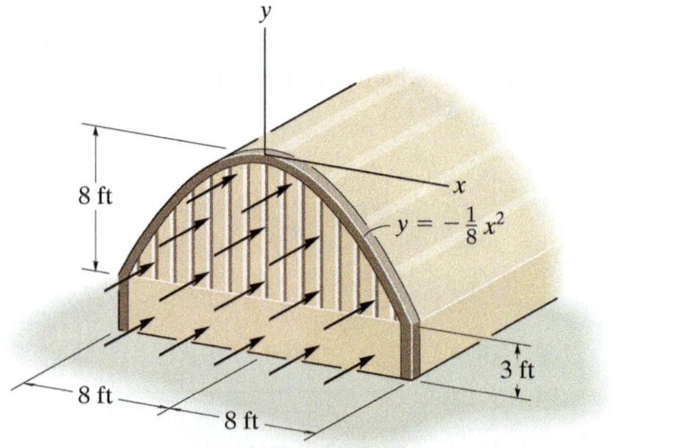

Prob. 9–119

***9–120.** The tank is filled with water to a depth of $d = 4$ m. Determine the resultant force the water exerts on side A and side B of the tank. If oil instead of water is placed in the tank, to what depth d should it reach so that it creates the same resultant forces? $\rho_o = 900\ \text{kg/m}^3$ and $\rho_w = 1000\ \text{kg/m}^3$.

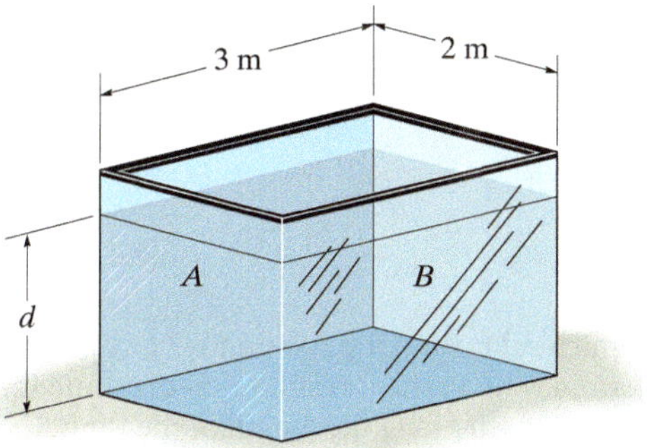

Prob. 9–120

9–121. When the tide water A subsides, the tide gate automatically swings open to drain the marsh B. For the condition of high tide shown, determine the horizontal reactions developed at the hinge C and stop block D. The length of the gate is 6 m and its height is 4 m. $\rho_w = 1.0\ \text{Mg/m}^3$.

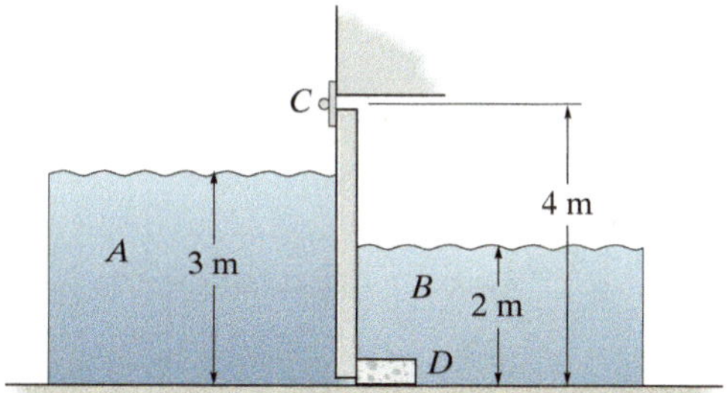

Prob. 9–121

9–122. Determine the resultant horizontal and vertical force components that the water exerts on the side of the dam. The dam is 25 ft long and $\gamma_w = 62.4\ \text{lb/ft}^3$.

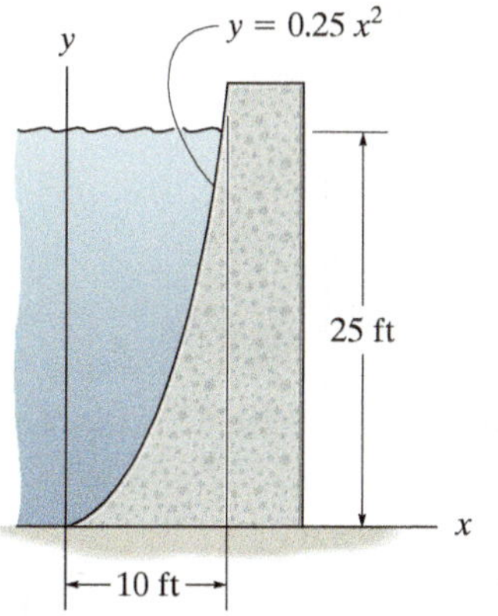

Prob. 9–122

9–123. Determine the magnitude of the resultant hydrostatic force acting on the dam and its location, measured from the top surface of the water. The width of the dam is 8 m; $\rho_w = 1.0\ \text{Mg/m}^3$.

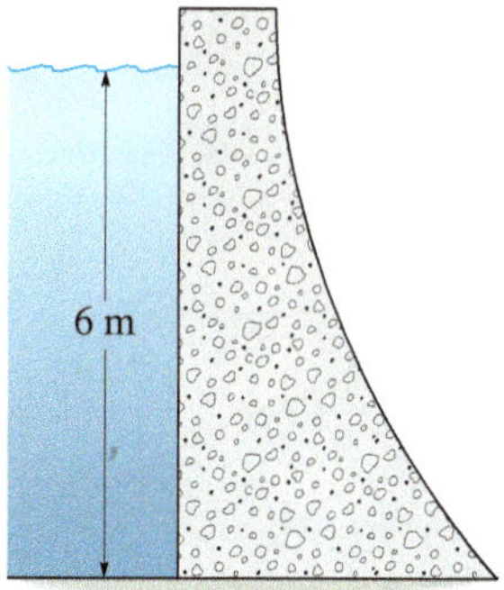

Prob. 9–123

***9–124.** The concrete dam in the shape of a quarter circle. Determine the magnitude of the resultant hydrostatic force that acts on the dam per meter of length. The density of water is $\rho_w = 1\ \text{Mg/m}^3$.

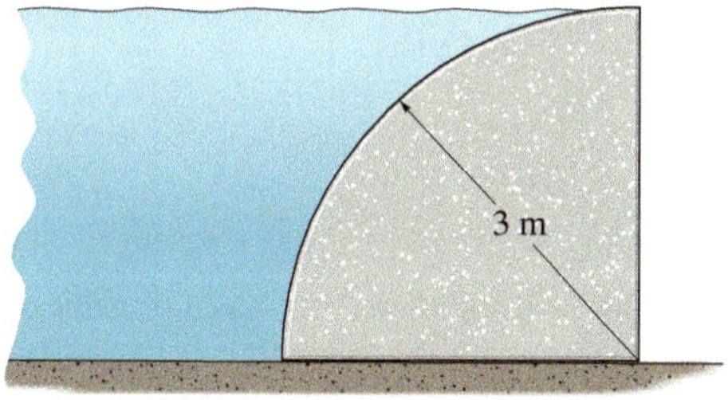

Prob. 9–124

9–125. The storage tank contains oil having a specific weight of $\gamma_o = 56\ \text{lb/ft}^3$. If the tank is 6 ft wide, calculate the resultant force acting on the inclined side BC of the tank, caused by the oil, and specify its location along BC, measured from B. Also compute the total resultant force acting on the bottom of the tank.

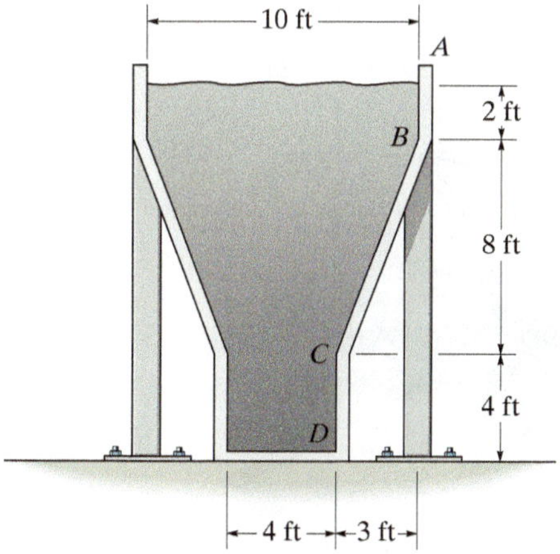

Prob. 9–125

9–126. The tank is filled to the top ($y = 0.5$ m) with water having a density of $\rho_w = 1.0\ \text{Mg/m}^3$. Determine the resultant force of the water pressure acting on the flat end plate C of the tank, and its location, measured from the top of the tank.

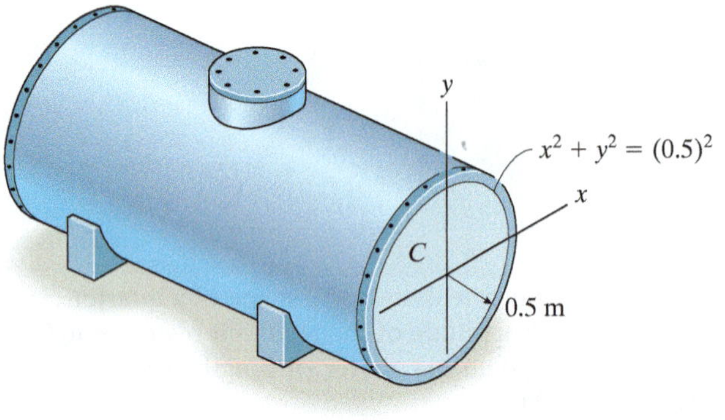

Prob. 9–126

9–127. The tank is filled with a liquid that has a density of $900\ \text{kg/m}^3$. Determine the resultant force that it exerts on the elliptical end plate, and the location of the center of pressure, measured from the x axis.

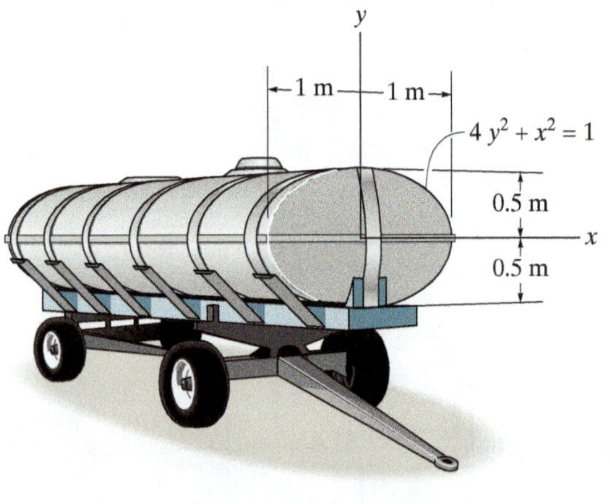

Prob. 9–127

***9–128.** The semicircular drainage pipe is filled with water. Determine the resultant horizontal and vertical force components that the water exerts on the side AB of the pipe per foot of pipe length; $\gamma_w = 62.4\ \text{lb/ft}^3$.

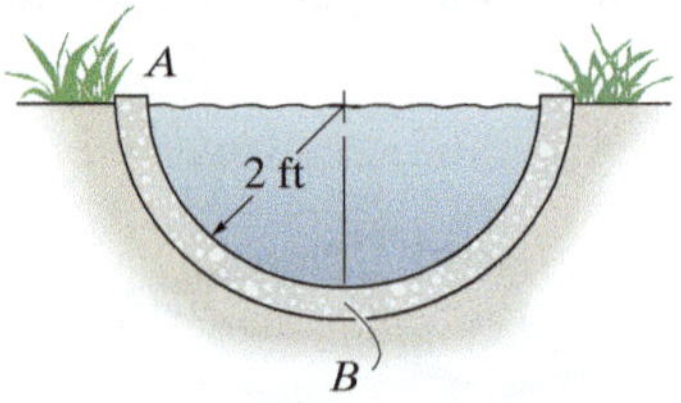

Prob. 9–128

9–129. The semicircular tunnel passes under a river that is 9 m deep. Determine the vertical resultant hydrostatic force acting per meter of length along the length of the tunnel. The tunnel is 6 m wide; $\rho_w = 1.0\ \text{Mg/m}^3$.

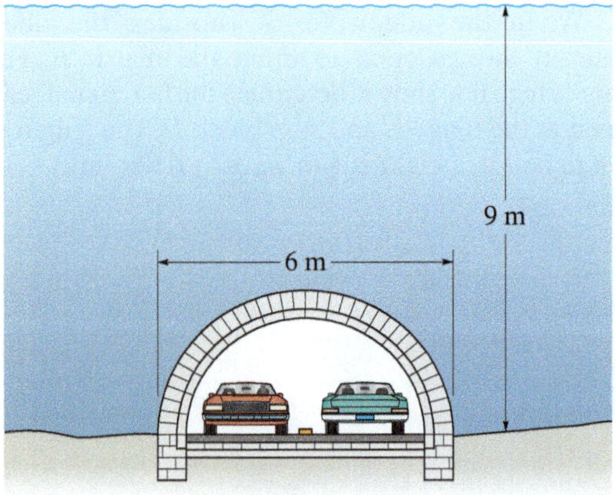

Prob. 9–129

9–130. The arched surface AB is shaped in the form of a quarter circle. If it is 8 m long, determine the horizontal and vertical components of the resultant force caused by the water acting on the surface; $\rho_w = 1.0\ \text{Mg/m}^3$.

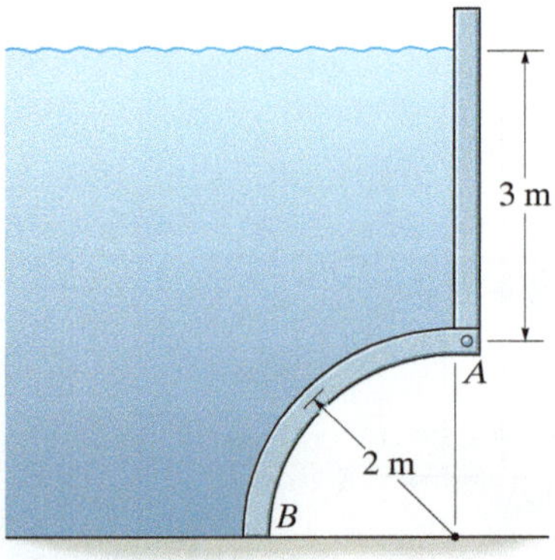

Prob. 9–130

9

CHAPTER REVIEW

Center of Gravity and Centroid

The *center of gravity* G represents a point where the weight of the body can be considered concentrated. The distance from an axis to this point can be determined from a balance of moments, which requires that the moment of the weight of all the particles of the body about this axis must equal the moment of the entire weight of the body about the axis.

$$\bar{x} = \frac{\int \tilde{x}\, dW}{\int dW}$$

$$\bar{y} = \frac{\int \tilde{y}\, dW}{\int dW}$$

$$\bar{z} = \frac{\int \tilde{z}\, dW}{\int dW}$$

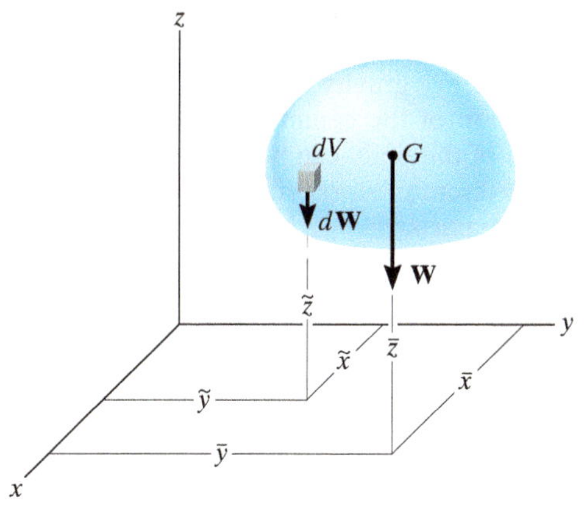

The center of mass will coincide with the center of gravity provided the acceleration of gravity is constant.

$$\bar{x} = \frac{\int_L \tilde{x}\, dL}{\int_L dL} \qquad \bar{y} = \frac{\int_L \tilde{y}\, dL}{\int_L dL} \qquad \bar{z} = \frac{\int_L \tilde{z}\, dL}{\int_L dL}$$

$$\bar{x} = \frac{\int_A \tilde{x}\, dA}{\int_A dA} \qquad \bar{y} = \frac{\int_A \tilde{y}\, dA}{\int_A dA} \qquad \bar{z} = \frac{\int_A \tilde{z}\, dA}{\int_A dA}$$

The *centroid* is the location of the geometric center for the body. It is determined in a similar manner, using a moment balance of geometric elements such as line, area, or volume segments. For bodies having a continuous shape, moments are summed (integrated) using differential elements.

$$\bar{x} = \frac{\int_V \tilde{x}\, dV}{\int_V dV} \qquad \bar{y} = \frac{\int_V \tilde{y}\, dV}{\int_A dV} \qquad \bar{z} = \frac{\int_V \tilde{z}\, dV}{\int_V dV}$$

The center of mass will coincide with the centroid provided the material is homogeneous, i.e., the density of the material is the same throughout. The centroid will always lie on an axis of symmetry.

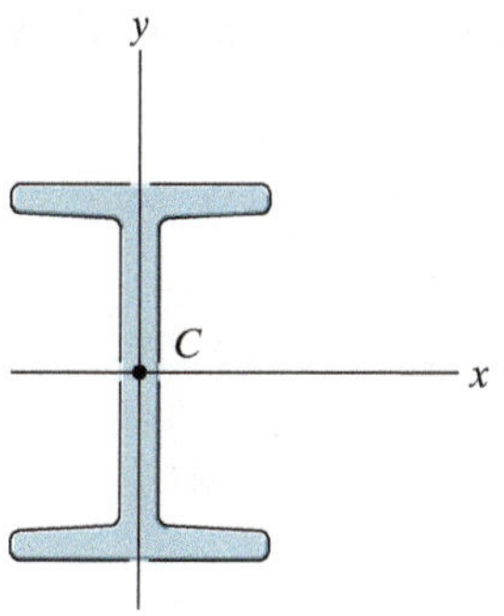

9

Composite Body If the body is a composite of several shapes, each having a known location for its center of gravity or centroid, then the location of the center of gravity or centroid of the body can be determined from a discrete summation using its composite parts.	$\bar{x} = \dfrac{\Sigma \tilde{x} W}{\Sigma W}$ $\bar{y} = \dfrac{\Sigma \tilde{y} W}{\Sigma W}$ $\bar{z} = \dfrac{\Sigma \tilde{z} W}{\Sigma W}$	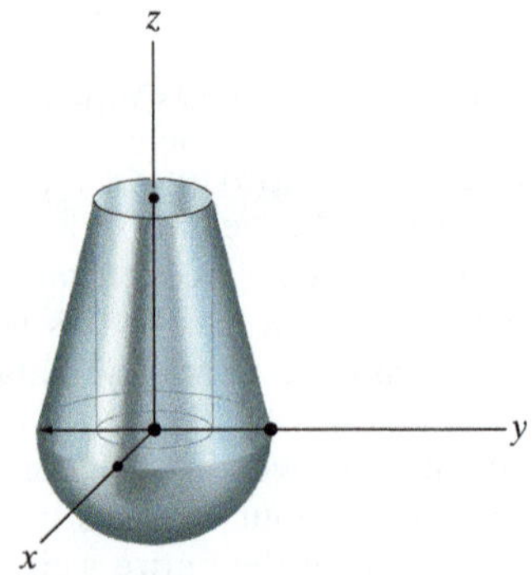
Theorems of Pappus and Guldinus The theorems of Pappus and Guldinus can be used to determine the surface area and volume of a body of revolution.		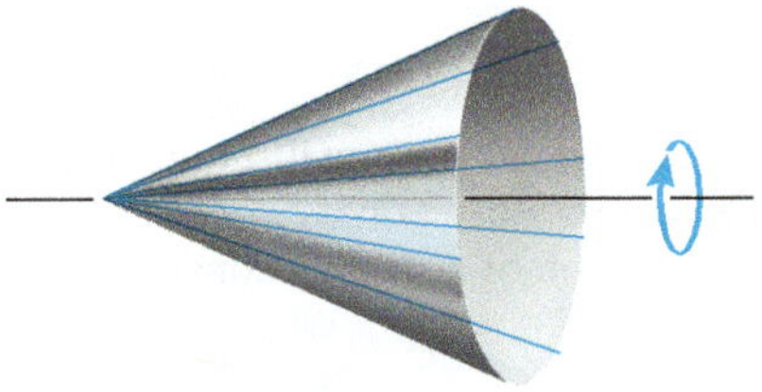
The *surface area* equals the product of the length of the generating curve and the distance traveled by the centroid of the curve needed to generate the area.	$A = \theta \bar{r} L$	
The *volume* of the body equals the product of the generating area and the distance traveled by the centroid of this area needed to generate the volume.	$V = \theta \bar{r} A$	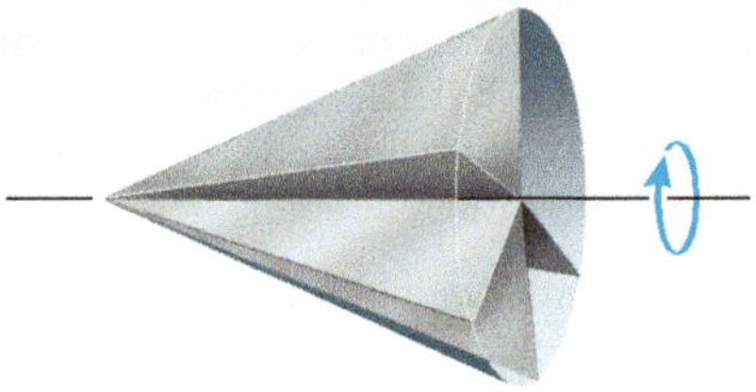

9

General Distributed Loading

The magnitude of the resultant force is equal to the total volume under the distributed-loading diagram. The line of action of the resultant force passes through the geometric center or centroid of this volume.

$$F_R = \int_A p(x, y)\, dA = \int_V dV$$

$$\bar{x} = \frac{\int_V x\, dV}{\int_V dV}$$

$$\bar{y} = \frac{\int_V y\, dV}{\int_V dV}$$

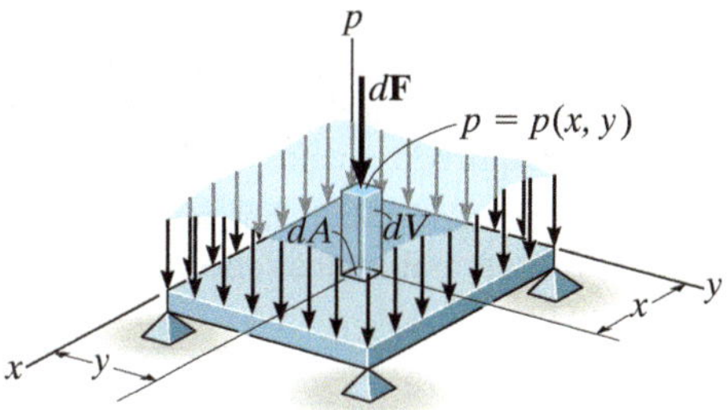

Fluid Pressure

The pressure developed by a liquid at a point on a submerged surface depends upon the depth of the point and the density of the liquid in accordance with Pascal's law, $p = \rho g h = \gamma h$. This pressure will create a *linear distribution* of loading on a flat vertical or inclined surface.

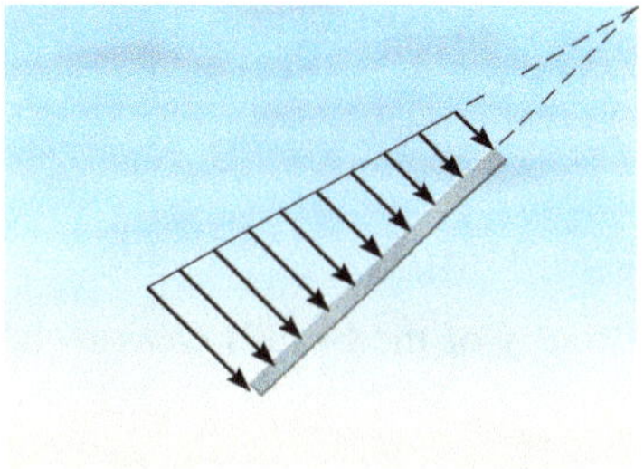

If the surface is horizontal, then the loading will be *uniform*.

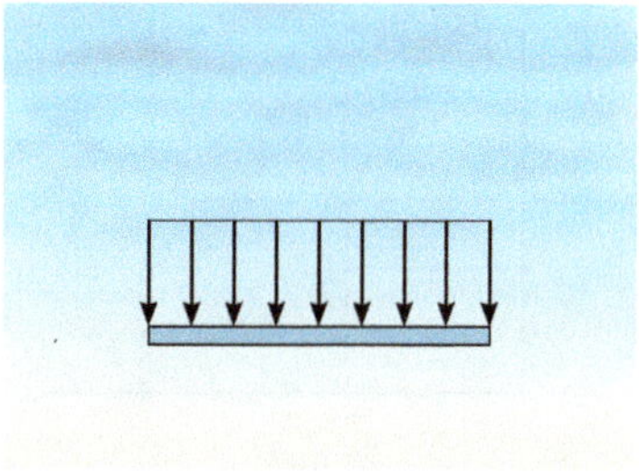

In any case, the resultants of these loadings can be determined by finding the volume under the loading curve or using $F_R = \gamma \bar{z} A$, where $\bar{z}$ is the depth to the centroid of the plate's area. The line of action of the resultant force passes through the centroid of the volume of the loading diagram and acts at a point P on the plate called the center of pressure.

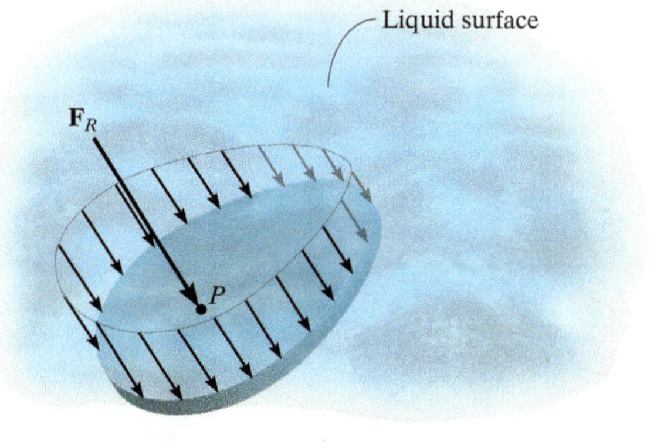

REVIEW PROBLEMS

9–131. A circular V-belt has an inner radius of 600 mm and a cross-sectional area as shown. Determine the volume of material required to make the belt.

***9–132.** A circular V-belt has an inner radius of 600 mm and a cross-sectional area as shown. Determine the surface area of the belt.

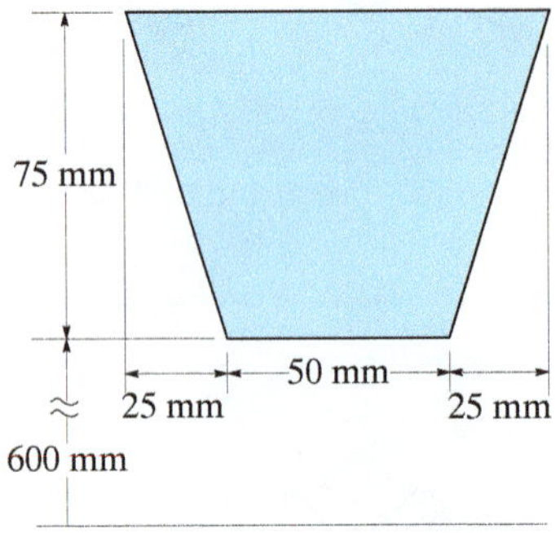

Probs. 9–131/132

9–133. Locate the centroid $\bar{y}$ of the beam's cross-sectional area.

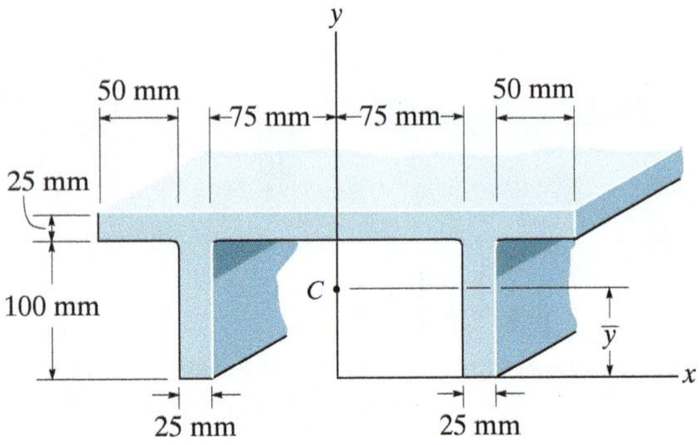

Prob. 9–133

9–134. Locate the centroid of the solid.

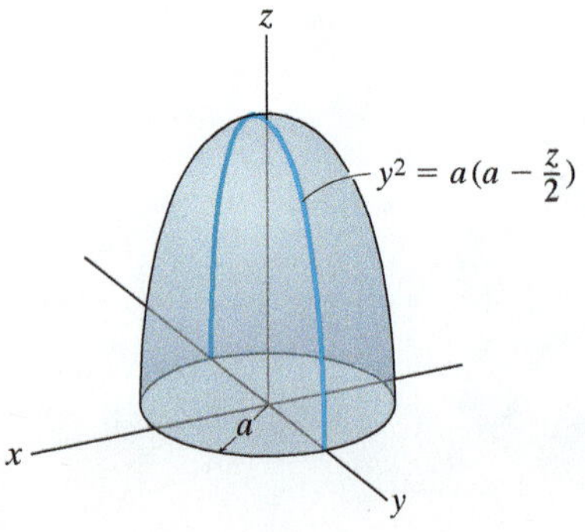

Prob. 9–134

9–135. Determine the magnitude of the resultant hydrostatic force acting per foot of length on the sea wall; $\gamma_w = 62.4\ \text{lb/ft}^3$.

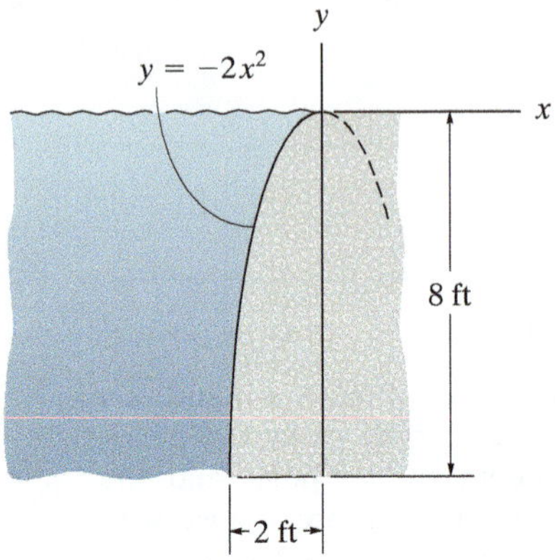

Prob. 9–135

***9–136.** The rectangular bin is filled with coal, which creates a pressure distribution along wall A that varies as shown, i.e., $p = 4z^{1/3}\ \text{lb/ft}^2$, where z is in feet. Compute the resultant force created by the coal, and its location, measured from the top surface of the coal.

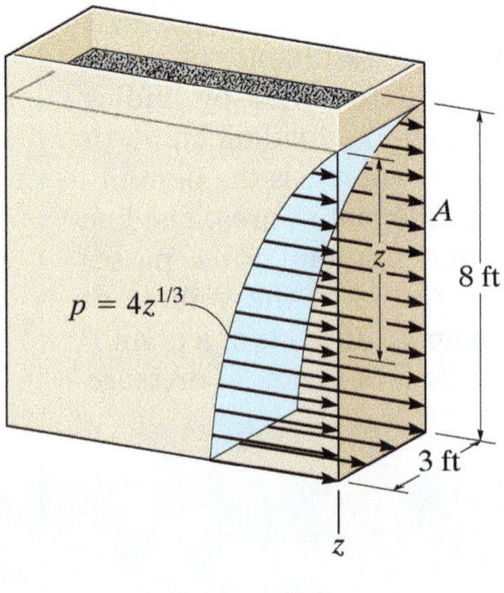

Prob. 9–136

9–137. The thin-walled channel and stiffener have the cross section shown. If the material has a constant thickness, determine the location ($\bar{y}$) of its centroid. The dimensions are indicated to the center of each segment.

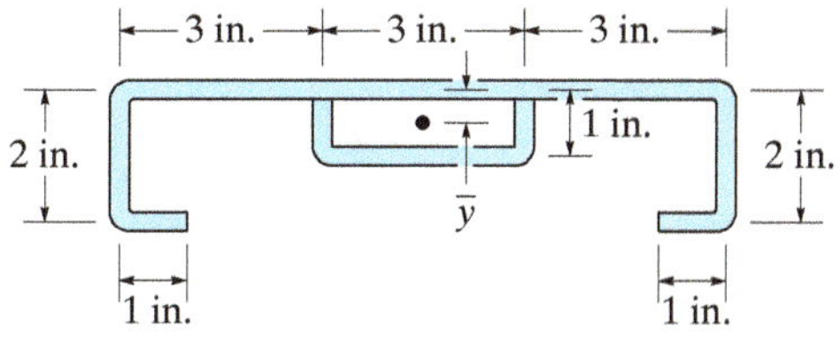

Prob. 9–137

9–138. Locate the center of gravity of the homogeneous rod. The rod has a weight of 2 lb/ft. Also, compute the x, y, z components of reaction at the fixed support A.

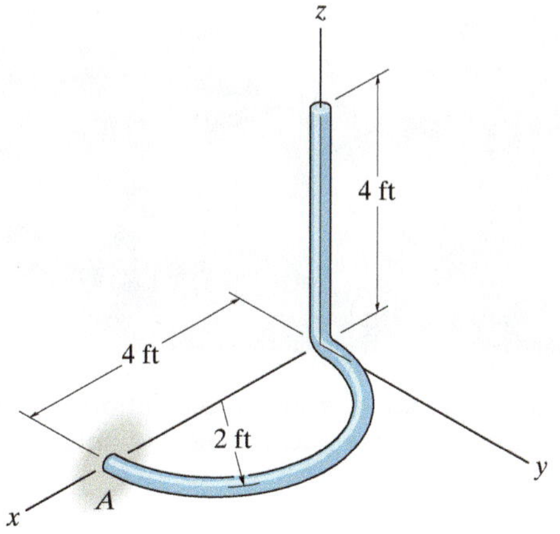

Prob. 9–138

9–139. The gate AB is 8 m wide. Determine the horizontal and vertical components of force acting on the pin at B and the vertical reaction at the smooth support A; $\rho_w = 1.0 \text{ Mg/m}^3$.

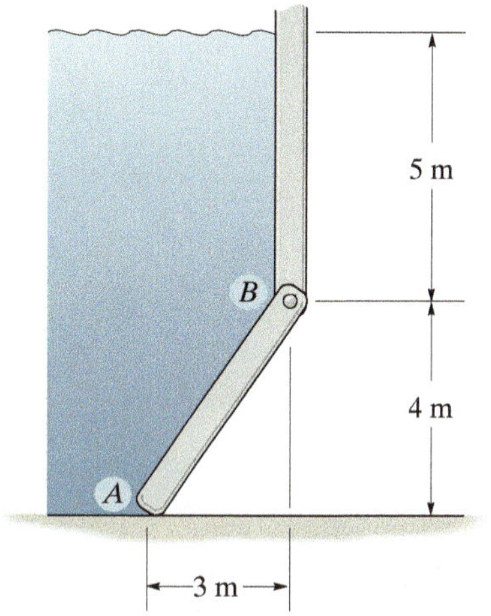

Prob. 9–139

***9–140.** The pressure loading on the plate is described by the function $p = \{-240/(x + 1) + 340\}$ Pa. Determine the magnitude of the resultant force and coordinates of the point where the line of action of the force intersects the plate.

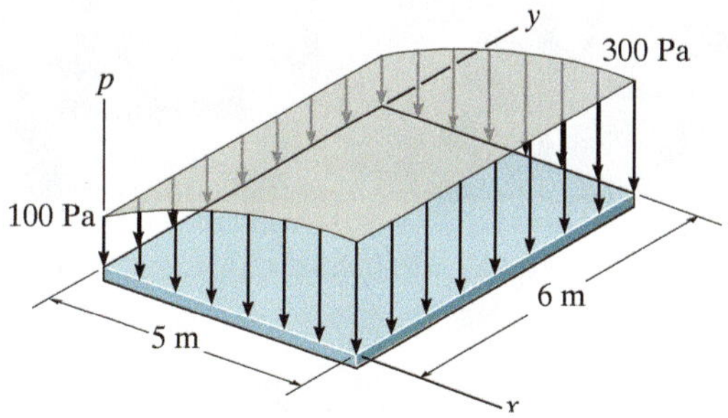

Prob. 9–140

Chapter 10

The design of these structural members requires calculation of their cross-sectional moment of inertia. In this chapter we will discuss how this is done.

Moments of Inertia

CHAPTER OBJECTIVES

- To develop a method for determining the moment of inertia for an area.
- To introduce the product of inertia and show how to determine the maximum and minimum moments of inertia for an area.
- To discuss the mass moment of inertia.

10.1 Definition of Moments of Inertia for Areas

Whenever a distributed loading acts perpendicular to an area and its intensity varies linearly, the computation of the moment of the loading distribution about an axis will involve a quantity called the *moment of inertia of the area*. For example, consider the plate in Fig. 10–1, which is subjected to a fluid pressure p. As discussed in Sec. 9.5, this pressure p varies linearly with depth, such that $p = \gamma y$, where γ is the specific weight of the fluid. Thus, the force acting on the differential area dA of the plate is $dF = p\, dA = (\gamma\, y)dA$. The moment of this force about the x axis is therefore $dM = y\, dF = \gamma y^2 dA$, and so integrating dM over the entire area of the plate yields $M = \gamma \int y^2 dA$. The integral $\int y^2 dA$ is called the *moment of inertia* I_x of the area about the x axis. Integrals of this form often arise in formulas used in fluid mechanics, mechanics of materials, structural mechanics, and mechanical design, and so the engineer needs to be familiar with the methods used for their computation.

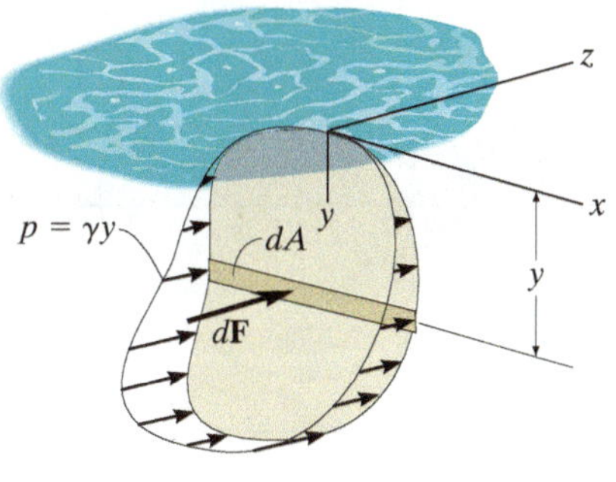

Fig. 10–1

Moment of Inertia. By definition, the moments of inertia of a differential area dA about the x and y axes are $dI_x = y^2\,dA$ and $dI_y = x^2\,dA$, respectively, Fig. 10–2. For the entire area A the *moments of inertia* are determined by integration; i.e.,

$$I_x = \int_A y^2\,dA \qquad I_y = \int_A x^2\,dA \tag{10–1}$$

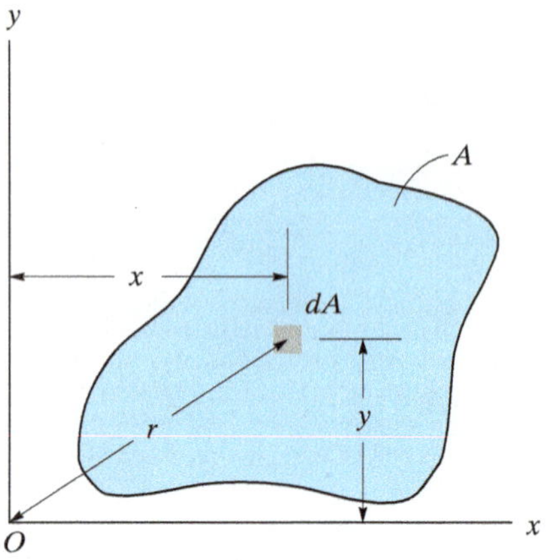

Fig. 10–2

We can also formulate this quantity for dA about the "pole" O or z axis, Fig. 10–2. This is referred to as the *polar moment of inertia*. It is defined as $dJ_O = r^2\,dA$, where r is the perpendicular distance from the pole (z axis) to the element dA. For the entire area the *polar moment of inertia* is

$$J_O = \int_A r^2\,dA = I_x + I_y \tag{10–2}$$

This relation between J_O and I_x, I_y is possible since $r^2 = x^2 + y^2$, Fig. 10–2.

From the above formulations it is seen that I_x, I_y, and J_O will *always* be *positive* since they involve the product of distance squared and area. Furthermore, the units for moment of inertia involve length raised to the fourth power, e.g., m^4, mm^4, or ft^4, $in.^4$.

10.2 Parallel-Axis Theorem for an Area

The *parallel-axis theorem* can be used to find the moment of inertia of an area about *any axis* that is parallel to an axis passing through the centroid and about which the moment of inertia is known. To develop this theorem, we will consider finding the moment of inertia of the shaded area shown in Fig. 10–3 about the x axis. To start, we choose a differential element dA located at an arbitrary distance y' from the *centroidal* x' axis. If the distance between the parallel x and x' axes is d_y, then the moment of inertia of dA about the x axis is $dI_x = (y' + d_y)^2\,dA$. For the entire area,

$$I_x = \int_A (y' + d_y)^2\,dA$$

$$= \int_A y'^2\,dA + 2d_y \int_A y'\,dA + d_y^2 \int_A dA$$

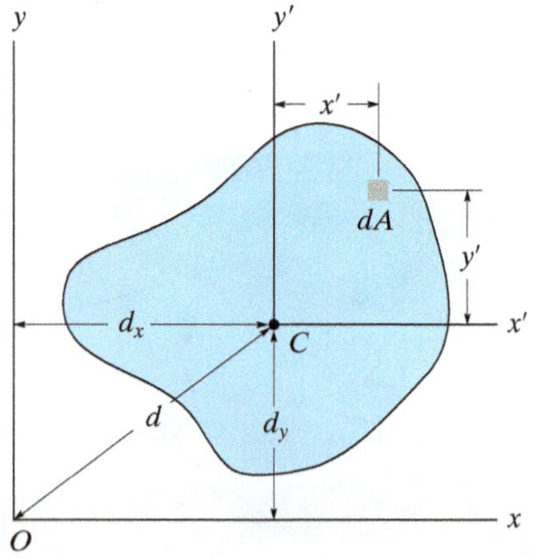

Fig. 10–3

10

The first integral represents the moment of inertia of the area about the centroidal axis, $\bar{I}_{x'}$. The second integral is zero since the x' axis passes through the area's centroid C; i.e., $\int y'\,dA = \bar{y}'\int dA = 0$ since $\bar{y}' = 0$. Since the third integral represents the total area A, the final result is therefore

$$I_x = \bar{I}_{x'} + Ad_y^2 \tag{10–3}$$

A similar expression can be written for I_y; i.e.,

$$I_y = \bar{I}_{y'} + Ad_x^2 \tag{10–4}$$

And finally, for the polar moment of inertia, since $\bar{J}_C = \bar{I}_{x'} + \bar{I}_{y'}$ and $d^2 = d_x^2 + d_y^2$, we have

$$J_O = \bar{J}_C + Ad^2 \tag{10–5}$$

The form of each of these three equations states that *the moment of inertia for an area about an axis is equal to its moment of inertia about a parallel axis passing through the area's centroid plus the product of the area and the square of the perpendicular distance between the axes.*

In order to predict the strength and deflection of this beam, it is necessary to calculate the moment of inertia of the beam's cross-sectional area.

10.3 Radius of Gyration of an Area

The *radius of gyration* of an area about an axis has units of length and is a quantity that is often used for the design of columns in structural mechanics. Provided the areas and moments of inertia are *known*, the radii of gyration are determined from the formulas

$$k_x = \sqrt{\frac{I_x}{A}}$$

$$k_y = \sqrt{\frac{I_y}{A}} \tag{10–6}$$

$$k_O = \sqrt{\frac{J_O}{A}}$$

The form of these equations is easily remembered since it is similar to that for finding the moment of inertia for a differential area about an axis. For example, $I_x = k_x^2 A$; whereas for a differential area, $dI_x = y^2\,dA$.

10

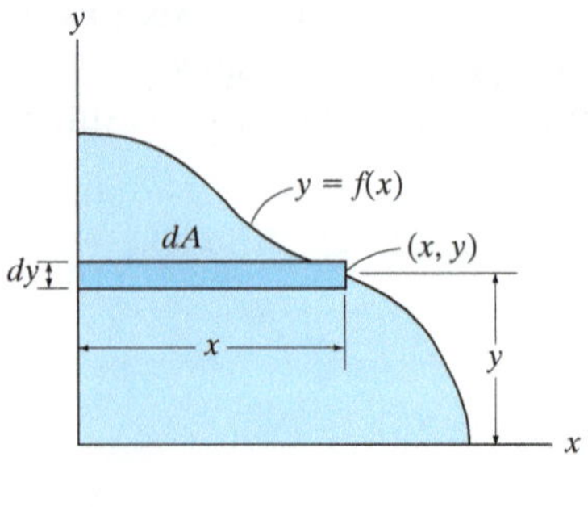

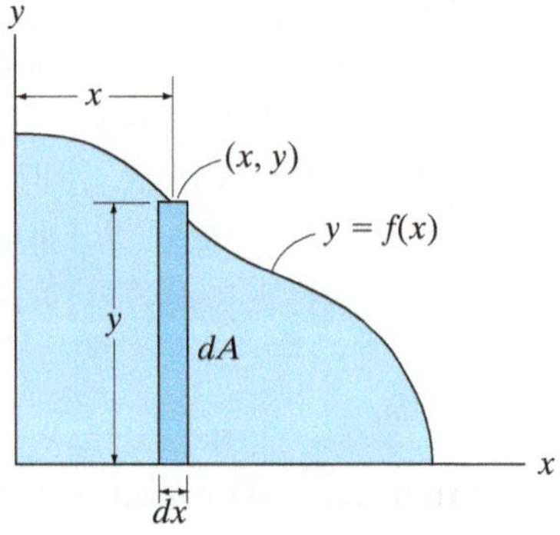

Fig. 10–4

Procedure for Analysis

In most cases the moment of inertia can be determined using a single integration. The following procedure shows two ways in which this can be done.

- If the curve defining the boundary of the area is expressed as $y = f(x)$, then select a rectangular differential element such that it has a finite length and differential width.
- The element should be located so that it intersects the curve at the *arbitrary point* (x, y).

Case 1.

- Orient the element so that its length is *parallel* to the axis about which the moment of inertia is computed. This situation occurs when the rectangular element shown in Fig. 10–4*a* is used to determine I_x for the area. Here the entire element is at a distance y from the x axis since it has a thickness dy. Thus $I_x = \int y^2 dA$. To find I_y, the element is oriented as shown in Fig. 10–4*b*. This element lies at the *same* distance x from the y axis so that $I_y = \int x^2 dA$.

Case 2.

- The length of the element can be oriented *perpendicular* to the axis about which the moment of inertia is computed; however, Eq. 10–1 *does not apply* since all points on the element will *not* lie at the same moment-arm distance from the axis. For example, if the rectangular element in Fig. 10–4*a* is used to determine I_y, it will first be necessary to calculate the moment of inertia of the *element* about an axis parallel to the y axis that passes through the element's centroid, and then determine the moment of inertia of the *element* about the y axis using the parallel-axis theorem. Integration of this result will yield I_y. See Examples 10.2 and 10.3.

10

EXAMPLE 10.1

Determine the moment of inertia for the rectangular area shown in Fig. 10–5 with respect to (a) the centroidal x' axis, (b) the axis x_b passing through the base of the rectangle, and (c) the pole or z' axis perpendicular to the x'–y' plane and passing through the centroid C.

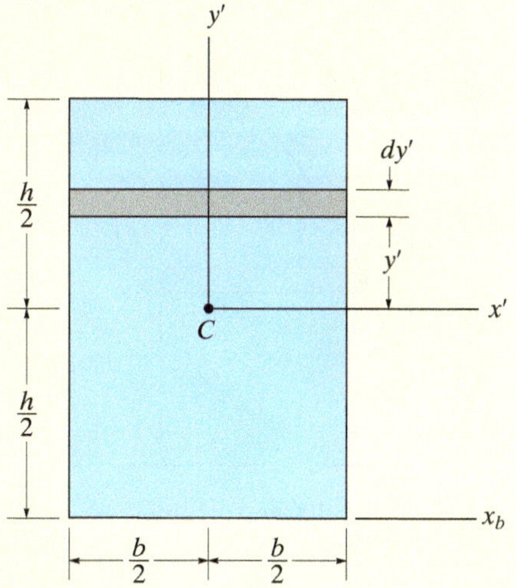

Fig. 10–5

SOLUTION (CASE 1)

Part (a). The differential element shown in Fig. 10–5 is chosen for integration. Because of its location and orientation, the *entire element* is at a distance y' from the x' axis. Here it is necessary to integrate from $y' = -h/2$ to $y' = h/2$. Since $dA = b\,dy'$, then

$$\bar{I}_{x'} = \int_A y'^2\,dA = \int_{-h/2}^{h/2} y'^2(b\,dy') = b\int_{-h/2}^{h/2} y'^2\,dy'$$

$$\bar{I}_{x'} = \frac{1}{12}bh^3 \qquad \textit{Ans.}$$

Part (b). The moment of inertia about an axis passing through the base of the rectangle can be obtained by using the above result of part (a) and applying the parallel-axis theorem, Eq. 10–3.

$$I_{x_b} = \bar{I}_{x'} + Ad_y^2$$

$$= \frac{1}{12}bh^3 + bh\left(\frac{h}{2}\right)^2 = \frac{1}{3}bh^3 \qquad \textit{Ans.}$$

Part (c). To obtain the polar moment of inertia about point C, we must first obtain $\bar{I}_{y'}$, which may be found by interchanging the dimensions b and h in the result of part (a), i.e.,

$$\bar{I}_{y'} = \frac{1}{12}hb^3$$

Using Eq. 10–2, the polar moment of inertia about C is therefore

$$\bar{J}_C = \bar{I}_{x'} + \bar{I}_{y'} = \frac{1}{12}bh(h^2 + b^2) \qquad \textit{Ans.}$$

10

EXAMPLE 10.2

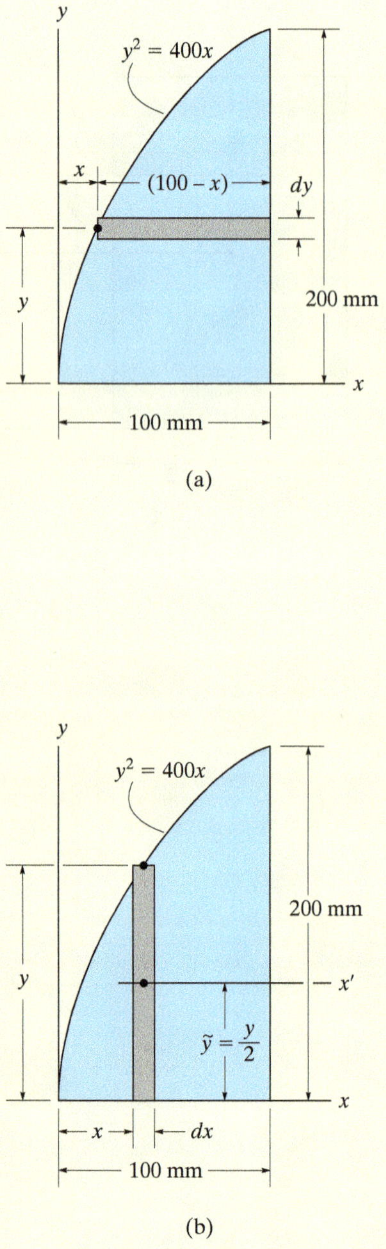

Fig. 10–6

Determine the moment of inertia for the shaded area shown in Fig. 10–6*a* about the *x* axis.

SOLUTION I (CASE 1)

A differential element of area that is *parallel* to the *x* axis, as shown in Fig. 10–6*a*, is chosen for integration. Since this element has a thickness *dy* and intersects the curve at the *arbitrary point* (x, y), its area is $dA = (100 - x)\,dy$. Furthermore, the element lies at the same distance y from the x axis. Hence, integrating with respect to y, from $y = 0$ to $y = 200$ mm, yields

$$I_x = \int_A y^2\,dA = \int_0^{200\text{ mm}} y^2(100 - x)\,dy$$

$$= \int_0^{200\text{ mm}} y^2\left(100 - \frac{y^2}{400}\right) dy = \int_0^{200\text{ mm}} \left(100y^2 - \frac{y^4}{400}\right) dy$$

$$= 107(10^6)\text{ mm}^4 \qquad \textit{Ans.}$$

SOLUTION II (CASE 2)

A differential element *parallel* to the *y* axis, as shown in Fig. 10–6*b*, is chosen for integration. It intersects the curve at the *arbitrary point* (x, y). In this case, all points of the element do *not* lie at the same distance from the x axis, and therefore the parallel-axis theorem must be used to determine the *moment of inertia of the element* with respect to this axis. For a rectangle having a base b and height h, the moment of inertia about its centroidal axis has been determined in part (a) of Example 10.1. There it was found that $\bar{I}_{x'} = \frac{1}{12}bh^3$. For the differential element shown in Fig. 10–6*b*, $b = dx$ and $h = y$, and thus $d\bar{I}_{x'} = \frac{1}{12}dx\,y^3$. Since the centroid of the element is $\tilde{y} = y/2$ from the x axis, the moment of inertia of the element about this axis is

$$dI_x = d\bar{I}_{x'} + dA\,\tilde{y}^2 = \frac{1}{12}dx\,y^3 + y\,dx\left(\frac{y}{2}\right)^2 = \frac{1}{3}y^3\,dx$$

10

(This result can also be concluded from part (b) of Example 10.1.) Integrating with respect to x, from $x = 0$ to $x = 100$ mm, yields

$$I_x = \int dI_x = \int_0^{100\text{ mm}} \frac{1}{3}y^3\,dx = \int_0^{100\text{ mm}} \frac{1}{3}(400x)^{3/2}\,dx$$

$$= 107(10^6)\text{ mm}^4 \qquad \textit{Ans.}$$

EXAMPLE 10.3

Determine the moment of inertia with respect to the x axis for the circular area shown in Fig. 10–7a.

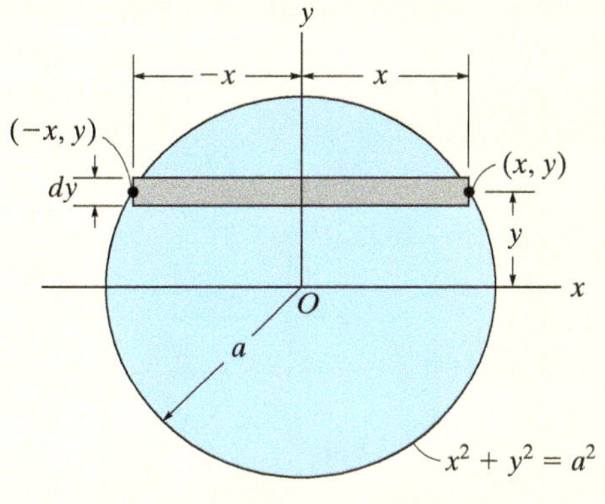

(a)

SOLUTION I (CASE 1)

Using the differential element shown in Fig. 10–7a, since $dA = 2x\,dy$, we have

$$I_x = \int_A y^2\,dA = \int_A y^2(2x)\,dy$$

$$= \int_{-a}^{a} y^2\left(2\sqrt{a^2 - y^2}\right)dy = \frac{\pi a^4}{4} \qquad \textit{Ans.}$$

SOLUTION II (CASE 2)

When the differential element shown in Fig. 10–7b is chosen, the centroid for the element happens to lie on the x axis, and since $\bar{I}_{x'} = \frac{1}{12}bh^3$ for a rectangle, we have

$$dI_x = \frac{1}{12}dx(2y)^3$$

$$= \frac{2}{3}y^3\,dx$$

Integrating with respect to x yields

$$I_x = \int_{-a}^{a}\frac{2}{3}(a^2 - x^2)^{3/2}\,dx = \frac{\pi a^4}{4} \qquad \textit{Ans.}$$

NOTE: By comparison, Solution I requires much less computation. Therefore, if an integral using a particular element appears difficult to evaluate, try solving the problem using an element oriented in the other direction.

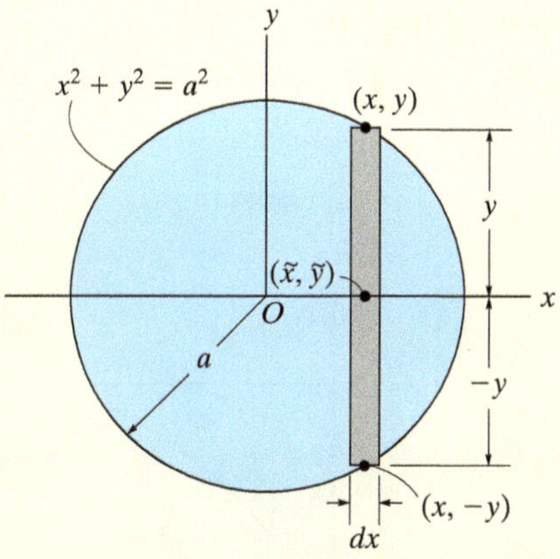

(b)

Fig. 10–7

10

FUNDAMENTAL PROBLEMS

F10–1. Determine the moment of inertia of the shaded area about the x axis.

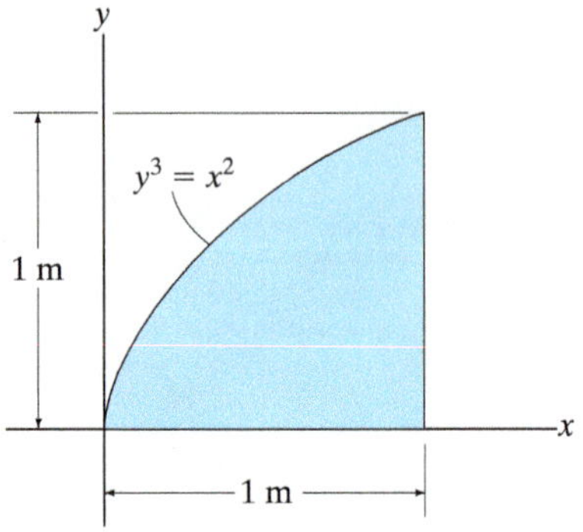

F10–1

F10–2. Determine the moment of inertia of the shaded area about the x axis.

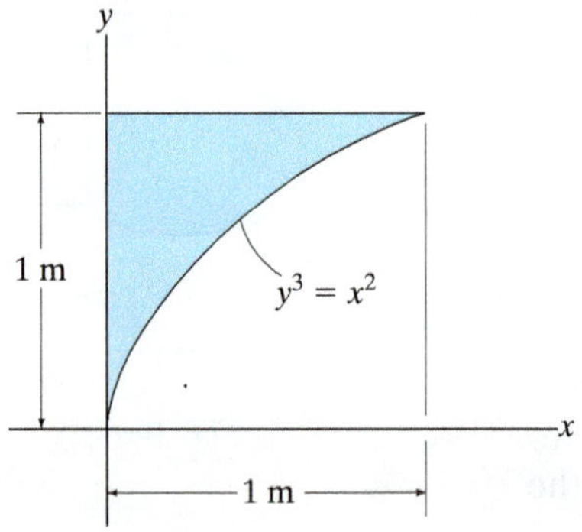

F10–2

F10–3. Determine the moment of inertia of the shaded area about the y axis.

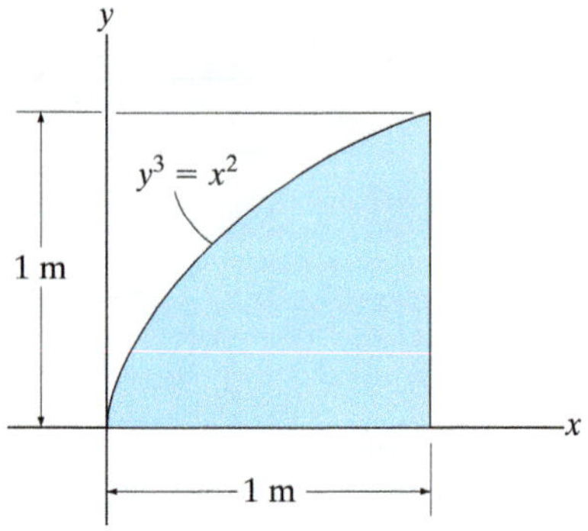

F10–3

F10–4. Determine the moment of inertia of the shaded area about the y axis.

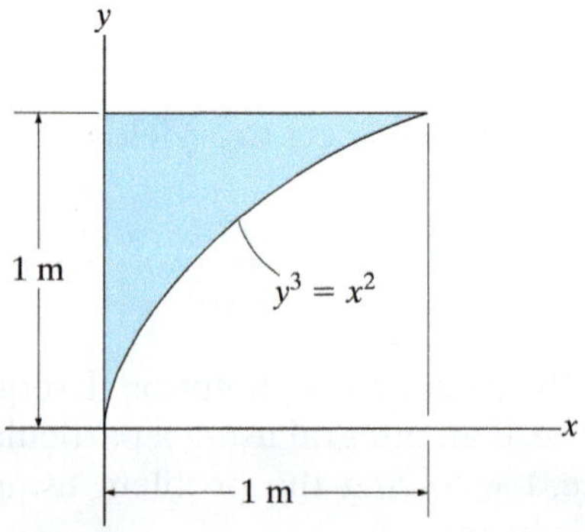

F10–4

10

PROBLEMS

10–1. Determine the moment of inertia of the shaded area about the x axis.

10–2. Determine the moment of inertia of the shaded area about the y axis.

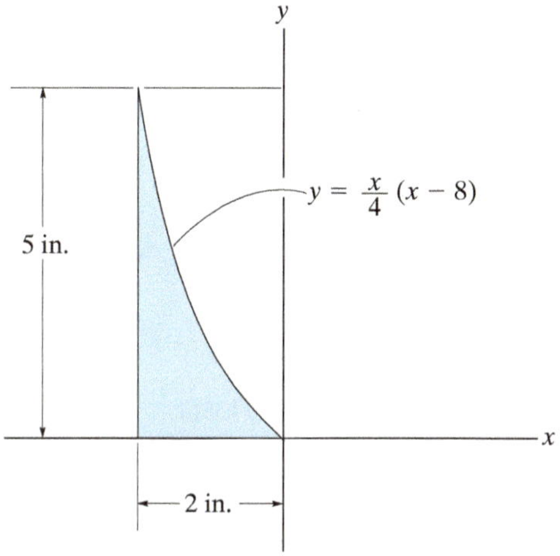

Probs. 10–1/2

10–3. Determine the moment of inertia of the area about the x axis.

***10–4.** Determine the moment of inertia of the area about the y axis.

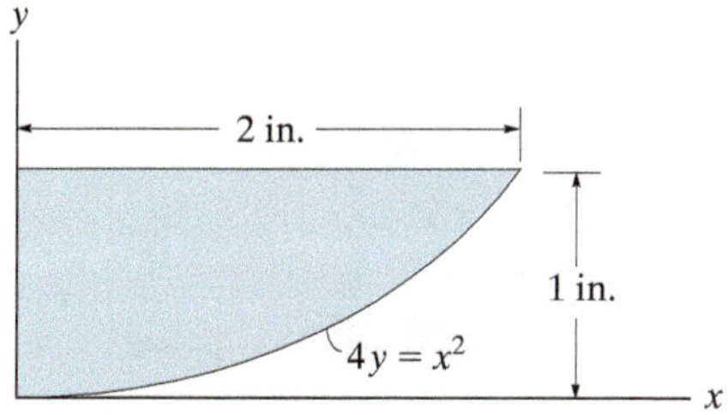

Probs. 10–3/4

10–5. Determine the moment of inertia of the area about the x axis.

10–6. Determine the moment of inertia of the area about the y axis.

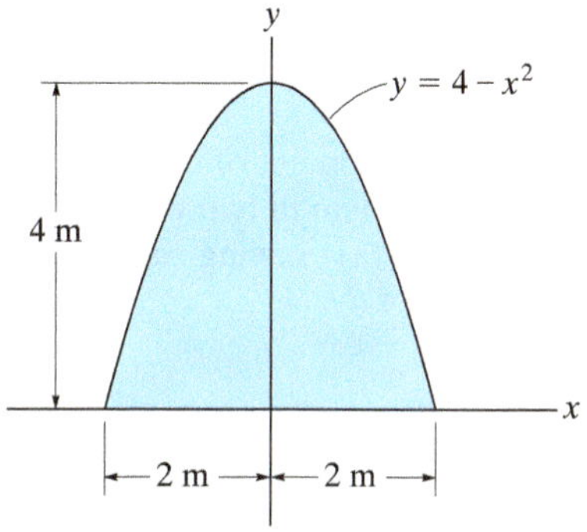

Probs. 10–5/6

10–7. Determine the moment of inertia of the area about the x axis.

***10–8.** Determine the moment of inertia of the area about the y axis.

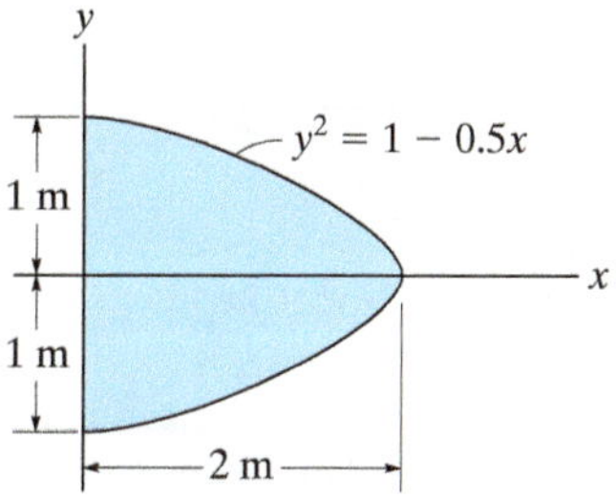

Probs. 10–7/8

10–9. Determine the moment of inertia of the area about the x axis.

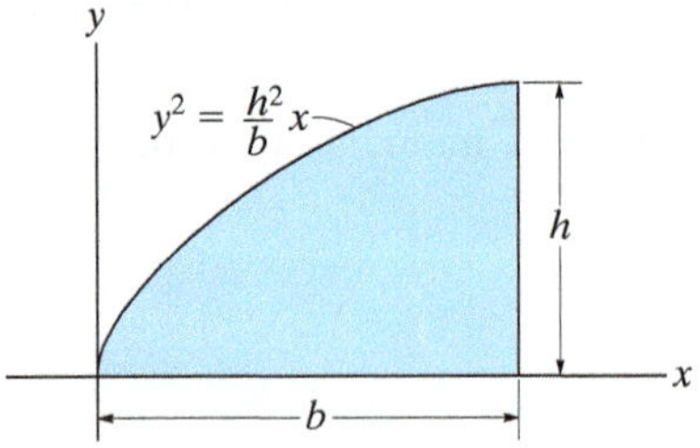

Prob. 10–9

10–10. Determine the moment of inertia of the area about the x axis. Solve the problem in two ways, using rectangular differential elements: (a) having a thickness dx and (b) having a thickness of dy.

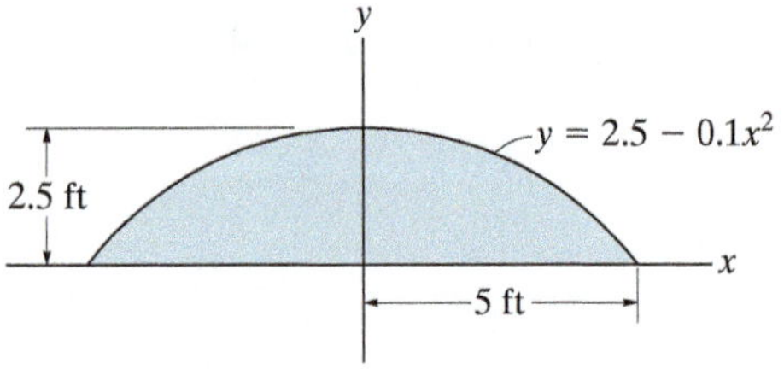

Prob. 10–10

10–11. Determine the moment of inertia of the area about the x axis.

***10–12.** Determine the moment of inertia of the area about the y axis.

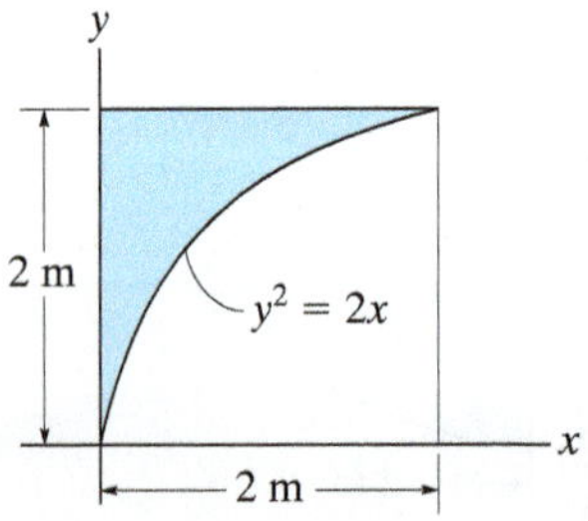

Probs. 10–11/12

10–13. Determine the moment of inertia of the area about the x axis. Solve the problem in two ways, using rectangular differential elements: (a) having a thickness of dx, and (b) having a thickness of dy.

10–14. Determine the moment of inertia of the area about the y axis. Solve the problem in two ways, using rectangular differential elements: (a) having a thickness of dx, and (b) having a thickness of dy.

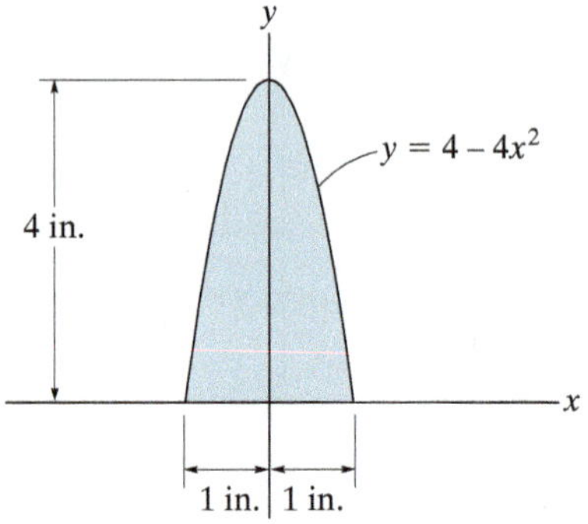

Probs. 10–13/14

■**10–15.** Determine the moment of inertia of the shaded area about the y axis. Use Simpson's rule to evaluate the integral.

*■**10–16.** Determine the moment of inertia of the shaded area about the x axis. Use Simpson's rule to evaluate the integral.

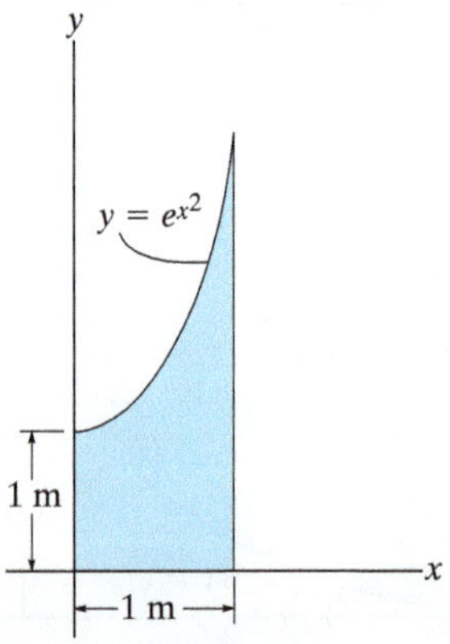

Probs. 10–15/16

10–17. Determine the moment of inertia of the shaded area about the x axis.

10–18. Determine the moment of inertia of the shaded area about the y axis.

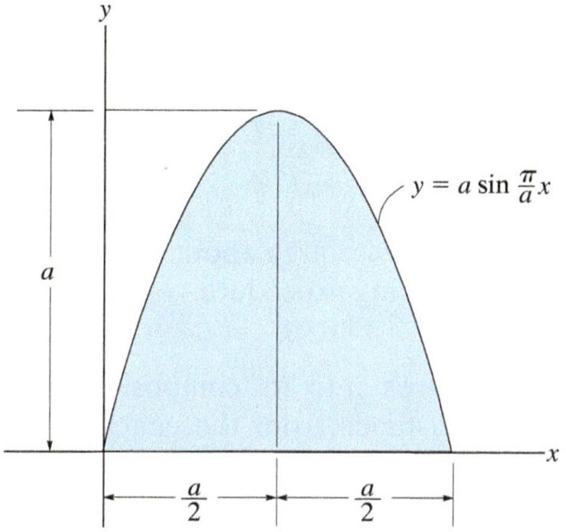

Probs. 10–17/18

10–19. Determine the moment of inertia of the shaded area about the x axis.

***10–20.** Determine the moment of inertia of the shaded area about the y axis.

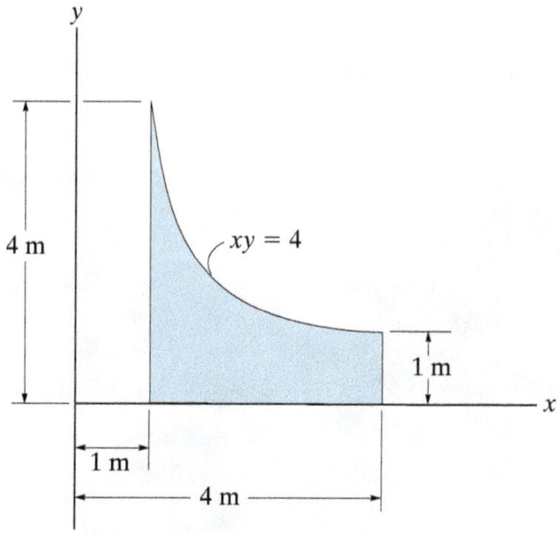

Probs. 10–19/20

10–21. Determine the moment of inertia of the shaded area about the x axis.

10–22. Determine the moment of inertia of the shaded area about the y axis.

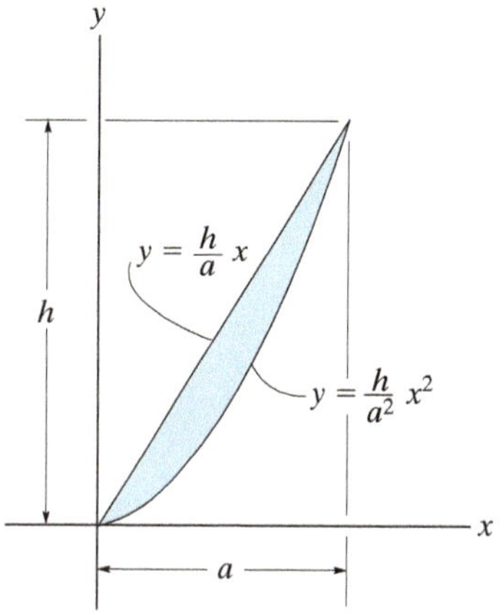

Probs. 10–21/22

10–23. Determine the moment of inertia of the shaded area about the x axis.

***10–24.** Determine the moment of inertia of the shaded area about the y axis.

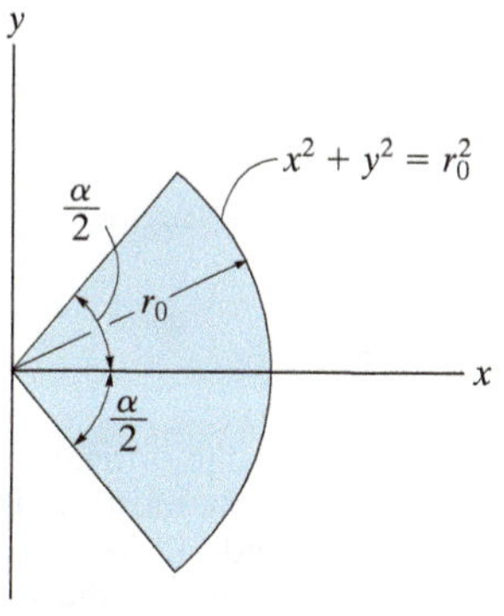

Probs. 10–23/24

10.4 Moments of Inertia for Composite Areas

A composite area consists of a series of connected "simpler" parts or shapes, such as rectangles, triangles, and circles. Provided the moment of inertia of each of these parts is known or can be determined about a common axis, then the moment of inertia for the composite area about this axis equals the *algebraic sum* of the moments of inertia of all its parts.

Procedure for Analysis

The moment of inertia for a composite area about a reference axis can be determined using the following procedure.

Composite Parts.

- Using a sketch, divide the area into its composite parts and indicate the perpendicular distance from the centroid of each part to the reference axis.

Parallel-Axis Theorem.

- If the centroidal axis for each part does not coincide with the reference axis, the parallel-axis theorem, $I = \bar{I} + Ad^2$, should be used to determine the moment of inertia of the part about the reference axis. For the calculation of $\bar{I}$, use the table on the inside back cover.

Summation.

- The moment of inertia of the entire area about the reference axis is determined by summing the results of its composite parts about this axis.
- If a composite part has a "hole", its moment of inertia is found by "subtracting" the moment of inertia of the hole from the moment of inertia of the entire part including the hole.

For design or analysis of this T-beam, engineers must be able to locate the centroid of its cross-sectional area, and then find the moment of inertia of this area about the centroidal axis.

EXAMPLE 10.4

Determine the moment of inertia of the area shown in Fig. 10–8*a* about the *x* axis.

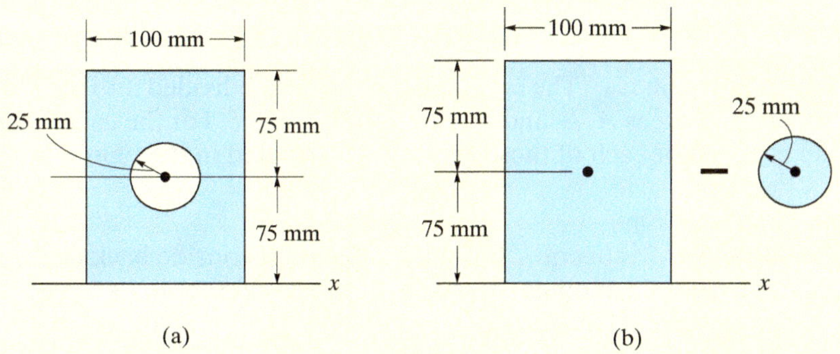

Fig. 10–8

SOLUTION

Composite Parts. The area can be obtained by *subtracting* the circle from the rectangle shown in Fig. 10–8*b*. The centroid of each area is located in the figure.

Parallel-Axis Theorem. The moments of inertia about the *x* axis are determined using the parallel-axis theorem and the geometric properties formulae for circular and rectangular areas $I_x = \frac{1}{4}\pi r^4$; $I_x = \frac{1}{12}bh^3$, found on the inside back cover.

Circle

$$I_x = \bar{I}_{x'} + Ad_y^2$$

$$= \frac{1}{4}\pi(25)^4 + \pi(25)^2(75)^2 = 11.4(10^6)\text{ mm}^4$$

Rectangle

$$I_x = \bar{I}_{x'} + Ad_y^2$$

$$= \frac{1}{12}(100)(150)^3 + (100)(150)(75)^2 = 112.5(10^6)\text{ mm}^4$$

Summation. The moment of inertia for the area is therefore

$$I_x = -11.4(10^6) + 112.5(10^6)$$

$$= 101(10^6)\text{ mm}^4 \qquad \textit{Ans.}$$

EXAMPLE 10.5

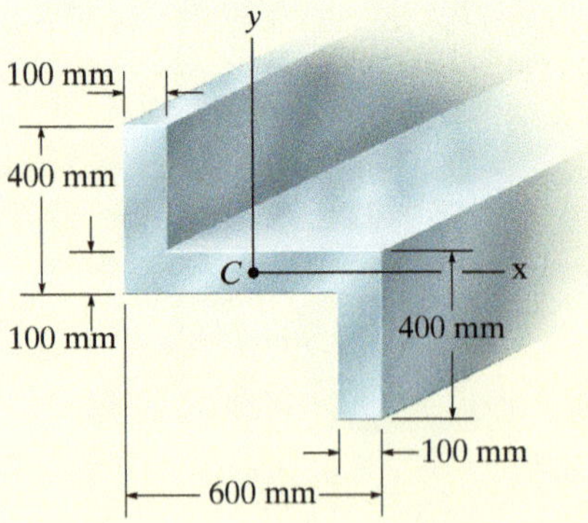

(a)

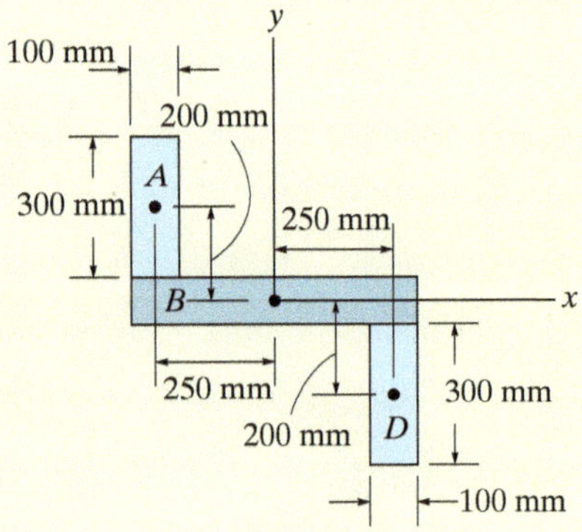

(b)

Fig. 10–9

Determine the moments of inertia for the cross-sectional area of the member shown in Fig. 10–9*a* about the x and y centroidal axes.

SOLUTION

Composite Parts. The cross section can be subdivided into the three rectangular areas A, B, and D shown in Fig. 10–9*b*. For the calculation, the centroid of each of these rectangles is located in the figure.

Parallel-Axis Theorem. From the table on the inside back cover, or Example 10.1, the moment of inertia of a rectangle about its centroidal axis is $\bar{I} = \frac{1}{12}bh^3$. Hence, using the parallel-axis theorem for rectangles A and D, the calculations are as follows:

Rectangles A and D

$$I_x = \bar{I}_{x'} + Ad_y^2 = \frac{1}{12}(100)(300)^3 + (100)(300)(200)^2$$

$$= 1.425(10^9)\text{ mm}^4$$

$$I_y = \bar{I}_{y'} + Ad_x^2 = \frac{1}{12}(300)(100)^3 + (100)(300)(250)^2$$

$$= 1.90(10^9)\text{ mm}^4$$

Rectangle B

$$I_x = \frac{1}{12}(600)(100)^3 = 0.05(10^9)\text{ mm}^4$$

$$I_y = \frac{1}{12}(100)(600)^3 = 1.80(10^9)\text{ mm}^4$$

Summation. The moments of inertia for the entire cross section are thus

$$I_x = 2[1.425(10^9)] + 0.05(10^9)$$

$$= 2.90(10^9)\text{ mm}^4 \qquad \textit{Ans.}$$

$$I_y = 2[1.90(10^9)] + 1.80(10^9)$$

$$= 5.60(10^9)\text{ mm}^4 \qquad \textit{Ans.}$$

FUNDAMENTAL PROBLEMS

F10–5. Determine the moment of inertia of the beam's cross-sectional area about the centroidal x and y axes.

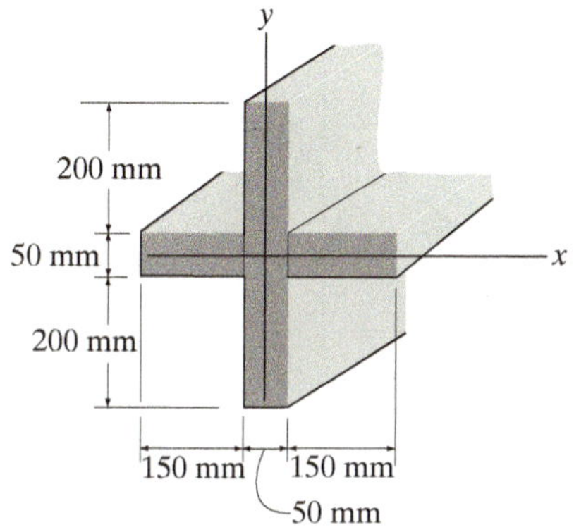

F10–5

F10–6. Determine the moment of inertia of the beam's cross-sectional area about the centroidal x and y axes.

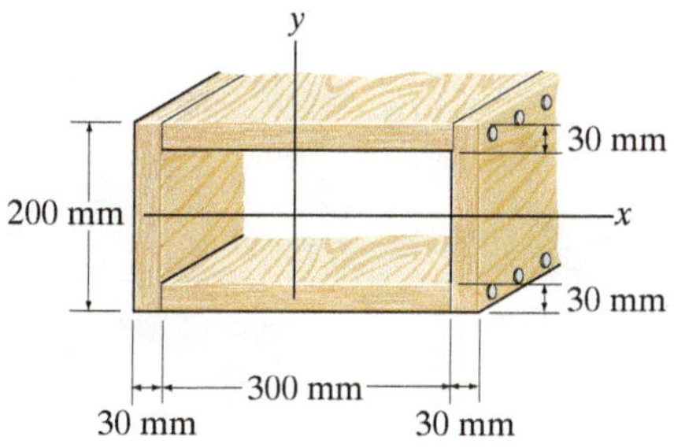

F10–6

F10–7. Determine the moment of inertia of the cross-sectional area of the channel with respect to the y axis.

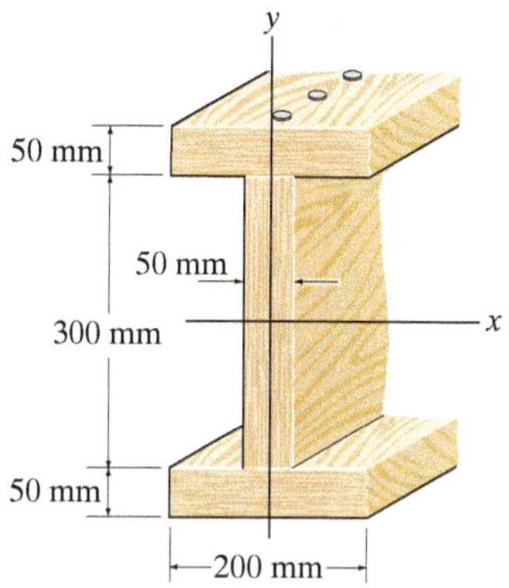

F10–7

F10–8. Determine the moment of inertia of the cross-sectional area of the T-beam with respect to the x' axis passing through the centroid of the cross section.

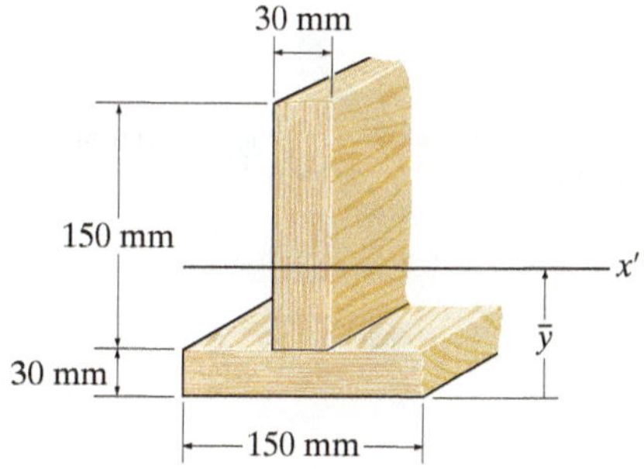

F10–8

PROBLEMS

10–25. Determine the moment of inertia of the composite area about the x axis.

10–26. Determine the moment of inertia of the composite area about the y axis.

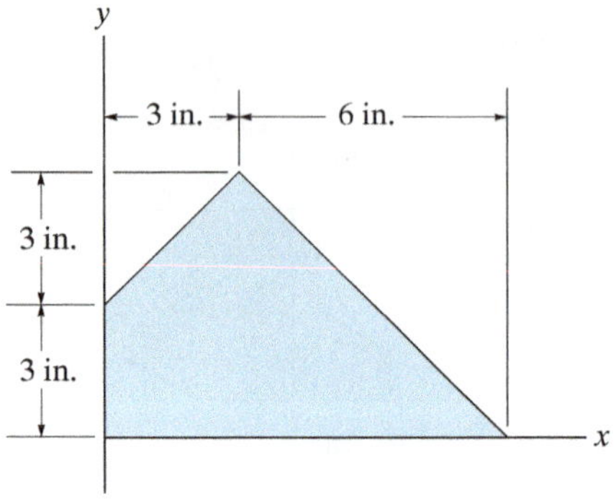

Probs. 10–25/26

10–27. Determine the radius of gyration k_x for the column's cross-sectional area.

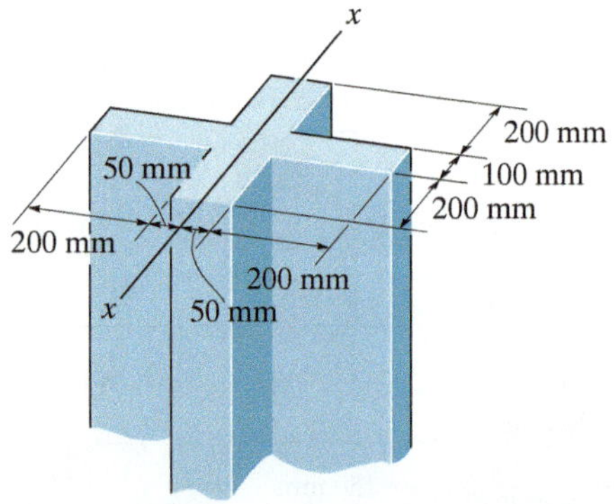

Prob. 10–27

***10–28.** Determine the moment of inertia of the beam's cross-sectional area about the x axis.

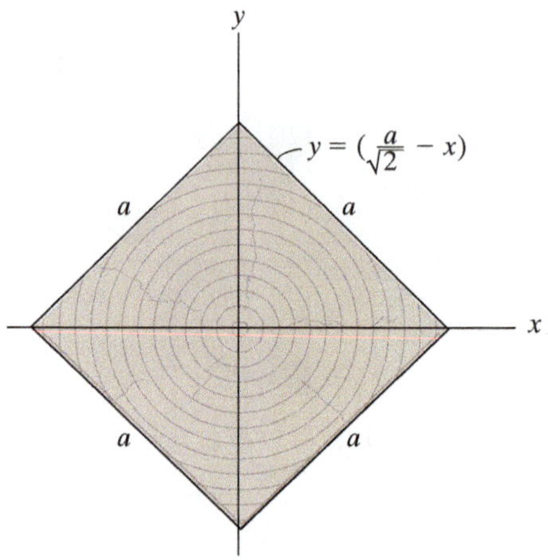

Prob. 10–28

10–29. Locate the centroid $\bar{y}$ of the channel's cross-sectional area, and then determine the moment of inertia with respect to the x' axis passing through the centroid.

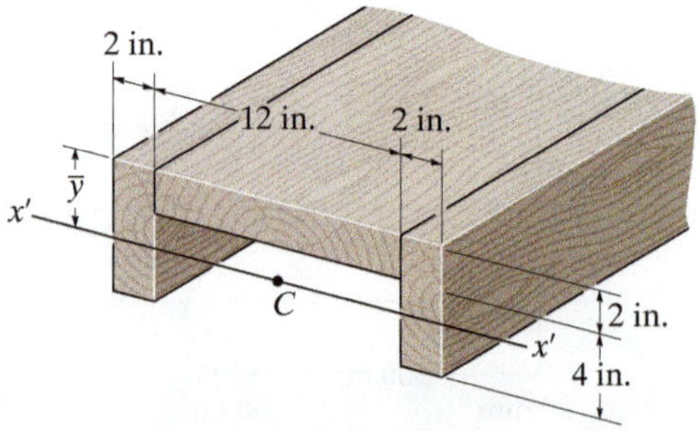

Prob. 10–29

10–30. Determine the distance $\bar{x}$ to the centroid of the beam's cross-sectional area, then find the moment of inertia about the y' axis.

10–31. Determine the moment of inertia of the beam's cross-sectional area about the x' axis.

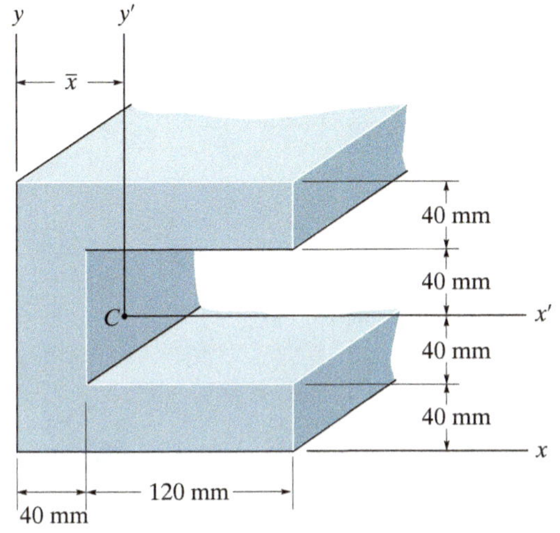

Probs. 10–30/31

10–34. Determine the moment of inertia of the beam's cross-sectional area about the y axis.

10–35. Determine $\bar{y}$, which locates the centroidal axis x' for the cross-sectional area of the T-beam, and then find the moment of inertia about the x' axis.

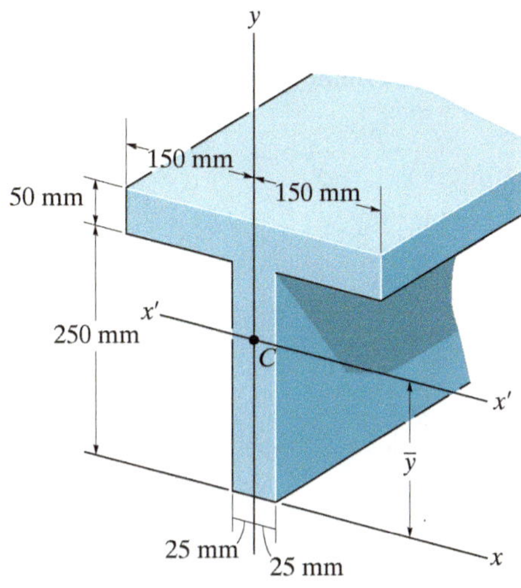

Probs. 10–34/35

***10–32.** Determine the moment of inertia of the shaded area about the x axis.

10–33. Determine the moment of inertia of the shaded area about the y axis.

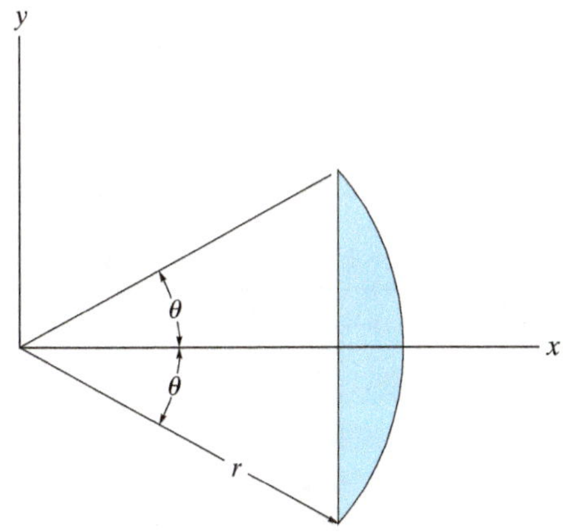

Probs. 10–32/33

***10–36.** Determine the moment of inertia I_x of the shaded area about the x axis.

10–37. Determine the moment of inertia I_y of the shaded area about the y axis.

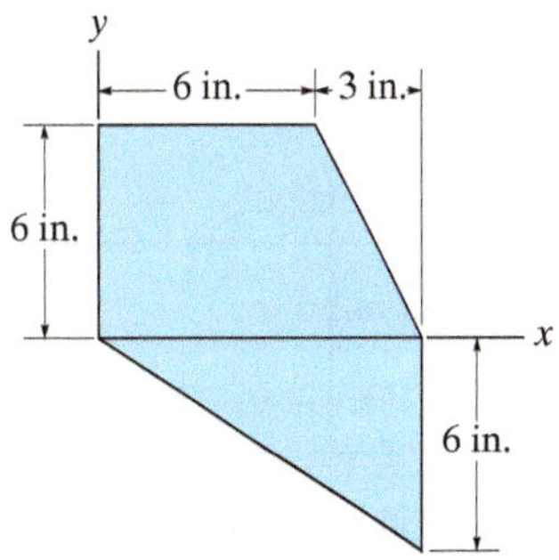

Probs. 10–36/37

10

10–38. The beam is constructed from the two channels and two cover plates. If each channel has a cross-sectional area of $A_c = 11.8\ \text{in}^2$ and a moment of inertia about a horizontal axis passing through its own centroid, C_c, of $(\bar{I}_{\bar{x}})_{C_c} = 349\ \text{in}^4$, determine the moment of inertia of the beam about the x axis.

10–39. The beam is constructed from the two channels and two cover plates. If each channel has a cross-sectional area of $A_c = 11.8\ \text{in}^2$ and a moment of inertia about a vertical axis passing through its own centroid, C_c, of $(\bar{I}_{\bar{y}})_{C_c} = 9.23\ \text{in}^4$, determine the moment of inertia of the beam about the y axis.

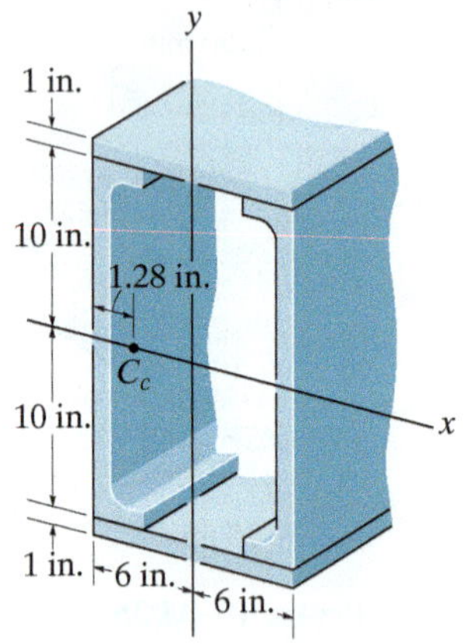

Probs. 10–38/39

10–42. Determine the moment of inertia of the beam's cross-sectional area about the x axis.

10–43. Determine the moment of inertia of the beam's cross-sectional area about the y axis.

***10–44.** Determine the distance $\bar{y}$ to the centroid C of the beam's cross-sectional area and then compute the moment of inertia $\bar{I}_{x'}$ about the x' axis.

10–45. Determine the distance $\bar{x}$ to the centroid C of the beam's cross-sectional area and then compute the moment of inertia $\bar{I}_{y'}$ about the y' axis.

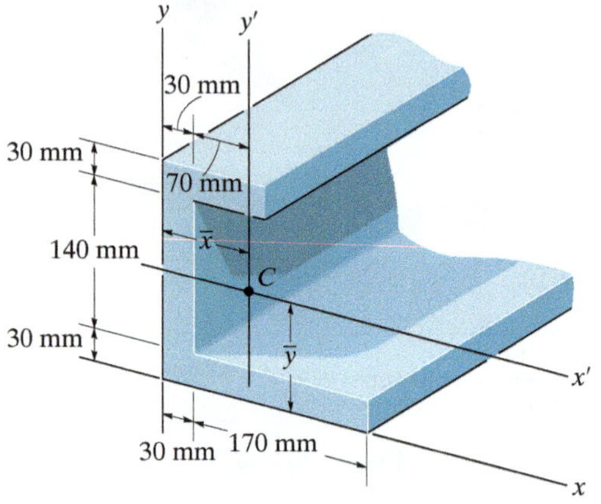

Probs. 10–42/43/44/45

***10–40.** Locate the centroid $\bar{y}$ of the composite area, then determine the moment of intertia of this area about the centroidal x' axis.

10–41. Determine the moment of inertia of the composite area about the centroidal y axis.

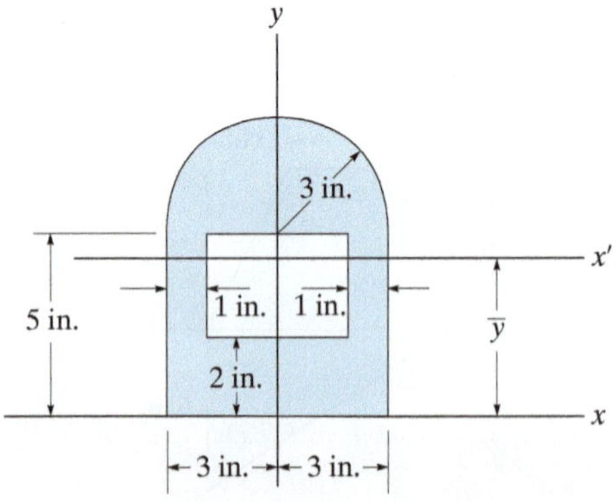

Probs. 10–40/41

10–46. Determine the distance $\bar{y}$ to the centroid for the beam's cross-sectional area; then determine the moment of inertia about the x' axis.

10–47. Determine the moment of inertia of the beam's cross-sectional area about the y axis.

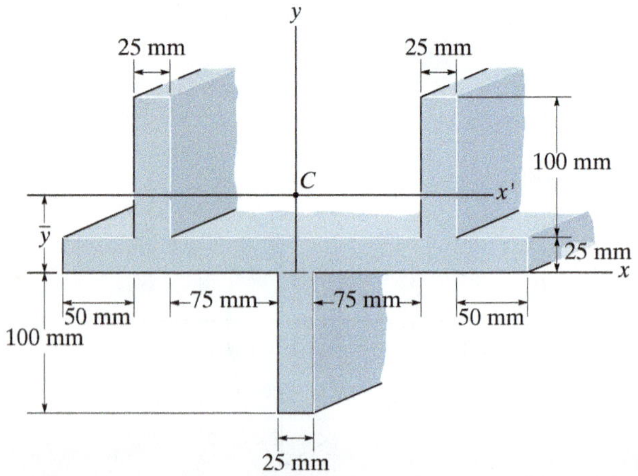

Probs. 10–46/47

***10–48.** Determine the moment of inertia of the area about the x axis.

10–49. Determine the moment of inertia of the area about the y axis.

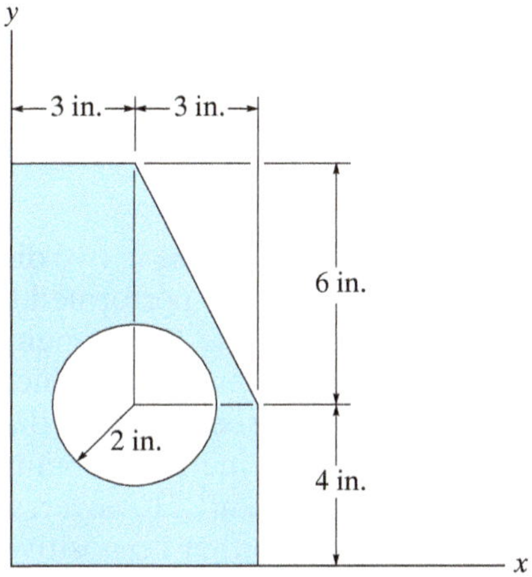

Probs. 10–48/49

10–51. Determine the moment of inertia of the beam's cross-sectional area with respect to the x' centroidal axis. Neglect the size of all the rivet heads, R, for the calculation. Handbook values for the area, moment of inertia, and location of the centroid C of one of the angles are listed in the figure.

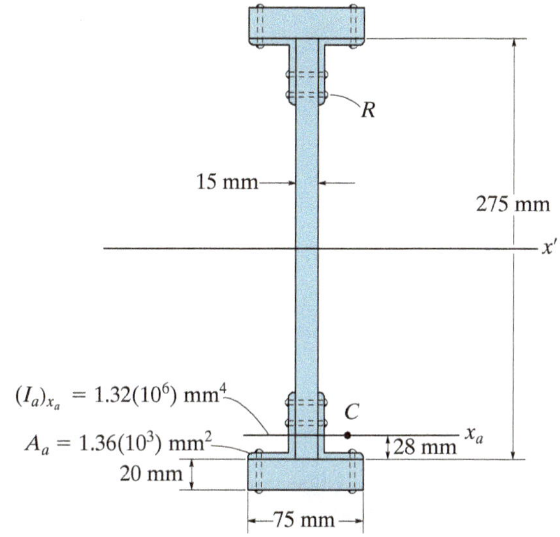

Prob. 10–51

***10–52.** Determine the moment of inertia of the parallelogram about the x' axis, which passes through the centroid C of the area.

10–53. Determine the moment of inertia of the parallelogram about the y' axis, which passes through the centroid C of the area.

10–50. Locate the centroid $\bar{y}$ of the cross section and determine the moment of inertia of the section about the x' axis.

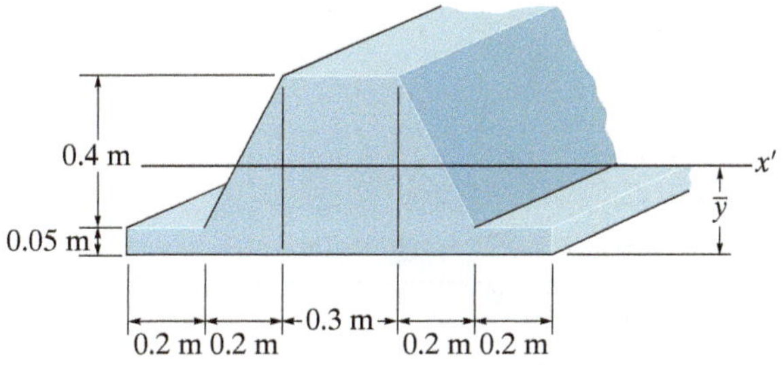

Prob. 10–50

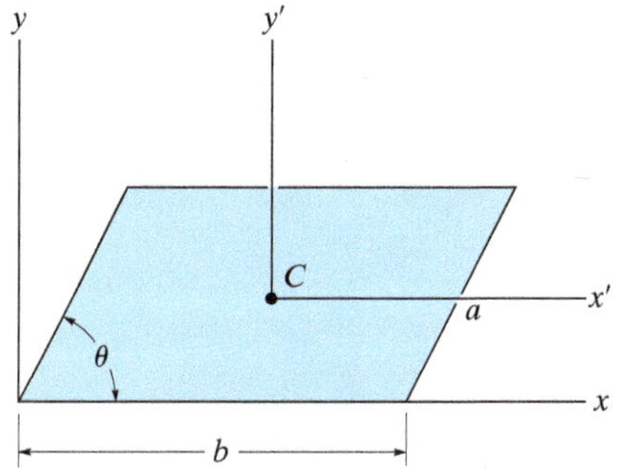

Probs. 10–52/53

10

*10.5 Product of Inertia for an Area

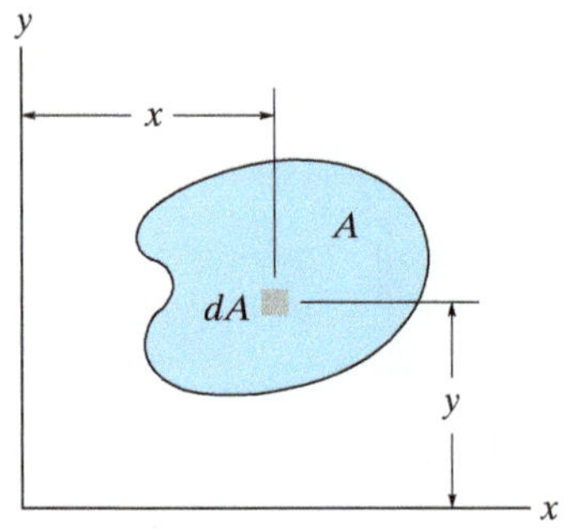

Fig. 10–10

It will be shown in the next section that the property of an area, called the product of inertia, is required in order to determine the *maximum* and *minimum* moments of inertia for the area. These maximum and minimum values are important properties needed for designing structural and mechanical members such as beams, columns, and shafts.

The *product of inertia* of the area in Fig. 10–10 with respect to the x and y axes is defined as

$$I_{xy} = \int_A xy\, dA \qquad (10\text{–}7)$$

If the element of area chosen has a differential size in two directions, as shown in Fig. 10–10, a double integration must be performed to evaluate I_{xy}. Most often, however, it is easier to choose an element having a differential size or thickness in only one direction in which case the evaluation requires only a single integration (see Example 10.6).

The effectiveness of this beam to resist bending can be determined once its moments of inertia and its product of inertia are known.

Like the moment of inertia, the product of inertia has units of length raised to the fourth power, e.g., m^4, mm^4 or ft^4, in^4. However, since x or y may be negative, the product of inertia may either be positive, negative, or zero, depending on the location and orientation of the coordinate axes. For example, the product of inertia I_{xy} for an area will be *zero* if either the x or y axis is an axis of *symmetry* for the area, as in Fig. 10–11. Here every element dA located at point (x, y) has a corresponding element dA located at $(x, -y)$. Since the products of inertia for these elements are, respectively, $xy\, dA$ and $-xy\, dA$, the algebraic sum or integration of all the elements that are chosen in this way will cancel each other. Consequently, the product of inertia for the total area becomes zero. It also follows from the definition of I_{xy} that the "sign" of this quantity depends on the quadrant where the area is located. As shown in Fig. 10–12, if the area is rotated from one quadrant to another, the sign of I_{xy} will change.

10

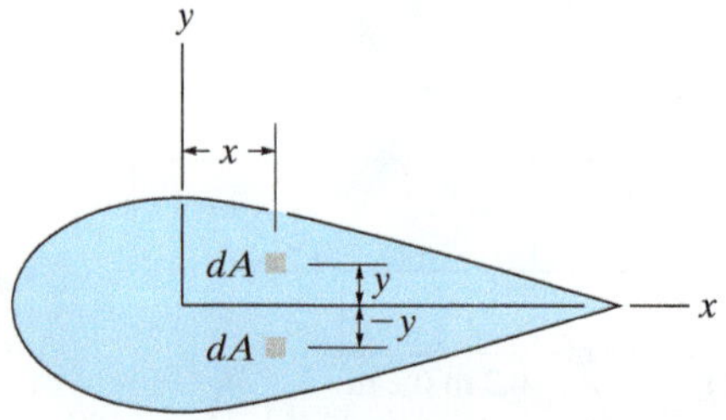

Fig. 10–11

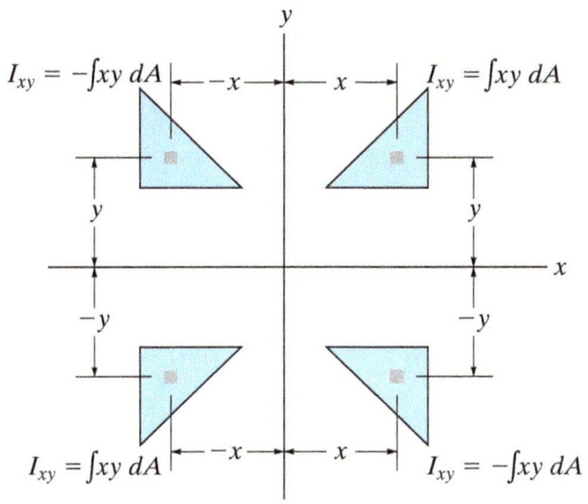

Fig. 10–12

Parallel-Axis Theorem. Consider the shaded area shown in Fig. 10–13, where x' and y' represent a set of axes passing through the *centroid* of the area, and x and y represent a corresponding set of parallel axes. Since the product of inertia of dA with respect to the x and y axes is $dI_{xy} = (x' + d_x)(y' + d_y)\, dA$, then for the entire area,

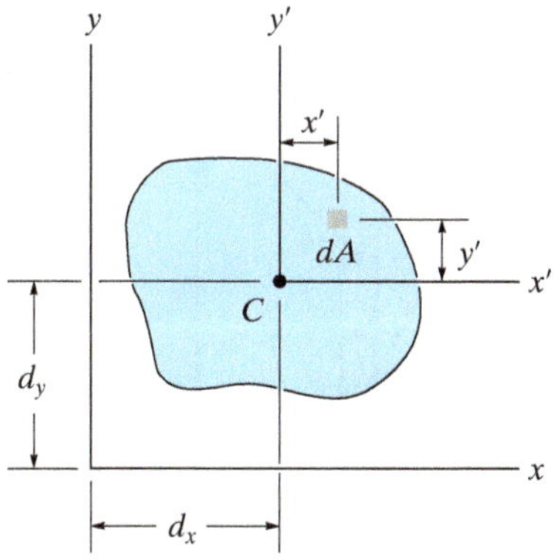

Fig. 10–13

$$I_{xy} = \int_A (x' + d_x)(y' + d_y)\, dA$$

$$= \int_A x'y'\, dA + d_x \int_A y'\, dA + d_y \int_A x'\, dA + d_x d_y \int_A dA$$

The first term on the right represents the product of inertia for the area with respect to the centroidal axes, $\bar{I}_{x'y'}$. The integrals in the second and third terms are zero since the moments of the area are taken about the centroidal axis. Realizing that the fourth integral represents the entire area A, the parallel-axis theorem for the product of inertia becomes

$$I_{xy} = \bar{I}_{x'y'} + Ad_x d_y \qquad (10\text{–}8)$$

It is important that the *algebraic signs* for d_x and d_y be maintained when applying this equation.

10

EXAMPLE 10.6

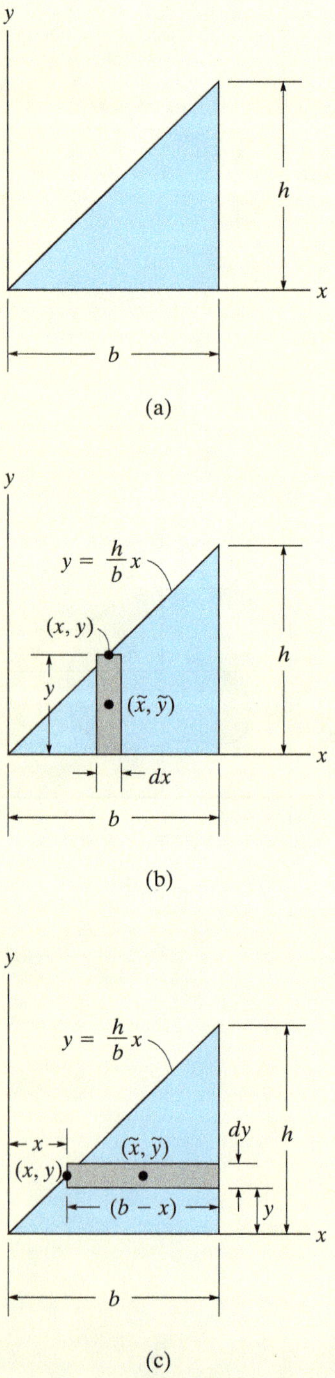

Fig. 10–14

Determine the product of inertia I_{xy} for the triangle shown in Fig. 10–14*a*.

SOLUTION I

A differential element that has a thickness dx, as shown in Fig. 10–14*b*, has an area $dA = y\,dx$. The product of inertia of this element with respect to the x *and* y axes is determined using the parallel-axis theorem.

$$dI_{xy} = d\bar{I}_{x'y'} + dA\,\tilde{x}\,\tilde{y}$$

where $\tilde{x}$ and $\tilde{y}$ locate the *centroid* of the element or the origin of the x', y' axes. (See Fig. 10–13.) Since $d\bar{I}_{x'y'} = 0$, due to symmetry, and $\tilde{x} = x$, $\tilde{y} = y/2$, then

$$dI_{xy} = 0 + (y\,dx)x\left(\frac{y}{2}\right) = \left(\frac{h}{b}x\,dx\right)x\left(\frac{h}{2b}x\right)$$
$$= \frac{h^2}{2b^2}x^3\,dx$$

Integrating with respect to x from $x = 0$ to $x = b$ yields

$$I_{xy} = \frac{h^2}{2b^2}\int_0^b x^3\,dx = \frac{b^2h^2}{8} \qquad \textit{Ans.}$$

SOLUTION II

The differential element that has a thickness dy, as shown in Fig. 10–14*c*, can also be used. Its area is $dA = (b - x)\,dy$. The *centroid* is located at point $\tilde{x} = x + (b - x)/2 = (b + x)/2$, $\tilde{y} = y$, so the product of inertia of the element becomes

$$dI_{xy} = d\bar{I}_{x'y'} + dA\,\tilde{x}\,\tilde{y}$$
$$= 0 + (b - x)\,dy\left(\frac{b + x}{2}\right)y$$
$$= \left(b - \frac{b}{h}y\right)dy\left[\frac{b + (b/h)y}{2}\right]y = \frac{1}{2}y\left(b^2 - \frac{b^2}{h^2}y^2\right)dy$$

Integrating with respect to y from $y = 0$ to $y = h$ yields

$$I_{xy} = \frac{1}{2}\int_0^h y\left(b^2 - \frac{b^2}{h^2}y^2\right)dy = \frac{b^2h^2}{8} \qquad \textit{Ans.}$$

EXAMPLE 10.7

Determine the product of inertia for the cross-sectional area of the member shown in Fig. 10–15*a*, about the *x* and *y* centroidal axes.

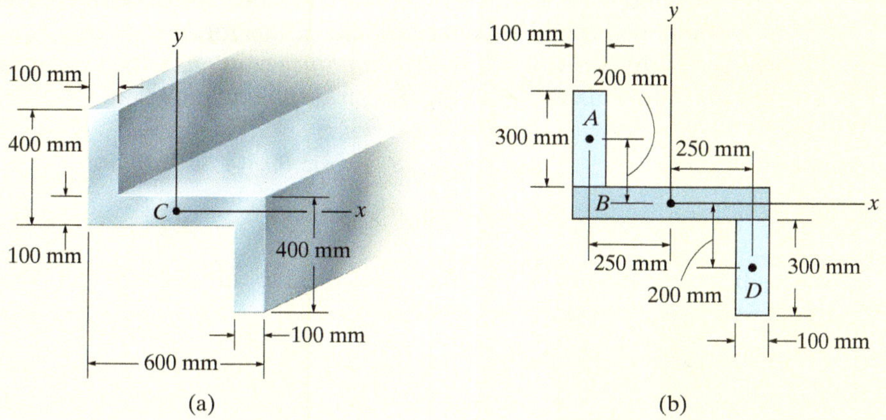

Fig. 10–15

SOLUTION

As in Example 10.5, the cross section can be subdivided into three composite rectangular areas *A*, *B*, and *D*, Fig. 10–15*b*. The coordinates for the centroid of each of these rectangles are shown in the figure. Due to symmetry, the product of inertia of *each rectangle* is *zero* about a set of x', y' axes that passes through the centroid of each rectangle. Using the parallel-axis theorem, we have

Rectangle A

$$I_{xy} = \bar{I}_{x'y'} + Ad_xd_y$$
$$= 0 + (300)(100)(-250)(200) = -1.50(10^9)\text{ mm}^4$$

Rectangle B

$$I_{xy} = \bar{I}_{x'y'} + Ad_xd_y$$
$$= 0 + 0 = 0$$

Rectangle D

$$I_{xy} = \bar{I}_{x'y'} + Ad_xd_y$$
$$= 0 + (300)(100)(250)(-200) = -1.50(10^9)\text{ mm}^4$$

The product of inertia for the entire cross section is therefore

$$I_{xy} = -1.50(10^9) + 0 - 1.50(10^9) = -3.00(10^9)\text{ mm}^4 \quad \textit{Ans.}$$

NOTE: This negative result is due to the fact that rectangles *A* and *D* have centroids located with negative *x* and negative *y* coordinates, respectively.

*10.6 Moments of Inertia for an Area about Inclined Axes

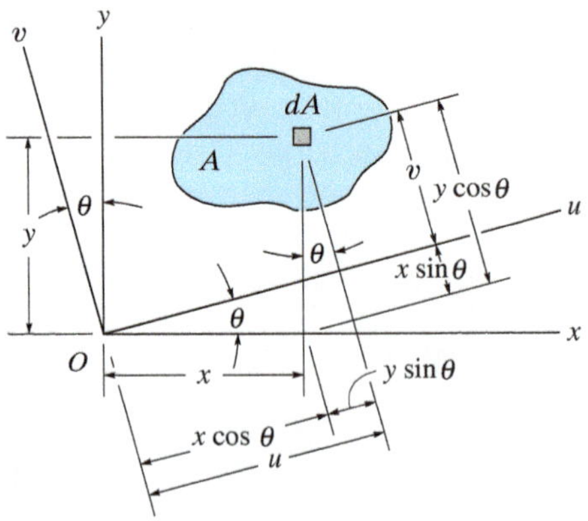

Fig. 10–16

In structural and mechanical design, it is sometimes necessary to calculate the moments and product of inertia I_u, I_v, and I_{uv} for an area with respect to a set of inclined u and v axes when the values for θ, I_x, I_y, and I_{xy} are *known*. To do this we will use *transformation equations* which relate the x, y and u, v coordinates. From Fig. 10–16, these equations are

$$u = x\cos\theta + y\sin\theta$$
$$v = y\cos\theta - x\sin\theta$$

With these equations, the moments and product of inertia of dA about the u and v axes become

$$dI_u = v^2\,dA = (y\cos\theta - x\sin\theta)^2\,dA$$
$$dI_v = u^2\,dA = (x\cos\theta + y\sin\theta)^2\,dA$$
$$dI_{uv} = uv\,dA = (x\cos\theta + y\sin\theta)(y\cos\theta - x\sin\theta)\,dA$$

Expanding each expression and integrating, realizing that $I_x = \int y^2\,dA$, $I_y = \int x^2\,dA$, and $I_{xy} = \int xy\,dA$, we obtain

$$I_u = I_x\cos^2\theta + I_y\sin^2\theta - 2I_{xy}\sin\theta\cos\theta$$
$$I_v = I_x\sin^2\theta + I_y\cos^2\theta + 2I_{xy}\sin\theta\cos\theta$$
$$I_{uv} = I_x\sin\theta\cos\theta - I_y\sin\theta\cos\theta + I_{xy}(\cos^2\theta - \sin^2\theta)$$

Using the trigonometric identities $\sin 2\theta = 2\sin\theta\cos\theta$ and $\cos 2\theta = \cos^2\theta - \sin^2\theta$ we can simplify the above expressions, in which case

$$I_u = \frac{I_x + I_y}{2} + \frac{I_x - I_y}{2}\cos 2\theta - I_{xy}\sin 2\theta$$
$$I_v = \frac{I_x + I_y}{2} - \frac{I_x - I_y}{2}\cos 2\theta + I_{xy}\sin 2\theta \qquad (10\text{–}9)$$
$$I_{uv} = \frac{I_x - I_y}{2}\sin 2\theta + I_{xy}\cos 2\theta$$

Notice that if the first and second equations are added together, we can show that the polar moment of inertia about the z axis passing through point O is, as expected, *independent* of the orientation of the u and v axes; i.e.,

$$J_O = I_u + I_v = I_x + I_y$$

10

Principal Moments of Inertia. Equations 10–9 show that I_u, I_v, and I_{uv} depend on the angle of inclination, θ, of the u, v axes. We will now determine the orientation of these axes about which the moments of inertia for the area are maximum and minimum. This particular set of axes is called the *principal axes* of the area, and the corresponding moments of inertia with respect to these axes are called the *principal moments of inertia*. In general, there is a set of principal axes for every chosen origin O. However, for structural and mechanical design, the origin O is located at the centroid of the area.

The angle which defines the orientation of the principal axes can be found by differentiating the first of Eqs. 10–9 with respect to θ and setting the result equal to zero. Thus,

$$\frac{dI_u}{d\theta} = -2\left(\frac{I_x - I_y}{2}\right)\sin 2\theta - 2I_{xy}\cos 2\theta = 0$$

Therefore, at $\theta = \theta_p$,

$$\tan 2\theta_p = \frac{-I_{xy}}{(I_x - I_y)/2} \tag{10–10}$$

The two roots θ_{p_1} and θ_{p_2} of this equation are 90° apart, and so they each specify the inclination of one of the principal axes. In order to substitute them into Eq. 10–9, we must first find the sine and cosine of $2\theta_{p_1}$ and $2\theta_{p_2}$. This can be done using these ratios from the triangles shown in Fig. 10–17, which are based on Eq. 10–10.

Substituting each of the sine and cosine ratios into the first or second of Eqs. 10–9 and simplifying, we obtain

$$I_{\substack{max\\min}} = \frac{I_x + I_y}{2} \pm \sqrt{\left(\frac{I_x - I_y}{2}\right)^2 + I_{xy}^2} \tag{10–11}$$

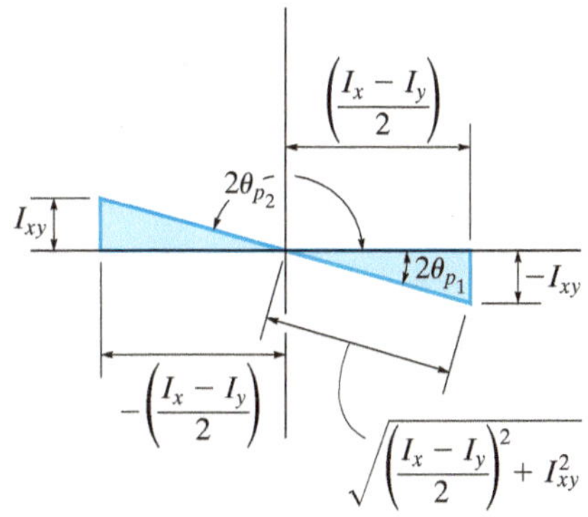

Fig. 10–17

Depending on the sign chosen, this result gives the maximum or minimum moment of inertia for the area. Furthermore, if the above trigonometric relations for θ_{p_1} and θ_{p_2} are substituted into the third of Eqs. 10–9, it can be shown that $I_{uv} = 0$; that is, the *product of inertia with respect to the principal axes is zero*. Since it was indicated in Sec. 10.6 that the product of inertia is zero with respect to any symmetrical axis, it therefore follows that *any symmetrical axis represents a principal axis of inertia for the area.*

10

EXAMPLE 10.8

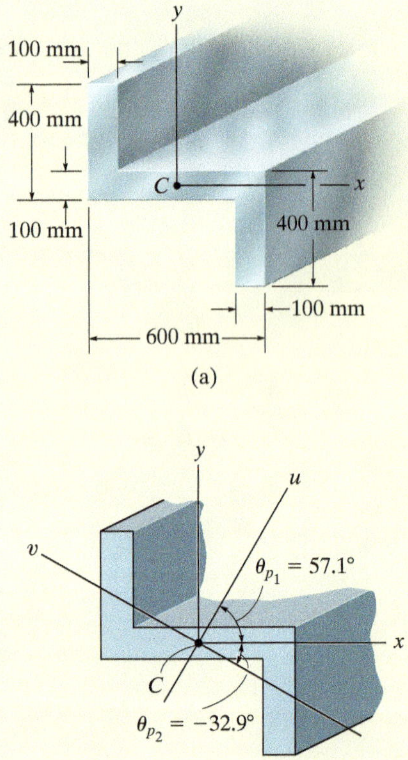

Fig. 10–18

Determine the principal moments of inertia and the orientation of the principal axes for the cross-sectional area of the member shown in Fig. 10–18*a* with respect to an axis passing through the centroid.

SOLUTION

The moments and product of inertia of the cross section with respect to the *x, y* axes have been determined in Examples 10.5 and 10.7. The results are

$$I_x = 2.90(10^9)\text{ mm}^4 \quad I_y = 5.60(10^9)\text{ mm}^4 \quad I_{xy} = -3.00(10^9)\text{ mm}^4$$

Using Eq. 10–10, the angles of inclination of the principal axes *u* and *v* are

$$\tan 2\theta_p = \frac{-I_{xy}}{(I_x - I_y)/2} = \frac{-[-3.00(10^9)]}{[2.90(10^9) - 5.60(10^9)]/2} = -2.22$$

$$2\theta_p = -65.8° \text{ and } 114.2°$$

Thus, by inspection of Fig. 10–18*b*,

$$\theta_{p_2} = -32.9° \quad \text{and} \quad \theta_{p_1} = 57.1° \qquad \textit{Ans.}$$

The principal moments of inertia with respect to these axes are determined from Eq. 10–11. Hence,

$$I_{\substack{\max\\ \min}} = \frac{I_x + I_y}{2} \pm \sqrt{\left(\frac{I_x - I_y}{2}\right)^2 + I_{xy}^2}$$

$$= \frac{2.90(10^9) + 5.60(10^9)}{2}$$

$$\pm \sqrt{\left[\frac{2.90(10^9) - 5.60(10^9)}{2}\right]^2 + [-3.00(10^9)]^2}$$

$$I_{\substack{\max\\ \min}} = 4.25(10^9) \pm 3.29(10^9)$$

or

$$I_{\max} = 7.54(10^9)\text{ mm}^4 \quad I_{\min} = 0.960(10^9)\text{ mm}^4 \qquad \textit{Ans.}$$

NOTE: The maximum moment of inertia, $I_{\max} = 7.54(10^9)\text{ mm}^4$, occurs with respect to the *u* axis since *by inspection* most of the cross-sectional area is farthest away from this axis. Or, stated in another manner, $I_{\max}$ occurs about the *u* axis since this axis is located within $\pm 45°$ of the *y* axis, which has the larger value of I $(I_y > I_x)$. Also, this can be concluded by substituting the data with $\theta = 57.1°$ into the first of Eqs. 10–9 and solving for I_u.

*10.7 Mohr's Circle for Moments of Inertia

Equations 10–9 to 10–11 have a graphical solution that is convenient to use and generally easy to remember. Squaring the first and third of Eqs. 10–9 and adding, it is found that

$$\left(I_u - \frac{I_x + I_y}{2}\right)^2 + I_{uv}^2 = \left(\frac{I_x - I_y}{2}\right)^2 + I_{xy}^2$$

Here I_x, I_y, and I_{xy} are *known constants*. Thus, the above equation may be written in compact form as

$$(I_u - a)^2 + I_{uv}^2 = R^2$$

When this equation is plotted on a set of axes that represent the respective moment of inertia and the product of inertia, as shown in Fig. 10–19, the resulting graph represents a *circle* of radius

$$R = \sqrt{\left(\frac{I_x - I_y}{2}\right)^2 + I_{xy}^2}$$

and having its center located at point $(a, 0)$, where $a = (I_x + I_y)/2$. The circle so constructed is called *Mohr's circle*, named after the German engineer Otto Mohr (1835–1918).

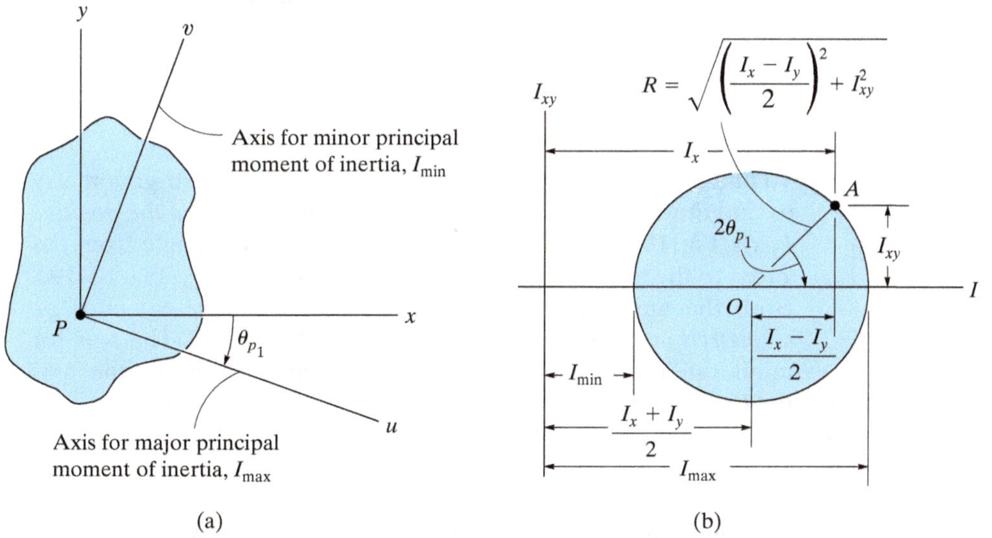

Fig. 10–19

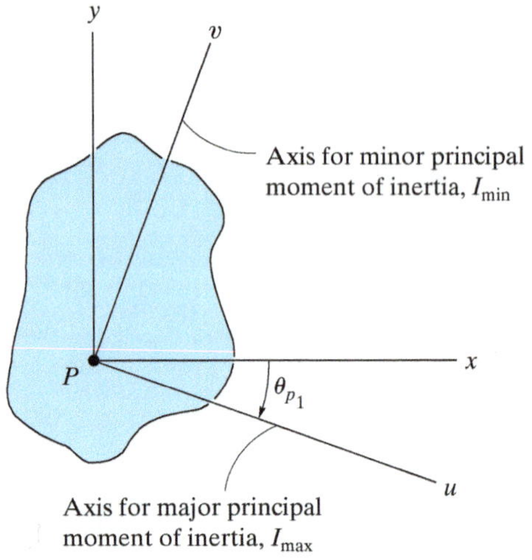

(a)

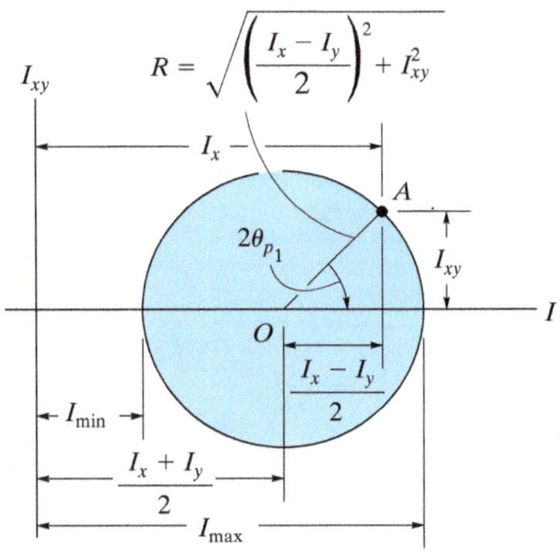

(b)

Fig. 10–19 (Repeated)

Procedure for Analysis

The main purpose in using Mohr's circle here is to have a convenient means for finding the principal moments of inertia for an area. The following procedure provides a method for doing this.

Determine I_x, I_y, and I_{xy}.

- Establish the x, y axes and determine I_x, I_y, and I_{xy}, Fig. 10–19*a*.

Construct the Circle.

- Construct a rectangular coordinate system such that the abscissa represents the moment of inertia I, and the ordinate represents the product of inertia I_{xy}, Fig. 10–19*b*.
- Determine the center of the circle, O, which is located at a distance $(I_x + I_y)/2$ from the origin, and plot the reference point A having coordinates (I_x, I_{xy}). Remember, I_x is always positive, whereas I_{xy} can be either positive or negative.
- Connect the reference point A with the center of the circle and determine the distance OA by trigonometry. This distance represents the radius of the circle, Fig. 10–19*b*. Finally, draw the circle.

Principal Moments of Inertia.

- The points where the circle intersects the I axis give the values of the principal moments of inertia I_{min} and I_{max}. Notice that, as expected, the *product of inertia will be zero at these points*, Fig. 10–19*b*.

Principal Axes.

- To find the orientation of the major principal axis, use trigonometry to find the angle $2\theta_{p_1}$, *measured from the radius OA to the positive I axis*, Fig. 10–19*b*. This angle represents *twice* the angle from the x axis to the axis of maximum moment of inertia I_{max}, Fig. 10–19*a*. Both the angle on the circle, $2\theta_{p_1}$, and the angle θ_{p_1} *must be measured in the same sense*, as shown in Fig. 10–19. The axis for minimum moment of inertia I_{min} is perpendicular to the axis for I_{max}.

Using trigonometry, the above procedure can be verified to be in accordance with the equations developed in Sec. 10.6.

EXAMPLE 10.9

Using Mohr's circle, determine the principal moments of inertia and the orientation of the major principal axes for the cross-sectional area of the member shown in Fig. 10–20*a*, with respect to an axis passing through the centroid.

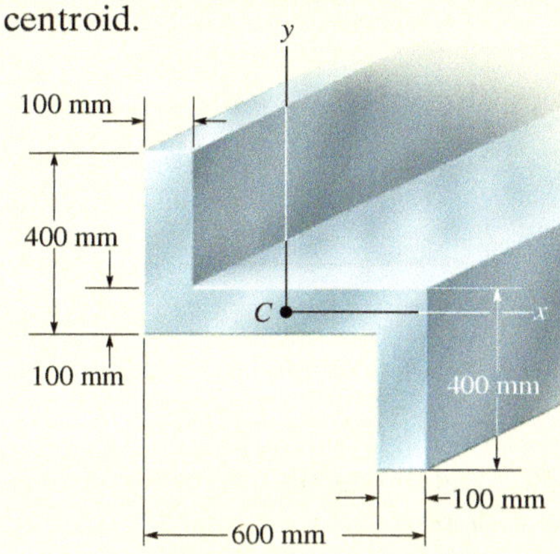

(a)

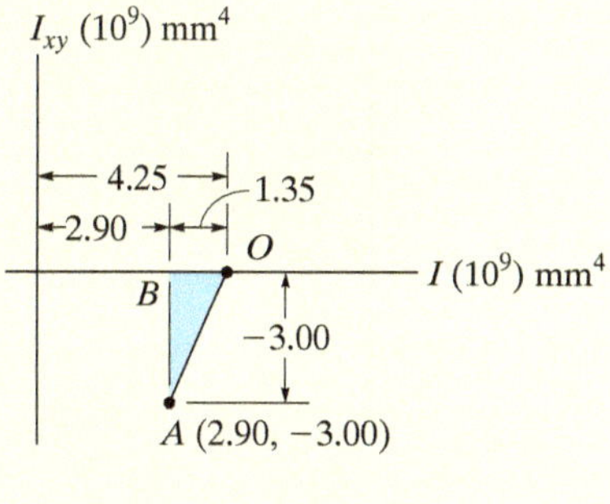

(b)

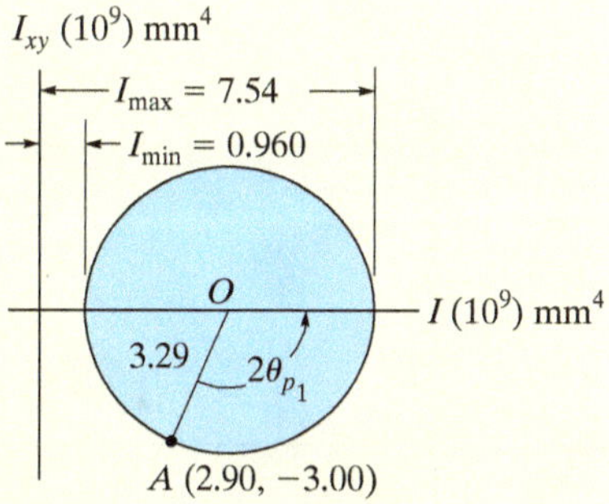

(c)

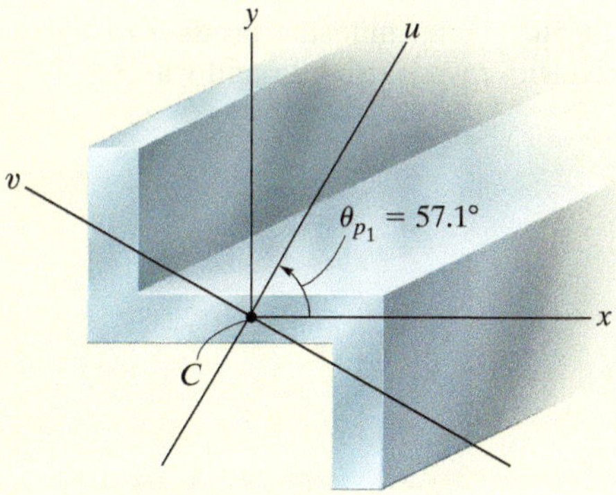

(d)

Fig. 10–20

SOLUTION

Determine I_x, I_y, I_{xy}. The moments and product of inertia have been determined in Examples 10.5 and 10.7 with respect to the *x*, *y* axes shown in Fig. 10–20*a*. The results are $I_x = 2.90(10^9)$ mm^4, $I_y = 5.60(10^9)$ mm^4, and $I_{xy} = -3.00(10^9)$ mm^4.

Construct the Circle. The I and I_{xy} axes are shown in Fig. 10–20*b*. The center of the circle, O, lies at a distance $(I_x + I_y)/2 = (2.90 + 5.60)/2 = 4.25$ from the origin. When the reference point $A(I_x, I_{xy})$ or $A(2.90, -3.00)$ is connected to point O, the radius OA is determined from the triangle OBA using the Pythagorean theorem.

$$OA = \sqrt{(1.35)^2 + (-3.00)^2} = 3.29$$

The circle is constructed in Fig. 10–20*c*.

Principal Moments of Inertia. The circle intersects the I axis at points (7.54, 0) and (0.960, 0). Hence,

$$I_{max} = (4.25 + 3.29)10^9 = 7.54(10^9)\text{ mm}^4 \qquad \textit{Ans.}$$

$$I_{min} = (4.25 - 3.29)10^9 = 0.960(10^9)\text{ mm}^4 \qquad \textit{Ans.}$$

Principal Axes. As shown in Fig. 10–20*c*, the angle $2\theta_{p_1}$ is determined from the circle by measuring counterclockwise from OA to the direction of the *positive* I axis. Hence,

$$2\theta_{p_1} = 180° - \sin^{-1}\left(\frac{|BA|}{|OA|}\right) = 180° - \sin^{-1}\left(\frac{3.00}{3.29}\right) = 114.2°$$

The principal axis for $I_{max} = 7.54(10^9)$ mm^4 is therefore oriented at an angle $\theta_{p_1} = 57.1°$, measured *counterclockwise*, from the *positive* x axis to the *positive* u axis. The v axis is perpendicular to this axis. The results are shown in Fig. 10–20*d*.

10

PROBLEMS

10–54. Determine the product of inertia of the thin strip of area with respect to the x and y axes. The strip is oriented at an angle θ from the x axis. Assume that $t \ll l$.

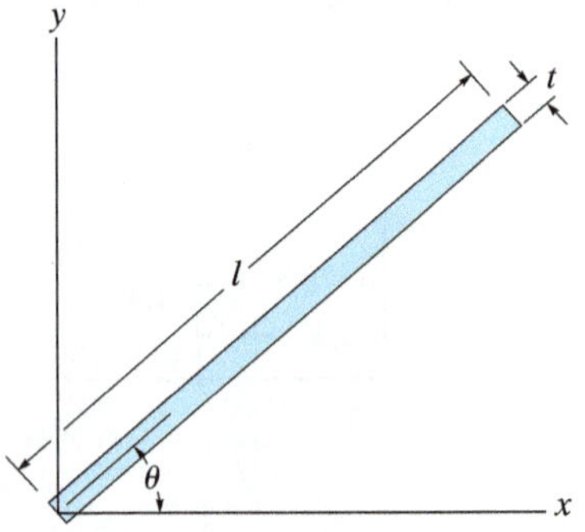

Prob. 10–54

10–55. Determine the product of inertia of the shaded area with respect to the x and y axes.

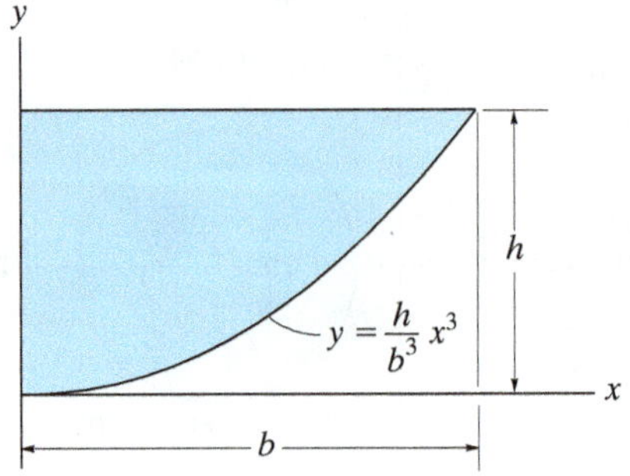

Prob. 10–55

***10–56.** Determine the product of inertia of the shaded area with respect to the x and y axes.

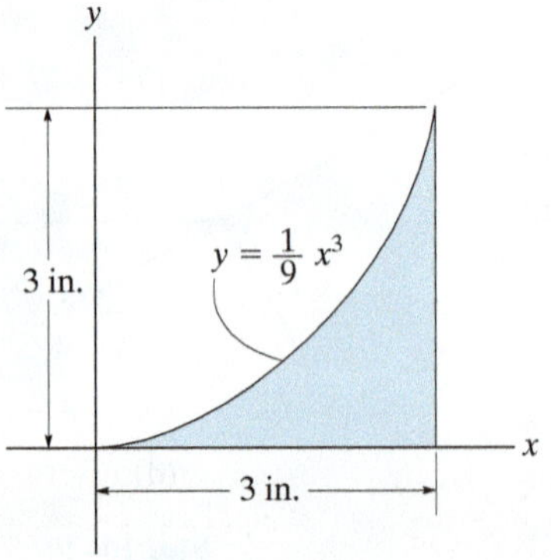

Prob. 10–56

10–57. Determine the product of inertia of the shaded area with respect to the x and y axes.

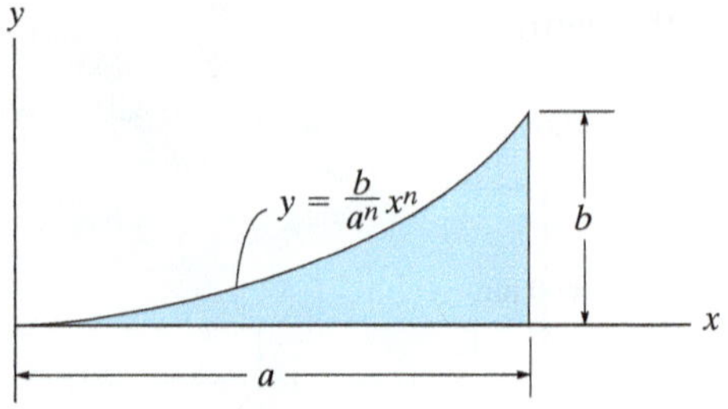

Prob. 10–57

10–58. Determine the product of inertia of the shaded area with respect to the x and y axes, and then use the parallel-axis theorem to find the product of inertia of the area with respect to the centroidal x' and y' axes.

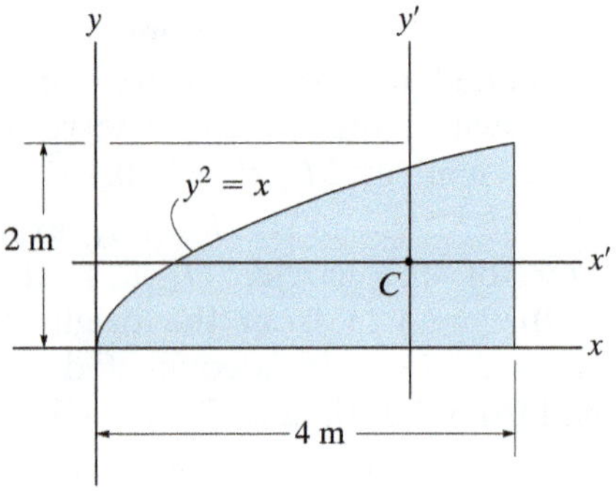

Prob. 10–58

■**10–59.** Determine the product of inertia of the shaded area with respect to the x and y axes. Use Simpson's rule to evaluate the integral.

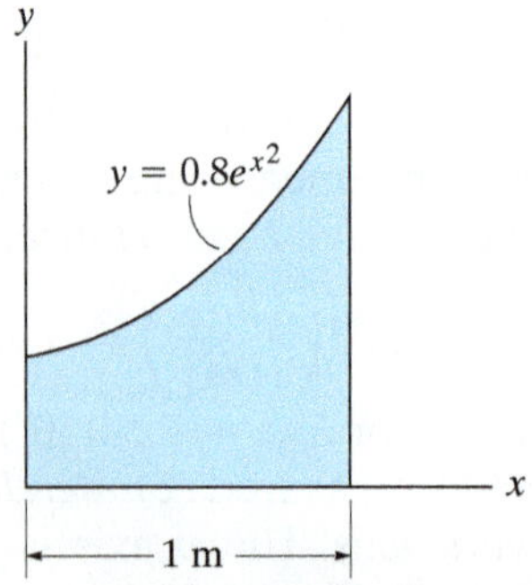

Prob. 10–59

***10–60.** Determine the product of inertia of the shaded area with respect to the x and y axes.

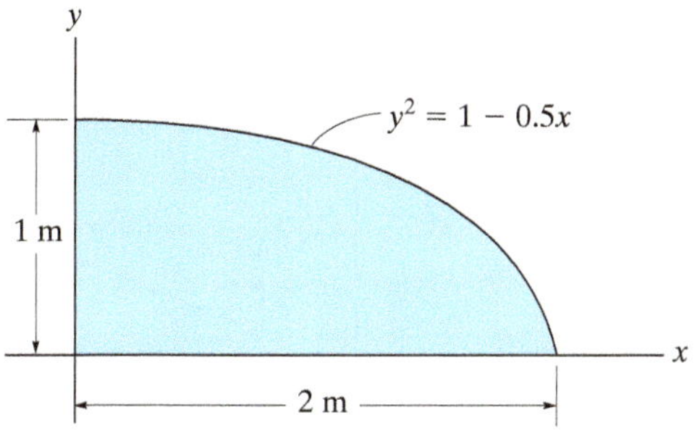

Prob. 10–60

10–61. Determine the product of inertia of the parallelogram with respect to the x and y axes.

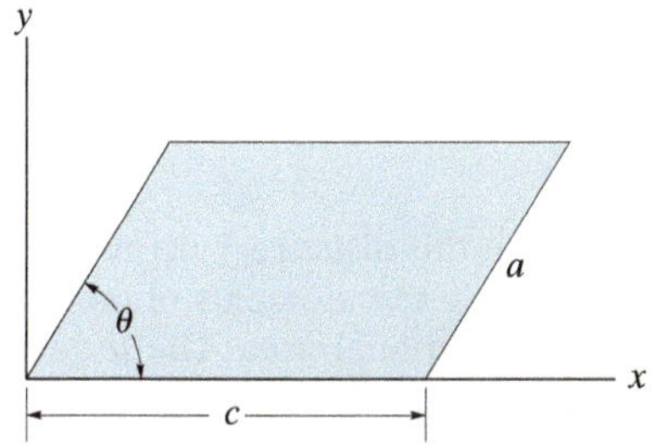

Prob. 10–61

10–62. Determine the product of inertia of the shaded area with respect to the x and y axes.

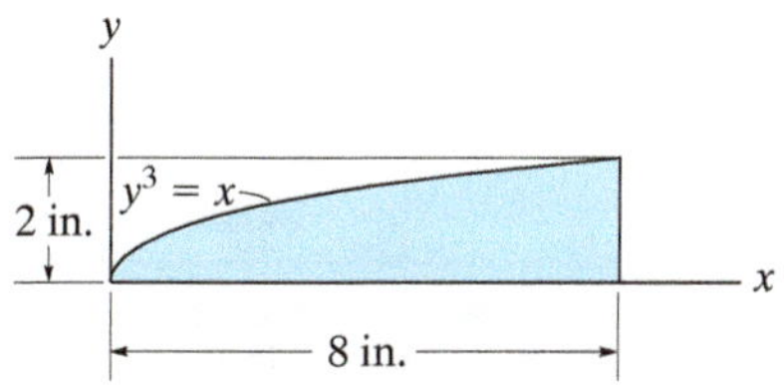

Prob. 10–62

10–63. Determine the product of inertia for the beam's cross-sectional area with respect to the u and v axes.

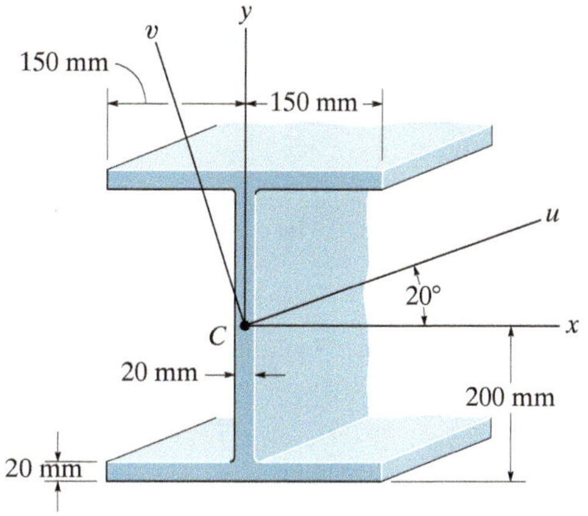

Prob. 10–63

***10–64.** Determine the moments of inertia of the shaded area with respect to the u and v axes.

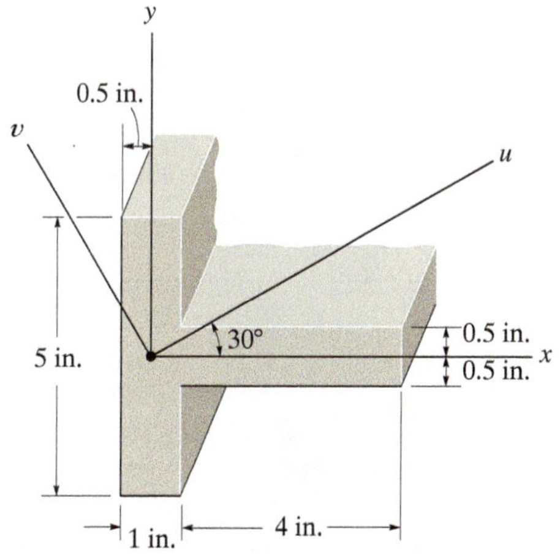

Prob. 10–64

10–65. Determine the product of inertia of the beam's cross-sectional area with respect to the x and y axes that have their origin located at the centroid C.

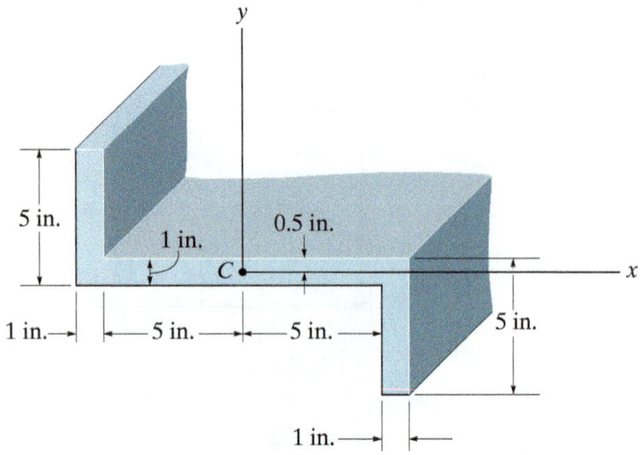

Prob. 10–65

10–66. Determine the product of inertia of the cross-sectional area with respect to the x and y axes.

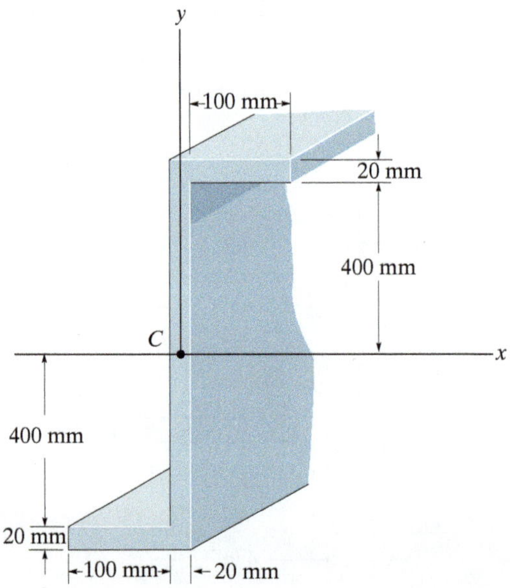

Prob. 10–66

10–67. Determine the product of inertia for the angle with respect to the x and y axes passing through the centroid C. Assume all corners to be square.

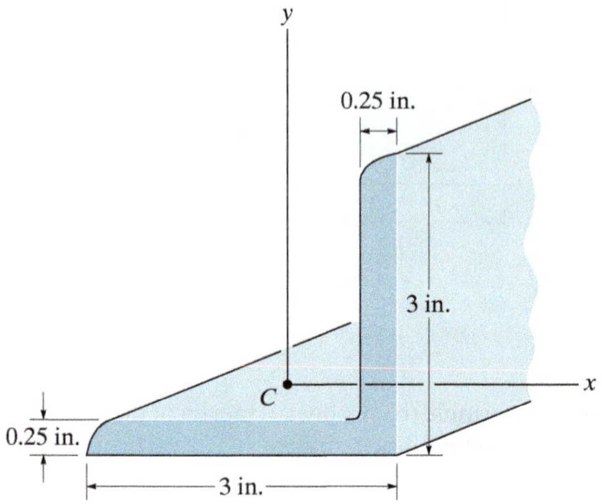

Prob. 10–67

***10–68.** Determine the distance $\bar{y}$ to the centroid of the area and then calculate the moments of inertia I_u and I_v of the channel's cross-sectional area. The u and v axes have their origin at the centroid C. For the calculation, assume all corners to be square.

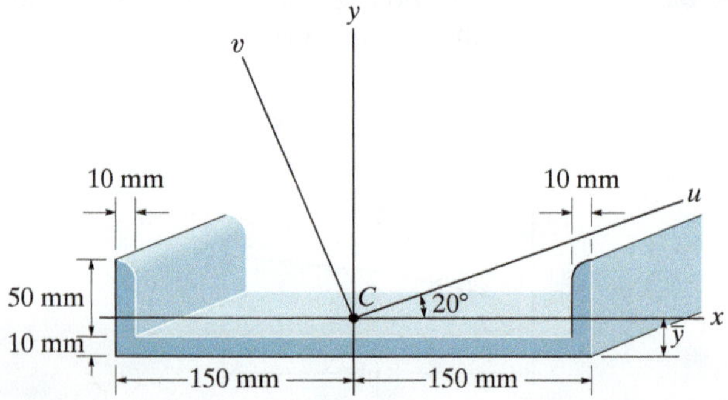

Prob. 10–68

10–69. Determine the moments of inertia I_u and I_v of the shaded area.

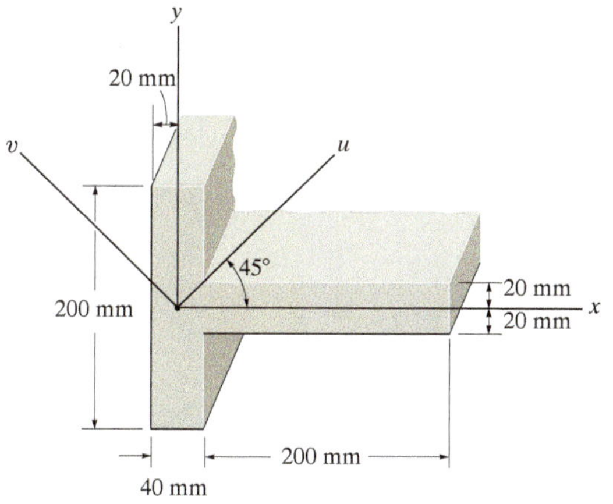

Prob. 10–69

10–70. Determine the moments of inertia and the product of inertia of the beam's cross sectional area with respect to the u and v axes.

10–71. Solve Prob. 10–70 using Mohr's circle. *Hint:* Once the circle is established, rotate $2\theta = 60°$ counterclockwise from the reference OA, then find the coordinates of the points that define the diameter of the circle.

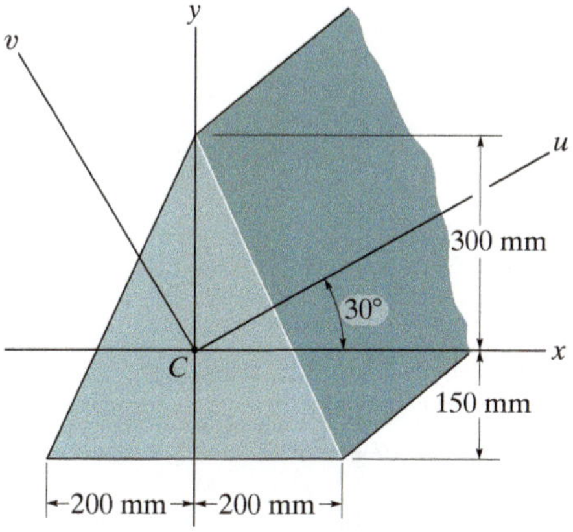

Probs. 10–70/71

***10–72.** Locate the centroid $\bar{y}$ of the beam's cross-sectional area and then determine the moments of inertia and the product of inertia of this area with respect to the u and v axes.

10–73. Solve Prob. 10–72 using Mohr's circle. *Hint:* Once the circle is established, rotate $2\theta = 120°$ counterclockwise from the reference OA, then find the coordinates of the points that define the diameter of the circle.

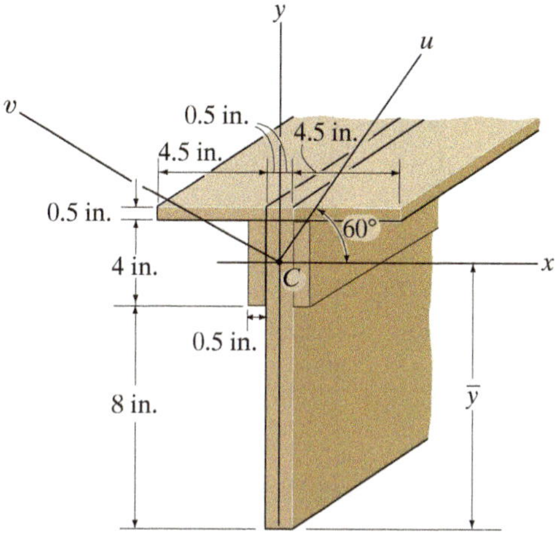

Probs. 10–72/73

10–74. Determine the principal moments of inertia of the beam's cross-sectional area about the principal axes that have their origin located at the centroid C. Use the equations developed in Section 10.7. For the calculation, assume all corners to be square.

10–75. Solve Prob. 10–74 using Mohr's circle.

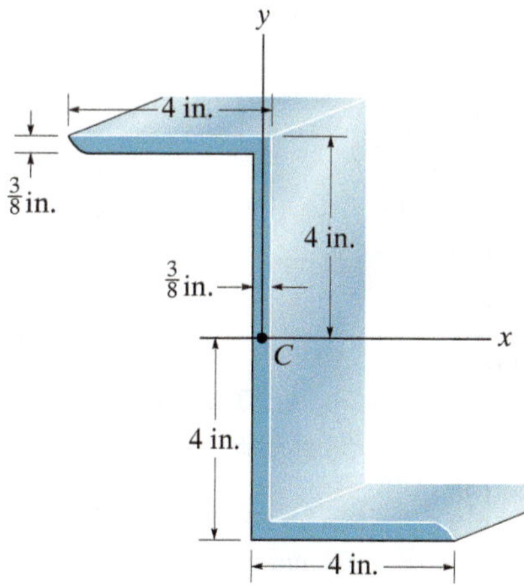

Probs. 10–74/75

*10–76. Determine the directions of the principal axes with origin located at point O, and the principal moments of inertia for the rectangular area about these axes.

10–77. Solve Prob. 10–76 using Mohr's circle.

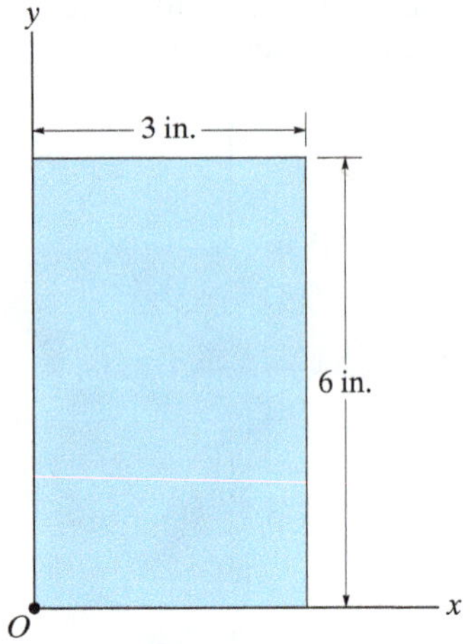

Probs. 10–76/77

*10–80. Determine the directions of the principal axes with origin located at point O, and the principal moments of inertia of the area about these axes.

10–81. Solve Prob. 10–80 using Mohr's circle.

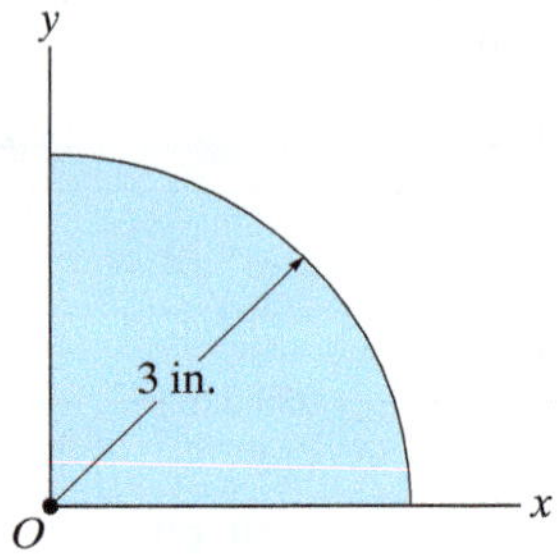

Probs. 10–80/81

10–78. Determine the principal moments of inertia for the angle's cross-sectional area with respect to a set of principal axes that have their origin located at the centroid C. Use the equation developed in Section 10.7. For the calculation, assume all corners to be square.

10–79. Solve Prob. 10–78 using Mohr's circle.

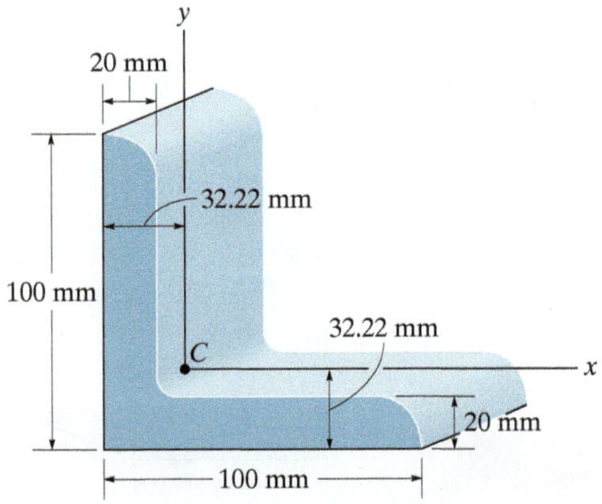

Probs. 10–78/79

10–82. The area of the cross section of an airplane wing has the following properties about the x and y axes passing through the centroid C: $\bar{I}_x = 450\ \text{in}^4$, $\bar{I}_y = 1730\ \text{in}^4$, $\bar{I}_{xy} = 138\ \text{in}^4$. Determine the orientation of the principal axes and the principal moments of inertia.

10–83. Solve Prob. 10–82 using Mohr's circle.

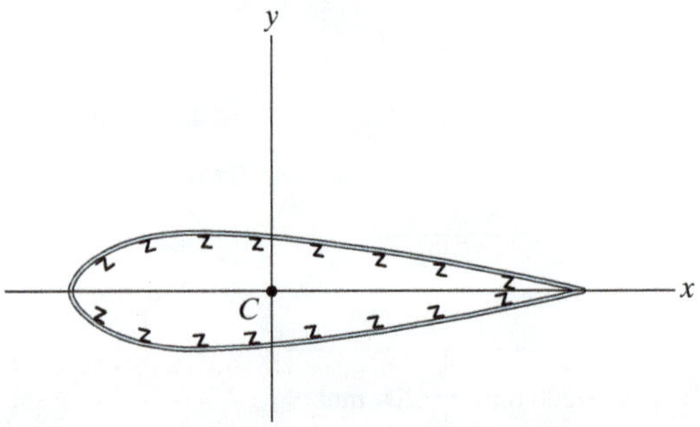

Probs. 10–82/83

10.8 Mass Moment of Inertia

The mass moment of inertia of a body is a measure of the body's resistance to angular acceleration. Since it is used in dynamics to study rotational motion, methods for its calculation will now be discussed.*

Consider the rigid body shown in Fig. 10–21. We define the *mass moment of inertia* of the body about the z axis as

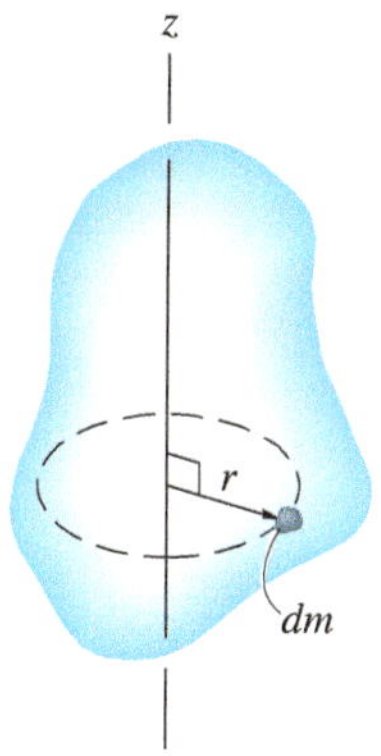

Fig. 10–21

$$I = \int_m r^2\, dm \tag{10–12}$$

Here r is the perpendicular distance from the axis to the arbitrary element dm. Since the formulation involves r, the value of I is *unique* for each axis about which it is computed. The axis which is generally chosen, however, passes through the body's mass center G. Common units used for its measurement are $kg \cdot m^2$ or $slug \cdot ft^2$.

If the body consists of material having a density ρ, then $dm = \rho\, dV$, Fig. 10–22*a*. Substituting this into Eq. 10–12, the body's moment of inertia is then computed using *volume elements* for integration; i.e.

$$I = \int_V r^2 \rho\, dV \tag{10–13}$$

For most applications, ρ will be a *constant*, and so this term may be factored out of the integral, and the integration is then purely a function of geometry.

$$I = \rho \int_V r^2\, dV \tag{10–14}$$

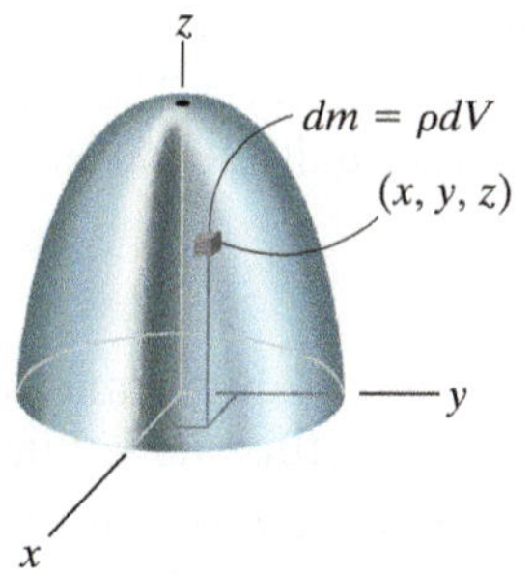

(a)

Fig. 10–22

*Another property of the body, which measures the symmetry of the body's mass with respect to a coordinate system, is the mass product of inertia. This property most often applies to the three-dimensional motion of a body and is discussed in *Engineering Mechanics: Dynamics* (Chapter 21).

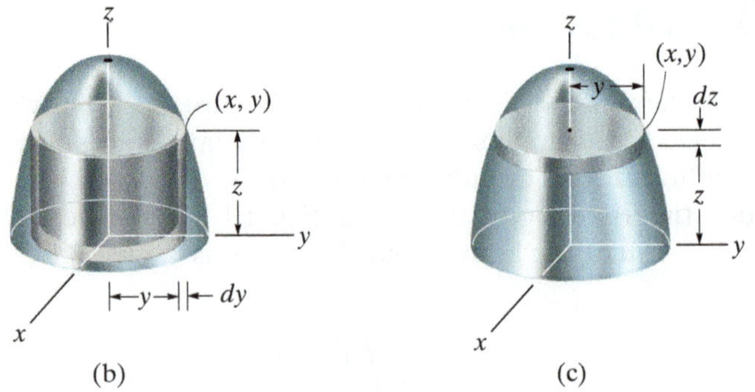

Fig. 10–22 (cont'd)

Procedure for Analysis

If a body is symmetrical with respect to an axis, as in Fig. 10–22, then its mass moment of inertia about the axis can be determined by using a single integration. Shell and disk elements are used for this purpose.

Shell Element.

- If a *shell element* having a height z, radius y, and thickness dy is chosen for integration, Fig. 10–22*b*, then its volume is $dV = (2\pi y)(z)\,dy$.
- This element can be used in Eq. 10–13 or 10–14 for determining the moment of inertia I_z of the body about the z axis since the *entire element*, due to its "thinness," lies at the *same* perpendicular distance $r = y$ from the z axis (see Example 10.10).

Disk Element.

- If a disk element having a radius y and a thickness dz is chosen for integration, Fig. 10–22*c*, then its volume is $dV = (\pi y^2)\,dz$.
- In this case the element is *finite* in the radial direction, and consequently its points *do not* all lie at the *same radial distance* r from the z axis. As a result, Eqs. 10–13 or 10–14 *cannot* be used to determine I_z. Instead, to perform the integration using this element, it is first necessary to determine the moment of inertia *of the element* about the z axis and then integrate this result (see Example 10.11).

10

EXAMPLE 10.10

Determine the mass moment of inertia of the cylinder shown in Fig. 10–23*a* about the *z* axis. The density of the material, ρ, is constant.

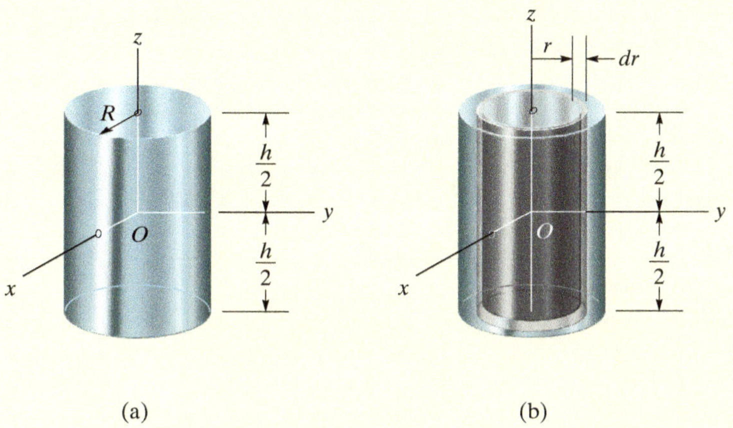

Fig. 10–23

SOLUTION

Shell Element. This problem will be solved using the *shell element* in Fig. 10–23*b* and thus only a single integration is required. The volume of the element is $dV = (2\pi r)(h)\,dr$, and so its mass is $dm = \rho\,dV = \rho(2\pi hr\,dr)$. Since the *entire element* lies at the same distance *r* from the *z* axis, the moment of inertia *of the element* is

$$dI_z = r^2\,dm = \rho 2\pi h r^3\,dr$$

Integrating over the entire cylinder yields

$$I_z = \int_m r^2\,dm = \rho 2\pi h \int_0^R r^3\,dr = \frac{\rho\pi}{2}R^4 h$$

Since the mass of the cylinder is

$$m = \int_m dm = \rho 2\pi h \int_0^R r\,dr = \rho\pi h R^2$$

then

$$I_z = \frac{1}{2}mR^2 \qquad \textit{Ans.}$$

EXAMPLE 10.11

A solid is formed by revolving the shaded area shown in Fig. 10–24*a* about the *y* axis. If the density of the material is 5 slug/ft^3, determine the mass moment of inertia about the *y* axis.

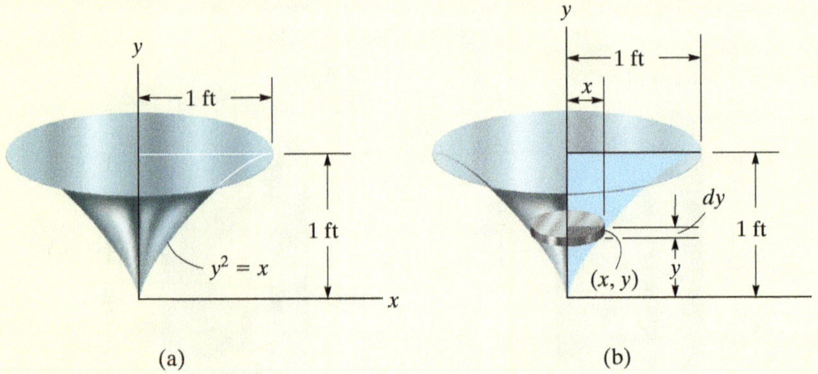

Fig. 10–24

SOLUTION

Disk Element. The moment of inertia will be determined using this *disk element*, as shown in Fig. 10–24*b*. Here the element intersects the curve at the arbitrary point (x, y) and has a mass

$$dm = \rho\, dV = \rho(\pi x^2)\, dy$$

Although all points on the element are *not* located at the same distance from the *y* axis, it is still possible to determine the moment of inertia dI_y *of the element* about the *y* axis. In the previous example it was shown that the moment of inertia of a homogeneous cylinder about its longitudinal axis is $I = \frac{1}{2}mR^2$, where m and R are the mass and radius of the cylinder. Since the height of the cylinder is not involved in this formula, we can also use this result for a disk. Thus, for the disk element in Fig. 10–24*b*, we have

$$dI_y = \frac{1}{2}(dm)x^2 = \frac{1}{2}[\rho(\pi x^2)\, dy]x^2$$

10

Substituting $x = y^2$, $\rho = 5$ slug/ft^3, and integrating with respect to y, from $y = 0$ to $y = 1$ ft, yields the moment of inertia for the entire solid.

$$I_y = \frac{5\pi}{2}\int_0^{1\text{ ft}} x^4\, dy = \frac{5\pi}{2}\int_0^{1\text{ ft}} y^8\, dy = 0.873 \text{ slug}\cdot\text{ft}^2 \qquad \textit{Ans.}$$

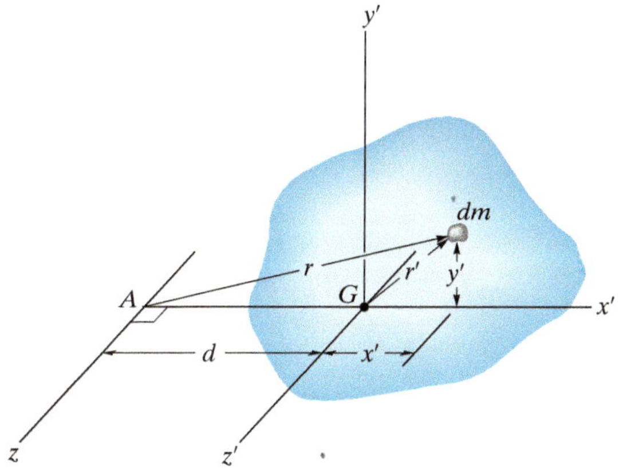

Fig. 10–25

Parallel-Axis Theorem. If the moment of inertia of the body about an axis passing through the body's mass center is known, then the moment of inertia about any other *parallel axis* can be determined by using the *parallel-axis theorem*. To derive this theorem, consider the body shown in Fig. 10–25. The z' axis passes through the mass center G, whereas the corresponding *parallel z axis* lies at a constant distance d away. Selecting the differential element of mass dm, which is located at point (x', y'), and using the Pythagorean theorem, $r^2 = (d + x')^2 + y'^2$, the moment of inertia of the body about the z axis is

$$I = \int_m r^2\,dm = \int_m [(d + x')^2 + y'^2]\,dm$$

$$= \int_m (x'^2 + y'^2)\,dm + 2d\int_m x'\,dm + d^2\int_m dm$$

Since $r'^2 = x'^2 + y'^2$, the first integral represents I_G. The second integral is equal to *zero*, since the z' axis passes through the body's mass center, i.e., $\int x'\,dm = \bar{x}\int dm = 0$ since $\bar{x} = 0$. Finally, the third integral is the total mass m of the body. Hence, the moment of inertia about the z axis becomes

$$I = I_G + md^2 \qquad (10\text{–}15)$$

10

where

I_G = moment of inertia about the z' axis passing through the mass center G

m = mass of the body

d = distance between the parallel axes

Radius of Gyration. Occasionally, the moment of inertia of a body about a specified axis is reported in handbooks using the *radius of gyration, k*. This value has units of length, and when it and the body's mass m are known, the moment of inertia can be determined from the equation

$$I = mk^2 \quad \text{or} \quad k = \sqrt{\frac{I}{m}} \tag{10–16}$$

Note the *similarity* between the definition of k in this formula and r in the equation $dI = r^2\,dm$, which defines the moment of inertia of a differential element of mass dm of the body about an axis.

Composite Bodies. If a body is constructed from a number of simple shapes such as disks, spheres, and rods, the moment of inertia of the body about any axis z can be determined by adding algebraically the moments of inertia of all the composite shapes calculated about the same axis. Algebraic addition is necessary since a composite part must be considered as a negative quantity if it has already been included within another part—as in the case of a "hole" subtracted from a solid plate. Also, the parallel-axis theorem is needed for the calculations if the center of mass of each composite part does not lie on the z axis. For calculations, a table of some simple shapes is given on the inside back cover.

This flywheel, which operates a metal cutter, has a large moment of inertia about its center. Once it begins rotating it is difficult to stop it and therefore a uniform motion can be effectively transferred to the cutting blade.

10

EXAMPLE 10.12

If the plate shown in Fig. 10–26*a* has a density of $8000\ \text{kg/m}^3$ and a thickness of 10 mm, determine its mass moment of inertia about an axis perpendicular to the page and passing through the pin at *O*.

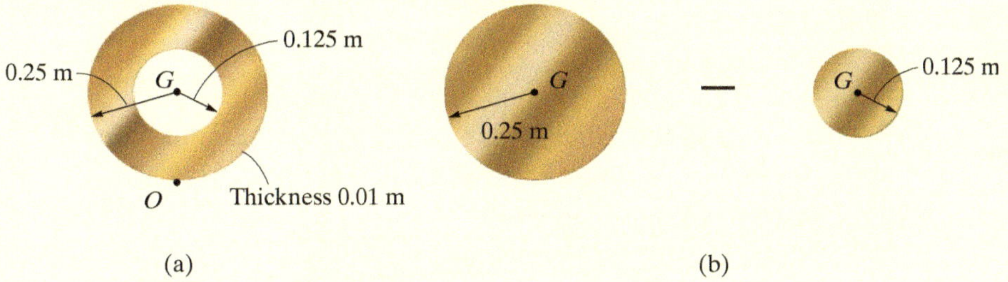

Fig. 10–26

SOLUTION

The plate consists of two composite parts, the 250-mm-radius disk *minus* a 125-mm-radius disk, Fig. 10–26*b*. The moment of inertia about *O* can be determined by finding the moment of inertia of each of these parts about *O* and then *algebraically* adding the results. The calculations are performed by using the parallel-axis theorem in conjunction with the mass moment of inertia formula for a circular disk, $I_G = \frac{1}{2}mr^2$, as found on the inside back cover.

Disk. The moment of inertia of a disk about an axis perpendicular to the plane of the disk and passing through *G* is $I_G = \frac{1}{2}mr^2$. The mass center of both disks is 0.25 m from point *O*. Thus,

$$m_d = \rho_d V_d = 8000\ \text{kg/m}^3\ [\pi(0.25\ \text{m})^2(0.01\ \text{m})] = 15.71\ \text{kg}$$

$$\begin{aligned}(I_O)_d &= \tfrac{1}{2}m_d r_d^2 + m_d d^2\\ &= \tfrac{1}{2}(15.71\ \text{kg})(0.25\ \text{m})^2 + (15.71\ \text{kg})(0.25\ \text{m})^2\\ &= 1.473\ \text{kg}\cdot\text{m}^2\end{aligned}$$

Hole. For the smaller disk (hole), we have

$$m_h = \rho_h V_h = 8000\ \text{kg/m}^3\ [\pi(0.125\ \text{m})^2(0.01\ \text{m})] = 3.93\ \text{kg}$$

$$\begin{aligned}(I_O)_h &= \tfrac{1}{2}m_h r_h^2 + m_h d^2\\ &= \tfrac{1}{2}(3.93\ \text{kg})(0.125\ \text{m})^2 + (3.93\ \text{kg})(0.25\ \text{m})^2\\ &= 0.276\ \text{kg}\cdot\text{m}^2\end{aligned}$$

The moment of inertia of the plate about the pin is therefore

$$\begin{aligned}I_O &= (I_O)_d - (I_O)_h\\ &= 1.473\ \text{kg}\cdot\text{m}^2 - 0.276\ \text{kg}\cdot\text{m}^2\\ &= 1.20\ \text{kg}\cdot\text{m}^2\end{aligned}$$ *Ans.*

10

EXAMPLE 10.13

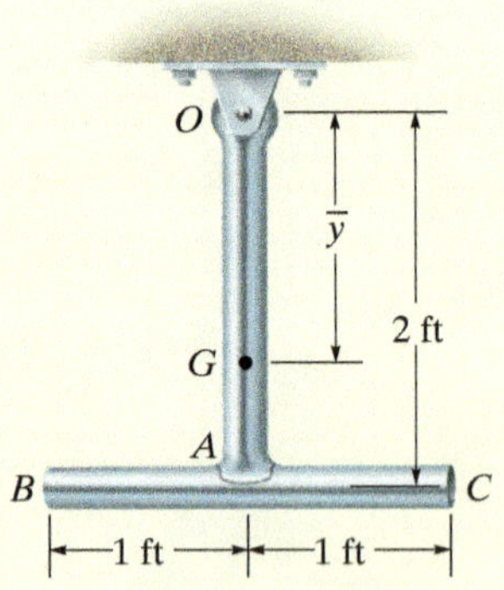

Fig. 10–27

The pendulum in Fig. 10–27 consists of two thin rods each having a weight of 10 lb. Determine the pendulum's mass moment of inertia about an axis passing through (a) the pin at O, and (b) the mass center G of the pendulum.

SOLUTION

Part (a). Using the table on the inside back cover, the moment of inertia of rod OA about an axis perpendicular to the page and passing through the end point O of the rod is $I_O = \frac{1}{3}ml^2$. Hence,

$$(I_{OA})_O = \frac{1}{3}ml^2 = \frac{1}{3}\left(\frac{10 \text{ lb}}{32.2 \text{ ft/s}^2}\right)(2 \text{ ft})^2 = 0.414 \text{ slug} \cdot \text{ft}^2$$

Realize that this same value may be determined using $I_G = \frac{1}{12}ml^2$ and the parallel-axis theorem; i.e.,

$$(I_{OA})_O = \frac{1}{12}ml^2 + md^2 = \frac{1}{12}\left(\frac{10 \text{ lb}}{32.2 \text{ ft/s}^2}\right)(2 \text{ ft})^2 + \frac{10 \text{ lb}}{32.2 \text{ ft/s}^2}(1 \text{ ft})^2$$
$$= 0.414 \text{ slug} \cdot \text{ft}^2$$

For rod BC we have

$$(I_{BC})_O = \frac{1}{12}ml^2 + md^2 = \frac{1}{12}\left(\frac{10 \text{ lb}}{32.2 \text{ ft/s}^2}\right)(2 \text{ ft})^2 + \frac{10 \text{ lb}}{32.2 \text{ ft/s}^2}(2 \text{ ft})^2$$
$$= 1.346 \text{ slug} \cdot \text{ft}^2$$

The moment of inertia of the pendulum about O is therefore

$$I_O = 0.414 + 1.346 = 1.76 \text{ slug} \cdot \text{ft}^2 \qquad \textit{Ans.}$$

Part (b). The mass center G will be located relative to the pin at O. Assuming this distance to be $\bar{y}$, Fig. 10–27, and using the formula for determining the mass center, we have

$$\bar{y} = \frac{\Sigma \tilde{y} m}{\Sigma m} = \frac{1(10/32.2) + 2(10/32.2)}{(10/32.2) + (10/32.2)} = 1.50 \text{ ft}$$

The moment of inertia I_G may be computed in the same manner as I_O, which requires successive applications of the parallel-axis theorem in order to transfer the moments of inertia of rods OA and BC to G. A more direct solution, however, involves applying the parallel-axis theorem using the result for I_O determined above; i.e.,

$$I_O = I_G + md^2; \qquad 1.76 \text{ slug} \cdot \text{ft}^2 = I_G + \left(\frac{20 \text{ lb}}{32.2 \text{ ft/s}^2}\right)(1.50 \text{ ft})^2$$

$$I_G = 0.362 \text{ slug} \cdot \text{ft}^2 \qquad \textit{Ans.}$$

10

PROBLEMS

***10–84.** Determine the moment of inertia of the thin ring about the z axis. The ring has a mass m.

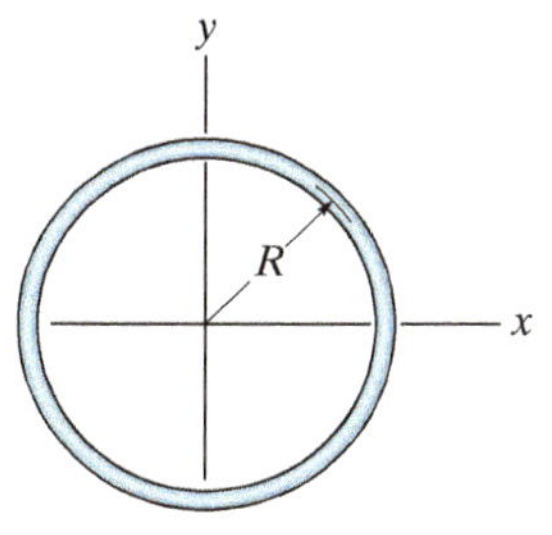

Prob. 10–84

10–85. Determine the moment of inertia of the semi-ellipsoid with respect to the x axis and express the result in terms of the mass m of the semiellipsoid. The material has a constant density ρ.

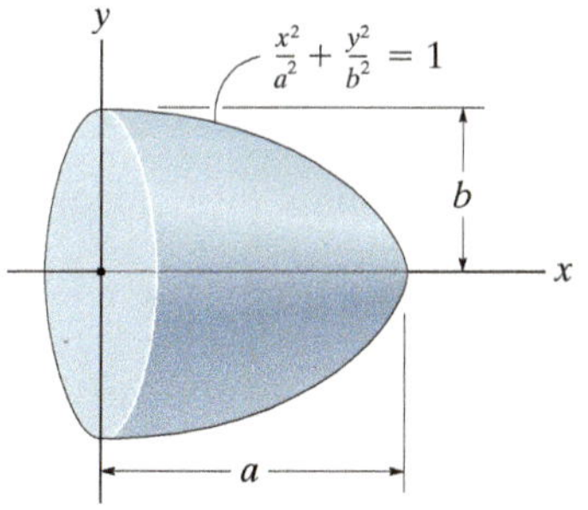

Prob. 10–85

10–86. Determine the radius of gyration k_x of the body. The specific weight of the material is $\gamma = 380 \text{ lb/ft}^3$.

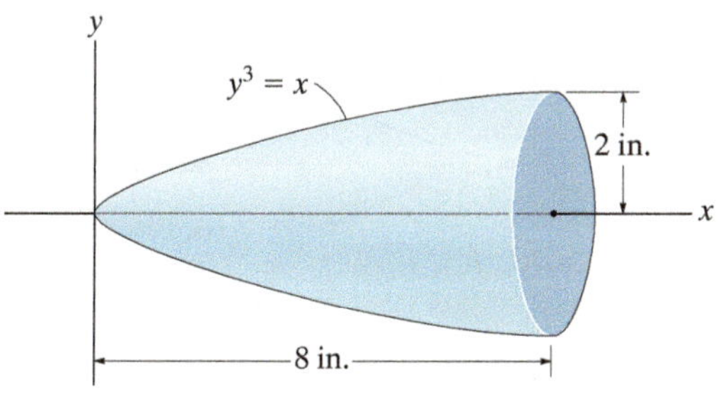

Prob. 10–86

10–87. Determine the radius of gyration k_x of the paraboloid. The density of the material is $\rho = 5 \text{ Mg/m}^3$.

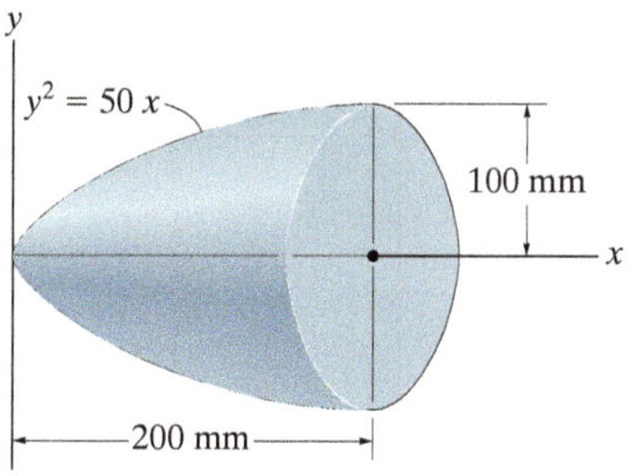

Prob. 10–87

***10–88.** Determine the moment of inertia of the ellipsoid with respect to the x axis and express the result in terms of the mass m of the ellipsoid. The material has a constant density ρ.

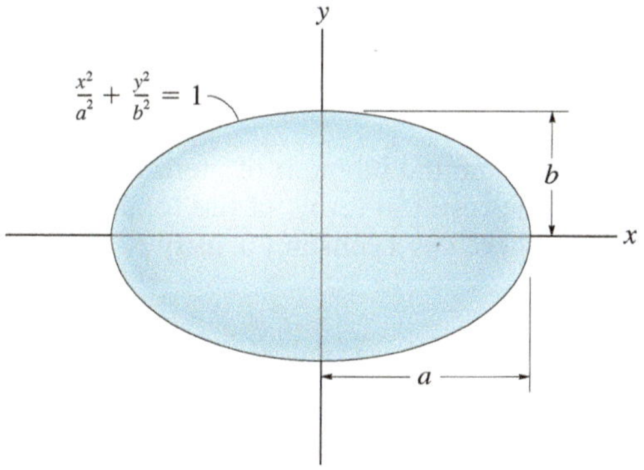

Prob. 10–88

10–89. Determine the moment of inertia of the homogenous triangular prism with respect to the y axis. Express the result in terms of the mass m of the prism. *Hint:* For integration, use thin plate elements parallel to the x–y plane having a thickness of dz.

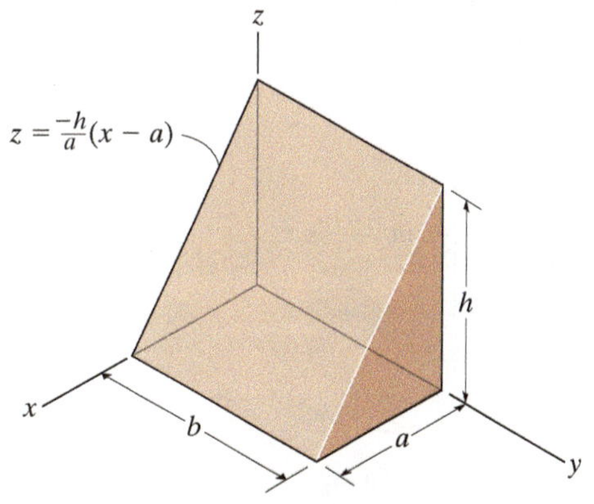

Prob. 10–89

10–90. Determine the mass moment of inertia I_z of the solid formed by revolving the shaded area around the z axis. The density of the materials is ρ. Express the result in terms of the mass m of the solid.

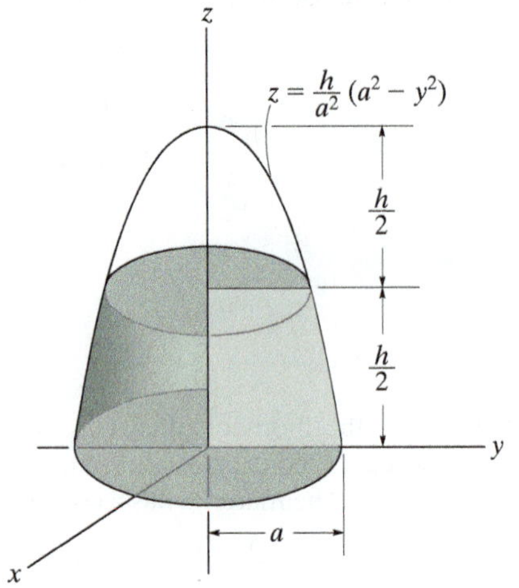

Prob. 10–90

10–91. Determine the moment of inertia I_x of the sphere and express the result in terms of the total mass m of the sphere. The sphere has a constant density ρ.

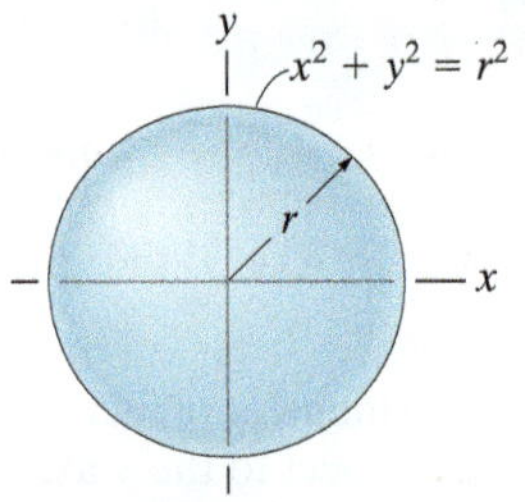

Prob. 10–91

***10–92.** The concrete shape is formed by rotating the shaded area about the y axis. Determine the moment of inertia I_y. The specific weight of concrete is $\gamma = 150\ \text{lb/ft}^3$.

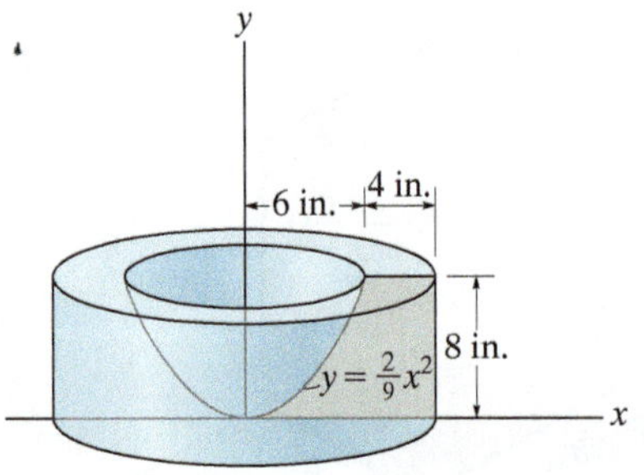

Prob. 10–92

10–93. Determine the mass moment of inertia I_y of the solid formed by revolving the shaded area around the y axis. The density of the material is ρ. Express the result in terms of the mass m of the solid.

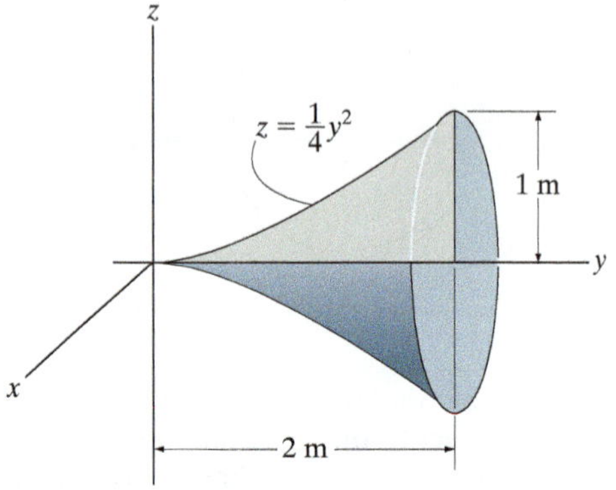

Prob. 10–93

10–94. Determine the mass moment of inertia I_y of the solid formed by revolving the shaded area around the y axis. The total mass of the solid is 1500 kg.

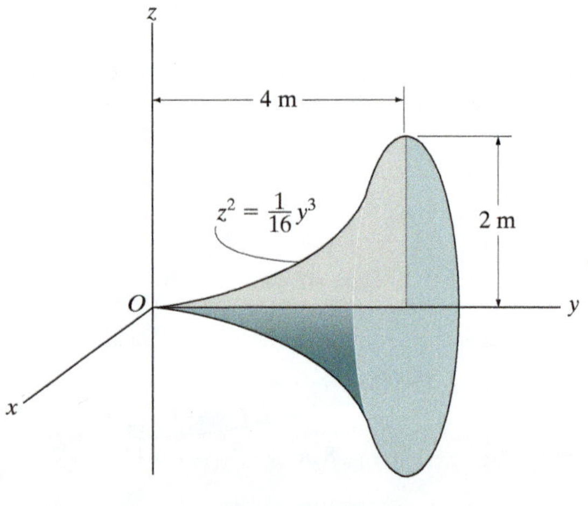

Prob. 10–94

10–95. The slender rods have a weight of 3 lb/ft. Determine the moment of inertia of the assembly about an axis perpendicular to the page and passing through point A.

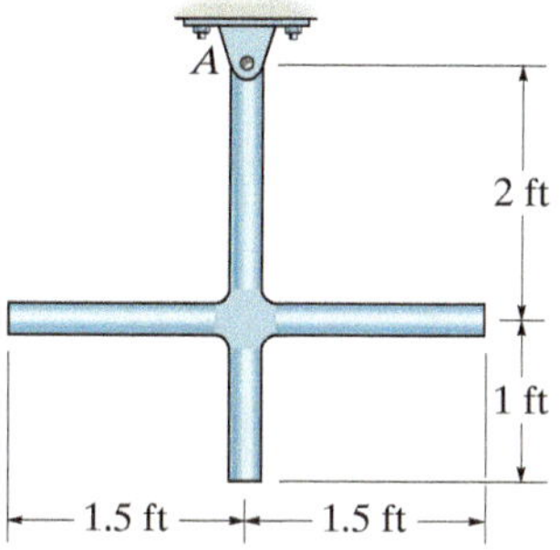

Prob. 10–95

10–97. The pendulum consists of the 3-kg slender rod and the 5-kg thin plate. Determine the location $\bar{y}$ of the center of mass G of the pendulum; then find the mass moment of inertia of the pendulum about an axis perpendicular to the page and passing through G.

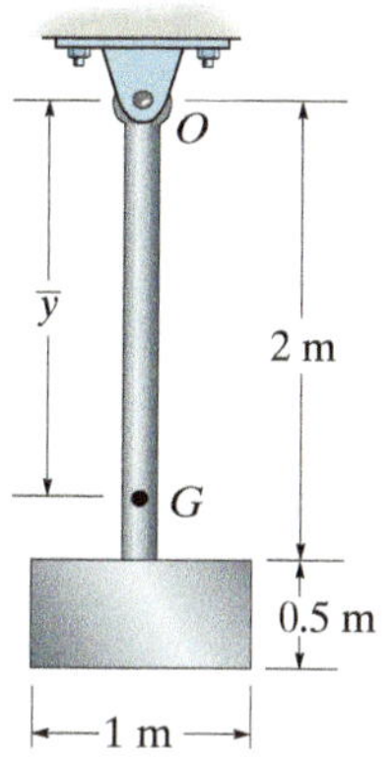

Prob. 10–97

***10–96.** The pendulum consists of a disk having a mass of 6 kg and slender rods AB and DC which have a mass of 2 kg/m. Determine the length L of DC so that the center of the mass is at the bearing O. What is the moment of inertia of the assembly about an axis perpendicular to the page and passing through point O?

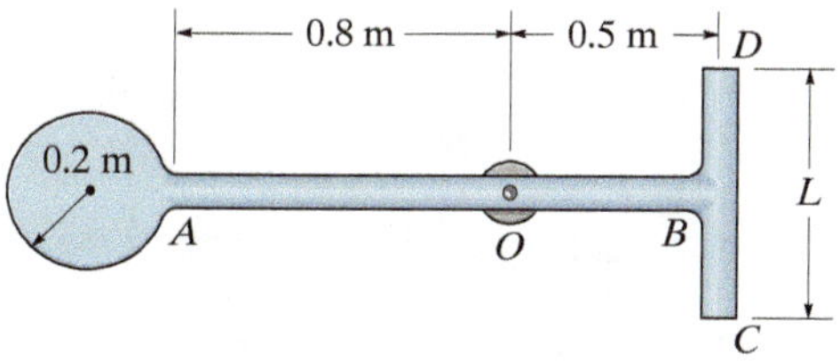

Prob. 10–96

10–98. Determine the location $\bar{y}$ of the center of mass G of the assembly and then calculate the moment of inertia about an axis perpendicular to the page and passing through G. The block has a mass of 3 kg and the mass of the semicylinder is 5 kg.

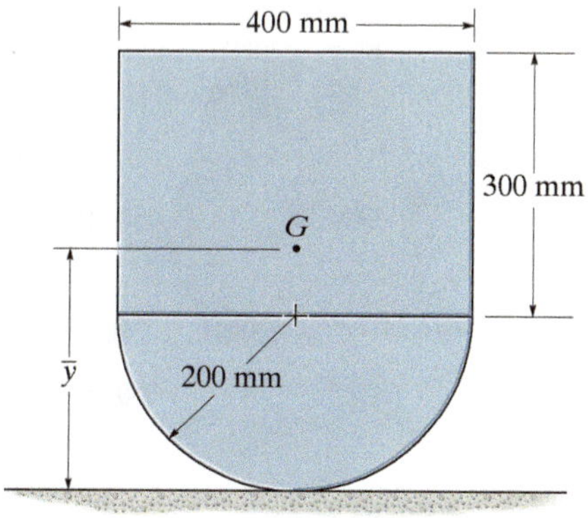

Prob. 10–98

10–99. If the large ring, small ring and each of the spokes weigh 100 lb, 15 lb, and 20 lb, respectively, determine the mass moment of inertia of the wheel about an axis perpendicular to the page and passing through point A.

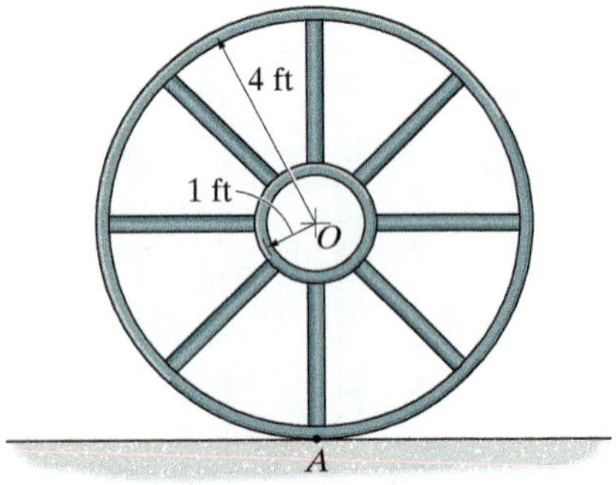

Prob. 10–99

***10–100.** Determine the mass moment of inertia of the assembly about the z axis. The density of the material is $7.85\ \text{Mg/m}^3$.

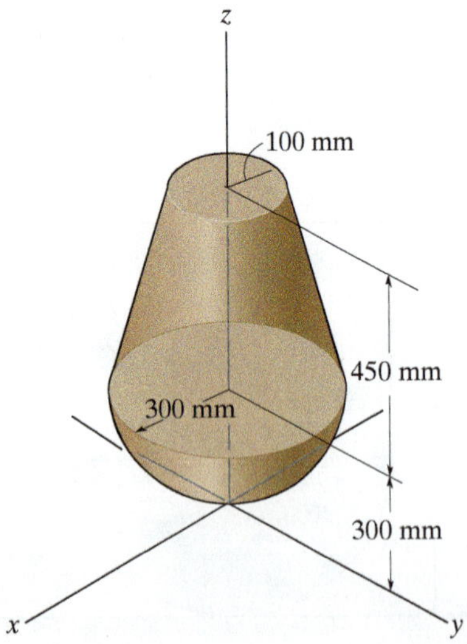

Prob. 10–100

10–101. Determine the moment of inertia I_z of the frustum of the cone which has a conical depression. The material has a density of $200\ \text{kg/m}^3$.

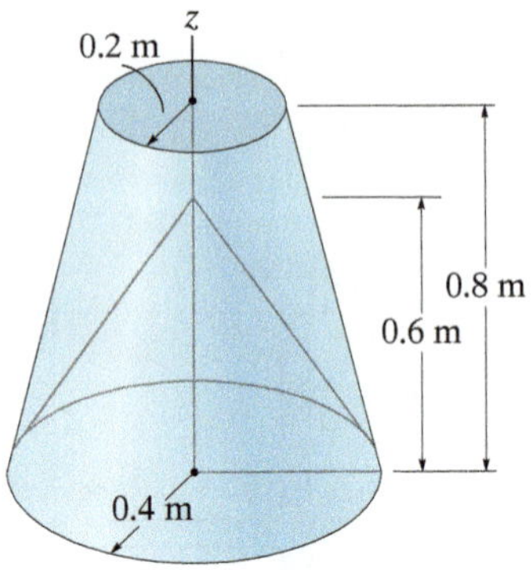

Prob. 10–101

10–102. The pendulum consists of a plate having a weight of 12 lb and a slender rod having a weight of 4 lb. Determine the radius of gyration of the pendulum about an axis perpendicular to the page and passing through point O.

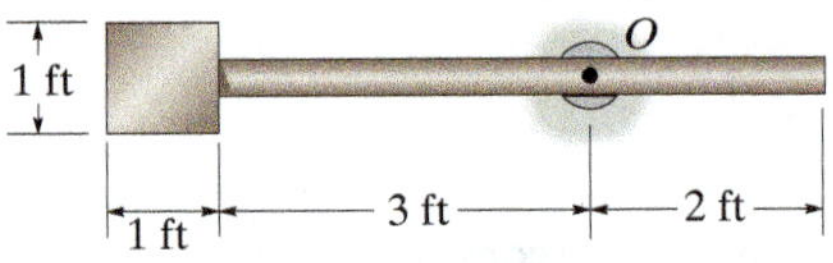

Prob. 10–102

10–103. The slender rods have a weight of 3 lb/ft. Determine the moment of inertia of the assembly about an axis perpendicular to the page and passing through point A.

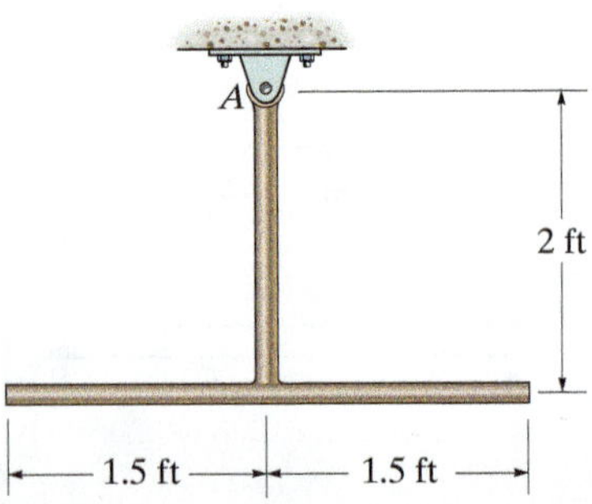

Prob. 10–103

10

*10–104. Determine the moment of inertia I_z of the frustrum of the cone which has a conical depression. The material has a density of 200 kg/m^3.

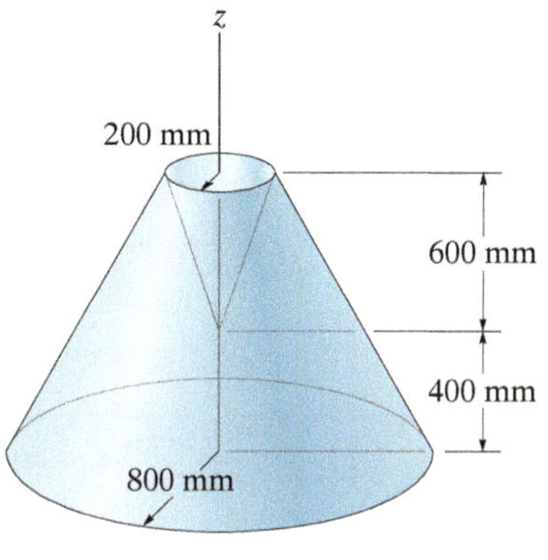

Prob. 10–104

10–106. The thin plate has a mass per unit area of 10 kg/m^2. Determine its mass moment of inertia about the y axis.

10–107. The thin plate has a mass per unit area of 10 kg/m^2. Determine its mass moment of inertia about the z axis.

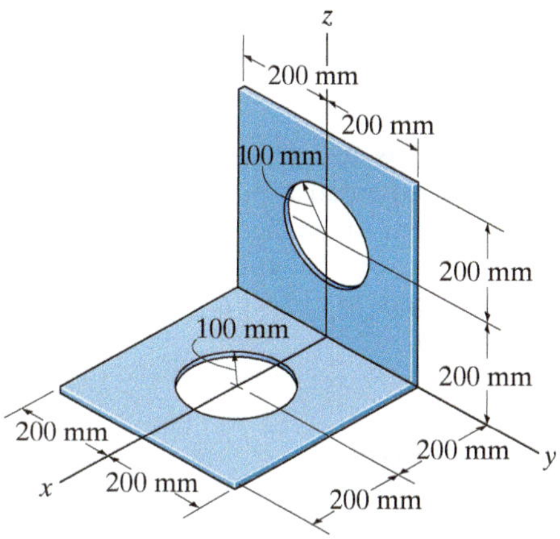

Probs. 10–106/107

10–105. Determine the moment of inertia of the wire triangle about an axis perpendicular to the page and passing through point O. Also, locate the mass center G and determine the moment of inertia about an axis perpendicular to the page and passing through point G. The wire has a mass of 0.3 kg/m. Neglect the size of the ring at O.

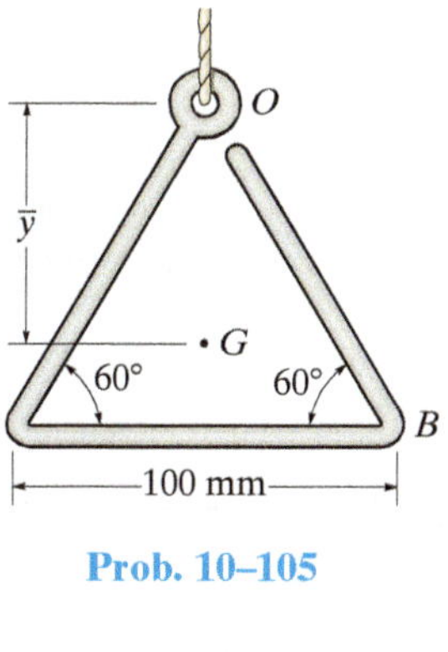

Prob. 10–105

*10–108. Determine the moment of inertia of the wheel about the x axis that passes through the center of mass G. The material has a specific weight of $\gamma = 90$ lb/ft^3.

10–109. Determine the moment of inertia of the wheel about the x' axis that passes through point O. The material has a specific weight of $\gamma = 90$ lb/ft^3.

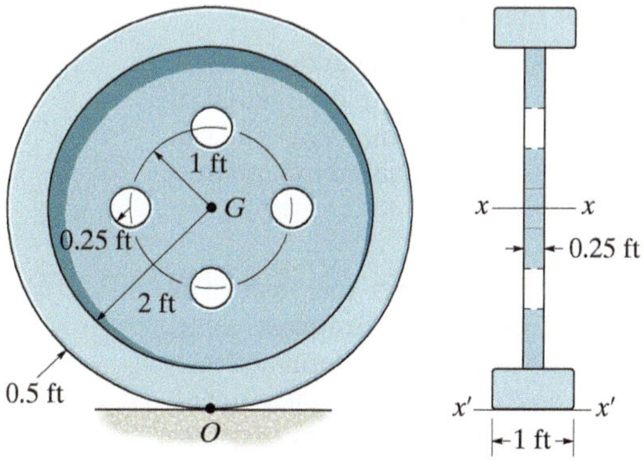

Probs. 10–108/109

10

CHAPTER REVIEW

Area Moment of Inertia

The *area moment of inertia* represents the second moment of the area about an axis. It is frequently used in formulas related to the strength and stability of structural members or mechanical elements.

$$I_x = \int_A y^2\,dA$$

If the area shape is irregular but can be described mathematically, then a differential element must be selected and integration over the entire area must be performed to determine the moment of inertia.

$$I_y = \int_A x^2\,dA$$

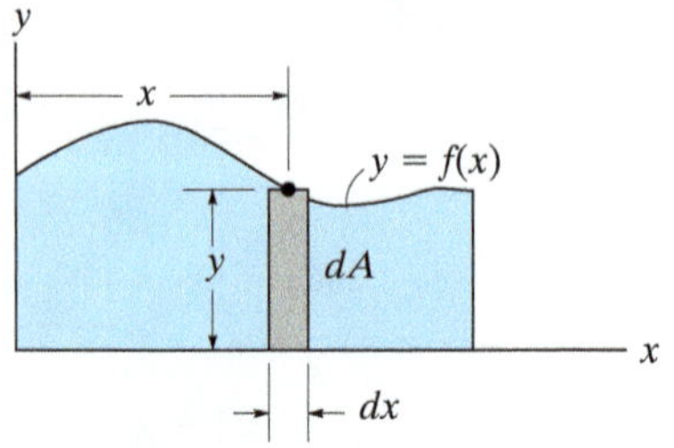

Parallel-Axis Theorem

If the moment of inertia for an area is known about a centroidal axis, then its moment of inertia about a parallel axis can be determined using the parallel-axis theorem.

$$I = \bar{I} + Ad^2$$

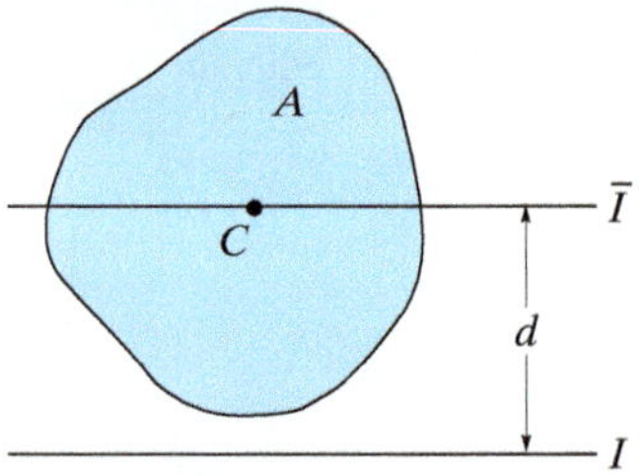

Composite Area

If an area is a composite of common shapes, as found on the inside back cover, then its moment of inertia is equal to the algebraic sum of the moments of inertia of each of its parts.

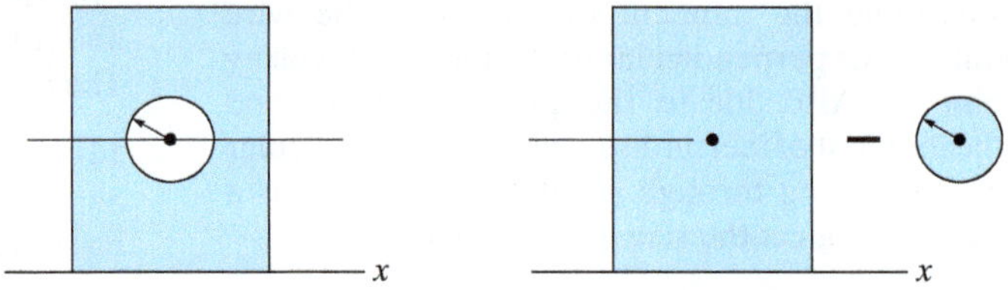

Product of Inertia

The *product of inertia* of an area is used in formulas to determine the orientation of an axis about which the moment of inertia for the area is a maximum or minimum.

$$I_{xy} = \int_A xy\,dA$$

If the product of inertia for an area is known with respect to its centroidal x', y' axes, then its value can be determined with respect to any x, y axes using the parallel-axis theorem for the product of inertia.

$$I_{xy} = \bar{I}_{x'y'} + Ad_xd_y$$

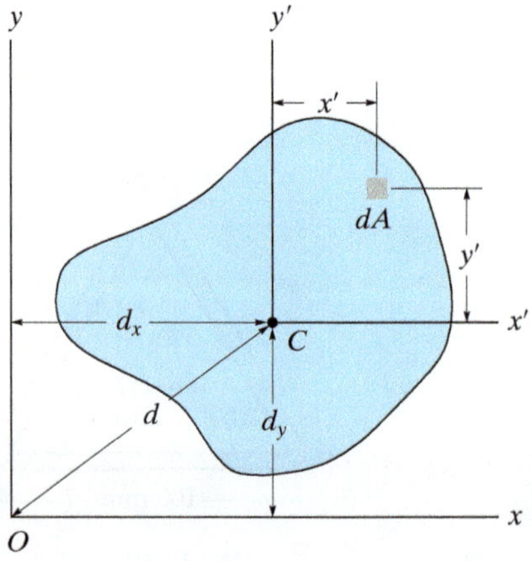

10

Principal Moments of Inertia

Provided the moments of inertia, I_x and I_y, and the product of inertia, I_{xy}, are known, then the transformation formulas, or Mohr's circle, can be used to determine the maximum and minimum or *principal moments of inertia* for the area, as well as finding the orientation of the principal axes of inertia.

$$I_{\substack{\max\\\min}} = \frac{I_x + I_y}{2} \pm \sqrt{\left(\frac{I_x - I_y}{2}\right)^2 + I_{xy}^2}$$

$$\tan 2\theta_p = \frac{-I_{xy}}{(I_x - I_y)/2}$$

Mass Moment of Inertia

The *mass moment of inertia* is a property of a body that measures its resistance to a change in its rotation. It is defined as the "second moment" of the mass elements of the body about an axis.

$$I = \int_m r^2\, dm$$

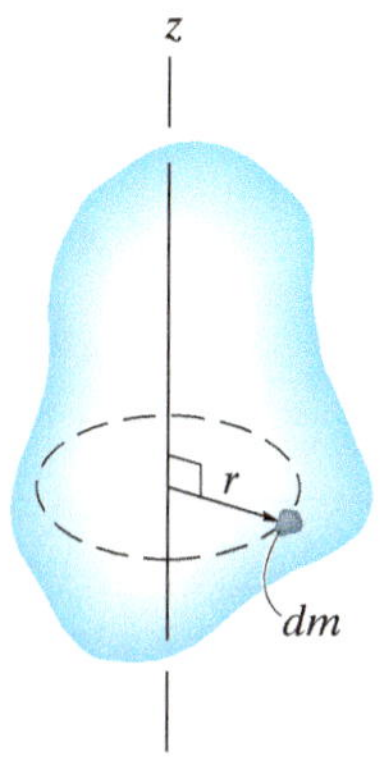

For homogeneous bodies having axial symmetry, the mass moment of inertia can be determined by a single integration, using a disk or shell element.

$$I = \rho \int_V r^2\, dV$$

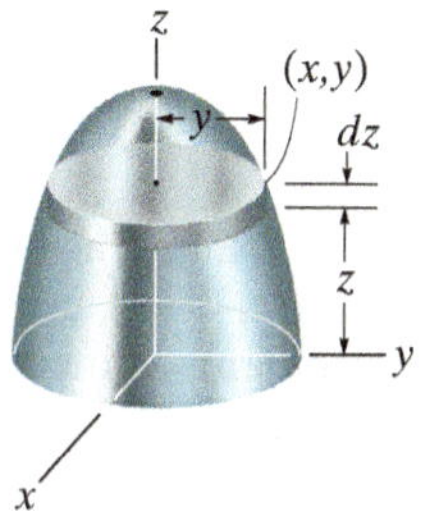

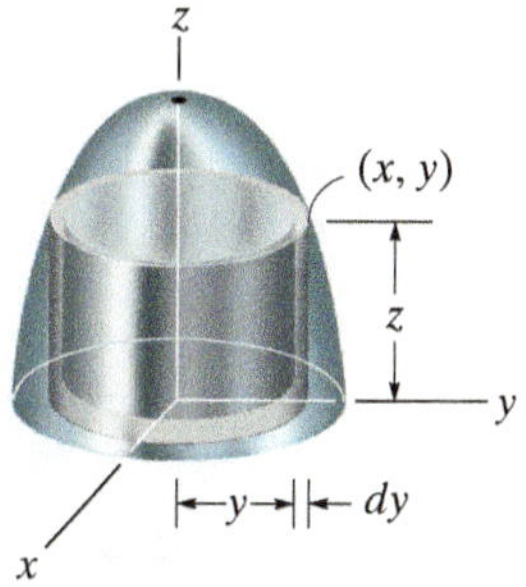

The mass moment of inertia of a composite body is determined by using tabular values of its composite shapes, found on the inside back cover, along with the parallel-axis theorem.

$$I = I_G + md^2$$

REVIEW PROBLEMS

10–110. Determine the moment of inertia for the shaded area about the x axis.

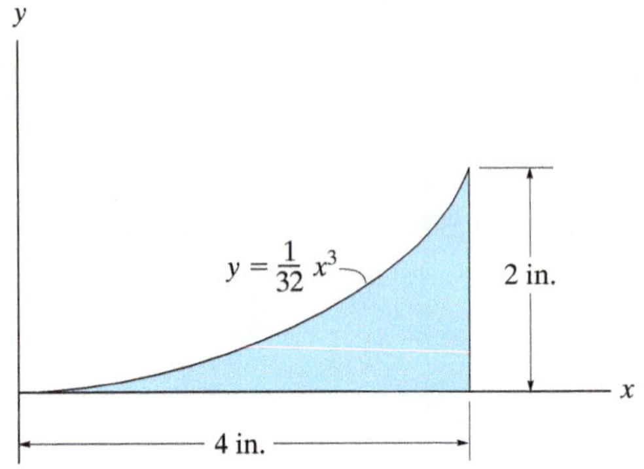

Prob. 10–110

10–111. Determine the area moment of inertia of the area about the x axis. Then, using the parallel-axis theorem, find the area moment of inertia about the x' axis that passes through the centroid C of the area. $\bar{y} = 120$ mm.

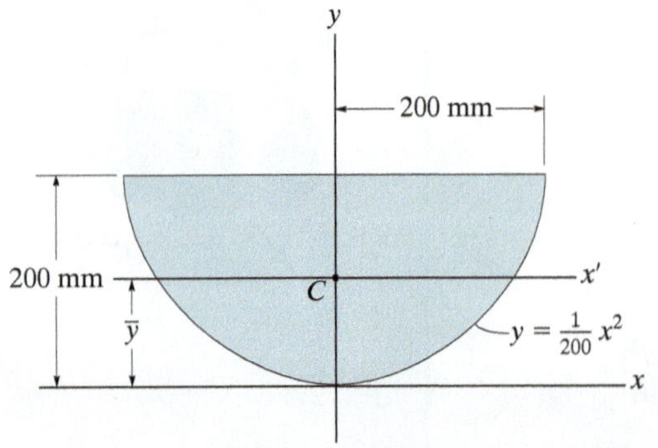

Prob. 10–111

***10–112.** Determine the product of inertia of the shaded area with respect to the x and y axes.

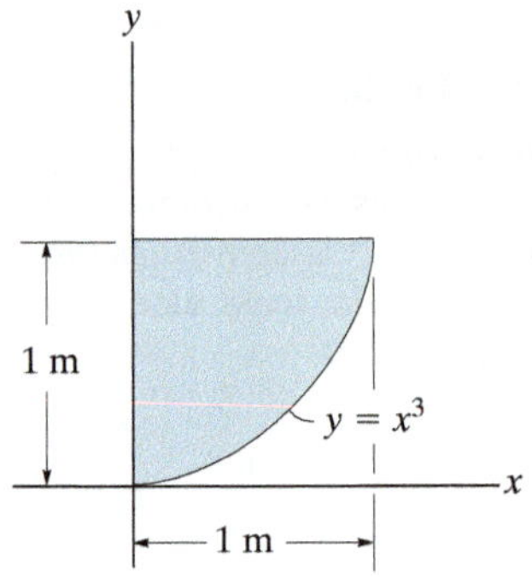

Prob. 10–112

10–113. Determine the area moment of inertia of the triangular area about (a) the x axis, and (b) the centroidal x' axis.

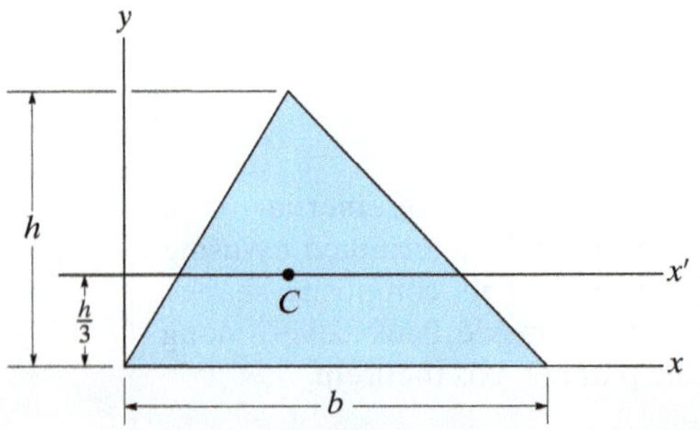

Prob. 10–113

10–114. Determine the mass moment of inertia I_x of the body and express the result in terms of the total mass m of the body. The density is constant.

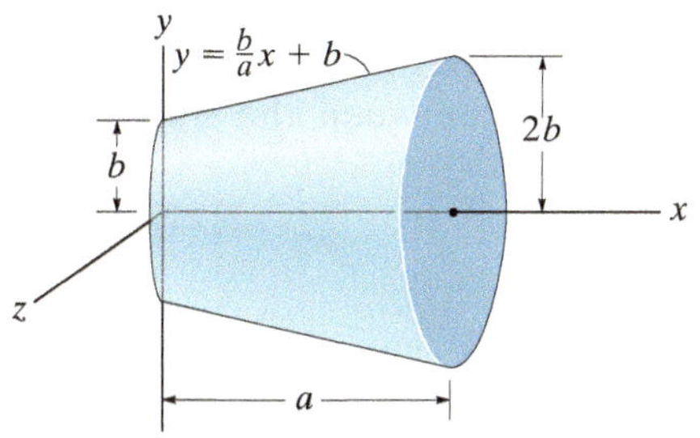

Prob. 10–114

10–115. Determine the area moment of inertia of the shaded area about the y axis.

***10–116.** Determine the area moment of inertia of the shaded area about the x axis.

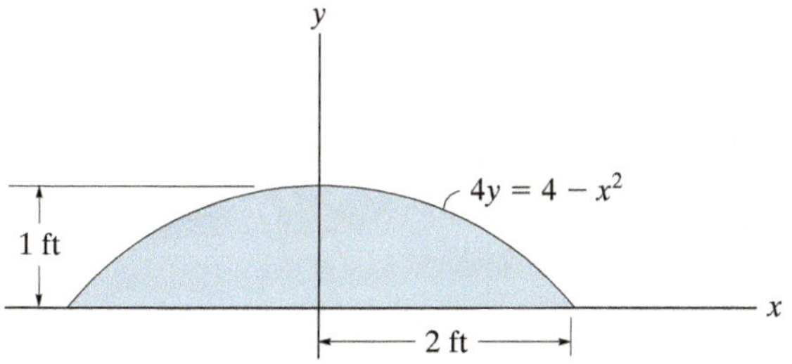

Probs. 10–115/116

10–117. Determine the area moments of inertia I_u and I_v and the product of inertia I_{uv} for the semicircular area.

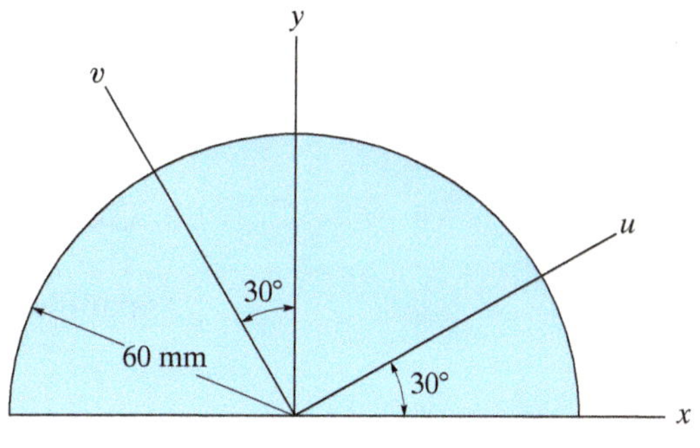

Prob. 10–117

10–118. Determine the area moment of inertia of the beam's cross-sectional area about the x axis which passes through the centroid C.

10–119. Determine the area moment of inertia of the beam's cross-sectional area about the y axis which passes through the centroid C.

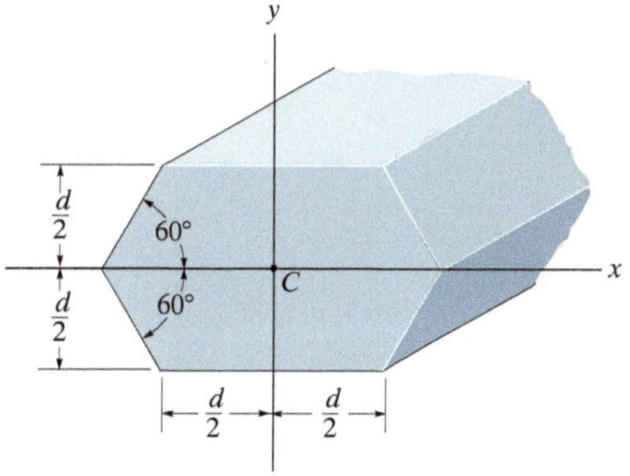

Probs. 10–118/119

APPENDIX

A Mathematical Review and Expressions

Geometry and Trigonometry Review

The angles θ in Fig. A–1 are equal between the transverse and two parallel lines.

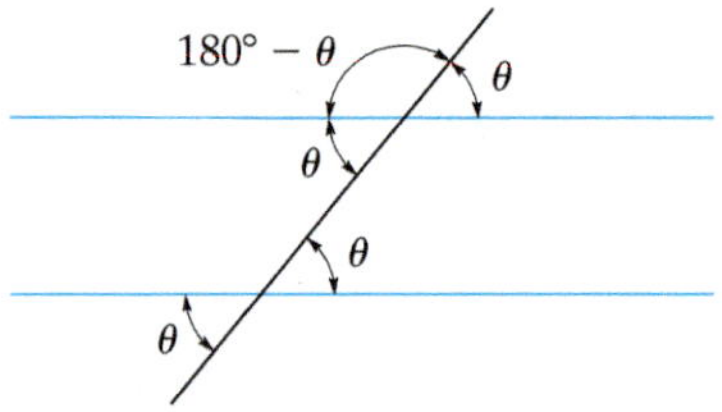

Fig. A–1

For a line and its normal, the angles θ in Fig. A–2 are equal.

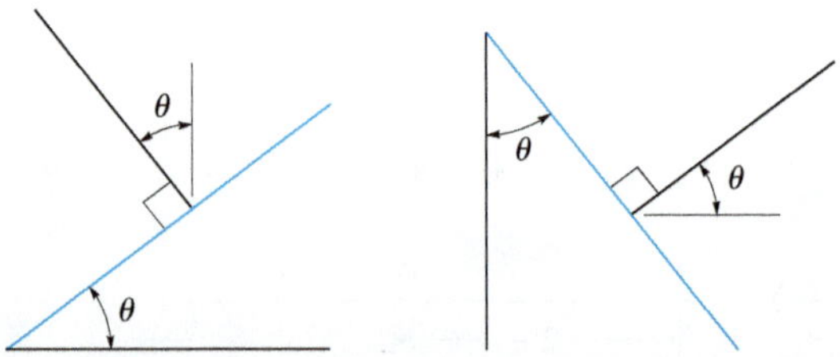

Fig. A–2

For the circle in Fig. A–3 $s = \theta r$, so that when $\theta = 360° = 2\pi$ rad then the circumference is $s = 2\pi r$. Also, since $180° = \pi$ rad, then $\theta\,(\text{rad}) = (\pi/180°)\theta°$. The area of the circle is $A = \pi r^2$.

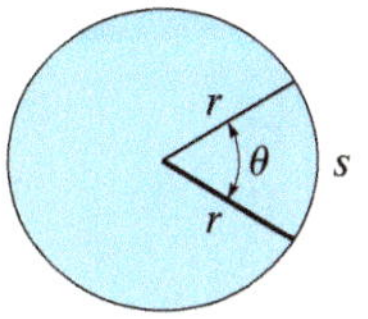

Fig. A–3

The sides of a similar triangle can be obtained by proportion as in Fig. A–4, where $\frac{a}{A} = \frac{b}{B} = \frac{c}{C}$.

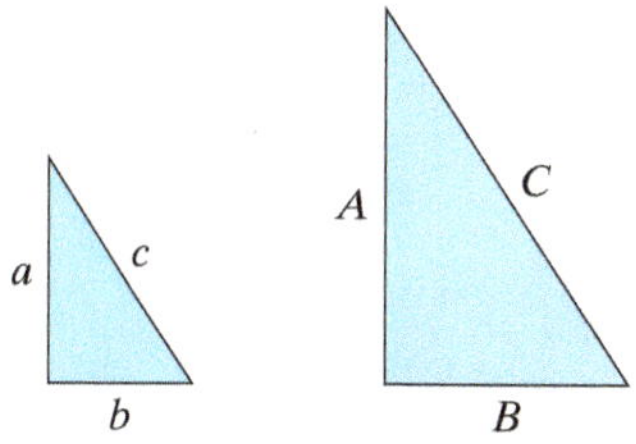

Fig. A–4

For the right triangle in Fig. A–5, the Pythagorean theorem is

$$h = \sqrt{(o)^2 + (a)^2}$$

The trigonometric functions are

$$\sin\theta = \frac{o}{h}$$

$$\cos\theta = \frac{a}{h}$$

$$\tan\theta = \frac{o}{a}$$

This is easily remembered as "soh, cah, toa", i.e., the sine is the opposite over the hypotenuse, etc. The other trigonometric functions follow from this.

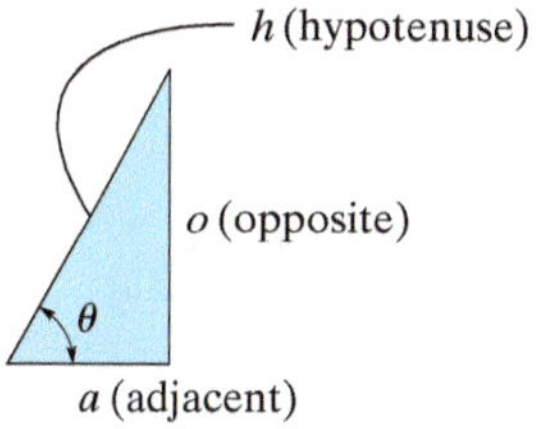

Fig. A–5

$$\csc\theta = \frac{1}{\sin\theta} = \frac{h}{o}$$

$$\sec\theta = \frac{1}{\cos\theta} = \frac{h}{a}$$

$$\cot\theta = \frac{1}{\tan\theta} = \frac{a}{o}$$

A

Trigonometric Identities

$$\sin^2\theta + \cos^2\theta = 1$$

$$\sin(\theta \pm \phi) = \sin\theta\cos\phi \pm \cos\theta\sin\phi$$

$$\sin 2\theta = 2\sin\theta\cos\theta$$

$$\cos(\theta \pm \phi) = \cos\theta\cos\phi \mp \sin\theta\sin\phi$$

$$\cos 2\theta = \cos^2\theta - \sin^2\theta$$

$$\cos\theta = \pm\sqrt{\frac{1+\cos 2\theta}{2}},\ \sin\theta = \pm\sqrt{\frac{1-\cos 2\theta}{2}}$$

$$\tan\theta = \frac{\sin\theta}{\cos\theta}$$

$$1 + \tan^2\theta = \sec^2\theta \qquad 1 + \cot^2\theta = \csc^2\theta$$

Quadratic Formula

If $ax^2 + bx + c = 0$, then $x = \dfrac{-b \pm \sqrt{b^2 - 4ac}}{2a}$

Hyperbolic Functions

$$\sinh x = \frac{e^x - e^{-x}}{2},$$

$$\cosh x = \frac{e^x + e^{-x}}{2},$$

$$\tanh x = \frac{\sinh x}{\cosh x}$$

Power-Series Expansions

$$\sin x = x - \frac{x^3}{3!} + \cdots,\ \cos x = 1 - \frac{x^2}{2!} + \cdots$$

$$\sinh x = x + \frac{x^3}{3!} + \cdots,\ \cosh x = 1 + \frac{x^2}{2!} + \cdots$$

Derivatives

$$\frac{d}{dx}(u^n) = nu^{n-1}\frac{du}{dx} \qquad \frac{d}{dx}(\sin u) = \cos u\frac{du}{dx}$$

$$\frac{d}{dx}(uv) = u\frac{dv}{dx} + v\frac{du}{dx} \qquad \frac{d}{dx}(\cos u) = -\sin u\frac{du}{dx}$$

$$\frac{d}{dx}\left(\frac{u}{v}\right) = \frac{v\dfrac{du}{dx} - u\dfrac{dv}{dx}}{v^2} \qquad \frac{d}{dx}(\tan u) = \sec^2 u\frac{du}{dx}$$

$$\frac{d}{dx}(\cot u) = -\csc^2 u\frac{du}{dx} \qquad \frac{d}{dx}(\sinh u) = \cosh u\frac{du}{dx}$$

$$\frac{d}{dx}(\sec u) = \tan u\sec u\frac{du}{dx} \qquad \frac{d}{dx}(\cosh u) = \sinh u\frac{du}{dx}$$

$$\frac{d}{dx}(\csc u) = -\csc u\cot u\frac{du}{dx}$$

Integrals

$$\int x^n\,dx = \frac{x^{n+1}}{n+1} + C,\ n \neq -1$$

$$\int \frac{dx}{a+bx} = \frac{1}{b}\ln(a+bx) + C$$

$$\int \frac{dx}{a+bx^2} = \frac{1}{2\sqrt{-ab}}\ln\left[\frac{a+x\sqrt{-ab}}{a-x\sqrt{-ab}}\right] + C,\quad ab < 0$$

$$\int \frac{x\,dx}{a+bx^2} = \frac{1}{2b}\ln(bx^2+a) + C$$

$$\int \frac{x^2\,dx}{a+bx^2} = \frac{x}{b} - \frac{a}{b\sqrt{ab}}\tan^{-1}\frac{x\sqrt{ab}}{a} + C,\ ab > 0$$

$$\int \sqrt{a+bx}\,dx = \frac{2}{3b}\sqrt{(a+bx)^3} + C$$

$$\int x\sqrt{a+bx}\,dx = \frac{-2(2a-3bx)\sqrt{(a+bx)^3}}{15b^2} + C$$

$$\int x^2\sqrt{a+bx}\,dx = \frac{2(8a^2-12abx+15b^2x^2)\sqrt{(a+bx)^3}}{105b^3} + C$$

$$\int \sqrt{a^2-x^2}\,dx = \frac{1}{2}\left[x\sqrt{a^2-x^2} + a^2\sin^{-1}\frac{x}{a}\right] + C,\quad a > 0$$

$$\int x\sqrt{a^2-x^2}\,dx = -\frac{1}{3}\sqrt{(a^2-x^2)^3} + C$$

$$\int x^2\sqrt{a^2-x^2}\,dx = -\frac{x}{4}\sqrt{(a^2-x^2)^3} + \frac{a^2}{8}\left(x\sqrt{a^2-x^2} + a^2\sin^{-1}\frac{x}{a}\right) + C,\ a > 0$$

$$\int \sqrt{x^2 \pm a^2}\,dx = \frac{1}{2}\left[x\sqrt{x^2 \pm a^2} \pm a^2\ln\left(x+\sqrt{x^2 \pm a^2}\right)\right] + C$$

$$\int x\sqrt{x^2 \pm a^2}\,dx = \frac{1}{3}\sqrt{(x^2 \pm a^2)^3} + C$$

$$\int x^2\sqrt{x^2 \pm a^2}\,dx = \frac{x}{4}\sqrt{(x^2 \pm a^2)^3} \mp \frac{a^2}{8}x\sqrt{x^2 \pm a^2} - \frac{a^4}{8}\ln\left(x+\sqrt{x^2 \pm a^2}\right) + C$$

$$\int \frac{dx}{\sqrt{a+bx}} = \frac{2\sqrt{a+bx}}{b} + C$$

$$\int \frac{x\,dx}{\sqrt{x^2 \pm a^2}} = \sqrt{x^2 \pm a^2} + C$$

$$\int \frac{dx}{\sqrt{a+bx+cx^2}} = \frac{1}{\sqrt{c}}\ln\left[\sqrt{a+bx+cx^2} + x\sqrt{c} + \frac{b}{2\sqrt{c}}\right] + C,\ c > 0$$

$$= \frac{1}{\sqrt{-c}}\sin^{-1}\left(\frac{-2cx-b}{\sqrt{b^2-4ac}}\right) + C,\ c < 0$$

$$\int \sin x\,dx = -\cos x + C$$

$$\int \cos x\,dx = \sin x + C$$

$$\int x\cos(ax)\,dx = \frac{1}{a^2}\cos(ax) + \frac{x}{a}\sin(ax) + C$$

$$\int x^2\cos(ax)\,dx = \frac{2x}{a^2}\cos(ax) + \frac{a^2x^2-2}{a^3}\sin(ax) + C$$

$$\int e^{ax}\,dx = \frac{1}{a}e^{ax} + C$$

$$\int xe^{ax}\,dx = \frac{e^{ax}}{a^2}(ax-1) + C$$

$$\int \sinh x\,dx = \cosh x + C$$

$$\int \cosh x\,dx = \sinh x + C$$

Fundamental Problems Partial Solutions And Answers

Chapter 2

F2–1.

$F_R = \sqrt{(2\text{ kN})^2 + (6\text{ kN})^2 - 2(2\text{ kN})(6\text{ kN})\cos 105^\circ}$

$= 6.798\text{ kN} = 6.80\text{ kN}$ *Ans.*

$\frac{\sin \phi}{6\text{ kN}} = \frac{\sin 105^\circ}{6.798\text{ kN}}, \quad \phi = 58.49^\circ$

$\theta = 45^\circ + \phi = 45^\circ + 58.49^\circ = 103^\circ$ *Ans.*

F2–2. $F_R = \sqrt{200^2 + 500^2 - 2(200)(500)\cos 140^\circ}$

$= 666\text{ N}$ *Ans.*

F2–3. $F_R = \sqrt{600^2 + 800^2 - 2(600)(800)\cos 60^\circ}$

$= 721.11\text{ N} = 721\text{ N}$ *Ans.*

$\frac{\sin \alpha}{800} = \frac{\sin 60^\circ}{721.11}; \quad \alpha = 73.90^\circ$

$\phi = \alpha - 30^\circ = 73.90^\circ - 30^\circ = 43.9^\circ$ *Ans.*

F2–4. $\frac{F_u}{\sin 45^\circ} = \frac{30}{\sin 105^\circ}; \quad F_u = 22.0\text{ lb}$ *Ans.*

$\frac{F_v}{\sin 30^\circ} = \frac{30}{\sin 105^\circ}; \quad F_v = 15.5\text{ lb}$ *Ans.*

F2–5. $\frac{F_{AB}}{\sin 105^\circ} = \frac{450}{\sin 30^\circ}$

$F_{AB} = 869\text{ lb}$ *Ans.*

$\frac{F_{AC}}{\sin 45^\circ} = \frac{450}{\sin 30^\circ}$

$F_{AC} = 636\text{ lb}$ *Ans.*

F2–6. $\frac{F}{\sin 30^\circ} = \frac{6}{\sin 105^\circ} \quad F = 3.11\text{ kN}$ *Ans.*

$\frac{F_v}{\sin 45^\circ} = \frac{6}{\sin 105^\circ} \quad F_v = 4.39\text{ kN}$ *Ans.*

F2–7. $(F_1)_x = 0 \quad (F_1)_y = 300\text{ N}$ *Ans.*

$(F_2)_x = -(450\text{ N})\cos 45^\circ = -318\text{ N}$ *Ans.*

$(F_2)_y = (450\text{ N})\sin 45^\circ = 318\text{ N}$ *Ans.*

$(F_3)_x = \left(\frac{3}{5}\right)600\text{ N} = 360\text{ N}$ *Ans.*

$(F_3)_y = \left(\frac{4}{5}\right)600\text{ N} = 480\text{ N}$ *Ans.*

F2–8. $F_{Rx} = 300 + 400\cos 30^\circ - 250\left(\frac{4}{5}\right) = 446.4\text{ N}$

$F_{Ry} = 400\sin 30^\circ + 250\left(\frac{3}{5}\right) = 350\text{ N}$

$F_R = \sqrt{(446.4)^2 + 350^2} = 567\text{ N}$ *Ans.*

$\theta = \tan^{-1}\frac{350}{446.4} = 38.1^\circ$ ⦟ *Ans.*

F2–9.

$\overset{+}{\rightarrow}(F_R)_x = \Sigma F_x;$

$(F_R)_x = -(700\text{ lb})\cos 30^\circ + 0 + \left(\frac{3}{5}\right)(600\text{ lb})$

$= -246.22\text{ lb}$

$+\uparrow(F_R)_y = \Sigma F_y;$

$(F_R)_y = -(700\text{ lb})\sin 30^\circ - 400\text{ lb} - \left(\frac{4}{5}\right)(600\text{ lb})$

$= -1230\text{ lb}$

$F_R = \sqrt{(246.22\text{ lb})^2 + (1230\text{ lb})^2} = 1254\text{ lb}$ *Ans.*

$\phi = \tan^{-1}\left(\frac{1230\text{ lb}}{246.22\text{ lb}}\right) = 78.68^\circ$

$\theta = 180^\circ + \phi = 180^\circ + 78.68^\circ = 259^\circ$ *Ans.*

F2–10. $\overset{+}{\rightarrow}(F_R)_x = \Sigma F_x;$

$750\text{ N} = F\cos\theta + \left(\frac{5}{13}\right)(325\text{ N}) + (600\text{ N})\cos 45^\circ$

$+\uparrow(F_R)_y = \Sigma F_y;$

$0 = F\sin\theta + \left(\frac{12}{13}\right)(325\text{ N}) - (600\text{ N})\sin 45^\circ$

$\tan\theta = 0.6190 \quad \theta = 31.76^\circ = 31.8^\circ$ ⦟ *Ans.*

$F = 236\text{ N}$ *Ans.*

F2–11. $\overset{+}{\rightarrow}(F_R)_x = \Sigma F_x;$

$(80\text{ lb})\cos 45^\circ = F\cos\theta + 50\text{ lb} - \left(\frac{3}{5}\right)90\text{ lb}$

$+\uparrow(F_R)_y = \Sigma F_y;$

$-(80\text{ lb})\sin 45^\circ = F\sin\theta - \left(\frac{4}{5}\right)(90\text{ lb})$

$\tan\theta = 0.2547 \quad \theta = 14.29^\circ = 14.3^\circ$ ⦟ *Ans.*

$F = 62.5\text{ lb}$ *Ans.*

F2–12. $(F_R)_x = 15\left(\frac{4}{5}\right) + 0 + 15\left(\frac{4}{5}\right) = 24\text{ kN} \rightarrow$

$(F_R)_y = 15\left(\frac{3}{5}\right) + 20 - 15\left(\frac{3}{5}\right) = 20\text{ kN} \uparrow$

$F_R = 31.2\text{ kN}$ *Ans.*

$\theta = 39.8^\circ$ *Ans.*

F2–13. $F_x = 75\cos 30^\circ\sin 45^\circ = 45.93\text{ lb}$

$F_y = 75\cos 30^\circ\cos 45^\circ = 45.93\text{ lb}$

$F_z = -75\sin 30^\circ = -37.5\text{ lb}$

$\alpha = \cos^{-1}\left(\frac{45.93}{75}\right) = 52.2^\circ$ *Ans.*

$\beta = \cos^{-1}\left(\frac{45.93}{75}\right) = 52.2^\circ$ *Ans.*

$\gamma = \cos^{-1}\left(\frac{-37.5}{75}\right) = 120^\circ$ *Ans.*

F2–14. $\cos\beta = \sqrt{1 - \cos^2 120° - \cos^2 60°} = \pm 0.7071$

Require $\beta = 135°$.

$\mathbf{F} = F\mathbf{u}_F = (500\text{ N})(-0.5\mathbf{i} - 0.7071\mathbf{j} + 0.5\mathbf{k})$

$= \{-250\mathbf{i} - 354\mathbf{j} + 250\mathbf{k}\}\text{ N}$ *Ans.*

F2–15. $\cos^2\alpha + \cos^2 135° + \cos^2 120° = 1$

$\alpha = 60°$

$\mathbf{F} = F\mathbf{u}_F = (500\text{ N})(0.5\mathbf{i} - 0.7071\mathbf{j} - 0.5\mathbf{k})$

$= \{250\mathbf{i} - 354\mathbf{j} - 250\mathbf{k}\}\text{ N}$ *Ans.*

F2–16. $F_z = (50\text{ lb})\sin 45° = 35.36\text{ lb}$

$F' = (50\text{ lb})\cos 45° = 35.36\text{ lb}$

$F_x = \left(\frac{3}{5}\right)(35.36\text{ lb}) = 21.21\text{ lb}$

$F_y = \left(\frac{4}{5}\right)(35.36\text{ lb}) = 28.28\text{ lb}$

$\mathbf{F} = \{-21.2\mathbf{i} + 28.3\mathbf{j} + 35.4\mathbf{k}\}\text{ lb}$ *Ans.*

F2–17. $F_z = (750\text{ N})\sin 45° = 530.33\text{ N}$

$F' = (750\text{ N})\cos 45° = 530.33\text{ N}$

$F_x = (530.33\text{ N})\cos 60° = 265.2\text{ N}$

$F_y = (530.33\text{ N})\sin 60° = 459.3\text{ N}$

$\mathbf{F}_2 = \{265\mathbf{i} - 459\mathbf{j} + 530\mathbf{k}\}\text{ N}$ *Ans.*

F2–18. $\mathbf{F}_1 = \left(\frac{4}{5}\right)(500\text{ lb})\,\mathbf{j} + \left(\frac{3}{5}\right)(500\text{ lb})\mathbf{k}$

$= \{400\mathbf{j} + 300\mathbf{k}\}\text{ lb}$

$\mathbf{F}_2 = [(800\text{ lb})\cos 45°]\cos 30°\,\mathbf{i}$

$+ [(800\text{ lb})\cos 45°]\sin 30°\mathbf{j}$

$+ (800\text{ lb})\sin 45°\,(-\mathbf{k})$

$= \{489.90\mathbf{i} + 282.84\mathbf{j} - 565.69\mathbf{k}\}\text{ lb}$

$\mathbf{F}_R = \mathbf{F}_1 + \mathbf{F}_2 = \{490\mathbf{i} + 683\mathbf{j} - 266\mathbf{k}\}\text{ lb}$ *Ans.*

F2–19. $\mathbf{r}_{AB} = \{-6\mathbf{i} + 6\mathbf{j} + 3\mathbf{k}\}\text{ m}$ *Ans.*

$r_{AB} = \sqrt{(-6\text{ m})^2 + (6\text{ m})^2 + (3\text{ m})^2} = 9\text{ m}$ *Ans.*

$\alpha = 132°,\quad \beta = 48.2°,\quad \gamma = 70.5°$ *Ans.*

F2–20. $\mathbf{r}_{AB} = \{-4\mathbf{i} + 2\mathbf{j} + 4\mathbf{k}\}\text{ ft}$ *Ans.*

$r_{AB} = \sqrt{(-4\text{ ft})^2 + (2\text{ ft})^2 + (4\text{ ft})^2} = 6\text{ ft}$ *Ans.*

$\alpha = \cos^{-1}\left(\frac{-4\text{ ft}}{6\text{ ft}}\right) = 131.8°$

$\theta = 180° - 131.8° = 48.2°$ *Ans.*

F2–21. $\mathbf{r}_{AB} = \{2\mathbf{i} + 3\mathbf{j} - 6\mathbf{k}\}\text{ m}$

$\mathbf{F}_{AB} = F_{AB}\mathbf{u}_{AB}$

$= (630\text{ N})\left(\frac{2}{7}\mathbf{i} + \frac{3}{7}\mathbf{j} - \frac{6}{7}\mathbf{k}\right)$

$= \{180\mathbf{i} + 270\mathbf{j} - 540\mathbf{k}\}\text{ N}$ *Ans.*

F2–22. $\mathbf{F} = F\mathbf{u}_{AB} = 900\text{N}\left(-\frac{4}{9}\mathbf{i} + \frac{7}{9}\mathbf{j} - \frac{4}{9}\mathbf{k}\right)$

$= \{-400\mathbf{i} + 700\mathbf{j} - 400\mathbf{k}\}\text{ N}$ *Ans.*

F2–23. $\mathbf{F}_B = F_B\mathbf{u}_B$

$= (840\text{ N})\left(\frac{3}{7}\mathbf{i} - \frac{2}{7}\mathbf{j} - \frac{6}{7}\mathbf{k}\right)$

$= \{360\mathbf{i} - 240\mathbf{j} - 720\mathbf{k}\}\text{ N}$

$\mathbf{F}_C = F_C\mathbf{u}_C$

$= (420\text{ N})\left(\frac{2}{7}\mathbf{i} + \frac{3}{7}\mathbf{j} - \frac{6}{7}\mathbf{k}\right)$

$= \{120\mathbf{i} + 180\mathbf{j} - 360\mathbf{k}\}\text{ N}$

$F_R = \sqrt{(480\text{ N})^2 + (-60\text{ N})^2 + (-1080\text{ N})^2}$

$= 1.18\text{ kN}$ *Ans.*

F2–24. $\mathbf{F}_B = F_B\mathbf{u}_B$

$= (600\text{ lb})\left(-\frac{1}{3}\mathbf{i} + \frac{2}{3}\mathbf{j} - \frac{2}{3}\mathbf{k}\right)$

$= \{-200\mathbf{i} + 400\mathbf{j} - 400\mathbf{k}\}\text{ lb}$

$\mathbf{F}_C = F_C\mathbf{u}_C$

$= (490\text{ lb})\left(-\frac{6}{7}\mathbf{i} + \frac{3}{7}\mathbf{j} - \frac{2}{7}\mathbf{k}\right)$

$= \{-420\mathbf{i} + 210\mathbf{j} - 140\mathbf{k}\}\text{ lb}$

$\mathbf{F}_R = \mathbf{F}_B + \mathbf{F}_C = \{-620\mathbf{i} + 610\mathbf{j} - 540\mathbf{k}\}\text{ lb}$ *Ans.*

F2–25. $\mathbf{u}_{AO} = -\frac{1}{3}\mathbf{i} + \frac{2}{3}\mathbf{j} - \frac{2}{3}\mathbf{k}$

$\mathbf{u}_F = -0.5345\mathbf{i} + 0.8018\mathbf{j} + 0.2673\mathbf{k}$

$\theta = \cos^{-1}(\mathbf{u}_{AO} \cdot \mathbf{u}_F) = 57.7°$ *Ans.*

F2–26. $\mathbf{u}_{AB} = -\frac{3}{5}\mathbf{j} + \frac{4}{5}\mathbf{k}$

$\mathbf{u}_F = \frac{4}{5}\mathbf{i} - \frac{3}{5}\mathbf{j}$

$\theta = \cos^{-1}(\mathbf{u}_{AB} \cdot \mathbf{u}_F) = 68.9°$ *Ans.*

F2–27. $\mathbf{u}_{OA} = \frac{12}{13}\mathbf{i} + \frac{5}{13}\mathbf{j}$

$\mathbf{u}_{OA} \cdot \mathbf{j} = u_{OA}(1)\cos\theta$

$\cos\theta = \frac{5}{13};\quad \theta = 67.4°$ *Ans.*

F2–28. $\mathbf{u}_{OA} = \frac{12}{13}\mathbf{i} + \frac{5}{13}\mathbf{j}$

$\mathbf{F} = F\mathbf{u}_F = [650\mathbf{j}]\text{ N}$

$F_{OA} = \mathbf{F} \cdot \mathbf{u}_{OA} = 250\text{ N}$

$\mathbf{F}_{OA} = F_{OA}\,\mathbf{u}_{OA} = \{231\mathbf{i} + 96.2\mathbf{j}\}\text{ N}$ *Ans.*

F2–29. $\mathbf{F} = (400\text{ N})\dfrac{\{4\,\mathbf{i} + 1\,\mathbf{j} - 6\,\mathbf{k}\}\text{ m}}{\sqrt{(4\text{ m})^2 + (1\text{ m})^2 + (-6\text{ m})^2}}$

$= \{219.78\mathbf{i} + 54.94\mathbf{j} - 329.67\mathbf{k}\}\text{ N}$

$\mathbf{u}_{AO} = \dfrac{\{-4\,\mathbf{j} - 6\,\mathbf{k}\}\text{ m}}{\sqrt{(-4\text{ m})^2 + (-6\text{ m})^2}}$

$= -0.5547\mathbf{j} - 0.8321\mathbf{k}$

$(F_{AO})_{\text{proj}} = \mathbf{F}\cdot\mathbf{u}_{AO} = 244\text{ N}$ *Ans.*

F2–30. $\mathbf{F} = [(-600\text{ lb})\cos 60°]\sin 30°\,\mathbf{i}$

$+ [(600\text{ lb})\cos 60°]\cos 30°\,\mathbf{j}$

$+ [(600\text{ lb})\sin 60°]\,\mathbf{k}$

$= \{-150\mathbf{i} + 259.81\mathbf{j} + 519.62\mathbf{k}\}\text{ lb}$

$\mathbf{u}_A = -\frac{2}{3}\mathbf{i} + \frac{2}{3}\mathbf{j} + \frac{1}{3}\mathbf{k}$

$(F_A)_{\text{par}} = \mathbf{F}\cdot\mathbf{u}_A = 446.41\text{ lb} = 446\text{ lb}$ *Ans.*

$(F_A)_{\text{per}} = \sqrt{(600\text{ lb})^2 - (446.41\text{ lb})^2}$

$= 401\text{ lb}$ *Ans.*

F2–31. $\mathbf{F} = 56\text{ N}\left(\frac{3}{7}\mathbf{i} - \frac{6}{7}\mathbf{j} + \frac{2}{7}\mathbf{k}\right)$

$= \{24\mathbf{i} - 48\mathbf{j} + 16\mathbf{k}\}\text{ N}$

$(F_{AO})_{\parallel} = \mathbf{F}\cdot\mathbf{u}_{AO} = (24\mathbf{i} - 48\mathbf{j} + 16\mathbf{k})\cdot\left(\frac{3}{7}\mathbf{i} - \frac{6}{7}\mathbf{j} - \frac{2}{7}\mathbf{k}\right)$

$= 46.86\text{ N} = 46.9\text{ N}$ *Ans.*

$(F_{AO})_{\perp} = \sqrt{F^2 - (F_{AO})_{\parallel}} = \sqrt{(56)^2 - (46.86)^2}$

$= 30.7\text{ N}$ *Ans.*

Chapter 3

F3–1. $\overset{+}{\rightarrow}\Sigma F_x = 0;\ \frac{4}{5}F_{AC} - F_{AB}\cos 30° = 0$

$+\uparrow\Sigma F_y = 0;\ \frac{3}{5}F_{AC} + F_{AB}\sin 30° - 550 = 0$

$F_{AB} = 478\text{ lb}$ *Ans.*

$F_{AC} = 518\text{ lb}$ *Ans.*

F3–2. $+\uparrow\Sigma F_y = 0;\ -2(1500)\sin\theta + 700 = 0$

$\theta = 13.5°$

$L_{ABC} = 2\left(\frac{5\text{ ft}}{\cos 13.5°}\right) = 10.3\text{ ft}$ *Ans.*

F3–3. $\overset{+}{\rightarrow}\Sigma F_x = 0;\quad T\cos\theta - T\cos\phi = 0$

$\phi = \theta$

$+\uparrow\Sigma F_y = 0;\quad 2T\sin\theta - 49.05\text{ N} = 0$

$\theta = \tan^{-1}\left(\frac{0.15\text{ m}}{0.2\text{ m}}\right) = 36.87°$

$T = 40.9\text{ N}$ *Ans.*

F3–4. $+\nearrow\Sigma F_x = 0;\ \frac{4}{5}(F_{sp}) - 5(9.81)\sin 45° = 0$

$F_{sp} = 43.35\text{ N}$

$F_{sp} = k(l - l_0);\ 43.35 = 200(0.5 - l_0)$

$l_0 = 0.283\text{ m}$ *Ans.*

F3–5. $+\uparrow\Sigma F_y = 0;\quad (392.4\text{ N})\sin 30° - m_A(9.81) = 0$

$m_A = 20\text{ kg}$ *Ans.*

F3–6. $+\uparrow\Sigma F_y = 0;\quad T_{AB}\sin 15° - 10(9.81)\text{ N} = 0$

$T_{AB} = 379.03\text{ N} = 379\text{ N}$ *Ans.*

$\overset{+}{\rightarrow}\Sigma F_x = 0;\quad T_{BC} - 379.03\text{ N}\cos 15° = 0$

$T_{BC} = 366.11\text{ N} = 366\text{ N}$ *Ans.*

$\overset{+}{\rightarrow}\Sigma F_x = 0;\quad T_{CD}\cos\theta - 366.11\text{ N} = 0$

$+\uparrow\Sigma F_y = 0;\quad T_{CD}\sin\theta - 15(9.81)\text{ N} = 0$

$T_{CD} = 395\text{ N}$ *Ans.*

$\theta = 21.9°$ *Ans.*

F3–7. $\Sigma F_x = 0;\quad \left[\left(\frac{3}{5}\right)F_3\right]\left(\frac{3}{5}\right) + 600\text{ N} - F_2 = 0 \quad (1)$

$\Sigma F_y = 0;\quad \left(\frac{4}{5}\right)F_1 - \left[\left(\frac{3}{5}\right)F_3\right]\left(\frac{4}{5}\right) = 0 \quad (2)$

$\Sigma F_z = 0;\quad \left(\frac{4}{5}\right)F_3 + \left(\frac{3}{5}\right)F_1 - 900\text{ N} = 0 \quad (3)$

$F_3 = 776\text{ N}$ *Ans.*

$F_1 = 466\text{ N}$ *Ans.*

$F_2 = 879\text{ N}$ *Ans.*

F3–8. $\Sigma F_z = 0;\quad F_{AD}\left(\frac{4}{5}\right) - 900 = 0$

$F_{AD} = 1125\text{ N} = 1.125\text{ kN}$ *Ans.*

$\Sigma F_y = 0;\quad F_{AC}\left(\frac{4}{5}\right) - 1125\left(\frac{3}{5}\right) = 0$

$F_{AC} = 843.75\text{ N} = 844\text{ N}$ *Ans.*

$\Sigma F_x = 0;\quad F_{AB} - 843.75\left(\frac{3}{5}\right) = 0$

$F_{AB} = 506.25\text{ N} = 506\text{ N}$ *Ans.*

F3–9. $\mathbf{F}_{AD} = F_{AD}\left(\frac{\mathbf{r}_{AD}}{r_{AD}}\right) = \frac{1}{3}F_{AD}\mathbf{i} - \frac{2}{3}F_{AD}\,\mathbf{j} + \frac{2}{3}F_{AD}\,\mathbf{k}$

$\Sigma F_z = 0;\quad \frac{2}{3}F_{AD} - 600 = 0$

$F_{AD} = 900\text{ N}$ *Ans.*

$\Sigma F_y = 0;\quad F_{AB}\cos 30° - \frac{2}{3}(900) = 0$

$F_{AB} = 692.82\text{ N} = 693\text{ N}$ *Ans.*

$\Sigma F_x = 0;\quad \frac{1}{3}(900) + 692.82\sin 30° - F_{AC} = 0$

$F_{AC} = 646.41\text{ N} = 646\text{ N}$ *Ans.*

F3–10. $\mathbf{F}_{AC} = F_{AC}\{-\cos 60°\sin 30°\,\mathbf{i}$

$+ \cos 60°\cos 30°\,\mathbf{j} + \sin 60°\,\mathbf{k}\}$

$= -0.25F_{AC}\,\mathbf{i} + 0.4330F_{AC}\mathbf{j} + 0.8660F_{AC}\,\mathbf{k}$

$\mathbf{F}_{AD} = F_{AD}\{\cos 120°\,\mathbf{i} + \cos 120°\,\mathbf{j} + \cos 45°\,\mathbf{k}\}$

$= -0.5F_{AD}\,\mathbf{i} - 0.5F_{AD}\,\mathbf{j} + 0.7071F_{AD}\,\mathbf{k}$

$\Sigma F_y = 0;\quad 0.4330F_{AC} - 0.5F_{AD} = 0$

$\Sigma F_z = 0;\quad 0.8660F_{AC} + 0.7071F_{AD} - 300 = 0$

$F_{AD} = 175.74\text{ lb} = 176\text{ lb}$ *Ans.*

$F_{AC} = 202.92\text{ lb} = 203\text{ lb}$ *Ans.*

$\Sigma F_x = 0;\quad F_{AB} - 0.25(202.92) - 0.5(175.74) = 0$

$F_{AB} = 138.60\text{ lb} = 139\text{ lb}$ *Ans.*

F3–11.
$$\mathbf{F}_B = F_B\left(\frac{\mathbf{r}_{AB}}{r_{AB}}\right)$$
$$= F_B\left[\frac{\{-6\mathbf{i} + 3\mathbf{j} + 2\mathbf{k}\}\text{ ft}}{\sqrt{(-6\text{ ft})^2 + (3\text{ ft})^2 + (2\text{ ft})^2}}\right]$$
$$= -\tfrac{6}{7}F_B\mathbf{i} + \tfrac{3}{7}F_B\mathbf{j} + \tfrac{2}{7}F_B\mathbf{k}$$
$$\mathbf{F}_C = F_C\left(\frac{\mathbf{r}_{AC}}{r_{AC}}\right)$$
$$= F_C\left[\frac{\{-6\mathbf{i} - 2\mathbf{j} + 3\mathbf{k}\}\text{ ft}}{\sqrt{(-6\text{ ft})^2 + (-2\text{ ft})^2 + (3\text{ft})^2}}\right]$$
$$= -\tfrac{6}{7}F_C\mathbf{i} - \tfrac{2}{7}F_C\mathbf{j} + \tfrac{3}{7}F_C\mathbf{k}$$
$$\mathbf{F}_D = F_D\mathbf{i}$$
$$\mathbf{W} = \{-150\mathbf{k}\}\text{ lb}$$
$$\Sigma F_x = 0;\ -\tfrac{6}{7}F_B - \tfrac{6}{7}F_C + F_D = 0 \quad (1)$$
$$\Sigma F_y = 0;\ \tfrac{3}{7}F_B - \tfrac{2}{7}F_C = 0 \quad (2)$$
$$\Sigma F_z = 0;\ \tfrac{2}{7}F_B + \tfrac{3}{7}F_C - 150 = 0 \quad (3)$$
$$F_B = 162\text{ lb}$$ *Ans.*
$$F_C = 1.5(162\text{ lb}) = 242\text{ lb}$$ *Ans.*
$$F_D = 346.15\text{ lb} = 346\text{ lb}$$ *Ans.*

Chapter 4

F4–1.
$$\circlearrowleft +M_O = -\left(\tfrac{4}{5}\right)(100\text{ N})(2\text{ m}) - \left(\tfrac{3}{5}\right)(100\text{ N})(5\text{ m})$$
$$= -460\text{ N}\cdot\text{m} = 460\text{ N}\cdot\text{m} \circlearrowright$$ *Ans.*

F4–2.
$$\circlearrowleft +M_O = [(300\text{ N})\sin 30°][0.4\text{ m} + (0.3\text{ m})\cos 45°]$$
$$- [(300\text{ N})\cos 30°][(0.3\text{ m})\sin 45°]$$
$$= 36.7\text{ N}\cdot\text{m}$$ *Ans.*

F4–3.
$$\circlearrowleft +M_O = (600\text{ lb})(4\text{ ft} + (3\text{ ft})\cos 45° - 1\text{ ft})$$
$$= 3.07\text{ kip}\cdot\text{ft}$$ *Ans.*

F4–4.
$$\circlearrowright +M_O = 50\sin 60°\,(0.1 + 0.2\cos 45° + 0.1)$$
$$- 50\cos 60°(0.2\sin 45°)$$
$$= 11.2\text{ N}\cdot\text{m}$$ *Ans.*

F4–5.
$$\circlearrowleft +M_O = 600\sin 50°\,(5) + 600\cos 50°\,(0.5)$$
$$= 2.49\text{ kip}\cdot\text{ft}$$ *Ans.*

F4–6.
$$\circlearrowleft +M_O = 500\sin 45°\,(3 + 3\cos 45°)$$
$$- 500\cos 45°\,(3\sin 45°)$$
$$= 1.06\text{ kN}\cdot\text{m}$$ *Ans.*

F4–7.
$$\circlearrowleft +(M_R)_O = \Sigma Fd;$$
$$(M_R)_O = -(600\text{ N})(1\text{ m})$$
$$+ (500\text{ N})[3\text{ m} + (2.5\text{ m})\cos 45°]$$
$$- (300\text{N})[(2.5\text{ m})\sin 45°]$$
$$= 1254\text{ N}\cdot\text{m} = 1.25\text{ kN}\cdot\text{m}$$ *Ans.*

F4–8.
$$\circlearrowleft +(M_R)_O = \Sigma Fd;$$
$$(M_R)_O = \left[\left(\tfrac{3}{5}\right)500\text{ N}\right](0.425\text{ m})$$
$$- \left[\left(\tfrac{4}{5}\right)500\text{ N}\right](0.25\text{ m})$$
$$- [(600\text{ N})\cos 60°](0.25\text{ m})$$
$$- [(600\text{ N})\sin 60°](0.425\text{ m})$$
$$= -268\text{ N}\cdot\text{m} = 268\text{ N}\cdot\text{m} \circlearrowright$$ *Ans.*

F4–9.
$$\circlearrowleft +(M_R)_O = \Sigma Fd;$$
$$(M_R)_O = (300\cos 30°\text{ lb})(6\text{ ft} + 6\sin 30°\text{ ft})$$
$$- (300\sin 30°\text{ lb})(6\cos 30°\text{ ft})$$
$$+ (200\text{ lb})(6\cos 30°\text{ ft})$$
$$= 2.60\text{ kip}\cdot\text{ft}$$ *Ans.*

F4–10.
$$\mathbf{F} = F\mathbf{u}_{AB} = 500\text{ N}\left(\tfrac{4}{5}\mathbf{i} - \tfrac{3}{5}\mathbf{j}\right) = \{400\mathbf{i} - 300\mathbf{j}\}\text{ N}$$
$$\mathbf{M}_O = \mathbf{r}_{OA} \times \mathbf{F} = \{3\mathbf{j}\}\text{ m} \times \{400\mathbf{i} - 300\mathbf{j}\}\text{ N}$$
$$= \{-1200\mathbf{k}\}\text{ N}\cdot\text{m}$$ *Ans.*

or
$$\mathbf{M}_O = \mathbf{r}_{OB} \times \mathbf{F} = \{4\mathbf{i}\}\text{ m} \times \{400\mathbf{i} - 300\mathbf{j}\}\text{ N}$$
$$= \{-1200\mathbf{k}\}\text{ N}\cdot\text{m}$$ *Ans.*

F4–11.
$$\mathbf{F} = F\mathbf{u}_{BC}$$
$$= 120\text{ lb}\left[\frac{\{4\mathbf{i} - 4\mathbf{j} - 2\mathbf{k}\}\text{ ft}}{\sqrt{(4\text{ ft})^2 + (-4\text{ ft})^2 + (-2\text{ ft})^2}}\right]$$
$$= \{80\mathbf{i} - 80\mathbf{j} - 40\mathbf{k}\}\text{ lb}$$
$$\mathbf{M}_O = \mathbf{r}_C \times \mathbf{F} = \begin{vmatrix} \mathbf{i} & \mathbf{j} & \mathbf{k} \\ 5 & 0 & 0 \\ 80 & -80 & -40 \end{vmatrix}$$
$$= \{200\mathbf{j} - 400\mathbf{k}\}\text{ lb}\cdot\text{ft}$$ *Ans.*

or
$$\mathbf{M}_O = \mathbf{r}_B \times \mathbf{F} = \begin{vmatrix} \mathbf{i} & \mathbf{j} & \mathbf{k} \\ 1 & 4 & 2 \\ 80 & -80 & -40 \end{vmatrix}$$
$$= \{200\mathbf{j} - 400\mathbf{k}\}\text{ lb}\cdot\text{ft}$$ *Ans.*

F4–12.
$$\mathbf{F}_R = \mathbf{F}_1 + \mathbf{F}_2$$
$$= \{(100 - 200)\mathbf{i} + (-120 + 250)\mathbf{j}$$
$$+ (75 + 100)\mathbf{k}\}\text{ lb}$$
$$= \{-100\mathbf{i} + 130\mathbf{j} + 175\mathbf{k}\}\text{ lb}$$
$$(\mathbf{M}_R)_O = \mathbf{r}_A \times \mathbf{F}_R = \begin{vmatrix} \mathbf{i} & \mathbf{j} & \mathbf{k} \\ 4 & 5 & 3 \\ -100 & 130 & 175 \end{vmatrix}$$
$$= \{485\mathbf{i} - 1000\mathbf{j} + 1020\mathbf{k}\}\text{ lb}\cdot\text{ft}$$ *Ans.*

F4–13. $M_x = \mathbf{i}\cdot(\mathbf{r}_{OB}\times\mathbf{F}) = \begin{vmatrix} 1 & 0 & 0 \\ 0.3 & 0.4 & -0.2 \\ 300 & -200 & 150 \end{vmatrix}$

$= 20\ \text{N}\cdot\text{m}$ *Ans.*

F4–14. $\mathbf{u}_{OA} = \dfrac{\mathbf{r}_A}{r_A} = \dfrac{\{0.3\mathbf{i} + 0.4\mathbf{j}\}\ \text{m}}{\sqrt{(0.3\ \text{m})^2 + (0.4\ \text{m})^2}} = 0.6\,\mathbf{i} + 0.8\,\mathbf{j}$

$M_{OA} = \mathbf{u}_{OA}\cdot(\mathbf{r}_{AB}\times\mathbf{F}) = \begin{vmatrix} 0.6 & 0.8 & 0 \\ 0 & 0 & -0.2 \\ 300 & -200 & 150 \end{vmatrix}$

$= -72\ \text{N}\cdot\text{m}$ *Ans.*

$|M_{OA}| = 72\ \text{N}\cdot\text{m}$

F4–15. Scalar Analysis

The magnitudes of the force components are

$F_x = |200\cos 120°| = 100\ \text{N}$

$F_y = 200\cos 60° = 100\ \text{N}$

$F_z = 200\cos 45° = 141.42\ \text{N}$

$M_x = -F_y(z) + F_z(y)$

$= -(100\ \text{N})(0.25\ \text{m}) + (141.42\ \text{N})(0.3\ \text{m})$

$= 17.4\ \text{N}\cdot\text{m}$ *Ans.*

Vector Analysis

$M_x = \begin{vmatrix} 1 & 0 & 0 \\ 0 & 0.3 & 0.25 \\ -100 & 100 & 141.42 \end{vmatrix} = 17.4\ \text{N}\cdot\text{m}$ *Ans.*

F4–16. $M_y = \mathbf{j}\cdot(\mathbf{r}_A\times\mathbf{F}) = \begin{vmatrix} 0 & 1 & 0 \\ -3 & -4 & 2 \\ 30 & -20 & 50 \end{vmatrix}$

$= 210\ \text{N}\cdot\text{m}$ *Ans.*

F4–17. $\mathbf{u}_{AB} = \dfrac{\mathbf{r}_{AB}}{r_{AB}} = \dfrac{\{-4\mathbf{i} + 3\mathbf{j}\}\ \text{ft}}{\sqrt{(-4\ \text{ft})^2 + (3\ \text{ft})^2}} = -0.8\mathbf{i} + 0.6\mathbf{j}$

$M_{AB} = \mathbf{u}_{AB}\cdot(\mathbf{r}_{AC}\times\mathbf{F})$

$= \begin{vmatrix} -0.8 & 0.6 & 0 \\ 0 & 0 & 2 \\ 50 & -40 & 20 \end{vmatrix} = -4\ \text{lb}\cdot\text{ft}$

$\mathbf{M}_{AB} = M_{AB}\mathbf{u}_{AB} = \{3.2\mathbf{i} - 2.4\mathbf{j}\}\ \text{lb}\cdot\text{ft}$ *Ans.*

F4–18. Scalar Analysis

The magnitudes of the force components are

$F_x = \left(\frac{3}{5}\right)\left[\frac{4}{5}(500)\right] = 240\ \text{N}$

$F_y = \frac{4}{5}\left[\frac{4}{5}(500)\right] = 320\ \text{N}$

$F_z = \frac{3}{5}(500) = 300\ \text{N}$

$M_x = -F_y(z) + F_z(y)$

$= -320(3) + 300(2) = -360\ \text{N}\cdot\text{m}$ *Ans.*

$M_y = -F_x(z) - F_z(x)$

$= -240(3) - 300(-2) = -120\ \text{N}\cdot\text{m}$ *Ans.*

$M_z = F_x(y) - F_y(x)$

$= 240(2) - 320(-2) = -160\ \text{N}\cdot\text{m}$ *Ans.*

Vector Analysis

$\mathbf{F} = \{-240\mathbf{i} + 320\mathbf{j} + 300\mathbf{k}\}\ \text{N}$

$\mathbf{r}_{OA} = \{-2\mathbf{i} + 2\mathbf{j} + 3\mathbf{k}\}\ \text{m}$

$M_x = \mathbf{i}\cdot(\mathbf{r}_{OA}\times\mathbf{F}) = -360\ \text{N}\cdot\text{m}$

$M_y = \mathbf{j}\cdot(\mathbf{r}_{OA}\times\mathbf{F}) = -120\ \text{N}\cdot\text{m}$

$M_z = \mathbf{k}\cdot(\mathbf{r}_{OA}\times\mathbf{F}) = -160\ \text{N}\cdot\text{m}$

F4–19. $\circlearrowleft +M_{C_R} = \Sigma M_A = 400(3) - 400(5) + 300(5)$

$+ 200(0.2) = 740\ \text{N}\cdot\text{m}$ *Ans.*

Also,

$\circlearrowleft +M_{C_R} = 300(5) - 400(2) + 200(0.2)$

$= 740\ \text{N}\cdot\text{m}$ *Ans.*

F4–20. $\circlearrowleft +M_{C_R} = 300(4) + 200(4) + 150(4)$

$= 2600\ \text{lb}\cdot\text{ft}$ *Ans.*

F4–21. $\circlearrowleft +(M_B)_R = \Sigma M_B$

$-1.5\ \text{kN}\cdot\text{m} = (2\ \text{kN})(0.3\ \text{m}) - F(0.9\ \text{m})$

$F = 2.33\ \text{kN}$ *Ans.*

F4–22. $\circlearrowleft +M_C = 10\left(\frac{3}{5}\right)(2) - 10\left(\frac{4}{5}\right)(4) = -20\ \text{kN}\cdot\text{m}$

$= 20\ \text{kN}\cdot\text{m}\ \circlearrowright$ *Ans.*

F4–23. $\mathbf{u}_1 = \dfrac{\mathbf{r}_1}{r_1} = \dfrac{\{-2\mathbf{i} + 2\mathbf{j} + 3.5\mathbf{k}\}\ \text{ft}}{\sqrt{(-2\ \text{ft})^2 + (2\ \text{ft})^2 + (3.5\ \text{ft})^2}}$

$= -\frac{2}{4.5}\mathbf{i} + \frac{2}{4.5}\mathbf{j} + \frac{3.5}{4.5}\mathbf{k}$

$\mathbf{u}_2 = -\mathbf{k}$

$\mathbf{u}_3 = \frac{1.5}{2.5}\mathbf{i} - \frac{2}{2.5}\mathbf{j}$

$(\mathbf{M}_c)_1 = (M_c)_1\mathbf{u}_1$

$= (450\ \text{lb}\cdot\text{ft})\left(-\frac{2}{4.5}\mathbf{i} + \frac{2}{4.5}\mathbf{j} + \frac{3.5}{4.5}\mathbf{k}\right)$

$= \{-200\mathbf{i} + 200\mathbf{j} + 350\mathbf{k}\}\ \text{lb}\cdot\text{ft}$

$(\mathbf{M}_c)_2 = (M_c)_2\mathbf{u}_2 = (250\ \text{lb}\cdot\text{ft})(-\mathbf{k})$

$= \{-250\mathbf{k}\}\ \text{lb}\cdot\text{ft}$

$(\mathbf{M}_c)_3 = (M_c)_3\,\mathbf{u}_3 = (300\ \text{lb}\cdot\text{ft})\left(\frac{1.5}{2.5}\mathbf{i} - \frac{2}{2.5}\mathbf{j}\right)$

$= \{180\mathbf{i} - 240\mathbf{j}\}\ \text{lb}\cdot\text{ft}$

$(\mathbf{M}_c)_R = \Sigma \mathbf{M}_c;$

$(\mathbf{M}_c)_R = \{-20\mathbf{i} - 40\mathbf{j} + 100\mathbf{k}\}\ \text{lb}\cdot\text{ft}$ *Ans.*

F4–24. $\mathbf{F}_B = \left(\frac{4}{5}\right)(450\text{ N})\mathbf{j} - \left(\frac{3}{5}\right)(450\text{ N})\mathbf{k}$

$= \{360\mathbf{j} - 270\mathbf{k}\}\text{ N}$

$$\mathbf{M}_c = \mathbf{r}_{AB} \times \mathbf{F}_B = \begin{vmatrix} \mathbf{i} & \mathbf{j} & \mathbf{k} \\ 0.4 & 0 & 0 \\ 0 & 360 & -270 \end{vmatrix}$$

$= \{108\mathbf{j} + 144\mathbf{k}\}\text{ N}\cdot\text{m}$ *Ans.*

Also,

$\mathbf{M}_c = (\mathbf{r}_A \times \mathbf{F}_A) + (\mathbf{r}_B \times \mathbf{F}_B)$

$$= \begin{vmatrix} \mathbf{i} & \mathbf{j} & \mathbf{k} \\ 0 & 0 & 0.3 \\ 0 & -360 & 270 \end{vmatrix} + \begin{vmatrix} \mathbf{i} & \mathbf{j} & \mathbf{k} \\ 0.4 & 0 & 0.3 \\ 0 & 360 & -270 \end{vmatrix}$$

$= \{108\mathbf{j} + 144\mathbf{k}\}\text{ N}\cdot\text{m}$ *Ans.*

F4–25. $\overset{+}{\leftarrow} F_{Rx} = \Sigma F_x;\ F_{Rx} = 200 - \frac{3}{5}(100) = 140\text{ lb}$

$+\downarrow F_{Ry} = \Sigma F_y;\ F_{Ry} = 150 - \frac{4}{5}(100) = 70\text{ lb}$

$F_R = \sqrt{140^2 + 70^2} = 157\text{ lb}$ *Ans.*

$\theta = \tan^{-1}\left(\frac{70}{140}\right) = 26.6°$ *Ans.*

$\circlearrowleft +M_{A_R} = \Sigma M_A;$

$M_{A_R} = \frac{3}{5}(100)(4) - \frac{4}{5}(100)(6) + 150(3)$

$M_{R_A} = 210\text{ lb}\cdot\text{ft}$ *Ans.*

F4–26. $\overset{+}{\rightarrow} F_{Rx} = \Sigma F_x;\quad F_{Rx} = \frac{4}{5}(50) = 40\text{ N}$

$+\downarrow F_{Ry} = \Sigma F_y;\quad F_{Ry} = 40 + 30 + \frac{3}{5}(50)$

$= 100\text{ N}$

$F_R = \sqrt{(40)^2 + (100)^2} = 108\text{ N}$ *Ans.*

$\theta = \tan^{-1}\left(\frac{100}{40}\right) = 68.2°$ *Ans.*

$\circlearrowleft +M_{A_R} = \Sigma M_A;$

$M_{A_R} = 30(3) + \frac{3}{5}(50)(6) + 200$

$= 470\text{ N}\cdot\text{m}$ *Ans.*

F4–27. $\overset{+}{\rightarrow} (F_R)_x = \Sigma F_x;$

$(F_R)_x = 900\sin 30° = 450\text{ N} \rightarrow$

$+\uparrow (F_R)_y = \Sigma F_y;$

$(F_R)_y = -900\cos 30° - 300$

$= -1079.42\text{ N} = 1079.42\text{ N}\downarrow$

$F_R = \sqrt{450^2 + 1079.42^2}$

$= 1169.47\text{ N} = 1.17\text{ kN}$ *Ans.*

$\theta = \tan^{-1}\left(\frac{1079.42}{450}\right) = 67.4°$ *Ans.*

$\circlearrowright +(M_R)_A = \Sigma M_A;$

$(M_R)_A = 300 - 900\cos 30°(0.75) - 300(2.25)$

$= -959.57\text{ N}\cdot\text{m}$

$= 960\text{ N}\cdot\text{m} \circlearrowright$ *Ans.*

F4–28. $\overset{+}{\rightarrow} (F_R)_x = \Sigma F_x;$

$(F_R)_x = 150\left(\frac{3}{5}\right) + 50 - 100\left(\frac{4}{5}\right) = 60\text{ lb} \rightarrow$

$+\uparrow (F_R)_y = \Sigma F_y;$

$(F_R)_y = -150\left(\frac{4}{5}\right) - 100\left(\frac{3}{5}\right)$

$= -180\text{ lb} = 180\text{ lb}\downarrow$

$F_R = \sqrt{60^2 + 180^2} = 189.74\text{ lb} = 190\text{ lb}$ *Ans.*

$\theta = \tan^{-1}\left(\frac{180}{60}\right) = 71.6°$ *Ans.*

$\circlearrowright +(M_R)_A = \Sigma M_A;$

$(M_R)_A = 100\left(\frac{4}{5}\right)(1) - 100\left(\frac{3}{5}\right)(6) - 150\left(\frac{4}{5}\right)(3)$

$= -640 = 640\text{ lb}\cdot\text{ft} \circlearrowright$ *Ans.*

F4–29. $\mathbf{F}_R = \Sigma\mathbf{F};$

$\mathbf{F}_R = \mathbf{F}_1 + \mathbf{F}_2$

$= (-300\mathbf{i} + 150\mathbf{j} + 200\mathbf{k}) + (-450\mathbf{k})$

$= \{-300\mathbf{i} + 150\mathbf{j} - 250\mathbf{k}\}\text{ N}$ *Ans.*

$\mathbf{r}_{OA} = (2 - 0)\mathbf{j} = \{2\mathbf{j}\}\text{ m}$

$\mathbf{r}_{OB} = (-1.5 - 0)\mathbf{i} + (2 - 0)\mathbf{j} + (1 - 0)\mathbf{k}$

$= \{-1.5\mathbf{i} + 2\mathbf{j} + 1\mathbf{k}\}\text{ m}$

$(\mathbf{M}_R)_O = \Sigma\mathbf{M};$

$(\mathbf{M}_R)_O = \mathbf{r}_{OB} \times \mathbf{F}_1 + \mathbf{r}_{OA} \times \mathbf{F}_2$

$$= \begin{vmatrix} \mathbf{i} & \mathbf{j} & \mathbf{k} \\ -1.5 & 2 & 1 \\ -300 & 150 & 200 \end{vmatrix} + \begin{vmatrix} \mathbf{i} & \mathbf{j} & \mathbf{k} \\ 0 & 2 & 0 \\ 0 & 0 & -450 \end{vmatrix}$$

$= \{-650\mathbf{i} + 375\mathbf{k}\}\text{ N}\cdot\text{m}$ *Ans.*

F4–30. $\mathbf{F}_1 = \{-100\mathbf{j}\}\text{ N}$

$$\mathbf{F}_2 = (200\text{ N})\left[\frac{\{-0.4\mathbf{i} - 0.3\mathbf{k}\}\text{ m}}{\sqrt{(-0.4\text{ m})^2 + (-0.3\text{ m})^2}}\right]$$

$= \{-160\mathbf{i} - 120\mathbf{k}\}\text{ N}$

$\mathbf{M}_c = \{-75\mathbf{i}\}\text{ N}\cdot\text{m}$

$\mathbf{F}_R = \{-160\mathbf{i} - 100\mathbf{j} - 120\mathbf{k}\}\text{ N}$ *Ans.*

$(\mathbf{M}_R)_O = (0.3\mathbf{k}) \times (-100\mathbf{j})$

$$+ \begin{vmatrix} \mathbf{i} & \mathbf{j} & \mathbf{k} \\ 0 & 0.5 & 0.3 \\ -160 & 0 & -120 \end{vmatrix} + (-75\mathbf{i})$$

$= \{-105\mathbf{i} - 48\mathbf{j} + 80\mathbf{k}\}\text{ N}\cdot\text{m}$ *Ans.*

F4–31. $+\downarrow F_R = \Sigma F_y;\quad F_R = 500 + 250 + 500$

$= 1250\text{ lb}$ *Ans.*

$\circlearrowleft +F_R x = \Sigma M_O;$

$1250(x) = 500(3) + 250(6) + 500(9)$

$x = 6\text{ ft}$ *Ans.*

F4–32. $\xrightarrow{+} (F_R)_x = \Sigma F_x;$

$(F_R)_x = 100\left(\frac{3}{5}\right) + 50 \sin 30° = 85 \text{ lb} \rightarrow$

$+\uparrow (F_R)_y = \Sigma F_y;$

$(F_R)_y = 200 + 50 \cos 30° - 100\left(\frac{4}{5}\right)$

$= 163.30 \text{ lb}\uparrow$

$F_R = \sqrt{85^2 + 163.30^2} = 184 \text{ lb}$

$\theta = \tan^{-1}\left(\frac{163.30}{85}\right) = 62.5°$ *Ans.*

$\circlearrowleft +(M_R)_A = \Sigma M_A;$

$163.30(d) = 200(3) - 100\left(\frac{4}{5}\right)(6) + 50 \cos 30°(9)$

$d = 3.12 \text{ ft}$ *Ans.*

F4–33. $\xrightarrow{+} (F_R)_x = \Sigma F_x;$

$(F_R)_x = 15\left(\frac{4}{5}\right) = 12 \text{ kN} \rightarrow$

$+\uparrow (F_R)_y = \Sigma F_y;$

$(F_R)_y = -20 + 15\left(\frac{3}{5}\right) = -11 \text{ kN} = 11 \text{ kN}\downarrow$

$F_R = \sqrt{12^2 + 11^2} = 16.3 \text{ kN}$ *Ans.*

$\theta = \tan^{-1}\left(\frac{11}{12}\right) = 42.5°$ *Ans.*

$\circlearrowleft +(M_R)_A = \Sigma M_A;$

$-11(d) = -20(2) - 15\left(\frac{4}{5}\right)(2) + 15\left(\frac{3}{5}\right)(6)$

$d = 0.909 \text{ m}$ *Ans.*

F4–34. $\xrightarrow{+} (F_R)_x = \Sigma F_x;$

$(F_R)_x = \left(\frac{3}{5}\right) 5 \text{ kN} - 8 \text{ kN}$

$= -5 \text{ kN} = 5 \text{ kN} \leftarrow$

$+\uparrow (F_R)_y = \Sigma F_y;$

$(F_R)_y = -6 \text{ kN} - \left(\frac{4}{5}\right) 5 \text{ kN}$

$= -10 \text{ kN} = 10 \text{ kN}\downarrow$

$F_R = \sqrt{5^2 + 10^2} = 11.2 \text{ kN}$ *Ans.*

$\theta = \tan^{-1}\left(\frac{10 \text{ kN}}{5 \text{ kN}}\right) = 63.4°$ *Ans.*

$\circlearrowleft +(M_R)_A = \Sigma M_A;$

$5 \text{ kN}(d) = 8 \text{ kN}(3 \text{ m}) - 6 \text{ kN}(0.5 \text{ m})$

$- \left[\left(\frac{4}{5}\right)5 \text{ kN}\right](2 \text{ m})$

$- \left[\left(\frac{3}{5}\right)5 \text{ kN}\right](4 \text{ m})$

$d = 0.2 \text{ m}$ *Ans.*

F4–35. $+\downarrow F_R = \Sigma F_z;\quad F_R = 400 + 500 - 100$

$= 800 \text{ N}$ *Ans.*

$M_{Rx} = \Sigma M_x;\ -800y = -400(4) - 500(4)$

$y = 4.50 \text{ m}$ *Ans.*

$M_{Ry} = \Sigma M_y;\quad 800x = 500(4) - 100(3)$

$x = 2.125 \text{ m}$ *Ans.*

F4–36. $+\downarrow F_R = \Sigma F_z;$

$F_R = 200 + 200 + 100 + 100$

$= 600 \text{ N}$ *Ans.*

$\circlearrowleft +M_{Rx} = \Sigma M_x;$

$-600y = 200(1) + 200(1) + 100(3) - 100(3)$

$y = -0.667 \text{ m}$ *Ans.*

$\circlearrowright +M_{Ry} = \Sigma M_y;$

$600x = 100(3) + 100(3) + 200(2) - 200(3)$

$x = 0.667 \text{ m}$ *Ans.*

F4–37. $+\uparrow F_R = \Sigma F_y;$

$-F_R = -6(1.5) - 9(3) - 3(1.5)$

$F_R = 40.5 \text{ kN}\downarrow$ *Ans.*

$\circlearrowleft +(M_R)_A = \Sigma M_A;$

$-40.5(d) = 6(1.5)(0.75)$

$- 9(3)(1.5) - 3(1.5)(3.75)$

$d = 1.25 \text{ m}$ *Ans.*

F4–38. $F_R = \frac{1}{2}(6)(150) + 8(150) = 1650 \text{ lb}$ *Ans.*

$\circlearrowright +M_{A_R} = \Sigma M_A;$

$1650d = \left[\frac{1}{2}(6)(150)\right](4) + [8(150)](10)$

$d = 8.36 \text{ ft}$ *Ans.*

F4–39. $+\uparrow F_R = \Sigma F_y;$

$-F_R = -\frac{1}{2}(6)(3) - \frac{1}{2}(6)(6)$

$F_R = 27 \text{ kN}\downarrow$ *Ans.*

$\circlearrowleft +(M_R)_A = \Sigma M_A;$

$-27(d) = \frac{1}{2}(6)(3)(1) - \frac{1}{2}(6)(6)(2)$

$d = 1 \text{ m}$ *Ans.*

F4–40. $+\downarrow F_R = \Sigma F_y;$

$F_R = \frac{1}{2}(50)(6) + 150(6) + 500$

$= 1550 \text{ lb}$ *Ans.*

$\circlearrowright +M_{A_R} = \Sigma M_A;$

$1550d = \left[\frac{1}{2}(50)(6)\right](4) + [150(6)](3) + 500(9)$

$d = 5.03 \text{ ft}$ *Ans.*

F4–41. $+\uparrow F_R = \Sigma F_y;$

$-F_R = -\frac{1}{2}(3)(4.5) - 3(6)$

$F_R = 24.75 \text{ kN}\downarrow$ *Ans.*

$\circlearrowleft +(M_R)_A = \Sigma M_A;$

$-24.75(d) = -\frac{1}{2}(3)(4.5)(1.5) - 3(6)(3)$

$d = 2.59 \text{ m}$ *Ans.*

F4–42. $F_R = \int w(x)\,dx = \int_0^4 2.5x^3\,dx = 160\text{ N}$

$\circlearrowleft +M_{A_R} = \Sigma M_A;$

$$x = \frac{\int xw(x)\,dx}{\int w(x)\,dx} = \frac{\int_0^4 2.5x^4\,dx}{160} = 3.20\text{ m} \quad \textit{Ans.}$$

Chapter 5

F5–1. $\xrightarrow{+} \Sigma F_x = 0; \quad -A_x + 500\left(\frac{3}{5}\right) = 0$

$A_x = 300\text{ lb}$ *Ans.*

$\circlearrowleft +\Sigma M_A = 0; \quad B_y(10) - 500\left(\frac{4}{5}\right)(5) - 600 = 0$

$B_y = 260\text{ lb}$ *Ans.*

$+\uparrow \Sigma F_y = 0; \quad A_y + 260 - 500\left(\frac{4}{5}\right) = 0$

$A_y = 140\text{ lb}$ *Ans.*

F5–2. $\circlearrowleft +\Sigma M_A = 0;$

$F_{CD} \sin 45°(1.5\text{ m}) - 4\text{ kN}(3\text{ m}) = 0$

$F_{CD} = 11.31\text{ kN} = 11.3\text{ kN}$ *Ans.*

$\xrightarrow{+} \Sigma F_x = 0; \quad A_x + (11.31\text{ kN}) \cos 45° = 0$

$A_x = -8\text{ kN} = 8\text{ kN} \leftarrow$ *Ans.*

$+\uparrow \Sigma F_y = 0;$

$A_y + (11.31\text{ kN}) \sin 45° - 4\text{ kN} = 0$

$A_y = -4\text{ kN} = 4\text{ kN} \downarrow$ *Ans.*

F5–3. $\circlearrowleft +\Sigma M_A = 0;$

$N_B[6\text{ m} + (6\text{ m}) \cos 45°]$

$- 10\text{ kN}[2\text{ m} + (6\text{ m}) \cos 45°]$

$- 5\text{ kN}(4\text{ m}) = 0$

$N_B = 8.047\text{ kN} = 8.05\text{ kN}$ *Ans.*

$\xrightarrow{+} \Sigma F_x = 0;$

$(5\text{ kN}) \cos 45° - A_x = 0$

$A_x = 3.54\text{ kN}$ *Ans.*

$+\uparrow \Sigma F_y = 0;$

$A_y + 8.047\text{ kN} - (5\text{ kN}) \sin 45° - 10\text{ kN} = 0$

$A_y = 5.49\text{ kN}$ *Ans.*

F5–4. $\xrightarrow{+} \Sigma F_x = 0; \quad -A_x + 400 \cos 30° = 0$

$A_x = 346\text{ N}$ *Ans.*

$+\uparrow \Sigma F_y = 0;$

$A_y - 200 - 200 - 200 - 400 \sin 30° = 0$

$A_y = 800\text{ N}$ *Ans.*

$\circlearrowleft +\Sigma M_A = 0;$

$M_A - 200(2.5) - 200(3.5) - 200(4.5)$

$- 400 \sin 30°(4.5) - 400 \cos 30°(3 \sin 60°) = 0$

$M_A = 3.90\text{ kN} \cdot \text{m}$ *Ans.*

F5–5. $\circlearrowleft +\Sigma M_A = 0;$

$N_C(0.7\text{ m}) - [25(9.81)\text{ N}]\,(0.5\text{ m}) \cos 30° = 0$

$N_C = 151.71\text{ N} = 152\text{ N}$ *Ans.*

$\xrightarrow{+} \Sigma F_x = 0;$

$T_{AB} \cos 15° - (151.71\text{ N}) \cos 60° = 0$

$T_{AB} = 78.53\text{ N} = 78.5\text{ N}$ *Ans.*

$+\uparrow \Sigma F_y = 0;$

$F_A + (78.53\text{ N}) \sin 15°$

$+ (151.71\text{ N}) \sin 60° - 25(9.81)\text{ N} = 0$

$F_A = 93.5\text{ N}$ *Ans.*

F5–6. $\xrightarrow{+} \Sigma F_x = 0;$

$N_C \sin 30° - (250\text{ N}) \sin 60° = 0$

$N_C = 433.0\text{ N} = 433\text{ N}$ *Ans.*

$\circlearrowleft +\Sigma M_B = 0;$

$-N_A \sin 30°(0.15\text{ m}) - 433.0\text{ N}(0.2\text{ m})$

$+ [(250\text{ N}) \cos 30°](0.6\text{ m}) = 0$

$N_A = 577.4\text{ N} = 577\text{ N}$ *Ans.*

$+\uparrow \Sigma F_y = 0;$

$N_B - 577.4\text{ N} + (433.0\text{ N})\cos 30°$

$- (250\text{ N}) \cos 60° = 0$

$N_B = 327\text{ N}$ *Ans.*

F5–7. $\Sigma F_z = 0;$

$T_A + T_B + T_C - 200 - 500 = 0$

$\Sigma M_x = 0;$

$T_A(3) + T_C(3) - 500(1.5) - 200(3) = 0$

$\Sigma M_y = 0;$

$-T_B(4) - T_C(4) + 500(2) + 200(2) = 0$

$T_A = 350\text{ lb}, \; T_B = 250\text{ lb}, \; T_C = 100\text{ lb}$ *Ans.*

F5–8. $\Sigma M_y = 0;$

$600\text{ N}(0.2\text{ m}) + 900\text{ N}(0.6\text{ m}) - F_A(1\text{ m}) = 0$

$F_A = 660\text{ N}$ *Ans.*

$\Sigma M_x = 0;$

$D_z(0.8\text{ m}) - 600\text{ N}(0.5\text{ m}) - 900\text{ N}(0.1\text{ m}) = 0$

$D_z = 487.5\text{ N}$ *Ans.*

$\Sigma F_x = 0; \quad D_x = 0$ *Ans.*

$\Sigma F_y = 0; \quad D_y = 0$ *Ans.*

$\Sigma F_z = 0;$

$T_{BC} + 660\text{ N} + 487.5\text{ N} - 900\text{ N} - 600\text{ N} = 0$

$T_{BC} = 352.5\text{ N}$ *Ans.*

F5–9. $\Sigma F_y = 0;\ \ 400\text{ N} + C_y = 0;$

$C_y = -400\text{ N}$ *Ans.*

$\Sigma M_y = 0;\ \ -C_x(0.4\text{ m}) - 600\text{ N}(0.6\text{ m}) = 0$

$C_x = -900\text{ N}$ *Ans.*

$\Sigma M_x = 0;\ \ B_z(0.6\text{ m}) + 600\text{ N}(1.2\text{ m}) + (-400\text{ N})(0.4\text{ m}) = 0$

$B_z = -933.3\text{ N}$ *Ans.*

$\Sigma M_z = 0;$

$-B_x(0.6\text{ m}) - (-900\text{ N})(1.2\text{ m}) + (-400\text{ N})(0.6\text{ m}) = 0$

$B_x = 1400\text{ N}$ *Ans.*

$\Sigma F_x = 0;\ \ 1400\text{ N} + (-900\text{ N}) + A_x = 0$

$A_x = -500\text{ N}$ *Ans.*

$\Sigma F_z = 0;\ \ A_z - 933.3\text{ N} + 600\text{ N} = 0$

$A_z = 333.3\text{ N}$ *Ans.*

F5–10. $\Sigma F_x = 0;\ \ B_x = 0$ *Ans.*

$\Sigma M_z = 0;$

$C_y(0.4\text{ m} + 0.6\text{ m}) = 0\ \ C_y = 0$ *Ans.*

$\Sigma F_y = 0;\ \ A_y + 0 = 0\ \ A_y = 0$ *Ans.*

$\Sigma M_x = 0;\ C_z(0.6\text{ m} + 0.6\text{ m}) + B_z(0.6\text{ m}) - 450\text{ N}(0.6\text{ m} + 0.6\text{ m}) = 0$

$1.2C_z + 0.6B_z - 540 = 0$

$\Sigma M_y = 0;\ -C_z(0.6\text{ m} + 0.4\text{ m}) - B_z(0.6\text{ m}) + 450\text{ N}(0.6\text{ m}) = 0$

$-C_z - 0.6B_z + 270 = 0$

$C_z = 1350\text{ N}\ \ B_z = -1800\text{ N}$ *Ans.*

$\Sigma F_z = 0;$

$A_z + 1350\text{ N} + (-1800\text{ N}) - 450\text{ N} = 0$

$A_z = 900\text{ N}$ *Ans.*

F5–11. $\Sigma F_y = 0;\ \ A_y = 0$ *Ans.*

$\Sigma M_x = 0;\ \ -9(3) + F_{CE}(3) = 0$

$F_{CE} = 9\text{ kN}$ *Ans.*

$\Sigma M_z = 0;\ \ F_{CF}(3) - 6(3) = 0$

$F_{CF} = 6\text{ kN}$ *Ans.*

$\Sigma M_y = 0;\ \ 9(4) - A_z(4) - 6(1.5) = 0$

$A_z = 6.75\text{ kN}$ *Ans.*

$\Sigma F_x = 0;\ \ A_x + 6 - 6 = 0\ \ A_x = 0$ *Ans.*

$\Sigma F_z = 0;\ \ F_{DB} + 9 - 9 + 6.75 = 0$

$F_{DB} = -6.75\text{ kN}$ *Ans.*

F5–12. $\Sigma F_x = 0;\ \ A_x = 0$ *Ans.*

$\Sigma F_y = 0;\ \ A_y = 0$ *Ans.*

$\Sigma F_z = 0;\ \ A_z + F_{BC} - 80 = 0$

$\Sigma M_x = 0;\ (M_A)_x + 6F_{BC} - 80(6) = 0$

$\Sigma M_y = 0;\ 3F_{BC} - 80(1.5) = 0\ \ F_{BC} = 40\text{ lb}$ *Ans.*

$\Sigma M_z = 0;\ (M_A)_z = 0$ *Ans.*

$A_z = 40\text{ lb}\ \ \ (M_A)_x = 240\text{ lb}\cdot\text{ft}$ *Ans.*

Chapter 6

F6–1. *Joint A.*

$+\uparrow\Sigma F_y = 0;\ \ 225\text{ lb} - F_{AD}\sin 45° = 0$

$F_{AD} = 318.20\text{ lb} = 318\text{ lb (C)}$ *Ans.*

$\overset{+}{\rightarrow}\Sigma F_x = 0;\ \ F_{AB} - (318.20\text{ lb})\cos 45° = 0$

$F_{AB} = 225\text{ lb (T)}$ *Ans.*

Joint B.

$\overset{+}{\rightarrow}\Sigma F_x = 0;\ \ F_{BC} - 225\text{ lb} = 0$

$F_{BC} = 225\text{ lb (T)}$ *Ans.*

$+\uparrow\Sigma F_y = 0;\ \ F_{BD} = 0$ *Ans.*

Joint D.

$\overset{+}{\rightarrow}\Sigma F_x = 0;$

$F_{CD}\cos 45° + (318.20\text{ lb})\cos 45° - 450\text{ lb} = 0$

$F_{CD} = 318.20\text{ lb} = 318\text{ lb (T)}$ *Ans.*

F6–2. *Joint D.*

$+\uparrow\Sigma F_y = 0;\ \tfrac{3}{5}F_{CD} - 300 = 0;$

$F_{CD} = 500\text{ lb (T)}$ *Ans.*

$\overset{+}{\rightarrow}\Sigma F_x = 0;\ -F_{AD} + \tfrac{4}{5}(500) = 0$

$F_{AD} = 400\text{ lb (C)}$ *Ans.*

$F_{BC} = 500\text{ lb (T)},\ F_{AC} = F_{AB} = 0$ *Ans.*

F6–3. $A_x = 0,\ A_y = C_y = 400\text{ lb}$

Joint A.

$+\uparrow\Sigma F_y = 0;\ -\tfrac{3}{5}F_{AE} + 400 = 0$

$F_{AE} = 667\text{ lb (C)}$ *Ans.*

Joint C.

$+\uparrow\Sigma F_y = 0;\ -F_{DC} + 400 = 0;$
$F_{DC} = 400\text{ lb (C)}$ *Ans.*

F6–4. *Joint C.*

$+\uparrow\Sigma F_y = 0;\quad 2F\cos 30° - P = 0$

$F_{AC} = F_{BC} = F = \frac{P}{2\cos 30°} = 0.5774P\ (C)$

Joint B.

$\xrightarrow{+}\Sigma F_x = 0;\ 0.5774P\cos 60° - F_{AB} = 0$

$F_{AB} = 0.2887P\ (T)$

$F_{AB} = 0.2887P = 2\text{ kN}$

$P = 6.928\text{ kN}$

$F_{AC} = F_{BC} = 0.5774P = 1.5\text{ kN}$

$P = 2.598\text{ kN}$

The *smaller value* of P is chosen,

$P = 2.598\text{ kN} = 2.60\text{ kN}$ *Ans.*

F6–5. $F_{CB} = 0$ *Ans.*

$F_{CD} = 0$ *Ans.*

$F_{AE} = 0$ *Ans.*

$F_{DE} = 0$ *Ans.*

F6–6. *Joint C.*

$+\uparrow\Sigma F_y = 0;\quad 259.81\text{ lb} - F_{CD}\sin 30° = 0$

$F_{CD} = 519.62\text{ lb} = 520\text{ lb (C)}$ *Ans.*

$\xrightarrow{+}\Sigma F_x = 0;\quad (519.62\text{ lb})\cos 30° - F_{BC} = 0$

$F_{BC} = 450\text{ lb (T)}$ *Ans.*

Joint D.

$+\nearrow\Sigma F_{y'} = 0;\quad F_{BD}\cos 30° = 0\quad F_{BD} = 0$ *Ans.*

$+\searrow\Sigma F_{x'} = 0;\quad F_{DE} - 519.62\text{ lb} = 0$

$F_{DE} = 519.62\text{ lb} = 520\text{ lb (C)}$ *Ans.*

Joint B.

$\uparrow\Sigma F_y = 0;\quad F_{BE}\sin\phi = 0\quad F_{BE} = 0$ *Ans.*

$\xrightarrow{+}\Sigma F_x = 0;\quad 450\text{ lb} - F_{AB} = 0$

$F_{AB} = 450\text{ lb (T)}$ *Ans.*

Joint A.

$+\uparrow\Sigma F_y = 0;\quad 340.19\text{ lb} - F_{AE} = 0$

$F_{AE} = 340\text{ lb (C)}$ *Ans.*

F6–7. $+\uparrow\Sigma F_y = 0;\ F_{CF}\sin 45° - 600 - 800 = 0$

$F_{CF} = 1980\text{ lb (T)}$ *Ans.*

$\circlearrowleft+\Sigma M_C = 0;\ F_{FE}(4) - 800(4) = 0$

$F_{FE} = 800\text{ lb (T)}$ *Ans.*

$\circlearrowleft+\Sigma M_F = 0;\ F_{BC}(4) - 600(4) - 800(8) = 0$

$F_{BC} = 2200\text{ lb (C)}$ *Ans.*

F6–8. $\circlearrowleft+\Sigma M_A = 0;\quad G_y(12\text{ m}) - 20\text{ kN}(2\text{ m}) - 30\text{ kN}(4\text{ m}) - 40\text{ kN}(6\text{ m}) = 0$

$G_y = 33.33\text{ kN}$

$+\uparrow\Sigma F_y = 0;\ F_{KC} + 33.33\text{ kN} - 40\text{ kN} = 0$

$F_{KC} = 6.67\text{ kN (C)}$ *Ans.*

$\circlearrowleft+\Sigma M_K = 0;$

$33.33\text{ kN}(8\text{ m}) - 40\text{ kN}(2\text{ m}) - F_{CD}(3\text{ m}) = 0$

$F_{CD} = 62.22\text{ kN} = 62.2\text{ kN (T)}$ *Ans.*

$\xrightarrow{+}\Sigma F_x = 0;\quad F_{LK} - 62.22\text{ kN} = 0$

$F_{LK} = 62.2\text{ kN (C)}$ *Ans.*

F6–9. From the geometry of the truss,

$\phi = \tan^{-1}(3\text{ m}/2\text{ m}) = 56.31°.$

$\circlearrowleft+\Sigma M_K = 0;$

$33.33\text{ kN}(8\text{ m}) - 40\text{ kN}(2\text{ m}) - F_{CD}(3\text{ m}) = 0$

$F_{CD} = 62.2\text{ kN (T)}$ *Ans.*

$\circlearrowleft+\Sigma M_D = 0;\quad 33.33\text{ kN}(6\text{ m}) - F_{KJ}(3\text{ m}) = 0$

$F_{KJ} = 66.7\text{ kN (C)}$ *Ans.*

$+\uparrow\Sigma F_y = 0;$

$33.33\text{ kN} - 40\text{ kN} + F_{KD}\sin 56.31° = 0$

$F_{KD} = 8.01\text{ kN (T)}$ *Ans.*

F6–10. From the geometry of the truss,

$\tan\phi = \frac{(9\text{ ft})\tan 30°}{3\text{ ft}} = 1.732\quad \phi = 60°$

$\circlearrowleft+\Sigma M_C = 0;$

$F_{EF}\sin 30°(6\text{ ft}) + 300\text{ lb}(6\text{ ft}) = 0$

$F_{EF} = -600\text{ lb} = 600\text{ lb (C)}$ *Ans.*

$\circlearrowleft+\Sigma M_D = 0;$

$300\text{ lb}(6\text{ ft}) - F_{CF}\sin 60°(6\text{ ft}) = 0$

$F_{CF} = 346.41\text{ lb} = 346\text{ lb (T)}$ *Ans.*

$\circlearrowleft+\Sigma M_F = 0;$

$300\text{ lb}(9\text{ ft}) - 300\text{ lb}(3\text{ ft}) - F_{BC}(9\text{ ft})\tan 30° = 0$

$F_{BC} = 346.41\text{ lb} = 346\text{ lb (T)}$ *Ans.*

F6–11. From the geometry of the truss,

$\theta = \tan^{-1}(1\text{ m}/2\text{ m}) = 26.57°$

$\phi = \tan^{-1}(3\text{ m}/2\text{ m}) = 56.31°.$

The location of O can be found using similar triangles.

$$\frac{1\text{ m}}{2\text{ m}} = \frac{2\text{ m}}{2\text{ m} + x}$$

$$4\text{ m} = 2\text{ m} + x$$

$$x = 2\text{ m}$$

$\circlearrowleft + \Sigma M_G = 0;$

$26.25 \text{ kN}(4 \text{ m}) - 15 \text{ kN}(2 \text{ m}) - F_{CD}(3 \text{ m}) = 0$

$F_{CD} = 25 \text{ kN (T)}$ *Ans.*

$\circlearrowleft + \Sigma M_D = 0;$

$26.25 \text{ kN}(2 \text{ m}) - F_{GF} \cos 26.57°(2 \text{ m}) = 0$

$F_{GF} = 29.3 \text{ kN (C)}$ *Ans.*

$\circlearrowleft + \Sigma M_O = 0;\ 15 \text{ kN}(4 \text{ m}) - 26.25 \text{ kN}(2 \text{ m})$

$- F_{GD} \sin 56.31°(4 \text{ m}) = 0$

$F_{GD} = 2.253 \text{ kN} = 2.25 \text{ kN (T)}$ *Ans.*

F6–12. $\circlearrowleft + \Sigma M_H = 0;$

$F_{DC}(12 \text{ ft}) + 1200 \text{ lb}(9 \text{ ft}) - 1600 \text{ lb}(21 \text{ ft}) = 0$

$F_{DC} = 1900 \text{ lb (C)}$ *Ans.*

$\circlearrowleft + \Sigma M_D = 0;$

$1200 \text{ lb}(21 \text{ ft}) - 1600 \text{ lb}(9 \text{ ft}) - F_{HI}(12 \text{ ft}) = 0$

$F_{HI} = 900 \text{ lb (C)}$ *Ans.*

$\circlearrowleft + \Sigma M_C = 0;\ F_{JI} \cos 45°(12 \text{ ft}) + 1200 \text{ lb}(21 \text{ ft})$

$- 900 \text{ lb}(12 \text{ ft}) - 1600 \text{ lb}(9 \text{ ft}) = 0$

$F_{JI} = 0$ *Ans.*

F6–13. $+\uparrow \Sigma F_y = 0;\quad 3P - 60 = 0$

$P = 20 \text{ lb}$ *Ans.*

F6–14. $\circlearrowleft + \Sigma M_C = 0;$

$-\left(\tfrac{4}{5}\right)(F_{AB})(9) + 400(6) + 500(3) = 0$

$F_{AB} = 541.67 \text{ lb}$

$\overset{+}{\rightarrow} \Sigma F_x = 0;\ -C_x + \tfrac{3}{5}(541.67) = 0$

$C_x = 325 \text{ lb}$ *Ans.*

$+\uparrow \Sigma F_y = 0;\ C_y + \tfrac{4}{5}(541.67) - 400 - 500 = 0$

$C_y = 467 \text{ lb}$ *Ans.*

F6–15. $\circlearrowleft + \Sigma M_A = 0;\ 100 \text{ N}(250 \text{ mm}) - N_B(50 \text{ mm}) = 0$

$N_B = 500 \text{ N}$ *Ans.*

$\overset{+}{\rightarrow} \Sigma F_x = 0;\quad (500 \text{ N}) \sin 45° - A_x = 0$

$A_x = 353.55 \text{ N}$

$+\uparrow \Sigma F_y = 0;\ A_y - 100 \text{ N} - (500 \text{ N}) \cos 45° = 0$

$A_y = 453.55 \text{ N}$

$F_A = \sqrt{(353.55 \text{ N})^2 + (453.55 \text{ N})^2}$

$= 575 \text{ N}$ *Ans.*

F6–16. $\circlearrowleft + \Sigma M_C = 0;$

$F_{AB} \cos 45°(1) - F_{AB} \sin 45°(3)$

$+ 800 + 400(2) = 0$

$F_{AB} = 1131.37 \text{ N}$

$\overset{+}{\rightarrow} \Sigma F_x = 0;\ -C_x + 1131.37 \cos 45° = 0$

$C_x = 800 \text{ N}$ *Ans.*

$+\uparrow \Sigma F_y = 0;\ -C_y + 1131.37 \sin 45° - 400 = 0$

$C_y = 400 \text{ N}$ *Ans.*

F6–17. Plate A:

$+\uparrow \Sigma F_y = 0;\ 2T + N_{AB} - 100 = 0$

Plate B:

$+\uparrow \Sigma F_y = 0;\ 2T - N_{AB} - 30 = 0$

$T = 32.5 \text{ lb},\ N_{AB} = 35 \text{ lb}$ *Ans.*

F6–18. Pulley C:

$+\uparrow \Sigma F_y = 0;\ T - 2P = 0;\ T = 2P$

Beam:

$+\uparrow \Sigma F_y = 0;\quad 2P + P - 6 = 0$

$P = 2 \text{ kN}$ *Ans.*

$\circlearrowleft + \Sigma M_A = 0;\ 2(1) - 6(x) = 0$

$x = 0.333 \text{ m}$ *Ans.*

F6–19. Member CD

$\circlearrowleft + \Sigma M_D = 0;\quad 600(1.5) - N_C(3) = 0$

$N_C = 300 \text{ N}$

Member ABC

$\circlearrowleft + \Sigma M_A = 0;\quad -800 + B_y(2) - (300 \sin 45°)\,4 = 0$

$B_y = 824.26 = 824 \text{ N}$ *Ans.*

$\overset{+}{\rightarrow} \Sigma F_x = 0;\quad A_x - 300 \cos 45° = 0;$

$A_x = 212 \text{ N}$ *Ans.*

$+\uparrow \Sigma F_y = 0;\quad -A_y + 824.26 - 300 \sin 45° = 0;$

$A_y = 612 \text{ N}$ *Ans.*

F6–20. AB is a two-force member.

Member BC

$\circlearrowleft + \Sigma M_c = 0;\ 15(3) + 10(6) - F_{BC}\left(\tfrac{4}{5}\right)(9) = 0$

$F_{BC} = 14.58 \text{ kN}$

$\overset{+}{\rightarrow} \Sigma F_x = 0;\quad (14.58)\left(\tfrac{3}{5}\right) - C_x = 0;$

$C_x = 8.75 \text{ kN}$

$+\uparrow \Sigma F_y = 0;\quad (14.58)\left(\tfrac{4}{5}\right) - 10 - 15 + C_y = 0;$

$C_y = 13.3 \text{ kN}$

Member CD

$\overset{+}{\rightarrow} \Sigma F_x = 0;\quad 8.75 - D_x = 0;\ D_x = 8.75 \text{ kN}$ *Ans.*

$+\uparrow \Sigma F_y = 0;\quad -13.3 + D_y = 0;\ D_y = 13.3 \text{ kN}$ *Ans.*

$\circlearrowleft + \Sigma M_D = 0;\ -8.75(4) + M_D = 0;\ M_D = 35 \text{ kN}\cdot\text{m}$ *Ans.*

F6–21. Entire frame

$\circlearrowleft +\Sigma M_A = 0; \quad -600(3) - [400(3)](1.5) + C_y(3) = 0$

$C_y = 1200 \text{ N}$ *Ans.*

$+\uparrow \Sigma F_y = 0; \quad A_y - 400(3) + 1200 = 0$

$A_y = 0$ *Ans.*

$\overset{+}{\rightarrow} \Sigma F_x = 0; \quad 600 - A_x - C_x = 0$

Member AB

$\circlearrowleft +\Sigma M_B = 0; \quad 400(1.5)(0.75) - A_x(3) = 0$

$A_x = 150 \text{ N}$ *Ans.*

$C_x = 450 \text{ N}$ *Ans.*

These same results can be obtained by considering members AB and BC.

F6–22. Entire frame

$\circlearrowleft +\Sigma M_E = 0; \quad 250(6) - A_y(6) = 0$

$A_y = 250 \text{ N}$

$\overset{+}{\rightarrow} \Sigma F_x = 0; \quad E_x = 0$

$+\uparrow \Sigma F_y = 0; \quad 250 - 250 + E_y = 0; \quad E_y = 0$

Member BD

$\circlearrowleft +\Sigma M_D = 0; \quad 250(4.5) - B_y(3) = 0;$

$B_y = 375 \text{ N}$

Member ABC

$\circlearrowleft +\Sigma M_C = 0; \quad -250(3) + 375(1.5) + B_x(2) = 0$

$B_x = 93.75 \text{ N}$

$\overset{+}{\rightarrow} \Sigma F_x = 0; \quad C_x - B_x = 0; \quad C_x = 93.75 \text{ N}$ *Ans.*

$+\uparrow \Sigma F_y = 0; \quad 250 - 375 + C_y = 0; \quad C_y = 125 \text{ N}$ *Ans.*

F6–23. AD, CB are two-force members.

Member AB

$\circlearrowleft +\Sigma M_A = 0; \quad -[\frac{1}{2}(3)(4)](1.5) + B_y(3) = 0$

$B_y = 3 \text{ kN}$

Since BC is a two-force member $C_y = B_y = 3 \text{ kN}$ and $C_x = 0 \ (\Sigma M_B = 0)$.

Member EDC

$\circlearrowleft +\Sigma M_E = 0; \quad F_{DA}(\frac{4}{5})(1.5) - 5(3) - 3(3) = 0$

$F_{DA} = 20 \text{ kN}$

$\overset{+}{\rightarrow} \Sigma F_x = 0; \quad E_x - 20(\frac{3}{5}) = 0; \quad E_x = 12 \text{ kN}$ *Ans.*

$+\uparrow \Sigma F_y = 0; \quad -E_y + 20(\frac{4}{5}) - 5 - 3 = 0;$

$E_y = 8 \text{ kN}$ *Ans.*

F6–24. AC and DC are two-force members.

Member BC

$\circlearrowleft +\Sigma M_C = 0; \quad [\frac{1}{2}(3)(8)](1) - B_y(3) = 0$

$B_y = 4 \text{ kN}$

Member BA

$\circlearrowleft +\Sigma M_B = 0; \quad 6(2) - A_x(4) = 0$

$A_x = 3 \text{ kN}$ *Ans.*

$+\uparrow \Sigma F_y = 0; \quad -4 \text{ kN} + A_y = 0; \quad A_y = 4 \text{ kN}$ *Ans.*

Entire Frame

$\circlearrowleft +\Sigma M_A = 0; \quad -6(2) - [\frac{1}{2}(3)(8)](2) + D_y(3) = 0$

$D_y = 12 \text{ kN}$ *Ans.*

Since DC is a two-force member $(\Sigma M_C = 0)$ then

$D_x = 0$ *Ans.*

Chapter 8

F8–1. $+\uparrow \Sigma F_y = 0; \quad N - 50(9.81) - 200(\frac{3}{5}) = 0$

$N = 610.5 \text{ N}$

$\overset{+}{\rightarrow} \Sigma F_x = 0; \quad F - 200(\frac{4}{5}) = 0$

$F = 160 \text{ N}$

$F < F_{max} = \mu_s N = 0.3(610.5) = 183.15 \text{ N},$

therefore $F = 160 \text{ N}$ *Ans.*

F8–2. $\circlearrowleft +\Sigma M_B = 0;$

$N_A(3) + 0.2N_A(4) - 30(9.81)(2) = 0$

$N_A = 154.89 \text{ N}$

$\overset{+}{\rightarrow} \Sigma F_x = 0; \quad P - 154.89 = 0$

$P = 154.89 \text{ N} = 155 \text{ N}$ *Ans.*

F8–3. Crate A

$+\uparrow \Sigma F_y = 0; \quad N_A - 50(9.81) = 0$

$N_A = 490.5 \text{ N}$

$\overset{+}{\rightarrow} \Sigma F_x = 0; \quad T - 0.25(490.5) = 0$

$T = 122.62 \text{ N}$

Crate B

$+\uparrow \Sigma F_y = 0; \quad N_B + P \sin 30° - 50(9.81) = 0$

$N_B = 490.5 - 0.5P$

$\overset{+}{\rightarrow} \Sigma F_x = 0;$

$P \cos 30° - 0.25(490.5 - 0.5P) - 122.62 = 0$

$P = 247 \text{ N}$ *Ans.*

F8–4. $\overset{+}{\rightarrow}\Sigma F_x = 0;\quad N_A - 0.3N_B = 0$

$+\uparrow\Sigma F_y = 0;$

$N_B + 0.3N_A + P - 100(9.81) = 0$

$\zeta + \Sigma M_O = 0;$

$P(0.6) - 0.3N_B(0.9) - 0.3\,N_A(0.9) = 0$

$N_A = 175.70\text{ N} \qquad N_B = 585.67\text{ N}$

$P = 343\text{ N}$ *Ans.*

F8–5. If slipping occurs:

$+\uparrow\Sigma F_y = 0;\quad N_c - 250\text{ lb} = 0;\ N_c = 250\text{ lb}$

$\overset{+}{\rightarrow}\Sigma F_x = 0;\quad P - 0.4(250) = 0;\ P = 100\text{ lb}$

If tipping occurs:

$\zeta + \Sigma M_A = 0;\ -P(4.5) + 250(1.5) = 0$

$P = 83.3\text{ lb}$ *Ans.*

F8–6.

$\zeta + \Sigma M_A = 0;\quad 490.5(0.6) - T\cos 60^\circ(0.3\cos 60^\circ + 0.6) - T\sin 60^\circ\,(0.3\sin 60^\circ) = 0$

$T = 490.5\text{ N}$

$\overset{+}{\rightarrow}\Sigma F_x = 0;\quad 490.5\sin 60^\circ - N_A = 0;\quad N_A = 424.8\text{ N}$

$+\uparrow\Sigma F_y = 0;\quad \mu_s(424.8) + 490.5\cos 60^\circ - 490.5 = 0$

$\mu_s = 0.577$ *Ans.*

F8–7. A will not move. Assume B is about to slip on C and A, and C is stationary.

$\overset{+}{\rightarrow}\Sigma F_x = 0;\quad P - 0.3(50) - 0.4(75);\quad P = 45\text{ N}$

Assume C is about to slip and B does not slip on C, but is about to slip at A.

$\overset{+}{\rightarrow}\Sigma F_x = 0;\quad P - 0.3(50) - 0.35(90) = 0$

$P = 46.5\text{ N} > 45\text{ N}$

$P = 45\text{ N}$ *Ans.*

F8–8. A is about to move down the plane and B moves upward.

Block A

$+\nwarrow\Sigma F_y = 0;\quad N = W\cos\theta$

$+\nearrow\Sigma F_x = 0;\quad T + \mu_s(W\cos\theta) - W\sin\theta = 0$

$T = W\sin\theta - \mu_s W\cos\theta \qquad (1)$

Block B

$+\nwarrow\Sigma F_y = 0;\quad N' = 2W\cos\theta$

$+\nearrow\Sigma F_x = 0;\quad 2T - \mu_s W\cos\theta - \mu_s(2W\cos\theta) - W\sin\theta = 0$

Using Eq.(1);

$\theta = \tan^{-1} 5\mu_s$ *Ans.*

F8–9. Assume B is about to slip on A, $F_B = 0.3\,N_B$.

$\overset{+}{\rightarrow}\Sigma F_x = 0;\quad P - 0.3(10)(9.81) = 0$

$P = 29.4\text{ N}$

Assume B is about to slip on A, $x = 0$.

$\zeta + \Sigma M_O = 0;\quad 10(9.81)(0.15) - P(0.4) = 0$

$P = 36.8\text{ N}$

Assume A is about to slip, $F_A = 0.1\,N_A$.

$\overset{+}{\rightarrow}\Sigma F_x = 0\quad P - 0.1[7(9.81) + 10(9.81)] = 0$

$P = 16.7\text{ N}$

Choose the smallest result. $P = 16.7\text{ N}$ *Ans.*

Chapter 9

F9–1. $$\bar{x} = \frac{\int_A \tilde{x}\,dA}{\int_A dA} = \frac{\frac{1}{2}\int_0^{1\text{ m}} y^{2/3}\,dy}{\int_0^{1\text{ m}} y^{1/3}dy} = 0.4\text{ m} \quad \textit{Ans.}$$

$$\bar{y} = \frac{\int_A \tilde{y}\,dA}{\int_A dA} = \frac{\int_0^{1\text{ m}} y^{4/3}\,dy}{\int_0^{1\text{ m}} y^{1/3}dy} = 0.571\text{ m} \quad \textit{Ans.}$$

F9–2. $$\bar{x} = \frac{\int_A \tilde{x}\,dA}{\int_A dA} = \frac{\int_0^{1\text{m}} x(x^3\,dx)}{\int_0^{1\text{ m}} x^3\,dx}$$

$$= 0.8\text{ m} \quad \textit{Ans.}$$

$$\bar{y} = \frac{\int_A \tilde{y}\,dA}{\int_A dA} = \frac{\int_0^{1\text{ m}} \frac{1}{2}x^3(x^3\,dx)}{\int_0^{1\text{ m}} x^3\,dx}$$

$$= 0.286\text{ m} \quad \textit{Ans.}$$

F9–3. $$\bar{y} = \frac{\int_A \tilde{y}\,dA}{\int_A dA} = \frac{\int_0^{2\text{ m}} y\left(2\left(\frac{y^{1/2}}{\sqrt{2}}\right)\right)dy}{\int_0^{2\text{ m}} 2\left(\frac{y^{1/2}}{\sqrt{2}}\right)dy}$$

$$= 1.2\text{ m} \quad \textit{Ans.}$$

F9–4. $$\bar{x} = \frac{\int_m \tilde{x}\,dm}{\int_m dm} = \frac{\int_0^L x\left[m_0\left(1 + \frac{x^2}{L^2}\right)dx\right]}{\int_0^L m_0\left(1 + \frac{x^2}{L^2}\right)dx}$$

$$= \frac{9}{16}L \quad \textit{Ans.}$$

F9–5. $\bar{y} = \dfrac{\int_V \tilde{y}\, dV}{\int_V dV} = \dfrac{\int_0^{1\text{ m}} y\left(\frac{\pi}{4} y dy\right)}{\int_0^{1\text{ m}} \frac{\pi}{4} y\, dy}$

$= 0.667\text{ m}$ *Ans.*

F9–6. $\bar{z} = \dfrac{\int_V \tilde{z}\, dV}{\int_V dV} = \dfrac{\int_0^{2\text{ ft}} z\left[\frac{9\pi}{64}(4-z)^2\, dz\right]}{\int_0^{2\text{ ft}} \frac{9\pi}{64}(4-z)^2\, dz}$

$= 0.786\text{ ft}$ *Ans.*

F9–7. $\bar{x} = \dfrac{\Sigma \tilde{x} L}{\Sigma L} = \dfrac{150(300) + 300(600) + 300(400)}{300 + 600 + 400}$

$= 265\text{ mm}$ *Ans.*

$\bar{y} = \dfrac{\Sigma \tilde{y} L}{\Sigma L} = \dfrac{0(300) + 300(600) + 600(400)}{300 + 600 + 400}$

$= 323\text{ mm}$ *Ans.*

$\bar{z} = \dfrac{\Sigma \tilde{z} L}{\Sigma L} = \dfrac{0(300) + 0(600) + (-200)(400)}{300 + 600 + 400}$

$= -61.5\text{ mm}$ *Ans.*

F9–8. $\bar{y} = \dfrac{\Sigma \tilde{y} A}{\Sigma A} = \dfrac{150[300(50)] + 325[50(300)]}{300(50) + 50(300)}$

$= 237.5\text{ mm}$ *Ans.*

F9–9. $\bar{y} = \dfrac{\Sigma \tilde{y} A}{\Sigma A} = \dfrac{100[2(200)(50)] + 225[50(400)]}{2(200)(50) + 50(400)}$

$= 162.5\text{ mm}$ *Ans.*

F9–10. $\bar{x} = \dfrac{\Sigma \tilde{x} A}{\Sigma A} = \dfrac{0.25[4(0.5)] + 1.75[0.5(2.5)]}{4(0.5) + 0.5(2.5)}$

$= 0.827\text{ in.}$ *Ans.*

$\bar{y} = \dfrac{\Sigma \tilde{y} A}{\Sigma A} = \dfrac{2[4(0.5)] + 0.25[(0.5)(2.5)]}{4(0.5) + (0.5)(2.5)}$

$= 1.33\text{ in.}$ *Ans.*

F9–11. $\bar{x} = \dfrac{\Sigma \tilde{x} V}{\Sigma V} = \dfrac{1[2(7)(6)] + 4[4(2)(3)]}{2(7)(6) + 4(2)(3)}$

$= 1.67\text{ ft}$ *Ans.*

$\bar{y} = \dfrac{\Sigma \tilde{y} V}{\Sigma V} = \dfrac{3.5[2(7)(6)] + 1[4(2)(3)]}{2(7)(6) + 4(2)(3)}$

$= 2.94\text{ ft}$ *Ans.*

$\bar{z} = \dfrac{\Sigma \tilde{z} V}{\Sigma V} = \dfrac{3[2(7)(6)] + 1.5[4(2)(3)]}{2(7)(6) + 4(2)(3)}$

$= 2.67\text{ ft}$ *Ans.*

F9–12. $\bar{x} = \dfrac{\Sigma \tilde{x} V}{\Sigma V}$

$= \dfrac{0.25[0.5(2.5)(1.8)] + 0.25\left[\frac{1}{2}(1.5)(1.8)(0.5)\right] + (1.0)\left[\frac{1}{2}(1.5)(1.8)(0.5)\right]}{0.5(2.5)(1.8) + \frac{1}{2}(1.5)(1.8)(0.5) + \frac{1}{2}(1.5)(1.8)(0.5)}$

$= 0.391\text{ m}$ *Ans.*

$\bar{y} = \dfrac{\Sigma \tilde{y} V}{\Sigma V} = \dfrac{5.00625}{3.6} = 1.39\text{ m}$ *Ans.*

$\bar{z} = \dfrac{\Sigma \tilde{z} V}{\Sigma V} = \dfrac{2.835}{3.6} = 0.7875\text{ m}$ *Ans.*

F9–13. $A = 2\pi \Sigma \tilde{r} L$

$= 2\pi\left[0.75(1.5) + 1.5(2) + 0.75\sqrt{(1.5)^2 + (2)^2}\right]$

$= 37.7\text{ m}^2$ *Ans.*

$V = 2\pi \Sigma \tilde{r} A$

$= 2\pi\left[0.75(1.5)(2) + 0.5\left(\tfrac{1}{2}\right)(1.5)(2)\right]$

$= 18.8\text{ m}^3$ *Ans.*

F9–14. $A = 2\pi \Sigma \tilde{r} L$

$= 2\pi\left[1.95\sqrt{(0.9)^2 + (1.2)^2} + 2.4(1.5) + 1.95(0.9) + 1.5(2.7)\right]$

$= 77.5\text{ m}^2$ *Ans.*

$V = 2\pi \Sigma \tilde{r} A$

$= 2\pi\left[1.8\left(\tfrac{1}{2}\right)(0.9)(1.2) + 1.95(0.9)(1.5)\right]$

$= 22.6\text{ m}^3$ *Ans.*

F9–15. $A = 2\pi \Sigma \tilde{r} L$

$= 2\pi\left[7.5(15) + 15(18) + 22.5\sqrt{15^2 + 20^2} + 15(30)\right]$

$= 8765\text{ in.}^2$ *Ans.*

$V = 2\pi \Sigma \tilde{r} A$

$= 2\pi\left[7.5(15)(38) + 20\left(\tfrac{1}{2}\right)(15)(20)\right]$

$= 45\,710\text{ in.}^3$ *Ans.*

F9–16. $A = 2\pi \Sigma \tilde{r} L$

$= 2\pi\left[\frac{2(1.5)}{\pi}\left(\frac{\pi(1.5)}{2}\right) + 1.5(2) + 0.75(1.5)\right]$

$= 40.1\text{ m}^2$ *Ans.*

$V = 2\pi \Sigma \tilde{r} A$

$= 2\pi\left[\frac{4(1.5)}{3\pi}\left(\frac{\pi(1.5^2)}{4}\right) + 0.75(1.5)(2)\right]$

$= 21.2\text{ m}^3$ *Ans.*

F9–17. $w_b = \rho_w g h b = 1000(9.81)(6)(1)$
$= 58.86\text{ kN/m}$
$F_R = \frac{1}{2}(58.76)(6) = 176.58\text{ kN} = 177\text{ kN}$ *Ans.*

F9–18. $w_b = \gamma_w\, hb = 62.4\,(4)(4) = 998.4\text{ lb/ft}$
$F_R = 998.4(3) = 3.00\text{ kip}$ *Ans.*

F9–19. $w_b = \rho_w g h_B b = 1000(9.81)(2)(1.5)$
$= 29.43\text{ kN/m}$
$F_R = \frac{1}{2}(29.43)\left(\sqrt{(1.5)^2 + (2)^2}\right)$
$= 36.8\text{ kN}$ *Ans.*

F9–20. $w_A = \rho_w g h_A b = 1000(9.81)(3)(2)$
$= 58.86\text{ kN/m}$
$w_B = \rho_w g h_B b = 1000(9.81)(5)(2)$
$= 98.1\text{ kN/m}$
$F_R = \frac{1}{2}(58.86 + 98.1)(2) = 157\text{ kN}$ *Ans.*

F9–21. $w_A = \gamma_w h_A\, b = 62.4(6)(2) = 748.8\text{ lb/ft}$
$w_B = \gamma_w h_B\, b = 62.4(10)(2) = 1248\text{ lb/ft}$
$F_R = \frac{1}{2}(748.8 + 1248)\left(\sqrt{(3)^2 + (4)^2}\right)$
$= 4.99\text{ kip}$ *Ans.*

Chapter 10

F10–1.

$$I_x = \int_A y^2\, dA = \int_0^{1\text{ m}} y^2\left[\left(1 - y^{3/2}\right)dy\right] = 0.111\text{ m}^4 \quad \textit{Ans.}$$

F10–2.

$$I_x = \int_A y^2\, dA = \int_0^{1\text{ m}} y^2\left(y^{3/2}\, dy\right) = 0.222\text{ m}^4 \quad \textit{Ans.}$$

F10–3.

$$I_y = \int_A x^2\, dA = \int_0^{1\text{ m}} x^2\left(x^{2/3}\right)dx = 0.273\text{ m}^4 \quad \textit{Ans.}$$

F10–4.

$$I_y = \int_A x^2\, dA = \int_0^{1\text{ m}} x^2\left[(1 - x^{2/3})\, dx\right] = 0.0606\text{ m}^4 \quad \textit{Ans.}$$

F10–5. $I_x = \left[\frac{1}{12}(50)\left(450^3\right) + 0\right] + \left[\frac{1}{12}(300)\left(50^3\right) + 0\right]$
$= 383\left(10^6\right)\text{ mm}^4$ *Ans.*
$I_y = \left[\frac{1}{12}(450)\left(50^3\right) + 0\right]$
$+ 2\left[\frac{1}{12}(50)\left(150^3\right) + (150)(50)(100)^2\right]$
$= 183\left(10^6\right)\text{ mm}^4$ *Ans.*

F10–6. $I_x = \frac{1}{12}(360)\left(200^3\right) - \frac{1}{12}(300)\left(140^3\right)$
$= 171\left(10^6\right)\text{ mm}^4$ *Ans.*
$I_y = \frac{1}{12}(200)\left(360^3\right) - \frac{1}{12}(140)\left(300^3\right)$
$= 463\left(10^6\right)\text{ mm}^4$ *Ans.*

F10–7. $I_y = 2\left[\frac{1}{12}(50)\left(200^3\right) + 0\right]$
$+ \left[\frac{1}{12}(300)\left(50^3\right) + 0\right]$
$= 69.8\,(10^6)\text{ mm}^4$ *Ans.*

F10–8.

$$\bar{y} = \frac{\Sigma \tilde{y} A}{\Sigma A} = \frac{15(150)(30) + 105(30)(150)}{150(30) + 30(150)} = 60\text{ mm}$$

$\bar{I}_{x'} = \Sigma(\bar{I} + Ad^2)$
$= \left[\frac{1}{12}(150)(30)^3 + (150)(30)(60 - 15)^2\right]$
$+ \left[\frac{1}{12}(30)(150)^3 + 30(150)(105 - 60)^2\right]$
$= 27.0\,(10^6)\text{ mm}^4$ *Ans.*

Answers to Selected Problems

Chapter 2

2–1. $F_R = 393$ lb
$\phi = 353°$

2–2. $F_R = 497$ N
$\phi = 155°$

2–3. $F = 960$ N
$\theta = 45.2°$

2–5. $F_{1u} = 205$ N
$F_{1v} = 160$ N

2–6. $F_{2u} = 376$ N
$F_{2v} = 482$ N

2–7. $F_{AB} = 448$ N
$F_{AC} = 366$ N

2–9. $F_{1v} = 129$ N
$F_{1u} = 183$ N

2–10. $F_{2v} = 77.6$ N
$F_{2u} = 150$ N

2–11. $F_a = 30.6$ lb
$F_b = 26.9$ lb

2–13. $F = 917$ lb
$\theta = 31.8°$

2–14. $F_{BC} = 434$ lb
$\phi = 56.5°$

2–15. $F_R = 10.8$ kN
$\phi = 3.16°$

2–17. $\theta = 53.5°$
$F_{AB} = 621$ lb

2–18. $\phi = 38.3°$

2–19. $F_R = 19.2$ N
$\theta = 2.37° \measuredangle$

2–21. $\theta = 75.5°$

2–22. $\phi = \frac{\theta}{2}$
$F_R = 2F\cos(\frac{\theta}{2})$

2–23. $\theta = 36.9°$
$F_R = 920$ N

2–25. **a.** $F_n = -14.1$ lb
$F_t = 14.1$ lb
b. $F_x = 19.3$ lb
$F_y = 5.18$ lb

2–26. $F_A = 439$ N
$F_B = 311$ N

2–27. $\theta = 60°$
$F_A = 520$ N
$F_B = 300$ N

2–29. $F_R = 4.01$ kN
$\phi = 16.2°$

2–30. $\theta = 90°$
$F_B = 1$ kN
$F_R = 1.73$ kN

2–31. $F_{R1} = 264.6$ lb, $\theta = 10.9°$
$F_{min} = 235$ lb

2–33. $F_R = 546$ N
$\theta = 253°$

2–34. $\mathbf{F}_1 = \{200\mathbf{i} + 346\mathbf{j}\}$ N
$\mathbf{F}_2 = \{177\mathbf{i} - 177\mathbf{j}\}$ N

2–35. $F_R = 413$ N
$\theta = 24.2°$

2–37. $F_R = 1.96$ kN
$\theta = 4.12°$

2–38. $\mathbf{F}_1 = \{90\mathbf{i}, -120\mathbf{j}\}$ lb
$\mathbf{F}_2 = \{-275\mathbf{j}\}$ lb
$\mathbf{F}_3 = \{-37.5\mathbf{i}, -65.0\mathbf{j}\}$ lb
$\mathbf{F}_R = 463$ lb

2–39. $\mathbf{F}_1 = \{400\mathbf{i} + 693\mathbf{j}\}$ N
$\mathbf{F}_2 = \{-424\mathbf{i} + 424\mathbf{j}\}$ N
$\mathbf{F}_3 = \{600\mathbf{i} - 250\mathbf{j}\}$ N

2–41. $F_R = 111$ lb
$\theta = 202°$

2–42. $\theta = 68.6°$
$F_B = 960$ N

2–43. $F_R = 839$ N
$\theta = 14.8°$

2–45. $F_1 = 143$ N
$F_R = 91.9$ N

2–46. $\phi = 29.2°$ $F_1 = 528$ N

2–47. $F_R = \sqrt{F_1^2 + F_2^2 + 2F_1F_2\cos\phi}$

$$\tan\theta = \frac{F_1\sin\phi}{F_2 + F_1\cos\phi}$$

2–49. $\phi = 42.4°$ $F_1 = 731$ N

2–50. $\theta = 29.1°$ $F_1 = 275$ N

2–51. $F_R = 1.03$ kN
$\theta = 87.9°$

2–53. F = 2.03 kN
$F_R = 7.87$ kN

2–54. $\theta = 37.0°$
$F_1 = 889$ N

2–55. $F_R = 717$ N
$\phi = 37.1°$

2–57. $F_R = 161$ lb
$\theta = 38.3°$

2–58. $\phi = 10.9°$ $F_1 = 474$ N

2–59. $F_R = \sqrt{(0{,}5F_1 + 300)^2 + (0.866F_1 - 240)^2}$
$F_R = 380$ N, $F_1 = 57.8$

2–61. $F_R = 114$ lb
$\alpha = 62.1°$
$\beta = 113°$
$\gamma = 142°$

2–62. $\mathbf{F}_1 = \{53.1\mathbf{i} - 44.5\mathbf{j} + 40\mathbf{k}\}$ lb
$\alpha_1 = 48.4°$
$\beta_1 = 124°$
$\gamma_1 = 60°$
$\mathbf{F}_2 = \{-130\mathbf{k}\}$ lb
$\alpha_2 = 90°$
$\beta_2 = 90°$
$\gamma_2 = 180°$

2–63. $F_x = 40$ N
$F_y = 40$ N
$F_z = 56.6$ N

2–65. $\mathbf{F} = \{217\mathbf{i} + 85.5\mathbf{j} - 91.2\mathbf{k}\}$ lb

2–66. $\mathbf{F}_1 = [480\mathbf{i} + 360\mathbf{k}]$ lb
$\mathbf{F}_2 = [200\mathbf{i} + 283\mathbf{j} - 200\mathbf{k}]$ lb

2–67. $F_R = 754$ lb
$\alpha = 25.5°$
$\beta = 68.0°$
$\gamma = 77.7°$

2–69. $F_R = 733$ N
$\alpha = 53.5°$
$\beta = 65.3°$
$\gamma = 133°$

2–70. $\mathbf{F}_1 = \{176\mathbf{j} - 605\mathbf{k}\}$ lb
$\mathbf{F}_2 = \{125\mathbf{i} - 177\mathbf{j} + 125\mathbf{k}\}$ lb
$F_R = 496$ lb
$\alpha = 75.4°$
$\beta = 90.0°$
$\gamma = 165°$

2–71. $\alpha = 121°$
$\gamma = 53.1°$
$F_R = 754$ N
$\beta = 52.5°$

2–73. $\mathbf{F}_1 = \{14.0\mathbf{j} - 48.0\mathbf{k}\}$ lb
$\mathbf{F}_2 = \{90\mathbf{i} - 127\mathbf{j} + 90\mathbf{k}\}$ lb

2–74. $\mathbf{F}_R = \{90\mathbf{i} - 113\mathbf{j} + 42\mathbf{k}\}$ lb

2–75. $\alpha = 46.1°$
$\beta = 114°$
$\gamma = 53.1°$

2–77. $\mathbf{F}_1 = \{225\mathbf{j} + 268\mathbf{k}\}$ N
$\mathbf{F}_2 = \{70.7\mathbf{i} + 50.0\mathbf{j} - 50.0\mathbf{k}\}$ N
$\mathbf{F}_3 = \{125\mathbf{i} - 177\mathbf{j} + 125\mathbf{k}\}$ N
$F_R = 407$ N
$\alpha_x = 61.3°$
$\beta_y = 76.0°$
$\gamma_z = 32.5°$

2–78. $F_3 = 166$ N
$\alpha = 97.5°$
$\beta = 63.7°$
$\gamma = 27.5°$

2–79. $\alpha_{F_1} = 36.9°$
$\beta_{F_1} = 90.0°$
$\gamma_{F_1} = 53.1°$
$\alpha_R = 69.3°$
$\beta_R = 52.2°$
$\gamma_R = 45.0°$

2–81. $F_R = 1.60$ kN
$\alpha = 82.6°$
$\beta = 29.4°$
$\gamma = 61.7°$

2–82. $\alpha_3 = 139°$
$\beta_3 = 128°$ $\gamma_3 = 102°$ $F_{R1} = 387$ N
$\beta_3 = 60.7°$ $\gamma_3 = 64.4°$ $F_{R2} = 1.41$ kN

2–83. $\mathbf{F}_1 = \{86.5\mathbf{i} + 186\mathbf{j} - 143\mathbf{k}\}$ N
$\mathbf{F}_2 = \{-200\mathbf{i} + 283\mathbf{j} + 200\mathbf{k}\}$ N
$\mathbf{F}_R = \{-113\mathbf{i} + 468\mathbf{j} + 56.6\mathbf{k}\}$ N
$F_R = 485.30$ N $= 485$ N
$\alpha = 104°$
$\beta = 15.1°$
$\gamma = 83.3°$

2–85. $F = 2.02$ kN
$F_y = 0.523$ kN

2–86. $\mathbf{r} = \{-2.35\mathbf{i} + 3.93\mathbf{j} + 3.71\mathbf{k}\}$ ft
$r = 5.89$ ft
$\alpha = 113°$
$\beta = 48.2°$
$\gamma = 51.0°$

2–87. $r_{AD} = 1.50$ m
$r_{BD} = 1.50$ m
$r_{CD} = 1.73$ m

2–89. $x = 5.06$ m
$y = 3.61$ m
$z = 6.51$ m

2–90. $\mathbf{F}_B = \{-400\mathbf{i} - 200\mathbf{j} + 400\mathbf{k}\}$ N
$\mathbf{F}_C = \{-200\mathbf{i} + 200\mathbf{j} + 350\mathbf{k}\}$ N

2–91. $F_R = 960$ N
$\alpha = 129°$
$\beta = 90°$
$\gamma = 38.7°$

2–93. $F_R = 1.17$ kN
$\alpha = 68.0°$
$\beta = 96.8°$
$\gamma = 157°$

2–94. $F_R = 1.50$ kN
$\alpha = 77.6°$
$\beta = 90.6°$
$\gamma = 168°$

2–95. $d = 6.71$ km
2–97. $F_R = 110$ lb
$\alpha = 35.4°$
$\beta = 68.8°$
$\gamma = 117°$
2–98. $\mathbf{F} = \{13.4\mathbf{i} + 23.2\mathbf{j} + 53.7\mathbf{k}\}$ lb
2–99. $F_R = 316$ N
$\alpha = 60.1°$
$\beta = 74.6°$
$\gamma = 146°$
2–101. $\mathbf{F}_A = \{169\mathbf{i} + 33.8\mathbf{j} - 101\mathbf{k}\}$ lb
$\mathbf{F}_B = \{97.6\mathbf{i} + 97.6\mathbf{j} - 58.6\mathbf{k}\}$ lb
$F_R = 338$ lb
$\alpha = 37.8°$
$\beta = 67.1°$
$\gamma = 118°$
2–102. $\mathbf{F}_{EA} = \{12\mathbf{i} - 8\mathbf{j} - 24\mathbf{k}\}$ kN
$\mathbf{F}_{EB} = \{12\mathbf{i} + 8\mathbf{j} - 24\mathbf{k}\}$ kN
$\mathbf{F}_{EC} = \{-12\mathbf{i} + 8\mathbf{j} - 24\mathbf{k}\}$ kN
$\mathbf{F}_{ED} = \{-12\mathbf{i} - 8\mathbf{j} - 24\mathbf{k}\}$ kN
$\mathbf{F}_R = \{-96\mathbf{k}\}$ kN
2–103. $F_R = 240$ lb
$\alpha = 90°$
$\beta = 90°$
$\gamma = 180°$
2–105. $\mathbf{F} = \{-6.61\mathbf{i} - 3.73\mathbf{j} + 9.29\mathbf{k}\}$ lb
2–106. $\mathbf{F}_A = \{28.8\mathbf{i} - 16.6\mathbf{j} - 49.9\mathbf{k}\}$ lb
$\mathbf{F}_B = \{-28.8\mathbf{i} - 16.6\mathbf{j} - 49.9\mathbf{k}\}$ lb
$\mathbf{F}_C = \{33.3\mathbf{j} - 49.9\mathbf{k}\}$ lb
$F_R = 150$ lb
$\alpha = 90°$
$\beta = 90°$
$\gamma = 180°$
2–107. $F = 52.1\ P$
2–109. $F_C = 442$ N; $F_B = 318$ N; $F_D = 866$ N
2–110. $\mathbf{F} = \{143\mathbf{i} + 248\mathbf{j} - 201\mathbf{k}\}$ lb
2–111. $r_{AB} = 10.0$ ft
$\mathbf{F} = \{-19.1\mathbf{i} - 14.9\mathbf{j} + 43.7\mathbf{k}\}$ lb
$\alpha = 112°$
$\beta = 107°$
$\gamma = 29.0°$
2–113. $\theta = 82.0°$
2–114. $\theta = 74.2°$
2–115. $r_{BC} = 5.39$ m
2–117. $(F_{AC})_z = 569$ lb
2–118. Proj $F = 0.667$ kN
2–119. $\theta = 70.5°$
2–121. $F_u = 246$ N
2–122. $\theta = 46.7°$
2–123. $\phi = 44.1°$
2–125. $(\mathbf{F}_{AC})_{AO} = \{-40.8\mathbf{j} - 13.6\mathbf{k}\}$ lb
2–126. $|(\mathbf{F}_1)_{F_2}| = 5.44$ lb
2–127. $\theta = 100°$
2–129. $F_{AC} = -25.87$ lb
$\mathbf{F}_{AC} = \{-18.0\mathbf{i} - 15.4\mathbf{j} + 10.3\mathbf{k}\}$ lb
2–130. $\theta = 132°$
2–131. $\theta = 74.4°$
$\phi = 55.4°$
2–133. $\theta = 97.3°$
2–134. $[(F)_{AB}]_{\parallel} = 63.2$ lb
$[(F)_{AB}]_{\perp} = 64.1$ lb
2–135. $F_1 = 19.4$ lb
$F_2 = 53.4$ lb
2–137. $F_{\parallel} = 82.4$ N
$F_{\perp} = 594$ N
2–138. $F = -300 \sin 30° \sin 30°\ \mathbf{i} +$
$300 \cos 30°\ \mathbf{j} + 300 \sin 30° \cos 30°\ \mathbf{k}$
$F_x = 75$ N
$F_y = 260$ N
2–139. $F = -300 \sin 30° \sin 30°\ \mathbf{i} +$
$300 \cos 30°\ \mathbf{j} + 300 \sin 30° \cos 30°\ \mathbf{k}$
$F_{OA} = 242$ N
2–141. $F_{1x} = 141$ N
$F_{1y} = 141$ N
$F_{2x} = -130$ N
$F_{2y} = 75$ N
2–142. $F_R = 217$ N
$\theta = 87.0°$
2–143. $F_{1x} = -200$ lb, $F_{1y} = 0$
$F_{2x} = 320$ lb, $F_{2y} = -240$ lb
$F_{3x} = 180$ lb, $F_{3y} = 240$ lb
$F_{4x} = -300$ lb, $F_{4y} = 0$
2–145. $F_R = 25.1$ kN
$\theta = 185°$
2–146. $\mathbf{F} = \{98.1\mathbf{i} + 269\mathbf{j} - 201\mathbf{k}\}$ lb
2–147. $\mathbf{F}_R = 178$ N
$\theta = 85.2°$
2–149. $r_{AB} = 17.0$ ft
$\mathbf{F} = \{-160\mathbf{i} - 180\mathbf{j} + 240\mathbf{k}\}$ lb

Chapter 3

3–1. $F_2 = 9.60$ kN
$F_1 = 1.83$ kN
3–2. $\theta = 4.69°$, $F_1 = 4.31$ kN
3–3. $F_{AC} = F_{AB} = \{2.45 \csc \theta\}$ kN
$l = 1.72$ m
3–5. $T = 13.3$ kN, $F = 10.2$ kN
3–6. $T = 14.3$ kN, $\theta = 36.3°$
3–7. $F_{BC} = 2.99$ kN, $F_{AB} = 3.78$ kN
3–9. $W = 76.6$ lb
3–10. $F_{CA} = 80.0$ N, $F_{CB} = 90.4$ N

3–11. $\theta = 64.3°$
$F_{CB} = 85.2$ N
$F_{CA} = 42.6$ N
3–13. $s = 5.33$ ft
3–14. $W = 6$ lb
3–15. $F_{BD} = 171$ N
$F_{BC} = 145$ N
3–17. $m = 2.37$ kg
3–18. $x_{AD} = 0.4905$ m
$x_{AC} = 0.793$ m
$x_{AB} = 0.467$ m
3–19. $m = 8.56$ kg
3–21. $d = 1.56$ m
3–22. $F_s = 2k(\sqrt{5 - 4\cos\theta} - 1)$,
$\theta = 35.0°$
3–23. $l' = \sqrt{12}$ ft; $\angle BCA = 30°$,
$l = 2.66$ ft
3–25. $F = 39.3$ lb
3–26. $d = 7.13$ in.
3–27. $k = 6.80$ lb/in.
3–29. $m = 20.4$ kg
3–30. $\tan\theta = dy/dx = 5.0(0.4)$
$m_B = 3.58$ kg
$N = 19.7$ N
3–31. $F_{ED} = 30.2$ lb
$F_{EB} = 43.6$ lb
$F_{BC} = 69.8$ lb
$F_{BA} = 86.6$ lb
3–33. $F_{ED} = 28.9$ lb
$F_{CE} = 28.9$ lb
$F_{CA} = 35.4$ lb
$F_{CD} = 10.6$ lb
$F_{DB} = 35.4$ lb
3–34. $W = 56.6$ lb
3–35. $(3.5 - x)/\cos\phi + x/\cos\phi = 5$
$x = 1.38$ m
$T = 687$ N
3–37. The attachment of the cable to point C and D will produce the least amount of tension in the cable.
$T = 106$ lb
3–38. $T = \{50 \sec\theta\}$ lb
3–39. $W_B = 18.3$ lb
3–43. $P = 1.61$ kN
$\alpha = 136°$
$\beta = 128°$
$\gamma = 72.0°$
3–45. $F_1 = 800$ N
$F_2 = 147$ N
$F_3 = 564$ N
3–46. $F_{DA} = 10.0$ lb
$F_{DB} = 1.11$ lb
$F_{DC} = 15.6$ lb
3–47. $s_{OB} = 327$ mm
$s_{OA} = 218$ mm
3–49. $F = 843$ N
3–50. $F_{AO} = 319$ N
$F_{AB} = 110$ N
$F_{AC} = 85.8$ N
3–51. $W = 138$ N
3–53. $F_{AB} = 520$ N
$F_{AC} = 260$ N
$F_{AD} = 260$ N
$d = 3.61$ m
3–54. $F_{AB} = 274$ lb, $F_{AC} = 295$ lb
$F_{AD} = 547$ lb
3–55. $F_{AD} = 557$ lb
$W = 407$ lb
3–57. $F_{AB} = 1.47$ kip
$F_{AC} = 0.914$ kip
$F_{AD} = 1.42$ kip
3–58. $F_{AB} = 79.2$ lb
$F_{AC} = 119$ lb
$F_{AD} = 283$ lb
3–59. $W_C = 265$ lb
3–61. $F_{AB} = 808$ lb
$F_{AC} = 538$ lb
$F_{AE} = 2.35$ kip
3–62. $F_{AE} = 2.91$ kip
$F = 1.61$ kip
3–63. $F_{AB} = F_{AC} = F_{AD} = 426$ N
3–65. $F_{AB} = 35.9$ lb
$F_{AC} = 25.4$ lb
$F_{AD} = 25.4$ lb
3–66. $W = 267$ lb
3–67. $h = 1.64$ ft
3–69. $F = 40.8$ lb
3–70. $y = 2.46$ ft
3–71. Yes, Romeo can climb up the rope.
Yes, Romeo and Juliet can climb down.
3–73. $\theta = 90°$ $F_{AC} = 160$ lb
$\theta = 120°$ $F_{AB} = 160$ lb
3–74. $F_{AB} = 175$ lb
$l = 2.34$ ft
3–75. $W = 240$ lb
3–77. $F_1 = 0$
$F_2 = 311$ lb
$F_3 = 238$ lb

Chapter 4

4–5. $(M_{F_1})_B = 4.125$ kip · ft (*Counterclockwise*)
$(M_{F_2})_B = 2.00$ kip · ft (*Counterclockwise*)
$(M_{F_3})_B = 40.0$ lb · ft (*Counterclockwise*)

4–6. $M_A = \{1.18 \cos \theta(7.5 + x)\}$ kN · m (*Clockwise*)
The maximum moment at A occurs when $\theta = 0°$ and $x = 5$ m.
$\circlearrowleft +(M_A)_{\max} = 14.7$ kN · m (*Clockwise*)

4–7. $\circlearrowleft +(M_{F_1})_A = 433$ N · m (*Clockwise*)
$\circlearrowleft +(M_{F_2})_A = 1.30$ kN · m (*Clockwise*)
$\circlearrowleft +(M_{F_3})_A = 800$ N · m (*Clockwise*)

4–9. $\circlearrowleft +M_B = 90.6$ lb · ft $\circlearrowleft$
$\circlearrowleft +M_C = 141$ lb · ft $\circlearrowleft$

4–10. $\circlearrowleft +M_A = 195$ lb · ft $\circlearrowleft$

4–11. $(M_R)_A = 2.08$ kN · m (*Counterclockwise*)

4–13. $(M_{F_A})_C = 162$ lb · ft $\circlearrowright$
$(M_{F_B})_C = 260$ lb · ft $\circlearrowleft$
The gate will rotate counterclockwise.

4–14. $F_A = 28.9$ lb

4–15. $(M_R)_A = 2.09$ N · m (*Clockwise*)

4–17. $\circlearrowleft +M_A = 0.418$ N · m (*Counterclockwise*)
$\circlearrowleft +M_B = 4.92$ N · m (*Clockwise*)

4–18. $(M_R)_A = 76.0$ kN · m (*Counterclockwise*)

4–19. $M_C = 4.97$ Mg

4–21. $F = 27.6$ lb

4–22. $M_A = 13.0$ N · m
$F = 35.2$ N

4–23. $\circlearrowleft +(M_O)_{max} = 80$ kN · m $\circlearrowright$
$x = 24.0$ m

4–25. $(M_{AB})_A = 3.88$ kip · ft (*Clockwise*)
$(M_{BCD})_A = 2.05$ kip · ft (*Clockwise*)
$(M_{man})_A = 2.10$ kip · ft (*Clockwise*)

4–26. $(M_R)_A = 8.04$ kip · ft (*Clockwise*)

4–27. $M_A = 100$ N · m (*Clockwise*)

4–29. $F = 239$ N

4–30. $\mathbf{M}_A = \{-16.0\mathbf{i} - 32.1\mathbf{k}\}$ N · m

4–31. $\mathbf{M}_O = \{-720\mathbf{i} + 120\mathbf{j} - 660\mathbf{k}\}$ N · m

4–33. $\mathbf{M}_O = [1080\mathbf{i} + 720\mathbf{j}]$ N · m

4–34. $\mathbf{M}_O = \{-720\mathbf{i} + 720\mathbf{j}\}$ N · m

4–35. $\theta_{max} = 90°$
$\theta_{min} = 0, 180°$

4–37. $\mathbf{M}_B = \{-37.6\mathbf{i} + 90.7\mathbf{j} - 155\mathbf{k}\}$ N · m

4–38. $\mathbf{b} = \mathbf{r}_{CA} \times \mathbf{r}_{CB} = 12\mathbf{i} + 9\mathbf{j} + 12\mathbf{k}$
$\mathbf{F} = F(\mathbf{b}/b) = \{249.88\mathbf{i} + 187.41\mathbf{j} + 249.88\mathbf{k}\}$ N
$\mathbf{M}_A = [1.56\mathbf{i} - 0.750\mathbf{j} - 1.00\mathbf{k}]$ kN · m

4–39. $\mathbf{b} = \mathbf{r}_{CA} \times \mathbf{r}_{CB} = 12\mathbf{i} + 9\mathbf{j} + 12\mathbf{k}$
$\mathbf{F} = F(\mathbf{b}/b) = \{249.88\mathbf{i} + 187.41\mathbf{j} + 249.88\mathbf{k}\}$ N
$\mathbf{M}_B = [1.00\mathbf{i} + 0.750\mathbf{j} - 1.56\mathbf{k}]$ kN · m

4–41. $\mathbf{M}_B = \{10.6\mathbf{i} + 13.1\mathbf{j} + 29.2\mathbf{k}\}$ N · m

4–42. $\mathbf{M}_O = \{373\mathbf{i} - 99.9\mathbf{j} + 173\mathbf{k}\}$ N · m
$\mathbf{r}_{AB} = -0.5 \sin 30°\mathbf{i} + (1 \cos 30° - (0.5 + 0.5 \cos 30°)\mathbf{j} + (1 \sin 30°)\mathbf{k}$

4–43. $\mathbf{M}_C = \{-35.4\mathbf{i} - 128\mathbf{j} - 222\mathbf{k}\}$ lb · ft

4–45. $y = 2$ m
$z = 1$ m

4–46. $y = 1$ m
$z = 3$ m
$d = 1.15$ m

4–47. $(M_{AB})_1 = 72.0$ N · m
$(M_{AB})_2 = 0$
$(M_{AB})_3 = 0$

4–49. $P = 8.50$ lb

4–50. $M_x = 44.4$ lb · ft

4–51. $\mathbf{M}_y = \{-78.4\mathbf{j}\}$ lb · ft
$M_y = 4.00$ lb · ft
$M_z = 36.0$ lb · ft

4–53. $\mathbf{M}_{AC} = \{11.5\mathbf{i} + 8.64\mathbf{j}\}$ lb · ft

4–54. $M_x = -21.7$ N · m

4–55. $F = 139$ N

4–57. $\mathbf{M}_x = \{77.1\mathbf{i}\}$ N · m
$\mathbf{M}_z = \{-91.9\mathbf{k}\}$ N · m

4–58. $M_{CD} = 432$ lb · ft

4–59. $F = 162$ lb

4–61. $M_x = -286$ lb · ft
$M_y = 165$ lb · ft
$M_z = 0$

4–62. $M_{OA} = 132$ lb · ft

4–63. $W = 56.8$ lb

4–65. $F_B = 192$ N
$F_A = 236$ N · m

4–66. $M_y = 282$ lb · ft

4–67. $F = 133$ N
$P = 800$ N

4–69. $F = 625$ N

4–70. $\circlearrowleft +(M_R)_C = 435$ lb · ft $\circlearrowleft$

4–71. $F = 139$ lb

4–73. $M_C = 22.5$ N · m $\circlearrowright$

4–74. $F = 83.3$ N

4–75. $M = 900$ lb · ft
$R_B = 500$ lb

4–77. $\circlearrowleft +(M_R)_C = 240$ lb · ft $\circlearrowleft$

4–78. $F = 167$ lb
Resultant couple can act anywhere.

4–79. $(M_c)_R = 260$ lb · ft (*Clockwise*)

4–81. $(M_c)_R = 5.20$ kN · m (*Clockwise*)

4–82. $F = 14.2$ kN · m

4–83. $M_C = \{-5\mathbf{i} + 8.75\mathbf{j}\}$ N · m

4–85. $M_R = \{52.0\mathbf{i} + 50\mathbf{j} + 30\mathbf{k}\}$ N·m
$M_R = 78.1$ N·m
$\alpha = 48.3°$
$\beta = 50.2°$
$\gamma = 67.4°$

4–86. $M_R = 59.9$ N·m
$\alpha = 99.0°$
$\beta = 106°$
$\gamma = 18.3°$

4–87. $\mathbf{M}_R = \{11.0\mathbf{i} - 49.0\mathbf{j} - 40.0\mathbf{k}\}$ lb·ft
$M_R = 64.2$ lb·ft
$\alpha = 80.1°$
$\beta = 140°$
$\gamma = 129°$

4–89. $\mathbf{M}_R = \{-12.1\mathbf{i} - 10.0\mathbf{j} - 17.3\mathbf{k}\}$ N·m

4–90. $d = 342$ mm

4–91. $M = 40.8$ N·m
$\alpha = 11.3°$
$\beta = 101°$
$\gamma = 90°$

4–93. $\mathbf{M}_C = \{7.01\mathbf{i} + 42.1\mathbf{j}\}$ N·m

4–94. $F = 35.1$ N

4–95. $(M_C)_R = 71.9$ N·m
$\alpha = 44.2°$
$\beta = 131°$
$\gamma = 103°$

4–97. $F_R = 2.10$ kN
$\theta = 81.6°$ $\measuredangle$
$M_O = 10.6$ kN·m ↻

4–98. $F_R = 2.10$ kN
$\theta = 81.6°$ $\measuredangle$
$M_P = 16.8$ kN·m ↻

4–99. $F_R = 5.93$ kN
$\theta = 77.8°$ $\measuredangle$
$M_{R_A} = 34.8$ kN·m (*Clockwise*)

4–101. $F_R = 416$ lb
$\theta = 35.2°$ $\measuredangle$
$(M_R)_A = 1.48$ kip·ft (*Clockwise*)

4–102. $F_R = 29.9$ lb, $\theta = 78.4°$ $\measuredangle$
$M_{R_O} = 214$ lb·in. ↺

4–103. $F_R = 26.4$ lb
$\theta = 85.7°$ $\measuredangle$
$M_{R_O} = 205$ lb·in. ↺

4–105. $F_R = 8.27$ kN
$\theta = 69.9°$ $\measuredangle$
$(M_R)_A = 9.77$ kN·m (*Clockwise*)

4–106. $F_R = 650$ N
$\theta = 72.0°$ $\measuredangle$
$(M_R)_A = 431$ N·m (*Counterclockwise*)

4–107. $\mathbf{F}_R = \{270\mathbf{k}\}$ N
$\mathbf{M}_{RO} = \{-2.22\mathbf{i}\}$ N·m

4–109. $\mathbf{F}_R = \{400\mathbf{i} + 300\mathbf{j} - 650\mathbf{k}\}$ N
$\mathbf{M}_{R_A} = \{-3100\mathbf{i} + 4800\mathbf{j}\}$ N·m

4–110. $\mathbf{F}_R = \{-40\mathbf{j} - 40\mathbf{k}\}$ N
$\mathbf{M}_{RA} = \{-12\mathbf{j} + 12\mathbf{k}\}$ N·m

4–111. $\mathbf{F}_R = \{-28.3\mathbf{j} - 68.3\mathbf{k}\}$ N
$\mathbf{M}_{RA} = \{-20.5\mathbf{j} + 8.49\mathbf{k}\}$ N·m

4–113. $F_R = 10.75$ kip ↓
$d = 13.7$ ft

4–114. $F_R = 10.75$ kip ↓
$d = 9.26$ ft

4–115. $F = 798$ lb
$\theta = 67.9°$ $\measuredangle$
$x = 7.43$ ft

4–117. $F = 1302$ N
$\theta = 84.5°$ $\measuredangle$
$x = 7.36$ m

4–118. $F = 1302$ N
$\theta = 84.5°$ $\measuredangle$
$x = 1.36$ m (to the right)

4–119. $F_R = 462$ lb
$\theta = 39.1°$ $\measuredangle$
$d = 3.07$ ft

4–121. $F_R = 991$ N
$\theta = 63.0°$ $\measuredangle$
$x = 2.64$ m

4–122. $F_R = 65.9$ lb
$\theta = 49.8°$ $\measuredangle$
$d = 2.10$ ft

4–123. $F_R = 65.9$ lb
$\theta = 49.8°$ $\measuredangle$
$d = 4.62$ ft

4–125. $F = 542$ N
$\theta = 10.6°$ $\measuredangle$
$d = 2.17$ m

4–126. $F = 922$ lb
$\theta = 77.5°$ $\measuredangle$
$x = 3.56$ ft

4–127. $F_C = 600$ N
$F_D = 500$ N

4–129. $F_B = 163$ lb
$F_C = 223$ lb

4–130. $F_R = 140$ kN
$y = 7.14$ m
$x = 5.71$ m

4–131. $F_R = 140$ kN
$x = 6.43$ m
$y = 7.29$ m

4–133. $F_A = 30$ kN
$F_B = 20$ kN
$F_R = 190$ kN

4–134. $\mathbf{F}_R = \{141\mathbf{i} + 100\mathbf{j} + 159\mathbf{k}\}$ N
$\mathbf{M}_{R_O} = \{122\mathbf{i} - 183\mathbf{k}\}$ N·m
4–135. $z = 0.586$ ft
$\mathbf{F}_w = \{-10\mathbf{j}\}$ lb
$\mathbf{M}_w = \{-14.1\mathbf{j}\}$ lb·ft
4–137. $F_R = 990$ N
$M_R = 3.07$ kN·m
$x = 1.16$ m
$y = 2.06$ m
4–138. $F_{RO} = 13.2$ lb ↓
$x = 0.340$ ft
4–139. $F_R = 6.75$ kN ↓
$\bar{x} = 2.5$ m
4–141. $F_R = 3.25$ kip,
$\theta = 67.2°$ ⦣
$x = 3.86$ ft
4–142. $F_R = 10.6$ kip ↓
$x = 0.479$ ft
4–143. $F_R = 0.525$ kN ↑
$d = 0.171$ m
4–145. $F_R = 18.0$ kip ↓
$x = 11.7$ ft
4–146. $F_R = \frac{1}{2}w_0L$ ↓
$\bar{x} = \frac{5}{12}L$
4–147. $b = 4.50$ ft
$a = 9.75$ ft
4–149. $w_1 = 660$ lb/ft
$w_2 = 720$ lb/ft
4–150. $F_R = 51.0$ kN ↓
$M_{R_O} = 914$ kN·m (*Clockwise*)
4–151. $F_R = 51.0$ kN ↓
$d = 17.9$ m
4–153. $F_R = 577$ lb
$\theta = 47.5°$ ⦣
$M_{RB} = 2.80$ kip·ft ↻
4–154. $F_R = 1.35$ kN
$\theta = 42.0°$ ⦣
$y = 0.1$ m
4–155. $F_R = 1.35$ kN
$x = 0.556$ m
4–157. $F_R = 36$ kN ↓
$\bar{x} = 2.17$ m
4–158. $F_R = 2.96$ kip
$\bar{x} = 9.21$ ft
4–159. $F_R = 107$ kN ←
$h = 1.60$ m
4–161. $F_R = 1.87$ kip ↓
$\bar{x} = 3.66$ ft
4–162. $F_R = 14.9$ kN
$\bar{x} = 2.27$ m
4–163. $\mathbf{M}_{CR} = \{63.6\mathbf{i} - 170\mathbf{j} + 264\mathbf{k}\}$ N·m
4–165. $\mathbf{M}_O = \{1.06\mathbf{i} + 1.06\mathbf{j} - 4.03\mathbf{k}\}$ N·m
$\alpha = 75.7°, \beta = 75.7°, \gamma = 160°$
4–166. $\mathbf{M}_{RP} = \{-26\mathbf{i} + 357\mathbf{j} + 127\mathbf{k}\}$ N·m
4–167. $\mathbf{F}_R = \{14.3\mathbf{i} + 21.4\mathbf{j} - 42.9\mathbf{k}\}$ lb,
$\mathbf{M}_A = \{-1.93\mathbf{i} + 0.429\mathbf{j} - 0.429\mathbf{k}\}$ kip·ft
4–169. $\mathbf{M}_O = \{298\mathbf{i} + 15.1\mathbf{j} - 200\mathbf{k}\}$ lb·in
4–170. $\mathbf{M}_A = \{-59.7\mathbf{i} - 159\mathbf{k}\}$ N·m
4–171. $M_{a-a} = 59.7$ N·m
4–173. $P = 23.8$ lb

Chapter 5

5–10. $N_B = 3.46$ kN
$A_x = 1.73$ kN
$A_y = 1.00$ kN
5–11. $B_y = 586$ N
$F_A = 413$ N
5–13. $B_y = 736$ N
$A_x = 920$ N
$B_x = 920$ N
5–14. $F_{BC} = 3.92$ kN
$A_x = 3.13$ kN
$A_y = 950$ N
5–15. $F_B = 67.6$ lb
$A_x = 25$ lb
$A_y = 111$ lb
5–17. $F_B = 105$ N
5–18. $T = 74.6$ lb
$A_x = 33.4$ lb
$A_y = 61.3$ lb
5–19. $B_x = 989$ N, $A_x = 989$ N, $B_y = 186$ N
5–21. $F_A = 30$ lb, $F_B = 36.2$ lb, $F_C = 9.38$ lb
5–22. $N_A = 250$ lb, $N_B = 9.18$ lb, $N_C = 141$ lb
5–23. $P_{max} = 210$ lb
5–25. $F_A = 2.06$ kip
5–26. $C_x = 333$ lb, $C_y = 722$ lb
5–27. $F_B = 3.68$ kN
$x_A = 53.2$ mm
$x_B = 50.5$ mm
5–29. $N_A = 1.85$ kip, $N_B = 1.15$ kip
5–30. $W = 5.34$ kip
5–31. $\theta = 26.4°$
5–33. $F_{BD} = 628$ N
$C_y = 68.2$ N, $C_x = 432$ N
5–34. $F_{CD} = 195$ lb, $A_x = 97.5$ lb, $A_y = 31.2$ lb

5–35. $l = 0.67664$ m, $\dfrac{\sin\theta}{0.3} = \dfrac{\sin 150^\circ}{0.67664}$, $F_s = 2.383$ N, $N_B = 2.11$ N, $F_A = 2.81$ N
5–37. $F_{CB} = 782$ N
$A_x = 625$ N
$A_y = 681$ N
5–38. $F_2 = 724$ N, $F_1 = 1.45$ kN, $F_A = 1.75$ kN
5–39. $N_B = 3.33$ kip
$A_y = 5.00$ kip
$A_x = 3.33$ kip
5–41. (a) $N_A = 2.19$ kip, $N_B = 1.16$ kip,
(b) $W = 4.74$ kip
5–42. $N_A = 975$ lb
$B_x = 975$ lb, $B_y = 780$ lb
5–43. $A_x = 1.46$ kip, $F_B = 1.66$ kip
5–45. $F_B = 6.38$ N, $A_x = 3.19$ N, $A_y = 2.48$ N
5–46. $d = 3a/4$
5–47. $\Delta_A = \dfrac{533.33}{5(10^3)}$, $\Delta_B = \dfrac{266.7}{5(10^3)}$, $\alpha = 1.02^\circ$
5–49. $P = 71.3$ N
$N = 446$ N
5–50. $\theta = \tan^{-1}\dfrac{b}{a}$
5–51. $\dfrac{L}{\Delta_B} = \dfrac{2L}{\Delta_C}$, $\dfrac{F_C}{k} = \dfrac{2F_B}{k}$, $F_B = 0.3P$
$F_C = 0.6P$
$x_C = 0.6P/k$
5–53. $T = \dfrac{W\cos\theta}{2\sin(\phi - \theta)}$

$A_x = \dfrac{W\cos\phi\cos\theta}{2\sin(\phi - \theta)}$

$A_y = \dfrac{W(\sin\phi\cos\theta - 2\cos\phi\sin\theta)}{2\sin(\phi - \theta)}$
5–54. $d = \dfrac{a}{\cos^3\theta}$
5–55. $N_A = 346$ N
$N_B = 693$ N
$a = 0.650$ m
5–57. $w_1 = 83.3$ lb/ft, $w_2 = 167$ lb/ft
5–58. $w_1 = \dfrac{2P}{L}$

$w_2 = \dfrac{4P}{L}$
5–59. $a = \sqrt{(4r^2\,l)^{\frac{2}{3}} - 4r^2}$
5–61. $h = \sqrt{\dfrac{s^2 - l^2}{3}}$
5–62. $T = 1.84$ kN, $F = 6.18$ kN
5–63. $N_A = 28.6$ lb
$N_B = 10.7$ lb, $N_C = 10.7$ lb
5–65. $R_D = 22.6$ kip, $R_E = 22.6$ kip, $R_F = 13.7$ kip
5–66. $W = 750$ lb, $x = 5.20$ ft, $y = 5.27$ ft
5–67. $F_A = 663$ lb, $F_C = 569$ lb, $F_B = 449$ lb
5–69. $P = 75$ lb, $A_y = 0$, $B_z = 75$ lb, $A_z = 75$ lb
$B_x = 112$ lb, $A_x = 37.5$ lb
5–70. $A_z = 100$ lb
$B_x = 0$, $B_y = 37.5$ lb
$A_x = 0$
$A_y = 37.5$ lb
5–71. $N_B = 1000$ N
$(M_A)_z = 0$
$A_y = 0$
$(M_A)_y = -560$ N·m
$A_z = 400$ N
5–73. $A_x = 0$, $A_y = 1.50$ kip, $A_z = 750$ lb, $T = 919$ lb
5–74. $F = 1.31$ kip, $A_x = 0$, $A_y = 1.31$ kip, $A_z = 653$ lb
5–75. $A_y = 0$
$T = 1.23$ kN
$B_x = 433$ N
$B_z = 1.42$ kN
$A_x = 867$ N
$A_z = 711$ N
5–77. $P = 100$ lb, $B_z = 40$ lb, $B_x = -35.7$ lb,
$A_x = 136$ lb, $A_z = 40$ lb, $B_y = 0$
5–78. $F_{BC} = 0$, $A_y = 0$, $A_z = 800$ lb,
$(M_A)_x = 4.80$ kip·ft, $(M_A)_y = 0$, $(M_A)_z = 0$
5–79. $A_x = 633$ lb, $A_y = -141$ lb, $B_x = -721$ lb
$B_z = 895$ lb, $C_y = 200$ lb, $C_z = -506$ lb
5–81. $F_{BD} = 294$ N
$F_{BC} = 589$ N
$A_x = 0$
$A_y = 589$ N
$A_z = 490.5$ N
5–82. $T_{DE} = 32.1$ lb, $T_{BC} = 42.9$ lb, $A_x = 3.57$ lb,
$A_y = 50$ lb, $(M_A)_x = 0$, $(M_A)_y = -17.9$ lb·ft
5–83. $T = 58.0$ N, $C_z = 87.0$ N, $C_y = 28.8$ N, $D_x = 0$,
$D_y = 79.2$ N, $D_z = 58.0$ N
5–85. $F = 354$ N
5–86. $N_A = 8.00$ kN, $B_x = 5.20$ kN, $B_y = 5.00$ kN
5–87. $N_B = 400$ N, $F_A = 721$ N
5–89. $\theta = \tan^{-1}\left(\dfrac{1}{2}\cot\psi - \dfrac{1}{2}\cot\phi\right)$
5–90. $A_x = 0$, $A_y = 0$, $A_z = 5.33$ lb, $B_z = 5.33$ lb,
$C_z = 5.33$ lb
5–91. $A_x = 0$,
$A_y = -200$ N, $A_z = 150$ N, $(M_A)_x = 100$ N·m
$(M_A)_y = 0$, $(M_A)_z = 500$ N·m

Chapter 6

6–1. $F_{AB} = 286$ lb (T)
$F_{AC} = 571$ lb (C)
$F_{BC} = 808$ lb (T)

6–2. $F_{AB} = 286$ lb (T)
$F_{AC} = 271$ lb (C)
$F_{BC} = 384$ lb (T)

6–3. $F_{CD} = 5.21$ kN (C)
$F_{CB} = 4.17$ kN (T), $F_{AD} = 1.46$ kN (C)
$F_{AB} = 4.17$ kN (T), $F_{BD} = 4$ kN (T)

6–5. $F_{DC} = 400$ N (C)
$F_{DA} = 300$ N (C)
$F_{BA} = 250$ N (T)
$F_{BC} = 200$ N (T)
$F_{CA} = 283$ N (C)

6–6. $F_{DE} = 1.00$ kN (C)
$F_{DC} = 800$ N (T)
$F_{CE} = 900$ N (C)
$F_{CB} = 800$ N (T)
$F_{EB} = 750$ N (T)
$F_{EA} = 1.75$ kN (C)

6–7. $F_{JD} = 33.3$ kN (T)
$F_{AL} = F_{GH} = F_{LK} = F_{HI} = 28.3$ kN (C)
$F_{AB} = F_{GF} = F_{BC} = F_{FE} = F_{CD} = F_{ED} = 20$ kN (T)
$F_{BL} = F_{FH} = F_{LC} = F_{HE} = 0$
$F_{CK} = F_{EI} = 10$ kN (T)
$F_{KJ} = F_{IJ} = 23.6$ kN (C)
$F_{KD} = F_{ID} = 7.45$ kN (C)

6–9. $F_{AB} = 7.5$ kN (T), $F_{AE} = 4.5$ kN (C)
$F_{ED} = 4.5$ kN (C), $F_{EB} = 8$ kN (T),
$F_{BD} = 19.8$ kN (C), $F_{BC} = 18.5$ kN (T)

6–10. $F_{AB} = 196$ N (T), $F_{AE} = 118$ N (C)
$F_{ED} = 118$ N (C), $F_{EB} = 216$ N (T)
$F_{BD} = 1.04$ kN (C), $F_{BC} = 857$ N (T)

6–11. $F_{DE} = 16.3$ kN (C), $F_{DC} = 8.40$ kN (T)
$F_{EA} = 8.85$ kN (C), $F_{EC} = 6.20$ kN (C)
$F_{CF} = 8.77$ kN (T), $F_{CB} = 2.20$ kN (T)
$F_{BA} = 3.11$ kN (T), $F_{BF} = 6.20$ kN (C)
$F_{FA} = 6.20$ kN (T)

6–13. $F_{GB} = 30$ kN (T),
$F_{AF} = 20$ kN (C), $F_{AB} = 22.4$ kN (C)
$F_{BF} = 20$ kN (T), $F_{BC} = 20$ kN (T)
$F_{FC} = 28.3$ kN (C), $F_{FE} = 0$
$F_{ED} = 0$, $F_{EC} = 20.0$ kN (T)
$F_{DC} = 0$

6–14. $F_{AB} = 330$ lb (C), $F_{AF} = 79.4$ lb (T)
$F_{BF} = 233$ lb (T), $F_{BC} = 233$ lb (C)
$F_{FC} = 47.1$ lb (C), $F_{FE} = 113$ lb (T)
$F_{EC} = 300$ lb (T), $F_{ED} = 113$ lb (T)
$F_{CD} = 377$ lb (C)

6–15. $F_{AB} = 377$ lb (C), $F_{AF} = 190$ lb (T)
$F_{BF} = 267$ lb (T), $F_{BC} = 267$ lb (C)
$F_{FC} = 189$ lb (T), $F_{FE} = 56.4$ lb (T)
$F_{ED} = 56.4$ lb (T), $F_{EC} = 0$, $F_{CD} = 189$ lb (C)

6–17. Maximum force:
$F_{DC} = F_{CB} = F_{CE} = F_{BE} = F_{BA} = 1.1547P$
$P = 5.20$ kN

6–18. $F_{CD} = 3.61$ kN (C), $F_{CB} = 3$ kN (T)
$F_{BA} = 3$ kN (T), $F_{BD} = 3$ kN (C)
$F_{DA} = 2.70$ kN (T), $F_{DE} = 6.31$ kN (C)

6–19. $F_{CD} = 467$ N (C), $F_{CB} = 389$ N (T)
$F_{BA} = 389$ N (T), $F_{BD} = 314$ N (C)
$F_{DE} = 1.20$ kN (C), $F_{DA} = 736$ N (T)

6–21. $P = 1.25$ kN

6–22. $F_{FE} = 0.667P$ (T), $F_{FD} = 1.67P$ (T)
$F_{AB} = 0.471P$ (C), $F_{AE} = 1.67P$ (T)
$F_{AC} = 1.49P$ (C), $F_{BF} = 1.41P$ (T),
$F_{BD} = 1.49P$ (C), $F_{EC} = 1.41P$ (T),
$F_{CD} = 0.471P$ (C)

6–23. $F_{CD} = 0.577P$ (C), $F_{DB} = 0.289P$ (T)
$F_{CE} = 0.577P$ (T), $F_{BC} = 0.577P$ (C)
$F_{BE} = 0.577P$ (T), $F_{AB} = 0.577P$ (C)
$F_{AE} = 0.577P$ (T)

6–25. $F_{BA} = P \csc 2\theta$ (C), $F_{BC} = P \cot 2\theta$ (C)
$F_{CA} = (\cot\theta \cos\theta - \sin\theta + 2\cos\theta)P$ (T)
$F_{CD} = (\cot 2\theta + 1)P$ (C)
$F_{DA} = (\cot 2\theta + 1)(\cos 2\theta)(P)$ (C)

6–26. Maximum force: $F_{CA} = 2.732P$ (T),
$F_{CD} = 1.577P$ (C), $P_{max} = 732$ N

6–27. $F_{HG} = 1125$ lb (T), $F_{DE} = 3375$ lb (C)
$F_{EH} = 3750$ lb (T)

6–29. $F_{GF} = 671$ lb (C), $F_{GB} = 671$ lb (T)

6–30. $F_{EF} = 729$ lb (C), $F_{FC} = 76.9$ lb (T)
$F_{EC} = 729$ lb (T)

6–31. $F_{KJ} = 11.2$ kip (T), $F_{CD} = 9.38$ kip (C),
$F_{CJ} = 3.12$ kip (C), $F_{DJ} = 0$

6–33. $F_{GJ} = 2.00$ kip (C)

6–34. $F_{GJ} = 2.00$ kip (C), $F_{GC} = 1.00$ kip (T)

6–35. $F_{BC} = 10.4$ kN (C), $F_{HG} = 9.15$ kN (T),
$F_{HC} = 2.24$ kN (T)

6–37. $F_{GF} = 1.80$ kip (C), $F_{FB} = 693$ lb (T)
$F_{BC} = 1.21$ kip (T)

6–38. $F_{FE} = 1.80$ kip (C)
$F_{EC} = 693$ lb (C)

6–39. *AB, BC, CD, DE, HI,* and *GI* are all zero-force members.
$F_{JE} = 9.38$ kN (C), $F_{GF} = 5.625$ kN (T)

6–41. $F_{CB} = 3.60$ kN (T), $F_{GC} = 1.80$ kN (C)
$F_{FG} = 4.02$ kN (C)

6–42. $F_{LK} = 10.0$ kip (C), $F_{LC} = 2.50$ kip (C)
$F_{BC} = 10.0$ kip (T)

6–43. $F_{JI} = 7333$ lb(C), $F_{DE} = 9000$ lb(T)
$F_{JE} = 3005$ lb (C)

6–45. $F_{AB} = P$ (T), $F_{EF} = P$ (C), $F_{BF} = 1.41P$ (C)

6–46. ***Method of Joints***: By inspection, members *BN, NC, DO, OC, HJ, LE* and *JG* are zero-force members.
$F_{CD} = 5.625$ kN (T), $F_{CM} = 2.00$ kN (T)

6–47. ***Method of Joints***: By inspection, members *BN, NC, DO, OC, HJ, LE* and *JG* are zero-force members.
$F_{EF} = 7.88$ kN(T), $F_{LK} = 9.25$ kN(C)
$F_{ED} = 1.94$ kN(T)

6–49. $F_{BG} = \{-600 \csc \theta\}$ N
$F_{BC} = -200L$ N
$F_{HG} = 400L$ N

6–50. $F_{CA} = 122$ lb(C), $F_{CD} = 173$ lb(T)
$F_{BD} = 86.6$ lb(T), $F_{BA} = 0$
$F_{DA} = 86.6$ lb (T)

6–51. $F_{AB} = 6.46$ kN(T), $F_{AC} = F_{AD} = 1.50$ kN(C)
$F_{BC} = F_{BD} = 3.70$ kN(C), $F_{BE} = 4.80$ kN(T)

6–53. $F_{BD} = 896$N (C), $F_{DC} = 554$N (T)
$F_{DA} = 146$N (C), $F_{AB} = 52.1$N (T)
$F_{AC} = 31.25$N (T), $F_{CB} = 448$N (C)

6–54. $F_{DB} = 474$N (C), $F_{DC} = 146$N(T)
$F_{DA} = 1.08$kN (T), $F_{AB} = 385$N (C)
$F_{AC} = 231$N (C), $F_{CB} = 281$N (T)

6–55. $F_{BC} = F_{BD} = 1.34$ kN (C), $F_{AB} = 2.4$ kN (C)
$F_{AG} = F_{AE} = 1.01$ kN (T), $F_{BG} = 1.80$ kN (T)
$F_{BE} = 1.80$ kN (T)

6–57. $F_{CE} = 721$N (T), $F_{BC} = 400$N (C)
$F_{BE} = 0$, $F_{BF} = 2.10$ kN (T)

6–58. $F_{AB} = 1.50$ kN (C), $F_{AF} = 1.08$ kN (C)
$F_{AD} = 600$ N (T), $F_{FD} = 0$, $F_{ED} = 1.40$ kN (C)
$F_{BD} = 361$ N (C)

6–59. $F_{BF} = 0$, $F_{BC} = 0$, $F_{BE} = 500$ lb (T)
$F_{AB} = 300$ lb (C), $F_{AC} = 583$ lb (T)
$F_{AD} = 333$ lb (T), $F_{AE} = 667$ lb (C)
$F_{DE} = 0$, $F_{EF} = 300$ lb (C)
$F_{CD} = 300$ lb (C), $F_{CF} = 300$ lb (C)
$F_{DF} = 424$ lb (T)

6–61. **a.** $P = 25.0$ lb
b. $P = 33.3$ lb
c. $P = 11.1$ lb

6–62. $T = 100$ lb, $\theta = 14.6°$

6–63. $P = 40.0$ N, $x = 240$ mm

6–65. $C_x = 75$ lb, $C_y = 100$ lb

6–66. $A_x = 4.20$ kN, $B_x = 4.20$ kN, $A_y = 4.00$ kN
$B_y = 3.20$ kN, $C_x = 3.40$ kN, $C_y = 4.00$ kN

6–67. $C_x = 160$ lb, $C_y = 107$ lb, $B_y = 26.7$ lb
$B_x = 80.0$ lb, $E_x = 0$, $E_y = 26.7$ lb,
$A_x = 160$ lb

6–69. $A_x = 120$ lb, $A_y = 0$, $N_C = 15.0$ lb

6–70. $C_y = 34.4$ lb
$C_x = 16.7$ lb
$B_x = 66.7$ lb
$B_y = 15.6$ lb

6–71. $N_E = 5$ kN
$D_x = 0$
$N_C = 16.7$ kN
$A_x = 0$
$A_y = 2.67$ kN
$M_A = 21.5$ kN·m

6–73. $C_y = 0$
$B_y = 7.50$ kN
$M_A = 45.0$ kN·m
$A_y = -7.50$ kN
$A_x = 0$

6–74. $T = 350$ lb
$A_y = 700$ lb
$A_x = 1.88$ kip
$D_x = 1.70$ kip
$D_y = 1.70$ kip

6–75. $T = 350$ lb
$A_y = 700$ lb
$D_x = 1.82$ kip
$D_y = 1.84$ kip
$A_x = 2.00$ kip

6–77. $m = 366$ kg, $F_A = 2.93$ kN

6–78. $N_A = 3.67$ kN
$M_A = 5.55$ kN·m
$C_x = 2.89$ kN
$C_y = 1.32$ kN

6–79. $F_E = 3.64F$

6–81. $F_{AB} = 9.23$ kN, $C_x = 2.17$ kN, $C_y = 7.01$ kN
$D_x = 0$, $D_y = 1.96$ kN, $M_D = 2.66$ kN·m

6–82. $T_{AI} = 2.88$ kip, $F_H = 3.99$ kip

6–83. $N = 41.7$ lb
This normal force **does not** stop the wheel from turning. A frictional force (See Chapter 8), which acts along on the wheel's rim stops the wheel.

6–85. $N_A = 130$ N

6–86. $N_B = N_C = 49.5$ N

6–87. $1.75 \text{ ft} \leq x \leq 17.4 \text{ ft}$

6–89. $m = 1.11$ Mg

6–90. $F_C = 19.6$ kN

6–91. $M = 314$ lb·ft

6–93. $N_A = 490.5$N, $N_B = 294.3$N, $T = 353.7$ N,
$\theta = 33.7°$, $x = 177$ mm

6–94. **a.** $F = 175$ lb
$N_C = 350$ lb
b. $F = 87.5$ lb
$N_C = 87.5$ lb

6–95. **a.** $F = 205$ lb
$N_C = 380$ lb
b. $F = 102$ lb
$N_C = 72.5$ lb

6–97. $F = 9.42$ lb

6–98. $N_C = 20$ lb, $B_x = 34$ lb, $B_y = 62$ lb, $A_x = 34$ lb
$A_y = 12$ lb, $M_A = 336$ lb · ft

6–99. $F = 370$ N

6–101. $F = 120$ lb

6–102. $F_{AB} = 981$ N, $F_E = 2.64$ kN, $F_{CD} = 16.3$ kN
$F_F = 14.0$ kN

6–103. $A_x = 0$
$A_y = 175$ lb
$B_x = 200$ lb
$C_x = 0$
$C_y = 200$ lb

6–105. $p = 3000$ psi

6–106. $F = 66.1$ lb

6–107. $F_{sp} = 150(2d - 1)$, $F_{CD} = (150d - 75)d$,
$d = 0.638$ ft

6–109. $E_x = 6.79$ kN, $E_y = 1.55$ kN
$D_x = 981$ N, $D_y = 981$ N

6–110. $\theta = 6.38°$

6–111. $\theta = \sin^{-1}\left(\frac{8W}{kL}\right)$

6–113. $T_{AB} = 300$ lb

6–114. $T = W$

6–115. $F_D = 20.8$ lb, $F_F = 14.7$ lb, $F_A = 24.5$ lb

6–117. $F_{ED} = 270$ lb, $B_z = 0$,
$B_x = -30$ lb, $B_y = -13.3$ lb

6–118. $M_{Cx} = 0$, $C_x = 0$, $F_{BA} = 1.54$ kip
$C_z = -0.18$ kip, $C_y = -1.17$ kip
$M_{Cz} = -4.14$ kip · ft, $A_x = 0$, $A_y = 1.44$ kip
$A_z = 0.540$ kip

6–119. $F_{BE} = 1.53$ kip
$F_{CD} = 350$ lb

6–121. $A_y = 250$ N, $A_x = 1.40$ kN, $C_x = 500$ N,
$C_y = 1.70$ kN

6–122. $F_{AD} = 2.47$ kip (T), $F_{AC} = F_{AB} = 1.22$ kip (C)

6–123. $m = 3.86$ kg

6–125. $A_x = 117$ N, $A_y = 397$ N, $B_x = 97.4$ N, $B_y = 97.4$ N

6–126. $F_{AG} = 471$ lb (C)
$F_{AB} = 333$ lb (T), $F_{BC} = 333$ lb (T)
$F_{GB} = 0$, $F_{DE} = 943$ lb (C)
$F_{DC} = 667$ lb (T), $F_{EC} = 667$ lb (T)
$F_{EG} = 667$ lb (C), $F_{GC} = 471$ lb (T)

Chapter 7

7.1. a) $M = 3688$ ft-lb
$V = 737.5$ lb
b) $M = 6850$ ft-lb
$V = -262.5$ lb
c) $M = 1312.5$ ft-lb
$V = -262.5$ lb

7.2. $M_{max} = 30{,}000$ N-m
$V_{max} = -6000$ N

7.3. $0 < x < 10$
$V = 453$ lb
$M = 453x$ ft-lb
etc

7.4. $10 < x < 25$
$V = -200 - 30x$ lb
$M = -15x^2 - 200x + 1000$ ft-lb
etc

7.5. $16 < x < 20$
$V = -1510$ N
$M = 4290x - 5000(x - 3) - 800(x - 9) + 8000$ N-m
etc

7.6. —

7.7. $8 \le x < 13$
$V = 24.4$ lb
$M = -100x - 40(x - 4) + 164.4(x - 5)$ ft-lb
etc

7.8. $9 \le x < 12$
$V = -650$ N
$M = 250x - 800(x - 4.5)$ N-m
etc

7.9. $3 \le x < 6$
$V = 750 - \frac{250}{3}(x - 3)^2$ N
$M = 750x - 6750 - \frac{250}{9}(x - 3)^3$ N-m
etc

7.10. $14.14' \le s < 29.14'$
$V = 52.5$ lb
$M = 52.5s - 218$ ft-lb
etc

7.11. $\frac{\pi}{4} < \theta < \frac{\pi}{2}$
$V = -29.3 \sin\theta$ lb
$M = 176 \cos\theta$ ft-lb
etc

7.12. $M_y = \frac{1}{8}x^3 - 15x^2 + 400x$ N-m

etc

7.13. $M = -.85x^2 + 34x - 350$ ft-lb

etc

7.14. $10 \le x \le 12$

$V_y = -1000$ N

$H = 0$

$M_x = -100$ N-m

$M_z = -1000x$ N-m

etc

7.15. $M_x = -5.68s^2 + 262s - 2194$ ft-lb

$M_y = 410$ ft-lb

etc

7.16. —

7.17. $22 < x < 30$

$V = -46.7$ lb

$M = 453.3x - 800(x - 10) + 300(x - 22)$ ft-lb

etc

7.18. $10 \le x < 25$

$V = -500 - 30(x - 10)$ lb

$M = -500(x - 5) - \frac{30(x - 10)^2}{2}$ ft-lb

etc

7.19. $13 \le x < 16$

$V = -1510$ N

$M = 4290x - 5000(x - 3) - 800(x - 9)$ N-m

etc

7.20. —

7.21. $13 < x < 23$

$V = 24.4$ lb

$M = -100x - 40(x - 4) + 164.4(x - 5) + 100$ ft-lb

etc

7.22. $12 < x < 15$

$V = -1650$ N

$M = 250x - 900(x - 4.5) - 1000(x - 12)$ N-m

etc

7.23. $6 \le x < 8$

$V = 0$

$M = -6750 + 750x - 750(x - 5)$ N-m

etc

7.24. $15 < x < 20$

$V = -900$ lb

$M = 600x - 500(x - 5) - 1000(x - 15)$ ft-lb

etc

7.25. $6 < x \le 10$

$V = 1000$ N

$M = 2000x - 23{,}000 - 1000(x - 3) + 5000$ N-m

etc

7.26. —

7.27. $10 < x < 15$

$V = 50x + 250$

$M = 25x^2 + 250x - 9375$ ft-lb

etc

7.28. $13 < x < 17$

$V = 0$ N

$M = -43{,}329 + 9333x + 10{,}000(x - 5) + 667(x - 13)$ N-m

etc

7.29. $30 < x < 40$

$V = -262.5$ lb

$M = 737.5x - 1000(x - 10) + 500$ ft-lb

$M_{max} = 5439$ ft-lb

etc

7.30. $8 < x \le 10$

$V = 112.5$ N

$M = -8900 + 1112.5x - 1000(x - 8)$ N-m

etc

7.31. $.8 \le x < 1.4$

$V = 15 - 50(x - .8)$ N

$M = -1112 + 1015x - 1000(x - .4) + 700 - 25(x - .8)^2$ N-m

etc

7.32. —

7.33. —

7.34. $R_1 = 20$ K

$R_2 = 10$ K

etc

7.35. $\bar{x} = 13.6$ ft

$M(\bar{x}) = 43{,}000$ ft-lb

7.36. —

7.37. $M_{max} = \frac{400}{\pi^2}$ ft-lb

7.38. —

7.39. —

7.40. —

7.41. —

7.42. —

7.43. $M_{max} = 121.65$ kN-m

7.44. $b \approx 35.4$ ft

7.45. $a = 11.13$ ft

7.46. $\bar{x} = .707h$

7.47. $|M_{max}| = 500$ N-m

7.48. —

7.49. —

7.50. —

7.51. —

7.52. —

7.53. $10 \leq x < 19$

$V = -175.3$ lb

$M = 77.47 - 450\,(x - 7.5) - 500\,(x - 11.67)$ ft-lb

etc

7.54. $7 < x \leq 13.2$

$V = 2480 - 64.5\,(x - 7)^2$ N

$M = -32{,}911 - 3580x - 600\,(x - 3) - 500\,(x - 7) - 21.5\,(x - 7)^3$ N-m

etc

7.55. $5 < x < 10$

$V = -930$ N

$M = 1070x - 1000\,(x - 2.5) + 1800 - 1000\,(x - 5)$ N-m

etc

7.56. —

7.57. $0 \leq x \leq 8$

$H = -50x$ lb

etc

7.58. $90 < x < 120$

$V = 0$

$M = -60{,}000$ ft-lb

7.59. $0 < x < 5$

$V = -.596x^{3/2}$kN

$M = -.238x^{5/2}$kN-m

etc

7.60. $0 < x < 3$

$V = 60 - 5x^2$ N

$M = 60x - \frac{5}{3}x^3$ N-m

etc

7.61. $6 < x < 11$

$V = 1000$ N.

$M = 1000x - 11{,}000$ N-m.

etc.

7.62. —

7.63. —

7.64. —

7.65. —

Chapter 8

8–1. $N_A = 16.5$ kN, $N_B = 42.3$ kN, the **mine car does not move.**

8–2. $P = 12.8$ kN

8–3. $T = 3.67$ kip

8–5. The ladder will not slip.

8–6. $\theta = 46.4°$

8–7. **a.** No
b. Yes

8–9. $P = \dfrac{M_0}{\mu_s ra}(b - \mu_s c)$

8–10. $\mu_s \geq \dfrac{b}{c}$

8–11. $P = \dfrac{M_0}{\mu_s ra}(b + \mu_s c)$

8–13. $P = 147$ N

8–14. **a.** $W = 318$ lb,
b. $W = 360$ lb

8–15. $\theta = 21.8°$

8–17. $F = 30.4$ lb, $\mu_m = 0.195$

8–18. $\mu = 0.354$

8–19. $d = 537$ mm

8–21. $\theta = 2\tan^{-1}\mu_s$

8–22. Slip at clamp, $n = 9.17$, slip between end boards; $n = 8.11$, use $n = 8$.

8–23. $P = 182$ N

8–25. $F_f = 10$ lb

8–26. $F_f = 70.7$ lb

8–27. $P = \dfrac{0.3W}{\cos\theta + 0.3\sin\theta}$, set $dP/d\theta = 0$, $\theta = 16.7°$,
$P = 0.287\,W$

8–29. $\theta = 11.0°$

8–30. $\mu_s = 0.577$

8–31. $\theta = 31.0°$

8–33. $\mu_s = 0.268$

8–34. Thus, **he can move the crate.**

8–35. $\mu_s' = 0.376$

8–37. $A_y = 474$ lb
$B_x = 36.0$ lb
$B_y = 232$ lb

8–38. $F_D = 36.9$ lb, $A_y = 468$ lb, $B_x = 34.6$ lb
$B_y = 228$ lb

8–39. $\theta = 7.50°$
$T = 452$ N

8–41. $\theta = 8.53°$, $F_A = 1.48$ lb, $F_B = 0.890$ lb

8–42. $\mu = 0.176$

8–43. $\mu_s = 0.509$

8–45. If A and B move together, $P = 69.3$ lb. If only block A moves, then $P = 63.5$ lb.
8–46. $P = 355$ N
8–47. $\mu_C = 0.0734$
$\mu_B = 0.0964$
8–49. $P = 60$ lb
8–50. $P = 90$ lb
8–53. $P = 13.3$ lb
8–54. $P = \frac{1}{2}\mu_s W$
8–55. $\theta = 33.4°$
8–57. $\mathbf{N} = (-a\mathbf{j} + h\mathbf{k}) \times (2a\mathbf{i} - a\mathbf{j}), n = \mathbf{N}/N,$
$h = \frac{2}{\sqrt{5}}a\mu$
8–58. $\theta = 33.4°$
8–59. $P = 5.53$ kN
Since a force P (> 0) is required to pull out the wedge, **the wedge will be self-locking when** $P = 0$.
8–61. $P = 34.5$ N
8–62. $x = 18.3$ mm
8–63. $P = 2.39$ kN
8–65. $P = 304$ N
8–66. $x = 32.9$ mm
8–67. $P = 617$ lb
8–69. $W = 7.19$ kN
8–70. Since $\phi_s > \theta_p$, **the screw is self-locking.**
8–71. $F = 66.7$ N
8–73. $F_{AB} = 1.38$ kN (T), $F_{BD} = 828$ N (C)
$F_{BC} = 1.10$ kN (C), $F_{AC} = 828$ N (C)
$F_{AD} = 1.10$ kN (C), $F_{CD} = 1.38$ kN (T)
8–74. $M = 4.53$ N·m
8–75. $M = 48.3$ N·m
8–77. $F_E = 72.7$ N
$F_D = 72.7$ N
8–78. $P = 880$ N
$M = 352$ N·m
8–79. $T = 4.02$ kN
$F = 11.6$ kN
8–81. $P = 1.98$ kN
8–82. $M = 0.202$ N·m
8–83. **a.** $F = 1.31$ kN
b. $F = 372$ N
8–85. Use $n = 2$ turns.
8–86. $P = 1.54$ kN
8–87. $\mu_s = 0.0583$
8–89. Approx. 2 turns (695°)
8–90. $\theta = 99.2°$
8–91. $T_1 = 73.3$ lb
8–93. $P = 17.1$ lb
8–94. $W = 9.17$ lb
8–95. $P = 78.7$ lb
8–97. $P = 736$ N
8–99. $M = 187$ N·m
8–101. $m_C = 136$ kg, $M = 134$ N·m
8–102. $M = 3.37$ N·m
8–103. $F = 2.49$ kN
8–105. $M = 50.0$ N·m, $x = 286$ mm
8–106. $F_s = 85.4$ N
8–107. $\mu_s = 0.145$
8–109. $F = 10.7$ lb
8–110. $M = 46.7$ N·m
8–111. $\mu_s = 0.321$
8–113. $P = 118$ N
8–114. $M = \frac{2\mu_s PR}{3\cos\theta}$
8–115. $M = 0.521\,P\mu R$
8–117. $d = 0.140\,L$
8–118. $P = 13.8$ lb
8–119. $P = 29.0$ lb
8–121. $F = 6$ lb, $\mu_k = 0.696$
8–122. $T = 289$ lb, $N = 479$ lb, $F = 101$ lb
8–123. $\mu_s = 0.0407$
8–125. $r = 20.6$ mm
8–126. $P = 2.45$ lb
8–127. $P = 1.88$ lb
8–129. $P = 1333$ lb
8–130. $P = 245$ N
8–133. $d = 4.60$ ft
8–134. $n = 30, M = 150$ lb·ft
8–135. $M = 270$ N·m
8–137. $P = 625$ lb
8–138. Check if crate slips on dolly, if crate tips, or if dolly tips. $P = 196$ N.
8–139. $P = 15$ lb
8–141. $M = 145$ lb·ft
8–142. $P = 140$ N
8–143. $P = 474$ N

Chapter 9

9–1. $\bar{x} = 124$ mm
$\bar{y} = 0$
9–2. $\bar{x} = \frac{4}{\pi}$ ft, $A_x = B_x = 1$ lb, $A_y = 3.14$ lb
9–3. $\bar{x} = 0.531$ ft, $O_x = 0$, $O_y = 0.574$ lb,
$M_O = 0.305$ lb·ft
9–5. $\bar{x} = 0.299a$
$\bar{y} = 0.537a$
9–6. $\bar{y} = \frac{2}{5}$ m
9–7. $\bar{x} = \frac{3}{8}a$
9–8. $\bar{y} = \frac{3h}{5}$
9–9. $\bar{x} = 5$ in.
9–10. $\bar{y} = 1.43$ in.

9–11. $\bar{x} = \frac{3b}{4}$

9–13. $\bar{x} = 6$ m

9–14. $\bar{y} = 2.8$ m

9–15. $\bar{x} = \frac{b - a}{\ln\frac{b}{a}}$

9–17. $\bar{x} = \frac{n + 1}{2(n + 2)}a$

9–18. $\bar{x} = \frac{a(1 + n)}{2(2 + n)}$

9–19. $\bar{y} = \frac{hn}{2n + 1}$

9–21. $\bar{x} = \frac{3a}{8}$

9–22. $\bar{y} = \frac{3a}{5}$

9–23. $\bar{x} = \frac{4a}{3\pi}$

9–25. $\bar{x} = 3.20$ ft
$\bar{y} = 3.20$ ft
$T_A = 384$ lb
$T_C = 384$ lb
$T_B = 1.15$ kip

9–26. $\bar{x} = \left(\frac{\pi - 2}{2\pi}\right)a$

9–27. $\bar{y} = \frac{\pi}{8}a$

9–29. $\bar{y} = \frac{25}{56}$ m

9–30. $\bar{x} = 1$ m

9–31. $\bar{y} = 0.4$ m

9–33. $\bar{y} = \frac{\pi a}{8}$

9–34. $\bar{x} = 1.26$ m
$\bar{y} = 0.143$ m
$N_B = 47.9$ kN
$A_x = 33.9$ kN
$A_y = 73.9$ kN

9–35. $\bar{x} = 1.61$ in.

9–37. $dm = \rho dV = \rho_o xy\,(tydx),\ m = \frac{1}{8}\rho_0 r^4 t$
$\bar{x} = \frac{8}{15}r$
$\bar{y} = \frac{8}{15}r$

9–38. $\bar{r} = 0.833a$

9–39. $\bar{y} = 2.67$ m

9–41. $\bar{z} = \frac{3}{8}a$

9–42. $\bar{y} = 2.61$ ft

9–43. $\bar{z} = 12.8$ in

9–45. $\bar{z} = 2.50$ ft

9–46. $\bar{x} = \frac{8}{15}r$

9–47. $\bar{z} = \frac{h}{4},\ \bar{x} = \bar{y} = \frac{a}{\pi}$

9–50. $\bar{z} = \frac{c}{4}$

9–51. $d = 3$ m

9–53. $\bar{x} = 77.3$ mm
$\bar{y} = 121$ mm

9–54. $\bar{x} = 0, \bar{y} = 58.3$ mm

9–55. $\bar{x} = 1.60$ ft, $\bar{y} = 7.04$ ft,
$A_x = 0$, $A_y = 149$ lb,
$M_A = 502$ lb · ft

9–57. $\bar{x} = \frac{W_1}{W}b$
$\bar{y} = \frac{b(W_2 - W_1)\sqrt{b^2 - c^2}}{cW}$

9–58. $\bar{y} = 154$ mm

9–59. $\bar{x} = 3$ in., $\bar{y} = 2$ in.

9–61. $\bar{x} = 77.2$ mm, $\bar{y} = 31.7$ mm

9–62. $\bar{y} = 291$ mm

9–63. $\bar{x} = 4.74$ in., $\bar{y} = 2.99$ in.

9–65. $\bar{y} = 10.2$ in.

9–66. $\bar{y} = 272$ mm

9–67. $\bar{y} = 53.0$ mm

9–69. $x = \frac{\frac{2}{3}r\sin^3\alpha}{\alpha - \frac{\sin 2\alpha}{2}}$

9–70. $y = \frac{\sqrt{2}(a^2 + at - t^2)}{2(2a - t)}$

9–71. $\bar{y} = 85.9$ mm

9–73. $\bar{z} = 1.625$ in.

9–74. $\bar{x} = -1.14$ in., $\bar{y} = 1.71$ in., $\bar{z} = -0.857$ in.

9–75. $\theta = 53.1°$

9–77. $\bar{x} = 2.81$ ft, $\bar{y} = 1.73$ ft, $N_B = 72.1$ lb
$N_A = 86.9$ lb

9–78. $\bar{x} = 1.47$ in.
$\bar{y} = 2.68$ in.
$\bar{z} = 2.84$ in.

9–79. $\bar{x} = 22.7$ mm
$\bar{y} = 29.5$ mm
$\bar{z} = 22.6$ mm

9–81. $\bar{z} = 359$ mm

9–82. $h = 323$ mm

9–83. $\bar{z} = 128$ mm

9–85. $h = 2.00$ ft

9–86. $\bar{z} = 463$ mm

9–87. $\bar{x} = 8.22$ in.

9–89. $\bar{z} = 754$ mm
9–90. $A = 3.56\ (10^3)\ \text{ft}^2$
9–91. $V = 22.1\ (10^3)\ \text{ft}^3$
9–93. $V = 20.5\ \text{m}^3$
9–94. $V = 4.25\ (10^6)\ \text{mm}^3$
9–95. $A = 4856\ \text{ft}^2$
9–97. $A = 188\ \text{m}^2$
9–98. $V = 207\ \text{m}^3$
9–99. $A = 141\ \text{in}^2$
9–101. $A = 8\pi ba$
$V = 2\pi ba^2$
9–102. $A = 118\ \text{in}^2$
9–103. $W = 84.7$ kip
9–105. $W = 0.377$ lb
9–106. $V = \frac{2}{3}\pi ab^2$
9–107. $A = 1.33\ \text{m}^2$
$\bar{x} = 0.6$ m
$V = 5.03\ \text{m}^3$
9–109. $V = 36.8\ \text{ft}^3$
9–110. $V = 1.40(10^3)$
9–111. $h = 29.9$ mm
9–113. $m = 138$ kg
9–114. $A = 1365\ \text{m}^2$
9–115. $F_R = 678$ lb, $\bar{x} = 0.948$ ft,
$\bar{y} = 1.50$ ft
9–117. $\theta F_R = \frac{2}{3}(xdx)[(4 - y)\,dy]$, $F_R = 24.0$ kN,
$\bar{x} = 2.00$ m, $\bar{y} = 1.33$ m
9–118. $F_R = \frac{4ab}{\pi^2}p_0$
$\bar{x} = \frac{a}{2}\quad \bar{y} = \frac{b}{2}$
9–119. $F_R = 4.00$ kip, $\bar{y} = -6.49$ ft
9–121. $D_x = 101$ kN
$C_x = 46.6$ kN
9–122. $F_{R_v} = 260$ kip, $F_{R_h} = 487.5$ kip
9–123. $F = 1.41$ MN, $h = 4$ m
9–125. $F_R = 17.2$ kip,
$d = 5.22$ ft, $F_R = 18.8$ kip
9–126. $F = 3.85$ kN
$d' = 0.625$ m
9–127. $F_R = 6.93$ kN, $\bar{y} = -0.125$ m
9–129. $F = 391$ kN
9–130. $F_x = 628$ kN
$F_y = 538$ kN
9–131. $V = 22.7(10)^{-3}\ \text{m}^3$
9–133. $\bar{y} = 87.5$ mm
9–134. $\bar{x} = \bar{y} = 0, \bar{z} = \frac{2}{3}a$
9–135. $F_R = 2.02$ kip
9–136. $\bar{z} = 4.57$ ft
9–137. $\bar{y} = 0.600$ in.
9–138. $\bar{x} = 1.22$ ft, $\bar{y} = 0.778$ ft, $\bar{z} = 0.778$ ft,
$M_{Ax} = 16.0$ lb·ft, $M_{Ay} = 57.1$ lb·ft, $M_{Az} = 0$,
$A_x = 0$, $A_y = 0$, $A_z = 20.6$ lb
9–139. $A_y = 2.51$ MN, $B_x = 2.20$ MN, $B_y = 859$ kN

Chapter 10

10–1. $I_x = 18.5\ \text{in}^4$
10–2. $I_y = 9.6\ \text{in}^4$
10–3. $I_x = 0.571\ \text{in}^4$
10–5. $I_x = 39.0\ \text{m}^4$
10–6. $I_y = 8.53\ \text{m}^4$
10–7. $I_x = 0.533\ \text{m}^4$
10–9. $I_x = \frac{2}{15}bh^3$
10–10. $I_x = 23.8\ \text{ft}^4$
10–11. $I_x = 3.20\ \text{m}^4$
10–13. $I_x = 19.5\ \text{in}^4$
10–14. $I_y = 1.07\ \text{in}^4$
10–15. $I_y = 0.628\ \text{m}^4$
10–17. $I_x = \frac{4a^4}{9\pi}$
10–18. $I_y = \left(\frac{\pi^2 - 4}{\pi^3}\right)a^4$
10–19. $I_x = 10\ \text{m}^4$
10–21. $I_x = \frac{ah^3}{28}$
10–22. $I_y = \frac{a^3h}{20}$
10–23. $dA = (rd\theta)dr$, $y = r\sin\theta$, $I_x = \frac{r_0^4}{8}(\alpha - \sin\alpha)$
10–25. $I_x = 209\ \text{in}^4$
10–26. $I_y = 533\ \text{in}^4$
10–27. $k_x = 109$ mm
10–29. $\bar{y} = 2$ in., $\bar{I}_{x'} = 128\ \text{in}^4$
10–30. $\bar{x} = 68.0$ mm, $\bar{I}_{y'} = 36.9\ (10^6)\ \text{mm}^4$
10–31. $I_{x'} = 49.5(10^6)\ \text{mm}^4$
10–33. $I_y = \frac{r^4}{4}(\theta + \frac{1}{2}\sin 2\theta - 2\sin\theta\cos^3\theta)$
10–34. $I_y = 115\ (10^6)\text{mm}^4$
10–35. $\bar{y} = 207$ mm
$\bar{I}_{x'} = 222\ (10^6)\ \text{mm}^4$
10–37. $I_y = 1971\ \text{in}^4$
10–38. $I_x = 3.35(10^3)\ \text{in}^4$
10–39. $I_y = 832\ \text{in}^4$
10–41. $\bar{y} = 3.79$ in. $I_{x'} = 198\ \text{in}^4$
10–42. $I_x = 154\ (10^6)\ \text{mm}^4$
10–43. $I_y = 91.3\ (10^6)\ \text{mm}^4$
10–45. $\bar{x} = 61.6$ mm, $\bar{I}_{y'} = 41.2(10^6)\ \text{mm}^4$

10–46. $\bar{y} = 22.5$ mm, $\bar{I}_{x'} = 34.4(10^6)$ mm^4
10–47. $\bar{I}_{y'} = 122(10^6)$ mm^4
10–49. $I_y = 365$ in^4
10–50. $\bar{y} = 0.181$ m, $\bar{I}_{x'} = 4.23(10^{-3})$ m^4
10–51. $\bar{I}_{x'} = 162(10^6)$ mm^4
10–53. $\bar{I}_{y'} = \dfrac{ab \sin\theta}{12}(b^2 + a^2\cos^2\theta)$
10–54. $I_{xy} = \dfrac{1}{3}tl^3 \sin 2\theta$
10–55. $I_{xy} = \dfrac{3}{16}b^2h^2$
10–57. $I_{xy} = \dfrac{a^2b^2}{4(n+1)}$
10–58. $I_{xy} = 10.7$ m^4, $\bar{I}_{x'y'} = 1.07$ m^4
10–59. $I_{xy} = 0.511$ m^4
10–61. $I_{xy} = \dfrac{a^2c\sin^2\theta}{12}(4a\cos\theta + 3c)$
10–62. $I_{xy} = 48$ in^4
10–63. $I_{uv} = 135(10)^6$ mm^4
10–65. $I_{xy} = -110$ in^4
10–66. $I_{xy} = 98.4(10^6)$ mm^4
10–67. $I_{xy} = 0.740$ in^4
10–69. $I_u = 85.3(10^6)$ mm^4
$I_v = 85.3(10^6)$ mm^4
10–70. $I_u = 909(10^6)$ mm^4
$I_v = 703(10^6)$ mm^4
$I_{uv} = 179(10^6)$ mm^4
10–71. $I_u = 909(10^6)$ mm^4
$I_v = 703(10^6)$ mm^4
$I_{uv} = 179(10^6)$ mm^4
10–73. $\bar{y} = 8.25$ in. $I_u = 109$ in^4, $I_v = 238$ in^4,
$I_{uv} = 111$ in^4
10–74. $I_{max} = 64.1$ in^4, $I_{min} = 5.33$ in^4
10–75. $I_{max} = 64.1$ in^4, $I_{min} = 5.33$ in^4
10–77. $\theta = -22.5°$, $I_{max} = 250$ in^4, $I_{min} = 20.4$ in^4
10–78. $I_{max} = 4.92(10^6)$ mm^4, $I_{min} = 1.36(10^6)$ mm^4
10–79. $I_{max} = 4.92(10^6)$ mm^4, $I_{min} = 1.36(10^6)$ mm^4
10–81. $I_{max} = 26.0$ in^4
$I_{min} = 5.78$ in^4
$\theta = -45°$
10–82. $\theta = 6.08°$, $I_{max} = 1.74(10^3)$ in^4, $I_{min} = 435$ in^4
10–83. $\theta = 6.08°$, $I_{max} = 1.74(10^3)$ in^4, $I_{min} = 435$ in^4
10–85. $I_x = \dfrac{2}{5}mb^2$
10–86. $k_x = 1.20$ in.
10–87. $k_x = 57.7$ mm
10–89. $I_y = \dfrac{m}{6}(a^2 + h^2)$
10–90. $I_z = \dfrac{7}{18}ma^2$
10–91. $I_x = \dfrac{2}{5}mr^2$
10–93. $I_y = \dfrac{5}{18}m$
10–94. $I_y = 1.71(10^3)$ kg · m^2
10–95. $I_A = 2.17$ slug · ft^2
10–97. $\bar{y} = 1.78$ m, $I_G = 4.45$ kg · m^2
10–98. $\bar{y} = 203$ mm, $I_G = 0.230$ kg · m^2
10–99. $I_A = 222$ slug · ft^2
10–101. $I_z = 1.53$ kg · m^2
10–102. $k_O = 3.15$ ft
10–103. $I_A = 1.58$ slug · ft^2
10–105. $I_O = 0.450(10^{-3})$ kg · m^2
$\bar{y} = 57.7$ mm
$I_G = 0.150(10^{-3})$ kg · m^2
10–106. $I_y = 0.144$ kg · m^2
10–107. $I_z = 0.113$ kg · m^2
10–109. $\bar{I}_{x'} = 293$ slug · ft^2
10–110. $I_x = 1.07$ in^4
10–111. $I_x = 914(10^6)$ mm^4
$\bar{I}_{x'} = 146(10^6)$ mm^4
10–113. **a.** $I_x = \dfrac{bh^3}{12}$
b. $\bar{I}_{x'} = \dfrac{bh^3}{36}$
10–114. $I_x = \dfrac{93}{70}mb^2$
10–115. $I_y = 2.13$ ft^4
10–117. $I_u = 5.09(10^6)$ mm^4, $I_v = 5.09(10^6)$ mm^4, $I_{uv} = 0$
10–118. $0.0954\,d^4$
10–119. $0.187\,d^4$

Geometric Properties of Line and Area Elements

Centroid Location	Centroid Location	Area Moment of Inertia

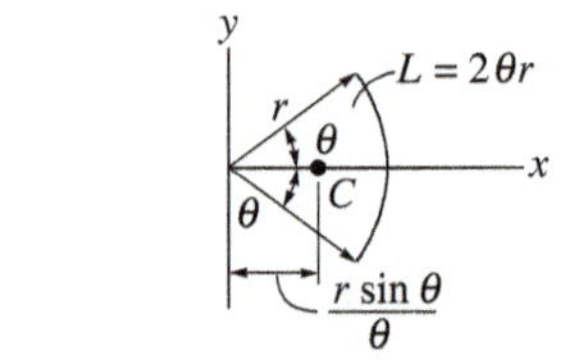

Circular arc segment

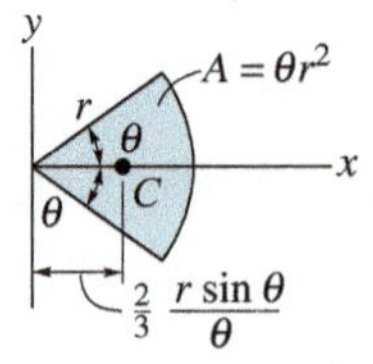

Circular sector area

$I_x = \frac{1}{4} r^4 (\theta - \frac{1}{2} \sin 2)\theta$

$I_y = \frac{1}{4} r^4 (\theta + \frac{1}{2} \sin 2)\theta$

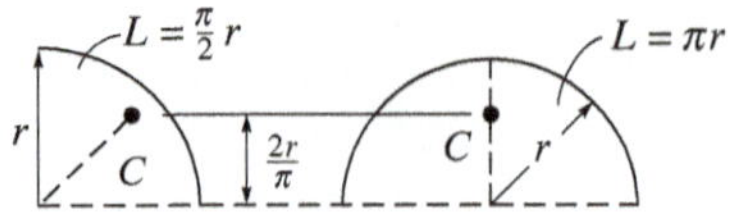

Quarter and semicircle arcs

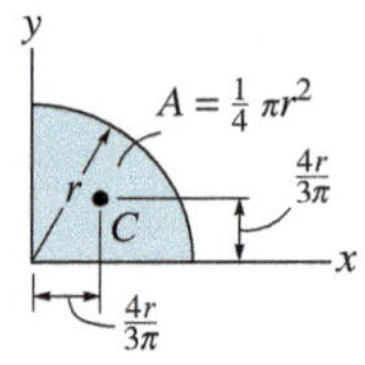

Quarter circle area

$I_x = \frac{1}{16} \pi r^4$

$I_y = \frac{1}{16} \pi r^4$

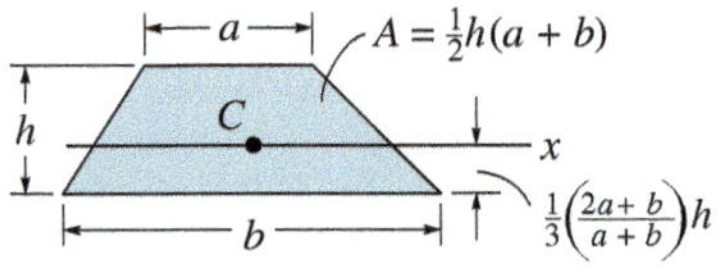

Trapezoidal area

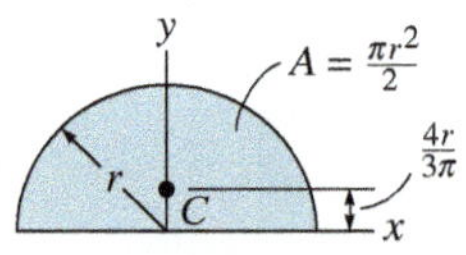

Semicircular area

$I_x = \frac{1}{8} \pi r^4$

$I_y = \frac{1}{8} \pi r^4$

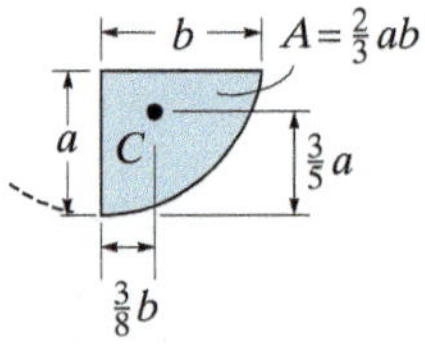

Semiparabolic area

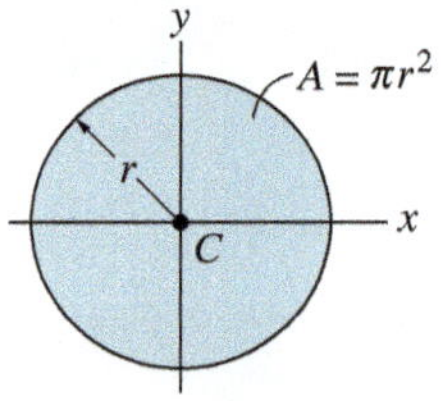

Circular area

$I_x = \frac{1}{4} \pi r^4$

$I_y = \frac{1}{4} \pi r^4$

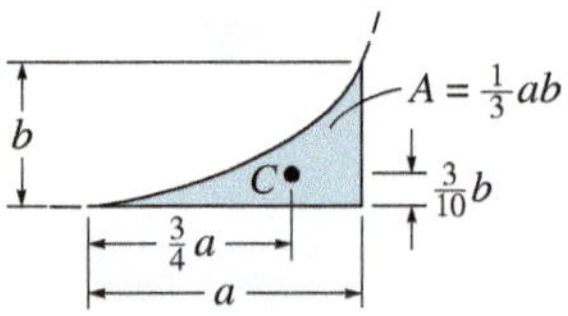

Exparabolic area

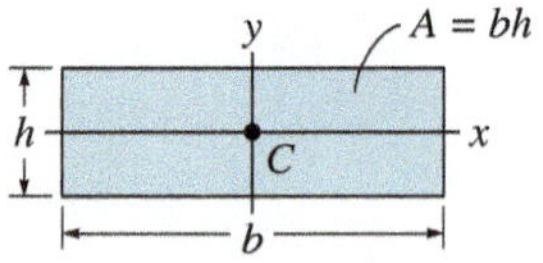

Rectangular area

$I_x = \frac{1}{12} bh^3$

$I_y = \frac{1}{12} hb^3$

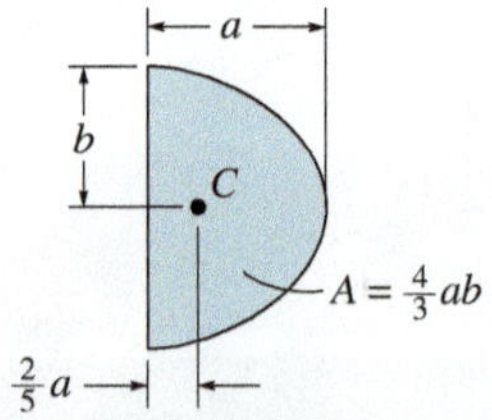

Parabolic area

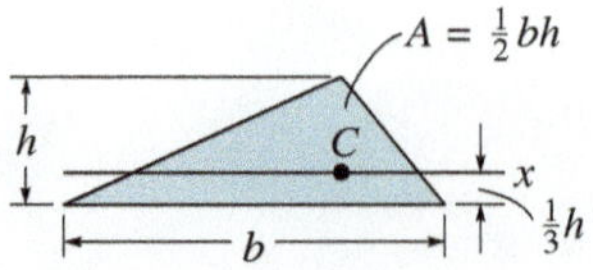

Triangular area

$I_x = \frac{1}{36} bh^3$

Center of Gravity and Mass Moment of Inertia of Homogeneous Solids

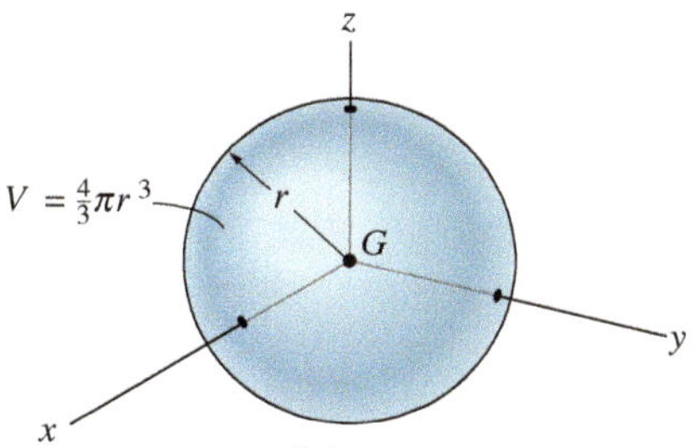

Sphere

$I_{xx} = I_{yy} = I_{zz} = \frac{2}{5} mr^2$

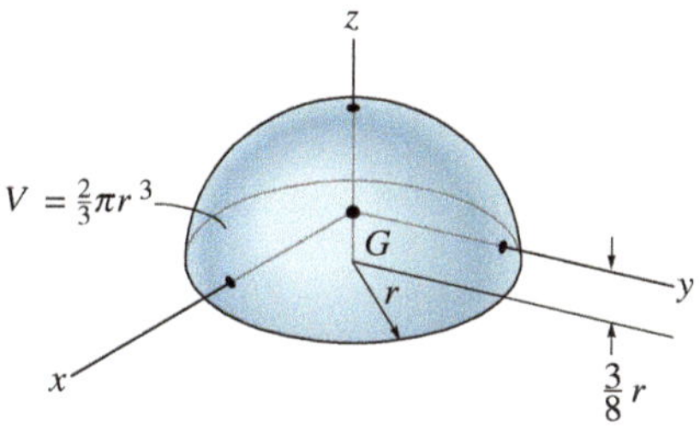

Hemisphere

$I_{xx} = I_{yy} = 0.259mr^2 \quad I_{zz} = \frac{2}{5}mr^2$

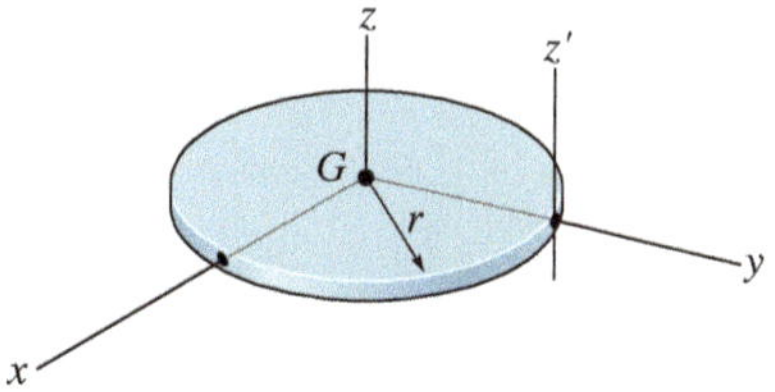

Thin Circular disk

$I_{xx} = I_{yy} = \frac{1}{4} mr^2 \quad I_{zz} = \frac{1}{2}mr^2 \quad I_{z'z'} = \frac{3}{2} mr^2$

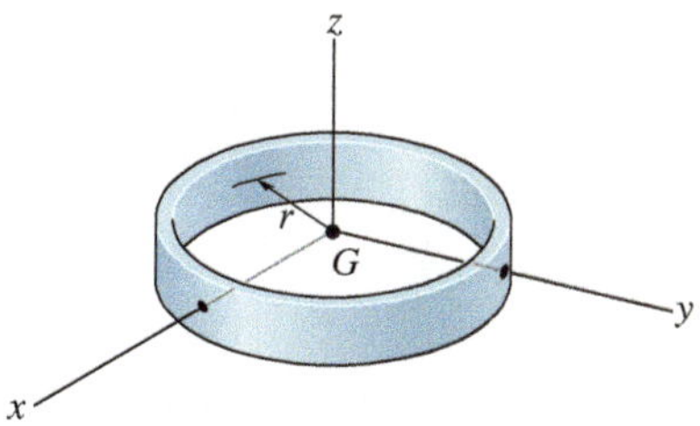

Thin ring

$I_{xx} = I_{yy} = \frac{1}{2} mr^2 \quad I_{zz} = mr^2$

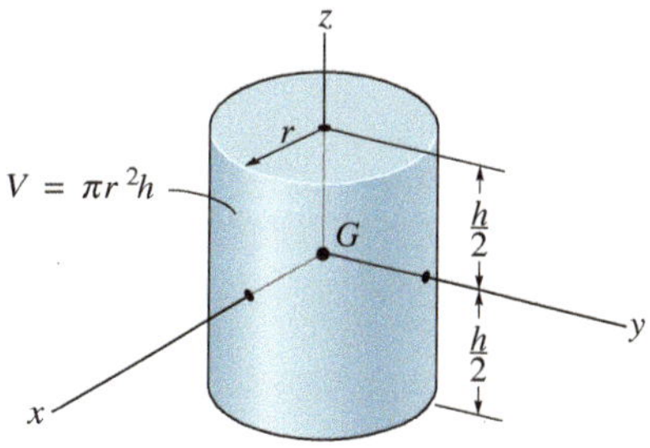

Cylinder

$I_{xx} = I_{yy} = \frac{1}{12} m(3r^2 + h^2) \quad I_{zz} = \frac{1}{2} mr^2$

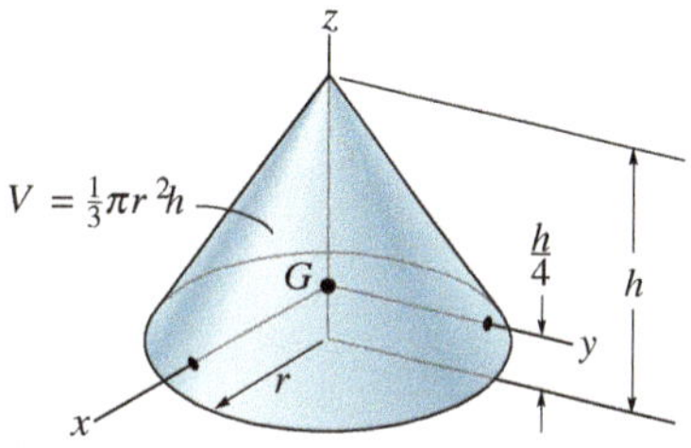

Cone

$I_{xx} = I_{yy} = \frac{3}{80} m(4r^2 + h^2) \quad I_{zz} = \frac{3}{10}mr^2$

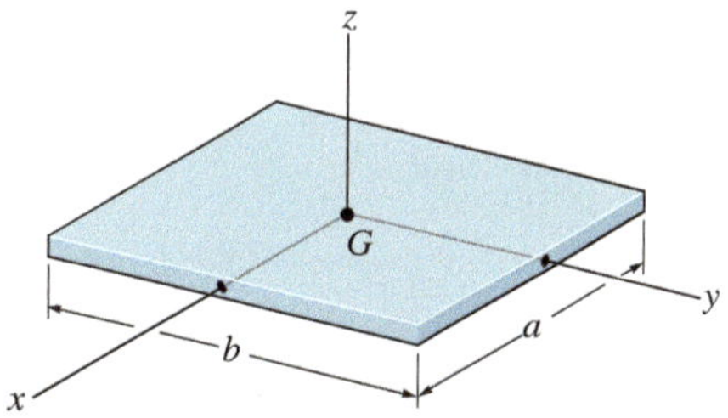

Thin plate

$I_{xx} = \frac{1}{12} mb^2 \quad I_{yy} = \frac{1}{12} ma^2 \quad I_{zz} = \frac{1}{12} m(a^2 + b^2)$

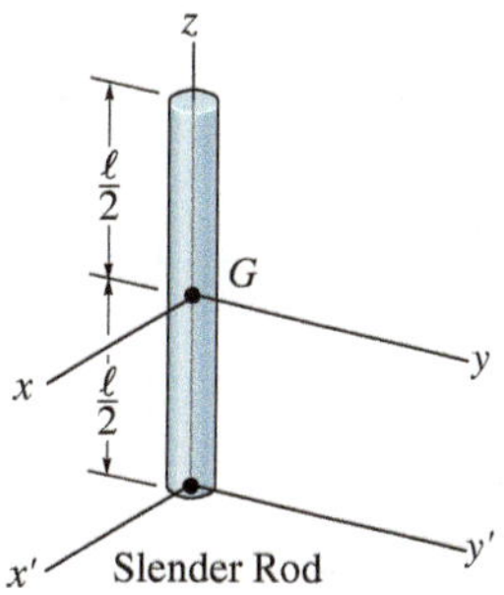

Slender Rod

$I_{xx} = I_{yy} = \frac{1}{12} m\ell^2 \quad I_{x'x'} = I_{y'y'} = \frac{1}{3} m\ell^2 \quad I_{z'z'} = 0$